IMAGINAL DISCS

With the elucidation of the complete fly genome, traditional fly genetics is in more demand than ever. Genetics will allow us to explain the role of each of the 14,000 genes, many of which are involved in the development of imaginal discs. These hollow sacs of cells make adult structures during metamorphosis, and their study is crucial to comprehending how a larva becomes a fully functioning fly.

This book examines the genetic circuitry of the well-known "fruit fly," tackling questions of cell assemblage and pattern formation, of the hows and the whys behind the development of the fly. The book first establishes that fly development relies primarily on intercellular signaling, and then discusses how this signaling occurs. After an initial examination of the proximity versus pedigree imperatives, the book delves into bristle pattern formation and disc development, with entire chapters devoted to the leg, wing, and eye. Extensive appendices include a glossary of protein domains, catalogs of well-studied genes, and an outline of signaling pathways. More than 30 wiring diagrams, among 67 detailed schematics, clarify the text. The text goes beyond the Internet databases insofar as it puts these myriad facts into both a conceptual framework and a historical context. Overall, the aim is to provide a comprehensive reference guide for students and researchers exploring this fascinating, but often bewildering, field.

Lewis I. Held, Jr., is Associate Professor in the Department of Biological Sciences at Texas Tech University.

Developmental and Cell Biology Series

SERIES EDITORS

Jonathan B. L. Bard, *Department of Anatomy, Edinburgh University*
Peter W. Barlow, *Long Ashton Research Station, University of Bristol*
David L. Kirk, *Department of Biology, Washington University*

The aim of the series is to present relatively short critical accounts of areas of developmental and cell biology where sufficient information has accumulated to allow a considered distillation of the subject. The fine structure of cells, embryology, morphology, physiology, genetics, biochemistry and biophysics are subjects within the scope of the series. The books are intended to interest and instruct advanced undergraduates and graduate students and to make an important contribution to teaching cell and developmental biology. At the same time, they should be of value to biologists who, while not working directly in the area of a particular volume's subject matter, wish to keep abreast of developments relevant to their particular interests.

BOOKS IN THE SERIES

18. C. J. Epstein *The Consequences of Chromosome Imbalance: Principles, Mechanisms and Models*   0 521 25464 7
19. L. Saxén *Organogenesis of the Kidney*   0 521 30152 1
20. V. Raghavan *Developmental Biology of the Fern Gametophytes*   0 521 33022 X
21. R. Maksymowych *Analysis of Growth and Development in Xanthium*   0 521 35327 0
22. B. John *Meiosis*   0 521 35053 0
23. J. Bard *Morphogenesis: The Cellular and Molecular Processes of Developmental Anatomy*   0 521 43612 5
24. R. Wall *This Side Up: Spatial Determination in the Early Development of Animals*   0 521 36115 X
25. T. Sachs *Pattern Formation in Plant Tissues*   0 521 24865 5
26. J. M. W. Slack *From Egg to Embryo: Regional Specification in Early Development*   0 521 40943 8
27. A. I. Farbman *Cell Biology of Olfaction*   0 521 36438 8
28. L. G. Harrison *Kinetic Theory of Living Pattern*   0 521 30691 4
29. N. Satoh *Developmental Biology of Ascidians*   0 521 35221 5
30. R. Holliday *Understanding Ageing*   0 521 47802 2
31. P. Tsonis *Limb Regeneration*   0 521 44149 8
32. R. Rappaport *Cytokinesis in Animal Cells*   0 521 40173 9
33. D. L. Kirk *Volvox: Molecular Genetic Origins of Multicellularity and Cellular Differentiation*   0 521 45207 4
34. R. L. Lyndon *The Shoot Apical Meristem: Its Growth and Development*   0 521 40457 6
35. D. Moore *Fungal Morphogenesis*   0 521 55295 8
36. N. Le Douarin & C. Kalcheim *The Neural Crest, Second Edition*   0 521 62010 4
37. P. R. Gordon-Weeks *Neuronal Growth Cones*   0 521 44491 8
38. R. Kessin *Dictyostelium*   0 521 58364 0
39. L. I. Held, Jr. *Imaginal Discs: The Genetic and Cellular Logic of Pattern Formation*   0 521 58445 0

Thomas Hunt Morgan (3rd from right) and his associates at Columbia University. This luncheon was held in the "Chart Room" on 2 January 1919, to celebrate the return of Alfred Henry Sturtevant (foreground with beer and cigar) from his brief stint as a soldier in World War I [72, 651, 1556, 2283]. Calvin Bridges (center) is feigning a chat with a museum mannequin (*Homo erectus*) dressed in Sturt's uniform. Clockwise from this anthropoid "guest" are Hermann J. Muller, T. H. Morgan ("the Boss"), Frank E. Lutz, Otto L. Mohr, Alfred F. Huettner, A. H. Sturtevant, Franz Schrader, Ernest G. Anderson, Alexander Weinstein, S. C. Dellinger, and Calvin B. Bridges. Curt Stern (not shown) did not join the team until 1924 [3071]. This merry band of pioneers launched a great quest for the secrets of genetics, and they had a knack for solving mysteries that rivaled Sherlock Holmes [72, 650, 651, 2951, 4182, 4184]. Nevertheless, as the informality of this party indicates, these legendary heroes did not take themselves too seriously [72, 3903]. Indeed, their lightheartedness has suffused this field ever since [4696] and is reflected in the whimsical names of many fly genes [2561]. Most of the mutations they studied affect the adult's anatomy by altering the development of the larva's imaginal discs. Those discs are the subject of this book, one of whose aims is to celebrate the triumph of the quest. This picture is from Sturt's photo album. It was provided courtesy of the Archives, California Institute of Technology.

IMAGINAL DISCS

The Genetic and Cellular Logic of Pattern Formation

LEWIS I. HELD, JR.

Texas Tech University

CAMBRIDGE
UNIVERSITY PRESS

CAMBRIDGE UNIVERSITY PRESS
Cambridge, New York, Melbourne, Madrid, Cape Town, Singapore, São Paulo

Cambridge University Press
The Edinburgh Building, Cambridge CB2 2RU, UK

Published in the United States of America by Cambridge University Press, New York

www.cambridge.org
Information on this title: www.cambridge.org/9780521584456

First published 2002
This digitally printed first paperback version 2005

A catalogue record for this publication is available from the British Library

Library of Congress Cataloguing in Publication data
Held, Lewis I., 1951–
   Imaginal discs : the genetic and cellular logic of pattern formation / Lewis I. Held, Jr.
    p. cm.
   Includes bibliographical references and index.
   ISBN 0-521-58445-0
   1. Drosophila melanogaster – Genetics.   2. Drosophila melanogaster – Embryology.
   3. Drosophila melanogaster – Morphogenesis – Molecular aspects.   4. Cellular signal
transduction.   5. Cell interaction.   I. Title.
   QH470.D7 H45 2002
   576.5 – dc21   2001043553

ISBN-13 978-0-521-58445-6 hardback
ISBN-10 0-521-58445-0 hardback

ISBN-13 978-0-521-01835-7 paperback
ISBN-10 0-521-01835-8 paperback

Contents

Preface   *page* xi

CHAPTER ONE. **CELL LINEAGE VS. INTERCELLULAR SIGNALING**   1

Discs are not clones   1
No part of a disc is a clone, except claws and tiny sense organs   4
Cells belong to lineage "compartments"   4

CHAPTER TWO. **THE BRISTLE**   5

Numb segregates asymmetrically and dictates bristle cell fates   5
Delta needs to activate Notch, but not as a signal per se   9
Amnesic cells can use sequential gating to simulate a binary code   10
Notch must go to the nucleus to function   12
E(spl)-C genes are Su(H) targets but play no role in the SOP lineage   15
The transcription factor Tramtrack implements some cell identities   18
Hairless titrates Su(H)   20
Several other genes help determine the 5 cell fates   23
Pox neuro and Cut specify bristle type   27
Bract cells are induced by bristle cells   28
Macrochaetes and microchaetes differ in size but not in kind   29

CHAPTER THREE. **BRISTLE PATTERNS**   31

Surprisingly, different macrochaete sites use different signals   33
Prepatterns may contain hidden "singularities"   36
How Achaete and Scute control bristles was debated for decades   36
In 1989, Achaete and Scute were found to mark "proneural clusters"   43
In 1995, the old AS-C paradigm toppled and a new one emerged   44
Proneural "spots" shrink to SOP "dots"   45
The SOP uses a feedback loop to raise its Ac and Sc levels   48
Two other bHLH genes (*asense* and *daughterless*) assist SOPs   48

"Lateral" or "mutual" inhibition ensures one SOP per PNC 49
Notch-pathway and proneural genes are functionally coupled 50
Doses of Notch-pathway genes can bias the SOP decision 52
Extra SOPs could be inhibited by contact or diffusion (or both) 52
Scabrous may be the diffusible SOP inhibitor 53
Inhibitory fields dictate the spacing intervals of microchaetes 55
Microchaetes come from proneural stripes, not spots 57
Hairy paints "antineural" stripes on the legs 61
Leg bristles use extra fine-tuning tricks 62
Chemosensory leg bristles are patterned like notal macrochaetes 67
Extramacrochaetae superimposes an uneven antineural "mask" 68
Dose dependency implies that HLH proteins "compute" bristles 71
Robustness of patterning may be due to a tolerant time window 73
Atonal and Amos are proneural agents for other types of sensilla 74
Other (upstream) pathways govern bristle patterning 75

CHAPTER FOUR. **ORIGIN AND GROWTH OF DISCS** 76

Segmentation genes set the stage for disc initiation 76
Prepatterns and gradients clashed in trying to explain homeosis 80
Homeotic genes implement regional identities 84
Wing and haltere discs "grow out" from 2nd- and 3rd-leg discs 85
Thoracic discs arise at Wingless/Engrailed boundaries 87
Cell lineage within compartments is indeterminate 91
The Polar Coordinate Model linked regeneration to development 92
But regeneration has peculiarities that set it apart 96

CHAPTER FIVE. **THE LEG DISC** 97

The Molecular Epoch of disc research was launched in 1991 97
Bateson's Rule (1894) governs symmetry planes in branched legs 99
Meinhardt's Boundary Model deftly explained Bateson's Rule 100
The Boundary and PC Models jousted in a "Paradigm War" 103
Hh, Dpp, and Wg are the chief intercellular signaling molecules 105
P-type cells use Hh to "talk" to A-type cells nearby 107
Hh elicits expression of Dpp and Wg along the A/P boundary 109
Dpp dorsalizes and Wg ventralizes, or do they? 111
Dpp and Wg are mutually antagonistic 114
Dpp and Wg jointly initiate distal outgrowth 115
But Dpp seems more crucial than Wg as a growth factor 118
The A/P boundary can migrate when its "jailors" are "asleep" 119
Regeneration is due to a Hh spot in the peripodial membrane 122
The Polar Coordinate Model died in 1999 123
How Hh, Dpp, and Wg move is not known, nor is their range 124
Whether Dpp and Wg travel along curved paths is not known 125
Hairy links global to local patterning 128
Questions remain about the Hh-Dpp-Wg circuitry 128
Distal-less is necessary and sufficient for distalization 129

Proximal and distal cells have different affinities   132
Dachshund is induced at the Homothorax/Distal-less interface   132
Homothorax and Extradenticle govern the proximal disc region   133
Fasciclin II is induced at the BarH1/Aristaless interface   134
BarH1 and Bric à brac affect P-D identity, joints, and folds   135
Leg segmentation requires Notch signaling   135

CHAPTER SIX. **THE WING DISC**   137

The A-P axis is governed by Hh and Dpp but not by Wg   137
Dpp turns ON *omb* and *spalt* at different thresholds   140
Dpp regulates *omb* and *spalt* similarly despite clues to the contrary   143
Dpp does not regulate *tkv* in 3rd instar despite clues to the contrary   143
A vs. P identities might explain how a straight A/P line emerges   148
But the A/P line appears to straighten via a signaling mechanism   149
Intercalation is due to a tendency of Dpp gradients to rise   153
The variable height of Dpp gradients makes them appear seamless   155
A Wg gradient specifies cell fates along the wing's D-V axis   156
Perpendicular (Dpp × Wg) gradients suggest Cartesian coordinates   157
But cells do not seem to record positional values per se   158
Wg's repression of Dfz2 is inconsequential   158
Apterous's role along the D-V axis resembles Engrailed's A-P role   158
Chip cooperates with Apterous, and "Dorsal wing" acts downstream   160
Serrate and Delta prod Notch to evoke Wg at the D/V line   161
Fringe prevents Notch from responding to Serrate   164
The core D-V circuit plugs into a complex network   165
The wing-notum duality is established by Wg and Vein   167
But Vestigial and Scalloped dictate "wingness" per se   171
Straightening of the D/V border requires Notch signaling and Ap   173
Straightening of veins may rely on similar tricks   173
Two cell types predominate in the wing blade: vein and intervein   174
Veins come from proveins that look like proneural fields   175
But the resemblance is only superficial   175
All veins use the EGFR pathway   177
But interveins also use the EGFR pathway (at a later time)   184
Veins 3 and 4 are positioned by the Hh pathway   185
Veins 2 and 5 are positioned by the Dpp pathway   186
The Dpp pathway later implements the vein state   188
A cousin of Dpp (Gbb) fosters the A and P cross-veins   188
Vein 1 uses a combination of Dpp and Wg signals   189
Macrochaetes are sited by various "prepattern" inputs   190
How bristle axons get wired into the CNS is not known   191

CHAPTER SEVEN. **THE EYE DISC**   197

Compound eyes have ~750 facets, with 8 photoreceptors per facet   197
Unlike the bristle, the ommatidium is not a clone   201
The eye has D and V compartments (despite doubts to the contrary)   202

The Iroquois Complex controls D-V polarity via Fringe and Notch   203
A morphogenetic wave creates the ommatidial lattice   208
D-V polarity depends on a rivalry between R3 and R4 precursors   209
R1–R8 cells arise sequentially, implying a cascade of inductions   211
But the final cell (R7) is induced by the first one (R8)   213
Various restraints prevent more than one cell from becoming R7   213
The information content of the inductive signals may be only 1 bit   216
No transcription factor "code" has yet been found for R cells   218
The lattice is created by inhibitory fields around R8 precursors   224
The lattice is tightened when excess cells die   227
Eye bristles arise independently of ommatidia   228
The MF operates like a moving A/P boundary   229
Dpp and Wg control the rate of MF progress   234
The MF originates via different circuitry   234

CHAPTER EIGHT. **HOMEOSIS**   237
BX-C and ANT-C specify gross metameric identities along the body   237
Ubx enables T3 discs to develop differently from T2 discs   243
But Ubx does so by directly managing target genes in multiple echelons   244
Pc-G and Trx-G "memory" proteins keep homeotic genes ON or OFF   247
Homothorax, Distal-less, and Spineless specify leg vs. antennal fates   249
If a "master gene" exists for the eye, then it is also a micromanager   252
The manifold "enhanceosome" is a wondrous Gordian Knot   254
The deepest enigma is how evolution rewired the circuit elements   254

EPILOGUE   256

APPENDIX ONE. **Glossary of Protein Domains**   257

APPENDIX TWO. **Inventory of Models, Mysteries, Devices, and Epiphanies** 266

APPENDIX THREE. **Genes That Can Alter Cell Fates Within the (5-Cell)
    Mechanosensory Bristle Organ**   271

APPENDIX FOUR. **Genes That Can Transform One Type of Bristle Into
    Another or Into a Different Type of Sense Organ**   276

APPENDIX FIVE. **Genes That Can Alter Bristle Number by Directly Affecting
    SOP Equivalence Groups or Inhibitory Fields**   278

APPENDIX SIX. **Signal Transduction Pathways: Hedgehog,
    Decapentaplegic, and Wingless**   285

APPENDIX SEVEN. **Commentaries on the Pithier Figures**   297

References   307

Index   441

Preface

How embryos "self-assemble" has fascinated thinkers for millennia [2918, 3064, 3190]. Among the ancient Greeks, Aristotle (384–322 BCE) made copious observations and coined the term "morphogenesis," which is still in use today [2989, 4305]. For the past century, the science of "developmental mechanics" has hammered at this problem relentlessly, but it is only in the last decade that the core mysteries have finally cracked [1487]. The deepest secrets have come from a fairylike fly named *Drosophila melanogaster*, probably the same species of "gnat" that Aristotle himself noticed hovering over vinegar slime [217, 3361, 4184]. Unfortunately, these insights can only be fully appreciated in the arcane language of fly genetics. Hence this book full of runes and rules.

This book concerns cuticular patterns, the cellular machinery that makes them, and the genetic circuitry that runs the machinery. Although it is mainly a survey, it is also a narrative that traces the roots of our knowledge. The story that it tells – albeit in condensed form – rivals the *Iliad* in scope (legions of researchers devoting decades to attacking thousands of genes) and the *Odyssey* in wonderment (monstrous mutants posing riddles that challenge even the most clever explorer-heroes). Indeed, truth is often stranger than a fairy tale in the realm of the fly. Believe it or not, there are even remote islands where giant drosophilids with dappled wings and feathery legs have been spied dancing and fighting in the misty forests [668, 669].

Ever since 1910 when T. H. Morgan's first "fly paper" was published [2948], the field of fly genetics has brimmed with intriguing curiosities [820, 2951, 3673] and equally colorful human personalities [120, 327, 2283, 4183]. Added to these delights is a menagerie of recently discovered molecules – e.g., the midget "Bearded" (81 a.a.) [2499] and the giant "Dumpy" (3680 a.a.) [4668]. Now that the fly genome project is ending [14], the world is peering into this circus. What newcomers may not realize is that this field offers many diversions beyond its databases.

Like other holometabolous insects, flies live two lives – first as a grub, then as a flying adult [82]. During metamorphosis, 19 "imaginal discs" erupt from inside the maggot and are quilted together to form most of the adult skin. The gold-colored cuticle secreted by that skin is exquisitely ornate. The head is embossed with hundreds of domes that focus light onto bundles of photoreceptors, the thorax is sculpted into dozens of jointed parts that form a contraption for walking and flying, and the abdominal wall (built from non-disc tissue [2648]) is pleated into an expansible chamber for digestion and reproduction. Nearly everywhere, the body surface sprouts bristles whose patterns can be as orderly as soldiers on parade.

Why do only some cells make bristles? That is a problem of differentiation. Why do bristles arise only at certain sites? That is a problem of pattern formation, and these questions can be asked for structures in general. Beneath both problems is a coding enigma: how does the fly's 1-dimensional genome encode the 2-dimensional cuticular landscape? Once, it seemed that each body part might be governed by its own set of genes [4509, 4512], but this notion proved wrong [1094, 1114, 2410, 4643]. In fact, most patterns are built by the same ensembles of genes. These modules arose eons ago in the mythical common ancestor of insects and vertebrates [1439, 3840]. Since then, evolution has customized the circuitry by making new intra- and inter-modular links [968, 1440].

What is the nature of the circuitry, and how does it program cells to "compute" patterns? That is the subject of this book. Topics are arranged roughly in order of increasing complexity. Chapter 1 establishes one simple fact: in contrast to nematodes, flies rely primarily on intercellular signaling (vs. cell lineage) to assign cell fates. The rest of the book traces how signaling occurs. Chapter 2 delves into the 5-cell cluster that constructs a mechanosensory bristle. The bristle is an exception to the signaling rule: its cell fates are dictated almost entirely by lineage. Chapter 3 uses bristle *patterns* to show how cells communicate in populations larger than a bristle but smaller than a disc, and Chapter 4 sets the stage for a discussion of larger-scale patterning by reviewing how discs arise and grow. Chapters 5 to 7 explore how leg, wing, and eye discs use similar toolkits of genes in idiosyncratic ways. The other two major discs – haltere and genital – are excluded because their strategies so closely resemble wing [16, 51, 3875, 4683, 4684] and leg discs [679, 735, 1163, 2343, 2942, 3732], respectively. (Fly genitalia are evolutionarily modified appendages [1137, 1179, 1562].) Chapter 8 contemplates the phenomenon of homeosis in the context of evolution.

Overall, the book's quest is to understand cellular "epistemology" (what do cells know?) and "psychology" (how do they think?). Its approach involves de- and reconstruction: to cut through the jargon, tease out the facts, and then try to make sense of the models by piecing the clues back together using *a priori* reasoning.

The bad news is that there are so many pieces in the puzzle that persistence will be needed. The good news is that their interactions are so limited that no fancy math is required to learn the rules of the game [3588, 3841]. A recurrent theme in the saga is how cellular riddles were solved by molecular genetics. The abiding moral is that there is much more experimental work to be done if we are to comprehend how the fly's ~14,000 genes [14, 1559, 3618, 3674] – or a large portion thereof [280, 615, 963, 4273] – are orchestrated during patterning [698, 2162, 2237, 2845, 4084]. In short, the fly still holds many secrets, and genomics will need genetics to ferret them out [465].

Thus, the book is a sampler of case studies and gedanken exercises, not an encyclopedia. That function is served by the Internet databases, and readers should consult two main websites: *FlyBase* (*flybase.bio.indiana.edu*) [124, 279] and *The Interactive Fly* (*sdb.bio.purdue.edu*) [484]. Fly lore is best savored by browsing the classics: the 1993 Cold Spring Harbor 2-volume compendium on development [238], its gargan-

tuan 12-volume predecessor *The Genetics and Biology of Drosophila* [122], Mike Ashburner's huge "handbook" [118], Lindsley and Zimm's dictionary of fly genes [2561], Bridges and Brehme's Barnumesque catalog of freakish mutants [470], and the Morgan team's *magnum opus* of 1925 [2951]. However, the fun of fly research is best portrayed in the charming *Fly* by Martin Brookes (2001, Harper-Collins, N.Y.).

Despite this disclaimer about breadth, a few topics are covered in depth in the appendices. Appendix 1 is a glossary of protein domains. Appendix 2 lists most of the ideas that have guided research in this field. Appendices 3 to 5 catalog the well-studied genes that affect bristles, sensilla, or bristle patterns, and Appendix 6 outlines three of the key signaling pathways in disc development (Hedgehog, Wingless, and Decapentaplegic). The other two pathways are discussed in Chapters 2 (Notch) and 6 (EGFR). Appendix 7 contains additional comments about the figures.

Historically, disc research has been reviewed intermittently. Disc histology was codified by Dietrich Bodenstein in 1950 [377]. Disc development and genetics were surveyed by Gehring and Nöthiger (1973) [1421], Postlethwait and Schneiderman (1973) [3448], Bryant (1978) [526], Shearn (1978) [3881], Poodry (1980) [3422], and Oberlander (1985) [3165]. The first blush of molecular-genetic data was evaluated by Stephen Cohen in 1993 [834], and the fundamentals of signaling were summarized by Seth Blair in 1999 [358]. Two books that nicely bracket the last 30 years of investigation are *The Biology of Imaginal Discs* (1972, H. Ursprung and R. Nöthiger, eds.) [4426] and *Developmental Genetics of Drosophila* (1998, A. Ghysen, ed.) [1452].

Conventional nomenclature is used. Locations of genes are stated in terms of the salivary gland chromosome map [2561]: the 3-part code (e.g., "92E12–14") denotes the chromosome section (1–20 span the X, 21–60 the 2nd, 61–100 the 3rd, and 101–102 the tiny 4th chromosome), the lettered subdivision (A–F), and the band or range of bands. Genes are italicized, but gene complexes (e.g., Bar-C) are not. All proteins are in plain type. Mutations are superscripted (e.g., *numb*[LOF]), whereas wild-type alleles are not (*numb*) or are labeled with "+" (*numb*[+]). Null alleles are designated by a "null" or "−" superscript. Most gene names record the dominant (capital) or recessive (lowercase) nature of early mutations (e.g., *Notch* vs. *numb*). Capital "D" (*D̲rosophila*) is used for paralogs within the species (e.g., *Dfz2* [310] in the *frizzled* series), whereas lowercase "d" refers to orthologs of vertebrate genes (e.g., *dTcf* [692, 1517]). Proteins are always capitalized (e.g., Numb).

Given these rules, the normal symbols for *Hairless* (*H*) and *hairy* (*h*) are distinct for the genes but not for the proteins ("H" in both cases), so "H" will be used only for Hairless, while "Hairy" will be written out. Likewise, Beadex will be written out to avoid confusion with the protein encoded by *bithorax* (both would be "Bx"). Small capitals are employed for Boolean states (ON, OFF), conditions (IF, THEN, NOT), and conjunctions (AND, OR). Amino acid and nucleotide sequences are underlined. Boundaries are denoted by slash marks (e.g., "A/P") and axes by hyphens (e.g., "A-P"). Short gene names (≤5 letters) are not usually abbreviated.

Abbreviations include a.a. (amino acid), AEL (after egg laying), AP (after pupariation) a.k.a. (also known as), b.p. (base pair), h (hour), hs (heat shock), kb (kilobase), kD (kiloDalton), MC (macrochaete), mC (microchaete), St. (stage of embryogenesis), t.s. (temperature-sensitive), pers. comm. (personal communication), and unpub. obs. (unpublished observations). Times (h AEL or h AP) refer to a culture temperature of 25°C, unless stated otherwise. Polypeptide lengths are for the unprocessed (nascent) precursor. Genes that are usually called "neurogenic" (based on mutant phenotype) [436] are here termed "antineural" (based on function) [4387] to contrast them with "proneural" (based on function) genes [2018]. "Eye disc" refers to both the eye and antennal parts, and "wing disc" denotes the entire dorsal mesothoracic disc (wing, notal, and pleural parts). By tradition (quirky though it may be), fate maps employ *left* legs (Ch. 5), *right* wings (Ch. 6), and *left* eyes (Ch. 7) [185, 320, 526, 531], although right eyes are used by some authors [2962].

Readers must be familiar with the basics of fly development [358, 2434, 3517] and the methods of modern genetics [354, 4671], including (1) induction of cell clones by *flp*-mediated recombination [1530–1532, 3952, 4781] and the *flp*-out trick [4159], (2) regional misexpression of genes via *Gal4-UAS* constructs [435, 3857], (3) temporal misexpression via heat-sensitive alleles [4214] or heat-shock promoters [2953], (4) enhancer trapping using *lacZ* reporter genes [278, 329, 1286, 4687], and (5) two-hybrid screening for protein interactions [222, 763, 1228, 1229, 1316].

Wherever possible, circuits are formulated in terms of Boolean logic [399] because this format shows syntax better than the "spaghetti diagrams" of genetics, electronics, or neural networks [2870]. The temptation to compare fly circuits with vertebrate or nematode circuits is generally resisted here for the sake of conciseness. Such comparisons can be found in Eric Davidson's book *Genomic Regulatory Systems* [968] and at Tom Brody's website *The Interactive Fly*.

The term "link" is used in the sense of "causal linkage." Links are symbolized as "➡" (activation) or "⊣" (inhibition). When a gene is the object (e.g., "Dpp ➡ *omb*"), the effect is always at the transcriptional level, but pathways may be distilled in terms of either genes (*en* ⊣ *ci* ➡ *ptc*) or proteins (En ⊣ Ci ➡ Ptc), and any attendant ambiguities will be clarified by context. Epistatic links need not be direct. Thus, "*a* ➡ *c*" could reflect a longer chain such as "*a* ➡ *b* ➡ *c*" or "*a* ⊣ *b* ⊣ *c*." The reason for listing so many links in this book is to facilitate Aristotle's goal of delineating the entire chain of causes from the fertilized egg to the adult [1993, 2919, 4305]. Only by concatenating all the known fragments can we see the gaps that remain to be filled.

The terms "LOF" (Loss of Function) and "GOF" (Gain of Function) typically denote decreases or increases in levels of gene activity (i.e., under- or overexpression) [1117, 1455], but in the broader sense that will be used here, GOF also includes ectopic misexpression where the "gain" is regional (cf. Fig. 6.13). For example, clones of cells that express a wild-type allele of *engrailed* (*en*⁺) outside the territory where *en*⁺ is normally transcribed will be called "*en*^GOF" [4848]. Cases do arise where overexpressing a wild-type allele has effects that differ from expressing a constitutively active construct [3545], and these will be so indicated. Mutations that are neither LOF nor GOF (e.g., neomorphs and antimorphs) are rarer, and allele-specific superscripts will be retained for them (e.g., *ci*^D [3818] and *en*¹ [1636]).

LOF and GOF tests are used to assess the necessity (LOF) and sufficiency (GOF) of a specific gene for a particular process [173, 3643, 4333, 4671], and they are valuable tools. However, neither is foolproof. For example, if we delete gene "*a*" and see no effect on bristles (a negative LOF result), then *a* is clearly dispensable for bristle formation, but we cannot conclude that *a* is irrelevant because it might be acting redundantly with gene "*b*" [2845, 4584]: "*a* OR *b* ➡ bristle." GOF data can also be misleading [6, 682, 1329]. For instance, if we drive the expression of gene "*a*" in a region where it is not normally transcribed and find that it induces bristles (a positive GOF result), then *a* is clearly sufficient for evoking bristles [1458, 1854, 2019, 3267], but this does not mean that *a* promotes bristle formation *in wild-type flies* because GOF perturbations can saturate limiting components (e.g., bHLH partners [438, 918, 1854] or external ligands [421]) or provoke interactions with other pathways (e.g., converging RTK cascades [326, 1117, 2623] or branched Frizzled chains [3912, 4365, 4867]), resulting in all sorts of artifacts [6]. Researchers beware!

It is...unsafe to deduce normal gene function [when] the product is forced into inappropriate cells, perhaps in the absence of proteins with which it normally interacts and the presence of others that it does not normally encounter. [1304]

Results derived from mutant analyses or from utilizing ectopic expression of a gene product reveal the potential of a particular interaction to occur, not whether the interaction actually occurs during wild-type development. [3248]

Artifacts can be minimized by combining LOF and GOF tests [147, 3462]. Indeed, that is the only way to distinguish factors that are "instructive" for cell fates from those that are merely "permissive." Instructive agents have both LOF and GOF effects, whereas permissive agents have a LOF but no GOF effect [449, 1455]. Even this 2-pronged approach may not be able to resolve epistatic relations, however, where (1) interactions are cooperative as in multiprotein complexes, (2) pathways are nonlinear, (3) feedback obscures causality, or (4) the "upstream" vs. "downstream" ranking of genes contradicts the order of cellular actions in time. An example of the last difficulty involves *scute* and *Notch*. In general, *scute* is epistatic to *Notch* (i.e., *scute*^{LOF} *Notch*^{LOF} flies show the *scute*^{LOF} missing-bristle trait instead of the *Notch*^{LOF} extra-bristle trait) [918, 1797, 1802, 3270, 3983], so *scute* should be acting downstream of *Notch*, but in fact *scute* must endow cells with "proneural competence" before *Notch* can enforce any "lateral inhibition" (cf. Ch. 3). The situation is even more complex at certain sites where *Notch* also acts before *scute* during a "prepattern" (pre-proneural) stage [461, 886].

Not all the fly's circuitry is as inscrutable as the *Notch-scute-Notch* cascade, but our view of every subsystem is distorted by the imperfect lens of genetic dissection [2917, 3881, 4085, 4671]. Conclusions must therefore be qualified by layers of caveats about this or that alternative interpretation. The problem with such equivocation, of course, is that it can put readers to sleep.

How much of this blather can readers tolerate? Why not just present "best guess" models and avoid all the dithering? Good advice on this issue comes from a delightful little essay entitled, "Wingless signaling: The inconvenient complexities of life." Therein, Rachel Cox and Mark Peifer argue that cartoon-like abstractions are essential but must be tempered by critiques that convey the subtleties. Around every "gospel truth" there is a Talmudic aura of uncertainty. The author's goal should be to make the material as accessible as possible without hiding any ambiguities. This book will attempt to do just that.

Nature is a home handywoman. Constrained by evolution, she does the job with the tools at hand, using a screwdriver for a hammer if necessary.... This machinery is neither elegant nor simple, but consists rather of a complex set of interacting proteins that were cobbled together by evolution.... Models help to organize our thoughts and offer testable hypotheses. Of course, in constructing a model, some data may need to be hammered into place, and the inconvenient data that cannot be coaxed into place have to be left out. The models that are frequently illustrated in minireviews...thus cannot be viewed as the "truth," or they would narrow thought processes and squelch novel lines of research. We must be thoughtful iconoclasts, remembering that ultimately all models are wrong, fundamentally flawed or lacking the full complexity of systems shaped by evolution rather than intelligent design. We will thus use this forum to critique rather than prop up our model. It is increasingly clear that life is more complicated than portrayed there. [894]

Only by venturing into the ocean of literature can novices experience the richer Fly World beyond the Internet harbors. Alas, it is all too easy to get lost in those rougher seas. For that reason, an effort is made to supply the equivalents of charts and buoys. To wit, all key mysteries that have taunted investigators are set in boldface when introduced. So are the models and metaphors that have been contrived to explain the mysteries, plus the epiphanies encountered whenever great mysteries were slain. All these concepts are inventoried in Appendix 2. Some of the coined names for the concepts are whimsical, but no more so than the silly names of many fly genes. Indeed, working in this field has been so much fun *because* of its playful irreverence – a legacy of the neophyte pioneers in Morgan's team [119]. Even "the Boss" himself loved to clash ideas [2947] and smash idols [2946]. Ideas are contrasted here wherever possible, and the style is decidedly iconoclastic.

All statements are source-referenced, and cross-references that are not direct attributions are listed as "cf. such-and-such" – a style that is common in the humanities but rare in the sciences [1630]. The cf.'s mean to compare, confer, or just "see also." Due to space limitations, some citation strings had to be truncated. Those cases are flagged with a "Δ" superscript to alert readers who want to trace earlier sources thereby. An unabridged bibliography is posted at *The Interactive Fly*.

Esoterica are banished to tables, figures, and appendices wherever possible, and supportive evidence is crammed into indented blocks of text so that readers can skip them if they want. Even so, readers may find some sections of the text unnavigable without looking up the cited papers and tracing their lines of reasoning. Subheadings are worded as sentences so that the Table of Contents reads like a summary for each chapter. Gene abbreviations are defined wherever they are used in the text. Overall, the layout is designed to avoid boring the expert without confusing the novice. I still remember how hard it was to make my way into this field as an apprehensive apprentice.

This field has seen paradigm clashes of Promethean proportions, and those wars must be recounted to do the subject justice. For that reason, the modern facts have been woven into a historical tapestry, with a few homilies stitched in for good measure. Admitting past mistakes can help in spotting future pitfalls...even in the Olympian realm of molecular genetics, which surprisingly has more than a fair share of mortal foibles [1879, 2414, 3909, 4669, 4673]. The potential pitfalls include not only (1) the aforementioned LOF and GOF artifacts, but also (2) reporter anomalies (e.g., perdurance of β-gal [3764, 4188]), (3) antibody limitations (e.g., misleading epitopes on proteins that are cleaved [155, 3271] or reshaped [1980]), (4) confocal illusions [3293, 4760], and (5) *in vitro* infidelities relative to *in vivo* conditions [655, 871]. For the next generation of researchers, some of the parables may sound quaint, but for those of us who toiled through this period, they are a chronicle worth preserving.

Readers accustomed to color photos may bemoan the book's reliance on black-and-white diagrams. I am sorry for any disappointment. The latter style just seemed more fitting for an abstract analysis. All the figures were drawn in ADOBE *Illustrator* by me (a hopeless attempt to compete with my truly artistic siblings). They evolved from cartoons into montages. When many grew too big to fit the standard 6 × 9-in. size of this series, I tried breaking them into pieces but found that the surgery was lethal. The montages had acquired a life of their own. They tell whole stories (some of which spill over into App. 7). I thank Cambridge for approving a larger trim size and for letting me set my own deadline. The cusp of the millennium seemed an apt time to step back and take a wide-angle "snapshot" of this blossoming field. The last batch of citations came from the annual *Drosophila* Research Conference (in Washington, DC) entitled, "2001: A Fly Odyssey."

This project began in 1992 when Robin Smith (then Life Sciences Editor at Cambridge) asked me to write a book for this series at the behest of Paul Green (a series editor). The topic took shape gradually, and the contract was signed in 1996. By 1997, my other professional pursuits had to be sidelined as the writing became all-consuming. I thank Peter Barlow (another series editor) for calming my fears and Ellen Carlin (Assistant Life Sciences Editor) for trusting my judgment.

Encouragement was provided by my dear parents (Maj. Gen. Lewis I. Held and Minnie Cansino Held), siblings (Lloyd, a.k.a Grey, and Linda), other relatives and sundry friends – most of whom remain skeptical that any sane adult can adore flies. Maybe this book will change their minds? Probably not!

Critical comments on portions of the manuscript were kindly furnished by Seth Blair, Tom Brody, Ian Duncan, Matt Gibson, Robert Holmgren, Teresa Orenic, Grace Panganiban, Amy Ralston, Allen Shearn, David Sutherland, and Tanya Wolff. The idea about Notch and Argos in the Skeptic-Theorist debate (Ch. 6) was Seth's. I regret any overlooked errors.

As one foot soldier in the global army of fly pushers, I have met many "generals" over the years who figure prominently in this saga. By far the greatest – and humblest – was Curt Stern. His musings on the mysteries of patterning were the siren songs that lured me to this lovely fly. Those of us who heeded his call have long dreamt of finding insights one day. Little did any of us suspect, though, that the bounty of revelations in the last decade would go so far beyond merely sating our curiosity. As we sift the treasure, the sparkle of so many answers is fostering – even in the saltiest among us – a profound sense of awe.

Lewis I. Held, Jr.
Lubbock, Texas
April 2001

CHAPTER ONE

Cell Lineage vs. Intercellular Signaling

Imaginal discs are hollow sacs of cells that make adult structures during metamorphosis. They are so named because "imago" is the old term for an adult insect [4008], and their shape is discoid (i.e., flat and round like a deflated balloon) [377]. They arise as pockets in the embryonic ectoderm and grow inside the body cavity until the larva becomes a pupa, at which point they turn inside out ("evaginate") to form the body wall and appendages [3165]. In a *D. melanogaster* larva there are 19 discs (Fig. 1.1). Nine pairs form the head and thorax, and a medial disc forms the genitalia. The abdominal epidermis comes from separate cell clusters called "histoblast nests" [2301, 2648, 3647]. Unlike discs, histoblast nests remain superficial during larval life [927] and do not grow until the pupal stage [2650].

Given the diversity of cell types in the adult skin (e.g., bristles, sensilla, photoreceptors) and the commonality of their descent from one progenitor (the fertilized egg), it is natural to ask how cells specialize to adopt divergent roles. In principle, cells can acquire instructions from ancestors or contemporaries [1654]. More specifically, a cell can inherit predispositions from its mother ("intrinsic" mode), take cues from neighbors ("extrinsic" mode), or both [477, 1614, 2019, 2451, 3741]. The predispositions could be gene states, while the cues could be diffusible ligands [1144, 3182].

To the extent that fates are assigned intrinsically, there should be a rigid correspondence between (1) parts of the anatomy and (2) branches of the lineage tree [1362, 4086, 4087]. That is, a clone of cells descended from ancestral cell "x" should make structure "X", while another clone descended from ancestor cell "y" should make structure "Y". Moreover, these rules should be obeyed in every member of the species. *C. elegans* worms adhere closely to this strategy [1284, 4201, 4202], but flies do not [1839, 1881]. In *D. melanogaster*, the only adult structures that use an intrinsic mode are tiny sense organs [532, 1410, 3441]. All larger parts of the body use extrinsic mechanisms. Thus, the problem of how discs develop can be reduced to questions about how cells communicate [695]. Who signals to whom? Over what distance? With what molecules? To what end?

Discs are not clones

Proof that cell pedigrees are irrelevant for disc patterning was first provided in a 1929 paper [4180] by Alfred Henry Sturtevant (1891–1970) – a wunderkind of the Morgan lab [257, 2504, 2615]. Sturtevant studied a strain that produced freakish flies called "gynandromorphs" [2950]. Each such fly is a patchwork of purely male and female tissues (Fig. 1.2) [1715]. They begin life with two X chromosomes but typically lose an X during the first mitosis, so that one of the two zygotic nuclei becomes 1X [1695]. Because gender in flies is dictated by the number of Xs relative to the numbers of autosomes [817], the 1X nucleus – and the half of the body that it populates – becomes male. Sexual traits are expressed autonomously at a single-cell level because flies lack circulating sex hormones. The male/female boundary can be mapped throughout the cuticle (not just in dimorphic organs) by using recessive mutations to mark one of the Xs. The *yellow*[LOF] mutation is often used because it turns the normally brown bristles (and cuticle) yellow [4101]. Such flies are useful for cell lineage analysis because any body part that develops clonally must come

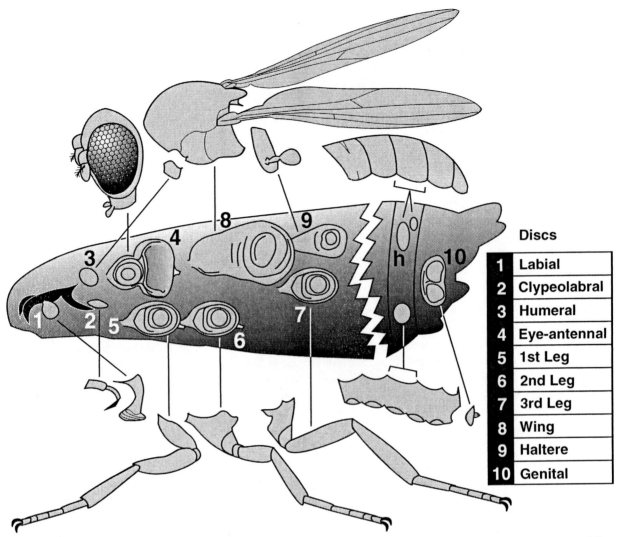

Discs	
1	Labial
2	Clypeolabral
3	Humeral
4	Eye-antennal
5	1st Leg
6	2nd Leg
7	3rd Leg
8	Wing
9	Haltere
10	Genital

FIGURE 1.1. Imaginal discs and their cuticular products. The fly exterior is assembled from separate parts (like an automobile). The epidermis of the head and thorax come from 9 bilateral pairs of discs (one of each kind is shown), and genitalia come from a medial disc, so there are 19 discs total. Abdominal wall comes from histoblast nests (**h**): tergites from dorsal nests, and sternites and pleurae from ventral nests.

Discs are drawn to the same scale, and are oriented to display their mature shapes and folding. Placements are approximate. Clypeolabral and labial discs are attached to the pharyngeal skeleton (black hooks) [3285], while other discs adhere to other larval organs (not shown) [527,834,4565]. "Humeral" is synonymous with "dorsal prothoracic" disc. Bristles are omitted for clarity, and flank sclerites are simplified.

An adult fruit fly is ~3 mm long. Full-grown larvae are roughly twice that length [3421]. About half the larval midsection is omitted here. Adapted from [1739,4565].

Discs look more alike than the structures they produce. The same is true at the cell level, where discs are nearly indistinguishable by ordinary histology [3165,4424]. Even at the molecular level, different discs make virtually identical suites of proteins [1459,1611,3625,3756,3865], although amounts vary. The reason for these common features – as later chapters show – is that all discs use the same basic "toolkit" of molecules for intercellular signaling [662], although the circuitry (i.e., how those molecules interact) is tailored to the disc-specific patterns [1440].

from one single male or female progenitor cell and hence be purely yellow or brown.

Sturtevant discovered that cuticular derivatives of all the larger discs can be bisected by a yellow/brown boundary. Hence, these discs do not develop as clones. Subsequent studies found mosaicism in the smaller discs as well [1370, 2026, 2029, 2828]. By implication, each disc must come from ≥2 cells [2411]. In fact, when discs are first

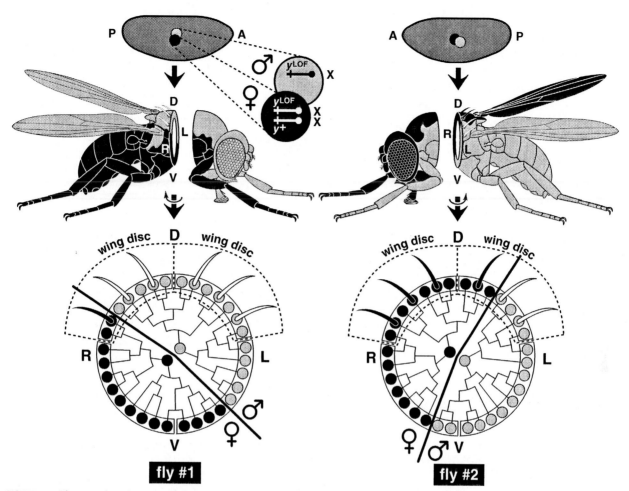

FIGURE 1.2. The nonclonal nature of fly development. The irrelevance of cell clones to pattern formation is seen in the piebald variegation of sexually mosaic "gynandromorphs" (middle panel) [1370,2026]. Such flies are typically half male (gray) and half female (black) [1715,2950]. They start life as a heterozygous female (2X) zygote but lose an X chromosome from one nucleus at the first mitosis to create a male (1X) clone (fly 1, top panel) [1695]. If the X that remains has the *yellow*^{LOF} (y^{LOF}) allele (enlarged gray circle), then the descendants of that nucleus will make yellow (instead of brown = wild-type) bristles or cuticle in the adult (fly 1, bottom panel).

The two embryos at the top of the figure (A, anterior; P, posterior) differ in the orientation of the first mitotic spindle [3274,4021]. This disparity causes the male/female boundary to trace different paths in the cuticle (middle panel) [4649,4652,4845].

The adults are bisected in the middle panel, and the cross-sections are turned ~90° to a frontal view in the bottom panel (D, dorsal; V, ventral; R, right; L, left). The outer ring of circles (nuclei) schematically represents the thoracic epidermis. The inscribed "tree" represents an imaginary series of mitoses (branch points) from the initial two nuclei to the adult epidermis. Bristle numbers and cell densities are drastically reduced for clarity.

If the wing disc (dashed outline) were a clone – i.e., derived from a single nucleus – then it should be purely yellow or brown because its progenitor nucleus must be one or the other. In actual gynandromorphs, however, the wing disc is often mosaic (R disc in fly 1 and L disc in fly 2), so it cannot be a clone. Moreover, the ability of the male/female boundaries to pass between any two landmarks (e.g., the different pairs of bristles in fly 1 vs. fly 2) argues that the patterns of cell lineage within the disc (inscribed trees) must also vary from fly to fly.

Overall, therefore, such flies reveal a fundamental uncoupling between pedigrees and patterning. This uncoupling is abstractly seen in the ability of the male/female "hour hands" to lie anywhere on the epidermal "clockface." The two flies shown here are only two examples from a large set of possibilities.

detectable histologically, each contains at least 10 cells (cf. Ch. 4). It is a quirk of history that the full import of Sturtevant's study was only realized 40 years later [119] when Antonio García-Bellido and John Merriam used Sturtevant's data to map the embryonic disc primordia [1370].

No part of a disc is a clone, except claws and tiny sense organs

Yellow/brown gynandromorphs are as eye-catching as a herd of Appaloosa horses because each individual has a unique pattern of colored patches (Fig. 1.2) [2026]. Their harlequin variegation is due to (1) the random orientation of the zygote's first mitotic division in all three dimensions from one individual to the next [3274, 4021] and (2) the tendency of sister nuclei to stay together during cleavage [4899]. The male/female line hence intersects the egg surface at random angles [4222, 4845], and the yellow/ brown boundary should bisect any given area of the adult surface if a sufficiently large population is examined – *unless that area is delimited clonally.* Among the 96 specimens that Sturtevant analyzed, many groups of cuticular landmarks were divided by such boundaries. This "indeterminate" cell lineage was epitomized by two pairs of bristles that belong to the wing disc: the dorsocentrals and postalars. From one fly to the next, Sturtevant found that

both dorsocentrals may be alike [i.e. both male or both female] but different from both postalars, or the posterior dorsocentral and posterior postalar may be alike but different from the corresponding anteriors, or any one of the four may be different from the other three. Such relations occur for any group of mesonotal bristles one examines. [4180]

Indeed, male/female boundaries meander relatively freely through every bristle array on the adult surface [1800, 2026, 3007, 3539, 4652]. Clearly, discs are not balkanized into subregions where individual cells obey commands such as *"Divide 'n' times and tell your descendants to make this part of the adult."* The only exceptions are (1) bristles and sensilla [3441] whose few component cells ($\leq$10) come from single "mother" cells and (2) claws [1356], which follow a similar developmental path [1587]. Additional instances are found in embryonic development – e.g., neural ganglia [627], muscle subtypes [250, 3684, 3698], and cardiac precursors [1339, 4194, 4547]. Wherever cell-type determination is uncoupled from cell lineage – as here in the case of large-scale patterning within discs – it must perforce rely on intercellular signaling [293, 354, 4727].

Cells belong to lineage "compartments"

Despite the rarity of rigid pedigrees in disc development, cells commonly obey looser edicts such as *"You may make any portion of region 'R', but nothing outside it"* [4671]. Regional limits of this kind were discovered in wing discs when marked cells were spurred to grow faster than background cells. Oversize anterior or posterior clones grew up to – but failed to cross – a boundary that roughly bisects the disc [1376, 1377], and analogous "compartments" were later found in halteres [1358, 1771], legs [1800, 2449, 4076], antennae [2931], genitalia [1107, 2028], and the proboscis [4144, 4145]. Compartments are essential for patterning (cf. Ch. 4 *ff*), but their lineage constraints per se are not [754, 2428, 2448, 2677, 4491]. Hence, the existence of these clans does not negate the **"Proximity vs. Pedigree Rule"** [3445] enunciated above. Put simply, this rule asserts that cells select fates based on input from peers, not parents [354, 526, 1808].

CHAPTER TWO

The Bristle

Tactile stimuli are hard for arthropods to detect through the armor of their rigid exoskeleton [1666, 3582]. To solve this problem, flies use bristles (Fig. 2.1). When a bristle is deflected, the pivoting of the shaft in its socket deforms the dendrite of a neuron attached to the shaft's base [789, 1352, 2174, 2787]. The resulting depolarization sends an action potential to the central nervous system (CNS) [1118, 2173, 2196, 4527]. Flies can pinpoint sensations because axons from different bristles get "wired" to different CNS target cells during metamorphosis, although much remains to be learned about the topology of these neurosensory maps (cf. Ch. 6).

Mechanosensory bristles are formed by 5 cells: 2 superficial cells that secrete cuticle (the shaft and socket cells) and 3 subepidermal cells that do not (the neuron, sheath, and glial cells) [2475, 3351, 3552, 3832, 4531]. These 5 cells descend from a "sensory organ precursor" (SOP). The SOP divides to produce one daughter (IIa) that yields the outer cells, and another (IIb) that yields the inner cells [1447, 1741, 1925]. The sheath cell wraps the neuron's dendrite [602, 789, 3351], while the glial cell wraps the axon [2173]. A sixth cell – the "bract cell" – is found in association with bristles on the distal leg and proximal wing [524, 1714, 1808]. It secretes a thickened hair ("bract") that is pigmented like the bristle shaft but much smaller [3362, 3421]. The bract cell is not part of the SOP clone [1808]. The way in which it is recruited from epidermal cells is discussed later.

Until 1999, the glial cell's origin was obscure [1463, 1465, 1741, 1925], and only the shaft, socket, sheath, and neuron were considered to comprise the SOP clone. In 1999, a debate about the sequence of bristle cell mitoses [2680, 3550] prompted a reinvestigation of the mitoses themselves [1447, 3549], whereupon a new mitosis was discovered. It had hitherto been overlooked because the glial cell is small and migrates away from its birthplace. Pre-1999 models are being revised to include this amendment [2382].

Chemosensory bristles have all the components of a mechanosensory bristle plus 4 extra neurons, whose dendrites project to a pore at the shaft's tip [1741, 3061, 3529, 4841] where they detect chemicals (Fig. 2.8) [3835]. Strangely, such bristles (on the legs and wings at least) are also photosensitive, with independently entrainable circadian clocks [2333, 3327, 3401]. Aside from sensory modality [3005], fly bristles also vary in size, shape, pigmentation, and pattern.

Bristles are intriguing not only because their stereotyped mitoses violate the general rule of indeterminate lineage (cf. Ch. 1), but also because they encapsulate the problem of differentiation (how do cells acquire differences?) [2424, 2577, 4658]. In theory, the instructions for assigning fates could be unequally inherited from the SOP, with no need for cross-talk among descendants. According to this "**Obey Your Mother! Model**," bristle cells adopt fates based on cues inherited from their mothers. The main cue appears to be the presence or absence of a membrane-associated protein called "Numb." Numb has all the features expected for a heritable determinant of cell fate.

Numb segregates asymmetrically and dictates bristle cell fates

The gene *numb* was isolated in a screen for mutations affecting the embryonic peripheral nervous system (PNS) [4417]. In a seminal 1994 article that provided the key to deciphering bristle differentiation, Michelle Rhyu *et al.*

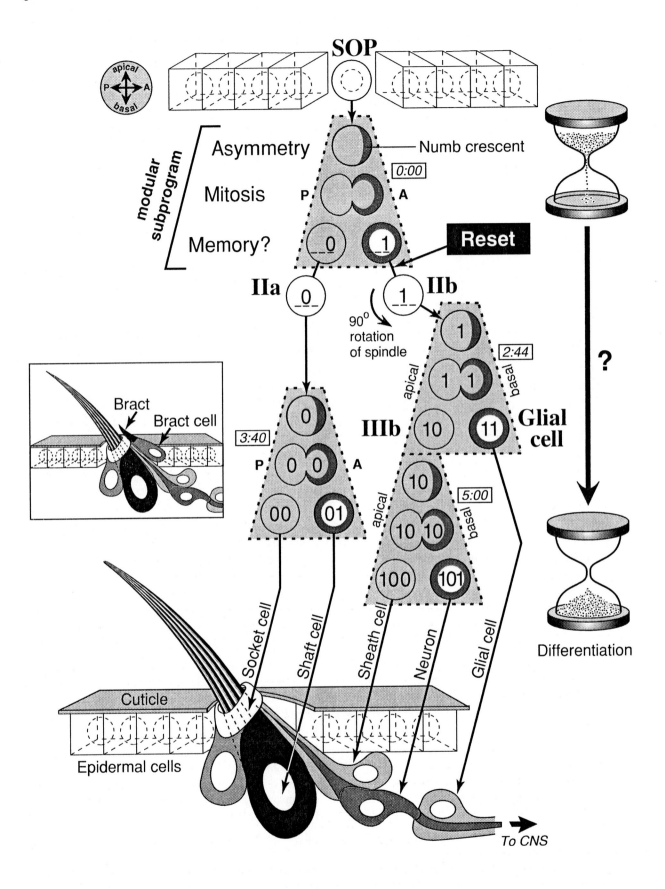

in San Francisco reported that *numb* mutations also affect adult bristles [3579] and, more important, that Numb protein is distributed unequally during SOP divisions. Indeed, this was the first gene product in flies ever shown to segregate asymmetrically in mitosis, although others soon followed [2021].

Within the SOP lineage, 4 cells inherit Numb (IIb, shaft cell, glial cell, neuron; Fig. 2.1) [1447, 3579, 4542] while 4 do not (IIa, socket cell, IIIb, sheath cell), and mutant defects are generally consistent with this parceling. Thus, *numb*^{null} mutations cause SOPs to produce 4 outer cells and no inner ones – implying that IIb adopts a IIa fate – and the outer cells are often all sockets, so a shaft-to-socket transformation must also be involved. A third conversion (neuron-to-sheath) occurs in hypomorphs [4542]. Collectively, these phenotypes imply a fate-assigning role for *numb* at every mitosis in the lineage, with the possible exception of the glia-producing IIb mitosis, which, as mentioned above, has only recently begun to be studied.

The history of a cell's Numb states can be denoted by the left-to-right order of digits in a binary code (Fig. 2.1), where "1" signifies Numb's presence and "0" its absence. Thus, the various sister cells in the SOP lineage would have the following paired codes:

IIa (0) vs. IIb (1).
Socket (00) vs. shaft (01).
IIIb (10) vs. glial cell (11).
Sheath (100) vs. neuron (101).

From the standpoint of a strict "**Coding Model**," the code would be causal. That is, a bristle cell's fate would be dictated by the series of Numb states (0 or 1) experienced by its ancestors. This code would explain the null phenotype where all cells assume a 00 (socket) state, and it would also explain the hypomorphic condition where neurons (101) switch to sheaths (100). To wit, leaky Numb levels might be high enough to let IIb attain its "1" state but not to push neurons into their later "1" state.

One test of this model would be to overexpress *numb*. Flooding the lineage with Numb protein should raise all "0" states to "1" and cause all cells to differentiate as glia (11). When *UAS-numb* is driven by a *Gal4* transgene expressed in SOPs, no clusters of 4 glial cells were reported [4542]. The most extreme defect was a 4-neuron trait where IIa likely became IIb (0 ➜ 1) and sheath cells became neurons (100 ➜ 101). Milder abnormalities were also seen, including "2 sheaths: 2 neurons" (0 ➜ 1 but not 100 ➜ 101) and duplicated shafts (00 ➜ 01 but not 0 ➜ 1). Overall, the data agree with the model, although the failure to force cells into a glial fate is problematic. Perhaps the excess Numb cannot prevent Numb's level from being reset to "0" in IIb (Fig. 2.1).

Additional support for the model comes from flies carrying a *hs-numb* construct (*numb* joined to a heat-shock promoter). When such flies are heat-shocked around the time of SOP mitoses, they display "2 sheaths: 2 neurons" as well as "2 shafts, sheath, neuron" (socket-to-shaft conversion) and "socket, shaft, 2 neurons" (sheath-to-neuron). These defects are explicable by the forced presence of Numb in the IIa (0 ➜ 1), socket (00 ➜ 01), or sheath (100 ➜ 101) cell [3579]. Four-neuron

---

FIGURE 2.1. Development of a mechanosensory bristle from a sensory organ precursor (SOP). Compass (upper left) gives initial directions (A, anterior; P, posterior). Times (hours: minutes at 23°C) are for microchaete mitoses on the notum but are similar for other bristles [1447].

The SOP arises from an ordinary epithelial cell. It starts to divide (at ~16 h after pupariation) to form IIa and IIb. IIa's daughters will make a socket and shaft. IIb's daughters are IIIb and a glial cell. The glial cell is smaller and buds off basally in the manner of a CNS neuroblast division [1073, 1740]. IIIb divides to form a sheath cell and neuron. Some bristles have a thick hair ("bract") atop their sockets (inset), which is made by a clonally unrelated cell.

Each mitosis obeys stereotyped steps (dashed trapezoid) that comprise a modular subprogram: (1) Numb localizes to one side of the cell cortex (crescent), (2) segregates to one daughter, and (3) alters cell fate. Letting 1 and 0 signify Numb's presence or absence, each cell can acquire a unique code if it "remembers" its former Numb states. Imaginary memory registers (underlined spaces) are shown for a few cells, with left-to-right order recording successively later states.

For such a binary code to work, IIb must eliminate ("reset") Numb before dividing. When SOPs are prevented from dividing, they become neurons [1743]. This result has been interpreted as a default condition, but it may instead reflect persistence of Numb: the continual presence of Numb should lead to a "nonsense" code (111) that might be interpreted as "neuron" (101). The mechanism whereby cells remember former Numb states is unknown.

Timing and branching of the pedigree are as per [1447, 3549]. Other details are based on [1449, 1741, 1808, 3579]. See [3195] for lineage comparisons with other sensory organs.

N.B.: Grooves are absent from some bristles (e.g., sex comb teeth [1714]). Epidermal cells are sometimes aligned with this degree of precision [2388], although they need not be. Chemosensory bristles have 4 additional neurons (cf. Fig. 2.8) [4125], and their SOPs obey a different lineage [3529]. See also App. 7.

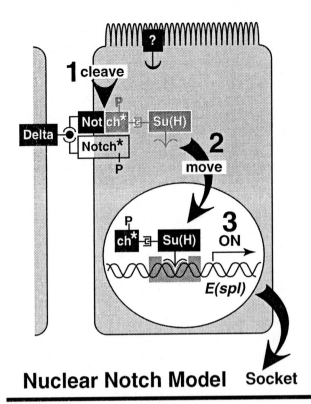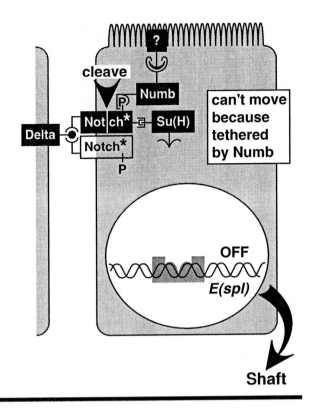

Nuclear Notch Model

Catalysis Model

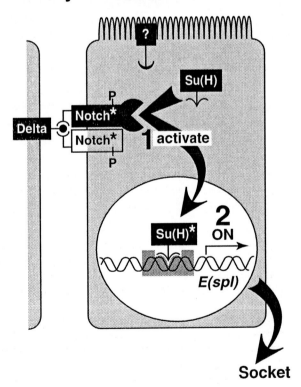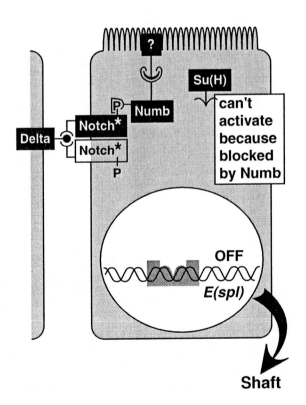

phenotypes were not observed, probably because pulses were too short to affect all three rounds of (asynchronous) mitoses.

If Numb were a traditional "cytoplasmic determinant," then it would specify only one type of tissue or cell [1904]. On the contrary, it marks 4 different cells in the SOP lineage. Moreover, it plays similar roles in sense organs of the larval PNS [3579, 4417], in neuroblasts of the embryonic CNS [2451, 3579, 4028, 4523], in cardiac cell progenitors [1339, 4194, 4547], in sibling founder cells of larval vs. adult muscles [653, 3684], and in muscle subtype determination [251, 910, 3263, 3687]. Thus, its role transcends histotype.

Evidently, Numb functions as a versatile switch that enables daughter cells to become different from one another, regardless of what those differences may be [831, 1761, 3263, 3579, 4875]. As a binary digit ("bit"), Numb is the best example ever adduced that flies can use abstract symbols for instructions just as computers employ machine language. As explained below, this **"Numb Epiphany"** of 1994 is not only helping to elucidate how genes can work as switches, but it is also revealing how an intrinsic mechanism of fate specification can dovetail with an extrinsic pathway of intercellular signaling.

Delta needs to activate Notch, but not as a signal per se

Although the Numb code should be sufficient for assigning all fates, some cell interactions have also been implicated. The 4-neuron trait that is caused by gain-of-function (GOF) *numb* manipulations is also seen with loss-of-function (LOF) mutations in *Delta* (*Dl*) and *Notch* (*N*) [1742, 3272]. Because *Dl* and *N* mediate "sibling rivalries," whereby equivalent cells become different (cf. Fig. 3.6) [2222, 3022], they could – in theory – create binary codes by refereeing a series of bouts (winner = 1; loser = 0) without relying on cell pedigrees at all [1614].

Might fates be computed by either lineage (via Numb) or signaling (via Dl and N), with one agent assuring success if the other fails? No, because such redundancy would imply that phenotypes should be wild-type unless both strategies fail, but (as stated above) fates can be altered by single LOF mutations in *numb*, *Dl*, or *N*. Rather, it seems that the two devices are connected in series, not in parallel.

Dl and N are transmembrane proteins that interact as ligand (signal) and receptor (receiver), respectively [1204, 2626]. When a N-expressing cell contacts a Dl-expressing cell, N is activated by dimerization [3022] or oligomerization [2209, 2299]. Activation causes N's intracellular domain ("N-intra") to detach from the membrane and go to the nucleus, where it stimulates transcription of target genes [4155]. Numb may block signaling by tethering N-intra to the cortex (Fig. 2.2), thus keeping it from reaching its targets. Enough Numb would normally be present to sequester all N, although an artificial excess of N could escape Numb's grasp and cause the kinds of N^{GOF} phenotypes that are seen [1307, 1651].

The need for ligand may suggest extrinsic signals, but there cannot be any instructive (ON/OFF) signaling per se [1433] because Dl is expressed at equal levels in IIa, IIb, and surrounding nonbristle cells [3270]. Evidently, Dl plays only a permissive role, essentially like a seaman sending Morse code by using a shutter (Numb) to blink a light (Dl-N) that stays ON. This **"Blinker Model"** supposes that Dl's job is merely to keep N active so that the nucleus only gets a "N = OFF" signal when Numb is present. Mosaic analyses suggest that the SOP descendants themselves supply one another with the ligands for N stimulation, with no reliance on surrounding epidermal cells [4859]. This intrabristle cross-talk has been confirmed in an interesting experiment. When Dl is overexpressed in the neuron, the adjacent shaft cell

---

FIGURE 2.2. Models for Notch signaling and its blockage by Numb. Black rectangles are proteins, and connecting "wires" are binding sites. Contact with Delta ligand on a neighbor's surface activates (asterisk) the Notch receptor, possibly by dimerization (partner outlined) [3022]. Cells that lack Numb (left) can relay the signal to its nucleus, while those that express Numb (right) cannot.

The models differ in how Numb stops the signal. In the **Nuclear Notch Model** (above the line) [1307, 1448, 1651, 2299, 4027, 4244, 4542], Numb stops Notch from leaving its roost (ghost image) by anchoring it to the membrane [2267] via an unknown linker ("?" = possibly Partner of Numb [2609]). In the **Catalysis Model** (below the line) [112, 132, 1131, 3022, 4244], Numb blocks an active site for Su(H) activation (covalent modification?).

Numb is shown binding Notch at a phosphotyrosine (P), but Numb's PTB domain is unusual and may not need a phosphate [2530, 4789], and Notch is only known to have phosphoserines [2209]. Notch resides in the apicolateral membrane [184, 1203, 1448, 2070]. The cell's apex is carpeted with microvilli. Su(H) can activate transcription (right-angle arrow) of *E(spl)* (a.k.a. "*m8*"; cf. Fig. 2.4) by binding its promoter (gray rectangle), but *E(spl)* may not dictate bristle cell fates, nor is *Su(H)* needed for signal relay in neurons or sheath cells (see text). Estimates are that a signal at the membrane takes ~20–90 min to cause detectable changes in target gene expression [184]. See also App. 7.

transforms into a socket cell (Fig. 2.7c) [2008]. Clearly, Numb's lock on the Notch pathway can be artificially overridden by excess Dl.

In contrast to the Blinker Model, the popular view has been that Numb merely biases Dl-N contests [1613, 2019, 2021, 2222, 3437], rather than being the sole deciding factor. Yuh Nung Jan and Lily Yeh Jan, who pioneered this field, advocated this "**Bias Model**" but recognized an inherent paradox: because one sister cell should win every contest (with or without a Numb handicap) the *numb*[null] phenotype should be wild-type, but it is not (and the same dilemma applies in the CNS [552]). To explain why, they invoked time constraints [2020]:

We think that…an intrinsic mechanism utilizing numb protein is superimposed on the Notch/Delta system to bias the competition….We speculate that this Notch/Delta system is not sufficiently reliable to ensure that the two cells always acquire two different fates in the allotted time. (In the case of IIa vs. IIb fates, the time window is less than 2 hr.)…This hypothesis could explain the variable phenotype resulting from complete loss of *numb* function. In *numb*[null] clones some sensory bristles show the severe phenotype of having four socket cells, whereas other sensory bristles develop normally. Our interpretation is that, in the absence of *numb*, the Notch/Delta system still operates, but is not sufficiently reliable….Some sensory bristle cells were able to finish the competition and form normal sensory bristles with four distinct fates, whereas others were unable to do so.

The Bias Model predicts that contests will end in Dl-rich/Dl-poor (winner/loser) cell pairs. On the contrary, only Dl-equivalent pairs are detected in wild-type flies [3270]. Rejecting the Bias Model in favor of the Numb-dictated "Obey Your Mother!" Model still leaves the question of why all bristle sites in *numb*[null] clones do not have a 4-socket phenotype [2019]. Perhaps the normally dormant Dl-N rivalry mechanism has been awakened in these clones, in which case they should manifest Dl-rich/Dl-poor cell pairs (a testable prediction). Alternatively, unknown asymmetries may be augmenting Numb's function (i.e., a *partial* redundancy). Either way, Numb's control over N begs the evolutionary question: how did a heritable determinant (Numb) "hijack" an intercellular signaling pathway (Dl-N)?

Amnesic cells can use sequential gating to simulate a binary code

If Numb is the bit in the bristle formula, then how do cells interpret 2- and 3-bit "words" for the various cell types? A simple ratcheting mechanism, whereby cells count how many times they have been "1," cannot suffice because in that case "01" and "10" would be synonyms. It would seem that cells must use some sort of combinatorial code where genes aside from *numb* are used for recording previous Numb states. Figure 2.1 illustrates such a Coding Model.

Do any known genes behave like a primary memory register – namely, their mutant alleles convert IIa into IIb (or vice versa) without switching any subsequent states? Among the genes whose mutant phenotypes connote a IIa-IIb switch, only *Bearded* (*Brd*) lacks later effects (App. 3) [2500]. Its GOF phenotype consists of neurons and sheath cells without shafts or sockets – indicative of a transformation of IIa into IIb. Thus, *Brd* could store the outcome of the first mitosis. (N would turn *Brd* OFF in IIa.) *Brd*[null] mutants look wild-type [2500], but this impotence is attributable to redundant paralogs [2382].

There is another way of thinking about Numb's mode of action that does not involve memory genes per se. To wit, Numb's first state (0 or 1) might simply "gate" IIa and IIb into divergent signal transduction pathways (STPs), so that the second Numb signal (0 or 1) is interpreted differently by IIa daughters (STP 1) vs. IIb daughters (STP 2). According to this "**Gating Model**," genes that act only in the IIa STP should interconvert shafts and sockets when mutated, but should have no effect on neurons, sheath, or glial cells (IIb descendants), and separate sets of STP genes would operate exclusively in the IIb and IIIb sublineages.

Indeed, *Suppressor of Hairless* appears to be a IIa-specific STP gene. Null *Su(H)* mutations suppress only part of the phenotype caused by *numb*[LOF] – namely, the shaft-to-socket switch but not the neuron-to-sheath switch – implying that *Su(H)* is only needed in the IIa lineage [4542]. This conclusion is bolstered by the ability of excess Su(H) to transform IIa (shaft-to-socket) but not IIIb daughters [200, 3827, 4542]. Su(H) is detected in both the IIa and IIIb lineages, but its level is highest in the socket cell (as are *Su(H)* transcripts [3826]) – a IIa daughter [1448]. Su(H) moves from the socket cell's cytoplasm to its nucleus when N is activated [1448] – precisely the behavior expected for a messenger molecule [1269, 1307]. Su(H) can bind both to N (signal acquisition?) [1269] and to DNA sites (signal delivery?) upstream of genes in the *Enhancer of split* Complex [1131, 2453], which may control bristle cell fates (but see below). Thus, Su(H) has not only the phenotypic properties of a IIa STP agent, but also the histological hallmarks.

When homozygous *Su(H)* [null] clones are induced near the time of SOP mitoses (3rd instar or pupal period) they transform sockets into shafts [3827, 4542], but when they are induced earlier (1st or 2nd instar) most sockets and shafts are missing and replaced by neurons and possibly sheath cells [3820]. This switching of IIa to IIb implies that Su(H) mediates not only the socket/shaft decision but also the IIa/IIb decision. If so, then the signal relays in STP 1 would be as follows, where States 0 and 1 indicate Numb's absence or presence, "➡" and "⊣" denote activation or inhibition, and "default" refers to the state of a cell in the absence of extrinsic signals:

1. Decision to become IIa vs. IIb (first mitosis):
   State 0 (N = ON): Dl ➡ N ➡ Su(H) ➡ {genes in Set 1} ➡ **IIa.**
   State 1 (N = OFF): Numb ⊣ N. {Genes in Default Set 2} ➡ **IIb.**

2. Decision to become socket vs. shaft cell (second mitosis):
   State 0 (N = ON): Dl ➡ N ➡ Su(H) ➡ {genes in Set 3} ➡ **socket.**
   State 1 (N = OFF): Numb ⊣ N. {Genes in Default Set 4} ➡ **shaft.**

Although it is easy to imagine how N could toggle cells from one set of genes to another, it is unclear why cells that do not receive a signal should switch their sets of default genes (Set 2 to Set 4) from one generation to the next. Conceivably, the latter is caused by a separate mitosis-counting mechanism.

Counting of mitoses could be accomplished by a digital ratchet or an analog quantifier [1880, 2762]. An example of an analog device exists in the fly CNS, where descendants of the NB4-2 neuroblast apparently count mitoses of ganglion mother cells via the quantity of two homeodomain proteins (Pdm-1 and Pdm-2) whose concentration slowly falls [313, 314, 3713, 4804, 4816]. Alternatively, SOP progeny might get their timing cues from the cascade of genes that respond to ecdysone [247, 2111, 2134, 3059, 3814]. Indeed, SOP mitoses malfunction at specific macrochaete sites when ecdysone levels are depressed [3990], possibly due to this hormone's management of disc cell cycles [1185, 1599].

It should be possible to distinguish intrinsic vs. extrinsic clocks by letting SOPs differentiate *in vitro*, as has been done for neuroblasts [485]. With a "schedule" strategy (i.e., events A, B, and C are triggered independently at clock times 1, 2, and 3), later steps should occur when early ones are blocked; whereas with a "dominoes" strategy (A licenses B which licenses C), later steps should not proceed in the absence of earlier ones [1805, 2943, 3010]. In either case, the phenotype expected for LOF mutations in "counting genes" would be completion of SOP mitoses but no differentiation because descendants will wrongly think that they are still in the first generation [916].

One gear in the escapement mechanism could be the p69 Tramtrack protein, which helps specify bristle cell fates (see below). Overexpression of Tramtrack during part of the cell cycle can halt bristle differentiation [3502], but it also blocks mitosis. Prospero has similar effects in the CNS [2522]. Intriguingly, LOF mutations in *bazooka* cause extra cell divisions in the SOP lineage [3629]. Bazooka is needed for neuroblast asymmetry and for adherens junctions [2985, 3803, 4709].

Other possible cogs in the clockwork include (1) Dacapo, a cyclin-dependent kinase inhibitor [6, 1002, 1004, 2404]; (2) Twins, the regulatory subunit of a mitosis affecting phosphatase (see below) [6, 1535, 2761]; and (3) String, a Cdc25 phosphatase that is rate-limiting for mitosis [6, 2740, 3081]. Interestingly, *string* has separate *cis*-enhancers for epidermal vs. SOP mitoses [2480]. In nematodes, some Notch-dependent fates are indeed gated by cell-cycle phases [79], and a few recently recovered fly genes that are suspected of being SOP gating agents also interact with Notch [6].

Assuming that a gene in Set 2 diverts IIb into a new transduction pathway, a second messenger other than Su(H) would need to mediate the sheath/neuron decision. Su(H)-independent Notch pathways operate elsewhere [2353, 2453, 2540, 2549, 2747, 2839, 3694, 3695], so this idea is plausible [444, 4542, 4581].

In fact, a IIb-specific transducer called "Sanpodo" (Spdo) has been identified for the larval PNS, and it plays a similar role in the CNS [3985]: (1) *spdo* acts downstream of *numb*, (2) *spdo* [null] alleles transform sheaths into neurons but have minimal effects elsewhere in the SOP lineage, and (3) *spdo* may act likewise in adult bristle development [6, 1128, 3728]. Spdo is a homolog of tropomodulin – a protein associated with microfilaments. Two conjectures have been offered as to how Spdo might aid signal transduction [1128]: (1) Spdo could be part of a multiprotein complex (including N) at the cortex, or (2) Spdo could act in a cytoskeletal process that transports ligand-activated (truncated) N to the nucleus. (Both ideas find support in Canoe, which colocalizes with N [1203, 4236] and has kinesin- and myosin-like domains that could mediate transport [2882, 3417, 3418].) Thus, STP 2 may use Spdo in the IIIb portion of the SOP lineage:

<u>Decision to become sheath cell vs. neuron:</u>
State 0 (N = ON): Dl ➔ N ➔ Spdo? ➔ {genes in Set 5} ➔ **sheath.**
State 1 (N = OFF): Numb ⊣ N. {Genes in Default Set 6} ➔ **neuron.**

If the Gating Model is correct, then Numb would be acting less like a croupier (dealing cards that are stored by sentient cells and interpreted collectively after a full hand is dealt) and more like a gondolier (steering forgetful cells to destinations by choosing a path at each intersection).

Notch must go to the nucleus to function

The issue of whether Notch acts directly in the nucleus was debated for years [1891, 2222, 3153, 4244, 4823] and has recently been settled in favor of a nuclear role [443, 446, 711, 1613, 2514], although some skeptics remain unconvinced [113, 4809] and recent data support their contention that Su(H) can act without N [218, 2255].

The history of the debate is worth recounting because (1) its grittiness shows the power of scientific dialectic and (2) its reliance on imperfect techniques shows the fallibility of scientific deduction. However, readers may safely skip this section without losing the thread of the main story.

The controversy began in 1993, when N-intra (N's intracellular domain) was found to be able to enter the nucleus and activate the Notch pathway [1271, 2542, 4161]. It was unclear whether these correlated events were causally interrelated.

Based on these findings, a "**Nuclear Notch Model**" was formulated (Fig. 2.2). It postulated that N-intra detaches (by proteolysis) from the extracellular part of N upon binding of ligand and goes to the nucleus where it relays the signal. Why else, the model's advocates asked, would N have nuclear localization sequences (NLSs) [132, 4161]? Skeptics stressed that N is undetectable in nuclei of wild-type flies [112, 368], but defenders countered that this invisibility could be due to rapid degradation triggered by N's PEST motif [2299, 2542]. Recent work indeed confirms that (1) N-intra is needed in only small amounts to carry out its nuclear function [3805] and (2) N-intra accumulates when the proteasome is disabled [3821].

Because N has no obvious DNA-binding motif of its own, the presumption has been that N-intra functions as a "co-activator" that binds Su(H) or another transcription factor to activate downstream genes [1448, 2542].

Although now considered proven (see below), the co-activator idea is still hard to reconcile with the genetics of *Su(H)* and *deltex*:

1. The ability of excess Su(H) to transform shafts into sockets [3827] is puzzling if Su(H) needs N-intra, because shaft cells should lack nuclear N-intra (due to Numb). Conceivably, excess Su(H) compensates for the dearth of N-intra by behaving as an activator all by itself. This notion is consistent with the fact that N and Su(H) elevate transcription synergistically [750, 1131], but it is contradicted by evidence that Su(H) recruits co-repressors when N-intra is absent [449, 1330, 2129, 2941]. Perhaps, a sufficiently high titer of Su(H) swamps the pool of co-repressors, leaving "naked" Su(H) as the predominant species, and the solo Su(H) is then free to turn ON certain target genes. Other target genes, however, are turned OFF, so its effects must be context dependent [1329, 1330].

2. The gene *deltex* (*dx*) was named for the resemblance of its LOF phenotype to that of *Delta* – namely, wing veins that widen into deltas at their tips [2561, 4778]. It is one of only four genes recovered in an extensive screen for modifiers of N^{GOF} lethality [4778, 4780]. When Deltex binds N's ankyrin repeats, it apparently displaces Su(H), which goes to the nucleus [1448, 2746]. Such displacement implies that excess Deltex should (1) sequester N in N-Deltex heterodimers, (2) prevent N-intra from joining Su(H) as a co-activator, and (3) result in a N^{LOF} phenotype [132]. Instead, overexpression of *deltex* yields a N^{GOF} phenotype [2746]. Strangely (like *Brd*), *deltex*^{LOF} mutations rarely affect bristle cell fates, so either *deltex* plays no role in the SOP clone or its function is redundantly shared by another gene [1570, 4778].

N's ability to bind Su(H) [1269, 2747, 4244] suggested that inactive N might keep Su(H) out of the nucleus by restraining it at the membrane [1269, 1448, 4244]. Indeed, Su(H) colocalizes at the membrane with inactive N in cultured cells [1269, 1307], and movement of Su(H) to the nucleus depends on N activation both in cultured cells [1269, 1307] and *in vivo* in socket cells [1448]. However, the restraint idea made no sense genetically. If it were true, then deleting N should liberate Su(H) to go to the nucleus (just like activating N) [711, 1131, 1613, 2299, 2746]. In fact, N^{null} mutations disable the pathway (App. 3).

Until 1997, the Nuclear Notch Model (sans the restraint amendment) fended off criticism fairly well. Active N constructs were shown to go to the nucleus, bind Su(H), and increase transcription of Su(H)-responsive genes [750, 1914, 2043, 2611]. Still, it was unclear how the facts fit into a causal chain *in situ* [4580].

In 1997, several findings were reported that seemed to contradict the model, although each challenge had a technical loophole:

1. Activated N constructs that are anchored to the membrane (in fly or mammal cells) are as effective at stimulating transcription of Su(H)-dependent reporter genes as constructs that demonstrably go to the nucleus [132, 1131]. (However, proteolytic conversion of anchored N to free N-intra could not be ruled out, and minute amounts of nuclear N-intra might be enough for reporter activation.)

2. Disabling both NLSs of hNotch1 (the human Notch homolog) – by replacing 2 a.a. and deleting 49 a.a., respectively – reduces nuclear localization but does not significantly decrease transactivation of Su(H)-responsive genes [132]. (However, nuclear localization was not abolished, and residual N-intra in the nucleus may suffice for reporter activation.)

3. Eliminating hNotch1's only stable binding site for the Su(H) homolog (in its RAM23 domain [2747, 4244]) does not significantly decrease its ability to transactivate Su(H)-dependent genes [132]. (However, Su(H) can also bind hNotch1's ankyrin repeats, so the reduction in binding may not be enough to stop co-activation.)

Doubts raised by these results made an alternative "**Catalysis Model**" look more attractive (Fig. 2.2) [1613]. In that model, transient contact with N supposedly changes Su(H) to an activated state, and "Su(H)*" then enhances transcription of downstream genes with no further need for N [112, 1448, 1651, 2299, 4244]. Because N itself has no known enzymatic activity [1613], N might use a binding partner as the catalyst [1131]. Conceivably, N alters Su(H)'s shape to unveil a domain without covalent modification [132, 1131]. This "hit-and-run" scenario agreed with (1) the instability of N-Su(H) binding and (2) the fact that N and Su(H) do not always colocalize [4244].

In 1998, the tide turned for the last time. Incisive studies provided direct evidence for the Nuclear Notch Model. They showed ligand-dependent proteolysis and transit of N-intra to the nucleus, as well as transcriptional stimulation by amounts of nuclear N below histological detection [443, 711, 2514, 4582]:

1. Upon binding of ligand, mNotch1 (the mouse Notch homolog) is cleaved on the cytoplasmic side of its transmembrane domain – between residues that in

the fly correspond to Met-1762 and Val-1763 [3805]. Mutating the Val to Leu or Lys virtually blocks cleavage, but even the residual amount of cleaved product is able to activate transcription of a reporter gene. The quantity of nuclear N needed for such activation is so small that it is barely detectable immunologically, even when sensitivity is enhanced with multiple Myc tags on N-intra. This result explains why nuclear N had hitherto been "invisible" *in vivo*, and it validates the idea that trace amounts of cell-surface receptors can directly influence nuclear events [2030, 2031].

2. Inserting a Gal4 DNA-binding domain into the intracellular tail of full-length N produces a receptor that can activate *UAS*-regulated reporter genes in a ligand-dependent manner *in situ* [2454, 4155], whereas similar insertions into the extracellular domain do not. Hence, cleavage and nuclear transit of N-intra are natural consequences of ligand binding.

3. Inserting domains for transcriptional activation (Gal4 or VP16) or repression (WRPW or an Ala-rich piece of Engrailed) into N's tail, respectively, rescues or fails to rescue N^{null} embryos – implying that N-intra directly (in the nucleus) regulates targets of the Notch pathway *in situ* [4155].

4. A cleaved, Su(H)-bound form of N that is soluble and phosphorylated was identified in fly embryos in addition to full-length N [2211], and the ratio of processed to full-length N was shown to depend on (1) the presence of N's binding domain for Dl and (2) the amount of Dl both below and above the wild-type level.

5. The cytoplasmic domain of N was found to have an intrinsic ability (85% that of Gal4) to activate transcription at a heterologous promoter in yeast and *in vivo* [2211]. To confirm that N is a co-activator for Su(H), a foreign activation domain was substituted for N. When VP16 was fused to Su(H), the Su(H)-VP16 product activated transcription of *m8* (a natural target of Notch signaling) in N^{null} embryos [2211].

The inescapable conclusion is that a piece of the N molecule stimulates transcription directly in the nucleus after ligand-dependent cleavage of full-length N. The implied stoichiometry of "1 N molecule: 1 unit of activation" is consistent with the dose sensitivity of the Notch pathway [118, 1797]. This situation contrasts with other transduction pathways, where the quantity of second messenger is catalytically amplified relative to the input signal [3297].

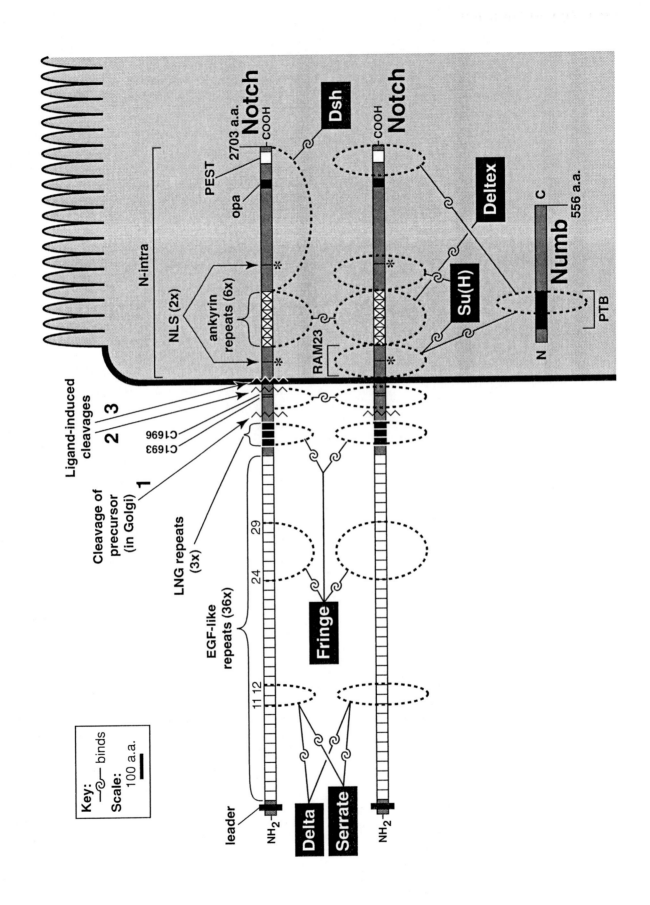

14

How does Numb fit into this scheme? An obvious possibility (stated above) is that Numb stops Notch signaling by tethering N-intra to the cortex, and Numb does bind N-intra *in vitro* [1307, 1651]. If this were the whole story, then freeing Numb from the cortex should let N-intra function in the nucleus (hence shifting State 1 cells to State 0), but as the following experiments demonstrate, this prediction is not fulfilled.

1. Deleting Numb's "mooring cleat" (at its N terminus) releases it into the cytoplasm but does not affect its ability to switch cell fates (by inhibiting N) when overexpressed [2269].
2. The mouse protein "Numblike" resembles Numb except that it is cytoplasmic [4876], and expressing a *Numblike* mouse transgene in *numb*^null flies shuts off the Notch pathway, despite Numblike's freedom from the cortex.
3. Numb can suppress N in cultured cells even when both overexpressed proteins are cytoplasmic [1307]. Indeed, Numb appears to block N function by sequestering N-intra in cytoplasmic heterodimers, thus keeping it out of the nucleus.

While Numb's normal function could still be to tether N, it must be able to stifle N by other means. N's Numb-binding domain contains a nuclear localiza-

tion sequence (Fig. 2.3), so Numb could block nuclear entry of N-intra by masking this site. Alternatively, because N's Numb- and Su(H)-binding sites overlap [1307, 1651, 2747, 4244], Numb-N heterodimers may be unable to dock with Su(H) even if some of them enter the nucleus. (Numb does not bind Su(H) directly [4542].) A third possibility is that Numb obstructs access of the ligand-dependent protease [4523]. One clue to this mystery is that microtubules must be intact for Numb to muzzle N [4523].

In summary, the physical basis for the "Numb ⊣ N" and "Dl → N → Su(H)" interactions appears to be as follows. Numb masks one or more of N's functional domains. In Numb's absence, binding of Dl to N releases N-intra from the membrane by proteolytic cleavage, allowing N-intra to enter the nucleus, where it functions as a co-activator with Su(H) or other transcription factors to modulate the expression of downstream genes.

E(spl)-C genes are Su(H) targets but play no role in the SOP lineage

In vitro, Su(H) protein binds the nucleotide heptamer $\underline{GTG}^G/_A\underline{GAA}$ (where "$^G/_A$" means a G or A at this position) [492, 1131, 2453, 4408]. This heptamer resides in the promoters of at least 4 of the 13 genes in the *Enhancer of split* Complex (E(spl)-C) at 96F11–14 – namely, *mγ*, *m4*, *m5*,

---

FIGURE 2.3. Notch, its domains, and some of its binding partners.

Part of a cell is shown (gray = cytoplasm; microvilli at top) with two Notch (middle) and one Numb molecule (below) drawn as bars (cf. scale at upper left). Domains (variously shaded or hatched) are mapped (cf. App. 1). Other proteins (black rectangles) are not to scale. Partners are linked by hooks (cf. key), and binding sites are delimited by dashed ovals or half ovals.

Despite their regularity, the various EGF-like repeats play different roles [459, 4601, 4823], and no two are identical [2182, 2542]. Delta and Serrate both bind EGF-like repeats 11 and 12 [459, 985, 3544], although they also rely on repeats 24–26 (not shown) [2419], and binding alone may not suffice for activation [1943, 1944]. Fringe (cf. Ch. 6) binds repeats 24–29 [990] and at the LNG domain [2096]. Scabrous requires repeats 19–26 for its association with N (not shown) [3456], although any binding must be indirect [4601]. The famous allele *split* (whence the link to the E(spl)-C was deduced [3028]) is due to a missense mutation (Ile-to-Thr) in repeat 14 [1747, 2182], and the widely used heat-sensitive allele *N*^ts1 has a missense mutation (Gly-to-Asp) in repeat 32 (a.a. 1272) [4779].

Notch is cleaved at three sites (zigzag lines) [491, 2995, 4809]. Cleavage at site 1 occurs during maturation in the *trans*-Golgi along the secretory pathway [368, 2299], and the fragments stay together, possibly by disulphide bridges at C1693 and C1696 [3153] (but see [446, 578]). Cleavage at site 2 occurs after ligand binding [491, 2995], releasing an ectodomain that is "swallowed" by the ligand-bearing cell [2263, 3271] (cf. the reverse with Boss [601, 2318, 4211]). The smaller size of the extracellular vestige (now <300 a.a.) stimulates a Presenilin-dependent protease to make the next (final?) cut [3533, 4156, 4582]. Cleavage at site 3 occurs in the transmembrane domain (a.a. 1746–1765) between M1762 and V1763 [2995, 3805], releasing an intracellular fragment (N-intra) that goes to the nucleus and turns ON target genes as a co-activator with Su(H) [4155] (see text for evidence). For review, see [2994].

Numb, when present, apparently tethers N-intra to the membrane (cf. Fig. 2.2). Half of Numb's PTB domain binds Notch at RAM23, where Su(H) also binds [1307, 1651, 2747, 4244], and at a less crucial C-terminal site [1139, 1307, 1651]. Deltex may displace Su(H) from Notch when Delta is absent [112, 1448, 2746, 3022], although binding sites for Deltex and Su(H) do not overlap [2747, 4244]. (Deltex switches cell fates when overexpressed but has no LOF effect on bristles [2746].) Notch is thought to dimerize via cysteines (C1693 and C1696?) [2209, 2542] but may also do so via its ankyrin repeats [2747] or opa motif (links not shown) [3353].

Binding of Dishevelled (Dsh) to the C-terminal tail [151] short circuits the Notch and Wingless signaling pathways [356, 3690]. Notch's other binding partners include Wingless itself [4601] (see App. 3 for other modulators). Sites of phosphorylation [368, 2209, 2211, 4582] and glycosylation [2070, 4611] are not shown. Fringe is presumed to glycosylate Notch at O-linked fucose sites in EGF-like repeats 3, 20, 24, 26, and 31 [2904]. Adapted from [368, 446, 2210, 2542, 4611]. See [4602, 4603] for apparently heretical modes of Notch signaling. See also App. 7.

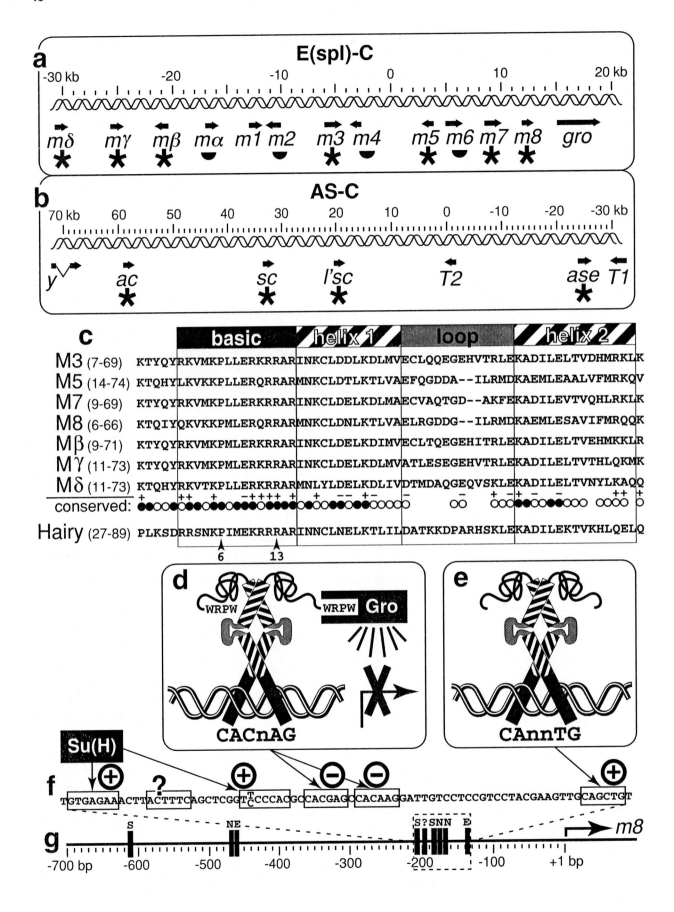

a E(spl)-C

-30 kb   -20   -10   0   10   20 kb

mδ mγ mβ mα m1 m2 m3 m4 m5 m6 m7 m8 gro

b AS-C

70 kb   60   50   40   30   20   10   0   -10   -20   -30 kb

y   ac   sc   l'sc   T2   ase   T1

c

	basic	helix 1	loop	helix 2
M3 (7-69)	KTYQYRKVMKPLLERKRRAR	INKCLDDLKDLMV	ECLQQEGEHVTRLE	KADILELTVDHMRKLK
M5 (14-74)	KTQHYLKVKKPLLERQRRAR	MNKCLDTLKTLVA	EFQGDDA--ILRMD	KAEMLEAALVFMRKQV
M7 (9-69)	KTYQYRKVMKPLLERKRRAR	INKCLDELKDLMA	ECVAQTGD--AKFE	KADILEVTVQHLRKLK
M8 (6-66)	KTQIYQKVKKPMLERQRRAR	MNKCLDNLKTLVA	ELRGDDG--ILRMD	KAEMLESAVIFMRQQK
Mβ (9-71)	KTYQYRKVMKPMLERKRRAR	INKCLDELKDIMV	ECLTQEGEHITRLE	KADILELTVEHMKKLR
Mγ (11-73)	KTYQYRKVMKPMLERKRRAR	INKCLDELKDLMV	ATLESEGEHVTRLE	KADILELTVTHLQKMK
Mδ (11-73)	KTQHYRKVTKPLLERKRRAR	MNLYLDELKDLIV	DTMDAQGEQVSKLE	KADILELTVNYLKAQQ

conserved:

Hairy (27-89) PLKSDRRSNKPIMEKRRAR INNCLNELKTLIL DATKKDPARHSKLE KADILEKTVKHLQELQ

6   13

d

WRPW   WRPW Gro

CACnAG

e

CAnnTG

f

Su(H)

⊕   ?   ⊕   ⊖   ⊖   ⊕

TGTGAGAA ACTTACTTTCAGCTCGGT[T/C]CCCAC CACGAG CACAAG GATTGTCCTCCGTCCTACGAAGTTGCAGCTGT

g

S   NE   S?SNN E   m8

-700 bp   -600   -500   -400   -300   -200   -100   +1 bp

and *m8* (Fig. 2.4) [171, 2453]. *In vivo*, reporter genes attached to these promoters are activated by Su(H), and the activation requires the heptamers [171, 2453]. The implication is that E(spl)-C genes are targets of Su(H) during Notch signaling (i.e., Dl ➡ N ➡ Su(H) ➡ "Transcribe E(spl)-C genes"). Indeed, these same promoters are responsive to activated N *in vivo* [171]. Some connection between *E(spl)* (a.k.a. *m8*) and *N* had long been presumed because a mutant allele of *E(spl)* enhances the phenotype of *split* – an allele of *N* (hence the gene's name) [3028, 3896].

Seven genes in the E(spl)-C contain a "basic helix-loop-helix" (bHLH) motif (Fig. 2.4) [2246, 2274]. For bHLH proteins in general, the "basic" domain mediates DNA binding and the helices mediate dimerization [1152, 2634]. Some bHLH proteins form homodimers, others form heterodimers with preferred partners, and still others can do both [68, 245, 2039, 3028, 3029, 3375].

A separate cluster of four bHLH genes – the *achaete-scute* Complex (AS-C) at 1B1–7 – also governs bristle development (cf. Ch. 3), and certain proteins from the two complexes can heterodimerize [1484, 3171]. Whereas AS-C bHLH proteins have canonical bHLH features [348, 1152,

3375], E(spl)-C bHLH proteins are unorthodox in several respects (Fig. 2.4) [135, 3171, 4318]:

1. Their dimers typically bind an "N box" consensus nucleotide sequence CACnAG (where "n" denotes a variable nucleotide), instead of the CAnnTG "E box" that characterizes other bHLH proteins. Nevertheless, E(spl)-C dimers can compete with AS-C dimers at E boxes [2054], and flanking bases are also relevant.

2. Their basic domain contains a conserved proline, although substituting a different amino acid does not alter their affinity for N boxes [3171].

3. Their carboxy terminus invariably ends with the sequence WRPW (tryptophan, arginine, proline, tryptophan).

These features define a distinct bHLH subfamily that also includes the genes *hairy* and *deadpan* [331, 1017, 2274, 3697, 3759]. The WRPW tetrapeptide binds the ubiquitous protein Groucho [2064, 3278] – a "co-repressor" (i.e., an inhibitory factor that "piggybacks" on DNA-binding proteins). Groucho reduces transcription by recruiting a histone deacetylase [737]. Interestingly, the *groucho* gene resides in the E(spl)-C. By bringing Groucho to N boxes,

FIGURE 2.4. Clusters of bHLH genes and the roles of their proteins as transcriptional regulators.

a, b. *Enhancer of split* (E(spl)-C; **a**) and *achaete-scute* (AS-C; **b**) complexes. Genes are depicted as arrows, indicating the direction of transcription. E(spl)-C spans ~50 kb on the 3rd chromosome; AS-C spans ~100 kb near the tip of the X. Asterisks mark genes whose products have a bHLH motif. Oddly, all these genes are devoid of introns. Within each complex, the bHLH genes exhibit partial functional redundancy [2548, 4674]. AS-C bHLH proteins are transcriptional activators, whereas E(spl)-C bHLH proteins are repressors. Half circles mark E(spl)-C genes that belong to a separate ("Bearded") gene family [2383].

c. Amino acids in bHLH motifs of E(spl)-C proteins and Hairy (AS-C proteins not shown; for code, see App. 1). Dashes are inserted to aid alignment [2274]. Numbers are residues counted from the N-terminus. E(spl)-C bHLH proteins are ~200 a.a. long, with ~60 a.a. in their bHLH domain. The "conserved" row pertains to E(spl)-C proteins: filled circles (invariance); unfilled circles (>50% but <100% identity); +s or −s (charged residues). The many +s define the "basic" domain, which, in this subfamily, has proline at position 6 and arginine at position 13 [3179].

d, e. bHLH proteins dimerize via helical domains (striped), adopt a scissors shape, and bind DNA via basic domains (black). Within each subunit, the upper and lower helices touch (unlike in this cartoon). **d.** E(spl)-C bHLH dimers bind an "N box" consensus sequence "CACnAG" ("n" is a dimer-specific nucleotide) [3171, 3759], which Hairy can also bind, although it prefers CACGCG [3179, 4451]. Their C-terminal "tails" (top) end in "WRPW" (as does Hairy's [4524]) [2274, 3697], which recruits Groucho (Gro, black rectangle). Gro is a co-repressor (X'd arrow) [2064, 3278] that in turn binds a chromatin-modifying histone deacetylase [737], although E(spl)-C proteins also have Gro-independent effects [3029]. The *gro* gene resides in the E(spl)-C (**a**). **e.** Most other bHLH dimers (including AS-C) bind an "E box" hexamer "CAnnTG" [348, 3375], do not end in WRPW, and activate transcription (cf. the bHLH-PAS subgroup [4022, 4612]).

f, g. Promoter region of the *m8* (a.k.a. *E(spl)*) gene. **f.** Details. Binding sites (S = Su(H); E = E box; N = N box) are boxed (**f**) or shown as bars (**g**). "?" denotes a hexamer (function unknown) found between invertedly repeated "S" sites [171, 3075], whose interrepeat distance is constant among promoters. +s and −s signify stimulation vs. inhibition of transcription. **g.** "Wide-field" view of the *m8 cis*-enhancer region. Dashed rectangle marks the section shown in **f** – viz., bases −133 to −211 b.p. from the transcription start site ("+1 b.p."; right-angle arrow). Negative feedback of *m8* onto its own N boxes [871] may help to stabilize output and minimize noise [262, 1383, 3347]. The crowding of binding sites suggests steric competition or "quenching" [1984] that mediates "either/or" (vs. "both/and") logic [2350]. Additional N boxes and "S" sites (of varying affinity) map in the 700 b.p. span (**g**) [2317, 3075], although we do not know whether they are all needed [449]. The base at −185 (**f**) may be T [171] or C [2246]. *Cis*-regulatory regions for all E(spl)-C genes (except *gro*) have been analyzed similarly [871, 3075].

Maps of loci were compiled from Fig. 3 of [3806] amended as per [2383,4767] (**a**) and from [636, 1538] (**b**). Sequences in **c** are from [2274]. Panels **d** and **e** are based on [1152, 2634], and panels **f** and **g** are adapted from [171, 2246, 2317, 3171]. See [2054] for an exegesis of E vs. N boxes and an exploration of target gene preferences for AS-C vs. E(spl)-C proteins. See also App. 7.

E(spl)-C bHLH proteins should repress the transcription of any genes that have N boxes in their promoters. One such gene is *m8*, which hence should repress itself, although the proximity of its N boxes to Su(H)-binding sites (Fig. 2.4g) suggests that autorepression is prevented by competitive binding during Notch signaling [171]. If so, it seems odd that SOPs express a *lacZ* reporter linked to an *m8* promoter (ectopically) when its Su(H)-binding sites are deleted [2453], but the promoter could still respond to AS-C bHLH proteins that bind its E boxes [171, 3171].

Phrasing the logic at *m8* promoter as an imperative, the rule would be roughly as follows: "Turn ON IF you are occupied by Su(H) OR AS-C proteins, but NOT IF there is too much of your own gene product around, in which case you should remain OFF." Another way of thinking about the output of *m8* is as an equation whose inputs – Su(H), AS-C, and M8 – have weighting factor coefficients. This glance at *m8* serves to show (in microcosm) the intricacy of the control system.

We do not yet know which (if any) E(spl)-C genes are expressed in the SOP progeny [991]. Genetic analysis has been difficult [2561, 4885] due mainly to the functional redundancy of the bHLH genes [991, 1017, 2548, 3806]. Indeed, deletion of *m8* (whose neomorphic *E(spl)^D* allele had long implied a key role for this gene [4318]) has no detectable phenotypic effect [1018, 3806].

Overexpression of wild-type E(spl)-C bHLH genes can cause various changes in cell fates [2317, 4256], but these abnormalities may be misleading because some of them persist when conserved domains (bHLH and WRPW) are deleted [68, 1475]. The biologically relevant question is: Do E(spl)-C bHLH proteins *normally* assign fates in the bristle organ? Given the following data, the answer seems to be "No":

1. Deletion of all 7 E(spl)-C bHLH genes in somatic cell clones increases the density of bristles (for reasons discussed in Ch. 3), but has virtually no impact on bristle cell identities [991, 3027].
2. Deletion of various E(spl)-C bHLH genes rescues the bristle-loss phenotype of null mutations in *Hairless* (an antagonist of the Notch pathway, see below), but does not alter the shaft-to-socket transformation also caused by these mutations [197].
3. Replacing the WRPW (Groucho-binding) site of M7 with a transcriptional activator produces a chimeric M7 protein that should bind N boxes (since it still has its bHLH domain) but *activate* (not inhibit) its gene targets (due to the C-terminal substitution). Expres-

sion of the chimeric transgene (via *Gal4-UAS* control) increases bristle density, but fails to switch cell fates in the bristles themselves [2063], implying that *m7*'s downstream targets are irrelevant to the SOP lineage.
4. Overexpressing the bHLH genes *mδ*, *mγ*, or *m7* suppresses SOP initiation at many sites, but the remaining bristles have a normal set of constituent cell types [3027]. When either of the non-bHLH genes *mα* or *m4* is overexpressed, it increases bristle density (by inhibiting bHLH E(spl)-C genes) but fails to alter the SOP lineage [92].

Only one bHLH gene remains a viable candidate for the setting of cell fates within the bristle organ during normal development – the AS-C gene *asense*. Asense is found in SOPs, and its expression persists into IIa and IIb, although no *asense* mRNA has been detected beyond this point [438, 1079, 2039]. Given that all SOPs make Asense, it is surprising that only one group of bristles is affected when the gene is deleted: a row of stout bristles along the wing margin [2039]. In *asense*^{null} flies, this row manifests empty sockets, missing bristles, stunted shafts, and twinned shafts. Twinning could indicate a socket-to-shaft or IIb-to-IIa conversion, although neither is certain because the SOP lineage here (in wild-type flies) can violate the rule of fixed fates (e.g., socket and shaft cells do not need to be sisters) [1741]. When the *asense* deletion also removes certain AS-C enhancers, a sheath-to-neuron switch is seen in the giant sensillum of the radius [1079]. The importance of *asense* inferred from this LOF data is not upheld by the GOF data. Overexpression of Asense (via a heat-shock promoter) induces extra bristles, but it fails to alter any cell fates within the bristles themselves [438, 1079].

The transcription factor Tramtrack implements some cell identities

The gene *tramtrack* (*ttk*) was recovered in screens for proteins that bind enhancers at the *ftz* pair-rule locus [502, 1731]. Its transcripts are spliced to produce two protein isoforms – "p69" (69 kD) and "p88" (88 kD) [3537]. Both isoforms typically function as transcriptional repressors [503, 1483, 4596, 4773], but p69, unlike p88, becomes an activator in late eye development [2391]. Both proteins have an N-terminal "BTB" (protein interaction) domain and two DNA-binding zinc fingers, but the fingers of p69 and p88 differ in amino acid sequence (and hence target DNA sequence) because they come from different exons (Fig. 2.5) [3537].

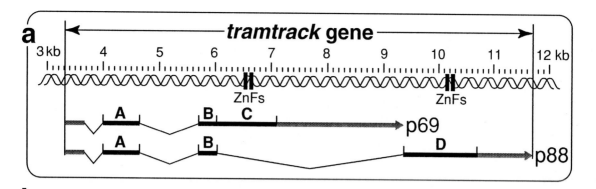

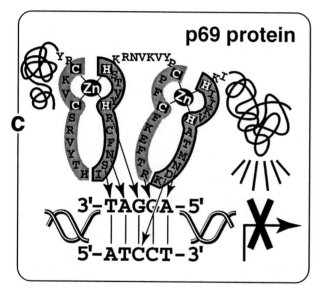

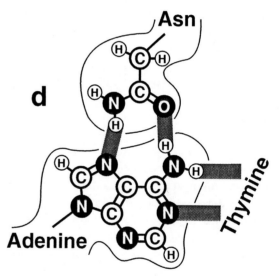

FIGURE 2.5. Alternative splicing of *tramtrack* yields different DNA- binding proteins.

a. Transcripts (fragmented arrows) of the ~9-kb gene *tramtrack* (*ttk*, 100D) encode two proteins, each of which has different zinc fingers ("ZnFs") [3537, 4773]. Kinked lines are spliced-out pieces, and lightly shaded bars are untranslated sequences. The 69-kD "p69" isoform (641 a.a.) has sections A, B, and C. The 88-kD "p88" isoform (811 a.a.) has sections A, B, and D. Shared segments (A, B) encode a "BTB" protein-interaction domain [211, 1516, 4891], and unshared ones (C, D) encode zinc fingers.

b. Amino acid sequences of the fingers. White letters are Cys and His residues that coordinate Zn^{2+} ions, dashes are gaps to aid alignment, and black circles mark identical residues.

c. Zinc fingers (shaded) of p69 are drawn flat, although each outlined half is actually an α-helix, and the other half forms β-strands. p69 binds DNA as a monomer [1186], inserting the α-helices of its fingers consecutively into DNA's major groove. Arrows indicate a.a.-base contacts. The DNA oligomer used for structural analysis was devised from p69's enigmatic set of binding sites [1186]. p88's fingers (not shown) bind other sequences [3537, 4775] and have not been studied crystallographically. Transcriptional repression (lines radiating to X'd arrow) might be due to a C-terminal domain (BTB is N-terminal). In some unknown way, p69 (but not p88) is converted from a repressor to an activator in eye development, where the two isoforms act differently in different cell types [2391, 2529]. Both isoforms are regulated by proteolysis [1051, 2529, 4249].

d. Hydrogen bonds (shaded) between asparagine and adenine (cf. N-to-A arrows in **c**) vs. bonds within the A-T base pair (at right) revealed by crystallography [1187]. Such interlocks have fueled hopes for a "finger code" that could allow fingers to be designed for desired DNA sequences [296, 2265].

This diagram is adapted from [4773] (**a**), [3537] (**b**), [1187, 2123] (**c**), and [296] (**d**).

This use of alternative splicing to send transcription factors to different gene destinations is a clever sort of grammar in cell programming [345, 2599, 3996]. The imperative "Turn OFF!" (or in some cases "Turn ON!") can thus act on various genes via different "Go to" addresses (the fingers of p69 vs. p88) [4339]. This "**Finger Shuffling Trick**" is used by at least two other fly genes: (1) the "puffing cascade" gene *Broad-Complex*, whose BTB domain is spliced to one of four pairs of zinc fingers [247, 248, 1044, 2974] in a spatially regulated manner [458], and (2) a (non-BTB) chorion transcription factor [1519, 1918]. Alternative splicing is also used for other types of "Go to" commands, which (1) add or delete NLSs that send proteins to the nucleus vs. cytoplasm [957, 2032], (2) insert a peptide that sends proteins to intercellular junctions [4574], and (3) may even direct the wiring of sensory axons to specific CNS neurons [4763].

Deleting *ttk* function in somatic clones causes the same 4-neuron trait seen in N^{LOF} flies, but overexpressing *ttk* (via a heat-shock promoter) only rarely yields the 4-socket trait of N^{GOF} [1650]. Instead, the major GOF phenotype is "2 shafts, 2 sockets" [1650, 3502] – indicative of a conversion of IIb into IIa. (Strangely, the GOF phenotypes for p69 and p88 are similar despite their different binding-site preferences.) Hence, *ttk* may implement the IIa/IIb decision. Its null phenotype suggests that *ttk* also mediates the sheath/neuron decision because two switches must occur for an SOP to make 4 neurons (IIa-to-IIb and sheath-to-neuron). Consistent with this reasoning, p69 is detectable in IIa (not IIb) and in sheath cells (not neurons) [3502]. The rarity of 4-socket or 4-shaft defects implies that *ttk* plays no role in the shaft/socket decision – a conclusion bolstered by Ttk's presence in both shaft and socket cells [3502].

Thus, as a rule, *ttk* is expressed in non-neural cells (or their ancestors) in the SOP lineage [3502]. Similar rules govern *ttk* function in the CNS and eye, so *ttk* may be a generic repressor of neural fates [1483, 4773]. Ttk may be kept OFF in IIb by proteolysis [3550] because (1) in the eye Ttk is targeted for degradation when it binds Seven in absentia (Sina) and Phyllopod (Phyl) [4249], and (2) LOF mutations in *sina* and *phyl* cause doubled bristles [671, 717] that suggest a IIb-to-IIa conversion [3550].

If *ttk* toggles cell fates in the IIIb sublineage (sheath vs. neuron) in the same way that *Su(H)* behaves in the IIa sublineage (socket vs. shaft), then is Ttk the counterpart of Su(H)? Probably not, because Ttk has not been shown to bind N (whereby it could form a factor-cofactor com-

plex). Other candidate transducers for the IIIb sublineage have been recovered [6].

Hairless titrates Su(H)

Given the involvement of both Su(H) and Ttk in the IIa/IIb decision, but their separate roles in the IIa and IIb sublineages, there may be at least three transduction modes. *Hairless* (*H*) [200, 2659] also fits into this scheme. Until 2000, its effects were thought to be confined to the socket/shaft decision, but a detailed LOF-GOF analysis proved its participation in all three decisions [3027]. Removing H forces cells into State 0, while overexpression forces them into State 1 (App. 3).

STP 1 : Decision to become IIa vs. IIb:
  State 0 (N = ON): Dl ➡ N ➡ Su(H) ➡ {*ttk*, etc., but not E(spl)-C} ➡ **IIa.**
  State 1 (N = OFF): Numb ⊣ N. H and {genes in Default Set 2} ➡ **IIb.**

STP 2 : Decision to become socket vs. shaft cell:
  State 0 (N = ON): Dl ➡ N ➡ Su(H) ➡ {genes? [not *ttk* or E(spl)-C]} ➡ **socket.**
  State 1 (N = OFF): Numb ⊣ N. H and {genes in Default Set 4} ➡ **shaft.**

STP 3 : Decision to become sheath cell vs. neuron:
  State 0 (N = ON): Dl ➡ N ➡ Spdo? ➡ {*ttk*, etc., but not E(spl)-C} ➡ **sheath.**
  State 1 (N = OFF): Numb ⊣ N. H and {genes in Default Set 6} ➡ **neuron.**

Moderate H^{LOF} alleles delete shafts (hence its name), either by blocking SOP initiation ("bristle loss") or by converting shaft cells into socket cells ("double-socket" phenotype) [198]. Macrochaetes (large bristles) are affected more than microchaetes (small bristles), and certain sites are consistently affected more than others (Fig. 2.6a) [997, 2690].

Why should similar bristles differ in their need for the same gene product? This "**Nonequivalence Riddle**" has nagged theorists not only for H^{LOF} [3053, 3054] but also for other "missing bristle" mutants [3405] – especially *scute*^{LOF} whose subpatterns are allele specific [650, 767, 1108, 1328, 4100] – and for "extra bristle" mutants [733, 3028]. Topological models (wherein particular genes control particular bristles in 1:1 correspondence) were discounted because (1) phenotypes are so easily modifiable by temperature or overcrowding [767–769, 1997, 1998, 3405] and (2) double-mutant combinations can remove bristles outside the areas affected by either mutant alone

[3405, 4187]. For the AS-C, the spatial heterogeneity is due to each bristle's reliance on the sum of (functionally redundant) Achaete and Scute, whose amounts are set by position-specific enhancers (cf. Ch. 3). *Hairless* may likewise have a partially redundant "partner" in some body regions because leg bristles remain virtually normal even in H^{null} mutants [198]. Although the reason for H's spatial nonuniformity is unclear, the basis for its quantitative effects is understood, thanks to a peculiar feature of this locus – namely, its "haplo-insufficiency."

Only 21 genes in the fly genome (aside from *Minutes*) manifest haplo-insufficiency (i.e., display a mutant phenotype when present in haploid dose as a deficiency/+) [118]. Presumably, the products of such genes are "limiting" for the rate of particular reactions [117] or for the stoichiometry of particular partnerships [2582, 3458]. In the case of *Hairless*, the haplo-insufficiency is attributable to a physical coupling between the H and Su(H) proteins.

As the name indicates, *Suppressor of Hairless*^{LOF} mutations suppress H^{LOF} phenotypes. The suppression can be mimicked by lowering the dose of *Su(H)*. Thus, heterozygous $H^{LOF}/+$ flies that are also heterozygous for a *Su(H)* deficiency look nearly wild-type [117]. Conversely, increasing the dose of *Su(H)* proportionally enhances the phenotype of $H^{LOF}/+$ flies, as expected for a stoichiometric interaction (Fig. 2.6c). This titratable antagonism is illustrated most dramatically by the creation of a H^{LOF} phenotype in a wild-type (H^+/H^+) background by simply raising the dose of *Su(H)* to 8 (from its normal 2) [3827]. The identity of the Su(H) "substitute" that H opposes in STP 3 (sheath cell vs. neuron choice) remains to be determined [3027]. H is detectable in all SOP descendant cells [3027], but subtle differences in its levels may still exist in State 0 vs. State 1 cells. Alternatively, H may be modified posttranscriptionally by auxiliary factors in one or the other cell state.

Hairless is a weird protein. It is extremely basic (overall pI = 9.5; pI in parts = 11, based on a.a. composition; Fig. 2.6d), although it has no known DNA-binding motif. It is 40% alanine, serine, or proline [200, 2657]. Its function in flies seems to be to stifle Notch signaling during bristle development (SOP lineage and SOP initiation) [3027, 3824], but less so during the other processes where Notch is deployed [113, 444]. Su(H) and H bind each other in yeast two-hybrid assays [171, 1332, 2657], and their interfaces have been identified [492, 2695]. Because the H- and DNA-binding domains within Su(H) overlap, H might block Su(H) by masking its DNA-binding site [492]. How-

ever, truncated H transgenes that contain the Su(H)-binding domain do not fully rescue H phenotypes *in vivo* [2657], so H must do more than just dock with Su(H) [1329, 2658]. Indeed, it has recently been shown that the Su(H)-H complex binds DNA and that H recruits the co-repressor "dCtBP" (see Fig. 3.12e and [2129] sequel).

Why should any antagonist for Su(H) be needed – titratable or otherwise? Shouldn't shaft cells have a silent Notch pathway after Numb has gagged N? Perhaps, but the pathway may not be silent enough. Residual N-intra or Su(H) might linger from the mitosis that created IIa. H's role could be to "mop up" excess Su(H) to ensure that prospective shaft cells do not hear any N signal whatsoever and hence become shafts with 100% fidelity. H could set a constitutive threshold above which Su(H) must rise in order for a cell to become a socket [197, 3826]. A similar damping step has been proposed for gating N-Su(H) heterodimer entry into the nucleus [2211]. In principle, thresholds are an easy way to convert messy (analog) signals into discrete (digital) states [3573, 4513] – as Charles Plunkett, a pioneer of bristle research, argued in 1926:

The only way in which I can conceive of an "all-or-nothing" reaction, such as the production of a bristle, being determined by the concentration of a continuously varying substance, is that the reaction occurs if, and only if, that concentration equals or exceeds a certain "threshold" value. [3405]

Thresholds also help to explain "fluctuating asymmetry" [2693, 2901, 4460], where the left and right sides of single individuals show different phenotypes. Such asymmetries are rare in wild-type flies [2866, 4746, 4747] but common in mutants such as H^{LOF} [1961, 3405, 3571, 3572, 4514, 4674]. For H^{LOF}, the reason may be that the amount of H sinks to the level of Su(H) in some prospective shaft cells. In the close contests that ensue, the outcome will depend on random local fluctuations (a.k.a. "noise" [1383, 2259, 2518, 4513]) in both values (cf. error bars in Fig. 2.6e). Thus, a prospective shaft cell on the fly's left side might sense more H than Su(H) and make a shaft (resulting in a normal bristle), while its counterpart on the right senses more Su(H) than H and makes a socket (yielding a "double-socket" phenotype). Because the same is true for H-deficiency heterozygotes, such asymmetries cannot be dismissed as being due to "leakiness" of alleles [767, 1998, 3067]. Rather, they are symptomatic of a breakdown in the "robust buffering" of the fate-assignment mechanism itself [1440, 3845, 4513].

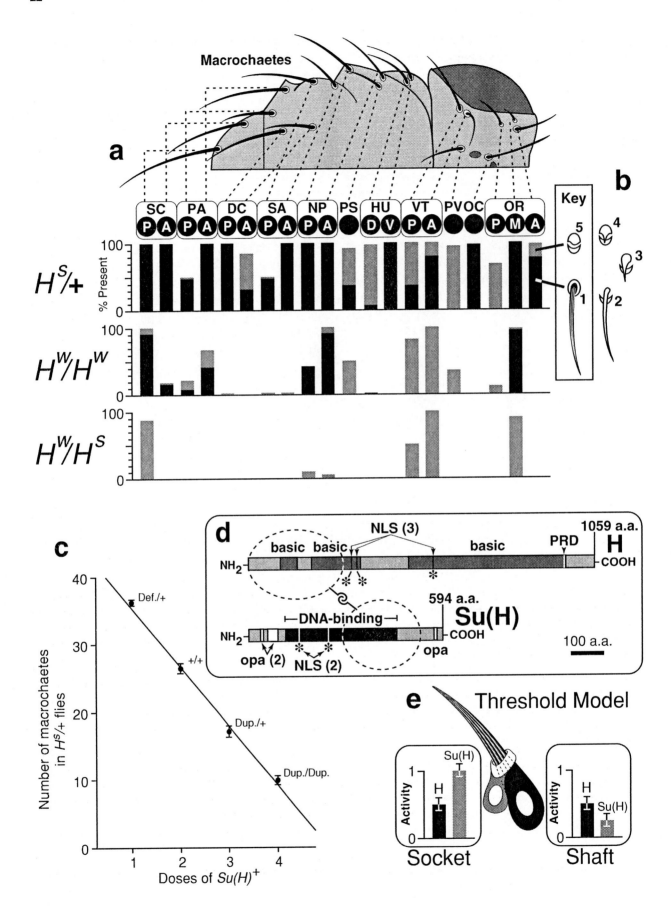

Interestingly, shaft cells in H hypomorphs can display any morphology on the spectrum from shaft to socket (Fig. 2.6b) [198]. Likewise, overexpressing H during shaft morphogenesis (via a *hs*-promoter) can drastically deform the shaft:

Unexpectedly, heat shocks after bristle determination affected morphogenesis of shaft and socket cells of macro- as well as microchaetae. At an early stage, very frequently shaft trifurcations were observed at the expense of the socket, the central shaft always being the longest. Later shocks caused a thickened shaft base with one to several spines. . . . If shocks were applied even later, shafts were completely absent, but instead, misshapen sockets were observed. This bald phenotype is inducible over a very long period until rather late stages of [the pupal period]. [2657]

Conversely, overexpressing N [3270] or Su(H) [3827] during that period partially transforms shafts into sockets, making them shorter and thicker. Intermediate shapes such as these are probably due to close contests where cells cannot select fates decisively. The intergradations also suggest that the "obvious" dissimilarity of shafts and sockets is deceptively superficial. They may actually be outputs of the same dendrite-wrapping algorithm [1741, 2173]. If so, then shafts and sockets would actually differ *quantitatively* (e.g., in their extent of elongation). Little is known about the molecular machines that cells use to modulate their shapes, and even less is known about the genetic circuitry that runs them (see below).

Several other genes help determine the 5 cell fates

The aim of this chapter is to chart as straight a path as possible from Numb, which ordains cell fates from its perch at the cell membrane, to the genes that must implement those fates. Six players have been profiled (Numb, Delta, Notch, Su(H), Hairless, Tramtrack), a contender has been considered (Asense), and one putative set of implementing genes has been ruled out (the E(spl)-C). Below, this same route is retraced from surface to nucleus, with five more actors added, whose roles are less central: Serrate, Nak, Twins, Musashi, and Shibire.

Until 1998, Dl was thought to be the only ligand for N in bristle development. This notion stemmed from the similarity of phenotypes in heat-sensitive Dl^{LOF} and N^{LOF} mutants. Both kinds of mutants exhibit the same "balding" (superficial bristle loss) when exposed to pulses of high temperature around the time of SOP mitoses, and in both cases the 4-neuron clusters at many denuded sites indicate switching of IIa to IIb and sheath to neuron [1742, 3272]. However, when null alleles are made homozygous in somatic clones, Dl and N do not give the same phenotype: N^{null} clones are bald, whereas Dl^{null} clones make tufts of relatively normal bristles

---

FIGURE 2.6. Effects of *Hairless* mutations on bristle cell fates (**a, b**), and interactions between *Hairless* and *Suppressor of Hairless* (**c–e**).

a. Half head and thorax of a wild-type fly (above), with all bristles omitted except macrochaetes. Beneath, macrochaetes (black circles) are seriated in posterior-to-anterior order, except that some are grouped [2560] as scutellars (SC), postalars (PA), dorsocentrals (DC), supra-alars (SA), notopleurals (NP), humerals (HU), verticals (VT), or orbitals (OR). Relative positions of intragroup members are posterior (P), anterior (A), dorsal (D), ventral (V), or middle (M). Presutural (PS), postvertical (PV), and ocellar bristles (OC; oval is an ocellus) are unpaired. The histograms show how each site is affected by *Hairless* genotypes. "H^s" and "H^w" are strong and weak LOF alleles of *Hairless*, respectively ("+" = wild-type). Black bars are frequencies (percent) of "normal" bristles (shaft and socket); shaded bars are frequencies of "double sockets," which likely arise from a shaft-to-socket cell transformation (key at right). H function is evidently more limiting in shaft cells than in incipient SOPs because mild reductions in H levels (uppermost histogram) cause more double sockets than absent bristles [198]. The variation in sensitivity among sites makes no sense in terms of their seriation.

b. Intermediate phenotypes between a normal bristle (type 1) and a "double socket" (type 5). In type 2, the shaft has abnormal fluting and pigmentation, and the socket fails to form a complete circle around the base. In types 3 and 4, the shaft is a vestige that ultimately (in type 5) resembles a stunted socket.

c. Number of "normal" (shaft and socket both present) macrochaetes on the head and thorax of $H^{LOF}/+$ flies (H^1 allele) carrying varying doses of the wild-type allele of *Su(H)*. "Def." and "Dup." are a deficiency and duplication for the *Su(H)* locus.

d. Domains in the H and Su(H) proteins (cf. App. 1). Dashed ovals connect regions needed for H-Su(H) binding [492]. "Basic" here means a segment of the protein where the average pI is 11 (*sic!*) [2657]. The DNA-binding domain of Su(H) is probably bipartite (i.e., same limits but centrally inert), given the properties of its mouse homolog [788].

e. Threshold Model of Bang *et al.* [197]. In this model, H and Su(H) titrate one another, and whichever protein remains undimerized determines cell fate. That is, excess Su(H) causes a cell to become a socket, and excess H causes it to become a shaft. *N.B.:* In other contexts, Hairless may act independently of Su(H) [725].

Data in **a** are from Table 2 of [198], **b** is a schematic of that article's Fig. 5, **c** is replotted from [117], **d** incorporates data culled from [492, 1331, 2657, 3826], and **e** is based on [197].

[4859]. (*N.B.:* In the original study [1797], *Dl*^{9P39} patches made tufts, whereas *Dl*^{RevF10} patches – like *N*^{null} – were bald, but a later study found no such allelic difference [4859], and both alleles may be nulls [3272].) In the absence of either N or Dl, all SOPs should have made neurons instead of bristles.

How can normal bristles develop without Dl? The answer is a redundant ligand. *Serrate* (*Ser*) was known to encode a N ligand but was not suspected of acting in the SOP lineage because none of its mutant phenotypes (including null ones [4029]) involve bristle defects [2561]. However, Ser and Dl are related proteins [1252, 4302], and Ser (when artificially expressed) can substitute for Dl in embryonic neurogenesis [1634], implying an overlap in functional ability. Whereas *Ser*^{null} clones look wild-type and *Dl*^{null} clones make tufts, doubly mutant *Ser*^{null} *Dl*^{null} clones are bald like *N*^{null} and have extra neurons at denuded sites [4859]. Evidently, Dl and Ser both activate N in the SOP lineage, although Dl's contribution is greater (and hence less dispensable) than Ser's. (Why t.s. *Dl*^{LOF} defects should be stronger than *Dl*^{null} clonal defects remains unclear.) Consistent with this conclusion, shaft cells can be transformed into socket cells by overexpressing either Dl or Ser in neurons [2008] (Fig. 2.7c).

A yeast two-hybrid screen [763, 1228] for Numb-binding proteins netted *numb-associated kinase* (*nak*), which encodes a putative serine/threonine kinase [765]. Nak interacts with Numb's PTB domain, which is interesting from several standpoints:

1. Numb can localize properly without this domain [1307, 2269], so Nak probably does not escort or anchor Numb to the cortex [765].
2. Like Nak, N binds Numb's PTB domain [1307], so Nak might displace N from Numb, in which case it should antagonize Numb. Indeed, it does: overexpression of Nak alters cell fates in the opposite way from *numb*^{GOF} [765].
3. The PTB domain is a docking motif that binds phosphotyrosines [765, 2530], so Nak might hinder Numb indirectly by phosphorylating substrates that directly displace N.

Nak's true role will not be known until LOF alleles can be isolated, but the Numb-N mechanism must involve phosphorylation somehow because LOF mutations in *twins* can switch IIb to IIa just like *numb*^{LOF} mutations [3904]. Twins is the regulatory subunit of serine/threonine phosphatase PP2A in *D. melanogaster* [4418]. Other *twins*^{LOF} alleles interfere with mitosis [1535, 2761]. Such

effects are consistent with PP2A's role in other species [830, 3352, 3904], and they suggest that Twins might be responsible for the cell-cycle dependence of Numb's cortical localization [2267].

LOF mutations in *musashi* also produce *numb*^{LOF}-like phenotypes [3035]. This gene encodes an RNA-binding protein. Because Musashi protein is nuclear, it probably does not localize mRNA (like Staufen [478, 1321, 2526]) or regulate translation. It might control RNA processing (splicing? [193, 1100, 2186]) in a positive manner for *numb*, *nak*, *twins*, or *Hairless*, or in a negative manner for *Dl*, *N*, *Su(H)*, or *tramtrack*. Indeed, Musashi has recently been found to bind and muzzle *tramtrack* transcripts ([3035] sequel).

Ever since Numb's cardinal role in dictating bristle cell fates was revealed in 1994 [3579], many pieces of the genetic puzzle have fallen into place. One remaining mystery [2268] is: What upstream cell-polarity cues cause Numb's cortical asymmetry?

During asymmetric mitoses, the Numb crescent always overlies one pole of the spindle [2267, 4028]$^\Delta$. This coordination makes sense because otherwise the division plane could bisect the crescent and dispense equal amounts to both daughters [3580, 4875]. Embryonic neuroblasts (NBs) and ganglion mother cells also parcel Numb unequally. In their case, spindle orientation and Numb localization are both controlled by Inscuteable (Insc) and Miranda [4323]$^\Delta$. Miranda is an adaptor protein that interacts with Insc and Numb [1322, 3893]$^\Delta$. Numb's N-terminal 227 a.a. suffice for proper crescent formation in NBs [2269] and may be hooked to Insc via Miranda [3893]. However, *insc*^{LOF} mutants do not show defects in bristles [3550] or larval sense organs [2327] (although IIb mitoses are misoriented [3630]), so Insc seems dispensable [2607]. Miranda is present in the IIb cell [3630] (despite an earlier report to the contrary [2680]), but its function there (if any) is unknown. Other links in the NB chain of command may also be irrelevant [2017], such as Bazooka (upstream of Insc [3803, 4709]) and "Partner of Numb" (downstream [2606]), which binds Numb and Miranda [2609]. This conclusion seems odd given the many similarities between SOPs and NBs [587, 1805]$^\Delta$. For example, (1) IIb and IIIb divide along the apical-basal axis [1447, 3630] like NBs, which localize Insc apically [2328, 2607, 2609]), and (2) the homeodomain protein Prospero segregates basally in IIb and IIIb [1447], as in the embryonic CNS [477]$^\Delta$, and helps decide cell fate in both systems [2680]$^\Delta$.

The current working hypothesis for Numb localization (in both SOPs and NBs) is that Numb is recruited to the cortex by the WD-repeat protein Lethal giant larvae and then driven to its final destination by a polarized

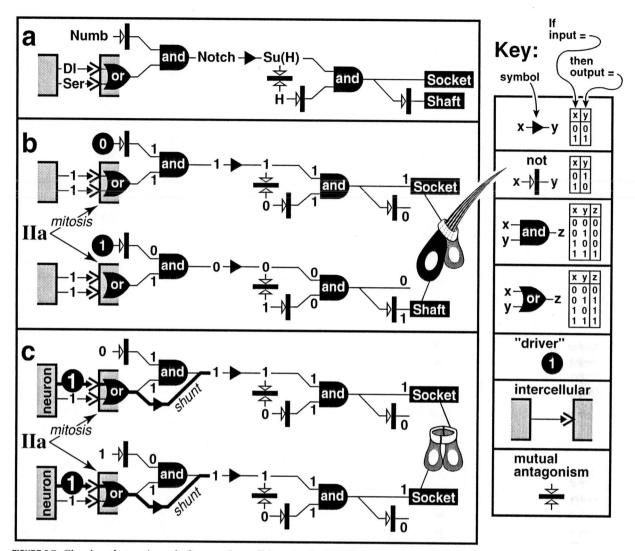

FIGURE 2.7. Circuitry that assigns shaft vs. socket cell fates in the SOP lineage. Logic symbols (key at right; cf. [2284, 4735]) denote activation, inhibition, combinatorial criteria ("AND" gate), and redundancy ("OR" gate). The "NOT" symbol is a blend of genetic and electronic icons [1751, 2770], and the reciprocal "NOT" condition ("mutual antagonism," "flip-flop," or "seesaw") means that if one partner is ON ("1"), then the other must be OFF ("0"). These relations are summarized in the truth tables, where "1" means "true" and "0" means "false" [399, 1433]. The "driver" is the element that chiefly dictates the outcome, and "intercellular" connotes a ligand (arrowhead) and receptor ("V").

a. Core pathway of proteins that decide shaft vs. socket cell identities (cf. Fig. 2.2 and text for details). Abbreviations: Dl (Delta), H (Hairless), Ser (Serrate), Su(H) (Suppressor of Hairless). In plain English, this circuit means: "Activate the Notch receptor if it binds either Dl or Ser on an adjacent cell (but go no farther if Numb is present!), then let it bind Su(H) and (as long as there's not too much Hairless around!) let the Notch-Su(H) complex turn ON genes for socket identity; otherwise, turn ON genes for shaft identity."

b. The circuit as it functions in daughters of the IIa cell (Fig. 2.1) in wild-type flies. Numb is normally the decisive factor in setting cell fate (see text). The daughter that lacks Numb (above) has an active Notch pathway and hence forms a socket, while the daughter that inherits Numb (below) has an inactive Notch pathway and hence forms a shaft.

c. The circuit as it malfunctions when neurons are forced (via neuron-specific *Gal4-UAS* driver "*31-1*") to make excess Delta [2008]. In this situation, the prospective shaft cell makes a socket instead, presumably because hyperstimulation of Notch receptors supersaturates the limited amount of Numb. The ability of such excesses to "short circuit" the system (thick line) implies that care must be exercised in interpreting GOF phenotypes (in general) whenever proteins "flood" a network.

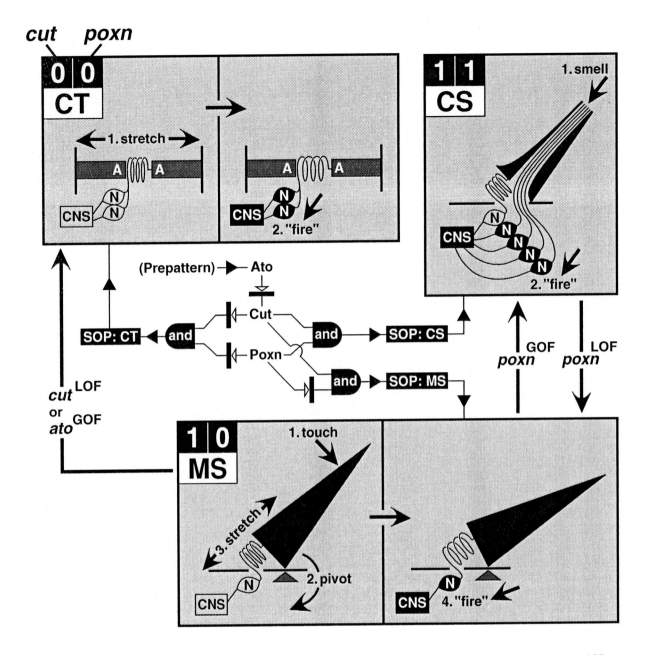

actomyosin motor [468,3205,3326], where it is anchored by Gαi (inhibitory subunit of GTP-binding protein) [3776].

Near Numb's C terminus is a tripeptide – NPF (asparagine-proline-phenylalanine) – that binds "EH" (Eps15 Homology) domains [3725, 4417]. Because NPF or EH motifs are found in proteins that control clathrin-mediated endocytosis [2699, 2756], Numb's inhibition of N might entail effects on endocytosis. Indeed, Numb can influence the endocytic internalization of membrane receptors [3757].

Evidence that Notch signaling requires endocytosis comes from *shibire* (*shi*). Heat-sensitive shi^{LOF} mu-

tations cause the same anatomical defects as N^{LOF} mutations and share the same sensitive periods [1057, 3423, 3425, 3863, 3891]. Among the defects are socket-to-shaft transformations [1803, 3425]. The protein encoded by *shi* is the fly homolog of the mammalian GTPase "dynamin" [740, 4443]. In neurons, dynamin self-assembles around the necks of clathrin-coated pits, and a conformational change in these "collars" appears to pinch off vesicles from the membrane, although the mechanism may involve pushing instead of pinching [2795]$^\Delta$. Detachment of vesicles halts in t.s. shi^{LOF} flies at high temperature [1016]$^\Delta$, which explains why *shi* was recovered in a screen

FIGURE 2.8. Circuitry (cf. Fig. 2.7 for key) controlling the identity of mechanosensory (MS) vs. chemosensory (CS) bristles vs. chordotonal (CT) organs. Those organs are shown with their neurons (N) but without their support cells. They are created from single-cell SOPs (black rectangles) via differentiative mitoses (not shown; cf. Fig. 2.1). In the upper left of each panel are transcriptional states (1 = ON; 0 = OFF) of *cut* and *poxn* depicted as a binary code.

The circuit (middle area) is simple. "Prepattern" genes (cf. Ch. 3) activate *atonal* (*ato*) at certain spots in the skin, and Ato suppresses Cut. If the SOP also lacks Paired box-neuro (Poxn), then it makes a CT organ. If Ato is absent, but other proneural proteins (not shown) are present (cf. Ch. 3), then Cut will be expressed and the cell will make a bristle. The bristle type will be CS if *poxn* is ON and MS if *poxn* is OFF. The state of *poxn* (like *ato* and *cut*) must be set by region-specific genes upstream (not shown). Thus, LOF or GOF mutations in *poxn* interconvert MS and CS bristles, and *cut* ^{LOF} or *ato* ^{GOF} transforms MS bristles into CT organs. Transformations of bristles into olfactory sensilla (not shown) are elicited by *amos* ^{GOF} [1587].

CT organs lack protrusions [378, 1454, 2787, 2831] and arise from the SOP via a different lineage tree [1756]. Stretching of their dendrites (springs attached at two anchor points "Λ") causes the neurons to fire an action potential to the CNS [789, 2018]. CT sensilla on the fly femur have 2 neurons [3867], whose cell bodies are embedded along the stretching axis (unlike depicted here).

MS bristles transduce touch stimuli by lever action [2174]: deflection of the rigid shaft (triangle) toward the body surface (horizontal line) causes it to pivot. Neural depolarization appears to be due to stretching of the dendritic membrane (vs. compression of the tubular body) [789, 874, 2787] (but see [4527]).

CS bristles should technically be termed "chemomechanosensory" because they use one neuron to sense touch like MS bristles, and they have four extra neurons whose dendrites extend up the shaft to a pore at the tip [2502, 4125], where they "taste" salts or sugars [111, 2955, 3061, 3529, 4841].

This schematic is adapted from [2174] (MS), [3529] (CS), and [378, 2128] (CT), with logic based on [1455]. Many authors have conjectured that cells differentiate by making a series of binary choices. A favorite metaphor for decision trees has been a rail yard of bifurcating railroad tracks [3063]. *N.B.:* Expression of *cut* at the wing margin goes through two phases; only the later phase controls SOP identity [353].

---

for conditional paralysis (viz., it stops synaptic vesicle recycling) [4759]$^{\Delta}$. Shibire may also play other roles in vesicular traffic [2232, 2900].

Endocytosis typically attenuates signaling by removing activated receptors from the surface [2031]$^{\Delta}$ and sending them to lysosomes for destruction [3217]$^{\Delta}$. If true here, then blockage of endocytosis by *shi* ^{LOF} should enhance Notch signaling [702, 3863, 4482] but, as stated above, *shi* ^{LOF} actually causes a N^{LOF} (not a N^{GOF}) phenotype. This paradox was finally solved in 2000 by using separate antibodies to follow the fates of N's extracellular (N-extra) vs. intracellular (N-intra) domains [3271]. It turns out that N-intra cannot be released from the membrane of the "listener" cell until N-extra is engulfed by the "speaker" cell as a ligand-receptor complex. The engulfing process requires Shi [3271], and Notch signaling requires release [4156] – hence the similarity of *shi* ^{LOF} and N^{LOF} phenotypes. Because Shi seems to not play a critical role in the listener cell, Numb is probably not acting via Shi in that cell.

Notch may be the only ligand that needs endocytosis for transduction per se in discs because other ligands can signal without it [3863]. Nevertheless, endocytosis does appear to propagate certain signals within the epidermis (viz., Dpp [1169, 1546] and Wg [273, 2900, 2984]), perhaps via a "transcytosis" mechanism that uses separate sets of (nontransducing) receptors (cf. Ch. 5) [811, 1154, 2843, 4379, 4694].

Pox neuro and Cut specify bristle type

The gene *paired box-neuro* (a.k.a. *pox neuro* or *poxn*) was isolated in a search for DNA-binding motifs of the "paired box" class [401] and was named "neuro" because it is transcribed in neural precursors of the CNS and PNS [950]. In imaginal discs, Poxn is expressed in SOPs of chemosensory (CS), but not mechanosensory (MS), bristles [149, 950]. CS bristles of flies have many features that differ from MS bristles [4125]. For example, they have a shaft that is thin and curved (vs. thick and straight), a characteristic axonal CNS projection, behavioral reactivity to airborne chemicals, and a different lineage that produces 5 neurons (vs. 1) – 4 of whose dendrites extend to the bristle tip (Fig. 2.8) [1741, 1802, 3004, 3061, 3529]. On the dorsal tibia and wing margin, *poxn* ^{null} converts CS into MS bristles [149], while *poxn* overexpression has the opposite effect [149, 3141, 3142]. Evidently, the role of the wild-type *poxn* gene is to divert SOPs from a MS to a CS mode of differentiation. One aspect of CS bristles that fails to be transformed in *poxn* ^{null} mutants is the (earlier) times when SOPs arise [149], which may explain the larger bristle sizes (see below).

Null mutations at the *cut* locus convert external sensilla into "chordotonal" (CT) organs in embryos [378, 2831], and a similar transformation (albeit often partial) is seen for adult bristles in *cut*-null areas of genetic mosaics. Overexpression of *cut* causes a reciprocal phenotype in embryos [372], although effects in adults have not been reported. CT organs are stretch receptors that fasten to the cuticle (or muscle) at two points and normally lack an external projection (Fig. 2.8) [1454, 2787]. Aside from a sensory neuron (femoral sensilla have 2 [3867]) they have at least 2 support cells: a "cap" cell and a sheath cell whose "scolopale" abuts the neuron's dendrite [378, 789, 2831]. In this regard, they resemble bristles, which may well have evolved from CT-like structures [378, 3437]. Chordotonal SOPs express *atonal* (*ato*) [4898], a bHLH gene located outside the AS-C (at 84F) [2040], but bristle SOPs do not (they express *cut* instead [149, 2000]). When bristle SOPs are forced to express *ato*, they stop expressing *cut* and transform into chordotonal organs [2038]. Evidently, the normal duty of *ato* is to ensure that SOPs develop into chordotonal organs by suppressing *cut* (*ato* ⊣ *cut*). SOPs "know" whether to turn ON *ato* due to *cis*-regulatory enhancers beside the *ato* transcription unit that respond to region- specific cues [4208]. Atonal also governs the development of photoreceptor cells (cf. Ch. 7).

Based on the normal expression of *poxn* in *cut* [null] embryos [4481] and the normal expression of *cut* in *poxn* [null] discs [149], *poxn* and *cut* appear to act independently, despite initial reports (ubiquitous expression and reporter-gene studies) that suggested otherwise (cf. App. 4) [4481]. Apparently, SOPs "read" the state (ON or OFF) of these genes and adopt a particular pathway of differentiation (MS vs. CS vs. CT) accordingly (Fig. 2.8).

The failure of *cut* [null] to achieve a complete transformation implies that other genes may overlap *cut* functionally in setting organ-type identity. Candidates include the paired genes *BarH1* and *BarH2* [3763], which, like *cut*, are expressed in embryonic external sensilla [1843]. Mutations in these genes switch one type of sensillum to another: deletion of both genes converts "campaniform" to "trichoid" sensilla (see [2128, 2174, 2787] for normal anatomy and [2473, 4309] for transitional intermediates), and overexpression of *BarH1* induces the reverse [1843]. Like *cut*, the *Bar* genes contain a homeobox [1842, 2286], which suggests that SOPs adopt fates via the same kind of logic that operates on a larger scale through *bona fide* homeotic genes (cf. Ch. 8) [282].

Olfactory sensilla are governed by yet another bHLH gene named "*amos*" (*absent solo-MD neurons and olfactory sensilla*, where MD stands for multiple dendritic) [448, 1587, 1928]. Amos (at 36F) and Atonal (at 84F) are more closely related to each other than to the AS-C family (at 1B) [1756]. Indeed, their basic (DNA-binding) domains are identical, except for a conservative Arg-to-Lys change. Nevertheless, they have dissimilar LOF and GOF phenotypes – presumably because different sets of cofactors bind idiosyncratic motifs outside their bHLH domains and steer them to different target genes [1587].

Bract cells are induced by bristle cells

The bract is a noninnervated hair made by a single cell [3552, 4531] (Fig. 2.1). It differs from other "trichomes" [2876, 2877, 3362] insofar as it is thick, pigmented, and found only on legs and wings [1361, 1714, 4338]. On the legs, bracts reside next to MS bristles on segments distal to the trochanter [1883]. On the wing, they reside next to MS bristles on the proximal costa [524, 793].

Because the bract cell is not part of the SOP clone [544, 1356, 1800, 4334, 4344], it has long been thought to arise by induction [1808, 3421]. When disc cells are dissociated and reaggregated, isolated bracts are never seen [4334], implying a dependence on bristles. Indeed, bracts do not develop unless both the shaft and socket cells are present [1357, 1808, 3448]:

1. Injecting larvae with Mitomycin C [4335, 4531] or nitrogen mustard [4337] often suppresses bristle sockets, and bracts are missing wherever shafts lack sockets. Likewise, bracts fail to form when sockets are transformed into shafts by N [LOF] or *shi* [LOF] [1803, 3425].
2. In the mutants *Hairless* [2] and *shaven* [de] (a LOF allele of *dPax2* [1319]), bristle shafts are often absent or vestigial [2168], and bracts are missing wherever sockets lack shafts [4338].

Why do some bristles have bracts, while others – only a few cells away – do not? Bristles that normally lack bracts seem to be those that develop earliest (CS bristles and macrochaetes) [149, 1598, 1803, 3142, 3628]. Thus, it is possible that every SOP emits a signal at one stage of its differentiation, but if this stage fails to overlap the period when epidermal cells are responsive, then the bristle will lack a bract. Consistent with this conjecture, CS bristles can acquire bracts when t.s. N [LOF] or *shi* [LOF] mutants are heat-pulsed [1803] (because delayed signals now enter the competence window?), and MS bristles can lose bracts when wild-type pupae are heat-shocked [1803] (because signals are delayed beyond the end of the window?). The Notch pathway is also involved in allocating bracts to particular parts of the wing margin [793, 3092].

Recruitment of cells by induction is also instrumental in the development of ommatidia in the eye (cf. Ch. 7), olfactory sensilla in the antenna [3531, 3548], and CT organs in the femur [4898]. The latter case is intriguing because it involves iteration (old SOPs inducing new ones) that could theoretically go on forever. In fact, the process is limited by the duration of the competence period and by the extent of the competent region [4898].

The CT inductive signal is transduced by the EGF receptor pathway [4898] (cf. Fig. 6.12). Evidently, this pathway also mediates bract induction because (1) $Star^{LOF}$ causes missing bracts, (2) $Star^{LOF}$ $Ras1^{LOF}$ double mutants are missing even more bracts, (3) all bracts can be eliminated by heat-treating t.s. $Egfr^{LOF}$ mutants, and (4) ectopic bracts can be induced throughout the epidermis by heat-shocking hs-$Ras1^{GOF}$ pupae (L. Held, unpub. obs.). Either the receptor (Egfr) or the ligand (unknown) must be localized in accord with proximal-distal cell polarity because bracts are only found on the side of the socket opposite the shaft, and this rule is obeyed even when bristles are misoriented in mutant phenotypes [1357, 1810]. Implementation of the received signal requires the homeobox gene $Distal$-$less$ (Dll): (1) Dll is expressed in bract, but not bristle, cells on the femur [618]; (2) Dll^{LOF} mutations suppress bracts [618, 4212]; and (3) Dll^{null} clones lack bracts [618, 1561].

Macrochaetes and microchaetes differ in size but not in kind

Unlike the sense organs interconverted by mutations in $poxn$, cut, $BarH1$, and $BarH2$, macrochaetes (MCs) and microchaetes (mCs) do not differ in kind. Regardless of size, all tactile bristles share the same cellular composition, lineage, and anatomy. Indeed, a full spectrum of bristle lengths is seen on the legs [1714, 1808, 1883], and mutations can cause notal bristles to have sizes between MCs and mCs [3067, 3373]. Thus, it is not surprising that no "selector" genes have been found that dictate MC vs. mC identity in the same way, for example, that $poxn$ specifies CS vs. MS bristles.

Bristle size is a function of polytenization [2475, 4641] – a cyclin-dependent process of DNA replication [961, 1246, 1262, 2551, 4376, 4820] that enlarges the cell's nucleus and cytoplasm without subsequent cell division [216, 1336, 1764, 3030, 3301] (cf. lepidopteran scales [2338, 4014]). In notal mCs (whose cells are initially diploid), the shaft cell undergoes two rounds of endoreplication, and the socket cell undergoes one [1741]. The degree of polyploidy achieved by MC shaft and socket cells is much greater [2475]. In

wild-type males, dorsocentral MCs average 336 μm (posterior) or 250 μm (anterior) in length, while nearby mCs are only ~80 μm long [3067] and tarsal bristles are ~50 μm long [1803, 2544].

When SOP mitoses end, the shaft cell is packed with rough endoplasmic reticulum as protein synthesis starts for shaft elongation and cuticle deposition [3216, 3552]. Indeed, ribosomes are probably working at top speed because reducing the supply of any one component slows the entire process. Thus, there are more than 50 sites in the genome where halving the gene dose (i.e., haplo-insufficiency) causes the same short-and-thin "$Minute$" bristle phenotype [118, 2561], and the majority of these sites appear to encode ribosomal proteins [3718]. Likewise, bristle size depends on the number of ribosomal RNA genes (at $bobbed$ loci on the X and Y) [1520, 1909, 3604, 4091, 4104]. The situation is analogous to the reduced wing size seen with LOF alleles of $rudimentary$: this "housekeeping" gene encodes a pyrimidine-synthesizing enzyme, which becomes rate-limiting during wing growth [1314, 1315, 1865, 3838]. Bristle length also depends on two bHLH-PAS genes – $spineless$ and $tango$ – but their role is unclear [1119, 1166].

Larger bristles typically have more longitudinal ridges (up to ~10) [2474, 3362, 3832, 4639], probably because ridge interval is constrained by an invariant molecular feature such as the diameter of an actin filament bundle [1645, 1898, 3216, 3351, 3552]. The bundles (up to ~20 around each shaft [94]) are organized by $chickadee$ [4474], $forked$ [3356], $rotundRacGAP$ [1642], $sanpodo$ [1128], $singed$ [640, 3283], $Stubble$ [94], and $tricornered$ [1430] around the time of cuticle secretion [1642, 3552]. When seen from the side, the grooves in wild-type bristles seem to wind helically around the shaft [3583, 4096], but in fact they meet along a top seam [4639, 4641]. These chevrons suggest that bristles evolved from triangular scales, which rolled up into a cone [1340, 4640]. Shaft differentiation is enforced by $dPax2$ [1319]: LOF alleles remove shafts, and excess dPax2 elicits two shaft-like spikes from each socket cell [2168]. However, $dPax2$ probably operates more like a switch because $dPax2^{GOF}$ can also convert neurons into glial cells [4303].

Bristle cells must divide, endoreplicate, and differentiate before cuticle deposition starts. The larger the bristle, the earlier it should have to begin these events in order to finish on time. This rationale explains why MC SOPs divide before mC SOPs [1741, 1925], and in general why bristle lengths are inversely correlated with the times of onset of differentiative events (cf. Fig. 3.4c) [806, 2837, 3420]. Similar reasoning explains why SOP mitoses commence earlier for multiply innervated CS bristles than for singly innervated MS bristles within each body region [149, 1598, 1741, 1803, 3142]. They simply need more time to make more cells.

If the number of replication endocycles is rate-limiting for bristle size, then delaying the time of SOP initiation should reduce the number of endocycles and cause a smaller bristle. Indeed, tiny bristles are often seen in flies carrying subnormal doses of *achaete* or *scute* [801, 802, 4095, 4098, 4108], and the stunting may be due to late onset of differentiation [1462].

Conversely, prodding an SOP to arise prematurely should allow more endocycles and cause a larger bristle. MS bristles that develop from transformed CS SOPs in *poxn*[null] flies are probably larger for this reason: the SOPs still arise at an early (unaltered) stage and hence have more time than neighboring MS SOPs to endoreplicate. SOP initiation can also apparently be accelerated by augmenting AS-C gene function [1563, 2690]:

1. Notal mCs can become MCs when *lethal at scute* is overexpressed (via a *UAS-l'sc* transgene driven by *scabrous-Gal4*) [3027, 3037].
2. Derepression of the AS-C in the dorsocentral notum (in *polychaetoid*[LOF] mutants [733]; Ch. 3) changes mCs into MCs [3067], while the opposite transformation (MCs into mCs) is seen when *spalt* (an AS-C regulator) is overexpressed [987].
3. When combined with mutations that derepress the AS-C, *ac*[GOF] (*Hairy wing*) alleles transform certain leg bristles into MCs [1802].
4. Ectopic overexpression of *sc* in the eye can convert interommatidial mCs to MCs [4209].
5. Extra doses of *sc* (in a background where *sc* is dere-

pressed) convert tergital mCs to MCs [2690], and similar enlargements occur when *sc* is overexpressed using a heat-shock promoter [3628]. (See [3905] for how *wingless* may be setting AS-C levels for this purpose.)

When cells are forced to divide by artificially expressing the mitosis-licensing agent String, unusually small bristles often arise at sites where SOPs are being recruited [2225]. This effect has been attributed to a delay in SOP initiation (consistent with the above argument), rather than to precocious entry into the SOP differentiation pathway [2225]. In summary, it appears that bristle size is controlled by rheostats, rather than by switches.

A bizarre bristle transformation occurs in flies with a construct of the male-specific cDNA of *doublesex* (a gene involved in sex determination) driven by a heat-shock promoter. Whereas most leg bristles in wild-type flies are thin, tapered, and brown, most leg bristles in these flies are thick, blunt, and black, like bristles of the sex comb [2106]. Whether this trait involves changes in the timing of SOP events remains to be determined.

Evolutionarily, bristle development has been entrained by various anatomic gradients. For example, bristle lengths increase from anterior to posterior among notal MCs [805, 2837, 3067], tergital mCs [801, 2690], and sternital bristles [804]. They increase from dorsal to ventral for tergital MCs [801], from ventral to dorsal for tarsal bristles [1803], and from distal to proximal for leg bristles in general [1714, 1883]. Bristle intervals typically increase with bristle lengths. Various aspects of bristle *patterns* have also been tailored to fit certain features of the cuticular landscape. Explanations for some of these correlations are presented in Chapter 3.

CHAPTER THREE

Bristle Patterns

The epidermis of a *D. melanogaster* adult has on the order of 500,000 cells, ~5,000 of which (~1%) make bristles [1804]. *A priori*, it would seem reasonable to expect bristles to sprout as randomly as the hairs on a human arm. However, even the most scattered bristles – the tergite microchaetes (mCs) – have fairly uniform spacing [801, 2301-2303]. At the other extreme of precision are the 40 macrochaetes (MCs) on the head and thorax, whose basic layout has been conserved for 50 million years (Fig. 3.1) [1625, 3966, 3980, 4185].

Except for the MCs, the bristles of each body region tend to vary in number and position from one fly to the next. Interestingly, most bristles are organized in rows that run parallel or perpendicular to axes of the body or limbs [1808, 2883, 4015]. Within such rows, the bristles are aligned more or less accurately and are spaced more or less evenly. Different rules govern different patterns. Thus, notal mCs form jagged rows along the anterior-posterior axis [805, 4428], while wing bristles form straight rows along the margin [1741], eye bristles arise at alternating vertices of each ommatidium [3539], and belly (sternital) bristles are spaced at intervals proportional to their shaft lengths [804].

Why do such patterns exist? Surely, some are adaptive. For example, flies use "brushes" (parallel transverse rows) on the legs to wipe dust from their eyes [4225], and other patterns appear to map air currents, prevent wetting, or act as shock absorbers [1116, 3206]. However, many may simply be accidents of evolution [3966]. Algorithms that evolved to align neuroblasts in the embryo's central nervous system (CNS) may later have acquired the duty of building an adult peripheral nervous system (PNS) [3981, 3984]. Old tendencies to obey particular cues may

have been retained in the new disc-specific contexts [1804, 1807], causing bristles to sprout at certain spots, interfaces, or contour lines (cf. Chs. 5 and 6). Moreover, bristle patterns in different species may manifest whimsical idiosyncrasies. Variations within the genus [376, 1361, 1801] imply a frivolity that rivals the pastiches of colored scales on butterfly wings [210, 434, 3119].

Regardless of why bristle patterns arose, the designs themselves epitomize the general problem of pattern formation in development. Namely, what causes the correlation of particular cell types with particular positions? We are far from a complete explanation, but it is now clear that the diversity of the patterns is deceptive. In fact, all of them use the same genetic tools to enable each epidermal cell to make the Hamletian choice of whether to be, or not to be, a bristle.

Given the two dimensionality of the cuticle and the single-cell origin of each bristle, the riddle posed by all these patterns is reducible to a simple Euclidean problem. To wit, what kind of geometry does the genome use to place points in a plane?"

Throughout the twentieth century, many of the intellectual giants of genetics tried to solve this "**Bristle Plotting Puzzle**" by clever experimentation or deft theorizing. Most of the speculation focused on mutants with abnormal numbers of MCs. T. H. Morgan toyed with the genetics of extra-bristle mutants as early as 1915 [2640]. Morgan's student, A. H. Sturtevant, was more intrigued with *missing*-bristle mutants, whose asymmetric patterns are as eye-catching as a gap-toothed smile. In 1918, he used *Dichaete* in artificial selection trials [4178]; in 1931, he toppled a popular theory about *scute* [4187]; and in 1970 (just before his death), he showed a dose

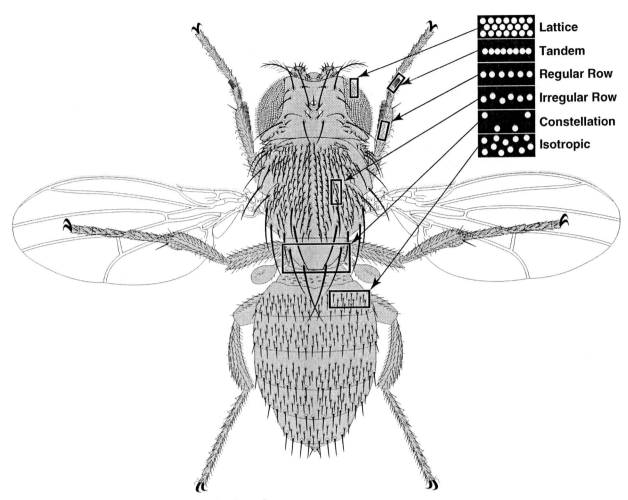

FIGURE 3.1. Diversity of bristle patterns on the fly surface.

In this adult male, the eye bristles occupy alternating vertices of ommatidia in a hexagonal lattice. The sex comb typifies a "tandem" array where bristles touch. Other leg bristles are aligned in straight ("regular") rows and spaced evenly. On the notum the rows tend to be irregular, especially laterally. The four scutellar macrochaetes form an aperiodic ("constellation") pattern that is constant from one fly to the next, as are the arrangements of other macrochaetes. Most tergite bristles (except those at the posterior edges) are arranged randomly ("isotropically") but spaced uniformly. Tergite pigmentation has been omitted for clarity. Adults are ~3 mm long [3421]. Adapted from [1804]. For anatomical nomenclature, see [1224]. See also App. 7.

What sort of geometry does the fly use to create these patterns? Do all the arrays arise via the same algorithm? Do bristles know where they are? The mysteries of this "golden fleece" have intrigued investigators for decades [1799, 1804, 1808, 2434, 4096].

---

dependency of *achaete* and *hairy* that later proved prescient [4185]. Another family member of the Morgan lab, H. J. Muller, inspired a team of Russian researchers to dissect the *achaete-scute* Complex (AS-C) [650] – an enterprise that later became a cottage industry for Antonio García-Bellido and his school in Madrid [1354, 1452]. British patriarchs John Maynard Smith [2768, 2769] and C. H. Waddington [4517] invented cute schemes for categorizing mutant patterns, while Vincent Wigglesworth and Peter Lawrence speculated about the chaotic origins of natural patterns [2427, 2439, 4657]. Lawrence and García-Bellido used bristle patterns as a platform for

digging into deeper issues of development, and these two mentors trained many talented students who became leaders in this field.

Indeed, bristle pattern research is so interesting from a historical standpoint because of these crosscurrents. Its parochial models have often influenced the larger fields of development and genetics and vice versa. For instance, Curt Stern's ideas about bristle patterning (discussed below) colored Lewis Wolpert's thinking about patterning in general [4724], and Wolpert's theory of positional information in turn revolutionized the bristle subfield via the team founded by Howard Schneiderman and Peter Bryant in California [530, 1485]. Most of the old questions have now been answered by modern molecular techniques, but the story of how we got here is as charming as the answers themselves.

Surprisingly, different macrochaete sites use different signals

Alan Turing, the British mathematician who cracked Germany's Enigma Code in World War II [1863], is also famous for helping to found the modern field of computer science [872, 4409]. It is less widely known that Turing also dabbled in developmental biology. In 1952, he proposed a new theory to explain how biological patterns arise [4410]. He argued that chemical reactions among randomly diffusing substrates should cause the products to accumulate at regular intervals. Given the right parameters, his formulae predict stable spots or stripes whose spacing depends on diffusion constants, and whose arrangement depends on the shape of the region where the reactions occur [1176, 2769, 3014]. Turing's ideas founded a flourishing school of thought about the abilities of reaction-diffusion systems to create patterns [1727, 2806, 3013].

In 1954, Curt Stern invoked comparable dynamic forces to explain how bristle patterns develop [4095, 4096]. Stern (1902–1981) had learned fly lore as a postdoc in Morgan's lab [2614, 3070], where he earned kudos for proving that crossing over occurs in somatic tissues [119, 4089, 4347]. Poetically, this very phenomenon allowed him to test his new ideas [4100, 4103]. He acknowledged that chemical interactions such as those in **Turing's Model** could mediate patterning, but he thought that physical forces might work just as well. For example, Stern conjectured that uneven growth in a sheet of cells might lead to stress points, which cells could then "read" as signals for making bristles.

The initiation of differentiation of specifically localized organs cannot be a purely locally caused process but is possible only on the basis of a prepattern which itself is the result of interrelation of parts of a developing whole. [4098]

Possibly bristles differentiate at places of specific strains or stresses set up in the folded embryonic tissue which later smoothly covers the thorax.... Of course, the physical scheme of tensions used in this train of thought is only one of many variants. Different general patterns of biochemical properties ... in the embryonic future epidermis could be considered equally well. [4096]

Stern coined the term "prepattern" for the array of signals that prefigures the final pattern. Initially, he imagined that the sites containing the cues must be interdependent. He called those sites "singularities." In the above example, the singularities were the stress points, but in general they could be any heterogeneity. This idea came to be known as the "**Prepattern Hypothesis**" [4346].

In a population of somatic cells which are genetically alike, the differential fate of some specially located ones must be due to a superimposed differential organization of the population.... The existence of a differential organization, a "prepattern," within any imaginal disc must precede and hence be responsible for the patterned origin of bristles. [4095]

Because the prepattern was deemed to be a gestalt, any change in one of its components should alter the remaining components in the same way that tugging one node of a spider's web will alter the whole web. To test whether influences spread beyond the point of an interference, Stern used mosaic flies with discrete regions of mutant and wild-type tissue. He created the genetic patchiness by somatic crossing over. His reasoning was straightforward. If a mutant fly differs from the wild-type in its network of force vectors, then mutant tissue in a wild-type background should cause novel webs of interactions, and strange new patterns should emerge that might yield clues to the mechanism.

Over the next ~20 years, Stern and his co-workers tested this "**Force Field Model**" by analyzing ~20 possible "prepattern" genes in mosaic settings. Disappointingly, they found virtually no new patterns [4100, 4346]. Nevertheless, some clues were uncovered that led to profound insights.

Ironically, the pithiest clues came from one of the first mutations that Stern studied – ac^1, an allele of *achaete* (*ac*), whose name means "no bristle" in Latin. This gene was mentioned earlier as part of the AS-C (Fig. 2.4). Homozygous ac^1 flies usually lack a notal MC called the "posterior dorsocentral" (PDC). In mosaics, this PDC bristle is typically absent if its site is occupied by ac^1 tissue and typically present if heterozygous tissue resides there [4095, 4096]. Evidently, cells at the PDC site use their own genotype to decide whether to make a bristle, regardless of the distribution of genotypes elsewhere in the disc. This local autonomy was not what Stern expected based on his Force Field Model, nor was it what Turing's Model would have predicted since diffusing chemicals should naturally affect nearby cells [2769, 4016].

Given the ability of ac^1 tissue to suppress the PDC bristle but virtually no other MCs when located at extraneous sites, Stern surmised [4100] that

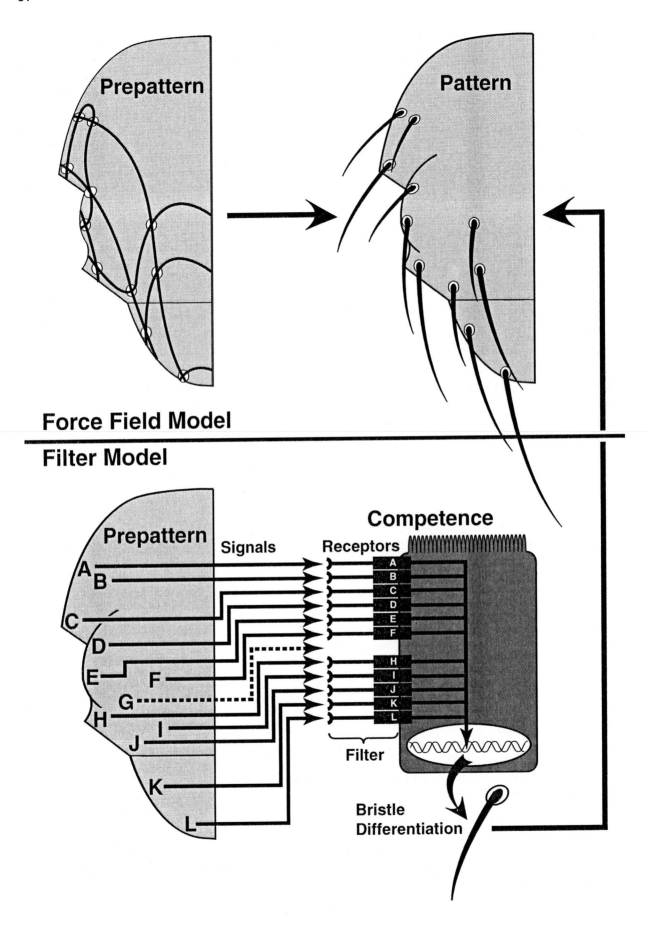

Force Field Model

Filter Model

1. *ac¹* must interfere with the "competence" of cells to respond to a specific signal at the PDC site.
2. Different MCs must use different signals.

His second deduction is a special case of the general **Principle of Nonequivalence** [2516, 4727, 4729] – the idea that identical structures are specified by nonidentical signals. This notion is counterintuitive because *a priori* Occam's Razor suggests a minimum number of signals for any given pattern. In a similar way, stripes of pair-rule gene expression in the embryo blastoderm were once thought to arise via spatially periodic signals in a Turing-like process [1728, 2159, 2377, 2379, 3031]. However, there, too, the Principle of Nonequivalence trumped Occam's Razor: most stripes are actually governed by unique combinations of transcription factors [104, 1908, 3247, 3587, 4834]. Thus, MCs – like pair-rule stripes – are made "inelegantly" [49, 2378, 2765, 3248].

Based on the above conclusions, Stern revised his model by inserting a new step (Fig. 3.2). For a prepattern signal to evoke a pattern element in the final array, this signal must first be transduced by the cell at that site. To illustrate this "competence" step, he used a prospecting analogy where signals differ qualitatively from one another:

This then is a picture of gene action in differentiation. A whole developing embryo, or a developing part of an embryo, possesses regional differences of varied nature. One gene (or its products) responds to one specific regional peculiarity, another gene responds to other regional characteristics. These responses, which are the necessary prerequisites for a complete series of processes, result in specific differentiations. One may compare the groups of genes present in all regions of the developing organism to groups of prospectors working in different regions of a country. Each group would be alike in consisting of different men looking for coal, gold, uranium, and so on. In some regions only the coal prospectors would respond to the coal deposits, at other places the gold prospectors would respond to gold veins, at still other places the uranium prospectors to ores of this metal. Each response would depend on the pre-pattern of geological differentiation. Each response would be the necessary prerequisite for subsequent differentiation in the form of the specific structures erected by mining engineers – perhaps deep shafts and elevators for coal mines, surface excavations and washing troughs for gold, shallow tunnels and radiosensitive installations for uranium. [4096]

In this new formulation, mutant genes that affect the pattern should be classifiable into those that do so by altering the prepattern ("prepattern genes") and those that do so by modifying cellular competence ("competence genes"). Signals that cannot be transduced due to a competence mutation like *ac¹* would be "filtered out" before they elicit bristles at those sites [4100].

One way to envision this "**Filter Model**" is with cell surface receptors serving as the filter (Fig. 3.2), although Stern never advocated any particular molecular device. If each bristle site has a unique signal and each signal has a unique receptor, then the failure of *ac¹* cells to make a PDC bristle could be due to a defective receptor.

In fact, the *ac* gene does not encode a surface receptor, nor does the *ac¹* mutation disable the Ac protein. Rather, *ac* encodes a nuclear protein, and *ac¹* deletes a region-specific *cis*-enhancer that lies 5 kb upstream from the *ac* gene's transcription start site. As explained below, it is the ensemble of these enhancers at the AS-C locus that actually "filters" the prepattern signals.

FIGURE 3.2. Curt Stern's Prepattern Hypothesis in its original (above) and final (below) forms, both of which were proposed in 1954 [4095, 4096].

Force Field Model. Initially, Stern envisioned patterns as the steady-state outcomes of dynamic physical or chemical forces. In this illustration, uneven growth deforms the epithelium so that each half mesonotum acquires 7 stress lines. Each of the 11 sites ("singularities") where the lines intersect then induces a MC. Because mutations can only change the pattern by altering the force vectors, any patches of mutant tissue in a wild-type background should distort the entire web of forces. Contrary to expectation, virtually every pattern-altering mutation that Stern studied (including *achaete¹*, Fig. 3.3c) affected differentiation autonomously in mosaics.

Filter Model. To explain why cells exhibit such insularity, Stern postulated that (1) each MC must use a different signal (A–L, although signals could differ quantitatively instead), and (2) a MC will only form at a site if the cell there is "competent" to respond to that specific signal. Inserted between the prepattern and the pattern is an extra step wherein each cell "filters out" signals that it cannot "hear." In this illustration, the competence of each cell (depicted with apical microvilli) depends on surface receptors. In reality, positional signals are "filtered" through enhancer elements within the *achaete-scute* Complex (Fig. 3.4) [2891]. Wild-type cells (like this one) cannot respond to Signal G, a "cryptic singularity" that Stern discovered [4097]. Similarly, *ac¹* cells would be "deaf" to Signal I (not shown).

N.B.: Stern described these variations of his theme in general terms, but never formalized or named them as specific models. Here they are made explicit to highlight their differences. The idea that tensions affect gene expression (top model) seemed farfetched at the time, but various examples are now known [301, 761, 762, 1942, 2580].

Prepatterns may contain hidden "singularities"

In male flies, the foreleg bears a "sex comb" [1716], so-called because its single file of stout bristles resembles a hair comb [4344]. In gynandromorphs, sex comb "teeth" can arise when even a small patch of male tissue (in an otherwise female leg) overlaps this site [3441, 4105]. Evidently, female forelegs have inductive signals for teeth but lack a comb because female cells are not attuned to them. This case is interesting because it shows that even wild-type cells routinely "filter out" certain signals in their own epidermis.

Extending this reasoning into the realm of evolution, Stern argued that signals may arise evolutionarily before the ability of cells to respond to them:

Many *Drosophila* species are known which do not possess a sex comb in either sex. *D. melanogaster* undoubtedly evolved from such sex-combless ancestors. Must we postulate that a basically new configuration of a foreleg had to originate in evolution before a sex comb made its appearance; or can we not more simply assume that the sex comb pre-pattern was long in existence in the legs of flies and had to wait only for a mutation toward a single new gene which could respond to the hitherto unheard call? [4096]

In other words, ancestral drosophilids may have possessed a signal for sex combs, but this signal failed to evoke comb development until *D. melanogaster* males evolved the ability to "hear" it.

Stern called such hidden signals "cryptic singularities" [4100], and he uncovered another instance in studying a certain duplication of the *yellow* gene, where AS-C enhancers reside. *D. melanogaster* normally lack a MC at the "interalar" site [3966, 4185] ("G" in Fig. 3.2) on the notum (dorsal thorax). However, an extra MC can arise there when the site is occupied by duplication-bearing cells, even if the rest of the notum is wild-type. Apparently, the duplication enables cells to "read" a subliminal signal that wild-type cells cannot [4097]. Because other (ancestral?) dipterans *have* an interalar bristle, *D. melanogaster* may have retained a vestige of the signal but lost the ability to transduce it, in which case any mutation that reveals it would be considered atavistic [1354].

Other authors have adduced evidence for cryptic singularities in line with the dorsocentrals [2640, 4507], between the ocellars and orbitals [2769, 4013], between the anterior and posterior scutellars [1282, 2413, 3890, 4517], and on the tarsus [1802]. Indeed, subthreshold bristle-forming tendencies have been found in nests of cells that normally do not form a bristle, but can be prodded to do so in certain mutant backgrounds [589].

There are two ways of thinking about singularities in general, regardless of whether they are overt or covert:

1. Singularities are instructive signals, tantamount to telling a cell to "Make a bristle!"
2. Singularities are not signals per se, nor are they inherently tied to any particular structure.

Stern favored the second view [4098], as illustrated by the following passage where he cites some classic experiments of Spemann's [4030].

There are some famous experiments in which embryonic tissues of a salamander have been transplanted into the mouth region of frog embryos, and reversely frog tissue transplanted into the mouth region of salamander embryos. These experiments have shown that both salamander and frog are endowed with a prepattern which distinguishes their mouth regions from the rest of the body.... In the evolution of the two branches of amphibians to which salamanders and frogs belong, no more than the appearance of new genes which responded in different ways to the previously established inductive capacity of the mouth region was required. The complexities of any organism must imply an inexhaustible array of prepatterns, a few of which are realized in accomplished differentiations and most of which are waiting for the response of genes not yet arisen in a specific organism. [4096]

The molecular-genetic analyses described below support this argument. Singularities appear to be *trans*-acting factors that elicit bristles only because the AS-C has *cis*-enhancers to bind them. In principle, any bristleless site could be a "cryptic singularity," because evolution could endow the AS-C with enhancers to "read" virtually any position-specific protein (cf. Ch. 8) [662]. Thus, each prepattern is an accidental result of the cross-linking (by evolution) of bristle differentiation to certain heterogeneities of gene expression in the epidermis.

Our understanding of prepattern mechanisms rests mainly on studies in the wing disc, where many *trans*-acting factors are known. Those agents will be examined in Ch. 6 after readers have been briefed on the key signaling pathways (cf. Table 6.2 and Fig. 6.14). The rest of this chapter focuses on the AS-C itself and its regulation by circuits that operate wherever bristles arise.

How Achaete and Scute control bristles was debated for decades

Among *ac¹* mosaics that lack a PDC due to homozygous *ac¹* tissue occupying that site, a MC often arises nearby in heterozygous territory (Fig. 3.3) near the normal site [802, 3613, 4095, 4098]. This apparent ability of the PDC to be displaced led Stern to infer that the PDC can come from any cell within a cluster of equivalent cells at this spot. When the cell at the "preferred" site within this "equivalence group" is prevented from becoming a MC, another cell can substitute. Why don't all of these cells make MCs? Stern guessed that whichever cell emerges as the SOP must block the others from adopting this fate. He likened the situation to embryonic fields, where the tissue area that can potentially make a structure (e.g., a limb [1955, 4589]) is typically larger than the region that actually does so.

This astonishing result fits perfectly well into existing concepts of the embryologist. He has discovered the existence of pre-patterns which he calls embryonic fields. These are areas in which a specific differentiation may occur anywhere. Actually, under normal circumstances, the differentiation takes place in only a limited part of the whole field, at a peak, figuratively speaking. Once differentiation has set in at the peak, no other differentiation occurs within the larger field. If, however, differentiation at the peak is suppressed, then a lower region of the field may differentiate. This differentiation itself will exclude other differentiation within the remainder of the field.... The pre-patterns of the embryonic tissue of *Drosophila*, which call forth the response of genes involving the differentiation of bristles, are embryonic fields of larger dimensions than the limited points of normal location of bristles. If the influence of the *achaete* gene at the normal point prevents differentiation of a bristle, other parts of the field may assume the properties of peaks and differentiate bristles. [4096]

Given the directionality of the displacements, Stern deduced that the PDC cluster must be elongated along the anterior–posterior axis, rather than having the circular shape expected if the sensory organ precursor (SOP) emits a diffusible inhibitor.

It seems that differentiation of a typically located [PDC] inhibits other nearby potential bristle differentiations in the line defined by the position of the dorsocentral bristles but that absence of the [PDC] permits the realization of some of these potencies. [4095]

The inhibitory mechanism is considered later. It does not appear to depend on *achaete* or any other genes in the AS-C.

The AS-C spans ~100 kb near the tip of the X chromosome (Fig. 3.4). Four of its genes – *achaete, scute, lethal-at-scute, asense* – are termed "proneural" because they promote neural development in the CNS or PNS [1454, 3637]. Innervated bristles depend almost exclusively on *achaete* (*ac*) and *scute* (*sc*), as shown by doubly mu-

tant flies with null alleles for both genes. Such flies lack virtually all bristles due to a failure of SOP inception [912] but otherwise look normal [1379, 1462, 1802, 3812]. Not only are *ac* and *sc* necessary, they are also sufficient for bristle initiation because overexpression of either gene can induce extra bristles at ectopic sites [191, 635, 1348, 3628] by evoking extra SOPs [912, 3982]. AS-C polymorphisms account for much of the naturally occurring variation in bristle number seen in this species [2593, 2645]. Historically, this gene complex was viewed as holding the key to deciphering bristle patterning [1354, 4766], and rightly so.

Partial-LOF *sc* mutants are striking because they create holes in an otherwise perfect pattern of MCs, and because the bristles that they remove depend strictly on the allele. Among the 11 notal MCs {PSC, ASC, PPA, APA, PDC, ADC, PSA, ASA, PNP, ANP, PS} (Fig. 3.4), the subsets that tend to be missing include sc^1 {PSC, ASC}, sc^4 {PSC, ASC, PPA, APA, ASA, ANP, PS}, sc^6 {ANP}, sc^7 {PSC, ASC, PPA, APA}, and sc^9 {PSC, ASC, ANP} [1360]. How does the AS-C encode bristle positions? This "**Bristle Coding Enigma**" baffled geneticists from the discovery of the first sc^{LOF} allele (sc^1) by Calvin Bridges in 1916 [470] until 1995 when the mystery was solved by Juan Modolell's team in Madrid [1538, 2890]. The intervening 80 years were punctuated by intriguing clues and colorful controversies [628, 650, 1354, 1799, 2891]. Some milestones in this saga are recounted below.

1916	On January 22, the first *scute* mutant was found by Calvin Bridges [470]. Twelve days later (Feb. 3), the first *achaete* mutant was found by Alexander Weinstein. Both men were then graduate students in Morgan's lab [2283], where the Great Quest for mutants had begun in 1910. Neither *achaete* nor *scute* could be given a one-letter abbreviation because "*a*" and "*s*" had been taken by *arc* and *sable* (in 1912) [470]. Hence, the two genes became known as *ac* and *sc*.
1920s	Following H. J. Muller's first sojourn in the Soviet Union, A. S. Serebrovsky used Muller's X-ray method to induce new *sc* mutations [650, 651]. His assistant, N. P. Dubinin, found partial complementation among alleles [650] and used these data to deduce a linear array of *sc* "subgenes." Each subgene was supposed to control a discrete subset of bristles [1108, 1109]. His inferred sequence of subgenes

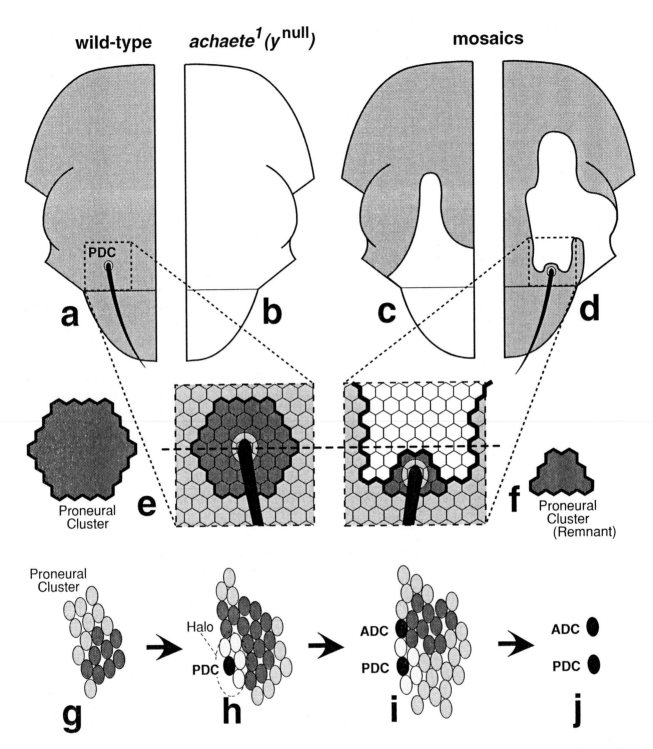

wild-type **achaete^1(y^{null})** **mosaics**

a b c d

Proneural
Cluster

e

f

Proneural
Cluster
(Remnant)

Proneural
Cluster

Halo

PDC

ADC

PDC

ADC ●

PDC ●

g h i j

agrees roughly with what we now call "en-hancers" along the AS-C [1360, 1453]. A similar mistake was made at the Bithorax Complex [2507, 2782], where enhancers were thought to be genes per se before DNA cloning clari-fied the situation [146, 1121, 2672, 4671].

1931 A. H. Sturtevant and Jack Schultz supported the "linear array" tenet of Dubinin's "**Sub-**

gene Hypothesis" but disproved the "dis-crete subset" tenet by showing that allele-specific bristle groups can be modified via changes in the genetic background [4187]. This malleability implied that the pattern-ing of bristles is "a system in which limiting factors are important and in which, there-fore, changes in the rest of the system make

FIGURE 3.3. Proneural clusters, whose existence was deduced by Stern in 1954 [4095] and confirmed histologically in 1989–91 [912, 3637, 3982].

a–f. Stern's reasoning, based on the nonautonomous formation of an ectopic MC by wild-type tissue in *achaete¹* (*ac¹*) mosaics. **a.** Heminotum (minus humerus) of a wild-type fly, with all bristles omitted except the PDC MC. **b.** Homozygous *ac¹* flies typically lack a PDC. Absence of shading indicates *yellow*^{null} (*y*^{null}, yellow cuticle), which was a linked marker for *ac¹* [636, 1442], as was *singed¹* (*sn¹*, gnarled bristles). **c, d.** Gynandromorphs arising from X-chromosome loss in *y*^{null} *ac¹* *sn¹* /+++ heterozygotes (redrawn from [4095]). Both heminota lack a PDC at its normal site, evidently because the tissue there is mutant. In some cases (e.g., **d**), a PDC-like MC arises in nearby heterozygous tissue [3613, 4095, 4096]. **e, f.** Illustrations based on Stern's hypothesis for bristle displacement [4095]. Squares show imaginary details for PDC areas in **a** and **d**. Dashed line marks the normal PDC site. A "proneural cluster" (dark shading) of cells (small hexagons) is "competent" to make a PDC. The central cell becomes the PDC (**e**) unless its *ac¹* genotype renders it unable to do so (**f**), whereupon a cell in the remaining part of the cluster substitutes, thus displacing the PDC. All SOPs, Stern argued, must inhibit their neighbors from making bristles.

g–j. Drawing of a real cluster on a right heminotum at ~25, 20, 10, and 0 h before pupariation, as revealed by antibodies to Scute [912, 3982]. Ovals are nuclei (diameters ≈ 4 μm). The actual dorsocentral cluster (1) has an irregular shape, (2) grows, and (3) yields two MCs whose SOPs arise at the cluster's edge. **g.** Some cells express more Scute (dark shading). **h.** The SOP for the PDC (black) arises in a high-Scute area and acquires a "halo" of low-Scute cells. **i.** As the cluster grows, the high-Scute area shifts and the ADC (no halo yet) arises. The reason for this shift is not known. **j.** In this cluster, non-SOP cells stop expressing Scute before SOPs, which do so later (before dividing). Further growth widens the gap between the ADC and PDC. As for why the ADC and PDC typically form on the medial (dorsal) side of their common cluster, the answer appears to be that a diffusible bristle-promoting signal (Dpp) encounters this cluster on its medial face and then wanes (cf. Fig. 6.14) [3373, 4368].

---

changes in threshold values." Dosage thresholds have since been shown to be decisive in bristle patterning and to depend on a balance between opposing groups of HLH genes (see below). Sturtevant and Schultz advocated a version of Plunkett's **Spreading Model** [3405], wherein a dorsocentral source ("center") emits a bristle-inducing substance, and different *sc* alleles impede its diffusion in different directions by acting at different times. In the same year, Richard Goldschmidt (Curt Stern's colorful mentor [4102]) proposed a similar diffusion-based theory [1520].

1932   Muller [2986] and Sturtevant [4181] independently disproved the Spreading Model by demonstrating the same autonomy for *sc* that Stern later showed for *ac* [4095, 4096]. To wit, they showed that *sc*^{LOF} tissue can remove bristles without affecting nearby wild-type tissue – a result that is inconsistent with the involvement of diffusible factors. Our modern view of AS-C function affirms Muller's conclusion that "the development of bristles...is not governed by one or a few centers, but is in its major features autonomous at the site of each bristle."

1935   Like Sturtevant and Schultz, George Child disproved Dubinin's "one-subgene-one-bristle-group" idea but did so by manipulating the environment instead of the genetic background. He showed that allele-specific traits can be changed by raising *sc*^{LOF} flies at different temperatures [767–769].

1954   Curt Stern proposed his Prepattern Hypothesis [4095, 4096]. The autonomy that he found in *ac* mosaics ruled out a Force Field Model for prepatterns. He concluded that cells respond to position-specific, bristle-inducing signals only if they are "competent" to receive them. Displacements of wild-type bristles led him to postulate that (1) each MC arises within a group of equipotent cells, any one of which can become a bristle; and (2) the nascent SOP inhibits its neighbors from adopting a bristle fate.

1975   Clifton Poodry used irradiation to probe the timing of bristle differentiation on the notum [3420]. He found that when MCs are grouped according to the time when they acquire insensitivity to irradiation, the groups do not match the subsets removed by *sc*^{LOF} alleles. Thus, those subsets cannot be explained by loss of *sc* function at allele-specific times (as Sturtevant and Schultz had imagined).

1978   The time when a bristle site becomes insensitive to irradiation was found to be correlated with bristle length [2837]. This trend is probably due to a need for larger bristles to begin differentiating earlier in order to

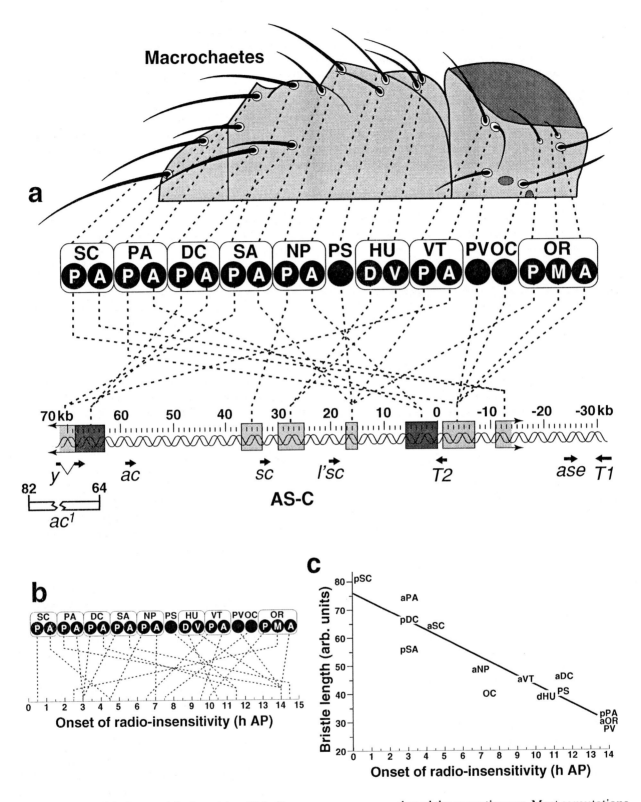

finish before cuticle deposition. This discovery (Fig. 3.4c) had no bearing on why different sc^{LOF} alleles affect specific bristles.

1985   Cloning of the AS-C was reported by Juan Modolell's team [636], which had previously cloned the sc portion [656]. Most sc mutations were found to map within a 50-kb region of DNA proximal to the sc gene. Nucleotide sequencing revealed similarities (and redundancies) of the four bHLH genes in the

FIGURE 3.4. Spatiotemporal control of macrochaete development on the head and thorax.

a. The genotype-phenotype "switchboard" for MCs. Half head and thorax of a wild-type fly (dorsal view; mCs omitted). Black circles below symbolize MCs seriated in posterior-to-anterior order, except that some are grouped (cf. Fig. 2.6 for key: P = posterior, A = anterior, D = dorsal, V = ventral, M = middle). The double helix spans the ~100-kb *achaete-scute* Complex (AS-C) near the tip of the X chromosome. Arrows denote transcription units (kink = intron), with distal to the left and zero as an arbitrary EcoR1 site. Four "proneural" genes (*achaete, ac*; *scute, sc*; *lethal-at-scute, l'sc*; *asense, ase*) encode bHLH proteins, although only *ac* and *sc* normally affect MCs. Enhancers (rectangles) were mapped genetically (light shading; double arrows mark unknown limits) [636, 3688] or by their ability to drive reporter genes in proneural clusters (dark shading) [1538, 2891]. The minimal effective sequence per rectangle is unknown except for the DC enhancer [1380]. Note that the MCs and enhancers are not colinear (dotted lines). Genetic loci for ANP and APA enhancers (5–7 kb and −1 to −7 kb, respectively; not shown) differ trivially (a few kb) from those depicted (as assessed by reporters). The *ac¹* allele affects DCs (Fig. 3.3) due to an 18-kb (64–82) deletion that overlaps the DC enhancer. The functional core of this 5.7-kb enhancer is really only $\leq$1.4 kb long [1380]. The DC enhancer may be looped into contact with the *sc* promoter (30 kb away) when Pannier (a transcription factor) binds this enhancer and forms a complex with bHLH dimers that bind the promoter [3504]. The *yellow* gene (*y*) has its own enhancers (not shown) [1442, 1534]. In Ref. [1538], ASA and PPA are (mistakenly?) ascribed to the same enhancer. See also App. 7.

b. Times (hours after pupariation) when sites become insensitive to γ-irradiation. Because the high dose that was used (10^4 Rad) kills dividing cells but spares postmitotic ones, onset of insensitivity probably signifies completion of SOP mitoses [3420]. Indeed, onset times agree roughly with histological detection of four SOP progeny at these various sites [1925]. More precisely, times may reflect the start of endoreplication because shaft cells become insensitive and endoreplicate ~1–3 h before socket cells [1741, 3420, 3421]. No colinearity exists between onset times and the overall posterior-anterior seriation (dotted lines), although posterior bristles precede anterior ones within most pairs – a correlation that may have less to do with location than with shaft length (see **c**). Times for mCs are 19–24 h AP (not plotted).

c. Inverse correlation between bristle length and onset of radio-insensitivity. This trend may stem from a need for SOPs to begin mitoses or endoreplication early enough to finish differentiating before cuticle deposition. Strangely, the earlier a MC SOP arises (cf. Fig. 3.7), the longer it waits before starting mitosis [1925]. Lengths (arbitrary units) are for wild-type flies raised at 17°C [2837], times are from [3420], and the line was plotted by best-fit regression. *N.B.*: For calibration purposes, the actual lengths for pDC and aDC bristles are 336 μm and 250 μm, respectively [2544].

---

complex [73, 258, 4485]. Surprises included the paucity of transcription units (only 6 genes in 100 kb of DNA) and the absence of introns [2891].

1987 Ruiz-Gómez and Modolell used 74 terminal deficiencies with breakpoints in the AS-C for a deletion analysis [3688] and tested two hypotheses: (1) the "**Differential Threshold Model**," where each MC needs a different amount of Ac or Sc, thus explaining why bristles are affected in a definite sequence as enhancers are whittled away; and (2) the "**Site-specific Enhancer Model**," where each enhancer controls *sc* expression at a certain site. They favored the latter model, which turned out to be correct.

1989 The groups of equipotent cells that Stern predicted in 1954 were documented histologically by *in situ* RNA hybridization and named "proneural clusters" (PNCs) [1454, 3637, 3957].

1991 *In situ* detection of Ac and Sc protein distributions allowed PNCs to be resolved in greater detail [912, 3982]. James Skeath and Sean

Carroll validated Stern's fanciful idea of cryptic singularities by identifying an "inactive cluster" anterior to the dorsocentrals and by showing that derepressing the AS-C induces MCs there (their Fig. 5) [3982].

1992 Skeath *et al.* found that when the AS-C is broken in half by an inversion, co-expression of Ac and Sc in (embryonic) PNCs ceases, and Ac and Sc become expressed in complementary subsets of clusters [3986]. The heretical implication was that *ac* and *sc* respond independently to PNC-specific enhancers, rather than controlling certain subsets of clusters via differences in their gene products [2721].

1995 José Luis Gómez-Skarmeta *et al.* described the first mutation that eliminates *sc* function without disabling any AS-C enhancers (*sc^{M6}*) [1538], thereby allowing the role of *sc* to be assessed unambiguously for the first time [628]. They also produced the first enhancer map compiled via reporter gene constructs (Fig. 3.5). By these means, they showed that the "classic" roles that had been assumed

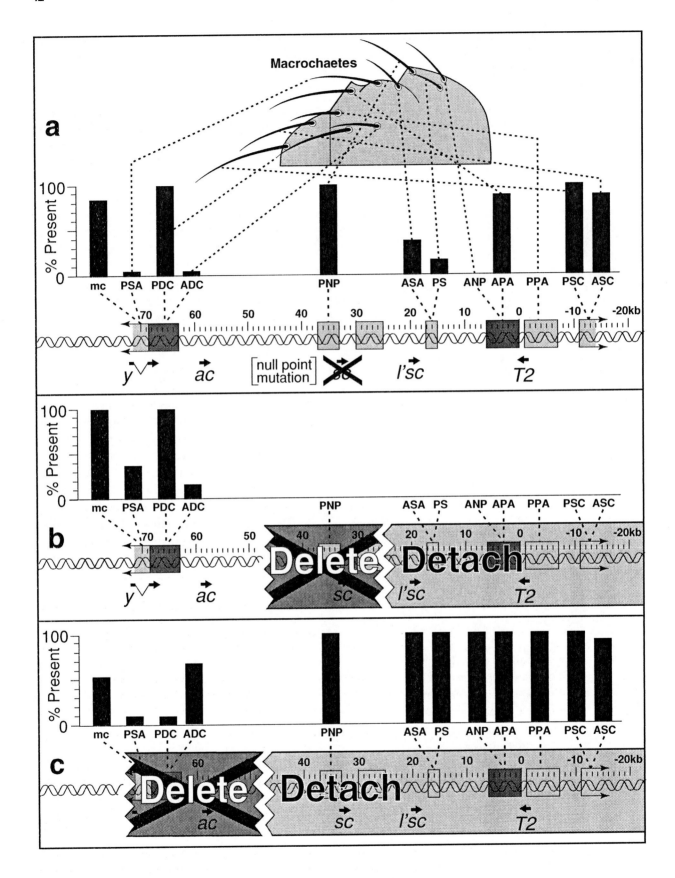

for *ac* and *sc* were wrong. The old AS-C paradigm toppled, and a new one emerged, as recounted below. In the ensuing "gold rush," many researchers have been picking apart each enhancer to ascertain what *trans*-acting factors influence it (e.g., see [1380]).

In 1989, Achaete and Scute were found to mark "proneural clusters"

As discussed above, Stern deduced that each SOP arises from a group of equipotent cells, whose competence to respond to bristle-inducing signals is conferred by *ac* or *sc*. Because prepattern signals should be congruent with the final pattern (except for cryptic singularities), the expression of *ac* and *sc* need not be confined to bristles sites. In theory, *ac* and *sc* could be transcribed ubiquitously. In fact, they are not. In 1989, Romani *et al.* [3637] found that *ac* and *sc* RNA is restricted to nests of cells at future bristle sites. They called these nests "proneural clusters" (PNCs) [3950]. PNCs fulfill Stern's 1954 prophecy. They were resolved more clearly when Ac and Sc protein distributions were seen in 1991 [912, 3982].

Each PNC has a characteristic size, shape, time of appearance, and time of disappearance [589, 912, 1538, 3637, 3982].

1. Ac and Sc are detectable in each cluster before the SOP(s) appears. This order is consistent with the idea that Ac and Sc enable a cell to adopt the SOP state. The number of Ac- and Sc-expressing cells increases until SOP emergence, as does the intensity of expression.

2. Most clusters yield only one SOP. In those cases, the nest is roughly circular. Cells near the center typically manifest more Ac and Sc than those near the periphery, and one of the intensely expressing central cells becomes the SOP. Stern had imagined such a central "peak" based on his embryonic field analogy [3613, 4096–4098].

3. At their zenith, most single-SOP nests contain ~20–30 cells. This size is roughly what was expected based on PDC shifts in *ac* mosaics. The PSA nest is exceptionally small (~10 cells).

4. The DC and SC clusters produce two SOPs each. In each case, the posterior SOP develops first. The PDC arises at a clearly eccentric location within its nest [3373], as does the APA. Eccentric SOPs rule out certain models for how the SOP is selected.

5. In each PNC, Ac and Sc reach higher levels in a subcluster of ~6 cells. Among those few cells one cell attains a still greater level, which evidently boosts it above some sort of "ignition threshold" because it inevitably becomes the SOP [637, 918]. In nests that produce two SOPs, the subclusters that precede the SOPs are spatially and temporally distinct.

6. The location of the SOP(s) within a PNC is virtually the same from fly to fly.

7. Ac and Sc fade below the limits of detection before the SOP begins dividing. Presumably, the levels of Ac and Sc can wane because they have triggered the

---

FIGURE 3.5. Roles of *achaete* and *scute* in bristle development on the notum.

a. Phenotype of *sc*^{M6}, a *sc*^{null} with a nonsense-codon mutation ("X") before the HLH domain [1538]. Unlike previous nulls, this allele is not associated with deletions or inversions that confound gene-phenotype analyses (e.g., **b, c**). Bars indicate frequencies of MC presence. Sites are listed in enhancer order (dashed lines; cf. Fig. 3.4 for details and Fig. 2.6 for MC key; mc = microchaetes). Surprisingly, *sc* is not needed for bristles that were thought to depend on it (**b**), including scutellars (PSC, ASC) whence *scute* got its name [470].

b, c. "Classical" (but incorrect) roles of *ac* and *sc* as deduced from nulls that involve major DNA alterations. In each case, deletion of a gene (*sc* or *ac*) was achieved by crossing over between inverted chromosomes with breakpoints at different AS-C sites [2561]. With such complex "knock outs," it is impossible to distinguish the effects of gene inactivation from the effects of disabling enhancers. Thus, the complementary sets of bristles affected by these constructs (**b** vs. **c**) give a false impression of the relative roles of *sc* and *ac* (cf. **a**). **b.** Phenotype of a *sc*^{null}, $In(1)sc^{8L}sc^{4R}$, where L and R denote left or right halves of inversions contributing to the recombinant. The deletion removes *sc* and the PNP enhancer, and the inversion moves proximal enhancers ASA, etc., away from *achaete* to the other (heterochromatic) end of the chromosome. Thus, the only enhancers still able to influence *ac* are mc, PSA, PDC, and ADC. Other PSA and ADC enhancers must exist in the deleted or detached domains (probably in the *sc* promoter region [2721]) because these bristles are often missing. **c.** Phenotype of an *ac*^{null}, $In(1)y^{3PL}sc^{8R}$, which removes *ac* and enhancers for PSA, PDC, and ADC. Microchaetes may be only partly affected because their enhancers extend past the deletion [3688]. Although proximal enhancers PNP, etc., are disconnected from *ac* by the inversion, they remain linked to *sc* and therefore can apparently still function normally.

All data are from [1538]. *N.B.:* The *yellow* (*y*) gene has its own set of *cis*-enhancers [1442, 1443], and mutations therein produce a menagerie of pigment patterns [1435, 3055] (as variegated as dog or cat breeds), suggesting that some of the same spatiotemporal *trans*-factors that control the AS-C may also regulate *yellow* [3056].

next step in SOP differentiation. The SOP must be able to regulate its levels of Ac and Sc separately from other cells of the nest because the SOP typically stops expressing Ac and Sc at a different time (earlier or later) from the remaining cells.

PNC cells are arrested in the G2 phase of the cell cycle [4427] and are recognizable by higher levels of Cyclin A [2225]. Mitotic quiescence appears essential for proper selection of SOPs because premature entry into mitosis (prompted by artificial activation of *string*) disturbs SOP emergence [2225]. As for why cells must stop dividing to select a SOP, it is possible that apical signaling is hindered when dividing cells retract basally [3823], and the same logic may apply in the morphogenetic furrow where synchrony is also imposed [1126, 1903].

Quiescence cannot be under AS-C control because *ac⁻ sc⁻* discs have G2-arrested clusters like the wild-type [4427]. Arrest is probably enforced by a parallel circuit: the same *trans*-acting factors that affect the AS-C may bind directly to a separate set of *cis*-enhancers at the *string* locus [2480], where they can suppress this mitosis-licensing agent without needing Achaete or Scute. The discovery of these pools of nonmitotic cells answered the old question of why bristle vs. non-bristle lineages appear to diverge so early in development [996, 1372, 1742]: the split is not due to compartment-like lineage restrictions, but rather to withdrawal of pre-bristle cells from the cell cycle long before cell types are finalized.

The above observations provide clues about how SOPs are selected. Before trying to fit these puzzle pieces together, however, additional clues at the genetic level are examined below.

In 1995, the old AS-C paradigm toppled and a new one emerged

The old AS-C paradigm was based mainly on breakpoint-associated *ac*^LOF and *sc*^LOF alleles [1360, 1453]. Given the complementarity of their phenotypes (Fig. 3.5), the expectation was that "*ac*-dependent" bristle sites would mainly use Ac, and "*sc*-dependent" sites would mainly use Sc. It thus came as a shock when *ac* and *sc* were found to be expressed together in PNCs [637, 3530, 3637]. To explain this correlation, a mutual activation (of *ac* by *sc* and of *sc* by *ac*) was invoked and a two-step process was imagined [2721, 3982, 4453]. In Step 1, separate enhancers would turn *ac* and *sc* ON in complementary sets of clusters, with each site using one or the other gene as its founder. In Step 2, Ac would *trans*-activate *sc* and Sc would *trans*-activate *ac* in each cluster. Although parsi-

monious, this idea turned out to be wrong with regard to PNCs (although it does apply to SOPs, as explained below).

In 1995, Gómez-Skarmeta *et al.* disproved this "**Mutual Activation Model**" and solved the Bristle Coding Enigma of the AS-C [1538]. To knock out *sc* function, they used a nonsense mutation (*sc*^M6) that differs from all prior *sc* and *ac* nulls insofar as it does not disrupt any surrounding AS-C enhancers. Thus, it revealed the actual bristles controlled by *sc* – a subset that happens to differ dramatically from the *sc*-dependent subset as classically defined (Fig. 3.5). The Mutual Activation Model predicts that Ac should only be found at *ac*-dependent sites in a *sc*^null. In fact, Ac accumulates in all PNCs of *sc*^M6 flies. Additional disproof of the model came from inversions that abolished *ac-sc* coexpression by merely severing the connections between certain enhancers and either *ac* or *sc* [1538, 3984, 3986]. Therefore, coexpression is due to shared enhancers that activate both genes per cluster (cf. point 3 below), not to stimulation of *ac* by *sc* or vice versa. Such a possibility had been suspected [2062] but not tested rigorously. E(spl)-C genes rely on the shared-enhancer strategy to a much more limited extent [871, 3075].

This "**AS-C Epiphany**" of 1995 has forced us to rethink how anatomy is controlled by *ac* and *sc*. The old and new conceptual frameworks are contrasted below:

1. The reason breakpoint-associated *ac*^LOF and *sc*^LOF alleles (vs. *sc*^M6) remove certain subsets of bristles is not because they affect one or the other gene per se. (Indeed, *ac* and *sc* are functionally interchangeable [191, 918, 948, 3628], so different bristles cannot rely on qualitatively different Ac vs. Sc proteins.) Rather, allelic specificity is due to deletion or detachment of specific AS-C enhancers, each of which affects a definite body region (Fig. 3.5). For example, *ac¹* removes the PDC because the PDC enhancer is deleted (Fig. 3.1), not because the *ac* gene is defective, and the *sc*^M6 phenotype shows that this enhancer does not need *sc*.

2. Categories of bristles [1802, 2561] and sensilla [948, 2519, 3637] that were formerly seen as being "*ac* dependent" or "*sc* dependent" must now be viewed as being governed by enhancers that happen to lie near *ac* or *sc*, respectively. For example, the scutellars were thought to be controlled by *sc*, but the *sc*^M6 phenotype indicates that they are more dependent on *ac* (Fig. 3.5).

3. Temporal differences that were ascribed to *sc* vs. *ac*

(e.g., *sc*-dependent MCs preceding *ac*-dependent mCs [1741], and *sc*-dependent chemosensory bristles preceding *ac*-dependent mechanosensory bristles [1802]) are probably due to "heterochronic" enhancers that are activated at different times [4584, 4844].

4. The reason no model ever succeeded in parsimoniously explaining the allele-specific bristle subsets is that the order of enhancers along the AS-C is scrambled relative to the MC pattern [1453, 1538] (Fig. 3.4). There is no "MC homunculus" analogous to the "homeobox homunculus" [1805] that obeys the order of body segments [1113, 2508, 2672, 3494]. In this regard, the AS-C resembles the *hairy* locus, where enhancers for pair-rule stripes 1–7 (anterior to posterior) are scrambled: (3 & 4), 7, (6 & 2), 5, 1 [2376] (cf. similar chaos at *even skipped* [1323, 3717, 3992] and the lack of a "muscle homunculus" in mice [2336]).

5. The phenomenon of competence does not involve reception of an external inductive signal at the cell surface (Fig. 3.2). Rather, it entails AS-C enhancers that directly respond to *intracellular* position-specific, *trans*-acting factors. Because the enhancers (not the genes) are responsible for the competence, *ac* and *sc* are more properly called "proneural" than "competence" genes. Operationally, this distinction is supported by the ability of generalized *ac* or *sc* overexpression to cause extra MCs at abnormal sites [191, 635, 1348, 3628]. Such a result would not be expected if the genes merely enabled cells to respond to pre-existing signals at fixed locations.

6. Now that the old puzzle of why specific *ac*^LOF or *sc*^LOF alleles remove certain bristles has been solved, the remaining question is how AS-C enhancers are controlled by spatial or temporal cues that are upstream in the genetic hierarchy [4584]. Much progress has been made in identifying these "prepattern" factors, and these findings are discussed in Chapters 5 and 6.

Not all enhancers activate *ac* and *sc* equally, because only certain MCs are missing in the *sc*^null point mutant (Fig. 3.5). This favoritism may be due to the kind of cognate lock-and-key affinities between enhancers and promoters that are found at the adjacent *yellow* locus [1534], in the Bithorax Complex [275, 1821, 3975], and elsewhere [349, 2194, 2537, 2827, 3180, 3833]. In other words, transcription factors that bind certain enhancers (for sites that lack bristles in *sc*^M6) would have greater affinity for the *sc* promoter than for the *ac* promoter. Indeed, the *sc* and *ac* promoters differ appreciably from one another [918, 3171]. Such specificity would also explain why *l'sc* is

silent in SOPs, despite its closer proximity than *ac* or *sc* to some MC enhancers (Fig. 3.4) [3504].

Alternatively, these preferences could be mediated by factors that bend AS-C DNA into knotlike configurations where contacts are sterically possible only at certain points [275]. One candidate for such a role is the product of *Dichaete*. Dominant mutations in this gene have long been known to cause *sc*^LOF-like patterns of missing bristles [3405, 4178]. Dichaete is a Sox-domain protein that kinks DNA into an 85° angle [2635, 3707], so it might constrain the contacts between *trans*-acting factors (bound at *cis*-enhancers) and the basal transcription apparatus (bound at core promoters) [317, 1480, 1629, 3360], depending on where its binding sites are located within the AS-C. However, Dichaete cannot be a general partner in AS-C function because it is not normally expressed in wing discs [2977]. A more likely candidate there is Chip [3504], which ties DNA into loops [3600, 4456]. Still another suspect is the gargantuan Mediator complex [134, 412, 4386], which is known to bridge enhancers to promoters in other systems [1703, 2488, 2665].

The *sc*^M6 phenotype suggests that some bristles need more Sc than Ac protein, but this cannot be true because *sc* and *ac* are functionally redundant [952, 2019, 2891]. Rather, the sum total of both proteins must dictate whether a bristle forms. If that total exceeds the threshold, then a bristle will arise. Otherwise, it will not.

Proneural "spots" shrink to SOP "dots"

How is a SOP chosen within a PNC? The deciding factor seems to be the amount of Ac and Sc. As stated above, the intensity of AS-C proteins is uneven within each cluster, and the cell with the most Ac and Sc inevitably becomes the SOP [912, 1538]. The dose dependence of the selection mechanism is confirmed by mosaics where *ac*^+ *sc*^+ (full-dose) cells become SOPs 2–4 times more often in a background of *ac*^− *sc*^− /++ (half-dose) cells than in a background of *ac*^+ *sc*^+ cells [912, 1794].

Why does one cell acquire more Ac and Sc than other cells in its cluster, and why does the SOP tend to arise at a certain site in each PNC? Genes that are hierarchically upstream of the AS-C (e.g., *u-shaped* and *pannier* [911, 1671]) are known to impose spatial biases [3966], but those biases can fit several sorts of scenarios:

1. **"Predestined SOP Model."** SOP sites could be fixed by antecedent patterns of expression by genes that are hierarchically upstream of the AS-C (cf. Ch. 6). Sharp expression boundaries could pinpoint SOP sites within PNCs by their intersections, but whether

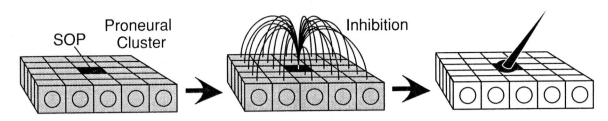

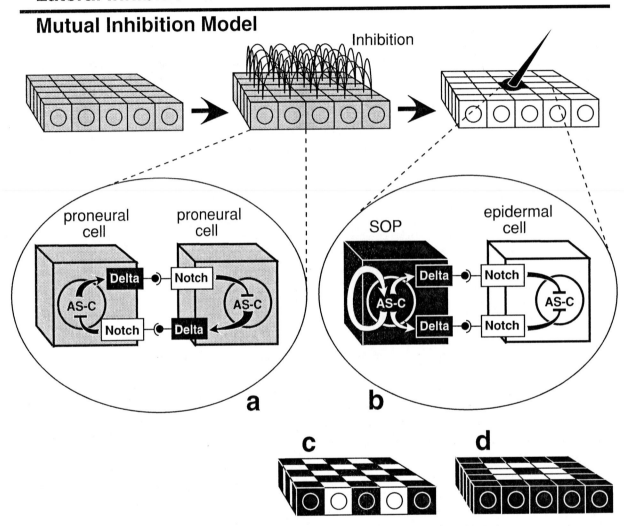

there are enough boundaries to designate all sites is not known. Also unknown is whether such precision is attainable by diffusible signals [210, 902, 903, 2515, 3584], given the formidable task of addressing a one-cell target in a "forest" of tall, thin epithelial cells.

2. **"Contest Model."** PNC cells might compete to become the SOP, perhaps via Dl-N "shouting matches" (Fig. 3.6; see below) [178]. Because SOPs arise within predictable subclusters, the competition must be biased [3954]. The bias could be spatial (e.g., proximity

to a diffusible activator or inhibitor of the AS-C) or temporal (e.g., starting to make Ac and Sc earlier). Accuracy need not be as great as in the previous model because a single victor will emerge as long as there are ≤6 contiguous contenders (i.e., the biasing process need only delimit 6-cell spots).

3. **"Diffusible Activator Model."** If all PNC cells were to secrete a diffusible activator of the AS-C, then its concentration should intensify in the middle of the PNC and a cell there would be first to cross the

FIGURE 3.6. Hypothetical mechanisms for ensuring a single bristle per proneural cluster.

In the **Lateral Inhibition Model** (above), the SOP prevents neighbors from becoming SOPs [4095]. Arcs denote inhibitory signals that could be transmitted by diffusion or contact. Contact between nonadjacent cells would require extensions like the filopodia seen on moth scale cells [3051].

In the **Mutual Inhibition Model** (below), every proneural cell inhibits adjacent cells, and the SOP can be chosen later [1563, 3022, 3270]. Uncommitted proneural cells are shaded, the SOP is black, and cells that succumb to inhibition (and become epidermal) are white.

a, b. Supposed interactions at different stages, according to the Mutual Inhibition Model. Cubes are individual cells, and inscribed circles are nuclei. **a.** Initially, all cells are equivalent in the extent to which they emit (Delta ligand) and receive (Notch receptor) inhibitory signals, but this balance is unstable because each cell's AS-C feeds back on itself in a loop that traverses the other cell [1458]. The feedback is positive because the loop has two negative steps. (Arrows denote activation; cross-barred "⊣" lines denote repression.) Any asymmetry will be amplified, so if one cell is slightly more inhibited, its AS-C and Delta activities will wane, as will its ability to inhibit. Amplification is damped by a separate loop that uses the EGFR pathway (not shown) [917]. Other factors may also prevent cells from stably expressing both Notch and Delta [990, 2839]. **b.** Eventually this cell shuts OFF its AS-C and becomes epidermal. Conversely, the cell that is less inhibited keeps increasing its AS-C activity and, at some point, crosses a threshold that triggers an autocatalytic loop, which keeps the AS-C switched ON. By this stage the "winner" has purged its Notch receptors, and the "loser" has purged its Delta ligands.

c, d. Two of the possible outcomes for a 5-by-5 array of cells undergoing mutual inhibition [1614, 1805, 4269]. (Another is parallel stripes.) The intended outcome, a single SOP among 25 cells, is actually impossible. One way out of this dilemma is to imagine that the inhibitory arc of the loop is functional, but the excitatory arc is disabled. In that case, cells would "damp" one another but would not become SOPs, based on how severely they themselves are damped. Another means of biasing the SOP decision would then be needed, perhaps involving underlying prepattern factors or an overlay of Extramacrochaetae protein [1458].

This montage is adapted from [1458, 3866, 4115] (**a, b**) and [1181, 1614, 1889, 4269] (**c, d**). *N.B.:* In the wing, Delta-Notch signaling manages to create 3-cell wide stripes that become veins [989] by a feedback mechanism that seems similar [1951] but may not be (cf. Ch. 6). Thus, the same circuit may be used in different ways in different contexts [444, 627, 1457, 2018]. Odd results in the eye have raised questions, however, about the circuit's versatility and validity [184]. See also App. 7.

"ignition threshold" to become the SOP [1462, 3584]. This scheme finds support in the intermediate level of Ac-Sc (a pre-SOP state? [912]) in subclusters where SOPs arise, but eccentric SOPs are problematic.

The first two schemes are actually components of larger hypotheses (the Lateral vs. Mutual Inhibition Models [4480]) that are discussed later. They are being treated here as stand-alone models to contrast their premises and predictions.

The Predestined SOP Model is hard to reconcile with the ability of *ac¹* clones to displace the PDC within its cluster, because only the ordained cell should be endowed with enough Ac or Sc. Perhaps it is the highest *relative* (vs. absolute) level of Ac or Sc that determines the SOP. Cells within a cluster might be comparing their levels of Ac or Sc via Dl-N signaling. However, the decision cannot be purely relative because raising the absolute level of Ac or Sc can evoke extra SOPs per cluster [191, 635, 3628].

The Contest Model implies that close matches should sometimes arise in which two rivals both attain an advanced state of SOP commitment before one retreats to the default epidermal state. Indeed, pairs of SOP-like cells are occasionally seen in PNCs that ultimately

make only one bristle [1925]. How such disputes might be resolved to yield a single SOP is discussed later.

The Diffusible Activator Model could explain eccentric SOPs if those nests have more of an AS-C inhibitor (e.g., Extramacrochaetae) centrally than peripherally, thus forcing the SOP to arise near the edge (like PDC displacements seen in *ac¹* mosaics). Some kind of diffusible factor is likely because mosaics for AS-C^{LOF} [807, 3613, 4095, 4181] and AS-C^{GOF} [1577, 4097] mutations often violate the rule of autonomy when the mutant/wild-type boundary skirts a bristle site. In those cases, mutant cells acquire the ability to make bristles, probably by diffusion of a rescuing agent from nearby wild-type cells [1462, 4098].

It is worth stressing that it is close neighborhood [which causes the nonautonomy] unrelated to total extent of the *ac¹* area which may appear as a small island just covering the bristle site or a large patch whose border happens to pass by it. This suggests a simple spread of *ac⁺* dependent material into the *ac¹* tissue patch. This material then nonautonomously endows the *ac¹* cells with the ability to respond to the prepattern by differentiation of a bristle if only of rudimentary size. It seems to be a case of substitution therapy equivalent to the presumed support in mosaics of gene-deficient cells by nondeficient tissue. [4098]

The rescuing molecule cannot be Ac or Sc, because they localize to the nucleus [912]. The soluble agent, which appears to be cleaved from a transmembrane precursor [3640], is probably Spitz [917] – a ligand for the EGFR pathway (cf. Ch. 6). That pathway participates in a positive feedback loop with the AS-C (see below): AS-C ➜ EGFR ➜ AS-C [917].

The SOP uses a feedback loop to raise its Ac and Sc levels

Although *ac-sc* cross-activation was disproven for PNCs, it apparently does occur in SOPs [918, 1538, 2721]. When a 3.7-kb piece of DNA immediately upstream of the *sc* transcription unit is joined to a *lacZ* reporter, this construct is found to be expressed in SOPs but not in any other cluster cells (except at DC and PSA sites, which must have enhancers in this fragment [2721]). The same is true for a construct of the 0.8-kb *ac* promoter joined to *lacZ*. Both of these transgenes are silenced in an *ac⁻ sc⁻* background unless ligated to PNC enhancers, implying that *ac* and *sc* genes of SOPs require Ac or Sc for auto- or cross-activation.

These feedback loops apparently must be primed to a threshold before they sustain themselves [917], and PNC enhancers supply the needed boost [918]. Earlier studies showed expression in all PNC cells (not just SOPs) of *lacZ* reporter genes driven only by the *ac* or *sc* promoter (sans PNC enhancers) [2721, 4453], but those results may have been complicated by staining artifacts or deletion of control sites that restrict expression *in vivo* [918, 1538].

Ac and Sc belong to the class of bHLH transcription factors that bind E boxes (Fig. 2.4). There are three E boxes in the *sc* promoter [918] and four in the *ac* promoter, although one is unresponsive to proneural proteins [3171, 4452]. The "SOP enhancer" for *sc* has been whittled down to a 356 b.p. piece (~2.7 kb from the transcription start site) [918], and only one E box appears essential for auto-activation.

These results imply the positive feedback loop "more {Ac AND Sc} ➜ bind E boxes ➜ more {Ac AND Sc}," but if this loop automatically raises Ac-Sc levels high enough to spark SOP differentiation [1538], then what prevents non-SOP cells in each PNC from doing the same? Conceivably, access of proneural proteins to E boxes in non-SOP cells is blocked by *trans*-acting repressors that are expressed in all PNC cells except the SOP [918, 1538]. One such candidate – Extramacrochaetae – is discussed later. Another candidate is encoded by the *klumpfuss* (*klu*) gene [2250]:

1. *klu* is switched ON in PNCs when Ac becomes detectable but is switched OFF in the SOP as soon as it arises. Thus, *klu*'s expression could allow it to distinguish non-SOP cells from SOPs.
2. *klu* is genetically downstream of the AS-C (*klu*^LOF is epistatic to *ac*^GOF in causing bristle loss), but its transcription in PNCs is independent (*klu* still turns ON in *ac⁻ sc⁻* flies). The agents that activate *klu* are unknown.
3. Klu has four zinc fingers and other hallmarks of a transcription factor, so it could implement SOP vs. non-SOP states directly at the DNA level.
4. Despite the above features, *klu* cannot be a simple selector gene because its effects appear to be nonautonomous: SOPs do arise in *klu*^LOF mutants (as judged by high Ac levels) but fail to complete bristle development. Thus, *klu* causes bristle loss via an effect on a cell (SOP) that does not express *klu*. However, there is an autonomy-based alternative: the cell that becomes the SOP might need *klu* to first turn ON (leaving a legacy of other gene settings) and then OFF to launch the SOP on its (*klu*-independent) differentiative path.

The AS-C must act differently in SOPs vs. non-SOPs because ubiquitous overexpression of an antagonist (*hairy*) quashes Ac in all proneural cells except SOPs [3982]. Further evidence for distinct (SOP vs. non-SOP) regulation is given below regarding the E(spl)-C.

Two other bHLH genes (*asense* and *daughterless*) assist SOPs

A third bHLH gene in the AS-C – *asense* (*ase*) – also participates in bristle development. Like *ac* and *sc*, it is transcribed in SOPs, but unlike them it is not transcribed in other PNC cells (except on the wing edge) [438, 1079, 2039]. Its SOP expression appears late in cluster development, suggesting that it is turned ON by high levels of Ac and Sc. Indeed, there are four E boxes in the 5′ untranslated region of the *ase* transcript [918, 2039], and *ase* expression disappears in an *ac⁻ sc⁻* background in all SOPs (except for the wing edge) [438, 1079].

However, *ase* is dispensable for most bristles, perhaps because other (AS-C?) genes compensate for its absence. In *ase*^null flies, only the wing [2039] and tergites [2690] seem affected. The stout bristles on the wing margin are often stunted, fused, or twinned, but rarely absent – implying a function for *ase* in bristle differentiation [1079, 2039], although the partial loss of tergital mCs suggests a role in SOP inception as well.

HLH proteins typically prefer to homo- or hetero-dimerize with only certain partners, which in turn dictates preferences for particular DNA-binding sites [68, 348, 2089, 3017, 3179]. For example, Ac and Sc were each found to require a binding partner from outside the AS-C in order to function [588, 589, 4452, 4453]. That partner is Daughterless.

As its name implies, the *daughterless* (*da*) gene was originally identified by its LOF effects on sex determination. Only later was its involvement in bristle development discovered. Flies that are doubly heterozygous for a *da*[null] allele and an AS-C deficiency lack some MCs, while heterozygotes for either lesion alone look wild-type [949]. This dose-dependent interaction is attributable to a need for Ac/Da and Sc/Da heterodimers, which have been shown to form *in vitro* [588, 589, 918, 4452].

Unlike Ac and Sc, Da is expressed ubiquitously in discs at a level that is no higher inside bristle-yielding PNCs than outside them [905, 4435]. Based on its role in the embryonic PNS, Da seems to be required for the SOP to divide [2019]. Specifically, *da* appears to activate cell-cycle genes in the SOP after it emerges from the cluster [949, 1755]. One downstream gene regulated by *da* may be *Cyclin A* (*CycA*): a *CycA*[LOF] allele that causes missing MCs was induced by a P-element insertion in *CycA*'s first intron, and that intron contains five E boxes [4415].

"Lateral" or "mutual" inhibition ensures one SOP per PNC

As mentioned earlier, Stern drew two inferences from PDC displacements in *ac¹* mosaics: (1) each MC SOP comes from a group of equipotent cells, and (2) the SOP inhibits its neighbors from adopting its fate [4095, 4098]. The first conjecture was validated by the discovery of PNCs. The second was later formalized as the "**Lateral Inhibition Model**" (Fig. 3.6) [112, 1458, 1808, 3207, 3950]. This model made several predictions that were also supported by subsequent observations:

Prediction 1: Killing an incipient SOP should remove the inhibition and allow a different cell to become the SOP.

Prediction 2: Disabling the inhibitory mechanism should allow all the cells in a PNC to become bristles.

The first prediction was tested in the embryonic CNS of grasshoppers, whose neuroblasts arise somewhat like SOPs of the fly PNS [587, 1071, 1073, 1745, 1804]. Ablating an incipient neuroblast does indeed cause a nearby

cell to change its fate (from ectodermal to neural) to replace the dead one [1072, 4231]. This ability of PNCs to regenerate a progenitor cell provides direct evidence for the Lateral Inhibition Model.

Consistent with the second prediction, LOF alleles of various genes cause tufts of bristles at existing MC sites (and high densities in mC zones). Most of these genes are in (or associated with) the Notch pathway [114], including *Notch* [1742, 1797, 3891], *Delta* [3277, 4859], *Su(H)* [3820, 3826], E(spl)-C bHLH genes [991, 1794], *groucho* [1794], *kuzbanian* [3640, 4025], *shibire* [1802, 1803, 3425], and *big brain* [1075, 3519]. Among the foregoing, *big brain*[LOF] has milder effects (even as a null), as do GOF alleles of *Hairless* [200, 2657] – a *Su(H)* antagonist.

GOF alleles [191, 635, 1348, 1802] and artificial overexpression [1854, 3027, 3628] of proneural genes also cause extra bristles, but these phenotypes differ in several ways from the *Notch*[LOF] tuft syndrome [2959, 3950]:

1. Ectopic bristles arise in areas that are normally bare.
2. MCs remain separated by intervening epidermal cells.
3. Increases in mC density are less severe.

The ability to put bristles into new areas (e.g., the flanks of the abdomen) underscores the role of proneural genes in conferring neural competence to specific body regions [1804]. Notch-pathway genes do not share this role, with one exception. LOF alleles of *groucho* cause extra bristles in the wing blade and scutellum – the same areas affected by LOF alleles of *hairy* [1794]. The explanation is that Groucho is a promiscuous corepressor that interacts with Hairy, but not via the Notch pathway [1794, 3278].

The milder MC and mC phenotypes seen in AS-C[GOF] mutants [191, 1577, 1802] imply that lateral inhibition still functions, whereas it is disabled in Notch-pathway LOF mutants. Herein lies the key difference between AS-C and Notch-pathway genes.

The idea that bristle cells prevent nearby cells from becoming bristles was first proposed in 1940 by Vincent Wigglesworth [1145, 4657]. The hemipteran insect that he studied – *Rhodnius prolixus* – bears evenly spaced bristles on its abdomen [4663]. By comparing bristle positions in successive instars of a single individual, he found that the new bristles are added in the largest gaps of the old pattern [4657]. To explain this trend, he proposed that bristle cells deplete their vicinity of a diffusible factor that is needed for bristle initiation. This depletion prevents any new bristles from arising within a certain range of any extant bristle. Only when growth pushes bristles far enough apart should an intervening cell become a bristle in the next instar. This "**Local Depletion Model**" is formally equivalent to the Lateral Inhibition Model, where bristle cells emit an inhibitor instead of consuming an inducer [3207, 4660]. These exclusionary ("inhibitory") fields are thought to serve two different roles in *Drosophila*:

1. *Fine-tuning:* Inhibitory fields ensure one bristle per potential site.
2. *Bristle spacing:* Inhibitory fields place bristles a minimum distance apart.

In more recent years, an alternative hypothesis – the "**Mutual Inhibition Model**" – has gained popularity [1563, 1797, 3022]. It asserts that *every* proneural cell (not just the SOP) inhibits its neighbors via Dl ligands. Inhibited cells reduce AS-C activity so low that they cannot become SOPs (Fig. 3.6), except for one cell (the future SOP), which somehow escapes inhibition.

Like the *C. elegans* scheme that inspired it [3866], this model is usually depicted as a contest between equivalent cells, which adopt alternative fates after one wins and the other loses [1794, 4115]. If groups of ~25 cells compete, then multiple SOPs should arise in checkerboard, stripe, or ring patterns (Fig. 3.6) [1889, 4269].

How can a *single*-SOP outcome be mandated? One way would be for the contests to occur in only a core subcluster (~6 out of the ~25 cells). Alternatively, the SOP could be shielded from inhibition by boosting its AS-C above a threshold for autocatalysis [443, 1458, 3022, 3823].

The two models ("LI" vs. "MI") are debated below. Like the Contest Model discussed above, these schemes invoke cellular chitchat, but they are more concerned with preventing excess SOPs than with SOP selection per se.

Notch-pathway and proneural genes are functionally coupled

The MI model envisions a feedback loop between Notch-pathway and proneural genes (Fig. 3.6). This same loop could mediate inhibition in the LI model. Some kind of cell-autonomous "N ⊣ Dl" connection must exist because cells with excessive N-intra signaling in their cytoplasm lose Dl from their surface [3270]. Moreover, proneural-Notch cross-talk must exist because haplo-insufficiency for *da* enhances N^{LOF} (*split*) eye phenotypes [437].

Genes wired to the end of the Notch pathway (Dl → N → Su(H) → E(spl)-C ⊣ targets?) will be repressed because E(spl)-C bHLH proteins recruit the Gro corepressor, whereas genes affected by Ac/Sc/Da should be activated (Ac/Sc/Da → targets?). If Dl were an AS-C target and the AS-C were an E(spl)-C target, then the loop would be complete. Both links appear to be functional, as the following evidence argues. However, it is important to keep in mind that showing a link under one set of circumstances does not prove that the link is used at a physiologically meaningful level in other cell types at other times [113, 871, 1330, 4584].

1. Evidence for an "AS-C → Dl" link:
   a. GOF: In wing discs, *Dl* is expressed in PNCs [2296], and ectopic expression of *sc* (or *l'sc*) activates *Dl* transcription in ectopic congruent areas [1854]. The transcription factor Senseless (zinc-finger class) has been implicated as a mediator of this link [3127].
   b. LOF: Deleting the AS-C abolishes *Dl* expression in PNCs of the embryonic CNS [1672].
   c. LOF: Mutating the binding sites for Ac/Da heterodimers in the *Dl* promoter reduces expression of a *lacZ* reporter gene (driven by that promoter) in PNCs of the embryonic neuroectoderm [2359].
2. Evidence for an "E(spl)-C ⊣ AS-C" link:
   a. GOF: In cultured cells, expression of *m8* and *m5* reduces the ability of Da to activate transcription of a *CAT* reporter gene (driven by an *m8* promoter fragment that has a Da-binding site) [3171].
   b. GOF: Overexpression of *m8* (via a *UAS-m8* transgene driven by *da*- or *hs-Gal4*) prevents SOPs from arising [3037, 4256] – an effect traceable to reduced *sc* transcription in SOPs (mild *m8*^{GOF}) [918] or PNCs (extreme *m8*^{GOF}) [991]. Also, bristles fail to develop when *m8* is overexpressed in SOPs (*UAS-m8* driven by *scabrous-Gal4*) [3037].
   c. Correlation: E(spl)-C bHLH proteins are naturally absent from the SOP (high AS-C level) but present in the rest of the PNC (low AS-C level) [2052].

In theory, the "E(spl)-C ⊣ AS-C" effect could be mediated in at least three ways [1458, 2063, 4318]:

1. Protein-protein binding via HLH domains. E(spl)-C bHLH proteins could trap Ac, Sc, or Da monomers in inert heterodimers [68, 348, 1484].
2. Transcriptional repression of AS-C genes by DNA binding.
3. Transcriptional repression of genes that are downstream targets of AS-C proteins.

The first idea was discounted because M8 does not block Ac/Da or Sc/Da dimer formation *in vitro* [4452], although M8 can bind Ac, Sc, and Da under other conditions [68, 1484].

The second route seems likely for "*m8* ⊣ *ac*." Although M8 does not bind E boxes [4452], it does bind N boxes (Fig. 2.4) [3171, 4318], and there is an N box in *ac*'s promoter. Indeed, *m8* reduces transcription of an *ac*-promoter *CAT* reporter [1794]. The "*m8* ⊣ *sc*" link should work the same way because there is an N box in *sc*'s "SOP enhancer" [918], but M8 still represses *sc* when this box is deleted. The reason seems to be that M8 can also get to the enhancer indirectly by binding an NF-κB-like factor that, in turn, binds a motif near the N box [918].

Convincing evidence for a transcriptional mode of action for M7 comes from a clever "chimera" experiment. E(spl)-C bHLH proteins are repressors mainly because the WRPW motif at their C-terminus recruits the co-repressor Groucho (Fig. 2.4) [1794]. When an artificial "*m7*^{ACT}" gene was built by replacing WRPW with a VP16 activator domain (from herpes virus [4394, 4433, 4483]) [2063] and put into flies with a *Gal4* driver, it caused extra bristles in various body regions – an effect opposite to the balding seen when native E(spl)-C bHLH proteins are overexpressed [918, 991, 3037, 4256]. It also caused ectopic activation of *lacZ* reporter genes linked to 4-kb promoter fragments from *ac* and *sc*. Apparently, *m7*^{ACT} – and, by inference, native *m7* – acts by influencing the AS-C at the transcriptional level (Possibility 2). The effect is apparently not exerted via E boxes because M8 does not displace AS-C/Da heterodimers from E boxes *in vitro* [4452].

The existence of E boxes in *m8*'s promoter (cf. Fig. 2.4) [171] suggests an "AS-C ➙ E(spl)-C" link [2317, 3171], and E boxes in *m7*'s promoter are required for reporter gene expression in PNCs [3974]. However, this connection is probably moot because

1. E(spl)-C bHLH genes are downregulated by N^{LOF}, despite a functional AS-C [2052].
2. E(spl)-C bHLH genes are activated by N^{GOF}, even in an *ac*⁻ *sc*⁻ background [2052].

Indeed, if an "AS-C ➙ E(spl)-C" link were to exist, then the SOP's abundant AS-C proteins should cause its level of E(spl)-C proteins to be the highest in the PNC, but the opposite is true: the SOP is the only PNC cell that lacks E(spl)-C bHLH proteins [2052]. The loss of these proteins from SOPs is attributable to a separate (indirect) path that is triggered by high levels of AS-C gene products: "AS-C ➙ *senseless* ⊣ E(spl)-C" [3127].

According to the MI model, the loop between the AS-C and Notch should allow the AS-C loci of adjacent cells to interact [436, 1458, 1794]. In tracing the circuit from one cell's AS-C into the other cell and back again (Fig. 3.6), there are two negative steps (E(spl)-C ⊣ AS-C) so the overall loop is positive [1458]. For equivalent cells, the AS-C levels will be unstable. Under these conditions,

any stochastic change should "boomerang" ("higher AS-C ➙ loop ➙ even more AS-C" or "lower AS-C ➙ loop ➙ even less AS-C"), breaking the symmetry and tipping the balance. At some point, the victor's AS-C level will cross a threshold, after which its autonomous SOP loop (Ac/Sc ➙ bind E boxes ➙ more Ac/Sc) will take over. Other facts are consistent with this scenario:

1. Heat-pulsing flies carrying a t.s. N^{LOF} allele upregulates *sc* in PNCs via an effect on *sc*'s "SOP enhancer" [918]. Similar stimulation of *ac* is seen in embryos that are defective in Notch-pathway genes [3983].
2. In *Su(H)*^{null} wing discs, *ac* is upregulated in PNCs [3820, 3827], supposedly via downregulation of the E(spl)-C. Interestingly, *Dl* is also upregulated – an effect consistent with the proposed feedback loop.

Nevertheless, the circuitry inside SOPs must differ from that of ordinary PNC cells.

1. Deleting all 3 Su(H)-binding sites in the *m8* promoter turns OFF *m8* in PNCs (assayed by an *m8* promoter-*lacZ* reporter) but turns it ON in SOPs [2453]. Is the E(spl)-C regulated by genes outside the Notch pathway, and, if so, why should *m8* be upregulated when deprived of activation by Su(H)? Conceivably, the deletions remove binding sites for unknown *trans*-acting repressors that are specific for SOPs (e.g., Senseless? [3127]). Deleting N boxes in *m8*'s promoter has the same effect in embryos: expression stops in PNCs and starts in neuroblasts [2317].
2. PNCs arise normally in H^{null} flies, but SOPs fail to appear [197]. This balding is suppressed when E(spl)-C bHLH genes are also deleted [197], which is strange because these genes are not normally expressed in SOPs [2052]. Possibly, H prevents Su(H) from turning E(spl)-C genes ON in SOPs, but (for unknown reasons) has no effect in other PNC cells. Disabling *H* would cause "Su(H) ➙ E(spl)-C ⊣ AS-C" (and stifling of SOPs), but additionally disabling E(spl)-C would restore AS-C function [197]. These results have been interpreted as showing that the SOP is sensitive to inhibition from non-SOP cells [197] – a conclusion that would invalidate the LI model. However, the H^{null} and double-mutant traits can also be explained by effects solely inside the SOP (given the circuitry above and the H ⊣ Su(H) antagonism discussed in Ch. 2), so the facts do not inherently favor either model. (A similar restoration of normalcy is seen in the H^{LOF} *Su(H)*^{LOF} double mutant [3827].) Moreover, *H* transcripts are uniform in the epidermis when SOPs emerge [200], so it is unclear how H could be acting differently in SOP vs. non-SOP cells [3827].

3. The zinc-finger transcription factor Senseless (Sens) accumulates only in SOPs [3127]. Because (1) E boxes reside in the *sens* promoter and (2) binding sites for Sens (AAATCA) exist in the *cis*-regulatory regions of *ac* and *sc* (as well as *m8*), Sens could be mediating the positive feedback loop that fosters the surge of proneural proteins in SOPs: "AS-C ➔ Sens ➔ AS-C". Indeed, *sens* exhibits strong synergy (i.e., dense tufts of bristles) with *scute* when they are co-expressed (via *dpp-Gal4*) in eye discs [3127].

4. Overexpression of the EGFR inhibitor Argos suppresses SOPs without affecting PNCs [917]. It does so by blocking an autocrine loop: "AS-C ➔ EGFR ➔ AS-C". The blockage prevents the SOP from exceeding the level of Ac or Sc that is present in the PNC. The external ligand in this loop appears to be Spitz [917].

Doses of Notch-pathway genes can bias the SOP decision

As stated above, SOP selection depends on AS-C dose in mosaics: a wild-type cell is 2–4 times more likely to become a SOP than a cell with half that dose ($ac^- sc^-$/++) [912, 1794]. When the boundary between high- and low-dose cells bisects a PNC, the two genotypes presumably compete within the mosaic PNC, and their abilities to attain the SOP state are reflected in the overall frequencies of differently marked bristles. Such contests are a key part of the MI model, but they can also fit the LI model. Maybe cells with more Ac and Sc are simply getting a head start in the race to become a SOP [917]. Such a temporal advantage might explain why female cells become SOPs 4 times more often than male cells at gender boundaries in gynandromorphs [1800].

Considering the ties between Notch-pathway genes and the AS-C, it is not surprising that doses of those genes also affect SOP choice. *Notch*, *Delta*, and *Hairless* are among the 21 fly genes (excluding *Minutes*) that are haplo-insufficient [118]. In haploid dose, both *Notch* and *Delta* cause more mCs [997, 4466], while a haploid dose of *Hairless* stifles some MCs, reduces the number of mCs, and transforms cell types within the bristle organ (cf. Ch. 2) [997].

Neuroblast commitment in the embryonic CNS was already known to depend on the doses of Notch-pathway genes [1000, 4467] when, in 1991, two labs reported that *N* dosage can bias SOP fates in discs. García-Bellido's team induced marked +/+ clones in N^{null}/+ vs. +/+ flies and found more +/+ bristles in the +/+ controls (*N* dosage of clone:background = 2:2) than in the N^{null}/+ flies (2:1) [996]. In other words, the more N a cell has relative to its neighbors, the *less* likely it is

to become a SOP [112, 1458, 4193]. This same conclusion was reached by Heitzler and Simpson: in *N* mosaics, the bristles typically develop from cells having less *N* wherever tissues of unequal dose (1:2 or 2:3) confront one another at a genotype boundary [1797]. In *Dl* mosaics, the border bristles tend to come from cells with more *Dl* (4 vs. 3 doses) [1797], while in *Su(H)* mosaics, they tend to arise from cells with less *Su(H)* (2:4 or 3:4) [3820]. In summary, PNC cells compete to become SOPs, and the outcome can be affected by virtually any component in the AS-C-Notch feedback loop.

The dosage sensitivity of the SOP decision may help explain an odd class of dominant *N* alleles called *Abruptex* (N^{Ax}) [997, 1274, 3435, 3436]. Like artificially expressed *N-intra*, the N^{Ax} mutations eliminate bristles by stifling SOP initiation [197, 1798, 2627, 3235, 3863]. Unlike N-intra, however, their effects are ligand dependent: N^{Ax} phenotypes are suppressed by Dl^{LOF} [992, 4780] or lower *Dl* dosage [460, 1798]. N^{Ax} mutations localize to EGF-like repeats 24–29 in *N*'s extracellular domain [1747, 2182], while N-Dl binding relies on repeats 11 and 12 (Fig. 2.3) [985, 3544]. Nevertheless, for some reason, N^{Ax} proteins have less affinity for Dl [2543]. Attempts to solve this "**Abruptex Paradox**" have invoked effects on N-N oligomerization [2182] or *cis* N-Dl interactions on the same cell [1798], but the only certain culprit is Fringe, an auxiliary ligand [992, 2543]. Fringe normally inhibits N's response to Serrate (cf. Ch. 6), but N^{Ax} mutations enable Fringe to *activate* the pathway [990, 994, 2096], possibly by hypersensitizing N to Dl [3245]. The ways in which N^{Ax} traits are modifiable by downstream genes (e.g., *H* or *gro* [992]) or by AS-C dosage [996] suggest a GOF effect of N^{Ax} on the pathway, as does the ability of excess N to mimic N^{Ax} traits [3235, 3863]. However, N^{Ax} cannot mimic N's ability to bypass *shibire*^{LOF} [3863], and certain N^{Ax} alleles must be LOFs based on their enhancement over a deficiency [460]. Surprisingly, the latter N^{Ax} alleles do not act via Su(H), but rather via the Wingless pathway [461]. This cross-talk (which precedes PNCs) may involve binding between N and Dishevelled – a transducer for Wingless [151, 356, 1054].

Dl appears to be the sole ligand for N in setting SOP fates in PNCs because Dl^{null} somatic clones exhibit a maximal tuft phenotype [4859]. Thus, Serrate is not an auxiliary ligand here, unlike its redundant role inside the 5-cell bristle organ (cf. Ch. 2).

Extra SOPs could be inhibited by contact or diffusion (or both)

PNCs that produce a single MC sometimes contain 2 or 3 SOP-like cells, as detected by low-level expression of a neural *lacZ* reporter [1925]. The implication is that these "pre-SOPs" are competing [918, 1458]. Presumably, one eventually wins and proceeds to form the bristle, while the other loses and regresses to an epidermal state [918]. Curiously, some pre-SOP rivals are nonadjacent (Fig. 4c of [1925]). How can they compete via a Dl-N mechanism if they do not touch?

This finding raises a key question. If Dl alone transmits inhibitory signals in PNCs, then inhibition should require direct contact, unless Dl can diffuse (see below). Given how epidermal cells are packed [4641], a SOP should touch no more than 5 or 6 cells [3641]. Until 1991 (when Ac and Sc were seen *in situ*), the number of PNC cells per MC site was thought to be in this range (based on the number of bristles per tuft in partial-LOF *N* and *Dl* mutants [3641, 3950] though shi^{LOF} is more extreme [3425]), but then it was found that a mature PNC actually contains 20–30 cells [912, 3982] – far too many for an ordinary cell to reach [1458, 3272]. Two possible ways out of this dilemma have been proposed:

1. The only PNC cells that are truly in contention to become a SOP are those that express Ac and Sc at a *higher* level [918, 1458, 3022]. The number of such cells per PNC is indeed in the 5 or 6 range [912, 913, 3982]. In PNCs that yield two MCs (e.g., the dorsocentrals), the posterior SOP arises before the anterior SOP, and the PNC area that stains intensely for Ac and Sc appears to shift accordingly (Fig. 3.3). "Heterochronic pairs" [1925] are governed by single AS-C enhancers [1538]. The timing mechanism is unknown, but it too may rely on the AS-C [1925] because less Ac can cause the dorsocentral PNC to yield a single bristle that occupies an intermediate position [4095], and other odd spatial and temporal shifts affect SOPs when AS-C function is derepressed [1926].

2. The SOP inhibits other PNC cells via filopodial extensions of its cell body [1802, 1810, 1813]. Filopodia indeed extend from SOPs before they divide [2387, 3630], and the filopodia of neighboring SOPs on the notum can touch (F. Roegiers, pers. comm.). Anti-Dl antibodies do not detect such extensions *in situ* [2296, 3270], but their visibility may be limited by their size. Intriguingly, when Dl- and N-expressing cells are cocultured *in vitro*, filopodia sprout from the former but not the latter:

We consistently noted a morphological difference between Delta[+] and Notch[+] cells in mixed aggregates that were incubated overnight. Delta[+] cells often had long extensions that completely surrounded adjacent Notch[+] cells, while Notch[+] cells were almost always rounded in appearance without noticeable cytoplasmic extensions. [1204]

In 1999, the old dogma of contact-mediated Dl-N signaling [2296, 3022] was challenged with the discovery of a soluble *in vivo* form of Dl that is snipped from full-length protein by a protease [3479]. The protease is encoded by *kuzbanian* (*kuz*). Kuz may also cleave N [2298, 3153, 3239, 3640] ([3479] sequel), but N appears to be cleaved by a different metalloprotease [491, 2995], as well as by Presenilin [4156] and by a Furin-like convertase (cf. Fig. 2.3) [113, 712, 1723, 2589, 3335].

Somatic clones of kuz^{null} cells exhibit bizarre phenotypes. The clone interior is devoid of bristles, but dense tufts of kuz^{null} bristles develop at the borders wherever a MC or mC site is encountered [3640]. The phenotype can be explained if Kuz plays both a proneural and an inhibitory role [3640]. Conceivably, (1) a Kuz-dependent proneural factor "X" diffuses several cell diameters from the wild-type tissue and (2) X prompts kuz^{null} SOPs to arise at the border, but (3) those SOPs cannot inhibit other PNC cells because they cannot make a Kuz-dependent inhibitor (Dl). Assuming that Kuz's main duty is to liberate Dl, it is easy to see why kuz^{LOF} mutations have no effect on cell fates inside the bristle organ (App. 3): in that case (unlike the PNC situation), Dl only needs to activate N receptors on an adjacent cell and hence would not need Kuz to help it diffuse.

Certainly, Dl is needed for inhibition, but it may not be the actual diffusible signal. Indeed, the notion of diffusible Dl is hard to reconcile with what has been learned about Dl-N interactions in wing veins and ommatidia. In those tissues, Dl cannot activate N unless cells express membrane-bound Dl, which grabs N's extracellular domain from apposed cell membranes and escorts it into the *Dl*-ON cell via endocytosis [3271]. The idea also seems incompatible with the antagonistic properties of artificially secreted Dl constructs [4207].

Scabrous may be the diffusible SOP inhibitor

If the AS-C-Notch-pathway loop were solely responsible for ensuring one SOP per PNC, then the same phenotypes should be attainable by all genes in the loop. As mentioned before, however, LOF mutations of Notch-pathway genes cause denser bristles than AS-C^{GOF} mutations. Indeed, even when AS-C genes are overexpressed using powerful promoters, no N^{LOF}-like tufts are ever seen [918, 3628]. Why don't high AS-C amplitudes drive the loop to saturation and transform every PNC cell into a SOP? Conceivably, even under these extreme conditions, the SOP still emits an inhibitor that prevents SOPs from arising closer than a few cell diameters. Indeed, this assumption was axiomatic in the LI model [3641].

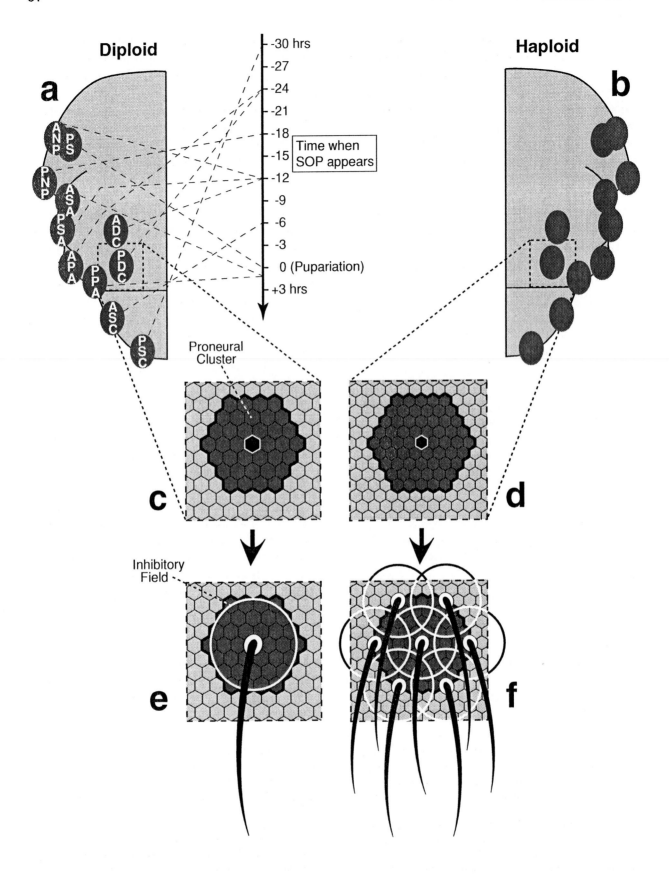

Diploid

a

ANP
PS
PNP
ASA
PSA
ADC
APA
PDC
PPA
ASC
PSC

-30 hrs
-27
-24
-21
-18
-15
-12
-9
-6
-3
0 (Pupariation)
+3 hrs

Time when
SOP appears

Haploid

b

Proneural
Cluster

c

d

Inhibitory
Field

e

f

The irrepressible inhibitor may be Scabrous (Sca). LOF alleles of *sca* cause extra bristles near normal sites [1921, 2885], and Sca has attributes (listed below) appropriate for this role [178]. Paradoxically, however, its GOF effects in the wing are opposite to what would be expected: excess Sca inhibits Notch signaling by interfering with Dl's activation of N [2462]. Sca does not appear to be a ligand for N [186, 2461]: when phage libraries of fly DNA sequences were screened for N-binding proteins, *Dl*, *Serrate*, *wingless*, *big brain*, and *N* were fished out, but *sca* was not [4823]. Nevertheless, Sca does form complexes with N [3456], and *sca* interacts genetically with both *Dl* and *N* [179, 437, 1921, 2885, 3490].

1. Sca is a secreted protein (soluble glycoprotein dimer) that travels at least 3 cell diameters from its source cells (R8 photoreceptor precursors) in the eye disc [2461]. A comparable diffusion range would suffice for it to act as an inhibitor for bristle SOPs.
2. Sca is first expressed in PNCs and strongly in SOPs, after which it fades from all PNC cells, except the SOP [197, 2885]. This time course mimics Ac and Sc.

Genetically, *sca* is downstream of the AS-C: (1) excess Ac (due to *Hairy wing* mutations) causes *sca* expression in the same (normally bare) areas as *ac* (e.g., posterior wing blade) [2885], and (2) *sca* transcription ceases in an *ac⁻ sc⁻* background [3974].

This "AS-C ➔ *sca*" link is due to 4 E boxes in *sca*'s promoter [3974]. As AS-C levels rise in the SOP, Ac/Sc/Da should turn ON *sca* by binding these E boxes. The "AS-C ➔ Dl" link (described earlier) should also cause upregulation of *Dl*, which likewise has E boxes ($\geq$12) in its promoter [2359]. However, incipient SOPs do not express

more Dl than other PNC cells [2296, 3270]. This uniformity was interpreted as favoring the MI over the LI Model [3022, 3270], but this fact is moot if a separate inhibitor is operating.

According to one report [1203], N fades from SOPs as they mature, and this "deafness" should allow AS-C levels in the SOP to climb without interference from its Dl-expressing neighbors. Is the SOP similarly "deaf" to its own Sca? The answer must await identification of the Sca receptor and determination of whether it is expressed in the SOP [2461, 3456].

One remaining mystery is why the *sca*ⁿᵘˡˡ phenotype is so mild. Conceivably, additional SOP inhibitors might be acting redundantly with Sca [1158, 2885]. Sca must be as important as Dl [2592, 2628] because much of the second chromosome's naturally occurring variation in bristle number (32% for abdominal and 21% for sternopleural bristles) maps to this locus [2172, 2381], and *sca* contributes heavily to the modification of bristle number in artificial selection trials [1658, 2638, 2643, 2644].

Inhibitory fields dictate the spacing intervals of microchaetes

If Sca is a SOP inhibitor, then it should suppress would-be SOPs within a ~3-cell radius. In a hexagonally packed epithelium, an "inhibitory field" of this size would cover 36 cells (6, 12, and 18 cells in concentric shells) – enough to reach all 20–30 cells in an average PNC (Fig. 3.7).

Regardless of how Sca acts, it is worth pondering how inhibition might work in general. If each SOP creates its own inhibitory field, then patterns could be constructed "dot by dot." That is, extant SOPs could be used as points of reference for nascent SOPs in the same way that a

---

FIGURE 3.7. Origin of macrochaete SOPs on the notum.

a, b. Heminota (minus humeri) composed of diploid (2n, **a**) vs. haploid (1n, **b**) cells should look grossly alike because ploidy does not alter body size in animal species [1190, 1191]. The chief effect of ploidy is a proportional change in cell volume [1801]. Ovals denote macrochaete PNCs, most of which arise (asynchronously) before the notal region acquires the shape shown (cf. Fig. 2.6 for acronyms and Fig. 3.3 for a realistic depiction of a PNC). Dashed lines mark when each SOP appears [1925, 3374]. This temporal sequence is essentially irrelevant to the pattern of axonal fasciculation within the disc [4431, 4432] and axonal projection within the CNS [1455] (not shown).

c, d. Area where the PDC macrochaete arises (enlarged). Ploidy uncouples mechanisms that depend on cell size (inhibitory field diameter?) from those that do not (PNC diameter?). Assuming that PNCs (dark shading) come from a disc's global prepattern, a 1n PNC will have the same area as a 2n PNC, but should contain more cells (small hexagons) because 1n cells are ~21% smaller in diameter ($0.79^3 = 0.5$; 1n wing cells have 40% less area [3750]). Central cell (black with white outline) is the SOP.

e, f. SOPs may use a diffusible inhibitor (circle = range) to prevent excess SOPs. If smaller cells secrete less inhibitor, then 1n SOPs may not be able to inhibit outlying cells, some of which will become SOPs and set up their own fields. Indeed, 1n-2n mosaics do have extra bristles at MC sites (and higher bristle densities overall) when occupied by 1n tissue [3750, 3754]. However, there is usually only one extra bristle per site and it tends to arise anteriorly or posteriorly, but not laterally. This result favors the Lateral Inhibition Model over the Mutual Inhibition Model (Fig. 3.6; see text).

geometer uses a compass for triangulation [800, 4657]. This supposition leads to two predictions:

Prediction 1:   The inhibitory radius should depend on the size of the SOP cell.

Prediction 2:   Inhibitory fields set the intervals between evenly spaced bristles.

Both expectations fit the LI model, but neither of them easily fits the MI model.

Prediction 1 is testable by using ploidy to uncouple cell size from body size. As a rule in animal species, alterations in ploidy cause ubiquitous (proportional) changes in cell volume but do not change body size [855, 1190, 1191]. Thus, the SOPs in haploid (1n) flies should be half the volume of SOPs in diploid (2n) flies, but 1n PNCs should be as large as 2n PNCs because PNC dimensions should depend only on body size. Under these conditions, a 1n SOP might not be able to secrete enough inhibitor to cover its PNC, in which case multiple bristles should arise (Fig. 3.7). Indeed, extra MCs develop at the normal MC sites when those sites are occupied by 1n tissue in 1n/2n chimeras [3750]. This result is consistent with Prediction 1.

Prediction 2 is supported by the observations on the spatial distribution of mCs under various conditions:

1. In the 1n/2n chimeras mentioned above, the mCs are more closely spaced in the 1n areas [3750, 3754], presumably because 1n SOPs exude less inhibitor than 2n SOPs.

2. Triploid flies have larger cells and larger bristle intervals than diploid flies [1801], presumably because 3n SOPs exude *more* inhibitor than 2n SOPs.

3. Starved flies have smaller cells and smaller bristle intervals [1801, 3617].

4. *Egfr*^{LOF} mutations cause smaller cells and denser bristles [1042].

5. On the abdominal sternites of wild-type flies, bristle lengths and spacing vary, and "nearest-neighbor distance" is proportional to bristle length [804]. Tergites exhibit a similar trend [801, 2083, 2690], and bristle interval varies with bristle length among bristle rows on the second-leg basitarsus (Fig. 5.1e) [1803].

6. Females, which are larger than males, tend to have more bristles (except MCs) on the notum [1741, 4428], wing [1741], legs [1714], tergites [801, 1357, 2690], sternites [3522, 3555], and sternopleura [3522].

7. Sternopleurae exhibit extra bristles (within a gender) when cell number increases [1471, 4036].

The above trends argue that mCs are spaced by a "bottom-up" process ("$\rightarrow$" denotes "causes") [862]:

SOP size $\rightarrow$ inhibitory field radius $\rightarrow$ distance to next bristle $\rightarrow$ pattern.

In contrast, MCs appear to be patterned by a "top-down" strategy:

Disc-wide coordinate system(?) $\rightarrow$ pattern of PNCs $\rightarrow$ pattern of SOPs.

Here, we encounter a "**MC vs. mC Paradox**." How can MCs and mCs be patterned differently if they each come from the same sort of PNC? The answer is: They don't.

In fact, mCs tend to come from proneural areas that are larger than the PNCs for MCs (see below) [1453], and the locations of the SOPs within those areas are not preset [913, 3954]. This indeterminacy explains the variability of mC numbers and positions (1) from fly to fly [3522], (2) on the two sides of the same fly [432, 850, 1714, 3553, 3554], and (3) from one row to the next [1808, 1813, 1883]. In contrast, MCs are notoriously constant in pattern [3206, 3966].

Our insights into why MC and mC patterns behave differently gelled in the 1990s, but Curt Stern guessed the answer in 1956 from his studies of *Minute*^{LOF} mosaics. As mentioned in Ch. 2, *Minutes* are genes whose LOF alleles cause small bristles [118, 2561]. Stern found several mosaics where a "not-*Minute*" (+/+) clone formed an extra MC near a "*Minute*" (*Minute*^{LOF}/+) dorsocentral MC [4097] (cf. his Fig. 4). His interpretation resembles the argument above for 1n/2n chimeras insofar as he attributes the duplicate MCs to inadequate inhibition. In this same passage, he explains how asynchrony – if the hiatus is long enough – could permit the inhibition from one SOP to fade before a new bristle is initiated at virtually the same site (italics are author's):

All cases of duplicate or supernumerary bristles in mosaics may be interpreted by the assumptions (1) that *the prepattern defines not sharp points but larger regions* characterized by gradients in strength, (2) that the differentiation of a bristle tends to inhibit the differentiation of other bristles in its surroundings and (3) that under equal circumstances *the inhibiting effect of a differentiating large not-Minute bristle is stronger than that of a Minute bristle*. Therefore, if a Minute bristle develops in a mosaic and the more competent not-Minute tissue covers nearby parts of the prepatterned region, then the inhibition exerted by the differentiating Minute bristle is insufficient to suppress the formation of a twin or supernumerary bristle in the [clone]. . . . [But] difficulties remain. Inhibition of bristles should be a mutual phenomenon. Why then, when it is assumed that a Minute bristle is not strong enough to inhibit a not-Minute twin, should not a not-Minute bristle inhibit the formation of its Minute twin? One might speculate that *twin bristles arise when the Minute partner happens to begin differentiation earlier than the not-Minute duplicate and once having started on its developmental path is not subject to inhibition any more.*

A similar rule governs the PNCs at MC sites where two SOPs normally develop (e.g., the ADC and PDC bristles within the dorsocentral PNC). Such SOPs are called "heterochronic pairs" because one member always arises before the other. Huang *et al.* argued that the interval between the two MCs is fixed by the radius of the inhibitory field that emanates from the "first-born" SOP [1925]:

In each case, the late SOP of the pair appears 3–4 cells away from the early one. The relief from inhibition of the second SOP could simply result from the intercalary growth of the disc, leading to an extension of AS-C expression beyond the radius of inhibition.... If our interpretation of the heterochronic pair is correct, it indicates that, in the notum and wing veins, lateral inhibition plays a dual role in pattern formation: first, ensuring that only one SOP will form at each site; second, allowing late SOPs to form at regular distances from early ones.

The "halo" of lower Ac-Sc expression around incipient SOPs (Fig. 3.3) fits this scenario, because it implies an inhibitory influence from the SOP [912], although it could be an artifact of changes in cell size and nuclear position [1458]. ADC-PDC idiosyncrasies in *ac¹* mutants had long ago intimated an interdependence of paired SOPs:

The two central bristles [ADC, PDC] together form an interdependent system. If the PDC develops, the ADC is also always present, but if the PDC is absent the ADC may or may not develop. Another property links the two central bristles together. Their positions on the thorax, as measured by the distance from some rather fixed landmarks, are variable in a peculiar way. When both bristles occur, the ADC is about where it would be in normal flies, but the PDC arises much closer to the ADC than a normal PDC would. When the PDC is absent, then the ADC does not form at its accustomed place but appears farther back. The apparent shift in position toward the posterior is clear-cut, though usually not extreme. [4096]

In summary, the PNCs at MC sites can contain more than one SOP (either naturally or under abnormal circumstances). In those cases, the interbristle distance appears to be set by the radius of an inhibitory field that is produced by whichever member of the SOP pair arises first. The use of an inhibitory field in this way by a MC is unusual, but it is commonplace for mCs. The

histological evidence for these conclusions is reviewed below.

Microchaetes come from proneural stripes, not spots

Notal mCs (and indeed most small bristles) develop after pupariation, whereas most MCs originate during the 3rd instar. For this reason, the same area can be patterned twice by separate mechanisms with no interference whatsoever [2427, 3067]. This "**Heterochronic Superposition Trick**" is used not only for MCs vs. mCs on the notum but also for (1) chemo- vs. mechanosensory (early vs. late) bristles on the legs [1803, 3628] and (2) photoreceptors vs. bristles (early vs. late) in the eyes [602, 603].

Between the stages of MC (early) and mC (late) patterning, the intervals between MC SOPs increase due to cell proliferation [1363, 3951] and epithelial stretching (during evagination). Therefore, the inter-SOP distances for MCs and mCs may not be as different at the outset as they appear in the adult. If so, then the radii of the inhibitory fields used by these different sets of SOPs might also be comparable. The inhibitory fields of the MC SOPs probably vanish before the patterning of mC SOPs begins [1458].

During the "classical" era of bristle-pattern research from ca 1954 (Stern's work on *ac¹*) to 1985 (Modolell's cloning of the AS-C), most of the theorizing revolved around MCs, and this trend continued through 1991 when MC PNCs [912, 3982] and SOPs [1925] were documented. By that point, a dogma had emerged whose tenets were as follows:

1. Every bristle has its own PNC (2-SOP nests are exceptions).
2. AS-C action (conferring competence) always precedes Notch-pathway action (ensuring one SOP per PNC).

In 1993 and 1997, analogous studies were made on SOPs [4428] and PNCs [3270] of notal mCs and leg bristles [3193]. These bristles, which are aligned in rows, arise differently from most MCs, although they share some features with 2-SOP MC nests. The heretical findings are listed below and summarized in Figs. 3.8 and 3.9 (cf. Fig. 3.7).

1. SOPs within a notal mC row do not originate in a fixed sequence from fly to fly. Rather, late SOPs emerge in the largest gaps between SOPs that happen to develop first [4428], just as Wigglesworth observed in

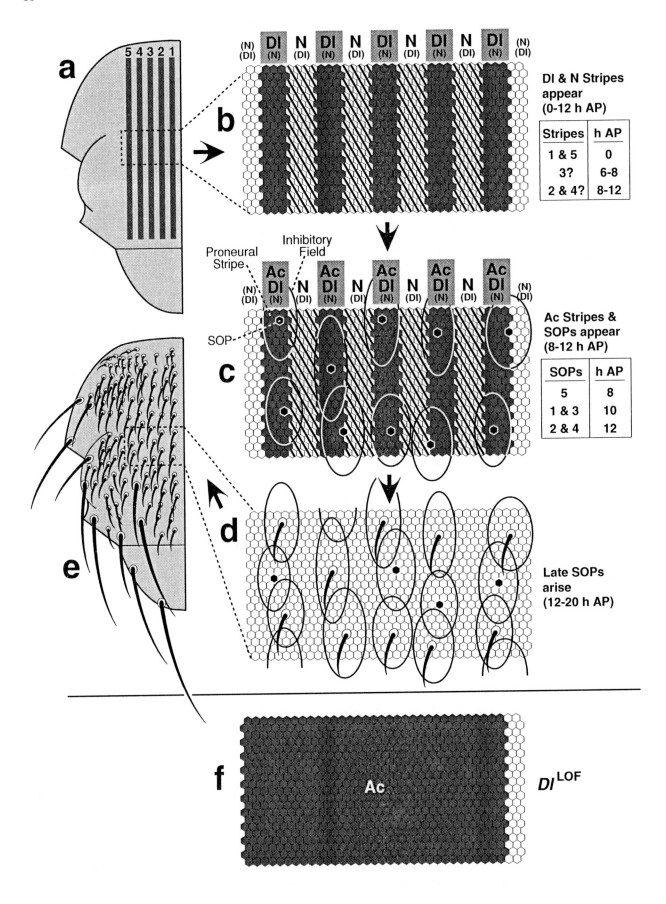

DI & N Stripes appear (0-12 h AP)

Stripes	h AP
1 & 5	0
3?	6-8
2 & 4?	8-12

Ac Stripes & SOPs appear (8-12 h AP)

SOPs	h AP
5	8
1 & 3	10
2 & 4	12

Late SOPs arise (12-20 h AP)

Dl^LOF

hemipterans [4657]. These vagaries contradict any notion of prepatterned PNCs or predestined SOPs.

2. Unlike MCs, where Ac is expressed in PNC "spots," notal and leg bristle rows are preceded by Ac stripes [3193, 3270]. The stripes could be chains of fused spots [3951], but other evidence argues that the distinction is real [1804, 3193, 4428].

3. In the notal mC region, Dl and N are expressed in alternating stripes ~8 h before Achaete is detectable [3270]. This "backward" sequence rules out an "AS-C → Dl" link, and this conclusion was confirmed when Dl stripes were shown to develop normally in an ac^- sc^- background.

4. Each Dl or N stripe is 3–5 cells wide [3270]. (Dl and N are both expressed outside their own stripes at low levels.) If Dl-N contests (Fig. 3.6) only occur between adjacent cells, then Dl and N must be using some other mechanism to form alternating stripes that are more than one cell wide.

Hints of "proneural stripes" had previously come from heat-sensitive N^{ts1} and shi^{ts1} hypomorphs. Their notal rows 1 and 5 swell with extra bristles to stripe-like widths when pulsed at 6–12 h (N^{ts1}) or 14–20 h (shi^{ts1}) after pupariation [1742, 3863, 3891]. The N^{GOF} allele N^B suppresses all rows except row 1 [102, 2626, 2627]. Somatic clones that are null for all 7 bHLH E(spl)-C genes cause simi-

lar stripes of densely packed mCs [1794]. If each mC had its own PNC, then the LOF defects should have been punctate clumps (cf. Brd^{GOF} flies [2499, 2500]), rather than uniformly dense bands of bristles.

The fact that Dl precedes Ac in stripes 1–5 suggests a "Dl → AS-C" link (cf. proneural roles for N [460, 2540]), but the old "Dl ⊣ AS-C" rule must still hold because (1) reducing Dl function during this period upregulates ac, and (2) raising N function suppresses ac [3270]. The "Dl ⊣ AS-C" effect may initially be blocked within Dl stripes because Dl is seen first on cell surfaces and only later in cytoplasmic vesicles (as is N).

Derepression of ac in a Dl^{LOF} background causes Ac to fill the row 1–5 area (Fig. 3.8f) [3270]. Evidently, Ac stripes in wild-type flies arise by a geometric subtraction. (The term "antineural" denotes inhibition of SOP initiation [1440, 2293, 4387, 4898].)

> 1 large Ac proneural area (Fig. 3.8f)
> minus 4 high-N antineural stripes (Fig. 3.8b)
> _____
> equals 5 Ac proneural stripes (Fig. 3.8c)

Extra bristles in N^{ts1} flies are mostly within the rows, which makes sense because an overall drop in N function should affect low-N (high-Dl) stripes first, where

---

FIGURE 3.8. Origin of microchaete SOPs on the notum.

a. Heminotum (minus humerus), showing proneural stripes where Achaete (Ac) protein (**c**) foreshadows the 5 medial rows of microchaetes (**e**). Stripes 1 and 5 converge and fuse at their posterior tips (not shown), and other stripes develop laterally. Stripe 5 overlaps the ADC and PDC macrochaete sites, whose SOPs arise ≥12 h before pupariation (Fig. 3.7). These SOPs may still exude inhibition because no microchaetes form between them.

b. Enlarged swath, showing cells (hexagons). Delta (Dl) protein appears at ~0 h after pupariation (AP), is expressed in all 5 stripes (shaded) by 12 h AP (cf. table), and is detectable at low levels ("(Dl)") elsewhere. Notch (N) protein is also ubiquitous but more intense (hatched) between Dl stripes. Whether these stripes arise via the kinds of AS-C *cis*-enhancers that govern macrochaetes is unknown [1538]. Also unknown is whether the early stripes constrain the later ones [3966].

c. By ~8 h AP, Ac is congruent with Dl but is confined to incipient SOPs (black cells outlined in white) by 12 h AP. SOPs appear in stripe 5, then 1 and 3, then 2 and 4 (cf. table). Inside each stripe, SOPs arise asynchronously.

d. Most SOPs are evident by 12 h AP, but some emerge as late as 20 h AP. By that time, the earlier SOPs have not formed bristles as shown here, but have divided and appear as 4-cell tetrads. Late SOPs arise in the largest spaces between extant SOPs, as Wigglesworth found in hemipterans. He imagined that SOPs stifle would-be SOPs nearby by emitting an inhibitor [4660] or absorbing an inducer [4657]. Only when a competent (Ac-expressing) cell finds itself outside inhibitory fields can it become a SOP. These rules produce a roughly even spacing of bristles in the available area. Proneural stripes will yield zigzag bristle rows if the fields are as wide (3–5 cells) as the stripes. Depending on how the signal is transmitted, inhibitory fields need not be circular. Here they are drawn as ovals (**c, d**).

e. Heminotum, showing all bristles.

f. Effect of lowering Dl function on Ac expression. When t.s. Dl^{LOF} mutants are heat-pulsed at 5–10 h AP, Ac is expressed ubiquitously except along the midline (the same band that remains bare in N^{LOF} and shi^{LOF} mutants [1742, 3863]). The implication is that Dl normally represses ac between the "Dl stripes." Apparently, the interstripes default to an ac-ON state in the absence of N signaling.

This diagram is based on data from [3270, 4428]. *N.B.:* It makes sense that bristles would be positioned late in development (after proliferation has ceased) because any pattern that is created earlier would be "messed up" by scattered mitoses [695]. This system thus obeys the general rule that proliferation and differentiation are mutually exclusive in eukaryotic cells [4882].

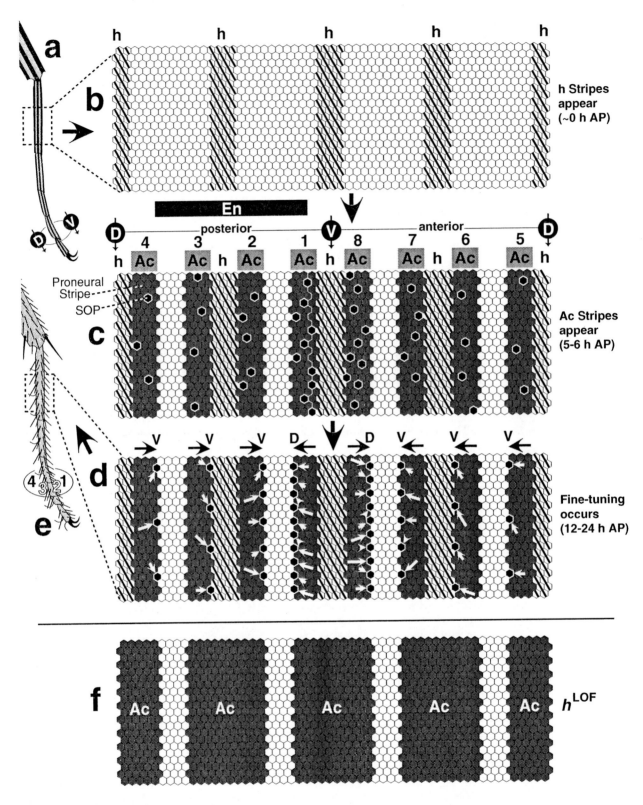

N is the limiting factor [1742, 3270]. Likewise, reducing *Dl* function derepresses *ac* in the low-Dl (high-N) stripes where Dl is the limiting factor, but there appears to be only minimal Ac intensification in the high-Dl stripes [3270].

A priori, one would expect a signal-receiver system to be silenced to the same extent, whether the signal or receiver is reduced [184]. Clearly, however, the Notch pathway is not muzzled as much in high-N–low-Dl as high-Dl–low-N stripes: the former suppress *ac*, whereas

FIGURE 3.9. Origin of bristle SOPs on the leg.

    a. Distal half of a left second leg (mid-tibia to claws), showing stripes of Hairy protein that appear at pupariation [665, 3193]. Only 3 of the 4 stripes are visible from this posterior view. Transverse stripes are omitted. Dorsal (D) and ventral (V) midlines are marked. At this stage, leg segments are actually shorter and wider than drawn here.

    b. Entire 360° surface of an enlarged swath (filleted at D midline), showing cells (hexagons) and all 4 Hairy stripes (hatched). Black bar (below) shows the extent of Engrailed (En) expression, which characterizes the posterior compartment.

    c. Same swath, showing 8 stripes of Achaete (Ac) protein (shaded) that all appear at 5–6 h AP (after pupariation) [3193]. These proneural stripes (each 3–4 cells wide) foreshadow the tarsal bristle rows (**e**). The number of bristles per row (SOPs per stripe) decreases with distance from the V midline [1714]. Positions of SOPs (black cells outlined in white) are hypothetical. Their emergence has not been examined histologically. Lineage studies indicate that SOPs in stripe 1 (at least) originate from ≥2 columns of cells because row 1 bristles can come from both compartments [1800, 2449]. SOPs in stripe 1 are depicted posterior (6 at left) or anterior (4 at right) to the boundary (white zigzag line). Inhibitory fields around SOPs are omitted.

    d. Inferred movements (white arrows). Lineage studies suggest that (1) bracts and bristles in each row come from separate columns of cells and (2) bristle cells move ventrally (V arrows; stripes 2–7) or dorsally (D arrows; stripes 1 and 8) to one edge of their stripe [1800, 2449]. Adjustments of this kind (previously seen with scale precursor cells on moth wings [3051]) would explain why leg bristles are more aligned than notal bristles. Other "fine-tuning" shifts may occur vertically, based on abnormal spacing in mutants with misoriented bristles [1810]. The horizontal and vertical shifts have been combined into one vector per SOP, although they probably happen at different times [1803] (Fig. 3.11).

    e. Adult left leg (posterior view), showing bristle rows 1–4.

    f. Effect of eliminating h function on Ac: Ac now appears in 4 broad stripes, implying that the 8 Ac stripes arise by repression of ac within Hairy stripes. This "subtractive" logic is similar to the strategy used on the notum (cf. Fig. 3.8). Dl and N protein distributions have not yet been assessed for the legs. Notch signaling might be repressing ac in the other 4 interstripes.

---

the latter do not. The reason could reside in (1) subtleties of Dl-N stoichiometry or (2) restricted distribution of other Notch-pathway components [3270]. For example, the $m8$ promoter drives reporter gene expression in 5 stripes per heminotum [2453], so ac could be OFF in high-N stripes because $m8$ can only be turned ON there. However, Dl might be diffusing into the high-N stripes [3479] and activating the pathway there and *only* there. In that case, high-Dl cells would be "deaf" to their own signals [988], perhaps due to *cis* interactions between Dl and N molecules on their surfaces [2008].

Hairy paints "antineural" stripes on the legs

LOF mutations in *hairy* (*h*) cause extra bristle rows on the notum [803] and ectopic mCs on the scutellum [1981, 2959], where Hairy is strongly expressed in wild-type flies [665]. Extra bristles also arise on the head, wings, legs, and flanks [1802, 2561]. The added bristles vanish when h^{LOF} is combined with ac^{LOF}, indicating an antagonism between h and ac. In fact, many of the excess bristles disappear even when ac^{LOF} is heterozygous [2959, 4185]. This sensitivity implies a delicate balance of proneural (ac) and antineural (h) dosages in certain regions [665].

    Curiously, the bHLH protein encoded by *hairy* does not bind Ac, Sc, or Da [68, 4452], so *hairy* cannot be inhibiting the AS-C via heterodimers. Hairy belongs to the same HLH subclass as E(spl)-C bHLH proteins. Like them, it should be acting as a repressor because its WRPW terminus recruits the Gro co-repressor (Fig. 2.4) [1242, 4865]. However, (1) h^{LOF} and gro^{LOF} phenotypes differ

[991], (2) *hairy*'s interaction with sc depends on a different ("Orange") domain [976], and (3) still another Hairy domain binds a regulatory C-terminal Binding Protein [3377, 3430, 4865] (App. 5). Hairy must operate by DNA binding because it is converted from a repressor to an activator when its WRPW tip is replaced by a VP16 activator domain [2065]. Proof that it functions exclusively in this capacity was provided by two studies in 1994 [3179, 4451] that showed the following:

1. Hairy specifically binds a GGCACGCGAC sequence in the ac promoter (∼300 b.p. upstream of the transcription start site). This same consensus was fished out from a pool of random oligonucleotides by Hairy "bait," so it is Hairy's optimal binding site. At least two E(spl)-C bHLH proteins (M5 and M7) can also bind the core hexamer (CACGCG) with high affinity.

2. In cultured cells, Hairy represses transcription of a transfected reporter gene linked to four of these sequences. The effect is abolished when the core binding site is mutated, so Hairy must exert its repression via this site.

3. In ac^- sc^- [4451] or wild-type [3179] flies, an ac^+ transgene (with its own promoter) is repressed by the endogenous h^+ but is derepressed (as assayed by extra bristles) when h^+ is replaced by h^{LOF}. This result argues that no redundant gene (besides h) represses ac via this site (despite the ability of E(spl)-C proteins to bind there *in vitro*). The same phenotype is obtained when the core binding site is mutated,

so all *hairy*'s repressive effects must be mediated by this *cis*-silencer.

On the legs, Hairy stripes play the same antineural role as Notch-pathway stripes on the notum. They appear before *ac* turns ON (Fig. 3.9b), and their removal derepresses *ac* (Fig. 3.9f) [665, 3193]. Unlike the notum, however, there are only half as many antineural stripes relative to proneural stripes (4 vs. 8). How is *ac* repressed in the four other interstripes? The Notch pathway must be involved somehow because heat pulses to N^{ts1} flies cause extra bristles between bristle rows as well as within them [1802, 1803].

Single Ac stripes on the notum and legs can produce more than a dozen SOPs without any obvious fragmentation of the stripes into proneural "clusters" *sensu stricto*. Hence, MC PNCs must be special cases of a more general regional identity that has been called the proneural "field" [913, 1804, 2890, 3127, 3193]. Proneural fields can assume various shapes and yield various numbers of SOPs. Both sorts of variations are seen to a limited extent among PNCs themselves [912]. Field shapes include not only irregular spots (PNCs) and smooth stripes (leg zones), but also the expansive rectangles that Wigglesworth had envisioned for the tergite epidermis 60 years ago [4657].

Is it necessary to visualize the prepattern only as a series of constantly located, sharply defined peaks rising from a flattened valley floor? Could not rounded domes, ridges or even plateaus be other forms in the prepattern landscape? If so, then the inhibition . . . in the neighborhood of a developing structure may conceivably be less extensive than the particular boundary imposed by the prepattern, and more than one and perhaps many structures might appear in a single element of the prepattern. The arrangement of structures within each of these regions could then be controlled by the mechanism envisaged in Wigglesworth's . . . model. [800].

In light of what has been learned about notal and leg bristles, the old MC-based models can now be reassessed. Just as haploid tissue makes extra bristles at MC sites (Fig. 3.7), it also makes more bristles per row [3750, 3754], so the number of SOPs per stripe is probably dictated by how many inhibitory fields can fit therein. In wild-type flies, the inhibitory field would cover any PNC wherein a single SOP arises, whereas inhibitory fields would be dwarfed by proneural areas that foster multiple SOPs. Within the latter areas, any synchrony of SOP inception could lead to SOPs developing too close together. Thus, asynchrony may have evolved to guarantee a certain minimum spacing interval (by forcing later SOPs to arise outside inhibitory fields of earlier

ones). That interval would precisely equal the inhibitory radius wherever SOPs develop in linear sequence [805], but only the eye appears to use a trick of this sort in the creation of its photoreceptor array (cf. Fig. 7.7) [1804, 1808].

Apparently, the Notch pathway has the same duty inside as outside the proneural notal stripes (and in proneural fields generally) – to downregulate the AS-C. Dl is probably not an SOP inhibitor here (as conjectured for MCs) because it is not elevated in mC SOPs [3270]. Notch signaling must be low enough (in the high-Dl stripes) that the AS-C can spark SOPs, which then escape N-dependent inhibition and emit a N-independent inhibitor (Scabrous?) [3270]. This working hypothesis is an amalgam of the MI and LI models in which mutual inhibition sets the stage for lateral inhibition.

Leg bristles use extra fine-tuning tricks

Leg bristles are intriguing because of their precise alignment and spacing. Moreover, they are organized into diverse subpatterns, unlike the homogeneous notal mCs. The basitarsi are especially rich in modular motifs [1714], and their development has been analyzed genetically [1799, 1802, 1807, 1808, 4346]. Their features are illustrated in Fig. 3.10.

The 2nd-leg basitarsus has 8 bristle rows symmetrically disposed about the dorsal-ventral plane. The number of bristles per row varies from fly to fly, and adjacent rows vary independently [1813, 1883]. Bristle length and interval increase with distance from the ventral midline (cf. Fig. 5.1e) [1801, 1803], implying that dorsal SOPs have larger inhibitory fields. Higher ventral density may be adaptive for finer resolution in detecting stimuli where the leg grasps objects (like the human palm). The 8-row trait has been retained in the genus over a 3-fold span of basitarsal circumference [64, 4120], implying that it is under disc-wide (vs. local) control, and this conclusion is affirmed by its invariance in starved flies [1801]. Superimposed upon the 8 rows are 5 chemosensory bristles, whose inter-row sites are also symmetric [2544]. A sixth bristle arises in the "vacancy" (distally between rows 2 and 3) in ~30% of legs [1802].

The 1st- and 3rd-leg basitarsi have transverse rows of pale bristles on the front or rear face, respectively. These "t-rows" are used as brushes where the legs contact the eyes (1st leg) or wings (3rd leg) during the cleaning ritual [3376, 4225, 4462]. The 2nd and 3rd legs are not dimorphic, but the male 1st leg has a "sex comb" that females lack.

Because the 2nd leg is simplest, its pattern may be most primitive (cf. Ch. 8) [1807, 1883, 4095]. If so, then evolution may have cobbled together the 1st- and 3rd-leg basitarsal patterns via the following "**Basitarsal Elaboration Scenario**" (Fig. 3.10), starting with the basic array of 8 longitudinal rows. Figuring out how genes were rewired to make such patterns might lead to clues about how evolution tinkers with development [1361]. The following steps are largely speculative (as is the rest of this section), and readers may safely skip to the next section:

1. *Evolution inserted t-rows into 1st legs (between rows 7 and 8) and 3rd legs (between rows 1 and 2).* T-rows are in the anterior lineage compartment on 1st legs, but they are in the posterior compartment on 3rd legs [4076]. First legs apparently kept row 7, but 3rd legs widened their t-rows to replace row 2 [4349]. *N.B.:* Homeotic mutations that transform A regions of 3rd legs to resemble A regions of 1st legs fail to affect P rows [1713, 4349], thus creating legs that have two symmetric sets of t-rows.

2. *Evolution suppressed part of row 1 on 3rd legs.* On the 3rd leg, row 1 seems to have lost its middle bristles at the compartment boundary. The proximal few bristles and the distalmost "orphan" bristle (all of which tend to arise anterior to the A/P line [1800]) persisted, but the orphan bristle moved across the midline to abut row 8, while the ventral campaniform sensillum made a similar move (cf. misplaced sensilla in *sple¹* legs [1810]; Fig. 5.12g). *N.B.:* In *Ubx^{bxLOF}* flies, the A region of the 3rd leg mimics the A region of the 2nd leg: the long-lost middle bristles reappear and the orphan bristle moves back to join a restored row 1 (cf. Fig. 3c of [2449]).

3. *The sex comb originated as a modified t-row.* Before this step, the 1st legs of both sexes probably had the modern female-type pattern (cf. *D. ananassae* [4095]), wherein the proximal ends of rows 1 and 8 converge, and the second bristle in row 8 mimics the corresponding bristle of row 1 (cf. the big second and third bristles on the 3rd leg). Various types of sex combs exist in the genus [1361, 4095]. Some are merely t-rows whose bristles are thick, blunt, dark, and curved. In others, the combs are oriented longitudinally. Cell lineage studies in *D. melanogaster* show that its comb begins as a distal t-row, which then turns ~90° (Fig. 3.11g–i) [4344]. The following conjectures pertain mainly to how the sex comb rotates during development, but they also have implications about how it evolved:

   a. The comb has more bristles than any t-row, but distal rows are also broader than proximal ones, so a growth gradient may exist along the segment. Indeed, the comb probably rotates via growth [3446], rather than by cell migration [1357]. The rotation may push the t-row just above it since this row often bends away dorsally.

   b. When flies carrying a t.s. sex-transforming mutation are shifted to high temperature, the sequence of their comb defects reveals that bristle number is set before bristle type [281].

   c. One t-row seems to be suppressed to make room for the rotation [2978, 2979, 3566, 3987, 4344], although two t-rows might be merging to form a single comb [4109]. Intersexual basitarsi may reveal a "missing link" transitional stage [1845].

   d. Cell death may be involved in the rotation because this region hypertrophies (an overreaction to expanded apoptosis? [3442]) in cell-lethal mutants [3442, 3964].

   e. The "central bristle" [1714] seems homologous to the 2nd leg's distalmost row-8 bristle. It probably belonged to the sex comb but detached and migrated to its central site in the bare triangular area. Similar detachments of row 8's middle bristles from the t-rows may have occurred that led to their elimination (cf. the male vs. female patterns) [4344].

If this evolutionary scenario is even partly correct, then it connotes a thorough reprogramming of cell behaviors. We know virtually nothing about the genetics of those alterations. In contrast, we are beginning to understand how SOPs rearrange within incipient bristle rows (Fig. 3.11). By heat-shocking wild-type pupae, abnormal phenotypes can be induced [1803], and the sequence of their sensitive periods suggests the following stages:

1. *Lateral movements refine SOP alignment.* One odd phenotype in *h^{null}* legs is bristle misalignment [3193]. Conceivably, SOPs on wild-type legs use Hairy/Ac boundaries as guidelines for alignment. If so, then the loss of these lines in *h^{null}* legs might be causing the disorder [3193]. Additional evidence for transverse SOP movements comes from cell lineage: in certain rows, the bristle SOPs arise ventral to their prospective bract cells, while in others the bristles arise relatively dorsally (Fig. 3.9) [1800, 2449]. Because each bristle eventually lines up with its bract, the SOPs must undergo directed lateral movements within each proneural stripe (Fig. 3.11a–c). Stripes 1, 2, 4, 5, 7, and 8 lack a Hairy/Ac boundary at the edge that their

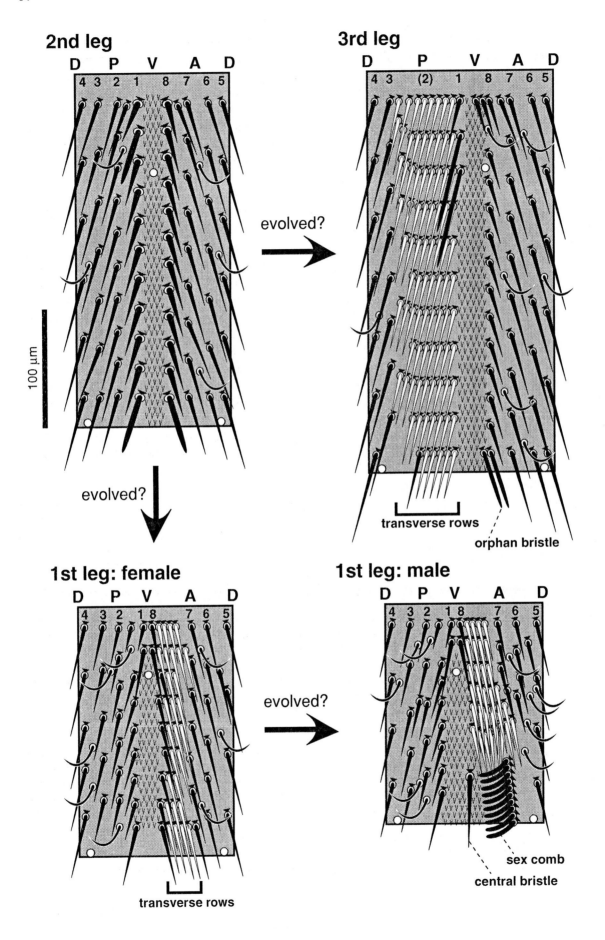

2nd leg

100 µm

3rd leg

evolved?

evolved?

transverse rows

orphan bristle

1st leg: female

evolved?

transverse rows

1st leg: male

sex comb

central bristle

SOPs supposedly seek. It is thus unclear what "homing" cues these SOPs use.

2. *Proximal-distal movements refine SOP spacing.* Although no SOP movements have yet been witnessed in flies, scale precursor cells have been shown to rearrange on moth wings [3051]. Those shifts appear to be mediated by filopodial extensions from the precursor cells, and the same filopodia may help the cells within each scale row acquire their regular spacing. Evidence for a filopodium-mediated spacing mechanism in *Drosophila* comes from phenotypes of mutants that have abnormally oriented bristles. In the wild-type, each row's bristles face in the same direction. When mutant bristles face one another, they tend to have an unusually large space between them, while bristles that point away from one another tend to have an unusually small space. These correlations are explicable if SOPs are repelling one another using filopodia that are longer on one side than the other (Fig. 3.11d–f) [1810]. Such filopodia have indeed been found [2387, 3630].

Short-range adjustments of these kinds could explain why bristle rows are neater on the legs than on the notum [805, 1813, 4428]. Mutant legs that fail to elongate fully have bristle rows that are as jagged as notal rows [4348], so fine-tuning may require proper evagination. Notal SOPs probably differentiate *in situ* (*i.e.*, wherever they arise within Ac stripes), without any subsequent corrective movements to reduce the scatter. Leg SOPs might become mobile after delaminating [1745] since they (or their descendants) could then crawl freely on the underside of the epidermis without needing to displace other cells within the crowded monolayer. Consistent with this reasoning, marked bristles are often isolated from somatic clones induced on legs [544] and tergites [1371], but not the notum [521, 1373, 1797].

Within each t-row the bristle sockets touch one another, but adjacent t-rows are separated by several cell diameters. It is therefore unlikely that t-row SOPs use isodiametric inhibitory fields [4641]. Diffusion of the inhibitor could be channeled along one axis, but it seems more plausible to imagine that the t-rows and sex comb "self-assemble" from scattered SOPs that join into tandem files by homophilic adhesion. Homophilic adhesion could also explain the tandem bristles in the medial row of the wing margin [1741]. Some sort of cellular cooperation must be occurring in both cases because sex combs and marginal rows can reform when leg or wing discs are dissociated and reaggregated [1356, 3424, 4334].

FIGURE 3.10. The four types of basitarsal bristle patterns in *D. melanogaster*. Each segment is drawn as if slit along its dorsal (D) midline and spread flat (P, posterior; V, ventral; A, anterior), with its proximal edge at the top. Only the 1st leg is sexually dimorphic.

The 2nd-leg pattern is simplest and maybe most primitive. Its basitarsus has 8 rows of mechanosensory bristles (numbered at top), each of which bears a bract (triangle) at its base. The 5 curved "bractless" bristles reside between the rows and are chemosensory [3061]. Also shown are sensilla campaniformia (stretch detectors [4342, 4841, 4887], white circles) and trichoid hairs (small "v"s). Between rows 1 and 8 almost every cell makes a hair, so the hair density indicates cell size on the segment. There are ~2,000 epidermal cells on the 2nd-leg basitarsus [1801].

The 1st- and 3rd-leg basitarsi augment this basic pattern with transverse rows of lighter-colored bristles. Similar rows decorate the tibias of those legs, but most of the tibial transverse-row bristles lack bracts [1714]. Transverse rows serve as brushes for cleaning the eyes (1st leg) or wings (3rd leg), and they are ideally located (A vs. P) for this role [4462]. Bristle lengths and intervals tend to increase from V to D in all patterns.

In males, the most distal transverse row rotates ~90° during development to form the "sex comb" [4344], which may function during courtship [1692, 4039]. Males also differ from females in having extra bractless bristles between rows 5 and 6 [1714].

In the hypothetical scheme indicated by the arrows [1883, 4096], all three leg pairs originally resembled the 2nd-leg pattern. Transverse rows then evolved in 1st and 3rd legs of both sexes, and males acquired a sex comb. The central and orphan bristles may be vestiges of these changes (see text). Alternatively, evolution might have gone the other way (i.e., simplifying an initially complex pattern). In *Polycomblike*[LOF] mutants, the 2nd leg acquires two sets of transverse rows (not shown) [1122]: one set on its A side (plus a sex comb in males) where it becomes a 1st leg and another set on its P side where it becomes a 3rd leg (see [1713]). This "homeotic schizophrenia" is a default state (cf. Table 8.1) where the Polycomb Group of "memory genes" malfunctions, thereby allowing certain homeotic genes to stay ON where they should be OFF [1508, 1509]. Such defaults can be equivocal, however. Indeed, an opposite type of default is seen when the latter homeotic genes are disabled [4149] – i.e., 1st and 3rd legs look like 2nd legs.

Drawings are for left legs from actual wild-type flies but are idealized insofar as (1) segments are not perfect cylinders, (2) hairs are not so neatly arranged, and (3) color shadings *in vivo* make it difficult to discern where transverse rows end and longitudinal rows begin, and (4) segment widths are exaggerated to avoid bristle overlaps (lengths are accurate). See also App. 7.

Observation

Inferred cell movements

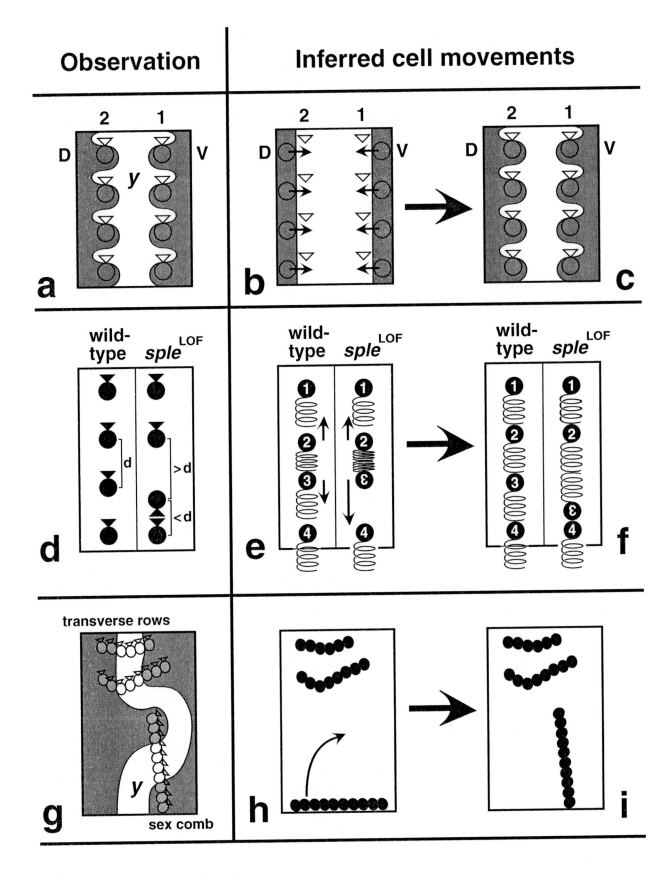

The chimeric composition of the reconstituted patterns proves that physically separate (differently marked) cells can align, but it is not known whether this process requires regrowth and repatterning. Hawaiian *Drosophila* have a tandem file of alternating "bristle-bract-bristle-bract..." cells in basitarsal row 8 [1801], and this peculiar row (bracted bristles are typically separated by several ordinary epidermal cells [1598]) may have evolved via heterophilic association of bristle and bract cells.

Chemosensory leg bristles are patterned like notal macrochaetes

As mentioned above, the 2nd-leg basitarsus has 8 rows of mechanosensory (MS) bristles and typically 5 chemosensory (CS) bristles between the rows (Fig. 3.10). Each MS bristle bears a bract on the proximal side of its socket, whereas CS bristles are "bractless" [1714]. In many ways, basitarsal CS bristles develop like notal MCs, while the MS bristles develop like notal mCs:

1. There is an order of magnitude fewer CS than MS bristles on the 2nd-leg basitarsus (5 vs. ~70) [1802], analogous to the disparity between notal MCs and mCs (26 vs. ~225) [1741].

2. Like notal MCs [3966], the number and positions of CS bristles are constant, whereas MS bristles vary in number and position from fly to fly (and on the left vs. right sides of individual flies) [1714, 4344].

3. CS bristles originate before MS bristles [1803, 3628], just as MC SOPs precede mC SOPs [1925, 4428]. Thus, patterning is essentially a two-act play on both the leg and notum: in Act I a few (CS or MC) SOPs arise in a fixed array, and in Act II many more (MS or mC) SOPs arise relatively randomly inside parallel proneural zones [3954]. The ability of the later kinds of patterns to form independently of (and be superimposed upon) the earlier ones is also seen on the wing margin [353, 882, 1741] and has been proposed for the integument of other insects [2425].

4. Under certain circumstances, both CS bristles and MCs can form rows. Although CS bristles are not arranged periodically on the basitarsus, rows of such bristles are manifest on the tarsus as a whole [1811], the tibia [149, 3141, 4481], and the wing margin [1741], and other *Drosophila* species have strikingly regular rows of evenly spaced CS bristles on the legs [667]. Likewise, related genera have rows of evenly spaced MCs on the notum [1354], and these rows emerge (atavistically?) in *D. melanogaster* when proneural potential increases [1354]. Bristle intervals in CS or MC rows are larger than in adjacent MS or mC rows, presumably because later intercalary growth pushes SOPs apart after their initial interval is created by (standard size?) inhibitory fields [1354, 2959, 3950].

5. CS bristle sites are immune to changes in cell size or organ size (and hence appear to be part of a global

---

FIGURE 3.11. Movements of bristle cells inferred from various studies. Bristles are denoted by circles and bracts by triangles. Three types of hypothetical movements are shown – none of which have actually been witnessed yet.

a–c. Lateral shifts that may align SOPs in longitudinal rows [1800]. **a.** Sections of rows 1 and 2 from a left leg (D, dorsal; V, ventral). Wavy lines indicate the kinds of boundaries seen in gynandromorphs. Male tissue is marked with *yellow*[LOF] (*y*). Curiously, bristles within a row tend to be more closely related to each other than to their own (physically closer) bracts. **b, c.** These affiliations are explicable if SOPs in each proneural stripe (Fig. 3.9) migrate in one direction (dorsally for row 1, ventrally for row 2) to a different column of cells where they induce neighbors to become bracts [1800, 1808]. Shaded zones in **b** and **c** denote *y*+ tissue as in **a** (they are not Ac stripes), and triangles in **b** mark future bracts.

d–f. Proximal-distal shifts that may fine-tune SOP intervals within those rows [1803, 1810, 1813]. **d.** LOF mutations of *spiny legs* (*sple*) cause bristles to be misoriented along with their bracts, usually a 180° polarity reversal (third bristle). Strangely, intervals between bristles that face one another tend to be larger (">d") than wild-type distance ("d"), while spaces between bristles that face away from each other tend to be smaller ("<d"). **e, f.** These correlations are explicable if SOPs repel one another via filopodia (springs) [1809, 1810]. If SOPs 2 and 3 happen to be unusually close in a wild-type fly (left half of **e**), then they would be pushed apart by the cramped filopodium of SOP 2. Such repulsions would make spaces more uniform (left half of **f**). In contrast (right half of **e**), reversed-polarity SOPs (SOP 3) should feel much more force from the proximal neighbor (due to 2 intervening springs) than from the distal one (0 springs), resulting in the observed correlations (right half of **f**). Filopodia have indeed been seen on bristle SOPs [2387, 3630], and they are longer in one direction. Similar filopodia appear to mediate rearrangements of scale SOPs in moth wings [3051]. (Lepidopteran scales are believed to be homologous to dipteran bristles [1340, 2173, 2422, 4640, 4641].)

g–i. Sex comb rotation [4344]. **g.** Clones that pass through the centers of the transverse rows (two shown) often include the center of the sex comb, despite its perpendicular orientation. **h, i.** This zigzag is explicable if the sex comb arises as a transverse row and rotates 90° [4344]. Conceivably, movements of this sort might also be guided by filopodia: apical "cytonemes" have been seen in leg and wing discs [3507], and basal "epidermal feet" [2175, 2585–2588] have been seen in leg, wing, and eye discs [1133, 2863, 3426] and on leg and wing disc cells in culture [3303]. *N.B:* The boundary in **g** is not actually traceable in such smooth detail because *y*[LOF] does not affect the background cuticle, and the same is true for **a**.

positioning system of some kind), whereas MS bristle spacing and number depend on cell size and cell number, respectively [1801]. The same is true on the notum [3966], where a prepattern seems to dictate positions for MC PNCs and mC stripes but not to "micromanage" mC SOP sites within those stripes.

6. Like notal MCs, CS bristles are more affected by sc^{LOF} than ac^{LOF} mutations [1802], while the opposite is true for notal mCs and MS leg bristles [1360, 1453].

For these reasons, "constellation" patterns (notal MCs and CS leg bristles) and "row" patterns (notal mCs and MS leg bristles) were thought to be fundamentally different (Fig. 3.1). It therefore came as a surprise when MCs were coaxed into a mC-like pattern by gene dose manipulations (see below), implying that both types of patterns are divergent outputs of a common mechanism.

Extramacrochaetae superimposes an uneven antineural "mask"

The *extramacrochaetae* gene (*emc*) was isolated in a screen for mutations that interact with the AS-C in a dose-dependent way [408]: LOF mutations were sought that produce extra bristles when the ratio of the tested gene to the AS-C was 1:3 or 1:4, instead of the normal 2:2. Despite the large scale of the screen, the only lesions recovered were in *emc* and *hairy*, although Notch-pathway genes have similar properties [733, 997, 2690]. Thus, all genes of this kind may now have been identified. Although *emc* and *hairy* both behave as *trans*-repressors of the AS-C, emc^{LOF} alleles interact mainly with *sc* and affect notal MCs, whereas h^{LOF} alleles interact mainly with *ac* and affect notal mCs [2959, 2960]. In other body regions their effects overlap more, and their interactions with *ac* and *sc* are less restricted [1349]. Indeed, *emc*'s link to MCs (whence its name) is not inherent in the protein because mCs can be removed by overexpressing *emc* just before their SOPs arise [4453].

Like Hairy and the E(spl)-C effectors of the Notch pathway, Emc also has an HLH motif, but it lacks a basic domain for binding DNA [1156, 1388]. This feature alone suffices to explain its mode of action [213, 3083] because (1) any dimers that Emc forms with Ac, Sc, and Da will have only one basic domain, but (2) two basic domains are needed to bind DNA [286]. Indeed, Emc does block DNA binding by Ac, Sc, and Da *in vitro* [589, 3179, 4452, 4453], and this "inert decoy" effect can been mimicked by a Da construct whose basic domain has been deleted [1754]. Hence, Emc silences the AS-C post-translationally

by sequestering Ac, Sc, and Da in inactive heterodimers. Curiously, when a basic domain from L'sc is substituted for the corresponding piece of Emc, the chimeric protein can still block E-box binding by L'sc/Da dimers [2722]. Conceivably, the inserted domain cannot dock properly with DNA because the rest of the Emc protein has evolutionarily (through disuse?) lost the ability to orient this region at a suitable angle [2722].

By adding extra doses of *sc* in an emc^{LOF} background, the number of notal MCs can be increased far above the wild-type condition. Interestingly, as new MCs arise, they form rows, as if they are now obeying patterning rules that govern mCs [2959]. Indeed, the aligned MCs shift positions as new ones are added so as to maintain even spacing, so they cannot merely be sprouting from fixed "cryptic singularity" sites.

How can MCs acquire organizational attributes of mCs, given that MCs and mCs come from distinct proneural fields (PNC vs. stripe)? Because some MCs reside where mC rows later develop, emc^{LOF} mutations might just be accelerating the process of stripe formation so that it intrudes into the time window for MC SOP initiation [1926]. The following facts support this hypothesis:

1. In emc^{LOF} mutants, two rows of ectopic MCs develop along the same dorsocentral and presutural lines [1349, 1926, 2959] where mC SOPs emerge first [4428]. Extra MCs can also be induced in these zones by overexpressing *sc* ubiquitously [3628].
2. Most MC SOPs arise earlier in emc^{LOF} mutants than in wild-type flies [1926] – indicating a general acceleration of bristle development. Indeed, the plethora of "MCs" may just be an illusion resulting from a premature initiation of SOPs. If mC shaft cells start endoreplication early enough, then they could grow to the size of MCs (cf. Ch. 2), as has been inferred for mutant tergites [2690].
3. "Excess function" (*Hairy wing*) mutations in the AS-C put extra MCs along the dorsocentral line [191, 635, 1577], indicating a proneural proclivity along these corridors prior to mC SOP emergence [1926]. (See [3890] for evidence of a similar tendency on the scutellum.)
4. The overall level of Sc protein increases in emc^{LOF} mature wing discs, and the pattern of accumulation changes [912] such that the dorsocentral PNC expands to form a vague stripe along the future anterior-posterior axis. In wild-type flies that stripe does not appear until ~8 h later [3270]. These effects, and the reduction of Sc in emc^{GOF} discs [912], are surprising, considering that Emc is supposed to affect the AS-C

post-transcriptionally, but they may be indirect (see below).

Heterochronic changes of this sort may have been instrumental in bristle pattern evolution, because the emc^{LOF} array resembles MC designs of more primitive dipterans [1354, 2959].

Emc is expressed ubiquitously [1388], so one of its functions appears to be to set a threshold which AS-C activity must exceed to evoke SOPs [637, 997, 1926, 3982] – the same sort of argument proposed earlier for the Notch pathway's role inside PNCs (the MI model). Essentially, Emc would form an inhibitory "ocean" that lets peaks (PNCs) and ridges (stripes) of a submerged prepattern landscape rise above "sea level" at certain times to make proneural territories [918]. This metaphor is emblematic of Turing-like models that appear to be irrelevant here since they invoke *diffusible* molecules [1727, 2806]. However, the same rules of damped autocatalysis can govern cellular automata [1033, 1437, 2651, 2694, 4699] that may be applicable.

The ocean analogy is an oversimplification because the amount of *emc* RNA actually varies in a landscape of its own throughout the disc epithelium. Its density tends to be low wherever Ac and Sc are expressed [913, 4453]. Although the complementarity is only approximate, it suggests a second function for Emc: to restrict the regions where proneural fields can arise by suppressing AS-C transcription [997]. Indeed, many of the extra bristles in emc^{LOF} mutants are located in areas that are normally bare [1349], and the AS-C is derepressed at those sites [589, 3982, 4452, 4453]. In wild-type flies, *emc* and AS-C expression are both high in some naturally nude regions, so Emc's antineural effect there must only be post-transcriptional [589, 913].

Heterogeneous transcription of *emc* inside some PNCs implies a third role (aside from its global "baseline" and "carving" duties) – to confine proneural potential to a subset of cells within certain PNCs [637, 913, 2062]. In such cases, the mask (of high-level *emc* expression) partly overlaps the PNC, rather than skirting it. The dynamics of *emc* and AS-C expression inside the dorsocentral PNC are revealing in this regard:

After the PDC precursor has emerged, strong *emc* and *ac-sc* expression co-exist in the anterior part of the cluster, an area in which the ADC precursor later appears. When this SOP arises, the *emc* expression has been extinguished in that area. [913]

Consistent with the idea that Emc helps "fine-tune" MC sites [1458], MCs are indeed positioned less accurately

in emc^{LOF} mutants than in the wild-type. Particular bristles are displaced by as much as several cell diameters [1926] – as if Emc's absence allows the area of maximal competence to broaden and shift within the PNC. Such shifts probably also occur in mosaics wherever a PNC is bisected by the boundary separating lower from higher AS-C doses [912, 1794] because the low-dose areas would be artificially mimicking the natural role envisioned for Emc [913]. Another implication from those mosaic studies is that Emc need only reduce proneural activity by a factor of two in order to perform its spatial sharpening function. Gradients of *emc* transcription within certain PNCs appear steep enough to span at least a 2-fold range [913].

In summary, *emc* affects the AS-C in time (by damping its protein accumulation) and in space (by overlaying an uneven antineural "filter" on the epidermis). This antineural *emc* mask (plus the prohibitive stripes painted by *N* and *hairy*) helps explain the amazing ability of bristles to still form in normal patterns, even when AS-C genes are uncoupled from all their position-specific enhancers [1454, 1458, 1804].

1. When ac^- sc^- flies carrying a *hs-sc* [3530, 3628] or *hs-ase* [1079] transgene are heat shocked, bristles develop mainly at normal sites, despite the proneural competence that is supposedly thereby induced throughout the epidermis.
2. Bristles still arise mostly in normal areas when these same transgenes are overexpressed in a normal (ac^+ sc^+) background [438, 1079, 3628].
3. Similar results are obtained in both genetic backgrounds (and even if the entire AS-C is deleted) when *sc* or *l'sc* is overexpressed under *Gal4-UAS* control [1854].

Clearly, topographic cues outside the AS-C must be constraining proneural potential. Emc provides at least some of these cues because emc^{LOF} (in a background where *ac* is derepressed) decreases the positional precision of normal MCs and increases the scatter of extra bristles in general [913]. Because *emc* transcription is normal in ac^- sc^- discs [913, 4453], its peaks and valleys must be dictated by prepattern factors at a higher echelon in the chain of command. Competence is probably also gated by mitotic quiescence [637, 4427], and the dosage of *da* becomes limiting when AS-C monomers supersaturate the system [438, 918, 1854]. All *trans*-regulators of the AS-C thus far discussed (Notch pathway, *hairy*, *emc*) exert negative control in defining (or refining)

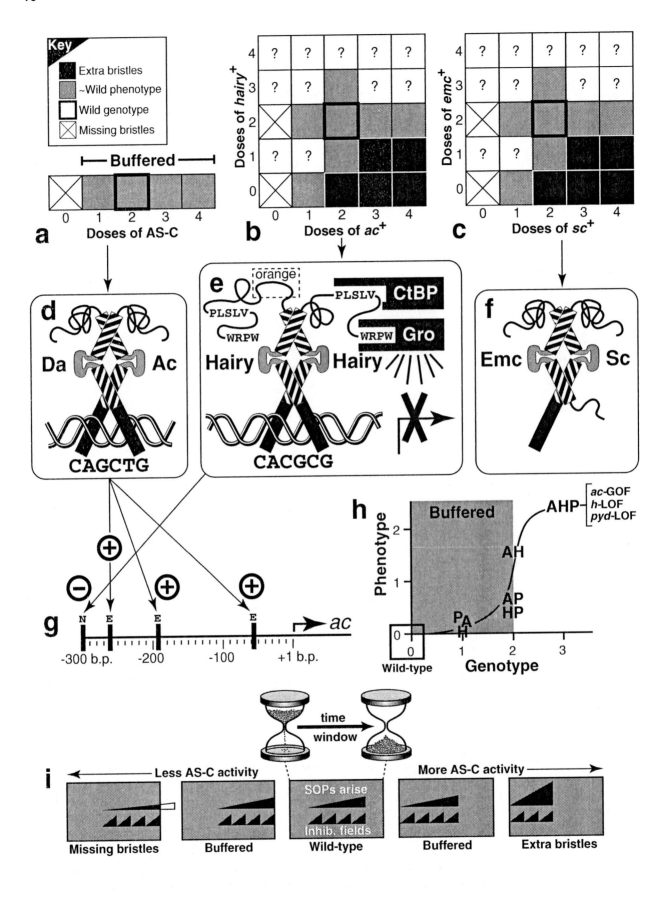

AS-C expression. Upstream factors also exist which establish spots or stripes of proneural expression by activating the AS-C, but they tend to belong to pathways that are tailored to the needs of individual discs (see below).

Dose dependency implies that HLH proteins "compute" bristles

When the amount of a proneural or antineural protein deviates too far from its wild-type level, the resulting pattern abnormalities can be "cured" by artificially adjusting the amount of an antagonist until the deviation is "titrated" (Fig. 3.12) [408]. (Da also has dose-dependent interactions with AS-C proteins, but they are positive [949].) Examples include

1. Higher doses of *emc* [1348] or *h* [1354] erase extra bristles caused by *ac*^{GOF} mutations.

2. Heterozygosity for AS-C deletions erases extra bristles caused by *h*^{LOF} or *emc*^{LOF} or both [2960].

3. Heterozygosity for *ac*^{LOF} erases extra bristles caused by *h*^{LOF} [2959, 4185].

4. Halving the AS-C dose erases extra bristles caused by half doses of *N* or *Dl* [997, 2690].

Imbalances among pro- and antineural gene products may account for the bald patches seen in interspecific *Drosophila* hybrids [318, 3980, 4238, 4239]. In such flies, certain SOPs emerge but then abort differentiation. Hybridization may be disturbing the relative amounts of HLH agonists vs. antagonists at various sites [3980]. Those amounts would be co-adapted within species but incompatible across species [2901]. The incompatibility presumably arose via genetic drift in the parallel lineages of flies after their divergence from a common ancestor [4259].

FIGURE 3.12. How HLH proteins "compute" bristles. Proneural proteins (Ac, Sc, Da) prod cells to become SOPs, whereas antineural proteins (Hairy, Emc) restrain them.

a–c. Dose-dependent interactions (see key), based on data from [408, 1348, 1354, 2959, 2960]. **a.** Wild-type females have two AS-Cs (one on each X), and "dosage compensation" rectifies the level in males to parity by a chromatin-remodeling trick (not shown) [1633, 2183, 3760] that is poorly understood [307, 308, 3249, 3997]. The phenotype resists change when AS-C dose deviates by up to a factor of two (halved to 1 or doubled to 4), and such "buffering" is also seen with *hairy*$^+$ (**b**) or *emc*$^+$ (**c**). However, extra bristles arise when increases in Ac or Sc combine with decreases in Hairy or Emc.

d–f. HLH dimers that affect a cell's ability to become a SOP. Shading conforms to Fig. 2.4: basic region (black), helices (hatched), loop (gray), and remainder ("noodle"). **d.** Ac/Da (and Sc/Da) heterodimers stimulate transcription of *ac* by binding "E boxes" ("class A" sites [1242, 3179]) in the *ac* promoter (**g**): CAGCTG (-262 and -195) or CAGGTG (-58) [2722, 4452, 4453]. They also activate *sc* [918] (not shown). These effects are blocked by Hairy and Emc, but in different ways. **e.** Hairy does not dimerize with Ac, Sc, or Da [68, 4452]. Its homodimers bind upstream of *ac* at CACGCG (a variant "N box" or "class C" site) and repress *ac* transcription (**g**) [3179, 4451], possibly via the "orange" domain [976], but more likely by recruiting Groucho (Gro, a long-range silencer that binds WRPW) [1242, 4524, 4865], C-terminal binding protein (CtBP, a short-range quencher that binds PLSLV) [2180, 3377, 3430], or other co-repressors such as dDrap (not shown) [3648]. **f.** Emc sequesters Ac, Sc, and Da in dimers that are inert because Emc lacks a basic (DNA-binding) domain [589, 3179, 4452, 4453]. As a result of this "broken pliers" configuration, *ac* and *sc* cannot auto- or cross-activate in Emc's presence. Emc's effects *in vivo* may be greater on *sc* due to when (vs. where) *emc* is expressed. This "disabled dimer" trick is also used in the LIM-HD and POU-HD families, where Beadex [2854, 3908] and I-POU [4380] play roles analogous to that of Emc. This process whereby an inhibitor sequesters an activator is called "squelching" [1984, 4806] to distinguish it from "quenching" (i.e., the blocking of an activator by an inhibitor after both have bound a DNA site) [1600, 1984, 2350].

g. Promoter region upstream of the *ac* transcription start site (crooked arrow). E and N boxes are marked; "+" = activation; "−" = repression.

h. An example of buffering, where three mutations were studied singly or combined: *ac*^{GOF} (A, a.k.a. *Hairy wing*), *hairy*^{LOF} (H), and *pyd*^{LOF} (P). All of them increase activity of the AS-C [635, 733, 3179, 4451]. The number of mutations per fly is along the *x* axis; the number of extra scutellar macrochaetes is along the *y* axis [3069]. Phenotypes remain nearly wild-type until the double-mutant threshold (cf. **a–c**), whereupon synergism ensues.

i. Time Window Model for pattern buffering. In each panel, the *x* axis is time, the upper triangle denotes gradual production of SOPs in a proneural field, and the lower triangles denote the time needed for SOPs to extend inhibitory fields (they are not really sequential). In wild-type flies (middle rectangle), the time available for patterning (window width) exceeds the time needed (upper triangle width), so slight changes in AS-C activity (2nd and 4th rectangles) may merely shift the timing of events. In contrast, more extreme changes (1st and 5th rectangles) may additionally alter the rate of accumulation of Ac and Sc (gradient slope). High-AS-C flies should make SOPs so quickly that they cannot fully extend inhibitory fields before new SOPs arise nearby (leading to extra bristles), whereas low-AS-C flies will make SOPs so slowly that time expires before the wild-type number is attained. Whether HLH proteins also help terminate this time window is unknown, but a HLH Hourglass has been found in vertebrates [2293, 2294] that uses proneural and antineural bHLH antagonists. See also App. 7.

Evolutionary drift in "volume settings" of interacting HLH genes might also explain the intersexuality of many hybrids in this genus [1174, 1882, 1883] and other insect groups [1520, 1524, 3398]. Gender-specific traits in *Drosophila* appear to be decided by the ratio of X chromosomes to autosomes [1173, 4077]△, as measured by relative doses of gene products [3266]. Remarkably, the numerator and denominator elements include the proneural bHLH proteins Scute [1028]△ and Daughterless [3268]△, the panneural bHLH protein Deadpan [209], and the antineural HLH protein Extramacrochaetae [1155]△, suggesting that evolution may have commandeered a cassette of HLH neural-patterning genes for secondary duty in sex determination [1173, 2018, 4824].

HLH genes decide cell fates not only in PNS neurogenesis and sex determination, but also in photoreceptor initiation [182, 1076]△, CNS neurogenesis [485, 3981]△, myogenesis [2177]△, and gliogenesis [1464]△, as well as in the development of midline mesectoderm [899]△, Malpighian tubule tip cells [4672]△, appendage tips [1166]△, tracheal tubes [4548]△, salivary ducts [899]△, endodermal tissues [4269], etc. [914, 1343, 1744].

Like Numb, therefore, HLH transducers function at an abstract level that transcends histotype (cf. Ch. 2). Collectively, they act like analog-to-digital transducers to ensure all-or-none outcomes [816, 2958, 3186, 4584]. In short, cells use HLH proteins to do arithmetic.

The virtuosity of HLH input/output devices is illustrated by the sex-determining genes, which reliably produce discretely male or female (vs. intermediate) bristles in sex combs despite X:A ratios that range between 0.5 (male) and 1.0 (female) [4370]△. The choice of bristle type is made on a cell-by-cell basis [3734, 4105] a few hours after the number of sex comb SOPs is set [281, 2106]. What *kind* of math is being used to convert quantitative signals into qualitative states?

The "**SOP Computer**" seems to operate by combining positive and negative scalar inputs (Ac, Sc, Da, Emc) as HLH heterodimers. For the scheme outlined below to work, Ac and Sc must be rate limiting for dimer formation with Da and Emc. If Da/Da dimers form too readily, then Da's presence before Ac and Sc appear [905, 4435] could be problematic (although dimer preferences could be imposed biochemically or sterically by third-party agents [245, 286, 2046, 3083]). Indeed, Da dimerizes only weakly with itself [588, 4453].

1. Upstream *trans*-activators of *ac* and *sc* intensify at certain epidermal sites, bind site-specific enhancers near *ac* and *sc*, and start to evoke Ac and Sc at time "t_1." Ac and Sc form heterodimers with Da or Emc [68, 589, 4452], instead of binding each other or homodimerizing [588, 918, 1484, 3974, 4453].

2. In these proneural areas, the number of Ac/Da and Sc/Da heterodimers per cell will increase at a rate "r_1" that is damped by diversion of Ac, Sc, and Da monomers into inert heterodimers with Emc [4453]. Indeed, all these heterodimers form readily [68, 589, 4452], and Emc has been shown to inhibit E-box binding [4452] and transcriptional stimulation [589, 4453] by Ac-Sc-Da combinations in a competitive, dose-dependent manner.

3. Eventually, Emc is depleted (unless Emc is resupplied via transcription), and this event marks the first titration threshold "T_1" [1387, 4584]. Thereafter, the rate of Ac/Da and Sc/Da accumulation will increase to "r_2" (assuming there was initially more Da than Emc).

4. At some point, the concentration of Ac/Da and Sc/Da heterodimers crosses a second threshold "T_2" that triggers autocatalytic production of Ac and Sc [918] via (1) direct feedback of the heterodimers on E boxes near *ac* and *sc* [2722, 4452], and (2) indirect feedback via *sens*, whose promoter has E boxes and whose product binds near *sc* and *ac* [3127]. Levels of Ac and Sc rise exponentially to saturation (limited by Da) [997]. The cell becomes a SOP (via target genes whose E boxes bind Ac/Da or Sc/Da) and emits a signal that prevents nearby cells from making more Ac and Sc.

5. SOPs continue to arise within the proneural area (in either a stochastic or a spatially biased manner) until time "t_2," when all competent cells have either become SOPs or are inhibited by nearby SOPs.

This algorithm does not explain why Sc transcription rises in the wing disc when Emc is reduced [912], because Emc only acts post-transcriptionally and autoregulation of AS-C loci is only supposed to occur in SOPs. Conceivably, some *trans*-activators of the AS-C may also be bHLH proteins that could be disabled by Emc [1387]. The only gene so far identified upstream of *emc* itself is *polychaetoid* (*pyd*) [4237]. Pyd's presence in intercellular junctions suggests that *emc* might be activated by intercellular signaling.

The HLH scheme shows how easy it is to do math with molecules by titrating monomers. It uses addition (Ac + Sc = Total proneural), subtraction (Total proneural − Emc = Net proneural), and multiplication (feedback in Step 4), but not division. Indeed, it now appears that even the famous X:A "ratio" is an illusion of a procedure that strictly relies on counting [816]. The dose sensitivities of Notch-pathway components are probably also due to dimer interactions – the most obvious of which is the docking of Dl and N in *trans* [1204, 3544] and possibly also in *cis* [2008] at cell surfaces.

Titration-based arithmetic may also operate at the RNA level. A motif exists in the 3'-UTRs (untranslated regions) of transcripts from *hairy*, *emc*, the E(spl)-C genes *m3*, *m4*, *m5*, and *mγ*, and the *m4*-related genes *Bearded*, *Bob* (*Brother of Brd* = actually 3 genes A, B, and C), and *Tom* (*Twin of m4*) – all of which act antineurally [2382, 2386, 2499]. This "GY" box reads <u>GUCUUCC</u> (except for *emc*'s <u>GUUUUCC</u>). The complementary <u>CAGAAGG</u> appears in 3'-UTRs of transcripts from the proneural genes *ac*, *l'sc*, and *atonal* [2386]. Hence, mRNAs from these pro- and antineural genes could form heteroduplexes that could conceivably modulate transcription, mRNA localization, processing, turnover, or translation [233, 1219, 1702, 2686, 2920, 2965, 3400]. Docking between *hairy* and *ac* mRNAs might explain why interactions between these genes are so dose sensitive (Fig. 3.12), despite the failure of their proteins to dimerize [68, 4452]. Other motifs in these antineural-class UTRs include "Brd" (<u>AGCUUUA</u>) and "K" (<u>UGUGAU</u>) boxes, both of which decrease accumulation of transcripts [92, 2382, 2384, 2385].

Robustness of patterning may be due to a tolerant time window

For Threshold T_1, what presumably matters are the absolute amount of Emc and the rate of production of Ac and Sc, and the latter rate should depend on the amount of stimulatory "prepattern" factors [2891]. Because the variables are all constitutive (with no feedback until Step 4), the stoichiometry easily explains why the process is sensitive to gene dosage.

What is not so easy to understand is why phenotypes remain wild-type when the AS-C dose is halved or doubled, or when the dose of *emc* is halved (Fig. 3.12). Why don't such changes affect bristle number like greater alterations do (e.g., halving *emc* dose plus doubling *sc* dose)? Historically, the ability of animals to maintain phenotypic constancy in the face of perturbations (genetic or environmental) led to the related concepts of "buffering" [1748, 3709]$^\Delta$, "canalization" [1470]$^\Delta$, and "robustness" [2565, 3032, 4646]$^\Delta$.

Certainly, flies do use a variety of error-correction strategies at every echelon (e.g., genes [4584]$^\Delta$, proteins [2760]$^\Delta$, cells [1292]$^\Delta$, and tissues [12, 1940, 2527, 3042]), and many of these "quality control" tricks are undoubtedly sophisticated. However, the buffering of SOP number may

instead have a trivial explanation. To wit, there may be enough tolerance in the system to accommodate minor changes in AS-C activity by letting events shift along a time line (Fig. 3.12). If the slack is finite, then greater changes in AS-C activity will eventually push events to the edges of this time window, and no further slippage will be possible. At that point, the phenotype should begin to overtly reflect added deviations in genotype. Specifically,

1. Two-fold increases in AS-C dose (or halving the *emc* dose) might merely hasten Step 5 (before t_2). Further flooding of the system with Ac and Sc (i.e., changes greater than a factor of two) might accelerate the process so much that more cells can cross T_2 and become SOPs before SOP inhibitory fields can extend to their maximal diameter – hence leading to greater bristle density.

2. Halving the AS-C dose should postpone the time at which Step 5 is reached. However, if upstream *trans*-activators (Step 1) are still present, then the process might continue beyond t_2 until it reaches completion. Further decreases might slow the process so much that the prepattern factors disappear before all the gaps between SOPs are filled in.

This "**Time Window Model**" is buttressed by timing shifts of *emc*^{LOF} SOPs [912, 1926, 2690]. It is also consistent with temporal changes in (1) N^{null} SOPs [3689], (2) the CNS and PNS of N^{null} embryos [379, 1563], and (3) SOPs in *Egfr*^{LOF} mutants [917]. On the wing, SOPs of ectopic sensilla (caused by *h*^{LOF} or *ac*^{GOF}) can arise within a broad time span and still develop normally [362], and the same is true for extra SOPs in PNCs of *pyd*^{LOF} mutants [733]. Like late passengers missing a train that always leaves the station on schedule, these delayed *pyd*^{LOF} SOPs make shorter bristles [3067] because cuticle secretion starts on time, and a similar constraint explains the fewer bristles that come from depleted histoblast nests [581]. Indeed, AS-C and E(spl)-C genes may be short and intronless so that their proteins can be made quickly and timed precisely [92]. Figure 3.13 summarizes the core circuit for bristle SOP selection.

The idea that SOP inception is time sensitive may explain why it is temperature sensitive in hypomorphs [767, 769] not only during the PNC-SOP window [768], but also much earlier [770, 1997, 1998, 3405]. As Goldschmidt argued long ago [1520], shifts in the relative rates of reactions can uncouple key processes, leading to a later failure of the system to reach a critical threshold. The

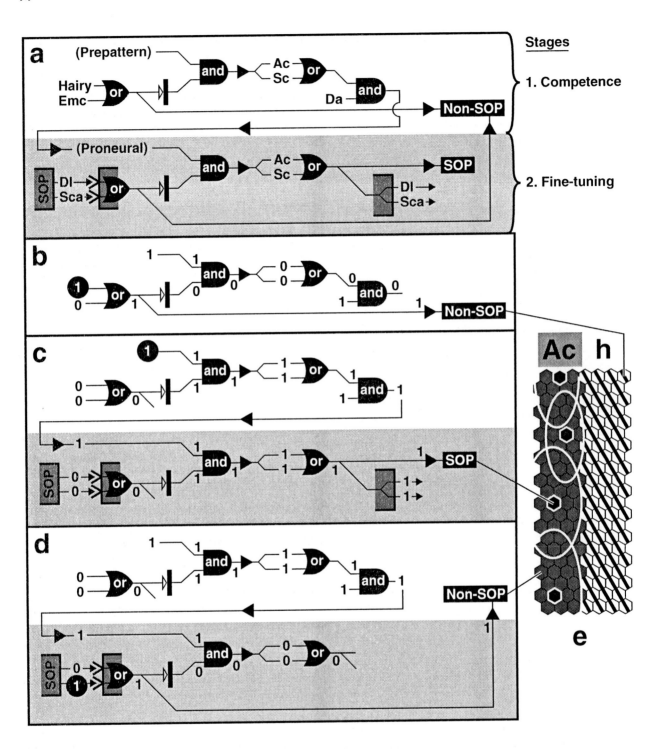

antecedent reactions for bristle patterning remain to be identified.

A time window also constrains the number of SOPs that form the cluster of chordotonal receptors in the femur. In contrast to how bristle SOPs arise, chordotonal SOPs emerge from their PNC as a tandem chain via reiterative induction of new SOPs by old SOPs. The process starts and stops at definite times, and the number of SOPs can be increased or decreased by raising or lowering the perceived volume of the inductive (EGFR pathway) or inhibitory (Notch pathway) signals during that window [4898].

Atonal and Amos are proneural agents for other types of sensilla

Chordotonal PNCs differ from bristle PNCs insofar as they rely on the bHLH gene *atonal* instead of *ac* or *sc* [3181, 4897], and the same is true for photoreceptors in

FIGURE 3.13. Circuitry that assigns SOP (bristle) vs. non-SOP (smooth cuticle) cell fates in the epidermis (cf. Fig. 2.7 for key to symbols).

    a. Core pathway of proteins that decide SOP vs. non-SOP identities (see text for further details). Abbreviations: Ac (Achaete), Da (Daughterless), Dl (Delta), Emc (Extramacrochaetae), h (hairy), Sc (Scute), Sca (Scabrous). In the first stage, "prepattern" factors (cf. Chs. 5 and 6) prompt the synthesis of Ac and Sc in certain regions by acting through *cis*-enhancers in the AS-C. When Ac and Sc (which act redundantly) heterodimerize with Da (which is ubiquitous), they endow a cell with the "competence" to become a SOP. This "proneural" state can be prevented by the antineural agents Hairy (which primarily affects Ac) and Emc, both of which act intracellularly (Fig. 3.12). In the second stage (shaded), one or more SOPs are selected by "fine-tuning" within the proneural field. Continued accumulation of Ac and Sc prods a cell to become a SOP, unless it is blocked by external antineural signals emitted by SOPs themselves. Dl and Sca are thought to be short- and long-range agents, respectively (see text), that create an "inhibitory field" around each SOP.

    b–e. Schematics of cells (**b–d**) that are selecting various fates at different tissue locations (**e**). States of components (cf. **a**) are recorded as "1" (present) or "0" (absent), and black circles denote determining factors. **b.** The circuit as it functions in areas where Hairy is expressed (e.g., leg stripes, **e**). The cell never becomes competent because Hairy is the controlling factor. **c.** If a cell is impelled by prepattern factors (circled "1") and no antineural agents are present, then it will complete the gauntlet and become a SOP. **d.** Any cell inside the inhibitory field of a SOP cannot become a SOP. Sca (circled "1") is here assumed to be the diffusible inhibitor. **e.** Rectangular piece of leg skin (Fig. 3.9c) containing one Ac stripe (shaded) and the adjoining Hairy interstripe (hatched). Among these ~150 cells (hexagons), only 4 SOPs (black cells, **c**) will form. Remaining cells are prevented from becoming SOPs because they contain Hairy (**b**) or are inhibited by a nearby SOP that adopted this state (stochastically) before they could do so (**d**). Ovals mark limits of inhibitory fields, which are assumed to be anisotropic because the inhibitor (Sca?) does not diffuse freely.

    N.B.: In *D. melanogaster*, non-SOP cells look alike, regardless of whether they were previously proneural, but other species convert some proneural fields into pigment stripes [615]. Although the circuitry depicted here is digital, the "SOP computer" actually operates partly in an analog (threshold) mode (Fig. 3.12; see text). Also, it is not known whether Sca acts via the same pathway as Dl [113], and other questions remain about the core logic [178]. In principle, the "AND" gate preceding Ac and Sc should cancel out the subsequent "OR" gate, but this part of the circuit is included because Ac and Sc can be independently changed by mutations (Fig. 3.5) or various LOF-GOF manipulations (App. 5).

---

the eye disc [182, 1076, 2041, 2042]. Still another bHLH gene – *amos* – establishes PNCs for olfactory sensilla and leg claws [1587]. LOF and GOF studies have shown that these different bHLH genes not only set up PNCs, but also determine the identities of the sensilla that arise therein (cf. Ch. 2). For example, overexpressing Scute induces bristles only [912, 918, 2038], whereas overexpressing Atonal induces chordotonal organs [764, 2038, 2040], and overexpressing Amos elicits both chordotonal and olfactory sensilla [1587, 1928]. Thus, no clear distinction exists between proneural genes on the one hand and sensillum-identity genes on the other [448, 1928].

Other (upstream) pathways govern bristle patterning

Polychaetoid is intriguing because it may link the HLH circuitry to the Notch pathway [4811] and adherens junctions [533]. The involvement of *pyd* in HLH affairs is indicated by (1) reduced transcription of *emc* in *pyd*[LOF] wing discs [4237], and (2) synergy of *pyd*[LOF] with *emc*[LOF], *h*[LOF], and *ac*[GOF] [733, 1802, 3068, 3069]. The dose-dependent synergy of *pyd*[LOF] with both *N*[LOF] and *Dl*[LOF] is equally dramatic [733], and other clues also point to an affiliation

of *pyd* with the Notch pathway – viz., (1) Pyd's location at junctions [4236] in the same apicolateral ring where N is concentrated (Pyd is homologous to mammalian Zonula Occludens-1) [1203], and (2) Pyd's ability to bind the product of a gene (*canoe*) that also interacts with *N* (App. 5) [4236].

    If *pyd* serves such a key role, then why don't *pyd*[LOF] mutations affect all bristles equally? Why, for example, are extra MCs found in the dorsocentral area ~10 times more often than in the notopleural area [733]? The spatial heterogeneity cannot simply be due to LOF allele quirks (a bugbear of the AS-C) [533], because it also occurs with null alleles [733]. This same question was raised for *Hairless* in Ch. 2, where it was argued that a second gene might share its function. Although a redundant agent might also solve the *pyd* problem, it is possible that *pyd* is part of a separate pathway that is upstream of the entire SOP siting program [2363]. Pyd may instigate certain PNCs (via "Pyd ➜ AS-C"), but genes in other pathways may establish the other ones. Such genes are the elusive prepattern factors that Stern tried so hard to find. How they act is discussed in subsequent chapters.

CHAPTER FOUR

Origin and Growth of Discs

Insect imaginal discs are barely visible to the naked eye, so detailed observations had to await the invention of adequate magnifying lenses [1266, 3064]. Discs were first described by the great naturalist Jan Swammerdam (1637–1680) [821, 3422, 4586], a contemporary of Leeuwenhoek's, who applied his training in human anatomy to the study of insect morphology [3133]. In his *Book of Nature* (printed in English in 1758), Swammerdam waxes lyrical about the metamorphosis (which he calls "mutation") of appendages ("horns" are antennae) in hymenopteran larvae ("worms" of bees):

The wings, horns, and other parts which worms without legs seem to acquire about their chests at the time of their mutation are not truly produced during the period of mutation, or, to speak more agreeably to truth, during the time of the limbs shooting or budding out, but . . . have grown there by degrees under the skin, and as the worm itself has grown by a kind of accretion of parts, and will make their appearance in it upon breaking the skin on its head or its back, and thereby give it the figure of a nymph, which it would afterwards of itself assume.

Hence it is, that we can with little trouble produce [by dissection] the legs, wings, horns, and other parts of an insect, which lie hid under its skin while in the shape of a naked worm, which has neither legs nor any other limbs. . . .

The nymph . . . is nothing more than a little worm, which, the growth of legs, wings, and other limbs hid under its skin being perfected by time, at last bursts that skin, and casting it off, gives us a clear and distinct view of all those parts. This change . . . is no more mysterious or surprising than what happens when one of the meanest plants, despised and trodden under foot, gradually swells on every side, and after producing a bud, by bursting the little case containing it, presents an elegant and beautiful flower. [4218]

The first account of discs in dipterans appeared in 1864 when August Weismann (renowned for his germ plasm theory [1487]) analyzed the stages of development in *Musca* and *Sarcophaga* [4586]. In 1936, after *D. melanogaster* had become the darling of genetics, a former student of Spemann's, Charlotte Auerbach [2215, 4696], traced the development of leg, wing, and haltere discs back to the newly hatched larva [142]. Dietrich Bodenstein studied the development of all the discs for Demerec's *Biology of Drosophila* (1950) [1019], which served as the standard reference for a generation of fly researchers.

The precursors of discs within the embryonic blastoderm were mapped in the 1970s by microcautery [425], microbeam irradiation [2591], and cell-lineage techniques [1695, 2026, 4652]. Each disc was found to come from a distinct cluster of blastoderm cells (Fig. 4.1).

Segmentation genes set the stage for disc initiation

Not until the genetic basis of segmentation began to be analyzed ca 1980, however, was it possible to discern the factors that *cause* discs to originate. In that year, Christiane Nüsslein-Volhard and Eric Wieschaus defined the hierarchy of "segmentation genes" that divides the embryo's anterior-posterior (A-P) axis into metameres [2178, 3151, 4069]. The hierarchy has four echelons (Fig. 4.2) [2717, 3248, 3843, 4671]:

1. *Axis genes* (a.k.a. "maternal-effect" genes). Three proteins form gradients along the A-P axis: "Bicoid" is made from mRNA maternally deposited at the anterior pole [1104]$^\Delta$, "Nanos" comes from posterior

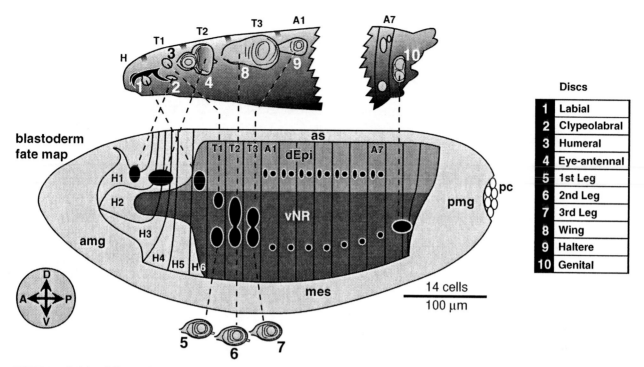

FIGURE 4.1. Origin of discs as "islands" within the embryonic ectoderm. Left side of an embryo (below) at the cellular blastoderm stage (D, dorsal; V, ventral; A, anterior; P, posterior). Spots indicate where discs (key at right) or histoblast nests later form. In this fate map, segments (straight or warped rectangles) are numbered in each region (Head, Thorax, Abdomen). Areas shaded medium (dEpi, dorsal epidermis) or dark (vNR, ventral neurogenic region) form larval skin, except that ~25% of vNR cells ingress as neuroblasts. Gastrulation internalizes the midgut (amg, pmg), mesoderm (mes), pole cells (pc), and virtually the entire head [4421]. Amnioserosa (as) is extraembryonic.

This fate map is based on Hartenstein's atlas [1739], with the following exceptions: (1) D midline is not tilted toward viewer; (2) clypeolabral, eye, labial, and humeral sites are as per gynandromorph maps [1695, 1777, 2026, 4146], although the actual humeral site may be more ventral because its spiracle and the 1st-leg disc can share blastoderm cell ancestors [237, 928, 2819, 4565]; (3) eye anlage is shown as one oval instead of the three spots that Volker Hartenstein used (pers. comm.) as an arbitrary way to connote the eye disc's mysteriously diffuse origin [1147, 1224, 2103] and its patchy apoptosis [3058, 4825]; (4) thoracic discs are more dorsal as per cell transplant data [2820]; (5) wing disc and 2nd-leg disc are fused as per histologic [827, 834] and lineage data [2819, 4076], as are haltere disc and 3rd-leg disc; and (6) genital disc spans A8–A10 [429, 735, 2343].

N.B.: H1–H3 are stylized [1739] and probably not contiguous [832, 1300, 1631, 2104, 4825]. See [2103, 3058, 3631] for head and tail details and [631, 3791, 3792] for other nuances. In the schematic drawing of the larva (above), discs are spread out and the midsection is omitted. See also App. 7.

---

mRNA [939]$^\Delta$, and "Caudal" is made from uniformly distributed mRNA whose translation is repressed by Bicoid [3116]$^\Delta$. Among these three factors, only Bicoid appears to be a *bona fide* morphogen (see below) [1945, 2886, 4154, 4163]. The division of labor between maternal and zygotic (gap, pair-rule, and segment polarity) genes [2105, 3152, 4653] makes sense evolutionarily, given the time constraints of embryogenesis [4650].

2. *Gap genes.* During the syncytial blastoderm stage, genes in the "gap" class are expressed in broad (~10–50% egg length) bands [1398]. They are so-named because LOF mutants are missing large parts of the segment series. Some initial overlaps between ex-pression bands are erased by mutual repression [2329, 3567].

3. *Pair-rule genes.* Spatial periodicity first appears when genes of the "pair-rule" class are transcribed. Their name comes from an absence of alternating segments in their LOF mutants. This echelon, which came as a surprise [659, 3248], is divisible into primary and secondary tiers [1750, 3248]. Primary pair-rule genes integrate nonperiodic cues from axis and gap genes (with some *inter se* inputs), and most of their expression stripes (~7 per gene) are separately regulated by a combination of upstream factors [242, 1722, 1750]. Secondary pair-rule genes rely more on the periodic outputs of primary genes. For example, *fushi tarazu*

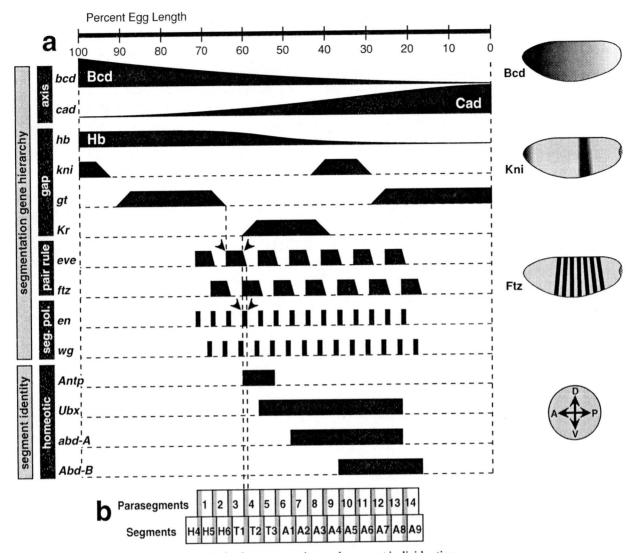

FIGURE 4.2. Categories of genes involved in body segmentation and segment individuation.

a. Protein levels (Bcd, Cad, Hb) or mRNA levels (remainder) detected or inferred [3605] for a few members of each class in the segmentation gene hierarchy (axis, gap, pair rule, segment polarity), or (below) expression patterns of key homeotic genes at the time when their activity is instrumental in patterning [3248]. Gene abbreviations: *abd-A* (*abdominal-A*), *Abd-B* (*Abdominal-B*), *Antp* (*Antennapedia*), *bcd* (*bicoid*), *cad* (*caudal*), *en* (*engrailed*), *eve* (*even skipped*), *ftz* (*fushi tarazu*), *gt* (*giant*), *hb* (*hunchback*), *kni* (*knirps*), *Kr* (*Krüppel*), *Ubx* (*Ultrabithorax*), *wg* (*wingless*). Embryos at right depict realistic Bcd, Kni, and Ftz protein distributions [105, 661, 1102, 1103, 4055]. The Cad gradient arises via Bcd repression of *cad* mRNA translation [1487, 3116]. Vertical lines show a few other interactions: (1) boundaries of *eve* stripe 2 (arrowheads) are defined by *gt* and *Kr* [4764]; (2) even-numbered *en* stripes (arrowheads) emerge due to a trailing gradient of Eve that turns OFF *odd skipped* (not shown) in the front cells of each *ftz* stripe (thus activating *en* by repressing its repressor) [877, 1324, 2681, 2682]. See [1605] for a comparable chain of negative regulators and [2002] for *Deformed*, where the control is more combinatorial than hierarchical.

b. Embryos make 14 parasegments and 15 segments, with an *en* stripe (shaded) at the front or rear of each, respectively.

Panel **a** is adapted from [162, 1416, 1946, 3605], with *eve* stripe 2 and *en* stripe 4 data from [4764] and [1324]; **b** follows [3843] ("A9" stands for A9–A10).

N.B.: As a "wiring diagram," this illustration is woefully inadequate because it omits so many key genes [47, 380, 1962, 3248], spatial details (e.g., uneven levels within homeotic gene domains [682, 2717]), and dynamic modulations [242, 1790, 2181, 3770, 4834]. Nevertheless, at least it conveys a feeling for the flow of control. See [1487, 2121] for excellent exegeses and [2869, 4893] for evolutionary context.

has a compact "zebra" *cis*-enhancer that governs all 7 stripes [4834]$^\triangle$.

4. *Segment polarity genes.* At the cellular blastoderm stage, the foregoing factors turn ON the "segment polarity" genes. Their name comes from the segmental periodicity of LOF defects and associated reversals in cuticular polarity. Among them, *engrailed* and *wingless* are instrumental in setting metamere boundaries.

Bicoid and Caudal are transcription factors [1011, 1709], as are all gap and pair-rule gene products [3768]$^\triangle$, whereas Nanos acts on mRNA processing [4756]. These roles make sense because early mitoses are syncytial, which allows for free diffusion of signals among nuclei [1962]. By the time pair-rule stripes sharpen, however, plasma membranes are partitioning the cortical nuclei into a cellular blastoderm [47, 657, 3843], and direct internuclear communication is no longer possible [1429]. Not surprisingly, segment-polarity genes encode all sorts of proteins involved in transducing signals from the cell surface to the nucleus [3192, 3341]. Only two signals are critical here: Wingless and Hedgehog [1010, 1167, 4499]$^\triangle$.

Ultimately, the segmentation gene hierarchy regulates expression of "Hox" genes, which implement segmental identities (cf. Ch. 8 and App. 1) [2002]$^\triangle$. Control of Hox genes stems mainly from the gap gene level, but inputs come from other levels as well (cf. Fig. 8.1) [683]$^\triangle$.

The overall circuitry is complex because many of the ~50 core segmentation genes interact with ≥5 other such genes in the same or different echelons [1487, 2121]$^\triangle$. Moreover, in some cases, the links skip a level or go back up the chain of command [3815]$^\triangle$. Details are still being worked out [3383]$^\triangle$, but the main design features are now clear [968]$^\triangle$:

1. Both analog and digital controls are employed, and the driving inputs can be positive or negative [657, 2446]$^\triangle$.

2. Binary (ON or OFF) states are stabilized by positive feedback [2057]$^\triangle$ and arise, at the DNA level, via cooperative binding of *trans*-acting factors at *cis*-enhancers of target genes [1708]$^\triangle$.

3. Inputs are typically processed by combinatorial logic (as "AND" gates) at these *cis*-enhancers [414]$^\triangle$, while other aspects of Boolean syntax are mediated by competitive binding [3993]$^\triangle$, dimerization [3768]$^\triangle$, quenching [1600]$^\triangle$, physical spacers [3992]$^\triangle$, and "dual control" [657]$^\triangle$.

These elements mesh to form a dynamic and robust system [4499]$^\triangle$ that orchestrates thousands of target genes [2541]$^\triangle$. Despite its virtuosity, however, this "gene machine" is not optimized [4670] since certain pairs of components (*nanos* and *hunchback* [1945, 1986, 4154] or *odd skipped* and *engrailed* [878]) can be removed with little or no effect on the patterning of denticle belts. Aspects of the system have been modeled mathematically [2811, 3704], but computer models are best for tracking its behavior in real time. Simulations by John Reinitz [3567]$^\triangle$ and others [380, 2200, 2201] are yielding more testable predictions as additional parameters of the components become known [1010].

By 1987, the segmentation gene network was coming into focus, and seminal reviews were authored by Philip Ingham and Michael Akam [47, 1962]. Tellingly, these normally reserved British dons could barely contain their excitement at humankind's first glimpse into the molecular clockwork of pattern formation. Indeed, the insights galvanized the whole field [125, 3669]. Akam realized that the embryo's strategy of making segments from overlapping stripes fulfilled Curt Stern's vision of prepatterns [47]:

The pattern generated by the gap and pair-rule genes is transient and can appropriately be described as a prepattern.

As discussed in Ch. 3, Stern thought that prepatterns are transient networks whose interdependent nodes evoke elements in the final pattern [4100, 4346]. In this case, the nodes would be the boundaries that delimit metameric regions along the A-P axis [657, 2433], and their interdependence would be attributable to the circuitry of the segmentation gene hierarchy [3248].

Ironically, Stern's Prepattern Hypothesis had been toppled as the field's reigning paradigm in 1969 by Lewis Wolpert's "**Positional Information Hypothesis**" [4724]. Wolpert argued that patterns are organized by global coordinate systems instead of by skeletal frameworks (see below).

Akam's remark heralded the resurrection of the prepattern concept as a useful way of thinking about embryos and discs [2446, 3953], although its full rehabilitation took several more years. To contrast these distinct schools of thought, the history of their conflict is briefly recounted below. Although retrospective, the next section is not antiquarian, and readers would do well to study it. It constitutes the bedrock for the rest of the book.

Prepatterns and gradients clashed in trying to explain homeosis

Stern's reasoning was straightforward. If prepattern genes establish the prepattern's nodes and the nodes are interdependent, then disabling one or more nodes within a LOF mutant clone should cause the remaining nodes outside the clone (in wild-type terrain) to redistribute themselves. In other words, any prepattern gene would have to behave nonautonomously in genetic mosaics [4100]. Today we know many such genes, most of which encode diffusible signals, but at that time (1954) none were known. Stern and his associates tested one mutation after another to see whether they could alter patterns nonautonomously in mosaics. By 1967, they had assayed ~20 genes but had not yet found a *bona fide* case of pattern reorganization [4346].

In 1967, Stern's long quest for a "prepattern mutant" was rewarded when he and Chiyoko Tokunaga tested *eyeless-Dominant* (ey^D) in mosaics and found that it acted nonautonomously [4100, 4346]. Although its name comes from its eye defects (a result of cell death [1309]), this mutation also causes extra rows (~2–6 rows vs. one in the wild-type) of sex comb bristles (~39 vs. ~10 bristles in the wild-type) on male forelegs. In mosaics, wild-type cells were found to be able to participate in extra rows when they intrude into the sex comb area of an otherwise mutant leg [4109].

Apparently, ey^D augments the sex comb node or "singularity" in the leg prepattern, and wild-type cells respond accordingly – i.e., "ey^D mutation ➡ bigger sex comb singularity ➡ phenotype" ("➡" denotes "causes"). However, other aspects of the mutant syndrome are visible in forelegs of *both* sexes and hence cannot be specific for the sex comb: swelling of the distal basitarsus, extra bristles of other types, disruption of pattern and polarity, and fusions of 1st and 2nd tarsal segments. These additional defects suggested other etiologies [1421, 1799]:

1. *Overgrowth Scenario:* ey^D mutation ➡ overgrowth ➡ more sex comb positional coordinates ➡ phenotype [3213, 3448, 4348, 4724]. Conceivably, ey^D induces overgrowth by disrupting *Sex combs reduced* [3332, 3333] – a Hox regulator at this spot.
2. *Segmentation Scenario:* ey^D mutation ➡ gap in intersegmental membrane ➡ distorted gradient ➡ more sex comb positional coordinates ➡ phenotype [523, 2427, 3427, 4100, 4517]. This argument is predicated on the assumption that the fly leg uses a sawtooth series of segmental gradients to specify positions along its

proximal-distal axis [385, 386, 1801, 2425]. Interrupting the barrier between the low end of one gradient and the high end of the next one (i.e., intersegmental membrane) could cause a "backflow" of morphogen, with the consequences listed above [2423].
3. *Apoptosis Scenario:* ey^D mutation ➡ cell death ➡ compensatory overgrowth ➡ phenotype [1314, 3441, 3442, 3964, 4346] (see [4336] for a striking phenocopy).

Despite the fact that we still do not know which of the above explanations (if any) is right, ey^D gave a fitting ending to the prepattern saga, and Stern featured this story in his 1968 opus *Genetic Mosaics and Other Essays* [4100]. The celebration was short lived, however, because Wolpert proposed his model in the next year.

The coup de grace that killed Stern's model was not the paucity of prepattern mutants, which he rightly attributed to lethal side effects that would preclude their survival [4100]. Rather, it was the fact that homeotic mutations act autonomously [526, 1357, 4346].

"Homeotic" mutations, by definition, transform particular body parts to resemble other body parts – e.g., a leg into a wing (cf. Ch. 8) [2509, 3214, 4486]. Assuming that each organ has its own prepattern, Stern expected mosaic organs to display signs of jousting prepatterns – nonautonomous influences of homeotic tissue on nearby wild-type cells or vice versa [4097, 4098].

By 1968, however, several contradictory cases had been documented. For example, *extra sex combs*^{LOF} partly converts 2nd and 3rd legs into 1st legs [1713], yet acts autonomously in mosaics [4349]. This result was not too unsettling for the hypothesis because all leg cells might "speak" a common language. No such rationalizing was possible, however, for *bithorax*^{LOF} (anterior haltere into anterior wing) [2506] or *aristapedia*^{LOF} (distal antenna into distal leg) [3444, 3614] because it seemed inconceivable that the same scaffolding could be used for organs as different as halteres and wings or legs and antennae (Fig. 4.3) [3214].

In his 1968 book, Stern acknowledged this paradox but did not offer a solution [4100]. In a review of Stern's book, Peter Bryant argued that stretching the prepattern concept to cover these situations undermined its usefulness to the point of absurdity [520].

Some workers, including Stern, have employed the concept [of prepattern] as specifying, in *Drosophila*, individual bristle positions. However, experiments using mosaics for homeotic mutants tend to indicate that the prepattern for antennal structures is identical to that for leg structures. In that case, the prepattern could not be specifying individual bristle positions since the bristle patterns are entirely different in these two appendages. If the concept must be generalized to such an extent that the prepattern for leg is identical to that for antenna, then it seems to lose much of its usefulness as a working hypothesis. We are left, in fact, with the rather hackneyed interpretation of pattern formation as resulting from a gradient of some hypothetical morphogenetic influence. [520]

A new conceptual framework emerged in 1969 when Lewis Wolpert blended Hans Driesch's old notion of coordinate systems [1101, 3741, 3742] with modern information theory [95, 2547, 2766, 2989] to create the idea of positional information (PI) [4724, 4731, 4734]. It was not this homeosis problem that PI aimed to solve, but rather the "**Regulation Riddle**": how can patterns robustly regenerate? Nevertheless, an explanation for homeosis followed naturally. Indeed, Wolpert cited the autonomy of *aristapedia*^{LOF} as support for the "universality" corollary of his theory – namely, that the same PI coordinates should be readable by cells of any histotype.

Coincidentally, in that same year Wolpert was handed an even better example. Antenna-to-leg homeosis in *Antennapedia*^{GOF} (*Antp*^{GOF}) affects not just the arista but the entire antenna, and the patchiness of the transformation in nonmosaic mutant flies revealed a striking correspondence between antennal and leg regions (Fig. 4.3a) [3445, 4728]. Such different organs could not possibly share a common prepattern (or so it was thought), but they might easily use a common coordinate system. The Prepattern Hypothesis seemed doomed.

For me, the most significant contributions to the study of pattern formation over the last 30 years come from the work of Stern on genetic mosaics and the concept of prepattern.... This work provides excellent evidence for the concept of positional information and polarity potential, and the best evidence for the postulate of universality, at least within the same animal. As will be seen, the concept of positional information gets over some of the difficulties associated with the concept of prepattern.... Along similar lines one can interpret homeotic mutants which involve genes such as *aristapedia* which cause antennae to form legs. Once again the positional information may be the same and only the interpretation different. This is in line with the postulate of universality and it is thus again encouraging to find genetic mosaics of *aristapedia* with normal tissue behaving [autonomously] according to position and genome. [4724]

Wolpert invoked three stages for patterning and used a French Flag to abstractly represent three distinct cell types [4723]:

1. *Specification of PI.* A scalar variable changes linearly or exponentially along an axis [3989] (e.g., the activin gradient [1656, 4220]). In the archetypal model, the variable is a "morphogen" (diffusible signaling molecule) whose concentration describes a "gradient" across the "field" of cells whose fates it "specifies" [4732]. Ambiguities in these terms can be minimized by operational definitions [967, 4673] (e.g., a "morphogen" is a signal that elicits expression of different genes at different concentrations [865, 1655]). Morphogens were imagined to emanate from a terminal source, with or without a sink where the morphogen is degraded [210]. Gradients are usually assumed to reach equilibrium before being "read" [3074], but this need not be true [2780, 3225]. Also unclear is whether cells "ratchet" to a particular level of response whence they are unable to retreat [1657]. In theory, cells can tell the absolute size of a field and their orientation therein by sensing slopes of gradients [2434, 2448]. Finally, there is the issue of resolution (signal-to-noise ratio) as a function of distance. For example, in fly eyes, the signal for ommatidial polarization is sensed over distances of >100 cells with 100% fidelity (cf. Fig. 7.5) [2883], but how?

2. *Recording of PI.* To gauge its position, each cell measures the local height of the PI variable – like a person guessing the nearness of an ambulance by the loudness of its siren. (How finely cells can measure concentration differences is unknown [1142, 1655, 4366].) Cells then record this information (e.g., as ON or OFF states of "memory" genes) [1608, 1964, 2417, 2515, 2803, 4730] as a function, perhaps, of receptor occupancy on the cell surface [1129, 1655, 2779] or of sensitivity thresholds in target gene promoters [1284, 1923, 2678, 3087, 3406]. The gradient itself can then disappear since it is no longer needed. How long such memories can persist in dividing cells is unclear.

3. *Interpretation of PI.* At a later time, cells translate the memories of their old locations ("positional values" [4729, 4730, 4732, 4734]) into differentiated states (blue, white, or red) through some sort of downstream gene circuitry [98, 2646] – like a letter carrier deducing city names from postal codes (e.g., see [472]). Whereas Step 2 converted analog to digital information, Step 3 must convert quantitative to qualitative states [1655], although memories may also control analog traits such as adhesivity [2436] or mitotic rate [1263]. One abiding mystery is the extent to which neighboring cells interact to sharpen the interpretation zones [1143, 1609, 4691].

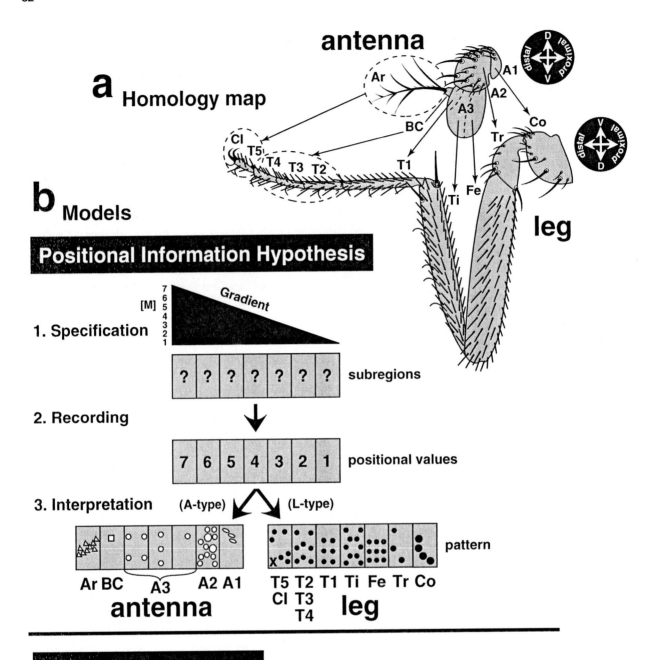

a Homology map

b Models

Positional Information Hypothesis

1. Specification

2. Recording

3. Interpretation

antenna

leg

Prepattern Hypothesis

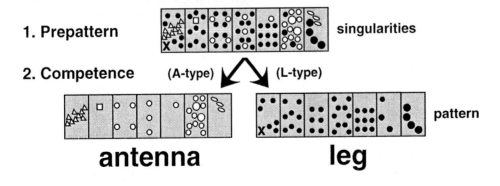

1. Prepattern   singularities

2. Competence

antenna   leg

PI and prepatterns are often contrasted in terms of how they might create an array of evenly spaced bristles within a line of cells [526, 2764, 4671]. PI would use a gradient to assign a number to each cell, and certain numbers would signify "bristle" because the genome is "wired" accordingly. A prepattern strategy, on the other hand, would use some sort of device (e.g., reaction-diffusion) to put bristle-inducing signals at regular intervals, and any cell would be able to respond (above a certain threshold). With PI, every cell knows where it is and differs from every other cell, whereas prepatterned cells would know their state (bristle vs. nonbristle) but not their location [4725, 4726, 4729].

In the PI scheme, patterns are essentially encoded as bitmaps [1805, 3610], with each cell (≈pixel) having a separate genetic representation (≈ON/OFF state). The huge coding capacity needed for intricate patterns (even with combinatorial encoding) strained credulity [862, 2142, 2646], as did the paucity of plausible molecular mechanisms [1805, 2809, 3087, 4730, 4732], but neither quibble slowed the PI model's rise to popularity [864, 3988, 4734]. It quickly attained paradigm status. Stern's camp formally "surrendered" in 1978 [4346].

Soon thereafter, direct support for PI came from the segmentation genes [4733]. They furnished the first definitive morphogen (Bicoid) [1103]$^\Delta$, plus insights into

---

FIGURE 4.3. Demise of the Prepattern Hypothesis in 1969 and ascendancy of the Positional Information Hypothesis, due in part to their clumsy or elegant explanations for homeotic autonomy.

a. Correspondence (arrows) between antennal and leg parts, as deduced from *Antennapedia^R* (*Antp^R*) flies, whose antennae are patchily transformed into 2nd legs [3445]. Abbreviations: A1–3 (antennal segments 1–3), BC (basal cylinder), Ar (arista), Co (coxa), Tr (trochanter), Fe (femur), Ti (tibia), T1–5 (tarsal segments 1–5), Cl (claws). Compasses denote adult polarities (D = dorsal, V = ventral; the inversion is explained below). Antennae of *Antp^R* flies (not shown) have patches of leg tissue. Whenever leg tissue arises at a particular proximal-distal level, it makes leg structures appropriate for that level. The implication is that these leg cells can somehow ascertain their location along this axis of the antenna. Wolpert and Stern agreed that the antenna and leg probably share some kind of "ground plan" that both types of cells can "read." They disagreed over the *kind* of ground plan. In Wolpert's scheme, it is a gradient; in Stern's, it is a prepattern.

b. The two models. Each of the 7 gray rectangles represents a segmental region (eventually containing 10^2–10^3 cells) of the antennal or leg rudiment along the proximal-distal axis. Certain segments are subdivided or grouped.

Lewis Wolpert's **Positional Information (PI) Hypothesis** explained antenna-to-leg homeosis in terms of a shared coordinate system [4724]. Antennal and leg cells are supposed to use the same positional signal – a diffusible molecule "M." This "morphogen" is produced at one end of the organ (left) and diffuses to form a concentration gradient (triangle). The gradient could be linear (as shown) but is more likely exponential. In Step 1, cells assess their distance from the source by measuring the amount of M: the more M that a cell "tastes" at its site, the nearer it "thinks" that it must be to the source. In Step 2, cells record these levels (1–7) as "positional values" (ON or OFF states of memory genes?) that persist after M vanishes. In Step 3, cells translate these values into structures that suit the disc to which they belong. For example, antennal cells interpret the number 7 as "arista," whereas leg cells interpret it as "T5 and claws." The interpretation mode that a cell uses (A vs. L) is predetermined by the turning ON or OFF of a particular homeotic gene in one of the cell's ancestors. Mutations in that gene cause homeosis: mutant cells "think" they are leg cells, but they can still sense M properly, so they make leg structures that are appropriate for their proximal-distal level in the antenna. Finer-grain patterning within subregions may rely on a second echelon of (segmental) gradients (not shown) [386]. The elegance of Wolpert's model has been its universality: the same gradients can theoretically pattern all the discs [20, 523, 3701].

Curt Stern's **Prepattern Hypothesis** had a more difficult time explaining homeosis. Its basic assumption was that every pattern element is preceded by a discrete signal ("singularity") [4095]. Unlike the morphogen, these signals (circles, ovals, square, triangles, and "x") differ *qualitatively*: they are not graded. In Step 1, the disc lays out an array of these signals (the "prepattern"). The autonomy of the *Antp^R* mosaicism implied that antenna and leg must express one another's signals. (How else could nouveau-leg cells that find themselves "stranded" in the antenna figure out what structures to make?) In other words, both prepatterns must coexist in each disc [4346]. Antennal cells would normally read one subset of signals (white symbols), while leg cells read the other (black symbols). The filtering of these subsets (cf. Fig. 3.2) would depend on a cell's "competence," which in turn would be determined early in development when one of the cell's ancestors turned a particular homeotic gene ON or OFF. Mutations in such genes would lead to interdisc transformations (cf. Ch. 8). This argument was tolerable for a single type of transformation (e.g., antenna-to-leg), but the large number of interdisc homeoses [3214] implied that every disc must contain the prepattern for virtually every other disc. Although not impossible, this notion seemed implausibly clumsy, especially when contrasted with how easily the PI Hypothesis handled this same phenomenon.

Panel **a** is redrawn from [3447], which depicts a right antenna and left leg (see [1587] for A3 subregions, [1516, 1561, 2287] for molecular "homology," and [4522] for a different map). Panels **b** and **c** are based on the ideas of Stern (Fig. 58 in [4100]) and Wolpert [4734], as contrasted *inter se* by Bryant [526] and Tokunaga [4346]. Black or white symbols in the antennal and leg schematics are merely abstract representations of the actual pattern elements. See Figure 8.3 for data on how antennal vs. leg identities are actually controlled and [2956, 3744] for the origin of the gradient concept. See also App. 7.

(1) how gradients can specify zones (e.g., gap gene en-hancers varying in Bicoid-binding affinity) [1105]$^\Delta$, (2) how boundaries can be sharpened (e.g., competitive and cooperative *trans* interactions) [575]$^\Delta$, and (3) how interpretation can be implemented (e.g., morphogen sensitivity being dictated by distance from *cis*-enhancer to core promoter) [1835, 2421].

The first step involves the localization of a cytoplasmic de-terminant, in this case the messenger RNA of ... *bicoid* in the unfertilized egg. On fertilization the localized RNA be-comes translated into bicoid protein, which spreads by dif-fusion and forms a morphogenetic gradient. The bicoid pro-tein concentration provides positional information to the nuclei in which it is taken up. The third step involves the conversion of the graded pattern into a repetitive pattern of stripes under the control of the segmentation genes. Finally, the repetitive pattern is converted into a sequential pattern; the segments differentiate and each acquires its own identity [via] the homeotic genes. [1416]

Gradients and threshold responses appear to be used not only in Bicoid's control of gap genes [1284]$^\Delta$, but also in cross-talk among gap genes [4764]$^\Delta$, in gap gene control of pair-rule genes [1323]$^\Delta$, in cross-talk among pair-rule genes [4553]$^\Delta$, and in pair-rule gene control of segment-polarity genes [1324]$^\Delta$. Despite the network's reliance on gradients, the way that it processes infor-mation is more reminiscent of prepatterns [2448]. To wit, cells only use certain Bicoid levels to establish zones of gap gene expression, and the zone borders act like singularities to delimit pair-rule stripes. The remaining levels are ignored. In a pure PI scheme, *all* levels should be recorded as heritable positional values [3087], but Stage 2 of the PI scenario seems to be bypassed. Thus, the process seems more like a PI-prepattern hy-brid than strictly one or the other. Peter Lawrence, like Michael Akam, realized that these theories could dove-tail nicely into a unified explanation [2446]:

[The prime function of pair rule genes is] to locate bound-aries that delimit fields or gradients of positional information. [2433]

Within each segment, cell positions were thought to be specified by a "segmental gradient" [2426]$^\Delta$, based on a long history of surgical experiments [1300, 2430]$^\Delta$. The segment-polarity echelon fulfilled this prophecy inso-far as its stripes have a segmental periodicity [2717]$^\Delta$, although the initial metamerism consists of "paraseg-ments" [1978, 2438, 2724] that are out of phase with segment boundaries (Fig. 4.2b) [1772, 4152].

Two protein products of segment polarity genes – Hedgehog and Wingless – act like PI morphogens in embryonic body segments [1789, 2448, 2716, 3748] and adult abdominal segments [2301, 2303, 3905, 4157, 4158] (see [2649] for a prescient model). Early experiments seemed to re-fute Wingless as a morphogen in the embryo [3730], but this conclusion was overturned by later studies which showed that *sloppy paired* imposes an uneven "land-scape" of cellular competence to respond to Wg [596, 3129].

Homeotic genes implement regional identities

How could *identical* gradients (segmental or paraseg-mental) produce *different* structures? According to PI theory, they must use different "code books" for in-terpreting their coordinates. In terms of Wolpert's flag metaphor, each segment would use the same coor-dinates, but Segment T1 might employ a British Flag interpretation mode, while Segment T2 uses an Italian Flag mode, etc. Only a few bits of information would be required to establish 15 or so interpretation modes along the A-P axis (one per segment). The process whereby segments become different from one another ("individuation") is genetically separable from segmen-tation, because certain homeotic mutants undergo nor-mal segmentation with virtually no individuation [2507, 4147].

What would happen if the interpretation mode of a particular segment were to malfunction? The segment should develop like another segment (i.e., undergo homeosis), and the same is true for interdisc transfor-mations [1418]. Thus, the phenomenon of homeosis fit neatly into the PI paradigm [523, 3701]. Presumably, ho-meotic genes would serve three PI-related functions: (1) use their ON or OFF states to record various levels of the whole-egg gradient, (2) perpetuate these mem-ories during disc growth, and (3) set the interpretation mode for each segment or parasegment [1365, 2755, 3882]. These expectations turned out to be basically correct (cf. Ch. 8).

As Wolpert's ideas were rippling through the research community in the early 1970s, a complementary model of Stuart Kauffman's was also gaining notoriety [2155, 2156]. Its formulation for disc identities was based on fate switches observed during "transdetermination" – a spo-radic metaplasia seen during long-term culture of disc tissue *in vivo* [1668–1670] or *in vitro* [3883, 3885]. Kauffman was struck by the fact that each disc only transforms into a few other types (cf. Fig. 6.9d). He argued that (1) these limitations are indicative of an identity code, (2) the code is binary, and (3) the "transition rules" (i.e., which disc can transform into which other discs) ap-ply equally well to homeosis [2153, 2164, 3214]. According to his "**Binary Code Conjecture**," each disc follows a preferred sequence of transdeterminations because only one bit changes at a time (i.e., 0 to 1, or 1 to 0). For example, if the code for 2nd leg is "1110," then single switches could produce 0110, 1010, 1100, or 1111, but not other states (e.g., 1001) [2158, 2159]. Transdetermination and homeosis share other features [2082, 2159, 2754, 3788, 4142], although differences do exist [2138, 3448, 3883], and there are other possible explanations aside from coding (cf. Ch. 8) [1472, 2411].

In 1975, Antonio García-Bellido proposed his "**Selector Gene Hypothesis**" [1358]. Like Kauffman's conjecture, it invoked switch genes for regional identities [701, 1635, 2930, 3457], but the regions were lineage compartments within discs rather than entire discs [2431]. At the time, it seemed conceivable that a succession of compartments might subdivide discs down to the single-cell level [1358, 2142, 2440], so that each cell would acquire an "area code" tantamount to a Wolpertian coordinate [1369, 1477, 3914]. However, that idea proved false [354, 494, 1375, 1639, 4671].

The discovery of the homeobox, announced by several labs in 1984 [2783, 2785, 3844], provided dramatic support for the selector gene idea [1412–1414, 2421]. This ~180 b.p. sequence encodes a ~60 a.a. DNA-binding domain (cf. App. 1) [1417, 1420]. As homeoboxes kept showing up in one cloned homeotic gene after another, it became clear that these genes constitute a distinct class structurally and functionally [1418, 3556] as contrasted with "housekeeping" genes whose products (e.g., metabolic enzymes) are used ubiquitously [293, 1865]. Since then, more evidence has surfaced for "master genes," but their roles defy all the early models. Because it is not possible to understand how homeotic genes act without delving into each disc's idiosyncrasies, further discussion is deferred until Ch. 8.

One quirk of homeosis itself, though, merits mention here – the "**Collective Amnesia Conundrum**" [1805] (cf. "homeogenetic induction" [1005] and the "community effect" [4063]). Most models presume that disc states are carried by individual cells. A leg disc cell, for example, would know that it belongs to a leg disc and not to some other disc, regardless of its environment. This axiom was based on autonomy in genetic mosaics (see above) [2922, 4346] and retention of histotypes by dissociated disc cells in mixed aggregates [1374, 1411, 1421, 3138]. However, the islands of leg tissue in *Antp*^GOF antennae arise nonclonally [3445], as if neighboring cells *jointly* forget their disc of origin, and transdetermination involves the same sort of process [1353, 1405, 1406]. Social cell behavior like this might depend on the nonclonal clus-

ters of synchronously dividing cells that are often seen in normal discs [23, 532, 2744, 2848] if (1) disc cells normally forget their identity at some stage of the cell cycle [4882], (2) they recover a sense of who they are by "reading" the identities of neighbors when they exit mitosis [533], and (3) too many cells simultaneously undergo mitosis, in which case groups of cells would remain amnesic and be forced to adopt a default state. Excess growth is indeed correlated with transdetermination [1421, 1670, 3883, 4334] and with "homeotic regeneration" [1421, 3448, 4142], but this "**Cyclic Amnesia Scenario**" is too simple [3448] because it predicts

1. Only one default state should exist per disc; however, each disc can actually transform into several other types [1421, 3214, 3881].
2. Tumorous discs should exhibit frequent fate switches *in situ*; however, they typically do not [541, 1386, 2107, 3887].
3. Fate switches should occur in normal discs by chance alone. On the contrary, homeotic outgrowths are virtually never seen in wild-type flies [3883].

Observed cases of communal conversion may arise from diffusion [4063, 4563]: if single cells mistakenly start expressing a morphogen, then it could diffuse and activate inappropriate gene circuits so that unrelated neighbors jointly switch their fate [2082, 2754] (cf. Ch. 8). The dependence of transdetermination on wounding is consistent with this notion [3887] because wounding can evoke morphogens [489], and some homeoses are associated with tissue loss [839].

Wing and haltere discs "grow out" from 2nd- and 3rd-leg discs

Rudiments of the thoracic discs become morphologically recognizable when they invaginate after germ-band shortening (St. 13) at 9–10 h AEL (After Egg Laying, 25°C) [237, 2717, 2820]. At this time, the wing and haltere discs contain ~24 and ~12 cells, respectively [237, 664, 827, 2335], excluding mesoderm cells [57, 1886, 2819]. In a newly hatched larva (~24 h AEL), the major discs contain 20–70 cells (wing ~40; eye ~70; leg ~40; haltere ~20; genital ~60) [2650]. In mid-to-late 1st instar, all disc cells resume mitosis [474, 528, 2650] (they had stalled at ~10 h AEL, St. 13 [1142, 1259]) and start multiplying exponentially [3139, 3441]. During the 3rd instar, cell numbers double every ~10 h (wing 8 h; eye 11 h; leg 8 h; haltere 13 h) [2081, 2935], so that by pupariation each disc has 10–50×10^3 cells (wing ~49,000; eye ~44,000; leg ~17,000; haltere ~10,000) [542, 1123, 1185, 2710] and measures 50–300 μm across [524, 3139, 3507]. Indeed, one reason for studying discs is that they must pattern themselves *during growth* – a constraint that is typical of most developing systems (e.g., vertebrates) – whereas fly embryos employ a quirky syncytial phase whose circuitry may be less useful in trying to understand common (ancient) patterning mechanisms [1303, 3074, 3747].

When do disc cells become determined? That is, when do they irrevocably decide to make (1) adult vs. larval structures [1717] and, more narrowly, (2) structures appropriate to their own disc type vs. some other type [523, 1411, 2411, 3139, 3795]? Both decisions must occur after 3 h AEL (St. 5) because single blastoderm cells can contribute to both imaginal and larval tissues when transplanted homotopically at that time [2819]. Leg determination probably happens by 5.5 h AEL (St. 11) because the early enhancer for *Distal-less* (*Dll*) is activated then [827, 1571, 1572]. Dll is a marker for distal leg, although its expression is not so restricted initially [1572]. The first transcription of *vestigial* (*vg*), a marker for wing and haltere cells, is not detected until ~2 h later [827, 4681]. Strangely, however, neither *Dll* nor *vg* is required for disc development at this stage [827, 3969, 4681], so neither gene provides ironclad evidence for determination [884, 4682]. Nevertheless, the ~5–7 h AEL (St. 11) estimate is probably close [2335], given that (1) disc fates can be changed until that time (but not afterward) by misexpressing homeotic genes [684, 3930] and (2) X-rays can induce pattern duplications in discs (implying that they have become embryonic fields) after 6 h AEL (but not at 3 h AEL) [4651].

Remarkably, the Vg-expressing cells in Segments T2 and T3 originate as subsets of Dll-expressing cells that move dorsally in the extended germ band [827, 1571, 1572]. Because this movement is correlated with locally increased mitosis [237], it is probably driven by growth, although cell migration is likely involved later when wing and haltere cells cling to tracheal branches [237, 2335]. Evidently, wing and haltere progenitor cells arise within the same nests that form 2nd and 3rd legs, respectively (Fig. 4.4). This conclusion fits with the visible "budding" of prospective wing from 2nd-leg rudiments (and haltere from 3rd leg) in the dipteran *Dacus tryoni* [81], and it affirms the controversial idea that the insect wing evolved as a branch from the arthropod leg [145, 2348, 2349, 3911, 4662]. Also, it clarifies why single blastoderm cells can contribute to both wing and 2nd leg (or to haltere and 3rd leg) [2442, 4076, 4651].

Moreover, this recent finding settles an old dispute between two titans from the Morgan era: A. H. Sturtevant and Curt Stern. In his 1929 gynandromorph study (cf. Ch. 1), Sturtevant noticed that sternopleu-ral bristles are clonally linked with thoracic bristles as often as with leg bristles [4180]. From this result, he inferred that the sternopleura (a flank sclerite; cf. Figs. 4.4i and 5.1d) can come from either the wing or 2nd-leg disc, depending on which disc happens to spread into this region first during metamorphosis (cf. other claims for indeterminacy [3250] that were refuted by Stern [343, 3006]). Stern argued (correctly) that sternopleurae *always* come from 2nd-leg discs, and he marshaled much evidence to prove his point [4090, 4099]. Sturtevant remained stubbornly unconvinced [4182]. To explain Sturtevant's data, Stern guessed that leg and wing primordia must abut one another in the ectoderm [4090, 4652], and Dll and Vg now validate his 1940 conjecture:

If the division line between [male and female areas] of a developing gynandromorph has an equal chance of falling either between the ventral [leg] and dorsal [wing] anlage or between the two subregions [sternopleura vs. remainder] of the ventral [leg] anlage, then the results of Sturtevant can be explained without recourse to the hypothesis of indeterminate overgrowth of imaginal disc ectoderm during metamorphosis. [4090]

By ~7 h AEL (St. 12) when disc fates appear to be set, the leg and wing primordia contain ~20 and ~30 cells, respectively [827, 828, 2819] (including mesoderm cells [1886]), and these figures agree with indirect estimates from cell lineage studies [2828, 3139, 4649], which put the number of prospective epidermal cells per disc in the 10–40 cell range (excluding labial and clypeolabral discs). Because discs are not clones (cf. Ch. 1), the disc-initiating factors (whatever they may be) need not be so precise as to pinpoint single cells in the embryonic ectoderm.

Discs retain their identities during larval life [2411], although their cells look embryonic and do not differentiate until metamorphosis [82, 526, 3795, 3797]. Thus, it is not possible to distinguish wing- from leg-disc cells, for instance, just by looking at them [3165, 4424]. This prolonged period of "determination sans differentiation" [1654, 3062, 3448] made it possible to study the stability of determined states [1384, 1406, 1668] and growth regulation [543], which led to the discovery of transdetermination and regeneration, respectively [2755, 3794]. The independence of larval from imaginal development is vividly illustrated by discless larvae, which feed and grow and molt apparently normally and only die when attempting to undergo metamorphosis [3428, 3880, 3886, 4221].

Disc cells remain diploid [499, 1201, 1764, 1765, 2151], whereas the larval skin cells that surround them become polytene [1126, 1259, 1346] via endocycles of DNA replication

[1261, 3551, 3927, 3995, 4587]. Historically, differences like these made sense in terms of the once-popular view that separate batteries of genes control imaginal vs. larval development [1384, 4686].

To my mind the simplest explanation of such a clear cut dichotomy brought about by a single kind of stimulus is to be found in the analogy of the locked door to which the juvenile hormone is the key. In other words we are concerned with two sets of genes so disposed that in the presence of a small amount of juvenile hormone one set takes precedence, while in the presence of a large amount of juvenile hormone the other set is brought into action. [4659]

That view was primarily buttressed by "disc-specific" mutations recovered in screens for pupal lethals [3880, 3881, 3886, 3888, 4121] (especially ones that cause a discless phenotype), but subsequent work showed that defects can be disc-specific for a trivial reason – viz., the egg has enough maternally supplied gene products to build the larva (in 1 day) but not the adult (requiring 4 more days) [834, 4221, 4257, 4671]. Regardless of whether the "**Battery Dichotomy Hypothesis**" has any validity in other areas (e.g., physiology), it is certainly defunct in the realm of patterning: many of the same genes that pattern the larval epidermis also pattern discs [834, 3341, 4675] and histoblast nests [4158].

Except for the humeral disc (which lacks a lumen), discs become hollow sacs via invagination or delamination [1421, 1544]. The apices of their cells face the lumen [142, 2650, 4424] and ruffle to form a carpet of microvilli prior to cuticle secretion [194, 3421, 3422]. The disc epithelium is one-cell thick [3165, 3426, 3539] like the ectoderm whence it comes [1259] and like insect skin in general [2584, 3421, 4661]. One side of each disc thickens into a columnar epithelium that will secrete adult cuticle [1133, 1315, 1431, 3564], while most of the other side forms a thin "peripodial membrane" whose squamous cells make few cuticular parts in the thorax [500, 2862, 2863] but significant parts of the head [1777]. Curiously, mitoses may be synchronized across the lumen [2744]. Coordination of this sort (as well as other sorts of signals) is probably mediated by the filopodia ("translumenal extensions") that stretch from peripodial cells to the columnar surface [773, 1473, 3508]. During 3rd instar, the columnar epithelium looks "pseudostratified" because its nuclei occupy many levels [2650, 4715].

In early 3rd instar, the wing portion of the wing disc (as distinct from the notal part) is initiated via a separate genetic circuit [885, 2219, 2254, 4683, 4684]. Characteristic folds emerge there (eventually forming a "pouch") and in the leg disc (eventually forming a comparable "endknob") [142, 377, 2287], presumably due to region-specific mitotic

rates or orientations [544, 2848, 3422, 4427] or histotypic affinities [351, 1060, 2570]. Cell death cannot play a major role in morphogenesis or patterning [532, 3422] because it is so rare [12, 1501, 2015, 2848, 2849], except in eyes [396, 397, 489, 1309, 1763] where it serves to tighten the ommatidial lattice [4713, 4715].

Aside from epithelial cells, most discs contain adepithelial cells [239, 476, 497, 3104, 3660] (= mesodermal myoblasts [236, 1214, 1886, 2820, 3609, 3661]), neurons [4330], and tracheal cells – all of which reside between the epithelium and the basal lamina [3422, 3426]. Genetic evidence suggests that adepithelial cells in leg discs induce patterning in the tarsal epithelium [3332, 3333]. The male genital disc is freakish insofar as it recruits mesodermal cells *into* the epithelium [38]. Those cells go on to form the paragonia and vas deferens.

During metamorphosis, each disc everts through its stalk [1311, 3422, 4429]. Historically, the planarity and insularity of discs (the latter feature being uncommon among embryonic fields [4516]) proved useful for analyzing mechanisms [523, 2429, 3448], as did the durability and intricacy of the cuticle, which indelibly records the final state of virtually every epidermal cell [526, 3421, 4663].

Thoracic discs arise at Wingless/Engrailed boundaries

Although blastoderm cells are not restricted to larval vs. imaginal fates [2819], nor to ventral (leg) vs. dorsal (wing/haltere), nor even to left vs. right disc fates [4076, 4651], they are confined to anterior vs. posterior fates within each segment [4076]. Why do A/P compartments exist? One early clue was that A/P boundaries precede disc formation, so they might help decide where discs arise [2812, 3135]. In fact, they do.

The wing's A/P border traverses a featureless field of hairs (Fig. 4.4) [354, 1376] – a finding that seemed odd because PI axes were thought to be "seams" for anatomic designs [904, 1409]. This invisible line was previously revealed in the phenotypes of *bithorax*[LOF], *postbithorax*[LOF], and *engrailed^1* mutants [4671]. The first two mutations reside in the *cis*-regulatory region of *Ultrabithorax* (*Ubx*) [254, 3940] and hence are written more correctly as "*Ubx^{bxLOF}*" and "*Ubx^{pbxLOF}*," whereas *en^1* is an idiosyncratic allele of *engrailed* [2356] that does not behave as a simple LOF [852, 1138, 1636, 1837, 2447]. *Ubx^{bxLOF}* and *Ubx^{pbxLOF}* transform the A or P part of the haltere into the corresponding part of the wing [2924] and A or P 3rd leg into 2nd leg [4324], whereas *en^1* transforms P into A compartments in the wing [493, 1378, 2306, 2441] and foreleg [439, 852, 4343]. Thus, while the *Ubx* alleles might be acting via interdisc

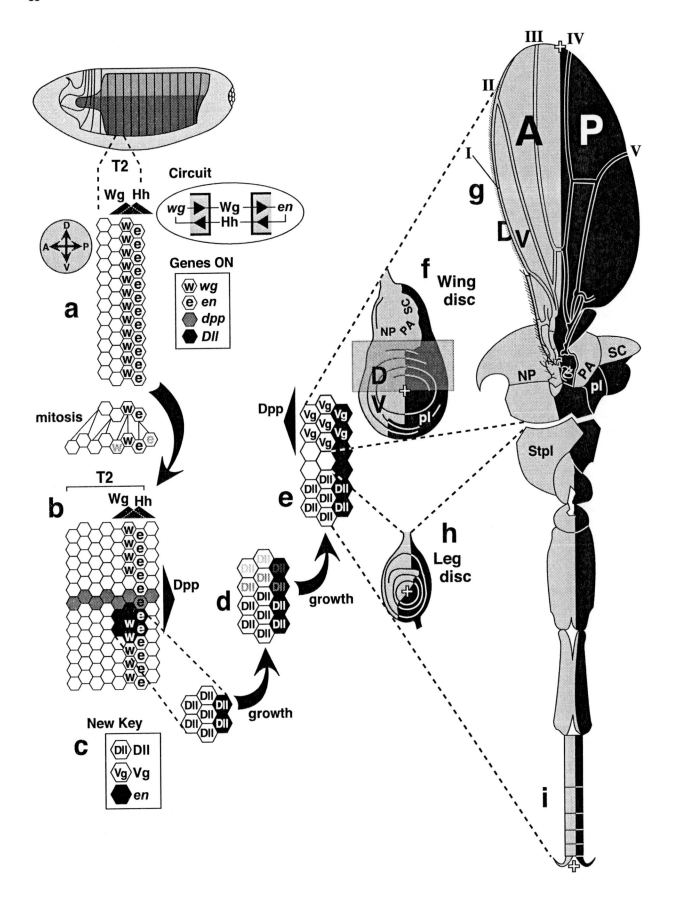

serial homologies, *en¹* cannot be doing so. Nevertheless, all three alleles fit the Selector Gene Hypothesis since they imply an abstract (histotype-independent) code of some sort [1355, 1375].

Because *en* is ON in the P part of each disc and OFF in the A part [495, 1697, 1963, 2307], it could serve as a PI "mode selector" gene to allow A and P cells to use different interpretation modes for reading the same PI gradient [2441, 2930]. If so, then the A and P gradients should be "back to back" [904, 4734], given the mirror-image phenotype of *en¹* in the wing and leg. Comparable duplications caused by other mutations are attributable to cell death followed by compensatory growth [523, 526, 539, 1499, 2015, 3441], but *en¹* cannot be acting thusly because it mainly behaves cell autonomously [1378, 2441, 2928, 4343]. Indeed, its autonomy led to Morata and Lawrence to postulate that

---

FIGURE 4.4. Initiation of thoracic discs at the Wingless/Engrailed interface and the later role of the A/P boundary during disc development. Gene abbreviations: *dpp* (*decapentaplegic*), *Dll* (*Distal-less*), *en* (*engrailed*), *hh* (*hedgehog*), *vg* (*vestigial*), *wg* (*wingless*).

a, b. Magnified view of the embryo's T2 (second thoracic segment, flank area only) at two stages after egg laying (AEL). Axes (A-P, anterior-posterior; D-V, dorsal-ventral) are indicated by compass at left. **a.** T2 at ~5 h AEL (extended germ band, St. 10/11). The posterior two rows of cells (hexagons) transcribe *wg* or *en*; key at right) and secrete Wg or Hh. Wg and Hh diffuse (black triangles = concentration gradients) and sustain one another's synthesis by a feedback circuit that is abridged here (inset at right; see Fig. 2.7 for key and [2816] for theory). Because the gradients are symmetric, each should create *two* flanking counterparts, but the response is asymmetric due to other genes (not shown) [596, 1628, 3129, 3748]. Before the stage in **b**, all ectoderm cells divide at least once. The pedigree (lines connect mother to daughter cells) depicts mitoses in one file of cells along the A-P axis, although divisions are not really so orderly. The important point is this: whether a cell expresses Wg or En is *not* heritable. To wit, as cells move away from the Wg/En interface, they stop making En [4491] and Wg [3363] (fading letters). **b.** T2 at ~8 h AEL (germ band shortening, St. 12). Wg and En domains do not grow in proportion to segment width (although En stripes widen more than depicted [272, 1064, 2717]) because diffusion ranges of Wg and Hh are fixed [1063]. An A-P stripe of *dpp*-ON cells crosses a gap in the Wg stripe. Dpp is also a morphogen (gradient at right). Around the tip of the Wg stripe remnant, cells transcribe *Dll* [834]. The upper limit of the *Dll*-ON spot is set (at ~St. 11) by Dpp (Dpp ⊣ *Dll*). The lower limit is set by Spitz (Spitz ⊣ *Dll*) – another morphogen (not shown; cf. Fig. 6.12) from the V midline [1572, 2335, 3536].

c–e. Subsequent changes in the T2 Dll cluster. Reshaping is apparently due to mitosis (vs. active cell migration) [827, 1571], but cells also rearrange to some extent [1572, 2335]. **c.** Note the new symbol key. **d.** As cells move dorsally with the Dpp stripe (see **e**), they stop making Dll [1571] (fading letters). **e.** Eventually a dorsal cohort starts transcribing *vg* – a wing (and haltere) marker. Thus, wing and leg anlagen arise from a common pool of cells [827]. The pool is partitioned (St. 15) based on how much Dpp (D signal) the cells perceive relative to Spitz (V signal; not shown) [2335]:

High $^{Dpp}/_{Spi}$ ratio ➡ "wing" state (*vg*-ON/*Dll*-OFF)
Low $^{Dpp}/_{Spi}$ ratio ➡ "leg" state (*vg*-OFF/*Dll*-ON)

f–i. Mature (3rd-instar) discs and their adult derivatives, showing regions that do (black) or do not (gray) make En (oriented as per compass in **a**). Within both discs, the *en* ON/OFF boundary becomes a reference line for specifying cell positions along the A-P axis, using either Wg or Dpp as a morphogen (not shown). In the leg disc, a remnant of *wg*-ON cells is retained ventrally, and a stripe of *dpp*-ON cells is induced dorsally (cf. Fig. 5.4). In the wing disc, a stripe of *dpp*-ON cells is induced along the entire boundary (cf. Fig. 6.3). Plusses mark the appendage tips, which would come out of the page during eversion and then flop down. **f.** Left wing disc. Tissue in the translucent rectangle (D wing surface) is not visible in **g** because it folds behind (cf. Fig. 6.1). Eversion brings the notum (NP, PA, and SC = notopleural, postalar, and scutellar areas) into contact with the pleura (pl). **g.** Adult wing (V surface only) and heminotum in side view. Bristles are omitted except at wing margin. A and P denote lineage compartments. The A edge of the *en*-ON region per se is actually closer to vein III due to a shift during late 3rd instar (cf. Fig. 6.7d) [350]. **h.** Left 2nd-leg disc. **i.** Adult leg, drawn as if filleted along the ventral A/P boundary and spread flat so its whole surface is visible. Thoracic structures also come from the leg disc, including StPl (sternopleural sclerite). Holes in joints are cartographic artifacts due to odd shapes of proximal leg segments.

Panel **a** is based on [1063]; **b–e** are adapted from [834, 884, 1571]; **f** is traced from [2467]; **g** is modified from [1376, 4076]; **h** is sketched from stained discs pictured in [495, 2754]; and **i** is amalgamated from [1800, 2449, 4076].

N.B.: Cells are not usually packed so neatly [1064], nor are they so constant in shape or size [4743]. Indeed, they may not even stay in a monolayer [1572]. In **b**, *dpp* transcription is actually greater posteriorly along its stripe (possibly due to upstream control by segment-polarity genes) [2003]. In **c–e**, a ring of *escargot*-ON cells (not shown; future pleura and coxa) surrounds the *Dll*-ON core (all other leg segments) [1542, 1572, 1573]. Unlike the leg disc whose axes persist from the blastoderm, the wing disc appears to rotate ~90° during development (not shown) [4652]. As shown in **h**, the A/P (*en*-ON/OFF) line of the leg disc is diagonal near the center but zigzags toward the stalk above [495, 3747, 4254]. This irregularity has led to confusion because some authors schematize the A/P line vertically (D pole at 12 o'clock) like the outer part, while others slant it (D pole at 1 or 2 o'clock) like the central part (cf. Fig. 5.1).

"the state of activity of the selector gene is permanent and . . . propagated by cell heredity" [2440], and the discovery that *en* has a homeobox [1031, 1247, 3429] corroborated this conjecture [354, 1979]. Subsequent research, however, showed that at certain stages a cell only expresses *en* as long as its neighbors force it to do so [1839, 1968, 3310, 4491].

In general, the following terms help distinguish between these autonomous (independent) and nonautonomous (dependent) states [523, 3448, 3795, 3796]:

1. "Determined" states are intrinsic, permanent, and heritable [1411, 1421, 2448]. They are "firm biases" [3794].
2. "Specified" states are extrinsic (imposed by intercellular signals), transient, and not heritable [2111, 2716, 4645]. They are "transitory biases" [3794] that depend on the persistence of the external signals.

En and Wingless (Wg) stripes first appear at the cellular blastoderm stage (~3 h AEL, St. 5) [67, 176, 271, 2146, 3157]. When a dye was injected into single cells and their descendants were examined after several mitoses (~5–8 h AEL), 15 (out of 37) clones were found to contain a mixture of En-expressing and nonexpressing cells [4491]. Clearly, En states are not irrevocably fixed in the blastoderm because if they were, then only pure (*en*-ON or *en*-OFF) clones should exist.

A second major finding in this study was that none of the 15 clones straddled a Wg/En (parasegment) boundary [1963]. The inability of blastoderm clones to cross this line confirmed cell-lineage studies of leg and wing discs [2432, 4076] and showed that A/P lineage restrictions apply not only to discs, but also to the ectodermal fabric from which they are cut [2444, 4488]. In theory, segregation of A and P cells could be enforced by killing trespassers, but it seems that cells never even venture into alien territory. The reason appears to be an incompatibility of cell affinities [2437]. That is, the *en*-ON state endows a cell with (as yet unidentified) surface molecules (cadherins? [3308]) that prevent its immersion in a field of *wg*-ON cells, and vice versa.

The idea that disc cells seek "their own kind" came from old experiments where cells from marked discs were mixed together, and the aggregates were found to form chimeric patterns. The harmonious integration in these patterns was attributed to homophilic cell sorting [1356, 4423, 4425], but it could equally have been due to repatterning during growth [526, 1357, 1775, 3424, 3448]. To distinguish between these alternatives, short-term cell movements (sans growth) were monitored and disc-type preferences were thereby proven [1202, 1205]. A vs. P

preferences had been suspected from the behavior of *en¹* clones in mosaics [1358], and recent evidence supports this general notion (cf. Ch. 6) [354, 2437, 2993].

From ~3 to 5 h AEL (St. 5–10), En states are sustained by signals that come from the adjacent Wg stripes (Fig. 4.4) [271, 1093, 2719, 3921, 4489]. Manifestations of this dependence include (1) decay of En expression in cells as they move away from the Wg/En border during segment growth [4491] and (2) premature decay of En stripes in *wg*^null embryos [1065, 2719]. Interestingly, the dependence is mutual [1063, 1836, 1841, 4743]. Wg expression is sustained by a Hedgehog (Hh) signal emitted by the En-expressing cells [1969, 1977, 1982, 2494, 2719]. Wg and Hh are both secreted proteins [334, 574, 1789, 4440, 4455], whose modes of action are discussed later. For now, the essential point is that the diffusion ranges of Wg [271, 920, 1541, 1779, 4442] and Hh [1969, 4262] are a only few cell diameters in the embryo [3748], so any cell that moves outside this range will cease expressing whatever genes were dependent on that morphogen [2810, 2814]. The third morphogen that is instrumental in disc development is Decapentaplegic (Dpp).

Cells whose states depend on morphogens can be thought of as revelers in a nightclub. If the colored lights ($\approx$ morphogen ranges) are aimed at fixed spots on the floor ($\approx$ epithelium), then dancers can jostle ($\approx$ divide or rearrange) into or out of a colored area, but each area will be occupied by roughly the same number of people at any given time. The system is in a steady state, despite a flux of its parts. After ~5 h AEL (St. 10) both *en*-expressing cells [1093, 1790, 3129] and *wg*-expressing cells [1894, 2013, 2534, 2536, 2683] adopt determined states that are analogous to *indelible* skin colors [2717].

This "**Cabaret Metaphor**" also pertains to how discs arise. By ~5 h AEL (when germ band extension is completed), Wg (but not En) stripes split into ventral and dorsal pieces [176, 271, 880, 1093] with an intervening gap (Fig. 4.4). A perpendicular (A-P) stripe of Dpp-expressing cells intersects the tips of the ventral (D-V) Wg stripe remnants [2003, 3833], forming a ladder of struts (Dpp) and rungs (Wg) along the trunk [827, 3122]. At each Dpp-Wg intersection in T1–T3, clusters of cells make Dll. This coincidence implied that *Dll* is only transcribed in cells that receive both signals [827, 833]:

$$\{Wg \text{ AND } Dpp\} \rightarrow Dll?$$

Along the A-P axis, Dll is activated by Wg (see below) and limited by Wg's range of diffusion. However, *dpp*^null embryos still express Dll [1572], so *dpp* cannot be needed for activation. On the contrary, Dpp must be stifling *Dll* dorsally because Dll spots elongate dorsally in *dpp*[null] embryos [1572] (cf. similar effects on salivary gland anlagen [1820]). Ventrally, *Dll* is confined by Spitz, another secreted signal [1572, 2335]. Together, Dpp and Spitz confine Dll along the D-V axis. Thus, the actual rule is

{Wg AND NOT-Dpp AND NOT-Spitz} ➜ *Dll*.

Still unexplained is how Dll can coexist with Dpp at Dpp-Wg intersections if Dpp inhibits Dll. Conceivably, some factor (Wg?) overrides Dpp's inhibition of Dll at those points.

In each thoracic segment (of wild-type embryos), the Dll clusters straddle the Wg/En border [1571]. Hence, both states (Wg and En) get incorporated into the founder population of prospective disc cells. As described above, these cells are not yet segregated into ventral (leg) vs. dorsal (wing/haltere) fates. Soon thereafter, subgroups split off (under the influence of Dpp [1572]) and migrate dorsally (tracking the Dpp stripe [1571]). Dll expression wanes in the cells as they move away from the Wg rung tips [827, 1571], so Wg is probably sustaining Dll expression (the Cabaret Scenario). Dependence of Dll on Wg was confirmed by showing that Dll disappears when Wg is artificially inactivated (by heat-pulsing a t.s. mutant) at this stage [827]. The turning ON of the wing-haltere marker Vg in the dorsally migrating subgroups in T2 and T3 [1571, 1572] may require that Dll be turned OFF. That is, while Dll confers the potency to make Vg, such "competent" cells can only make Vg when they subsequently shut OFF Dll. This "ON-**then**-OFF **Licensing Trick**" is also used elsewhere (e.g., Dll-Hth's launching of *spineless* expression in the antenna, cf. Fig. 8.3).

Leg vs. wing states are imprinted at ~6 h AEL (St. 11) by the relative doses of dorsal (Dpp) and ventral (Spitz) signals that the cells perceive [2335]. Inside the nascent discs, cells cannot be undergoing any drastic rearrangements because their blastodermal fate maps roughly match the layout of adult structures [3602, 4652]. Nevertheless, leg cells must reorganize to some extent because the future proximal cells are initially located dorsal to the future distal cells, whereas they later surround them. This situation arises because these fates are assigned by differing intensities of Dpp signal [1572].

Wing disc initiation is also more complicated (in a genetic, not a topological, sense). It involves two phases. In Phase 1, Vg (a transcriptional co-activator [1686]) is expressed (as described above) together with Escargot (Esg) and Snail (Sna) in response to extrinsic inductive cues ("{Dpp AND NOT-Wg} ➜ Vg"?). Esg and Sna

are zinc-finger proteins whose preferences for DNA sequences are virtually identical, and so are their downstream effects (as deduced genetically) – a redundancy remarkably like the bHLH proteins Achaete and Scute [1333]. In Phase 2, Esg and Sna establish an intrinsic "wing" or "haltere" state by auto- and cross-activation (again like Ac and Sc), and Vg then comes under their control [1333].

The above scenarios do not apply to genital [679, 735, 1170, 2028, 3817] or eye discs [2933, 3041, 4146], each of which comes from several segments [2103, 4825]. Indeed, the eye disc does not acquire an A/P boundary until long after it invaginates from the ectoderm (i.e., during 2nd instar) [1635, 1637].

Cell lineage within compartments is indeterminate

Nowhere is the Cabaret Metaphor more apt than within each compartment as a disc grows. Cells are free to jostle relative to the adult regions (colored areas) that they will eventually occupy (when the music stops) [2142, 3947]. This fluidity was revealed by Sturtevant's studies of gynandromorphs [4180] (cf. Ch. 1), and it was confirmed at a fine-grained level in all major discs by randomly marking cells at various stages of development [3441].

Because descendants of marked cells tend to stay together [3441], most cell movements must be due to passive displacements [3, 1610, 1888, 1890] (newborn cells pushing extant ones? [1525, 3515, 3518]), rather than to active migration, which would fragment the clone [532, 3422, 4671]. Clones tend to have irregular outlines [3441], and when outlines from different individuals are superimposed, they tend to overlap in virtually every region of the body [532], including the eye [189, 260, 261], antenna [3446], leg [544], wing [521, 1545], notum [521, 3007], and genitalia [1107]. The only exceptions are the compartment borders.

The flexibility of cell lineage is epitomized by *Minute* mosaics [354]. When single cells are spurred to grow faster than their neighbors (by expulsion of a retarding *Minute*[LOF] allele), their descendants can occupy more than 10 times the area that they normally would within the A or P compartment without disturbing the pattern [497, 2935].

One consequence of this plasticity is robustness [543, 2375, 2448]. Amazingly, 75% of a disc's cells can be killed without incurring *any* cuticular defects [1776]. The dead cells are simply replaced by compensatory growth. Apparently, the goal of disc growth is to generate enough tissue to fill all subregions, regardless of cell pedigrees: "the developing pattern seems to control proliferation, rather than the other way around" [532].

Strictly speaking, it is not proliferation per se that is being controlled, but rather tissue mass [4579]. Cell sizes can be altered drastically without affecting disc size, gene expression zones, or cuticular pattern [2081, 2771, 3081, 3750, 4576]. Any PI gradients that exist in discs must therefore be heedless of cell boundaries [546, 3862]. In other words, the "graininess" [1822, 2016] that cells impose on the overall "image" is irrelevant to the production of the image itself.

Disc cell proliferation is sustained by circulating growth factors from the larval fat body [2707]$^\Delta$ and possibly the brain [1346]. Endocrine signals like these are transduced by the insulin receptor pathway [486, 3450]$^\Delta$, which responds to insulin in cultured cells [919] and has significant effects on cell and organ size *in vivo* [2476]$^\Delta$.

Within the leg and wing discs, the chief paracrine stimulant appears to be Dpp [569, 619, 643, 1674, 1839], and the same is true for the eye disc during its early stages [570]. Nevertheless, many other pathways participate [543, 1126, 1142, 4667, 4886], including dActivin [511]$^\Delta$, Hh [2214]$^\Delta$, Wg [734]$^\Delta$ (although, oddly, *not* in the wing pouch [2254]$^\Delta$), EGFR [3470]$^\Delta$ (*including* the wing pouch [3025]$^\Delta$ and eye [185]$^\Delta$), JAK-STAT [1906]$^\Delta$, and Notch [2362]$^\Delta$. Indeed, Dpp can only elicit outgrowth of the leg [617, 4490] or wing [558, 559, 617, 643, 3862] by cooperating with an unknown factor whose expression and regulation mimic those of Wg [2254] (cf. Chs. 5 and 6). A common conduit for multiple growth-regulating pathways is the Ras-MAPK cascade [2132]$^\Delta$, which promotes dMyc stability [1342, 2081, 3846, 4124] and blocks apoptosis [2793]$^\Delta$.

Discs also use *anti*-mitogenic signals that evidently rely on junctions [533, 1907, 4745]$^\Delta$ and cadherin-like adhesive proteins [1386]$^\Delta$ because LOF defects in these components can cause overgrowth [4742]$^\Delta$. One such retardant is nitric oxide [2370]. There must not be many such restraints on growth, however, because no single or compound LOF mutants have ever been found that rival the drosophilid titans of Hawaii [669]. Indeed, no such giants have yet been produced by any GOF manipulations either [2561].

Until about mid-3rd instar, mitoses are distributed evenly in leg discs [1599], and the same is true for wing discs [20, 2015, 2848, 4427]. The uniformity is attributable to a compact (≤ 0.7 kb) *cis*-enhancer that ensures ubiquitous expression of String – a rate-limiting (Cdc25) regulator of mitosis [2480]. Around mid-3rd instar, there is a transition from random to patterned mitoses. The change is probably due to a switch from the previous (cell-size responsive?) enhancer to other (prepattern responsive?) enhancers at the *string* locus [2480]. Proneural areas be-

come quiescent [1925], including the future wing margin [2079, 2081, 3154, 3374]. So does a band of cells just anterior to the A/P compartment boundary [543, 3813] (possibly *dpp*-ON cells [2739]) and a spot at the center of the leg disc [1599, 4666] (possibly *aristaless*-ON cells [620, 2287]). In the early pupal period, most SOPs undergo their differentiative divisions, and mitoses occur in a belt around the notum, along wing veins, and then in the interveins [1545, 1741, 3813]. Proliferation ceases one day after pupariation [1312].

Cessation of growth must be controlled intrinsically because wild-type discs do not exceed their normal size when given extra time to grow (via pupariation delay [3955] or transplantation [19, 542, 546, 1357]). In theory, terminal size could be targeted in terms of total cell number [99, 480, 1548, 3483, 3492], but in fact, as stated above, growth is instead shut off at a certain mass [2078, 3081, 4576].

The unresolved "**Growth Cessation Mystery**" is: what local cues do cells use to monitor such a global property [3416, 4114, 4196, 4886]? One possible answer is that some morphogens also act as growth (or survival) factors [847, 3862]. In that case, organ dimensions could be limited by the range [855, 2291, 4265] and/or slope [1142, 2448, 2852] of a gradient that is steeper in young discs [977, 1169, 3213, 3862]. Indeed, discs with extra Dpp-producing zones can triple in breadth [4229, 4418]. Whatever the mechanism, it evidently operates separately in different compartments because their sizes can be manipulated independently in the same disc [1169, 3947, 3961, 4265].

The Polar Coordinate Model linked regeneration to development

When mature discs are cut into pieces and forced to metamorphose, each piece forms a certain subset of elements. In this way, fate maps were charted [526, 3165]. In 1971, experiments were described where parts of leg discs were allowed to grow before metamorphosis. Regardless of whether postsurgical growth occurred *in situ* [522] or in an adult host [3808], the upper (stalk-end) half typically regenerated (i.e., made a whole leg), while the lower half "duplicated" (i.e., made its fated structures but in two copies).

This outcome can be restated as "ABC makes ABC<u>DEF</u>, and DEF makes <u>FED</u>DEF," where letters are identities (~ dorsal to ventral) and underlined elements are added postsurgically. Lateral and medial halves also obey this "Reciprocity" Rule [3808]. That is, one fragment (medial) regenerates while its complement (lateral) duplicates.

To explain why leg discs behave in this way, Peter Bryant proposed that cells at a cut edge can only adopt lower values in a PI-like gradient [522, 523]. If 654321 denotes the gradient (each digit = $^1/_6$ of the disc along an axis), then this "Down the Slope" constraint means that both halves must make DEF because they share the same (~3.5) wound edge [20]: 654 (ABC) makes 654<u>321</u>, and 321 (DEF) makes <u>123</u>321. This "**Gradient of Developmental Capacity (GDC) Model**" argues that a cell's positional value dictates its regenerative repertoire [525].

Research soon shifted to the wing disc because its larger size offered much higher resolution. Bryant bisected wing discs at 5 longitudinal and 8 transverse levels, and found that the fragment pairs obey the same Reciprocity Rule as leg discs [524, 525]. These orthogonal series suggested a Cartesian coordinate system, but diagonal cuts behaved similarly. Thus, regeneration proceeds in all directions away from a peak near the center of the disc, indicating a *cone*-shaped gradient (Fig. 4.5c). Tests of the model, however, produced confounding results:

1. *Paradox 1 (Fig. 4.5d): the central fragments.* Any piece that contains the high point should regenerate, but central pieces (bearing the gradient's peak) actually duplicate [536]. Bryant initially guessed that loss of the peripodial membrane from these 4-cut ("cookie cutter") fragments might be responsible for this odd result [524].
2. *Paradox 2 (Fig. 4.6a): the quadrants.* The regenerative power of a fragment should only depend on its wound edge(s). However, all four quadrants duplicate, and complementary pieces regenerate, so the behavior of an edge actually seems to depend on whether it belongs to a $^1/_4$ or $^3/_4$ piece [526].
3. *Paradox 3 (Fig. 4.6b): intercalary regeneration.* Because the gradient's peak is central, marginal pieces must duplicate. When such pieces are cultured singly or intermixed with identical fragments this result is indeed observed. However, when pieces from opposite sides of the disc are co-cultured as a scrambled mass, they regenerate the midsection, including the peak [1773, 1775].

Intercalary regeneration had been previously documented in cockroach legs [386, 539], and the Reciprocity Rule had been found to govern amphibian limbs [20]. A concerted effort (among Peter Bryant, Susan Bryant, and Vernon French) to reconcile these diverse systems led to the "**Polar Coordinate (PC) Model**" in 1976 [1303].

In contrast to the GDC Model, the PC Model explains the Reciprocity Rule by postulating that growth depends on *interactions* between wound edges, rather than on the free edges themselves [354, 536]. It departs from Wolpert's idea of morphogen gradients insofar as it relies exclusively on communication between *adjacent* cells [4682]. Cells are presumed to execute only two cardinal instructions:

1. *"Shortest Intercalation" Rule.* When cells are confronted at wound edges, they assess one another's coordinates. Disparities provoke mitosis, and newborn cells adopt values between those of the flanking cells. In so doing, the cells pursue the shorter of the two possible arcs (on a circle) that connect the points. How cells might compute or compare alternate path lengths was never clear [1805, 2291, 2808, 3862]. Because the intercalated span of values is identical for complementary fragments, only the context dictates whether the new growth is "regeneration" or "duplication," but this distinction is moot to the cells themselves. Typically, the model is illustrated as a clock face, but cells are not supposed to perceive the "12/0" meridian as a discontinuity. Hence, the PC system is seamless, except for a singularity at the origin where opposite angular values converge like a contracted purse string. What happens to the rules at this point (the "Central Degeneracy Problem") was never solved [3915, 3916, 4698].
2. *"Distalization" Rule.* In the original version of the model, a complete circle of circumferential values was thought to be essential for newborn cells to "distalize" – i.e., adopt fates at the next level (ring) in a distal (central) direction. (How cells sense completeness was unclear.) Exceptions [3811, 4141] led to a revision [547]. Newborn cells were now supposed to distalize whenever their intended coordinates are already taken by cells at that radial level (regardless of whether a whole circle is present). The revised Rule 2 conforms with Rule 1 insofar as it also depends solely on nearest-neighbor interactions.

The geometry advocated by the PC Model was not new (Driesch had toyed with polar coordinates in 1894 [1101, 3742]), but the synthesis that it achieved was historic because it deftly explained a huge mass of confusing data. Also, it was the first real hint that arthropods and vertebrates might build their limbs in similar ways. Over the ensuing decades, the model's popularity waxed and waned [354, 531, 863, 2420]. One lemma, at least, endures to this day: when the normal chain of events is diverted by

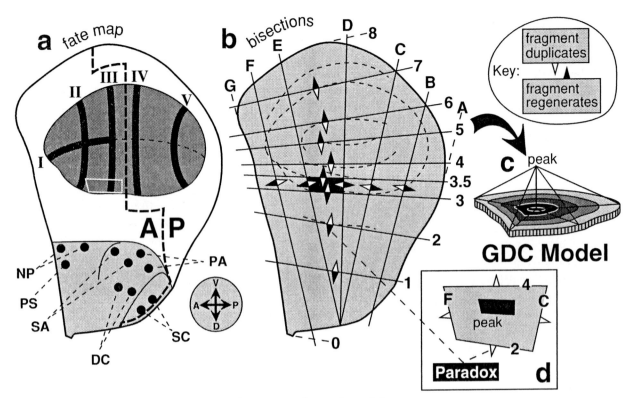

FIGURE 4.5. Regenerative potency of wing-disc fragments and an early model.

a. Fate map (abridged) of a mature right wing disc (notum, light shading; arc = suture; wing, dark shading) as per Bryant [524], except that dots are actual SOP sites [1925] and thick lines are prevein zones (I–V) [4189]. Vein I is the bristled part of the margin [1741] (thin dashed line = bristleless part). Thick dashed line is the A/P compartment boundary (cf. Fig. 4.4). Directions (A, anterior; P, posterior; D, dorsal; V, ventral) are given in the compass at right. In all bristle pairs except scutellars (SC; bounded area = scutellum), the P partner is on the right (cf. Fig. 3.4 for abbreviations). White trapezoid is a unique spot (cf. **b**).

b. Results of bisecting a disc lengthwise (B–F) or transversely (1–7) along folds (dashed lines) or elsewhere and letting each piece grow in isolation inside the body of a host adult [524]. As indicated in the key, an unfilled arrowhead means that the fragment to its rear "duplicates" (i.e., makes two copies of its fated structures as mirror images), whereas a solid arrowhead means that the fragment behind it "regenerates" (i.e., restores the whole). Note that every cut line has arrowheads of opposite type. Thus, the wing disc obeys a "Reciprocity Rule": when one piece regenerates, the reciprocal fragment duplicates. The fact that regeneration proceeds away from D, E, 3, and 3.5 lines suggests that the disc uses *x-y* coordinates, but diagonal cuts (not drawn) showed that potency actually declines *radially* from the center piece (black trapezoid).

c. The **Gradient of Developmental Capacity (GDC) Model** was based on the findings shown in **b** [524]. It proposed that the center piece contains the peak of a conical gradient of regenerative potency. The gradient is schematized here as triangles (side view) above the columnar epithelium (inscribed with imaginary contour lines). When the disc is cut, the gradient is supposed to force cells at each edge to grow down its slope.

d. According to the GDC Model, any piece that includes the peak should regenerate, but the "CF24" piece actually duplicates (as does BF16; not shown) [526]. This and other paradoxes (cf. Fig. 4.6) led to the GDC model being abandoned in favor of the Polar Coordinate Model [1775].

miscues, it can produce global deformities (e.g., trefoil limbs [539, 1805]) that appear freakish to human observers but look natural to the cells themselves because no local rules are broken.

The PC Model was theoretically able to encompass normal development as well as regeneration because disc growth *in situ* might also be regulated by intercalation [1301, 1303]. If nascent discs contain only a subset

of a mature disc's positional value spectrum – say "1" and "5" of a final "12345" array – then juxtaposition of 1 and 5 could lead to birth of a 3, and growth would stop with insertion of 2 and 4. PI (and pattern) would thus be elaborated steadily through such "intercellular negotiation" [834, 4682]. Growth is probably not limited in this way, however, since final disc mass is independent of cell density [4576].

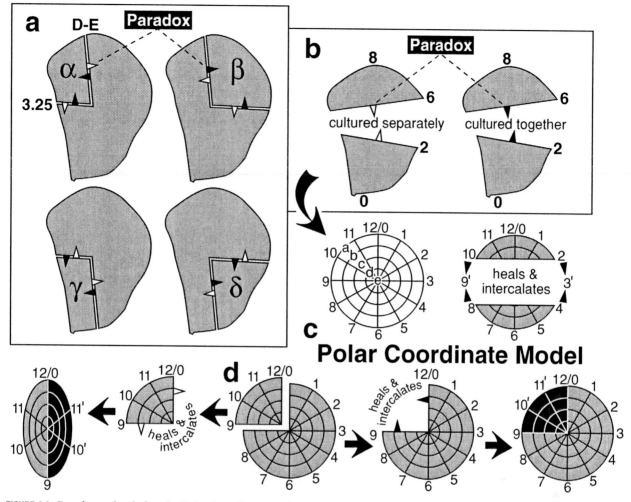

FIGURE 4.6. Paradoxes that led to the Polar Coordinate Model (cf. Fig. 4.5 for symbol key and context).

a. According to the GDC Model, a fragment's decision to regenerate vs. duplicate should depend only on its edges, but quadrants (cornered at the inferred peak of regenerative potency) violate this axiom. For example, the α quadrant duplicates, but a piece with the same vertical edge (3/4 complement to β) regenerates.

b. Dorsal ("02") fragments only make dorsal structures, and ventral ("68") fragments only make ventral ones. However, when 02 and 68 pieces are scrambled together, they regenerate middle ("26") structures. (See Fig. 4.5 for the numbering scheme.) This result is also hard to explain in terms of free edges.

c. The **Polar Coordinate (PC) Model** avoided both these paradoxes. Based on the phenomenon of intercalation (**b**), this model argued that growth depends on *interactions* between wound edges, not on *free* edges. Every cell is supposed to have a circumferential (clockface) coordinate (= angle from a 12/0 reference line) and a radial (letter) coordinate (= distance from center). The "Shortest Intercalation Rule" asserts that when cells touch (via healing), they assess each other's coordinates and fill in missing values by the shorter route (no discontinuity at 12/0). For example (at right; cf. **b**), when "10" abuts "8," a "9" fills the gap, instead of "11, 12/0, 1... 7" (the longer route). Prime marks denote new values.

d. The PC Model obeys the Reciprocity Rule (see α in **a**). For example, when a fragment contorts to heal "9" and "12/0" edges together, this contact creates "10" and "11" (black tissue), regardless of whether the edges belong to a quadrant (left) or its 3/4 partner (right). Thus, quadrants duplicate because none has enough angular values (i.e., ≥ half) to regenerate.

The various panels are adapted from [524–526,1775].

The idea of a common cellular logic for development and regeneration was supported by (1) the ability of immature disc tissues to provoke regeneration in mature disc fragments [1183], and (2) the ability of cell-killing agents (e.g., γ-rays [3440], X-rays [3449], UV-rays [1424], or microcautery [425]) to evoke the same kinds of mirror-image duplications from *developing* discs as are obtained with fragments of *mature* discs. In the latter case, the presumption was that necrosis-induced duplications arise via the same route as surgical tissue removal [539]. Indeed, wing-disc structures that maximally resist γ-ray-induced duplication were mapped to precisely the same spot as the GDC cone peak that was localized via surgery (Fig. 6.1) [3440].

Subsequent studies confirmed some of the model's assumptions, including the topology of wound healing [537, 947, 3564, 3565], the need for contact between cut edges [1184], and the confinement of growth to a blastema at the wound interface [5, 489, 537, 946, 3155]. *De novo* formation of wing margin between dorsal- and ventral-type cells validated the concept of intercalation [354, 1038], and the ability of different discs to trigger intercalation *inter se* in a position-specific manner [18, 535, 1774, 4665] affirmed the idea of a universal PI "language" [3448].

But regeneration has peculiarities that set it apart

Contrary evidence, however, also accumulated. Below is a partial list of findings that defied the PC scheme (see [1301, 2161] for critiques):

1. Violations of the Reciprocity Rule (e.g., fragments regenerating that should only duplicate) were found after tissue maceration [3788, 4141], during long-term ($\geq$1 week) culture [1125, 2144, 2230], and as a function of the fragment's proximity to – or straddling of [2139, 2143] – the A/P compartment boundary [2140, 2141].

2. Complementary fragments should grow equally because they intercalate alike. However, duplicating fragments hardly grow at all despite adding more bristles [3808]: they neither double in mass nor grow as much as their regenerating partners [19, 20]. These heretical results challenge the dogma [525] that discs only use "epimorphosis" (patterning that re-

quires mitosis) vs. "morphallaxis" (repatterning sans growth) [2945, 4724, 4736].

3. The eye-antenna disc should have the coordinate system of a single disc because one part (eye) regenerates and the other (antenna) duplicates [1404, 1406, 1423, 3810]. However, in other contexts (homeosis [3439, 3443, 3447], transdetermination [3788], or ubiquitous *Antp* expression [1416]) it behaves like a compound leg-and-wing disc. This "**One Disc or Two? Paradox**" [1805, 2933, 3788] seemed to have been settled in favor of two discs in 1997 based on the homeotic effects of *Distal-less* [1561], but the one-disc idea rallied in 2000 based on the homeotic effects of the Iroquois Complex [697]. The rule that "antenna never regenerates" [1404, 1411, 1423, 3810] is broken when antennal tissue is macerated before culturing: in ~20% of the implants, structures form that normally come from the eye part of the disc (ocelli, ptilinum, frons, and head bristles, but not facets) [3788]. Such regeneration should be impossible if the antenna has fewer than half the angular positional values.

4. Blastemas can form at both edges of a fragment [2213] and can regenerate structures in a correct circumferential sequence (as per the fate map) [2144] *before healing together*. This behavior is more consistent with the GDC Model [523] than with the PC Model.

One challenge that the PC Model initially seemed to deflect was the sequestering of regenerative potential in a single quadrant of the leg disc [3808]. As explained in Ch. 5, the model attributed this anomaly to a clustering of angular values [1303, 4140], but a recent molecular analysis of wound healing [1472] reveals *trans*-lumenal interactions that have mortally wounded the PC Model itself.

CHAPTER FIVE

The Leg Disc

In an article entitled "Pattern formation in the embryo and imaginal discs of *Drosophila:* What are the links?" [4675], Adam Wilkins and David Gubb posed a question that was on the minds of many researchers at that time (1991). The embryo's segmentation hierarchy was basically understood, but it was unclear what these various genes might be doing in discs [4682]. Wilkins and Gubb argued that segment-polarity genes supply the angular values of the Polar Coordinate (PC) Model, which until then had only been an abstract formalism.

Several predictions of this "**Angular Values Conjecture**" were soon put to the test by molecular genetics. Chief among them was the expectation that LOF and GOF alterations in segment-polarity genes should reorganize disc anatomy. This prophecy was indeed fulfilled, and the experimental probing uncovered a trove of insights into the machinery of disc patterning.

The conjecture itself, however, turned out to be wrong. Segment-polarity genes do not paint a pinwheel on each disc. Rather, they draw a few important lines – the compartment boundaries.

The Molecular Epoch of disc research was launched in 1991

The Wilkins-Gubb paper, with its clarion call for a molecular assault on discs, provides a convenient demarcation between the Cellular and Molecular Epochs of disc research. In the 1990s, many links were patiently forged between the blastoderm and the adult. During this process, several old puzzles at the cellular level were solved by clever experiments at the molecular level. Among those puzzles were the following [1807]:

1. "**Gaps Mystery**" (Fig. 5.1e). *Cellular Problem:* Each leg's bristle pattern is roughly symmetric about the D-V plane [1883], as are afferent projections of the bristle neurons [3005]. It therefore came as a surprise when the A/P compartment boundary was found not to coincide with this plane, but rather to be offset from it by several cell diameters [1800, 2449]. Shouldn't they be congruent if this line is used as a reference axis for specifying mirror-symmetric positional information [904]? *Molecular solution:* The reference axis turns out not to be the A/P boundary itself but a morphogen-producing zone just anterior to it [1807]. There are no gaps between that zone and the mirror plane.

2. "**Quadrant Regeneration Mystery**" (Fig. 5.2a). *Cellular Problem:* In contrast to the wing disc, where every quadrant duplicates, one quadrant of the leg disc regenerates [3808, 4140]. Why is this upper medial (UM) quadrant special? *Molecular solution:* Its peripodial membrane turns out to have a spot of gene expression that dictates regeneration vs. duplication [1472]. The dogma that regeneration and development always use the same rules is false.

3. "**Dorsal Remnant Mystery**" (Fig. 5.2b). *Cellular Problem:* When embryos are exposed to X-rays [3449] or UV-rays [1424] at doses that are sufficient to kill cells [2774], the legs that manage to complete development often manifest a "duplication-deficient" anatomy, where part is missing and the rest is duplicated in mirror image. Similar defects can be produced by microcautery [425], by heating t.s. cell-lethal mutants [101, 2102, 3702, 3964], and by *in situ* excisions [522]. First legs tend to fuse rather than duplicate [3442] because their

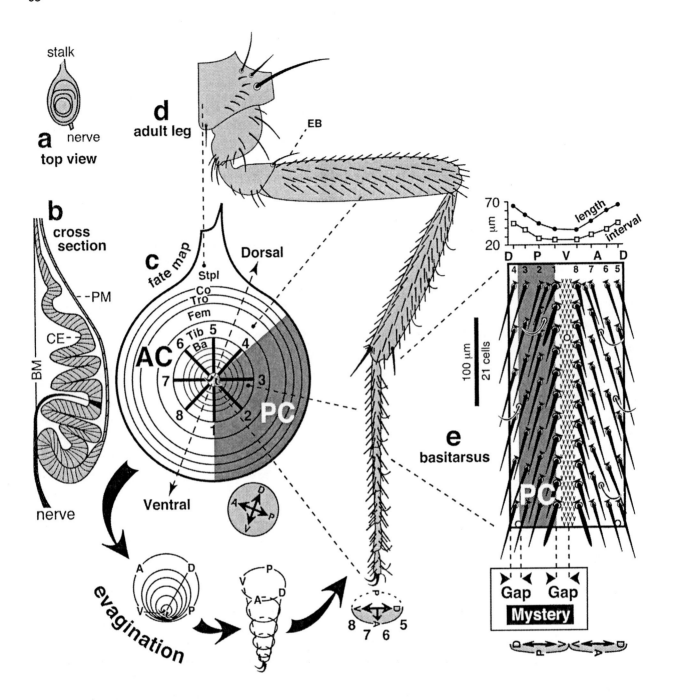

discs are conjoined [377, 3426], but regardless of how the symmetry arises, the remnants tend to retain dorsal structures at the expense of ventral ones [425, 2102, 3442, 3449, 3705]. Why is the dorsal region (outside the UM quadrant) more robust? *Molecular solution*: It is the source of a diffusible growth factor that stimulates mitoses throughout the disc [529, 1807], and this growth factor is upregulated in response to trauma [2856].

4. **"Triplications Mystery"** (Fig. 5.2c). *Cellular Problem:* Triplications can also be induced in cell-lethal

mutants during certain sensitive periods [1495, 2102]. Typically, such legs have their normal tip (stem) plus an outgrowth (branch) that either diverges or converges (i.e., gains or loses circumferential elements distally). Outgrowths that branch from the stem's ventral face tend to diverge, while dorsal ones tend to converge [1495] – a bizarre trend also documented in cockroach legs [384, 547, 2804]. Why? *Molecular solution*: The deciding factor appears to be the mixture of morphogens on one side of the disc vs. the other that is created under these conditions [2856].

FIGURE 5.1. Geometry of leg development.

 a. Mature, left 2nd-leg disc (~150 μm across [3140]). Lines are folds. The central circle is the "endknob." Connections to larval skin (stalk) and CNS (nerve) have been cut.

 b. Longitudinal section ("BM" and "PM" = basement and peripodial membranes; "CE" = columnar epithelium). Transverse lines are cell boundaries. The black cell is one of several larval neurons that innervate Keilin's organ [834, 2287]. These neurons invaginate with the disc (a piggybacking that is ancient in dipterans [2824]). Their dendrites (not shown) stay attached to larval skin through the stalk [4330], while their axons project to CNS (a route later taken by adult axons). Adepithelial cells are omitted.

 c. Idealized fate map of left 2nd-leg disc. Segment primordia are the coxa (Co), trochanter (Tro), femur (Fem), tibia (Tib), basitarsus (Ba), and tarsal segments 2–5 (unmarked), with claws mapping centrally. The sternopleura (Stpl) is part of the body wall (cf. Fig. 4.4). AC and PC (shaded) are A and P compartments. Spokes denote bristle rows, whose spacing is estimated: it may not be so uniform (cf. **e**). As shown below for the tarsus, the annuli telescope out during evagination (toward the viewer) as the covering PM (not shown; cf. **b**) peels away [1311]. In this schematic, the tarsus also rolls clockwise ~90°. Compass directions (D, V, A, P) refer to prospective or actual adult axes (cf. **d, e**).

 d. Anterior face of left 2nd leg (~2 mm long). There are 2 macrochaetes (distal tibia) and numerous chemosensory (curved) bristles on the tibia and tarsus. "EB" (edge bristle) is at the D midline. Tarsal rows 5–8 are marked on the compass below.

 e. Basitarsus (whole surface) drawn as if slit along D midline and spread flat. Most bristles have a bract (triangle) and align in rows (numbers at top). Five chemosensory bristles lack bracts and reside between rows, as do 3 sensilla (circles at 3.5, 5.5, and 8.5). Between rows 1 and 8 is a lawn of hairs. Row-1 and -8 bristles are peg shaped and darker; others are tapered and lighter. Aside from bristle thickness and pigmentation, other useful landmarks include bristle lengths and intervals – both of which increase linearly from V to D (graph at top). Indeed, the symmetry is so precise [1883] that it was surprising when the A/P compartments were found to be offset from this D-V mirror plane by a few cells ("gaps" below) [1800, 2449]. Row-1 bristles can reside in either AC or PC (cf. Fig. 3.9c).

 Panels **a** and **b** were traced from [377] and [3426], respectively (see [1774, 2144] for nerve-stalk asymmetry, [1516, 2287] for folding details, and [1311] for region-specific cell shapes); **c** is adapted from [1807, 3807]; and **e** is based on data in [1801, 1803].

 N.B.: For reasons explained in Fig. 4.4h legend, some authors orient the axes (**c**) with the D pole pointing into the stalk (at 12 o'clock vs. ~1 o'clock here). Claws (**c**) actually map more dorsally [1812, 3807, 4140]. Numbering of rows (**e**) follows original nomenclature [1714]. The numbers are backward in [837, 1039, 2533, 4159].

Before examining how these and other riddles were demystified, it may help to review some of the clues that accumulated before the Molecular Epoch. Those clues tantalized two generations of researchers, and the older analyses were at least as interesting as the studies of the recent past.

Bateson's Rule (1894) governs symmetry planes in branched legs

The latter three mysteries listed above are interrelated insofar as they all involve mirror-symmetric partial patterns. Near the end of the nineteenth century, William Bateson collected numerous examples of such abnormalities as naturally occurring "sports" in various species of arthropods [240]. In virtually every 3-tipped leg, he found that (1) the three tarsi lie in one plane (vs. a tripod), and (2) the outer two are mirror images of the inner one [2809]. These constraints constitute "Bateson's Rule." Although the tips always manifest R-L-R or L-R-L symmetry (R and L = right vs. left-handed), the inner member can face the flanking ones with any surface (D, V, A, or P). To illustrate this rule, Bateson built a cute wooden model (Fig. 153 in his 1894 tome) where the bases of three tarsi are geared together. When the in-

ner one rotates clockwise, the outer ones turn counterclockwise, thus maintaining mirror symmetry.

 Bateson's Rule must reflect a deep law of cellular "psychology" because the syndrome that it describes transcends the mode of tissue removal (irradiation, surgery, or apoptosis) [547]. The PC Model invoked a cellular imperative to "*Always maintain continuity!*": when distant cells are apposed, they multiply to fill the gap in positional values by the shortest possible route [539, 2513]. With this basic edict and one *ad hoc* assumption, the PC model could explain why the UM quadrant regenerates (Fig. 5.3b). To wit, this quarter might contain >50% of the disc's angular coordinates [1303, 4140]. However, the idea of crowded coordinates undermined the model's other proposal that discs stop growing when a certain density of coordinates is reached (like people filling seats in a stadium) [546, 1303].

 The PC Model was also able to account for duplications, triplications, and other outgrowths [538, 1496]. Those conjectures were supported by clonal analyses [1497, 1503, 1504, 2102], apoptosis studies [797, 1501], and phenocopy simulations [1498]. In short, the model was robust.

 Even so, there were doubts. Some of arguments that were contrived to defend the model strained credulity

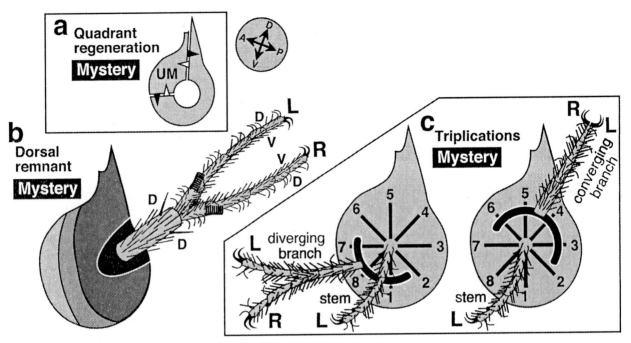

FIGURE 5.2. Mysteries of leg development (cf. text and Fig. 5.1e).

a. Quadrant Regeneration Mystery. Unlike the wing disc, where every quadrant duplicates, the upper medial (UM) quadrant of the 1st leg disc regenerates (black triangles; cf. key in Fig. 4.5). UM regeneration is seen either with or without the endknob (center region) [3808, 4140].

b. Dorsal Remnant Mystery. When cells are killed by X-rays [3449] or cell-lethal mutations [2102, 3705], D structures tend to remain, whereas V ones are lost (except near branch tips). First-leg discs, which are conjoined [3426], fuse medially instead of duplicating, and losses are symmetrical on left (L) vs. right (R). In this contour map, darker zones denote higher frequencies (0–30; –60; –90; –100%) of elements present in 65 pairs of fused 1st legs.

c. Triplications Mystery. Triplicated legs possess a "stem" [1875] and a symmetric outgrowth. V-side outgrowths tend to diverge, whereas D-side ones converge, and the (2/6.5) line between cohorts is sharp.

Maps in **a**, **b**, and **c** are redrawn from [3808], [3442], and [1495], respectively.

N.B.: "Medial" and "lateral" mean toward or away from the larval midline, whereas A, P, D, and V refer to prospective or actual directions in the adult.

---

because they demanded odd regions of cell loss and contorted angles of healing (Fig. 5.3c–e) [1495, 3442]. Moreover, much of the later data came from a cell-lethal mutation [1502, 3945] whose gene product was then unknown. (Suppressor of forked turned out to be a polyadenylation protein that affects mRNA stability [2879, 4235] and licenses metaphase [140, 141].) Skeptics felt that the model relied too heavily on *ad hoc* fantasies and molecular phantoms [863, 2161].

Meinhardt's Boundary Model deftly explained Bateson's Rule

In 1980 (four years after the PC Model's debut), Hans Meinhardt, a mathematically gifted theoretician, offered a clever new solution to Bateson's Rule (Fig. 5.3) [2804]. His "**Boundary Model**" was based on the leg's A/P compartment boundary and a supposed D/V bound-

ary that subdivides the A region into D and V parts [4076]. Meinhardt inferred that distal tips grow out wherever P, V, and D areas convene [2807]. Normally, this condition is only met at the distal tip, but extra tip "organizers" could arise if P, V, or D cells were to appear in the wrong place due to wounding or apoptosis. The idea that trauma could trigger ectopic gene expression seemed farfetched at the time, but it has since been confirmed [489, 1472, 2856]. As for how three territories might cooperate to induce distalization, Meinhardt conjectured that each cell type secretes an ingredient needed for synthesis of a tip morphogen "T." That is, the P, V, and D molecules react chemically to make T.

Each compartment may be responsible for a particular step in the synthesis of the morphogen or . . . may produce a diffusible cofactor which is required for morphogen production. [2804]

The fictional P, V, and D molecules should not diffuse far beyond their source sectors since, in that case, tips would form everywhere. The model thus invokes three short-range inducers and one long-range morphogen. T forms a cone-shaped gradient that could encode radial distance like the PC system's radial coordinate [2805, 2813].

In Meinhardt's original model (1980), there is no cellular imperative about continuity. Rather, Bateson's Rule is an emergent property of the model's criterion for distal outgrowth [617, 2706]. The geometry is easy to visualize (Fig. 5.3g–i) [2808, 2809] but hard to describe in words. Imagine that the random switching of cell states is simulated by spattering P-, V-, or D-colored "paint" onto a tricolored pie chart. Wherever the "P + V + D = T" criterion is satisfied, a tip will grow out. Among the possible configurations are

1. A dab lands wholly inside a sector. In this case, nothing happens because there would only be a bipartite reaction (P + V, V + D, or D + P) at the spot's edge.
2. A dab lands at the perimeter and straddles a boundary (Fig. 5.3g). If the spot's color matches a flanking color, then nothing happens, but if it differs from both, then T will be made where the three colors meet, and an extra tip will emerge in mirror image to the endogenous tip (for reasons described in 3).
3. A dab straddles a boundary but does not reach the perimeter (Fig. 5.3h). If the spot's color differs from both flanking colors, then T will be made wherever its edge crosses the border. Rounded spots will have two T points, and the line connecting them must intersect the disc's center because they lie on a radius (P/V, V/D, or D/P). Thus, the first part of Bateson's Rule is satisfied: the three tips will lie in one plane. Because each T point has a pinwheel (PVD) handedness, a person in the P zone near a juncture must walk either clockwise (L) or counterclockwise (R) to step into V and then into D. Along a radius crossed by a dab, the handedness changes at every T point, thus fulfilling the second part of Bateson's Rule (i.e., R-L-R or L-R-L alternation).
4. A dab's edge touches, but does not cross, a boundary (Fig. 5.3i). T will be produced on either side of the intersection, and a pair of tips begins to grow out but eventually converges (vs. diverges as in 2) because (1) "the distance between the two [tips decides] . . . how many elements are missing . . . due to overlap of the two gradients," or (2) "the normal maximum concentration is not achieved [because . . . too few

cells of a particular compartmental specification are available or if they are not close enough" [2804].

The leg's bristle pattern is too rich to be encoded by three broad angular values (P, V, D sectors) [3253]. Meinhardt recognized this problem and in a 1983 paper sketched a "**Hybrid PC-Boundary Model**" (Fig. 5.3k) where the P/V, V/D, and D/P lines provide three angular values in the nascent disc [2808], with remaining values emerging later via the Shortest Intercalation Rule. One difficulty with this remedy is that all three spokes lie outside the coordinate-dense UM quarter, so it should be unable to regenerate.

In the same article, Meinhardt concocted an alternative solution, which will here be called "**Boundary Model II**" (Fig. 5.4). Its key amendment was that "the compartment borders must act as a frame for the finer circumferential subdivision, [with the] distance from a particular compartment border [being] measured by a diffusible morphogen" – an idea earlier suggested by Francis Crick and Peter Lawrence [904].

Echelon 1 (at the boundaries)

P + V = A

V + D = B

D + P = C

Echelon 2 (at the center of the disc)

A + B + C = T

T would thus be created in two steps (vs. one in Boundary Model I), and its source point would be defined by the intersection of three lines (vs. the juncture of three areas). More important, an angular coordinate could be cobbled from the new factors. A, B, and C are supposed to be long-range morphogens (like T but unlike inducers P, V, and D), so they could form tent-shaped gradients [2804]. Cells could then figure out where they are by measuring concentrations of A, B, and C. To encode angular PI, each gradient must curve around the circumference (Fig. 5.4a), a constraint that *a priori* seems implausible [618, 2456]. Conceivably, the molecules could be channeled or actively transported (see below). Cartesian and polar systems could easily be interconverted [922, 1298, 2157, 3252, 3703], but their coordinates cannot be swapped directly *inter se* [548, 2512, 2881, 3915, 4697].

Boundary Model II also explains why some disc fragments duplicate (Fig. 5.4b). Thus, for example, if the D sector is excised, then the P and V edges should synthesize A (when they heal together) at a site opposite to the extant A source [2808]. If the new A gradient respecifies cells by morphallaxis, then it will cause a mirror-image

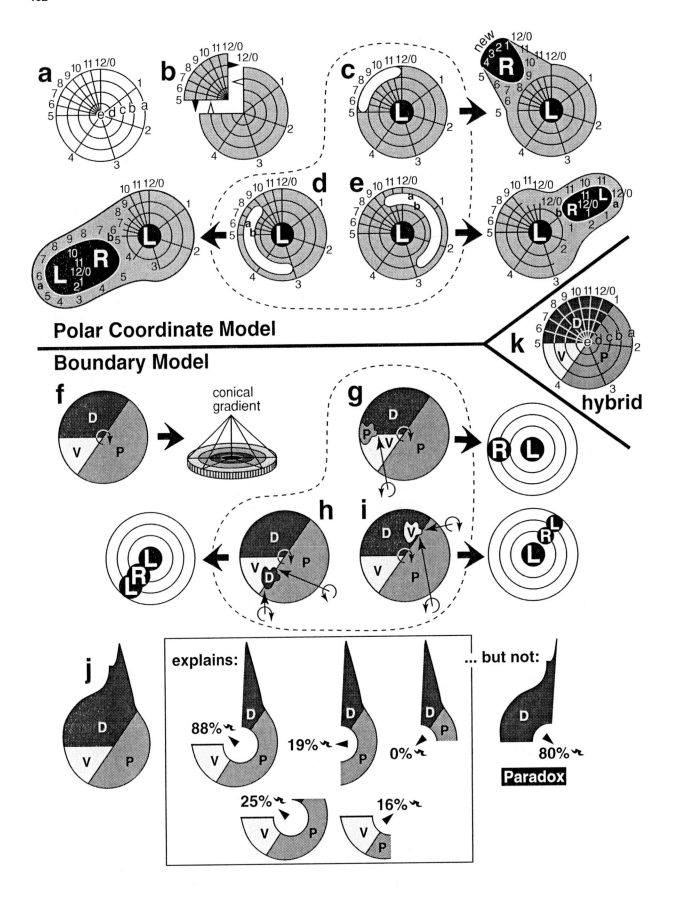

Polar Coordinate Model

Boundary Model

duplication. In accord with this scenario (but not with the epimorphosis required by the PC Model), duplicating fragments grow negligibly and lose structures near wound edges [20].

The Boundary and PC Models jousted in a "Paradigm War"

Given the reasoning in the original Boundary Model, distal outgrowths should only sprout from dorsal, ventral, or anteroventral faces of the leg (P/D, P/V, and D/V radii). In this regard, it surpassed the PC Model (where constraints are looser) at explaining why outgrowths tend to lie in the D-V plane [617, 4490]. It also afforded a simpler view of how a disc's coordinate system might be "geared" to the segmentation gene machine of the embryo [2812, 2813]. To wit, the A/P boundary in each thoracic segment would be recruited as a reference axis for each leg or wing disc [2808] – a conjecture verified by later molecular studies (Fig. 4.4) [834]. Ironically, Boundary Model II was stumped by the same Quadrant Regeneration Mystery that the PC Model handled relatively easily (Fig. 5.3j) [2804]: how could D-type cells alone be able to regenerate a tip?

Despite their ability to "cross-hybridize" (Fig. 5.3k) [1807, 2814, 3253], no two models ever differed so fundamentally as the Boundary and PC Models [834]. Their rivalry centered on several pivotal issues [354, 1807].

1. *Who's talking and who's listening?* According to the PC Model, every cell identifies itself to its neighbors and receives such input in return. In the Boundary Model, only a few cells emit signals; most are silent receivers.

2. *Do cells shout or whisper?* For the PC system to work, cells need only converse with their neighbors, and they are thought to do so by direct contact [1301],

---

FIGURE 5.3. Models for pattern formation that were prominent in the 1980s, and how they explained regeneration-duplication phenomena in leg discs (cf. Fig. 5.2). In the **Polar Coordinate (PC) Model (a–e)** [1303], growth is supposedly stimulated when distant cells abut via healing, whereas in the **Boundary Model (f–j)** [2804, 2808] distal growth is provoked when three kinds of cells convene.

a. Polar coordinates in a left leg disc (cf. Fig. 4.6). Each cell assesses its position as (1) radial distance (lettered rings) from the center and (2) angular declination (numbered spokes) from an arbitrary radius (12/0). Unlike the wing disc where angular values are presumed to be spaced evenly, here they must be crowded because the UM quadrant regenerates.

b. In the PC Model, fragments (e.g., UM) with >50% of the angular values regenerate (black triangles), whereas pieces (e.g., 3/4 part) with <50% duplicate (hollow triangles). In each case, the edges (5 and 12/0) intercalate via the shortest route (cf. Fig. 4.6d).

c–e. Distal outgrowths can occur if tissue is lost from certain regions, and the edges of the hole (white "sausages") heal as shown. **c.** If cells are lost from the UM perimeter, then an extra "clockface" arises when new tissue (black) fills the 5–12/0 gap by the shorter route (viz. 1–4). Because the new clockface is backward, the outgrowth will be right handed (R vs. L). Comparable losses from other peripheral parts should not form whole circles, which is problematic because mirror planes of twinned legs tend to be at right angles to spoke 4, not 8 [3442]. **d.** Loss of internal coordinates can generate two whole new circles if wound edges reorient (during evagination?) to abut values from ≥50% declination (e.g., 9 and 3; "a" and "b" = outer two rings) [1495]. **e.** Confrontations of <50% declination will form partial circles and stunted outgrowths. L-R-L handedness (**d, e**) obeys Bateson's Rule (see text), but the model does not explain why converging outgrowths (**e**) occur near angular value 1 (cf. Fig. 5.2c) and not near 2–4.

f. In Meinhardt's Boundary Model, the anterior compartment is subdivided into ventral (V) and dorsal (D) parts, whereas the posterior (P) compartment is not subdivided. The PVD sequence is clockwise in left legs (curved arrow). Each cell identity – P, V, or D – furnishes one key ingredient for making a morphogen. The morphogen is created only at the nexus, and its diffusion forms a conical gradient (rings = contours; epithelium in side view).

g–i. The model explains outgrowths. **g.** Duplications arise when a peripheral patch that straddles a boundary (e.g., D/V) transforms into the third type (P), thus creating a new PVD nexus (arrow) that grows out to form a right limb. **h.** If such a patch is internal, then limbs arise at two new PVD points. **i.** If the patch merely touches a border, then too little morphogen may be made to sustain two tips, in which case the outgrowth converges. As for why ventral outgrowths diverge and dorsal ones converge (cf. Fig. 5.2c), the model must assume (*ad hoc*) that D spots tend to straddle P/V (**h**), whereas V spots only touch D/P (**i**).

j. The model (in a leg disc at left) explains most surgical outcomes. When the boxed fragments are cultured, they regenerate claws (bicurved symbol) and other distal structures that map in the excised endknob at frequencies indicated. Regenerative ability is maximal when all 3 domains remain (88%), as expected. However, the model does not explain why the UM quadrant (sans endknob) also replaces claws often [2804].

k. **Hybrid PC-Boundary Model**, where P/V, V/D, and D/P radii specify angular values (4, 5, 1) in a PC system [2808].

Panels **a** and **b** are redrawn from [1303] (angular spacing was later revised [4140]); **c–e** are adapted from [1495, 1496, 3442]; **f–k** are based on [2804, 2808] (he uses AD and AV, not D and V); data in **j** are from [3811, 4140]. See also App. 7.

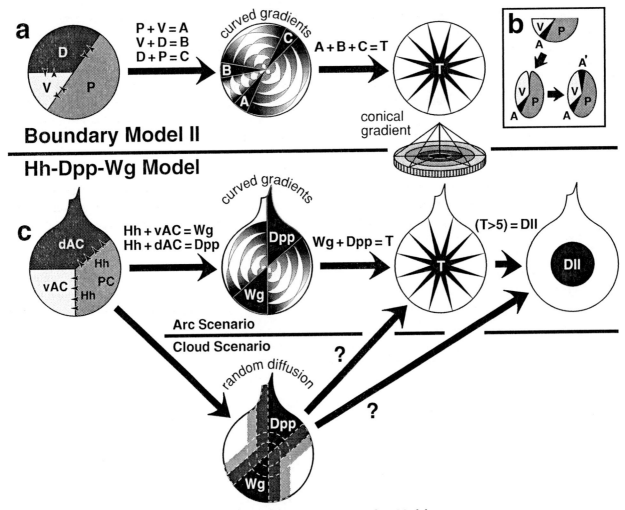

Boundary Model II

Hh-Dpp-Wg Model

FIGURE 5.4. Molecular validation of the basic premises in Meinhardt's Boundary Model.

a. The **Boundary Model II** (as revised by Meinhardt in 1983 [2808]). In this version (cf. Fig. 5.3), there are 3 new morphogens (A, B, C). Each of them arises at an interface (P/V, V/D, D/P) where region-specific molecules (P, V, D) interact after diffusing a short distance into the adjacent region. A, B, and C then react at the center to form the tip morphogen T. Finally, T diffuses to form a conical gradient that provides the radial coordinate of a polar system, while the angular coordinate is specified by the graded concentrations of A, B, and C. The gradients must overlap (not shown) to encode every angle uniquely.

b. According to this model, a VP fragment should duplicate because V and P will create a second source of A (A') when they heal together.

c. Hh-Dpp-Wg Model. Developing legs actually use a strategy like Meinhardt's model, although the first induction involves one-way (not bidirectional) diffusion (arrowheads), and there are only 2 (not 3) long-range morphogens. Cells in the Posterior Compartment (PC) make Hedgehog (Hh) – a short-range signal. When Hh diffuses across the A/P border, cells in the ventral AC (vAC) respond by making Wingless (Wg), while dorsal (dAC) cells make Decapentaplegic (Dpp).

How Wg and Dpp travel is not known. One idea (Arc Scenario, above) is that they move in arcs to form curved gradients, with subsequent steps (e.g., specification of angular values) as in Boundary Model II [1807]. The gene *Distal-less* (*Dll*) is shown being activated above a threshold (5) in the T gradient (assumed to be 10-to-1 centrifugally). Another idea (Cloud Scenario, below) is that Wg and Dpp diffuse randomly [618, 2456], in which case they could activate downstream genes directly (rightmost arrow) where their clouds overlap (via the rule "Wg + Dpp = Dll") without any intermediate tip morphogen [2813]. Indeed, one Wg target gene (*Dfz3*) does appear to be expressed in a parabola-shaped domain [3977]. However, defect arcs in genetic mosaics [1811] are more easily explained by the former scheme (see text). The Arc Scenario may also apply to the developing notum (cf. Fig. 6.14), where the *dpp*-ON stripe is at a ~30° angle relative to where its target gene (*wg*) is turned ON [4369].

Panels **a** and **b** are based on [2808] (p. 381, para. 2), and **c** is adapted from [618, 1807, 1991, 2456].

whereas in Meinhardt's scheme, cells must broadcast signals over distances of tens or hundreds of cells using diffusible molecules as messengers.

3. *Is the coordinate system seamless?* This issue has been debated ever since compartments were discovered in 1973 [2161, 2513]. In the PC Model, the circumference has no discontinuities. In Boundary Model II, compartment boundaries are distinctive as reference axes (as in a Cartesian system) [2808].

As the Cellular Epoch drew to a close in the late 1980s, evidence was mounting for the Boundary Model. Peculiarities of the A/P border (e.g., the two below) were difficult to reconcile with the PC Model's notion that all angular values are equivalent. The leg disc offered the ideal battlefield because its geometry fits a polar coordinate system so neatly (leg segments = rings, and bristle rows = spokes; Fig. 5.1c) [1311, 1807], and its A/P line is as inviolate as the wing's [4076, 4254]. Nevertheless, the wing disc was better known in terms of growth and lineage, so some of the skirmishes took place there instead.

Certain properties of the A/P border were problematic for the PC Model. For example, regenerative ability was found to be correlated with the cut's location relative to the A/P line. Compartments had originally seemed irrelevant for PI because they can be reestablished during growth of wing- and leg-disc fragments [5, 354, 2213, 4223] or during apoptosis-induced duplication of legs *in situ* [1504, 3701], and Bryant's map of regenerative potency revealed no discontinuity at the A/P line (Fig. 4.5). However, when Jane Karlsson reinvestigated using other cuts, she found a clustering of angular values near the wing's A/P boundary [2140]. No such correlation is seen in the *Drosophila* leg disc [4140], but circumferential intercalation in the legs of other (hemimetabolous) insects is constrained by A/P lineage boundaries [1297, 1299, 1301, 1302].

The PC Model was also unable to explain why the A/P border region is a haven for slower growing cells in the wing blade. In the developing wing (although perhaps not elsewhere in that disc [1431, 2935, 4703]), cells that divide more slowly than their neighbors are winnowed from the population [2229, 2935, 3949, 3961, 3962], except near the A/P and D/V compartment borders [3947, 3948]. "It is as if cells at a compartment boundary form a competitive pool that is separate from that of the internal cells" [1639]. In the case of the leg disc, the gynandromorph fate map is compressed on either side of the A/P boundary [1800] – indicating less mixing of unrelated cells

there – but this trend might not involve competition per se [3961].

At the dawn of the Molecular Epoch, many genes were found to be expressed in sectors (Fig. 5.8) or rings (Fig. 5.11) within the leg disc. These patterns were hailed as confirmation of the PC model [531, 880, 1304, 1574, 1970], but subsequent analyses of the gene circuitry tipped the scale in favor of Meinhardt's model [617, 2814]. The awkward truth, however, is that much of the new data does not fit neatly into either scheme. Chief among these "ugly facts" are the mirror-image duplications manifest by wg^{LOF} and dpp^{LOF} legs (Fig. 5.5), which imply a bipolar coordinate system [1812].

Hh, Dpp, and Wg are the chief intercellular signaling molecules

Hh, Dpp, and Wg are all secreted signaling proteins. Hh and Wg are founding members of the Hedgehog [1701] and "Wnt" (Wingless and Int-1 [3148]) families [3923], while Dpp is a member of the TGF-β family [3099]. During the 1990s the molecular players in each transduction chain were fitted together, like clues in a Sherlock Holmes mystery, to reveal the routes from the ligands to the nucleus. Among the surprises in these pathways were

1995 Discovery (in several labs) that Hh transduction involves a cAMP-dependent kinase (*DC0* encodes its catalytic subunit) whose only known developmental role until then was in slime molds [355, 2115, 3343].

1996 Identification of the long-sought Wg receptor (Dfz2) [310, 1971, 3656] and Hh co-receptor (Smo) [61, 4441], which were unexpectedly alike [3203, 3344, 4708].

1997 Identification of Pan as the final link in the Wg chain [519, 1088, 4439], which solved one mystery but led to a deeper one: why should the *pan* (Wg pathway) and *ci* (Hh pathway) genes be adjacent (~10 kb apart)?

1998 Realization that the Hh and Wg pathways also share an unsuspected Slimb-dependent proteolysis of their transcription factors (Ci and Arm) [2060, 4279].

1999 Revelation that the proteoglycan Dally is a co-receptor for both Wg [641, 2556, 3309, 4400] and Dpp [2004, 4400] – hence challenging the idea that Wg or Dpp are freely diffusible, as had earlier been disputed for Hh [277, 4275].

Figure 5.6 sketches the Hh, Dpp, and Wg transduction pathways. The players within each pathway are

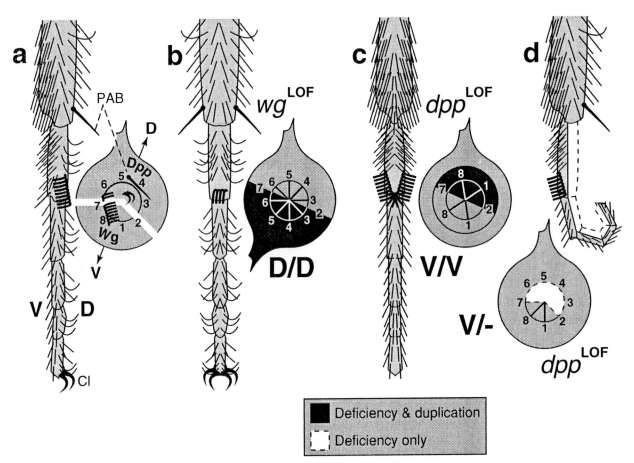

FIGURE 5.5. Phenotypes of wg^{LOF} and dpp^{LOF} legs that argue for a bipolar coordinate system instead of Meinhardt's 3-azimuth model. For gene abbreviations, see App. 6.

a. Anterior face of a wild-type left 1st leg from midtibia to the distal tip. Ventral (V) and dorsal (D) sides face left and right. Landmarks include pre-apical bristle (PAB), sex comb, and claws. The apical bristle (not labeled) is less prominent on 1st (vs. 2nd) legs. At right is a fate map (cf. Fig. 5.1c). Inner circle is the tarsus (1–8 are bristle rows). The sex comb begins as a ventral transverse row; claws and PAB are dorsal [1812]. Crooked white line delimits Dpp (~225° sector) and Wg (~135°) "domains of influence" (bounded by rows 2 and 7; cf. Fig. 5.2c), based on LOF defects (**b–d**). Because these sectors are so much larger than the slivers where dpp and wg are transcribed (cf. Fig. 5.4c), Dpp and Wg must act as secreted signals [834], but whether they are *bona fide* morphogens is debatable. These sectors may actually be more equal in size because D bristle rows are closer to one another than are V rows (cf. Fig. 5.1e).

b. In wg^{LOF} legs, V structures are missing and replaced (cf. key at bottom) by a set of D structures arranged in mirror symmetry [880, 1812]. This D/D ("Janus") symmetry extends through the sternopleura [177]. D/D legs also arise in arm^{LOF} [3317], dsh^{LOF} [2262], gam^{LOF} [599], pan^{LOF} [519], $porc^{LOF}$ [2110], and sgl^{LOF} [1673] flies and are inducible by wg^{null} [1811], arm^{null} [3317], and probably arr^{null} [4570] V clones (cf. Fig. 5.6). D/D legs can also be created by heat-pulsing t.s. N^{LOF} larvae [886] – a perplexing result because there is no known role for the Notch pathway in Wg signaling in legs. The few sex comb bristles that remain fail to rotate. A milder D/D duplication is seen in $Egfr^{LOF}$ gro^{LOF} double mutants [3465]. Finally, various kinds of treatments that cause trauma (microcautery, irradiation, etc.) can produce D/D phenotypes [425, 2102, 3442, 3449, 3705], presumably because the dpp-ON D sector has more robust growth potency than the wg-ON V sector (cf. Dorsal Remnant Mystery, Fig. 5.2b) [1807].

c, d. In dpp^{LOF} legs, D structures tend to be missing from the tibia and tarsus [1812]. Often they are replaced with V structures (**c**), but sometimes not (**d**). **c.** V/V phenotype. When a second set of V structures arises, the first legs have U-shaped sex combs. Janus V/V legs are also seen in $punt^{LOF}$ [3932, 4277] and tkv^{LOF} flies [753] and are inducible by dpp^{null} [1811] or $punt^{LOF}$ D clones [3329]. **d.** "V/-" phenotype. Missing structures in dpp^{LOF} legs are commonly not replaced. (No analogous "D/-" phenotypes are seen with wg^{LOF} legs.) In such cases, the V remnant tends to shorten and curl dorsally.

These figures are based on [177, 1811, 1812, 3317].

outlined in App. 6. Some familiarity with these molecular components is essential to grasp the tissue-level narrative that follows.

P-type cells use Hh to "talk" to A-type cells nearby

Based on detailed analyses of the above pathways, we now know the raison d'être for A and P compartments in most discs: they establish separate groups of "signaler" and "receiver" cells [354, 2448, 4488, 4848]. The P group emits a signal (Hh) to which the A group can respond [571, 1078, 4229]. Limited diffusion of the Hh signal ensures that only a subset of A cells (those at the border) will actually respond [358]. This "**Short-range Inducer Model**" thus relies on a principle that henceforth will be called the "**Deaf-Speakers/Mute-Listeners Trick.**"

A vs. P identities in thoracic discs are implemented by two different transcription factors: En (homeodomain class) [1031, 3719, 3860] and Ci (zinc finger class) [65, 155, 1827, 3194, 4503]. In P cells, En somehow activates *hh* transcription [1647, 1659, 4227, 4848] so P cells secrete Hh. Simultaneously, En represses *ci* transcription [1135, 1647, 2581, 3818] so P cells are refractory to their own signal (because Ci is needed for Hh transduction). All A cells can detect the signal because they express every transduction component (Ptc, Ci, etc.), but most A cells should be unable to emit Hh because Ci-75 represses *hh* [155, 1078, 2832].

The asymmetric roles of A vs. P cells in this discourse [354, 2448, 4848] contrast with the mutual speaker–listener roles envisioned by Boundary Model II [2808], but the two models are similar beyond this point [1427, 3497]. When secreted Hh spreads across the A/P line [1472, 4228], it rescues the activator form of Ci (Ci-155) from the proteasome guillotine so Ci-155 can convey its signal to target genes. In summary, three distinct zones are maintained along the A-P axis of each thoracic disc as a result of this circuitry (Fig. 5.7):

1. P-type cells express En, which activates *hh* but suppresses *ci* and *dpp* [4229]. En's inhibition of *dpp* occurs by two routes: En directly binds *dpp*'s *cis*-enhancers [2980, 3747], and En blocks *ci* expression so Ci-155 cannot activate *dpp* [1647, 1827, 2832].
2. A-type cells at the border receive Hh signal that diffuses from the P side. They respond (via PKA and other agents) by rescuing Ci-155 from being cleaved so it can activate *dpp* and *ptc*.
3. A-type cells beyond the range of Hh express *ptc* at a basal level [2115], and Ptc (by a double-negative route) relieves repression of PKA. PKA then phosphorylates Ci to provoke proteolysis of Ci's transcription-

activating (dCBP-binding) domain, and the remnant (Ci-75) keeps *hh*, *dpp*, and other target genes OFF.

The *en*-OFF state of A cells is inherited from the embryo [1093, 1124, 1790, 3129] (cf. Fig. 4.4). It is maintained by *Polycomb* [582, 2889, 4327], *polyhomeotic* [2728, 3512, 3753], *dRaf* [2532], *fat-head* [4607], and *groucho* (D/V boundary only) [998], so no further control by Ci-75 is needed. Hence, the Ci-vs.-En circuit is asymmetric: En keeps *ci* OFF in P cells, but Ci-75 does not keep *en* OFF in A cells (*ci*^null clones in the A region do not turn *en* ON [1078, 2832]).

The *en*-ON state of P cells is maintained by *trithorax* [452], *moira* [475], *polyhomeotic* [2728], and the EGFR pathway [207], rather than by the autoregulatory "En ➔ *en*" loop seen in embryos [1790]. Indeed, En ⊣ *en* in discs [207, 1647, 4229], although this GOF effect may be spurious relative to the circuit's normal range of operation [156].

It is possible to artificially drive *ci* expression (via *act5c-Gal4:UAS-ci*) in *en*-ON (P-type) cells so as to make them co-express both Ci and En [943]. Such schizophrenic cells "hear themselves talk" (via "Ci-155 ➔ Ptc" and via "En ➔ Hh") and behave like border cells [943]. In contrast, deaf-mute (*en*^null *ci*^null) cells behave uniquely (vs. A or P or border cells) in terms of adhesive affinity (cf. Ch. 6) [943].

Other questions pertain to Ci-75 vs. Ci-155, which have identical DNA-binding domains [2980]:

1. Because *hh* is a target gene of Ci-75 [2832], Ci-155 should turn *hh* ON at the border [155], but it does not. Evidently, *hh* is more sensitive to inhibition by Ci-75 than to activation by Ci-155 (both of which exist there) [2832], but how this selectivity is achieved physically is unclear [2980]. Oddities of this zone suggest that other factors might override these controls [2728, 4478, 4539]. For instance, ectopic Ci-75 can repress *ptc* there but not *dpp* [155].
2. A minimal amount of *ptc* transcription in A cells is needed to keep *dpp* OFF beyond the border zone [642], but if this baseline were due to a constitutive promoter, then *ptc* should also be expressed in the P region. So why isn't it? Apparently, *ptc*'s expression in A cells stems not from its promoter per se but from an unknown A-type activator aside from Ci-155. Because anterior *ptc* expression does not increase when *ci* is removed [2832], *ptc* must be selectively "deaf" to Ci-75 (just as *hh* is deaf to Ci-155). Again, the basis for such selectivity is obscure [2980].

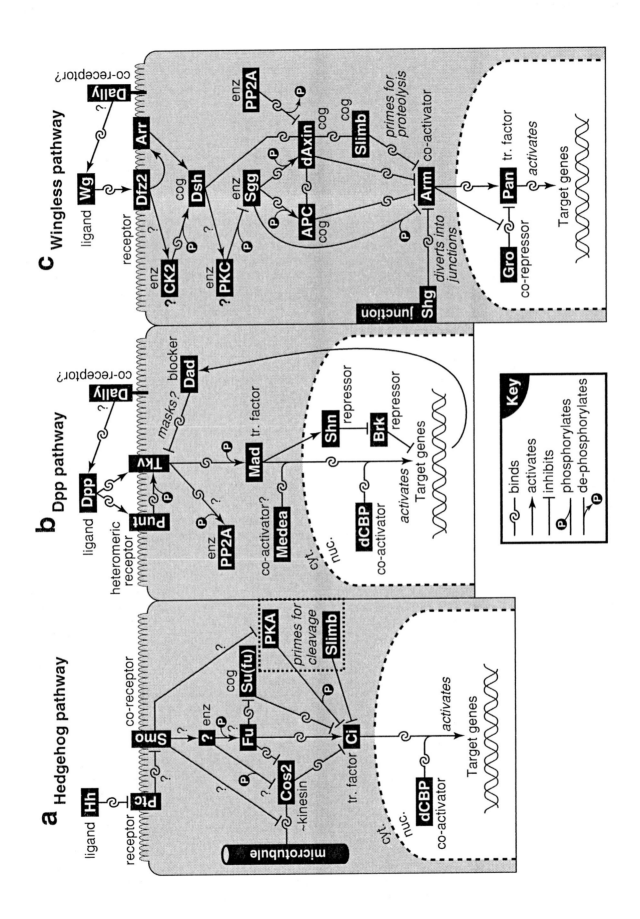

a Hedgehog pathway

ligand Hh
receptor Ptc
co-receptor Smo
? enz
Fu
cog Su(fu)
Cos2
~kinesin
microtubule
PKA
primes for cleavage
Slimb
Ci
tr. factor
cyt.
nuc.
dCBP co-activator
activates
Target genes

b Dpp pathway

ligand Dpp
heteromeric receptor Punt
co-receptor? Dally
Tkv
masks? blocker Dad
enz PP2A
Mad tr. factor
co-activator? Medea
cyt.
nuc.
Shn repressor
Brk repressor
dCBP co-activator
activates
Target genes

c Wingless pathway

ligand Wg
receptor Dfz2
co-receptor? Dally
Arr
? CK2 enz
Dsh cog
? PKC enz
Sgg enz
PP2A enz
APC cog
dAxin cog
Slimb cog
Arm co-activator
primes for proteolysis
Shg junction
diverts into junctions
Pan tr. factor
Gro co-repressor
activates
Target genes

Key

— binds
→ activates
⊣ inhibits
P phosphorylates
P de-phosphorylates

Hh elicits expression of Dpp and Wg along the A/P boundary

In 1994, the Short-range Inducer Model was invoked to explain Hh's effects in leg discs. Hh was shown to turn ON *wg* in the ventral half of the border, and to turn ON *dpp* in both the dorsal and ventral halves (Fig. 5.8) [231, 1037, 2466]. Ventral *dpp* transcription, which is relatively meager [2739], turned out to be a red herring from the standpoint of patterning since *dpp*^null^ V clones, unlike D ones, are anatomically normal [1811].

For some unknown reason, the *dpp* [2739, 2954] and *wg* [177, 3317] transcription zones are shaped like sectors, rather than bands. The *dpp*-ON domain is narrower than the *wg*-ON wedge and hence is often called a "stripe" [487, 3747]. The width disparity could trivially be due to *dpp* turning ON at a higher Hh threshold, but it might exist for other reasons (e.g., *dpp*-ON cells adhering more tightly or dividing more slowly). The sectors coincide roughly with the leg's prospective D (*dpp*) and V (*wg*) midlines. If, as argued below, Dpp and Wg are morphogens for specifying positions along the D-V axis, then this state of affairs would explain the Gaps Mystery. To wit, the reference line would not be the A/P compartment boundary per se as first thought [904], but rather the slightly offset line that bisects the *dpp* and *wg* transcription sectors (Fig. 5.4c).

Embryos also manifest Hh-dependent Wg expression (cf. Ch. 4) [3386], but Dpp is not under Hh control at that time. In comparing a Wg blastoderm stripe to the Wg sector of the leg disc (and doing the same for En [3747]), the leg disc seems equivalent to a body segment (or part thereof) that has curled itself into a circle [880, 1039]. This basic topology agrees with both the PC and Boundary Models, as well as with the Angular Values Hypothesis [2291].

Viewing the nascent disc as a circularized segment is appealing because it provides a simple way to co-opt the established intercellular signaling machinery of the segment polarity genes for circumferential patterning in the discs. [834]

The leg disc also resembles a body segment insofar as Dpp apparently acts as a dorsalizing agent in both systems (see below) [1807]: in the embryo, a Dpp gradient specifies diverse fates within the dorsal 40% of the ectoderm [127, 1091, 1211, 4613].

Why should Hh turn ON different genes in different parts of the A compartment? The answer is that dorsal (dAC) and ventral (vAC) cells of the A compartment differ in competence (Fig. 5.4c). These biases are revealed when Hh is expressed ubiquitously: Hh then activates *dpp* throughout dAC and *wg* throughout vAC [231, 617, 1037, 2466], rather than just along the A/P boundary (Fig. 5.9c). Activation of *dpp* in dAC (and weakly in vAC) and *wg* in vAC is also seen in somatic clones where the Hh pathway is turned ON downstream of Hh by LOF mutations in *patched* [2058, 2491] or *DC0* [2058, 2059, 2491, 2533, 3238]. For *slimb*^LOF^ clones, the situation is more complex because *slimb* also functions in the Wg pathway [2060, 2856, 4279].

FIGURE 5.6. Some key signaling pathways in disc development. In the working models depicted here, certain components may be replaced by others (not shown) in different tissues, regions, or time periods. Three cells are depicted, with apical microvilli at top, although receptor-ligand interactions may actually take place elsewhere on the surface. Black rectangles = proteins; gray area = cytoplasm ("cyt."); dashed line = nuclear membrane. Arrows (activation) and blunt "⊣" lines (repression) indicate epistatic relations. Noncovalent binding is symbolized by interlocking hooks (cf. key). Question marks concern how or whether the interactions occur (see text). A "cog" (a.k.a. "adaptor") is a component that plays a steric (nonenzymatic) role [441, 1954, 3299, 3676] (e.g., Slimb [2668]). Enz = enzyme, CK2 = Casein Kinase 2, PKA = Protein Kinase A (likewise PKC), Shg = Shotgun (a.k.a. dE-cadherin). For other abbreviations, see App. 6. For perspective, see App. 7.

a. Hedgehog pathway [1554, 1974, 2075, 3685, 3831]. Binding partners (listed from surface to DNA) are based on fly studies or vertebrate homologs (vH): Hh-Ptc [755]^Δ^, Ptc-Smo (vH [4130]), Fu-Cos2 [4068]^Δ^, Fu-Ci [4068], Fu-Su(fu) [4068]^Δ^, Cos2-microtubules [3612, 3976], Cos2-Ci [4068, 4536]^Δ^, Su(fu)-Ci [4068]^Δ^, dCBP-Ci [54], Ci-DNA [751]^Δ^. Ci's interaction with dCBP may be inhibited by PKA (link not shown) [54].

b. Dpp pathway [959, 1983, 3498]. Dally is placed upstream of Dpp, based on the ability of extra doses of *dpp*^+^ to suppress *dally*^LOF^ phenotypes [2004], and a similar argument applies to Wg [2556]^Δ^. Evidence for binding: Dpp-Punt [2495], Dpp-Tkv [2495]^Δ^, Dpp-proteoglycans (e.g., Dally) [2004]^Δ^, Punt-Tkv [2495], Tkv-Mad [1983], Tkv-Dad [1983], Tkv-PP2A (vH [1627]), Mad-Medea [1983, 4703], Mad-dCBP [4534], Mad-DNA [2217]. Mad may also bind microtubules (not shown) [1084]. The "Shn ⊣ *brk*" link does not actually occur at the protein level, but rather occurs via inhibition of *brk* transcription [2727].

c. Wingless pathway [422, 3919]. Arr is thought to be a co-receptor for Dfz2, but it does not "present" Wg in the same way as Dally [4570]. Conceivably (not shown), Arr may act via CK2: Dfz2 ➔ Arr ➔ CK2 ➔ Dsh, etc. The APC homolog here (labeled simply APC) is likely E-APC [4832]. Sgg is thought to be Arm's natural kinase [3683] (but see [3228, 4062]). Evidence for binding (see [3316] for details): Wg-Dfz2 [310], Wg-proteoglycans like Dally [598]^Δ^, CK2-Dsh [3149, 4676], Dsh-dAxin (vH [2238]^Δ^), Sgg-APC (vH [3675]), Sgg-dAxin [3683], APC-dAxin [1698], PP2A-dAxin (vH [1919]), dAxin-Slimb (vH [2241]), APC-Arm [4832], dAxin-Arm [1698], APC-Arm [4832], Slimb-Arm (vH [4701]^Δ^), Shg-Arm [3749, 4268, 4416]^Δ^, Arm-Pan [519, 4439], Gro-Pan [692, 2496], Pan-DNA [519, 4439].

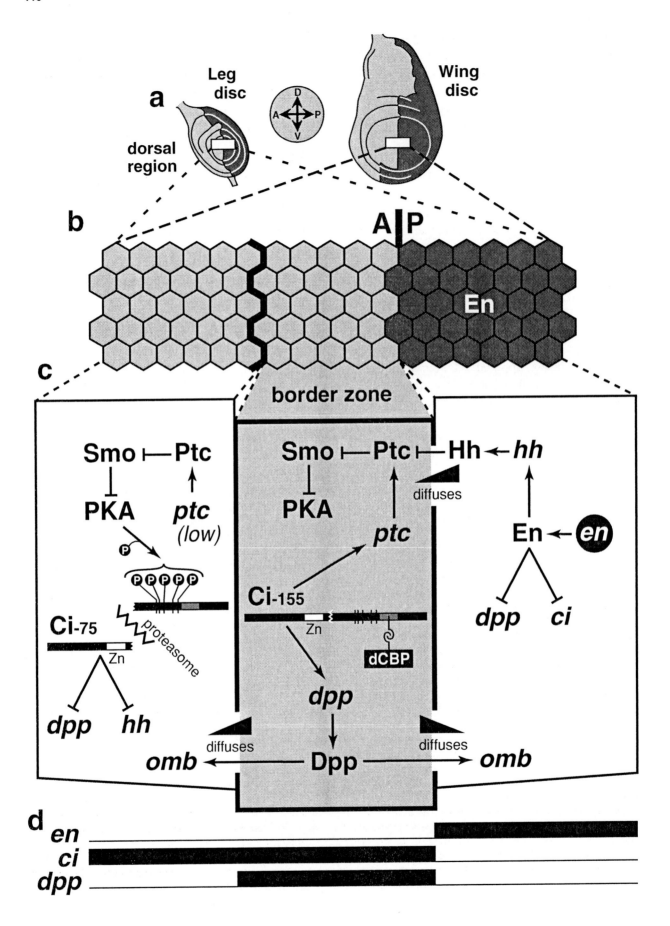

The dAC/vAC border is not a strict lineage boundary [487, 1800, 4076], so dAC-type cells can presumably jostle into vAC territory and vice versa. No genes have been identified that are expressed strictly within dAC or vAC [3919], nor are any genes transcribed along the dAC/vAC border (as expected based on Boundary Model II) [4277], so the nature of these areas is not obvious [617].

Dpp dorsalizes and Wg ventralizes, or do they?

Conceivably, dAC and vAC states are maintained by the Dpp and Wg signals themselves [487, 2954, 4277]. That is, AC cells within range of the Dpp signal adopt dAC competence, while those within range of Wg adopt vAC competence. Such a system could be stable without cell-lineage restrictions (cf. the Cabaret Metaphor). Support for this idea comes from an experiment where cells are tricked into switching states: when dAC cells are duped

(via sgg^{LOF}) into thinking they are receiving a maximal Wg signal, they stop making Dpp and activate their own *wg* gene whenever they are within range of Hh [1833, 2059] (notwithstanding a report to the contrary [1039, 4487]).

Additional evidence is provided by the fact that the dAC/vAC border (which runs along row 7 in the fate map; cf. Figs. 5.1c and 5.4c) coincides with the line that separates the Dpp and Wg "spheres of influence" (Fig. 5.5a): in dpp^{LOF} mutants the tarsus lacks structures on the D side of "$L_{2\&7}$" (the line that runs along rows 2 and 7), whereas, in wg^{LOF} mutants the whole leg loses structures on the V side of $L_{2\&7}$ (Fig. 5.5b-d) [1812, 3317]. $L_{2\&7}$ is also defined by leg defects caused by dpp^{null} or wg^{null} clones [1811]. It divides the disc into two sectors that would measure ~225° (D) and ~135° (V) if bristle rows were evenly spaced around the leg (cf. a femoral septum at this angle [3867]), but in fact there is a larger space

---

FIGURE 5.7. Effects of Hh diffusion into the A compartment of thoracic discs. For gene abbreviations, see text.

a. Left leg and wing discs, with D-V axes of the central regions oriented vertically (cf. compass). A and P compartments are lightly or darkly shaded, respectively. The boxed areas of both discs obey the "Hh ➜ Dpp" link, but the ventral part of the leg disc obeys "Hh ➜ Wg" instead (not shown).

b. Part of the disc epithelium that straddles the A/P compartment boundary (magnified from **a**). Hexagons are cells (although packing is not really so orderly). Cells in the P region express the transcription factor En, while A cells do not. Zigzag line marks the anterior limit of the *dpp*-ON zone.

c. Circuitry of genes and proteins (abridged; cf. Fig. 5.6) that operates in each of the cell types delineated in **b**. The chain of events (right to left) starts with expression of *en* in all P cells. (P cells have Smo but not Ci or Ptc [1022, 2832].) En activates *hh* but represses *ci* and *dpp* [1647, 2832, 2980, 3747, 4277]. Hh diffuses ~8–10 cell diameters into the A region [4136]. Binding of its receptor Patched (Ptc) triggers synthesis and release of Dpp, which turns ON genes like *omb* (cf. Fig. 6.3) [2455, 3074]. The "Hh ➜ Dpp" link is mediated by the transcription factor Ci (horizontal bar). Full-length Ci (155 kD, 1396 a.a.) has 5 tandem zinc fingers (Zn), a C-terminal domain that binds the co-activator dCBP, a proteasome cleavage site (zigzag line), 5 residues capable of phosphorylation by Protein Kinase A (vertical lines) and other domains (cf. text) [54, 2832]. A-type cells do not express *dpp* in the absence of Hh (left) because basal transcription of *ptc* [4136] supplies enough Ptc to derepress PKA (via Smo). Phosphorylation of Ci by PKA leads to Ci's proteasome-dependent cleavage, and the N-terminal fragment Ci-75 acts as a repressor (due to an alanine-rich section [65]) of *dpp* and *hh* [2832]. Dpp and Hh are both diffusible (triangles denote gradients), but Dpp's range is greater (cf. Fig. 6.3). Omitted here is the enigmatic transcriptional regulation of *ci* by the zinc-finger protein Combgap (Cg) [621, 4017, 4217]: Cg ➜ *ci* throughout the A region of wing and leg discs, but Cg ⊣ *ci* in the P region of wing (but not leg) discs. Curiously, the En ➜ *hh* link is inoperative in the developing eye [4168].

d. Summary of ON/OFF states (thick vs. thin lines) of *en*, *ci*, and *dpp*. The *dpp*-ON stripe is evidently "painted" in two ways: (1) directly via "Hh ➜ *dpp*" and (2) indirectly by a figure-ground "illusion" [657, 1585, 3747], where "Ci-75 ⊣ *dpp*" anteriorly and "En ⊣ *dpp*" posteriorly. Both actions stem from the disc's *en*-ON/OFF duality.

Validation of this logic comes from larvae carrying an *en* transgene driven by a heat-shock promoter. When such larvae are heat-shocked, every cell transiently becomes *en*-ON (P-type) and the *dpp*-ON stripe disappears [3747]. A similar shut-OFF is expected for discs devoid of En, but this condition has not yet been created [1839]. Not shown is a trespass of the *en*-ON edge into the border zone that occurs in late 3rd instar when a "Hh ➜ *en*" link pervades the disc [1647]. (Its ubiquity is shown by the ability of *ptc*^null A clones to turn ON *en* [4136].)

N.B.: Events in the boundary zone are more complex than depicted here (cf. Figs. 6.3–6.5), especially for Ci [4540]. En's inhibition of *dpp* (En ⊣ *dpp*) occurs both directly (by binding of En to *dpp*'s *cis*-enhancers [2980, 3747]) and indirectly via En's suppression of *ci* (because Ci-155 ➜ *dpp* [1827, 2832]). Suppression of *ci* is also partly mediated by Polyhomeotic [3512] – a member of the Polycomb Group of repressors (cf. Ch. 8) [1012, 1867, 2374]. En's activation of *hh* (En ➜ *hh*) is probably indirect (En ⊣ X? ⊣ *hh*) [224] because En normally acts a repressor [2047, 2048, 3175, 3860, 4003] by recruiting Groucho [1243, 2064, 4351] (but see [1706, 2642, 2729, 3481]). "X" is probably not Ci-75 because *ci*^null clones in the A region do not turn ON *hh* at P-type levels [1078, 2832, 2834]. Alternatively, En might turn ON *hh* by using a co-activator, such as Blistered [2729], Extradenticle [3324, 3325, 3861], or the Trx-G chromatin-remodeling complex (cf. Ch. 8) [451]. En is potentially secretable [811, 2085, 2661] but is not used thusly here. Slimb helps keep *wg* OFF in the P region (not shown) via a Hh-independent route [2856]. After [2832, 4479, 4539]. See [2834] for further nuances.

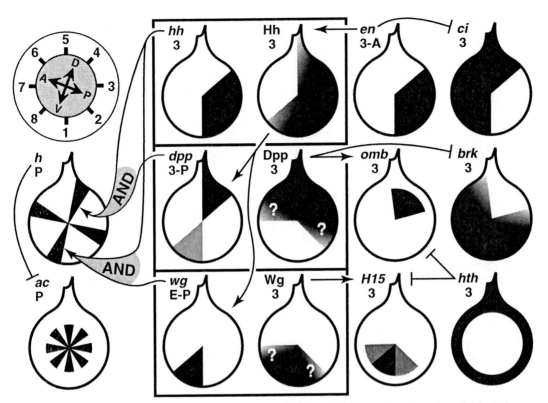

FIGURE 5.8. Sectorial domains of gene expression in the leg disc (which become stripes along the leg) plus one annular domain (*hth*). The flow of control is from *en* (upper right), which establishes the A-P axis, through *dpp* and *wg*, which polarize the D-V axis, to *ac* (lower left), which allocates cells to bristle vs. nonbristle fates.

Black areas chart mRNA (transcribed from endogenous or reporter gene) or protein levels during embryogenesis (E), late 3rd instar (3), early pupal period (P), or adult (A) stage. Protein levels are given for Hh (estimated based on data from wing discs [4136]$^\Delta$), Dpp (range is not exactly known [1169, 4265]), and Wg [2456]$^\Delta$. Shading denotes degree of expression. Areas are mapped onto a disc, although some genes turn ON only after 3rd instar (e.g., *ac*). The span of stages (e.g., "3-P") indicates when expression has been seen but does not connote confinement to that period. Wiring (➡ activation; ⊣ inhibition) illustrates a few genetic interactions (see text or sources cited below). "AND" means that both inputs are needed to activate the target gene. Thus, the dorsal *hairy* stripe is caused jointly by Dpp and Hh (neither alone is sufficient), while the ventral stripe is caused jointly by Wg and Hh [1778]. Interactions between Dpp and Wg (not shown) involve mutual antagonism along the D-V axis and cooperation along the proximal-distal axis (see text).

Genes ("DO" = details omitted): *ac* (*achaete*; DO: expression areas proximal to the tibia) [3193], *brk* (*brinker*) [2049], *ci* (*cubitus interruptus*) [1135], *dpp* (*decapentaplegic*) [2069, 3122]$^\Delta$, *en* (*engrailed*) [2954]$^\Delta$, *h* (*hairy*) [3193]$^\Delta$, *H15* (a T-box relative of *omb*) [3762]$^\Delta$, *hh* (*hedgehog*) [4254]$^\Delta$, *hth* (*homothorax*) [4760]$^\Delta$, *omb* (*optomotor-blind*) [2049]$^\Delta$, *wg* (*wingless*, whose stripe in pupal legs is ~5 cells wide [1039]) [4277]$^\Delta$.

All these genes are known to have LOF or GOF effects on pattern, except *H15* (whose null phenotype is wild-type [3087]) and *omb*. Two genes whose expression resembles that of *H15* – *nkd* [4863] and *Dfz3* [3977] (not shown) – also have no LOF defects. Enhancer-trap line c409 (not shown) matches *ci*'s pattern [2684], as does aldehyde oxidase [2340, 4051]. See [489] and *FlyView* (*flyview.uni-muenster.de*) [2027] for other enhancer-trap expression patterns.

Areas of expression that are not wedge shaped (omitted) include arcs (e.g., *GAL30A* [487], *grain* [509], *Sex combs reduced* [2393]$^\Delta$, *Ultrabithorax* [3397]), parabolas [2977], a spot at the femoral chordotonal organ [1235], etc. [834, 1574, 2684, 4522]. Augmented from [531].

In some specimens the *dpp*-ON and *wg*-ON sectors appear as wide as shown [2954, 4277, 4849]$^\Delta$, but often they look narrower [231, 487, 3762]$^\Delta$. Likewise, the *en* domain sometimes displays a central acute angle as shown [2954, 3329]$^\Delta$, but it can also seem straight [1472, 4254] or even obtuse [315]. These vagaries arise from variations in folding, viewing angle, and focal plane, as well as from inherent irregularities in the A/P (and other?) boundaries [1833, 4254]. In early pupae, a gap arises between the *en* domain and the dorsal *h* sector [1778], although it may not be as large as implied here.

N.B.: Hh, Dpp, and Wg only seem to define D-V sectors of *h* expression [1778]. Therefore, one remaining mystery is: what factors define *h*-ON sectors along the A-P axis? Another is: what factors define the edges of *ac*-ON stripes that are not created by "*h* ⊣ *ac*" inhibition?

between V rows 1 and 8 than between D rows 4 and 5 (Fig. 5.1e). Thus, the areas flanking $L_{2\&7}$ may actually be closer in size – a detail that is important in assessing how far Dpp or Wg signals travel.

Reduction of *dpp* or *wg* function does not just result in loss of structures from one or the other side of the leg. Commonly, the missing structures are replaced by a mirror-image copy of the remaining part (Fig. 5.5) [177, 1812, 3317, 4032]. The V/V *dpp*[LOF] phenotype is like a hand with two palms and no back, while the D/D *wg*[LOF] phenotype is like a hand with two backs and no palm. Both D/D [560, 1812] and V/V [2082] legs have uneven girths, suggesting that leg segments vary in the fraction of their circumference that is under *wg* or *dpp* control.

According to the PC Model, such "Janus" oddities should, in principle, be easy to explain. The deficiency portion of the deficiency-duplication phenotype would be due to tissue loss, and the duplication would arise via juxtaposition of cells with disparate angular coordinates from around the necrotic zone, followed by circumferential intercalation by the shortest route. In fact, however, although necrosis is seen in *dpp*[LOF] discs [529, 2739, 2849], it is negligible in *wg*[LOF] discs [2015, 2929, 4683]. An even more serious problem for the model pertains to intercalation. According to PC logic, the V/V *dpp*[LOF] phenotype arises because loss of the D (*dpp*-dependent) sector removes ≥50% of the angular coordinates. If so, then the shortest route for row 2 and 7 cells to take when confronted must be the V arc that includes rows 1 and 8. However, the D/D *wg*[LOF] phenotype implies a different shortest route (i.e., the D arc encompassing rows 3–6) for the same row 2/7 confrontation. The PC Model is stumped by this paradox.

When *wg* is artificially expressed in GOF clones on the D side of the leg, the cells in and around the clone make ventrolateral (row 2- or 7-type) bristles [837, 1039, 4159]. This result proved that Wg can act as a ventralizing agent [3923], but it raised the question of why the cells do not become fully ventralized. The failure to induce ventral (row 1- or 8-type) bristles was initially attributed to low output of the *wg* transgene that was used [4159], but two later findings revived doubts as to whether Wg is really a morphogen in the leg disc (cf. proof for such a role in the wing disc [4849]):

1. Activating the Wg pathway (cf. Fig. 5.6) downstream of the receptor (via *sgg*[null] clones) can induce 1/8-type bristles in the very heart of the Dpp domain or anywhere else around the tarsus [1039, 1833, 4666]. This

fact refutes the excuse that Dpp is too strong in its own realm to permit V-type states there. Instead, it suggests that Wg only specifies ventrolateral fates, with fates nearer the V midline being specified by a second ligand (another Wnt? [3919]) that is also transduced via Sgg [4666]. (Similar logic may govern the embryo's Dpp gradient [2730].)

2. If Wg were a morphogen, then it should specify fates in a concentration-dependent way. However, when the disc is flooded with Wg (via *Gal4* drivers that enforce nearly ubiquitous expression of *UAS-wg*), the V region sprouts a forest of 2/7-type bristles but no extra 1/8-type bristles [4666]. Again, the implication is that Wg merely controls ventrolateral fates. Midventral identities might be dictated by Hh (cf. D rows [2533]), because extra 1/8-type bristles arise when the Hh pathway is turned on in the V area by *DC0*[LOF] (Figs. 3d of [2491] and 6h of [3238]), although row 1 lies in the P compartment (deaf to Hh?) [1800, 2449].

Dpp's role as a dorsalizing morphogen in the leg is equally dubious. Excess Dpp (*UAS-dpp* driven by *dpp-Gal4*) can induce expression of *omb* (*optomotor-blind*) – a marker gene for D identity – in the V region of the leg [487], but overexpression of Dpp within its own domain fails to induce extra (4/5-type) D pattern elements [2954]. The lack of a significant GOF phenotype for a gene that has a strong LOF phenotype was seen before in *scute* (cf. Ch. 3). In that situation, the explanation was a combinatorial interaction between Scute and Daughterless – a dimerization partner that may be limiting under excess-Scute conditions. Here, too, with Dpp, there could be a simple reason why "shouting" achieves no more than "whispering." Namely, another signal might be needed in addition to Dpp. Indeed, Hh might be the hidden player in the both the Dpp and Wg stories:

{Hh AND Dpp} ➤ row 4/5-type bristles?
{Hh AND Wnt?} ➤ row 1/8-type bristles?

Dpp does display some dose-dependent properties (see below) [2954], but they are irrelevant to Dpp's role as a dorsalizing agent per se. This confusing state of affairs stands in stark contrast to the embryo, where Dpp's morphogen properties are so well documented in terms of fate specification [1281, 3532, 3693, 4058, 4492], gradient distribution [1212, 1927, 2003], and dose sensitivity that is manifest not only in discrete thresholds [127, 1091, 1211, 4613], but also in haplo-insufficiency [1872, 1987, 3458, 3693]. It also contrasts with the wing, where Dpp's role as a morphogen has been well established (cf. Ch. 6) [2455, 3074].

Dpp and Wg are mutually antagonistic

In 1996, a flurry of articles revealed a mutual antagonism between Dpp and Wg signaling. By manipulating *dpp* or *wg* expression directly [487, 2059, 2082, 2954, 4277] or by interfering indirectly with components of their pathways [2059, 3329, 4277], these experiments uncovered a flip-flop circuit that was suspected from earlier clues [1812, 2058]. Evidence that Wg inhibits Dpp is summarized below.

1. Under *wg*^{GOF} conditions (*dpp-Gal4:UAS-wg*), where *wg* is forced to be expressed in the Dpp sector (i.e., *dpp*-expression zone), the level of *dpp* transcription decreases [487, 2082, 4277].

2. In *wg*^{LOF} mutants, *dpp* expression increases in the Wg sector (as *wg* expression decreases) [487]; this substitution is noticeable within 24 h after reducing *wg* function [4277].

3. When *wg* expression is extinguished in the Wg sector (due to *dpp-Gal4* activating *UAS-dpp* along the entire A/P border), expression of *dpp-lacZ* increases in that V area [487, 2954]. The same effect is seen in Wg-sector somatic clones whose Wg pathway is blocked (by *dsh*^{null} [1833, 2059]).

4. Somatic clones that are both *DC0*^{null} (which makes cells think they have received a Hh signal) and *wg*^{null} express *dpp* when they arise anywhere within the vAC (or dAC) [2059].

Evidence for a reciprocal inhibitory effect of Dpp on Wg is as follows:

1. Overexpressing *dpp* along the A/P boundary (*dpp-Gal4:UAS-dpp*) suppresses Wg in the Wg sector [487] except at the disc center (endknob) [2954, 4277]. Concomitantly, expression of the enhancer trap *H15-lacZ* (a V marker) vanishes in the V region, and *omb* expression (a D marker) is now detected [487] (see Fig. 5.8 for expression domains). A similar effect is achieved when a constitutively active (ligand-independent) Dpp receptor (*tkv*^{GOF}) is used instead of excess Dpp (*dpp-Gal4:UAS-tkv*^{GOF}) [4277].

2. When the level of *dpp* transcription decreases in the Dpp sector (due to *dpp-Gal4:UAS-wg*), the endogenous *wg* (distinguishable from the *UAS-wg* transgene by a *lacZ* insert at the *wg* locus) is activated [487]. (Concomitantly, *H15-lacZ* expression appears and *omb* expression disappears [487].) Upregulation of *wg* also occurs (although only rarely) when *dpp*^{null} clones reside in the Dpp sector [2059] or when Dpp transduction is blocked in dorsal somatic clones (by *punt*^{LOF} [3329] or *tkv*^{LOF} [2059, 3329]).

3. In *dpp*^{LOF} mutants, Wg is now detected in the Dpp sector (as *dpp* expression decreases) [487]. The ectopic Wg is restricted to the tarsal region – the only area converted to V identity in these hypomorphic genotypes [1812]. A similar substitution of *wg* for *dpp* expression occurs when the Dpp pathway is blocked ubiquitously (by *punt*^{LOF} [4277]), but in that case *wg* is transcribed in the entire dorsal sector [4277].

4. Somatic clones that are both *DC0*^{null} and *dpp*^{null} express *wg* when they arise anywhere within the dAC (or vAC) [487, 2059]. Similar effects are seen when *DC0*^{null} is coupled with *sgg*^{null}, which inhibits *dpp* [2059].

In wild-type discs, Dpp's ability to suppress *wg* transcription must be greater than Wg's ability to suppress *dpp* transcription because *dpp* is transcribed at a low level in the Wg sector, whereas *wg* is not transcribed at all in the Dpp sector [231, 2466]. Dpp has been shown to also block Wg's downstream effects: when the same group of (dorsal femur) cells is forced to express both *dpp* and *wg* (by using *30A-Gal4* to drive *UAS-dpp* and *UAS-wg*), *wg* is unable to activate its puppet gene *H15* [487]. Whether Wg can similarly block Dpp's downstream actions is not known.

The Dpp-Wg antagonism means that the D and V sectors are each like a seesaw that must tilt one way (Dpp) or the other (Wg), but it does not elucidate why each sector tilts a specific way. The answer may be found in the embryo [487, 3862]: the nascent leg disc incorporates part of an ectodermal Wg stripe as its V sector (Fig. 4.4b), and henceforth Wg predominates there [880]. Why the D sector tilts to a Dpp state is unclear, but abundant evidence (see below) indicates that the alternative (i.e., two symmetric Wg sectors) is incompatible with distal outgrowth. The mutual exclusivity of Dpp or Wg activity may sharpen the boundary between their domains so that other circuits can come into play [2954]. Oddly, ventral suppression of *dpp* also requires Notch during 2nd and early 3rd instars [886]: "{N AND Wg} ⊣ *dpp*". Notch enforces a comparable bipolarity in wing discs (wing vs. notum; cf. Fig. 6.9) [886] and eye discs (eye vs. antenna) [2353, 2362].

The ability of *dpp-Gal4:UAS-wg* to convert the D side of the leg into a purely V phenotype (with attendant *H15-lacZ* [487]) has been construed as evidence that Wg can specify midventral identities after all [2082], despite the skepticism voiced above. However, the objection raised by the *sgg*^{null} clones still stands because they fully ventralize without "cheating" (i.e., using the mutual-antagonism circuit to cripple the opponent

(Dpp) prior to specifying cell fates). Another attempt to resurrect Wg as a leg-disc morphogen was based on a dose dependence of Wg's effects on the D side of the leg. With increasing dose, Wg successively (1) removes D structures, (2) induces leg-to-wing homeosis (cf. Ch. 8), and (3) achieves total repression of *dpp*, whereupon the D side vanishes and is replaced by a V duplicate [2082]. Again though (as with *dpp*), none of the thresholds is really pertinent to whether Wg is a morphogen along the D-V axis. What is needed is proof that greater doses of Wg directly induce increasingly ventralized bristles, but no such evidence exists [3087].

Dpp and Wg jointly initiate distal outgrowth

In the Boundary Model, distal outgrowth is supposed to be triggered by a combination of morphogens [2804]. Given that the leg tip comes from the disc center, which is the sole point of contact between the Dpp and Wg sectors, it was reasonable to suppose that the combination of these two diffusible signals might jointly cause outgrowth [617, 4490]. This "{Dpp AND Wg} ➔ distalize!" imperative was proposed by Campbell *et al.* [620] in 1993. In 1994, the upstream step "Hh ➔ {Dpp AND Wg}" was added by several teams (Short-range Inducer Model) [231, 1037, 2466]. Thus was born the "**Hh-Dpp-Wg Model**" for distal outgrowth (Fig. 5.4c), which invoked a dependence of the proximal-distal axis on the D-V axis:

Hh ➔ {Dpp AND Wg} ➔ distalize!

Many proteins in the TGF-β [63, 2733] and Wnt [2261, 3150, 3910] families act as growth factors, and the data below argue that Dpp and Wg also do so. Some of these experiments use *Distal-less* (*Dll*) and *aristaless* (*al*) as indicators for prospective distal tips. Both *Dll* and *al* are homeobox genes [836, 3798]. *Dll* is expressed in the disc's tarsal region (plus trochanter and tibia) [881, 1037, 1561], while *al* is expressed where the claws arise (plus dorsal coxa and ventral tibia) [620, 881, 3798]. The facts below support the Hh-Dpp-Wg Model:

1. Ubiquitous expression of Hh expands the solid circle of Dll into a broad band straddling the dAC/vAC border (as Dpp and Wg spread to fill dAC and vAC) [1037] and similarly stretches the Al spot into a stripe (Fig. 5.9c) [617]. Both responses imply a combinatorial "{Dpp AND Wg} ➔ {*Dll* AND *al*}" control mechanism.
2. Ubiquitous expression of Dpp extends the Dll circle into the ventral sector where Wg signal is naturally received (as per a "{Dpp AND Wg} ➔ *Dll*" rule)

and provokes overgrowth in that same area [1037]. Outgrowths from the Wg sector (along with an extra Al spot) are also induced when *dpp-Gal4* drives *UAS-dpp* in that region [2954].

3. Ubiquitous expression of Wg lengthens the Al spot into a stripe along the Dpp sector (Fig. 5.9d) [620] (as per a "{Dpp AND Wg} ➔ *al*" rule), and concentric oval folds (indicating overgrowth) form around this stripe. The same effect (plus overproliferation of dorsal cells) is seen when Wg is misexpressed in the Dpp sector (*dpp-Gal4:UAS-wg*) [4666].
4. *hh*[GOF] clones only induce outgrowths when they straddle the dAC/vAC border (where both Dpp and Wg would be induced) [231, 617], whereas AC *hh*[GOF] clones that miss the border cause partial AC duplications and local overgrowth (cf. Fig. 6.6) but no distal outgrowths. The same is true for *DC0*[LOF] [2058, 2491, 3238] or *slimb*[LOF] clones [2060, 4279], although the outgrowths tend to be distally incomplete [2533]. Such boundary dependence recalls the Boundary Model (cf. the paint-spattering scenario).
5. *dpp*[GOF] clones only induce outgrowths ventrally [1037, 2059] as per a "{Dpp AND Wg} ➔ distalize!" imperative. The branches are typically stunted (1–2 segments long), but truncation could be due to low output of the transgene. Longer branches are seen when the endogenous *dpp* gene is turned ON along the ventral A/P border (via the Dpp-Wg antagonism circuit) in somatic clones whose Wg pathway is turned OFF (by *dsh*[null] [1833, 2059] or by the compound *DC0*[null] *wg*[null] [2058, 2533]).
6. *wg*[GOF] clones only induce outgrowths when they reside along the Dpp sector as per the "{Dpp AND Wg} ➔ distalize!" rule [4159], in which case they induce ectopic Al [617, 620] and Dll [1037]. Similar effects are seen when clones in this area have a hyperactive Wg pathway due to intrinsic malfunction (by *sgg*[null] [1039, 1833, 2059] or *dAxin*[null] [1698]) or a seesaw response (Dpp-down causes Wg-up) to blockage of Dpp signal transduction (by *punt*[LOF] [3329] or *tkv*[LOF] [2059, 3329, 4277]). Because *sgg*[null] clones induced after mid-2nd instar fail to trigger outgrowths [4666], such clones may need to exceed a certain size to "knock out" the *dpp*-ON target area and tip the seesaw. A bizarre twist on this story may explain *in situ* duplication of 3rd-leg discs in *lethal (2) giant discs*[LOF] (*l(2)gd*[LOF]) overgrowth mutants. Over a period of several days, a tongue of Wg expression arcs around anteriorly to meet the Dpp sector, where an extra disc then arises [558], but duplication is prevented when *l(2)gd*[LOF] flies are also made homozygous for *dpp*[LOF] or *wg*[LOF].

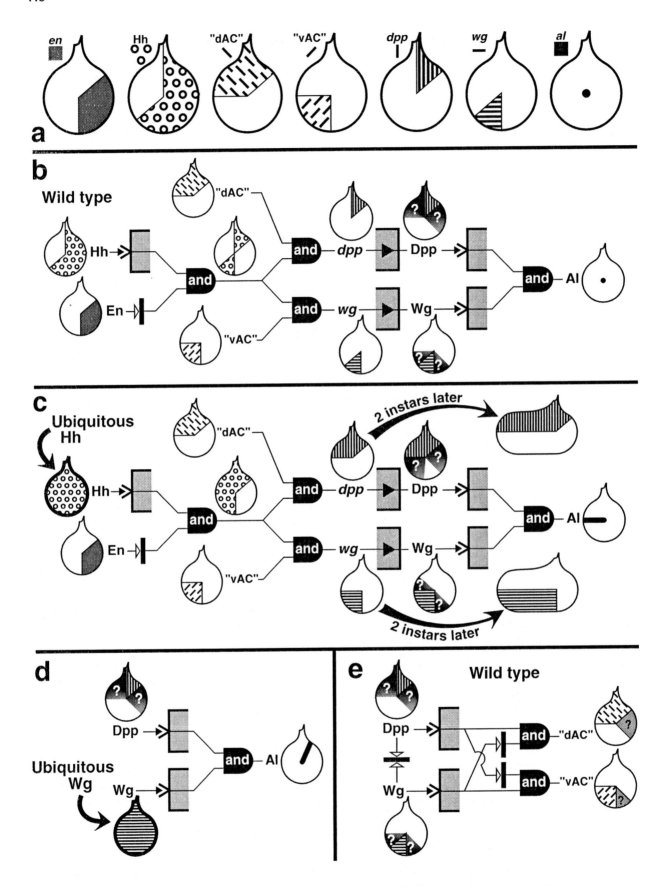

Given the "{Dpp AND Wg} ➝ *al*" link and Al's expression in the claw region, it is odd that no extra claws arise when the Al spot widens as a result of ubiquitous Hh or Wg [4666]. Conceivably, other factors elicited by "Dpp AND Wg" may be suppressing extra claws (e.g., a central spot of mitotic quiescence [1599, 1811] that expands along with Al [4666]). Equally perplexing is why Al is expressed at the inception of the disc (in a pattern similar to Dll) and during 3rd instar [620, 881, 1304], but not during the intervening 1st and 2nd instars. One last riddle about Al is: why do outgrowths at the dAC/vAC interface (e.g., due to DCO^{null} clones) tend to be distally incomplete [2533], because they should express Al and other distal markers? Indeed, al^{GOF} clones fail to induce leg duplication at a significant frequency [620]. Attempts to figure out where *al* acts within the distalization circuitry were stymied until recently by the lack of a null allele [620, 4682]. In 1998, it was shown that *al* function is not needed for distal leg development [618], so *al* must act downstream. These quibbles about Al are minor compared with the issues about Dpp.

The presence of endogenous *dpp* mRNA in the Wg sector was initially problematic for this model because distalization should occur wherever Dpp and Wg signals are both received [617, 620]. What prevents extra tips from growing out of the V region? Evidently, Dpp's concentration is insufficient to let it interact with Wg to trigger outgrowths. When the amount of Dpp is increased (via *dpp-Gal4:UAS-dpp*), extra tips indeed emerge [2954]. When the Dpp level is raised still farther (by using other strains or temperatures), the V sector virtually disappears due to suppression of Wg. These results indicate that the threshold for Dpp-Wg antagonism ("T_a") is higher than the threshold for Dpp-Wg cooperation in distalization ("T_d") [2954]. Another threshold is evident in the prospective claw area ("T_c"): slight reductions in Dpp signaling prevent claws from forming (but legs are otherwise normal) [4033], and a tongue of Wg extends into the dorsal endknob (where claws arise) as Dpp expression retreats [2954]. At the highest level of Dpp expression achieved by *dpp-Gal4:UAS-dpp*, a final threshold ("T_t") is seen: the only vestige of the Wg wedge that still

---

FIGURE 5.9. Logic of the Hh-Dpp-Wg circuitry that governs leg patterning (cf. Fig. 5.4c).

a. Regions (cf. Figs. 5.1c, 5.8, 5.11) where key genes are transcribed (*en, engrailed*; *dpp, decapentaplegic*; *wg, wingless*; *al, aristaless*), where protein is detected (Hh, Hedgehog), or where discrete types of competence prevail ("dAC" vs. "vAC" are dorsal and ventral parts of the anterior compartment). Shading or hatching symbol is shown under each name.

b. Core circuit (cf. Fig. 2.7 for symbol key), whose logic is as follows. Like *en*, *hh* is transcribed in the posterior compartment (PC), but unlike En, Hh protein diffuses some distance into the AC. PC cells are "deaf" to Hh signal because they express En (cf. Fig. 6.4) [4229], but AC cells within Hh's range can respond. Hh's diffusion range in the AC is depicted as two opposing sectors (after the first "AND"). Cells in the dAC and vAC areas respond differently to Hh: dAC cells transcribe *dpp*, while vAC cells transcribe *wg*. Dpp and Wg (like Hh) are diffusible signals, but their ranges (graded shading) and activity levels are uncertain. Cells at the center (where *dpp* and *wg* sectors touch) make Al. From this diagram, it would seem that Dpp and Wg need Hh continually, but in fact, turning *hh* OFF during the last day of larval life causes no notable defects aside from missing claws [1472].

c. Evidence for the "Dpp + Wg = Al" circuit. Ubiquitous expression of Hh during mid-late 3rd instar expands *dpp* and *wg* transcription into the entire dAC or vAC [231, 2466] (assessed by *lacZ* reporters). As a result, cells bordering the dAC/vAC boundary are exposed to the same high-Dpp, high-Wg conditions as cells at the center. Hence, the Al spot extends into a stripe along this line. When *hh* is turned ON (transiently) during 1st instar, discs become grossly misshapen, and Dpp antibody (above) or a *wg* reporter-gene (below) reveals that Dpp and Wg continue to be expressed, hence suggesting that each acts as a mitogen within its own (dAC or vAC) territory. The PC remains normal in size (despite a ~3-fold larger AC), while the Al stripe elongates to span the enlarged AC (not shown).

d. Further evidence for the "Dpp + Wg = Al" circuit. Ubiquitous expression of Wg causes the Al spot to lengthen into a stripe along the Dpp sector.

e. Mutual antagonism between Dpp and Wg. The act of synthesizing Dpp apparently precludes synthesis of Wg within the same cell, and vice versa. AC cells appear to adopt different states of competence based on whether they receive a Dpp or a Wg signal: cells that receive Dpp (dAC plus "?" sector) remain competent to make Dpp in response to Hh, while cells that receive Wg (vAC plus "?" sector) remain competent to make Wg in response to Hh. This "seesaw" can be forced the wrong way by (1) blocking Dpp [1811, 1812] or its transduction [753, 3329, 3932, 4277] in D cells, (2) blocking Wg [177, 880, 1673, 1811, 2110] or its transduction [519, 2262, 3317] in V cells, (3) forcing D cells to make Wg [487, 2082, 4277], or (4) forcing V cells to make Dpp [487, 2954, 4277].

Data for panel **a** comes from refs. in Figs. 5.8 and 5.11. Sources for other panels are **b** [231, 620], **c** [231, 617, 2466], **d** [620], and **e** [1811].

N.B.: When Wg is ubiquitously expressed (**d**) during 1st instar, discs also deform (not shown) but do not grow as much as with excess Hh (**c**) [617]. Distal-less behaves like Al (**c**) but in a broader area [1037]. Ventral sector of *dpp* transcription is omitted (see text), as are coxal and tibial regions of Al expression [620].

expresses Wg is the tip (just ventral to the claw spot) [2954, 4277]. In summary, the following events occur in sequence as the amount of Dpp rises from a slightly subnormal level: (1) Dpp displaces Wg from the claw area, (2) an extra distal outgrowth arises in the V region, (3) Wg totally vanishes except at the tip of the Wg sector (i.e., $T_c < T_d < T_a < T_t$).

But Dpp seems more crucial than Wg as a growth factor

The experiments listed above prove that Dpp and Wg are sufficient for inducing extra distal tips, but they do not address whether both are necessary. The ability of wg^{LOF} legs (branched or unbranched) to fully distalize [177, 880, 1812, 3317] raises the question of whether Wg is needed. Such legs can be D/D symmetric from the sternopleura to the claws with only a slight change in leg length [177, 519, 1812, 3317, 4278], while V/V dpp^{LOF} legs (which express wg on both sides) are severely stunted [2082, 4277]. Likewise, when wg is transiently silenced before 1st instar (12–24 h AEL), 60% of the D/D legs distalize fully (vs. 80% when wg is stifled for 12 h in early 3rd instar) [880]. These effects are not just due to leakiness of the LOF alleles used because wg^{null} clones (induced in 1st instar) can also yield distally complete (albeit only partly D/D) legs [1811]. Other clues also point to an inequality in the roles played by Dpp and Wg:

1. Unlike (distally complete) wg^{LOF} legs, dpp^{LOF} legs tend to lose distal elements as a function of the strength of the LOF allele [529, 1812, 4033]. In fact, dpp was once thought to primarily govern the proximal-distal axis (instead of the D-V axis) for this reason [33, 834, 1427, 4675, 4682].

2. Janus phenotypes are manifest by both wg^{LOF} (D/D) and dpp^{LOF} (V/V) legs, but dpp^{LOF} legs can also exhibit a "V/-" phenotype where the deficiency is not associated with a duplication [1812]. In such cases, the D side of the tarsus is missing but is not replaced by V-type tissue (Fig. 5.5d). No analogous "D/-" defects are seen in wg^{LOF} legs. Why? Unlike Wg, Dpp may not only be serving as a morphogen, but also as a trophic survival factor [4080]. Among the facts that support this conjecture are (1) dpp^{LOF} discs display appreciable apoptosis [529, 2739, 2849], whereas wg^{LOF} discs manifest relatively little [2015, 2929, 2932, 4683]; (2) the tissue loss and poor growth of dpp^{LOF} wing discs are rescuable by co-culturing with wild-type discs, implying a diffusible anti-apoptosis factor [529]; and (3) apoptosis in wing and leg discs is triggered (via the JNK pathway [971]) when genes in the Dpp pathway are suppressed [12],

whereas milder wg^{LOF} effects occur via omb. Dpp's link to apoptosis may be at its receptor because Tkv binds proteins that inhibit apoptosis [3170].

3. Unlike wg^{LOF} legs, which commonly branch [177, 1812, 3317, 4278], dpp^{LOF} legs bifurcate only rarely [529] and when they do, the side branch dwindles distally (L. Held, unpub. obs. of 60 dpp^{d2} legs). Evidently, branches easily grow when the seesaw circuit causes dpp to be derepressed (e.g., in V clones that are dsh^{null} [1833, 2059, 2262, 4278], DCO^{null} wg^{null} [2058, 2533], or $dAxin^{GOF}$ [1698]), but dorsal outgrowths are pitifully small by comparison when wg is derepressed (e.g., in D clones that are $punt^{LOF}$ [3329], tkv^{LOF} [2059, 3329, 4277], or sgg^{null} [1039, 1833]). This disparity may partly be due to Dpp's greater ability to snuff out Wg nearby than vice versa [1833, 2059, 4277].

4. A similar imbalance is seen when comparing dpp-$Gal4$:UAS-dpp legs to dpp-$Gal4$:UAS-wg legs: the former (excess Dpp) often have outgrowths from the V surface [2954], whereas the latter (excess Wg) rarely display outgrowths from either side [2082, 4666]. Why then can wg^{GOF} clones induce outgrowths on the D side [4159]? Perhaps because the ones that arise near, but not in, the Dpp sector spark a "Dpp + Wg" reaction without downregulating dpp. When wg^{GOF} (or sgg^{null}) clones reside laterally (far from the Dpp sector), they evoke only tiny outgrowths [1039, 1833], indicating that Wg alone is a poor mitogen.

5. Distal outgrowths can be induced ventrally by doubly mutant DCO^{null} wg^{null} clones (which express Dpp), but no outgrowths (dorsal or elsewhere) are seen for DCO^{null} dpp^{null} clones (which express Wg) [2058, 2533], implying a greater role for Dpp than Wg in distal outgrowth.

6. Frequencies and sizes of somatic clones are reduced when the clones are prevented from transducing Dpp (by $punt^{LOF}$ [3329] or tkv^{null} [3329, 4277]) but not when they are prevented from transducing Wg (by dsh^{null} [2059]). This rule is broken by arm (arm^{LOF} clones die if they arise ventrally or distally [3317]), but Arm plays a vital role (as a component of junctions) aside from signaling.

7. Discs whose whole A/P border is dominated by Dpp (dpp-$Gal4$:UAS-dpp) widen perpendicularly to the border [487, 2954, 4277] (by expanding the A compartment?), whereas discs whose A/P border is ruled by Wg (dpp-$Gal4$:UAS-wg) lengthen along the A/P line instead [487, 2082, 4277].

Dpp cannot be the sole agent in distalization because tips would arise all along the D midline. Thus, if Wg is superfluous [4666], then another agent "X" must overlap Dpp centrally (see the argument above for why wg^{GOF} does not fully ventralize cells) and the rule must be "{Dpp AND X} ➡ distalize!". (X might override the Dpp-Wg antagonism to enforce the T_t threshold [2954].) As for why ectopic Wg triggers extra tips, the leg disc circuitry must have a "Wg ➡ X" link, but other factors may redundantly keep X active centrally [2533].

Heating t.s. wg^{LOF} mutants during 2nd instar abolishes Dll centrally and causes truncations [1037] like what happens when discs are deprived of Dpp or Hh. However, shift studies with t.s. alleles show that neither Wg nor Dpp is needed for distalization after early 3rd instar (although they are still required for the D-V axis) [883].

The conjecture that Dpp is more important than Wg for distalization would help explain two enigmas from the Cellular Epoch: the "Dorsal Remnant" and "Triplications" Mysteries (see the start of this chapter). When cells are randomly killed in a young disc, whole sectors can vanish. Judging from the structures made by surviving cells, virtually all parts of a disc except its Dpp sector are dispensable (Fig. 5.2b) [3449]. The reason may be that a disc's growth factors come mainly from there. Identical D-remnant legs can be produced by overexpressing Dpp in the Wg sector (which then fails to develop due to "Dpp ➡ wg") [2954]. The Triplications Mystery [1495] also implies an asymmetry of mitogenic function along the D-V axis. Perhaps, D patches of apoptosis that overlap the Dpp sector reduce the Dpp level so much that D outgrowths starve for mitogens and hence converge, whereas V patches of apoptosis have no such effect and thus allow V outgrowths to diverge. The ability to elicit convergent vs. divergent outgrowths by nicking D vs. V sides of cockroach legs [384, 547, 2804] reveals a similar asymmetry that is consistent with this argument.

These etiologic scenarios are seemingly contradicted by the fact that dpp upregulates, rather than downregulates, when discs undergo widespread cell death [489, 2856], and indeed, apoptosis is induced by the same t.s. allele of *suppressor of forked* that causes triplications [1496, 1502]. However, too much Dpp has been shown to reinforce apoptosis as severely as too little Dpp (in wing discs) [12], so this objection may be unwarranted. One fact at least is certain: the dpp-ON and wg-ON sectors respond differently to trauma. The dpp-ON sector broadens, whereas the wg-ON sector is unaffected [2856]. This asymmetry probably holds the key to both mysteries, although important links in the connecting circuitry remain obscure.

If Dpp is the major mitogen in a leg disc, then it should be able to reach every cell. At one time it was thought that cells on the V side of the disc might receive sustenance from the low level of Dpp in the Wg sector [2954], but ventral dpp expression can be eliminated (via dpp^{null} cell clones) without any detectable effect on growth or patterning [1811]. If ventral cells must depend on dorsally produced Dpp, then the question of Dpp's diffusion range becomes pivotal (see below).

The mitogenic capacity of at least one element in the Hh-Dpp-Wg circuit is made obvious by a grotesque deformity seen when hh is transiently turned ON ubiquitously in 1st-instar discs (Fig. 5.9c) [617]. Under these conditions, dpp and wg are transcribed throughout dAC and vAC, respectively, but oddly (given the Hh-Dpp-Wg Model) there seem to be no outgrowths at the dAC/vAC boundary. The AC of these discs broadens tremendously, becoming ≥3 times wider than the P compartment by the 3rd instar. Because dAC and vAC expand equally, it would seem that Dpp and Wg must both be mitogens, but each might be suffusing the whole disc. When Wg alone is overexpressed via dpp-$Gal4$:UAS-wg, some excess growth occurs [4666], but it is meager by comparison. Overexpressing Dpp via dpp-$Gal4$:UAS-dpp is also relatively ineffectual in provoking extra growth in its own domain [2954], implying that a Dpp threshold for hyperplasia is crossed in one (hh^{GOF}) but not the other (dpp^{GOF}) experiment.

Another peculiarity of the Hh-Dpp-Wg circuitry is seen with DCO^{null} clones, which commonly cause distal outgrowths but are prevented from doing so when they are also null for dpp or wg [2533, 3238]. This result is heretical because the (Wg-expressing) DCO^{null} dpp^{null} clones should have induced outgrowths whenever they were near the dpp-ON sector, and the (Dpp-expressing) DCO^{null} wg^{null} clones should have induced outgrowths whenever they were near the wg-ON sector.

The A/P boundary can migrate when its "jailors" are "asleep"

When Wg is forced into the Dpp domain (via dpp-$Gal4$:UAS-wg [4277] or ptc-$Gal4$:UAS-wg^{ts} [2082]), it causes a mysterious shift in the boundary that separates en-ON from en-OFF cells: this boundary migrates anteriorly, but only in the dorsal half of the disc where dpp is downregulated (Fig. 5.10d):

en is normally expressed in cells of the posterior compartment of each disc. In 15°C ptc-$Gal4$:UAS-wg^{ts} discs [the permissive temperature for wg^{ts}], en expression has broadened into anterior cells.... In leg discs the en expansion is primarily restricted to dorsal cells and occurs in approximately 80%

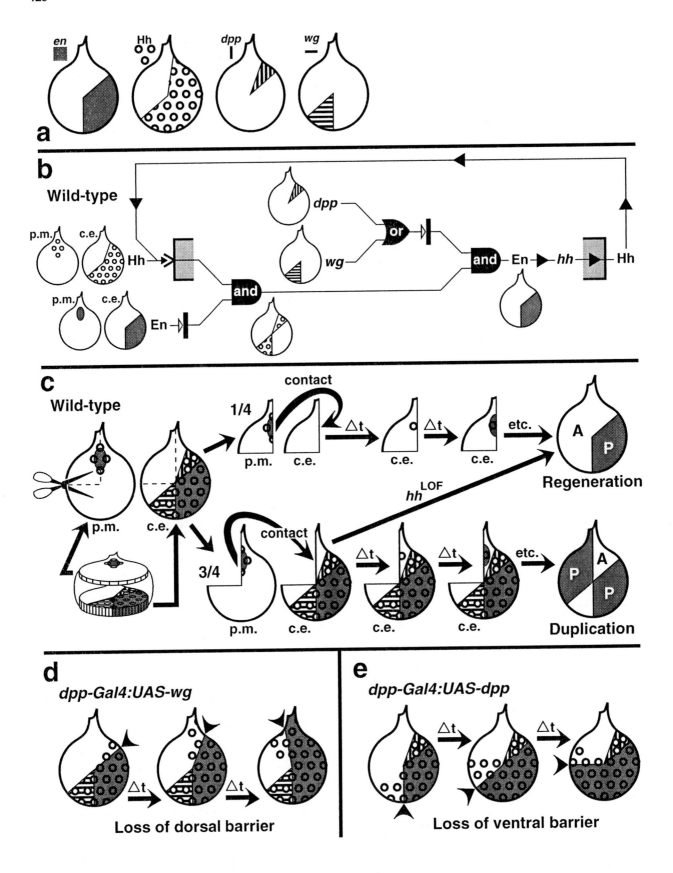

($n = 60$) of leg discs. The anterior *en* expression is dependent upon ectopic Wg.... However, anterior *en*-expressing cells do not take on posterior fates in the adult.... The mechanism behind the *en* expansion is not clear. [2082]

A similar phenomenon was independently discovered when Dpp is forced into the Wg domain (via *dpp-Gal4:UAS-dpp*) [2954, 4277]: the *en* ON/OFF boundary shifts anteriorly, but only in the ventral half of the disc where *wg* is downregulated (Fig. 5.10e):

In addition to the reduction in ventral-anterior *wg* expression, the leg discs with increased *dpp* expression exhibited an altered pattern of *en* expression. Cells expressing *en* were present across the region that would normally make up the ventral-anterior region of the leg disc. The mechanism for this pattern change is not known. We propose that the increased expression of dpp along the ventral A/P boundary reduces wg expression and that this destabilizes the A/P boundary.

Two mechanisms seem plausible. In one, the destabilized A/P boundary migrates across the anterior-ventral region of the disc and as the boundary passes, cells begin to express *en*. Ectopic *hh* overexpression in the A compartment can induce *en* expression [1647]. We postulate that expression of *wg* in the anterior-ventral [area] normally prevents cells on the A side of the boundary from inducing *en* expression in response to *hh* from the P compartment. However, when increased *dpp* reduces the level of *wg* expression in the anterior-ventral [area], the cells on the A side of the boundary might respond

to the *hh* coming from the P compartment by activating *en* expression. This in turn would permit these cells to induce *hh* expression, which could signal *en* expression to occur in the next most adjacent anterior cells. This would result in a processive movement of the compartment boundary across the anterior-ventral region that is normally blocked by *wg*. This proposed migration is reminiscent of the processive migration of the morphogenetic furrow across the eye imaginal disc, a process which is also blocked by *wg* expression [4390]. [2954]

As the above quote asserts, the correlation of boundary migration with reduced *dpp* or *wg* transcription may be causal. To wit, cells that express either *dpp* or *wg* may be unable to turn ON *en* in response to Hh. Because Hh activates *dpp* or *wg* in each A cell that it reaches (as per the Hh-Dpp-Wg Model), it behaves like a minotaur that builds its own cage. If the cage dissolves due to insufficient Dpp or Wg, then the *en*-ON state can invade the A compartment. The wavefront ratchets itself stepwise by a Hh-En-Hh feedback loop (Fig. 5.10b; cf. the eye [571, 1784, 4169]):

Step 1. Hh diffuses ahead of the *hh*-ON (= *en*-ON) territory and turns ON *en*. It is this "Hh → *en*" link that appears to be disabled by Dpp or Wg. This circuit element is dormant until late 3rd instar [350, 364, 3177, 4228] when it

---

FIGURE 5.10. Hh-En-Hh feedback loop and how it solved the Quadrant Regeneration Mystery.

a. Regions where key genes are transcribed (*en*, *engrailed*; *dpp*, *decapentaplegic*; *wg*, *wingless*) or protein (Hh, Hedgehog) is detected. Symbols (shading or hatching) are below names.

b. Hh-En-Hh loop (cf. Fig. 2.7 for icons). The logic begins at left with 4 maps of a 1st-leg disc: upper vs. lower are Hh vs. En; left vs. right are peripodial membrane (p.m.) vs. columnar epithelium (c.e.). In the p.m. of 1st-leg discs, *en* and *hh* are transcribed in a dorsal spot [1472]. In the c.e. (of all leg discs), *en* and *hh* are expressed in the P compartment, and Hh diffuses into two sectors (map below first "and") where *en*-OFF (A-type) cells can "hear" it (P cells are "deaf"). In theory, these cells should respond by turning ON *en* because a "Hh → *en*" link emerges in late 3rd instar (cf. Fig. 6.4) [1472, 4136], and indeed a sliver of cells within these sectors does so (not shown) [350]. However, remaining cells in the border zone cannot because (1) Hh causes them to express Dpp or Wg (cf. Fig. 5.9b), and (2) Dpp or Wg blocks this "Hh → *en*" step. Thus, the En map (below second "and") does not change perceptibly. In this steady state (remainder of the loop), En continues to activate *hh*, and Hh continues to diffuse out of the P compartment.

c. As shown in the side-view schematic (lower left), the leg disc is a flat sac (cf. Fig. 5.1b): p.m. cells (upper layer) are squat, whereas c.e. cells (lower layer) are tall. (Elsewhere layers are drawn side by side.) After a cut, both tissues heal together ("contact"), at least briefly. If the cut bisects the *en* spot, then Hh diffuses (via contact) from p.m. to c.e., where it engenders some *en*-ON cells. (Remaining drawings depict c.e. only.) Such interactions should occur in both 1/4 (upper series) and 3/4 (lower series) pieces. In the former, the *en*-ON patch leads to regeneration (by restoring A-P duality), whereas in the latter it leads to duplication (by instigating a second P compartment). Without Hh activity (*hh*^LOF arrow), the 3/4 piece fails to acquire an extra P domain, so it regenerates instead, as do 3/4 pieces of wild-type 2nd- or 3rd-leg discs, which lack an *en* spot [1472]. "Δt" denotes time periods on the order of hours (after healing), whereas "etc." connotes an extended growth period of several days.

d, e. Invasion of the A compartment by the En wavefront (arrowhead) when the Hh-En-Hh loop is unleashed. In these cases, the Dpp-Wg antagonism (cf. Fig. 5.9e) is used to downregulate *dpp* dorsally (**d**) or *wg* ventrally (**e**).

Domains in **a** are from sources listed in Fig. 5.8. The *dpp* sector is rendered narrower (cf. Fig. 5.9) so that the ectopic *en* patch (**c**) can fit into a *dpp*-OFF part of the 3/4 piece, because otherwise the circuit will not work. (Distance from vertical cut line to *dpp* sector is not known exactly.) Circuitry in panel **b** is deduced from data in [1472, 2082, 2954, 4136]; **c** is adapted from figures in [1472]; and **d** and **e** are based on pictures in [2082, 2954, 4277], whence wavefront stages are inferred.

prods the *en* ON/OFF line to creep a few cells ahead of the A/P compartment boundary (Fig. 6.7d) [78, 642, 2832, 3818, 4136].

Step 2. As a result of Step 1, the *en*-ON edge has stepped ahead of the *hh*-ON edge. The newly created *en*-ON "border" cells should then activate *hh* via an "*en* ➡ *hh*" link that was used in the embryo [3747, 4227]. This link should not work in *ci*-ON cells [2992] because *hh* is more sensitive to repression by Ci-75 than to activation by Ci-155 [2980], so removal of Dpp or Wg must somehow derepress it. (See [703] for how a promoter can be wired for such inputs.) At this stage, the *hh*-ON edge catches up with the *en*-ON edge.

Step 3. As soon as the border cells (with *en* and *hh* both ON) begin making Hh, the new Hh will diffuse and rekindle the loop at Step 1. (A direct "Hh ➡ *hh*" shunt is precluded because *hh* is insensitive to Ci-155 [2980].)

This loop can be sparked by ectopic *hh*^GOF virtually anywhere in A compartments of wing [998, 1647, 3739, 4136] or leg discs [1472] (cf. tergites [2436]), but again the *en*-ON state (due to "Hh ➡ *en*") does not spread far beyond the *hh*^GOF area (clone or region) because of the "**Minotaur Scenario.**" That is, the *hh*^GOF tissue creates a cage around itself (Hh ➡ {Dpp OR Wg}) that breaks the feedback loop ({Dpp OR Wg} ⊣ {Hh ➡ *en*}). Nevertheless, any *dpp* expression that is thereby evoked could sustain the area's growth and thus amplify the response, as if the minotaur henceforth grows by feeding itself rather than by escaping (cf. Fig. 6.6).

In theory, Dpp and Wg could be interfering with a different step (i.e., {Dpp OR Wg} ⊣ {*en* ➡ *hh*}). However, this possibility seems less likely because the *dpp*-ON zone remains an effective barrier even when *polyhomeotic*'s blockage of "*en* ➡ *hh*" at the A/P border is alleviated [2728]. Oddly, in that case, the *en*-OFF edge migrates posteriorly.

Regeneration is due to a Hh spot in the peripodial membrane

The observations quoted above were made in 1996. In 1999, Matt Gibson and Gerold Schubiger in Seattle invoked a similar scenario [1472] to explain the same Quadrant Regeneration Mystery that, poetically, Schubiger first articulated in 1971 [3808]. As mentioned at the outset of this chapter, that mystery had baffled theorists for decades [354, 1807]. How could one quarter of a disc regen-

erate the other three quarters? What is so special about the upper medial (UM) quadrant?

To accommodate this odd behavior, the PC Model had imagined a warped coordinate system where more than half the leg's angular values are crowded into the UM area. However, this *ad hoc* assumption undermined the model's conjecture that a disc stops growing when the density of coordinates attains a preset limit [546, 1301, 1303]. How could the UM quadrant end up with a higher density? The Boundary Model was even more bedeviled because the tip morphogen that it demanded for outgrowth is only made when several compartments convene. How could the UM quadrant, which lies entirely in the A compartment [4076], regenerate distally (Fig. 5.3j) [2804, 2808]? Equally puzzling was why certain fragments that are composed of both A and P cells should duplicate [354].

The key to solving the mystery was Gibson and Schubiger's discovery of a Hh spot in the peripodial membrane of the 1st-leg disc [1472]. Like the other discs, each leg disc is a flat hollow sac with two surfaces (Fig. 5.1b). The thick columnar layer is separated from the thin "peripodial" layer by a fluid-filled lumen, and columnar cells are packed ≥10x more densely than squamous peripodial cells [2862, 3426]. The peripodial side has usually been ignored in model building [1008] because it makes so few cuticular structures [526, 2862]. Nevertheless, its cells can interact with the other layer when wound edges heal together after surgery [3564, 3565], and this interaction turns out to be pivotal for leg disc regeneration.

One of the two cuts that liberates the "1/4" piece passes through the Hh spot (Fig. 5.10c). When this peripodial edge adheres to the columnar edge, some of the Hh evidently diffuses into the columnar epithelium because a few of its cells begin expressing *en* (due to the "Hh ➡ *en*" link) and later *hh* (due to the "*en* ➡ *hh*" link). The activation of *en* can be blocked by using a t.s. *hh*^LOF allele to turn *hh* OFF during this period (thus disabling the "Hh ➡ *en*" link).

Spreading of the *en*-ON state should cease due to the Minotaur Scenario (Hh ➡ Dpp ⊣ {Hh ➡ *en*}), but the few *en*-ON cells that are thus created are evidently enough to "seed" the columnar layer with a new P compartment because the 1/4 piece regenerates circumferentially and distally. The same sort of interaction occurs in the "3/4" piece, but its extra P compartment leads to duplication instead (in P/A/P symmetry), thus obeying the Reciprocity Rule that one fragment regenerates while its complement duplicates (cf. Ch. 4). In both cases, subsequent growth at the wound site is probably sustained by the new *dpp*-ON zone (Hh ➡ *dpp*), regardless of whether the peripodial and columnar surfaces later detach from one another (as they often do [3564, 3565]). Other facts support this hypothetical sequence of events [1472]:

1. Whereas the 1st-leg disc has an appreciable Hh spot (20–30 cells), the 2nd-leg disc does not (0–5 cells). Consistent with the above argument, the UM quadrant of the 2nd-leg disc behaves like that of a hh^{LOF} 1st-leg disc: it typically fails to regenerate. First legs are peculiar insofar as the left and right discs are conjoined [377, 3426] (cf. eye [1777] and genital discs [540]), and it is conceivable that the Hh spot is a vestige (and/or mediator) of that fusion event [1472].

2. In contrast to the 3/4 piece of the 1st-leg disc (which duplicates), the 3/4 piece of the 2nd-leg disc regenerates. Its Dpp and Wg signaling centers evidently suffice for this "Type 2" kind of regeneration (where A and P compartments are both present at the outset) vs. the "Type 1" regeneration exhibited by 1/4 pieces of the 1st-leg disc (where only the A compartment is initially present). The 3/4 fragment of the 1st-leg disc can likewise be coaxed to regenerate (instead of duplicate) by using a t.s. hh^{LOF} allele to silence all hh activity during the postsurgical growth period. The implication is that Hh itself is not needed for Type 2 regeneration (cf. fragments that can duplicate without an A/P boundary [354]).

3. The idea that a few en-ON cells can spark a discwide duplication may seem farfetched, but clonal analyses show that the new halves of duplicated legs can indeed arise from as few as 5–10 cells [1472, 1503]. Moreover, tiny hh^{GOF} clones have a similar David-vs.-Goliath power: they can cause huge wing duplications (cf. Fig. 6.6). Even colossal changes such as homeosis and transdetermination may stem from small errors in diffusible signals (e.g., Hh) because they are nonclonal in origin [1405, 3445] (cf. Ch. 8).

However, not all the details of this process have been worked out. For example, both Dpp and Wg must supposedly be present for distal outgrowth to occur, and an incursion of Hh into the dAC (as described above) should only induce Dpp, not Wg. Nevertheless, Wg is detected in precisely this wound area [1472]. What circuit activates Wg in the dAC? Another question concerns the endogenous Hh spot itself. If the same "Hh ➔ en" link that is awakened in the columnar epithelium is also activated in the peripodial layer in late 3rd instar, then what prevents this spot from expanding (via the Hh-En-Hh loop)? Is there a "cage" comparable to the *dpp* and *wg* sectors?

The implications of this "**Regeneration Epiphany**" of 1999 are only beginning to emerge [2698]. One minor lesson is the quirkiness of the three legs. Although they share the same basic circuitry, each has special traits. For example, most studies of *in situ* leg duplication have used an allele ("*ts726*") of *suppressor of forked* that induces apoptosis at high temperature [1501]. When homozygotes are heated during the sensitive period, the frequency of duplications is higher in 2nd legs (ratio for 1st:2nd:3rd legs = 1:32:2) [3442], but the original 1915 study of branched legs (in Morgan's lab) used an allele of *reduplicated* [2561] that causes more branching in 1st legs [1875].

Another, more general lesson is that regeneration can follow paths apart from normal development [354] (despite evidence for harmony [548, 1183, 2998, 2999]), so regeneration-based theories must be reexamined. For instance, it is now clear how some fragments can duplicate morphallactically [19, 20, 695] – viz., extant tissue can be repatterned by newly synthesized morphogens. For the same reason, cells in the wound blastema can switch their compartmental affiliation [5, 354, 2213, 4223].

The Polar Coordinate Model died in 1999

Because the events described above for leg disc fragments require contact between opposite surfaces, the results affirm the PC Model's contention that growth is stimulated by such contact. However, in no sense do cells "compute" a "shortest route" to "decide" whether to regenerate vs. duplicate [1472], nor is the growth necessarily "intercalary" *sensu stricto*. Hence, the PC Model, despite its yeoman's service as a heuristic aid, seems moribund in light of these findings. In fairness, a final burial should await comparable studies with wing disc fragments, although that corpus of data may also yield to a simple molecular explanation. The wing disc's peripodial membrane has *en*-ON and -OFF areas like the columnar side [2072], but its *en*-ON area is larger, so the zone where Hh ➔ {*dpp* AND *ptc*} is shifted anteriorly [2867, 3831, 4404]. Interestingly, the wing disc's peak of regenerative potency lies squarely in the gap between the A/P lines of the two layers (Fig. 6.1e), and each quadrant (and its 3/4 partner) has one edge in this gap. Although these geometrical clues alone do not solve the puzzle, the problem now seems tractable, and the probes that deciphered leg disc regeneration may soon accomplish the same for the wing disc.

Discarding the PC Model also implies dismissing the Angular Values Conjecture (see the start of this chapter). It had tried to reconcile the PC Model with the wedge-shaped expression domains of segment-polarity genes (*en*, *hh*, *ptc*, *ci*, and *wg*) by postulating that the latter encode circumferential coordinates [531, 880, 4675]. The ability of some of those genes to cause outgrowths when expressed on the "wrong" side of the disc [231, 1806, 4159, 4848] agreed with the notion that xenotopic confrontations should elicit intercalary growth [354, 532, 1304, 1807, 3862]. However, such growth is apparently not stimulated by confrontations per se because cells can be duped into thinking that they are in the wrong place without causing them to intercalate. Neither the Angular Values Conjecture nor the PC Model satisfactorily explains these findings. A fitting eulogy for both ideas is provided by the following quote from a paper by Serrano and O'Farrell entitled, "Limb morphogenesis: Connections between patterning and growth." (See [569, 2455, 3074] for data on which their argument is based; italics are author's.)

Clones expressing a constitutively active form of a Dpp receptor demonstrate that the growth response is an autonomous feature of Dpp signaling. Although cells within these clones over-proliferate in response to the activation of the Dpp signaling pathway, no proliferation – that is, no intercalary growth – is induced in the surrounding cells, despite the incongruous juxtapositioning of cell fates at the border of the clone. The unique feature of the transformed cells within these clones is that they have a fate normally associated with a high concentration of Dpp, but lack high Dpp levels. *The failure to activate intercalary growth in this circumstance suggests that it is not the juxtapositioning of incongruous cell fates, but rather the juxtapositioning of cells with discordant levels of Dpp that induces this proliferation.* Similar data argue that Wg also has a fairly direct role in stimulating intercalary growth. Thus, the discordance that induces intercalary growth appears to be a discontinuity in Dpp and/or Wg morphogen concentrations.... The growth factors will diffuse across the junction to stimulate the growth of the cells with low inhibitor levels. [3862]

How Hh, Dpp, and Wg move is not known, nor is their range

Hh, Dpp, and Wg signals could theoretically be integrated in many ways to specify fates. For example, in the "**Dpp-Wg Growth Potential Model**" [3862] cells are supposed to combine Dpp and Wg inputs to compute growth rates. Such a mechanism could be used not only for growth control, but also for patterning.

This idea is appealing for the leg disc where Dpp and Wg gradients have opposing slopes but can be forced to coincide, in which case some bizarre phenotypes arise [1812, 2954]. Indeed, both ligands have effects beyond the $L_{2\&7}$ boundary that supposedly separates their domains of influence (Fig. 5.5a): (1) tkv^{null} clones (whose cells think they are receiving no Dpp) on the V

side of the tarsus produce excess V-type (1/8) bristles and evoke autonomous outgrowths [4277], and (2) dsh^{null} clones (whose cells think they are receiving no Wg) autonomously exhibit excess bristles as far dorsally as rows 3 and 6 [1833]. If, as these data suggest, the Dpp and Wg gradients overlap and are arc shaped, then leg cells could assess their azimuth by computing a ratio of the two inputs. Such "**Double-Gradient Models**" have historically been appealing because (1) the opposing gradients afford a greater level of precision (especially at low morphogen levels) than single-gradient schemes [91, 1142, 4725] and (2) they let cells figure out when the field reaches final size so they can stop growing [2448, 4724].

Such models are hard to evaluate without knowing how far Hh, Dpp, and Wg actually move within the epithelium [641, 783, 1554, 3305, 4275]. Unfortunately, we do not even know *how* they move [1649, 2448, 3507, 3919, 4275, 4276]. Three chief modes of transit are theoretically possible, not to mention potential routes through the extracellular matrix [277, 2466, 3774, 4275, 4375]:

1. *Passive diffusion.* Morphogens were originally thought to diffuse randomly through extracellular space [902, 2199, 3087, 3989], and some may in fact do so [674, 2780]. However, free diffusion is unlikely for Hh_N [1972, 2466, 3432, 4275, 4283], Dpp_C [2004, 3244], or Wg [598, 3561] because they preferentially bind cell-surface proteoglycans and/or lipids [299, 3345, 3854] (cf. glycosylation of Notch, which potentiates its binding of Delta [510]). The epithelial folds in discs should force freely diffusing molecules to skip from crest to crest, whereas the gradients of Hh, Dpp, and Wg deduced genetically imply that the ligands are tracking the surface contours [1910, 4138]. Also, if Hh were diffusing freely, then it is hard to see why changes in cell shape should affect its movement [287].

2. *Active transport via transcytosis.* In some systems, proteins can move within the plane of an epithelium, rather than through the extracellular space [1154, 2843]. How they enter and exit cells varies: this "transcytosis" may involve classical endo- or exocytosis or both [1604, 3364, 4562], plus intracellular vesicular transport [811, 4379, 4694] along the same route, perhaps, as recycled receptors [654, 1450] so as to avoid degradation [702, 3217]. Dpp moves via endocytosis in wing discs (dependent on clathrin [1546] and dynamin [1169]), and so may Hh because Ptc (its receptor) resides mainly in lateral membranes and is actively internalized [644, 731, 1022, 1554, 3506]. Hh can traverse ptc^{null} tissue [277, 754, 755, 4136] but may use an unnatural route to do so. Hh_N's

cholesterol tail suggests a "**Raft Scenario,**" where Hh$_N$ is tethered to "lipid rafts" [1896, 1960, 3595, 3946] that are transferred intact from cell to cell [2297, 4275] in a polarized manner. Alternatively, the sterol-sensing domains in Ptc [255] and Dispatched (a protein that releases Hh$_N$ from *hh*-ON cells) [572] may be used like hands to grip Hh$_N$'s tail long enough to pass it to the next cell [572, 1975] – a "**Tiger-by-the-Tail Scenario.**" Anchoring of some kind must be involved because a "tailless" Hh construct escapes control by Ptc and Dispatched to travel 5 times farther than normal and turn ON at least one target gene (*dpp*) throughout the entire A compartment [572]. For Wg, the case is equivocal [1054]. In the embryonic epidermis, Wg moves directionally via a dynamin-dependent (endocytic) route [2900] that is uncoupled from transduction [273, 1055, 1779]. In wing discs, however, Wg is also detected basolaterally outside cells, and this extracellular Wg (signaling ability unknown [1910]) moves independently of dynamin [4138]. Retention (during growth) of Wg-filled secretory vesicles (in cells whose *wg* used to be ON) may also broaden its range [3363].

3. *Active transport via "cytonemes."* In 1999, the old concept of disc cells as inert "cobblestones" in an epithelial plaza was shattered with the discovery of long filopodia that extend centripetally from outlying cells to the A/P compartment boundary in wing and leg discs [3507]. Might disc cells be communicating like neurons – via axon-like "railways" [4496]? (Cytonemes seem to use microfilaments instead of microtubules.) If so, then morphogens would be taken up at the tips and transported to the cell body [534] (in lipid rafts? [1975, 4275]). This scenario may not be as fanciful as it seems because retrograde transport of morphogens evidently can occur in axons [562, 1933], and the degree of actin polymerization has been shown to affect Hh movement in eye discs [287]. For Dpp, however, this mode of movement is unlikely because Dpp is not detected along cytoneme routes [4265].

A further complication is that ligand movement can be constrained by receptor density [674, 775], and all three ligands can regulate the amount of receptor under certain circumstances.

1. Hh ➜ *ptc*. Hh's direct effects (as assayed genetically in leg and wing discs) extend over a shorter span (~10 cells) than those of Dpp or Wg (~50 cells; cf. Table 6.1), but this limitation is not inherent in the Hh molecule. Hh's range can be doubled by removing its receptor (Ptc) from the cells that it traverses: wherever a *ptc*^{null} A clone abuts the A/P line, the *dpp*-ON stripe is displaced to the anterior side of the clone [277, 754, 755, 4136]. This ability to detach the border zone from the P compartment (cf. Fig. 6.6i) confirms the Hh-Dpp-Wg Model's logic. The reason that Hh bothers to upregulate *ptc* transcription [642, 1841, 1982, 2074] must therefore be, at least in part, to reduce its own diffusion range [277, 754, 755, 2074].

2. Dpp ⊣ *tkv*. Dpp can reduce the expression of its receptor (Tkv) [2457], but the level of Dpp that is needed to do so exceeds the physiological range so greatly that this effect is probably irrelevant to normal development [1674]. The virtual complementarity of Tkv and Dpp profiles must therefore arise indirectly (i.e., not by a "Dpp ⊣ *tkv*" link). Tkv must impede Dpp transit because (1) the slightly higher endogenous level of Tkv in the P compartment is correlated with a steeper response to Dpp [1327], (2) overexpressing Tkv in the dorsal half of the wing pouch shrinks the band where Dpp's target gene *spalt* is expressed [4251], and (3) the "wall" of concentrated Tkv outside the wing pouch's *omb* domain prevents that domain from expanding when the output of Dpp increases (in *dpp-Gal4:UAS-dpp* discs) [1674].

3. Wg ⊣ *Dfz2*. The distribution of Wg and its receptor (Dfz2) are roughly complementary in wing discs because Wg represses *Dfz2* transcription [595]. Strangely, however, the density of Dfz2 receptors has no effect on Wg movement, possibly due to a low Wg-Dfz2 binding affinity. Indeed, other Wg-binding molecules must import Wg into recipient cells because uptake of Wg into vesicles is seen even in receptor-less (*Dfz2*^{null} *fz*^{null}) mutants [2984]. Thus, the "Wg ⊣ *Dfz2*" link does not illuminate how Wg travels. In contrast to how Wg regulates *Dfz2*, Wg ➜ *Dfz3* [3977], although it is unclear to what extent *Dfz3* serves as a Wg receptor because *Dfz2*^{null} *fz*^{null} clones in the wing evince no sign of any residual Wg signaling [734].

Whether Dpp and Wg travel along curved paths is not known

A priori, diffusion of Hh into the A region would be expected to turn ON target genes in zones parallel to the A/P line. The *dpp*-ON "stripe" roughly fits this expectation [487, 3747], but the *wg*-ON domain does not [177, 3317]. Its wedge shape implies that Hh travels farther in the periphery than near the center. As mentioned above, such wedges were once thought to be manifestations

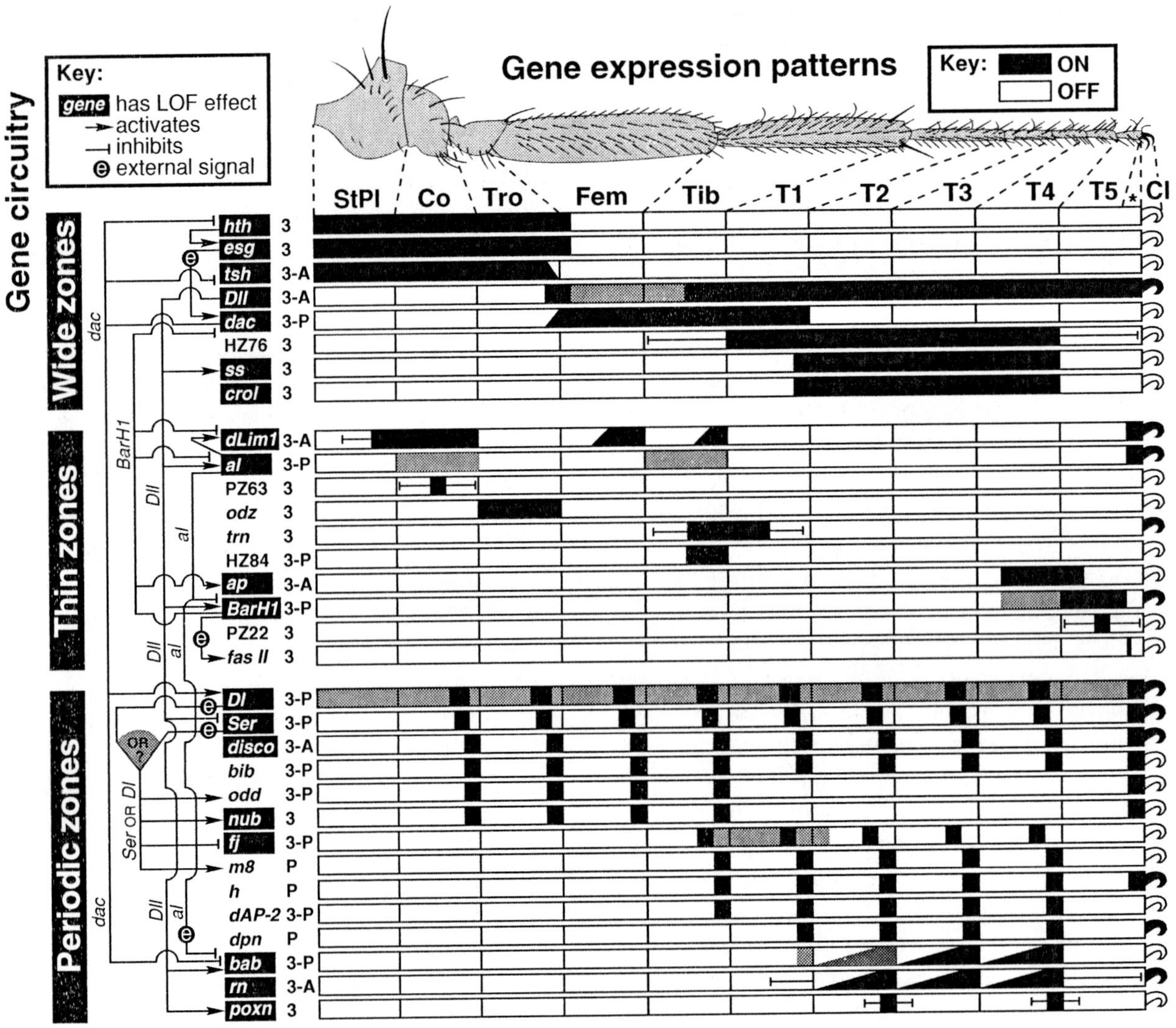
Gene circuitry
Key:
gene has LOF effect
activates
inhibits
external signal
Gene expression patterns
Key: ON
OFF
StPl
Co
Tro
Fem
Tib
T1
T2
T3
T4
T5
Cl
Wide zones
hth 3
esg 3
tsh 3-A
Dll 3-A
dac 3-P
HZ76 3
ss 3
crol 3
Thin zones
dLim1 3-A
al 3-P
PZ63 3
odz 3
trn 3
HZ84 3-P
ap 3-A
BarH1 3-P
PZ22 3
fas II 3
Periodic zones
Dl 3-P
Ser 3-P
disco 3-A
bib 3-P
odd 3-P
nub 3
fj 3-P
m8 P
h P
dAP-2 3-P
dpn P
bab 3-P
rn 3-A
poxn 3
dac
BarH1
Dll
al
OR ?
Ser OR Dl

of angular coordinates [531], but they could be due to the confining of morphogens to circular paths.

According to this "**Arc Scenario**" [358, 1807, 4159], Dpp or Wg molecules stay within the ring (future segment) where they are secreted, although diffusion around the ring would be unimpeded. Different rings could allow different diffusion ranges (e.g., farther at the periphery). Restraints on Wg diffusion have been found at certain folds in the wing disc [595, 3088], but no analogous boundaries are known in the leg disc. An alternative scheme is the "**Cloud Scenario**" [618, 2456], which assumes random diffusion of Dpp and Wg from their sites of synthesis (Fig. 5.4c). One key difference between these proposals is that in the latter case Dpp and Wg would spread across the center to invade the other's territory, whereas in the former no "Dpp AND Wg" overlap would need to exist.

The debate cannot be settled at present because Wg gradients cannot be resolved finely enough to assess curvature [595, 4666], and no suitable antibodies exist for Dpp [1546, 1807, 3087, 4251]. Anti-Dpp antibodies used thus far [617, 2457] recognize mainly N-terminal epitopes of the uncleaved precursor rather than C-terminal epitopes of the diffusible ligand [2954, 3244]. GFP-tagged Dpp circumvents this problem [1169, 4265], but the mobility and turnover of the chimeric molecule may be abnormal. Antibodies to phosphorylated Mad reflect levels of Dpp transduction [4265], but the only published photo for the leg disc is equivocal [4251]. Hence, the argument has been waged mainly with genetic data.

Many genes are known to be expressed in concentric rings within the leg disc (Fig. 5.11) [881]. By itself, this fact could be dismissed as irrelevant to D-V patterning because such domains might reflect thresholds

FIGURE 5.11. Annular domains of gene expression, plotted along the leg's proximal-distal axis (= peripheral to central in the disc). Genes are grouped based on whether expression is broad, narrow, or iterated, and they are seriated within each group based on proximal limits.

Black bars are regions that express the indicated genes (mRNA or reporter) during late 3rd instar (3), early pupal period (P), or adult (A) stage. Degrees of expression are indicated by shades of gray or by graded slopes. All areas are plotted on an adult template, but not all genes are expressed then. The span of stages (e.g., "3-P") indicates when expression has been seen, but it does not mean that expression is confined to that time. When the claw icon is filled in, this means that two dots (or one solid spot) are detected where claws arise [1587, 2287], although they could be apodemes [33, 3971], and the "spot" might actually occupy the pretarsus [4401]. See [851, 2287] for histology and [3523] for overview of circuitry. See App. 7 for discussion.

Genes whose LOF alleles affect anatomy are in white letters on black background. Wiring at left (➔ activation; ⊣ inhibition) shows some genetic interactions (see text or references below at upstream or downstream gene). "External signal" means that the upstream gene emits a signal that affects the downstream gene in nearby cells by diffusion (*esg* ➔ ? ➔ *dac*) or contact (*BarH1* ➔ ? ➔ *fasII*). Horizontal "T" extensions within the template are confidence limits (not inhibitor icons), although their absence does not imply certainty. Charting is more precise in pupal or adult legs than in discs [999]. Sternopleura (Stpl) comes from the leg disc but is not part of the leg per se (Co, coxa; Tro, trochanter; Fem, femur; Tib, tibia; T1-5, tarsal segments; * = "pretarsus" [2287, 4008]; Cl, claws).

Genes ("DO" = details omitted): *al* (*aristaless*; DO: Co expression is dorsal only and Tib expression is ventral only) [618, 881]$^\Delta$, *ap* (*apterous*) [3474]$^\Delta$, *bab* (*bric à brac*) [1516], *BarH1* [2287, 3474, 3762], *bib* (*big brain*) [999], *crol* (*crooked legs*; DO: later spots arise in Tib and Fem) [935], *dac* (*dachshund*; DO: dorsal patch outside Tro ring) [4761]$^\Delta$, *dAP-2* (*Drosophila AP-2*) [2905], *disco* (*disconnected*; DO: squares should actually be bell shaped) [344]$^\Delta$, *Dl* (*Delta*) [344, 3525]$^\Delta$, *dLim1* [3474, 4401], *Dll* (*Distal-less*) [4761]$^\Delta$, *dpn* (*deadpan*) [331], *esg* (*escargot*) [1573]; *fas II* (*fasciclin II*; DO: the ring's perimeter is one cell wide) [2287], *fj* (*four jointed*; DO: stripes arise asynchronously) [4852]$^\Delta$, *h* (*hairy*) [3193]$^\Delta$, *hth* (*homothorax*) [4761]$^\Delta$, *m8* (a.k.a. *E(spl)*) [344], *nub* (*nubbin*) [3525]$^\Delta$, *odd* (*odd skipped*; DO: Fem band may be more proximal) [834] (his Fig. 11d), *odz* (*odd Oz*; DO: 2nd leg has more bands in Co and Stpl) [2497], *poxn* (*paired box-neuro*) [150], *rn* (*rotund*; DO: claws are unstained in some preparations) [881]$^\Delta$, *Ser* (*Serrate*) [344, 3525]$^\Delta$ (cf. LOF effects [1944]$^\Delta$), *trn* (*tartan*) [721], *ss* (*spineless*; DO: expression shifts to Stpl in late 3rd instar; cf. *tango*) [1166]$^\Delta$, *tsh* (*teashirt*) [4761]$^\Delta$. HZ76, HZ84 (DO: spot in dorsal T5), PZ22, and PZ63 are enhancer-trapped transposons [1574]. Dll fills the adult tibia in some stained preparations (not shown) [1561] but only the distal half in others [3242, 4761].

N.B.: Most expression domains are dynamic (e.g., see Fig. 5.12b). In late 3rd instar, the *hth* ring overlaps the *dac* ring less than the outer *Dll* ring [4760]. Depending on proximity to the *dac*-ON region, Hth either inhibits or activates *tsh* (not shown) [4761]. Adult defects imply that *dac* is needed in T2–3 [2689], BarH1 and BarH2 are needed in T3–5 [2287], and *al* is needed in T4–5 [618]. In *pleiohomeotic*^{LOF} flies, the pretarsus looks transformed into a 6th tarsal segment [1500].

---

in a radial gradient. However, some of these "annulus genes" are first transcribed in the D part of their realms, whence the expression spreads laterally and ventrally to close each ring (cf. *BarH1* [2287] and *dachshund* [8, 4575]). In the most extensive screen to date, four "annular" *lacZ*-tagged transposon inserts were recovered among ~1200 enhancer-trap lines. Three of the four (HZ84, PZ22, PZ63) showed D-V spreading, and two of these manifested a D-V gradient of expression intensity within the ring [1574]. The following quotes serve not only to document this trend, but also to convey the dynamics of the phenomenon:

HZ84: Expression in the tibial segment expanded gradually to the ventral side of the segment, and just prior to puparium formation, it formed a gradient; the density of expression was higher on the dorsoanterior side, but lower on the ventroposterior side. This gradient was maintained through the early pupal stage, during which it was localized at the distal half of the tibial segment.

PZ22: In the early third instar, lacZ was expressed in a single cell at the center of the leg disc.... The area of lacZ expression spread from this cell to neighboring cells, but the highest expression remained in the single cell at the presumptive dorsal side of the disc. By the late third instar, the pattern was expressed as a circle in the fused tarsal area.... This circular expression formed a gradient with the density of the expression being higher on the dorsoanterior side and lower on the ventroposterior side, as with HZ84. However, this gradient was short-lived. As soon as the larva [pupariated], the expression became uniform and was restricted to the fifth tarsal segment.

PZ63: In the leg disc, the expression appeared in the mid third instar at the dorsal side of the coxal segment.... The expression extended to the ventral side of the segment, becoming circular and uniform in the late third instar.

One of the three genes associated with these inserts has been cloned and found to contain a zinc finger motif. Interestingly, another zinc-finger protein, Klumpfuss (Klu) [4803], also shows D-V spreading [2250].

Expression in the leg discs starts early during the third larval instar. At this time, the *klumpfuss* expression domain occupies a wedge-like sector encompassing roughly one third of the circumference of the leg disc. Rings of expressing cells successively become visible underneath a knob-like central structure during the third larval stage. The rings correspond to the anlagen of the leg segments and the order of their appearance reflects the developmental pattern of the leg disc. Around the time of puparium formation, *klumpfuss* expression seems to be restricted to the distal half of each leg segment in concentric domains, spreading later over the whole leg.

In klu^{LOF} homozygotes, all three leg pairs manifest fusions of the trochanter and femur, as well as of tarsal segments 3–5 [2250]. With severe alleles, the basitarsus is totally missing, but all these losses appear to arise via apoptosis after a normal phase of leg segmentation.

Stronger evidence for the Arc Scenario comes from mosaics where parts of the *dpp*-ON or *wg*-ON domains are erased by dpp^{null} or wg^{null} clones. When the source of Dpp or Wg vanishes in a leg segment, the effects (missing or duplicated structures) do not spread beyond the confines of that segment [1811]. According to the Cloud Scenario, the effects should have spread across segmental annuli. Another fact favoring the Arc Scenario is that V/V and D/D mirror planes in Janus dpp^{LOF} and wg^{LOF} legs coincide with bristle rows [1812] (Fig. 5.5). According to the Cloud Scenario, the planes should have arisen at the cloud perimeter, which would not coincide with bristle rows.

Hairy links global to local patterning

The geometry of the leg disc seems to demand some version of a polar coordinate system, even if the Polar Coordinate Model per se is invalid. In the fate map, the prospective leg segments are nested as concentric rings, and the future bristle rows occupy radial spokes (Fig. 5.1c). It is easy to see how the Arc Scenario could create spokes as thresholds in semicircular morphogen gradients, but the Cloud Scenario's contour lines would be parabolas that have no obvious relation to spokes (Fig. 5.4c) [1807].

As described in Chapter 3, bristle rows arise within stripes of Achaete-expressing cells. In the tarsus, there are 8 such stripes (Fig. 3.9c). Hairy-expressing cells reside between Ac stripes 4–5 (h-ON$_D$), 1–8 (h-ON$_V$), 6–7 (h-ON$_A$), and 2–3 (h-ON$_P$). This alternation has a causal basis, which is made manifest by disabling *hairy*. Under these conditions, all the cells that normally express Hairy now express Ac, so 4 broad Ac bands arise (Fig. 3.9f) [3193]. Evidently, Hairy's duty is to define Ac stripes by negative regulation (Hairy $\dashv$ *ac*). How Ac is suppressed in the other 4 interstripes (1–2, 3–4, 5–6, and 7–8) remains unknown.

Two of the 4 Hairy stripes are visible in the disc, whereas the other two arise after pupariation. The early stripes – h-ON$_D$ and h-ON$_V$ – abut the A/P line and are regulated by Hh, as shown by LOF (*h* turns OFF in smo^{null} clones) and GOF (*h* turns ON where *ci* is misexpressed) studies [1778]. These stripes are also under Dpp and Wg control (*h* turns OFF in Mad^{LOF} wg^{null} clones), and separate *cis*-enhancers at the *h* locus respond to Dpp vs. Wg. Because *dpp* and *wg* are activated by Hh, Hh's control of *h* could theoretically be indirect (Hh → {Dpp OR Wg} → *h*), but the link is actually direct (Hh → *h*) because (1) *h* stays OFF in every cell of a smo^{null} clone even when Dpp or Wg is present, and (2) ectopic expression of Dpp or Wg can turn *h* ON only where Ci-155 is also present [1778]. Thus, the circuitry must be combinatorial: the h-ON$_D$ enhancer responds to {Ci-155 AND Mad}, while the h-ON$_V$ enhancer responds to {Ci-155 AND Arm-Pan}. Overall, therefore, the logic is: "{Hh AND [Dpp OR Wg]} → *h* $\dashv$ *ac*."

One unresolved issue is why a gap arises between the *en*-ON region and the h-ON$_D$ stripe in pupal legs [1778, 3193] if En → Hh and diffusion of Hh → *h*. The gap is 3–4 cells wide and accommodates Ac stripe 4. No such gap should exist if Hh turns *h* ON at the edge of the *en*-ON domain.

Questions remain about the Hh-Dpp-Wg circuitry

Despite the Hh-Dpp-Wg Model's success in explaining so many previously enigmatic aspects of leg development and regeneration, additional questions remain:

1. **"Yin Yang Paradox"**: Why do some cells express supposedly incompatible genes? Given that *en* $\dashv$ *ci*, it seems odd that cells ahead of the A/P boundary can express both *en* and *ci* [350, 4136]. Such overlaps may be trivially due to one gene turning ON as its inhibitor turns OFF in a "fuzzy" transition zone [364, 1561, 2728]. The problem seems more serious for *en* $\dashv$ *dpp*, although here too the shut-OFF may simply require a threshold of inhibitor [998]. In mature wing discs, the *dpp*-ON stripe overlaps the *en*-ON area by 1–2 cells along much of the border but by 3–4 cells at the dorsal edge of the wing pouch [3497] (cf. overlaps of *wg*-ON and *en*-ON sectors in leg discs [1472, 4159]). No such overlap is seen in young discs [4228], so the Hh → *en* link may evoke agents (*polyhomeotic*? [2728, 3512]) that override *en* $\dashv$ *dpp* at certain places [364]. Much of the zone has a gap (largest in the notal area [3497]) – not an overlap – between the *dpp*-ON and *en*-ON domains [4136, 4479], so thresholds may be region specific. Intergradations

(or lack thereof) may also involve nuances in Ci processing [4478, 4536, 4539] or cross-talk between Hh and Dpp pathways [4251].

2. **"Fragmented Stripe Dilemma":** Why are different parts of the leg's *dpp*-ON stripe controlled by different enhancers? If Hh turns ON *dpp* in the dAC (cf. Fig. 5.4) by the usual Hh pathway (cf. Fig. 5.7), then wouldn't it be simplest to use a single *cis*-enhancer as the "Ci-155 ➜ *dpp*" link [3747]? Indeed, the wing pouch seems to use this sort of strategy [2980]. In contrast, the leg's stripe is created in pieces – each regulated by a distinct control element [347, 2739, 4056]. We have been fooled before in expecting stripes to be made neatly [49]. Maybe subdividing a stripe into subunits allows each piece's output to be adjusted to suit each leg segment. The piecemeal strategy makes sense in terms of the Arc Scenario: morphogen output per unit of stripe must increase with distance from the center in order to pump the morphogen around a larger arc. Consistent with this logic, the fly's distal leg segments are most sensitive to an overall reduction in Dpp signaling [4277], and a centrifugal intensity gradient is seen for the *wg*-ON stripe in grasshopper legs [2069]. Indeed, a preliminary analysis of *cis*-enhancers at the *wg* locus shows that the *wg*-ON stripe in fly legs may be subdivided like the *dpp*-ON stripe [2687].

3. **"Leg Stump Riddle:"** Why do tarsi tend to be truncated in *dpp*LOF, but not *wg*LOF, mutants [528, 529, 1812, 1871]? The rationale in 2 does not help here because the *dpp*-ON and *wg*-ON stripes should behave alike. Nor can the biased loss be ascribed to a need for *dpp* in only the distal region: *dpp* must be needed along the entire A/P boundary because a total LOF (via *dpp-Gal4:UAS-wg* ➜ *dpp*) deletes all D elements and creates legs that are perfectly V/V symmetric from top to bottom [2082]. The answer may lie in how *cis*-enhancers are arrayed at the *dpp* locus [347, 2739, 4056]. To wit, the breakpoints of mutants studied thus far might mainly affect tarsal enhancers [834]. Alternatively, a second TGF-β-like gene may act redundantly with *dpp* proximally. (It would also have to be antagonized by Wg to explain the total V/V phenotype [2082].) Although *punt*LOF and *tkv*null somatic clones only affect the distal tibia and tarsus [3329, 4277], repression of *punt* function for ≥40 h can cause Wg to replace Dpp throughout the entire Dpp sector [4277]. Distal segments are preferentially affected when *punt* is suppressed for shorter (23 h) periods.

Evidently, Punt receives dorsalizing signals along the whole proximal-distal axis (regardless of any redundancy), but the signals may inherently be weaker distally due to the geometry of the disc (cf. 2 above).

Distal-less is necessary and sufficient for distalization

As indicated above, *Distal-less* (*Dll*) encodes a homeodomain protein [836, 4434] whose expression turns ON at the inception of the leg disc [827, 828, 833] and stays ON centrally thereafter (Fig. 5.11) [834, 1037, 1544]. LOF alleles cause recessive loss of distal leg structures (hence its name), plus a dominant transformation of distal antenna to distal 2nd leg [1085, 3765, 4212].

Dll function must be necessary for distal cell survival during 3rd instar because (1) chilling a cold-sensitive *Dll*LOF mutant during this period causes truncations [4212], (2) *Dll*null clones induced before this time do not survive distal to the coxa [838, 1561], and (3) tarsus-specific genes (*al* and *bric à brac*) are not expressed in mature *Dll*LOF discs [618]. LOF alleles can be roughly seriated in terms of the severity of distal loss [839, 4212], but the most sensitive spot is actually the basitarsus (not the tip) [839], and vestiges have gaps that defy ordering [836]. The seriation suggested a proximal-distal (P-D) gradient [839], but Dll expression is uniform in the tibial-tarsal region [881, 1037, 1561, 3243], thus dashing any hope that *Dll* might execute the elusive P-D gradient (e.g., as a "memory" gene) [618]. Besides the disc's central Dll circle, there is faint expression in the femur and a proximal (~trochanter) *Dll*-ON ring that arises in 3rd instar in response to other stimuli (including *homothorax* [4760]).

In 1997, Lecuit and Cohen proposed a scheme (here called the "Cloud Scenario"; Fig. 5.4) for P-D patterning [2456] based on how Dpp and Wg affect *Dll* and *dachshund* (*dac*). The *dac* gene encodes a novel nuclear protein that is expressed in the femur, tibia, and basitarsus [2456], although it is also needed in tarsal segments 2–3 [2689]. In early 3rd instar, the Dll and Dac domains abut one another (solid circle of Dll surrounded by a Dac ring), but by mid-3rd instar they overlap considerably (Fig. 5.12).

Because Dll depends on both Dpp and Wg (see above), the overlap of these signals at the disc center might evoke a tip morphogen ("Dpp + Wg = TipM") whose cone-shaped gradient could specify Dll and Dac at different "T" thresholds ("*Dll* turns on if TipM > T_1"; "*dac* turns ON if T_2 > TipM > T_3"). This general idea, the **"Distal Organizer Scenario"** [488, 618], was implicit

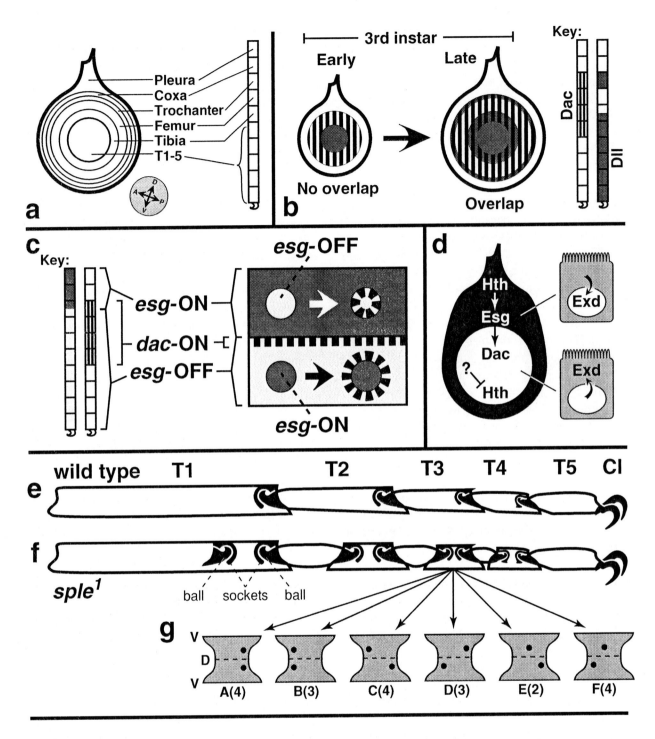

in Meinhardt's Boundary Model [2804]. Lecuit and Cohen argued instead that Dpp and Wg diffuse randomly, so Dll and Dac arise in regions of overlap: *Dll* turns ON at high {Dpp AND Wg}, while *dac* turns ON at lower {Dpp AND Wg}. In that case, Dpp and Wg would act directly.

The critical test involved clones whose Dpp (tkv^{GOF}) or Wg (sgg^{LOF}) transduction pathway is activated in "alien" territory (V or D region, respectively). Such clones (unlike clones that emit Dpp or Wg) fail to induce Dll or Dac in surrounding cells [2456]. Ergo, the Distal Organizer scheme must be wrong. (TipM should have seeped out and induced Dll and Dac.) TipM could still be an instructive signal if cells also need permissive levels of Dpp and Wg to turn ON *Dll* or *dac* (e.g., "*Dll* turns ON only if TipM > T_1, Dpp > T_{DPP}, and Wg > T_{WG}") – a condition met inside, but not outside, the clone.

FIGURE 5.12. Proximal-distal axis of the leg and how it is subdivided into discrete domains.

a. Blank templates for charting zones of gene expression: disc (left) and derivatives (right) consisting of pleura (body wall) and the leg proper (9 segments; cf. Figs. 5.1 and 5.11). Abbreviations: Dac (Dachshund), Dll (Distal-less), Esg (Escargot), Exd (Extradenticle), Hth (Homothorax).

b. Areas where Dac and Dll are expressed during 3rd instar. The Dll circle initially abuts the Dac ring but later expands to overlap it. (Note also the emergence of an outer ring.) An opposite situation occurs more distally (not shown) where an Aristaless circle and a Bar ring first overlap and later abut [2287].

c. Areas where Dac and Esg are expressed (left) and intercalation of a *dac*-ON zone wherever *esg*-ON and *esg*-OFF cells abut (right). Dac and Esg zones normally overlap slightly (omitted at right). Two clones (circles) are cartooned. When an *esg*^null clone (above) is induced in *esg*-ON territory (where Esg and Dac overlap [1573]), its peripheral cells make Dac. When an *esg*^GOF clone (below) arises in *esg*-OFF territory, Dac is expressed by cells around the clone. Evidently, *esg*-ON cells emit a signal that causes nearby *esg*-OFF cells to turn *dac* ON (Esg ➤ ? ➤ *dac*). An analogous induction (Bar ➤ contact? ➤ *fasciclin II*) occurs more distally (not shown) [2287].

d. Proximal vs. distal domains of the leg disc. In the proximal area (black; dorsal region exaggerated), *hth* and *esg* are expressed (Hth ➤ *esg*), and Hth enables Exd to enter the nucleus (Hth ➤ Exd$_{nuc}$) [9, 130, 677, 2673, 3589]. In the distal area (white), *hth* is inhibited jointly by Dpp and Wg so Exd cannot enter the nucleus [8]. The inhibition involves high doses of Dpp and Wg but apparently is not mediated by Dac or Dll [4760], so its route is unclear. The likely agent until early 3rd instar is *Antennapedia* (Antp ⊣ *hth*), but Antp fades from the endknob by mid-3rd instar [677].

e, f. Extra tarsal joints caused by mutations in "planar polarity" genes. "T1–T5" are tarsal segments, and "Cl" denotes claws (dorsal is at top, ventral at bottom). **e.** Tarsus of wild-type 2nd leg (bristles omitted) showing normal ball-and-socket joints (solid black). Within each joint, the ball belongs to the distal segment, whereas the socket belongs to the proximal one. Pivoting of balls inside sockets lets the tarsus curl ventrally when an apodeme (not shown) is pulled by tibial muscles (like a marionette) [236, 2857, 3421]. The joints arise dorsally [1812] and constrain movement to the D-V plane (like an elbow), although the details of their morphogenesis are unclear [4009]. Note unequal lengths of T1–T5 and flared tip of T4/T5 socket. **f.** 2nd-leg tarsus from a fly homozygous for the *spiny legs*^{1} (*sple*^{1}) mutation (a.k.a. *pk*^{sple1}) at the *prickle* locus [1641, 1795, 1810, 2884]. In each segment (except T5, which is immune), there is an extra joint. The extra joint resides near each segment's midpoint (except in T1, where the joint lies distally and only makes a partial intersegmental membrane). All extra joints are upside down (balls proximal to sockets; cf. the embryo's "segment-polarity" phenotypes [3151]. The extra joint in T4 has a flared socket like the T4/T5 joint, so joints are probably copies of their segment's distal (vs. proximal) end. Mutations in other planar-polarity genes (e.g., *fz* or *dsh*) cause milder phenotypes [1810, 4716]. One clue to the etiology is that ommatidia, like joints, are ensembles controlled by the same polarity circuit. In that case, the whole proneural group is oriented by a few key cells (cf. Fig. 7.5) [4366], so maybe the same is true for joints.

g. "Fickle Sensilla Mystery" associated with the double-joint syndrome (cf. Fickle Bristle Mystery; Fig. 7.8). Hourglass shapes are full-surface views of the distal half of T3 from legs of various *sple*^{1} individuals. Dorsal (D) and ventral (V) midlines are marked. Black dots are sensilla campaniformia (stretch receptors [4342]). Types of sensillar patterns in a sample of *sple*^{1} legs are shown, with numbers of legs per class (A-F) in parentheses (total = 20). In wild-type flies, two sensilla always arise at the distal end of T3 (as in A) and T1 (not shown) [3705]. In the mutants, sensilla must only arise near the normal joints (not near the ectopic ones) because otherwise there would be more than two sensilla. However, they must then be able to move to the extra joint. Normal and extra joints seem to "attract" sensilla equally (cases of A ≈ B, and cases of C ≈ D), with occasional cases (E or F) where a sensillum cannot seem to decide and hence has been frozen in midsegment.

Panel **b** is adapted from [2456, 4497, 4575], **c** depicts data from [1573], **d** is based on [8], **e** and **f** are traced from [1810] (although the *sple*^{1} tarsus is actually ~10% shorter than the wild-type one), and **g** depicts data from [1810]. See [883, 1341] for an intercalation scheme.

Additional experiments confirmed that Dpp and Wg turn ON *Dll* and *dac* at different thresholds, consistent with the Cloud Scenario. The Arc Scenario seems defunct because it does not explain how Dpp and Wg can achieve stepped plateaus of combined activity coincident with the Dll and Dac domains. However, as mentioned above, each part of the *dpp*-ON and *wg*-ON sectors seems to be governed by separate *cis*-enhancers [347, 2687, 2739], and thus the central output of Dpp and Wg could be higher than in the periphery. Admittedly, the "Dpp AND Wg" profile would be uneven (low near D and V

midlines and high at A and P midlines), but it might still exceed a basal threshold for *Dll* activation throughout the central area. Thus, the Arc Scenario is still viable, although now laden with *ad hoc* assumptions.

Given the dependence of Dll and Dac on Dpp and Wg, it came as a surprise when *Dll* and *dac* were found to stay ON after the Wg (*dsh*^null clones) or Dpp (*tkv*^null or *Mad*^LOF clones) pathways are shut OFF after 2nd instar [883, 2456]. The inference was that Dpp and Wg are needed to turn ON *Dll* and *dac* but not to maintain the ON states. Lecuit and Cohen thought that *Dll*'s ON state might be perpetuated by cell lineage, but subsequent studies showed that the *Dll*-on state is not heritable [618, 1561, 4575]. The reason for *Dll*'s and *dac*'s endurance remains unclear.

Interpretation of these results is complicated by some quirks of the circuitry. First, overexpressing *Dll* in its own area – a GOF manipulation – causes a LOF phenotype (leg truncation) [1561]. This effect implies downregulation of the endogenous gene via "Dll ⊣ Dll" autoregulation. More important, ectopic clones of *Dll*-expressing cells induce leg duplications with expression of *dpp* and *wg* inside the clone [1561]. This "Dll → {*dpp* AND *wg*}" link seems to violate the rules of normal development (*dpp* and *wg* are only on in subregions of the Dll circle), although it may only operate at Dll levels above the physiologic range. Evidently, *Dll* is not only necessary for distalization, but also sufficient to induce it. *Dll* also shows some biases along the D-V axis: (1) *Dll* function is essential in the ventral, but not dorsal, femur [618]; and (2) Dll^{LOF} interacts strongly with wg^{LOF} but weakly with dpp^{LOF} [1812].

Proximal and distal cells have different affinities

Dll^{null} clones induced during 3rd instar grow (albeit slowly) in segments distal to the coxa and form vesicles that detach from surrounding Dll^+ tissue, except in the femur and upper tibia (where they integrate normally) [618, 1561]. The bristles in the vesicles differ from those in the overlying epidermis, suggesting a switch in regional identity [1561]. Evidently, *Dll*-on and -off cells have different affinities [3242, 4760]. P-D affinity differences were discovered long ago in moth wings by transplanting epidermis along this axis [3047, 3049, 3050]. *Dll* does not affect affinities in fly wings [618], where Wg (or another Wnt) instead must preside because fz^{null} $Dfz2^{null}$ clones in the hinge are round with smooth edges [734].

Further evidence for P-D cell sorting comes from studies with other transcription factors. From the inception of the leg disc when Dll appears distally [1572], the proximal cells express *homothorax* (*hth*, homeobox class) [2361, 3226, 3589] and its downstream target *escargot* (*esg*, zinc-finger class) [1542, 1573, 1765]. When hth^{GOF} or esg^{GOF} clones are induced in distal territory (where both genes are normally off), the clones detach from the epidermis and form vesicles [1573, 4760].

What role, if any, does this immiscibility play in the wild-type leg disc? One idea is that it prevents cells from crossing a boundary near the future trochanter [4760]. This line appears to be a legacy of how the arthropod

leg evolved [1544, 2937]. Even if this exclusionary rule is true, however, cells must still be able to change their affinity states (cf. the Cabaret Metaphor; Ch. 4) because the P-D axis has no lineage restriction at the trochanter or elsewhere [544, 618, 1561, 4076, 4575] (cf. a similar situation in the notum [614]). Consistent with this reasoning, cells that are permanently unable to turn on *Dll* (Dll^{null} clones) apparently fail to cross the P/D line [4760].

Like the P/D line, the D/V line that divides the disc into *dpp*- and *wg*-dependent sectors (Fig. 5.5a) is also maintained without lineage barriers. Might it also separate immiscible (D vs. V) populations [4760]? Indeed it may, because sgg^{null} clones (whose Wg pathway is permanently on) often form vesicles when they are induced in 1st instar outside the V region (cf. similar behavior in the wing and notum [3603, 3956]).

All sgg^{null} clones in dorsal positions cause ectopic growths in which the mutant tissue rounds up and appears to minimize contact with non-mutant cells. This correlates with the adult phenotypes of dorsal sgg^{null} clones where the mutant cells are often displaced distally in ectopic outgrowths. [1833]

sgg^{null} clones in lateral positions cause simple outgrowths of leg. The outgrowths consist entirely of genetically marked sgg^{null} mutant cells, which adopt ventral-most identity.... Since the fates of the mutant cells are inappropriate for their position in the leg, the clones are apparently unable to integrate into the normal pattern and the mutant cells are extruded from the leg as outgrowths.... We have also observed clones of sgg^{null} mutant cells, which segregate internally to form vesicles. The vesicles are composed of mutant cells, which produce genetically marked ventral bristles projecting into the lumen. Since the vesicles are often detached from the leg, it is not possible to know where these clones of ventral cells originate; however, it is possible that they form in lateral positions and segregate internally. [1039]

Vesicles are also seen with dsh^{null} clones [4278] so the effect is not gene specific, although it may only be manifest by the Wg (vs. Dpp) pathway along the D-V axis. The D/V immiscibility must eventually wane because sgg^{null} clones induced in 2nd instar integrate normally at virtually any point around the disc circumference [4666].

Long before the P-D and D-V axes were suspected of having affinity subtypes, such differences were implicated along the A-P axis [2441, 2928]. Because most investigations of A-P affinities have been pursued with the wing disc, that corpus of work is discussed in Ch. 6.

Dachshund is induced at the Homothorax/ Distal-less interface

Despite the absence of true P/D compartmentalization, the segregation of (albeit transient) proximal and distal cell types could allow other genes to be turned ON at the interface, analogous to how Hh turns on *dpp* and *wg* at the A/P interface. Indeed, such signaling does seem to occur. Both *hth*[GOF] and *esg*[GOF] distal clones elicit expression of *dac* in nearby wild-type cells (Fig. 5.12) and disturb their polarity [1573, 2825]. These nonautonomous effects are probably mediated by a diffusible signal, which – given the polarity defects – could be a Wnt: Hth ➞ Esg ➞ Wnt? ➞ Dac. This inference is hard to reconcile with Lecuit and Cohen's surmise that *dac* responds autonomously to Dpp and Wg [2456]. Conceivably, as argued above, Dpp and Wg may only act permissively (albeit in a threshold-dependent manner).

This process whereby a third state (*dac*-ON) emerges at the interface between two others offers a way of understanding intercalation [1341] without the need to invoke coordinates per se (e.g., the Polar Coordinate Model [1303]). Similar logic might explain intercalary regeneration in hemimetabolous insect legs because polarity reversals arise there when a donor leg is severed proximally and grafted to the stump of a leg amputated more distally (e.g., see [3878]). Certainly, the results help demystify the old "**Ends-Before-Middle Riddle**" of why young discs fail to make mid-level P-D structures when they undergo premature metamorphosis [3809]. Namely, intermediate identities would emerge stepwise as the pattern is elaborated during growth [1573]. In fact, Dac only begins to be expressed at the interface between *hth*-ON (outer) and *Dll*-ON (inner) regions at the end of the 2nd instar [8, 2287, 4575, 4760]. Presumably, the "Esg ➞ Dac" link is only activated at that time (i.e., a temporal trigger). Alternatively, a critical mass of signaling cells may be needed to overcome a response threshold (i.e., a scalar trigger).

Other peculiarities of the leg's P-D axis have been uncovered over the years. Within each leg segment, for example, distal bristle SOPs tend to divide before proximal ones [1598, 1803, 1808, 3142]. Mitoses, in general, orient along the P-D axis during 3rd instar [544] – a trend that explains the narrow shapes of wild-type clones [3441, 4344], which can be 100 cells long but only 3 cells in width [544]. In conjunction with a higher rate of distal growth (spurred by more Dpp?), this radial pattern of growth may also explain why the epithelium buckles into concentric folds during this period.

Homothorax and Extradenticle govern the proximal disc region

As mentioned earlier, the proximal part of the leg disc (pleura to trochanter) expresses *hth*, while the distal part (trochanter to tip) does not [677, 1573, 3589]. These areas also roughly delimit the action of *extradenticle* (*exd*, it is also needed in the proximal femur [1544]), another homeobox gene [1248, 3526]. Although *exd* is transcribed uniformly in the disc [1248, 3527], Exd's subcellular location differs proximally (nuclear) vs. distally (cytoplasmic) [130, 2673] as does its amount [1543, 3226]. Exd's distribution is dictated by Hth [9, 2045, 3716]: Exd localizes to the nucleus ("Exd$_{nuc}$") wherever Hth is ectopically expressed but localizes to the cytoplasm ("Exd$_{cyt}$") when Hth is removed [8, 302, 677, 3226, 3589]. In turn, Exd sustains Hth: Hth vanishes in *exd*[null] clones, although *hth* is still transcribed, so Exd may block Hth degradation [8, 2045]. These links make sense because Hth binds Exd [9, 2045, 3589, 3716] and these heterodimers act as transcription factors to activate downstream genes [2045, 3716]. One of those targets is *esg*: *esg* is turned ON in distal *hth*[GOF] clones [1573]. The first three links below would operate proximally, while the fourth would evoke Dac (via a paracrine signal) in nearby distal cells.

Link 1: Hth ➞ Exd$_{nuc}$.
Link 2: Exd ➞ Hth.
Link 3: {Hth AND Exd$_{nuc}$} ➞ *esg*.
Link 4: Esg ➞ Dac (3rd instar only).

Thus, the leg disc has a proximal ring where Hth-Exd$_{nuc}$ heterodimers govern one set of target genes and a distal core where Dac and Dll might orchestrate a different set (Fig. 5.12d). Dac and Dll, in turn, are controlled by Dpp and Wg – the ultimate sovereigns of the distal province (although Dpp [4277] and Wg [177, 519, 1812, 3317, 4278] are also needed proximally). Indeed, some targets of Dpp alone (*omb*) or Wg alone (*H15*) are expressed in sectors that abut the *hth*-ON ring, even though the *dpp*-ON and *wg*-ON sectors extend into that ring (Fig. 5.8) [8,1542, 1544]. Could Hth-Exd$_{nuc}$ be overruling the commands of Dpp and Wg? Yes, because (1) expression in the *omb* or *H15* sectors vanishes wherever *hth*[GOF] clones overlap them, (2) *omb* and *H15* are derepressed in proximal *hth*[LOF] clones (in D or V regions, respectively) [8], and (3) *omb* is turned ON by proximal *exd*[LOF] D clones and OFF by distal *exd*[GOF] D clones [1542]. Hth's veto of Dpp's (and Wg's?) authority is not mediated by *esg* because *esg*[GOF] clones do not turn OFF *omb* distally [1573].

Link 5: Hth ⊣ {Dpp ➞ *omb*}.
Link 6: Hth ⊣ {Wg ➞ *H15*}.

An initial report suggested that Dac and Dll are snuffed out in *hth*[GOF] clones [8] (excess Exd alone has no effect [1542, 1544]), in which case Hth would also block

Dpp and Wg's joint targets. However, more recent data refutes the "Hth ⊣ *Dll*" link and possibly the "Hth ⊣ *dac*" as well [4760]. Rather, *Dll* and *dac* may be OFF proximally simply because peripheral Dpp and Wg signals are too weak to prod them into an ON state [2456].

Based on the earlier work, a **Mutual P-D Antagonism Model**" was proposed wherein (1) distal states are supposedly suppressed in the proximal area by the disputed "Hth ⊣ *Dll*" and "Hth ⊣ *dac*" links, and (2) proximal states are supposedly inhibited distally by the reciprocal "Dll ⊣ *hth*" and "Dac ⊣ *hth*" links [8, 1542]. However, the latter two links also appear to be false [4760] due at least in part to illusions involving the folds in the disc epithelium. One consistent finding is that proximal cells turn OFF *hth* [8, 4760] and hence replace Exd_{nuc} with Exd_{cyt} [1542] whenever they "think" they are receiving high doses of both Dpp and Wg (viz., V clones that are tkv^{GOF} or D clones that are wg^{GOF} or arm^{GOF}). Thus, the only confirmed inhibitors of proximal states in the distal region are Dpp and Wg, although it is unclear how these signals cooperate and whether they use proxies (excluding Dac and Dll).

Link 7:   {Dpp AND Wg} ⊣ *hth*.

Given the reciprocity of Links 5 and 6 on the one hand (i.e., Hth's damping of Dpp and Wg transduction) and Link 7 on the other hand (i.e., Dpp and Wg's damping of Hth), the basic premise of the Mutual Antagonism Model may still be correct, and the immiscibility described above could make sense as follows. If *hth*-ON cells are less adhesive than *hth*-OFF cells, then they would automatically sort to the periphery [4071-4074] and that outer ring would be desensitized to Dpp and Wg [4760]. The inner cells would then be left to form a central "plateau" where the {Dpp AND Wg} combination turns ON {Dll AND Dac} in neat circles despite its saddle shape. The humps of the saddle could be effectively flattened during transduction by damping factors such as Dad and Nkd.

Do Hth and Exd control proximal vs. distal cell identities per se? Dramatic evidence that they do so comes from proximal shifts in the identities of distal cells when they are forced to express *hth*: tarsal segment T4 makes basitarsal (T1) sex comb teeth in *ap-Gal4:UAS-hth* flies, and *hth*^{GOF} clones make femoral bristles in the tibia [2825] (cf. large bractless bristles in distal *hth*^{GOF} clones [1573]). These are the kinds of shifts expected if Hth muffles a cell's ability to "hear" (or act upon) Dpp and Wg signals [4760]. Moreover, the muffler would confine Dpp's mitogenic effects to the distal region and thereby fa-

cilitate appendage outgrowth from the body wall [2937]. Proximal *exd*^{LOF} clones cause reciprocal shifts: they tend to make distal-type (small bracted) bristles (e.g., tibial bristles in the trochanter) [1543, 1544]. No P-D shifts have been reported for proximal *hth*^{LOF} (or *esg*^{LOF}) clones or distal *exd*^{GOF} clones, while distal *esg*^{GOF} clones only make smooth cuticle [1573]. Curiously, *exd*^{LOF} discs that fail to form coxa or trochanter still express the proximal marker genes *teashirt* and *odd skipped* during 3rd instar [3527]. Like *esg*, *teashirt* (*tsh*) encodes a zinc-finger protein [66, 1197] that could control the transcription of identity-implementing genes (a.k.a. "realizator" genes [1358]), but distal *tsh*^{GOF} clones do not alter P-D identities [1175].

Fasciclin II is induced at the BarH1/Aristaless interface

Subtler shifts in P-D identity are seen when *BarH1* is overexpressed. The "*Bar*" gene – made famous by Sturtevant's studies of unequal crossing over [4179, 4186] – turned out to be two redundant homeobox genes *BarH1* and *BarH2* [1842]. During 3rd instar, both genes are strongly expressed in tarsal segment T5 and weakly in T4 [2287]. When *BarH1* is overexpressed in T4 (by an *ap-Gal4* driver), T4 bristles now resemble those of T5. In contrast, LOF clones homozygous for a deficiency of the *Bar* locus show a transformation to T3 [2287]. The implication is that the *Bar*-OFF "ground state" dictates T3, medium amounts of Bar dictate T4, and a high level dictates T5 [2287].

The leg disc's T5 ring surrounds densely packed "pretarsal" cells that express *aristaless* [851, 2287]. The BarH1 and Al domains initially overlap, but later (by mid-3rd instar) they abut along a sharp boundary. This line seems (based on LOF-GOF data) to be maintained by mutual repression: "BarH1 ⊣ *al*" and Al ⊣ *BarH1*" [2287, 4401]. Both domains are subsets of the Dll territory, and both genes are permissively kept ON by *Dll*: "Dll → *BarH1*" [2287] and "Dll → *al*" [618]. Moreover, BarH1 helps ban Dac from part of the Dll domain because BarH1 ⊣ *dac* [2287].

In late 3rd instar, the outermost pretarsal cells (those that touch *BarH1*-expressing cells) begin expressing Fasciclin II (Fas II) [2287], a homophilic adhesion molecule [1620, 4758]. Might BarH1 be inducing Fas II? If so, then the induction would likely involve a BarH1 target gene that encodes a ligand, with the ligand fitting a receptor on *al*-ON cells: BarH1 → Ligand? → Receptor? → Fas II. Support for some sort of induction comes from both LOF data (*BarH1*^{null} clones erase Fas II from its normal domain) and GOF data (*Gal4:UAS*-driven expression of *BarH1* in a stripe within the Al circle elicits Fas II in two flanking lines of cells) [2287]. Because the responding cells (endogenous or ectopic) always occupy single files, the induction mechanism is probably mediated by direct cell contact (i.e., a cell-surface ligand), rather than by diffusion (cf. the "Esg ➡ Dac" induction).

Fas II-expressing "border" cells have a unique rectangular shape apically [851, 2287, 2408]. Conceivably, the homophilic Fas II molecules might be "zippering" these cells into a file by maximizing contact. These tensile forces might also help the Fas II file keep its hoop shape, although the hoop is probably not stressed by growth because (1) it arises so late that there is little further proliferation, and (2) inner cells are mitotically quiescent [1599, 4666]. Al may play an executive role in cell adhesion or polarity [4350], as may dLim1 – a LIM-homeodomain protein that is co-expressed with Al [4401].

BarH1 and Bric à brac affect P-D identity, joints, and folds

As discussed above, BarH1 seems to specify T4 and T5 states relative to T3 by its level (high in T5, lower in T4, absent from T3) [2287]. A similar role seems to be played by the gene *bric à brac* (*bab*, named after ovary defects [3722]), although *bab*'s "ground state" is T1. Bab (a BTB-domain protein) localizes to the nucleus and may influence target genes [1516]. Bab is expressed strongly in T4 and T3, less in T2, and minimally in T1. Strong LOF alleles of *bab* transform T2, T3, and T4 to resemble T1 [1516] – an effect that is dramatized by T1-like sex combs on T2, T3, and T4 of the male foreleg. Weaker alleles only affect T2. The sensitive period occurs shortly after pupariation [1516].

A T1 ground state makes sense because so many *Drosophila* species have combs not only on T1, but also on distal segments [376, 1361, 3752, 4095]. Flies can be coaxed to make combs on segments distal to T1 by the following agents:

1. Treatments that cause cell death: X-irradiation [4343, 4349], nitrogen mustard [4336], or surgery [4143].
2. Apoptosis-inducing mutations in *suppressor of forked* [3441, 3442].
3. Growth-perturbing mutations in *comb gap* [962], *dachs* [4348, 4515], *four jointed* [4348, 4515], *hyperplastic discs* [2685, 2709], *l(1)ts504* [3964], or *Notch* [3525].
4. Homeotic mutations in *cramped* [3752, 4793], *dachshund* [3752], *Enhancer of zeste* [3752], *extra sex combs* [4095], *multi sex combs* [3721, 3755], *pleiohomeotic* [1500],

poils aux pattes (whose product is part of the Mediator complex) [412], *Polycomb* [645, 1024], *proboscipedia* [3333], *sex comb distal* [411, 3752], *spineless* [4095, 4515], and other genes [2561].

Evidently, T2, T3, and T4 of the male foreleg are predisposed to make sex combs (i.e., have a "prepattern"), but only T1 normally does so because *bab* and other genes block the realization of this potential in the T2–T4 region (i.e., suppress "competence") [4095]. This surmise is supported by the fact that several of the above genes belong to the Polycomb Group of repressors (cf. Ch. 8) [3752].

Aside from their different ground states (T1 vs. T3), Bab also differs from BarH1 in that it is expressed in a gradient within its respective segments (Fig. 5.11). Iterated gradients of positional information have long been suspected for insect leg segments based on (1) regeneration outcomes seen after grafting legs of cockroaches or other hemimetabolous insects, and (2) polarity disturbances when fly leg segments partially fuse in various mutants [386, 3427]. However, Bab probably is not serving this function because no intrasegmental displacements of sex combs (away from its normal distal site) were reported for *bab*^LOF mutants. (They should have occurred if Bab were specifying P-D position within each segment.)

Bab and BarH1 both must govern the T4/T5 joint (where their most intense zones abut) because T4 often fuses with T5 in *bab*^LOF mutants, *Bar*^null clones, and *BarH1*^GOF (*ap-Gal4*) tarsi [1516, 2287]. The ability of these genes to induce a joint invagination may be related to their ability to induce folds in 3rd-instar leg discs at their domain boundaries [1516, 2287]. LOF alleles of *bab* also affect the T2/T3 and T3/T4 joints, but the effects are milder than for T4/T5.

Leg segmentation requires Notch signaling

Mutations in various Notch pathway genes cause fusions of tarsal segments [881]: *Notch* [344, 999, 3525, 3891], *Delta* [344, 999, 3277, 4466, 4467], *deltex* [1570], *fringe* [999, 3525], *kuzbanian* [4025], *Serrate* [344, 4029], *strawberry notch* [895], *Suppressor of Hairless* [999], etc. [2195, 4467]. Many of these same mutations also prevent the detachment of the distal femur from the proximal tibia [999, 3525] (cf. *groucho*^LOF [998]), which normally occurs during the early pupal period [1311].

The involvement of the Notch pathway in leg segmentation was analyzed in 1998–99 [344, 999, 3525]. By turning ON the pathway (via *Notch*^GOF, *Delta*^GOF, or *Serrate*^GOF) in unusual locations, extra joints could be induced [344, 999, 3525] (cf. *Bar*^null [2287]), implying that the pathway is sufficient as well as necessary for creating joints. For example, when Serrate (a ligand) was artificially expressed in the middle of the tibia (using a *Gal4* driver), two new joints arose there – one at either end of the ectopic Serrate domain [344] – implying that joints are made at *Serrate* ON/OFF boundaries. Normally Serrate is only expressed at the distal end of each leg segment (Fig. 5.11), so this experimental result begs the question: why doesn't a joint *normally* arise on the proximal side of each Serrate band as well as on the distal side? There is no obvious answer because all the needed components of the transduction pathway are present on both sides.

One clue to this mystery is a "double-joint" syndrome where an extra joint does indeed arise on the proximal side of each Serrate band in T1–T4 [344, 1810] in addition to the normal joint, and each extra joint is upside-down (Fig. 5.12f). The syndrome is caused by LOF mutations in *frizzled* (*fz*), which encodes a receptor for Wg and maybe other Wnts, and *dishevelled* (*dsh*), which encodes a transducer for Wg and maybe other Wnts. Conceivably, a Wnt signal normally polarizes cells so that they can tell whether they are on the "right" or "wrong" side of the Serrate band. When that signal fails, cells on both sides think they are on the right side and hence make mirror-symmetric joints [344]. This idea is at least plausible because cross-talk between the Notch and Wingless pathways occurs in the wing disc [459, 886, 2718, 2839, 3690], and Dsh can bind Notch [151, 356]. A separate pathway involving the LIM domain protein Prickle also suppresses extra joints [1641, 1810], but the nature of that circuitry remains to be elucidated.

The annular expression domains categorized in Figure 5.11 are reminiscent of the hierarchy of embryonic segmentation genes (Fig. 4.2) [1341, 1988] – the "wide zone" class corresponding to gap genes and the "periodic zone" class to segment-polarity genes [3523]. Body and leg segmentation are similar processes of "slicing" a cylinder into rings, but does this analogy have any real mean-

ing? Could the leg be a scaled-down version of the body [2868]? Unlike body segmentation, leg segmentation does not involve cell-lineage restrictions [2852]: clones can cross segment boundaries virtually whenever and wherever they are induced [544]. Nevertheless, the double-joint syndrome may have an etiology related to segment-polarity phenotypes because *dsh*^LOF causes both traits [344, 1810]. Two pair-rule genes (*hairy* and *odd skipped*) are in the leg's periodic class and another (*odd Oz*) has a single ring, but there is no evidence of any two-segment periodicity in the anatomy or sequence of disc folds [834, 1516, 2287]. As a rule, wide-zone genes are expressed earlier than periodic-zone genes (echoing the embryo sequence). The Notch pathway is not involved in body segmentation in flies [3695, 4601] but may have been used ancestrally [1009, 3525]. A final answer regarding body-leg homology will be impossible until we know much more about how both body and leg circuitry evolved [1807, 1988, 2069, 2868, 3122, 4230].

The analogy at least has heuristic value. For example, if combinatorial inputs from the wide-zone class dictate periodic-gene expression patterns [3525], then the latter's *cis*-enhancers should bind *trans*-effectors from the former. The fact that several wide- and thin-zone genes encode putative transcription factors (e.g., homeobox genes *hth*, *Dll*, *al*, *BarH1*, and zinc-finger genes *esg* and *tsh*) is compatible with this scheme. The periodic-zone gene *disco* could mediate joint formation: it encodes a zinc-finger protein that controls intercellular adhesion [1791, 4079], perhaps the kind of adhesion needed for joint invagination [1311, 2468]. Regardless of its circuitry, this system is worth studying for its robustness [4515], especially its ability to reliably make exactly 9 segments.

An equivalent robustness is evident in the wing disc, where two symmetric surfaces (each with 5 longitudinal and 2 cross veins) are created separately and then plied together with virtually no mismatching of dorsal vs. ventral elements. Accuracy is of the utmost importance in that case because a defective airfoil is a debilitating handicap for a fly.

CHAPTER SIX

The Wing Disc

The A-P axis is governed by Hh and Dpp but not by Wg

Like the leg disc, the wing disc uses Hedgehog to set up a border zone just ahead of the A/P compartment boundary [231, 4228] (cf. Fig. 5.7), but its zone emits only one long-range morphogen – namely, Dpp [231, 354, 2448]. Wingless is irrelevant for the wing's A-P axis and instead functions along its D-V axis [4849]. Both morphogens are essential: wings fail to develop when the disc is deprived of either Dpp [529, 3438, 4032, 4848] or Wg [884, 4683].

Topologically, the wing is like a squashed leg (Fig. 6.1). Its D and V faces are apposed, and its veins run along its length like the leg's bristle rows. However, while the prospective bristle rows converge centrally in the leg disc (cf. Fig. 5.1), the primordia of veins 2–5 are parallel to one another and intersect a perpendicular line (the future margin). Thus, it is unclear whether the wing has a true "tip" like the leg [4682]. Certainly, the expression of Dll in a band along the wing margin (Fig. 6.2) differs from the circle of Dll in the leg disc (cf. Fig. 5.4) [2254].

The stripe where *dpp* is expressed in a mature disc is ~5 cells wide [754, 2739, 4188, 4251, 4479]. That width is the net result of (1) activation of *dpp* by Hh (Hh ➡ *dpp*) in a larger zone (8–10 cells wide; cf. Fig. 5.7) and (2) repression of *dpp* by En (En ➞ *dpp*) in the rear 1/3 (3–4 cells) of that zone [350, 2728, 2992, 4136] in the same way that En represses *dpp* in the P compartment throughout larval life [1647, 3747, 4229] (but see Fig. 6.3).

Suppression of *dpp* in the border zone is only seen in late 3rd instar [350, 364, 4228] when a "Hh ➡ *en*" link turns *en* ON in A cells that receive Hh [78, 642, 3818]. As was argued in Ch. 5 (the Minotaur Scenario), this new link could join the old "*en* ➡ *hh*" link [1647, 1659, 4227, 4848] to spark a chain

reaction ("Hh ➡ *en* ➡ Hh"), but is prevented from doing so in the leg disc by Dpp or Wg. In the wing disc (Fig. 6.4a), the "circuit breaker" is Dpp alone (Dpp ➞ {Hh ➡ *en*}). Overall, therefore, the logic consists of the 5 links listed below, but until recently it was unclear how Hh-responsive cells decide whether to activate (Link 1) or repress (Links 2 and 3) *dpp*.

Link 1: Hh ➡ *dpp*.
Link 2: Hh ➡ *en* (late 3rd instar only).
Link 3: En ➞ *dpp*.
Link 4: *en* ➡ *hh*.
Link 5: Dpp ➞ {Hh ➡ *en*}.

In 1997, Strigini and Cohen showed that this circuitry works not just in the border zone but anywhere in the A compartment [4136]. When they induced *hh*^{GOF} clones in the A region, *en* turned ON in every clonal cell and in nearby wild-type cells that got Hh by diffusion (Link 2). At a still greater radius ($r_{dpp} > r_{en} > r_{hh}$) were *dpp*-ON cells (Link 1). Unlike the *hh*-ON and *en*-ON "biscuits," however, the *dpp*-ON domain was a "doughnut" with a *dpp*-OFF hole in the middle where Link 3 evidently overrules Link 1. These clones could presumably have pushed their *en*-ON state out farther into the A region by alternating Links 2 and 4 ("Hh ➡ *en* ➡ Hh"), but they did not, evidently because the doughnut acts like a cage to block them (Link 5) [2082]. Given this policing ability of Dpp, how does the *en*-ON state normally creep (albeit weakly [350]) from the P into the A region at all? Maybe the intrusion reflects a balance of forces, with Link 2 barely subduing Link 5.

Strangely, Zecca *et al.* had obtained different results from similar experiments in 1995 [4848]. They saw *dpp*

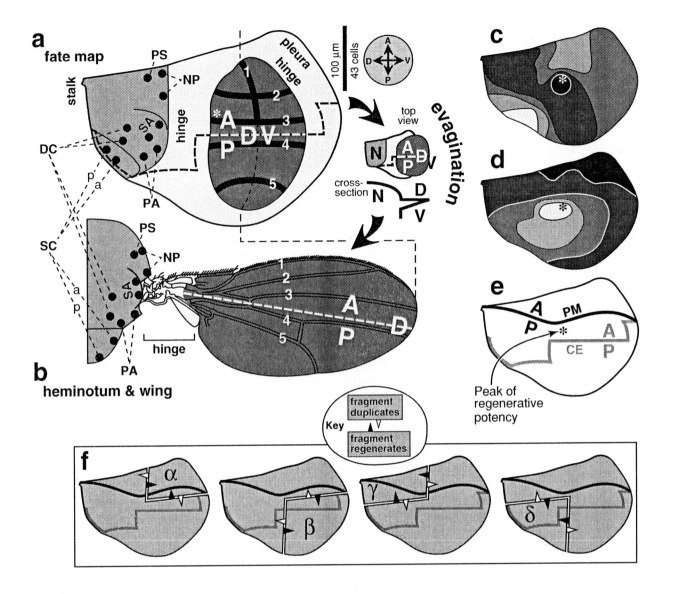

doughnuts around en^{GOF} A clones, but their hh^{GOF} A clones expressed *dpp* inside as well as outside each clone (i.e., biscuits vs. doughnuts), thus disobeying Link 3. This malfunction in Link 3 is traceable to a glitch in Link 2 because their *hh* transgene did not turn ON *en* as Strigini and Cohen's transgene did. Why not? Apparently, the constitutive promoter they used (*Tubulinα1* vs. *Actin5C*) was too weak to force Hh to turn ON *en*, although enough Hh was produced to turn ON *dpp*. Indeed, ptc^{LOF} clones (which make cells think they are receiving Hh) are able to turn *en* ON anywhere in the A region when a null *ptc* allele is used, but weaker LOF alleles only manage to turn *en* ON near the A/P boundary [2834, 4136].

The idea that more Hh is needed to activate *en* than *dpp* is supported by the seriation of strata in the border zone (*hh*-ON, *en*-ON, *dpp*-ON) and of radii at hh^{GOF} A sites ($r_{hh} < r_{en} < r_{dpp}$). Because concentration-dependent gene expression is the hallmark of a morphogen [865, 1655], it looks like Hh is a *bona fide* morphogen after all and not just a short-range inducer [2448, 4848]. A gentle nudge from Hh would therefore coax a cell into Link 1, whereas a firmer push would force it into Links 2 and 3.

The **"Hedgehog Gradient Model"** postulates that Hh activates different genes (e.g., *en* or *dpp*) at different thresholds ($T_{en} > T_{dpp}$) as a function of distance (near vs. far) from its source (usually the P compartment but potentially at ectopic A sites) [154, 4136]. Additional support for this model comes from varying the level of Hh in the P compartment [4136]: as the amount of Hh drops (when t.s. hh^{LOF} flies are raised at different temperatures), the band of *en* expression disappears before

FIGURE 6.1. Fate map of the wing disc and the regenerative potency of its fragments.

a. Fate map (abridged) of a mature right wing disc as per Bryant [524], except that dots are actual SOP sites [1925] and thick lines are prevein zones (1–5) [4189]. (See [834, 987, 1368, 3372, 4429] for further details.) The wing is darkly shaded, the heminotum (a.k.a. body wall) medium shaded, and the hinge and pleura lightly shaded. The wing disc is ~300 μm across × ~450 μm long [524] – roughly twice the dimensions of a leg disc (cf. Fig. 5.1). Like the 1st- and 2nd-leg discs (vs. the 3rd [3422]), the wing disc has a neural connection to the CNS [2128], but the nerve fiber enters along with the trachea (not shown) [3565]. Based on cell diameters in the wing pouch (mean of 10 counts along 25-μm transects in Fig. 8b of [350]; cf. scale bar), the disc would be 130 cells across × 190 cells long, but these are gross underestimates because the epithelium is highly folded. Vein 1 is the bristled part of the anterior wing margin [1741] (thin dashed line = posterior margin). The A/P compartment boundary is drawn as a thick dashed line. Directions (A, anterior; P, posterior; D, dorsal; V, ventral) are given in the compass at right. Bristle names ("a" and "p" = anterior or posterior members): DC (dorsocentrals), NP (notopleurals), PA (postalars), PS (presutural), SA (supraalars), SC (scutellars). The layout of SOPs prefigures the adult bristle pattern, except that scutellar SOPs rearrange ("p" moves posterior to "a"). In **a** and **c–e**, the asterisk marks the peak of regenerative potency as mapped surgically (cf. Fig. 4.5).

b. Derivatives of the wing disc (shaded as per fate map). During evagination (illustration between **a** and **b**) the wing pouch expands, folds along the D/V line, and tucks its V side underneath ("N" = notum) [1311, 3374, 4189]. When the hinge region contracts [4509], its cells pack densely [3374] with some apoptosis [2849] to form a menagerie of elements [526, 1866, 4065]. This contraption allows the wing to twist so as to maximize the lift per stroke [1045, 1115, 4748]. Distal edge of the hinge is approximate [678]. The hinge region is genetically distinct from the blade per se because (1) it overgrows in response to excess Wg [459, 2252, 2254, 3088], and (2) it does not depend on Dpp signaling [157]. Indeed, *wg* expression in the hinge is induced by *vestigial*-ON cells in the blade [2570]. Bristles (except along front edge) are omitted.

c, d. Regenerative potency, assessed by γ-irradiation of young wing discs [3440]. **c.** Darker shading indicates higher frequencies (50–60, –70, –80, –90, –100%) of remaining markers in discs that exhibit deficiencies but no duplications. **d.** Darker shading denotes higher incidence (0–5, –10, –20, –60, –80%) of duplicated markers in discs that manifest duplications and deficiencies. Note that the same (central) region resists becoming deficient (**c**) or duplicated (**d**).

e. Path of the A/P compartment boundary in the peripodial membrane (PM, solid line) and columnar epithelium (CE, shaded line), as assessed by *engrailed* and *patched* expression [1133, 2072, 2216]. The regenerative peak (asterisk) falls between the A/P lines of the two layers (in top view). The D/V boundary in the peripodial membrane has not been charted so precisely [498]. The signaling role of this membrane in normal development is only starting to be explored [773, 1473, 3508].

f. Superposition of the entire A/P boundary (**e**) on the 1/4 (α, β, γ, δ) and 3/4 fragments whose regenerative behavior helped found the Polar Coordinate Model (cf. Fig. 4.6a). As indicated in the key (cf. Fig. 4.5), an unfilled arrowhead means that the fragment to its rear duplicates, whereas a solid arrowhead means that the fragment behind it regenerates. Given recent insights into leg disc regeneration (cf. Fig. 5.10c), it may be possible to figure out why quadrants duplicate and 3/4 pieces regenerate, but no unifying hypothesis is obvious based on the geometry alone.

---

the band of *dpp* expression, implying that *en* requires more Hh than does *dpp*. Similar studies indicate that Hh turns ON two other genes – *patched* (which encodes its receptor) and *knot* (which evokes vein 3) – at thresholds between T_{en} and T_{dpp} (Fig. 6.3) [2894, 4136, 4478, 4479, 4539]. This "rainbow" of different qualitative states could emerge (as per Wolpert's French Flag [4723]) from "contour lines in a gradient landscape" [3074], and the thresholds might stem from variable affinities of target promoters for activator vs. repressor forms of Ci [154]. Disparate affinities might explain why the Dpp band persists for as long as 24 h after *hh* function is shut OFF (by pulsing heat-sensitive hh^{LOF} mutants), while the Ptc band vanishes within only 45 min [642].

These facts do not prove a morphogen mechanism, however, because seriations of states can arise in other ways [1805]. In the "**Signal Relay Scenario**" [713, 714, 2816, 2817, 4591], for example, Hh would induce a second signal (En?), which then induces a third (Dpp?), and so on [1807, 2076, 2466, 3923, 4159]. The key difference between morphogen and relay models is that the former demand direct action on target genes while the latter eschew it [3074, 3087, 3091, 4137].

A signal-relay mode of action [1607] (a.k.a. "signaling cascade" [91, 1144, 2192], "sequential induction" [2409, 2716, 3644], or "bucket brigade" [355, 2448]) was ruled out for Hh in the boundary zone by replacing the normal Hh with a membrane-tethered Hh construct [572, 4136]. Under these conditions, *en* and *dpp* both turn ON only in A-type cells that touch *hh*-ON cells – i.e., a single row of cells. (Coexpression of *en* and *dpp* is somehow tolerated despite the "En ⊣ *dpp*" link.) If Hh were using En (or some other proxy) to create the *dpp*-ON state, then two (or more) separate rows of cells should have appeared at the edge of the *hh*-ON domain – the inner row expressing *en* and the outer row expressing *dpp*. Thus, Hh must be directly activating *en* and *dpp*.

Ironically, a signal-relay scenario does apply from this point on. After Hh switches ON *dpp* in neighboring cells, Dpp diffuses more extensively (as a second messenger) to organize pattern on a discwide scale.

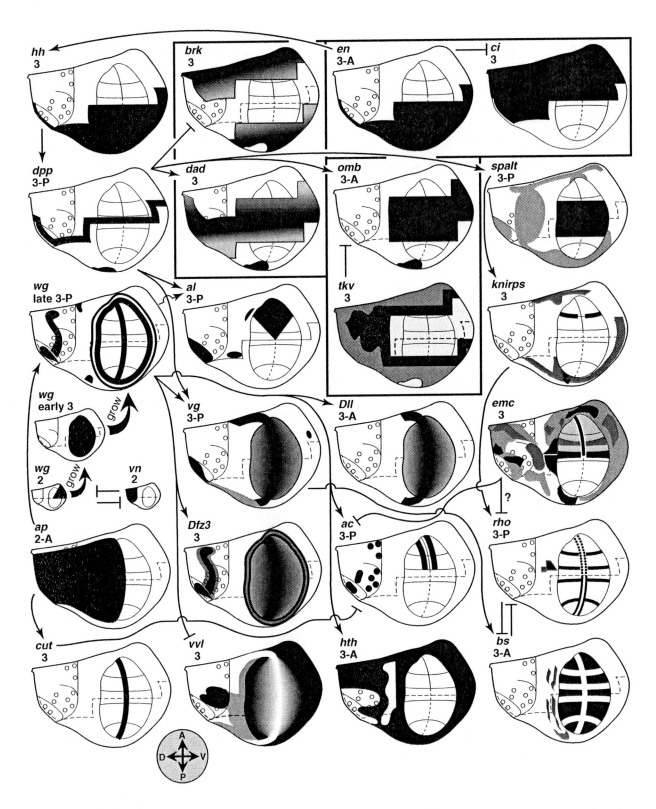

Dpp turns ON *omb* and *spalt* at different thresholds

While Hh specifies fates over a small range (~10 cells from its source), Dpp is evidently a morphogen for the wing's entire A-P axis (~50 cells in either direction within the wing pouch) [2115, 2455, 3074, 4251]. Suppressing *dpp* causes small or missing wings [529, 3438, 4033].

The argument in favor of this "**Dpp Gradient Model**" rests largely on two genes that are activated by different levels of Dpp. Both genes appear to encode transcription

FIGURE 6.2. Patterns of gene expression in the wing disc. The flow of control is from the selector genes *en* (upper right) and *ap* (lower left), which define compartments, through *dpp* and *wg* (left), which specify orthogonal coordinates, to *rho* and *bs* (lower right), which dictate vein vs. intervein identity.

Black areas chart mRNA (transcribed from endogenous or reporter gene) during 2nd instar (2), late-3rd instar (3), early pupal period (P) or adult stage (A). Shades of gray denote degrees of expression. Boxed pairs have roughly complementary patterns. Genetic interactions are indicated by connecting wires (➜ activation; ⊣ inhibition; see text or sources cited at the relevant genes below). See also App. 7.

Genes ("DO" = details omitted): *ac* (*achaete*; DO: other spots, asynchronies, and shape irregularities) [3689]△, *al* (*aristaless*) [618]△, *ap* (*apterous*) [361]△, *brk* (*brinker*) [3978]△, *bs* (*blistered*) [3145]△, *ci* (*cubitus interruptus*) [1135], *cut* (DO: *cut*-ON stripe is slightly thinner than *wg*-ON stripe [2839]△) [988]△, *Dad* (*Daughters against dpp*) [2867]△, *Dfz3* (*Drosophila frizzled3*) [3977], *Dll* (*Distal-less*; DO: pinching at edges of the wing pouch [618]△) [3242]△, *dpp* (*decapentaplegic*; DO: gap in notal area [3764]△ and tapering to a point at future wing tip [4575]△) [4188]△, *emc* (*extramacrochaetae*) [913], *en* (*engrailed*) [2954]△, *hh* (*hedgehog*) [4228]△, *hth* (*homothorax*) [157, 678]△, *knirps* [2617], *omb* (*optomotor-blind*; DO: indentation at wing margin, depression at A/P border [4251], fade-out at edges, and proximal tapering) [3978]△, *rho* (*rhomboid*) [4189]△, *spalt* (DO: indentation at wing margin, depression in the 3–4 intervein [984, 4251], fade-out at edges, A/P asymmetry, and pupal details [4188]) [2867]△, *tkv* (*thick veins*; DO: depression at A/P border [1327]△ and late-3rd instar expression in P compartment) [1674, 2457]△, *vg* (*vestigial*) [678]△ (see [353] for pupal stages), *vn* (*vein*) [3928]△, *vvl* (*ventral veinless*; a.k.a. *drifter*; DO: faint spot in notum) [703, 990]△, *wg* (*wingless*) [2954]△ (cf. details about early stages [207]△, pupal stage [834], hinge rings [678]△, notum [1380, 3373]△, and margin [3689]). Note the nesting of *spalt's* major domain inside *omb's* domain, which in turn is a subset of *Dad's* zone [619, 2867]. This sort of successive subtraction may involve the layering of regional repressors.

---

factors that could convey Dpp's signal to other echelons of genes downstream [3074].

1. The *optomotor-blind (omb)* gene was found in a screen for flies unable to use visual cues to navigate a maze [3367], and its LOF trait has been traced to neural defects in the optic lobes [3366, 3408]. Omb (974 a.a.) shares a T-box DNA-binding motif [3254, 3255, 4000] with vertebrate *Brachyury* [3365, 3366, 4001] and has been shown to bind DNA [3365]. While Dpp and Wg control *omb* expression in the wing [1626], *omb* is regulated by Hh in abdominal segments (sans Dpp or Wg mediation) [2302]. In both cases, the actions of *omb* are partly redundant with those of *Scruffy* [2302]. This redundancy may explain why *omb*^LOF clones only cause defects when they cross the wing margin [1626], despite (1) *omb's* endogenous expression in a wide swath of the wing pouch (Fig. 6.2) and (2) *omb's* GOF ability to induce ectopic wings when expressed in the notum. Although *omb* assists Dpp in patterning, it seems not to help Dpp in promoting growth [1626] (cf. a reciprocal case [511]).

2. The *spalt* gene ("spalt" is German for "split") was recovered based on its LOF embryonic lethality and head transformation [3152], which involves homeosis [672, 2101, 3737]. Along with its paralog *spalt-related* [221] located ~65 kb away, *spalt* is activated by Dpp [986] starting in 2nd instar [4188] via a *cis*-regulatory region that is packed with discrete enhancers [220, 987, 2345] (cf. the *en-inv* paralog pair [1659]). Spalt (1355 a.a.) has 7 zinc fingers (2 pairs and 1 triplet), 2 opa motifs, and C-terminal stretches of serine, alanine, and proline

[2346], while Spalt-related (1263 a.a.) has 8 zinc fingers and similar homopolymer stretches. A fragment of Spalt-related containing 3 conserved zinc fingers binds the DNA sequence TTATGAAAT [221], although its endogenous target genes remain unknown [987].

In 1996, two teams – Thomas Lecuit *et al.* in Heidelberg [2455] and Denise Nellen *et al.* in Zürich [3074] – showed that *omb* and *spalt* are targets of Dpp in the wing disc. The "Dpp ➜ *omb*" and "Dpp ➜ *spalt*" links were demonstrated by LOF-GOF manipulations of the Dpp pathway: (1) *omb* and *spalt* expression vanish in *dpp*^LOF discs, and similar effects are seen with *tkv*^LOF or *Mad*^LOF clones; (2) *omb*-ON and *spalt*-ON domains expand to fill the A-P axis when *dpp* or *tkv*^Q253D (encoding a constitutively active Tkv receptor) is expressed widely (viz., ubiquitously or along the future margin). Dpp's effect on *omb* and *spalt* must be direct (vs. via signal relay) because although both genes can be activated outside their normal areas by *dpp* or *tkv*^Q253D, only *dpp*^GOF clones force their expression into surrounding wild-type tissue [2455, 3074]: *tkv*^Q253D clones turn *omb* and *spalt* ON exclusively within the confines of the clones. If diffusible second messengers existed, then they should have evoked *omb-spalt* expression outside the clones (nonautonomy vs. autonomy) [3999].

Dpp appears to turn *omb* and *spalt* ON at different concentrations because the endogenous stripes are of different widths: the *omb*-ON stripe is broader than the *spalt*-ON stripe (Fig. 6.2). The inferred ranking of their thresholds ($T_{spalt} > T_{omb}$) is confirmed by the concentric circles of target gene expression that attend ectopic

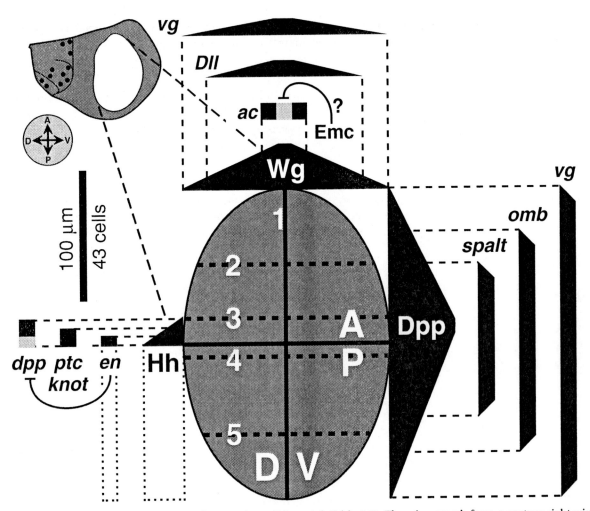

FIGURE 6.3. Morphogen gradients that dictate wing cell fates (cf. Table 6.1). The wing pouch from a mature right wing disc (above left) is idealized as an oval (cf. Fig. 6.1), with solid lines (A/P or D/V) between compartments. Dashed lines denote veins 2–5 (vein 1 = part of the margin). Scale bar applies to the enlarged oval. Black triangles around the oval are gradients of Hh, Wg, or Dpp, although Hh is also present in the P region (dotted rectangle). Outer bars indicate extents of target gene transcription, although *en* is also ON in P cells (dotted bar) and *ptc* is ON at a basal level in remaining A cells (not shown). For gene abbreviations, see Fig. 6.2 or below. For additional details and perspective, see App. 7.

Hh diffuses into the A region [4228] where it turns ON (1) *dpp* [4136, 4479], (2) *patched* (*ptc*) [2728, 4136, 4479, 4539], (3) *knot* (a.k.a. *collier*) [1840, 2894, 3077, 4479], and (4) *en* [4136] (late-3rd instar only [350, 4228]) at successively higher thresholds (although thresholds for *ptc* and *knot* are comparable). Because *en* ⊣ *dpp* [1647, 2980, 3747, 4229], *dpp* only remains ON in the front 2/3 of the Hh gradient *during late-3rd instar* [4136] (but see caveats in App. 7). Light shading denotes suppression. Ptc's level should also drop where it overlaps *en*-ON [277, 350, 2728, 3372] because *en* ⊣ *ci* and Ci is needed for *ptc* transcription [2832, 3747, 4229] (cf. Figs. 5.6 and 5.7). Why it does not is unclear [4539]. At the same threshold as *dpp*, Hh also turns ON *master of thickveins* (not shown) [1327]. At a lower threshold than any of those listed above, Hh turns ON the Iro-C genes *ara* and *caup* (not shown; cf. Fig. 6.11) [320, 1537, 2993].

Cells in the *dpp*-ON stripe make Dpp, which diffuses to turn ON (1) *vg* [2217, 2219], (2) *omb* [2455, 3074], and (3) *spalt* [2455, 3074] at successively higher thresholds.

Cells at the D/V boundary make Wg, which diffuses to turn ON (1) *vg* [678, 2217, 3091, 4849], (2) *Dll* [166, 1040, 3089, 3091, 4849], and (3) *achaete* (*ac*; also *scute*, not shown) [166, 912, 3689, 3982] at successively higher thresholds, though *Dll* and *vg* respond in a graded manner. So do *Dfz3* (activated) [3977], *vvl* (repressed) [703, 990, 995], and the Iro-C genes *ara* and *caup* (repressed) [1537] (not shown). The *ac* stripe splits into two bands (each ~4 cells wide; cf. Fig. 6.8) [2079, 3689, 3690] apparently due to Cut ⊣ *ac* [988], possibly via Emc [913] or Mβ [871] or both [205]. Cut was once thought to be a Wg target [352, 353], but the link is indirect [2839, 3089] via Vg [3936] and Scalloped [2940]. Oddly, Wg also turns ON *omb* (not shown) [1626] via the same *cis*-enhancer that is under Brinker-mediated Dpp control [3978].

The overall diagram is adapted from [3087, 4137], but amended for *vg* regulation [703, 2217], with specific pathway data from Hh [2992, 2993, 4136, 4478, 4479, 4539], Dpp [2455, 3074, 3406, 3972], and Wg [4849]. Cell dimensions (scale bar) reflect the mean of 10 counts along 25-μm transects in Fig. 8b of [350].

dpp^{GOF} clones: *omb* and *spalt* are activated at different ranges ($r_{dpp} < r_{spalt} < r_{omb}$) [3074]. Lecuit *et al.* verified this finding after initially obtaining contradictory results [2455, 3999] that were attributable to not allowing enough time for the "halos" to materialize [2457].

When the *dpp*-ON stripe and the *en*-ON P compartment are also taken into consideration, the endogenous ON or OFF states of *omb* and *spalt* would subdivide the A-P axis into 7 zones [3074] – each with a unique set of transcription factors (Ci-155 rules the *dpp*-ON zone). Using a binary code for the sets of ON/OFF states (En, Ci-155, Spalt, and Omb from left to right), the zones (A to P) are 0000, 0001, 0011, 0111, 1011, 1001, and 1000 (cf. Figs. 5.7 and 6.3).

Dpp concentrations evidently specify different structures along the A-P axis because the weaker *spalt* expression in tkv^{LOF} (or sax^{null}) clones is associated with shifts in fate (in terms of vein location or marginal bristles) [3972]. The autonomy of the shifts implies that the cells are less able to "hear" the Dpp signal, so they "think" they are farther from the Dpp source than they actually are (cf. a similar case for Hh [2993]).

Nellen *et al.* make a clever case against any other A-P gradients of fate-determining factors (e.g., Gbb [1674, 2203, 4610]) in the wing disc. If such factors existed, they argue, then the *omb*-ON and *spalt*-ON "halos" around ectopic dpp^{GOF} clones should be skewed in the direction of those other gradients. In fact, however, the halos are uniform in width all around the clones, regardless of where the clones are located in the disc.

Dpp regulates *omb* and *spalt* similarly despite clues to the contrary

Lecuit *et al.* examined *omb* and *spalt* expression in *dpp-Gal4:UAS-dpp* wing discs, where *dpp* is upregulated in its usual stripe. These discs are grossly (~4x) widened along the A-P axis (except in the notal area) because Dpp is a potent mitogen (cf. Ch. 5). On the simple assumption that *spalt* and *omb* are direct read-outs of high or low Dpp levels, respectively, both bands (*spalt*-ON and *omb*-ON) should widen to the same extent (viz. ~4x), but neither band does so. The *spalt*-ON band widens by ~1.3x, whereas the *omb*-ON band retains normal size. The net result is that the two bands become congruent. On this basis, Lecuit *et al.* devised an "**Omb Memory Hypothesis**." Expression of *spalt*, they argued, is a true read-out for Dpp (ignoring the 1.3x vs. 4x discrepancy) because *spalt* is continually sensitive to Dpp (cf. the Cabaret Metaphor, Ch. 4). In contrast, expression of *omb* would be sensitive to Dpp only early in development.

The *omb*-ON state would "ratchet up" early [1657, 3999] and stay on heritably in all descendant cells, regardless of their later locations in the Dpp gradient [4575]. (Assuming that *Gal4*-driven Dpp arises after *omb*-ON is recorded, *spalt*'s domain could theoretically expand without affecting *omb*.) This hypothesis cannot be true, however, because *omb* needs Dpp continuously [571, 4849]: *omb* expression vanishes in tkv^{null} clones until extremely late in 3rd instar [3074] (cf. similar *omb* downregulation in late Mad^{LOF} clones [2455]).

In 1998, a simpler explanation was proposed by Haerry *et al.* [1674] and Lecuit and Cohen [2457] based on the fact that Tkv is expressed intensely just outside the *omb* domain but weakly inside it until late 3rd instar (Fig. 6.2) [512, 980]. This complementarity could explain *omb*-*spalt* congruence if high levels of Tkv were acting as a barrier for Dpp diffusion. In that case, the excess Dpp in *dpp-Gal4:UAS-dpp* discs would push the edges of the *spalt*-ON domain out to the Tkv barrier but no farther, whenceforth the *spalt*-ON and *omb*-ON domains would coincide. Several lines of evidence support this "**Tkv Barrier Hypothesis**":

1. In *dpp-Gal4:UAS-dpp* discs, the *tkv*-ON domain continues to "hug" the *omb*-ON domain just as it does in wild-type discs (and just as it should if it is a barrier), despite excess tissue growth outside the *omb*-ON area [1674].

2. Overexpression of a wild-type *tkv* transgene in the wing pouch causes A-P shrinkage of both the *omb*-ON and *spalt*-ON bands [1674], as if the excess Tkv is impeding the outward movement of Dpp. Consistent with this inference, the *omb*-ON band becomes nonuniform, with intense expression along the *dpp*-ON stripe and lower expression peripherally.

3. Dramatic evidence for an effect of Tkv on Dpp diffusion is seen in discs where *en-Gal4* drives a wild-type *UAS-tkv* transgene [1674]. Because the *en*-ON domain in wild-type discs abuts the rear edge of the *dpp*-ON stripe, *en-Gal4:UAS-tkv* discs should acquire a new Tkv barrier just behind the *dpp*-ON stripe. Indeed, expression of *omb* in these discs is intense precisely there, as if posteriorly directed Dpp is hitting a wall and increasing in concentration.

Dpp does not regulate *tkv* in 3rd instar despite clues to the contrary

The complementarity between *omb*-ON and *tkv*-ON domains in wild-type discs (Fig. 6.2) could easily arise

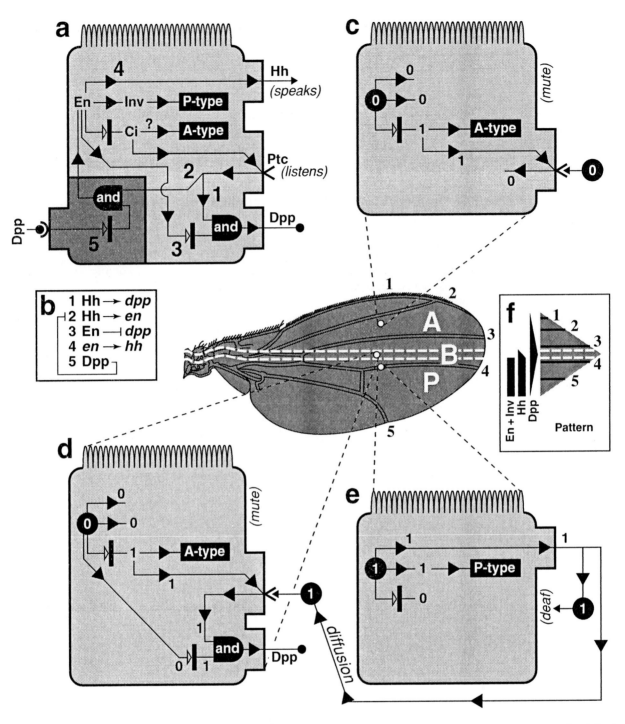

a

4

Hh *(speaks)*

En → Inv → P-type

Ci ? → A-type

Ptc *(listens)*

and

2

1

Dpp

and

Dpp

5

3

b

1 Hh → *dpp*
2 Hh → *en*
3 En ⊣ *dpp*
4 *en* → *hh*
5 Dpp ⊣

c *(mute)*

0

0

0

1 → A-type

1

0 ← 0

d *(mute)*

0

0

0

1 → A-type

1

1

1

0 1 and Dpp

1

e

1   1

1 → P-type

1

1

0

(deaf) 1

diffusion

1

f

1 2

3

4

5

En + Inv
Hh
Dpp

Pattern

from a "Dpp ➡ *omb* ⊣ *tkv*" cascade. The "*omb* ⊣ *tkv*" suppression cannot be total, however, because a minimum amount of Tkv is needed for *omb* expression in its endogenous domain [3074], and Tkv is also required there for cell proliferation [569, 3972]. Lecuit and Cohen argue that some sort of "Dpp ⊣ *tkv*" link (with or without *omb* as an intermediary) must exist because *tkv* expression is suppressed near clones that ectopically express *dpp* [2457]. However, the *Gal4* driver they used

("*C765*") cranks *dpp* expression far above physiological levels, and the discs were not examined until 48 h after clone induction. Thus, it is possible that the reduced Tkv is an artifact of the "industrial strength" Dpp or a distant side effect of prolonged exposure to Dpp (e.g., induced expression of *omb* or other genes that could confer new cell identities). Indeed, other experiments argue against any regulation of *tkv* by either *dpp* or *omb*.

FIGURE 6.4. Circuitry that enforces A vs. P identities of wing cells and causes expression of *dpp* at the A/P interface. Abbreviations: Ci (Cubitus interruptus), Dpp (Decapentaplegic), En (Engrailed), Hh (Hedgehog), Inv (Invected), Ptc (Patched).

a. Generic wing disc cell (cf. Fig. 2.7 for icons) showing the main circuit that operates throughout disc development (light gray), plus a "plug-in" module that works only in late-3rd instar (dark gray) [1647]. Numbers refer to links in **b**.

b. Major links, redrawn using epistasis symbols (cf. Figs. 5.6 and 5.7). Link 2 (receipt of Hh signal activates *en*) only becomes operative in late-3rd instar, when the *en*-ON state spreads into the A compartment to fill the zone marked "B" in the wing diagram (cf. Fig. 6.7d) [350]. Because Link 2 requires more Hh than Link 1 [4136], the band of *dpp*-ON cells induced by Hh extends farther anteriorly than the band of *en*-ON cells, and Link 3 then turns OFF *dpp* where these bands overlap (cf. Fig. 6.3). Link 5 is a "safety switch" that prevents a runaway cycle of *en* (Link 2) and *hh* (Link 4) activation that would convert all cells to a P-type state (cf. Minotaur Scenario; Fig. 5.10c-e; Ch. 5). Link 5 can be overridden by excess Hh (not shown) [2992], so an analog pressure clamp would be a more apt analogy than a digital switch. Literally, the clamp is probably the receipt of a maximal Dpp signal by Tkv-Punt receptors, but it might instead be an inability of *dpp*-ON cells to turn ON *en*. All we really know for sure is that *en*-ON states cannot traverse *dpp*-ON tissue. Link 4 is disabled in the B zone [2992] because *hh* is more sensitive to repression by Ci-75 than to activation by Ci-155 [2980]. In any event, the plug-in module is incidental to the grand scheme and is omitted from further consideration.

c–e. Schematics of cells in 3 different districts of gene expression along the A-P axis *before late-3rd instar*. "A" denotes the anterior A compartment beyond the range of diffusible Hh, "B" is the border zone (of A-type cells) that responds to Hh by forming a stripe of *dpp*-ON cells, and "P" is the P compartment. States of variables (cf. **a**) are recorded as "1" (present) or "0" (absent), although thresholds do matter (discussed above). Black circles indicate decisive factors. In plain English, the logic of the main circuit is as follows. If *en* is OFF (**c** and **d**), then the cell takes the Ci route, deploys Ptc (Hh's receptor), and adopts A-type identity, although an agent other than Ci may mediate the latter. If *en* is ON (**e**), then the cell secretes Hh and turns ON *inv*, which causes P-type identity. One-way signaling between Hh-secreting "speakers" (**e**) and Ptc-competent "listeners" (**d**) establishes a *dpp*-ON stripe in the B zone, which then secretes Dpp (cf. Fig. 6.5). Although P cells are bathed in Hh, Link 1 cannot work there because (1) they lack Ptc and so are deaf to Hh [328, 4229], (2) they lack Ci (due to "En ⊣ *ci*") and so cannot activate *dpp* ("Ci → *dpp*" link not shown) [1827, 2832, 4229], and (3) En represses *dpp* directly by binding its *cis*-enhancers [2980, 3747]. Because cell states are set by whether *en* is ON and whether Hh is received, these variables (black circles in **c–e**) act like a binary code, where the ordered pairs (*en* ON?; Hh bound?) are 00, 01, and 10. These codes define the A, B, and P identities.

f. Profiles of protein distributions (black bars or triangles) *before late-3rd instar* relative to districts of gene expression, which are demarcated on a realistic wing at left and a schematic at right. Rear dashed line is A/P boundary (after [497]). Invected (under En control) lets A vs. P cells interpret symmetric Dpp gradients differently to make an asymmetric pattern (i.e., veins 1–3 vs. 4–5) [3934, 3935].

Circuitry in **a** is based on [643, 1647, 1659, 3935, 4229, 4848], and the gradients in panel **f** are adapted from [4229, 4734]. The logic of this circuit is dramatically confirmed by using *dpp-Gal4* to drive *UAS-hh*. These conditions entrap the system in an "infinite loop": every time an A-type cell is instructed by Hh to turn ON *dpp* (**d**), it must now also turn ON *hh*, leading to another round of Hh signaling that enlarges the *hh*-ON area still farther. The outcome of this robotic iteration is a grotesque multiplication of bristles and sensilla that are normally limited to the A/P boundary [2992].

N.B.: Input/output ports (raised rectangles) on the cell surface are drawn at arbitrary apical-basal levels (apex denoted by microvilli). The "*en* → *hh*" link is written in gene symbols because – unlike "En ⊣ *dpp*" – it may be indirect [224] since En is normally a repressor (cf. Fig. 5.7 legend). The role of *inv* is unclear because null alleles have no apparent effect [358]. For other examples of "safety switches" that override morphogen inputs, see [2176, 3733].

1. Ubiquitous overexpression of *dpp* (using a moderate *Gal4* driver) does not downregulate *tkv*, nor does a constitutively active Tkv receptor (Tkv^{Q199D}) when it is expressed similarly [1674]. These results argue against a "Dpp ⊣ *tkv*" link, and another aspect of these same discs refutes an "*omb* ⊣ *tkv*" link. To wit, when the *omb*-ON band expands, it overlaps the normal *tkv*-ON domain in the distal half of the disc, but does not affect *tkv*.

2. During late 3rd instar, the wild-type *tkv*-ON pattern changes so there is now extensive overlap between *omb*-ON and *tkv*-ON domains in the P compartment [2457]. This overlap again refutes an "*omb* ⊣ *tkv*" link.

3. Clones of *dpp*^{GOF} cells show no lower Tkv-dependent phosphorylation of Mad within or around the clone, whereas *hh*^{GOF} clones do (implying that "Hh ⊣ *tkv*") [4251].

4. The wings of flies whose Hh is replaced by a membrane-tethered Hh construct have remarkably few defects (~50% linear dimensions, missing 3/4 intervein; Fig. 6.6h), considering that their *dpp*-ON stripe is only 1 cell wide (vs. the normal ~5-cell width; Table 6.1) [572, 4136]. Robustness of this sort implies a circuit wherein the Dpp output per cell varies as an inverse function of the number of Dpp-secreting cells.

TABLE 6.1. GRADIENTS THAT ESTABLISH CELL FATES IN THE WING PRIMORDIUM*

Feature	Hedgehog Gradient	Decapentaplegic Gradient	Wingless Gradient
Ligand (mature secreted fragment)	Hh = glycosylated polypeptide (214 a.a. piece of 471 a.a. precursor that cleaves itself) covalently linked to cholesterol [255, 2705, 3432, 3434].	Dpp = glycosylated polypeptide (132 a.a. piece of 588 a.a. precursor), which forms S-S linked dimers [1427, 3224, 3244].	Wg = glycosylated polypeptide (451 a.a. piece of 468 a.a. precursor) [175, 3561, 3599].
Source of ligand ("AC" and "PC" are anterior and posterior compartments)	PC [2467, 2832, 4227, 4228, 4254].	AC stripe that abuts the A/P boundary [754, 2739]. On average, the stripe is ~5 cells wide [4136, 4479], but it varies from ~2–8 cells in width and even dwindles to zero at the wing margin [3497].	Stripe that straddles the D/V compartment boundary [163, 177, 882, 4683]. Wg is expressed intensely in a band ~4 cells wide [882, 3091, 3689, 4138].
Mode of ligand movement (possibilities including free diffusion [2780], transcytosis [2843], and cytonemes [3507])	Unknown [786]. Needs proteoglycans (but not Dally) [277, 4275]. Hh's cholesterol tail suggests that Hh may ride "lipid rafts" [1896, 3946], but it could be a "handle" for passing Hh from cell to cell via Ptc [572]. Ptc colocalizes with Hh along the apical-basal span of each cell as punctate dots that suggest vesicles [572]. Ptc is mainly in lateral membranes and undergoes endocytosis [644, 731, 1554].	Unknown. Needs proteoglycans (including Dally) [2004, 2556, 3038] and clathrin (implying endocytosis) [1546].	Unknown [1910]. Needs proteoglycans (including Dally) [2556, 3038, 3561]. Different parts of the Wg molecule cause receptor activation vs. intercellular transport [273, 1055, 1779]. Uptake into vesicles by receiving cells occurs in the absence of Fz and Dfz2 (implying unknown nontransducing binding proteins) [2984]. Dynamin (and hence endocytosis) is needed for Wg transport in embryos [2900], but Wg can move extracellularly in discs without it [4138].
Effect of receptor on ligand movement	Ptc impedes Hh movement [277, 754, 755, 2074].	Tkv impedes Dpp movement [1674].	None [595]. Dfz2 retards Wg degradation and increases cells' sensitivity to Wg signal but does not affect Wg movement.
Effect of ligand on receptor gene expression	Hh ➡ *ptc* [642, 1841, 1982, 2074].	A "Dpp ⊣ *tkv*" link can be forced artificially [2457] but is unlikely to operate in nature [1674].	A "Wg ⊣ *Dfz2*" link operates in wing discs [595] but is not likely to play any role in patterning [734].
Orientation of gradient	A-P axis [4136, 4479, 4539].	A-P axis [3074].	D-V axis [4138, 4849].
Farthest effects (deduced from LOF-GOF studies and probes of target gene expression)	Direct effects (ignoring secondary consequences of Dpp) extend ~10 cells [277, 786, 2728, 2993, 4136] from A/P boundary to ~ vein 3 [572, 1837, 2992, 4136, 4188] (although Hh itself does not induce vein 3 [364]). A similar impact radius is seen around ectopic *hh*-GOF clones [572].	Entire A-P axis of the future wing (~50 cells in either direction from A/P line) [3438, 4032, 4251, 4265].	Entire D-V axis of the future wing (~40 cells in either direction from D/V line) [1910, 3977, 4849].

Range of ligand detected (by antibody staining) away from its source**	2–4 cells [277, 572, 4228].	~35 cells (80 μm) from A/P stripe of *dpp*-ON cells, as assessed by a GFP-tagged Dpp construct [1169, 4265]. This range nearly spans the pouch. Dpp traverses ~4 cells per hour [1169].	~25 cells from D/V boundary (≈ the entire breadth of the pouch) is the maximum reported [595, 3691].
Target genes ("high," "medium," or "low" refer to thresholds for activation)	*en* (high [4136], late-3rd instar only [4228]), *ptc* (lower than *en* [4136] but debatable as to whether it is high [572, 2728, 4479, 4539] or medium [277] or low [4136]), *kn* (medium? [2894, 4479]), *dpp* (low [4136, 4479]), Iro-C genes *ara* and *caup* (minimal [320, 1537, 2993]).	*spalt* (high [2455, 3074]), *optomotor-blind* (medium [2455, 3074]), *vestigial* (low [2219]).	*achaete* and *scute* (high [882, 912, 3374, 3689, 3982]), *Delta* (high [595, 988]), *Distal-less* (medium [3091, 4849]), *vestigial* (low [3091, 4849]).

*See Figure 6.3 for disc geometry. Ranges refer to mature discs. Gene abbreviations: *dpp* (*decapentaplegic*), *en* (*engrailed*), *hh* (*hedgehog*), *kn* (*knot*), *ptc* (*patched*), *wg* (*wingless*). N.B.: The ability of molecules to exert effects at concentrations that are undetectable by antibodies (e.g., Hh [4136]) is not unique to these morphogens [4489, 4849]. Historically, such negative results confounded the analysis of Notch (cf. Ch. 2), Ci [155, 4539, 4540], and Mad [3095, 3498], as well as the axis and gap gene products [1946, 3248]. A current example is *neuralized*, which is not detectably expressed in PNC cells but must directly affect them because its null allele autonomously transforms them into SOPs [2387] (but see [2387] sequels for evidence to the contrary).

**Until 2000, the diffusion range of Dpp was not directly known [3406] due to lack of suitable antibodies [1546, 1807, 3087] The use of Green Fluorescent Protein (GFP) to tag Dpp has been a breakthrough [1169, 4265], and the Dpp-GFP construct is biologically active. Nevertheless, its molecular weight exceeds that of the native Dpp ligand, so its concentration profile and turnover rate may be misleading. Diffusion ranges for other ligands are crudely estimated to be ~10 cells for Argos [1294, 2331, 4035], ~3 cells for Scabrous [186, 2461], and ~4 cells for Spitz [1288, 1337].

Indeed, cells in the *dpp*-ON stripe must be using Tkv to monitor the intensity of Dpp signal because (1) the *dpp*-ON stripe disappears when a *tkv*[GOF] transgene is expressed ubiquitously (see also the small wings in *dpp-Gal4:UAS-tkv* flies [4251]), and (2) the *dpp*-ON stripe doubles in width when a *tkv*[LOF] state is created along the stripe (via a *ptc-Gal4* driver and a dominant-negative *UAS-tkv*[DN]) [1674]. This sort of modulation could not work with a "Dpp ⊣ *tkv*" link because the latter would prevent cells from sensing the absolute amount of Dpp. Indeed, all cells in the wing pouch (not just the ones in the *dpp*-ON stripe) must be sensing an absolute level of Dpp because clones with excess Tkv ratchet their identities (e.g., from *omb*-ON to *spalt*-ON), as if they are "hearing" a stronger Dpp signal [2457]. Tkv-mediated autoregulation may also explain the ability of the *dpp*-ON state to sustain itself for ≥12 h when deprived of Hh [2992].

Despite the above caveats, the idea that Dpp affects its receptor's expression has become a gospel truth [1327, 1546, 3347, 3406, 4137, 4251], and derivative models are starting to be based on it [977, 3225]. The "Dpp ⊣ *tkv*" link is seductive because it would explain why mitoses can be uniform throughout each disc [1134, 1142], even though the major

mitogen (Dpp) is concentrated along one (A/P) line [1839]. To wit, if ligand and receptor gradients are opposed to one another, then all cells could "hear" the same level of Dpp regardless of their distance from the source – like a rock concert where fans standing closest to the band tend to be the deafest.

Two additional facts refute this "**Rock Concert Scenario.**" If a "Dpp ⊣ *tkv*" link were operating, then increasing the amount of Dpp via *dpp-Gal4:UAS-dpp* should have no effect because all cells should become deafer (i.e., reduce their Tkv) as the *dpp*-ON stripe raises its volume (i.e., increases Dpp output). In reality, however, the disc quadruples in width. One way out of this corner would be to assert that the "Dpp ⊣ *tkv*" link only works up to a certain level, above which cells hear more Dpp and multiply faster. In that case, proliferation should still be uniform throughout the disc, but here the second contradiction is encountered. The area where *omb* and *spalt* are expressed is virtually unaffected, while the outlying areas balloon to huge proportions [2455]. Such behavior implies a step function rather than a gradient, and Tkv does have a step function profile [4265] (Fig. 6.2; cf. the Tkv Barrier Hypothesis). Thus, although Tkv mediates both the mitogen and morphogen effects of Dpp, it does not suffer the kind of modulation that Dpp exerts on *Dad* or *brk* (cf. App. 6). Putting *tkv* under the control of "prepattern genes" that are independent of *dpp* makes sense because it would allow evolution to orchestrate the shapes of adult organs by targeting more or less Tkv to specific parts of a disc at specific times, thereby tinkering with growth rates in a balkanized way. The "programming language" of morphogenesis undoubtedly involves such tricks [1165, 4114].

In 2001, one of the prepattern genes that controls *tkv* was confirmed. It is *hedgehog*. When Hedgehog diffuses into the A compartment, it turns ON a gene called *master of thickveins (mtv)* [1327], which suppresses *tkv*. The "Hh ➔ *mtv* ➔ *tkv*" chain operates in parallel with the "Hh ➔ *dpp*" link. How *tkv* is regulated in the rest of the pouch is unclear because *mtv*^null clones have relatively little effect on *tkv* levels beyond the A/P border zone [1327].

Many other riddles persist about how Dpp functions as a mitogen. For example, if each section of the *dpp*-ON stripe is needed for normal growth [3438, 4032], then why do outgrowths arise only when ectopic *dpp*-ON stripes cross the D/V line [1839, 4848] (see below)?

A vs. P identities might explain how a straight A/P line emerges

The *dpp*-ON stripe establishes two gradients along the A-P axis. To a first approximation, the A gradient spans the A compartment and the P gradient spans the P compartment. In reality, the A/P gradient interface ("AG/PG") is offset from the A/P compartment boundary ("AC/PC") by ~6 cells because the peak expression of *dpp* is in the middle of the 5-cell wide *dpp*-ON stripe, and that stripe is separated from the AC/PC line by the 3–4-cell wide band of *en*-ON A cells (Fig. 6.3). The implications of this imperfect registration are addressed later.

How can back-to-back gradients that use the same morphogen produce different patterns of veins, bristles, etc., in the adult wing [904, 4734]? The obvious answer (cf. Ch. 4) is that *engrailed* acts a "selector gene" [1358]. To wit, *en*-ON P cells would adopt P-type competence that makes them interpret Dpp levels in one way, while *en*-OFF A cells adopt (by default) A-type competence that makes them interpret Dpp levels in a different way (cf. Fig. 4.3b) [643, 1628, 1838, 2441, 2930]. Enforcement of competence states would be a second duty of compartments [231, 1807, 3862], aside from their creation of a morphogen stripe via the Deaf-Speakers/Mute-Listeners Trick (cf. Ch. 5) [328, 4848]. By using this genetic gadgetry, an asymmetric pattern can thus be built on a symmetric scaffold.

Indeed, the effects of *en*^1 were a cornerstone for García-Bellido's Selector Gene Hypothesis [1355, 1358]. This allele partly transforms P compartments to look like A ones and thus causes asymmetric patterns to become symmetric (e.g., the foreleg acquires a second sex comb in mirror image to the normal one [439]). The *en* gene was long suspected of being involved in transducing rather than sending positional information because it essentially behaves autonomously in mosaics. For instance, small patches of *en*^1 tissue can form a sex comb on the P (wrong) side of the leg [4343] and triple-row bristles on the P (wrong) edge of the wing [1378]. Once *en* was cloned in 1985 [1031, 1247, 2356, 3429] and shown to encode a DNA-binding transcription factor [2047, 3719], the case for its involvement as a "switch gene" was bolstered.

The case was further strengthened when *en* was found to be expressed coextensively with P compartments (as defined by cell-lineage restrictions) of thoracic discs [495, 1697, 1963, 2307]. This congruence was expected because *en* offered a way to neatly explain the straightness of the wing's A/P boundary. The A/P line in the adult wing runs for hundreds of cells without wavering more than one cell diameter [350, 1078, 3441]. It is also straight across entire wing and leg discs through early 3rd instar [569, 2728, 3764], although it later manifests kinks [495, 497, 1022, 1833], possibly due to being "pushed" by uneven growth [497].

Speculation about *en* as a boundary-straightening agent began in 1975. Gines Morata and Peter Lawrence proposed that A and P cells have different affinities due to expression (P) or nonexpression (A) of *en* [2441, 2928]. They imagined that a line forms between A and P cells in the same way that oil and water segregate to form an interface (viz., via homophilic sorting). The evidence for this **"Selector Affinity Model"** is reminiscent of leg clones with improper P-D states (e.g., *Dll* [1561] or *hth* [1573]; cf. Ch. 5).

1. "Xenotopic" (i.e., located in alien territory) *en* clones can cross the A/P line: (1) crossing (P-to-A) by A-type *en*^1 clones [2441, 2927, 2928] or *en*^LOF clones [2306, 2447] and (2) crossing (A-to-P) by P-type *en*^GOF clones [4848] (cf. *Polycomb*^LOF [4327]). This strategy could correct sporadic mistakes so as to maintain separate pools of signaling vs. receiving cells [2677], although it is unclear how lineage restraints are enforced in late-3rd instar when *en*-ON cells exist on both sides of the A/P line [354].

2. Xenotopic *en* clones far from the A/P line minimize contact with surrounding tissue by adopting a circular shape and smooth outline: (1) A-type *en*^{null} *inv*^{null} (but not *en*^{1} [1639, 2441]) clones in the P region [4229], (2) P-type *en*^{GOF} clones (which have 2x normal cell density [4848]) in the A region [1078, 4848], and (3) *polyhomeotic*^{LOF} (*en*-ON) clones in the A region [3512].

3. Xenotopic *en* clones far from the A/P line can form vesicles in an apparent attempt to leave the epithelium (cf. sorting of wing and haltere cells [2924, 3875]): (1) A-type *en*^{LOF} clones in the P region [2447, 2927] and (2) *polyhomeotic*^{LOF} (*en*-ON) clones in the A region [3512]. However, A-type *en*^{1} bristles in secondary (P-side) sex combs do not leave the P territory in mosaics [4343].

Homophilic sorting implies A- vs. P-specific adhesion molecules [493], but none has so far been found [942, 1839], although En does control expression of Connectin and Neuroglian in the CNS [3925]. Generally speaking, cell identities must be implemented by echelons of "realizator" genes that are downstream of selector genes [1225, 1226, 1358, 1367, 1924], but only ~5 possible realizators for *en* have surfaced thus far [3457].

But the A/P line appears to straighten via a signaling mechanism

Notwithstanding all the supportive evidence, *en*'s status as a selector gene was controversial for many years [354, 1635, 1636, 1837] due to (1) neomorphic idiosyncrasies of *en*^{1} [852, 1138, 1636, 2306], (2) the inability of *bona fide* LOF alleles to mimic its mirror-image syndrome [852, 1636, 2447], and (3) expression of *en* in A-type cells at the A/P border [4539]. A hidden confounding factor all along, it turned out, was an adjacent paralog – *invected* (*inv*) [842] – whose function partly overlaps that of *en* [1659, 3925]. Only when both of these homeobox genes are inactivated is it possible to achieve complete P-to-A transformations (Fig. 6.6) [4229].

To the extent that *en* and *inv* act redundantly, they should have similar effects when misexpressed in the A region. In fact, their GOF phenotypes differ considerably: a viable *inv*^{GOF} mutant has nearly perfect P/P symmetric wings (A-to-P homeosis) [3934, 3935], while *en*^{GOF} A clones cause milder A-to-P conversions [4848]. Moreover, *inv*^{GOF} has virtually no effect on the expression of *hh*, *dpp*, *ci*, or *ptc* [3935], whereas *en*^{GOF} alters them dramatically [1647, 4848]. Evidently, a division of labor exists: *inv*'s duty is to implement P identity downstream of *en* ("*en* ➡ *inv* ➡ P identity" [1647, 4229] but see [358]), whereas *en* has two duties: (1) to activate *inv*, and (2) to

activate *hh* so as to trip the Hh-Dpp relay that patterns the wing ("*en* ➡ *hh* ➡ *dpp*-ON stripe at A/P boundary") [1837, 4229]. Other genes aside from *en* must keep *inv* ON in the P region because *en*^{null} clones do not mimic the P-to-A homeosis of *en*^{null} *inv*^{null} clones [4229].

In 1997, the orthodoxy of the Selector Affinity Model was shaken by a heretical discovery. Clones of *smo*^{null} cells can cross from A to P without ever changing their *en* state [364, 2435, 3627], and a similar A-to-P crossing was later seen with *fu*^{LOF} clones [3177]. Some involvement of the Hh pathway with cell affinities had earlier been suggested by the round shape of A clones (far from the A/P boundary) that secrete Hh (*ci*^{null} [1078]) or "think" they are receiving Hh signal (*DC0*^{LOF} [2058, 2072, 2491, 2533, 3238], *ptc*^{LOF} [642, 754, 1078, 2834, 3372, 3627], or *slimb*^{LOF} [2856]), and the A-to-P shift of *smo*^{null} clones can be halted by activating the Hh pathway downstream of Smo (via *DC0*^{LOF}) [3627].

These results imply that the confinement of A-born cells to A compartments and P-born cells to P compartments has nothing to do with A-type (*en*-OFF) vs. P-type (*en*-ON) states per se. Rather, the segregation appears to be enforced by a cohort of *dpp*-ON A cells near the A/P line. Any A cell can join this "ΔΠΠ fraternity" if it turns *dpp* ON, which will happen naturally as A cells jostle close to the A/P line and encounter enough Hh (cf. the Cabaret Metaphor, Ch. 4). Likewise, A cells can depart ΔΠΠ by turning *dpp* OFF, which happens as growth pushes *dpp*-ON cells too far from the P compartment, so the *dpp*-ON stripe itself is not a cell-lineage compartment. As the examples below illustrate, this "**Border Guard Model**" [364, 1839] can explain the old phenomena as well as the Selector Affinity Model, plus it accommodates the new results that the latter cannot (Fig. 6.5):

1. *Crossing the A/P border from P to A.* Given that a *dpp*-ON state is a passport into the *dpp*-ON stripe, while *dpp*-OFF means banishment thence, P cells never cross the border because their *en*-ON identity prevents them from turning *dpp* ON (*en* ⊣ *dpp*). In an *en*^{LOF} clone, the P-born cells acquire passports because the turning OFF of their *en* gene lets them hear the Hh signals all around them, whereupon they turn ON *dpp*. If they are close enough to the A/P line, then they can enter ΔΠΠ and move farther into the A region by the exit route that A cells normally use.

2. *Crossing the A/P border from A to P.* Banishment from ΔΠΠ is imposed on any member whose *dpp* gene gets turned OFF. Such a switch will happen in border clones whose *en* gene is artificially turned ON (*en*^{GOF} ⊣ *dpp*), but it can also happen when the Hh transduction pathway is shut OFF (e.g., *smo*^{null})

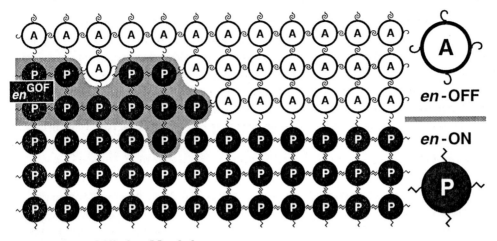

Selector Affinity Model

Border Guard Model

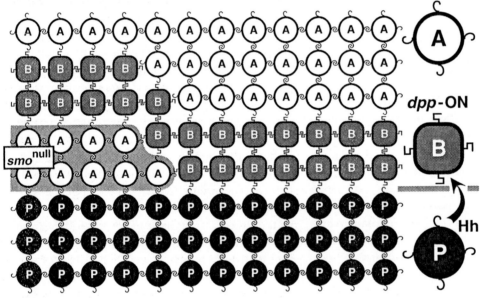

FIGURE 6.5. Alternative explanations for why cells cannot normally leave the compartment where they are born. Part of the A/P interface is shown in each panel, and cell types are enlarged at right.

According to the **Selector Affinity Model** (above) [904, 1369, 2441, 2928], the ON or OFF state of a cell's *engrailed* (*en*) gene "selects" whether it adopts P- or A-type identity. A-type cells use one kind of homophilic adhesion protein (round hook), P-type cells use another (zigzag hook), and A-P binding is disfavored. Only when cells switch identities – e.g., the shaded *en*^GOF A clone – can they cross the A/P line (gray bar at right). In this schematic, one of the clone's cells has already crossed.

The **Border Guard Model** [364, 1839, 3627] postulates a unique adhesion protein (square hook) for A (*en*-OFF) cells within a certain range (~5–10 cells vs. the 2 cells shown here; cf. Fig. 6.3) of the A/P line. These border ("B") cells turn ON *decapentaplegic* (*dpp*) in response to Hedgehog (Hh) diffusing from the P region, and they join into chains [4848]. No P or A cell can enter the B zone unless it turns ON *dpp*. A-type cells can do so when they approach the A/P boundary, but P cells cannot because *en* inhibits *dpp*. Hence, the B/P line restricts cell lineage, while the A/B line does not. If a B clone (shaded) stops transducing Hh due to *smo*^null (*smoothened*^null), then it forfeits its B identity but not its *en*-OFF state and so remains A-type. The B zone then shifts anteriorly because (1) *smo*^null suppresses the Hh receptor Ptc (Patched), and (2) Hh diffuses farther over cells having less Ptc (cf. Fig. 6.6i) [754, 755].

Interestingly, homotypic mingling (which both models presume) occurs when fast-growing clones cross the body midline from one foreleg disc to the other (A-to-A or P-to-P) after they fuse (not shown) [4076]. Segregation of cell types might not actually use qualitatively different linkers (as depicted here) but could instead rely on different linker densities [4070–4074]. Indeed, studies of *ptc*^GOF clones suggest that A and P affinities differ quantitatively because these partly deaf clones behave as if they have an intermediate affinity [2993]. Other molecules (not shown) could maintain epithelial integrity by nonspecific (or extracellular matrix) binding [1957, 3937]. Adapted from [942].

because the *dpp*-ON state cannot persist for long in cells that are deaf to Hh.

3. *Rounding and vesiculation of xenotopic clones.* ΔΠΠ members who find themselves outside the walls of their fraternity (the *dpp*-ON stripe) may still attempt to bind together as a satellite group. Ectopic ΔΠΠ identity is bestowed on A cells around *en*^{GOF} clones and on P cells just inside the edge of *en*^{LOF} clones (Fig. 6.6). Both types of clones would "round up" if *dpp*-ON cells adhere strongly, and further constriction of their *dpp*-ON annuli could lead to invagination, evagination, or detachment as a vesicle, although some of the puckering might arise from growth stimulation due to locally increased Dpp. (Reports do not state whether A-type *en*^{LOF} vesicles in P areas are totally detached [2447, 2927].)

According to this rationale, the rear edge of the *dpp*-ON stripe restricts mixing of A and P cells because (1) A cells at the edge cannot turn OFF *dpp* (Hh signal is too high), and (2) P cells cannot turn ON *dpp* (their *en*-ON state precludes it). The capacity of *dpp*-ON cells to build a "wall" is obvious in mosaics where a *dpp*^{GOF} clone partly restores wing formation in *dpp*^{LOF} flies that lack wings [1839]. Each rescued "winglet" bears a single median clone that is straight and narrow, as if the *dpp*-ON cells were organizing a stripe *de novo* by aligning and adhering tightly [4848]. When winglets consist of purely A or P tissue, they have A*A or P*P symmetry (where * is the clone). In one P*P case, a *dpp*^{GOF} clone (Fig. 6D of [4848]) was hundreds of cells long (D and V sides) but only ~5 cells wide.

Remarkably, such *dpp*^{GOF} clones create a straight *dpp*-ON ribbon without any guidance from an A/P line or Hh asymmetry, and similar "magic" is seen in *dpp*$^+$ wings where the *dpp*-ON zone is pushed far into the A region by a *smo*^{null} clone (Fig. 6.6i) [786]. Evidently, the auto-affinity is polarized along the axis of each cell so that *dpp*-ON cells align rather than clump (i.e., form a band vs. a biscuit). Polarized aggregation in other systems (e.g., fingerprints [1159, 1606] and slime molds [1492, 3594, 4783]) may offer clues to this process in flies. Suspiciously, *dpp*-ON cells are often found where tissues of the same type come together – e.g., tracheal branches [4082], dorsal closure in the embryo [1510, 1515, 3137, 3592, 4854], and midline fusion of discs during metamorphosis [34, 2714, 4429, 4853] (cf. *pannier* [614, 615, 3504]). Dpp may also help to align the eye's morphogenetic furrow [724, 930, 1616, 4648] (cf. the chain of cells therein and its buckling properties [4364]). At the tarsus-pretarsus border in leg discs, there is a ring of cells that expresses the homophilic adhesion molecule Fasciclin II [2287] but *fasII* is not under *dpp* control (cf. *dpp*-ON rings in grasshopper legs [2069]).

Rounding of *en*^{GOF} A clones and *en*^{LOF} P clones (Fig. 6.6) could thus be due to an inability of circularized *dpp*-ON stripes to break open and linearize. It is less clear why *DC0*^{LOF} and *ptc*^{LOF} clones (which express *dpp* in a solid circle instead of an annulus) behave likewise: under certain circumstances, *ptc*^{LOF} clones can straighten as keenly as *dpp*^{GOF} clones [1837]. Overgrowing clones can deform the A/P boundary but apparently cannot break it [497].

Whether a cell can join the "border guard" may actually depend not on its *dpp* state but rather on receipt of a saturating dose of Dpp, because rounding is also seen for *dpp*-OFF clones whose Dpp pathway is activated by a Tkv^{Q253D} receptor [569]. Receipt of an autocrine *dpp*-like (BMP-4) signal is thought to similarly maintain the border between neural and non-neural ectoderm in chick neurulation [4135]. Strangely, cells that both emit and receive Dpp at high doses grow slowly: (1) the endogenous *dpp*-ON stripe seems to slow its mitotic rate during 3rd instar [543, 3813], and (2) cells whose *dpp* transcription is cranked far above the normal range appear to stop dividing [2457]. Quiescence of this sort should limit the shedding of border cells into the *dpp*-OFF A population. Nevertheless, departures must be common because most of the cells in the mature A compartment have expressed *dpp* at an earlier time in wing disc development [4575].

The chain-forming ability of border cells might be expected to influence cell morphology. Cells at the wild-type A/P line are aligned and do adopt odd (rectangular vs. polygonal) shapes, but these features only extend one cell diameter on either side and thus fail to reach the *dpp*-ON stripe in late-3rd-instar discs [350]. Conceivably, some combination of Dpp with other inputs fine-tunes this line in wild-type flies. Hh can affect adhesivity directly in tergites [2437], and *en* may also participate because certain combinations of *en*^{LOF} alleles transform the wing's A/P border stripe into haltere tissue [1138, 1635]. Regardless of the gadgetry, the existence of some device for aligning the *dpp*-ON stripe makes sense because the precision of all subsequent patterning rests on this slender reference axis [942].

The quandary of A/P affinities was reinvestigated in 2000 by Dahmann and Basler [943]. They found rounding of *hh*^{GOF} *smo*^{null} A clones and *ci*^{GOF} P clones – both of which are consistent with the Border Guard Model because the *dpp*-ON rings that they induce (outside or inside the clone, respectively; cf. Fig. 6.6b) could cinch the clone into a circle. However, rounding was

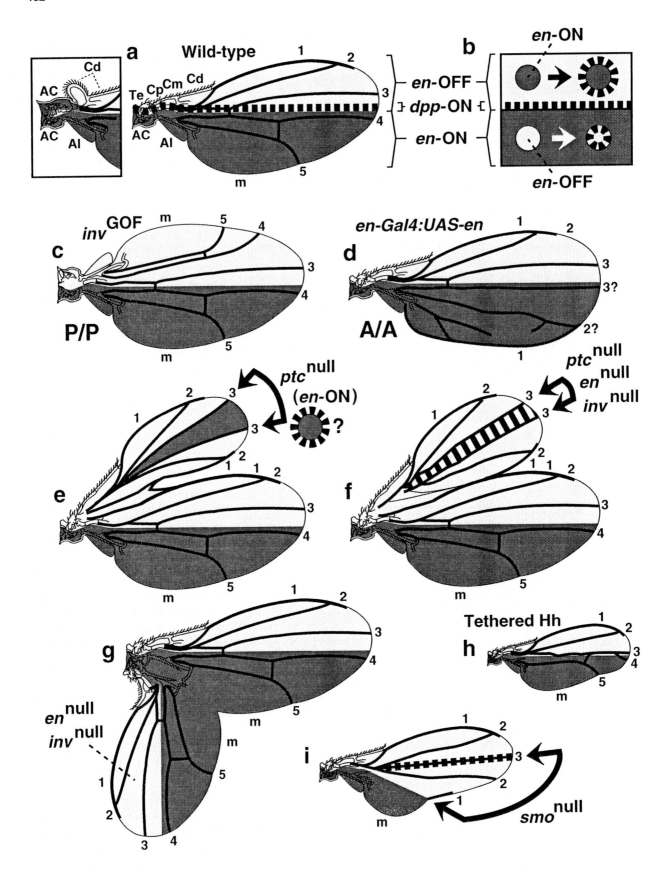

a Wild-type

Cd

AC
AC   Al

Te Cp Cm Cd
                    1   2
                         3
                    AC   Al   4
                              m   5

b   *en*-ON
        en-OFF

        dpp-ON

        en-ON

        en-OFF

inv GOF   m   5   4
c   3
                               4
P/P   m   5

en-Gal4:UAS-en   1   2
d   3
                               3?
A/A   2?
                      1

e   2   3   *ptc* null
    1   3   (*en*-ON)
         2   ?
    1   1   2
              3
                   4
         m   5

f   2   3   *ptc* null
    1   3   *en* null
                   inv null
    1   1   2
              3
                   4
         m   5

g   1   2
                        3
                        4
    en null   m   5
    inv null
         1   m
         2
         3   4

h Tethered Hh   1   2
                          3
                          4
                   m 5

i   1   2
                        3
              2
         1
         m   *smo* null

also seen with ci^{null} hh^{null} A clones and en^{null} inv^{null} ci^{null} P clones – neither of which has a dpp-ON ring. Through a series of LOF-GOF tests, the "antisocial" behavior of such "deaf-mute" clones was traced to their lack of Ci and En, rather than the deaf-mute condition per se. Clones lacking Ci and En (ci^{null} A clones, ci^{null} en^{null} A or P clones, etc.) shun both A and P cells and straddle the A/P border. Thus, another type of affinity must exist apart from the intraborder type [364]. We do not know what cell-surface molecules enforce either sort of affinity. E-cadherin is unlikely because dE-$cadherin^{null}$ clones respect the A/P line, although null or GOF dE–$cadherin$ clones do round up and GOF clones from opposite sides of the A/P line can fuse [943].

In sum, the available data favor the Border Guard Model but also argue for other tricks. If cells outside the dpp-ON stripe have a common affinity that transcends their A or P affiliation, then decreases in this stickiness might explain the old "**Cell Competition Conundrum**" [4884] – namely, why do slow-growing clones persist at A/P and D/V borders but not elsewhere [1639, 2132, 3947, 3948]. To wit, if the mitotic rates of cells were to alter their adhesivity, then retarded clones could be pushed to the edges of compartments, and those that fail to reach an edge might be squeezed out of the epithelium altogether.

Intercalation is due to a tendency of Dpp gradients to rise

Given our current picture of the circuitry (Fig. 6.4), it is now possible to devise plausible etiologies for how venation goes awry in various mosaic wings (Fig. 6.6). In the following cases, the wing's A-P pattern is formulated as

---

FIGURE 6.6. Weird wings that reveal the logic of A-P patterning. Gene abbreviations: dpp (*decapentaplegic*), en (*engrailed*), hh (*hedgehog*), inv (*invected*), ptc (*patched*), smo (*smoothened*). All panels show the dorsal side of a right wing (most bristles omitted for clarity), light vs. dark shading denote en-OFF vs. en-ON states, thick hatching (omitted in some panels) means dpp-ON, and arrows mark clone limits.

a. Wild-type wing. Landmarks: tegula (Te); proximal, medial, and distal costa (Cp, Cm, Cd); veins 1–5; posterior margin (m); alula (Al); and axillary cord (AC). Inset at left shows the base of a *30A-Gal4:UAS-en* wing where en is ectopically expressed in the Te-Cm region [1647]. This region has transformed to AC, and a mirror-image copy of Cd has grown out, possibly due to Dpp elicited at the en-ON/OFF interface. Evidently, this interface does not enforce a vein 3/4 identity immediately, though the dpp-ON cells may "bootstrap" themselves to such a level eventually if they have sufficient time to do so (e.g., see **g**).

b. How en controls dpp. Cells whose en gene is ON secrete Hh (not shown), and diffusion of Hh into en-OFF territory turns ON dpp (cf. Fig. 6.4). "Xenotopic" clones that express en in the A region (above) thus acquire a "halo" of dpp-ON cells, whereas en-OFF clones in the P region (below) acquire an inner dpp-ON "lining." Only clones that cross the D/V boundary reorganize pattern or grossly alter growth, and they are actually elongated (e.g., **e** and **f**) vs. round [1837].

c, d. Symmetries that arise when inv is misexpressed. **c.** Phenotype of an inv^{GOF} (dominant) mutant that expresses inv throughout the wing, but which still expresses en only in the P region [3935]. Unlike en, inv affects how cells respond to Dpp (i.e., A- vs. P-type competence) but does not affect dpp or its regulators (hh, ptc, ci, etc.). Thus, inv acts downstream of en (cf. Fig. 6.4). P/P wings are also seen when a $Gal4$ transgene forces en – and hence inv – ON in the A part of the blade (not shown) [1647], but in that case vein 1 stays intact. Likewise, *fat-head*^{LOF} wings manifest a P/P phenotype because *fat-head* normally represses en in the A compartment [4607]. **d.** Partial A/A symmetry (akin to en^1 [1369, 1837]). Surprisingly, *en-Gal4:UAS-en* represses en and inv in response to initial excess of En [4229], implying negative autoregulation (En $\dashv$ en) [207]. A/A wings are also seen with LOF changes in EGFR signaling: $dRaf^{LOF}$ [207], $pointed^{LOF}$ [3804], $Ras1^{LOF}$ [207], and $vein^{LOF}$ [3929]. The etiology of the EGFR effects is obscure.

e, f. Effects of inducing ptc^{null} (**e**) vs. ptc^{null} en^{null} inv^{null} (**f**) clones in the vein 1 area. Disparity in clone breadth (3-3 distance) may depend on activation (**e**) vs. failure to activate (**f**) en within the clone, which dictates whether dpp turns ON outside or inside (zone in **f** is conjectural) [4229]. In both cases, the ectopic dpp-ON tissue induces a symmetric outgrowth, and the R-L-R handedness obeys Bateson's rule (cf. Fig. 5.3). Weaker ("1221" vs. "123321") outgrowths arise with dpp^{GOF} clones (not shown) [4848].

g. Effect of an en^{null} inv^{null} clone at the P margin. This A-type clone evidently formed a new dpp-ON zone (not shown) at its interface with P-type tissue, and that zone fostered the outgrowth of a whole new A and P compartment. The R-L mirror plane (cf. **e**, **f**) illustrates an "edge effect" predicted by Meinhardt [2808].

h. Effect of substituting a membrane-tethered Hh (which cannot diffuse) for the endogenous Hh. The wing is small, and veins 3 and 4 are fused, but other features are relatively normal (cf. Fig. 6.13c).

i. Effect of a smo^{null} clone of A-type cells at the A/P border (cf. Fig. 6.5). Hh diffuses farther over smo^{null} cells but is not transduced by them, so Hh only turns ON dpp when it reaches competent tissue beyond the clone. Evidently, this journey diminishes Hh's strength because the amount of Dpp fails to foster (1) normal growth, (2) patterning of the P region, or (3) full "123321" duplication of the A region (cf. **e**, **f**).

Wings (all at same scale) were traced from pictures in [1647] (inset in **a**), [3935] (**c**), [4229] (**d–g**), [4136] (**h**), and [754, 786] (**i**). Schematic (**b**) is based on [354, 1647, 4136, 4229, 4848]. Another "brainteaser" phenotype (not shown) whose etiology was figured out using known circuits is *dpp-Gal4:UAS-tkv* [4251] (their Fig. 6b). How duplications arise in $Egfr^{LOF}$ gro^{LOF} wings (not shown) is unknown [3465].

"123/45m" [4848], where numbers are longitudinal veins, "/" is the A/P compartment boundary (inserted at every 3–4 juxtaposition), and "m" is the part of the margin behind vein 5. All structures derived from the clone (* = initial site) are underlined, new structures (clonal and wild-type) are bracketed, and *dpp*-ON zones are denoted by "z." All the analyzed clones run parallel to the A/P line and cross the D/V boundary.

1. *en*^{GOF} *clones in the A region.* When *en* is turned ON between veins 2 and 3 (12*3/45m), the final pattern is "12{3/4*4/3}3/45m" or "12{3/*/32}3/45m" [4848]. Evidently, a *dpp*-ON zone arises at each *en* ON/OFF interface ("2z*" and "*z3"), and the extra Dpp evokes growth inside (vein 4 = P-type response) and outside (vein 3 = A-type response) the clone, leading to a roughly symmetric {3z4*4z3} or {3z*z32} subpattern at the initiation site.

2. *hh*^{GOF} *clones in the A region.* When *hh* is turned ON between veins 2 and 3 (12*3/45m), a "12{3/4*4/32}3/45m" pattern appears to develop (the authors do not label vein 4 [231]). The etiology is likely the same as for *en*^{GOF}, except that the clone gets its *en*-ON (P-type) state indirectly via a "Hh → *en*" link. When *hh* is turned ON in the vein 1 area (1*123/45m), the resulting pattern seems to be "1{23/4*4/321}123/45m" [231] (but vein 4s can be missing [2058]). Although the whole {23/4*4/321} insert is symmetric, growth is asymmetric about each *dpp*-ON zone: for some reason, more growth occurs outside than inside the clone ("23z4" and "4z321").

3. *Clones in the A region that constitutively activate the Hh pathway (cf. Ch. 5).* If an A-type cell "thinks" that it "hears" Hh, then it should behave like a *hh*^{GOF} clone – i.e., turn *en* ON (Hh → *en*) and *dpp* OFF (En ⊣ dpp). However, as discussed above, this response requires a stronger signal than to turn ON *dpp* (Hh → dpp). The threshold for turning *en* ON is reached by *ptc*^{null} clones but not by weaker *ptc*^{LOF} clones [2834, 3177, 4136, 4229]. The greater width of *ptc*^{null} (vs. *ptc*^{null} *en*^{null}) clones in the vein 1 area (Fig. 6.6) [4229] may thus reflect an attempt to form a whole P compartment (vs. a *dpp*-ON stripe), but *ptc*^{null} clones also express *dpp* unevenly [642] and lack vein 4s ("1{23*321}123/45m") just like weak *ptc*^{LOF} clones [754, 3177]. Evidently, cells can cross the "Hh → *en*" threshold but fail to turn *dpp* OFF until they make enough En [998]. Indeed, *DC0*^{null} clones express *en* weakly [3177] but create the same "4-less" pattern as *ptc*^{LOF} clones [2491, 2533, 3177, 4538] via Dpp from

the clone itself (vs. flanking cells) because *DC0*^{null} *dpp*^{null} clones do not cause duplications [2533, 3238]. Only when *DC0*^{null} clones are also mutant for *Su(fu)* do they fully turn *en* ON [3177] (cf. *cos2*^{LOF} [3976]). Other components exhibit odd defects: *slimb*^{LOF} clones participate in making vein 3 and have a small 3*3 span [4538], and a vein 3 arises in the middle of *fu*^{LOF} *ptc*^{LOF} (but not *fu*^{LOF} *DC0*^{LOF}) clones [3177].

4. *dpp*^{GOF} *clones in the A region.* Clones whose dpp gene is switched ON should make a "1{23*32}123/45m" pattern [4229], but when a weak (*Tubulinα1*) promoter was used with a *dpp* transgene, the maximal effect was "1{2*21}123/45m" [4848]. Posterior clones (123/45m*m) also failed to make vein 3 (maximum = "123/45m{m54*45m}m") [4848], and so did winglets created via *dpp*^{GOF} "rescuer" clones in *dpp*^{LOF} discs ("1*1" or "m*m") [4848]. This limitation makes sense given that vein 3 depends on Hh directly, rather than on Dpp (cf. Fig. 6.13) [2992]. Ectopic *dpp*-ON clones that employ the *Tubulinα1* promoter turn ON *omb* (low threshold) but not *spalt* (high threshold) [3074].

The latter case is especially instructive. If *dpp*^{GOF} clones do not induce vein 3 (or *spalt*) when the *dpp* gene functions at submaximal capacity, then perhaps the *dpp*-ON condition does not impose any positional identity whatsoever. In other words, a cell does not think "If my *dpp* gene is ON, then I must be a 3–4 intervein cell" [4229]. Rather, a cell whose *dpp* gene is turned ON may adopt the A-P level of its environs and then try to "climb" to the 3–4 intervein peak. If its climb is hampered (e.g., by a weak *dpp* promoter) or aborted (e.g., by insufficient time), then its Dpp level – and hence the "vein state" of nearby wing cells – will never reach the "3–4" state goal. This **"Climbing Scenario"** contrasts with the **"Intercalation Scenario"** of the old Polar Coordinate Model because it does not rely on discontinuities per se to stimulate growth.

Indeed, the final nail in the PC Model's coffin may be a "123*21/m" small-wing phenotype that arises when *smo*^{null} A clones (123*/45m) displace the *dpp*-ON stripe (Fig. 6.6i) [754]. According to the Rule of Shortest Intercalation, the 1/m disparity should elicit growth [354], but it does not. Rather, the new growth (*21) occurs where there was no discontinuity – at the original 3/4 site. Subsequent changes in identities also make no sense in terms of polar coordinates but are explicable by the Climbing Scenario. The etiology seems to be as follows [754]. When A cells at the A/P border become

smo^{null}, they can no longer transduce Hh but remain A-type because their *en* gene is still OFF. As their Hh pathway shuts down, *ptc* stops being upregulated and the density of superficial Ptc proteins subsides. Removal of these Hh-binding sites lets Hh diffuse farther. Only when Hh reaches competent smo^+ cells can it induce *dpp* expression, so the *dpp*-ON stripe shifts ahead of the smo^{null} clone (Fig. 6.5) and continues to migrate anteriorly as the clone grows. As the gap widens between the Dpp source and the P compartment, Dpp's mitogenic stimulation of P-type cells dwindles, so their growth abates. Moreover, the falling levels of Dpp reaching the P region cannot sustain the formerly high A-P positional identities, so "vein states" ebb from "4" to "4–5 intervein" to "5" and finally to "5–m". (These identity shifts preclude a **"Ratchet Scenario"** [3225, 3999] because a ratchet should prevent such demotions [1657].) Within the A region, the smo^{null} clone's proximity to the *dpp*-ON stripe nurtures its growth until it spans two interveins [754] (wild-type clones are not usually so big [521, 1545]), but here too demotions are evident: cells within the *dpp*-ON stripe sink from a "3–4 intervein" identity to a "vein 3" state (again precluding a ratchet). The probable reason is that Hh's journey over the smo^{null} territory diminishes its strength so it induces a reduced level of *dpp* expression.

A similar apposition of opposite cell states ($\approx 1/m$) is seen in a *Gal4* line where *UAS-en* is expressed in a tiny part of the anterior wing (Fig. 6.6a). The full series of elements around a wild-type wing from A to P is tegula (Te); proximal, medial, and distal costa (Cp, Cm, Cd); 123/45m; allula (Al); and axillary cord (AC). When *en* is turned ON in the Te-Cp-Cm region, those elements are replaced by AC (a P-type response to the Dpp gradient), and Cd duplicates in mirror image to yield "AC{Cd}Cd-123/45mAlAC" [1647]. Apparently, the abutting of *en*-ON (AC) with *en*-OFF (Cd) cells creates a *dpp*-ON zone whose output is too small (or whose "climb" is too brief) to raise cells above a Cd identity. Larger outgrowths (with vein-1 identity) can arise here when *en* expression is evoked by $groucho^{LOF}$ [998] or $polyhomeotic^{LOF}$ [2728, 3512, 3753].

This climbing or "bootstrapping" tendency of Dpp gradients may be due to a simple law of mass action. To wit, "the range of signaling depends on the amount of signal generated, which in turn depends on the number of Dpp-secreting cells" [3074]. As the tissue around a Dpp source grows, so too does the source. The more it grows, the higher the Dpp concentration will rise, and the more thresholds for gene activation (*omb*, *spalt*) and vein induction (1, 2, 3) will be crossed [2457,

3225]. At some point, the growth must become self-limiting because discs stop growing even when given extra time (cf. Ch. 4) [542], but the terminal trigger for maturation is obscure [977, 2776, 4886] (cf. servomechanisms [40, 4725]).

The variable height of Dpp gradients makes them appear seamless

One of the appealing features of the Polar Coordinate Model was its seamlessness. No singularities were supposed to interrupt the smooth sequence of coordinates around a disc. Even the "12/0" line in the standard clockface depiction of the model was assumed to be invisible to the cells themselves because they were somehow able to interpret "12" and "0" as the same coordinate [1303, 1807]. Only a seamless mechanism, it was thought, could explain why the plane of symmetry in duplicated patterns can be situated anywhere along an organ's axes, and seamlessness was thought to be antithetical to cell-lineage compartments.

There seems to be no restriction on the possible positions or directions of symmetry lines in duplicated patterns. For example, Bryant [522, 524, 525] showed clear cases where the symmetry line divided organs such as sensilla groups, axillary sclerites, tegula, etc. This indicates that the pattern regulates as a whole and that there is no subdivision of the pattern-forming field into sub-fields during the process of duplication. [526]

In duplicated patterns . . . we found that the line of symmetry could pass through various recognizable structures such as axillary sclerites, notal wing processes, tegula, etc. In these cases, then, only part of the structure is present but that part is duplicated. . . . We conclude from cases such as these that the wing disc is not subdivided into separate morphogenetic fields corresponding to these components of the adult structure. The developmental behavior of cells during regeneration and duplication is defined by their relation to the overall pattern and not to any subdivision of it. [525]

The plane of symmetry forms without regard to the boundaries of the named recognizable structures. Furthermore . . . the demarcation between doubled and deficient areas on the disc is imposed without regard to the compartments described by Garcia-Bellido *et al.* [1376]. Evidently the plane of symmetry recognizes no natural barriers from the center of the disc to the periphery. It is somewhat surprising that sometimes the plane of symmetry goes through a single cell. For example, Fig. 7 shows the plane of symmetry going through the anterior post-alar bristle . . . and Fig. 17 illustrates the plane of symmetry going through a single sensillum. Unfortunately, both of these single cell structures are symmetrical even in normal flies, and so it cannot be proved that in these pattern alterations the symmetry of half a cell is reversed. [3440]

An example of this phenomenon is the ability of left and right 1st-leg discs to fuse together to varying extents when cells are randomly killed.

The precise mirror image patterns in fused prothoracic legs show either that exactly the same number of cells die in precisely the same location in each leg disc, or that pattern regulation in the left and right leg discs is coordinated to provide similar patterns after different degrees of cell death. [3442]

Interestingly, a similar spectrum of fusion phenotypes can be produced by adjusting the amount of ectopic Dpp that is expressed in the Wg sectors of 1st-leg discs [2954]. Clearly, the *dpp*-ON stripe can operate like a rheostat (vs. a digital switch) [1807].

Ironically, the earliest model for disc regeneration – the 1971 Gradient of Developmental Capacity Model (cf. Ch. 4) – resembles our current view about Dpp's role [522]. Where did the 1976 PC Model go wrong? Its most obvious mistake was discussed in Chapter 5 – namely, its premise that regeneration obeys the rules of normal development. In fact, repatterning in the foreleg disc proceeds via an abnormal contact between the peripodial and columnar layers (cf. Fig. 5.10c) [1472]. The ability of the upper medial quadrant to reform a whole P compartment after being "infected" with only a few *en*-ON cells must be due to its *dpp*-ON halo, which fosters the growth of the colony during the week needed for full regeneration [3808] ($\approx$ a self-feeding Minotaur).

In a similar way, perhaps, *en*^{null} *inv*^{null} clones in the P compartment instigate colonies of A-type cells that can grow to the size of a whole A compartment (Fig. 6.6g) [4229]. Whether the ectopic *dpp*-ON stripes (that arise at each interface between *en*^{null} *inv*^{null} and wild-type tissue) climb to a 3–4 intervein state depends on the time available: late-induced clones never attain a 3–4 level. Moreover, the height of the "valley" between adjacent Dpp peaks depends on the distance between them. For example, a "123/4*5m" clone can yield a "123/4 {4/$\underline{3}$*$\underline{3}$/4}5m" pattern (valleys = 4–5 intervein at "{" site and 2–3 intervein at "*" site), whereas a "123/45m*m" clone can produce a "123/45m{m54/$\underline{321}$*$\underline{123}$/45}m" pattern (valleys = m at "{" and 1 at "*") [4229]. The farther an ectopic "/" peak is from the endogenous peak, the deeper the intervening valley can sink. In these two examples, the valleys sank to "vein 1," "2–3 intervein," "4–5 intervein," or "m". Evidently, the valley can occupy any level of the gradient, as can the peak [3753]. Mirror planes between subpatterns can reside at valleys or peaks when ectopic *dpp*-ON stripes arise within compartments (e.g., *dpp*^{GOF} A clones [4848]).

This ability of mirror planes to "slide" freely along Dpp gradients reveals a subtler misstep in the history of the PC Model. It was erroneous to assume that compartment boundaries would have fixed cellular fates (e.g., 3–4 intervein) in reconstituted patterns [904]. In fact, Dpp gradients can produce patterns that are as seamless as the patterns that the PC Model can theoretically create by "ironing out" discontinuities. This scaffolding strategy is analogous to hanging clothes (structures) on a clothesline (gradient), where the poles can be raised to any height and the clothesline can be slackened to any depth. No one ever suspected that the scaffold itself could be so amazingly flexible or so utterly invisible.

If the PC Model is incorrect, then why do fragments obey the Rule of Shortest Intercalation (cf. Ch. 4)? A final answer must await further studies, but a guess may be attempted. Fragments might not "take the long way around" because any new *dpp*-ON zones that they acquire (via wounding [489, 2856] or A/P confrontation [1472]) only raise the gradient levels of nearby cells. For a cell's identity to "jump" to the opposite side of the fate map (cf. Fig. 6.6a), it would need to switch at least one selector gene, but Dpp does not alter those states. Fragments probably obey the Reciprocity Rule for a similar reason: wound edges are unlikely to replace structures on the opposite side of the disc unless they are assisted by unusual contacts. Indeed, the choice between regeneration and duplication for any given piece is critically dependent on the mode of wound healing [947].

In summary, the basic chain of events along the A-P axis is: En (P region) $\rightarrow$ Hh (diffuses across A/P line) $\rightarrow$ Dpp (diffuses in both directions) $\rightarrow$ target genes. Those targets of Dpp then cooperate with other targets of Hh to establish the pattern of veins (see below).

A Wg gradient specifies cell fates along the wing's D-V axis

In 1996, Wg was shown to act as a morphogen along the wing's D-V axis [4849] by the same sort of evidence that established Dpp's morphogen status in the same year [2455, 3074]. Wg's long-range action was monitored by the target genes *vestigial* (*vg*) and *Distal-less* (*Dll*) – both of which encode transcription factors [836, 1686, 4681] – and *neuralized* (*neu*), a marker for SOPs [417, 2387, 3374]. All three genes are positively regulated: their expression wanes when Wg signaling is suppressed (e.g., in *arm*^{LOF} clones or in t.s. *wg*^{LOF} larvae exposed to high temperature [4849]), and they turn ON when the Wg pathway is ectopically activated in the wing pouch (e.g., in *sgg*^{LOF} clones [351, 1040]).

Wg's effect on these target genes (Wg $\rightarrow$ *vg*, Wg $\rightarrow$ *Dll*, Wg $\rightarrow$ *neu*) must be direct (vs. a signal relay mode) because (1) ectopic expression of a membrane-tethered form of Wg induces *vg*, *Dll*, and *neu* expression only in adjacent cells, and (2) activation of the Wg pathway in ectopic *arm*[GOF] clones turns ON *vg*, *Dll*, and *neu* only within the clone [4849]. Moreover, each gene appears to respond to a different threshold level of Wg ($T_{neu} > T_{Dll}$ > T_{vg}) because (1) ectopic Wg-secreting clones induce *neu*, *Dll-lacZ*, or *vg* expression in nested circles (radii: $r_{neu} < r_{Dll} < r_{vg}$) when a strong promoter drives *wg*; and (2) *neu* fails to be expressed with a weaker promoter [4849]. As expected for a gradient mechanism, the seriation of the nested circles matches the endogenous rainbow of bands that straddle the *wg*-ON stripe at the D/V compartment boundary (Figs. 6.2 and 6.3), although the punctate *neu* transcription is excluded from *wg*-ON cells.

Perpendicular (Dpp × Wg) gradients suggest Cartesian coordinates

In 1997, the above results (from Zecca and Basler in Zürich and Struhl in New York) were corroborated by Neumann and Cohen in Heidelberg [3087, 3091]. One new insight came from using *dpp-Gal4:UAS-wg* to lengthen the point of overlap between *wg*-ON and *dpp*-ON domains into a solid A/P stripe. Dpp and Wg do not antagonize one another in the wing as they do in the leg [1833, 4277], so this sort of coexpression is stable. Such discs can double in length via expansion of the pouch [2252, 3091] or duplicate via inception of an extra pouch [2252]. This synergistic hyperplasia agrees with the commonly observed fact that *dpp*-activating clones must cross the (*wg*-ON) margin to cause ectopic outgrowths [1839, 4848].

1. Dpp pathway: *dpp*[GOF] [4848] and *brk*[LOF] [619] clones only cause outgrowths when they cross the margin.
2. Hh pathway: Outgrowths (123*y*321 where *y* is the clone) arise around the following anterior clones when they cross the margin: *hh*[GOF] [231, 4848], *ptc*[LOF] [754, 3077, 3177], *cos2*[LOF] [3976], *slimb*[LOF] [2060, 4279, 4538], and *DC0*[LOF] [2491, 2533, 3177, 3238, 4538], although *slimb*[LOF] and *DC0*[LOF] cells have identities closer to vein 3 [2058, 3238, 4538].
3. Engrailed ON/OFF boundaries: Doubly mutant *en*[null] *inv*[null] clones generally must reach the distal or posterior wing margin to reorganize the wing's vein pattern [4229]. This result confirms older studies using weaker alleles: *en*[1] [2441] and *en*[LOF] [2447] clones incite outgrowths only when they reach the P margin. The same is true for *en*[GOF] clones in the A compartment [4848]. The ultimate proof of this "Dpp-Wg

Intersection Rule" for outgrowth is a *Gal4* line that expresses *en* throughout the anterior wing pouch except the margin [1647]: the A region adopts a P pattern (veins 2 and 3 transform into veins 5 and 4, with an ectopic posterior cross-vein), but no outgrowths arise.

Because an intersection (or at least proximity) of *dpp*- and *wg*-expressing domains seems essential for wing outgrowth (natural or ectopic) [164, 643, 1976], wing outgrowth appears to obey the same "Dpp + Wg" rule that has been proposed for leg outgrowth [620, 3862]. The spot where the endemic *dpp*-ON and *wg*-ON stripes meet could be a reference point for positional information [1427]. Indeed, the *vg* gene's "quadrant" enhancer is first activated there [2219, 2254], and this spot is unique in how it responds to Notch [2462] and Dpp levels. Suppression of Dpp pathway genes causes apoptosis and nicking there [2455], whereas excessive Dpp signaling causes apoptosis elsewhere along the margin [12]. An extra "Dpp + Wg" spot appears to arise in the posterior margin of *twins*[LOF] wing discs [4418] and in the anterior margin of *lethal(2)giant discs*[LOF] wing discs [545, 558, 559, 2976]. The result in each case is a wing duplication. Twins is a regulatory subunit of protein phosphatase PP2A that is involved in both the Dpp and Wg signaling pathways (Fig. 5.6), so LOF mutations could be short-circuiting the system somehow. The etiology for *l(2)gd*[LOF] is obscure. The "world record" for multiple winglets from a single disc is 8: the extra 7 were induced by coexpressing large amounts of Dpp and Wg along the D/V boundary [2254]. All these clues are consistent with a "Dpp + Wg" trigger.

However, as in the case of the leg disc (cf. Ch. 5), the trigger for wing outgrowth seems to only depend on Wg indirectly, if at all [1074, 4682]. The formula may be "Dpp + X = distalize!," with factor "X" being elicited by circuit elements at the margin [2092] that happen to overlap the *wg*-ON band. The evidence against a direct role for Wg includes the following (see [2092, 2252] for more):

1. Wing growth proceeds virtually normally when *wg* function is repressed throughout the disc in 3rd instar via a t.s. *wg* allele [880] or when it is repressed at the tip via *dpp-Gal4:UAS-scabrous* [2462].
2. *DC0*[null] *wg*[null] double-mutant clones induce partial duplications of the blade as readily as *DC0*[null] (margin-crossing) clones [2533, 3238], implying that *Wg* is dispensable for ectopic tip formation.
3. *sgg*[null] clones (whose Wg pathway is activated) fail to induce outgrowths [1040, 2252].

4. wg^{GOF} cannot rescue wing growth in discs that are stunted due to ap^{null} or Ser^{null} or $Su(H)^{null}$ or vg^{null} [2092, 2252], but stunted wg^{LOF} discs (as well as ap^{null}, Ser^{null}, or $Su(H)^{null}$ discs) can be rescued by vg^{GOF} [2219, 2252].

If the "Dpp + Wg" rule is wrong, then how can duplications in *dpp-Gal4:UAS-wg* discs be explained? In those cases, the ectopic pouch always arises where the *dpp*-ON-*wg*-ON stripe meets a native *vg*-ON spot in the notum [2252], so the formula may actually be "Dpp + Vg + X = distalize!" where "Wg ➞ X" to provide the missing ingredient [2254]. In the absence of Vg, the main effect of excess Wg is overgrowth of the hinge (vs. the blade per se), with negligible impact on patterning [459, 2252, 2254, 3088]. This ability of wg^{GOF} to uncouple growth from patterning is undoubtedly relevant to the mystery of how Dpp and Wg gradients foster uniform (vs. concentration-dependent) growth throughout the epithelium (cf. Ch. 4) [3862], but its exact implications are unclear.

But cells do not seem to record positional values per se

Are the perpendicular Dpp and Wg stripes (Fig. 6.3) really reference axes in a *bona fide* Cartesian coordinate system (cf. Fig. 4.3b) [4682]? In theory, each cell could pinpoint its location by (1) deducing its quadrant from the states of its selector genes (*en* and *apterous*, see below), and (2) ascertaining its site within the quadrant by measuring the amounts of Dpp and Wg. However, there is no evidence that wing disc cells record their positions with any such precision. Rather, they switch ON only a few target genes per morphogen, based on preset threshold levels. The disc is thus subdivided into a checkerboard of districts ($\approx$ superimposed French Flags). Cells would know their district (as a binary address of ON/OFF "memory genes" *omb*, *spalt*, *vg*, *Dll*) but would not remember their exact location. The same conclusion was reached for how embryonic cells "interpret" the Bicoid gradient (cf. Ch. 4) – viz., as a code (of gap, pair-rule, and segment polarity gene states) whose resolution never attains the graininess of individual cells.

The upshot of this line of reasoning is that the PI Hypothesis is only partly correct here. Morphogen gradients do inform cells of their positions, but cells do not record "positional values" per se. Without any quantitative positional values, there is no need for an "interpretation" step. Cells go directly from gradients to a few qualitative states, thus bypassing a positional-value stage. These states are then used as anchors for pattern elements (e.g., veins, see below) and function more like

prepattern singularities. Overall, the mechanism would be a PI-prepattern hybrid [3087].

Wg's repression of Dfz2 is inconsequential

In 1998, through various LOF and GOF experiments, Cadigan *et al.* showed that Wg reduces the amount of Dfz2 (its receptor) [595]. They also found that higher densities of Dfz2 retard Wg degradation and sensitize cells to Wg signal. Based on these facts, they proposed a "gradient shaping" scheme like the Rock Concert Scenario discussed above for Dpp. The Wg gradient is steep near its source and shallow elsewhere, but nearby cells may sense a lesser slope due to fewer receptors (Wg ⊣ *Dfz2*), while outlying cells may sense a greater slope due to more receptors. The biphasic (spire-ramp) profile could hence be perceived as a smooth gradient with constant signal-resolving capacity per unit distance [2457], and the reciprocal profiles of Wg and Dfz2 would make sense in terms of the "Wg ⊣ *Dfz2*" reciprocity.

In 1999, however, this scheme was essentially disproven by the discovery that $Dfz2^{null}$ wings develop normally [734]. The smoothing idea could not be saved by invoking a redundancy of Dfz2 with Fz [734] because no "Wg ⊣ *fz*" link exists: Fz is distributed uniformly in wing discs [3264]. Nor can Dfz3 rescue the hypothesis because it manifests an opposite sort of regulation (Wg ➞ *Dfz3*) [3977]. Hence, whereas the "Wg ⊣ *Dfz2*" link clearly operates in a wild-type context (unlike "Dpp ⊣ *tkv*"), its ability to "shape" the Wg gradient appears to be essentially inconsequential.

Apterous's role along the D-V axis resembles Engrailed's A-P role

Because Dpp and Wg have similar diffusion ranges, the evolutionary decision as to which morphogen to use along each axis must have been arbitrary from an engineering standpoint. The only obvious constraint is that a single morphogen cannot be used for both axes unless they are patterned at different times [4724]. The same rationale applies to the compartments whose boundaries evoke Dpp and Wg. It turns out that separate selector genes (also arbitrary) are used in overlapping areas to delineate four quadrants.

Until 1993, García-Bellido's Selector Gene Hypothesis (cf. Ch. 4) [1358] rested mainly on a single gene (*engrailed*) and its idiosyncratic phenomenology. In that year, a similar suite of properties was found for a second gene (*apterous*) [361, 2445], thus bolstering the idea [354].

1. Like *engrailed*, *apterous* (*ap*) encodes a homeodomain transcription factor [290], but Ap also has two LIM (protein docking) domains [418, 826, 3159, 3601]. The

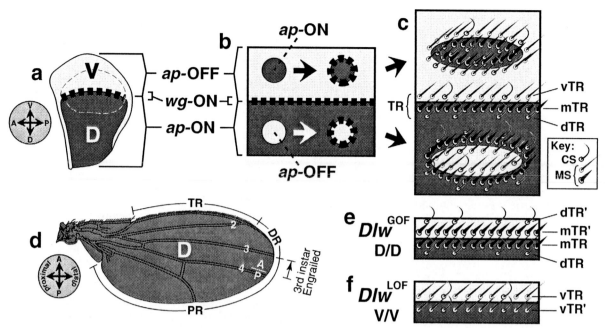

FIGURE 6.7. The role of *apterous* in establishing the wing margin.

a. Mature right wing disc. Directions (compass at left) are A, anterior; P, posterior; V, ventral; D, dorsal. Prospective D vs. V regions of the disc are indicated (dark vs. light shading), as are the wing pouch (thin dashed line) and future margin (thick dashed line). The D part expresses *apterous* (*ap*); the V part does not. The margin expresses *wingless* (*wg*), as do the pouch edge and a notal stripe (not shown). The Wg band straddles the D/V boundary (in contrast to the Dpp stripe, which abuts the A/P line; Fig. 6.6a).

b. Effects of xenotopic *ap* clones inferred from adult defects and GOF data. When *ap* is misexpressed ventrally (above), it turns *wg* ON at the edge of this GOF area [2851]. Based on their ectopic margins (c) [1038], *ap*[LOF] clones in the D region (below) must also turn *wg* ON peripherally.

c. The normal margin (center) has a triple row (TR) of MS (mechanosensory, straight) and CS (chemosensory, curved) bristles. The medial TR (mTR) has stout bristles, and the dorsal TR (dTR) has CS bristles, while the ventral TR (vTR) has a 4:1 ratio of slender MS to evenly spaced CS bristles. New TRs develop around *ap*[GOF] tissue in the V region (upper oval) [2851] and *ap*[LOF] clones in the D region (lower oval) [1038], whenever these spots are anterior to vein 2. The *ap*[GOF] V spots have a "dTR-mTR-E-vTR" polarity (E = clone edge) from inside to outside, whereas the *ap*[LOF] D clones have opposite polarity [1038].

d. Right adult wing (D surface). Only stout bristles are drawn. Along the front is TR. More posteriorly are a double row (DR) and a posterior row (PR). DR (like TR) has innervated bristles, but PR only has nonsensory hairs [1741, 1837]. The DR/PR transition (at vein 3) regresses to the A/P line (~ vein 4 [350]) if En expression in the A region (which arises in the 3rd instar [350]) is blocked [1837, 4229]. Conversely, it advances farther anteriorly if Hh is overexpressed [2992]. Why the anterior extent of the *en*-ON area is greater distally than proximally (not shown) is a mystery [350].

e, f. Symmetric (D/D or V/V) phenotypes caused by GOF (**e**) or LOF (**f**) alterations at the *Dorsal wing* (*Dlw*) locus [4325]. Strangely (via transvection?), the same mutation (*Dlw*[1]) switches cell fates in opposite ways (V-to-D vs. D-to-V), depending on whether it is in heterozygous (*Dlw*[1]/+ fly = GOF) or homozygous (*Dlw*[1]/*Dlw*[1] clones = LOF) condition. In each case, the effect is fully penetrant but partly expressed (i.e., not all bristles are transformed). V/V wings can also be created by using *ap-Gal4* with *UAS-Ser* and *UAS-fringe* to restore Notch signaling, a *wg*-ON stripe, and a wing margin (albeit a V/V one) in an *ap*[LOF] background [2850, 3158].

Panel **a** is after [361], **b** and **c** are based on [882, 1038, 1741] (although clones are typically more elongated along the proximal-distal axis than shown here [521, 1373, 1545]), **d** is adapted from [350, 524, 1224, 1741, 1837], and **e** and **f** sketch phenotypes from [4325]. See also App. 7.

gene is named for its missing-wing LOF phenotype [470, 3848, 4119, 4692, 4693]. Although *ap* is expressed in legs and antennae, these structures develop normally in *ap*[null] flies [826] so, unlike *en*, *ap* appears to function only in certain discs (wing and haltere). Part of the reason may lie in the leg's lack of D/V lineage restraints [354, 1800, 4076].

2. In the wing disc, Ap is expressed only in dorsal anlagen (notum, dorsal wing and hinge), and the distal edge of the *ap*-ON area aligns with the wing margin (Fig. 6.7) [352, 826, 1038, 4682, 4683]. This congruence implicated *ap* as a selector gene because the margin is a boundary that somatic clones fail to cross starting in mid-2nd instar [361]. Indeed, that is when

ap expression becomes detectable [1074, 2251, 4543, 4683], suggesting that the lineage restriction could be due to *ap*. At one time, it was thought that the restriction was due instead to a zone of mitotic quiescence [494, 498, 3154]. However, this notion proved false [352], partly because the (*wg*-dependent) quiescence arises too late to explain the (*ap*-dependent) lineage segregation [2079, 2846, 3374, 3979, 4683].

3. As with *en*, switching the *ap*-ON/OFF states of cells causes (1) compartmental switches in downstream genes (e.g., integrins) [361, 498, 4118], (2) homeotic changes in adult structures [1038, 3158], and (3) the emergence of boundary identities at ON/OFF interfaces [1038] (Fig. 6.7). The initiation of the *ap*-ON state dorsally depends on Hh (although the geometry makes no sense) [2251], while its OFF state ventrally is maintained (like that of *en* anteriorly) by *Polycomb* [3752, 4327].

4. Like *en*^{LOF} clones in the P (*en*-ON) region, *ap*^{LOF} clones in the D (*ap*-ON) region cross the compartment boundary when near it and round up when far from it [361], and *Polycomb*^{LOF} (*ap*-ON?) clones cross in the other direction [4327]. Thus, *ap* could control compartment-specific affinities (Selector Affinity Model), but it might be acting indirectly (Border Guard Model) [352, 361]. In the absence of Ap, a border still arises but is ragged [2251], possibly due to apoptosis [4683]. The ability of Fringe to rectify this border without any Ap activity rules out the simple Selector Affinity Model (see below) [3158].

The asymmetry of D vs. V bristles at the wing margin (Fig. 6.7) makes it possible to distinguish mirror-symmetric (D/D or V/V) and inverted (V/D) margins from the wild-type (D/V) condition. (See [1368, 3158, 3624, 4189] for vein asymmetry "corrugations.") When *ap*^{null} clones arise dorsally (and anterior to vein 2), they induce an ectopic V/D margin at their perimeter, with V-type bristles inside and D-type bristles outside [1038] (cf. similar spots in *ap*^{LOF} wings [4118]). Because the endogenous margin also arises at an *ap*-ON/OFF interface, the implication is that *ap* is responsible for instigating the margin. The ability of *ap*^{null} clones to autonomously convert cells from D- to V-type further implies that *ap* also controls regional (D vs. V) identities.

Chip cooperates with Apterous, and "Dorsal wing" acts downstream

Two other genes besides *ap* have been found to cause similar switches in cell identities – *Chip* [1217] and *Dor-*

sal wing (*Dlw*) [4325]. *Chip* was isolated independently by (1) a genetic screen for agents that bridge enhancers to promoters [2939, 2940], (2) a screen for mutations affecting the wing margin [1217], and (3) a two-hybrid screen for proteins that bind Ap [1217]. LOF-GOF analyses led to some odd results [1217] (cf. a similar paradox for *Dl* and *N* [75, 990, 1204]). These riddles are listed below, alongside solutions offered by a 1998 "**Ap-Chip Tetramer Model**" proposed by Fernández-Fúnez *et al.* [1217]. "Boxes" allude to graphics in Fig. 6.8b (see [2854, 3600] for further evidence).

Riddle	Explanation
1. *Chip*-LOF clones behave like *ap*-LOF clones (i.e., D clones induce margins, whereas V clones do not), despite the fact that *Chip* is ON ubiquitously (*ap* is only ON dorsally). The ectopic margins have V/D (inside/outside) polarity, indicating D-to-V transformation. Expression of *ap* is normal within the clone, so *ap* is not acting downstream of *Chip*.	1. *Chip*-LOF and *ap*-LOF have similar effects because the tetramer concentration drops when either Chip or Ap is in short supply (boxes 2 and 3). Defects arise when it drops below a threshold for target gene activation. Chip may normally be the limiting factor because *Chip* (not *ap*) has a haplo-insufficient (*Chip*-null/+) effect [118, 1217].
2. *Chip*-LOF and *Chip*-GOF have similar effects. Misexpression of *Chip* in the A/P zone (via *dpp-Gal4:UAS-Chip*) induces margin bristles at the P (but for some reason not A) edge of this band, and the new margin is confined to the D region, implying a D-to-V switch inside the *Chip*-GOF band.	2. Misexpression of *Chip* in the A/P zone has no effect ventrally because Ap is absent and Chip has no independent role. Dorsally, the extra Chip proteins sequester many Ap proteins in inactive trimers (box 4). Hence, the effects of *Chip*-GOF match those of *Chip*-LOF – a counterintuitive fact.
3. Overexpression of *Chip* in the D region (via *ap-Gal4:UAS-Chip*) suppresses wing development, but overexpressing *ap* with the same driver causes no such defect (although dorsal *ap*-GOF can have minor effects [2851]).	3. Excess Chip in the whole D region likewise diverts Ap into trimers (box 4), leading in this case to wing loss. In contrast, excess Ap is relatively harmless because Ap cannot distract Chip into any sort of inert complex (box 5).
4. The wing-loss defect of *ap-Gal4:UAS-Chip* can be rescued by simultaneously expressing *ap* in the same domain (*ap-Gal4:UAS-Chip:UAS-ap*).	4. Jointly raising the doses of *Chip* and *Ap* will make more tetramers (box 6), but this analog change may be moot if Ap's targets respond digitally (ON vs. OFF).

According to this model, which has since been confirmed [2851, 3159, 3600, 4456, 4457], Ap can only activate its target genes as a linear "Ap-(Chip-Chip)-Ap" tetramer. Within the tetramer, a Chip dimer links two Ap proteins. Chip proteins dovetail via a "DD" dimerization domain, while Ap's LIM domains bind Chip's "LID" (LIM-interaction) domains. Tetramers may serve a DNA-looping role by forming a "DNA-{Ap-(Chip-Chip)-Ap}-DNA" bridge (discussed below).

The Ap-Chip gadgetry recalls the AS-C story (cf. Ch. 3) insofar as a patterned agent (Ap ≈ Achaete) must dimerize with a ubiquitous partner (Chip ≈ Daughterless) to activate target genes (cf. Tango [899, 4022, 4548]). Might there exist an antagonistic interloper like Emc which diverts a partner(s) into inert dimers? Indeed, there is. Like Emc, Beadex is a protein with dimerization (LIM ≈ HLH) domains but no DNA-binding (homeo ≈ basic) domain [2854, 3908, 4858]. Consistent with the notion that Beadex is a competitive inhibitor of the Ap-Chip union, Beadex binds Chip (but not Ap) *in vitro* [2854], *Beadex*^{GOF} mutations counteract Ap's activation of target genes *in vivo*, and *Beadex*^{LOF} alleles augment Ap's effectiveness [2854, 3908, 4577].

One amusing twist in the Ap-Chip-Beadex operetta not seen in the Ac-Da-Emc imbroglio is a link that goes beyond the protein level. In early-mid 3rd instar discs, expression of Beadex is confined to the D compartment just like Ap, implying that Ap ➝ Beadex, and this link has been verified (*Beadex* turns ON wherever *ap* is misexpressed) [2851, 2854]. There are many reasons why genes evoke antagonists [3347, 3399, 3558]. However, in *ap*'s case the answer is not obvious because *ap* is a selector gene, and digitally operated genes should not need the kind of analog modulation that fine-tunes the AS-C's "SOP Computer" (cf. Ch. 3). The reason here appears to be that Ap's level must be lowered enough to allow installation of new 3rd-instar circuitry (involving Serrate), but not so low that Ap cannot execute its residual selector duties (e.g., D-V adhesion) [2853].

Another lingering mystery concerns the most intriguing of the screens that snagged *Chip*. Various lines of evidence [274, 1092, 1441, 4204, 4414] – mostly concerning the "insulators" *su(Hw)* [607, 1401, 2666], *scs-scs'* [2330], and *Fab-7-8* [212, 1680, 1681, 2844, 4877] – imply the existence of a "**Kilobase Spanner.**" This mythical device allows enhancers to control target promoters over large (multikb) distances [2188, 2511, 3002, 3577, 3975], perhaps by knotting the intervening DNA [275, 606, 1092, 2939, 3001]. Morcillo *et al.* hunted for genes that encode the elusive spanner's components and reported their findings in 1996 [2940]. To as-

say for spanner-disabling mutations, they used the *cis*-enhancer that turns the *cut* gene ON at the wing margin. This enhancer is 85 kb upstream of the *cut* promoter [2001] – a chasm almost as wide as the entire AS-C (cf. Fig. 3.4). *Chip* was one of three genes that they recovered. (The other two are enhancer-binding proteins.) Indeed, *Chip* was also recovered in a screen for factors that can bridge a 30-kb gap within the AS-C itself (between the DC *cis*-enhancer and the *scute* promoter) [3504]. Assuming that Chip is part of the Kilobase Spanner, why should Ap need such a tool? Perhaps the *cis*-enhancers that Ap binds (to control its target genes) are too far from their cognate promoters to be effective without help from Chip. At the *Ser* locus, for example, this span (~10 kb) might be big enough to pose such a handicap [163, 164].

Flies that are heterozygous for a LOF allele of *Dorsal wing* (*Dlw*^{LOF}/+) have D/D wings, and V/V patterns are induced by homozygous *Dlw*^{LOF}/*Dlw*^{LOF} clones, although both kinds of transformation are incomplete [4325]. Because the latter clones do not evoke margins at their borders, *Dlw* must be acting downstream of *ap* and *Chip* (in a nonsignaling pathway) to implement D-vs.-V identities. Thus, *Dlw*'s D-V role resembles *invected*'s role downstream of *en* along the A-P axis. In both cases, parallel pathways (diverging from *en* or *ap*) implement cell identity vs. *trans*-border signaling (cf. Figs. 6.4 and 6.8) [2448].

Serrate and Delta prod Notch to evoke Wg at the D/V line

Signaling between the D and V compartments relies on the Notch pathway, which mediates Ap's control of Wg (Fig. 6.8). The key players are as follows:

Wingless (Wg)	Morphogen for the wing's D-V axis. Slave of Notch.
Notch (N)	Activator for Wingless and Cut. Receptor for Delta and Serrate.
Serrate (Ser)	Dorsal ligand for Notch at the D/V line. Slave of Apterous.
Delta (Dl)	Ventral ligand for Notch at the D/V line. Slave of unknown masters.
Fringe (Fng)	Muzzler of Serrate. Creator of boundaries. Slave of Apterous.
Cut	Patron of Wingless. Killer of Delta and Serrate. Slave of Notch.

Wg's targets include *achaete* and *scute* [2079, 2839, 3374, 3690]. These proneural genes elicit bristles on either side

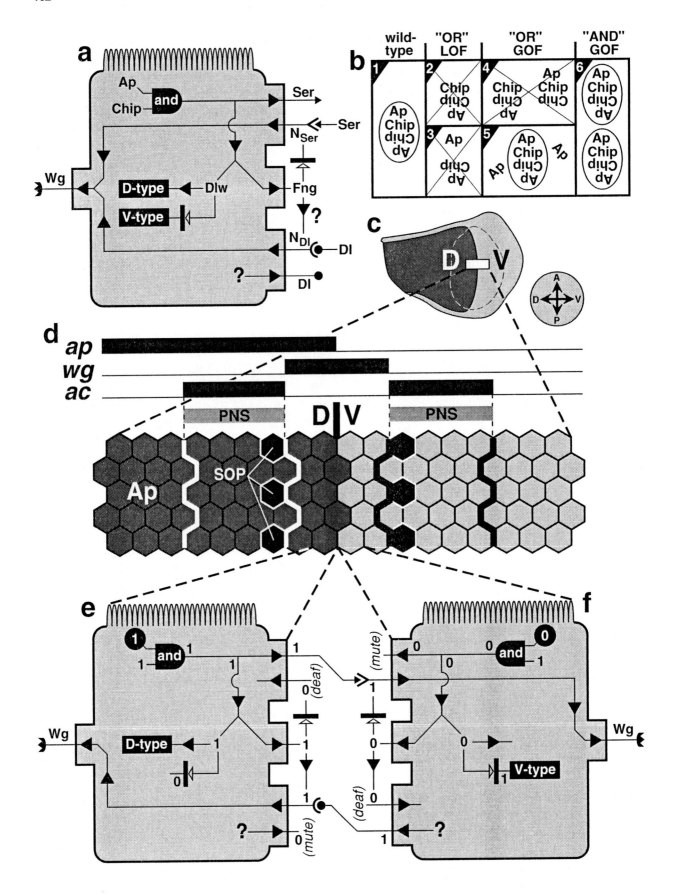

of the *wg*-ON stripe (Fig. 6.2) [882, 1741, 1851]. The *wg*-ON stripe spans the whole margin, but the flanking Ac-Sc stripes only arise in the A half [912, 2079, 3982], perhaps because En represses them in the P half (En ⊣ AS-C) [882]. The two Ac-Sc stripes express Dl and Ser strongly, while the center *wg*-ON stripe expresses N strongly [988, 1951]. This alternation of high-Dl with high-N resembles the notum [3270, 4428], where the stripes are also ~4 cells wide (cf. Fig. 3.8) [2079].

Marginal SOPs typically emerge at the inner edge of each high-Dl stripe (Fig. 6.8) [352, 882, 1851, 3689]. This trend suggests that Wg imposes a small gradient of proneural potential within the flanking stripes [356, 882]. Interestingly, when cells at the interface are prevented from transducing Wg (by *dsh*LOF or *arm*LOF clones), then cells farther out can become SOPs [882]. This result recalls the

*ac*LOF mosaics whose bristle shifts led Stern to postulate proneural clusters in the first place (cf. Fig. 3.3) [4095, 4096].

Most cells in the middle (*wg*-ON) stripe express *cut* [352, 353, 373, 882], and *cut*LOF discs have a single broad Dl-Ser (Ac-Sc?) stripe instead of two thin ones [882, 988]. This phenotype recalls the broad Ac-Sc stripes on *hairy*LOF legs (cf. Fig. 3.9) [3193] and suggests that Cut's role is to bisect the margin's Ac-Sc stripe, just as Hairy subdivides the leg's Ac-Sc stripes. Cut could "carve" the gap by repressing the AS-C (or Dl and Ser) as a transcription factor [370].

An antineural role for *cut* would explain why *wg*GOF [459, 2252, 3088], *sgg*LOF [351, 2252, 2839, 3958], and *dsh*GOF (to a lesser extent) [151] cause confluent lawns of bristles with no *cut*-like bald areas inside them: such clones fail to turn ON *cut*, which is directly under *N*–not *wg*–control

---

FIGURE 6.8. Circuitry that enforces D vs. V identities of wing cells and turns *wg* ON at the D/V line.

a. Generic wing disc cell (cf. Fig. 2.7 for icons) showing the main control circuit for Wg output. Abbreviations: Ap (Apterous), Dl (Delta), Dlw (Dorsal wing), Fng (Fringe), N (Notch), Ser (Serrate), Wg (Wingless). N_{Ser} and N_{Dl} are hypothetical Fng-modulated forms of Notch that are receptive to Ser or Dl [990, 1250]. Fng's damping of the Ser-N trigger has been well documented *in vivo* [1250, 3246], as has its facilitation of Dl-N binding *in vitro* [510]. The circuit works as follows. If *ap* is ON (**e**), then the cell expresses Ser and Fng and adopts D-type identity (via Dlw) but cannot receive Ser signals because Fng blocks the Ser-N interaction. If *ap* is OFF (**f**), then the cell expresses Dl (Dl's upstream controls are uncertain [2253]) and adopts V-type identity. Dlw (under Ap control) should allow D vs. V cells to interpret symmetric Wg gradients asymmetrically. The circuit allows two types of discourse: Ser-signal "speakers" with N_{Ser} "listeners" or Dl-signal "speakers" with N_{Dl} "listeners." Both dialogs should turn *wg* ON. Validation of the circuit's logic comes from (1) heating t.s. N^{LOF} larvae, which erases *wg* and *cut* expression in border cells [2839, 3689], and (2) expressing *Ser* or *fng* in the D region of *ap*LOF discs, which restores Notch signaling and a *wg*-ON stripe but not D identity (hence creating V/V symmetric wings) [2850, 3158].

b. Stoichiometry of Ap-Chip interactions and etiology of abnormal traits. Wings look wild-type when Ap and Chip are equimolar (boxes 1 and 6) or when Ap is in excess (box 5), but not otherwise (X'd boxes). Ap binds DNA (not shown) but can only activate target genes when it forms a tetramer (oval) with Chip (box 1). The tetramer is thought to be a Chip dimer with an Ap at each end. Reducing the amount of Ap (box 2) or Chip (box 3) lowers the number of functional complexes. Excess Chip (box 4) diverts Ap into inert trimers, whereas excess Ap (box 5) has no such effect. Raising the dose of both partners ("AND" GOF) maintains the balance and makes excess tetramers (box 6 shows a doubling). The counterintuitive LOF-GOF genetic data only make sense in terms of this sort of jigsaw-puzzle logic. See text for further details. Oddly, Ap and Chip show no such interdependence in leg discs [3474].

c. Mature right wing disc. Shading denotes D (dark) vs. V (light) compartments, and dashed line delimits the pouch.

d. Enlarged part of the anterior D/V border (box in **c**). Hexagons are cells. All cells in the D region express Ap. Zigzag lines delineate 3 zones (horizontal bars above): a median stripe where *wg* is ON (also *cut* and N: not shown) and flanking "proneural stripes" (PNS) where *achaete* (*ac*) is ON (also *scute*, with intense Dl and Ser but meager N: not shown) [988]. Sensory organ precursors (SOPs, black cells) arise in each PNS along the medial (high-Wg) edge [3689]. In *cut*LOF discs, the *wg*-ON cells express *ac* [882, 988], so Cut's normal role must be to split the otherwise broad *ac*-ON stripe into two parts – the same sort of subtractive logic that is used in the notum (cf. Fig. 3.8f).

e, f. Schematics of cells flanking the D/V line. The V cell is drawn as if flipped around to face the D cell. States of variables (cf. **a**) are recorded as "1" (present and active) or "0" (absent or inactive). Black circles indicate determining factors. "Mute" and "deaf" denote inability to send or receive particular signals (Ser above, Dl below). The functioning of the circuit is explained under **a** above. The net result is that both cells are induced to secrete Wg.

Circuitry in **a** is compiled from [1040, 2216, 2252, 3089, 3689] in general and [1217] for Chip, [1991, 1992, 2253] Fringe, and [4325] Dlw. Panel **b** illustrates data from [1217, 2851, 4456]. It is not known whether the tetramer's symmetry is rotational (as shown) or reflective. Panel **d** is adapted from [882, 2079, 3689, 3690, 3979]; and the dialog illustration (**e, f**) is based on [988, 994, 1074, 2839, 3246].

N.B.: Input/output ports (raised rectangles) on the cell surface are drawn at arbitrary apical-basal levels (apex laced with microvilli). The reciprocity of the D/V dialog is a vivid realization of Boundary Model II (cf. Fig. 5.4a) [2808], whereas signaling is unidirectional across the A/P line (cf. Fig. 6.4). Evolution might have opted for symmetric signaling here because the wing's D and V surfaces must match so precisely.

[353, 2839, 3089]. Similar reasoning might explain bristle densities in *sgg*[LOF] clones on the legs [1039, 1833, 4666], although the *cut*-like agent at the leg's V midline is unknown. In each case, Wg would foster a prepattern (i.e., 1 or 2 proneural stripes) [3374]. Thus, *wg* qualifies as the kind of prepattern gene sought by Stern (cf. Ch. 3) [886, 4095], while Wg's ability to bias proneural potential in a graded manner fits nicely with Wolpert's gradient scheme [3954]. In *cut*[LOF] discs, the marginal bristles become disorganized [886] probably because the interfaces that align the SOPs are gone (cf. Fig. 3.9d).

Given Wg's downstream links, it is not surprising that *wg*[GOF] clones induce ectopic margins when they arise anteriorly in the D or V region [1040] (albeit without turning ON *cut* [2839]; cf. *slimb*[LOF] clones [2060, 2856, 4279]). Although the *wg*-ON stripe straddles the D/V line [163, 3689], the D or V half of the stripe can be suppressed without appreciably affecting wing shape [1040, 2850]. A feedback loop apparently enables the remaining half-stripe to double its output of Wg [2839, 3690] (cf. Dpp [572, 1674, 4136]).

Unlike the leg disc, where the *wg*-ON state is inherited from the embryo [837, 4682], the nascent wing disc does not express *wg*. Only later does it activate and modulate *wg* expression in response to various spatial and temporal cues [834, 880, 4683]. At the margin, the chief cue for *wg* is the *ap*-ON/OFF interface [3689]. This stimulus is mediated by the Notch pathway.

Notch incites and sustains the marginal expression not only of *wg* [988, 994, 1074, 2839, 3689], but also of *cut* [988, 990, 994, 2839, 3089], *vg* [1250, 2216, 2254, 3089], and bHLH genes of the E(spl)-C [988] (cf. overall circuit [2252, 3089]). (The *vg* gene is regulated independently by Wg [2092, 3091, 4849] through a separate *cis*-enhancer [2219, 2254, 4684].) All pouch cells are competent to react in these ways [990]. For example, *N*[GOF] misexpression turns ON *wg* and *vg* anywhere in the pouch [1040, 2219] but not outside it.

The Notch links also explain why quashing Notch snuffs out Wg and Cut at the D/V line [459, 2548, 2839, 3689] and stunts pouch growth [1040, 3025]. Similar LOF effects are seen in *Su(H)*[null] discs [2253, 2453, 3089, 3826] and *Su(H)*[null] clones [994, 1040, 2079, 3089], implying that Su(H) relays the signal (N → Su(H) → target genes). Targets include bHLH genes at the E(spl)-C [991, 994], which may play an antineural role because deletions cause extra bristles [1794]. However, they seem irrelevant for the margin itself because deleting them causes negligible notching [991, 994, 1794], and *m8*[GOF] does not affect *vg* [2254]. (*N.B.*: Notching is the trait for which *Notch*, *Serrate*, *cut*, *Chip*, and *scalloped* were named, while *wingless*, *apterous*, and *vestigial* denote wing reduction [470, 2561, 2940]). Clones that lack all 7 bHLH genes repress the *cut* stripe slightly when they overlap it, but *groucho*[null] clones turn *cut* OFF completely [2548]. Evidently, Notch activates *cut* not only via the E(spl)-C bHLH proteins (*), but also redundantly via an unknown Gro-binding protein ("X"): "N → {[E(spl)-C* OR X] AND Gro} ⊣ Y ⊣ *cut*," where "Y" is an unknown intermediate repressor [2548]. Based on their role in lateral inhibition (cf. Ch. 3), E(spl)-C bHLH proteins would be expected to directly repress *ac* and *sc* (vs. acting through *cut*). However, inhibiting *cut* seems sufficient to totally derepress *ac* and *sc* [988], so such a parallel link is unlikely.

Fringe prevents Notch from responding to Serrate

Notch is expressed fairly uniformly in immature wing discs [988, 1203, 1851, 2070, 2209, 2296], but its ligands Ser and Dl are not.

1. *Ser* is activated (and kept ON) by *ap* in the D compartment in 2nd–3rd instar [885, 988, 1040] via a specific *cis*-enhancer [164]. Its downstream status (Ap → Ser) is confirmed by the ability of *Ser*[GOF] to rescue wing development in *ap*[LOF] discs [2251, 2253]. In mid–late-3rd instar, Ser fades from much of the D area and arises elsewhere (intensifying at the D/V border) [885, 2216, 4301] via different *cis*-enhancers that respond to other cues [163, 1250]. The only *Ser*[null] clones that affect the margin are those that reach it from the D side [885, 994, 1040], so Ser is only needed at the D edge of the D/V line. Misexpression of Ser in the V region induces "margin syndrome" (margin bristles, overgrowth, and activation of *wg*, *cut*, *vg*, and E(spl)-C genes) [885, 1040, 2092, 2216, 4029], implying that (1) all V-type cells can respond to Ser and (2) the only cells that normally do so abut the V edge of the D/V line because Ser is tied to D-cell surfaces [988, 1074] and is silenced by a co-expressed inhibitor. The inhibitor turns out to be Fringe (see below) [1250, 3246]. Ser can also cause the syndrome when expressed near the D/V line on either side [988], implying that Fringe's effects can be overridden near the border. Quasinormal triple rows (albeit V/V) only arise when Wg [2092] or Vg [2252] is also present.

2. *Dl* is transcribed throughout the disc as early as mid-2nd instar – more strongly ventrally, and most intensely at the future wing margin [1074, 2253, 3246]. In mature discs *Dl* is upregulated in two bands that flank the *wg*-ON D/V stripe [988, 994, 1074, 2296, 2839] (cf. N

[1951, 2070]). Thus, *Dl*'s expression pattern does not reveal as obvious a D-V signaling bias as *Ser*'s [357]. Nevertheless, Dl must be mediating polarized (V-to-D) signaling like Ser (D-to-V) because (1) Dl^{LOF} clones at the D/V border nick the margin only when they reach it from the V side [994, 1074], (2) Dl^{GOF} clones cause margin syndrome only in the D (and border) region [988], and (3) misexpression of *Dl* along the A-P axis (via *ptc-Gal4*) elicits ectopic margins, overgrowth, and *cut* and *vg* expression mainly on the D (vs. V) side [1250, 3246], although other aspects of the syndrome are induced on both sides [1074, 2092, 2253] and the effects with *UAS-Dl* lines are weaker than with *UAS-Ser* [1074, 2092]. Apparently, Dl is modulated somehow so that transduction of its signal is blocked ventrally.

The *fringe* (*fng*) gene was discovered ca 1991 [1990] via an enhancer-trapped *lacZ* at its locus that had an intriguing expression pattern in the wing disc [1992]. The inserted *lacZ* and the resident *fng* are expressed congruently with *ap* in the D compartment until late-3rd instar. Indeed, *ap* regulates *fng* [1992]. Like ap^{null} clones, fng^{null} clones in the D region that are near the endemic margin cause an ectopic margin along their edge (Fig. 6.7). However, the *fng*-ON/OFF margins contain only D-type bristles in D/D symmetry. Evidently, *fng* acts downstream of *ap* to control *trans*-border signaling [1992, 4762], whereas D-vs.-V identities are regulated separately by a parallel (*Dlw*) branch in the circuit (Fig. 6.8). Dramatic evidence for this role comes from *fng*'s ability to rescue ap^{LOF} winglets when it is artificially expressed dorsally [2850, 3158], resulting in full-size (albeit V/V symmetric) wings.

Fng's a.a. sequence suggested that it might be secreted [1992], and it is secreted in cell culture [2084] and to some extent *in vivo* [3246]. However, it only affects Notch in the Golgi [3000], and its effects *in vivo* are cell autonomous [2253, 3246], so its role here is intracellular. Indeed, Fng may be tethered to Notch during maturation [2096]. Fng turned out to be a toggle that switches the Notch receptor from one responsive state to another: Serrate cannot activate Notch in the presence of Fng [1250, 2253, 3246].

How does Fng interfere with the Ser-N interaction? Fng glycosylates Notch in the Golgi [510, 1268, 2904, 3000] before Notch reaches the surface (cf. earlier hints to this effect [1551, 2070, 2253, 4835]), and the added sugars increase Notch's affinity for Delta [510, 3246]. The same modification

may make Notch refractory to Ser [360, 3245]. The Fng-N interaction seems to be direct [357, 510, 1250]. Fng interacts with EGF repeats 24–29 in Notch's extracellular domain (cf. Fig. 2.3) [990].

There is no known counterpart of Fng in the V compartment that damps Dl-N signaling (while facilitating Ser-N signaling?) [1250], and it remains unclear whether such a doppelgänger must exist [360, 1944]. Scabrous can act in this way when artificially expressed [2462], but its normal expression at the margin occurs too late for it to be playing this role. The only sure conclusion is that the Deaf-Speakers/Mute-Listeners Trick is as instrumental for the D-V axis as it is for the A-P axis – albeit in a different way:

1. The D-V axis has two short-range signals (Ser and Dl) and hence two kinds of selective deafness, whereas the A-P axis has only one (Hh). In this regard, the D-V strategy is a vivid incarnation of Meinhardt's Boundary Model [2808].
2. The D-V axis enforces deafness at the cell surface (D cells cannot hear Ser because Fng modifies Notch), whereas the A-P axis does so at the genetic level (P cells cannot hear Hh because En prevents expression of the transducer Ci).

Although these strategies are equivalent developmentally, they must have had different origins evolutionarily. Regional suppression of a transducer like Ci should be trivial to evolve, but where on earth did a "deafness-promoting factor" like Fng come from? Like Numb, another regulator of Notch (cf. Ch. 2), Fng appears to have acquired an abstract switching function that is applicable in any context [1988]. Indeed, Fng also constrains Notch signaling to a compartment boundary in the eye (cf. Ch. 7), where it has other upstream regulators instead of *ap*. Hence, Fng's fame as a versatile "boundary gene" [1988, 1992] is well deserved.

Strangely, for Fng to block incoming Ser signals, the receiving cell itself must express Ser or Dl: when a *fng*-ON cell's Ser and Dl are both removed (dorsal Ser^{null} Dl^{null} clones), it can respond by turning ON *cut* [2839]. This mystery exposes our ignorance about how Notch interacts with its ligands (sterically and stoichiometrically) under normal conditions [1944].

The core D-V circuit plugs into a complex network

Figure 6.8 summarizes our basic understanding of D-V cross-talk and fate assignment. It incorporates the following links, all of which are well documented, except for 1 and 5:

Link 1: {Ap AND Chip} ➜ *Dlw*? *Dlw*'s spatial expression is unknown, but its LOF and GOF effects imply allegiance to the D (*ap*-ON) compartment [4325].

Link 2: {Ap AND Chip} ➜ *Ser*. Many other factors also regulate *Ser* in the wing [163, 1250].

Link 3: {Ap AND Chip} ➜ *fng*. No other regulators of *fng* are known in the wing [1988, 2851].

Link 4: Fng ⊣ {Ser ➜ N}. This link deafens D cells to their own signal [1250, 3246].

Link 5: Fng ➜ {Dl ➜ N}? Fng dramatically increases Dl-N binding *in vitro* [510], but the extent to which it potentiates Dl-N signaling *in vivo* is less certain [2253]. Normally, Dl can only turn *Ser* ON in the D area, but when Fng is put into the V area (by a *fng*GOF mutation or by *ptc-Gal4:UAS-fng*), Dl can activate *Ser* there as well (via *ptc-Gal4:UAS-Dl*) [3246], a result that clearly implies Link 5. However, this effect (especially as manifest in ectopic bristles) typically extends only a short distance into the V half of the pouch (*wg* activation can extend farther [510]) and hence may stem from border peculiarities (Links 7–11 and 16–18). If Link 5 were a more general feature, then excess Fng ventrally should enhance the response to Dl, but such wings actually grow less (e.g., *fng*D4 GOF heterozygotes [1992]), not more [2253]. Even more telling, perhaps, is the fact that excess Fng fails to exacerbate Dl-mediated signaling during CNS neurogenesis [1250].

Link 6: {Ser OR Dl} ➜ N ➜ *wg*. Notch signaling only activates *wg* within the confines of the wing pouch [994, 1074, 2253]. This restriction apparently stems from the same proximal vs. distal dichotomy that operates in the leg [157] (cf. Ch. 5), with Teashirt and Hth-Exd_nuc dimers governing the proximal domain [793, 2937, 3104] and suppressing key genes needed for distalization [678]. However, the boundary may instead be enforced by the Iroquois Complex [1060].

Omitted from the sketch in Fig. 6.8 are other components (e.g., *strawberry notch* [895, 2662]). Other connecting links (below) have been seen under various LOF and GOF circumstances. Why certain links only operate at specific stages remains a mystery [981].

Link 7: N ➜ {Ser AND Dl}. Notch stimulates transcription of its ligands [988, 3246] (but see [1951]).

Link 8: Ser ➜ *Dl*. Ser activates *Dl* transcription [2253, 3246].

Link 9: Dl ➜ *Ser*. Dl has a reciprocal effect on *Ser* [2253, 3246]. This reinforcement (Links 8 and 9 acting via Link 7) upregulates both ligands near the D/V border in 3rd instar [1988, 3246]. The two sides are interdependent because unilateral inactivation of N abolishes N activity on the other side [2838, 3689] (but see [2218] for D-V asymmetries and [2838, 3524] for exceptions).

Link 10: Dl ➜ *Dl*. Given that Dl can activate its own transcription in adjacent cells [2253], this link should cause Dl expression to spread like wildfire throughout the disc (cf. Minotaur Scenario; Ch. 5). Why it does not is unclear [2253].

Link 11: {Ser or Dl} ⊣ N. Excess ligand autonomously represses *N* (and hence *wg* and *cut*) in the signaling cells themselves [988, 1040, 2248, 2253, 2839, 4301] by an unknown (cell surface?) mechanism [357]. Dl is weaker in this regard [1074, 2092]. This link creates a single-cell version of the Deaf-Speakers/Mute-Listeners Trick that allows one-way signaling [2251, 2850, 3158]. Its normal role is to block N in cells flanking the *cut*-ON (*wg*-ON) stripe late in development [2839].

Link 12: Cut ⊣ {Ser AND Dl}. This link enables the *cut*-ON (high-N) stripe to create a gap in the wider *ac*-ON (high-Ser-high-Dl) band and thereby split it into two parts (Fig. 6.8c) [988].

Link 13: Cut ➜ *wg*. Expression of *wg* is maintained (but not initiated) by Cut [2839] (but see [882]). Not surprisingly, the *wg*-ON and *cut*-ON bands coincide [882]. The *wg*-ON band looks wider with a *lacZ* insert [353], but this illusion is attributable to perdurance of the β-gal product.

Link 14: Wg ⊣ *wg*. A paracrine feedback loop maintains the width of the *wg*-ON stripe [2839, 3690] (but see [2839]).

Link 15: Cut? ⊣ {Wg ➔ target genes}. Wg-secreting cells cannot respond to their own Wg [2839], possibly because they also express Cut.

Link 16: Wg ➔ {Ser AND Dl}. Wg has an autocrine-paracrine ability to activate *Ser* and *Dl* [988, 2839].

Link 17: Dsh ⊣ {N ➔ target genes}. Although Dsh belongs to the Wingless transduction pathway (cf. Fig. 5.6), it can interfere with Notch signaling, probably by directly binding Notch [151].

Link 18: N ⊣ {Wg ➔ target genes}. The ability of Dsh to "short circuit" the Wingless and Notch pathways implies a reciprocal antagonism [356], which may involve sequestration via N-Dsh heterodimers.

Link 19: Nubbin ⊣ {N ➔ target genes}. Nubbin is a POU protein that damps N target genes in the pouch [3092, 3103]. It may serve the sort of generalized "threshold-setting" function ascribed to Dad and Nkd for the Dpp and Wg pathways, respectively (cf. App. 6).

Link 20: Fng ➔ {Ser AND Dl}. When Fng is misexpressed ventrally, it activates *Ser* and *Dl* at its ON/OFF boundary, but there is no reciprocal activation of *fng* when Ser or Dl are misexpressed [2216, 2253, 3246]. This effect could be mediated by the loop of Links 8 and 9 [1250], but it likely just relies on loop 7 because Fng can activate the Notch pathway under steric circumstances that might be mimicked by supersaturation [2096].

Elements upstream of Wg are all expressed by mid-2nd instar [164, 1074, 4543, 4683], but *wg* does not turn ON at the margin until late-2nd [3104] or early-mid 3rd [2252, 3374, 3690, 4683] instar, and *cut* comes ON even later (mid-late 3rd instar) [2839]. The reason for these delays is not known.

A few other nuances should be mentioned (see [2251] for further riddles). Target genes (e.g., *wg* vs. *cut*) may need different thresholds of N activation [2839]. Responses may depend more on the relative amounts of ligand vs. receptor in conversing cells (i.e., a sharp vs. fuzzy boundary) than absolute amounts in either partner [1074, 2092, 2216, 2839]. Removal of *ap* function alone is not enough to fully convert a cell from D- to V-type identity (as assayed by *vg* induction) [4684], but it is unclear what other redundant factors help specify D-type

identity together with *ap* [2251]. Nor is it known whether V-type identity is a default state or is actively implemented by *ap* and *Dlw* counterparts [885].

If Dl and Ser are membrane bound (as generally assumed [1251]), then the dialog across the D/V line should create a *wg*-ON stripe that is only two cells wide [1988]. In fact, however, the stripe is 3–6 cells wide [353, 2079, 3689, 3690] (cf. *vg* [885]) and can broaden to ~20 cells when Nubbin is disabled [3092]. This greater-than-expected width could be due to (1) diffusion or transcytosis of N ligands [885, 3479], (2) perdurance of the *wg*-ON state [3363], (3) signal relay [3923], or (4) a larger-than-realized zone of productive ligand-receptor overlap [3092]. None of these possibilities can be excluded at present. The same quandary applies to wing veins [989, 1951] and to the activation of *E(spl)mβ* in intervein cells ≥5 diameters away from the source of Dl [871, 984].

In summary, the basic chain of events along the D-V axis is: Ap (D region) ➔ {Fng AND Ser} ➔ N (on V side of D/V line) ➔ Wg (diffuses in both directions) ➔ target genes, including Ac (➔ bristles) and Vg (➔ pouch), with N being activated differently on the D side (? ➔ Dl ➔ N). Metaphorically speaking, a "tent" (symmetric Wg gradients) is erected on a "pole" (activated N) that is planted at the edge of a territory (*ap*-ON domain), much like the process along the A-P axis. Viewing the process in this way elucidates the old "**Field Effect Mystery**" of molecular genetics: how can genes affect regions where they are not expressed? In this case, the *ap*^{null} phenotype involves loss of the whole wing – not just the D part where *ap* is expressed – because *ap* is needed to spark a chain reaction that builds the entire wing [4683]. The symmetric Wg gradients lead to nearly identical patterns in the D and V halves of the pouch, which makes sense because the two surfaces must match perfectly to make a flat airfoil. The slight differences that do exist (e.g., bristle types; cf. Fig. 6.7) are made possible by a gene (*Dlw*) that toggles the mode (D- or V-type) that cells use to interpret their gradient.

The wing-notum duality is established by Wg and Vein

What causes Ap to be expressed in the D region in the first place? Here, we encounter a queer irony. Whereas Wg is under Ap control during late stages of wing development, the opposite is true when the wing primordium first forms.

Unlike the leg disc where *wg* stays ON in one subregion throughout development, *wg* expression in the wing disc changes with time (Fig. 6.2) [880]. Wg is first

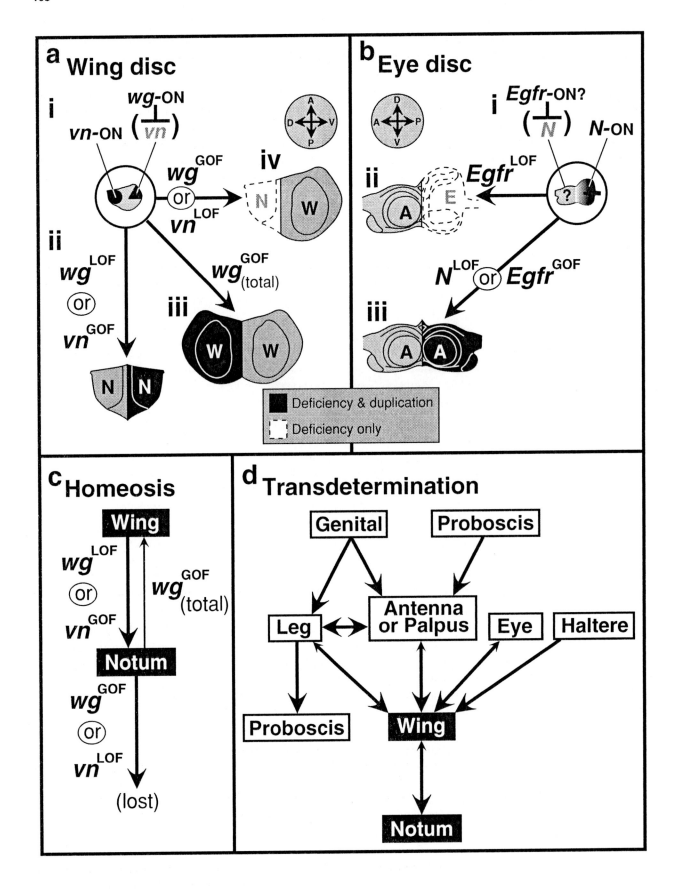

detectable in early-mid 2nd instar [880, 2252, 3104], when the wing disc contains ~100 cells [4683]. At this stage, *wg* is transcribed in ~10 cells that occupy an anterior-ventral sector (as in the leg disc) [207, 880, 4543, 4683]. Shortly thereafter (mid-late 2nd instar; ~200 cells) the *wg*-ON domain expands to span the ventral third of the disc [2252, 2253, 3089, 3104], while *ap* turns ON in the remaining D portion [1074, 2251, 3104, 4683]. This complementarity is not absolute (overlap exists), but the following facts argue that it is probably causal (i.e., Wg ⊣ *ap*).

When *wg* is suppressed during 2nd instar (by heating t.s. *wg*^{LOF} larvae), the wing is often replaced by a mirror-image copy of the heminotum (Fig. 6.9) [880]. This duplication arises by *in situ* transformation [158, 159, 2929, 4543], rather than by tissue loss and compensatory growth [2015, 4683]. When *wg*^{LOF} discs are examined around this time, they are found to express Ap throughout the disc [4683], implying that *ap* is normally repressed ventrally by Wg (Wg ⊣ *ap*) but can assert itself when Wg is removed.

FIGURE 6.9. Bipolar duality of the wing disc (i.e., wing vs. notum, **a**) and eye disc (i.e., eye vs. antenna, **b**), homeotic tendencies in the wing disc (**c**), and how these tendencies may explain the wing-to-notum trend in transdetermination (**d**). *N.B.:* The key for **a** (**ii–iv**) and **b** (**ii** and **iii**) straddles the **a/b** dividing line, and prospective adult axes are given in the compasses above (right wing disc, left eye disc). The letter "N" in plain type here represents the heminotum, whereas an italicized "*N*" signifies the gene *Notch*.

a. Genetic control of wing-vs.-notum identity in the wing disc. **i.** Wing disc (gray) showing regions (solid black) where *vein* (*vn*, dorsal spot) and *wingless* (*wg*, ventral sector) are transcribed in 2nd instar. Both *vn* and *wg* encode diffusible signaling molecules. Although Vn and Wg could probably diffuse into one another's territory and exert mutual inhibition, the chief link that appears critical is "Wg ⊣ *vn*". **ii–iv.** Abnormal discs resulting from LOF or GOF changes in *vn* or *wg* expression. **ii.** "N/N" double-heminotum phenotype resulting from *wg*^{LOF} mutations. The wing portion of the disc has transformed to notal identity. Evidently, *vn* is expressed by default when Wg is absent. N/N duplications can also be evoked by *skinhead*^{LOF} [3371], *vg*^{LOF} [2252], and *N*^{LOF} [886] (cf. Table 8.1). Although *vn*^{GOF} has not yet been tested, it is assumed to also behave this way because (1) Vn is the only essential ligand for Egfr in wing disc development [3025, 3928, 4604], (2) N/N defects can be created by misexpressing *Egfr* (via *omb-Gal4*) [4543], and (3) wing-to-notum transformations are seen in *Egfr*^{GOF} clones [4543]. **iii.** "W/W" double-wing phenotype resulting from misexpressing *wg* throughout the disc (via *T80-Gal4:UAS-wg*) [3025]. The notal portion of the disc has transformed to wing identity. Apparently, Wg suppresses *vn* within the dorsal domain (Wg ⊣ *vn*) and hence triggers the "wing subroutine" there as well as in *wg*'s native (ventral) domain. A partial W/W trait is rarely seen with *vn*^{LOF} alleles (omitted), which instead typically cause notal loss (as in **iv**) [3801, 3929]. Its rarity is likely due to a tendency of *wg* to remain OFF in the absence of Vn. **iv.** "W/-" (notumless) phenotype resulting from *vn*^{LOF} mutations [4543] or *wg* misexpression throughout the pouch (via *sd-Gal4:UAS-wg*) [2252]. The difference in *wg*^{GOF} effects (**iii** vs. **iv**) may be due to a threshold of Wg concentration below which *vn* can be turned OFF without inducing wing development. Forcing *wg* to be expressed along the A/P border (via *dpp-Gal4:UAS-wg*) can yield either outcome (**iii** or **iv**) [2252, 3025, 3104].

b. Genetic control of eye-vs.-antenna identity in the eye disc (cf. Fig. 7.1). **i.** Eye disc (gray) showing the region (dark gradient) where *Notch* is transcribed in 2nd instar [2353]. Egfr's transcriptional area is unknown; it may complement *Notch*'s. **ii.** "A/-" (eyeless) phenotype caused by LOF conditions of the EGFR pathway [2353] due to overexpressing the EGFR inhibitors *anterior open*, *tramtrack*, or Bar-C (also caused by *wg*^{GOF}). See Ch. 7 for "early eye" genes whose LOF alleles cause a similar phenotype. GOF conditions in the Notch pathway (*N*^{GOF}) can cause smaller, abnormal eyes (but see [2362]). **iii.** "A/A" double-antenna phenotype resulting from LOF effects ("DN" = dominant negative) in the Notch pathway (*N*^{DN}, *Dl*^{DN}, *Ser*^{DN}, or *mastermind*^{DN}, but not *Su(H)*^{DN} or *m8*^{DN}) or from GOF effects in the EGFR pathway (*Egfr*^{GOF}, *spitz*^{GOF}, *Ras1*^{GOF}, *dRaf*^{GOF}, or *pointed*^{GOF}, but not *dMAPK*^{GOF} or *dMEK*^{GOF}) [2353]. The eye portion of the disc has transformed to antennal identity. A comparably complete transformation of the opposite kind ("E/E") has never been found (but cf. Ch. 8).

c. Summary of data from **a.ii–iv** in terms of histotype identities. Cells can undergo a wing-to-notum switch in two ways (thicker arrow denotes greater likelihood) but a notum-to-wing switch in only one way.

d. Switches in histotype fate seen during long-term culture of disc fragments. Double-tipped arrows mean that switching can occur in either direction (size of arrowhead denotes relative frequency). Top-to-bottom order reflects the favored flow of events, so the notal fate is a "sink" for the system. The tendency of wing tissue to transdetermine to notum (**d**) may be due to the ease with which such a transformation can occur homeotically (**c**). "Proboscis" appears twice (as per original data [1421]), which poses a paradox for any scheme that tries to rank histotypes hierarchically. Although no arrow is shown from eye to antenna, such switches probably do occur during culture because (1) antennal fragments duplicate themselves while eye fragments regenerate antennae [1404, 1406, 1423, 3810] and (2) LOF mutations in various genes transform eye to antennal tissue (but not vice versa; cf. Table 8.1) [2362]. Third-instar *vn*^{LOF} leg discs can transdetermine to wing (not shown), even though mature *vn*^{LOF} larvae lack wing discs, because they bypass the 2nd-instar need for *vn* [4768].

Schematic in **a.i** is based on [3928, 3929, 4543] for *vn*-ON and [2252, 2253, 3089, 3104, 4683] for *wg*-ON. Illustrations in **a.ii–iv** and **b.iii** are sketches of discs pictured in [2252, 3104, 4543] (**a.ii**), [3025, 3104] (**a.iii**), [2252, 3929] (**a.iv**), and [2353] (**b.iii**). Panel **b.i** is based on [3382], and **d** is redrawn from [1421] with one arrow omitted (from unidentified tissues to a genital state). See also App. 7.

The heminotum is the most dorsal derivative of the wing disc (cf. Fig. 6.1), so the wg^{LOF} phenotype is analogous to D/D wg^{LOF} legs (cf. Fig. 5.5) [4683]. Conceivably, Wg might be playing a ventralizing role as in the leg disc (Wg ⊣ dpp), and wg does turn ON in a leg-like ventral sector [880, 2251, 4683]. The parallels between wg's action in wing vs. leg discs extend no farther, however. The leg disc has no *bona fide* D or V compartments (cf. Ch. 5) [4076], and the wing disc only uses the "Wg ⊣ ap" link briefly [4684]. The wing-notum dichotomy that Wg sets up may not stabilize until later when other circuits involving vg (see below) assume responsibility for its maintenance. To avoid confusion with D/D effects inside the wing per se (cf. Fig. 6.7), this phenotype will be termed "N/N" (i.e., double notum). ("N" can also mean Notch protein, but not when used here with a slash mark. "W" will be used to designate the wing.)

An N/N phenotype is also seen in vg^{LOF} strains [2252] and can be created by heat-pulsing t.s. N^{LOF} larvae during the same sensitive period as t.s. wg^{LOF} (2nd instar) [886]. Evidently, wg, N, and vg collaborate to initiate the wing anlage at this time [885, 2219, 2252–2254]. If they fail to do so, then the prospective wing tissue reverts to a notal fate. The nature of this 3-way partnership is obscure but probably involves some of the links that later operate at the D/V border [4463].

Aside from wg, N, and vg, another element shared in common by the "wing initiation" and "D/V border" subcircuits is $Egfr$ [207, 3025]. $Egfr$ encodes the *Drosophila* Epidermal growth factor receptor (a.k.a. DER, Torpedo, Ellipse, or Flb) [812, 3484, 3535, 3536]. Egfr is a canonical receptor tyrosine kinase (RTK) [2573, 3464, 3779, 3780]. The EGFR pathway (discussed in a later section) must dictate notal identity because ligand-independent $Egfr^{GOF}$ clones make notal tissue and induce ap expression even when located in the wing [4543].

Egfr has four known ligands [1291, 3830, 4556, 4604] – all of which have an EGF motif: Gurken (a TGF-α relative that only acts in oogenesis) [3086, 3120], Spitz (also a TGF-α cousin) [3711], Argos (a competitive inhibitor) [1294, 1911, 2067, 3828], and Vein (an agonist that has an Ig-like motif) [3801, 3929, 4808]. All four ligands function in the D-V axis of the embryo [1529, 3085, 3758, 4117, 4557], where Spitz sets the ventral limit of the leg disc [1572, 3536]. A fifth possible ligand (similar to Spitz) named "Gritz" has recently been identified [204, 2108, 4558].

Is any of the embryo's D-V circuitry incorporated into the D-V axis of the wing disc? Apparently not. The expression of Vein (Vn), the wing disc's chief Egfr ligand [3025, 3928], is not inherited from the embryo like en (cf. Fig. 4.4) [3929]. Rather, Vn is deployed in the notal region during 2nd instar [3928, 3929, 4543] – roughly the same stage when Wg arises in the wing area (Fig. 6.9) [880, 2252, 3104]. The "Egfr ⊣ wg" link (deduced from effects of $Egfr^{GOF}$ and $dRaf^{LOF}$ clones [207]) can thus be rewritten as "Vn ⊣ wg". This link could allow Vn to define the wg-ON area, but Vn evidently does not do so because wg is expressed normally in vn^{LOF} discs [4543]. These same discs do not express ap, so Vn is required for ap activation (Vn → ap) [4543].

Because Vn is secreted [3801], the Vn-Wg relationship could theoretically work like the Dpp-Wg "seesaw" in the leg disc (cf. Ch. 5). Indeed, Vn does "rebound" in the absence of Wg: 2nd-instar wg^{LOF} discs have an extra spot of Vn where the Wg spot would have been [4543], and they go on to show N/N symmetry in other features as well. Added to the previous "Vn ⊣ wg" link, this "Wg ⊣ vn" link completes the seesaw, or so it seems.

Strangely, however, Wg does not rebound in the absence of Vn: vn^{LOF} discs lack notal tissue, but no wing tissue develops in its place [3929]. This "notumless" ("W/-") phenotype is also seen in (1) $Egfr^{LOF}$ mutants [3025] (cf. the "V/-" subclass of dpp^{LOF} legs; Fig. 5.5), and (2) wg^{GOF} discs where wg is expressed throughout the pouch (via sd-Gal4:UAS-wg) [2252]. The sensitive period for this phenotype is 1st and early-2nd instar (based on a t.s. vn^{LOF} allele) [3929, 4543].

When the entire disc (vs. only the pouch) is flooded with Wg (via $T80$-Gal4:UAS-wg), a "W/W" phenotype (two pouches but no heminotum) arises (Fig. 6.9) [3025]. In this case, Wg cannot be totally suppressing Ap (Wg ⊣ ap) because enough Ap must persist to activate the D/V border circuit for pouch formation. A partial W/W phenotype is rarely observed in vn^{LOF} mutants [3801, 3929]. Its sensitive period is 2nd and early-3rd instar [3929]. The key facts presented thus far are summarized below:

1. Each ligand normally governs a separate region. Vn rules the notum, while Wg rules the wing.

2. The GOF effects are reciprocal. Forcing the Wg pathway into an ON state can shut Vn OFF within its own domain (Wg ⊣ vn), and vice versa (Vn ⊣ wg).

3. The LOF effects, however, are not reciprocal. Removing Wg from its normal domain results in Vn being expressed there, but removing Vn from its normal domain does not result in Wg being expressed there.

Why does Vn, but not Wg, rebound when its opponent is removed? Presumably, *vn* has a ubiquitous *cis*-enhancer (or basal promoter) stronger than that of *wg*. (In the leg disc, both *dpp* and *wg* must have strong controllers.) The "default state" of the wing disc would thus be *vn*-ON, with the *wg*-ON area being superimposed ON this baseline by (1) an upstream regulator turning *wg* ON ventrally and (2) Wg turning *vn* OFF when it reaches a certain level. An obvious candidate for *wg*'s initial regulator is Hh, although it is unclear why *wg* is not also activated dorsally and continually until pupariation. Interestingly, only the dorsal (wing, haltere, eye) discs require *vn* [80, 3291, 3928, 3931, 4768], while *wg* is needed in all the major discs [880, 1163, 1780, 1781, 3732].

But Vestigial and Scalloped dictate "wingness" per se

Vestigial is expressed in an even more restricted subset of discs than Vein – viz., the wing and haltere discs [4681]. Indeed, it is emblematic of their inception in the embryonic ectoderm (cf. Ch. 4) [827, 1571, 1572, 4681]. Both discs transcribe *vg* uniformly until 2nd instar, when transcription subsides outside the pouch [4683].

The launching of pouch development is tantamount to creating a proximal-distal axis for appendage outgrowth [4682, 4684], and *vg* has been suspected of encoding "wingness" because (1) it turns ON throughout the pouch [4683], and (2) it is required autonomously by wing cells for growth [3960]. Expression of *vg* in the pouch is regulated by two main *cis*-enhancers [982, 2254]:

1. The "boundary" enhancer (*vg*BE) is located in *vg*'s 2nd intron [2219]. It binds Su(H) and keeps *vg* ON at the D/V border in response to Notch [2218, 2219]. Initially, it also needs Wg [2254].

2. The "quadrant" enhancer (*vg*QE) is in *vg*'s 4th intron [2219]. It binds Mad [2217], Ventral veinless (Vvl) [703], and probably Arm-Pan [2254]. It turns *vg* ON in the rest of the pouch in response to Dpp, Wg, and Vvl [703, 2217, 3091, 4849], with input from Vg itself as well [2254, 3933].

Expression of *vg*QE vanishes in clones that are singly mutant for *Mad*[LOF] [2217], *arm*[LOF] [3091, 4849], or *vvl*[LOF] [703], so *vg*QE must use a combinatorial rule like the "{Dpp AND Wg} ➝ *al*" rule that governs *al* in the leg disc (viz., "{Dpp AND Wg AND Vvl} ➝ *vg*"). The *vg*QE is OFF at the margin because Vvl is absent there – a lack that is due to Wg (Wg ⊣ *vvl*) [703] and, ultimately, Notch (N ➝ *wg*) [994, 1074, 2253]. Interestingly, *vg*QE is first activated at the

center of the wing disc (analogous to *al*) before its expression broadens to fill the pouch [2219, 2254].

The suspicion that *vg* might be a "master gene" for wing development was confirmed in 1996 by showing that it can coerce cells in foreign (leg or eye) discs to make wing tissue when it is ectopically misexpressed [2219]. This "magic trick" can also be performed by *wg*[GOF] [2092, 2252], but *vg*[GOF] is more adept at turning ON downstream wing marker genes (e.g., *nubbin*) [1686, 2252]. Moreover, when *wg*[GOF] induces ectopic wing tissue, *vg* turns ON at those sites [2092], whereas the reverse is not necessarily true [2219] (cf. Table 8.1).

The *vg* gene is autoregulatory (*vg* ➝ *vg*) insofar as Vg activates both *vg*BE and *vg*QE [1686, 3291]. However, Vg is ineffective without its auxiliary protein Scalloped (Sd) [1686, 3291, 3936]. Vg and Sd are expressed congruently during wing disc development [4683], and *vg* and *sd* maintain one another's transcription: "Vg ➝ *sd*" and "Sd ➝ *vg*" [3291, 3936, 4463, 4683]. The former link also operates outside the wing disc, although the latter apparently does not [2219].

Vg is a nuclear protein without any known DNA-binding motif [4681]. Sd is a member of the TEA family of transcription factors [623, 1029, 2010], and as such it binds tandem 9 b.p. sequences that conform to the consensus of its human ortholog [1686, 2940]. Vg and Sd bind one another [3291] but do not form homodimers [3936]. In the absence of Sd, Vg is found primarily in the cytoplasm, but Vg localizes to the nucleus when Sd is present, implying that Sd escorts Vg or retains it after nuclear entry (Fig. 6.10) [1686, 3936]. The relative amounts of Vg vs. Sd are critical [3291, 3936, 4463]. For example, excess Sd reduces target gene activity [1686], and imbalances in either direction cause similar phenotypes [3936] (cf. the Ap-Chip story above [1217]).

When the Su(H)-binding site in the *vg*BE DNA is replaced with binding sites for Ci-155, the reporter gene switches from Notch to Hh control as expected [1686] and turns ON in a stripe at the A/P (vs. D/V) border. Oddly, however, it is not expressed in the leg disc, despite the presence of a similar A/P border. Why not? Conceivably, Vg-Sd binding sites elsewhere in the enhancer bind a repressor when Vg-Sd dimers are absent (as in the leg disc), and this repressor blocks activation by Ci-155. The repressor could be Sd itself, or a co-repressor recruited by Sd (Fig. 6.10f). Whatever its nature, the inhibitory agent must be powerful enough to prevent *trans*-activation by Ci. Any *cis*-enhancer having Vg-Sd binding sites would thus operate according

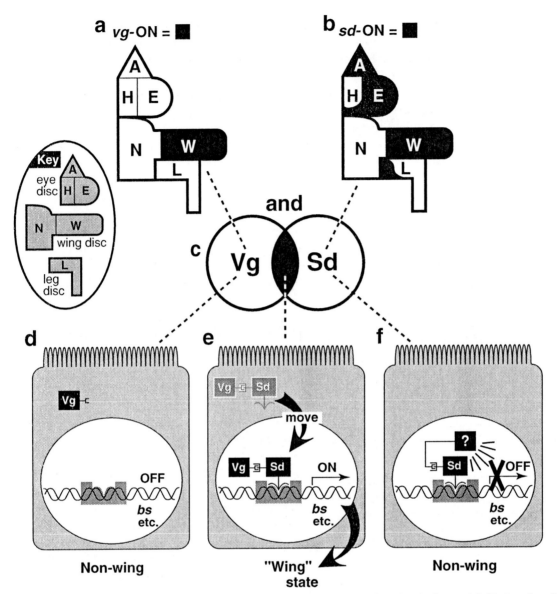

FIGURE 6.10. The "**Venn Overlap Rule,**" as manifest in the control of wing blade identity by *vestigial* (*vg*) and *scalloped* (*sd*).

a, b. Regions where *vg* and *sd* are normally transcribed (black areas). These regions are plotted on a schematic map (key in **a**) where the derivatives of eye, wing, and leg discs are outlined by thick lines, and subregions (A = antenna, H = head capsule, E = eye, N = notum, W = wing) are partitioned by thin lines. **a.** The *vg* gene is transcribed throughout the wing and in small areas of the hinge and notum (not shown; cf. Fig. 6.2) [4681]. **b.** The *sd* gene is transcribed throughout the *vg*-ON territory, as well as in eye and leg discs [623]. Although expression in eyes per se is well documented [623], the antennal, head, and leg areas plotted here are only guesses based on the ability of ectopic *vg* to induce wing tissue wherever *sd* is already ON [2219, 3936].

c. Venn diagram of the logic. Only when Vg and Sd are co-expressed (black area of overlap) will wing tissue be formed. Put more generally, the Venn Overlap Rule states that when two or more genes interact cooperatively, they only activate target genes where their domains of expression overlap [2353].

d–f. Illustrations that show how the rule is implemented at a molecular level. **d.** When a cell expresses *vg* alone, Vg resides mainly in the cytoplasm [1686, 3936]. Even those Vg molecules that do enter the nucleus are impotent because they cannot bind DNA without help from Sd. **e.** When both Vg and Sd are present, Vg binds Sd [3291] and the Vg-Sd complex is found mainly in the nucleus. Nuclear localization could be due to escorted import (as shown) or to selective retention after import. The Vg-Sd complex binds a DNA sequence (gray box) near target genes like *blistered* (*bs*) [1686]. Some of those genes (not *bs* itself) execute wing identity. Whether the Vg-Sd dimers activate transcription directly or merely allow other *trans*-activators to turn ON *bs*, etc., is unknown. **f.** When a cell expresses *sd* alone, Sd may bind some of those same sites (but see [1686] sequel). In the absence of Vg, *trans*-activators like Ci cannot turn ON such target genes [1686]. The reason for this "locked OFF" state may be that Sd recruits a strong co-repressor ("?"). See also Fig. 6.14f and App. 7.

to a Boolean rule: "Turn ON ONLY IF you are in the wing pouch AND you are prodded by the *trans*-activator of a signaling pathway; otherwise remain OFF, no matter what activators try to turn you ON!"

The "locked OFF" state is fundamentally different from the "passive OFF" state (Fig. 6.10), and it has different implications for evolution (cf. Ch. 8). This "locked OFF" trick is evident in the Pangolin story (cf. App. 6) [3147], although neither Pan nor its partners preside over a discrete body region as Vg does. Pan ($\approx$ Sd) locks its target genes in an OFF state by recruiting Groucho when Arm ($\approx$ Vg) is absent [692, 2496]).

Straightening of the D/V border requires Notch signaling and Ap

An old issue in embryology is the "**Euclidean Ruler Problem**": how do cells draw lines? Any system – be it sentient or cybernetic – that builds complex structures must be adept at drawing lines, squaring angles, and calibrating distances. Unlike humans, cells are handicapped as geometers because they cannot be surveyors. Except for neurons, cells have no "eyes" or their equivalent, so they must rely on local cues. Epithelial cells do not normally organize themselves into neat files [4641], but on occasion they must. Wing veins are a case in point. Somehow, wing cells have mastered the ability to form straight lines that become veins. Vein 1 forms along part of the D/V boundary, so this border may offer clues about vein formation in general.

Two possible alignment strategies were discussed before for the A/P boundary (Fig. 6.5): the Selector Affinity and Border Guard schemes. Which of them is used at the D/V boundary? In 1999, this question was analyzed by three different teams – Micchelli and Blair in Wisconsin [2838], Rauskolb *et al.* in New Jersey (Rutgers) [3524], and Milán and Cohen in Heidelberg [2850]. Overall, the results argue that both types of mechanism play a role. Notch signaling must be involved because

1. The D/V line becomes ragged when t.s. N^{LOF} larvae are heat-pulsed or a dominant-negative N allele is expressed [2838].
2. Some N^{null} clones cross (D to V or V to D) when they arise at the border [2838, 3524], and the same is true for clones that express a "deaf" N construct (truncated intracellular domain) [3524]. In contrast, clones whose N is constitutively activated never cross [3524].
3. $Dl^{LOF}\ Ser^{LOF}$ clones can cross the border in either direction [3524].

4. N^{null} clones only cross when N signaling ceases on both sides of the line [2838, 3524], and the same is true for $Dl^{LOF}\ Ser^{LOF}$ clones [3524].
5. Crossing also occurs with fng^{LOF} (D to V) and fng^{GOF} (V to D) clones that arise near the boundary [3524].

Because *ap* still functions in all these wings, *ap* cannot suffice for alignment. $Su(H)^{null}$ clones respect the D/V line and the line stays straight even when it traverses wg^{null} tissue, so neither Su(H) nor Wg is mediating N's effects [2838]. The above results suggest two types of affinity: (1) a "border" affinity for N-stimulated cells and (2) a "default" affinity for all other cells. N-activated cells evidently align to form a marginal band in the midst of an otherwise homogenous pool of blade cells [2838, 3524].

In 2001, this Border Guard Model was validated by O'Keefe and Thomas in San Diego [3158]. By expressing Fng and a D-specific integrin in the D part of ap^{LOF} wing discs, they enabled these discs to make a straight margin of *wg*-ON cells despite an apparent lack of any D-type cells. The rescued discs went on to make symmetric V/V wings that were otherwise remarkably normal in appearance.

In wild-type wings, Ap may control a subtler affinity within the border zone [2838]: *ap*-ON and *ap*-OFF cells segregate into $\sim$2-cell-wide bands within the $\sim$4-cell-wide N-activated zone [3524]. Cells can evidently drift into or out of the border zone by jostling during growth, but D-type cells cannot become V-type or vice versa [2838]. Evidently, this is how D vs. V lineages maintain the *ap*-ON/OFF compartment boundary [352].

Straightening of veins may rely on similar tricks

Similar arguments were made earlier for the *dpp*-ON zone at the A/P line. In each case, the border might be straightened by homophilic molecules that are downstream targets in the pathway. Interestingly, both pathways (Dpp and Notch) cooperate in vein development. This "double dose" of adhesivity could explain why extra vein tissue tends to (1) coalesce with existing veins to create branching networks, (2) form smooth curves [3372], (3) adopt narrow $\sim$4 cell widths [1368], and (4) acquire smooth edges [3624]. These traits are obvious with $blistered^{LOF}$ clones [1312, 2907, 3624] but are also evident in wings that are dpp^{GOF} [980], $Egfr^{GOF}$ [517, 2493, 4604], $heartless^{GOF}$ [3558], net^{LOF} [206, 320, 466, 4189, 4307] (cf. $daughterless^{LOF}$ [2741]), $plexus^{LOF}$ [206, 2742, 4307], rho^{GOF} [1643], $rolled^{GOF}$ [517,

2713, 3624], *spitz*^GOF [3800], *sprouty*^LOF [681], *vein*^GOF [4604], etc. [3076, 3558, 4794].

Whenever extra veins are elicited, they tend to arise midway between existing ones. These midway zones are distinctive in their timing of mitoses [3813], D-V annealing [431, 1312], and gene expression [4831] (cf. Fig. 3b of [4397]), so they may have special biases. The "**Paravein Hypothesis**" asserts that ectopic veins emerge there as an atavistic unveiling of ancient "paraveins" (old veins now lost) [320, 1368, 3237, 4307]. However, the channeling could be due instead to a mundane "**Stick-and-Straighten Scenario.**" To wit, if (1) excess vein tissue arises randomly and (2) contact with extant veins is likely within, say, one third of an intervein, then "rivulets" in the outer thirds will merge with extant veins [4189], while each leftover "puddle" in the middle will self-assemble into a new vein. This confluence idea does not preclude some corridors being less hostile to vein formation for other reasons [984], nor does it rule out other dormant relics from ancient circuits. However, the latter must be hard to awaken because *Drosophila*'s vein pattern has been stable for ≥23 million years [1625].

The inferred ability of vein tissue to cohere and linearize has its own sort of evolutionary significance. It is evidently a robust way of ensuring perfect veins from crude areas [3624]. Cell rearrangements have likewise been implicated in the straightening of bristle rows (cf. Ch. 3) [1800, 2449, 3603].

Intervein cells manifest a different kind of adhesion [430, 431, 3472, 4664]. They "zip" the D and V wing surfaces together by adhering with one another across the "transalar" lumen that is created during evagination [1312, 1313, 2907].

Two cell types predominate in the wing blade: vein and intervein

Wing veins serve as hydraulic pipes for wing expansion [1738, 4509], as spars for flapping flight [1116, 3015, 4749, 4750], and as conduits for nerves and tracheae [363, 1311, 3234, 4065]. The physical association of veins and tracheae [1368] is intriguing considering how many circuit elements are shared by vein patterning and tracheal branching [2576, 2679]: *blistered* [29, 268, 1646, 2907] (via separate *cis*-enhancers [3145]), *colt* [1738], *Delta* [766], *dpp* [4546], *dumpy* [4668], *knirps* [732], *Notch* [2574, 4082], *rhomboid* [4546], *spalt* [732, 2347], *sprouty* [681, 1665, 2321, 3558], *ventral veinless* (a.k.a. *drifter* [84, 85]) [337, 995, 2575], etc. Nevertheless, these parallels are perplexing because the styles of morphogenesis seem so different (e.g., polarization, movement, adhesion) [1312, 2679, 3729].

Excluding the margin and a few sensilla, there are only two salient cell types in the blade: vein (~10% of the surface) and intervein (~90%) [1312]. Vein cells differ from intervein cells in several key respects [1312, 2907].

1. *Density.* Vein cells are more densely packed than intervein cells [1368, 2847], with appreciably smaller apical diameters [1312, 3015]. They secrete a thicker cuticle [4188], which gives the veins more tensile strength. D vs. V convexities in the veins give the wing a "corrugation" [2847] that resists curling.

2. *Adhesion.* Intervein cells differentiate basal extensions that help anneal the D and V surfaces [431, 1312, 1313, 4407]. In contrast, vein cells manifest little affinity for the opposing surface, regardless of whether it contains vein (normal situation) or intervein (mosaic condition) cells [2907]. Instead, they form a lumen [1312, 1313, 3015]. This D-V cooperation during differentiation explains the transalar nonautonomy so often seen with vein-affecting clones [2847]. Nevertheless, veins can vanish [995, 1042, 1043] or form ectopically [4848] in one surface without affecting the other.

3. *Death.* Intervein cells die after the fly ecloses from its pupal case, but vein cells survive [2077, 3015].

4. *Upstream controls.* The vein state is the end result of a long cascade of signals (cf. Fig. 6.11). In theory, therefore, the intervein state could merely be a default [2742]. In fact, however, the selector gene that enforces it – *blistered* [320, 2907] – is actively turned ON by Hh and Dpp in different interveins via separate *cis*-enhancers [3145]. Other regulators might include Daughterless (bHLH) [2741], Extramacrochaetae (HLH) [206, 913, 983], Net (bHLH) [206, 467], and Plexus (nuclear matrix protein) [206, 2742].

5. *Downstream effectors.* The transalar adhesivity of intervein cells is attributable to higher levels of integrins [431, 504, 1313, 2907, 4664] and Dumpy (a huge adhesion protein) [4668]. In contrast, vein cells manifest higher levels of extracellular matrix components (laminin [1312, 1313, 3015] and collagen [3015]) and F-actin [1312, 1313], and have darker pigmentation [3624]. Vein cells express *Delta* but can differentiate without it [1951].

Because Blistered (Bs) has a MADS (DNA-binding) domain [29], it could establish the intervein state directly by activating "realizator" genes for cell shape, adhesivity, etc. [2907]. The vein state is controlled by the EGFR pathway (see below) and constrained by the distribution of *rhomboid* (*rho*) [320, 4556]. In the wing pouch, the expression patterns of *bs* and *rho* are complementary

(Fig. 6.2), except that *rho*-ON stripes are narrower than *bs*-OFF gaps [320]. LOF-GOF analyses reveal a mutual antagonism ("*rho* ⊣ *bs*" and "*bs* ⊣ *rho*") [320], although *rho*'s effects on *bs* are quantitative (except for L2 [320]), and initial expression of *rho* is normal in *bs*^{null} discs [3624].

Veins come from proveins that look like proneural fields

Veins widen to ~10 cell diameters [984] when the Notch pathway is suppressed by Dl^{LOF} [989, 1951, 2249, 3277, 4189] (cf. Dl^{DN} [1951, 4207]), *deltex*^{LOF} [1269], *gro*^{LOF} [1794], H^{GOF} [2657], *kuz*^{LOF} [4025], *mastermind*^{LOF} [437], N^{LOF} [981, 984, 2462, 2847, 3025, 4189], *shi*^{LOF} [3271], $Su(H)^{\text{LOF}}$ [989], or LOF alleles of bHLH genes in the E(spl)-C [989, 1794, 4256], although the E(spl)-C is dispensable for L1 [991, 1794]. In the case of Dl^{LOF}, the thickening is especially noticeable at the tips of the veins, which flare into "deltas" [76, 1269, 4466] (whence its name [2560]). Conversely, veins are suppressed (i.e., shortened, interrupted, or eliminated) when the pathway is hyperactivated by Dl^{GOF} [1951], H^{LOF} [117, 198, 989], N^{GOF} [1951, 4161], or GOF alleles of bHLH genes in the E(spl)-C [2063, 2548, 4256].

These hyper- and hypoplasias imply that Notch's role in vein formation is analogous to its role in SOP selection [989, 993, 4859]. That is, it might be whittling large areas to smaller size by preventing subsets of competent cells from adopting a desired fate. The competent cells in this case would occupy "proveins" (≈ proneural fields; cf. Ch. 3) [320, 1951, 3271] that are 4 to 5 times wider than the final veins. When the Notch pathway is disabled, its "lateral inhibition" process apparently fails, and entire proveins form veins that are thicker than normal. The provein may extend ~10 cell diameters on either side of each vein because N^{null} clones make vein tissue when they fall within this range [993] (cf. Dl^{LOF}, $Su(H)^{\text{LOF}}$, and E(spl)-C^{LOF} clones [989, 1074]), and veins of this width develop when N and *emc* are simultaneously incapacitated [205]. Moreover, these are the sorts of widths seen for S-phase zones (labeled with BrdU) at 12 h AP [3813], so proveins may be real entities [981].

Dl and N both have haplo-insufficient (N^{null}/N^+ or Dl^{null}/Dl^+) thick-vein phenotypes [3022], but the double heterozygote ($N^{\text{null}}/N^+; Dl^{\text{null}}/Dl^+$) looks wild-type [990], implying that Dl and N must balance for the pathway to function properly. This stoichiometry is starting to make sense now that the extracellular (N-extra) and intracellular (N-intra) parts of the Notch molecule can be tracked separately. When Dl binds N, the signaling cell ingests both its own Dl and its neighbor's N-extra, while the receiving cell transports N-intra from its surface to its nucleus to turn ON target genes (like two dogs sharing one bone; cf. Fig. 2.3) [3271]. Serrate may also serve as a ligand because removal of both Ser and Dl (in somatic clones) causes wider veins than the loss of either one alone [4859], but overexpressing Ser has no effect on the vein pattern [1951].

But the resemblance is only superficial

Despite the obvious parallels, the Dl-N circuitry cannot be operating here like it does in SOP selection. The reason is that veins are presaged by sharp stripes only ~1–2 cells wide before the Notch pathway is deployed in a refinement role [320]. For example, *rho* is expressed in a narrow stripe in the L2 vein area at pupariation [320, 2847, 4191], whereas Dl is not even detectable there until 22 h later [1951]. In fact, *rho* acts upstream of *Dl* (*Dl* is repressed in *rho*^{LOF} *vn*^{LOF} wings and turns ON ectopically in *rho*^{GOF} wings [320, 989]). How can any gene turn ON in a thin stripe before lateral inhibition has reduced proveins to a mature width?

There is no obvious solution to this "**Athena Enigma**" (referring to the goddess born in adult form), but there are a few clues [328]. In mature discs, *Dl* is expressed in stripes that are 3–6 cells wide (final veins are ~2–3 cells wide [984, 4831]) [320, 1951, 2296], while N is expressed in a complementary pattern (i.e., in interveins and lateral proveins) [1203, 1951, 2209, 2296]. The critical period when Dl and N are needed for width control is 22–30 h after pupariation [1951] (but see earlier effects [4189]: their Fig. 5k). At that time, the *Dl*-ON stripes are unchanged in width [1951], while the N-ON zones have shrunken to ~2–3 cell-wide flanking stripes (with a residual basal level elsewhere). The Notch pathway is activated in the latter stripes [3271]. Effectors such as *mβ* must stop lateral provein cells from implementing the vein state [989], but how? Physical forces are probably involved [1951] because D and V surfaces are annealing at this time [1313], and surgical operations can induce vein fragments [2472] or reroute existing veins just before then (11–21 h AP). Even wild-type wings manifest a wide-vein phenotype when their D and V surfaces are prevented from annealing (by transplantation of disc fragments) [2847].

At present, the only firm conclusion is that Notch is regulating differentiation here, rather than determination [1951]. Thus, unlike its role in proneural fields where Notch shrinks stripes to lines and points (cf. Ch. 3), Notch is charged here with maintaining pre-existing lines. It must keep the widths narrow by enforcing

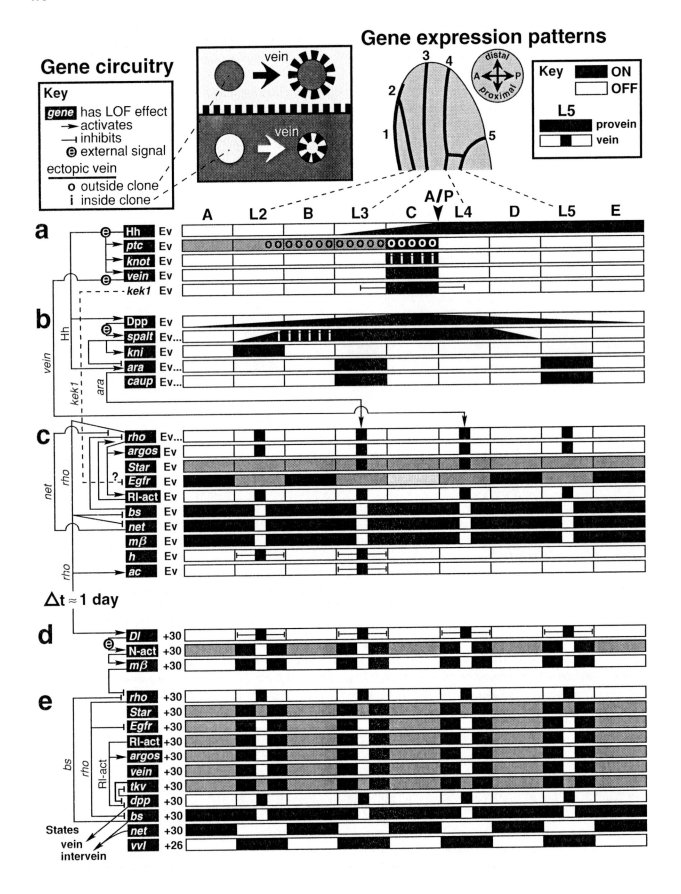

alignment in the face of mitoses, jostling, and other deviations within the plane and correcting any transalar mismatches that arise [320]:

1. *Straightening.* The Notch pathway is active in lateral provein, not vein, cells. Whatever homophilic adhesion molecules that it evokes would thus facilitate alignments on either side of the nascent vein. If vein cells themselves use a different adhesive protein, then each provein would be a 3-ply structure, where the flanking bands might help "sandwich" and reinforce the central one for added rigidity. This hypothesis could be tested by monitoring cell movements.

2. *Matching.* An engineering problem that wings face (which legs do not) is the matching of two patterns that arise independently [3052]. A possible reason for the provein strategy, therefore, is that it allows "last-minute" adjustments of identities if veins on the D and V surfaces are out of register [4189]. Each intervein "zips up" from the middle outward [431, 1312], and mistakes may be "ironed out" by revising states of determination that are kept labile for just this purpose [1312]. This lability may explain how extra crossveins arise when the level of integrins (≈"transalar glue" [3158]) drops too low [936, 4664]. Removal of a vein's D moiety typically blocks the V moiety from mak-

ing vein material [1359, 1643], although V/V wings have virtually normal venation [3158]. How transalar interactions cause fate switching at such a late stage is unclear. Again, cinematography might help, especially if D and V surfaces are contrived to have different vein patterns.

Notch finishes its fine-tuning chores long after the initiation phase has been completed. Vein initiation involves many different factors. As described below, each vein appears to have its own "board of directors" in the same sense that pair-rule stripes are managed by ensembles of different gap genes in the embryo [984, 4188] (cf. Ch. 4).

In the discussion below, particular veins will be designated by number (1–5) or by "L1–L5" [4189], alluding to their longitudinal (vs. cross-vein) orientation. Intervein areas will be denoted by "iv" plus the numbers of the flanking veins (e.g., "iv3–4").

All veins use the EGFR pathway

Another example of the Athena Enigma is found in the embryo [2617], where single files of mesectodermal cells turn on the bHLH-PAS gene *single-minded* (*sim*) on each side of the V midline [899, 4297, 4548] and then merge during gastrulation [3043, 3044, 4023]. Like *rho*-ON preveins, these *sim*-ON files (1) are perfected *ab initio* [1278], (2) also

---

FIGURE 6.11. Domains of gene or protein expression along the A-P axis of the wing pouch or pupal wing.

Data are organized in 5 panels based on stage (**a–c**: ~0–6 h after pupariation = "Ev"; **d** and **e**: ~30 h after pupariation = "+30") or pathway (**a** Hh, **b** Dpp, **c** EGFR, **d** Notch). "Ev" stands for evagination, while "Ev . . ." means ~0–24 h after pupariation. Certain genes are presented in more than one panel. Except for Dpp and Hh (where protein levels are plotted) and Rl and N (where activated states are mapped), black areas denote transcription (cf. key, upper left). Degrees of expression are indicated by shades of gray or by graded slopes.

Abbreviations: A/P (compartment boundary), A–E (intervein zones), L2–L5 (longitudinal veins 2–5). Full rectangles beneath L2–L5 (cf. key, upper right) are "proveins" [3271] (cf. preveins [377, 1312, 3015]), while squares denote mature veins (≈ 2–4 cells across [1312]), and flanking areas are termed "lateral provein." For genes whose LOF alleles cause ectopic veins, "i" or "o" denotes whether the vein is inside (autonomous) or outside (nonautonomous) the clone, and the span of symbols delimits the area where the effect is seen. Wiring at left (➔ activation; ⊣ inhibition) illustrates some genetic interactions (see text or references below). See also App. 7.

Genes or proteins ("Δ" = veins missing in LOF flies or clones; "≈" = mutations causing similar phenotypes; "DO" = details omitted): *ac* (*achaete*) [320], *ara* (*araucan*; ΔL3 [320]) [1536], *argos* [2715]Δ, *bs* (*blistered*) [984]Δ, *caup* (*caupolican*) [984]Δ (activated by Hh [2993], not shown), *Dl* (*Delta*) [205, 3271]Δ (but see [989]), *dpp* (*decapentaplegic*) [980]Δ, *Egfr* [1643]Δ, *h* (*hairy*) [665], Hh (Hedgehog) [1208]Δ, *kek1* (*kekkon1*) [3020], *kni* (*knirps*; ΔL2; compare *radius incompletus* [320, 2617]Δ and Fig. 4.2) [984]Δ, *knot* [2894]Δ, *mβ* (*E(spl)mβ*) [871, 984]Δ, N-act (activated intracellular fragment of Notch) [3271]Δ, *net* [466], *ptc* (*patched*) [2834]Δ, *rho* (*rhomboid*; a.k.a. *veinlet*) [1643]Δ, Rl-act (activated Rolled; a.k.a. MAP kinase [332]) [1643, 2715]Δ, *spalt* (ΔL2 [984]Δ) [2742]Δ, *Star* [1643], *tkv* (*thick veins*) [2457]Δ, *vein* (ΔL4 [320] ≈ *Egfr*^LOF [3624]; DO: gap at wing margin due to Wg) [4604]Δ, *vvl* (*ventral veinless*) [995]. Interestingly, *vvl* is the only generic "vein gene" that is OFF in L1 (not shown). Possibly, *vvl* continues to be repressed by Wg at the margin [703] (cf. Fig. 6.2), just as *ara* and *caup* are repressed there by Wg (and N) [1537].

The format is of this figure is based on [320]. Nomenclature follows [1366, 4307] (but see [4065]). For a review, see [328].

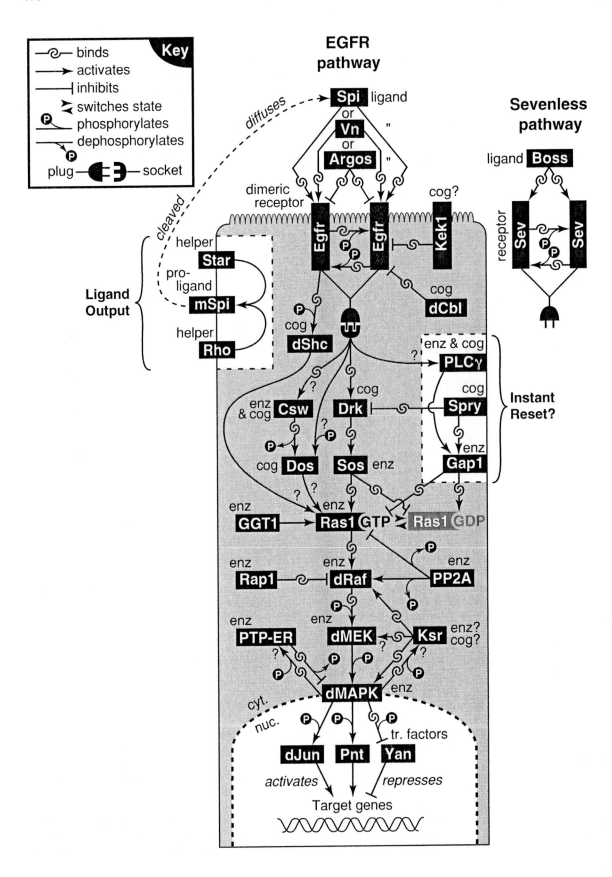

express *rho* [326, 330, 2405, 4192], and (3) widen or shrink like veins in response to LOF or GOF alterations in Notch signaling [2712, 2941]. Neither *sim* nor any other bHLH-PAS gene appears to be involved in wing development, but the coincident expression of *rho* suggests that the EGFR pathway, like the Dpp and Notch pathways, is a generic tool for line drawing.

The EGFR pathway is outlined in Fig. 6.12 [3348, 3672]. Egfr is one of 8 characterized RTKs in *Drosophila* [1338, 2321, 2660, 3342, 3558]. The others are Breathless [1507, 2245, 2469, 3560], Derailed [612], dTrk [3475], Heartless [267, 1505, 3906], Insulin Receptor [1215, 1216, 4815], Sevenless [226, 428, 1677], and Torso [809, 4045, 4046]. All these RTKs have been shown, or are assumed, to "plug into" the same basic Ras-MAPK engine [1081, 1098, 1117, 1337, 1338, 3340]. Domain-swap experiments reveal differences in the potency of RTK kinase domains [1446, 3559], but no qualitative differences downstream. However, some elements are receptor-specific [1086] (e.g., dShc is used by Egfr but not by Sev [2623]). Thus, there is no obvious answer to the "**Ras Specificity Riddle**" [1291, 2647, 3590, 3790, 4245]: how can the same Ras-MAPK chain evoke different cell states at different times and places? The likely answer, in some cases at least, is that cells respond differently because they have been primed with different transcription factors [1688, 3942] so the RTK signal can only affect key promoters if other factors are present [3874]. That is, cells use combinatorial codes (cf. Ch. 8) [3783].

Egfr is the most widely deployed RTK in flies [1337, 3342, 3348, 3830]. The core RTK transduction cascade was worked out for Sevenless [1048, 1678, 3294, 3671, 4555] and Torso [1117, 1167, 3340] in the early 1990s and was later shown to apply to Egfr [517, 1043, 1337, 1916, 3336]. However, the cascade is not a cassette *sensu stricto*: depending on the receptor or tissue or stage, certain elements may be replaced by others. For instance, *pnt* executes Ras-MAPK ON/OFF states in the eye [3162] and wing margin [3025], but not in veins 2–5 [3804, 4191] where the transcription factors are unknown [981], and *pnt* even acts as an EGFR antagonist during oogenesis [2952].

Cameos of the components involved (or implicated) in EGFR signaling are listed below by abbreviation or full name, with nonenzymatic domains given in braces (cf. App. 1) and sources for molecular-genetic data in brackets. Asterisks mark proteins omitted from Fig. 6.12 because their roles are uncertain (see [1922, 4233, 4286] for additional candidates). The length of the list reflects the baroque complexity of this pathway, which seems more like a network:

14-3-3ζ* {14-3-3}. Scaffold protein that regulates dRaf, etc. [244, 2279, 3638, 3639, 3672].

FIGURE 6.12. The EGFR and Sevenless signaling pathways. See text for proteins, abbreviations, and a summary of how the core chain operates. One cell is sketched with microvilli at top, although ligand contact may take place elsewhere on the surface. Black rectangles = proteins; gray area = cytoplasm ("cyt."); dashed line below = nuclear membrane. Arrows (activation) and blunt "⊣" lines (repression) denote epistatic relations. Noncovalent binding is symbolized by interlocking hooks (cf. key at top). Question marks concern how or whether interactions occur. A "cog" (a.k.a. "adaptor") is a component that plays a steric (vs. enzymatic) role [441, 1954, 3299, 3676] (e.g., Drk [3185]), while "enz" = enzyme. See also App. 7.

Egfr is depicted at the apex but it actually affiliates with apicolateral adherens junctions, as do Kek1 and dCbl [1001]. Conversely, the agents that process mSpi (white box at left) [199, 2244, 3385] are shown on the lateral membrane, but they may act apically [3692, 4192] via an ADAM-like protease [4558]. White box at right is a possible module for resetting Ras1 to an OFF state after signal transmission [3455]. Ras1 resides constitutively on the inner face of the membrane [781, 2653], as do Gap1 [3455], Csw [71], Dos [243], Sos [2137] (flies thus differ from mammals [106, 3485]), and Spry [681, 4340].

The Boss-Sev ligand-receptor complex is shown at right. It plugs into the same downstream circuitry as Egfr (sans Kek1, dShc, or the mSpi add-on) [1081, 3941], although effects of *sev*GOF and *Egfr*GOF can differ [1049]. The 7-pass transmembrane ligand Boss (Bride of sevenless) [1732] activates the RTK Sev (Sevenless) [1733, 2319, 2450, 3671, 4359]. Sev differs from Egfr in that its N-terminus is anchored to the membrane to form a big loop [226, 428, 1677] of unknown shape [3136]. Strangely, the entire Boss molecule (ripped from its own membrane?) is internalized by the Sev-expressing cell [601, 2318, 4211]. Sev is concentrated apically [194, 195] and at adherens junctions [4360]. Parochial players in the Sevenless and Torso pathways (e.g., dCdc37 [932]) are omitted.

Binding partners in the core chain (listed from surface to DNA) are based on fly studies or vertebrate homologs (vH): Egfr-Drk [3829] (cf. Sev-Drk [3488]△), Drk-Sos [3488]△, Sos-Ras1 (vH [402, 4705]), Ras1GTP-dRaf (vH [2913]△), and dRaf-dMEK (vH [898, 2912]).

Ancillary binding partners (listed alphabetically, upstream factor first): Csw-Dos [243]△, dCbl-Spry (not shown) [4738], Egfr-Csw (probably indirect based on Sev-Csw [71]), Egfr-dShc (vH [4445]), Gap1-Ras1 (vH [3778, 4044]), Kek1-Egfr [1445], Ksr-dMAPK [2007]△, Ksr-dRaf [4285], dMAPK-PTP-ER [2133], dMAPK-Yan [2007], Rap1-dRaf (vH [3057]), Spry-Drk [681], and Spry-Gap1 [681]. Star apparently binds mSpi [1915].

This diagram is adapted from [2131, 3348, 3672, 3782]. It is augmented with data from sources cited above.

Argos   (a.k.a. Giant lens or Strawberry) {EGF-like}. Diffusible ligand that inhibits Egfr [1294, 1528, 3183, 3772, 3828].

Canoe*   Binding partner for Ras1 [2748].

Clown*   Inhibitor that acts downstream of Argos and upstream of Ras1 [4233].

Cnk*   "Connector enhancer of ksr" {PDZ, PH}. Binding partner for dRaf [4288] and essential cofactor for Ras1 [4287].

Csw   "Corkscrew" {SH2}. Phosphatase for Dos [70, 3336, 3337].

dC3G*   Likely catalyst for Rap1 [1995].

dCbl   "dCable" {zinc finger}. Inhibitor of Egfr [2550, 2821, 3227] and binding partner for Spry [4738].

dHsp90* "*Drosophila* Heat shock protein 90". Crutch for dRaf [932, 4447].

dJun   {bZip}. Transcription factor regulated by dMAPK [383, 3359, 4381] but surprisingly not by JNK in this instance [2280, 2281, 3593].

dMAPK   "*Drosophila* Mitogen-Activated Protein Kinase" (a.k.a. Rolled). Kinase regulated by dMEK [332, 517].

dMEK   "*Drosophila* MAPK-ERK Kinase"(a.k.a. dSor1 or dMAPKK). Kinase regulated by dRaf [1916, 2700, 4399].

Dos   "Daughter of sevenless" {PH}. Adaptor for the receptor complex [243, 1829, 2639, 3489].

dRac1*   GTPase that increases the Ras-dependent activation of dMAPK [4024].

dRacGap* Negative regulator of dRac1 [4024].

dRaf   Kinase regulated by Ras1 [965, 1050, 2613, 2823, 3491].

Drk   "Downstream of receptor kinases" {SH2, SH3}. Adaptor for the receptor complex [3488, 3944].

dShc   {PTB, SH2}. Adaptor for the receptor complex [2389, 2623, 4445].

E(Elp)24D "Enhancer of *Ellipse*, 24D." E(Elp)24D binds Egfr (and is phosphorylated by it upon Egfr stimulation), resulting in a diminution of Egfr signaling [4034]. It thus may act like Kek1.

Ebi*   {F-box, WD40}. Slimb-like facilitator of protein degradation [1086], which may mediate cross-talk with the Notch pathway.

Egfr   "Epidermal growth factor receptor." Receptor for Spi or Vn (in discs) [2573].

Faf*   "Fat facets." Protease that cleaves ubiquitin from ubiquitin-conjugated proteins [1931, 4765]. The LOF phenotype of *faf* resembles that of integral EGFR genes (*argos*, *Gap1*, *yan*) [1932] in the eye (the only disc where *faf* is needed [1240]), and *faf* genetically interacts with various components of the EGFR/Sev pathway [1994].

Gap1   "GTPase-activating protein 1" (a.k.a. Mip or Sextra) {PH}. Antagonist for Ras1 [128, 550, 1399, 3455, 3633].

GGT1   "Geranylgeranyl transferase-1." Post-translational modifier of Ras1 [4284].

Gritz*   {EGF-like}. A Spitz-like protein (270 a.a.) that may be a ligand for Egfr [204, 2108, 4558]. Indeed, it may be the missing factor "X" inferred from various system analyses (e.g., see discussion of wing veins below).

Karst*   Cytoskeletal protein (β_{Heavy}-spectrin), which acts upstream of Sev. Karst may maintain the integrity of adherens junctions for juxtracrine (Boss-Sev) signaling [4296].

Kek1   "Kekkon1" {Ig-like}. Transmembrane inhibitor of Egfr [1445, 3020]. Possibly inhibits Egfr by sequestering it in inactive heterodimers [77].

Ksr   "Kinase suppressor of ras" {DEF}. Putative scaffold for a kinase complex [1096, 2007, 4284, 4285] with 14-3-3 proteins [592, 4122, 4772]. Ksr is phosphorylated by dMAPK [592].

Lilli*   "Lilliputian" {HMG} [4250, 4706]. Putative transcription factor that acts downstream of dRaf to regulate the levels of target gene transcription. Lilli also affects the Dpp pathway [3097, 4195].

Lqf*   "Liquid facets." Endocytosis-associated protein that appears to be a substrate for Faf [594, 1238]. Lqf is needed in the signaling cell, although the identity of the (antineural) signal remains unknown.

Mask*   "Multiple ankyrin repeats single KH domain" {ankyrin, KH, NLS, opa} [4002]. Isolated in a screen for modifiers of *csw*^{DN}, *mask* encodes a huge (~4000 a.a.) protein that acts positively in RTK signaling. It is especially needed in R7 photoreceptors [4002], implying that Mask may be more

involved in Boss-Sev than Egfr transduction.

mSpi   "<u>M</u>embrane-bound <u>Spi</u>." Spi precursor, which must be cleaved before it
can diffuse away from its source cell
[3829].

PLCγ   "<u>P</u>hospholipase <u>Cγ</u>" (a.k.a. Small wing)
{SH2/SH3/PH}. Activator for Gap1 [3455,
4272].

Pnt   "Pointed" {Ets}. Transcription factor
regulated by dMAPK [516, 745, 2242, 3162,
3804].

PP2A   "<u>P</u>rotein <u>P</u>hosphatase <u>2A</u>." Phosphatase for Ras1 and dRaf [2660, 4495, 4554].

PTP-ER   "<u>P</u>rotein <u>T</u>yrosine <u>P</u>hosphatase-ERK/
<u>E</u>nhancer of <u>R</u>as1." Phosphatase for
dMAPK [2133].

Rap1   (a.k.a. Roughened). Inhibitor of dRaf
[1724].

Ras1   GTPase ON/OFF switch [481, 1273, 3084, 3943].

RasGAP*   "<u>Ras</u> <u>G</u>TPase-<u>a</u>ctivating <u>p</u>rotein" {PH,
SH2, SH3}. Inhibitor of Ras1 [1207].

Rho   "Rhomboid." Facilitator for cleavage of
mSpi [330] (cf. a paralog that operates
during oogenesis [1644]).

Rin*   "Rasputin." Regulator of a RasGAP
other than Gap1 [3300].

RnRacGAP*   (RotundRacGAP). Regulator of the
actin cytoskeleton [1642].

Semang*   Isolated as a modifier of *Src42A* [4870],
semang functions downstream of, or
parallel to, Yan [4869].

Sos   "<u>S</u>on <u>o</u>f <u>s</u>evenless" {PH}. Activator for
Ras1 [391, 3635, 3943].

Spen*   "Split ends" {RNA-binding}. Positively
acting component that interacts synergistically with Pnt and antagonistically with Yan [736, 3542]. Spen also influences the Wg pathway [2554].

Spi   "Spitz" {EGF-like}. Diffusible ligand for
Egfr [3711, 3829].

Sprint*   "<u>S</u>H2, poly-proline-containing <u>Ras</u>
<u>int</u>eractor" {SH2, Pro-rich} [4224]. Prob-
able binding partner (and exchange
factor?) for Ras1-GTP.

Spry   "Sprouty" {Cys-rich}. Activator of Gap1
[681, 1665, 2321, 3558] and binding partner
for dCbl [4738]. Spry interacts synergistically with Argos [1999].

Src42A*   Part of a branch parallel to dRaf [2612, 4870].

Star   Facilitator for cleavage of mSpi [1782, 1785,
2288, 3385, 3678].

Svp*   "Seven-up." Transcription factor first ascribed to the Sevenless pathway [1856, 2322,
2888], and later shown to genetically interact
with *Egfr* [264] and *frizzled* (planar polarity)
[1193].

Vn   "Vein" {EGF-like, Ig-like}. Diffusible ligand
for Egfr [3801].

Yan   Transcription factor {Ets, DEF, DEJL} regulated by dMAPK [3467]$^\Delta$.

The trunk line of the EGFR pathway appears to
work as follows ("VH" indicates data from vertebrate
homologs):

1. Binding of ligand causes dimerization and reciprocal Tyr-phosphorylation of the receptor [1814, 3780, 3900,
4556, 4588].

2. The resulting Tyr^P recruits the SH2 domain of Drk
[3488, 3785].

3. Drk's N-terminal SH3 domain binds the Pro-rich C-
terminus of Sos [3488] to alleviate Sos's self-inhibition
[2137].

4. Sos binds Ras1-GDP, and catalyzes escape of GDP
and entry of GTP [391, 3943] (VH [402, 4705]).

5. Ras1-GTP breaks free of Sos (VH [4705]) and recruits dRaf to the cell membrane [167] (VH [1689, 2477,
4128]).

6. Ras1-GTP activates dRaf via dRaf oligomerization
(VH [1196, 2620, 2702]), plus an unknown agent [2531].

7. Ras1 switches back to its inactive state by hydrolyz-
ing GTP to GDP [2633, 2652, 2855, 4043] with or without the
help of Gap1 [2880, 3778, 4044], while dRaf phosphorylates dMEK [2700].

8. dMEK activates dMAPK by phosphorylating it [326,
638, 822, 2700].

9. dMAPK homodimerizes [823], moves to the nucleus [11, 2206, 2355], and phosphorylates transcription
factors that turn various target genes ON or OFF [4252,
4393]. The signal can travel from Steps 1 to 9 in as little
as 5 min [1338, 4035].

In the Sevenless pathway, which operates virtually
identically up to this point, MAPK phosphorylates dJun
[3359], PntP2 (an isoform of Pnt) [516], and Yan [516]. PntP2
is an effector for EGFR signaling in certain contexts

[2242, 2952, 3162, 3928, 4775] but not in wing veins [3804, 4191]. The following additional steps are used in the Sevenless pathway, and comparable ones may operate in the EGFR pathway [670, 3942], although the only disc where Yan is expressed is the eye disc [2392].

10. PntP2 cooperates with dJun [4381] and competes with Yan for cognate Ets-binding sites at the promoters of target genes [516, 1047, 3162, 4393, 4775]. In the R7 photoreceptor equivalence group, the first echelon of targets includes *phyllopod (phyl)* [717, 1047, 1052, 4381].

11. Phyl forms a complex with Sina (Seven in absentia [671]) and the transcriptional repressor Ttk88 (an isoform of Tramtrack [4773]; cf. Fig. 2.5) [2165, 2529, 4249]. Phyl must be rate-limiting because *sina*^{GOF} does not elicit extra R7 cells like *phyl*^{GOF} [717, 1052, 3082]. The complex leads to ubiquitination and degradation of Ttk88 [2529, 4249] via the conjugating enzyme UbcD1 [4249, 4382]. Like Ttk88, Ttk69 (the other isoform) also undergoes signal-dependent degradation [2529] but acts as a transcriptional activator at late stages [2391].

12. Degradation of Ttk88 alleviates repression of farther downstream genes [2165, 2390, 3082], including *prospero* [2165, 4775] (a direct target of Yan [670]), other pan-neural genes [4249], and *engrailed* [4773]. This strategy is reminiscent of the Wg pathway, where genes are controlled by proteolysis of Arm (cf. Fig. 5.6).

Among the target genes of the EGFR pathway is *Egfr* itself, which is downregulated after hours of signaling [4190], thus making cells refractory to further input [681]. Faster negative-feedback loops (via Spry, Kek1, or RasGAP [808, 1445, 3558]) may allow analog responses to graded input [1446, 1527, 1615, 3227, 4557, 4788]. Such dampers favor a rheostat mode, whereas "instant-reset" loops (via PLCγ?) favor an ON/OFF solenoid mode [3455]. One quirk of RTKs is that signaling can be ignited by excess receptor alone (sans ligand) [4556]. If signaling is not extinguished by the instant-reset module, then it will stop due to internalization and degradation of activated receptors [702, 3783, 4482, 4545].

LOF alleles of *rho* remove parts of veins (mostly distally) [1041, 1368], while mild overexpression (via a *hs*-promoter) induces extra veins [1643, 4191] as well as expanding vein widths. Stronger overexpression (via *Gal4* drivers) elicits even more vein material by switching intervein cells to a vein state throughout the wing

[1643, 3126]. Similar LOF and GOF effects are seen for other genes in the EGFR pathway [517, 1042, 2493, 3558, 4191], including *cnk* [4288], *csw* [1237, 3336], *dC3G* [1995], *dMAPK* [517, 1043, 1995], *dMEK* [1916, 2131, 3771], *dos* [243, 3489], *dRac1* [4024], *dRaf* [1043], *drk* [1043], *dShc* [2623], *Egfr* [181, 814, 3464, 3465, 3484], *ksr* [1995], *Ras1* [1043, 1995, 3470], *semang* [4870], *Sos* [1043], *spitz* (GOF only) [3800, 4604], *Star* [1643, 1782, 4191], and *vn* [1366, 3800, 3929], while opposite effects characterize the antagonists *argos* [3771, 3772, 3800, 3828, 4233], *dRacGap* [4024], *Gap1* [411, 550], *kek1* [4604], *PLCγ* [2669, 4272], and *spry* [681, 2321, 3558]. The only salient exception is *pnt* – an EGFR effector in the eye and elsewhere [2242, 2952, 3162], which has no LOF impact on veins [3804, 4191] (although it does affect other wing traits [3928]).

This syndrome implies that the EGFR pathway specifies the vein state. Based on heat-pulse studies with a t.s. *Egfr* allele and a *hs-rho* transgene, the pathway is required for vein formation at 0–30 h AP, with 6–18 h AP being especially critical [1643].

For years, *rho* and *Star* were known to act in the EGFR pathway [4191, 4192], but their functions were obscure. Both genes encode transmembrane proteins (7- or 1-pass [330, 2288], respectively). Recent experiments reveal that Rho cooperates with Star to promote the presentation, cleavage, and release of Spitz [199, 2244, 3385]. Without Rho and Star, Spitz remains bound to the membrane and hence cannot diffuse or activate Egfr on adjacent cells. Neither Vein nor Argos requires such processing because they are directly secreted (sans a transmembrane domain) [3830].

Unfortunately, these new facts do not help solve the problem at hand because Spitz is dispensable for vein development. Wing discs from *spitz*^{null} embryos develop normally [3928], and *spitz*^{null} clones do not affect veins [1643, 3025].

Might Rho and Star be assisting other Egfr ligands? The only other ligands known to be expressed in discs are Vn (Egfr activator) and Argos (Egfr inhibitor), but neither of them has a transmembrane domain, so the above talents of Rho and Star seem moot. Indeed, Vn can signal without Rho because overexpressing Vn induces vein tissue in *rho*-OFF interveins (Fig. 6.13j) [4604] as effectively as overexpressing Rho [1643]. Rho and Vn must interact somehow, though, because *rho*^{LOF} *vn*^{LOF} double mutants exhibit extreme synergy: neither *rho*^{ve} nor *vn*^{1} alone deletes more than a few vein sections, but *rho*^{ve} *vn*^{1} wings lack all veins except L1 (Fig. 6.13i) [1041, 1366, 2907] due to a failure of vein initiation [320, 3015, 3624]. Some of the major clues about *rho* and *Star* are reviewed below:

1. The *rho*-ON state is the earliest indicator of veins [1643, 4191], and *rho* is the only gene known to be expressed exclusively in veins throughout their development [1643]. Hence, the orthodox view has been that upstream agents funnel through the EGFR pathway [3558] by converging on *rho* to initiate all veins [1643]. However, the recently revealed *rho*^null^ phenotype (Fig. 6.13h) [1643] shows that *rho* is not needed in most of L3 and L4.

2. Rho and Star cooperate in vein development [3385]. When *Star* (which has little effect relative to *rho*^GOF^) is overexpressed along with a weak *rho*^GOF^ transgene, strong synergy (solid vein areas) occurs at 6–9 h AP, with weaker synergy (extra vein fragments) at 9–24 h AP [1643]. The roles of Rho and Star must differ, however, because *Star*^GOF^ does not rescue veins removed by *rho*^null^, nor can *rho*^GOF^ rescue veins removed by *Star*^null^ (or induce extra veins in a *Star*^null^ background) [1643].

3. Rho and Egfr probably act sequentially because no synergy exists between *rho*^GOF^ and *Egfr*^GOF^ transgenes that cause mild increases in vein material [1643].

4. Both Rho and Star act upstream of Egfr because wings that overexpress *rho* and *Star* in the presence of a dominant-negative *Egfr* (*Egfr*^DN^) show an *Egfr*^DN^ phenotype, rather than a *rho*^GOF^ (ubiquitous-vein) trait [1643].

5. The level of Egfr response can be crudely assessed by an antibody that recognizes the activated (doubly phosphorylated) state of dMAPK [1337, 1338]. This kinase – a key transducer – is encoded by *rolled (rl)* [332]. Activated Rl (Rl*) is expressed only in *rho*-ON vein cells at 0–24 h AP [1643], except for sporadic Rl*-ON-*rho*-OFF cells next to *rho*-ON stripes [1643]. This congruence suggests that Rho is involved in reception, rather than emission, of a ligand signal [326].

6. Unlike *Egfr*^LOF^ clones, which suppress veins cell autonomously [1042, 1043], *rho*^null^ or *Star*^null^ clones manifest nonautonomy. In regions of the wing where such clones erase veins, nearby wild-type cells may fail to make part of a vein, and mutant cells at the clone edge can be rescued to make part of a vein [1643]. Both types of nonautonomy imply involvement of Rho and Star in emission (vs. reception) of a vein-inducing signal, thus contradicting point 5 above.

In all cases of *rho*^null^ nonautonomy (point 6), the influence does not spread beyond a few cell diameters, so the Rho-dependent ligand – whatever its identity – must not diffuse very far. In contrast, Vn's range must include the whole wing because all veins (save the marginal one) are erased by *rho*^LOF^ *vn*^LOF^. During the main period when the EGFR pathway is needed by veins (6–18 h AP), Vn is expressed in the 3–4 intervein and margin. (After 18 h AP, Vn is expressed in all interveins [3929].) Therefore, Vn, or a secondary messenger, must be diffusing over large (Dpp-scale) distances [1643, 3830].

Conceivably, Vn is activating the pathway everywhere, but only at a basal level consistent with its weak potency [3800]. The level could then be raised above a threshold needed for vein induction by a Rho-dependent mechanism so that veins arise along *rho*-ON stripes. If this "**Rho-Vn Booster Model**" is correct, then why do the proximal parts of L3 and L4 persist in *rho*^null^ wings (Fig. 6.13h)? The apparent answer is that these zones experience the greatest dose of Vn because (1) they border iv3–4 where Vn is expressed, and (2) the iv3–4 cells themselves are relatively deaf to Vn because Egfr is downregulated there (Fig. 6.11) [1643, 4604] by Knot [2894] and possibly by Kek1 [3020]. Indeed, the proximal sections of L4 [3929, 4604] and (to a lesser extent) L3 [4189] are preferentially lost in *vn*^LOF^ mutants. This scenario thus invokes deaf speakers (iv3–4) and mute listeners (rest of the pouch) as per the Deaf-Speakers/Mute-Listeners Trick [328]. Although this logic explains L3 and L4, the Athena Enigma still applies to L2 and L5.

In theory, vein-specific activation of the EGFR pathway could be reinforced by ensuring that inhibitors of the pathway are expressed in interveins. It is thus surprising that the inhibitors Argos (diffusible [1294]) and Sprouty (intracellular [681]) are expressed not in interveins but mainly in veins during this period [2715, 3558]. A similar "**Autoinhibition Paradox**" has been encountered in other tissues where these inhibitors are deployed [3399, 4081]: why aren't the cells that produce an inhibitor sensitive to inhibition themselves? Conceivably, their deafness could be due to differences in thresholds or timing of the activator vs. inhibitor [1290, 1528, 1687, 2493, 3347, 4557]. However, neither of these explanations seems to help with wing veins because the pattern and time-course of *argos* expression tracks the pathway's activation too closely (Fig. 6.11).

Assuming that the Rho-dependent ligand "X" is expressed coincidently with Argos, the answer may hinge on differences in the diffusion rates of X and Argos [4556]. If Argos diffuses faster than X, then it will venture farther to form a flatter distribution [3347, 3830] – hence

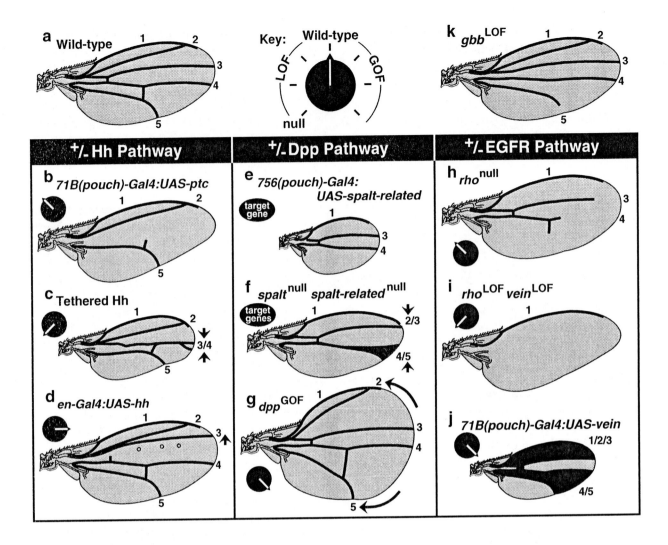

leaving ridges of X "high and dry" above a "sea" of Argos and sharpening the edges of incipient veins. This sort of scenario was devised in the 1970s by Gierer and Meinhardt in their "**Activator-Inhibitor Model**" [1478, 2802, 2806, 2815, 2818] (a.k.a. Lateral Inhibition [3207]) – a descendant of Turing's reaction-diffusion scheme [387, 390, 3014, 3208, 4410].

A gene that may encode X has been identified. The sequence of *gritz* is 49% identical to *spitz* [204, 2108, 4558].

But interveins also use the EGFR pathway (at a later time)

A drastic change occurs in EGFR signaling at ~24 h AP. Until then, the pathway is ON in future vein cells and OFF elsewhere, but henceforth the reverse is true [1643, 2715]. During this second phase (24–30 h AP), the EGFR pathway becomes most active in lateral provein bands that flank the veins (Fig. 6.11). Several of the pathway's

components undergo a similar transition in their expression around this time (viz., *argos* [2715], *Egfr* [4190], *spitz* [1643], and *Star* [1643]), but *rho* does not [1643]. Thus, the early congruence between *rho*-ON and Egfr-activated states gives way to a Phase 2 complementarity. The mechanism of this uncoupling is unclear. When the EGFR pathway is forced ON in veins (as well as interveins) during Phase 2, L2–L5 fail to form. Evidently, veins must shut OFF the pathway at ~24 h AP in order to differentiate.

The EGFR pathway is known to switch the allegiance of cells to different fates as a function of time in the egg chamber [2470, 3758] and eye disc (cf. Ch. 7) [1289]. Might a similar switch be happening here so that Egfr implements vein states in Phase 1 and intervein states in Phase 2? No, because the pathway's chronic LOF phenotypes (vein loss) reflect the earlier function only, and suppression of the pathway during Phase 2 only causes a mild increase in vein material [2715]. Evidently, the Phase 2

FIGURE 6.13. Dependence of vein pattern on signaling pathways. Black lines or areas mark veins or veinlike regions in all panels, and outer arrows denote vein displacements. For gene abbreviations, see below or legend of Fig. 6.11. The key to terms (above) shows a rheostat dial with LOF and GOF settings on either side of the wild-type level of gene expression. In the panels below, this knob is turned to indicate signal amplitudes for Hh, Dpp, and Vein (or other Egfr ligands). *N.B.:* When a gene is driven by *Gal4-UAS* transactivation in an unusual region (stated in parentheses), its transcription is typically higher than in its endogenous region (hence a GOF amplitude effect) [435, 3857].

a. Wild-type wing.

b–d. Hh pathway [4478]. This pathway governs veins 3 and 4. **b.** Overexpressing *patched* (*ptc*) in the pouch (via various *Gal4* drivers) suppresses L3 and L4 [2992]△ (but see [4604]). Presumably, the suppression occurs because excess Ptc impedes Hh diffusion [755] and inhibits Hh transduction [2992]△, although not severely enough to curtail Dpp-dependent growth. **c.** A stronger (~null) effect is obtained when the normal Hh is replaced with a tethered (nondiffusible) construct. This substitution eliminates the 3–4 intervein and collapses L3 onto L4 [4136]. Similar 3–4 fusions are seen with *fused*[LOF] [4479]△, *ptc*[GOF] [2993]△, *dispatched*[LOF] [572], *knot*[LOF] [2894]△, and *oroshigane*[LOF] [1171] (cf. *tout-velu*[LOF] [277]), as well as in transduction-defective *fu*[LOF] [3177] or *smo*[null] [364] clones that cross from A to P. Milder narrowing of the 3–4 span is seen with *Egfr*[GOF] [4604]. **d.** Overexpressing Hh in the 3–4 intervein region (via *en-Gal4:UAS-hh*) displaces L3 anteriorly, leaving behind its 3 sensilla campaniformia (circles) [2992]. Widening of the 3–4 intervein is also seen when cholesterol is added to the larval diet [341] – an effect attributable to the cholesterol tail that Hh acquires during its maturation [1960, 1975].

e–g. Dpp pathway [2992]. This pathway governs veins 2 and 5. **e.** Overexpressing the Dpp target genes *spalt* or *spalt-related* throughout the pouch eliminates L2 and L5 (and causes a smaller wing) [981, 984, 2617]. **f.** Deleting the entire *spalt*-Complex (i.e., both *spalt* and *spalt-related*) in large clones causes fusion of L2 with L3, and L4 with L5. This diagram is a composite of two wings – one with a large A clone and the other with a large P clone (Fig. 2b and c in [986]). Similar effects are seen with *dpp*[LOF] [4830]△, *punt*[LOF] [1674], *tkv*[LOF] [1674], and ubiquitously expressed "Supersog" constructs [4830]. **g.** Overgrowth occurs pervasively, except in the 3–4 intervein, when *dpp* is overexpressed by (1) using *MD-638-Gal4* or *71B-Gal4* to drive *UAS-dpp* throughout the wing pouch [1537, 2992], (2) using *dpp-Gal4* to drive *UAS-dpp* along the A/P line [2954], or (3) using *hs-flp*-out to induce scattered *dpp*-ON clones in 2nd instar [4188]. This uncoupling of L2 and L5 from L3 and L4 disproves the simple Dpp Gradient Model (cf. Fig. 6.3) [2992], wherein Dpp specifies all vein positions directly [2954, 4136].

h–j. EGFR pathway [320]. This pathway is instrumental in specifying all veins (except vein 1?), but at a lower level in the control hierarchy than Hh or Dpp (cf. Fig. 6.11). Two Egfr ligands are used redundantly: Vein and a second ligand regulated by *rho*. Eliminating one or the other (**h**) represses the pathway, but both must be disabled to shut it OFF entirely (**i**). **h.** When *rho* function is abolished (by null clones), certain veins (L1 and cross-veins) or fragments (L3 and L4) remain [1643]. Similar remnants are seen in *Star*[null] clones [1643]. **i.** All veins except L1 are absent in *rho*[LOF] *vein*[LOF] double mutants [1041, 1366, 2907] due to paucity of vein initiation [320, 3015, 3624], and wing shape is altered [1041]. This trait reflects extreme synergy because neither of these weak alleles – *rho*[ve] nor *vein*[1] – has much of an effect by itself [1041, 1368]. Near-total vein loss is also seen with *rho*[LOF] *argos*[GOF] [3772], *rho*[LOF] *Egfr*[LOF] [4191], *Egfr*[LOF] *rolled*[LOF] [1043], *Egfr*[LOF] *ast*[LOF] [1042], *dRaf*[GOF] [2715], *sprouty*[GOF] [681, 3558], and *dpp*[LOF] *tkv*[LOF] [980]. **j.** When *71B-Gal4* drives *vein* expression in the pouch, the wing is smaller, and 1–3 and 4–5 areas are veinlike [4604]. Similar effects are seen with *rho*[GOF] [3126], *rho*[GOF] *Star*[GOF] [1643], *spitz*[GOF] [4604], *net*[LOF] *N*[LOF] [4188], and *ara*[GOF] [1536]. The immunity of the 3–4 intervein is attributable to its shortage of Egfr [1643, 4604], which, in turn, is enforced by Knot [2894]. This immunity can be overcome at high (*Gal4*-driven) levels of Rho [1643]. Why the area behind L5 is immune is not known.

k. Cross-veins disappear when *gbb* (*glass bottom boat*) is repressed [4830], and similar defects are seen with *short gastrulation*[GOF] [4830] and *tolkin*[LOF] [1233, 3111]. Cross-veins are closer together in *approximated*[LOF], *dachs*[LOF], *dachsous*[LOF], and *four jointed*[LOF] (not shown) [4852]△.

Except for **g**, which is inferred from preveins in a mature disc (Fig. 2d in [4188]), all sketches (~ same scale) were traced from photographs of wings in [2072] (**b**), [4136] (**c**), [2992] (**d**), [984] (**e**), [986] (**f**), [1366] (**h**), [4604] (**i**), [4604] (**j**), and [4830] (**k**).

---

function of the pathway within the interveins involves differentiation more than determination.

Veins 3 and 4 are positioned by the Hh pathway

L3 shifts anteriorly when excess Hh is forced into iv3–4 (e.g., by *en-Gal4:UAS-hh* [320, 2992, 4479]), and it shifts posteriorly when Hh is prevented from entering iv3–4 (e.g., by *dispatched*[LOF] [572]). Such displacements indicate that Hh controls the position of L3 (Fig. 6.13) [320]. This inference also draws support from the extra L3-type veins that arise when Hh is overexpressed (with its receptor)

throughout the pouch (via *MD-638-Gal4:UAS-hh:UAS-ptc*) [2992].

L3 possesses 3 distal sensilla campaniformia [1368, 3234] – the only such stretch receptors in the distal blade [351]. These sensilla are ideally situated to monitor flexion during flight [1045, 1115, 4748] (cf. sensilla on the wing base [840] and haltere [4342]). They come from a proneural stripe along the L3 prevein (cf. Fig. 6.11) [320, 362], although it is unclear how that stripe transforms into 3 PNC spots. When L3 is pushed anteriorly (by *en-Gal4:UAS-hh* [2992, 4479] or *dpp-Gal4:UAS-knot* [2894]), the sensilla are often left behind, implying that they are sited independently of L3 (albeit coincidentally) in wild-type flies (cf. other instances of uncoupling [1042, 3624, 4308]). Indeed, the sensilla persist when L3 disappears entirely [320].

L4 often has gaps when Hh diffusion into iv3–4 is impeded (e.g., due to *en-Gal4:UAS-ptc*) [320], and similar gaps are also seen when vn^{LOF} clones abut the A/P boundary from the A side [1366]. The latter effect is nonautonomous because L4 is in the P compartment, and it suggests a 2-step (signal relay) process consistent with the Rho-Vn Booster Model described above [981, 2852, 2894, 3077]:

1. Hh diffuses anteriorly and activates *vn*.
2. Vn then diffuses posteriorly to induce L4 [320], which resides just posterior to the A/P compartment boundary [4188].

Both links ("Hh ➡ *vn*" and "Vn ➡ *rho*") are known to operate in the wing disc [320, 4189, 4604].

The iv3–4 area is kept "vein free" by Knot (a COE-domain transcription factor). Knot's expression there is induced by Hh [2894, 4479], and $knot^{LOF}$ A clones that abut the A/P line exhibit the same nonautonomy as vn^{LOF} clones (i.e., L4 loss) [3077]. When Knot is misexpressed in other parts of the wing, it is able to suppress veins there as well [2894], so evolution could have used it more broadly but did not (nor is Knot needed anywhere else in discs [3077]). Instead, Bs was granted dominion over all intervein states [2907]. Nevertheless, Knot rules Bs in iv3–4: *bs* transcription ceases there in $knot^{LOF}$ wings [4479]. Thus, iv3–4 is governed by ≥3 echelons of transcription factors: Hh ➡ Ci (zinc finger) ➡ Knot (COE) ➡ Bs (MADS) ➡ ? ➡ intervein. This control system rivals the segmentation gene hierarchy (cf. Fig. 4.2) in the extent of its "middle management."

Knot's untapped talent as a "vein suppressor" helps explain the perplexing loss of L2 (*sic!*) in ptc^{G20} flies [3077, 3372]. Given Ptc's role as a negative regulator of Hh transduction (cf. Fig. 5.6), this LOF allele should upregulate Hh target genes, including *knot*, throughout the A region where Ptc normally silences the pathway, and the resulting Knot could be suppressing L2 there. This reasoning is validated by the ptc^{G20} $knot^{LOF}$ double mutant, whose L2 is completely restored [3077].

Veins 2 and 5 are positioned by the Dpp pathway

L2 and L5 disappear when *spalt* or *spalt-related* is overexpressed throughout the pouch [984, 986, 2617]. Evidently, these veins rely on ON/OFF expression boundaries that vanish under these GOF conditions. These paralogs constitute the *spalt* Complex (Spalt-C). As discussed above, they are turned ON by Dpp above a certain threshold [986]. Each of them encodes a zinc-finger protein [221, 2346] that could directly regulate the transcription of downstream genes.

Four of those downstream genes are essential for L2 and L5 (Figs. 6.11 and 6.12) [984]: *knirps* (a gap gene) and *knirps-related* comprise the *knirps* Complex (Kni-C) [2617, 3653], while *araucan* (*ara*) and *caupolican* (*caup*) belong to the Iroquois Complex (Iro-C) [1536, 1537]. Iro-C also includes *mirror*, but that gene has virtually no expression in veins and instead acts in the alula [2170]. All genes in Kni-C and Iro-C encode transcription factors (zinc-finger [3060, 3199] or homeobox class [1536], respectively), thus permitting still another echelon of downstream (realizator?) genes.

The following facts argue that Spalt-C's ON/OFF boundaries establish the location of L2 by means of Kni-C:

1. When *spalt*^{null} clones arise in iv2–3, they can displace the native L2 posteriorly or induce ectopic L2 fragments [984, 986]. The fragments arise just inside the perimeter of *spalt*^{null} clones [2617, 4188].
2. When clones deficient for Spalt-C arise in iv4–5, they can displace the native L5 anteriorly or induce ectopic L5 fragments [984, 986]. In this case, the fragments arise at a distance (≤5–10 cell diameters) outside the clone edge [4188].
3. Knirps is suppressed autonomously wherever a Spalt-C^{null} clone overlaps its domain [984], implying "Spalt-C ➡ *knirps*". Conversely, a high level of Spalt-C expression throughout the pouch turns OFF Kni-C [2617]. Apparently, *knirps* is only activated at an intermediate Spalt-C level in the same way that genes turn ON at thresholds in gradients. Indeed, the *knirps*-ON stripe resides exactly where the native *spalt*-ON band fades out (Fig. 6.11).

A "*knirps* ➡ L2" link must exist because (1) *knirps* is expressed only in provein 2 [984, 2617], (2) Kni-C^{LOF} (associated with deletions or rearrangements) suppresses L2 [2617], and (3) expressing *knirps* transiently (via heat shock) induces an ectopic vein a few cells ahead of L2 [2617].

Collectively, the data imply that a medium level of Spalt-C ➡ *knirps* ➡ L2. In accordance with this linear scheme (essentially a gradient model), *knirps* can cause the entire dorsal half of the wing to form vein material when *MS1096-Gal4* drives *UAS-knirps* there [2617]. Because Knirps is normally a repressor [105], it may select the vein state indirectly by repressing the intervein state: Spalt-C ➔ *knirps* ⊣ *bs* (intervein state) ⊣ *rho* (vein state).

Alternatively, the first step could be using the Deaf-Speakers/Mute-Listeners Trick [328, 2617]. Spalt-C might (1) activate a short-range inducer "X" that diffuses anteriorly to evoke L2 and (2) deafen cells to "X" so that they cannot respond (i.e., Spalt-C ➔ X ➔ *knirps* ➔ L2).

L2 is intriguing because it specifically acquires bristles in *hairy*[LOF] or *ac*[GOF] mutants [995, 3068, 4189]. Evidently, Hairy normally keeps *ac* OFF along L2 (Fig. 6.11) [665], and *hairy*[LOF] lets *ac* rebound enough to support SOP initiation (cf. Ch. 3). The same sort of logic applies to L3, but there is not normally enough Hairy along L3 to keep *ac* OFF (thus allowing 3 sensilla), and L3 tends to acquire extra sensilla instead of bristles [362, 2473, 4309]. In some genotypes, the number of bristles along L2 rises to ~45 [3069]. At such densities, the spacing of bristles becomes strikingly regular. Despite this ability of wing veins to simulate leg bristle rows (cf. L1's rows), there may not exist any homology because *hairy* is ON between rows in legs, not within them [3193].

L5 also depends on Spalt, but the intermediary agent in this case is Iro-C. Ara and Caup are expressed exclusively in L5 and L3 (Fig. 6.11) [1537], and both veins disappear (L3 partly and L5 completely) when Ara and Caup are suppressed [1536]. The clues to L5's circuitry are listed below:

1. Ara and Caup are activated ectopically (and autonomously) whenever a Spalt-C[null] clone arises in iv2–3 or iv4–5 [984]. This response implies the link "Spalt-C ⊣ Iro-C". As for why iv3–4 is immune, the culprit might be Knot, which is endogenously expressed there: "Knot ⊣ Iro-C"?
2. Overexpressing Spalt-C throughout the pouch (via *756-Gal4*) erases expression of Iro-C in L5 but not in L3 [984].
3. Overexpressing Ci in the pouch activates Iro-C in the A region (ahead of L3) but not the P region (around L5) [1537]. Apparently, *en* blocks Iro-C activation: "En ⊣ {Ci ➔ Iro-C}". En cannot be repressing Iro-C directly, though, because Iro-C turns ON in L5 (squarely in *en*-ON territory).
4. Although Ara is not normally expressed in L4, it can rescue L4 when widely expressed (via *71B-Gal4*) in a pouch whose Hh pathway is repressed by excess

Ptc (driven by the same *Gal4*) [1537]. Thus, Ara has an untapped talent as a generic vein inducer.

Conceivably, Iro-C is activated within a certain range of Dpp concentrations (Dpp ➔ Iro-C) that extends beyond the Spalt-C band by the width of Iro-C provein 5 [984]. The resulting activation of *ara* and *caup* there could then evoke a vein via *rho* [1537]: "Iro-C ➔ *rho* ➔ L5".

L3 resembles L5 insofar as its immediate upstream regulator appears to be Iro-C [1536, 1537]. However, L3 cannot be obeying the same "Spalt-C ⊣ Iro-C" rule as L5 because it lies squarely in the Spalt-C expression zone. (L5 is outside it.) This defiance of Spalt-C inhibition explains L3's persistence (unlike L5) in the face of excess Spalt-C (point 2 above). Nor does L3 depend on Spalt-C (or *omb* [1626]) in a positive way because Spalt-C[null] clones still make L3 [986]. Nevertheless, L3 does need Dpp (in addition to Hh) because constitutive activation of the Hh pathway in *DCO*[null] *dpp*[null] clones cannot sustain Iro-C expression when the clones are too far from the endogenous source of Dpp (at the A/P line) [1537]. (Singly mutant *DCO*[null] clones show no such dependence because they manufacture their own Dpp via "Ci-155 ➔ *dpp*".) To a first approximation, the above facts can be explained by a simple rule (where sound is used as a metaphor for signal strength):

IF Hh is audible AND Dpp is somewhat loud AND you can make a vein, THEN make L3.

This rule accounts for why L3 can be pushed forward by excess Hh (it can still get enough Dpp) but not by excess Dpp (it cannot leave Hh's diffusion range). However, it fails to accommodate all the available data as the fanciful dialog below shows:

SKEPTIC: If this rule is correct, then why isn't L3 as wide as the entire Hh gradient?

THEORIST: L3 can't form if *knot* is ON, and *knot* turns ON when the Hh signal is loud.

SKEPTIC: Well, if that's true and you remove *knot*, then L3 should widen but it doesn't! It just moves closer to L4.

THEORIST: Quite right, but no one knows the diffusion range of Hh in the absence of Knot. Maybe Hh can't get very far without Knot and hence makes L3 closer to the A/P line.

SKEPTIC: Rubbish! Hh's speed depends on Ptc, and *knot* doesn't affect *ptc*! And don't try invoking Dpp because *knot* doesn't affect that either!

THEORIST: Very well, but maybe Knot does more than just enforce an intervein state. Maybe it helps amplify the Hh signal. In its absence the cell would be deafer and would only respond to higher levels of Hh, thus moving L3 closer to the A/P line.

SKEPTIC: There's no evidence for that, and anyway it can't be true. You'd expect all the cells in a *ptc*^null *knot*^null clone to make vein tissue (because *ptc*^null should activate their Hh pathway similarly, and *knot*^null should let them all make L3), but only the cells in the middle of the clone do so. How do you explain that?

THEORIST: Indeed, the intervein tissue in those clones is perplexing. Maybe the span of L3 tissue is limited by lateral inhibition (mediated by Notch or Argos)? I don't know. Why don't we make it an essay question for the grad students?

SKEPTIC: Good idea! Maybe they can figure it out!

A related issue concerns symmetry. Given that L2 and L3 are a crude mirror image of L4 and L5 (cf. truly symmetric wings [1647, 3935]), the *dpp*-ON stripe, which supposedly specifies L2 and L5, should coincide with the symmetry plane. Such a superposition may prevail before late-3rd instar [320], but not after the stripe shifts anteriorly along with the *en*-ON edge (Fig. 6.7d) [350]. How can a reference line specify a symmetric pattern if it is off center (closer to L3 than L4)? The solution to this "**Registration Riddle**" (cf. Gaps Mystery; Fig. 5.1) might be found in *en*'s role as a selector gene [3074]. If *en* enables cells on opposite sides of its ON/OFF boundary to interpret their identical Dpp gradients differently, then perhaps it can also make one group deafer to compensate for the displacement. Alternatively, the mismatch might be corrected by a shim strategy that relies on the rainbow of gene expression subzones inside iv3–4 (cf. Fig. 6.3): one or more of those states could be added or subtracted from Dpp inputs to modulate the perceived gradients.

The Dpp pathway later implements the vein state

Just as overexpressing *rho* in the pouch converts almost the whole wing to vein material [1643] (cf. *Egfr*^GOF [3558]), so does overexpressing *dpp* [980, 1674, 3109] (cf. clones that are *dpp*^GOF [4848], *tkv*^GOF [980, 1674], or *bs*^null [2907, 3624]). Con-

versely, reducing *dpp* activity (in clones or whole wings) suppresses veins [980, 3438, 3850] (cf. *tkv*^LOF and *schnurri*^LOF clones [569] but cf. *punt*^LOF clones [3329]). This late role for *dpp* in implementing vein identity is separate from its earlier gradient role in specifying vein locations [981]:

1. Prospective vein cells turn ON *dpp* after *dpp* expression fades from the A/P border. Estimates differ for the time of onset: 6–9 h AP [4831] or 12–16 h AP [980].
2. Expression of *dpp* in veins is controlled by a "short-vein" cluster of *cis*-enhancers [980, 3850, 4056], which is separate from the "disc" cluster that regulates expression at the A/P line [347, 2739].
3. Expression of *dpp* in veins is regulated by different upstream circuitry, which must include the EGFR pathway because dRaf ⊣ *dpp* [2715]. Indeed, as mentioned above, veins cannot keep *dpp* ON unless they turn OFF the EGFR pathway at ~24 h AP [2715].

Because *dpp* turns ON later than *rho* (or any other known gene) in vein cells, the Dpp pathway may execute the final stage in vein cell commitment. Indeed, the period when overexpression of Dpp maximally produces ectopic and thick veins is 16–28 h AP [4831], which is roughly coincident with the period (22–30 h AP) when the Notch pathway is required [1951], but later than the period (6–18 h AP) when the EGFR pathway is essential [1643]. Moreover, *dpp* must function downstream of the EGFR pathway because *dpp*^GOF rescues vein development in *rho*^LOF *vn*^LOF wings (without activating *rho*) [980].

The diffusion range of Dpp under these conditions is probably minimal because the lateral provein cells that flank each prospective vein express high levels of Thick veins [980]. This high receptor density should "sop up" most of the Dpp before it can travel very far [2457]. Thus, Dpp evidently functions exclusively in an autocrine mode to induce vein determination (and/or differentiation).

A cousin of Dpp (Gbb) fosters the A and P cross-veins

In embryos, the product of the *short gastrulation* (*sog*) gene is a diffusible Dpp antagonist with a distribution complementary to Dpp [96, 319, 1278]. Interestingly, *sog* is also expressed in a pattern that is complementary to *dpp* in pupal wings (i.e., interveins and lateral proveins like *tkv*; Fig. 6.11) [4831]. However, LOF-GOF studies show that Sog's role is minor: (1) *sog*^LOF veins are merely irregular in width and alignment, and (2) *sog*^GOF eliminates both

cross-veins. Thus, Sog's chief duty may be to trim *dpp*-
ON files of cells. The upstream factors that regulate *sog*
are unknown [2715] but likely include *rho* [4831].

Another agent that appears to function specifically
in cross-vein development is Glass bottom boat (Gbb)
[2203, 4610] – a cousin of Dpp in the TGF-β family [2226, 2734].
LOF mutations in *gbb* cause the same cross-veinless trait
as GOF alterations in *sog* (Fig. 6.13k) [4830]. Genetic interactions between *sog* and *gbb* are much stronger than
those between *sog* and *dpp* [4830]. An additional cross-
vein regulator belongs to the same class of BMP1-like
metalloproteases as Tolloid (the enzyme that cleaves
Sog [771, 1232, 2696, 2990, 3901]) – viz., Tolkin [1233, 3111]. Cross-
veins are removed by LOF alleles of still other genes [2561]
(e.g., *crossveinless* [469, 4583], *crossveinless 2* [854, 3501], and
crossveinless-like 6 [2897, 2898]) and by *dCul-1*GOF [1831]. Wg
is specifically expressed in the cross-veins (but not in
L2–L5) at 24–30 h AP [353, 3374], although it is relatively
unimportant for their formation [854].

Cross-veins are intriguing geometrically because (1)
their orthogonal orientation relative to L2–L5 implies a
different reference axis and (2) the anterior cross-vein
spans the A/P compartment boundary and hence reflects regional coordination [2260]. Ectopic cross-veins
form between L2 and L3 in backgrounds that exacerbate vein formation [936, 1431, 3484, 4189, 4228, 4664], and
similar conditions foster extra cross-veins at other
proximal-distal levels [936, 1431, 3484]. The sensitive period
for inducing extra cross-veins via RnRacGAP is 24–30 h
AP [1642] (just after the 18–24 h AP window when longitudinal veins can be suppressed), and cross-veins
can be eliminated by heat-shocking wild-type pupae at
16–28 h AP [2896]. Cross-veins do not express *rho* until
25 h AP (vs. L2–L5, which turn *rho* ON by 0 h AP) [854,
4189].

The abiding mystery is why cross-veins normally
only form at two fixed sites [2688]. The fact that extra
cross-veins appear between L3 and L4 when the iv3–4
span shrinks distally (e.g., due to *knot*LOF [3077]) suggests
that the proximity of adjacent longitudinal veins might
be crucial. According to this "**Jacob's Ladder Model**" (alluding to sparks that arc between antennae of a Van
de Graaff generator), cells would be programmed to
make lateral connections wherever the space between
longitudinal veins drops below some preset distance
at some predetermined time. Alternatively, cross-veins
might be specified by the Wg gradient. These contrasting views are relevant to the larger issue of the extent to
which patterning relies on physical (epigenetic) tricks –
an old [861, 3063, 3357, 3699, 4306] but ongoing dispute [858, 1557,

2154, 2765, 3494] in "evo-devo" biology [1489, 1552, 1691, 1876, 3495,
4521].

Vein 1 uses a combination of Dpp and Wg signals

The posterior wing margin shows vein markers early in
the pupal period but never forms a mature vein [1312].
The anterior margin forms a vein (L1) that resembles
the four blade-traversing veins (L2-L5) in histotype and
gene expression during the pupal period [1312, 1313, 3015],
but it differs from them in (1) its immunity to LOF defects in *Egfr* [3624] and other EGFR pathway genes [320, 1368]
and (2) its failure to express *vvl* [995].

More important, L1 is perpendicular to the *dpp*-ON
stripe along the A/P boundary, whereas all the other
veins run parallel to it. The significance of this difference can be appreciated by imagining yourself shrunk
to the size of a cell. If you were to walk along vein L2,
L3, L4, or L5 before evagination, the volume of the perceived Dpp signal would seem constant. In contrast, if
you were to stroll anteriorly from the wing tip along L1,
the intensity of the Dpp signal would gradually diminish
because the Dpp gradient decreases in this direction (cf.
Fig. 6.3).

The patterns of bristles along L1 define three basic sections: (1) a "tip" pattern in iv3–4 consisting of
long noninnervated hairs plus a few bristles near L3
[1837, 2894, 4479], (2) double row (DR) in iv2–3, and (3) triple
row (TR) anterior to vein 2 [1741, 1837] (cf. Fig. 6.7d). The
tip pattern is due to the creeping of *en* expression into
this region [1837, 4229], but the latter two sections probably depend instructively on the level of Dpp. Why don't
blade cells respond like L1 cells? It is likely that bristles are confined to the margin because a high level of
Wg is needed permissively (i.e., as a digital vs. analog
input) to induce *achaete* in the *en*-OFF area as a prerequisite for SOP initiation (cf. Ch. 3) [166, 912, 3689, 3982]. Evidence for this "**Instructive-Permissive Model**" is listed
below:

1. *Raising Dpp levels along the whole length of L1.* Suffusing the wing pouch with Dpp via *MD-638-Gal4*
   causes DR-like bristles along the whole span of L1
   [2992].
2. *Making cells think they hear Dpp more loudly.* Raising the perceived level of Dpp in *tkv*GOF (or *mtv*LOF)
   clones causes cells along L1 to autonomously shift
   their fates toward more posterior identities – hence
   making DR bristles in the TR region [1327].
3. *Making cells deafer to Dpp.* Reducing the perceived level of Dpp in *tkv*LOF or *sax*null clones

causes cells along L1 to autonomously shift their fates toward more anterior identities (DR to TR) [3972].

4. *Making cells think they hear Wg more loudly.* Increasing the perceived level of Wg throughout the wing pouch (by driving *UAS-Dfz2* with *1J3-*, *69B-*, or *71B-Gal4*) causes bristles to form sporadically throughout the blade [3691], and these extra bristles look like the marginal bristles at their A-P position (DR or TR). A stronger transformation is seen in *sgg*^{null} clones [351, 3958] (cf. *punt*^{LOF} clones [3329]). When such clones are far from the *wg*-ON zone, they tend to round up or form vesicles [351, 353, 3603] (cf. similar behavior on the notum [3956] and leg [1039] and its possible significance [3046, 3049, 3050]).

Macrochaetes are sited by various "prepattern" inputs

The heminotum develops as a sort of "island" within the wing disc [157, 793, 1060, 2937, 3104]. It has its own sources for Dpp and Wg, and is insulated from signals that govern the blade and hinge [678, 987, 994, 1074, 2253, 2570]. This island is ruled by an oligarchy of transcription factors or cofactors that evoke macrochaetes at fixed sites:

Iro-C	The Iroquois Complex (140 kb at 69D1-3) encodes 3 homeodomain proteins: Araucan (Ara), Caupolican (Caup), and Mirror (Mirr) [1536, 2170, 2794, 3079]. Iro-C^{LOF} causes an absence of bristles from the sides of the notum – a trait that resembles the "Mohawk" haircut of the Iroquois (a tribe of native Americans) [951, 2520]: hence, the names "Iroquois," "Araucan" (another tribe), and "Caupolican" (an Araucan hero) [1536].
Mad	Terminal effector of the Dpp pathway (cf. App. 6).
Pan	Terminal effector of the Wg pathway (cf. App. 6).
Pannier (Pnr)	A zinc-finger transcription factor [1671, 3503, 3504] (a.k.a. dGATAa [4700]) that binds GATA sequences [1380].
U-shaped (Ush)	A zinc-finger cofactor that heterodimerizes with Pnr [911, 1671] and represses transcription, probably by recruiting dCtBP [3764].

These regulators constitute the elusive "prepattern" agents [615, 918, 2890, 2891, 3953] foretold long ago by Stern [4095]. We have only begun to grasp how they control the SOP-making proneural machinery (Fig. 6.14, Table 6.2). Some of the key interactions identified so far are described below:

Link 1: Dpp (at high levels) → *ush* (cf. analogous regulation in the embryo [127, 2050]). When the Dpp pathway is constitutively activated (in *tkv*^{Q253D} clones), *ush* turns ON ectopically [3764]. When the pathway is suppressed (in *tkv*^{LOF} or *Mad*^{LOF} clones [3764] or in *punt*^{LOF} or *Dad*^{GOF} discs [4369]), *ush* turns OFF within its normal domain.

Link 2: Dpp → *pnr* (cf. the embryo [30, 127, 2050, 4700]). Like Ush, Pnr is expressed ectopically in *tkv*^{Q253D} clones and extinguished in *tkv*^{LOF} or *Mad*^{LOF} clones [3764, 4369].

Link 3: Dpp (at low levels) → *wg*. When notal cells are "duped" (by *tkv*^{LOF}, *punt*^{LOF}, or *Mad*^{LOF}) into thinking they are receiving less Dpp than they actually are, the *wg*-ON stripe shifts toward the *dpp*-ON stripe [3763, 3764, 4368], as if the *wg*-ON stripe is "tracking" a certain level of Dpp activity. Likewise, the *wg*-ON stripe moves away from the *dpp*-ON stripe when the perceived activity of Dpp rises (in *tkv*^{GOF} clones), and it vanishes in *dpp*^{LOF} discs [3763]. High Dpp levels (via *ap-Gal4: UAS-dpp* or *tkv*^{GOF}) exert an opposite effect (Dpp ⊣ *wg*) [3763, 4368], probably due to Links 1, 2, and 5: "Dpp → {Pnr AND Ush} ⊣ *wg*".

Link 4: {Pnr AND NOT Ush} → *wg*. Wg is ectopically expressed when *pnr* is artificially turned ON beyond its normal ventral edge (by *C765-Gal4*) [1380, 3764], and Wg stops being expressed in its normal (*pnr*-ON/*ush*-OFF) stripe when *pnr* is turned OFF (by *pnr*^{null} clones) [3764, 4369]. In *pnr*^{LOF} flies, the *wg*-ON stripe shifts dorsally [615, 1380, 4369] as it does when the Dpp pathway is repressed (cf. Link 3).

Link 5: {Pnr AND Ush} ⊣ *wg*. When *pnr* is turned OFF (in *pnr*^{null} clones or *pnr*^{LOF} flies), *wg* now turns ON in the *pnr*-ON/*ush*-ON (dorsal) region [3764, 4369]. This ability of *wg* to turn ON sans Pnr suggests that Link 3 works at all Dpp levels but is overridden at high levels by Link 5. However, *wg* also turns ON in this region sans Dpp [4368], so *wg* may be activated by an unknown gene [3764]. Ush behaves as a negative regulator of *wg* throughout the *pnr*-ON domain: (1) *wg* turns ON dorsally when *ush* is turned OFF there (by *ush*^null clones) [3764], and (2) the *wg*-ON stripe vanishes when *ush* is broadly overexpressed (by *C765-Gal4* [3764] or *pnr-Gal4* [4369]).

Link 6: Pnr ➔ AS-C. LOF mutations at these two loci interact synergistically, suggesting that the genes act in the same pathway [3503]. Pnr activates the AS-C directly by binding GATA sites within the *cis*-enhancer for the dorsocentral (DC) PNC [1380]. This enhancer may loop into contact with the *ac* and *sc* promoters via a bridging complex consisting of Pnr, Chip, and Ac/Da or Sc/Da dimers [3504].

Link 7: Ush ⊣ AS-C. Expression of *ac-lacZ* in the DC proneural cluster expands in *ush*^LOF mutant discs and fades under *ush*^GOF conditions [911, 1671]. These effects are evidently mediated by the DC *cis*-enhancer because the same defects arise when a *lacZ* reporter is driven by this enhancer alone [911].

Link 8: Wg ➔ AS-C. In *wg*^LOF mutants, *ac* fails to turn ON at the dorsocentral and scutellar PNC sites [3374]. Wg was once thought to bias the competition for SOP fate to the side of the PNC facing the *wg*-ON stripe [3374, 3954]. However, Wg's role here is only permissive: the site of the dorsocentral PNC is not altered when the distribution of Wg is modified [1380].

As a consequence of this circuitry, Pnr sets the ventral edge of the *wg*-ON stripe (Link 4), while Ush sets its dorsal edge (Link 5) [1380, 3764]. These same factors also set the V and D boundaries of the DC proneural cluster [1380]. How the A and P limits of this cluster are established remains a mystery, as do various aspects of its temporal dynamics (cf. Fig. 3.3). The other PNCs are even less well understood.

How bristle axons get wired into the CNS is not known

Ever since nervous systems were first described, the **"Position-Projection Mystery"** [1455] has been: how do peripheral axons find appropriate CNS targets [60, 1392, 2973, 4264]? For bristles on the fly's notum, the growing axons behave differently depending on their source site [4431, 4432]. Axons from medial bristles send branches to the contralateral side of the thoracic ganglion, while axons from lateral bristles remain ipsilateral [1624]. Within each of these two notal areas, the macrochaete (MC) and microchaete (mC) axons act alike. The areas coincide roughly with Iro-C ON or OFF regions (Fig. 6.14). When *sca-Gal4* forces *ara* or *caup* to be expressed in all SOPs, the axons of medial mCs now manifest an ipsilateral projection, although the axons of medial MCs retain their identity [1624].

The ability to switch mC-axon projections from one state to another by toggling a regionally expressed homeobox gene implies that sensory neurons give their axons a "birthplace code," which guides them to predestined targets in the CNS [1451, 1455, 1457, 1460, 3789, 4462]. Indeed, CNS neuroblasts seem to emerge from the embryonic ectoderm in just this way [3981]. One attractive idea is that the encoded addresses use cell-surface markers such as Dscam. This neuronal protein has 38,000 possible isoforms attainable by alternative splicing [700, 1597, 3793] (cf. cadherins [787, 3858, 3870] and neurexins [2874, 4420]).

Aside from the notum, axonal pathfinding has been examined in legs [3005, 4330] and wings [363, 1463, 4636], but the wiring strategies for these appendages remain obscure [4329]. The best clues so far have come from the antennae [2545] and eyes [1850, 2358, 2708, 3514, 4331] (cf. ocelli [1032, 1350]).

For antennae, the determining factor in PNS-CNS wiring is receptor subtype. Insects use various olfactory sensilla to detect odors [2955, 4840, 4841], and most of these sorts of sensilla in flies are found on the 3rd antennal segment [2221, 4469] and maxillary palps [818, 819, 4504, 4506]. Olfactory sensilla fall into three morphological classes: trichoid, basiconic, and coeloconic [1653, 4126, 4127]. Transduction of chemical signals is mediated by 7-pass (~380 a.a.) transmembrane receptor proteins [3389, 3835]. Receptors are encoded by 57 different genes – 17 of which make no detectable mRNA [4506]. The rules for how the remaining 40 genes operate are simple, as are the rules for how the sensory neurons (~1000 on each antenna and ~120 on each palp) connect to the 43 glomeruli in each of the brain's symmetric antennal lobes [2056, 3626, 4125, 4506]:

1. The gene *Or83b* is expressed in all neurons, implying that Or83b may serve a scaffolding (vs. ligand-binding) role.

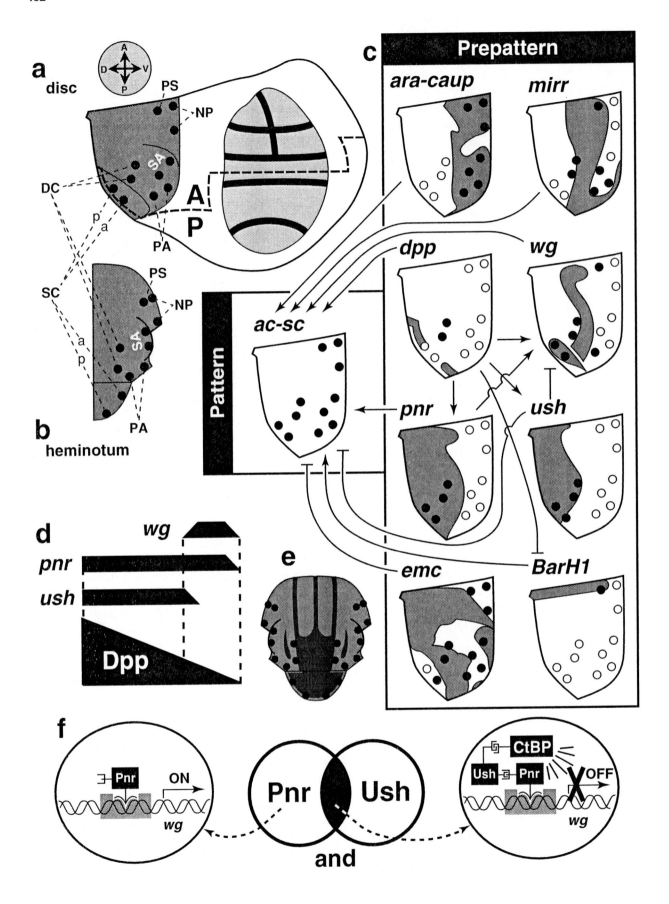

2. Each of the other 39 genes is expressed in only a subset of neurons.

3. The number of neurons per subset ranges from 2 to 50.

4. Each neuron expresses only 1 of the 39 specific genes. Expression of 32 genes is restricted to the antenna, while the other 7 are confined to the palp.

5. Neurons that express the same gene tend to be clustered on the antenna or palp. This clustering suggests that receptors might be acquiring "addresses" from a system of positional information, but subtypes are interspersed (i.e., the segregation is imperfect). Thus, a coordinate system *sensu stricto* is unlikely.

6. Neurons that express the same gene send axons to 1–2 specific glomeruli on the ipsilateral side and show a symmetric projection contralaterally.

FIGURE 6.14. Landscape of prepattern factors that evoke macrochaetes on the notum (cf. Fig. 3.4).

a. Fate map of a mature right wing disc (simplified from Fig. 6.1). Dots are actual SOP sites [1925]. Dashed line is the A/P compartment boundary. Directions (cf. compass above): A, anterior; P, posterior; D, dorsal; V, ventral. Within the notal region (dark shading), the future scutellum (posterior end) is demarcated by a thin solid line.

b. Heminotum, with macrochaete sites labeled. Abbreviations: "a" or "p" (anterior or posterior members per pair), DC (dorsocentrals), NP (notopleurals), PA (postalars), PS (presutural), SA (supra-alars), SC (scutellars) [2560]. The configuration of SOPs matches the adult pattern, except that scutellar SOPs rearrange ("p" moves posterior to "a").

c. Heminotal domains (shaded) where certain "prepattern genes" are expressed in mature discs. Boundaries are not as sharp as depicted, especially for *emc*, whose profile of expression has many subtle gradations (cf. Fig. 6.2), including a low background level (not shown). Solid circles are bristle sites affected by LOF mutations (cf. Table 6.2); remaining unfilled circles are unaffected. Genetic interactions are indicated by connecting wires (→ activation; ⊣ inhibition). Abbreviations (see references for expression patterns and circuitry): *ara-caup* (*araucan* and *caupolican*) [1536], *emc* (*extramacrochaetae*) [913] (see [205]$^\Delta$ for expression and [1349]$^\Delta$ for bristle effects), *BarH1* (postnotal expression domain not shown) [3763], *dpp* (*decapentaplegic*) [3764, 4369], *mirr* (*mirror*) [2170], *pnr* (*pannier*) [614, 3764, 4369]$^\Delta$, *ush* (*u-shaped*) [3764, 4369]$^\Delta$, *wg* (*wingless*) [3764, 4369]$^\Delta$. The DC cluster, which forms far from the *dpp*-ON stripe, depends on both Dpp and Wg signals [3373]. The "Dpp ⊣ *wg*" link inferred from LOF-GOF studies [3763]$^\Delta$ is mediated indirectly by Ush [3764, 4369]. Hence, it is omitted, as is a "Wg → *BarH1*" link that only works in part of the notum [3763]. The "*ush* ⊣ *wg*" link requires *pnr* (not shown; see below) [3764, 4369]. The genes *ara* and *caup* are each expressed like the depicted *iro*^{rF209} ("*ara-caup*") enhancer trap in the Iro-C [1536]. In co-stained specimens, their expression appears more complementary to that of *pnr* than depicted here [614]. Indeed, dorsal *pnr*^{LOF} clones turn ON Iro-C, indicating that "*pnr* ⊣ Iro-C" (not shown) [614]. Moreover, such clones sort out from the epithelium, implying a difference in affinity [614]. The overall circuitry suggests that Pnr sets the V limit of the *wg*-ON stripe (excluding the scutellar spot), while Ush sets its D limit [3764]. In fact, the *wg*-ON stripe extends a few cells beyond the *pnr*-ON area in late-3rd instar [614] due to a "Wg → *wg*" link (not shown) that is disabled in *wg*^{LOF} or *dsh*^{LOF} discs [4369]. The Gal4 insertion line *em462*, which exhibits a *wg*-like A-P stripe [615], is omitted because the patterning effects of its host gene are unknown.

d. Dpp gradient that supposedly governs the notal region. Dpp turns ON *ush* and *pnr* at different thresholds (high vs. low). Dpp also turns ON *wg*, but this activation is overridden where Pnr and Ush overlap. The DC cluster is regulated similarly but lies more dorsally than the *wg*-ON stripe [3373], presumably because the AS-C's DC *cis*-enhancer responds to slightly higher levels of Pnr and Ush than *wg* [1380].

e. "Trident" pattern of notal pigmentation, which is variably detectable in wild-type stocks [2982, 4511]. The pattern may reflect control elements at the *yellow* or *ebony* locus that rely on the kinds of prepattern agents shown in **c** [1435, 1442, 1443, 2968, 3055]. This pattern was the subject of Morgan's earliest (1909) artificial selection experiments with *Drosophila* [471, 2283]. The outer stripes of the trident are strikingly prominent in the dipteran genus *Zaprionus* [615]. See also the beautiful "picture wing" flies of Hawaii [669, 4397].

f. Model for how the "{Pnr AND Ush} ⊣ *wg*" link is implemented at a molecular level. Circles (center) are a Venn diagram (cf. Fig. 6.10). In the absence of Ush, Pnr is a transcriptional activator (left) [1380]. Ush is thought to mediate repression by recruiting dCtBP (right) [3764] as does Hairy (cf. Fig. 3.12). Direct binding of Pnr and Ush to the *wg* promoter has not yet been shown [1380]. This diagram is oversimplified because Pnr and Pnr-Ush probably do not compete for the same DNA sites (but see [4369]): excess Pnr (from *C765-Gal4:UAS-pnr*) apparently cannot displace Pnr-Ush heterodimers [1380].

Panel **c** is adapted from [3373, 3764, 3954, 4368]. Registration of the various domains follows [3763] (but see [615]). The thresholds in **d** summarize results of [3764, 4369], and the model in **f** is based on data in [1380, 1671, 3764].

N.B.: None of the boundaries in **c** restricts cell lineage [521, 615, 1373, 3007, 3143] (cf. the Cabaret Metaphor; Ch. 4): the sole constraint on lineage in the heminotum is the A/P boundary [1376, 1377]. Strangely, the zones where Dpp activates target genes are oriented at acute angles relative to the *dpp*-ON stripe itself, perhaps due to modulated diffusion rates (cf. the Arc Scenario; Fig. 5.4). Like *ac* and *sc* here (**c**), the proneural gene *atonal* is regulated by a zoo of upstream factors in the eye (Fig. 7.9e), but that menagerie is more like a traveling circus since it migrates across the disc.

TABLE 6.2.* GENES THAT CAN ADD OR DELETE MACROCHAETES ON THE NOTUM BY INDIRECTLY** AFFECTING PNCS OR INHIBITORY FIELDS

Pathway/complex: genes	SC (Scutellars)	DC (Dorsocentrals)	PA (Postalars)	SA (Supra-alars)	NP (Notopleurals)	PS (Presutural)
Hh pathway: *hh* [6, 2992], *cos2* [3373], *DC0* [3373], *ptc* [2992, 3373].	↑ by GOF in pathway (via *hh*-GOF or *ptc*-LOF).	↑ by GOF in pathway (via *cos2*-LOF, *DC0*-LOF, or *ptc*-LOF).	—	—	—	—
Dpp pathway: *dpp* [2954, 2992, 3373, 4831], *gbb* [4610], *punt* [3373, 4368], *spalt* [220, 987, 2455], *tkv* [4368].	↑ by *gbb*-LOF. ↓ by GOF in pathway (via *dpp*-GOF or *spalt*-GOF).	↓ by LOF in pathway (via *punt*-LOF). ↑ by GOF in pathway (via *dpp*-GOF or *tkv*-GOF) or by *gbb*-LOF.	↑ by GOF in pathway (via *dpp*-GOF).	↓ by GOF in pathway (via *dpp*-GOF or *spalt*-GOF).	↓ by *spalt*-LOF (although *spalt* may be acting independently of *dpp* here). ↓ by *spalt*-GOF.	↓ by *spalt*-GOF.
Wg pathway: *wg* [734, 880, 987, 3090, 3373, 3374], *arm* [3373], *dally* [6], *Dfz2* [734], *dsh* [151, 3373], *pan* [3591], *sgg* [3681, 3956].	↓ by LOF in pathway (via *wg*-LOF or *dally*-GOF), but not *Dfz2*-null. ↑ by GOF in pathway (via *sgg*-LOF, *dsh*-GOF, or *pan*-GOF).	↓ by LOF in pathway (via *wg*-LOF or *dally*-GOF), but not *Dfz2*-null. Also by *arm*-GOF. ↑ by GOF in pathway (via *sgg*-LOF or *pan*-GOF).	↓ pPA by LOF in pathway (via *wg*-LOF). ↓ aPA by GOF in pathway (via *wg*-GOF). ↑ by GOF in pathway (via *pan*-GOF).	↓ by GOF in pathway (via *wg*-GOF).	—	→ by LOF in pathway (via *wg*-LOF).
Notch pathway: *N* [461], *fng* [2253].	↓ by *N-Ax*.	↓ by *N-Ax*. ↓ aDC by *fng*-GOF.	—	—	—	—
EGFR pathway: *Egfr* [814, 917, 1042], *argos* [917], *dRaf* [917], *pointedP1* [917, 4604], *Ras1* [917], *rho* [917], *spitz* [917], *yan* [4604] (cf. *phyllopod* [717, 1052]).	↓ by LOF in pathway (via *yan*-GOF). ↑ by GOF in pathway (via *Egfr*-GOF or *pointedP1*-GOF).	↓ by LOF in pathway (via *Egfr*-LOF). ↑ by GOF in pathway (via *Egfr*-GOF) or by *dRaf*-DN.	↓ pPA by LOF in pathway (via *Egfr*-LOF). ↑ aPA by *Egfr*-LOF (but aPA is lost when PNC cells are flooded with Egfr-DN).	↓ by LOF in pathway (via *Egfr*-LOF). ↑ by GOF in pathway (via *Egfr*-GOF).	↓ by LOF in pathway (via *Egfr*-LOF). ↑ by GOF in pathway (via *Egfr*-GOF).	↓ by LOF in pathway (via *Egfr*-LOF). ↑ by GOF in pathway (via *Egfr*-GOF).
Iro-C: *ara* and *caup* [951, 987, 1624, 2520], *mirr* [2170] (cf. *Dichaete* [3405] and its breakpoints [3079]).	↑ aSC by *ara*-GOF. ↓ pSC by *ara*-GOF.	↓ by *mirr*-LOF (*Dichaete* effect only?).	↓ by *ara*-LOF *caup*-LOF or *mirr*-LOF.	↓ by *ara*-LOF *caup*-LOF. ↓ pSA by *mirr*-LOF. ↓ by *ara*-GOF.	↓ by *ara*-LOF *caup*-LOF.	↓ by *ara*-LOF *caup*-LOF or *mirr*-LOF. ↑ by *ara*-GOF.
Bar-C (*BarH1* and *BarH2*) [3763].	↓ by Bar-C-GOF.	↓ by Bar-C-GOF.	↓ by Bar-C-GOF.	↓ by Bar-C-GOF.	pNP by Bar-C-GOF.	↓ by Bar-C-LOF. ↓ by Bar-C-GOF.

pnr [615, 1380, 1671, 1796, 3503, 3504].	↓ by *pnr*-LOF (but *pnr*-null clones add bristles).	↓ by *pnr*-LOF (but *pnr*-null clones add bristles). ↑ by *pnr*-GOF.	—	—	—
ush [911].	↑ by *ush*-LOF (but *ush*-null clones never form bristles here).	↑ by *ush*-LOF (but *ush*-null can delete bristles). ↓ by *ush*-GOF.	—	—	—
emc [1349].	↑ aSC by *emc*-LOF.	↑ by *emc*-LOF.	↑ by *emc*-LOF.	↑ by *emc*-LOF.	↑ by *emc*-LOF.

*Column headings list notal macrochaetes as pairs (except PS) in a posterior-to-anterior sequence (cf. Fig. 3.4), except that DCs and SCs are abutted because they tend to be affected similarly. Symbols: ↓ (macrochaete loss), ↑ (extra macrochaetes at or near this site). Both bristles per pair are affected alike unless otherwise noted. LOF and GOF denote loss or gain of function. LOF effects are denoted by bold arrows (↓ or ↑) because they are more crucial for understanding the roles of genes in wild-type flies. GOF effects (↓ or ↑) can be misleading (cf. Preface). Gene abbreviations (see App. 6 for Dpp or Wg pathway genes): *ara* (*araucan*), *caup* (*caupolican*), *fng* (*fringe*), *mirr* (*mirror*), *N-Ax* (*Abruptex* is a type of *Notch* allele), *pnr* (*pannier*), *rho* (*rhomboid*), *ush* (*u-shaped*). Other abbreviations: DN (dominant negative); PNC (proneural cluster). Mutations in many other genes affect macrochaete number [2561], including some discussed in other contexts (e.g., *exd*^null (↓) [1543]).

**"Indirectly" here means that the gene affects SOP initiation via intermediate agents (e.g., Achaete or Notch) that create PNCs or inhibitory fields (cf. App. 5). This definition is blurred for the Wg pathway because Dsh (a cog in Wg transduction) affects Notch signaling (effectively short-circuiting the two pathways) by binding Notch [151, 3089, 3689]. The EGFR pathway may fall into a similar gray area because of its interactions with the Notch pathway [1291, 3465, 4898], so *Egfr* is listed both here and in App. 5. *Dichaete* (Iro-C) is also dually listed due to uncertainty about its breakpoint effects [3079], and Emc is dually listed because its spatial heterogeneity gives it a prepattern quality [1536, 4368]. The same is true for Hairy, although Hairy does not control MCs [2561]. Effects attributable to early function of Iro-C (i.e., specification of notum identity) are not shown [1060].

The eye's CNS projection is also dictated by receptor subtype [795, 1390, 1391, 1539, 1912, 2798], but arrival times of retinal axons in the brain's optic lobes govern the global order [2357, 3382, 3726] (cf. Ch. 7).

1. Axons from photoreceptors R1–R6 stop growing when they reach the lamina of the optic lobe.
2. Axons from R7 and R8 penetrate deeper into the medulla.
3. Axons that enter the optic lobes first innervate medial parts of the lobe, whereas those that enter later innervate successively more lateral domains.

Interestingly, the wiring process is directed by familiar morphogens. Ingrowing axons use Hedgehog and Spitz, respectively, to affect proliferation [1933, 1934] or differentiation [2970] of optic lobe neuroblasts [2799, 3726], while the optic lobe itself uses Dpp and Wg for neuropil maturation [2130].

CHAPTER SEVEN

The Eye Disc

Compound eyes have ~750 facets, with 8 photoreceptors per facet

A fly's face is dominated by its eyes (Fig. 7.1). Each of the two compound eyes is a honeycomb matrix of ~750 "ommatidial" subunits. Each subunit, in turn, has 8 photoreceptors or "R" cells (R1–R8) for a total of ~6,000 receptors per eye. At this pixel density, flies see grainier images than humans, who have ≥100,000 receptor cells in the fovea alone [925, 1720, 3415, 4737]. Because fly and human eyes appear to have had a common evolutionary origin [1130, 1407, 2037, 2835, 3093, 3557], the obvious "**One Eye or Many? Riddle**" is: Did our common ancestor have a simple or a compound eye? If the former, then why/how did insects multiply it [3290]? If the latter, then why/how did chordates reduce it to a solitary remnant [611, 1419, 1848, 3121, 4894]? Of course, there is a third possibility. Our common ancestor might have had only a primitive light detector [968], and we chordate or arthropod descendants then built our own versions of eyes based on the genes that were active at those spots on our face [3382].

The epithelium of the eye disc is a monolayer (as is true for all discs; cf. Ch. 4), but the epithelium of the adult eye is stratified. Above the bundle of 8 R cells, each adult ommatidium has 4 "cone" cells that secrete the lens (no relation to vertebrate cones). Between the bundles are pigment cells that prevent blurring by absorbing scattered photons: 2 PPCs, 6 SPCs, and 3 TPCs (primary, secondary, and tertiary pigment cells) per ommatidium. SPCs and TPCs are shared between neighboring ommatidia, as are 3 bristles at alternating vertices (Fig. 7.1c), so each ommatidium technically "owns" only 3 SPCs, 1 TPC, and 1 bristle, for a total of 23 unshared cells [3539] =

8 R cells + 4 cone cells + 2 PPCs + 3 SPCs + 1 TPC + 5 bristle cells, although the shaft and socket cells die before the adult stage [3351] (cf. Fig. 2.1 for bristle cells).

Photoreceptors R1–R6 share many features, whereas R7 and R8 exhibit certain idiosyncrasies (cf. Table 7.1) [1048].

1. *Rhabdomere* location. The rhabdomeres (light-gathering organelles) of R1–R6 form a trapezoid around the rhabdomeres of R7 and R8 [1726, 2354]. Hence, R1–R6 are termed "outer," while R7 and R8 are termed "inner" photoreceptors.

2. *Rhabdomere* size. R1–R6 have large rhabdomeres, whereas R7 and R8 have small ones [4715]. Thus, R1–R6 offer high sensitivity, while R7 and R8 provide high acuity [1720, 2044].

3. *Rhabdomere layering.* R1–R6 span the height of the epithelium, while the rhabdomere of R7 is stacked atop the rhabdomere of R8.

4. *Opsin subtypes.* R1–R6 express the rhodopsin Rh1 [2354, 3164, 4895]. R8 may weakly express Rh2 [889]. R7 can express either Rh3 or Rh4 [1272, 3381, 3414]. When an R7 has Rh3, its underlying R8 makes Rh5, and when an R7 has Rh4, its R8 makes Rh6 [779, 780, 3257]. In the absence of R7s, R8s make Rh6, implying that Rh6 is a default state that can be toggled to Rh5 only if the R8 gets an inductive signal (so far unknown) from an Rh3-type R7 [780].

5. *Spectral sensitivity.* As a consequence of their opsin composition, R1–R6 are primarily sensitive to light in the blue wavelengths, while R7 senses ultraviolet light and R8 responds to blue-green light [1272, 1726, 2235].

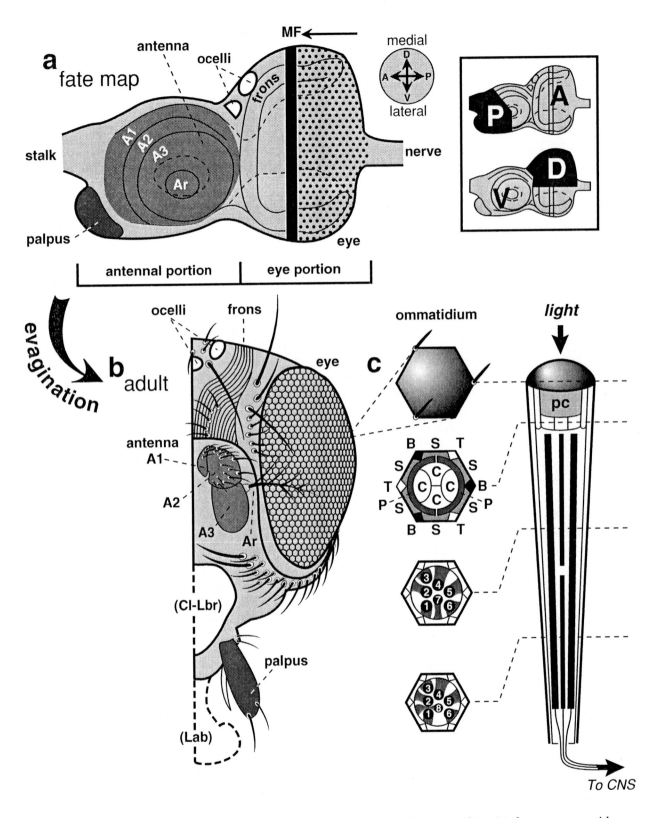

6. *Depth of axonal penetration into the brain's optic lobe.* The axons of R1–R6 terminate in the outer layer (lamina) of the optic lobe [2277, 3407, 4261], while those of R7 and R8 penetrate deeper to reach the medulla [1390, 1539, 1912, 2357, 2798, 4215].

The rhabdomeres of R1–R6 form a trapezoid [1305, 3877], and the trapezoids point in opposite directions in the dorsal (D) and ventral (V) halves of the eye [1059]. Thus, the eye is endowed with a D/V plane of mirror symmetry that defines its "equator" (Fig. 7.2). This

FIGURE 7.1. The eye disc, its adult derivatives, and the cellular components of an ommatidium.

a, b. Fate map (abridged) of an eye-antenna disc (**a**) and its cuticular derivatives (**b**). Abbreviations: A1–A3 (antennal segments 1–3), Ar (arista), MF (morphogenetic furrow, black bar; arrow shows direction of movement). The median hemi-ocellus (half oval) fuses with its counterpart from the other eye disc (not shown) [2864].

a. A left eye-antenna disc at maturity. In size and shape, the eye part crudely resembles a wing pouch, and the antenna part looks like a half-size leg disc. Dots are photoreceptor clusters. The nerve exits through the "optic stalk" (unlabeled). The disc's hind half is cup shaped (concave-side down) with two curled flaps (dashed lines on underside). The antennal part (medium shading) telescopes toward the viewer during eversion. Directions (compass at right) are prospective (**a**) or actual (**b**) axes of the adult. The vectors are confusing because eye discs rotate ~180° after they arise in the embryo, hence inverting their A-P and D-V axes [4146]. Thus, the P compartment ($\approx$ *en*-ON domain) winds up anterior [2931], and the original D compartment winds up ventral (compare the *fringe*-ON domains in Figs. 6.8 and 7.4) and is conventionally called the "V" compartment (as shown in the inset box). This inversion explains (1) why the *dpp*-ON and *wg*-ON sectors of the antenna are upside-down relative to those of the leg (compare Figs. 7.3 and 5.8) [2631, 4277] and perhaps also (2) why disabling the Dpp pathway (via *dpp*^{LOF} or *Mad*^{LOF}) removes the D half of the leg but the V half of the eye [723, 930, 1812, 4648] (but see [2752]). *N.B.:* The antennal D-V axis (mislabeled in [1037]) is defined by Wg and Dpp independently of the global D and V compartments. The A/P boundary is established by late-2nd instar (~72 h AEL) [2931, 2933], and the same is true for antennae transformed into legs [1635, 2926, 2933], whereas the D/V compartment boundary arises in early 1st instar [189]. These times contrast with the blastodermal onset of A/P restrictions in thoracic discs [4651], but they resemble the timing for the D/V line in the wing disc [354]. See Figure 6.9 for the origin of the eye vs. antenna dichotomy.

b. Left half of a head (frontal aspect). The frons manifests parallel grooves like a human fingerprint [1224, 1866]. Unlike the compound eyes, ocelli do not focus images, but rather may detect moving shadows [293]. The arista's branches may arise from hairs (vs. bristles) [3362]. Unlike the 3rd antennal segment, which senses odors [2221], the arista appears to sense temperature and humidity [1260]. Blank areas show where structures made by other discs would insert: "Cl-Lbr" (clypeus and labrum) from clypeolabral disc and "Lab" (labellum) from labial disc. Interommatidial bristles are omitted. Reshaping during evagination is complicated (not shown) [1311, 1777, 2864, 2865, 3516].

c. One facet from the lattice is shown in frontal view (left). At right is a side view of the entire conical ommatidium – a simple eye [2909, 3230]. Cross-sections are sketched (left) at 3 levels (dashed lines). Abbreviations: B (bristle), C (cone cell; no relation to "cone" photoreceptors in vertebrates); P, S, T are primary, secondary, and tertiary pigment cells; 1–8 are photoreceptors R1–R8. Nuclei are omitted. In the upper section, 4 C cells are embraced by 2 P cells, which are bordered by 6 S cells, with 3 B and 3 T cells at alternating vertices. Adjacent ommatidia share S, B, and T cells (cf. Fig. 7.2). Light is refracted by the cornea (dark shading) and underlying pseudocone (light shading) [1720, 4364]. The cornea is a cuticular secretion of the cone and primary pigment cells, while the pseudocone (pc) is a gel secreted by cone cells alone [602, 2034]. Photons are then transduced by the 8 rhabdomeres (black bars at right = numbered circles at left). Rhabdomeres are compact arrays of photosensitive microvilli [715, 844, 1817, 2323, 3568, 4458] on R1–R8 cells (shaded; intercellular spaces are exaggerated) [1270, 2594, 3350, 4270]. R7's rhabdomere lies atop R8's. Both are smaller than the other six, which nearly span the height of the ommatidium. Note (lower sections) that the cytoplasmic stalks (and hence rhabdomere gratings) of R7 and R8 are orthogonal [1318, 4355] – a cute trick for detecting polarized light [1720, 4398]. The transduction cascade culminates in electrical signals [374, 2908, 2909, 3229] that are sent to the CNS via axons (below) [1390, 2801]. Stray photons that might ricochet to adjacent ommatidia (and hence degrade the image) are absorbed by the intervening pigment cells [1048]. These pigment cells thus serve as insulating walls [3369].

Panel **a** is redrawn from [1777], with details added for ocelli [3663] and photoreceptors (Fig. 1b of [724]), **b** is adapted from [1777], with territories ascribed to different cephalic discs as per [1408], and **c** is simplified from [602, 1048, 1239, 1726, 4715]. See [3540] for a 3-dimensional rendition.

---

configuration, plus the wiring pattern of axons into the lamina, enhances visual acuity and makes the equator a sensitive "fovea" [433, 931, 1720, 2234, 2800].

The logic underlying this arrangement is a masterpiece of micro-optics and microcircuitry. The visual axes of the rhabdomeres in a single ommatidium diverge by exactly the same amount as the optical axes of neighboring ommatidia. Consequently, the axis of each rhabdomere actually coincides with those of six other rhabdomeres, each in a different ommatidium. The axons of each set of seven optically aligned rhabdomeres (in seven different ommatidia) converge onto a retinotopic unit in the first optic ganglion, the lamina. Here the signals of the six R1–6 cells are summed synaptically, while

the axons of R7 and R8 bypass the synapse and project to the second optic ganglion, the medulla. Therefore, for each point in space, light is collected through six different facets thus greatly enhancing the intensity of the retinal image, but without compromising spatial acuity. This is known as the neural superposition principle. [1721]

An intriguing exception to the rule that one point illuminates seven rhabdomeres is found at the so-called equator of the eye, about which the characteristic rhabdomere patterns are found to be mirror-symmetrical. In this region, the axes of eight or nine rhabdomeres in eight or nine ommatidia coincide. Remarkably, the projection pattern of the retinula cell axons follows this quite faithfully into the cartridges, and in the lamina the corresponding cartridges have crowns of

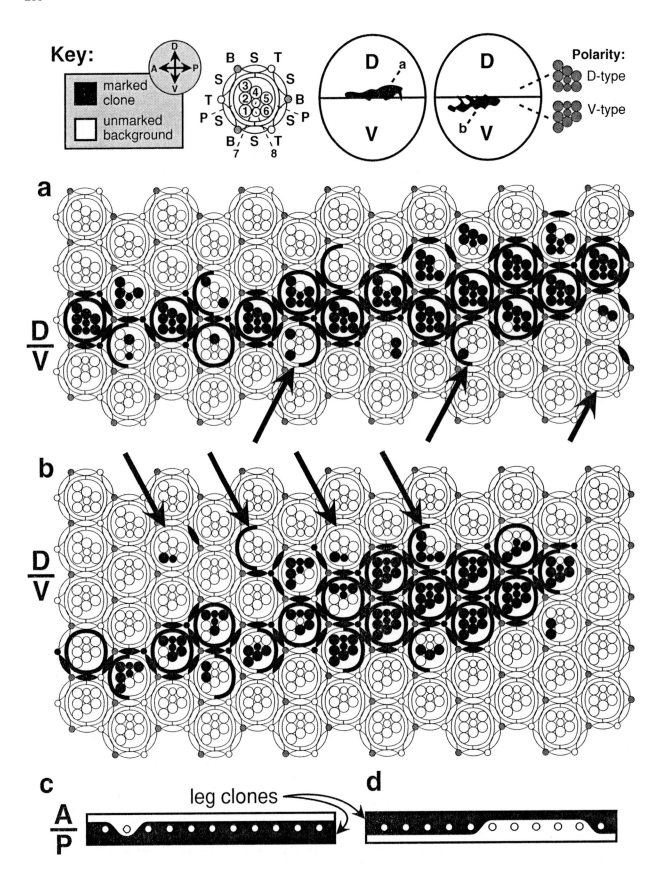

seven or eight receptor axons. Each neuro-ommatidium in this equatorial stripe can thus be expected to have an increased absolute sensitivity. [1720]

Unlike the bristle, the ommatidium is not a clone

A priori, the crystalline perfection of the ommatidial lattice and the geometric precision of its attendant trapezoids seem to demand an equally regimented process of assembly. One obvious solution would be a cell-lineage mechanism. An ommatidial "mother cell" could undergo a fixed number of differentiative mitoses to produce each of the needed cell types, just as bristle SOPs divide to make the 5 cells of the bristle organ (cf. Fig. 2.1). It was therefore not surprising when, in 1937,

Francis Bernard reported histological evidence (albeit circumstantial) for such a clonal mechanism in insect eyes [298]. "La rétinule" is the photoreceptor array: R1–R8.

Sur un matériel favorable, à savoir la nymphe de *Formicina flava* , j'ai pu retracer la filiation cellulaire des éléments de l'ommatidie: il y a, dès les premières divisions de l'ébauche hypodermique, séparation très précoce entre un groupe nucléaire interne et un groupe externe. La série interne conduit à la rétinule et aux cellules pigmentaires qui l'entourent, la série externe conduit, plus tardivement, aux cellules cornéennes et cristalliniennes. Ces faits n'apparaissent pas dans le texte des précédents auteurs, bien que les figures de Moroff (1912), relatives à l'ébauche optique de *Palaemon*, contiennent implicitement les mêmes séries nucléaires, dont la généralité reçoit ainsi un début de confirmation. [298]

FIGURE 7.2. Lineage restrictions at the eye's equator. Until recently, the eye disc was thought to lack a D/V compartment boundary because clones can cross the equator. However, such transgressions are trivial (**a, b**), and similar trespassing occurs at *bona fide* compartment lines in other discs (**c, d**). In both cases, the "noise" is attributable to patterning events that are superimposed upon – but only loosely linked to – the boundary. The key (top) shows (left to right): clone vs. background (black/white) markings; axes (D-V, dorsal-ventral; A-P, anterior-posterior); ommatidial template (B = 5-cell bristle organ; P, S, T = primary, secondary, and tertiary pigment cells; 1–8 = photoreceptor R1–R8 cells); conjectured locations of clones (enlarged in **a** and **b**) within their respective eyes (ovals); and trapezoid orientations in the D vs. V halves of the eye. Within the lattice, the ommatidial "repeat unit" has 23 cells, counting each bristle as 5 cells and excluding the cells that each ommatidium shares with its neighbors [3539].

a, b. Clones of cells (black) that are homozygous for *white*[LOF] as a result of somatic recombination induced by X-rays in late 1st-instar heterozygotes [3539] when the eye rudiment has ~20 cells. Because an adult eye has ~750 ommatidia and only one of the two cross-over segregants becomes homozygous [1695], a single cell that undergoes recombination should leave marked descendants in $750 \times \frac{1}{20} \times \frac{1}{2} \approx 20$ ommatidia. Indeed, each of these clones covers ~20 ommatidia, but most are mosaic. The mosaicism shows that the ommatidium is not a clone. In the living fly, each clone was a ragged white stripe across an otherwise red (left) eye (cf. key). Except for bristles (shaded), all the cell types shown are scorable phenotypically. The equator (D/V line at left) is defined by the symmetry planes of the columns (13 columns per eye are shown). The R1–R8 trapezoids in each column always point dorsally above a certain level and ventrally below it [2794], so that no D-type ommatidia ever get "stranded" in the V half or vice versa [4797]. The equator zigzags through the lattice, jogging by one ommatidium from column to column. Normally it alternates – jogging up then down (sawtooth mode) – but in places it jogs one or two rows in the same direction (staircase mode) before returning to the baseline [2794, 3539, 4715]. **a.** Clone that evidently originated in the D half but which includes a few cells in 3 adjacent V-type ommatidia (arrows). **b.** Clone that must have arisen in the V half but which includes some cells in 4 D-type ommatidia (arrows). All but one of these 7 ommatidia (rightmost arrow in **b**) have a majority of their cells from the "home" compartment. This same trend is seen in *Minute* mosaics where the marked clones (induced at various ages) have a growth advantage [633].

c, d. Clones (black) marked with *yellow*[LOF] on two left 2nd-leg basitarsi (proximal-distal axis is left to right). Only one row of 11 bristles (circles) – row 1 – is depicted. This row resides at the A/P compartment boundary. Its bristles can come from either compartment, depending on the individual leg. **c.** All bristles, except the second one, must have arisen from P cells because they are embraced by a clone of P provenance. **d.** Several of these same bristles must have arisen from A cells because they are in a clone of A provenance. Such vagaries probably reflect the stochastic way that SOPs are selected in proneural fields (cf. Fig. 3.9).

Panels **a** and **b** are redrawn from [3539]; panels **c** and **d** are redrawn from [1800] (6th leg in his Fig. 3b and 4th leg in Fig. 3a). In the key, the locations of the clones along the A-P axis (arbitrarily sited here) were not actually reported in the original paper. In **c** and **d**, the contour of the clone boundary through the background epidermis (drawn straight) is not actually known because the *yellow*[LOF] mutation does not affect ordinary cuticle.

N.B.: Clone edges (**a, b**) do not look this ragged in 3rd-instar discs, especially near the D/V border [696]. The choppiness seen here in the adult is due, at least in part, to 90° rotation of ommatidial clusters in late-3rd instar (cf. Fig. 7.5a–c), which must detach closely related cells from one another at cluster perimeters by breaking junctions. There is no A/P compartment boundary in the eye itself, which lies entirely in the A region (cf. Fig. 7.1) [4146].

Bernard's "**Ommatidial Lineage Hypothesis**" prevailed until 1973, when it was disproven in *Drosophila* by Seymour Benzer and his collaborators [292]. Benzer had been a hero in the Phage Genetics Era of molecular biology [291, 610, 1884, 4083, 4560], and he would go on to become a patriarch of modern fly genetics [4578]. In a popular article for *Scientific American* (1973), Benzer sketched their essential findings. Their refutation of the lineage model was based on somatically marked clones – a lineage-tracing method that is more accurate than histology alone.

Are the eight photoreceptor cells derived from one cell that undergoes three divisions to produce eight, or do cells come together to form the group irrespective of their lineage? This can be tested by examining the eyes of flies, mosaic for the *white* gene, in which the mosaic dividing line passes through the eye.... The result is clear: a single ommatidium can contain a mixture of receptor cells of both genotypes. This proves that the eight cells cannot be derived from a single ancestral cell but have become associated in their special group of eight irrespective of lineage. The same conclusion applies to the other cells in each ommatidium, such as the normally heavily pigmented cells that surround the receptors. [292]

A full account of the data came three years later in a *Developmental Biology* paper with the catchy title, "Development of the *Drosophila* retina, a neurocrystalline lattice" [3539]. Authored by Donald Ready, Thomas Hanson, and Benzer, this article was remarkable not only for its rejection of the clonal model [2128, 3233] but also for its pithy questions about eye development in general. It launched a brute-force mutational dissection of eye patterning that is arguably the greatest success story in modern genetics [293, 1048, 4471, 4715].

Despite its seminal contributions, however, the Ready *et al.* article had one unfortunate side effect. Like Bernard's paper it disseminated a myth of its own – viz., that the eye lacks D and V compartments (italics are author's) [3079].

The shape of a marked clone ... depends somewhat upon its position in the eye. The dorsal or ventral edge of a patch often runs horizontally near the middle of the eye. This suggests the possibility that the descendants of early cells tend to populate either the dorsal or ventral halves of the eye (Becker, 1957). ... [However,] while the cells of a white clone tend, by and large, to form a continuous patch, some marked cells lie removed from the main body of the clone. Such outlying cells are rarely more than one ommatidium removed from the main patch. ... Thus, although the equator often lies near a clonal boundary, *it is not determined by such a boundary.* [3539]

The history of this topic deserves scrutiny because it shows how paradigms can shift with the tides of opinion, not just with the facts on which the opinions are based.

The eye has D and V compartments (despite doubts to the contrary)

The equator was discovered by Wilhelm Dietrich in 1909 [1059]. In 1957, Hans Becker found that marked clones tend to respect the equator [260], and a similar conclusion was reached in 1967 by William Baker [188]. The exact path of the restriction line was uncertain, however, because neither Becker nor Baker assessed the orientations of the trapezoids [2128].

When A/P and D/V compartments were revealed in the wing disc in 1973 [1376], the eye's previously known D/V restriction boundary acquired, at least potentially, a new meaning. The suspicion was that the eye's D/V line might be acting like the wing's D/V border. This "**D/V Compartment Hypothesis**" was appealing because it offered the prospect of a universal coordinate system for discs.

In 1978, the "*Minute* technique" [2935] was used to assess whether this line behaves like a compartment boundary under "pressure" from overgrowing clones. Indeed, it does [189]. That is, it is not transgressed by faster-growing *Minute*[+] clones in a *Minute*[LOF] background. More important, clonal boundaries in the *Minute* mosaics were found to track the equator closely in the trapezoid array [633, 1080]. From one eye to the next, the discrepancy is typically no more than ±1 row of ommatidia (Fig. 7.2). That is, D clones "spill over" the equator by no more than one V-pointing trapezoid (and vice versa).

This degree of "noise" might seem negligible, but Ready *et al.* [3539] and others [1804, 1869, 2128, 4715] viewed *any* discrepancy as a sufficient reason to reject the hypothesis. At that time, this attitude seemed reasonable because all the then-known compartment lines were sharp and inviolate. The skepticism of these researchers was based on other facts as well [696]. The full list of attacks on the D/V Compartment Hypothesis is given below:

1. *"Trespass" Objection.* The ability of marked clones to cross the equator (even slightly) argues against the D/V line being a canonical compartment boundary. Shouldn't the edges of such clones be in perfect register with the equator [3539]? The lack of a D/V

boundary in the leg [1800, 4076] lent added plausibility to the notion that the eye might also lack such a boundary [1807].

2. *"Mosaicism" Objection.* The ability of marked clones to meander through the ommatidial array without respecting the boundaries between ommatidia [3539] ruled out a clonal mechanism (see above). It also seemed to preclude any strict D-vs.-V dichotomy. If ommatidial polarity is due to a difference in gene expression, then shouldn't ommatidia behave as units?

3. *"Selector Gene" Objection.* No mutations were known that convert one half of the eye into the other, although such a phenotype would be hard to detect in an ordinary mutant screen. Shouldn't the eye possess a selector gene for the D-V axis [633]?

The Trespass Objection was undermined in 1979, when a similar transgression was found at the A/P boundary of the leg [1800, 2449]: the bristles of tarsal row 1 can be embraced by either A or P clones (Fig. 7.2c, d). This variability from one fly to the next is attributable to the stochastic way that SOPs are selected. The proneural field for row 1 straddles the A/P line, so its SOPs can arise on either side (cf. Fig. 3.9). If bristle rows "don't care" about the compartmental provenance of their cells, then maybe ommatidia don't either. In that case, ommatidia would arise from "fields" that overlap the D/V line, and D/V-mixed ommatidia would still somehow manage to point dorsally or ventrally, rather than freezing at an intermediate angle.

The Mosaicism Objection has withered as we have learned more about how ommatidial polarity is controlled. We now know that ommatidia do not orient themselves based on the D-vs.-V states of their component cells. If that were true, then mosaic trapezoids (with foreign cells from across the equator) should point at intermediate angles, but they obviously do not (Fig. 7.2). Rather, polarity is superimposed nonautonomously on ommatidia by a signal [4366] that diffuses from a line between *mirror*-ON and *mirror*-OFF cells. It is thus possible for a trapezoid to straddle the *mirror*-ON/OFF line and still adopt a purely D or V orientation. The way in which this happens is discussed shortly.

The most serious criticism of the D/V Compartment Hypothesis was always the lack of a D/V selector gene in the eye (like *apterous* in the wing; cf. Ch. 6) [625]. One early indication that such a gene might exist came from *Ophthalmoptera* mutants. In these hideous flies, the

eyes are replaced by wings, and the margins of the wings grow out from the equator [3439] – precisely the site expected if the D and V parts of the eye and wing discs are homologous [696]. Another hint came from transplantation experiments. When fragments of the eye disc were transplanted to host larvae, an equator formed only from pieces that contained both D and V tissue [629]. Pieces with only D or V tissue made ommatidia with disordered polarity. Evidently, ommatidial polarity depends on the equator, which, in turn, depends on an interaction between D and V tissues.

The Iroquois Complex controls D-V polarity via Fringe and Notch

As mentioned in Chapter 6, the Iroquois Complex (Iro-C) has 3 homeobox genes: *araucan* (*ara*), *caupolican* (*caup*), and *mirror* (*mirr*). Their DNA sequences were described in 1996 (*ara* and *caup*) [1536] and 1997 (*mirr*) [2794]. All three genes are transcribed in the D half of the eye rudiment [483, 2170, 2794, 4210], beginning as early as late-1st/early-2nd instar (Fig. 7.3) [696, 2170]. Intriguingly, the ON/OFF boundary of Iro-C expression is a straight line that closely tracks the equator (± a few cell diameters) [696, 778, 2794].

In 1999, Florencia Cavodeassi *et al.* (in Madrid and Cambridge) showed that the Iro-C exhibits many of the properties expected for a D/V selector gene [696]. This evidence swept away the last remaining objection to the D/V Compartment Hypothesis. Hence, the eye does develop like a wing along its D-V axis (Fig. 7.4) [359]. This **"Iroquois Epiphany"** of 1999 has led to a rethinking of the rules for large-scale patterning in the eye.

1. Dorsal (but not ventral) Iro-C^{null} clones induce an ectopic equator at their edge. Wild-type trapezoids as far as 7 ommatidial rows away from the extra equator are repolarized (nonautonomously) so that they point away from this new line (Fig. 7.4) [696]. Weaker effects are seen near *mirr*^{LOF} D clones (which remain *ara*$^+$ *caup*$^+$) [2794, 4797], as well as with *fng*^{LOF} V clones [774, 1080, 3258], N^{GOF}, Dl^{GOF}, or Ser^{GOF} clones (see below) [3258].

2. Ectopic (GOF) expression of *ara*, *caup*, or *mirr* in a subset of the V region causes similar repolarization [696, 2794].

3. Dorsal (but not ventral) Iro-C^{null} clones have round shapes with smooth edges [696, 697], as do *mirr*^{LOF} D clones [4797]. This trait mimics null clones for D/V (*ap*) and A/P (*en*) selector genes in the wing disc

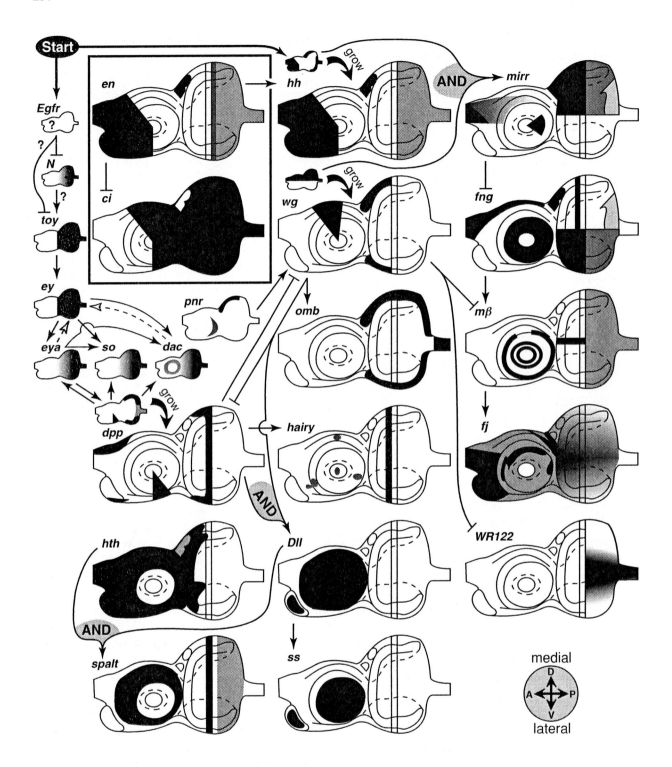

(cf. Ch. 6), and is presumably due to differences in cell affinities [2792].

4. Iro-C^{null} (but not wild-type) clones can cross the equator from D to V [696], again implying an effect on regional affinities.

5. A selector role is hard to prove in the eye because its D and V ommatidia look identical except for ori-entation. Stronger proof comes from the head cap-sule where (1) GOF misexpression of *caup* induces D-type bristles in the V area [1080] and (2) Iro-C^{null} (LOF) clones manifest V-type structures in the D area [697].

If Iro-C endows cells with D-vs.-V identities and those identities dictate ommatidial angles, then Iro-C^{null}

FIGURE 7.3. Patterns of gene expression in the eye-antenna disc during 3rd instar. Smaller drawings indicate earlier stages. Black areas chart mRNA (transcribed from the endogenous gene or a reporter). Shades of gray denote degrees of expression. Directions are shown in the compass at the lower right: A-P (anterior-posterior) and D-V (dorsal-ventral) refer to prospective axes of the adult, whereas "medial" and "lateral" denote directions within the larval body.

Genes ("DO" = details omitted; "ambiguous" means that conflicting reports have here been summed into a maximal-domain composite): *ci* (*cubitus interruptus*) [1135], *dac* (*dachshund*; early-3rd instar) [2353]△, *Dll* (*Distal-less*) [1085, 3242]△, *dpp* (*decapentaplegic*; above is early-3rd instar [930]△) [2004]△, *en* (*engrailed*) [4168]△, *ey* (*eyeless*; early-3rd instar) [930]△, *eya* (*eyes absent*; early-3rd instar) [2353]△, *fj* (*four jointed*; antennal expression is approximate) [4851]△, *fng* (*fringe*; ambiguous) [696]△, *hairy* [1616]△, *hh* (*hedgehog*; above is 2nd instar [1082]) [406]△, *hth* (*homothorax*) [1085, 3380]△, *N* (*Notch*; 2nd instar) [2353], *mβ* (*E(spl)mβ*) [871]△ (but see [991]), *mirr* (*mirror*; ambiguous) [4797]△, *omb* (*optomotor-blind*) [4849], *pnr* (*pannier*; mid-3rd instar) [2752], *so* (*sine oculis*; early-3rd instar) [2353]△, *spalt* [1085]△, *ss* (*spineless*) [1119], *toy* (*twin of eyeless*; 2nd instar) [2353]△, *wg* (*wingless*; above is early-2nd instar [696]) [3762]△, and the enhancer trap *WR122-lacZ* [1781].

Genes omitted (expression patterns is in parentheses): *araucan* and *caupolican* (≈ *mirr*) [3380]△, *extradenticle* (Exd is nuclear wherever Hth is present) [677], *orthodenticle* (ocellar region and ommatidia) [3663]△, *vein* (just posterior to the Hh ocellar spot) [80]. Annular expression domains of *BarH1* (not shown) and *dachshund* in the antenna appear identical to corresponding tarsal domains (Fig. 5.11) [3762].

Links (➔ activation; ⊣ inhibition) indicate genetic interactions; "AND" means that both inputs are needed to activate the target gene; dashed lines denote weaker effects. The flow of control begins at upper left in 2nd instar. Whether Egfr inhibits *toy* directly or via N is unknown [2353]. Egfr is probably activated in the antenna (cf. Fig. 6.9) [2353, 2362, 3382]. The antenna resembles the leg (cf. Fig. 5.8) insofar as "En ➔ Hh ➔ {*dpp* OR *wg*}" [1037]. For the eye's D-V axis, polarity is established by *mirr* (upper right; cf. Fig. 7.4). Wg may not actually inhibit *dpp* at the level of transcription [4390]. A "*dac* ➔ *eya*" link detected elsewhere does not work in the eye [554]. The "AND" logic for the early eye genes (middle left) has been omitted for clarity. For an exegesis of this nexus, see [1787]. Circuitry of these genes is based on [744, 930, 1780]. For ideas about how this network evolved, see [1419, 2034, 3895]. For enhancer-trap patterns, see [315, 3388, 4210] and *FlyView* (*flyview.uni-muenster.de*) [2027], and for expression of *hh*, *wg*, and *dpp* in the peripodial membrane, see [773].

N.B.: In comparing these expression patterns with those in the leg and wing discs (Figs. 5.8 and 6.2), it is important to recall that the eye disc rotates ~180° after emerging from the embryonic ectoderm [4146], thereby inverting its D-V axis relative to the D-V axes of the thoracic discs (cf. Fig. 4.3a).

---

clones in the D half of the eye should flip trapezoids from D- to V-type orientation. In fact, the trapezoids within such clones retain a D-type posture (Fig. 7.4), and inversions only occur nearby (point 1 above). This result suggests that ommatidia obey the following two rules:

Equator Rule: If you reside where Iro-C-ON and -OFF cells meet, then become an "equator-type" cell (nature unknown) and emit a diffusible signal.

Polarity Rule: If equator-type cells are nearby, then point your trapezoid away from them.

How might such rules be implemented? Conceivably, the eye's D-V axis operates like the wing's, with the Iro-C substituting for *apterous* [359]. With this new amendment, the D/V Compartment Hypothesis predicts that

1. A band of *wg*-ON cells should arise at the D/V line via the Fng-N-Dl-Ser (Fringe, Notch, Delta, Serrate) cassette.
2. Wg would diffuse away in both directions.

3. Trapezoids could then orient themselves relative to the slope of the gradient.

One aspect of this hypothesis does ring true. Trapezoid polarity is affected by some of the same Wg-pathway genes that control the planar polarity of bristles [24, 276, 1449, 3630], wing hairs [1638–1640], and leg joints [1641, 1810, 4716]: *arrow* (LOF) [4573], *fz* (LOF or GOF) [3912, 4170, 4365, 4873], and *dsh* (LOF or GOF) [367, 4278, 4365, 4573, 4873]. Similar changes can be achieved by genes in the branch of the pathway that is normally uncoupled from polarity [2883, 3563], although those alterations may be indirect [4573]: *arm* (LOF or GOF) [4573] and *sgg* (GOF) [4365, 4573]. However, Wg is not expressed at the equator [2631, 4390], so another Wnt may be dictating tissue polarity [25, 359, 2781, 4263]. Alternatively, the non-Wnt protein Four jointed (Fj) could be the eye's D/V polarizer [447, 483] (cf. the wing [4852]). Not only is Fj expressed in the expected gradient profile (Fig. 7.3) but it also seems to integrate inputs from all three pathways that are implicated in equator induction [4173]: Wg [4573], Notch [3258], and JAK-STAT [2618, 4851].

The notion that the eye uses the same Fng-N-Dl-Ser device as the wing was experimentally confirmed by

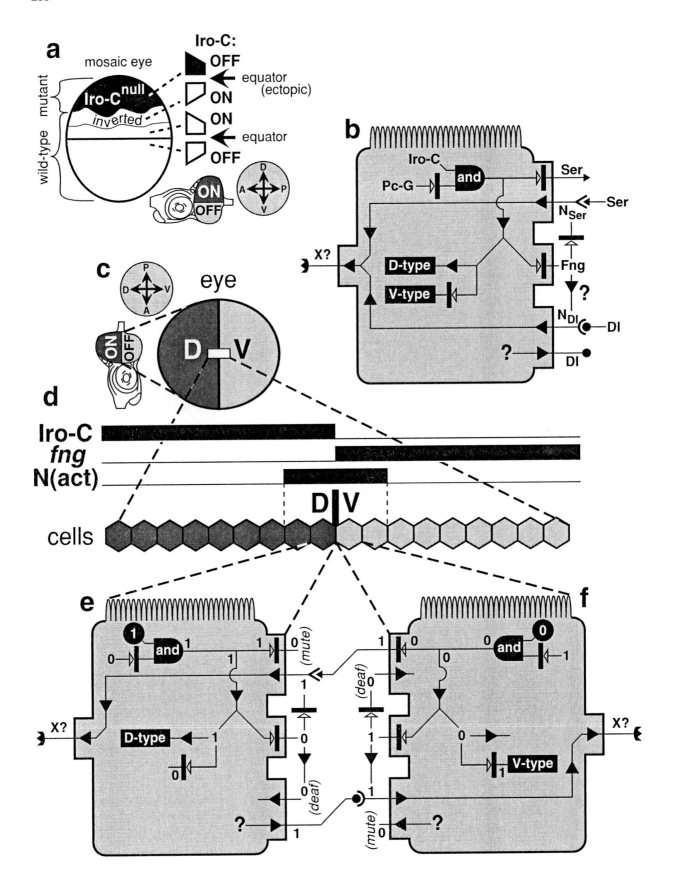

a spate of articles in 1998 [774, 1080, 3258]. The chief difference is that *fng* is regulated negatively (Iro-C ⊣ *fng*), instead of positively (Ap ➔ *fng*).

Link 1: Iro-C ⊣ *fng*. Within the eye anlage, *fng* is mainly expressed ventrally [3258], starting in 1st instar [774]. This expression vanishes when *caup* [1080] or *mirr* [774, 4797] is activated widely (GOF test). When Iro-C^{null} clones arise in the D area, these cells derepress *fng* [696] (LOF test; cf. similar effects in *mirr*^{LOF} discs [4797]). The derepression of this V marker affirms that Iro-C is a D-vs.-V selector switch. The downstream status of *fng* is supported by the fact that *mirr* must act via *fng* to induce equators: extra equators form around *mirr*^{LOF} D clones (which have a *fng* ON/OFF edge) but not around *mirr*^{LOF} *fng*^{LOF} D clones (which do not) [4797]. Mirr's ability to repress tar-

get genes may be due to recruitment of a chromatin-remodeling co-repressor [941].

Link 2: *fng* ON/OFF boundary ➔ N-act. Notch is normally activated (N-act form) at the D/V boundary because (1) the Notch target gene *mβ* is expressed there as early as 2nd instar [1080], and (2) a reporter gene with Su(H) binding sites is likewise activated there [1080]. Ubiquitous expression of *fng* or *caup* (via *eyeless-Gal4*) stifles eye development [774, 1080, 3258]. Eyes can be rescued by jointly expressing N-intra – Notch's constitutively active intracellular fragment [1080]. The *fng* ON/OFF boundary is essential for eye development from late-1st to mid-2nd instar [774]. As in the wing, Fng appears to disable N's reception of Ser signals and to enable reception of Dl signals [3245, 3258] – a bias that limits N-act to the D/V border [1988].

FIGURE 7.4. Control of ommatidial polarity by the Iroquois Complex (Iro-C).

a. Iro-C genes are expressed in the D, but not V, half of the eye field during normal development (disc schematic below). The equator (horizontal line) arises at this ON/OFF boundary, and all photoreceptor trapezoids (cf. Fig. 7.2) point away from it. When an Iro-C^{null} clone (black) is induced in the D area, it causes wild-type trapezoids within ~7 rows to invert their polarity. (The eye is ~33 rows high.) Hence, an ectopic equator emerges at the clone edge.

b. Generic eye cell (cf. Fig. 2.7 for icons) showing the main control circuit for emitting an unknown diffusible factor (X) that regulates polarity. Abbreviations: Dl (Delta), Fng (Fringe), N (Notch; "act" = activated form), Pc-G (Polycomb-Group genes), Ser (Serrate). N$_{Ser}$ and N$_{Dl}$ are hypothetical Fng-modulated forms of Notch that are receptive to Ser or Dl [990, 1250, 1988]. The circuit works as follows. If Pc-G products are sufficiently low, then Iro-C genes can be expressed (**e**) [3079]. In that case, the cell adopts D-type identity and expresses Dl (by an unknown route) but not Ser or Fng. D-type cells can receive Ser signals (from across the D/V line). If Iro-C genes are OFF (**f**), then the cell adopts V-type identity and expresses both Ser and Fng. In that case, Fng blocks the Ser-N interaction, so V-type cells can only receive Dl signals. The circuit allows two types of dialog: Ser-signal "speakers" with N$_{Ser}$ "listeners" or Dl-signal "speakers" with N$_{Dl}$ "listeners". Both conversations should turn ON Factor X. Factor X uses Frizzled as a receptor (cf. Fig. 7.5) [869, 1194, 4366].

c. Eye portion of a mature left eye disc. Shading denotes D (dark; Iro-C = ON) vs. V (light; Iro-C = OFF) compartments. Note that this eye was rotated 90° counterclockwise relative to the eye in **a** (cf. compasses and disc icons) to facilitate comparison with the wing disc's circuit diagram (Fig. 6.8).

d. Enlarged part of the D/V border (box in **c**). In this single file, cells on the D side express Iro-C genes, while cells on the V side express *fng*. Interactions between cells in **e** and **f** activate Notch at the D/V boundary. The width of the N(act) stripe (here shown spanning 4 cells) is not known exactly.

e, f. Schematics of cells flanking the D/V line. The V cell is drawn as if flipped around to face the D cell. States of variables (cf. **b**) are recorded as "1" (present and active) or "0" (absent or inactive). Black circles indicate determining factors. "Mute" and "deaf" denote inability to send or receive particular signals (Ser above, Dl below). The functioning of the circuit is explained under **b** above.

Panel **a** is a schematized adaptation of a photo in [696]. Circuitry in **b** is compiled from [696] (Iro-C), [3079] (Pc-G), and [774, 869, 1080, 1194, 3258] (Fng-N-Ser-Dl). Panel **d** incorporates data from the same sources.

N.B.: The D/V circuitry of the eye differs from that of the wing insofar as *fng* is expressed ventrally instead of dorsally (cf. Fig. 6.8). This difference is attributable to the 180° rotation that the eye disc undergoes after its inception [774, 4146] (cf. Fig. 7.1a legend). The "Iro-C ⊣ Ser" link is simplified from the "Iro-C ⊣ Fng ➔ Ser" route [696, 774, 3258], and the full path is probably even more complex. This model does not explain how equators arise from ectopic furrows that are triggered without altering Iro-C expression [722, 723, 4167, 4648]. Other key players in this scenario may be (1) *Lobe* (*Lobe*^{LOF} causes *mirr* to turn ON in the V half and hence suppresses ommatidia [758, 2560]) and (2) *teashirt* (*tsh*^{LOF} suppresses ommatidia in the D half) [3973].

Link 3:   Fng ➔ *Ser*. Like *fng*, *Ser* is expressed in the V half of the eye region [774, 1080, 3258], more intensely at the D/V border as development proceeds. *Ser* turns OFF in *fng*^{null} V clones [774] and turns ON in *fng*^{GOF} D clones when the latter are near the D/V line [3258].

Link 4:   Iro-C ➔ *Dl*? *Dl* is expressed in the D half of the eye region, most intensely at the D/V border [774, 1080]. No link with the Iro-C or any other upstream gene(s) has yet been shown.

Upstream of the Iro-C itself, several links are known to operate throughout most of larval life:

Link 5:   Wg ➔ Iro-C. In early 2nd instar, *wg* is expressed coincidentally with *ara-caup* in the D half of the eye rudiment [696]. Later this *wg*-ON zone shrinks to the edge of the D realm, while *ara-caup* persists throughout it (Fig. 7.3). When the Wg pathway is shut OFF by *dsh*^{LOF} clones during either period, expression of *ara* and *caup* ceases in the clones [696]. When the pathway is ectopically turned ON by *sgg*^{LOF} clones, *ara-caup* turns ON (except in the V region) [696]. Wg likewise controls *mirr*: *mirr* expression disappears from the D half of *wg*^{LOF} discs and arises ectopically in the V half in *wg*^{GOF} (*dpp-Gal4:UAS-wg*) discs [1781].

Link 6:   Hh ➔ Iro-C. Like *wg*, *hh* is expressed within a progressively shrinking D domain [696]. When the Hh pathway is shut OFF by *smo*^{LOF} clones, *ara* and *caup* turn OFF in the center of the clone [696]. When the pathway is constitutively activated by *ptc*^{LOF} clones that are induced in 2nd instar, *mirr* turns ON inside the clone and in nearby cells, although clones induced at later times have no such effect [696].

The investigations that established the above links also discovered something odd about *fng*: homozygous *fng*$^+$ clones affect ommatidial polarity in a heterozygous *fng*$^+$/*fng*^{null} background [1080]. This finding implies that differences in dosage (2 vs. 1 copy of *fng*$^+$) suffice to tilt ommatidia in one direction vs. another [4762]. Ergo, an analog step must exist somewhere in the polarity circuitry. That step was revealed in 1999 [869, 1194]. It also uses Notch, but in an entirely different way from

the equator module [359]. To explain how this gadgetry works, it is first necessary to briefly outline the process of ommatidial maturation.

A morphogenetic wave creates the ommatidial lattice

During 3rd instar, which lasts 2 days, a wave of cell differentiation sweeps across the eye disc epithelium from posterior to anterior. At its leading edge is a D-V groove called the "morphogenetic furrow" (MF) [3539]. Ahead of the MF are undifferentiated cells and scattered mitoses. Behind it are columns of nascent ommatidia. Columns emerge at a rate of one every ~1.5 h initially, but the rate accelerates to one per hour when the MF reaches the anterior region [228, 4715]. A mature disc has ~26 columns, and ~7 more are added during the first 10 h of the pupal period to complete the lattice [602, 1287, 4715]. The discussion here and in the next few sections focuses on the individual ommatidia. The way in which the lattice as a whole arises is considered subsequently.

Each incipient ommatidium undergoes a stereotyped series of changes after it leaves the MF (Fig. 7.5b) [186, 4355, 4364, 4712, 4715]. In the stages listed below, all columns are numbered starting at the MF (0) and counting posteriorly, with precursor cells being denoted by a "p" suffix:

1. **Rosette** (column 0): A ring of 10–15 cells around a core of 4–5 cells appears within the MF.
2. **Arc** (columns 1 and 2): The rosette's anterior cells are thought to return to the population of undifferentiated cells, leaving the posterior ones as a distinct arc. At the center of this chain is R8p, which is flanked by R2p and R3p on one side and R5p and R4p on the other, with "mystery" cells at the extremes [4366].
3. **Precluster** (columns 3 and 4): The arc zippers shut, so that R3p and R4p touch for the first time. Besides R2–5p and R8p, the incipient precluster retains a few mystery cells that merge into the background after R1p and R6p are added.
4. **Cluster** (column 5): Like R1p and R6p, R7p is recruited from cells that have just finished dividing in a "mitotic band" at ~ columns 3 and 4 [185, 625, 629, 1869, 3539, 4355]. At about this time, the clusters start to rotate (clockwise or counterclockwise) [4873].

Consistent with the above sequence, the first cell to express neural antigens is R8p, followed by the R2/5p pair, R3/4p pair, R1/6p pair, and finally R7 [4361, 4364, 4715]. Cone cells are incorporated just after the Cluster Stage, whereas pigment cells (primary, secondary, tertiary) and bristles are added later (after pupariation) [602, 605, 1287, 1289, 4355, 4715].

Until about the Cluster Stage all nascent ommatidia are bilaterally symmetric [4355], and R3p and R4p are equivalent in all discernable respects. Now these cells diverge: R4p loses its attachment to R8p, but R3p does not [4355, 4364, 4873], and both cells undergo a shift (dorsal in the disc's D half and ventral in the V half). Henceforth, each cluster has a handedness ("chirality") that dictates its direction of rotation [359, 1639]. Rotations occur in mirror symmetry so that D vs. V trapezoids later point in opposite directions (Fig. 7.5) [2883, 4715].

D-V polarity depends on a rivalry between R3 and R4 precursors

R3p and R4p are instrumental in steering the entire ommatidium one way vs. the other. This conclusion was reached by analyzing various kinds of *frizzled* (*fz*) mosaics. In virtually every case, the R3/4p cell that becomes R3 (the trapezoid tip) is the one with greater Fz activity, regardless of the disposition of genotypes in the other ommatidial cells [4366, 4873]. This rule was also obeyed regardless of absolute Fz levels. Thus, the same outcome was seen when the high-vs.-low discrepancy was created by (1) fz^+ vs. fz^{null}, (2) two copies of fz^+ driven by a *sevenless* (*sev*) enhancer vs. fz^{null}, or (3) two copies of *sev*-driven fz^+ vs. fz^+.

Similar studies with *Notch* and *Delta* mosaics affirmed the key role of the R3/4p pair. An uncommitted R3/4p cell can be forced to become R3 by suppressing its Notch pathway (via a chimeric Su(H)-Engrailed repressor) [4366]. Conversely, activating the Notch pathway (via a *sev*-driven N-intra) enforces the R4 state [4366], as does "muzzling" the cell by making it Dl^{null} while leaving its rival alone (Dl^+) [1194, 4366]. Evidently the two R3/4p cells compete. The one that emits more Dl becomes R3.

Epistasis experiments show that the Notch pathway acts downstream of the Fz receptor [869, 1194, 4366]. In fz^{null} flies whose R3/4p cells carry 2 or 3 doses of Notch, the R3/4p cell with 3 doses invariably trumps its 2-dose rival and becomes R4 (48/48 cases) [4366]. Amazingly, this factor of 1.5 suffices to decide the R3-vs.-R4 contest with ~100% fidelity. Evidently, a small analog imbalance can be amplified to produce an all-or-none binary output. The amplification is thought to involve a positive feedback loop ("N-act" is the activated N receptor): "Dl➔ N-act" between cells and "N-act ⊣ Dl" within cells (Fig. 7.5g) [869, 1194, 1195, 4366]. The net effect of this loop is that the more one cell signals, the more the other cell

is prevented from signaling [1613]. The dialog becomes a monolog, with one cell as the speaker and the other as the listener. A similar "**Delta-Notch Flip-Flop**" was invoked in the Mutual Inhibition Model of bristle patterning (cf. Fig. 3.6). Serrate is excluded from both circuits [1194, 4366, 4859].

In normal development, the R3/4p cell that is closer to the equator becomes R3. Putting this fact together with the above clues suggests the "**Scalar Model**" for ommatidial polarity (Fig. 7.5e) [359, 869].

1. Confrontation of Iro-C ON and OFF cells activates the Notch pathway (via the Fng-N-Dl-Ser module) in a stripe of equatorial cells: Iro-C-ON/OFF ➔ N-act.
2. The equatorial cells emit a diffusible molecule (factor X): N-act ➔ X.
3. Diffusion of X away from the equator forms a gradient in each half of the eye.
4. Fz senses the X ligand and represses Notch proportionally. Repression could be mediated by Dsh [869, 1194], which acts downstream of Fz [152, 422, 2326]. Dsh can bind and block the Notch receptor [151]: X ➔ Fz ➔ Dsh ⊣ N-act. The bias may also be due to a Notch-independent effect on *Dl* transcription [1194, 1195]: Dsh ➔ RhoA ➔ JNK ➔ dJun ➔ *Dl* [423, 2781, 4170, 4566]. Other agents (of uncertain linkage) include Dachsous [21], Expanded [367], Misshapen [3261], Prickle [1641], Starry night (a.k.a. Flamingo) [704, 4430], and Van Gogh (a.k.a. Strabismus) [25, 4716].
5. Within each cluster, the two R3/4p cells compete via the *trans*-cellular circuit "Dl ➔ N-act ⊣ Dl". Whichever cell expresses more Delta – and activates its Notch pathway less – "wins" and becomes R3. The "loser" becomes R4. Van Gogh is required for a cell to adopt the R4 fate [4716].
6. Within each pair of R3/4p cells the cell closer to the equator wins due to the bias from Step 4.
7. R3/4 asymmetry dictates each ommatidium's chirality and direction of rotation (but see [4716]).

One obvious question is whether Step 1 interferes with Steps 4 and 5, because they all use the same Notch pathway in different ways. Interference at the equator is apparently avoided by finishing Steps 1–4 before Steps 5–7 begin: the Notch target gene *mβ* is ON at the equator ahead of the MF, but is diffusely expressed behind it [871, 1080].

The bias introduced in Step 4 relies on the "decoding" logic of a canonical gradient model. That is, the amount of morphogen is measured and "interpreted" as a certain output (Fig. 7.5e). The initially large distance

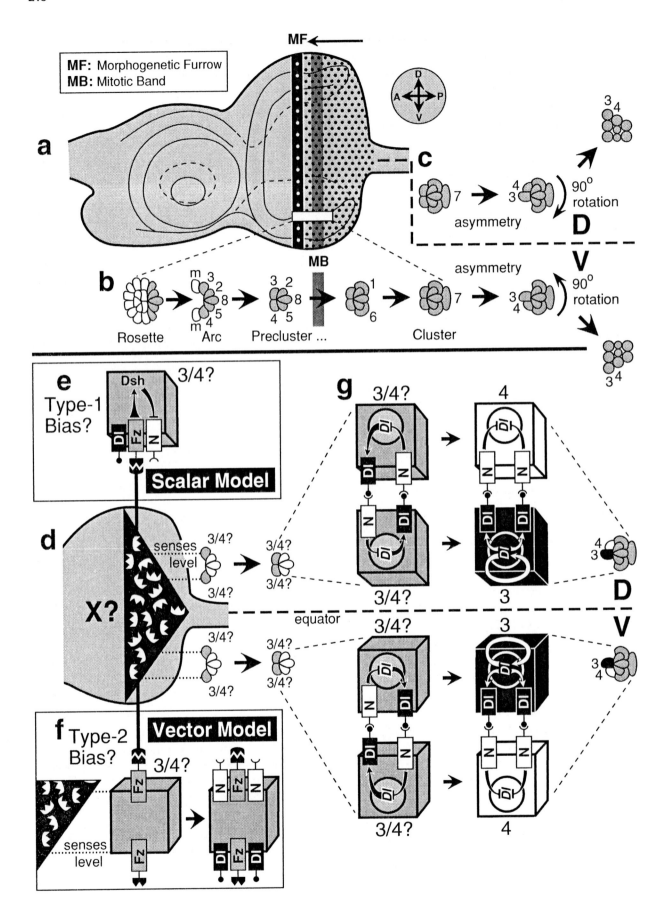

between the R3/4p cells (Arc Stage) would allow them to gauge local differences in X levels at a high signal-to-noise ratio [3563]. The levels could then be recorded in some more permanent "memory" before the two cells close the gap, touch, and interact (Precluster Stage).

Alternatively, Step 4 might use a different sort of bias (Fig. 7.5f: "Type 2" vs. "Type 1"). The **"Vector Model"** was proposed by Andrew Tomlinson and Gary Struhl [4366], who conducted the Fz-N-Dl experiments. Except for Step 4, it is like the Scalar Model. The new assumption is that differences in X levels can be sensed across the span of a single cell. For example, unoccupied Fz receptors might bind Dl in *cis*, while X-bound Fz receptors bind N in *cis*. As a result of this asymmetric "capping" (Fig. 7.5f), the equatorial side of every R3/4p cell would have more N and less Dl than the polar side.

When the two such R3/4p cells touch, the one closer to the equator will have more Dl and less N at the interface and hence should always become R3. This model agrees more closely with what is known about planar polarity in general [22], although Notch signaling only seems linked to cell polarity in the eye [4366].

R1–R8 cells arise sequentially, implying a cascade of inductions

In their 1976 "neurocrystalline" paper, which disproved Bernard's Lineage Hypothesis, Ready *et al.* proposed a model that relied on local cell interactions [3539]. Their **"Crystallization Model"** is a variation on the notion of sequential induction [91, 2409, 2448, 2716, 3645]. In this particular scheme, the MF acts as a template interface where naïve (anterior) cells "read" their location relative to

---

FIGURE 7.5. "Symmetry breaking" in nascent ommatidia. Precursor cells are denoted by a "p" suffix in the legend (omitted from figure) and equivalent pairs by a slash mark (e.g., "R3/4p"). Abbreviations: Dl (Delta), Fz (Frizzled), N (Notch). **a–c.** Histological appearance of photoreceptor clusters during development. **d–g.** Control of D/V chirality by Fz and N signaling. Cubes are individual cells, and inscribed circles are nuclei. Naïve R3/4p ("3/4?") cells are shaded (left). Committed R3p and R4p cells (far right) are black or white, respectively. See also App. 7.

a. Mature left eye disc. Dots are ommatidial sites. Vertical stripes are the morphogenetic furrow (MF) and mitotic band (MB; see [185, 4291, 4293, 4355] for details). Ahead of the MF are scattered mitoses (not shown). Compass (at right) gives axes in the fate map (cf. Fig. 7.1). "Rows" and "columns" run along A-P and D-V axes, respectively.

b. Stages in development of a cluster. Because the MF moves as a wave from P to A, this series is manifest in each row (magnified from box in **a**). Shading indicates future photoreceptors. Numbers denote the prospective fate of precursors ("3" = R3, etc.), *not* their current state of determination (which may be indefinite). Mystery ("m") cells leave the arc as it collapses into a precluster [186, 4364, 4715]. The sequence of depicted stages does not correspond exactly to the columns as numbered in the text: some transitional stages are omitted. For example, rotation is more gradual than implied here: it starts after the precluster stage (~ column 3–6) and ends by the 4-cone stage (~ column 15) [4361, 4364] (T. Wolff, pers. comm.).

c. Clusters become asymmetric at the 2-cone cell stage [4355, 4873] (≈ column 11 [4364]) when R4p severs its connection to R8p and undergoes a slight shift (D or V) jointly with R3p. At about this time, the clusters rotate clockwise or counterclockwise [777, 1639, 2794, 4873] on either side of the equator (thick dashed line). Trapezoid shapes arise later. (Only rhabdomeres really adopt this shape, not whole cells as shown here.)

d. Posterior (eye) part of the disc. The polarizing signal (X = E-shaped ligands) is assumed to be under N control (cf. Fig. 7.4) and to form 2 gradients by diffusing from the equator. Fz is the receptor for X. At the Arc Stage, the naïve R3/4p cell nearer the equator should sense a higher level of X (dotted line) and hence activate more of its Fz receptors [4366, 4873].

e. According to the **Scalar Model** [869], this higher Fz activity represses N (via Dsh) more strongly than in the rival cell. When the cells touch (Precluster Stage, **d**), this bias causes the cell nearer the equator to have less active N, so it wins the "shouting" contest (i.e., whoever has the most Dl) to become R3.

f. The **Vector Model** [4366] invokes a different kind of bias (Type 2 vs. 1). The side of the cell that faces the equator should have more of its Fz receptors occupied than the other side. If X-bound Fz receptors recruit N while vacant Fz receptors recruit Dl, then each R3/4p cell will have more N on its equatorial side and more Dl on its polar side. When the two such R3/4p cells meet (**d**), the one nearer the equator will have more Dl at the interface and so will always win and go on to become R3. (Other variations on this scheme are possible.)

g. Enlargement of R3/4p pairs after these naïve (gray) cells abut (Precluster Stage). The outcome of the contest (black "3" or white "4") is preordained by the bias introduced earlier (**e** or **f**), although no bias is depicted here (**g**). The positive feedback loop between the cells constitutes a **"Delta-Notch Flip-Flop"** device, whose instability eventually forces the cells to adopt binary (alternative) R3 vs. R4 fates (Cluster Stage at right) by amplifying the small initial bias [2883].

Relative locations of MF and MB in **a** are based on [604, 3539]. Panels **b** and **c** are adapted from [293, 604, 3539, 3563, 4355, 4715, 4873], although not all the depicted stages (**b**) occupy successive columns [605, 2887], and the mitotic band may actually span several columns [185, 228]. Models and schematics in **d–g** are modified from [359, 447, 869, 4173, 4366]. For details of rosette geometries, see [184, 4712]. For critiques of earlier models, see [3562, 3563]. See also App. 7.

already committed (posterior) cells and adopt fates accordingly.

Cell differentiation in the eye appears to be unrelated to cell lineage. The final cell type may thus be determined according to the lattice position into which the cell is recruited. This is reminiscent of the growth of a crystal, in which the leading edge of the pattern serves as a template on which new elements are incorporated. Rather than being required to assess its position independently, a cell joining the ensemble could use the information available at the growing edge to determine its role.

This assessment may be based on combinatorial cell contacts, the information being mediated by surface molecules, as in the "antigen-antibody" model of Tyler [4412] and Weiss [4590]. According to such a scheme, an undetermined cell ahead of the furrow might display a set of antigens. "Antibodies" on cells at the leading edge would bind a specific subset of these antigens, thus informing the newly added cell of its role and causing it to display an appropriate set of antigens in turn, propagating the pattern. [3539]

According to this model, the crystallization process would be initiated at the back of the eye by a special group of cells, and the remaining tissue should be unable to form ommatidia properly if this "seed crystal" is removed. In the 1960s, Richard White used this logic to prove the existence of such a nucleation center in the mosquito eye [4628, 4629]. He explanted prospective eye tissue with or without the "optic placode" (a thickening at the rear of the eye field) and found that ommatidia develop only when the placode is included in the explant. Hence, the placode behaves like a seed crystal. White was also able to prevent differentiation ahead of the MF *in situ* by implanting a strip of foreign epidermis in its path. In one case where the implant failed to reach the edge of the host's eye field, the MF traveled backward beyond the barrier, thus revealing an ability to move in an A-to-P as well as a P-to-A direction. This ability is also consistent with the Crystallization Model because the MF should be free to travel in any direction after being deflected (like a ripple in a pond [2036]).

When similar experiments were performed in *Drosophila*, different results were obtained. In 1986, Lebovitz and Ready reported that anterior pieces of eye discs (ahead of the MF) can make ommatidia when explanted [2452]. To rule out the possibility that the ectopic MF arises as a consequence of wounding, they cut the disc diagonally and showed that ommatidia tend to differentiate in a wave from P to A, rather than in a wave that is parallel to the wound edge. Their conclusion that wounding is irrelevant cannot be considered definitive, however, because *hedgehog* expression was not monitored. As discussed in Chapter 5, contact with

the peripodial membrane might be triggering patterning events *de novo* at particular sites [773, 1472].

The fact that killed the Crystallization Model was not the ability to incite ectopic MFs, but rather the wider-than-normal spacing of ommatidia in *Egfr*^{GOF} ("*Ellipse*") eyes [179–181, 2355, 2493, 3410] (cf. *ato*^{LOF} [4621] and *rap*^{LOF} [2145]). These lattice gaps are hard to reconcile with any template-based process [1048] because the ability of clusters to develop far from one another (i.e., in isolation) argues that ommatidia must be self-assembling units [185, 4360].

In 1986, Ready, Tomlinson, and Lebovitz [3540] salvaged the main premise of the Crystallization Model (viz., that naïve cells adopt fates based on the committed cells they happen to touch). Even if the entire MF does not act as a template, a template-like process might still be occurring within the confines of each nascent ommatidium. The **"Combinatorial Cascade Model"** that they devised focuses on intra-ommatidial signaling, while ignoring the issue of interommatidial patterning (Fig. 7.6e, f; see [3645, 4412, 4590] for earlier models of this kind):

The crystal-like regularity of the *Drosophila* compound eye thus does not arise like a crystal.... The orderly sequence of ommatidial assembly, together with the lack of cell migration revealed in the genetic marking experiments, suggest that the fates of undetermined cells in the retinal epithelium are determined in place by their interactions with neighboring cells that have already been incorporated into a precursor. At each stage of assembly, the stereotyped morphology of the precursor presents defined niches to surrounding epithelial cells which could supply the cues necessary for a cell to determine its appropriate fate.

Tomlinson and Ready expounded on this theme in 1987 [4364]:

The ommatidium appears to be an autonomously assembling unit. The stereotopy of the assembly process and the precise cellular contacts made, coupled with the autonomous assembly, point to local cues for pathway selection.... Taken together, the patterns of ommatidial development suggest that undetermined cells are directed into specific pathways by "reading" the identities of the cells they contact. A particular combination of cell types may specify a particular fate. For example, a cell in contact with R1, R6, and R8, would have an unambiguous cue to become R7. Once a cell has determined its fate, it must then display its identity so the next cells in line can make informed choices. A simple logic of cell contacts may direct pathway choice in the developing ommatidium.

In 1989, Cagan and Ready added cone cells and pigment cells (PPC, SPC, TPC) to this scheme based on geometric clues [602]: (1) a cell that contacts a cone cell becomes a

PPC, (2) a cell that contacts PPCs from two ommatidia becomes a SPC, and (3) a cell that contacts three PPCs becomes a TPC.

Order emerges in the eye as specific and stereotyped contacts are made between retinal cells.... The fact that these contacts are consistently seen in the early development of each pigment cell suggests that they are important for cell determination. ... Each cell's position appears to direct its pattern of gene expression, which in turn directs its fate. Ommatidial development progresses by evolving new determinative positions that allow for new cell fates. In this manner, positional information, encoded in stereotyped cell contacts, directs the patterns of gene expression which drive the self-assembly of the eye. [602]

As mentioned above, neural antigens are expressed by photoreceptor precursors in a definite order (Fig. 7.6b) [4361, 4364, 4715]. In the series below, "➡" refers to time. Thus, R8 manifests neural characteristics first and R7 manifests them last.

Neural features: R8p ➡ R2/5p ➡ R3/4p ➡ R1/6p ➡ R7.

A priori, it was natural to think that this sequence might reflect a causal chain of events [4769]. In the Combinatorial Cascade Model, each step (except the first) was thought to involve several signals, with different signals (and receptors) being used at successive steps because they come from different cell types. In each of the steps below, "➡" signifies "induces".

Step 1 (direct):   R8p ➡ R2/5p.
Step 2 (Boolean):   {R8p AND R2/5p} ➡ R3/4p.
Step 3 (Boolean):   {Some combination of R8p, R2/5p, AND R3/4p} ➡ R1/6p.
Step 4 (Boolean):   {Some combination of R8p, R2/5p, R3/4p, AND R1/6p} ➡ R7p.

Thus, two general kinds of mutations were expected: (1) mutations that block emission of signals, and (2) mutations that block receipt or transduction of signals.

Such a model predicts two classes of mutation, one class which interferes with the normal expression of cell-type identity signals and a second which prevents cells from properly reading these signals. [4363]

The gene *rough* fit neatly into this scheme [3670, 4356, 4358]. The only cells that need *rough* are the R2/5p pair. When they are *rough*[LOF], the R3/4p and R1/6p cells fail to mature as neurons [1783, 4361]. Evidently, R2/5p cells recruit neighbors into the cluster by emitting a signal. The

signal is probably not Rough itself because Rough has a homeodomain.

But the final cell (R7) is induced by the first one (R8)

Curiously, R7s were found to arise normally in *rough*[LOF] eyes [1783]. The implication was that R7 is not downstream of the *rough*-dependent signal but rather relies on a signal that comes from R8 alone. (Note that *rough*[LOF] R3/4p and R1/6p cells could still influence R7p despite their immaturity, and indeed they do, as discussed below [4367].) The model can easily be revised to incorporate two separate signals from R8p:

Step 1 (direct):   R8p (early signal) ➡ R2/5p.
Step 2 (Boolean):   {R8p AND R2/5p} ➡ R3/4p.
Step 3 (Boolean):   {Some combination of R8p, R2/5p, AND R3/4p} ➡ R1/6p.
Step 4 (direct):   R8p (late signal) ➡ R7p.

The "R8p ➡ R7p" induction does in fact occur, and its components manifest the expected specificity [4360]. Indeed, the ligand and receptor for this tête-à-tête are used nowhere else in the fly. The ligand is a 7-pass transmembrane protein encoded by *boss* (*bride of sevenless*) [1732], and the receptor is a tyrosine kinase encoded by *sev* (*sevenless*) [428, 1677]. As usual for RTK receptors (cf. Fig. 6.12), Sev is activated by Boss (vs. inhibited) [1733, 2319, 2450, 3671, 4359]. LOF mutations in either gene eliminate R7 [4890], but no other LOF effects are seen elsewhere [604].

Boss is peculiar insofar as it is a 7-span membrane protein that is totally engulfed by the receiving cell [601, 2318, 4890], and Sev is odd insofar as it is cleaved and rejoined to form an external loop that is anchored at both ends to the membrane [227, 2318, 2991, 3342]. Otherwise, these partners seem to work like an ordinary lock-and-key device (cf. Fig. 6.12), albeit customized for one cellular chat. Closer inspection, however, reveals nuances.

Genetic mosaics show that *boss* is only needed in R8 [3569] and that *sev* is only needed in R7 [632, 1726, 4363]. This result is not surprising for *boss* because R8 is the only *boss*-ON cell in the cluster [2319, 4459], but it is puzzling for *sev* because *sev* is ON in other cells that touch R8 [2450] — namely, R3/4p and weakly in R1/6p (but not at all in R2/5p) [194, 227, 230, 4360].

Various restraints prevent more than one cell from becoming R7

Why don't the other Sev-expressing cells that touch the Boss-expressing cell become R7s? Their unresponsiveness is attributable to the transcription factors Rough

Learning Resources

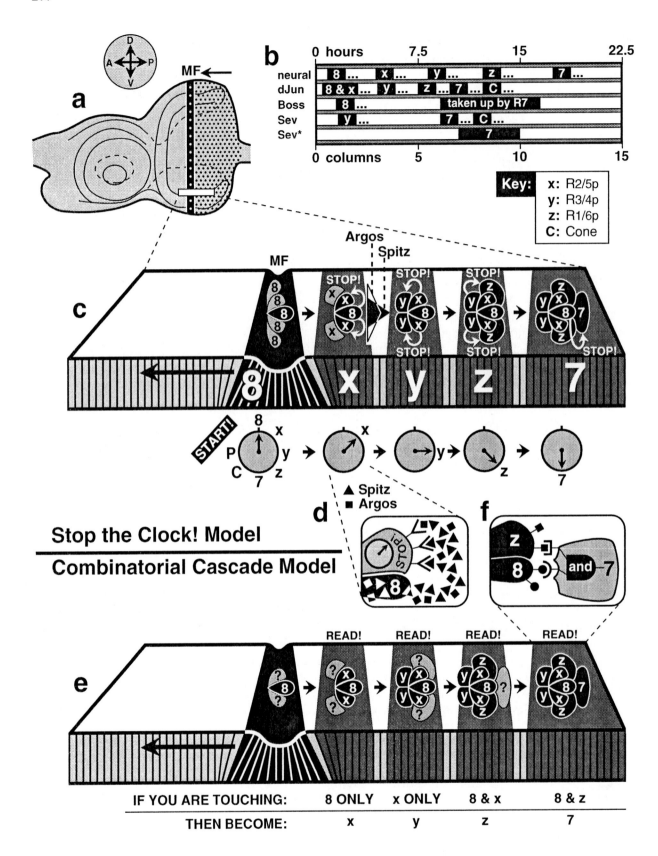

Stop the Clock! Model

Combinatorial Cascade Model

IF YOU ARE TOUCHING:	8 ONLY	x ONLY	8 & x	8 & z
THEN BECOME:	x	y	z	7

Key:
x: R2/5p
y: R3/4p
z: R1/6p
C: Cone

(in R2/3/4/5p) and Seven-up (Svp, in R1/3/4/6p) – neither of which is expressed in R7p [4890]. Thus, all outer R cells (R1–R6) can become R7s in doubly mutant *svp*^LOF *rough*^LOF clones [1783].

Cone cells normally express Sev but do not touch R8p [230, 4360] or turn ON *svp* [2888]. They can be induced to adopt an R7 fate by ubiquitously expressing *boss* (R1–R6 persist intact) [4459] or by artificially activating their Sev pathway [225, 229, 1049]. Hence, their failure to become R7s in wild-type flies is solely a consequence of their physical separation from R8p.

Overall, therefore, the data suggest that a cell must possess four traits to become an R7 [600]. Each trait is exhibited by a certain set of cells (in braces below: Cp = cone precursor, M = mystery cell) [897, 938, 4459].

Must have Sev:	Set 1:	{R1p, R3p, R4p, R6p, R7p, Cp, M}.
Must lack Rough:	Set 2:	{R1p, R6p, R7p, R8p, Cp, M}.
Must lack Svp:	Set 3:	{R2p, R5p, R7p, Cp, M}.
Must touch R8p:	Set 4:	{R1p, R2p, R3p, R4p, R5p, R6p, R7p}.

The overlap between Sets 1, 2, and 3 defines the "R7 equivalence group" {R7p, Cp, M}, whose members are competent to react to Boss [1049, 1678]. Within that pool, only a single cell satisfies the final criterion (Set 4) and hence becomes R7 [3540].

Additional prerequisites exist. For example, receipt of a Delta signal from R1/6p turns out to be essential for

---

FIGURE 7.6. Models for recruitment of photoreceptor cells into nascent ommatidia.

a. Mature left eye disc. Vertical stripe is the morphogenetic furrow (MF), which moves from P to A. Compass gives axes in the fate map (cf. Fig. 7.1). In **c** and **e**, an imaginary slice of the columnar epithelium (boxed in **a**) is magnified to show how each ommatidium acquires cells after arising as an arc in the MF. Stages are accentuated by stripes (black = MF; gray ≈ later columns) that correspond roughly to columns 0–4 in terms of cluster morphology, although the patterning cues (Stop! or Read!) may occur later. Cells are drawn as ovals (top view; black = determined; gray = uncommitted) or as narrow rectangles (side view).

b. Temporal aspects of photoreceptor development. Abscissa is calibrated in hours (25°C) and in columns of ommatidial clusters relative to the MF (origin). Bars followed by "..." signify onset (bar) and persistence (...), whereas bars alone indicate total duration. "Neural" means neuron-specific epitopes recognized by 22C10 and anti-HRP antibodies [1948] (but see [228] for a different pacing and [2322, 4712] for a different order); "dJun" is expression of dJun (duration ≈ 4 h, except cone "C" cells where it lasts longer); "Boss" and "Sev" are expression of these proteins; and "Sev*" is the period when Sev must be continuously activated for a prospective R7 cell to adopt the R7 fate [2991]. This period (columns 7–10) coincides roughly with the time when Boss is internalized ("taken up") by R7 [2319]. The mitotic band (not shown; cf. Fig. 7.9) roughly spans columns 3–5 [185, 228, 4712]. Omitted: Delta (timecourse ≈ neural) [184, 3273] (cf. Notch [603]). The R8p cell first becomes identifiable in column 0 when it expresses Scabrous and Atonal (not shown) [184, 2042]. Atonal is required for R8 differentiation [4621]. The "Irregular chiasm C-roughest" protein (not shown) is expressed by cells in the same order as they are thought to be induced (**c**) [3570].

c, d. The **Stop the Clock! Model** [1290, 3557] uses two variables: (1) the "state" of a cell and (2) a "Stop!" signal. Cells are supposed to automatically progress through a series of transcription factors (states), represented here as a clockface whose numbers and letters reflect cell fate: 7 and 8 refer to R7 and R8; x, y, and z to outer R cells (cf. key for **b**); and C and P to cone or pigment cells (**c**). All cells in each column change state synchronously at ~1.5-h intervals (gray bands), starting in the MF (Start!) with the R8 state, then x, y, z, R7, and finally (not shown) C and P. The "Stop!" signal forces cells to keep their current state. It is conveyed by Spitz (white arrows) – a ligand for Egfr. Each cell emits Spitz within ~1.5 h of receiving a Spitz signal, thus propagating the signal by "sequential induction" (although the exact "talking" order is unknown). R7p breaks the chain because (1) it is told to stop by the earliest cell in the series (R8p) and (2) it also requires a Boss signal (transduced by Sev). Because Spitz is diffusible, its concentration should dwindle at increasing distance from its source. This inferred gradient is depicted to the right of the "x" column (filled triangle), along with a gradient of the ligand Argos (unfilled triangle) – a competitive inhibitor. Spitz and Argos are both produced by whichever cell is signaling at any given time (e.g., R8p in **d**), but the Argos gradient is shallower because Argos diffuses faster [1288, 1290, 1294, 2331]. Thus, Spitz will be louder than Argos for cells next to the vocal cell (hence stopping their clock) but softer than Argos for cells farther away (hence letting them tick). See [2762, 2811, 3644–3646] for older models in this genre.

e, f. The **Combinatorial Cascade Model** [602, 3540, 4364] uses Boolean combinations of cell-surface ligands to elicit certain states in adjacent cells. The rules are given below. For example, the rule depicted in **f** is: "IF you are touching an *8* cell AND a *z* cell, THEN become a *7* cell." Each signaling cell is thought to exhibit a unique ligand (round vs. square lollipops on *8* vs. *z*), and each receiving cell is thought to have various receptors.

In both models, previously determined cells (black) talk to uncommitted cells (gray and "?"). Both models use "inductive signals," but the former uses them permissively while the latter uses them instructively [196].

Data in **b** are from [383, 2991, 4360, 4364, 4715]. Cross-sections in **c** and **e** are schematics simplified from [3540]. See [605, 4890] for a 3-dimensional perspective. The cartoon in **f** is adapted from [1290, 2351, 3538]. See also App. 7.

R7 identity (see below) [4367]. Moreover, members of the R7 equivalence group can adopt the R7 fate only during a definite time window [604]$^\Delta$, whose molecular basis is unknown.

The information content of the inductive signals may be only 1 bit

What happens to the would-be R7 cell when it is prevented (by sev^{LOF} or $boss^{LOF}$) from becoming R7? Oddly, it becomes a cone cell [1726, 2887, 4362]. (The total number of cone cells remains 4 because the transformed cell displaces a would-be cone cell [4363].) This non-neural default was surprising because *a priori* one might have expected R7p to start from – and revert to – a neural ground state [27, 1818, 4551].

The deeper question here is: what is Boss's information content [224, 1049, 3671]? Based on R7's default state, Boss might just be conveying the message, "Become neural!," with other factors (such as Notch [4367]) biasing the listener to an R7 identity [3671, 4551, 4790].

Curiously, circumstances were found where a cell can lack Sev and still become an R7. If the complement of R cells is represented as an ordered triplet (*a*, *b*, *c*) where the variables are the numbers of outer (R1–R6)-type photoreceptors (*a*), R7-like cells (*b*), and R8 cells (*c*), then a wild-type ommatidium would be "($a = 6$, $b = 1$, $c = 1$)." A typical svp^{null} ommatidium is "(3, 5, 1)," where the R3/4 and R1/6 pairs have become R7s, and there is an extra outer-type R cell of unknown origin. When svp^{null} clones are induced in sev^{null} eyes, the doubly mutant ommatidia are commonly "(3, 2, 1)" [2888], where R7 and two of the four R7 transformants (R3/4?) are missing, but the other two R7-like cells (R1/6?) manage to differentiate without Sev. Evidently, some other receptor aside from Sev can induce an R7 state. This conclusion was bolstered by the ability of ectopic Svp to transform cone cells (which have no access to Boss) to R7s [1856].

In 1995, the latter transformation was found to depend on Ras-MAPK pathway elements that are shared downstream of Sev and Egfr (cf. Fig. 6.12) [264]. Thus, Egfr was implicated as the accomplice in the *svp*-mediated conversions. The wider inference was that Egfr and Sev could substitute for one another elsewhere, as long as they exhibit the appropriate onset, duration, rate, and amplitude of signaling. The Combinatorial Cascade Model remained popular throughout this period [229, 605, 1048, 2961, 4714], despite a failure to find any receptors besides Sev that are dedicated to definite R-cell subsets [1288, 4890].

In 1996, Egfr and Sev were shown to be entirely interchangeable [1289], and this flexibility begat a new model [1290, 2243]. The experiments were performed by Matthew Freeman in Cambridge, England.

Egfr is naturally expressed throughout the developing eye epithelium (intensely in the MF) [2493, 4843]. Freeman sabotaged the normal Egfr by expressing a dominant-negative Egfr decoy (Egfr^{DN}) that has the ligand-binding and dimerization domains but lacks the cytoplasmic kinase domain, and he used a *hs-Gal4* driver (with *UAS-Egfr^{DN}*) to shut OFF the host Egfr at desired times. He found that virtually any R cell, except R8 [1083, 4799], can be eliminated by disabling its Egfr around the time of the R cell's inception [1289].

Surprisingly, the inactivation of Egfr also suppressed R7. R7 had been thought to only need a Sev-mediated signal. Apparently, R7 needs two signals – a "primer" via Egfr and a later "boost" via Sev [4322]. Actually, what a cell needs to become an R7 is a high stimulation of its Ras-MAPK pathway. The route that activates the pathway is moot. Thus, Freeman was able to rescue R7 cells in *sev^{null}* eyes by overexpressing a constitutively active Egfr^{GOF} construct (Tor^D-Egfr) that bypasses Sev entirely [1289]. Likewise, excess R7s (caused by *dMAPK^{GOF}*) can be suppressed by inactivating Egfr (via *argos^{GOF}*) [3771].

Another surprise was that later heat shocks stifle non-neural cells. The sensitive periods obey the same sequence as the appearance of these cell types in normal development: cone cells, PPCs, SPCs, and TPCs. Evidently, Egfr mediates the induction of both neural and non-neural cells. This deduction was affirmed by GOF studies: excess R cells, cone cells, or pigment cells can be elicited by expressing Tor^D-Egfr at successively later stages, and similar effects are seen with *spitz^{GOF}* [1289, 2493] (but not *vein^{GOF}* [2493]).

Freeman conjectured that Spitz is the only signal that is normally used to recruit all cell types into the ommatidium, except for Boss, which is needed for R7 but can be bypassed. (Delta was not then known to be instrumental [4367].) In each of the following steps, "➡" indicates that the foregoing cell uses Spitz to induce a naïve cell(s) to adopt the next state. For simplicity, the hypothetical series of inductions is written in a linear format to conform to the temporal sequence of inception in normal development.

Step 1:   R8p ➡ R2/5p.
Step 2:   R2/5p ➡ R3/4p.
Step 3:   R3/4p ➡ R1/6p.

Step 4:   R8p (late signal) ➡ R7p.
Step 5:   All R cells (?) ➡ Cone cells.
Step 6:   Cone cells ➡ PPCs.
Step 7:   PPCs ➡ SPCs.
Step 8:   SPCs ➡ TPCs.

This proposal differs from the Combinatorial Cascade Model insofar as a single inducer here acts as a 1-bit "STOP!" or "GO!" command. There are not supposed to be multiple signals acting in combination. Indeed, Freeman boldly entitled his 1996 paper, "Reiterative use of the EGF receptor triggers differentiation of all cell types in the *Drosophila* eye." The dependence of PPCs on cone cells (Step 6) has been proven directly by laser ablation [2859].

Here again, we encounter the Ras Specificity Riddle (cf. Ch. 6) [3638, 3783, 3942, 4245]: how can the same pathway achieve different outcomes? A plausible answer was put forward in 1994 by Thomas Reh and Ross Cagan [3557], and Freeman built on their premise to craft what may be called the "**Stop the Clock! Model**" (Fig. 7.6c, d) [1290]. The basic idea, as stated below in excerpts, is that cells progress synchronously through a series of states [1290]: A, B, C, etc. When a cell gets a Spitz signal, it keeps the state that it has at the time, and this state dictates its fate: A = R2/5, B = R3/4, C = R1/6, etc. The states thus furnish a context wherein 1-bit inputs can yield diverse outputs [224, 3287] (cf. Turing machines [872, 1034, 1897, 4409] and L-systems [2557–2559]).

The data suggest a model whereby external cues decide the developmental moment a cell will respond to its own evolving internal information. Each uncommitted cell appears to have an internal clock that changes its developmental potency with increasing developmental age. Perhaps this clock represents a progressive cascade of transcription factors. As the complement of factors change, the potency of the cell changes. [3557]

Upon activation by [Egfr], a cell would start to differentiate towards the fate appropriate to its developmental stage, indicating that it is the age or developmental history of a cell that is responsible for determining its ultimate fate. This implies that a cell passes through a series of "states," each representing a potential fate. Each cell state presumably derives from the subset of transcription factors that are present and which can be activated by the Ras pathway. [1290]

A ratcheting device of this kind has been found in neuroblasts. When isolated in a culture dish, embryonic neuroblasts autonomously express transcription factors in a fixed sequence, and their descendants stay in one of the 4 phases [485]: Hunchback (homeodomain),

Pdm (POU-homeodomain), Castor (zinc finger), or Grainyhead (bHLH). The "Stop!" signal in that case must linked to mitosis.

The classic example of a transcription factor hierarchy in flies is the "puffing" cascade [121, 123, 246, 2097, 3706], where successive batteries of genes trigger one another in response to ecdysone at metamorphosis [479, 1254, 3059, 4131, 4311, 4422]. Surprisingly, the puffing cascade may be more than just a metaphor here. It is directly harnessed to the MF: the proteins of the Broad-Complex (first tier of puffs) are required for ommatidial assembly behind the MF [457, 458], and Ultraspiracle (the ecdysone co-receptor) is needed for proper R-cell differentiation [3198, 4855]. Moreover, Ultraspiracle is inhibited by Seven-up [4856] – a fate-determining factor that also belongs to the steroid receptor superfamily [264, 1856, 2322, 2888]. These correlations suggest that the puffing cascade is literally furnishing the "clock" that Spitz supposedly stops in order to assign cell fates. However, other transcription factors that have no obvious endocrine associations are also candidates for this role and must be considered (see below).

Spitz is only active as a diffusible (vs. membrane-bound) ligand [3829]. It could easily go from R8p to R2/5p to R3/4p if it were to diffuse during the Arc Stage when these cells are in a single file (vs. later when they are clustered) [196]. Freeman argued that a diffusible inhibitor is limiting the range of Spitz's effects. If the inhibitor diffuses faster than the inducer, then cells far away will be "immunized" against stray inducer signals. This same sort of Activator-Inhibitor Model was discussed for wing veins (cf. Ch. 6) [3207]. In both cases, the supposed inhibitor is Argos. The following facts support the Stop the Clock! Model:

1. Egfr is required for the differentiation of all R cells except R8 [204, 1083, 1289, 2355, 4781]. R8 is also unique insofar as it requires the proneural gene *atonal* [2041, 2042], which enables it to found the R-cell cluster [4621]. In *Egfr*^null clones, the PNC spot of *atonal* expression shrinks to the R8p cell as it normally does, but this refinement is delayed [2493], apparently due to an "Egfr ➡ Notch ⊣ *atonal*" chain of control. Nevertheless, R8 does need Egfr later because it cannot retain its neural state without Egfr [1083, 2355].

2. Each R cell begins expressing Spitz when it becomes neural [4322], and a similar order is seen for Star [1782, 2288] and Rhomboid [1293, 4035]. Star and Rhomboid enable Spitz to be cleaved and released [3385]. For a normal facet to develop, Spitz is absolutely required in R8p, less so in R2/5p and R3/4p, and not at all in R1/6p

or R7p [1288, 4321] (cf. consistent data for Star [1785] and Rhomboid [1293]). Evidently, what matters is the local amount of Spitz: fewer than 8 R cells can make enough Spitz for facet completion (albeit perhaps delayed). *N.B.*: Some of the weak LOF effects might be attributable to Spitz's cousin Gritz [204, 2108, 4558], which could be acting redundantly here.

3. The first step (R8p → R2/5p) can be bypassed in *atonal*^{null} eyes by expressing an activated Egfr ahead of the MF [1083]. Under these conditions, ectopic R1–R7 cells arise but no R8 cells. Likewise, virtually all cells posterior to the MF can be converted to R1–R7 identity in wild-type eyes by expressing *spitz* there, but no excess R8s are seen [2493]. When R8p is *spitz*^{null}, it can still muster 4 other cells to form a "precluster," but it is the only cell therein that becomes neural [1083, 4322]. When R8p is *Egfr*^{null}, it can still induce neighbors to become neural (presumably by emitting Spitz), but only if they are wild-type (*Egfr*$^{+}$) [1083].

4. Like Spitz, Argos is expressed by each recruited R cell as it differentiates [1294, 2331, 3183]. (Argos is also detected in cone cells and PPCs, but not SPCs, TPCs, or bristles [1294].) Both *spitz* and *argos* may be targets of the EGFR pathway in the eye, with "Egfr → *spitz*" operating at a lower threshold than "Egfr → *argos*" [2493]. As expected, *argos*^{GOF} suppresses R cells (all types), cone cells, and pigment cells [515, 1287, 3772, 4233], while *argos*^{LOF} elicits extra R cells (R1–R6 only), cone cells, and pigment cells [1294, 2331, 3183, 3771, 3772]. The extra cells are recruited from the pool of uncommitted cells around nascent ommatidia [1294] – a conscription that rescues them from the doom of death [515].

5. Spitz and Argos do indeed have different effective ranges in the eye [1290]. Wild-type cells rescue *argos*^{null} clones over ~10–12 cell diameters [1294, 2331], but they rescue *spitz*^{null} clones over ~3–4 cell diameters at most [1288].

6. When an antibody to phosphorylated dMAPK is used to monitor the EGFR pathway, 1–3 central cells per cluster first show activity. This activity fades in older columns as staining emerges in the periphery of each cluster [1337]. Expanding rings of this sort are predicted by the model.

7. Timing is clearly important in assigning fates because cone cells can be converted to R7-type cells by activating their Ras-MAPK pathway prematurely [225, 516, 1049, 1273, 1399, 3633].

8. R7p continues to express an R7-specific marker gene (*klingon*) even after it transforms into a non-neural cone cell in *sev*^{LOF} flies [585, 2887]. Evidently, R7p has indelible qualities that precede receipt of the Boss signal. Early biases of this sort may constrain R7p to its correct fate [265, 1290].

9. Ubiquitously activated Notch is unable to summon the expression of specific E(spl)-C genes in certain tissues at certain times [871]. The selective inaccessibility of these genes to the Notch pathway offers support, albeit indirect, for the idea that specific genes might become available to the Ras-MAPK pathway at different times.

Some properties of the non-neural cells are difficult to reconcile with the above scheme. Cone cells do not seem to need their RTK pathway because they differentiate normally when the transcriptional repressor Ttk88 is expressed throughout the eye epithelium [670]. (See [4367] for an exegesis.) Also unclear is how cone and pigment cells can remain non-neural after receiving a Spitz signal because this signal should lead to the degradation of Ttk88 and hence to neural development [670].

No transcription factor "code" has yet been found for R cells

The Stop the Clock! Model envisions a succession of state-specific transcription factors. Disabling any of them should transform one type of ommatidial cell to another. In other words, there should exist genes that, when mutated, can switch cell fates within the ommatidium, and these genes should encode DNA-binding proteins. Many DNA-binding proteins have indeed been found to play roles in eye development (Table 7.1), but genetic screens have uncovered only two *bona fide* "homeotic" genes (aside from the effectors of the Boss-Sev pathway): *rough* and *svp* [4551].

When expressed ectopically, either Rough [232, 1676, 2223] or Svp [264, 1856, 2322] can cause R7p to adopt an outer-type (R1–R6) identity. If either of these transcription factors can enforce the outer-type state [4790], then both of them would have to be absent for a cell to become R7. This prediction has been tested. Rough and Svp are normally expressed in R2/3/4/5p, and R1/3/4/6p, respectively (Table 7.1). In *svp*^{null} eyes, R1/3/4/6p cells become R7s. This result is surprising because Rough should still be present in R3/4p, although mosaics reveal that Rough is not needed (functional?) there [4361]. In *rough*^{null} eyes, R2/5p resemble R1/3/4/6-type cells instead of R7s. The reason appears to be a "Rough ⊣ *svp*" link (Link 9 below), so Svp is derepressed in *rough*^{null} R2/5p cells. If this argument is correct, then *rough*^{null} *svp*^{null} clones should convert all outer cells to R7s (via contact with

R8p). The double-mutant phenotype agrees with this expectation [1783].

How are *rough* and *svp* turned ON or OFF in R-cell subsets during normal development? We know only part of the story: *rough* is kept OFF in R1/6 by *BarH1*, and *svp* is kept OFF in R7p by *lozenge*. These links (Links 8 and 4) and others are listed below. Each regulatory agent (on the left side of the control sign) is a transcription factor (see [1021, 1293] for other targets). Abbreviations include: *lz* (*lozenge*), *phyl* (*phyllopod*), *pros* (*prospero*), *sca* (*scabrous*), *sina* (*seven in absentia*), and *svp* (*seven-up*).

Link 1: Lz ➡ Bar-C. Ectopic expression of *lz* can activate the Bar-C [938] (= the *BarH1* and *BarH2* paralogs [1842]), whereas Bar-C gene expression in R1/6p cells is sharply reduced in lz^{null} eyes [897, 938]. Based on its effects here and elsewhere, *lz* is considered a prepattern gene [1255, 1587, 1653].

Link 2: Lz ➡ *dPax2*. In lz^{LOF} eyes, *dPax2* turns OFF in cone cells [1256]. The regulation of *dPax2* is direct: Lz binds 3 sites in the cone-cell enhancer of *dPax2* [1256]. For this 361 b.p. enhancer to turn *dPax2* ON, it must also bind the transcription factors PointedP2 (effector for EGFR) and Su(H) (effector for Notch) [1256]. Hence, this enhancer uses combinatorial logic: "IF Lz AND PointedP2 AND Su(H) are bound, THEN turn ON *dPax2*." How this logic is implemented sterically is not known.

Link 3: Lz ➡ *pros*. Ectopic expression of *lz* can activate *pros* [4775], whereas *pros* stops being expressed in R7p and cone cells in lz^{LOF} eyes [670, 4775]. The regulation of *pros* is direct: Lz binds 2 sites in the eye-specific enhancer of *pros* [670, 4775]. For this 1150 b.p. enhancer to turn *pros* ON, it must also bind PointedP2 [4775].

Link 4: Lz ⊣ *svp*. In lz^{LOF} eyes, *svp* turns ON in R7 and cone cells [897, 938], while in lz^{GOF} eyes, *svp* turns OFF in R3/4p [938]. This link must somehow be invalidated in R1/6p cells because they normally express both *lz* and *svp* [1255].

Link 5: Svp ➡ *BarH1*. In svp^{LOF} clones, the cells that normally would express *BarH1* (R1/6p) fail to do so [1856].

Link 6: BarH1 ➡ *svp*. (Converse of Link 5.) When *BarH1*-OFF (cone) cells are forced to express BarH1, they turn ON *svp* [1768].

Link 7: BarH1 ➡ *BarH2*. Ectopic expression of BarH1 (in cone cells) autonomously activates *BarH2* expression [1768].

Link 8: BarH1 ⊣ *rough*. When *rough*-ON (R3/4p) cells are forced to express BarH1, they turn OFF *rough* [1768].

Link 9: Rough ⊣ *svp*. In *rough*^{null} mutants, there are more *svp*-ON cells in the ommatidium [1783], apparently because R2/5p cells (*svp*-OFF) transform to resemble R1/3/4/6 (*svp*-ON). This link appears to be inactivated in R3/4p because those cells express both *rough* and *svp*.

Link 10: Rough ➡ *BarH1*. Expression of *BarH1* vanishes in *rough*^{LOF} eyes [1842].

Link 11: Glass ➡ *BarH1*. Expression of *BarH1* vanishes in *glass*^{LOF} eyes [1842].

Link 12: Glass ➡ *sca*. In *glass*^{LOF} eyes, *sca* is expressed at a subnormal level (although it stays ON longer) [4391].

Link 13: Glass ➡ *svp*. In *glass*^{LOF} eyes, *svp* fails to be expressed in its normal complement of photoreceptors [4391].

Link 14: Glass ➡ *boss*. In *glass*^{LOF} eyes, *boss* expression virtually disappears [4391], although *sev* expression persists [2963].

Link 15: Glass ➡ *glass*. In *glass*^{LOF} eyes, expression of an enhancer-trapped *lacZ* at the *glass* locus is greatly reduced [2964].

Link 16: dPax2 ➡ *cut*. Expression of *cut* vanishes temporarily in cone cells in *dPax2*^{LOF} eyes [1320].

Link 17: dPax2 ➡ Bar-C. Expression of Bar-C genes disappears in R1/6p cells in *dPax2*^{LOF} eyes [1320].

Link 18: Mδ ⊣ *BarH1*. Expression of *BarH1* vanishes in R1/6p cells under $m\delta^{GOF}$ conditions (*sev-Gal4:UAS-mδ*), while R7p expresses *BarH1* under N^{LOF} ($m\delta^{LOF}$) conditions [870].

The links among these transcription factors (excluding autoregulatory loops) are presented as a wiring diagram in Figure 7.9b, and their R-cell expression patterns are cataloged in Table 7.1. How the circuit operates is unclear. There is no obvious "receptor code" wherein each R subtype (e.g., R1/6) is encoded by a particular set of transcription factors. The only hint of such a code is the aforementioned enforcement of outer-type R-cell identity by Rough OR Svp. Undoubtedly, there are important links missing, and further screens will be necessary to

TABLE 7.1. CELL-TYPE SPECIFIC EXPRESSION OF *BOSS*, *SEVENLESS*, AND GENES THAT ENCODE TRANSCRIPTION FACTORS IN OMMATIDIA*

Gene	R8p Cell (Inner)	R2/5p Cells (Outer)	R3/4p Cells (Outer)	R1/6p Cells (Outer)	R7p Cell (Inner)	Cone Cells	PPC, SPC, TPC Cells	Bristle Cells
boss (ligand) [4890]△	**ON** & required. LOF: R7 (*sic*) becomes cone cell.	—	—	—	—	GOF: become R7s.	—	—
sev (receptor for Boss) [604, 1049, 1677, 4363]△	GOF: fails to develop?	GOF: fail to develop?	**ON** but not needed.	**ON** (weakly) but not needed.	**ON** & required. LOF: becomes cone cell (but # of cone cells remains 4).	**ON** but not needed. GOF: become R7s.	GOF: fail to develop.	—
atonal [bHLH] [4621].	**ON** & required. LOF: usually missing (no further cells recruited).	—	—	—	—	—	—	—
Bar-C [HD] genes (*H1* and *H2*) [1768]△.	—	—	—	**ON** & required. LOF: no obvious fate switch.	—	GOF: become outer-type R cells (R1/6?) or PPCs or disappear.	**ON** & required in PPCs. LOF: no obvious fate switch.	**ON** & probably required in neuron and sheath cells.
cut [HD] [373, 3659]△.	—	—	—	**ON**.	—	**ON**.	—	**ON**.
dachshund [?] [870].	—	—	—	**ON**.	**ON**.	—	—	—
dJun [bZip] [2280, 2281]△.	**ON** & required. LOF: missing?	**ON** & required. LOF: missing?	**ON** & required. LOF: missing (infrequently).	**ON** & required. LOF: missing (infrequently).	**ON** & required. LOF: missing (infrequently).	**ON**. GOF: become R7s.	—	—
dPax2 [paired & ~HD] [1256, 1319, 1320].	—	—	—	—	—	**ON** & required. LOF: arise but abn. diff.	**ON** & required in PPCs. LOF: arise but fail to form the lens properly.	**ON** in all cells of the bristle organ. LOF: arise but are malformed and misplaced.
eyes absent (*eya*) [?] [397, 930].	**ON**.	**ON**.	**ON**.	**ON**.	**ON**.	?	?	?
glass [ZF] [1157, 2963, 2964, 3161, 4391].	**ON** & required. LOF: arises but dies before maturing.	**ON** & required. LOF: arise but die before maturing.	**ON** & required. LOF: arise but die before maturing.	**ON** & required. LOF: arise but die before maturing.	**ON** & required. LOF: arises but dies before maturing.	**ON** (weakly) but not needed.	**ON** but not needed.	**ON** but not needed.

lozenge {runt} [241, 897, 938, 1255].	—	GOF: become R7s.	ON & required. LOF: absent or abn. dev.	ON & required. LOF: becomes outer-type R cell.	ON & required. LOF (weak): become R7s. Null: become outer-type R cells.	ON.	—
m8 {bHLH} [870, 1271].	—	ON in R4.	—	ON & required. LOF: becomes outer-type R cell.	—	—	—
muscleblind {ZF} [265].	ON & required. LOF: arises but abn. diff.	ON & required. LOF: arise but abn. diff.	ON & required. LOF: arise but abn. diff.	ON & required. LOF: arises but abn. diff.	ON (weakly) but not needed. LOF: suppresses transf. to R7s (or other R cells) caused by *svp*-GOF.	ON (weakly) but not needed.	ON (weakly) but not needed.
onecut {cut & HD} [3107].	ON & required. LOF: arise but abn. diff.	ON & required. LOF: arise but abn. diff.	ON & required. LOF: arise but abn. diff.	ON & required. LOF: arises but abn. diff.	—	—	ON in neuron.
orthodenticle {HD} [4461].	ON & required. LOF: arise but abn. diff.	ON & required. LOF: arise but abn. diff.	ON & required. LOF: arise but abn. diff.	ON & required. LOF: arises but abn. diff.	—		
phyllopod {?} [717, 1052, 4249].	—	ON & required. LOF: become cone cells.	ON & required. LOF: become cone cells.	ON & required. LOF: becomes cone cell.	GOF: become R7s.	GOF: become neuronal.	
pipsqueak (BTB-domain cofactor for Svp?) [4567].	—	ON & required. LOF: absent.	ON & required. LOF: absent.	—	—	—	—
pointed [Ets] [516, 3162, 4381].	ON & required. LOF: fails to arise.	ON & required. LOF: fail to arise.	ON & required. LOF: fail to arise (easily affected).	ON & required. LOF: fails to arise (easily affected). GOF: becomes cone cell (in *yan*-LOF background).	ON but not needed?	ON but not needed?	ON but not needed?
prospero {~HD} [1070, 1753, 2165, 4775].	—	—	—	ON (at higher level than cone cells) & required for repression of R8-type rhod.	ON.	—	—

(continued)

TABLE 7.1 (*continued*)

Gene	R8p Cell (Inner)	R2/5p Cells (Outer)	R3/4p Cells (Outer)	R1/6p Cells (Outer)	R7p Cell (Inner)	Cone Cells	PPC, SPC, TPC Cells	Bristle Cells
rough [HD] [232, 1676, 1783, 2223, 3723, 4361, 4459]△.	—	ON & required. LOF: lose R2/5 identity and mimic R3/4 or R1/6. R3/4 & R1/6 dev. abn., but R7 is ~normal, and an extra R7 often arises.	ON but not needed.	—	GOF: becomes outer-type R cell.	—	—	—
senseless (*sens*) [ZF] [1279, 3127].	ON & required. LOF: becomes R2/5. GOF: extra R8s.	—	—	—	LOF: absent.	—	—	—
seven in absentia (*sina*) [ZF] [671].	—	—	ON but not needed (except at late stage?).	ON but not needed (except at late stage?).	ON & required. LOF: becomes cone cell.	ON but not needed.	—	—
seven-up (*svp*) [ZF] [264, 1856, 2322, 2888].	—	GOF: remain neuronal but lose R2/5 traits.	ON & required. Null: become R7s. An extra outer-type R cell arises when R3 & R4 are both *svp*-null.	ON & required. Null: become R7s.	Extreme GOF: becomes outer-type R cell.	Mild GOF: become R7s. Extreme GOF: become outer-type R cells.	—	—
sine oculis (*so*) [HD] [759, 930].	ON & required. LOF: dies.	ON & required. LOF: dies.	ON & required. LOF: dies.	ON & required. LOF: dies.	ON & required. LOF: dies.	ON & required? LOF: dies.	ON & required? LOF: dies.	ON & required? LOF: dies.
spalt [ZF] [220, 221, 2902, 2903, 4855]△.	ON & required (late). LOF for Spalt-C: makes R8-like axon. proj., but ≈ outer-type R cell in rhabd. (large) & rhod. (Rh1).	ON (weakly?) but not needed.	ON but not needed.	ON (weakly?) but not needed.	ON & required (late). LOF for Spalt-C: makes R7-like axon. proj., but ≈ outer-type R cell in rhabd. (large) & rhod. (Rh1).	ON but not needed?	—	—
tramtrack [ZF] [670, 2391,2529, 4773]△. Isoforms: p69 (activator or repressor?) and p88 (repressor).	ON (p69 only) & required (late). LOF (p69): arises but fails to complete dev. GOF (p69 or p88): fails to arise (or dies thereafter).	ON (p69 only) & required (late). LOF (p69): arise but fail to complete dev. GOF (p69 or p88): fail to arise (or die thereafter).	ON (p69 only) & required (late). LOF (p69): arise but fail to complete dev. GOF (p69 or p88): fail to arise (or die thereafter).	ON (p69 only) & required (late). LOF (p69): arise but fail to complete dev. GOF (p69 or p88): fail to arise (or die thereafter).	ON (p69 only) & required (late). LOF (p69): arises but fails to complete dev. GOF (p69 or p88): fails to arise (or dies thereafter).	ON (p69 & p88) & required. LOF: become R7s (?) or fail to develop. GOF (p69 only): absent.	ON (p69 & p88).	ON (p69 & p88) in shaft, socket, and sheath cells; p69 is required in IIa and sheath cells (cf. App. 3).

gene						
ultraspiracle {ZF} [3198, 4855].	ON & required. LOF: arises but abn. diff.	ON & required. LOF: arise but abn. diff.	ON & required. LOF: arise but abn. diff.	ON & required. LOF: arise but abn. diff.	ON & required. LOF: arises but abn. diff.	—
yan {Ets} [2392, 3162, 3467, 4791]△.	ON (transient?). LOF: fails to arise (re-enters cell cycle). GOF (activated construct): lost.	ON (transient?). LOF: fail to arise (re-enters cell cycle). GOF (activated construct): lost.	ON (transient?). LOF: fail to arise (re-enters cell cycle). GOF (activated construct): lost.	ON (transient?). LOF: fail to arise (re-enters cell cycle). GOF (activated construct): lost.	ON (transient?). LOF: fail to arise (re-enters cell cycle). GOF (activated construct): lost.	ON (weakly). LOF: become R7s? GOF (activated construct): fail to arise.
					— ON.	ON in one cell only.

*"**ON**" means that the gene is transcribed in the indicated cell type; "—" means that no expression has been detected (or reported). Motifs (DNA-binding or other) are in braces, and roles (in signaling) are in parentheses. Domain abbreviations: HD (homeodomain), ZF (zinc finger). Phenotype abbreviations: abn. (abnormal), axon. proj. (axonal projection), dev. (development), diff. (differentiation), rhab. (rhabdomere), rhod. (rhodopsin), transf. (transformation). Gene abbreviations: *boss* (*bride of sevenless*), *sev* (*sevenless*). Cell types are listed left to right in order of their appearance (cf. Fig. 7.6). PPC, SPC, and TPC (Primary, Secondary, and Tertiary Pigment Cells) are distinct types but are grouped here to conserve space. The "p" suffix denotes a precursor cell, and slash marks indicate pairs of histologically similar cells (e.g., R1/6p = R1p and R6p). Considering the bristle as a single cell (the SOP generates a neuron, sheath, shaft, socket, and glial cell; cf. Fig. 2.1), but not counting R3-vs.-R4 asymmetry (cf. Fig. 7.5), the ommatidium has 10 basic cell types. LOF and GOF designate loss- or gain-of-function effects.

The so-called "mystery cells" (not listed) affiliate transiently with nascent clusters [186, 4364, 4715] and are known to express *sev* [225, 1049] and *dlun* (weakly after R2/5/8p and before R3/4p) [383]. These few cells are converted to photoreceptors by *argos*^LOF [1294], *boss*^GOF [225, 1049], *echinoid*^LOF [170], *fat facets*^LOF [746, 748, 1241,1932], *groucho*^LOF [1241], *liquid facets*^LOF [594], *rhomboid*^GOF [1293], *sev*^GOF [225, 229, 1049], *yan*^LOF [2392], and probably also by *Delta*^LOF [3273], *extramacrochaetae*^LOF [508], *PLCγ*^LOF [2669], *pointed*^GOF [516], *spitz*^GOF [1289], *string*^LOF [2971], *trantrack*^LOF [4773], and *ultraspiracle*^LOF [457, 4855], based on extra-R7 phenotypes. Mystery cells are normally blocked from adopting an R-cell fate by an unknown signal from neighboring (non-R) cells [594, 1241,1932]. Transmission of the signal appears to require specific intercellular adhesive contacts [170, 2492, 3106] and specific phasing of the cell cycle [747]. In the honey bee *Apis mellifera*, a second R7-type (UV-sensitive) cell apparently arises from a mystery cell via a contact with R8p that fruit flies lack [1049, 1149, 3538].

Several of these genes were named for effects on eye shape (*Bar*, *lozenge*), texture (*glass*, *rough*, *scabrous*), or R-cell composition (*boss*, *sevenless*, *seven-up*) [2561, 4357]. The transcription factors Eyelid (Bright family of DNA-binding domains) [4389], Eye gone (paired-HD) [2090, 2099], and Tailless (zinc-finger class) [955] are expressed in eye cells, but their roles are unclear. The same is true for the co-activators Blind spot and Kohtalo [4386].

Certain genes that do not encode transcription factors also regulate the differentiation of photoreceptors (e.g., *chaoptic* [2323, 3568, 4458] and the Broad-Complex isoforms [457]) or neurons in general (e.g., *abl* [289], *deadpan* [331,1164], *elav* [518, 938, 1887, 2312, 4805], and *scratch* [1164, 3611]).

Other genes are expressed or needed in subsets of photoreceptors [4600] – e.g., *dachshund* (R1/6 and R7) [2689], *daughterless* (R8, R2/5, and R3/4) [506], *Delta* (intense in R3/4 and R1/6) [870], *dead ringer* (R8 and R1–R6) [3917], *hairy* (expressed in R7 but not needed) [507], *hedgehog* (strongly in R2/5, weakly in R8, and later in all) [1077, 2632, 4621], *scabrous* (R8 only) [179, 182,186, 897, 2461], *retina aberrant in pattern* (R8 only) [2145, 3390], *rhomboid* (R8 and R2/5) [897, 4035], *semang* (R1/6 and R7) [4869], *Star* (R8 and R2/5) [1782, 1785], *vein* (R8 only) [4035], and the paralogs *Brother* and *Big brother* (needed for *lozenge* function in R1/6 and R7) [2524]. Cone cells specifically require *rugose* [4470].

This table is an expanded version of tables in [1047, 2351, 2961]. Genes whose products function in the Sev pathway between the receptor and its nuclear effectors are omitted. Their LOF-GOF effects are as expected, given the positive or negative roles of their products (cf. Fig. 6.12) [264, 1048, 4790] – e.g., *Ras1* [1273] and *dMAPK* [517]. Also omitted are genes in the Notch pathway (e.g., *mδ* in R4 [869]), which show transient expression in various precursors. See [897, 3574] for more cell-type markers and [1293, 2887, 2903] for enhancer-trap lines that express *lacZ* in subsets of R cells. See [1161, 3708] for *spalt*'s role in enforcing alternative cell identities in another PNS context. See [4052] for evolution of opsins and [968, 1272, 3895] for evolution of upstream circuitry.

N.B.: Ectopic overexpression of *scute* rescues *ato*^LOF eyes without restoring discernable R8 (or R7) cells [4209]. Conceivably, *scute* is enabling outer R cells to bypass their need to be induced by R8p, but a more prosaic explanation is that the mutant R8p progresses far enough in its maturation to induce R1–6. Indeed, *boss*-on R8p cells do arise in *scute*^GOF eyes, although their development is delayed and their numbers are reduced. The same explanation may hold for *senseless*^LOF eyes, whose ommatidia develop without mature R8 (or R7) cells [1279]. In the latter eyes, R8p cells do not express *boss* or *scabrous*, but they do express *atonal* faintly. Remarkably, R8 (*boss*-on) cells are rescued in *senseless*^LOF *rough*^LOF double mutants (attributable to a "*senseless*⊣*rough*" link) [1279]. Multiple R8 cells per ommatidium arise in *shattered*^LOF eyes [4247].

223

find the genes that can solve the puzzle. Fortunately, ingenious new screens are now underway [3381, 4132, 4292].

One of the many related riddles concerns *tramtrack* (*ttk*) [1051], which encodes two zinc-finger transcription factors (cf. Fig. 2.5). The p88 isoform prevents neural development wherever it is expressed [670, 2529, 4249]. Ttk88 is targeted for proteolysis when it binds Phyl and Sina (cf. Ch. 6) [4249, 4382]. Both Phyl and Sina are normally expressed in R1/6p and R7p cells, so it is easy to see how these cells become neural. What is not obvious, however, is how the other precursors (R8p, R2/5p, R3/4p) manage to stifle Ttk88 without the Phyl-Sina duo.

Phyl has its own secrets [1052]: *phyl* is supposed to be transcriptionally upregulated by the Ras-MAPK pathway [717, 1047, 1052, 4381], and this pathway is activated in all R cells [670, 1289, 1337], but *phyl* is only expressed in a subset thereof (R1/6/7p) [717, 1052]. Why?

Recently, two key clues to the **Photoreceptor Coding Enigma** have come to light. First, two teams (Cooper and Bray [870] and Tomlinson and Struhl [4367]) showed that R7p requires a Delta signal from R1/6p in addition to a Boss signal from R8p. This discovery has revived the Combinatorial Cascade Model, at least with regard to R7:

1. Removing Delta from R1p and R6p causes R7p to develop as an R1/6-like photoreceptor, whereas disabling Delta in R8p has no such effect.
2. Suppressing the Notch pathway in R7p (via a chimeric Su(H)-Engrailed repressor) likewise causes it to adopt an R1/6-like fate.
3. Activating the Notch pathway in R1/6p cells induces them to become R7s.

Second, Mollereau *et al.* reported that both R7 and R8 cells become outer-type (R1–R6) photoreceptors in Spalt-C^{null} eyes, although axonal projections are unchanged [2902]. Expression of *spalt* differs in early (R1p–R6p) vs. late (R7p–R8p) eye development, and this shift may signify a redeployment of *spalt* to encode certain aspects of the differentiated state (i.e., rhabdomere size and rhodopsin subtype). Prospero likewise suppresses R8-type rhodopsins in the R7 cell by directly binding the *rh5* and *rh6 cis*-enhancers [860].

The lattice is created by inhibitory fields around R8 precursors

For anyone interested in pattern formation the abiding allure of the fly eye has always been its "**Lattice Riddle**": how is the ommatidial lattice created? Ready *et al.*'s

Crystallization Model was a good first guess, but, as discussed above, it proved false. We now know that newly recruited cells do not "read" their fates from their predecessors. Rather, new cells assemble in shells around each R8p founder, with no interommatidial communication after the R8p "seed" is planted [4769]. The remaining question, of course, is how those seeds get planted so neatly in the first place.

Because the lattice acquires "nodes" (R8 sites) sequentially, old nodes could be dictating where new nodes arise. A phenomenon of this sort was described in hemipteran insects by Wigglesworth in 1940: bristles that arise in later instars are constrained to certain sites by bristles that arose in earlier instars [4657].

The Lattice Riddle was essentially solved in 1972 by Donald Ede [1140] – a disciple of the great theorist C. H. Waddington [1486, 4826]. Ede was familiar with Wigglesworth's work, and he was adept at building computer models. He combined Wigglesworth's notion of node-centered inhibitory fields [4657] with the concept of a node-inducing wavefront to craft the "**Inhibitory Field and Wavefront Model**" [1140]. This scheme was actually designed to explain feather lattices in bird skin, but it works equally well for the fly eye. The main feature that Ede was trying to simulate is the 0.5 internode offset between adjacent columns – an offset that automatically tessellates the plane with equilateral triangles (Fig. 7.7). (Italics are author's.)

One most tantalizing system which appears ripe for this sort of analysis ... is the establishment of the beautifully regular diamond lattice pattern of feather papillae on the back of the chick embryo. ... The condensations which give rise to feather papillae are initiated in single files, the first of which is immediately over the embryonic neural tube and vertebral column, and the back on either side is subsequently covered by condensations which appear file by file, rather like a line print-out from a computer. Experimental embryologists have shown that production of the feather papillae is dependent on induction by the neural tube and/or the vertebral column, that is the mesenchyme on either side of these structures only becomes capable of producing condensations as some substance diffuses outwards from the midline on each side. Thus, as a model, we may suppose that at first only a narrow band of skin is capable of undergoing mesenchymal condensation, or of initiating this process, but that this band becomes gradually broader. ... As each condensation is produced a zone of inhibition is generated around it, and *these zones will intersect at the apices of equilateral triangles whose bases extend from one condensation to the next.* As soon as the induced band has extended to these points, the next file of papillae will be initiated. In this way, as one file is produced after another, the lattice arrangement of feather papillae is marked out. [1140]

A similar argument was made by Patricia Renfranz and Seymour Benzer in 1989 [3574]. Like Ede, they stressed the half-phase offset between columns, but instead of a wave per se, they proposed that inductive signals (as well as inhibitory ones) emanate from the nodes (preclusters) themselves.

In the *Drosophila* eye disc at the morphogenetic furrow, negative and positive feedback loops between neighboring cells, like those acting in embryonic neurogenesis (reviewed in [626]), could control the spacing between nascent ommatidia at the furrow. Mathematical models of biological pattern formation, particularly formation of periodic structures, often invoke diffusible factors. These models postulate the presence of an activator and an inhibitor, the inhibitor diffusing more readily than the activator (for review, see [2806]). *Note that, in each successive column, the clusters are on diagonal centers, rather than immediately behind those in the preceding column.* An activator of cluster formation, coupled with an inhibitor, could provide a mechanism for regulating the spacing, and therefore the number, of ommatidia in a column. In the differentiating eye disc, such a mechanism would likely involve the cells that form "nucleation sites" at which the nascent ommatidia condense as five-cell preclusters at the furrow. Lateral inhibition between nucleation sites could control the number of ommatidia seeded in a column, and the activator could, in turn, influence the number of cells in each cluster. Such a mechanism, involving graded inhibitory signals, has been suggested for the regular spacing of sensory bristles on the insect cuticle [2959, 4657]. [3574]

In 1990, Nicholas Baker, Marek Mlodzik, and Gerald Rubin [179, 2885] proposed that the inhibitor is Scabrous (Sca) – a secreted glycoprotein that is distantly related to vertebrate fibrinogen [1921, 2461, 2463]. Sca is secreted by R8p cells at the trailing edge of the MF, and the forward diffusion of this molecule could easily constrain the next row of *sca*-ON cells via a "Sca ⊣ *sca*" feedback loop.

... the most straightforward model is that the *sca* protein itself is an extracellular regulator of ommatidial spacing. [179]

The observed eye disc phenotype indicates that *sca* plays a role in establishing the initial periodicity, because the spacing of preclusters and, thus, R8 precursors, in mutant eye discs is clearly irregular. Potentially, there may be interactions between forming preclusters and the assembled column immediately posterior. Indeed, the precise positioning of each new row one-half unit out of phase with respect to the last suggests that the existing pattern has an influence and that *sca* is involved in this regulation, as judged from its phenotype. Inhibition governed by *sca* expression in the R8 precursor might extend from the next posterior column, inhibiting *sca* expression in cells that are closer to the already developing precluster, and thus lead to the observed very regular array of developing ommatidia. [2885]

Sca has been thought to control not only the spacing of ommatidia in the eye but also the spacing of bristles throughout the entire body (cf. Ch. 3). Indeed, these two processes share other key features in common [229, 1804, 4621]:

Bristle Spacing	Ommatidial Spacing
1. Sca is transiently expressed in all cells of the proneural cluster (PNC), and it persists in the sensory organ precursor (SOP) [2885].	1. Sca is transiently expressed in "intermediate groups" [2042] (although this step is dispensable [1077]), and it persists in the R8p [179, 183, 186, 2461].
2. For a cell to become a SOP, it must express the ubiquitous proneural gene *daughterless* [905, 949, 4452] as well as the "regional" proneural genes *achaete* or *scute* [912, 3982].	2. For a cell to become an R8p, it must express *daughterless* [506] as well as the "regional" proneural gene *atonal* [2041, 2042].
3. The PNC is narrowed to a single SOP via the Notch pathway, using the ligand Delta [3022, 3270] but not Serrate [4859]. If that pathway is disabled, then many cells in the PNC become bristles [112, 3022].	3. The "R8 equivalence group" (i.e., the group of *atonal*-ON cells) is narrowed to one R8p via the Notch pathway using the ligand Delta [3273] but not Serrate [183, 2549]. If that pathway is disabled, then most cells in the group become R8s [179, 184, 603, 2461, 2549, 3273] (cf. a similar circuit in mice [4354]). Curiously, R1p-R7p can become neural without expressing *atonal* or any other bHLH factor [1616].

In the above table, the terms "intermediate group" and "R8 equivalence group" are used synonomously [184]. These groups of ~12 cells (henceforth called "IGs") express both *sca* [179, 1158, 1921] and the proneural gene *atonal* (*ato*) [2042, 4621]. They are spaced at uniform intervals near the front of the MF. Since they immediately precede the rosettes (column 0), they are considered to occupy column "−1" [184, 2042]. These nests of *sca*-ON-*ato*-ON cells dwindle to single R8p cells by column 0. If Sca is responsible for ommatidial spacing, then its secretion by R8p cells (column 0) must constrain new *sca*-ON cells in IGs (column −1). The constraint may even happen earlier [186, 2461] because groups of *ato*-ON cells ("initial clusters") have been detected in column "−2" [1077, 4208].

The main evidence for Sca as the R8p inhibitor is the *sca*[null] phenotype. Incipient ommatidia are closer together in *sca*[null] eyes [179]. However, the crowding is milder than it should be if the inhibitory mechanism

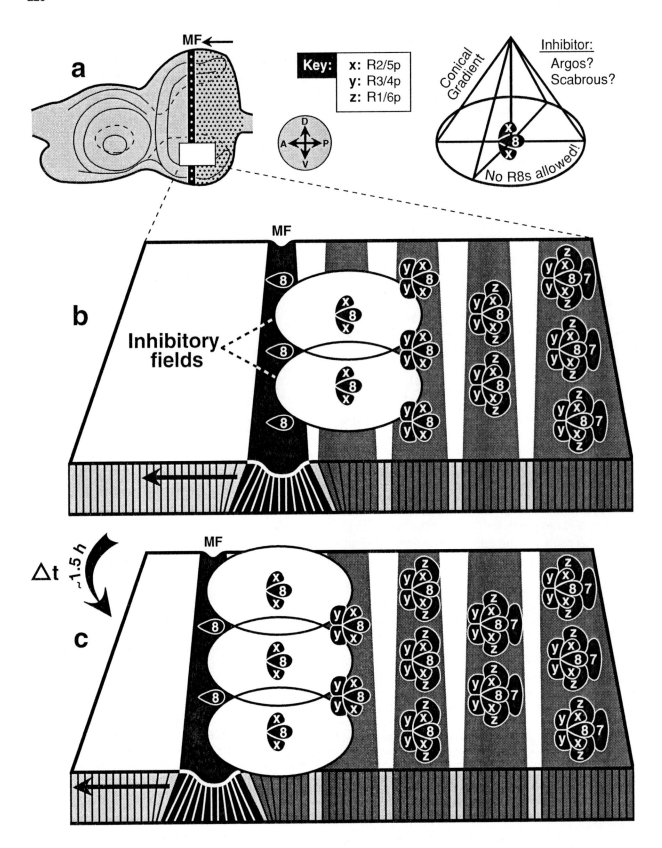

were totally disabled (cf. stronger N^{LOF} and Dl^{LOF} phenotypes [179, 184, 186, 603, 2461, 3273]): only rarely do R8p cells emerge as doublets (or higher multiples) from the MF [179, 186, 1158]. The implication is that *sca* acts redundantly with some other gene [179]. Genetic interactions between *sca* and *Notch* [1921, 2462, 2885, 3490] suggested that Sca might use Notch as its receptor [179], but Sca does not bind Notch [2460]. The actual Sca receptor remains unknown [178, 2461].

The main snag with the idea of Sca as the R8p inhibitor is the sca^{GOF} phenotype [186]. Overexpressing Sca at the front of the MF should extinguish ommatidia, but in fact, it causes ommatidial crowding just like sca^{LOF} [1158]. The LOF and GOF phenotypes also resemble one another insofar as nascent ommatidia are more irregularly spaced. This disorder suggests that cells default to a more random Delta-Notch contest when the amount of Sca is too low or too high [1158]. Conceivably, Sca biases these contests. R8p cells might arise wherever cells sense a particular slope (vs. absolute value) in Sca concentration. The ideal slope would exist near the edge of Sca's diffusion range, although it is unclear whether (and, if so, how) cells can measure such slopes [2434, 2448]. Unraveling this circuitry has been complicated because Notch is used twice during the process [182, 183, 2549, 3028]: first to activate *ato* in rosettes (N $\rightarrow$ *ato*) and later to suppress *ato* in all cells except R8p (N $\dashv$ *ato*). Sca apparently participates in the first phase but not the second [186]. (At the *ato* locus, there are separate *cis*-enhancers for expression in the rosettes vs. in the fainter band that precedes them [4208].)

In 1998, Spencer and Cagan argued that the diffusible R8p inhibitor is Argos, not Sca [4035]. They suspected the EGFR pathway in the bestowing of R8 competence because high levels of Argos [4035], Spitz [4322], and activated dMAPK [2355, 2493, 4035] are detected with Atonal [184, 4035] in the same 6–10 cells in each rosette. However, the EGFR pathway is not needed for ommatidial spacing: spacing is nearly normal in $Egfr^{null}$ [2355], $rhomboid^{null}$ [4035], and $spitz^{null}$ [1083, 4322] clones (although R8 cells fail to form inside doubly mutant $rhomboid^{null}$ $vein^{null}$ clones [4035]). Null alleles of *argos* do alter the lattice, but their effects are mild [204, 1294, 4035]. The absence of a strong LOF effect does not rule out Argos as the inhibitor since it might have a redundant partner. However, the hypothesis seems untenable in the face of a more troubling fact: $argos^{LOF}$ eyes can be restored to a wild-type phenotype by expressing *argos* with a *sev* enhancer [1287]. How important can Argos be for R8p spacing if it can be erased from the rosettes (by $argos^{LOF}$) and banished to clusters behind the MF (by *sevE-argos*) without disturbing the R8p pattern?

Thus, neither Scabrous nor Argos can claim "hero" status in this saga [4621]. The Lattice Riddle therefore endures as a daunting challenge in search of a molecular protagonist [204].

The lattice is tightened when excess cells die

Given that ommatidia self-assemble as separate islands by recruiting cells from the surrounding epithelium, the eye encounters a serious engineering problem. If the R8p founder cells are spaced too close together, then

---

FIGURE 7.7. The **Inhibitory Field and Wavefront Model** for creating a hexagonal lattice. See also Appendix 7.

a. Mature left eye disc. Vertical stripe is the morphogenetic furrow (MF). Compass (at right) gives axes in the fate map (cf. Fig. 7.1). The MF leaves a hexagonal lattice of ommatidial cell clusters (dots) in its wake. Oval (far right) denotes an inhibitory field wherein no R8 photoreceptors can arise. This field is supposed to be generated by a pre-existing R8 precursor ("8" or "R8p") and possibly other R-cell precursors (e.g., "x" = R2/5p) [2493]. The identity of the secreted inhibitor is unknown. Suspected candidates have included Argos [4035] and Scabrous (Sca) [179, 2352, 2885], but Argos cannot be acting alone because $argos^{null}$ clones have normal spacing of R8s [204]. Secretion implies a cone-shaped gradient.

b, c. Imaginary slices of the columnar epithelium (magnified from box in **a**) at two times separated by ∼1.5 h ("Δt"). The MF is a groove (black) that runs along the D-V axis. It moves from P to A (long arrows). R8 precursor cells (teardrop shapes) arise within "intermediate groups" (not shown; cf. Fig. 7.9) at intervals along the MF [184] and found the ommatidia (cf. Fig. 7.6) [1048, 1279]. Gray stripes mark columns of nascent ommatidia, which are older (and hence more complete) as distance (and time) from the MF increases. Successive columns are out of register (shifted vertically) by 0.5 R8p interval, thus forming a hexagonal lattice. The phase shift is attributable to rules that govern how new R8p cells arise: for a cell to become an R8p, it must be (1) in the MF and (2) not in an inhibitory field. Thus, as the MF moves anteriorly (compare **b** with **c**), the first cells that become competent for the R8 fate are those that lie in the crevices of the preceding inhibitory fields. Fields must arise within ∼1.5 h after R8p inception. How long they persist is not known. Here they are shown in only one column, although expression of Sca (and Ato) persists in R8p cells for ∼4 columns [183, 1158, 4621]. Also unknown is whether the signals come from single R8p cells [2885], from nests of a few cells as shown here [2632, 4035], or from the intermediate groups whence R8p cells arise [186, 2461, 4208]. The signaling cells are presumably deaf to the inhibitor because they probably disable their receptors (or transduction pathway) before they start signaling.

the epithelium will run out of cells before every ommatidium acquires its quota [1003, 3655]. However, if R8p cells are too far apart, then lots of unrecruited cells will be left over [2493]. Stationing R8p cells at exactly the right intervals *ab initio* is not an option because the spacing mechanism uses a sloppy diffusible signal. Evolution solved this "**Spacing Precision Problem**" by adding a "clean-up" phase after the spacing and accretion phases have run their course (cf. the "fine-tuning" tricks in bristle patterning; e.g., Fig. 3.9d). Thus, more cells are allocated to the epithelium than will be recruited, and the excess cells are pruned by cell death [602, 3509, 4715]. This "**Kill the Stragglers! Trick**" of producing an excess of competent cells and then destroying those that do not get incorporated is common in animal development [2009, 2840, 3493, 4037] – especially in nervous systems [887, 888, 1141, 3187, 3189, 3358] where it ensures parity of interconnected populations [798, 2149, 2150, 3188, 3478, 4685], as in the fly's retinolaminar projection [624, 1270, 4774].

At the beginning of pupation, the retina has more cells than it will ultimately use. As the initially disordered interommatidial cells are sorted into a lattice, ~2000 excess cells are eliminated by cell death.... Cell death appears to eliminate cells which fail to establish contacts appropriate to a specific cell type. As interommatidial cells are sorted into the lattice, some cells are left with no opportunity to make the contacts needed to survive. For example, a cell may be blocked from contacting a second primary pigment cell by the presence of an intervening cell. The blocked cell will be eliminated by cell death. As surplus cells are eliminated, the lattice tightens. At this stage two cells are often equally positioned to become the single secondary seen in the adult. The later phase of cell death generally eliminates one of these cells. [602]

This "Darwinian" strategy for tightening the lattice is clever, but it puts the eye in a precarious situation. To wit, slight mistakes in the timing or strength of induction could lead to runaway cell death. This instability may help explain why so many genes can mutate to give a small- or no-eye phenotype [396, 2561]. Other consequences of the strategy are seen in various features of eye development:

1. Rosettes in the MF are spaced more irregularly than the nodes in older columns [184, 4715].
2. The extra interommatidial cells can replace missing cone cells when the latter are forced by *sev*GOF to become R7s [225]. (Their transition to a cone-cell identity may be too late for *sev*GOF to also turn them into R7s.)
3. The number of deaths varies among normal ommatidia, as expected if there is "noise" in the initial spac-

ing of R8p cells [4769]. The average loss is 2 to 3 cells per ommatidium [4713].
4. More extensive cell death is observed in *Egfr*GOF discs whose clusters are widely separated [181], and the death occurs earlier (columns ~10–15 vs. a broad swath behind column 12 [4715]).

Stragglers commit suicide via the usual pathway of apoptosis (a.k.a. programmed cell death) [201, 1590, 2774, 2797, 4618]. In flies, this pathway is executed by *grim* [741, 3621], *reaper* [3134, 3621, 4078, 4619, 4620], and *hid* (*head involution defective*) [1621, 4402] via the caspases (cysteine proteases) DCP-1 [4018], drICE [1283], Dredd [742], etc. [1941]. The Notch pathway is also involved [396, 603, 605, 2859, 3273, 4248].

Doomed cells can be rescued by *Egfr*GOF, *Ras1*GOF, or *argos*LOF [515, 2859]. This result chimes nicely with the Stop the Clock! Model (see above) [2859], if one assumes that the final tick of the clock causes a cell to kill itself. Spitz would thus be a survival factor in addition to its fate-assigning role [185]. In this scenario, each epithelial cell would be brooding: *"If Spitz does not arrive before midnight, then I'll kill myself!"* Consistent with this surmise, more cells die when the Spitz signal is blocked in *argos*GOF or *Egfr*LOF eyes [201, 1287] or in *Star*null clones [1785, 1788]. Likewise, extensive apoptosis occurs anterior to the MF in "furrow-stop" mutants where the MF halts prematurely (Bar-CGOF [1309, 1310, 1788], *dpp*LOF [724], *Drop*LOF [1788, 2971], *hh*LOF [406, 724, 1788, 2632], *rough*GOF [725, 1788], and possibly *daughterless*LOF [506]) as well as in mutants where the MF never initiates properly (*dachshund*LOF [2689], *dpp*LOF [529, 2739], *eyeless*LOF [1309, 1684], *eyes absent*LOF [397, 2485], and *sine oculis*LOF [759] but apparently not *atonal*LOF [2042]). Finally, Spitz's rescuer role is confirmed by the nonautonomy of apoptosis in *spitz*LOF clones: core cells die, but cells near the edge survive apparently due to Spitz diffusing in from surrounding wild-type cells [185]. The level of EGFR signaling needed to keep a cell alive is less than the level needed to change its fate [1687].

Anterior cells can be rescued by agents that block apoptosis, whereupon they become SPCs or TPCs [1763, 4248]. Evidently, SPC and TPC are the default states when suicide is not an option.

Eye bristles arise independently of ommatidia

Apoptosis occurs maximally between 35 and 55 h AP (at 20°C) in wild-type eye discs [515, 1287, 1289, 4713, 4715]. By the start of this period, ommatidia have acquired all their photoreceptors (8), cone cells (4), and PPCs (2) but have not yet conscripted SPCs, TPCs, or bristle cells from the disorderly background of interommatidial cells [602, 4713].

Bristle development is well advanced by this time because (1) *ac*-ON proneural clusters appear just after pupariation [597], (2) bristle SOPs arise at ~12–24 h AP (20°C) [3273], and (3) differentiative mitoses ensue at ~14–28 h AP (20°C) [602, 1287, 4715]. Virtually every background cell can be forced to make a bristle (at the expense of SPCs and TPCs) by heating t.s. Dl^{LOF} mutants at 11–18 h AP (20°C) [3273] (cf. N^{LOF} [603]). This phenotype indicates that the Notch pathway selects bristle cells within the eye field, just as it does in other discs (cf. Ch. 3).

Because the same pathway also helps assemble ommatidial clusters [603], there is the potential for chaotic cross-talk between SOPs and R cells, which could wreak havoc on the lattice [1804]. The fly avoids such chaos by using the Heterochronic Superposition Trick (cf. Ch. 3). It times its SOP selection and R-cell recruitment so that these periods do not overlap. Indeed, SOPs do not rely on the MF at all [597]. They develop in a separate wave that spreads *radially* from the center to the periphery, not from P to A [602, 603, 605, 4715]. Moreover, bristles avoid using the EGFR pathway: $argos^{LOF}$ eyes have extra R cells, cone cells, and pigment cells, but the number of bristles (albeit disorganized) is normal [1294].

Bristles assume their final positions at ~35–40 h AP (20°C), as the lacework of background cells "shrink wraps" each ommatidium via apoptosis [4713]. When bristle initiation is prevented throughout the eye (by wg^{GOF}), the facet vertices where bristles would have sprouted are occupied by extra TPCs [597].

Because eye bristles arise independently of ommatidia, it is not surprising that the vertices to which they are relegated are not as orderly as the ommatidial lattice itself. These vagaries are apparent in the wild-type eye schematized in Figure 7.8. Most bristles reside at the anterior end of the horizontal edge between facets, but there are many exceptions to this trend [3539]. The exceptions obey a rule of their own: whenever a bristle is missing from the A end of the edge, a bristle is found at the P end (except at the periphery, which is bald). These A-to-P displacements are common in the anterior third of the eye but are sporadic elsewhere. Why are bristles near the front of the eye less choosy?

One clue to this "**Fickle Bristle Mystery**" (cf. Fig. 5.12g) comes from the eyes of t.s. N^{LOF} mutants that were heat-pulsed during the sensitive period for extra eye bristles. About 6 columns behind the vertical extra-bristle "scar," ≥30% of the bristles are shifted to the P end of the edge between facets [603]. Apparently, the

operational proneural cluster (PNC) stretches along each horizontal edge (SPC cell) [1804]. Within this PNC, only one SOP normally originates. Throughout most of the eye, the SOP might arise at the PNC's A end due to a "prepattern" agent that biases the Delta-Notch competition in that direction (cf. Fig. 3.6). If that agent is linked to the MF, then the fickleness of the anterior region could have something to do with the fact that the MF travels through this part of the eye field after pupariation [602, 1287, 4715]. During that period, the MF moves faster (~1.0 vs. 1.5 column per hour) [228, 4715], possibly accelerated by the higher titer of ecdysone [457]. Also, the proneural gene *ac* turns ON anterior to the MF for the first time [597]. Thus, the regions behind vs. ahead of the MF may be influenced differently by factors that control SOP initiation, but we do not yet know how.

The MF operates like a moving A/P boundary

The MF of the eye disc resembles the A/P boundary of the wing and leg discs insofar as Hh activates *dpp* at the border when it diffuses anteriorly from the P region [457, 2466, 2632, 3238]. Despite this shared "Hh ➤ *dpp*" link, *hh* expression behind the MF is not under *en* control [572, 4168], so the MF cannot be moving by a "Hh ➤ *en* ➤ *hh*" loop (cf. the Minotaur Scenario, Ch. 5). The obvious differences are that (1) the MF is not a compartment boundary [2933], and (2) the MF moves [571, 1784, 4387]. Thus, the MF is a state-altering machine wherein cells switch from one transient identity to another until they settle upon their final fate (cf. the Cabaret Metaphor, Ch. 4).

At first glance, Hh seems irrelevant for MF movement because the MF can traverse and "assimilate" (i.e., create ommatidia in) tissue that is deaf to Hh (e.g., smo^{null} clones [1077, 1616, 4169]; cf. hh^{null} clones [1082, 2632, 4169]). Similarly, Dpp seems irrelevant because the MF can traverse and assimilate tissue that is deaf to Dpp (e.g., clones that are Mad^{null} [4648], tkv^{null} [570, 1616], or $punt^{LOF}$ [570]; cf. large dpp^{null} clones [570, 1788, 4648]).

According to the "**MF-pushing Model I**," proposed in 1993 by Heberlein *et al.* [1788] and Ma *et al.* [2632], Hh is supposed to diffuse anteriorly to turn *dpp* ON in the MF, with Dpp somehow reciprocally enabling *hh* to turn ON in R cells. However, this "Hh ➤ *dpp* ➤ *hh*" loop is disproven by the ability of the MF to traverse singly deaf clones that should be disabling the loop.

Nevertheless, *hh* must be involved because the MF arrests in t.s. hh^{LOF} mutants when they are heated [2632]. Dpp was also thought to participate in MF movement because the MF at least slows down (if not stopping completely) when t.s. dpp^{LOF} mutants are heated [724].

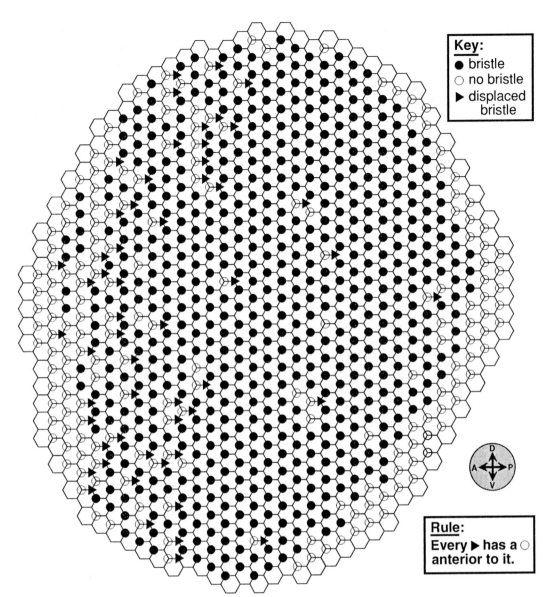

Key:
- ● bristle
- ○ no bristle
- ▶ displaced bristle

Rule:
Every ▶ has a ○ anterior to it.

FIGURE 7.8. "**Fickle Bristle Mystery**" of the fly's eye. The entire left eye of a wild-type female specimen is schematized (cf. key). Bristles (black symbols) occupy vertices in the lattice of facets (hexagons), but their pattern is less perfect. Directions (compass) are A (anterior), P (posterior), D (dorsal), V (ventral). The general rule is: "Bristles reside at the A end of each horizontal interface between ommatidia." Bristles that obey this rule are indicated by solid black circles. Wherever this rule is obeyed, eye bristles occupy alternating vertices around each facet, with tertiary pigment cells at the remaining vertices (cf. Fig. 7.2). The rule is violated at the A and P margins, where bristles are missing (unfilled circles). It is also broken by bristles that lie at P ends of horizontal interfaces (triangles). These sporadic "errors" occur most frequently near the A margin. They exhibit their own sort of rule (lower right): "Wherever a bristle arises at the P end of an interface, none forms at the A end." This correlation implies that (1) one bristle SOP arises per interface, and (2) this SOP is free to move to one end or the other [1804, 4715]. In other words, the errors are actually displacements (cf. Fickle Sensilla Mystery; Fig. 5.12g). Redrawn from [3539].

N.B.: The fly eye is nearly circular: there are 32–34 columns in a typical wild-type female, with ~32 ommatidia in the tallest column [2571, 3539, 4715]. The absence of bristles from the front of the eye is probably due to the remnant of *wg*-ON tissue that is left after the D and V margins of *wg*-ON cells have shrunk (as the MF encroaches anteriorly) [2554]. Wg's role as a diffusible "bristle suppressor" is shown by (1) *wg*^{GOF} clones which scar the eye and, in so doing, delete bristles from a swath of facets on either side of the scar [4390]; (2) *sevE-wg* eyes, which lack all bristles due to expression of *wg* in every facet under the control of the *sev* enhancer [518, 597, 3659]; and (3) *naked cuticle*^{LOF} eyes whose bristleless margin widens due to an increased cellular responsiveness to Wg [3659]. We do not know why bristles are normally absent from the rear of the eye. In other insects, the interommatidial bristles obey similar rules but vary in density [1853]. The number of bristles per vertex increases in *E(Elp)24D*^{LOF} eyes [4034].

Also, dpp^{LOF} flies have small or missing eyes [2739, 4033], although this trait has been traced to a separate role for Dpp in MF initiation (see below).

The aforementioned mosaics do offer one useful clue to solving this mystery: the MF slows down within the deaf patch, and the slowing is more pronounced at the center of the patch. This nonautonomy implies local rescue by an agent "X" that diffuses in from the surrounding wild-type tissue. Because the rescue is seen regardless of whether the patches are deaf to Hh [1616, 4169] or Dpp [570, 1616], X might be able to rescue either kind of tissue. Alternatively, Hh and Dpp might be redundantly serving as the rescuer for one another [1780]. Indeed, when cells are made simultaneously deaf for both Hh and Dpp (smo^{null} Mad^{null} clones [930] or smo^{null} tkv^{null} clones [1616]), they fail to support either MF movement or photoreceptor development. Evidently, therefore, the basic logic of MF progression is

{Hh OR Dpp} ➡ MF movement.

The notion that Hh and Dpp act redundantly is counterintuitive for several reasons. First, Hh and Dpp do not affect one another reciprocally: Hh activates dpp (albeit in only part of the anterior region) [457, 2466, 2632, 3238], but there is no evidence that Dpp activates hh. Second, Hh and Dpp are distributed quite differently relative to the MF (Fig. 7.9).

If Hh and Dpp are truly redundant, then they should behave similarly with respect to initiating extra MFs when they are ectopically expressed. In fact, however, they do not. Extra MFs (anterior to the endogenous MF) can be induced by goading either the Hh or Dpp pathway, but only the MFs induced by Hh-pathway stimulation then go on to express *atonal* and make photoreceptors. The full inventory of manipulations that can spark new MFs is listed below:

1. Locally activating the Hh pathway via hh^{GOF} clones [722, 1077, 1786], ptc^{LOF} clones [722, 723, 2631, 3562, 4167, 4572], or $DC0^{LOF}$ clones [1082, 1788, 3238, 4167, 4171, 4572].
2. Locally activating the Dpp pathway via dpp^{GOF} clones [724, 3388] or by overexpressing dpp using *dpp-Gal4:UAS-dpp* [724] or other *Gal4* drivers [3388].
3. Locally blocking Wg transduction via arr^{LOF} clones [4573] or ubiquitously blocking Wg signaling by exposing t.s. wg^{LOF} mutants to high temperature [2631, 3562, 4390].

Conceivably, dpp^{GOF} (unlike hh^{GOF}) is acting indirectly by suppressing wg and thereby enabling Hh to launch a MF from the dorsal anterior margin [1788] (i.e., "Dpp ⊣ Wg ⊣ Hh ➡ MF initiation" vs. "Dpp ➡ MF initiation"). The argument for an indirect role is bolstered by the aforementioned fact that tkv^{Q253D} (activated receptor) clones fail to undergo neuronal differentiation, in contrast to hh^{GOF} clones, which do [570, 1616, 1786]. It is also supported by several other findings:

1. Ectopic Dpp only induces extra MFs at the margins [3388] where hh is expressed in young discs [1082], implying a reliance on hh (although the "hot spot" for MFs at the A margin does not match Hh's high point and may involve a "Dpp ➡ hh" link [406]). This reliance is proven by the inability of MFs to arise in discs where dpp is overexpressed (via *dpp-Gal4:UAS-dpp*) sans hh function (due to hh^{LOF}) [406]. In the interior of the eye field, dpp^{GOF} clones behave like tkv^{Q253D} clones: they fail to make R cells [3388], presumably due to the paucity of Hh there.
2. Extra MFs in wg^{LOF} mutants also tend to arise at the margins [2631, 4390] – typically the D and A margins (but not the V margin where a redundant Wnt may be acting) [4387]. Again, the suspicion has been that removing Wg allows Hh to spark a new MF. This suspicion is confirmed by the inability of wg^{LOF} to induce new MFs when combined with hh^{LOF} [406].
3. Hh cannot be inducing MFs indirectly via Dpp (or Wg) because goading the Hh pathway via $DC0^{LOF}$ induces MFs even when triply mutant ($DC0^{LOF}$ dpp^{LOF} wg^{LOF}) clones are blocked from emitting Dpp [1082]. Nor do $DC0^{LOF}$ cells need to "hear themselves talk" because MFs are induced by $DC0^{LOF}$ smo^{LOF} clones [4169].

In 1999, Greenwood and Struhl stated the paradox clearly [1616]: how can Dpp rescue photoreceptor differentiation within smo^{null} clones if Dpp cannot evoke such differentiation ectopically? To solve it, they proposed a "**MF-pushing Model II**," which again invokes a diffusible factor X (cf. a similar model by Strutt and Mlodzik [4169]):

1. Hh ➡ {Dpp AND X}.
2. {Hh OR X} ➡ *atonal* ➡ photoreceptor differentiation and MF movement.

Their data suggests that X may act through dRaf via a RTK receptor other than Egfr. However, that receptor should still be present in smo^{null} tkv^{null} cells, so X should be able to rescue them in a paracrine manner. Because it does not, the model must be flawed. The flaw can be easily fixed though by supposing that X only acts in

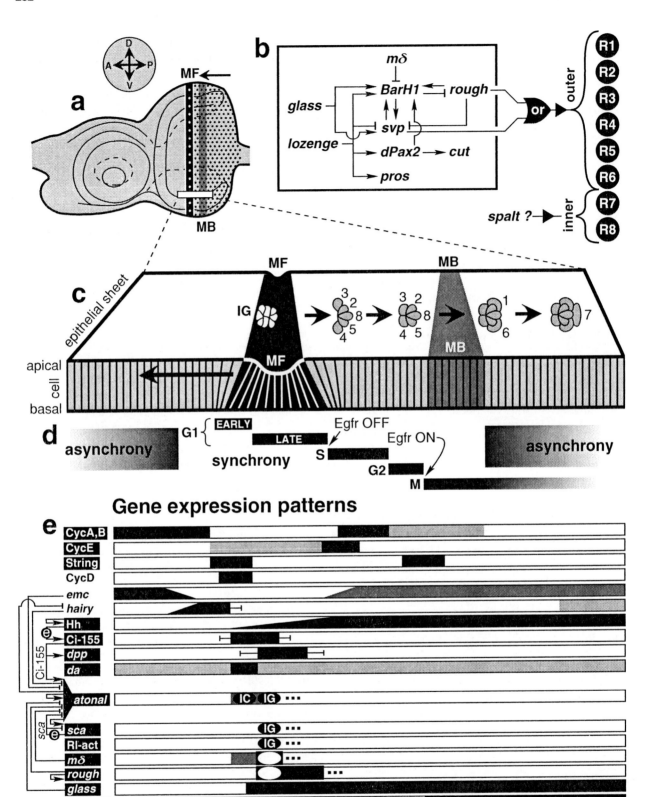

Gene expression patterns

conjunction with Dpp:

1. Hh ➡ {Dpp AND X}.
2. {Hh OR [Dpp AND X]} ➡ *atonal* ➡ etc.

Now, all the key facts can be explained:

1. A *smo*[null] clone is rescued because Dpp and X can both diffuse in.
2. A *tkv*[null] clone is rescued because Hh can diffuse in.
3. A *smo*[null] *tkv*[null] clone is not rescued because X alone is impotent.
4. MFs sparked by ectopic Hh can make ommatidia because they have Dpp and X.
5. MFs sparked by ectopic Dpp cannot make ommatidia because they lack X.

According to this reasoning, the cycle of events in the MF "engine" would be as follows:

Step 1.   Hh is produced by R cells behind the MF.
Step 2.   Hh diffuses anteriorly to induce Dpp, X, and *atonal*.
Step 3.   Dpp and X jointly activate *atonal*.
Step 4.   Atonal enables cells to adopt the R8 fate.
Step 5.   A new column of R8p cells is created, thus rekindling the process.

Other riddles remain, however. For instance, expressing Hh ubiquitously should induce MFs everywhere, but only one new MF appears when *hs-hh* larvae are exposed to high temperature [2466]. This extra MF lies parallel to the endogenous MF and ~1 MF-width ahead of it. Apparently, the "Hh ➡ *dpp*" link operates only in a certain part of the prospective eye region [1788, 3388] (cf. its restrictions in the leg disc, Ch. 5). Another odd finding is that overexpressing Hh in R8p cells (via

---

FIGURE 7.9. Genetic logic of eye development and dorsal-ventral stripes of gene expression in the eye disc epithelium during the 3rd-instar and early pupal period. See also Appendix 7.

a. Mature left eye disc. Vertical stripes are the morphogenetic furrow (MF) and mitotic band (MB). Compass (above) gives axes in the fate map (cf. Fig. 7.1). The MF traverses the epithelium from P to A, leaving a hexagonal lattice of nascent ommatidia (dots) in its wake. The MB is needed to supply enough cells to finish assembling the ommatidia [1003, 3655, 4712].

b. Circuitry among genes that encode transcription factors for photoreceptor cell fates (➡ activation; ⊣ inhibition; see text). Abbreviations: *pros* (*prospero*), *svp* (*seven-up*). *BarH1* is one of two genes (the other being *BarH2*) in the Bar Complex. A combinatorial code for R cells is thought to exist [1272], although there is little evidence for it except for *rough* and *svp*, which dictate outer-type identity, and even there the data is murky. Neither *rough* nor *svp* can convert R7p to an outer state in a *sev*[null] background [232, 1856], so R7p must require the Boss-Sev reaction to become neural [4551]. Also, Rough and Svp are not interchangeable: they have different effects when expressed in cone cells [264, 4551].

c. Imaginary slice of the columnar epithelium (magnified from box in **a**) at one instant of time. The MF is a groove (black) that runs along the D-V axis. It moves from P to A (long arrow). Each nascent ommatidium that it creates undergoes a series of transformations (short arrows). These stages can be seen in the (static) A-to-P sequence of nodes along each row because each successive column is older. The "intermediate group" stage (IG) precedes the rosette stage (not shown; cf. Fig. 7.5) [2042]. Within each IG, 2–3 R8p-like cells express *senseless* (not shown) [1279], which then turns OFF in all but one cell that becomes R8. Ovals (top view) are apical profiles of single cells (mystery cells not shown). Vertical lines (side view) are cell boundaries. The epithelium is actually thicker, and cell packing is less regular [4715].

d. Stages of the cell cycle. The span of each stage along the A-P axis is indicated by a black bar. As cells enter the furrow, they synchronize but lose this synchrony around the time that they begin mitosis [4715]. Arrows mean that a cell's EGFR pathway must be OFF for it to go from G1 to S but must be ON for it to go from G2 to M [185]. String is the downstream gating factor for the latter transition [185].

e. Domains of gene or protein expression relative to eye epithelium in **c**. Each stripe of gene expression along the D-V axis (parallel to the MF) is shown here in terms of its width along the A-P axis. Degrees of gene expression are indicated by shades of gray or by graded slopes. If LOF effects have been found, then the name of the gene/protein is in white letters on a black rectangle. Error bars indicate confidence limits. Black ovals denote expression in IGs or earlier "initial clusters" (ICs) [4208], whereas a black frame (hollow oval) denotes expression between IGs. Thus, *sca* exhibits a punctate pattern of spaced islands (IGs), while *atonal* displays a wider stripe with intense A (IC) and P (IG) foci and a P edge that describes a sine wave [183, 184, 1921, 2461]. ICs and IGs are out of phase relative to one another (not shown), and the IC phase is dispensable [1616]. Genes or proteins: *atonal* [1077, 2493]$^\Delta$ (autoregulation [4208]$^\Delta$; links to *glass* [2042], *hairy* [2042]$^\Delta$, and *sca* [1077]), Ci-155 (the activator form of Cubitus Interruptus) [1077, 1616]$^\Delta$, CycA–E (Cyclins A–E) [185]$^\Delta$, *da* (*daughterless*) [506]$^\Delta$, *dpp* (*decapentaplegic*) [725, 3380], *emc* (*extramacrochaetae*) [508], *glass* [2569]$^\Delta$, *hairy* [2540]$^\Delta$, Hh (Hedgehog) [287] (autoregulation [1786, 3238]), *lozenge* [639]$^\Delta$, *mδ* [2549]$^\Delta$, Rl-act (activated Rolled; a.k.a. MAP kinase [332]) [2493]$^\Delta$, *rough* [725]$^\Delta$, *sca* (*scabrous*) [183]$^\Delta$, and String [185]$^\Delta$.

Wiring at left (➡ activation; ⊣ inhibition; circled "e" = extracellular signal) shows some genetic interactions (see text or references above).

Links in panel **b** are discussed in the text (cf. Table 7.1). Panel **c** is adapted from [4715] and **d** is based on [185, 4291, 4293]. Panel **e** is styled after [1786, 1788, 4387], with additional data from the above references. For further circuit analysis, see [395, 4387]; for circuits that affect nonoptic head structures, see [80, 3380, 3663–3665, 4654]; and for comparable analyses in other systems, see [309, 2093, 4289].

109-68-Gal4:UAS-hh) does not perturb eye development, nor does overexpressing Hh in all cells behind the MF (via *GMR-Gal4:UAS-hh*) [4621].

Dpp and Wg control the rate of MF progress

Another observation made by Greenwood and Struhl indicates a role for Dpp in pacing the furrow: *tkv*^{Q253D} (activated receptor) clones cause the MF to move faster. This acceleration is attributable to a "Dpp → *hairy*" link uncovered in the same study. Hairy is one of two antineural regulators of *atonal* that are expressed just ahead of the MF [508] (cf. Fig. 3.12). The other is *extra-macrochaetae*: "{Hairy OR Emc} ⊣ *atonal*". The stripe of *emc*-ON cells precedes the *hairy*-ON stripe, which in turn precedes the *dpp*-ON stripe (Fig. 7.9). Although h^{LOF} and *emc*^{LOF} exhibit negligible effects in the adult eye, patches of doubly mutant h^{LOF} *emc*^{LOF} tissue allow the MF to outpace flanking wild-type tissue by as many as 8 columns [508]. Because Dpp is under Hh control, Hh has opposing effects on *atonal* at different distances [1784, 1786, 4387]:

Short-range activation: {Hh OR [Dpp AND X]} → *atonal.*

Long-range inhibition: Hh → *dpp* → *hairy* ⊣ *atonal.*

In contrast to its role as a brake far from the margins, Dpp acts an accelerator along the D and V margins, and Wg becomes the brake [2631, 3562, 4390]. (Wg also retards the MF per se [4390].) At each tip of the MF, the *dpp*-ON stripe extends a perpendicular arm anteriorly into the *wg*-ON patch (Fig. 7.3). The arms must be under separate genetic control because (1) they persist when hh^{LOF} causes the main *dpp*-ON stripe to disappear [406, 2632], (2) they are less sensitive to suppression by *rough*^{GOF} [4389], (3) they are the first *dpp*-ON domain to disappear when *dpp* expression is artificially reduced [724], and (4) they are the only *dpp*-ON cells to exhibit autostimulation (Dpp → *dpp*) [724]. The latter feedback loop pushes the *dpp*-ON domain into *wg*-ON territory independently of the MF engine described above:

Step 1.   Dpp spreads anteriorly from the "arm" cells into the *wg*-ON region.

Step 2.   Once there, Dpp turns OFF *wg* and turns ON *dpp*, thus extending the arm.

Given the nature of the feedback loop, Dpp diffusion should drive *dpp* expression into the interior of the eye field as well [724, 3388] (cf. the Minotaur Scenario, Ch. 5).

Because it does not, some unknown agent must keep the loop at bay.

The evidence for a mutual antagonism between Dpp and Wg in the arm regions – and elsewhere in the eye field – is as follows:

1. Overexpressing Dpp around the perimeter of the eye (via *dpp-Gal4:UAS-dpp*) abolishes the V patch of *wg*-ON cells and reduces the D patch to a tiny remnant in the ocellar region [724]. The same effect is seen with randomly induced *dpp*^{GOF} clones [3388]. Implication: "Dpp ⊣ *wg*".

2. Loss of Dpp from the P margin of the eye (in *dpp*^{LOF} mutants [724, 4648] or *Mad*^{LOF} clones [4648]) is associated with *de novo* expression of Wg there. Implication: "Dpp ⊣ *wg*".

3. When *wg* is suppressed by heat-treating t.s. *wg*^{LOF} mutants, the posterior arc of *dpp*-ON cells expands anteriorly to occupy what used to be the *wg*-ON areas along the D and V margins [2631, 4390]. Implication: "Wg ⊣ *dpp*".

4. Randomly induced *dpp*^{GOF} clones can stimulate the endogenous *dpp* gene in marginal regions, except where *wg* is expressed [3388]. Implication: "Wg ⊣ *dpp*".

Despite the similarity of this antagonism to the circuitry of leg development (cf. Ch. 5), Wg's effect on Dpp appears to be at a post-transcriptional level [1780, 4390]. Moreover, *dpp* is here playing only a permissive (vs. instructive) role: it coordinates the rates of movement of the tips with the MF proper [4387].

Dpp influences the MF rate by modulating the cell cycle [3330], specifically by promoting G1 arrest [1903]. Indeed, the famous groove that forms the MF may be an irrelevant side effect of mitotic synchrony. In the wing disc, two MF-like furrows flank the D/V line apically [498, 500, 1178] at zones of mitotic quiescence [2079, 2846, 3374, 3979, 4683] while a single groove runs basally, and an ectopic furrow has been artificially created along the wing's A/P border by using *ptc-Gal4* to drive a dominant-negative *dCdc42* allele [1133].

The MF originates via different circuitry

The MF arises where the D/V (*mirr*-ON/OFF) interface intersects the P margin [696]. This intersection point normally coincides with the optic stalk, but the stalk per se cannot be causal because the MF can be uncoupled from it. To wit, when the *mirr*-ON/OFF line is pushed ventrally relative to the stalk (due to *dpp-Gal4:UAS-wg* [1781] or *ey-Gal4:UAS-pannier* [2752]), the MF starts at the ventral site where this line touches the margin. The circuitry of MF initiation differs from the circuitry of MF progression in other features as well [1784, 4387].

During initiation, *atonal* is activated in four spaced clusters of cells just anterior to a *hh*-ON zone at the rear of the disc [1077, 1082]. The posterior two are unusual insofar as they never acquire all 8 R cells (one stops at 3, the other at 5), and both of them disappear without being incorporated into the final lattice [4364]. Thus, the next column (5 clusters) comes to occupy the rear edge of the final array [4364]. Nevertheless, the MF can arise (and move slightly) without either *atonal* expression or photoreceptor maturation [2042]. This ability is a salient difference from MF progression, which does require R cells [1786, 1788]. Other odd aspects of the eye's P edge include

1. The P margin is the only part of the eye where the Notch pathway is not required for proneural competence [183].
2. The EGFR pathway is required there for MF initiation [1083, 2355], whereas it is dispensable elsewhere for MF movement [2355].

Before MF initiation, *dpp* is expressed at the D and V margins and weakly at the P margin (cf. Fig. 7.3) [930, 2739]. At this same time, four "early eye" genes are also expressed [930] (HB = homeobox): *eyeless* (*ey*, HB), *eyes absent* (*eya*), *sine oculis* (*so*, HB), and *dachshund* (*dac*). Three of the latter genes are expressed in the P half of the eye region with an intensity that fades with distance from the P margin [397, 759, 930, 2689]: *eya*, *so*, and *dac*. The fourth (*ey*) is expressed throughout the eye region [930, 1684, 3486]. Four other HB genes – *eye gone* (*eyg*) [1780], *twin of eyegone* (*toe*) [2022], *optix* [3851], and *twin of eyeless* [933] – may also belong to this "early eye" group.

MF initiation requires Dpp signaling [723, 724]: whenever a clone that is deaf to Dpp (e.g., *Mad*^LOF [1780, 4648], *punt*^LOF [570, 1780], or *tkv*^null [570]) resides at the P margin, the MF fails to launch at that site. MF initiation also requires the early eye genes because no functional MF forms when any of them is disabled [1684, 2042, 2689, 3387]. What is the circuitry? Dpp evidently acts through *eya*, *so*, and *dac* because their expression vanishes in *Mad*^null clones that touch the margin [930], although expression of *ey* persists.

Dpp ➡ {*eya*, *so*, AND *dac*} ➡ MF initiation.
ey ➡ MF initiation.

The "Dpp ➡ {*eya*, *so*, AND *dac*}" link does not operate during MF movement because internal *Mad*^null clones

express *eya*, *so*, and *dac* [930]. (*Inter se* links among the early eye genes are considered in Ch. 8.)

MF initiation also requires *hh* signaling (despite prior evidence to the contrary [1788, 2632]): no MF originates when t.s. *hh*^LOF mutants are heated during the initiation stage [406] (cf. *hh*^null clones [1082]). Hh is normally detectable along the P edge in late-2nd/early-3rd instar [406, 1082, 3665] (despite a report to the contrary [2632]). Within this zone, *hh* may act upstream of *dpp* because *dpp* expression there vanishes when young t.s. *hh*^LOF larvae are heated [406, 3665], although *ptc*^LOF clones at the P margin can rescue aborted MFs in *dpp*^LOF eyes [723]. In turn, the Iro-C must act upstream of *hh* because erasure of the Iro-C ON/OFF border (by ubiquitous expression of *ara* via *ey-Gal4:UAS-ara*) prevents Hh levels from rising high enough to spark a MF [696]. Finally, LOF-GOF analysis shows that (1) *pannier* (*pnr*) regulates the Iro-C via Wg [696, 1781, 2752] (cf. the notum [1380]) and (2) the Iro-C ON/OFF border controls *hh* via the Notch pathway [696]. Thus, the basic chain appears to be

Pnr ➡ Wg ➡ Iro-C ON/OFF ➡ Notch ➡ Hh ➡ Dpp ➡ {*eya*, *so*, AND *dac*} ➡ MF initiation.

During the latter half of the 2nd instar, Notch is needed throughout the eye part of the disc (not just at the D/V border) to stop it from developing as an antenna (cf. Fig. 6.9) [2353]. At this time, the EGFR pathway acts antagonistically to Notch [2353]. Both pathways probably affect the above chain at the {*eya*, *so*, AND *dac*} node (cf. Fig. 7.3 and [1189]). Indeed, Eya cannot function without being phosphorylated by dMAPK [969].

When ectopic MFs are elicited anterior to the endogenous MF, they spread radially from the inception point like a ripple in a pond [723, 2036, 2631]. This independence from the global framework of the disc argues that the MF is a self-propagating wave. Evidently, the P-to-A motion that characterizes insects is arbitrary: evolution could have launched the MF from any point in the prospective eye field. Indeed, the only reason that MFs do not arise at the D and V margins is Wg's damping effects there [724, 1781, 2631, 3562, 4390].

The eye field thus constitutes the sort of excitable medium that has been simulated by a fascinating genre of models known as "cellular automata" [824, 1661, 4341]. Cellular automata are imaginary grids of "cells" that behave like parallel computers. Each cell adopts a state based on inputs from its neighbors, and its new state then influences their states in the next round of computation (cf. John Conway's popular game of "Life" [74, 1381, 1382]). A variety of dynamic patterns can be created by changing (1) the interactive rules and (2) the initial configuration of states [32, 1437, 1769, 2694, 4720]. These systems nicely illustrate how local rules can generate global patterns as emergent properties. No simulations of eye circuitry per se have been attempted since the Hh-Dpp-Wg rules were figured out, but they may be a useful way of testing models in the future [4005].

When an extra MF moves in an A-to-P direction, the ommatidia that it creates have a reversed A-P polarity [722, 4572]. D-V polarity, however, tends to be normal, as expected given the integrity of the Iro-C circuitry [3562, 3563]. Exceptions include (1) interior wg^{GOF} clones [4390] that activate $mirr$ [1781] and (2) marginal clones that may affect the early phase of equator initiation [722, 723, 1781, 4167, 4648]. (See [2631, 4573] for partial theories and [2752] for overview.) When opposing MFs collide, they typically merge to form an amazingly seamless ommatidial array [724, 3388].

CHAPTER EIGHT

Homeosis

Few phenomena are as entrancing as the transformation of one thing into another. The ancients believed that sorcerers had such powers, and modern magicians can still fool children with illusions of this sort. A special class of mutations can actually accomplish this feat.

"Homeosis" means a transformation of one body part into another [3214, 4486]. The term was coined by William Bateson to describe deformities that are occasionally found in nature. In his classic 1894 monograph, Bateson cataloged 886 abnormal biological specimens [240], many of which exhibited homeosis. His intent was to investigate how anatomy varies as a way of comprehending how evolution works [2509]. This goal was obvious from the book's overtly Darwinian title: "Materials for the Study of Variation Treated with Especial Regard to Discontinuity in the Origin of Species."

Variation has been supposed to be always continuous and to proceed by minute steps because changes of this kind are so common in variation. Hence it has been inferred that the mode of variation thus commonly observed is universal. That this inference is a wrong one, the facts will show.... The evidence of discontinuous variation suggests that organisms may vary abruptly from the definite form of the type to a form of variety which has also in some measure the character of definiteness. Is it not then possible that the discontinuity of species may be a consequence and expression of the discontinuity of variation?... For the word 'metamorphy' I therefore propose to substitute the term **homœosis**, which is also more correct; for the essential phenomenon is not that there has merely been a change, but that something has been changed into the likeness of something else. [240]

Ever since Bateson, the "**Homeosis Riddle**" has been: how do mutations cause such phenotypes? The ability of single mutations to drastically alter the anatomy led to speculation that species might have arisen saltationally via such "macromutations" [1177, 3619]. This idea was dramatized in Richard Goldschmidt's "**Hopeful Monster Hypothesis**" [1056, 1521, 1523]. (See [551, 1582, 1584, 4113, 4529] for critiques.) That hypothesis has experienced a revival [662, 4258, 4520] in the wake of Halder *et al.*'s 1995 finding that eyes can be induced nearly anywhere on the fly surface by misexpressing the gene *eyeless* (see below) [1685].

In the Fly World, the mystique of homeosis has been compounded by transdetermination (cf. Ch. 4) [1670] and by the ability of physical or chemical agents to "phenocopy" many of these monstrosities in wild-type flies [1522, 1524, 1667, 2398, 2399].

The weirdest example of phenotypic mimicry is the *bithorax* phenocopy: wild-type flies can acquire a Ubx^{bxLOF} phenotype (notum and winglets instead of halteres) if they are exposed to heat shocks [1826, 2636] or ether vapor [1511] as blastoderm stage embryos [427, 646, 1859]. The implication is that *Ubx* expression is transiently unstable at the front of the 3rd thoracic segment [647, 1469, 1860, 4258], but teratogenic effects there are more complicated [3751, 4595], and it is still unclear why other homeotic genes are not similarly affected elsewhere in the embryo.

BX-C and ANT-C specify gross metameric identities along the body

In *Drosophila* there are two major clusters of homeotic genes (i.e., genes that cause homeosis when mutated) [2421, 2561, 2656]. Both reside on the right arm of the 3rd chromosome, and together they control the identities of all embryonic parasegments (PS) [162, 2166, 2934].

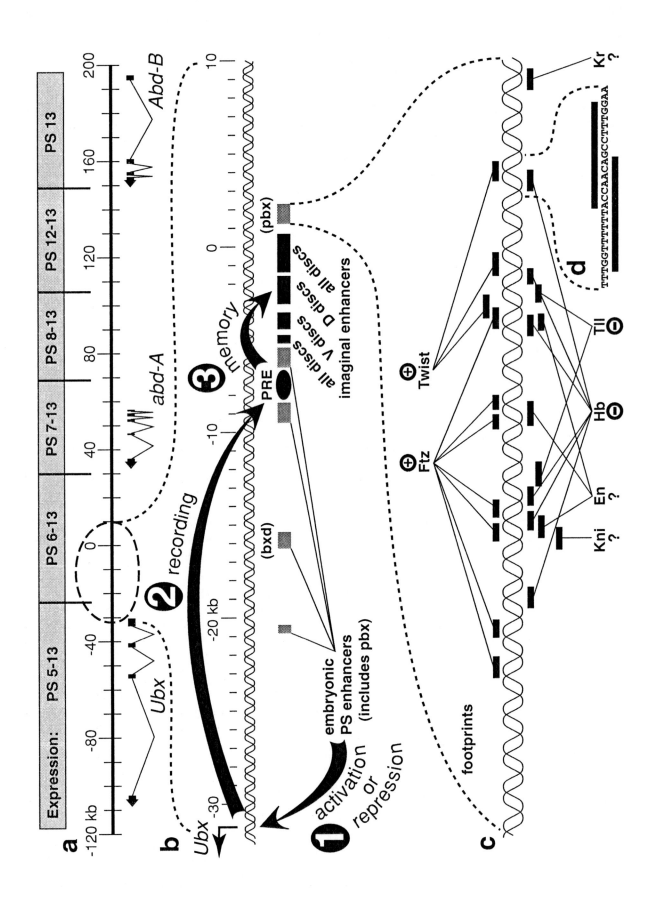

1. The Bithorax Complex (BX-C, 315 kb) has 3 homeotic genes (Fig. 8.1), which dictate fates in the abdomen and posterior thorax (PS 5–14) [2703]△: *Ultrabithorax (Ubx), abdominal-A (abd-A)*, and *Abdominal-B (Abd-B)*.

2. The Antennapedia Complex (ANT-C, ~330 kb) includes 5 homeotic genes that dictate fates in the head and anterior thorax (PS 1–5) [1023, 2167]△: *Antennapedia (Antp), Deformed (Dfd), labial (lab), proboscipedia (pb)*, and *Sex combs reduced (Scr)*. The ANT-C also has other types of genes [1025, 2717, 4671], including the segmentation genes *bicoid* and *ftz*, which seem to have acquired new functions during dipteran evolution [1467, 2590, 4067].

Until 1984, when the "homeo" box was revealed, researchers could only speculate about the nature of homeotic genes [3882]. The homeobox is a ~180 b.p. sequence that encodes a DNA-binding domain (cf. App. 1) [1417, 2498]. It is shared by genes in both the BX-C and ANT-C [2783, 2785, 3844]. Not all homeotic genes have a homeobox (Table 8.1) [1418], but the motif's widespread conservation within this group implies that evolution enlisted the homeobox family for administrative roles [45, 46, 1412, 1414, 4153] that were not bestowed on other families of transcription factors in animals (e.g., zinc finger) [3837], although the MADS box serves an analogous function in plants (cf. App. 1) [3105, 3477]. Most of the fly's homeotic genes fall into 3 functional classes [662]:

1. "Metamere identity" genes of the BX-C and ANT-C establish segmental or parasegmental identities (or groups thereof) along the body column [162, 1429]. The term "Hox" (contraction of <u>ho</u>meob<u>ox</u>) was coined to designate homologous complexes in vertebrates [48, 3677, 3839] (cf. reviews [658, 1223, 1236, 3871]), but it is also used more broadly to include the two clusters in flies [1429, 1591, 2778, 2937, 4563]. As explained below, two other

---

FIGURE 8.1. The Bithorax Complex (BX-C) and its regulation during development. See also Appendix 7.

a. The BX-C contains 3 protein-encoding homeotic genes: *Ultrabithorax (Ubx), abdominal-A (abd-A)*, and *Abdominal-B (Abd-B)* [3740]. Exons are drawn as black rectangles, with introns indicated as thin (kinked) connecting lines. When reporter genes (e.g., *lacZ*) insert into this 320-kb region, their expression is confined to a specific span of parasegments (PS, above) [284, 2773] due to nearby enhancers in the BX-C (cf. Fig. 4.2 for PS numbering) [2987, 3480, 3481, 3940, 4864]. The more rightward the reporter's insertion point (= distal on the right arm of the 3rd chromosome), the more the expression domain recedes toward PS13. This colinearity of enhancers (proximal to distal) relative to body regions (anterior to posterior) is also seen in the order of "*iab*" genetic control elements (not shown) [416, 699, 907, 3738]. The reason for the colinearity is unknown [162, 686, 3511, 4152, 4328, 4498], although clever guesses have been offered [1113, 1400, 2191, 2295, 2508, 2672, 3494]. A similar **"Homeobox Homunculus Mystery"** [1417, 1805] exists for Hox gene complexes of vertebrates [1111, 1112, 2266, 2292, 4446] – i.e., why are Hox genes and enhancers arranged like the parts of a miniature man or fly?

b. Enlarged view of the 40-kb section upstream of *Ubx*. Three stages in regulation are depicted. In Step 1, embryonic enhancers (shaded rectangles) turn *Ubx* ON or OFF in specific parasegments based on input that they receive from segmentation genes (**c**; cf. Fig. 4.2). In Step 2, the PRE (Polycomb Response Element, black oval) site indelibly "records" the cell's state by recruiting activator (Trithorax-Group) or repressor (Polycomb-Group) proteins [3196], depending on whether *Ubx* is actively transcribed (proteins not shown). Finally, in Step 3, the recruited proteins affect the imaginal disc enhancers (black rectangles) by blocking or aiding nucleosomal silencing of the DNA within ≥10 kb. Among the four disc enhancers, two drive expression in all discs, one affects mainly ventral (leg) discs, and the other affects mainly dorsal (wing and haltere) discs.

c. Magnified view of the *pbx* embryonic enhancer (619 b.p.), showing footprints (bars) of various regulatory proteins. Abbreviations: En (Engrailed), Ftz (Fushi tarazu), Hb (Hunchback), Kni (Knirps), Kr (Krüppel), Tll (Tailless). All these proteins participate in ectodermal segmentation (cf. Fig. 4.2 and [3680] for *tailless*, a terminal gap gene) except Twist, a bHLH transcription factor that activates mesodermal target genes [249, 1985, 2842, 4818] in the embryo's D-V patterning system [3576, 4290]. Plus signs indicate activation and minus signs denote repression, although individual footprint sites have not been tested for function *in vivo* [3397]. Kr and Kni can repress BX-C genes [680, 3902], but expression of a *pbx*-driven *lacZ* reporter is normal in *Kr*LOF or *kni*LOF embryos [4864], arguing against *pbx*-mediated regulation of *Ubx* by these agents. En is a repressor for *Ubx* in the haltere disc [1138, 1162, 1635, 4229], but in the embryo it can behave as either an activator [3481] or a repressor [684, 2670, 2725].

d. Nucleotide sequence in an enlarged section the *pbx cis*-enhancer, where a Twist binding site (bar above) appears to overlap a Hb binding site (bar below). It is important to remember that such sites are delimited by DNase protection assays, so the actual contact area with DNA may be smaller. Hence, these proteins might be able to bind simultaneously (i.e., without competing).

Demarcations of PS expression zones in **a** are only crude estimates due to a limited number of insertion sites [284]. Exons for BX-C genes are after [284, 2773, 3156, 3902], although each of the three genes has other splicing isoforms. Nevertheless, for *Ubx* at least, the isoforms act alike with regard to specifying parasegment identity [684]. Map coordinates obey convention [283]. Schematic in **b** is adapted from [3397]. See [3667, 4319] for finer dissections that identify separate Pc-G and Trx-G binding sites (PRE vs. TRE). Data in **c** and **d** are from [3397, 4864].

TABLE 8.1. GENES THAT CAN HOMEOTICALLY TRANSFORM LEG, WING, OR EYE DISC DERIVATIVES *INTER SE* *

FROM:	TO: Leg (L1, L2, L3)	Wing (W) or Notum (N)	Eye (E), Antenna (A), Head Capsule (H), or Palpus (P)
Leg (L1, L2, L3)	L1 → L2: Scr-LOF (a.k.a. *Msc*) [3286]△. Trx-G-LOF : See below**. {L2 or L3} → L1: Pc-G-LOF : See below**. *Ubx*-LOF (*Scr*-ON) [3311]△. *Antp*-GOF (a.k.a. *Scx*) [2190]△. Scr-GOF (a.k.a. *Msc*) [3285, 3286]△. L2 → L3: Pc-G-LOF : See below**. {*Pc*-LOF and *abd*-*A*-LOF and *Abd*-*B*-LOF} [582]. *Ubx*-GOF [1120, 2506]. L3 → L2: *Ubx*-LOF [4627]△. Trx-G-LOF : See below**.	L1 (dor.) → W: *vg*-GOF (*vg*-ON) [2082, 2753, 2754]. L2 (dor.) → W: *su(f)*-LOF [3442]. *vg*-GOF [2219, 2252, 4564], whereas L3 makes haltere tissue [4564]. *vg*-GOF (*vg*-ON) [2082, 2753, 2754]. {*vg*-GOF and *ug*-GOF} [2252]. {*vg*-GOF and *Ser*-GOF} (*vg*-ON) [885, 2092]. L3 (dor.) → W: *vg*-GOF (*vg*-ON) [2082, 2753, 2754].	L (dor.) → E: GOF in "early eye" genes: See below**. L (prox.) → A: *Dll*-GOF (*hth*-ON AND *spalt*-ON) [1085]. L (dist.) → A: *pb*-GOF [413]. *ss*-GOF [1119]. {L1 or L2} → A: {*Antp*-LOF AND *Scr*-LOF} [2641, 4149]. {L1 or L2 or L3} → A: {*Antp*-LOF AND *Scr*-LOF AND *Ubx*-LOF} [4149]. {*Dll*-GOF AND *hth*-GOF} [1085]. L2 → A: *Antp*-LOF [677]△. {L2 or L3} → A: {*Antp*-LOF AND *Ubx*-LOF} [4149]. L → P: *pb*-GOF [93].
Wing (W) or Notum (N)	W → L: *dLim1*-GOF [1218]. W (pouch) → L: *Dll*-GOF (*hth*-ON AND *spalt*-ON) [1085, 1561].	W → N: LOF in Wg pathway: See below**. LOF (misc.): *N*-LOF [886], *sd*-LOF [538, 2015], *tet*-LOF [2015], or *vg*-LOF [2570]△. GOF (misc.): *Egfr*-GOF [4543], *ara*-GOF [4543], or *pnr*-GOF (*vg*-OFF) [614]. N (pos.) → W: LOF in EGFR pathway: See below**. *hth*-LOF (*vg*-ON) [678]. GOF (misc.): *dpp*-GOF [2954], *sd*-GOF (*vg*-ON) [4463], {*vg*-GOF AND *ug*-GOF} (*vg*-ON) [2252], {*vg*-GOF AND *Dl*-GOF} (*vg*-ON) [2252], or {*vg*-GOF AND *Ser*-GOF} (*vg*-ON) [885, 2252]. N (unsp.) → W: GOF in Wg pathway: See below**. {*vg*-GOF AND *sd*-GOF} [1686].	W (hinge) → E: GOF in "early eye" genes: See below**. W (prox.) → A: *Dll*-GOF (*hth*-ON AND *spalt*-ON) [1085].

Eye (E), Antenna (A), Head Capsule (H), or Palpus (P)

A→L1:
Pc-G-LOF: *sxc*-LOF [1965].
Scr-GOF (*ss*-OFF [1119]) [4806]$^\Delta$.
A→L2:
Dll-LOF (*ss*-OFF) [1085, 3242]$^\Delta$.
exd-LOF [1543, 3527].
hth-LOF [3380]$^\Delta$ or *hth*-DN (minus HD) [3716].
pb-DN (altered HD) [3333].
Pc-G-LOF: See below**.
ss-LOF [1166]$^\Delta$ (cf. heat sensitivity [1623, 3810]).
tgo-LOF [1166].
Trx-G-LOF: See below**.
GOF in Hox genes: See below**.
Arp-GOF [513].
cut-GOF [2080].
tsh-GOF [3240].
A→L3:
Ubx-GOF [2676].
H (ant.-ven.) → L2:
exd-LOF [1543].

E→W:
opht (LOF?) [2561, 3210–3212].
{*Antp*-GOF and *N*-GOF} (*vg*-ON) [2362].
Dl-GOF [2092].
Opt (GOF?) [2561, 3439, 3788].
{*vg*-GOF and *Ser*-GOF} (*vg*-ON) [2092].
{E or A or H} → W:
vg-GOF [2219, 3291, 3936].
E→N:
trx-LOF [1966].
H (pos.-dor.) → N:
Dfd-LOF [2830].
exd-LOF [1543, 3527].
hth-LOF [3380].
lab-LOF [2829].
Antp-GOF [3802].
Scr-GOF [4872].
H (ven.) → W:
ug-GOF (*vg*-ON) [2082].

E→A:
LOF in Notch pathway: See below**.
ey-LOF (?) [3746].
GOF in EGFR pathway: See below**.
ci-GOF [2353].
Dll-GOF (*hth*-ON and *spalt*-ON) [1085]$^\Delta$.
E→H:
LOF (misc.): *eya*-LOF [1780], *Mad*-LOF [4648], {*Mad*-LOF and *DC0*-LOF} [4648], *sgg*-LOF [1833], *slimb*-LOF [2856], or *ttk*-LOF (?) [4773].
GOF in Hh pathway?: See below**.
hth-GOF [3380].
A→E:
GOF in "early eye" genes: See below**.
tsh-GOF [3240].
A→P:
pb-GOF [413]$^\Delta$.
ug-GOF [2082].
H (dor.) → E:
LOF (misc.): {*ara*-LOF and *caup*-LOF} [3380], *dsh*-LOF [1833], *ee*-LOF (?) [190], Iro-C-null [696,697], *pnr*-null [2752], or *tfd*-LOF (?) [2562].
tsh-GOF [3240].
H (ven.) → E:
exd-LOF [1543].
hth-LOF [3226, 3380].
GOF in "early eye" genes: See below**.
N-GOF [2362].
H (unsp.) → E:
{*Egfr*-LOF and *gro*-LOF} [3465].
GOF (misc.): *ara*-GOF (rare) [696], *caup*-GOF (rare) [696], or *dpp*-GOF [3388].
H→A:
LOF (misc.): {*ara*-LOF and *caup*-LOF} [3380], Iro-C-null [697], or *tfd*-LOF (?) [2562].
Dll-GOF (*hth*-ON and *spalt*-ON) [1085, 1561].
ss-GOF [1119].
H (ven.) → H (dor.):
ara-GOF or *caup*-GOF or *mirr*-GOF [697].
H (dor.) → P:
Iro-C-null [697].
P→A:
ss-GOF [1119],
ug-GOF (*pb*-ON) [2082].

(*continued*)

TABLE 8.1 (continued)

*The genes listed above are "homeotic" insofar as they can cause specific body parts to develop like other body parts, but only a subset has a homeobox [1418]. Conversely, there exist homeobox genes that do not cause homeosis [1467, 2934, 3836]. Arrows (x $\longrightarrow$ y) indicate transformations (from structure x to structure y), and letter codes for structures are given in row and column headings (e.g., N = notum). Within each category, LOF effects are listed first (alphabetically) followed by GOF effects. "AND" means "in combination with." Genes that are known to be on or OFF in the transformed regions are listed in parentheses. Homeotic switches between A and P compartments of the same disc or between D and V compartments are omitted (cf. Chs. 5 and 6), except for the eye disc, where different histotypes are involved. Directions: ant. (anterior), dist. (distal), dor. (dorsal), pos. (posterior), prox. (proximal), ven. (ventral), unsp. (unspecified subregion). "Weak" and "strong" refer to penetrance or expressivity or both.

Genes (excluding Hh, Dpp, and Wg pathways; cf. App. 6), with DNA-binding motifs (if any) in parentheses (HD, homeodomain; bHLH, basic helix-loop-helix; HMG, high mobility group; ZF, zinc finger): abd-A (abdominal-A; HD), Abd-B (Abdominal-B; HD), Antp (Antennapedia; HD), ara (araucan; HD), Arp (Aristapedoid), ash1 and 2 (absent, small, or homeotic discs 1 and 2), Asx (Additional sex combs), brm (brahma), caup (caupolican; HD), ci (cubitus interruptus, ZF), crm (cramped), cut (HD), dac (dachshund), Dl (Delta), Dll (Distal-less; HD), Dsp1 (Dorsal switch protein 1, HMG), ee (extra eye), Egfr (EGF receptor), esc (extra sex combs), exd (extradenticle; HD), ey (eyeless; paired/HD), eya (eyes absent), gro (groucho), hth (homothorax; HD), Iro-C (Iroquois Complex; HD), kis (kismet), lab (labial; HD), mirr (mirror; HD), moira, mxc (multi sex combs), N (Notch), opht (ophthalmoptera), Opt (Ophthalmoptera), optix (HD), pb (proboscipedia; HD), Pc-G (Polycomb Group), Pc (Polycomb), Pcl (Polycomblike), pco (polycombeotic), pho (pleiohomeotic), pnr (pannier), pnt (pointed, Ets), Psc (Posterior sex combs), Sce (Sex comb extra), Scm (Sex combs on midleg; ZF), Scr (Sex combs reduced; HD), sd (scalloped), Ser (Serrate), slimb, so (sine oculis; HD), ss (spineless; bHLH), su(f) (suppressor of forked), sxc (super sex combs), tet (tetraltera), tfd (two-faced), tgo (tango; bHLH), toy (twin of eyeless; paired/HD, trx (trithorax; ZF? [4066]), Trx-G (Trithorax Group), tsh (teashirt; ZF), ttk (tramtrack; ZF), Ubx (Ultrabithorax; HD), vg (vestigial). "Iro-Cnull" means null for ara, caup, and mirr. When alleles cannot be ascribed to LOF or GOF classes, they are labeled with original superscripts, except for DN (dominant-negative) alleles, which typically disable the endogenous gene and lead to a LOF effect. Some GOF effects may be misleading [50, 1418] due to promiscuity of DNA binding by HD proteins [1767, 1868, 2219] (cf. App. 1). The gene eye gone (Pax-6 family; not listed) causes extra eyes (locations not reported) when it is ectopically expressed [2090].

Some of the listed defects involve whole body segments or parasegments [1120, 2784]. For example, the following affect not only legs but also wing or haltere (T2 = mesothorax; T3 = metathorax): T2-to-T3 (PclLOF [4327], UbxGOF [591, 673, 1547, 2921, 4624]) and T3-to-T2 (UbxLOF [591, 2925, 4627]). In contrast, "field-specific" genes cause homeosis to a certain histotype virtually anywhere in the body when they are misexpressed [662]. By this definition, vg would be a master gene for wing identity [2219]. Other deformities defy classification. For example, duplicated antennae are seen in mirrLOF discs (where the fng-on area spreads dorsally into the ocellar region), but there is no compensatory loss of eye tissue [4797]. There is no simple correspondence between genes and body parts, except for the dorsal head, which appears to rely mainly on the Wg pathway [2631] under the control of Pannier [252].

For a comparable table of transdetermination frequencies, see [3883]. For further data, see Flybase and [2561, 3165, 3214, 3881].

**Pathway details:

L1 $\longrightarrow$ L2: Trx-GLOF: ash1LOF [3884]△, ash2LOF [757]△, brmLOF [4243], Dsp1LOF [1013], kisLOF [964], moiraLOF [475], or trxLOF [451]△.
[L2 or L3] $\longrightarrow$ L1: Pc-GLOF [2400]△: AsxLOF [3970], crmLOF [4793], escLOF [1509]△, mxcLOF [3721], PclLOF [3285]△, pcoLOF [3370], phoLOF [1500], PscLOF [2704]△, SceLOF [453], ScmLOF [405]△, or sxcLOF [1967]△.
L2 $\longrightarrow$ L3: Pc-GLOF: escLOF [1508], phoLOF [1500], or PclLOF [1122].
L3 $\longrightarrow$ L2: Trx-GLOF: ash1LOF [3884]△, ash2LOF [757]△, or trxLOF [1966]△.
L (dor.) $\longrightarrow$ E: GOF in "early eye" genes: dacGOF (ey-on) [3894], eyGOF (dac-on, eya-on, and so-on) [1684]△, eyaGOF [394], toyGOF (ey-on) [933], or [dacGOF AND eyaGOF] [743].
W $\longrightarrow$ N: LOF in Wg pathway: ugLOF [4543]△, armLOF [3317], dshLOF [3310], porcLOF [3310], skinheadLOF [3371], or NotumGOF [1494].
N (pos.) $\longrightarrow$ W: LOF in EGFR pathway: EgfrLOF (via argosGOF) [4543], or dRafLOF [207].
N (unsp.) $\longrightarrow$ W: GOF in Wg pathway: ugGOF [207]△ (but only rarely [252] unless strongly expressed [3025]), armGOF [3025], or osaLOF [849].
W (hinge) $\longrightarrow$ E: GOF in "early eye" genes: eyGOF (ss-OFF [1119]) [2517], or phoLOF [1500].
A $\longrightarrow$ L2: Pc-GLOF: PclLOF (ss-OFF [1119]) [2517], or phoLOF [1500].
A $\longrightarrow$ L2: Trx-GLOF: ash1LOF [3884]△, ash2LOF [757]△, laucLOF [4892], or trxLOF [1966].
A $\longrightarrow$ L2: GOF in Hox genes that turn OFF hth and hence block nuclear import of Exd [156, 3327, 4806]: abd-AGOF (ss-OFF [1119]) [4806]△, Abd-BGOF [676]△, AntpGOF (ss-OFF [1119]) [4241]△, or UbxGOF (ss-OFF [1119]) [4806]△.
E $\longrightarrow$ A: LOF in Notch pathway [2353]: NDN, DlDN, or SerDN.
E $\longrightarrow$ A: GOF in EGFR pathway [2353]: EgfrGOF, spitzGOF, Ras1GOF, dRafGOF, or pntGOF.
E $\longrightarrow$ H: GOF in Hh pathway?: hhGOF [722] or ptcLOF [722].
A $\longrightarrow$ E: GOF in "early eye" genes: eyGOF (eya-on) [394]△, eyaGOF [398]△, or optixGOF [3851]△.
H (ven.) $\longrightarrow$ E: GOF in "early eye" genes: dacGOF (ey-on) [743, 3894]△, eyaGOF [743], optixGOF [3851]△, {dacGOF AND eyaGOF} [743], eyaGOF [743], or {eyaGOF AND soGOF} [3387].

groups of genes (Pc-G and Trx-G) yield similar phenotypes because they propagate the states of Hox gene expression (ON or OFF) after embryogenesis.

2. "Field-specific" genes enforce identities in regions of the body that are not segments or parasegments [2410, 2677, 3716]. Examples include the Iro-C genes, whose ON/OFF states assign eye cells to D vs. V compartments (cf. Ch. 7) [696], and the *engrailed-invected* pair, whose ON/OFF states assign thoracic cells to A vs. P compartments (cf. Chs. 4–6) [4229]. Whether these outlying complexes were founded by "escapees" from the primordial Hox array is unclear. (See [153, 933, 1843, 2012, 2022] for more homeobox complexes in the fly genome, [567, 568, 2170] for homeobox phylogeny, and [490, 1236] for the issue of dispersal frequency.)

3. "Cell type" genes specify histotypes at the single-cell level [282, 853, 1069, 1901, 3033, 4630]. An example is the homeobox gene *cut*, which controls sensillar identity (cf. App. 4) [378].

The rest of this chapter explores how genes in the first two classes function. As for how genes in the third class work, a case study (the bristle) has already been presented in Chapter 2.

Ubx enables T3 discs to develop differently from T2 discs

The classic example of homeosis in *Drosophila* is the four-winged fly with two thoraxes [1214, 3608], hence the name "bithorax." This phenotype entails a conversion of T3 into T2 (i.e., 3rd into 2nd thoracic segment). Wild-type T3 discs (haltere and 3rd leg) express much more Ubx than T2 discs (wing and 2nd leg) [496, 782, 4625], and a maximal T3-to-T2 transformation is achieved when *Ubx* is mutationally shut OFF in T3 by disabling its *abx*, *bx*, and *pbx* enhancers [2510, 2564]. In that case, the haltere becomes a wing [2506, 2924], and the 3rd leg resembles a 2nd leg in its A compartment, although its P compartment adopts T1 identity [2198, 2925, 3311] due to (1) nullification of an embryonic "Ubx ⊣ *Scr*" link [162, 686, 1549, 1772, 3321], (2) consequent derepression of *Scr* [2564], and (3) enforcement of a T1 state by Scr [2190, 2561, 3285, 3286, 4345].

The four-wing trait is atavistic [664, 1429, 1466, 4550] because dipteran halteres evolved from paleopteran wings [664, 4750]. Ed Lewis cited this atavism in 1978 when he proposed his "**Ratchet Model**" for the evolution of metameric identity [2507]. In that model, the T2 segment represents a "ground state" that reflects the ancestral condition.

Flies almost certainly evolved from insects with four wings instead of two and insects are believed to have come from arthropod forms with many legs instead of six. During the evolution of the fly, two major groups of genes must have evolved: "leg-suppressing" genes which removed legs from abdominal segments of millipede-like ancestors followed by "haltere-promoting" genes which suppressed the second pair of wings of four-winged ancestors. . . . Each of the wild-type thoracic and abdominal segments has a unique pattern of differentiated structures which constitutes a morphologically defined state or "level of development". . . . The attainment of any level more advanced than [T2] is a stepwise process in which each step requires the presence of a specific BX-C substance. [2507]

According to this model, *Ubx* enables discs to raise their state from the T2 "baseline" to a T3 level, and *abd-A* and *Abd-B* would allow segments to "climb" to higher abdominal states (an idea that is counterintuitive because the latter segments seem simpler). This view of *Ubx* as a switch chimed with Kauffman's Binary Code Conjecture [2158, 2159] and García-Bellido's Selector Gene Hypothesis [1358] (cf. Ch. 4). García-Bellido rightly pointed out that, whatever its nature, the code must be abstract because halteres and legs look nothing alike yet are affected jointly by Ubx^{LOF} mutations. The ability of Ubx^{GOF} mutations to transform wings into halteres (the "Contrabithorax" effect) is consistent with this argument [673, 1547, 2921].

Given the presumption that every T3 cell uses *Ubx* in the same way (i.e., for its identity [1121]), it came as a surprise when Ubx protein was found to be distributed unevenly in T3 discs (Fig. 8.2) [591]. Ubx's expression is stronger in the P compartment of both discs, and the most intense subregions are the central (distal) part of the haltere disc and the tibial area of the leg disc [256, 496, 4626, 4627]. Conceivably, this analog heterogeneity might have no functional significance if a certain threshold of Ubx is needed for the T3 state. That threshold would be exceeded throughout T3 but not T2. This simplistic rationale is consistent with the fact that T3 subregions are differentially sensitive to *Ubx* dosage in their tendency to transform [4006].

Another possibility exists. Ubx might be unevenly deployed because it is needed more critically in those T3 areas that differ strongly from T2: (1) the haltere's bulbous "capitellum" (vs. the huge wing) and (2) the hind part of the 3rd-leg tibia where transverse rows arise (vs. the 2nd leg where there are no such rows; cf. Fig. 3.10). According to this "**Use-as-needed Scenario**" the anterior cells of the 2nd and 3rd legs would never

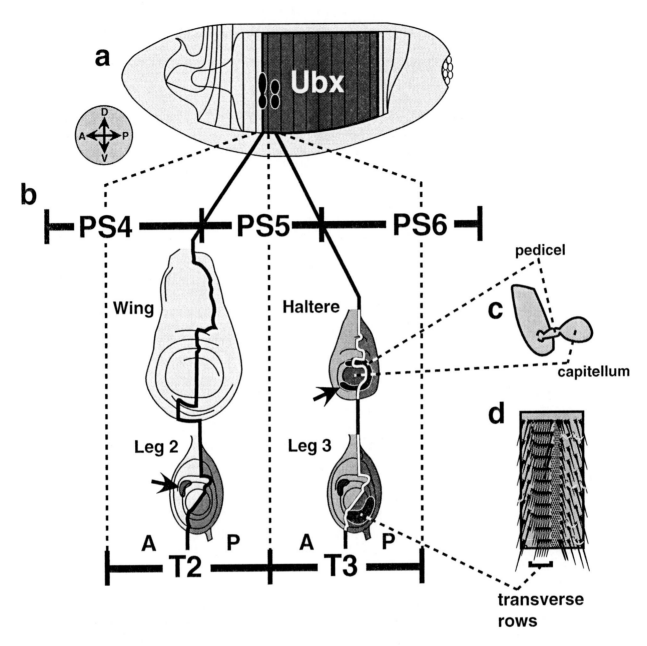

need to know that they belong to different metameres, so Ubx could safely disappear from those regions over evolutionary time (if it was ever there to begin with). As explained below, this scenario turned out to be correct.

But Ubx does so by directly managing target genes in multiple echelons

The first detailed model for eukaryotic gene circuitry was proposed in 1969 by Roy Britten and Eric Davidson [473]. They invoked 5 tiers of causally linked DNA or RNA elements called "sensors, integrators, activators, receptors, and producers." Ever since this "**Cascade**

Model," researchers have been predisposed to thinking of genetic control in terms of hierarchies [3773], where genes at one level only influence genes at the next lower level [98, 659, 974, 2410]. To use a military analogy, the basic idea is that generals would never talk to privates, but rather must give orders to subordinate soldiers through an inviolable chain of command. Ed Lewis made just such an assumption in his Ratchet Model. (Italics are author's.)

The various BX-C substances are presumed to act *indirectly* by repressing or activating other sets of genes which then directly determine the specific structures and functions that characterize a given segment. [2507]

FIGURE 8.2. Expression of Ubx in thoracic disc development. Abbreviations: PS (parasegments), T2 (2nd thoracic segment), T3 (3rd thoracic segment). Axes are indicated by the compass: A-P, anterior-posterior; D-V, dorsal-ventral. See also Appendix 7.

   a. Fate map of the left half of an embryo (cf. Fig. 4.1), showing the ectodermal region (PS 5–13) where Ubx is expressed (shaded) at Stage 11 (cf. Fig. 4.2). Black ovals mark where discs arise in T2 and T3.

   b. Expression of Ubx in T2 (wing and 2nd leg) and T3 (haltere and 3rd leg) discs from the left side a mature larva. Degrees of shading signify amounts. Each thick vertical line bisects the discs into A and P compartments (cf. Fig. 4.4). Note that Ubx expression has been altered since the embryonic stage. These changes are likely due to the facts that (1) embryonic and imaginal *cis*-enhancers are separate at the *Ubx* locus and (2) the hegemony of the Pc-G or Trx-G complex can be overturned by sufficiently powerful *trans*-acting factors (cf. Fig. 8.1) [3397]. In T2, Ubx persists in the leg disc's P compartment, but it has vanished from the wing disc except for the peripodial membrane (not shown). A new spot (arrow) arises in the anterior tibial area of the 2nd-leg disc. The 3rd-leg disc has a similar spot, along with an intense P patch that will form transverse rows of the tarsus (**d**) and tibia. In both T3 discs, Ubx expression is stronger than in the T2 discs, and it is stronger in the P than the A compartment. In the haltere disc, the intensity is also high in part of the A region (arrow) near the center (future capitellum) and in the surrounding fold (future pedicel).

   c. Derivatives of the haltere disc (pedicel and capitellum). Basal hinge (scabellum) and flank sclerite (hemi-metanotum) are not labeled.

   d. Bristle pattern of the 3rd-leg basitarsus, showing transverse rows that are unique to T3. Unlike the 1st leg, whose transverse rows are anterior, these rows are on the P side, whereas the 2nd leg lacks transverse rows altogether (cf. Fig. 3.10).

   Fate map in **a** is based on [1739]. Data on Ubx expression in the embryo is from [53, 324, 675, 684, 1990], which should be consulted for spatiotemporal subtleties. Ubx expression in the discs (**b**) is from [496, 4564, 4626]. The latter references are not entirely consistent with one another, and the sum total of the published domains is depicted here. For details of anatomy (**c** and **d**) and fate maps, see [17, 526, 1358, 1714, 1883].

In 1998, this paradigm was toppled by Scott Weatherbee *et al.* in Madison, Wisconsin [4564]. They studied *Ubx*'s target genes in the haltere disc. Their key findings are listed immediately below in terms of how *Ubx* achieves its evolutionarily assigned task to "rewire" the circuitry of T3 development so that a haltere is produced instead of a hindwing.

The ability of Ubx to exert opposite effects on *scute* in different parts of the haltere (row 3 vs. 4 in the table below) implies that it is binding different AS-C *cis*-enhancers jointly with regional co-activators or

co-repressors [4564]. Direct regulation is also suggested by its cell autonomy, which indicates that Ubx is necessary (LOF) and sufficient (GOF) for control of its target genes. Finally, the regional specificity of Ubx's effects within the haltere argues that it is affecting each gene independently of the others. In summary, the "general" in this case (Ubx) is apparently giving orders not only to colonels, but also to lieutenants, sergeants, and privates. This "**Micromanager Epiphany**" of 1998 [51] has sparked a rethinking not only about Ubx but also about Hox genes collectively:

Problem	Solution	Evidence
Reshape the hindwing into a balloon-like structure (the capitellum).	Ubx ⊣ *blistered* (one of the lowest genes in the venation hierarchy; cf. Fig. 6.11). Thus, Ubx prevents the D and V surfaces from annealing.	In *Ubx*-null haltere clones, *blistered* is autonomously derepressed. Conversely, *blistered* is turned ᴏꜰꜰ in *Ubx*-GOF wing cells.
Reduce the organ's size.	First, Ubx ⊣ {*spalt* ᴀɴᴅ *spalt-related* (*salr*)} so as to stifle the 2–3 and 4–5 intervein areas [986]. (*Ubx* does not affect *omb* – another Dpp target; cf. Fig. 6.3.) Second, Ubx ⊣ *vg*QE (cf. Ch. 6). Finally, Ubx ⊣ *wg* in the P compartment.	In *Ubx*-null clones in the A part of the haltere, *salr* is autonomously derepressed. Conversely, *salr* is turned ᴏꜰꜰ in *Ubx*-GOF wing cells. *Ubx*-null clones derepress *vg*QE in the "pouch" and *wg* along the P portion of the D/V boundary.
Eliminate the bristles along the margin.	Ubx ⊣ AS-C (a target of Wg). In the P region, *Ubx* blocks the Wg pathway at its source by preventing activation of *wg* by Notch: Ubx ⊣ *wg* (cf. Fig. 6.8). (*Ubx* does not affect *vg*BE – another Notch target.)	*Ubx*-null clones derepress *scute* autonomously in the A part of the haltere's D/V line, while they derepress *wg* in the P part. Conversely, *scute* is turned ᴏꜰꜰ in *Ubx*-GOF wing cells.
Put stretch-sensitive sensilla at the base (pedicel).	Ubx → AS-C in the pedicel.	*Ubx*-null clones in the pedicel fail to express *scute*.

1. *Dipteran halteres did not arise in a sudden ("hopeful monster") way* [52]. Rather, "the evolution of the haltere progressed through the accumulation of a complex network of Ubx-regulated interactions" [4564]. After Ubx suffused T3 of arthropod ancestors [144, 1618, 3431, 3911], mutations must have created Ubx-binding sites in the *cis*-regulatory regions of one gene after another [664]. Ultimately, Ubx infiltrated enough nodes in the circuitry to reconfigure the hindwing into a haltere [3875]. Ubx must have reprogrammed the 3rd leg from a T2 pattern in a similar way [4112], while Scr did the same for the 1st leg [3333], although only a few changes were needed in each case (cf. Scr's density in the sex comb [1508, 2393, 3285, 4243]; also cf. Fig. 3.10). (This process of "slaves" becoming entrained to new "masters" [1440] is also seen in endocrine systems where hormones evolved before receptors [219, 2969, 4530], and the idea that signals predate the ability of cells to respond was argued by Stern in 1954 [4096]; cf. Ch. 3.) Although T3 evolved gradually, some species could have arisen as "atavistic monsters" via a Ubx^{LOF} defect that reverted T3 to T2 in one step [4635].

2. *Ubx does not "label" the entire metamere* [682]. Because *Ubx* is not acting as a selector gene for every imaginal cell in T3, its ON-or-OFF state need not be clonally propagated like the state of a compartmental selector gene [591, 1547, 2921]. This insight may explain the Collective Amnesia Conundrum (cf. Ch. 4), and it sheds light on why transformed tissue can "revert" to an untransformed state during regeneration [16, 17, 4324, 4326]. Indeed, we can now understand why marginal Ubx^{null} haltere clones transform nearby Ubx^+ cells when the latter are goaded to proliferate [3875] – a heretical effect that violates the dogma of homeotic autonomy [2198, 2421, 2506, 2924, 2925]. Ironically, one compartmental selector gene can overrule Ubx: en^{GOF} clones transform anterior haltere to wing tissue at 100% frequency ($n = 65$) [1162] due to a paradoxical "En ⊣ Ubx" link [2670, 2725].

3. *There is nothing sacred about segmental or parasegmental Hox gene expression* [682]. Hox genes tend to be expressed parasegmentally in the embryonic ectoderm [2717], but these allegiances can later shift in discs to segmental [2421, 2723, 3930] or idiosyncratic patterns [1168, 2094, 2655, 3285], depending on the need to control target genes in specific subregions [682, 2717, 2937]. This Use-as-needed Scenario explains why disabling particular Hox genes can have different homeotic consequences at different stages of development [2934, 4671]. For example, haltere cells "read"

their *Ubx* state at least twice: once to choose a hair morphology for the capitellum and later to select a sensillar type for the pedicel [3623]. Likewise, leg vs. antennal cells read the *Antp* state during one time window for sensillar identity and during another for the type of limb morphology [2406, 3775, 3810] (cf. additional cases [281, 685]).

4. *Cell types are not necessarily encoded digitally* [3623]. Conventional wisdom has always asserted that cell types are distinct [3448, 4105, 4508, 4513] because they use sharply defined batteries of genes [473, 3115]. However, varying the dose of Ubx can convert haltere-type cells into wing-type cells through a spectrum of intermediate states [1735, 2921, 3623] (cf. intersexual bristles [1716, 1845, 1883, 2978, 2979]). These intergradations imply that analog logic may be used as often as digital logic in setting the outputs for "realizator" genes [1635, 1807, 2446].

This shift in our thinking about Hox genes actually began in 1995, when James Castelli-Gair and Michael Akam (Cambridge, UK) deftly solved the "**3 Genes vs. 9 Segments Paradox**" of the BX-C [684, 1121]. That paradox arose in 1985, when Ernesto Sánchez-Herrero *et al.* showed that the BX-C contains only 3 *bona fide* genes [3740], despite Ed Lewis's proof that the BX-C controls 9 different segment identities [2507]. Castelli-Gair and Akam focused on how *Ubx* specifies parasegment PS5 (= T2p and T3a, where "a" and "p" denote anterior vs. posterior) and PS6 (= T3p and A1a). Their survey of prior models is given below, along with their critiques:

1. "**Hox Code Model**" [661, 1705, 2156, 2507, 4149, 4695]: metamere identities are specified by particular combinations of Hox genes. Because only *Antp* and *Ubx* are expressed in PS4–PS6, their codes would be as follows, with *Antp*'s state listed first and *Ubx*'s second in each pair of digits (0 = OFF, 1 = ON): PS4 (10), PS5 (11), PS6 (01). Although consistent with the expression data [661], this model is contradicted by LOF-GOF data (switches are underlined):

   a. LOF test: $Antp^{\mathrm{null}}$ mutations should convert PS5 (1$\underline{1}$) to PS6 (0$\underline{1}$). Instead, they transform PS5 to antenna [4148] or PS3 [4525]. Also, they can cause non-homeotic defects [1834, 3766, 4148, 4149, 4151], and in subregions of the wing and leg they have no effect whatsoever [4149].

   b. GOF test: Ubiquitous expression of *Antp* should convert PS6 (0$\underline{1}$) to PS5 (1$\underline{1}$). Instead, it has little or no effect there [1468, 1550, 4862].

   c. GOF test: Ubiquitous expression of *Ubx* should convert PS4 (1$\underline{0}$) to PS5 (1$\underline{1}$). Instead, it transforms

PS4 to PS6 (<u>01</u>) [1549, 2676]. This effect is attributable to a "Ubx ⊣ *Antp*" link [1549], which fits two general models for how co-expressed Hox genes interact [648]: (1) the "**Posterior Prevalence Model**," wherein posterior Hox genes nullify the effects of more anterior ones in a functional hierarchy [1007, 1113, 1549, 1550, 2646], and (2) the "**Hox Competition Model**," wherein Hox proteins compete for DNA binding sites [89, 685, 2396, 4806]. While thoracic discs may not use a combinatorial Hox code, the labial and eye discs do appear to employ a heteromeric Hox protein complex for this purpose [3333].

2. "**Hox Isoforms Conjecture**" [256, 2371, 4197]: *Ubx* exons are spliced differently to dictate PS5 (isoform type 1) vs. PS6 (isoform type 2). The following facts refute this notion:

   a. Expression of the different *Ubx* isoforms is not metamere specific [115, 2308, 2600, 3156].

   b. LOF test: Mutations that eliminate specific *Ubx* isoforms affect metameres uniformly [583].

   c. GOF test: Ubiquitous expression of particular *Ubx* isoforms affects metameres uniformly [684, 2676, 4197].

3. "**Hox Threshold Conjecture**" [1549, 2421]: different amounts of the same Hox protein specify different metameres. The authors discuss the prior evidence for this idea and conclude that alternative explanations exist in each case. Their own data are most easily explained by assuming that (1) anlagen within each metamere respond to Ubx independently of one another and (2) they do so by using a single (ON/OFF) threshold. That is, the logic is digital.

In wild-type embryos, certain anlagen in PS5 develop like those in PS4 (where *Ubx* is OFF), while others mimic PS6 (where *Ubx* is ON). Ubx must be instrumental since disabling it (*Ubx*^{null} embryos) forces both PS5 and PS6 to look like PS4 [1772, 4152]. What Castelli-Gair and Akam found is that the PS6-like rudiments of PS5 express Ubx during the periods when their destiny is being decided. Evidently, Ubx is acting like a paintbrush to give them a PS6-like state. Ubx's dynamics are thus "not simply transitional stages to reach a mature pattern of relevant gene expression." Rather, "they are fundamental for the normal function of the gene." The importance of temporal regulation is obvious for the 3rd-leg disc, which is permitted to arise by the absence of Ubx at an early stage (*Ubx*-OFF allows *Dll*-ON) but then is modified by uneven deployment of Ubx at a later stage (Fig. 8.2). The authors proposed a model that is tantamount to the Use-as-needed Scenario:

According to our model, the unique character of a segment depends on its temporal and spatial pattern of Hox gene expression. This is its "Hox code". For a single cell, the Hox proteins are just like any other transcription factors. What they will do depends on the context in which they are expressed, and need not always have the same consequences in terms of "segment identity".... In limb or wing primordia, it may be more useful to think of Hox proteins as defining the difference between specific alternative developmental pathways (wing trichome vs. haltere trichome) than to think of them as specifying global segment identity. [684]

Elsewhere, Castelli-Gair has rightly pointed out how (1) this cellular (vs. segmental) role for Hox genes makes sense evolutionarily, and (2) we have been deluding ourselves with our own terminology.

Hox genes are not controlling segment identity, but cellular behavior that will result in a certain segment morphology. This is what Hox genes do in unsegmented organisms like *C. elegans* [3727] and is probably what they did in the common ancestor of all metazoans. Segment identity is a subjective concept that originates from the observation that in a particular species, a number of cell characteristics are always associated in a given segment. [682]

In conclusion, it seems that the Hox genes not only work like micromanagers but also have been assigned a fairly heavy "work load" during evolution [968, 1114, 4563]. They govern particular downstream genes [1591, 3457] on a cell-by-cell basis [682], although cells of like kind may be transiently organized metamerically during an early stage of development [4893]. As Akam poetically put it:

Some transcription factors appear to be specialists: they specify a particular fate or behavior whenever they are expressed in a cell; the myogenic factors might approximate this role, for example. The *Hox* gene products lie at the other extreme: they are versatile generalists. They operate in many different cell and tissue types, where they modulate, sometimes dramatically but more often subtly, a wide range of developmental processes. In each of these cell types, expression of a *Hox* gene means something different – to divide or not to divide, to make or not to make a bristle, to die or not to die. In any given lineage, that meaning probably changes several times during development, in response to hormonal and other developmental cues. [51]

Pc-G and Trx-G "memory" proteins keep homeotic genes ON or OFF.

Although the expression patterns of Hox genes are modulated in certain discs, the gross layout of their metameric domains is retained throughout development. How is this done? That is, when the segmentation gene hierarchy turns ON a specific Hox gene in a particular metamere (cf. Fig. 4.2), how does that gene manage to stay ON after the segmentation machinery fades away?

This question echoes the old "**Memory Riddle**" in embryology [2417, 2501] – viz., how do cells maintain particular states of determination from the stage when they acquire them until the stage when they express them [1365, 2755, 3882]? That riddle has been especially troubling for imaginal discs, where the bracketing stages (embryogenesis and metamorphosis) are separated by several days but can be extended (by serial transplantation) to months or even years [1668, 2755]. One solution proposed by Wolpert is that positional "values" of some kind (digital?) perpetuate the transient (analog) positional information (cf. Fig. 4.3) [4724].

Theoretically, the simplest way for a gene that encodes a transcription factor to remain ON is for it to activate itself in a feedback loop [1790, 2324]. Some Hox genes employ this strategy [1429, 2372], whereas others use an indirect route [321, 325, 1952, 4312]. However, autoregulation is the exception rather than the rule in the ectoderm [682].

In the imaginal discs, the ON or OFF states of most Hox genes are sustained by chromatin-remodeling proteins. Most of those proteins fall into two groups [3197, 3275, 3394] – one named after *Polycomb* and the other after *trithorax*. (See [482, 849, 1490, 2598, 3868] for shared factors.) These groups regulate not only the Hox family but also compartment selector genes, including *apterous* (D/V) [3752]$^\Delta$ and *engrailed* (A/P) [3512]$^\Delta$. LOF mutations in the Polycomb Group lead to the turning ON of homeotic genes wherever they are normally OFF, while LOF mutations in the Trithorax Group lead to the turning OFF of homeotic genes wherever they are normally ON. (*Ubx* may also regulate itself via a *negative* feedback loop [1989].)

Proteins in the Polycomb Group (Pc-G; ≥14 members [3395]$^\Delta$) form a variety of multiprotein complexes [3453, 4315]$^\Delta$. The complexes self-assemble via several types of interaction domains [2373, 2374, 4313]$^\Delta$. Pc-G complexes bind specific "PRE" (Pc-G Response Element) motifs [1902]$^\Delta$ at ≥80 polytene chromosome sites [3078]$^\Delta$, including the BX-C and ANT-C [2622]$^\Delta$, where they stabilize the nucleosomal structure of inactive chromatin [455, 3869]$^\Delta$. For this purpose, some Pc-G ensembles use histone deacetylation [4315]$^\Delta$, but others may not [3869]. Most important, Pc-G proteins stabilize the OFF states of BX-C [2772]$^\Delta$ and ANT-C genes [1491, 4026]$^\Delta$ during early embryogenesis [3452]$^\Delta$ and preserve those states throughout larval development [584]$^\Delta$. In other words, they ensure "the clonal transmission of a determined state" [1026], although the states are only "firm biases" [3794] that can be overruled later. Different subsets of Pc-G proteins govern different tissues [3453]$^\Delta$, due, in part, to regional feedback from target genes (e.g., "Ubx ⊣ Pc") [3276].

Pc-G genes can mimic position-effect variegation [3452]$^\Delta$, suggesting that they use the same silencing pathway as heterochromatin [388]$^\Delta$. Indeed, there are some shared components [2622, 2966], but in general the route is different [549, 1199, 3968]. Most Pc-G genes (e.g., *Polycomb*) are named for sex combs (cf. Table 8.1) [2189] because LOF mutants have extra combs. This phenotype stems from the fact that (1) Pc-G^{LOF} mutations lead to expression of *Sex combs reduced* (*Scr*) in subregions of T2 and T3 (in violation of the Posterior Prevalence Rule [1113]), and (2) Scr then forces a sex comb identity on the basitarsi. Such mutants thus have the normal number of legs, but all six legs look like forelegs – that is, like a man with arms where his legs should be.

Proteins in the Trithorax Group (Trx-G; ≥ 15 members [1490]$^\Delta$) also form a variety of chromatin-remodeling complexes [2114]$^\Delta$. Like Pc-G complexes, the member proteins associate via various interaction domains [3666]$^\Delta$. Trx-G complexes bind many of the same polytene chromosome bands as Pc-G complexes [1436]$^\Delta$, and "TREs" (Trx-G Response Elements) have been mapped to the same restriction fragments as PREs [3196]$^\Delta$. However, in the one case analyzed in detail (*Ubx* upstream region at ~10 b.p. resolution), the TREs and PREs do not overlap [4319], nor has any direct interaction between Pc-G and Trx-G complexes yet been found [3260]. Trx-G complexes sustain the ON states of genes in the BX-C [719]$^\Delta$ and ANT-C [4468]$^\Delta$ after embryogenesis [451]$^\Delta$ in a tissue-specific [454] and time-sensitive [3847] manner.

In contrast to Pc-G complexes, Trx-G complexes tend to stimulate transcription [848, 2114] by remodeling chromatin so as to make DNA more accessible to transcription factors [4892]. To do so, some Trx-G ensembles acetylate histones [203]$^\Delta$ – the same trick that is used by dosage-compensation complexes to double the transcriptional output of X-chromosome genes in males [1633]$^\Delta$. Other ensembles may displace nucleosomes [4377]$^\Delta$ via ATP-driven, mechanochemical reactions [1700]$^\Delta$.

Our current view of how Pc-G and Trx-G complexes work as memory devices [2981] can be illustrated by again considering the expression of *Ubx* in PS4 (OFF) vs. PS6 (ON). The differential regulation in this case is thought to involve least three distinct phases (Fig. 8.1b) [719, 3196, 3395, 3397]:

1. The segmentation gene hierarchy (SGH) turns *Ubx* ON in PS6 (by deploying an excess of *trans*-activators there) and OFF in PS4 (by deploying an excess of *trans*-repressors there). The impacts of these transcription

factors are local, transient, and interactive [323, 1712] (e.g., via quenching [1600, 1601]).

2. SGH activators recruit Trx-G complexes to a nearby TRE site, while SGH repressors recruit Pc-G complexes to a PRE site [2171, 3395, 3396], thus "bookmarking" the DNA as to which regions were being read or not read at that time. In both cases, the assembly process appears to be stepwise, with different subsets of proteins being deployed in turn to execute a series of chromatin-remodeling alterations [3101, 4315]. For example, Extra sex combs (Esc) is transiently needed when the Pc-G assumes command of the *Ubx* promoter at ~4 h AEL [3939, 4160], and Esc's 7-bladed propeller of WD loops may provide the hub for assembling the nascent Pc-G complex [1734, 3102, 3394, 4313, 4314, 4316]. Moreover, separate subsets of Pc-G proteins appear to implement (1) the bookmarks and (2) the transcriptional silencing machinery that is recruited to the bookmark [306].

3. SGH activators and repressors disappear as Pc-G and Trx-G assume command. In PS4, the Pc-G complexes catalyze a spreading of the "closed" state [4889] over ≥10 kb [3275, 4172], thereby silencing the nearby imaginal disc enhancers [3397, 3938], and those enhancers remain dormant unless awakened later by regional *trans*-activators [37, 690, 2227]. Spreading might occur via (1) iterative recruitment of histones or (2) looping of the DNA caused by adherence of Pc-G aggregates to one another [4448]$^\Delta$ or to the nuclear matrix [1393, 4690]$^\Delta$. In PS6, the Trx-G complexes tend to keep the nearby imaginal enhancers "open" to either positive [501] or negative [3100] *trans*-regulatory inputs [675, 3396, 4199, 4378, 4413]. Those later inputs would explain why Ubx's pattern in discs differs from the embryo (Fig. 8.2).

We still do not know how the open or closed states of chromatin are perpetuated so durably during repeated cycles of DNA replication [306, 1317]$^\Delta$. DNA methylation cannot be the answer here. It is used widely in mammals [1326]$^\Delta$ but less so by 1 to 2 orders of magnitude in flies [1589, 2625, 4411] (cf. its link to histone deacetylation [2521]). Also unclear is whether transdetermination is due to a "memory loss" of Pc-G or Trx-G proteins from homeotic gene promoters over prolonged periods of proliferation [849, 2082] (cf. Fig. 6.9d).

Homothorax, Distal-less, and Spineless specify leg vs. antennal fates

We may call one thing a wing and another a leg, but the key question is whether the cells themselves "know"

that they belong to one vs. the other type of organ [2411, 3794]. What does "wingness" or "legness" mean to them? Homeotic mutations have allowed us to interrogate fly cells in this regard, but their cryptic answers have taken awhile to decipher.

Wolpert argued that when two organs use the same morphogens (e.g., Hh, Dpp, or Wg), they must differ in their mode of interpretation [4724], and Stuart Kauffman proposed that those modes could be encoded by ON/OFF states of Hox genes [2155, 2156]. For three decades, this view prevailed [2755].

In 1998, the same Micromanager Epiphany that toppled the Cascade Model also challenged this Hox Code Model [51]. To wit, why should Ubx be unevenly expressed and variously employed in haltere cells when it is only supposed to label cells with T3 identity [51, 684, 3875, 4564]?

The antenna-leg dichotomy brings this issue into sharper focus. Like the haltere and wing, the antenna and leg are serially homologous [3447], and they use the Hh-Dpp-Wg signaling circuitry in similar ways [1037, 1833, 2082, 4277]. However, the antenna does not normally express any Hox gene [1168, 2103, 2167, 3527], so it is unclear how it manages to develop differently from a leg, which is considered to be the ground state [613, 677, 2936].

After the four-winged *Ubx*^{LOF} fly, the next most famous homeotic mutant has legs where its antennae should be, hence the name "*Antennapedia*" [1468]. In *Antp*^{GOF} flies, *Antp* is ectopically ON in antennae [2094, 4806] aside from its normal, albeit uneven, expression in leg discs [677]$^\Delta$. Based on this result alone, it would seem that the ectopic Antp is "instructing" the antennal tissue to adopt a leg fate [1416]. However, virtually any Hox gene can turn antennae into legs when forced ON there (Table 8.1), so *Antp*'s effect is actually nonspecific. This effect is due in part to disabling of the Eyeless protein [3402]. It also involves an inhibition of the genes *homothorax* (*hth*) [677, 4806] and *spineless* (*ss*) [1119], both of which encode non-Hox transcription factors. Hth is a homeoprotein [2360, 3226, 3589], and Ss is a bHLH-PAS protein [1119, 1166].

{*Antp* OR other Hox gene} ⊣ {*hth* AND *ss*}.

The etiology would be simple if *hth* and *ss* were expressed only in antennae (i.e., they could dictate antennal identity directly), but they are both expressed in leg discs also (Fig. 8.3; cf. Figs. 5.11 and 5.12) [9, 130, 677, 1119, 2673].

To state the dilemma more succinctly, how can genes *a* and *b*, expressed in both discs C and D, make disc C different from disc D? In theory, this "**Shared Genes**

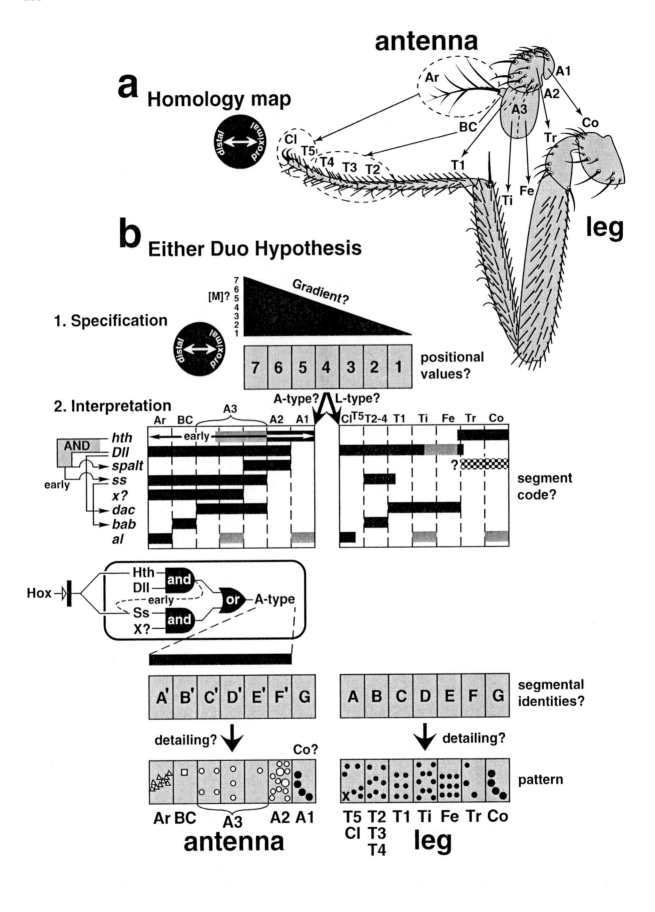

FIGURE 8.3. Our current understanding of what makes an antenna different from a leg. This problem was discussed earlier with regard to the 1969 debate between positional information (PI) and prepatterns (cf. Fig. 4.3). It is revisited here to see how much progress has been made since then. The answer, in short, is "a little."

   a. The old problem posed by the 1:1 correspondence of antennal and leg regions. These homologies were deduced from $Antp^{GOF}$ phenotypes [3445] and confirmed in other mutants (e.g., bab^{GOF} [1516]).

   b. The **"Either Duo Hypothesis"** for why antennae develop differently from legs. Antennae and legs are both thought to use an unknown morphogen (M) to specify radial positions relative to the future tip (cf. Fig. 5.4). In keeping with Wolpert's model, cells may convert this scalar signal into digital states ("positional values"), but those states are not propagated by cell lineage [544, 3446, 4575], nor are they stabilized by "memory" genes of the Pc-G or Trx-G (see text). Thus, no "recording" phase is interposed between specification and interpretation. (A and L denote antennal vs. leg types of development.) Eventually, the antennal and leg rudiments turn ON the same region-specific genes but express some of them differently. Horizontal bars denote expression domains; gray shading means weak or incomplete expression; checkered shading indicates heterogeneity. (*N.B.*: the axis is reversed relative to Fig. 5.11.) One code is well documented: the combination "hth-ON AND Dll-ON" (1) turns ON *spalt* and (2) evokes ectopic antennae in other discs (via *Gal4* drivers) [1085]. Also, hth^{LOF} or Dll^{LOF} clones turn OFF *spalt* and cause antennal-to-leg homeosis [677, 1085, 3226, 3380,3716]. The implication is that antennal identity is encoded by the Hth-Dll overlap, which leg discs lack (until late-3rd instar) [4760]. If so, then how do cells outside A2–A3 avoid making leg tissue? Distally, this role may be filled by *ss* because ss^{LOF} clones exhibit the same homeosis as hth^{LOF} or Dll^{LOF} [564, 1119, 1166, 2933, 4150, 4508]. However, ss^{GOF} cannot induce antennae outside the head (except for a minor claw-to-arista switch in the leg) [1119], so an unidentified factor "X" (in antenna but not leg) probably cooperates with Ss as Hth works with Dll. Either of these pairs of agents could enable antennal cells to deviate from the leg program in the same way that Ubx lets haltere cells deviate from the wing program (cf. Fig. 8.2). The logic is summarized in the circuit diagram (cf. Fig. 2.7 for symbols), which also shows how exogenous Hox proteins (e.g., Antp) cause homeosis by stifling Hth [4806] or Ss [1119]. The final stage of development involves creating patterns of small cuticular elements (e.g., bristles and sensilla, depicted abstractly here). We know a fair amount about how this is done [1653, 2055, 3531, 3548, 4125] (see Chs. 3 and 5 for the leg), but the links between the genes that act locally and the ones that act at a larger scale remain elusive. Those links may involve an intermediate level of prepatterns (cf. Fig. 6.14), so Stern's hypothesis is still as relevant as Wolpert's (cf. Fig. 4.3). Overall, this system is murkier than our picture of embryo segmentation (cf. Fig. 4.2). Remaining questions include: (1) What is the initial bias that steers tissue into an A- vs. L-type mode of interpretation? and (2) Is A1 actually a "coxa in disguise," or does it use a third duo of genes to veer away from a leg fate? Black or white symbols in the antennal and leg schematics are abstract renditions of the actual patterns.

   Genes ("DO" = details omitted): *al* (*aristaless*) [618, 881]$^\Delta$, *bab* (*bric à brac*) [1516], *dac* (*dachshund*) [4761], *Dll* (*Distal-less*; DO: expression above the tibia only appears in late-3rd instar) [1085, 3242, 4761]$^\Delta$, *hth* (*homothorax*) [1085, 4761]$^\Delta$, *spalt* [1085]$^\Delta$, *ss* (*spineless*; DO: expression in leg disc shifts to periphery in late-3rd instar; cf. *tango*) [1166]$^\Delta$. In the leg disc, *spalt* is expressed in "isolated cells scattered around the proximal region" [220].

   N.B.: During early antennal development, *hth* is expressed throughout the primordium [677], and *ss* expression is dependent on *hth* AND *Dll* (I. Duncan, pers. comm.). This link is drawn as a dashed line in the circuit diagram (near the bottom of the figure).

   The diagram in **a** is redrawn from [3445], and domains of gene expression (**b**) are modified from [3242] based on the above references.

Riddle" can be solved simply by combinatorial logic (cf. the Venn Overlap Rule, Fig. 6.10):

1. Genes *a* and *b* could jointly specify identity "C".
2. They would both be expressed throughout disc C.
3. However, they would be expressed in nonoverlapping subregions in disc D, thus leaving all of D's cells in a default state (e.g., leg vs. antenna).

   For *hth*, the riddle was solved in 2000 by Si Dong *et al.* in Madison, Wisconsin [1085]. There are two other homeobox (but non-Hox) genes that are important in the story: *extradenticle* (*exd*) [3319, 3526–3528, 3715] and *Distal-less* (*Dll*) [836, 4212]. Indeed, the original name for *Distal-less* was "*Brista*" – a contraction of "*Brista* on ar*ista*" (referring to the antenna-to-leg transformation) [2561, 4212].

1. Hth-Exd dimers are needed in the nucleus for antennae to develop differently from legs [677]. These dimers normally occupy the entire *hth*-ON domain because *exd* is ON in all *hth*-ON cells (indeed in all antennal and leg cells) [8, 9, 130, 677, 2673, 3589]. Thus, hth^{LOF} or exd^{LOF} clones cause an antenna-to-leg transformation on a cell-by-cell basis.

2. Ectopic expression of Hox genes inhibits transcription of *hth* (e.g., Antp ⊣ *hth*) [677, 4806] and hence prevents Extradenticle (Exd) from entering the nucleus [156, 3332, 3527, 3589].

3. In the antennal rudiment, the overlap between *hth*-ON (proximal) and *Dll*-ON (distal) domains is substantial, but in the leg disc it is nonexistent until late-3rd instar, and even then it is minimal (Fig. 8.3). The condition "*hth*-ON AND *Dll*-ON" must dictate

antennal identity because (1) Dll^{LOF} alleles partly transform antennae to legs [838, 1085, 4212] and so does hth^{LOF} [677, 3226, 3380, 3716], (2) ectopically expressed Dll induces antennae in places where hth is normally ON and vice versa [1085, 1561], and (3) co-expression of Dll and hth via $Gal4$ drivers induces ectopic antennae in legs, eyes, head capsule, and genitalia [1085]. The susceptibility of the genitalia makes sense because, like antennae, they evolved from legs [1179, 1562]. Interestingly, "hth-ON AND Dll-ON" turns ON a specific target gene – namely, $spalt$ [1085], although $spalt$ alone does not dictate antennal identity (cf. Ch. 6).

The Boolean logic of the hth-exd-Dll synergy could be implemented by a Hth-Exd-Dll complex that only activates antenna-specific genes when all three jigsaw pieces fit together [1085, 3242]. Indeed, the proteins do associate *in vitro* [784]. This complex must function like Ubx alone in the haltere (i.e., it allows antennal cells to depart from a leg fate in various ways).

For ss, the situation is less clear. LOF alleles transform the distal portion of the antenna into tarsal structures [426, 564, 1119, 1166, 2933, 4150, 4508], but ectopic GOF expression evokes only a paltry conversion of claws to aristae [1119]. This disparity may be due to an essential cofactor "X," whose gene is expressed throughout the ss-ON domain in the antenna but not the leg (Fig. 8.3). Tango is a dimerization partner for Ss [1166], but it does not appear to solve this puzzle because it is so broadly expressed in both antennae and legs. As a working hypothesis, the overall rule for antennal "identity" can be phrased as an "**Either Duo Hypothesis**":

{Hth-Exd AND Dll} OR {Ss AND X} ➔ antenna-specific cell behaviors.

This control circuit may be in two parts because it evolved that way [1119]. Moreover, evolution may have spared the basal "A1" segment, which expresses neither Dll nor Ss (Fig. 8.3). That annulus may thus retain a leg style of development, just as parts of the 3rd leg seem to have never been steered away from a 2nd-leg fate by Ubx (see above). Whatever the nature of these proximal-distal zones, they are certainly not clonal [544, 3446, 4575], so any biases they impose must be transitory and changeable via mitosis [3448, 3794] (cf. Cabaret Metaphor, Ch. 4).

The sensitive period when antennal cells can be diverted to a leg fate is early-3rd instar [1468, 1623, 3286, 3333, 3445, 3775], just after *spineless* is turned ON by Dll [1119]. Remarkably, the homeotic leg can come from as few as 10 transformed antennal cells [3445].

If a "master gene" exists for the eye, then it is also a micromanager

It is hard to imagine a stranger creature than one with eyes on its legs and wings, but flies with just this anatomy were reported in 1995 by Georg Halder *et al.* in Basel [1685]. Halder *et al.* created these monsters by turning ON the "Pax" (Paired box) gene *eyeless* (ey) in discs where it is normally not expressed. Although ey has a homeobox in addition to its paired box [3486], the ey^{LOF} phenotype is unlike any Hox^{LOF} fly insofar as it involves an absence, rather than homeosis, of the affected structure (hence the name "eyeless"). The ability of ey^{GOF} to turn ON all the downstream machinery needed to build an eye implied that ey must be the "master control gene" for eye development [214, 1415, 1416]. Ironically, Stern and Tokunaga had hailed an allele of this same gene (ey^D) as their long-sought "prepattern mutant" ~30 years earlier (cf. Ch. 4) [4109].

Further research, however, revealed several caveats that dampened the enthusiasm for the master gene concept [3658]:

1. The homeotic effect is dependent on the tissue context: ey can only induce ectopic eyes in certain dpp-ON zones in specific discs [394, 744, 1684, 1685, 3894]. Ectopic expression of dpp alone does not evoke homeosis in any disc [2755].

2. Other genes besides ey were found to also have the power to elicit ectopic eyes [1234, 1419, 4385], including *eye gone* (paired/homeobox) [1780, 2090] and *eyes absent (eya)* [394, 398]. Surprisingly, two such genes are normally ON in other discs as well: *dachshund (dac)* [3894] and *teashirt* (zinc finger) [3240] (cf. *sine oculis* in the leg disc [3118]). In theory, these other genes need not undermine ey's primacy because they could be acting downstream. However, several of them can act upstream (in a feedback loop) to turn ey ON at ectopic sites [743, 3387, 3894].

3. Still another homeobox gene *optix* can elicit extra eyes, and it does so completely independently of ey (i.e., in an ey^{null} background) [3851].

The growing disenchantment was summarized by Pan and Rubin in 1998 [3240]. (The homeobox gene *sine oculis* is abbreviated "so" [759, 3859].)

Although ey is required for the initial expression of *eya*, *so*, and *dac* in the eye primordium, the later genes are also involved in a positive feedback loop to activate the expression of ey. Therefore, ey does not function simply as a "master regulatory gene" to activate a linear pathway specifying the eye fate; rather, ey, *eya*, *so*, and *dac* form part of a regulatory network that together "locks in" the eye specification program. [3240]

Then, in 1999, an *ey* paralog called *"toy"* (*twin of eyeless*) was described [933], which acts upstream of *ey* (Toy ➜ *ey*) to induce ectopic eyes [1419, 1760]. Unlike *ey*, *toy* is not subject to any feedback regulation by *dac*, *eya*, or *so*. Hence, the title of "master gene" was transferred from *ey* to *toy*. The anointing of *toy* may have been premature, however, because we still do not know (1) its null phenotype, (2) its relationship to *optix*, (3) whether it can induce eyes with drivers other than *dpp-Gal4* (which may have its own idiosyncrasies) [1030, 4385], or (4) why it is able to induce ectopic eyes where *ey* cannot [3118].

If there were a hierarchical chain of command with *toy* or *ey* at the top, then co-expression of *ey* with one of its subordinates would not be expected to manifest much synergy because *ey* would have to act through them, but strong synergistic interactions (with regard to inducing extra eyes) have been documented for a number of combinations [930, 1030], including {*ey* AND *dac*} [744], {*ey*, *dac*, AND *dpp*} [744], {*ey* AND *dpp*} [744], {*ey* AND *eya*} [394, 744], {*eya* AND *dac*} [743], and {*eya* AND *so*} [553, 744, 3387].

The relationships among these genes at the inception of eye development are complicated [1787]. As discussed in Chapter 7, several are regulated by *dpp*, but *ey* is not [744, 930]. The links below have been demonstrated under endogenous or ectopic conditions, but there is no guarantee that the circuitry behaves similarly in the two contexts [4385]. Evidently, these genes operate as an interactive network rather than as a hierarchy (cf. Fig. 7.3) [3240, 3387].

Link 1: Dpp ➜ {*dac*, *eya*, AND *so*} during normal furrow initiation. In *Mad*^{null} clones at the P edge of the eye (where ommatidial patterning begins), the cells stop expressing *dac*, *eya*, and *so* [930], but this link ceases to operate once the furrow gets underway [930]. Dpp is not needed for *ey* to turn ON [744].

Link 2: Ey ➜ {*eya* AND *so*} during normal furrow initiation. Neither *eya* nor *so* is expressed in *ey*^{LOF} discs [1684]. Conversely, *ey* turns ON normally in *eya*^{LOF} or *so*^{LOF} discs [1684, 3387].

Link 3: {Ey AND Dpp} ➜ {*dac*, *eya*, AND *so*} when targeted to the wing disc [744]. Expression of *ey* in *dpp*-ON regions of leg and wing discs (via *dpp-Gal4:UAS-ey*) activates *dac* [3894], *eya* [394, 1684], and *so* [1684, 3118]. For the *eya* [554, 4888] and *so* [3118, 3476] loci, this activation has

been shown to occur via eye-specific *cis*-enhancers. The need for Dpp is shown by *ey*'s inability to activate *dac*, *eya*, or *so* in *dpp*^{LOF} eye discs when overexpressed there (via *ey-Gal4:UAS-ey*) [930]. Induction of eyes by *ey* also requires Hh [2126].

Link 4: Eya ➜ {*dac* AND *so*} during normal furrow initiation. When Dpp signaling is disabled (in *dpp*^{LOF} discs or *Mad*^{null} clones), expression of *dac* and *so* vanishes but can be rescued by overexpressing *eya* (via *dpp-Gal4:UAS-eya*) [930]. This talent is not shared by *dac* or *so*: neither gene (when expressed via *dpp-Gal4*) turns ON any other members of the *eya-dac-so* trio [930], nor is *so* alone able to induce extra eyes [930, 3387]. Likewise, *eya*^{LOF} eye discs fail to express *dac* above a trivial level, while *dac*^{LOF} eye discs still express *eya* at normal levels [743]. Indeed, the only hint of a "Dac ➜ *eya*" link is a spot of *eya* in *dpp-Gal4:UAS-dac* antennal anlagen [743] that is more diffuse than the *dac* spot induced by *dpp-Gal4:UAS-eya*. (Hence, no "*dac* ➜ *eya*" link is shown in Fig. 7.3.)

Link 5: Eya ➜ *dpp* during normal furrow initiation. Expression of *dpp* is reduced in *eya*^{LOF} discs and is eliminated entirely in clones that are homozygous for a strong LOF allele (*eya*^{1} [553]) [1780].

Link 6: {Eya AND Dpp} ➜ *dac* when targeted to the wing disc [744].

Link 7: {Eya AND Dpp AND So} ➜ *dac* when targeted to the wing disc [744]. Within the eye disc, *eya*^{LOF} and *so*^{LOF} clones fail to express significant amounts of *dac* [3387].

Link 8: {Eya OR Dac OR [So AND Eya]} ➜ *ey* in the antennal region of the eye disc [743, 3387, 3894]. However, this area appears to be the most sensitive to eye induction, so these effects are relatively weak and may not be physiologically significant.

Like Ubx, Toy and Ey appear to act as micromanagers for genes that are far downstream (as do Eya and So [3387]). Indeed, this fact alone refutes the notion of a hierarchy [4385]. The best example is the regulation of rhodopsin genes in photoreceptor cell types (cf. Ch. 7) [1272, 3895, 4600]. Within ~70 b.p. of the transcription start site, all rhodopsin gene promoters have a generic binding site for Ey or Toy (or both), which allows the genes to be activated throughout the eye [3256] (cf. the Glass-binding site [1157, 2963, 2964]). More distally, they each have *cis*-enhancers that are specific for each rhodopsin sub-type (e.g., Prospero represses these enhancers at *rh5* and *rh6* in R7 [860]). A promoter can only turn ON its gene if both the proximal AND distal sites are occupied. Thus, the promoter is like a door with two locks: only when both keys are turned will it open [1705].

Indeed, this sort of micromanagement makes even more sense for a field-specific gene like *ey* than it does for a metamere-specific gene like *Ubx*. Given that Hh, Dpp, Wg, Delta, and Spitz are used by virtually all discs (and in different parts of the same disc in some cases), the cells of each field need to have some way of knowing that a certain morphogen signal "means" a certain fate that is appropriate for their field [4385]. The meaning would be encoded grammatically by a combination of inputs from (1) a field-specific selector gene and (2) a particular signaling pathway [1477, 1686]. For example:

> IF you are in the eye
> AND your Ras is tickled at a certain time,
> THEN become such-and-such a cell type.

The former condition could be satisfied by *ey*. In general, field-specific genes appear to function as "licensing agents" for parochial pathways [871, 4563, 4857].

Ergo, the meaning of "wingness" or "legness" would be a certain transcription factor (e.g., Toy in eyes) or set of such factors (e.g., {Hth-Exd AND Dll} OR {Ss AND X} in antennae) that identifies each cell as belonging to that particular field. Cells would know their histotypic region of residence, although they might not know their disc per se. This surmise helps explain (1) why homeosis and transdetermination affect histotypes as well as metameres (Table 8.1; cf. Fig. 6.9d) and (2) why development can be perturbed to produce phenotypes that look grotesque to us (e.g., triplicated limbs; cf. Fig. 5.2) but feel natural to the cells themselves because no local rules are broken [617] (e.g., Bateson's Rule).

The manifold "enhanceosome" is a wondrous Gordian Knot

The synergy among early eye genes described above is partly attributable to a ≥3-protein complex wherein Eya binds Dac and So [553, 743, 3387]. Within the complex, So apparently functions to bind DNA [759, 3859], while Dac supplies a transcriptional activation domain [743], and Eya forms a bridge between them [4385], although a different architecture is used in other tissues [2367], and So recruits

other partners as well (including Groucho, Amos, and TAF250 [2193]).

The sad fact is that we cannot now solve either the Eya-Dac-So puzzle or the Hth-Exd-Dll puzzle discussed above, nor the riddle of how these aggregates dovetail with the transcription machinery so as to regulate their target genes. In general, humankind is still at a kindergarten level of understanding with regard to "promoter logic" [393, 968, 2836, 4165, 4755]. We would like to learn the following:

1. How do the various transcription factors and cofactors fit together to form "enhanceosomes" [649, 2109, 4274, 4304, 4454]? One example of a recent insight (in flies) is that Hox proteins use their hexapeptide motif to dock with the TALE domain of Extradenticle [879, 2207, 3281, 3393].

2. How does each enhanceosome dovetail with the basal transcription apparatus [1127, 1317, 2304, 2488, 2826], and to what extent does the gigantic Mediator complex assist this process [1703, 2665]? Virtually all ~20 subunits of the human Mediator complex are conserved in flies [412, 4386], and the fly complex binds a bewildering zoo of transcription factors – including Bicoid, dHSF, Dorsal, Ftz, Krüppel, and VP16 (but not Twist or Hunchback) [3262].

3. How do the steric rules implement "AND" or "OR" conjunctions [1477, 4042], "NOT" negations [1984, 2350], and "IF/THEN" contingencies [440, 464, 845, 2678, 4164]?

4. How are these logical conditions concatenated grammatically [693, 876, 960, 1046, 1444] so as to dictate criteria such as "IF factors A AND (C OR D) but NOT B are bound, THEN transcribe the gene at rate *r*" [103, 259, 2471, 3657, 4836, 4837]. It is at this level that proteins literally compute outputs, and it is here that we will undoubtedly find the key to how genes compute anatomy.

Arguably, this **"Enhanceosome Puzzle"** [1317, 1696, 2304, 2488, 4332] poses the greatest challenge for the postgenomic era of biology [968, 4847]. Will computational genomics alone succeed in untying this tangled Gordian Knot? No one knows. What we do know is that classical genetics has been incisive in cutting to the core of similar problems. Explorers of the future might therefore do well to study the past since those old tools may yet prove useful [410, 463, 1364, 2845].

The deepest enigma is how evolution rewired the circuit elements

It has become trite to say that development is like a computer program [2179]. Cliché or not, there is no better metaphor for gene circuitry [968, 4837]. In their 1997 book *Cells, Embryos, and Evolution*, John Gerhart and Marc Kirschner argued that evolution works like an electrician [1440]. By changing the wiring here or there, it reconfigures circuits to alter anatomies.

Collectively, the many "➞" and "⊣" links in the present book have sketched an elaborately ornate wiring diagram for disc development. The deeper question is: where did it all come from?

Over the eons of evolutionary time, the repertoire of genes and *cis*-enhancers must have been relatively stable because of their interdependence [968] (e.g., see [2610, 3601, 3895, 4777]; cf. the genetic code [1148]). Hence, they make up the genomic "hardware" [103]. In contrast, the links among these circuit elements were more changeable [3781] (e.g., see [1617]), so they make up the "software" [694, 2236, 3494]. For example, evolution has rearranged macrochaetes by adding or deleting *cis*-enhancers at the AS-C (cf. Fig. 3.4) [4766], and dipterans reflect this tinkering [662, 1354, 2959, 3966, 4096]. What has changed is not the genome's overall inventory of available genes or *cis*-enhancers, but rather the choice of which enhancers reside in the AS-C.

Linkage in metazoans represents a kind of easily modifiable software of development, which owes its success to the highly conserved hardware of basic cell biological mechanisms. [1440] (p.142)

Evolution is about teaching old modules new tricks. [1094]

Similar games were undoubtedly played within enhanceosomes themselves. For example, gene "*a*" encoding a *trans*-activator for gene "*b*" (*a* ➞ *b*) could mutate so that its protein acquires a WRPW motif that now recruits Groucho and turns it into a repressor (*a* ⊣ *b*). Pangolin may be a case in point because it can bind either Armadillo (Pan ➞ target gene) or Groucho (Pan ⊣ target gene). The implication is that evolution kept Pan in a schizophrenic state to serve as a toggle switch (cf. Fig. 5.6c). Other bipolar regulators may include Dorsal [1110, 1257], Mad [3377], NK-4 [776], Su(H) [1330, 2540], and Tramtrack [2391].

The making and breaking of connections is what nervous systems do as they learn, so neural networks may offer a useful way to simulate this process for gene circuits in the future [133, 297, 441, 1438]. One tantalizing question is whether living systems have "learned" to increase their "evolvability" [2154, 4520] by making certain types of genomic alterations easier than others [622, 1090, 3115, 4798] (e.g., transposition [420, 2212, 4632]).

For developmental biologists, the pithiest issue remains how the genes encode the anatomy. Do flies really have a "Build an eye!" program? In some sense the answer must be yes, but not at the cellular level. The cells themselves are just following a script of "IF this happens, THEN do that!" cues that evolution wrote in the DNA [825, 1094]. If we were to ask a cell, "What are you doing?," it could answer, but if we asked, "Where are you headed?," it could not.

Epilogue

Fly genetics in the 1900s succeeded in deciphering the logic of disc development. Its vaunted offspring – the field of fly genomics – is faster and sexier but no more powerful in its ability to solve the remaining riddles of circuitry and control. Curt Stern warned us about this irony in his essay, "The journey, not the goal":

One of the fundamental aspects of science is its lack of purpose.... Science, during the last one hundred or more years, has been in the dangerous position of a successful poet who started by composing songs of joy and sorrow to lighten the burden of his own soul only to find that they became best-sellers.... Science has become a profession.... The later comers [have] forgotten the beginnings of the highway. Dreamy followers of crooked paths [were] their predecessors.... We should encourage anew the roaming after knowledge for the sake of the joyful adventure. [4093]

Industrialization and commercialization notwithstanding, the Fly World still offers many mysteries for aimless explorers with curiosity alone to fill their sails.

Before launching the field of fly genetics, Thomas Hunt Morgan's passion was embryology [72, 3903]. In the preface to his 1934 book, *Embryology and Genetics*, Tom waxed lyrical about the promise of developmental genetics as a burgeoning hybrid field [2949]:

Since 1900, when the discovery of Mendel's work became known, one of the most amazing developments in the whole history of biology has taken place. The fundamental laws of heredity are now known, and since it is through the egg that the hereditary properties of the individual are carried on from generation to generation, the importance of an understanding of development to supplement the knowledge of the laws of heredity is apparent, and the interlocking of these two experimental branches of biology has become a subject of absorbing interest.... That much remains to be done will be only too obvious, but with the openings furnished by the experimental investigation of heredity and embryology there is promise that a great deal more is within our reach.

This hybrid has now borne fruit, and the fruit fly adorns the most glorious branch. Who could ever have guessed that this little "gnat" held such rich insights in its shimmering golden fleece?

Glossary of Protein Domains

All protein domains that were mentioned in the text or tables are inventoried below. For further information, consult *PROSITE* (*www.expasy.ch/prosite*) or the following reviews: domains in general [829, 1089, 3295], DNA-binding domains [3284], scaffolding domains [3299], extracellular domains [404], domain classification [4219], domain evolution [3021], protein-protein binding [2091], protein-peptide binding [3299], receptor-ligand binding [3080], signal transduction [3783], and an inventory of fly protein domains [3674].

D. melanogaster proteins vary in size from 21 a.a. (L38, a ribosomal protein) to 5201 a.a. (Kakapo, a cytoskeletal component needed for intercellular adhesion) [14]. The domains listed below vary from 4 a.a. (WRPW) to ~270 a.a. (PAS).

N.B.: Customarily, "domain" denotes a motif in proteins, while "box" refers to DNA. Thus, for example, the homeo*box* encodes the homeo*domain*. "Repeat" does not connote identity within a protein (e.g., only 6 of the 38 residues are invariant among Notch's 36 EGF-like repeats [2210]), nor does it imply interchangeability. For instance, the LIM domains of Lim3 can replace those of Apterous in wing development but not in the CNS [3159]. Likewise, Cactus's ankyrin repeats cannot functionally substitute for those of Notch [1053]. The specificity of repeats is epitomized by the "arm" domain:

Although individual repeats within a single protein are only about 30% identical, they are highly conserved during evolution. Thus, corresponding repeats of armadillo and β-catenin (which are direct homologs) are very similar (e.g., repeat 1 of armadillo is 90% identical to repeat 1 of β-catenin). Individual repeats, once free to diverge, are now fixed in sequence, presumably reflecting some well-conserved interaction with a target protein. [3312]

"Peptide-binding" and "protein-binding" both indicate mortise-and-tenon "docking" sites [1950, 3834], but the former fit ≤~10 contiguous amino acids, while the latter recognize larger or discontinuous epitopes [1730, 2364, 4871]. Abbreviations: "a.a." (amino acid), "a.k.a." (also known as), "n" (variable nucleotide), "^{P}" (phosphorylated a.a. such as S^P), "x" (variable a.a., with subscripts such as x_4 indicating the number of xs in a series). Standard 1- or 3-letter a.a. codes are given below. Sequences (a.a. or nucleotide) are underlined, and residues are listed in amino-carboxy or 5′-3′ order.

A	Ala	Alanine
C	Cys	Cysteine
D	Asp	Aspartic acid
E	Glu	Glutamic acid
F	Phe	Phenylalanine
G	Gly	Glycine
H	His	Histidine
I	Ile	Isoleucine
K	Lys	Lysine
L	Leu	Leucine
M	Met	Methionine
N	Asn	Asparagine
P	Pro	Proline
Q	Gln	Glutamine
R	Arg	Arginine
S	Ser	Serine
T	Thr	Threonine
V	Val	Valine
W	Trp	Tryptophan
Y	Tyr	Tyrosine
φ	(a hydrophobic a.a.)	

14-3-3 The numbers 14, 3, and 3 refer to chromato-graphic (14th of 15 fractions) and electrophoretic (3rd of 4 quartiles) mobilities of a founding class of mammalian brain proteins (band #3 in the 14-3 group) [2915, 2916], although such proteins are widespread among eukaryotes and participate in various cellular processes [42–44, 561]. Each mammalian isoform is named with a Greek letter (from α to η) based on its order in an HPLC elution sequence [44, 1958], and fly orthologs are named accordingly (e.g., 14-3-3δ). These proteins are ~260 a.a. long [44] and bind S^P-containing sequences that have the consensus RSxS^{P}xP or Rx(Y/F)xS^{P}xP [3023, 4787]. Although 14-3-3 is not a *bona fide* domain because it is not subsumed within larger proteins, it resembles PTB and SH2 domains insofar as it binds a cognate phosphopeptide. Monomers form 9 antiparallel α-helices [2567, 4770] and assemble into homo- or heterodimers [43, 2087]. Each 14-3-3 dimer is shaped like a cradle whose cavity is lined with negatively charged, highly conserved residues. The outer edges of the cavity bind (non-14-3-3) S^P-bearing guests [4787], with the cognate peptide in an extended main-chain conformation. In *Drosophila*, 14-3-3 isoforms serve scaffolding functions in different RTK signaling pathways [716, 2279, 4317] by binding to dRaf and Ksr [3639]. The details of how 14-3-3 dimers wrestle dRaf into one conformation vs. another are still being worked out [2957, 4116], but it is now clear that each half of the 14-3-3 dimer grips dRaf in a different way [43, 3638, 3639].

ADAM "ADAM" stands for a disintegrin and metalloprotease domain, and a synonym is "MDC" (metalloprotease-disintegrin-cysteine-rich) [346, 369]. It contains a putative integrin-binding "disintegrin" moiety and a protease that is typically zinc dependent [346, 369, 4721, 4722]. Some ADAM family members (e.g., Kuzbanian [3479] and TACE [491, 2995]) are "sheddases" that snip extracellular parts from other surface proteins on the same cell [3469, 4599], although the protease portions of others are nonfunctional [4721]. Other ADAM proteins may degrade extracellular matrix or assist in cell migration or adhesion [346, 3469].

ankyrin Named for a motif tandemly repeated 22 times in human erythrocyte Ankyrin, this ~33-a.a. protein-binding domain (a.k.a. "cdc10/SW16") [2624, 2841] is found in the Notch [1053, 2299, 2746] and IκB families [366, 2208, 3123, 4476], in various transcription factors [235, 403], in some yeast cell-cycle genes [2624], and in the putative ion channel (NompC, 29 repeats) that transduces touch stimuli in fly bristles [4527]. Its secondary structure is an L-shaped "girder" that stacks stably when tandemly repeated [1564] – a fact that explains why virtually all pro-teins that contain this domain have ≥4 consecutive copies [235, 403]. Some proteins use these modules to dimerize with other ankyrin-domain partners [1269, 2747], whereas others use them to dock with RH (Rel homology) domains of Rel-family transcription factors [168, 366, 2208, 3123] or other ankyrin-deficient partners [1870]. Surface residues dictate binding specificity [168]. Ankyrin itself "anchors" (hence the name) integral membrane proteins to the cytoskeleton and has one conserved (and 14 mostly conserved) residues in its repeats.

arm First identified in the fly gene *armadillo* (*arm*), where it is tandemly repeated 12 times (with one interruption) [892, 3320, 3598], the arm domain is a 42-a.a. protein-binding module [3146, 3312]. In β-catenin (Arm's vertebrate ortholog [3306]), each repeat forms an open triangle of 3 α-helices, and the (slightly nonplanar) triangles stack atop each other to build a cylindrical right-handed superhelix (110 Å long and 35 Å wide) [1593, 1936]. The (continuously wound) superhelix has a 95-Å groove that could fit a stretch of 25–30 a.a. (like a hot dog in a bun). Indeed, the groove is strongly positively charged and the binding partners of β-catenin (cadherins, Tcf-family transcription factors, APC) all have an excess of negative charges in their binding domains, so the partners likely adhere via charge complementarity [1936]. When the groove is not occupied by them, it may be filled by the acidic C-terminal tail of Arm/β-catenin itself [892]. The core of the superhelix is hydrophobic, which explains why no proteins have ≤6 tandem arm repeats (i.e., too few hydrophobic contacts for stability). Included in this family are importins α (8 repeats) and β (11 repeats) [2743], which escort NLS-bearing proteins into the nucleus [1568, 1569], and arm repeats may enable proteins to enter the nucleus (sans a NLS tag) [1180].

bHLH The ~60-a.a. "basic Helix-Loop-Helix" domain [2566] typically resides near a protein's N-terminus [899, 3375], although Atonal [2040] and Tap [576, 577, 1402] are exceptions. The basic part binds DNA via positively charged amino acids, while the amphipathic helices (separated by a loop of variable length) mediate dimerization [1152, 2634, 3016] or, when not bound to DNA, tetramerization [245]. Dimers have dyad symmetry and a scissors shape (Fig. 2.5), with the "blades" straddling DNA in its major groove [1152] and adopting an induced fit in some cases [1221]. Most bHLH dimers bind the consensus sequence CAnnTG [348, 3017], but dimers from the E(spl) Complex bind CACnAG [3171], and bHLH-PAS proteins bind ACGTG or GCGTG [899]. The identity of the variable ("n") nucleotides depends on the dimer [348], as do the preferences for flanking nucleotides [1245], with each member of a dimer dictating its own "ideal" half-site [899], although homodimers may bind asymmetric sites [348]. The basic part may also mediate interactions that do not involve DNA binding [764]. In bHLH subfamilies that have an auxiliary dimerization domain [135] – leucine zipper [245, 3375] or PAS [899, 900] – the other domain typically resides just C-terminal to the bHLH motif. Advantages of having two separate dimerization domains in the same protein include added affinity, combinatorial selectivity, or both [1221, 4857]. Such versatility is epitomized by the bHLH-PAS proteins that implement the circadian clock (see PAS below).

BTB This ~120-a.a. protein-binding domain is found in the fly proteins Broad-Complex, Tramtrack, and Bric à brac (hence the name; a.k.a. "POZ") [211, 1516, 3636, 4891]. Located usually at the N-terminus, the BTB motif reduces the DNA-binding ability of zinc fingers within the same protein, presumably because the oligomerization that it mediates obscures those DNA-binding surfaces [211]. (Analogous "self-defeatism" afflicts LIM-HD proteins [929].) In other cases, however, oligomerization facilitates DNA binding over a region large enough to bend DNA and exclude nucleosomes [2147]. Some transcription factors use their BTB domain to recruit a synergistic co-repressor [1036]. For example, the zinc-finger factor Tramtrack uses its BTB motif to recruit the co-repressor dCtBP [4596]. Tramtrack can form heterodimers, but most BTB proteins only homodimerize.

COE Named for the *Drosophila* gene *collier* (a.k.a. *knot*) and the mammalian transcription factor Olfactory-1 (a.k.a. Early B-cell factor) [909], this ~210-a.a. DNA-binding domain can mediate dimerization and transcriptional activation [1679]. COE family proteins also contain a C-terminal HLH domain [192] and resemble bHLH transcription factors. Collier helps specify muscle cell fate in embryos [910] and intervein identity in the wing [1840, 2894, 3077, 4479].

cut In addition to its homeodomain, the fly gene *cut* has 3 copies (sharing 55–68% a.a. identity) of a 60-a.a. motif termed the "*cut* repeat" [370]. In a mammalian *cut* counterpart, one repeat binds DNA independently of the homeodomain [2528], but none of the fly's repeats appears capable of doing so [87]. Rather, they may modulate the strength or sequence specificity of DNA binding by the homeodomain.

DEF DEF is a domain that recruits MAP kinase (a.k.a. ERK). "DEF" stands for "Docking site for ERK, FxFP" [2007]. The DEF peptide (FxFP) has two bulky hydrophobic (F) residues that are thought to fit into a pocket on MAP kinase. After MAP kinase docks with its DEF-bearing target (e.g., Yan), it phosphorylates S or T substrates near the DEF site.

DEJL Like DEF, DEJL (a.k.a. "D domain" [2601, 4800, 4801]) is a binding site for MAP kinase, but it can also bind JNK (Jun N-terminal Kinase). "DEJL" stands for "Docking site for ERK and JNK, LxL." It consists of an (L/I)x(L/I) tripeptide that has a cluster of basic residues on its N-terminal side [2007].

EGF-like Named for Epidermal Growth Factor (a 53-a.a. soluble fragment cleaved from a ~1200-a.a. transmembrane precursor), this ~40–50 a.a. protein-binding module has 6 conserved cysteines that pair to form 3 disulphide bonds. The 1st Cys binds to the 3rd, the 2nd to the 4th, and the 5th to the 6th, thus weaving a knot [970, 3521] whose loops dictate binding specificity. Notch's consensus is $Cx_4Cx_5Cx_8CxCx_8Cx_6$ [4611]. EGF-like repeats are found in extracellular matrix proteins or extracellular moieties of membrane-spanning proteins (ligands or receptors). Some EGF-like domains use 7 oxygen atoms from their side groups to form a cage that harbors Ca^{2+}. A canonical Ca^{2+}-binding sequence is seen in at least 6 of Notch's 36 repeats (including the two that bind Delta) and in one of Delta's 9 repeats [1707, 3521, 3544], and Notch-Delta binding is (for this reason?) Ca^{2+}-dependent [1204]. The world record for number of repeats in a single protein goes to Dumpy, with 308 EGF-like motifs in an extracellular domain that stretches to nearly 1 micron in length [4668]!

EH Named for "Eps15 Homology," this ~66-a.a. peptide-binding domain has 4 invariant and 21 mostly conserved residues [4740]. Human Eps15, which has 3 tandem repeats, is a substrate for the EGF receptor and other tyrosine kinases. EH domains use a hydrophobic pocket (braced by 4 α-helices) to bind a tripeptide "NPF" that is found in Numb near its C-terminus (but in few other fly proteins) [979, 3725]. Various EH- and NPF-bearing proteins aggregate to control clathrin-mediated endocytosis [739, 2699, 2756, 3596, 3856]. An EH partner of Eps15 has been implicated in the EGFR signaling pathway in flies [594].

Ets Named for the *ets* (E twenty-six) vertebrate proto-oncogene [3467], this ~85-a.a. domain functions in both DNA binding and protein-protein docking [2135, 4559]. It binds as a monomer to 10 b.p. sequences with a GGA core. Its C-terminal half is highly basic, whereas its N-terminal half has 3 conserved Trps at ~20-a.a. intervals. The first Trp is imbedded in a helix, while the third resides in a NLS motif.

F The F domain is a ~40-a.a. protein-binding sequence first identified in cyclin F [169]. It links proteins to ubiquitinating enzymes and thus can act as a destruction motif [169]. F-domain proteins that contain other kinds of protein-binding sites (e.g., Slimb, which has WD repeats [2060, 4279]) are evidently hubs for assembling ubiquitination complexes [3500, 4031, 4701].

GUK The ~160-a.a. guanylate kinase homologous ("GUK") domain is found near the C-terminus of MAGUK (membrane-associated GUK) proteins [533, 3419]. Given their enzyme homology, it is surprising that GUK domains appear catalytically inert [1062]. Instead, they tend to mediate protein-protein interactions, possibly via a leucine zipper subdomain [4237]. The role of this domain remains obscure, but at least one of its duties is to steer host proteins to neuromuscular synapses [4300]. MAGUK proteins typically form scaffolds for assembling complexes of membrane receptors [843, 1001, 1062, 4371] and ion channels [83, 896, 4811]. Some of those complexes are crucial for growth control [533, 4561]. At least one MAGUK (CASK) leaves its membrane perch (when a signal is received) and goes to the nucleus, where it functions as a co-activator [450, 1920]. A 28-a.a. MSR (MAGUK-specific repeat) motif (of unknown function) is found in the MAGUK family [3419].

HLH The Helix-Loop-Helix protein-binding motif usually adjoins a stretch of basic residues that mediate DNA binding (see bHLH above). Alternatively, it may adjoin a DNA-binding domain of the COE class [909]. Proteins that contain a HLH domain alone cannot bind DNA. Some of them (e.g., Extramacrochaetae [1156, 1388]) sequester bHLH partners in inactive heterodimers [213, 286, 3083]. This "kidnapping" strategy is also used by transcription factors in the LIM-HD family (Beadex disables Ap via Chip [2854, 3908]) and the POU-HD family (I-POU disables Cf1a [4380]). Other types of antagonists interfere with bHLH function by protein-protein binding outside the bHLH domain [2046].

HMG This ~80-a.a. DNA-binding (and DNA-*bending*) motif occurs in certain High Mobility Group (nonhistone chromatin) proteins [316] and in various other animal, plant, and yeast proteins [2035, 2402, 2415]. In one HMG subfamily, there are multiple HMG motifs per protein and DNA is recognized with little or no sequence specificity, whereas in the other subfamily (including Sox-type HMG domains [424]) there is a single HMG motif and DNA binding is sequence specific [810, 1629], although much of the specificity may depend on DNA-binding partners [2122]. HMG domains are L shaped, with two α-helices in one arm and one α-helix in the other.

They bend DNA [253, 4299, 4717] by inserting hydrophobic side chains (along DNA's minor groove) into the stack of bases [317, 1480, 2605]. Such insertions (plus a.a.-nucleotide hydrogen bonds) unwind the double helix [1481, 3008]. The "oregami" folds thus created in promoter DNA facilitate construction of an "enhanceosome" for initiating transcription [879, 1188, 4274] (e.g., the Pan-Arm complex [849]).

homeo The ~60-a.a. homeodomain (HD) binds DNA [1417, 1422, 1858] or, in some cases, RNA [710, 3541, 3606] or protein [2207, 3402]. It is encoded by the strongly conserved ~180 b.p. homeobox (HB) [1420] that is named for its prevalence among homeotic genes [1418] (cf. Ch. 8). HB genes appear to be "master control genes" for body patterning [1416]△. The term "Hox" (homeobox) is conventionally only used for homeobox genes within the clusters that govern body segment identities (i.e., BX-C and ANT-C in flies [1223]△).

The HD consists of 3 α-helices (#2–4) around a hydrophobic core, plus a flexible N-terminal arm containing a fourth (#1) helix [1417, 1420, 3284, 4198]. Residues ~30–50 within the HD (helices #2 and 3) form a helix-turn-helix (HTH) [335] that is superimposable on the HTH of prokaryotic repressors [3482] despite (1) a lack of appreciable sequence identity [1416, 3219] and (2) a different set of DNA contact sites [31, 1422, 1729, 2240, 4384]. One HD subclass called "TALE" (Three Amino Acid Loop Extension) has 3 extra a.a. between helices #1 and 2 [567, 568], whereas another has 10 extra a.a. between helices #2 and 3 [1335, 3033]. In Extradenticle, the TALE region forms a hydrophobic pocket that fits a hexapeptide motif (see below) [2207, 3281, 3393].

Particular HDs bind specific nucleotide sequences [857]△, as evidenced by "domain swap" experiments [2071, 4879]△. They function as transcriptional activators or repressors for a variety of downstream "realizator" genes [407, 1580, 2923, 2938, 3457]△. The universe of potential targets [2541, 4532] is reduced *in vivo* [707] (vs. *in vitro* [709]) due to (1) the masking of inactive genes by chromatin [655] and (2) the docking preferences of adjoining *trans*-acting factors in the *cis*-enhancer vicinity [1767, 2539].

HD proteins can bind DNA as monomers [1416], dimers [4688, 4689], or higher-order complexes [3716]. In some cases, binding is regulated by phosphorylation [300, 2011, 2677]. Many HD proteins have a second DNA-binding motif [1417] that alters the HD's DNA affinity or specificity [2098], or bends DNA [4477]. Auxiliary motifs include the cut [87]△, LIM [1862]△, paired [2098]△, and POU-specific [3713]△ domains, and one of these motifs (paired) can *trans*-complement the HD even when detached [2873].

The central two base pairs in the ∼10 b.p. bipartite binding site encode much of the specificity [709], which makes sense because the HD's N-terminal arm binds them, and it is also critical for specificity [1417, 2671, 4846, 4862]. Most HD proteins have NLSs that send them to the nucleus [1020, 4879] (e.g., Hth [2045]), but some (e.g., Exd [3716]) also have NESs that can divert them to the cytoplasm [2661].

DNA-binding affinity [3324, 3325], DNA-sequence specificity [706, 2675, 4450], and DNA curvature [3842] can be modulated by separate cofactors [683, 2421, 2539, 2674, 4623], including Extradenticle (Exd) [156]$^\Delta$ and Homothorax (Exd's escort for nuclear entry) [2360, 3716], both of which are also have a HD [2677]. Hox proteins capture Exd by inserting an N-terminal hexapeptide φ(Y/F)(D/P)WM(K/R) (a.k.a. YPWM [879]) into Exd's hydrophobic TALE pocket [2207, 2944, 3281, 3393, 3716, 3842]. Exd's chief role may be to coerce its partner into a new shape [708, 2675] that (1) alters the oligonucleotide specificity of the HD itself [2675] or (2) unmasks an activation [2535, 3391] or repressor [3715] domain elsewhere in the HD protein. The Hox-Exd tryst helps explain the astounding ability of some HD proteins to function without their HD [873, 2360]. To wit, the "jigsaw puzzle" of the HD protein and its cofactors can remain intact even with a few pieces missing. *N.B.:* Steric cooperativity wreaks havoc for other analyses in this book – e.g., the Cubitus interruptus complex (App. 6) and the Eya-Dac-So complex (Ch. 8).

Ig-like The immunoglobulin-like (Ig-like) domain defines a superfamily of cell-surface glycoproteins [4680]. The Ig motif folds into a "β-sandwich": two β-sheets (3 to 4 antiparallel strands per sheet) held together by one conserved Cys-Cys disulfide bond and by hydrophobic contacts. The sandwich (whose outer surfaces are hydrophilic) is a platform for displaying variable groups (peptide or sugar) on its faces or at the U-turns within its sheets. Ig-domain proteins commonly mediate adhesion or recognition by dimerizing (homo- or hetero-) with counterparts on opposing membranes [1296]. The vertebrate immune system may have evolved by giving these old molecules new duties in antigen binding and intercellular communication [4680]. The notorious diversity of antibodies in vertebrates may have a counterpart in flies [3793]: Dscam has 10 Ig-like domains – 3 of which (plus the transmembrane domain) can be varied (by alternative splicing) to produce 38,000 potential isoforms!

leucine zipper This ∼30-a.a. dimerization domain has a leucine every 7 a.a. [2401, 2790, 3219]. The leucines align along one face of an amphipathic α-helix, and 2 such helices (separate proteins) join by crossing (at a ∼20°

angle) and intertwining into a left-handed "coiled coil" [1864, 2621]. The coil is stabilized hydrophobic contacts between the leucines [1718, 1719, 3163]. The term "zipper" is thus misleading because the leucines do not interdigitate [1864, 2790]. At one end, the helices diverge into a "scissors grip" around DNA [3642, 3769, 4494]. The DNA-binding tips have a separate basic region [55, 915, 1153, 3160], hence the composite "bZip" domain [2197]. Some ("bHLH-Zip") proteins contain basic, leucine zipper, and HLH domains [245, 1594, 3375]. Whether a bZip protein forms homo- or heterodimers depends on its ability to make interhelical salt bridges with its partner [1513, 2334, 4493]. Because zippers can form higher order complexes (dimers of dimers), they can mediate binding to two separate DNA sites simultaneously, thus "tying" a DNA loop that may be instrumental in transcriptional activation [1221]. Their ability to bend DNA [3251] may facilitate the looping. Large-scale versions of leucine zippers appear to be used by transmembrane proteins (like glue) to join cell surfaces together [2323].

LIM Named for Lin11 (nematode), Isl-1 (vertebrate), and Mec-3 (nematode) proteins [160], the LIM domain has two tandem zinc fingers [973, 3736]. LIM repeats themselves tend to be tandemly clustered. Unlike orthodox zinc fingers, LIM domains bind proteins instead of DNA [97, 929, 972, 1862]. They can interact *inter se* (selectively or promiscuously, depending on developmental context) [3159, 4310] or with heterologous "LIM-binding" domains [161, 929, 3786, 4456]. In some LIM-HD proteins, the LIM domains prevent the HDs from binding DNA [929, 3735]. Apterous may be one such masochistic molecule, with Chip being its rescuer (cf. Ch. 6) [2854], but the story is not so simple [2851].

LNG This ∼31-a.a. motif is tandemly repeated 3 times in Notch, and in the nematode proteins Lin-12 and Glp-1 (hence the name) [2222]. Like EGF-like domains, the LNG domain has 6 conserved cysteines, but they are spaced differently ($Cx_4Cx_8Cx_3Cx_4Cx_6C$ instead of $Cx_4Cx_5Cx_8CxCx_8Cx_6$) [4611]. Removing these repeats causes a ligand-independent GOF phenotype [2542, 3543], so they may prevent activation of the receptor in the absence of ligand (by affecting oligomerization?) [2299].

MADS The MADS motif is named after Minichromosome maintenance-1 (yeast), Agamous (*Arabidopsis*), Deficiens A (*Antirrhinum*), and Serum Response Factor (mammal) [3819]. Mutations in *agamous* and *deficiens* cause homeotic changes in floral anatomy comparable to the cell-type (vein-intervein) switching seen with LOF defects in the fly's SRF gene *blistered* [2907]. Thus, animals and plants both use MADS-box genes to dictate histotype [3819], but to a different degree: evolution relied on MADS boxes for the selector-gene circuits in plants [1892, 3105, 3477], but it favored homeoboxes for wiring such circuits in animals [660, 1223, 2272]. The MADS domain is ~56 a.a. long, with 18 highly conserved residues, basic residues near the N-end and hydrophobic ones near the C-end [3907, 4392]. The N-terminus serves mainly for DNA binding, with the C-terminal part forming surfaces for protein interactions (dimerization plus accessory proteins). Dimers bind a palindromic decamer whose core 6 b.p. are typically A or T [3907], although each MADS factor binds a specific cognate target. Given these variations and the ability of MADS proteins to form heterodimers, different sites could be targeted by deploying certain subsets of MADS proteins just as bHLH proteins are mixed and matched during neurogenesis (cf. Ch. 3). However, there is no evidence that Blistered dimerizes with other partners [2907]. The 3-D structure of the SRF core bound to DNA [3323] reveals 4 levels in the dyad [4392] from the DNA outward: (1) N-terminal tails that line the minor groove, (2) antiparallel amphipathic 7-turn α-helices (one per monomer) that ride the DNA backbone along the groove and hold the dimer together with help from upper levels, (3) a β-sheet of 4 antiparallel ribbons whose hairpins touch the DNA, and (4) a cupola of two 3-turn α-helices that are outside the MADS domain.

NES "Nuclear Export Sequences" are typically ~10-a.a. leucine-rich peptides that cause proteins to leave the nucleus [1267, 2256, 3040]. The NES-binding protein Exportin escorts outbound cargo through the nuclear pore complex [953, 4057, 4419]. An interesting case study in the tug of war between NESs and NLSs is the heterodimer of Extradenticle and Homothorax [9, 28, 302, 3716]. *N.B.*: Certain "cytoplasmic localization" domains can keep proteins in the cytoplasm without resorting to NESs per se [1949, 1974, 3612, 3976].

NLS "Nuclear Localization Sequences" cause proteins to enter the nucleus [1566, 2032, 3040]. These ~5–10 a.a. (mono- or bipartite) peptides act like passwords for big proteins (≥ ~45 kD) to traverse the nuclear pore complex [4123]. (Smaller proteins traffic freely.) NLS-binding proteins (importins or transportin) dock with the NLS tag [1066, 2033], escort the cargo through the pore [1565, 1567, 3413, 3581, 4598], and then unload it [2637, 2750]. Unlike signal peptides, NLSs are not cleaved during transport. All NLSs contain clusters of hydrophilic residues, which must be superficial to function. NLSs are often rich in Lys and Arg, but not all such clusters serve as NLSs [3117],

and neutral residues [2663] or phosphorylation [1682] can be crucial. Whether a putative NLS truly acts as such can only be ascertained by mutating it or by putting it onto an inert "tester" protein. No consensus sequence exists in any of the 3 classes [3039], despite earlier indications [730].

opa Consisting almost entirely of glutamines (QQQQQ ...), the length of this ~30-a.a. motif (a.k.a. "strep" or "M") varies widely. It is named for *odd paired* (*opa*) – a pair-rule gene that has two repeats [285]. Q is the most often repeated a.a. in eukaryotic proteins having >16 tandemly repeated residues [1067], and opa-like stretches are found at over 300 sites in the fly genome. In some genes, the reading frame is shifted so that the iterated CAG or CAA codons yield pure strings of a different a.a. [4615]. The function of opa motifs is unclear. They might be transcriptional activation domains [1280, 2878, 4022], or they could be "polar zippers" (cf. leucine zippers) that mediate protein oligomerization via hydrogen-bonded, antiparallel β-strands [3353, 3354]. Runaway aggregation of opa-bearing proteins was thought to trigger various neurodegenerative diseases by recruiting caspases or clogging proteasomes. For example, expressing a 20-Q transgene in the fly eye does no harm, whereas a 127-Q transgene causes retinal degeneration [2169]. However, the etiology remains enigmatic [3201]$^\Delta$.

paired Named for the fly gene *paired* [456, 4785], this ~130-a.a. domain binds DNA sequences (e.g., in the *even skipped* promoter) [4383]. "Pax" denotes genes that contain a paired box [1632, 1757, 3125], and most Pax proteins serve as transcription factors in organogenesis [940, 1847, 1861, 4175]. In Paired and its relatives, the paired domain is associated with a homeodomain [303, 1325, 2098, 2873, 4383], whereas in others (e.g., Paired box-neuro) it is not [400, 401]. Only the N-terminal ~70-a.a. portion of the paired motif is essential for DNA binding by Paired [303, 608, 1325, 2873, 4383], while other members of the "Pox-Pax" family use both the N-terminal (PAI) and C-terminal (RED) subdomains [934, 2098, 2314]. The different strategies seem to hinge on a single residue in the RED subdomain [4776]. One fly gene (*eye gone*) has been found with a RED motif (and a homeodomain) but no PAI motif [2090, 2099].

PAS Period (circadian rhythm in flies), Arnt (aromatic hydrocarbon receptor nuclear translocator in mammals), and Single-minded (regulator of midline fates in fly CNS) share this ~270-a.a. domain, which contains two ~50-a.a. direct repeats (one near each end) [901, 1710, 3044]. The PAS motif folds into a β-sheet (5 or 6 strands) flanked by 2 α-helices [900]. It mediates homo- or heterodimer formation [1935], with or without an adjoining bHLH domain [899, 900]. Dimers can also form by binding of PAS to non-PAS domains [3575], and some of the partners that are recruited by a PAS domain can affect the DNA-binding specificity of the bHLH domain [4857]. Among bHLH-PAS proteins, Tango is a ubiquitous binding partner in flies [1166, 4022, 4548, 4857] (cf. Daughterless's role in the proneural bHLH subclass [588, 905, 4435]). The PAS domain is a critical gear in the clockwork that controls circadian rhythms (e.g., [1792]), so a brief libretto seems apt here. Although details are still being worked out [2486, 4821, 4822], this **"bHLH-PAS Circadian Clock"** clearly uses four key proteins: Clock and Cycle contain both PAS and bHLH domains [69, 165, 3710], whereas Period has a PAS but no bHLH domain [794, 3044], and Timeless has neither [3024]. Binding of Period and Timeless hides a cytoplasmic localization motif [1425, 2459, 3575, 3720] and lets an exposed NLS steer the dimer to the nucleus [926, 4505], where it indirectly activates transcription of *clock* [165]. Based on what happens in mammals, Clock presumably binds Cycle, and their dimer embraces E-boxes upstream of *period* and *timeless* to stimulate transcription [3710]. Timeless vanishes (degraded) at dawn [4860], leaving Period to snuff out its own gene's expression (and that of *timeless*) before being phosphorylated to death by the kinase Double-time [3654]. Whether Tango intrudes in these imbroglios is unknown [899]. Other questions concern why these events should take 24 h to complete [2395].

PDZ This ~90-a.a. globular domain (6 β-strands and 2 to 3 α-helices [586, 1097]) binds C-terminal tails of other proteins in a groove on one face [3419] but can also homo- or heterodimerize in a head-to-tail orientation by inserting a β-hairpin finger into this same groove [1192, 1849]. The archetypal PDZ subfamily binds E(S/T)x(V/I), whereas other subfamilies prefer different ligands [1062, 4019]. This versatility enables PDZ proteins to assemble complexes of signal-transducing molecules at plasma membranes [1385, 3513, 4236, 4405, 4811]. For example, Canoe and/or Pyd may tie components of Notch [4237], JNK [4236], or Ras-MAPK [2748] pathways to adherens junctions. The name PDZ (a.k.a. DHR = Discs-large homologous region) comes from three founding members of the family (Postsynaptic density protein-95, Discs-large, and Zona Occludens-1) [4744]. The PDZ backbone and groove resemble those of PTB [1730]. Quite literally, PDZ domains are the chief "nuts and bolts" of the cell [1954].

PEST Many rapidly degraded proteins (half lives ≤ a few hours) contain a ~10–50 a.a. domain that is enriched for the amino acids P, E, S, and T (hence the name) [3632]. Acidic residues tend to be clustered therein (uninterrupted by basic ones, which reside at the domain boundaries), but no consensus sequence exists. Whether PEST motifs are recognized directly by components of the ubiquitin pathway or require prior phosphorylation by kinases is not known in most cases [2572], but degradation apparently occurs in the 26S proteasome [3547]. Some proteins are quickly degraded without such motifs, and the PEST hypothesis in general remains controversial [382].

PH The ~120-a.a "Pleckstrin Homology" domain [1474, 2487, 3876] was first noticed in Pleckstrin (a substrate of Protein Kinase C) where it is present in two nonadjacent copies [2759]. This module is remarkable for (1) the constancy of its core scaffolding (4 β-strands in one plane, 3 β-strands in another, and an α-helix just outside this "sandwich"), despite great diversity in a.a. sequence; and (2) variability of its interstrand loops, where the ligand specificity of individual proteins often resides [1213, 2546, 3546]. This strategy (a constant core decorated with variable ligand-binding loops) resembles that of the Ig superfamily [2487]. Because most PH-family members assist in cell signaling or play cytoskeletal roles, PH domains may be protein- or lipid-binding "matchmaker" adaptors (cf. SH2 and SH3 domains) that bring together ≥2 proteins for signaling, anchorage, or catalysis (mainly at membranes) [1819, 3783]. Whether they function directly in signal transduction (by changing shape in response to ligand binding) is unclear [2478, 3546].

POU-specific The ~75-a.a. POU-specific domain is named for the mammalian genes *Pit-1*, *Oct-1*, and *Oct-2*, and the nematode gene *unc-86* [1832, 3713]. The domain folds into 4 interlaced α-helices with a hydrophobic core [131, 1014, 2258]. Two of the helices form a classic helix-turn-helix motif [1832]. Because POU-specific domains always adjoin a (~60-a.a.) homeodomain (HD), the term "POU domain" usually denotes the POU-HD compound [1832, 3713]. The POU and HD components are separated by a variable (15–56 a.a.) linker. They bind DNA cooperatively in its major groove [2257, 4177], with the POU portion bending the backbone [4477] and the linker constraining the overall configuration [409, 2525]. POU-HD motifs are found in various transcription factors [1160, 3649, 3712, 4569], which can bind DNA as monomers or dimers while interacting with a gaggle of cofactors [3713]. Recruitment of cofactors can depend on how the dimers are disposed along the DNA [4353]. In flies, examples include Nubbin (a.k.a. Pdm-1) [336, 793, 3092, 3103] and Ventral veinless (a.k.a.

Drifter or Cf1a) [703, 990, 995], both of which are critical for wing development.

PRD Consisting of HP dipeptide repeats, this motif is named for the fly gene *paired* (*prd*), which contains ~8 imperfect HP repeats in a 21-a.a. stretch of its sequence [1306]. (*N.B.:* The "paired" domain is likewise named for *paired* but is distinct [303].) Until recently, its role was unknown [4022]. One guess was a pH-sensitive chelation of metals (for DNA binding) because histidine's pK is close to physiological pH [2025]. In Paired itself, the PRD motif activates transcription [4785].

PTB The ~160-a.a. "Phospho-Tyrosine Binding" domain [4444, 4878] has 10 conserved residues. Its "β-sandwich" structure (two β-sheets plus an α-helix) resembles PH domains so exactly (despite dissimilar a.a. sequences) that it is considered a type of PH domain [2487]. Most PTB domains bind the peptide ϕxNPxY^P, wherein Y must be phosphorylated [3299, 4444]. Numb's PTB domain binds (1) YIGPY$^P\phi$, where Y can be unphosphorylated [2530, 4789]; (2) an NPxx sequence (9 of which exist in Partner of Numb [2609]) that is devoid of Y; and (3) an 11-mer that also lacks Y (GFSNMSFEDFP) [765]. The affinity of Numb for the latter peptide (from a natural mate) can be increased 15-fold (*sic!*) by replacing MS with AA (GFSNAA ...) [4900], implying that evolution settled for a tepid tryst rather than a stable union in this case. PTB motifs bind activated receptors, including Egfr [2623, 4444, 4445] and Notch [1307, 1651], and resemble PDZ domains insofar as their ligand docks onto the edge of a β-sheet (as an antiparallel β-strand) [1035, 1730].

PxDLSx(K/H) This motif is the consensus binding site for dCtBP (*Drosophila* C-terminal Binding Protein) in Knirps, Krüppel, and Snail [2350, 3112, 3113]. CtBP is a short-range, context-dependent co-repressor [2180, 3112, 3113] or co-activator [3377]. CtBP may act by recruiting chromatin-remodeling complexes of the Polycomb Group [3377, 3864] (cf. App. 5). Functional variants of this binding site include PLSLV in Hairy [3430] and PVNLA in Mδ (from the E(spl)-C) [3430].

RNA-binding Several kinds of RNA-binding motifs exist [1099, 3541]. The most common is the ~90-a.a. "RNP" (ribonucleoprotein) type (e.g., in Musashi [3035]). Its two consensus sequences are separated by ~30 a.a.: RNP1 (8 a.a.) and RNP2 (6 a.a.) occupy the central β-strands in a 4-strand antiparallel β-sheet that packs against two α-helices [1100, 2186]. Their basic and aromatic residues are implicated in salt-bridge and ring-stacking interactions with single-stranded RNA segments. Most of the speci-

ficity for target nucleotide sequences is thought to lie outside these oligopeptides.

runt The ~128-a.a. runt domain [2113] is named for the fly gene *runt* [2127, 3173], which encodes a transcriptional repressor [2065]. This motif binds DNA [1256, 2112, 2113, 2366, 3331] and mediates dimerization with non-runt partners [1518, 1533, 1929, 1930, 2113]. Those partners increase the dimer's DNA-binding affinity without binding DNA themselves [1533, 2524, 4234], although runt family members can bind DNA (as monomers) without such help [3172, 4541]. The greater affinity is due to a change in protein conformation [1930] and is reflected in stronger bending (~60°) of the DNA [1533]. The shape of the runt domain puts it in the Ig-like class of DNA-binding factors: its 10 antiparallel β-sheets form a 2-layer "β sandwich" whose two free loops insert into the major and minor grooves of DNA [294, 579, 3026].

SH2 The ~100-a.a. Src homology 2 (SH2) module binds phospho-Tyr residues that have a Y^{P}xxϕ consensus. ("Src" is from sarcoma virus.) SH2 domains are used in signal transduction by receptor tyrosine kinases (RTKs) [2910, 3298], in signal extinction [1209, 4320], and in ratcheting of substrates by non-receptor kinases [2758]. Adaptor proteins commonly use an SH2 to bind an activated RTK but an SH3 to relay the signal (cf. Fig. 6.12) [3185, 3488, 3944]. SH2s form two α-helices and two β-sheets, and the ligand fits a pocket on one face in lock-and-key fashion [2365, 3215, 4526]. The 3 a.a. on the carboxy side of the Y^P (xxϕ) usually dictate the fit [342, 3296, 3299, 4020], but flanking residues can also matter [3280]. A conserved Arg at the bottom of the pocket grips the Y^P phosphate. S^P or T^P would be too short to reach the Arg (hence SH2's fetish for Y^P), and Lys cannot reach the Y^P (ergo substitutions here are disabling) [3296]. The N- and C-ends of each SH2 converge – a feature that made it easy for evolution to insert or delete SH2s (like buttons) without disturbing nearby domains [3296]. Nevertheless, some SH2s are configured to nudge catalytic domains upon peptide binding [1200, 4535]. Aside from SH2 and PTB, another Tyr-binding domain probably exists [1575] because YYND is so rigidly conserved in the Shc family [2389, 2623].

SH3 The ~50–75 a.a. Src homology 3 (SH3) peptide-docking motif typically assigns proteins to certain sites within the cell [208, 3019, 3298], although the SH3 domain of Src itself also acts intramolecularly [1200]. Unlike SH2 domains, which bind Y^P, SH3 domains bind Pro-rich peptides (consensus = PxxPx), which are not modified during signaling [3299]. Hence, SH3 domains are often found (sans SH2) in adaptors not affiliated with phosphorylation cascades [2777, 3676] (e.g., MAGUKs [1062, 3419]). The ligand's variable and flanking residues dictate affinity for certain SH3 proteins [3298]. The orientation of the prolines is irrelevant for binding [1210, 2552, 2757] because the pocket holds each Pro by its nitrogen [3108], not by a lock-and-key fit. SH3 sequences form β-barrels (5–8 β-strands), with a hydrophobic ligand-binding pocket [3298, 4828] on the side opposite the N- and C-tips. Because the tips meet, SH3 domains have the same "button" modularity as SH2 domains [3298]. Yeast have SH3s but not SH2s, implying that SH3 evolved before SH2 [2365, 3296]. SH2s probably arose with RTK signaling in metazoans.

signal Signal sequences (a.k.a. leader peptides [172, 3510]) cause proteins to be (1) inserted in the plasma or organelle membranes [90] or (2) secreted [59, 1844, 3209, 4501, 4874]. They reside at the protein's N-terminus and have a ∼12-a.a. hydrophobic core with a relatively nonconserved sequence [3339, 3510]. The core is preceded by ∼1–10 a.a. and followed (C-terminally) by ∼6 a.a., at which point some proteins are marked for cleavage by a signal peptidase [4500].

Sox "Sox" denotes an "$\underline{S}R\underline{Y}$ b$\underline{ox}$" subgroup of HMG (∼80 a.a.) sequence-specific DNA-binding domains distinguished by ≥50% sequence identity with the HMG domain of the founding Sox member [2415, 3360] – the human gene *SRY* ($\underline{S}$ex-determining $\underline{R}$egion on the $\underline{Y}$) [3967]. All known Sox genes have only one HMG motif per gene, and most lack introns. Sox proteins bend DNA into an acute angle upon binding in the minor groove [1220] (cf. POU-specific domains [4477]). By such bending, it can enable nearby transcription factors to contact each other (or the basal transcription machinery) [317, 1480, 1629, 3360]. Besides this architectural role, some Sox-domain proteins (e.g., Dichaete) directly activate transcription [2635].

TEA This ∼68-a.a. DNA-binding domain is named after SV40 transcriptional enhancer factor $\underline{TE}$F-1, yeast TEC1, and the product of the *Aspergillus* $\underline{abaA}$ regulatory gene [566]. It is predicted to have 3 helices. Whatever its exact structure, the configuration must have been honed to perfection long ago by evolution because the TEA domain of *Drosophila*'s Scalloped protein is 98% identical (70/72 a.a.) to the TEA domain of its human TEF-1 ortholog [623]. Consistent with its mammalian TEA-toting cousins [2010], Scalloped cooperatively binds tandemly arranged 9 b.p. cognate sequences [1686]. It requires a tissue-specific co-activator (Vestigial) to stimulate transcription of target genes [1686, 3291, 3936], and there are hints of a co-repressor in humans [727, 728].

WD Named for Trp (W) and Asp (D), this ∼40-a.a. motif (a.k.a. "WD-40") typically ends in the dipeptide $\underline{WD}$ [3072, 4449]. Each repeat forms a β-sheet made of 3–4 antiparallel β-strands [3072, 4012]. Tandem copies (4–8 repeats/protein) can form a "β propeller" whose blades radiate symmetrically from a central core [3018, 4012, 4528]. WD-repeat proteins fall into ≥6 subfamilies [1006] and serve diverse regulatory (vs. enzymatic) functions, including signal transduction, RNA processing, gene regulation (e.g., Groucho [1243]), vesicular traffic, cytoskeletal assembly, and cell-cycle control [3072, 4004]. Many of these functions involve assembly of multiprotein complexes (e.g., Extra sex combs [1660, 3102, 3761, 3939, 4313]), so the main role of the repeats may be to provide docking interfaces [2289, 3072].

WRPW This tetrapeptide is found at the C-terminus of the bHLH transcription factors Hairy, Deadpan, and 7 proteins encoded by the E(spl)-C [331, 1017, 2274, 3697, 3759]. Like bait for a fish, $\underline{WRPW}$ serves to recruit Groucho – a ubiquitous transcriptional co-repressor [1243, 1244, 2064, 3278]. Groucho also binds (1) $\underline{WRPY}$ at the C-termini of proteins that have a Runt Homology domain [1243, 2496], (2) $\underline{FRPW}$ in Huckebein [1526], and (3) $\underline{FKPY}$ in Brinker (Dpp pathway) [4866], indicating a consensus of "aromatic-basic-Proline-aromatic" [4866].

zinc finger In this motif, Zn^{2+} ions pin together ≥4 Cys or His (or rarely Asp or Glu) residues [296, 841, 2265, 3578]. Among the ∼10 types of fingers, the most common is the "Cys_2His_2" DNA-binding module [296, 2878], which forms tandem arrays. Its ∼25-a.a. finger has a $\underline{Cx_{2-4}Cx_{12}Hx_{3-5}H}$ consensus sequence, wherein the central $\underline{x_{12}}$ "finger tip" is stabilized by hydrophobic interactions between (usually) Phe and Leu (= $x_3Fx_5Lx_2$). The Cys_2 side of the finger is a β-sheet (2 antiparallel strands), whereas the His_2 side is an α-helix that fits into DNA's major groove [4711]. (A variant type of zinc module binds in the minor groove [4883].) In each helix, 3 residues bind ∼3 nucleotides. Adjacent fingers (joined by ∼8 a.a. flexible links) function nearly independently (like modular "reading heads") [296, 1187, 2265]. Some zinc fingers read the same $\underline{CAnnTG}$ (E box) sequence as bHLH dimers, and co-expressing these different tribes of transcription factors in the same cells creates odd phenotypes [1334]. The "LIM" type of zinc finger [160, 3736] is used in protein-protein interactions [97, 3786] rather than for DNA binding [929, 973] (see LIM and homeo domains above). The "RING" type of finger [1295] can act as a ubiquitin ligase in some cases [2068, 2387, 2603].

Inventory of Models, Mysteries, Devices, and Epiphanies

Listed below are all the major models (a.k.a. hypotheses, metaphors, scenarios) and mysteries (a.k.a. enigmas, riddles, paradoxes) discussed in the text, with definitions in parentheses. Also listed are devices (a.k.a. gadgets, tricks) and epiphanies (a.k.a. principles, rules). Asterisks indicate names that were coined for convenience (cf. [1805, 1807]). Having a taxonomy of concepts – even a silly taxonomy – is a useful heuristic for thinking [1873, 2871, 3259]. Within each category, concepts are listed alphabetically. Numbers in bold are pages where the ideas are diagramed.

Models   **Pages**

Activator-Inhibitor Model (activation is local while inhibition occurs
    at longer range) 184
Angular Values Conjecture (segment polarity genes encode an
    angular coordinate)* 97
Ap-Chip Tetramer Model (a transcription factor is built like a jigsaw puzzle)... 160, **162**
Arc Scenario (Dpp and Wg are constrained to travel in arc-shaped paths)* **104**, 126
Basitarsal Elaboration Scenario (bristle patterns evolved as modules)*........... 63, **64**
Battery Dichotomy Hypothesis (larvae and adults use different sets of genes)*...... 87
Bias Model (Notch output is modulated by ligand input)*.......................... 10
Binary Code Conjecture (regional identities are specified by binary codes)*......... 85
Blinker Model (Notch output is modulated downstream of the ligand)* 9
Border Guard Model (segregation is achieved by preventing intermixing
    at the border)* 149, **150**
Boundary Model I and II (regional interfaces are used as coordinate axes)..... **102**, **104**
Cabaret Metaphor (transient states rely on intercellular signaling)* 90
Cascade Model (histotypic genes are regulated through a chain of command)* 244
Catalysis Model (Notch plays an indirect role in transcription regulation)*........ **8**, 13
Climbing Scenario (Dpp output tends to increase with time)* 154
Clock and Wavefront Model (cells change state cyclically until hit by a wave)....... 304
Cloud Scenario (Dpp and Wg diffuse randomly)*............................. **104**, 126
Coding Model (Numb plays an abstract role in specifying bristle-cell identity)* 7
Combinatorial Cascade Model (fates are induced sequentially by ≥ 2
    cell contacts)*... 212, **214**

Contest Model (SOP selection relies on Dl-N competition)*......................... 46
Crystallization Model (ommatidial cells adopt fates via contact
     with a template)... 211
Cyclic Amnesia Scenario (homeosis is due to memory loss)*...................... 85
Cyclops Scenario (signaling complexes can be "gated" by
     microtubule binding)*.. 287
Differential Threshold Model (bristles ≈ hills of various heights
     in a rising or falling tide)*.. 41
Diffusible Activator Model (SOP selection relies on accretion of inducer)*........... 46
Distal Organizer Scenario (the future tip of the leg is the origin for all
     radial coordinates)*..**104**, 129
Double-Gradient Models (two opposing gradients serve as dual inputs)............ 124
Dpp Gradient Model (Dpp is a morphogen that assigns cell fates across
     the disc)...140, **142**
Dpp-Wg Growth Potential Model (growth is dependent on both
     Dpp and Wg)*.. 124
D/V Compartment Hypothesis (the eye has *bona fide* D and V
     compartments)*...202, **206**
Either Duo Hypothesis (either of two pairs of factors dictates
     antennal identity)*..**250**, 252
Filter Model (bristle initiation relies on competence to varied signals)*..........**34**, 35
Force Field Model (bristle initiation relies on a web of physical forces)*..........33, **34**
Gating Model (bristle-cell identity relies on a series of decisions as in a maze)*...... 10
Gradient of Developmental Capacity (GDC) Model (discs regenerate
     via gradients)..93, **94**
Handcuff Scenario (Cos2 binds Ci to microtubules and thereby muzzles it)*....... 286
Hedgehog Gradient Model (Hedgehog is a short-range morphogen)...........138, **142**
Hh-Dpp-Wg Model (1 short-range and 2 long-range morphogens are used)*.. **104**, 115
Hopeful Monster Hypothesis (drastic mutations facilitate evolution).............. 237
Hox Code Model (metamere states are encoded by combinations of
     Hox genes).. 246
Hox Competition Model (Hox proteins compete for DNA-binding sites)........... 247
Hox Isoforms Conjecture (metamere identities rely on different
     splicing isoforms)*... 247
Hox Threshold Conjecture (different amounts of Hox protein dictate
     different fates)*.. 247
Hybrid PC-Boundary Model (polar coordinates cooperate with boundaries)*. 101, **102**
Inhibitory Field and Wavefront Model (these 2 variables create
     a hexagonal lattice)*...224, **226**
Instructive-Permissive Model (Dpp and Wg jointly dictate wing
     bristle identities)*... 189
Intercalation Scenario (growth is stimulated by xenotopic contact)*............**95**, 154
Jacob's Ladder Model (cross-vein positions are dictated by physical forces)*....... 189
Lateral Inhibition Model (SOP selection occurs via a winner-take-all strategy)....**46**, 49
Local Depletion Model (bristles are spaced apart because SOPs consume
     an inducer)*... 50
MF-pushing Model I (Hh and Dpp act in a loop that drives the MF)*.............. 229
MF-pushing Model II (Hh and Dpp act with agent X in a loop that drives
     the MF)*.. 231
Minotaur Scenario (the "crawling" of a boundary is self-limiting)*................ 122

Mutual Activation Model (*ac* turns ON *sc* and vice versa)*............................ 44

Mutual Inhibition Model (SOP selection occurs via egalitarian competition)..... **46**, 50

Mutual P-D Antagonism Model (proximal and distal domains compete)*.......... 134

Nuclear Notch Model (Notch regulates transcription directly)*.................... **8**, 12

Obey Your Mother! Model (bristle-cell fates are dictated by
    inherited instructions)*.. 5, **6**

Omb Memory Hypothesis (the *omb*-ON state is inherited after a
    certain stage)*.. 143

Ommatidial Lineage Hypothesis (ommatidia develop as clones)*.................. 202

Open for Business Model (more of the BX-C is accessible
    in posterior parasegments)... 305

Paravein Hypothesis (ectopic veins emerge atavistically between
    extant ones)*.. 174

Polar Coordinate (PC) Model (positions are specified by polar coordinates)..... 93, **95**

Positional Information (PI) Hypothesis (cell fates are dictated by
    coordinate systems).. 79, **82**

Posterior Prevalence Model (posterior Hox genes nullify the effects
    of anterior ones)... 247

Predestined SOP Model (SOP selection is based on prepatterned biases)*.......... 45

Prepattern Hypothesis (new states are spatially constrained by prior states)...... 33, **82**

Raft Scenario (Hh is transported by lipid rafts)*.................................. 125

Ratchet Model (BX-C genes enable segments to rise above a ground state)*........ 243

Ratchet Scenario (cells acquire states in an irreversible sequence)*............... 155

Rho-Vn Booster Model (wing veins are initiated by both Rho and Vn)*............ 183

Rock Concert Scenario (cells become deafer as morphogen volume rises)*........ 147

Scalar Model (ommatidial polarity arises via a standard
    gradient mechanism).. 209, **210**

Selector Affinity Model (cell assortment relies on compartment identity)...... 148, **150**

Selector Gene Hypothesis (regional identities are encoded by binary
    gene states).. 85

Short-range Inducer Model (a third state is induced at a two-state boundary)...... 107

Signal Relay Scenario (fates are assigned by a series of inductions)................ 139

Site-specific Enhancer Model (bristle initiation relies on unique
    cis-enhancers)*.. **40**, 41

Spreading Model (bristle patterning relies on diffusible inducers).................. 39

Stick-and-Straighten Scenario (cells align via polarized adhesion)*............... 174

Stop the Clock! Model (cells stop changing their state when they receive
    a signal)*.. **214**, 217

Subgene Hypothesis (heteroallelic phenotypes imply AS-C subdivisions)........... 38

Threshold Model (Hairless titrates Su(H) and the remaining one
    dictates cell fate)*.. **22**, 23

Tiger-by-the-Tail Scenario (Hh is passed from cell to cell by its
    cholesterol adduct)*... 125

Time Window Model (SOPs can only "percolate" during a limited period)*....... **70**, 73

Tkv Barrier Hypothesis (dense receptors block morphogen diffusion)*............. 143

Turing's Model (patterning depends on diffusible reactants)........................ 33

Use-as-needed Scenario (BX-C genes enable regions to deviate from
    a ground state)*... 243

Vector Model (ommatidial polarity arises via individually polarized cells)...... **210**, 211

Mysteries

Abruptex Paradox (what causes these odd *Notch* mutations?)* 52

Athena Enigma (how do stripes of cells arise so neatly *ab initio*?)* 175

Autoinhibition Paradox (how do cells avoid repression by their
own inhibitor?)* 183

Bristle Coding Enigma (how are bristle positions encoded?)* 37

Bristle Plotting Puzzle (how are bristles patterned?)* 31

Cell Competition Conundrum (why do sluggish cells end up
near boundaries?)* 153

Collective Amnesia Conundrum (how do clusters of cells switch fates?)* 85

Dorsal Remnant Mystery (how does dorsal leg tissue robustly resist insults?)* .. 97, **100**

Ends-Before-Middle Riddle (why do leg discs assign proximal and distal
fates early?)*, ... 133

Enhanceosome Puzzle (how do transcription factors sterically execute
Boolean logic?)* 254

Euclidean Ruler Problem (how do cells align themselves?)* 173

Fickle Bristle Mystery (why do bristles seek one or the other end of
a facet edge?)* 229, **230**

Fickle Sensilla Mystery (why do sensilla seek one or the other end
of a leg segment?)* **130**, 131

Field Effect Mystery (why are the effects of some mutations
nonautonomous?)* 167

Fragmented Stripe Dilemma (why are *dpp*-ON stripe enhancers "atomized"?)* 129

Gaps Mystery (why don't compartment lines coincide with symmetry axes?)* 97, **98**

Gearing Riddle (how do adjacent ommatidia rotate relative to one another?)* 304

Growth Cessation Mystery (how do organs "know" when to stop growing?)* 92

Homeobox Homunculus Mystery (why are Hox genes colinear with
the body?)* 239

Homeosis Riddle (how do genes transform the identities of body parts?)* 237

Lattice Riddle (how is the hexagonal lattice of ommatidia created?)* 224

Leg Stump Riddle (why are legs truncated by *dpp*^{LOF} but not by *wg*^{LOF}?)* 129

MC vs. mC Paradox (why do MC and mC patterns differ if their PNCs are
the same?)* 56

Memory Riddle (how do cells sustain their states of determination?)* 248

Nonequivalence Riddle (why do some mutations affect bristles in a spatially
heterogeneous way?)* 20

One Disc or Two? Paradox (is the eye-antenna disc a simple
or compound disc?)* 96

One Eye or Many? Riddle (what did the ancestral eye of insects and humans
look like?)* 197

Photoreceptor Coding Enigma (how are photoreceptor identities encoded?)* 224

Planar Polarity Puzzle (how do cells orient themselves within a plane?)* 304

Position-Projection Mystery (how do PNS axons find their CNS targets?)* 191

Quadrant Regeneration Mystery (why does only one
quadrant of the leg disc regenerate?)* 97, **100**

Ras Specificity Riddle (how does one signaling pathway manage to evoke
a variety of responses?)* 179

Registration Riddle (how does an eccentric stripe specify
a symmetric pattern?)* 188

Regulation Riddle (how do patterns regenerate and remain size invariant?)* 81
Shared Genes Riddle (how can the same genes expressed in two discs make
     them different?)* 249
Spacing Precision Problem (how are ommatidia positioned so regularly?)* 228
Three Genes vs. Nine Segments Paradox (how does the BX-C specify
     metamere identities?)* 246
Triplications Mystery (why do leg outgrowths converge vs. diverge?)* 98, **100**
Yin Yang Paradox (how can mutually inhibited genes be co-expressed?)* 128

Devices (Actual or Hypothetical)

bHLH-PAS Circadian Clock (a clock mechanism for daily cycles)* 263
Delta-Notch Flip-Flop (an amplifier for small disparities in signaling)* **46**, 209
Finger Shuffling Trick (the use of alternative splicing to change zinc
     finger "addresses") **19**, 20
Heterochronic Superposition Trick (a way to create overlapping patterns
     without interference)* 57
HLH Hourglass (a clock mechanism for triggering events)* 297
Kaleidoscope Toy (evolution's usage of Boolean logic to activate new genes)* 302
Kill the Stragglers! Trick (a way of robustly ensuring uniform periodicity)* 228
Kilobase Spanner (a fastener that ties DNA into loops)* 161
ON-then-OFF Licensing Trick (activation of a gene only after its upstream
     regulator turns OFF)* 91
POU Hourglass (a clock mechanism for counting mitoses)* 297
SOP Computer (an analog-to-digital HLH transducer)* 72
Stratification Device (a way to create a third state at a border between
     two others)* 302

Epiphanies

AS-C Epiphany of 1995 (the AS-C inputs regional cues and outputs bristles)* 44
Deaf-Speakers/Mute-Listeners Trick (inter-regional signaling
     is unidirectional)* 107
Iroquois Epiphany of 1999 (the eye has *bona fide* D and V compartments)* 203
Micromanager Epiphany of 1998 (Ubx directly controls multiple levels of
     target genes)* 245
Numb Epiphany of 1994 (bristles differentiate via a binary lineage code)* 9
Principle of Nonequivalence (identical elements are encoded differently) 35
Proximity vs. Pedigree Rule (most fates are assigned by
     intercellular signaling)* 4
Regeneration Epiphany of 1999 (regeneration does not recapitulate
     development)* 123
Venn Overlap Rule (certain genes only turn ON targets where their
     expression overlaps)* **172**, 251

Genes That Can Alter Cell Fates Within the (5-Cell) Mechanosensory Bristle Organ

In parentheses after each gene name are the salivary gland map location and the origin of the name, except where obvious. The **PHENOTYPES** section only lists effects on the SOP lineage. The **PROTEIN** section presents available data on function, length, subcellular location, domains (see App. 1; number of repeats in parentheses), and binding partners. For further information, see *FlyBase* and *The Interactive Fly*.

Abbreviations: "$\approx$" (resembles), a.k.a. (also known as), cyt. (cytoplasm), DS (downstream in causation; "+" activated or "−" inhibited by US gene), GOF (gain of function by mutant allele or transgene construct), LOF (partial loss of function), nuc. (nucleus), PNC (proneural cluster), SOP (sensory organ precursor), US (upstream), UTR (untranslated region). Evidence for the hierarchical order of genes in a pathway (US or DS) is given from a DS perspective only. Thus, the link "$a \rightarrow b$" would be entered under gene b as "DS(+) of a" followed by epistasis evidence, but under gene a as just "US(+) of b" without the data. Likewise, "$c \dashv d$" would be listed for d as "DS(−) of c" with evidence, but for c as "US(−) of d" without the data.

In a manner related to this class of genes, various subtypes of chemosensory neurons can be interconverted by LOF mutations in *gustB* [111, 3918] and *abnormal chemosensory jump 6* [818] (cf. *scalloped* [623]). Overexpressing the proneural gene *lethal at scute* has no effect on the SOP lineage [3027].

Some genes were not tabulated because the basis of their phenotypes has not been studied. For example, cellular etiology is unclear for LOF alleles of *seven in absentia* (*sina*) [671] and *phyllopod* (*phyl*) [717, 1052, 3378], which cause balding (absence of shaft and socket) or multi-

ple (2 or 3) shafts in what appears to be a single socket but may actually be multiple sockets [2680]. Sina and Phyl form a multimeric complex that promotes proteolysis of Tramtrack [2529, 4249]. Other genes were omitted because their roles are redundant (viz., the 6 other bHLH genes in the E(spl)-C [4767] and the $\geq$6 other Brd family genes [2382]) or seem incidental – viz., *big brain* (it transforms shafts to sockets when overexpressed, but only when *Notch* or *Delta* is also overexpressed [1075]), and *canoe* (App. 5; its LOF alleles dominantly enhance the N^{spl} 2-shaft trait, but it has no independent effects). Inactivation of Gαi by transgenic RNAi causes unspecified fate changes [3776]. Also excluded were *Suppressor 2 of zeste* and *Posterior sex combs*, whose GOF effects appear to arise by a trivially neomorphic route [514, 3872]. The *senseless* gene is needed for SOP differentiation but does not seem to function in fate assignment per se within the lineage [3127] (see App. 5). LOF mutations in proteasome subunit genes cause shaft-to-socket and neuron-to-sheath transformations indirectly by increasing the concentration of activated Notch [3821].

Genes whose LOF effects on the embryo CNS suggest allegiance to this group of bristle-affecting factors include *bazooka* [3803, 4709], *discs large* [3205, 3326], *inscuteable* [3195]$^{\triangle}$, *jumeaux* [729], *klumpfuss* [2250, 4803], *miranda* [3630]$^{\triangle}$, *partner of inscuteable* [3777, 3822, 4827], *partner of numb* [276, 3630]$^{\triangle}$, *sanpodo* [1128, 3263, 3985], *tap* [577]$^{\triangle}$, etc. [609]. The same goes for *glial cells missing*, which is expressed in the glial cell of the bristle complex [1447] and is required for its differentiation [4438].

Additional genes were isolated in recent screens [6, 295, 555, 2903, 3471], including *string*, which regulates mitosis (see text). "Neural precursor" genes (not listed) implement neural differentiation [2285]$^\Delta$ and hence are downstream of the genes tabulated here [2018]. A curious member of the precursor group is *cousin of atonal* (*cato*) [1586] – a bHLH gene outside the AS-C (53A3–5) that is expressed in SOPs after *asense*. Cato's chief role appears to be in the control of neuron morphology.

asense (*ase*, 1B1–7, AS-C, sensory defect). **PHENOTYPES: Null:** twinned bristles (anterior wing margin only) that may indicate socket-to-shaft or IIb-to-IIa transformations [1079, 2039], plus other defects unrelated to fate switching. **GOF:** no apparent effect on intra-bristle cell fates (although extra bristles are induced) [438, 1079]. **PROTEIN: Function:** pan-neural transcription factor (bHLH type; 486 a.a.) [1540]. **Location:** nucleus [438]. Detected in SOP, IIa, and IIb, but not in the final 5 cell types [1079]. **Domains:** bHLH, opa, PEST, Pro-rich region near N terminus, and acidic region centrally [1540]. **Partners:** binds as homodimer or Ase/Da heterodimer to "E box" DNA sequences [2039].

Bearded (*Brd*, 71A1–2, extra bristles) – one of 6 related genes in the Brd Complex that act alike [2382, 2383]. **Pathway:** same pathway as *N*: the extra-bristle phenotype of *Brd*^{GOF} is partly suppressed by an extra dose of *N*$^+$ [2500]. **PHENOTYPES: Null:** wild-type (i.e., no apparent defects) [2500], probably due to redundancy [2382]. **GOF:** switches IIa to IIb [2500]. **PROTEIN: Function:** unknown (81 a.a.) [2499]. May have no normal function in bristle cells, despite its ability to affect their fates when overexpressed via GOF mutations (cf. E(spl)-C). **Location:** unknown, although other members of the Brd family are cytoplasmic [92]. **Domains:** no obvious motifs except for a putative amphipathic α-helix [2383, 2499], although there are several intriguing motifs in the 3′ untranslated part of *Brd*'s mRNA [2384–2386, 2499]. **Partners:** none known.

Delta (*Dl*, 92A2, delta-shaped tips of wing veins). **Pathway:** US(+) of *N* [2542] (but see App. 5). **PHENOTYPES: Null** (assayed in mosaics): ~wild-type for fates within the bristle [4859], which is puzzling given its stronger partial-LOF effects. In double mutants with *Ser*^{null}, *Dl*^{null} switches IIa to IIb, socket to shaft, and (less often) sheath to neuron ($\approx N^{LOF}$) [4859]. **Severe LOF** (t.s. mutants): 4 neurons ($\approx$ severe N^{LOF}) [3272, 3273]. **LOF** (t.s. mutants): may switch IIa to IIb (cf. its "2-neuron-2?-sheath" trait), socket to shaft, and sheath to neuron (cf. its "2-shaft-2-neuron" trait) [218, 3272] ($\approx numb^{GOF}$). **GOF:** switches shaft to socket [2008]. **PROTEIN: Function:** ligand for Notch receptor (832 a.a.) [3022]. **Location:** transmembrane (single-pass) [2300, 4465] but also secreted under some conditions [2264, 3479]. Equal amounts in IIa vs. IIb (and shaft vs. socket

cell) [3270] argue against a "sibling rivalry" mode of fate determination here. **Domains:** signal peptide, EGF-like (9x), transmembrane [2300, 4465]. **Partners:** binds itself [2263] and N in *trans* (i.e., on adjacent cells) [1204, 2263, 3544] and possibly also N in *cis* (i.e., on the same cell) [2008].

deltex (*dx*, 6A, <u>Delta</u>-like gene on the <u>X</u>). **Pathway:** US(+) of *N* [2746] and US(−) of *H* [3027]. Suppressed by *Su(dx)* [1275, 1276], which encodes a ubiquitin ligase [129, 875]. **PHENOTYPES: Null:** unknown [580, 1053]. **LOF:** infrequent doubling or loss of eye bristles [1570, 4778]; main effects (thick wing veins, nicked wing margin, fused ocelli) are unrelated to bristles. **GOF** ($\approx N^{GOF}$): switches shaft to socket and may switch IIb to IIa (its "2-shaft-1?-socket" trait may actually be "2-shaft-2-socket") [2746]. Also switches sheath to neuron [3027]. **PROTEIN: Function:** unknown (737 a.a.) [580]. Although *deltex* overexpression alters bristle cell fates [2746], *deltex*^{LOF} mutations have virtually no effect [1570, 4778], so either (1) *deltex* plays no role in bristle development or (2) it acts redundantly [1053]. **Location:** cytoplasm [580, 1053]. **Domains:** zinc finger (ring-H2), opa (2x), and binding site for SH3 domains – all apparently nonessential. What's crucial is the N-terminal third, which binds Notch's ankyrin repeats [1053, 2746]. **Partners:** binds N [1053, 2746] but not H [3027].

Enhancer of split (*E(spl)*, a.k.a. *m8*, 96F11–14, enhances *split*, a *N*^{LOF} allele that causes double shafts) – one of 7 bHLH genes in the E(spl)-C that act alike in many contexts [3075] (but see [871, 2548, 2549]). **Pathway:** DS(+) of *N*: lethality of E(spl)-C^{null} is not rescued by *N*^{GOF} [2542]. DS(+) of *Su(H)*: Su(H)-binding sites in *m8* promoter are needed for expression of *m8* in PNCs (but not in SOPs) [2453]. **PHENOTYPES: Null:** wild-type, due to functional overlap with other E(spl)-C genes [1018, 1794, 3806]. Deletion of all 7 bHLH E(spl)-C genes (in somatic cell clones) fails to switch bristle cell fates, so E(spl)-C may have no normal role in assigning those fates [991] – a conclusion supported by the lack of any fate switches for *groucho*^{LOF} (Groucho encodes M8's co-repressor; see App. 5). **GOF:** various defects including double, missing, stunted, or deformed shafts and double sockets [2384, 4256, 4318]. In contrast, overexpressing the bHLH E(spl)-C genes *m7*, *mδ*, or *mγ* has no effect on the SOP lineage [3027]. **PROTEIN: Function:** transcription factor (bHLH type; 179 a.a.) [2246, 2274, 3171]. May have no normal function in bristle cells, despite its ability to affect their fates when overexpressed via GOF mutations [2384] (see text). **Location:** nucleus, but not in SOP [2052, 2053]. **Domains:** bHLH, <u>WRPW</u> (Groucho-binding) motif, PEST [1017, 2274]. **Partners:** binds Groucho [2064, 3278, 3430] and "N box" DNA sequences (CACnAG) [3171, 4318] but not "E boxes" [4452]. Also binds (and is phosphorylated by) Casein kinase II [4395], as are M5 and M7 but not M3 or Mγ (which may help explain subgroup idiosyncrasies within the E(spl)-C family).

Hairless (*H*, 92E12–14, bristle loss). **Pathway:** same pathway as *numb*: H^{LOF} synergistically enhances $numb^{\mathrm{LOF}}$ [4542]. Antagonistic (−) to *Su(H)*, but epistatic relationship is unclear: double mutant of H^{LOF} and $Su(H)^{\mathrm{null}}$ has mixture of H^{LOF} and $Su(H)^{\mathrm{null}}$ traits [3827]. DS(−) of *dx*: double-LOF $\approx H^{\mathrm{LOF}}$, and double-GOF $\approx H^{\mathrm{GOF}}$ [3027]. **PHENOTYPES: Null:** failure of SOP initiation at most sites [198]. Rare SOPs that do arise yield either a 4-socket cluster (implying a IIb-to-IIa and a shaft-to-socket switch) or a 2-sheath pair (implying a neuron-to-sheath switch and a loss of IIa) [3027]. **LOF** ($\approx Su(H)^{\mathrm{GOF}}$): switches shaft to socket [198, 200, 2475], but neurons transform only partially into sheath cells [3027]. **GOF** ($\approx Su(H)^{\mathrm{LOF}}$): switches socket to shaft [200, 218, 2657], IIa to IIb, and sheath to neuron [3027]. **PROTEIN: Function:** antagonist for Su(H). Thought to act by blocking Su(H)'s DNA-binding domain (1059 a.a.) [200, 492, 2659], H also recruits the co-repressor dCtBP [1329] ([2129] sequel). **Location:** nucleus [2657, 2658]. Expressed in all descendants of the SOP (glial cell is a possible exception) more strongly than in background epidermis [3027]. **Domains:** NLS (3x), PRD, and stretches of Ala [200, 2657]. Extremely basic overall (pI = 9.5), with 40% of its residues = Ala, Ser, or Pro [200], but also has conserved acidic sites [2695]. **Partners:** binds Su(H) [171, 492, 1332, 2657, 2658] and Numb [4542] but not Dx or Mam [3027].

lethal (2) giant larvae (*lgl*, 21A, neoplastic growth). **Pathway:** same pathway as *numb*: LOF phenotype is exacerbated when lgl^{LOF} is combined with $numb^{\mathrm{LOF}}$ [3205]. **PHENOTYPES: LOF** ($\approx numb^{\mathrm{LOF}}$): switches IIb to IIa [3205]. **PROTEIN: Function:** tumor suppressor (two splicing isoforms: 708 or 1161 a.a.) [2006, 2796] which maintains apical-basal polarity [333, 3205, 3326, 4707], cell shape [2667], epithelial integrity [1394–1396, 4742], and compartment boundaries [35, 2975]. **Location:** mainly in cytoskeletal matrix at the lateral cell cortex [3205, 4134]. **Domains:** WD repeats [3326], a basic region, 3 different homo-oligomerization domains, a matrix recognition tripeptide, and a Ser-Thr-rich tail in the larger isoform [2014, 2247, 4133]. **Partners:** binds itself, nonmuscle myosin II, and ~10 other matrix or junctional proteins [2120, 4133], probably including Discs large [3205, 3326, 4745].

mastermind (*mam*, 50C20–23, enlarged CNS). **Pathway:** Notch pathway [1000, 1514, 2353, 2394, 4780] possibly US

of Delta [1816]. **PHENOTYPES: LOF** (as deduced from constructs that act in a dominant-negative manner): 4 neurons, or switches socket to shaft ($\approx N^{\mathrm{LOF}}$) [1816]. **GOF:** unknown. **PROTEIN: Function:** possibly a transcription factor (1596 a.a.) [1816, 4007] or co-activator for Notch [3355, 4812]. **Location:** nuclear in SOP, IIa, IIb, and perhaps terminal cells [305, 4007]. **Domains:** a basic region (with one NLS) that resembles a bZip motif [305], 2 (separated) acidic regions, 3 (clustered) stretches of alternating Gly-Val, multiple opa motifs (Mam is 22% Gln!), and many homopolymeric stretches; very hydrophilic [1816, 4007, 4813]. **Partners:** binds Su(H) when Notch's ankyrin repeats are also present [3355]. Does not bind H [3027]. Localizes to polytene chromosomes (mainly at actively transcribing genes), but binding may not be directly to DNA [305].

musashi (*msi*, 96E1–4, double shafts reminiscent of 2-sword fighting style invented by samurai warrior Musashi Miyamoto). **PHENOTYPES: Severe LOF:** 4 sockets [3035, 3182]. **LOF** ($\approx numb^{\mathrm{LOF}}$): switches IIb to IIa, and shaft to socket [3035]. **GOF:** unknown. **PROTEIN: Function:** RNA-binding protein (606 a.a.) [3035]. Msi binds *tramtrack* transcripts ([3035] sequel). It may act like Elav [2563, 3620, 3731, 4805] to modulate distinct RNA isoforms [2312, 2313] and thereby toggle cell fates [1857]. **Location:** nucleus [3035]. Enhancer-trap *lacZ* reporter is expressed in SOP, IIa, IIb, and all bristle cells (more in neuron and sheath than in shaft and socket cells). **Domains:** RNA-binding (2x), plus N-terminal (a.a. 43–127) and C-terminal (a.a. 399–577) regions that are ~45% Ala or Gln [3035].

neuralized (*neu*, cf. its enhancer trap "A101," 85C, neural hyperplasia in the embryo). **Pathway:** Notch pathway, but not for wing veins or margin [2387]. US(+) of *N* [1000, 2542]. **PHENOTYPES: Null:** switches IIa to IIb, socket to shaft, and sheath to neuron ($\approx N^{\mathrm{LOF}}$) [2387, 4814]. **GOF:** no known effects on the lineage [2387] aside from suppression of SOPs (see App. 5). **PROTEIN: Function:** ubiquitin ligase (754 a.a.; see [2387] sequels) [417, 3463]. **Location:** cell membrane [2387, 4814]. Detectably expressed in SOPs [417, 1925], but must also function in SOP descendants based on the phenotypes of null clones [2387, 4814]. **Domains:** zinc (ring) finger near C terminus, NLS, opa (2x) [417, 3463].

Notch (*N*, 3C7, notched wing tip). **Pathway:** same pathway as *numb*: N^{LOF} synergistically enhances $numb^{\mathrm{GOF}}$ [1651]. DS(−) of *numb*: double-null $\approx N^{\mathrm{null}}$ [1651]; N^{null} does not alter Numb asymmetry [1651, 3579]; fate-switchable period extends beyond *numb*'s [1742, 3579]. DS(+) of *Dl*: double mutant of N^{GOF} and $Dl^{\mathrm{LOF}} \approx N^{\mathrm{GOF}}$ [2542]. DS(+)

of *dx*: N^{GOF} rescues dx^{LOF} phenotype [2746]. DS(+) of *neu*: N^{GOF} rescues neu^{LOF} [1000, 2542]. US(+) of *E(spl)* [2542] and *ttk* [1651]; US(−) of *pros* [3550]. **PHENOTYPES: Severe LOF:** 4 neurons [1742]; presumably same in bald areas of null clones [996, 1058] and in t.s. LOF mutants [603]. **LOF** (≈ $numb^{GOF}$): switches socket to shaft [218, 996, 1307, 1803, 3028]. **GOF** (≈ $numb^{LOF}$): switches IIb to IIa [3270], shaft to socket [1307, 1651, 2627, 3270, 3543], and possibly neuron to sheath [3270]. **Extreme GOF:** 4 sockets [1651]. **PROTEIN: Function:** receptor for Delta or Serrate ligands and signal transducer (2703 a.a.) [2210, 3022, 4611]. Intracellular part goes to nucleus and serves as transcriptional co-activator with Su(H) (cf. Fig. 2.2) [2211]. Notch was also thought to assist in cell adhesion [630, 1204, 1742, 1900] and in delaying differentiation [1271], but both ideas have been discounted [112, 994, 1612]. The significance of other isoforms is unresolved [4602]. **Location:** transmembrane glycoprotein (single-pass) [2210, 4611] in a subapical ring [1851]; cleaved (as precursor) in *trans*-Golgi at a prospective extracellular site [368]. Truncated construct (N-intra) goes to nuc. [2542, 3543, 4161], but not if Numb is added [1307]. Ubiquitous in disc epithelium, including SOPs [3270]. **Domains:** signal peptide, EGF-like (36x), LNG (3x), transmembrane, RAM23, ankyrin (6x), opa, PEST, and NLS (2x) [2210, 4244, 4611]. **Partners:** Binds itself [4601, 4823] (cf. Fig. 2.3), Big brain [4601, 4823], Delta [985, 1204, 2263, 3544, 4601], Deltex [1053, 2746], Disabled [1493], Dishevelled [151], Fringe [990, 2096, 4601], Notchless [3662], Numb [1307, 1651], Pecanex (high affinity) [4601], Presenilin [3533, 3534], Serrate [2263, 3544, 4601], Su(H) [1269, 2747, 4244], and possibly Su(deltex) [129]. N also binds Wingless [4601, 4823], but not as a receptor [310, 1971, 3689], despite suspicions based on *in vivo* cross-talk [459, 886, 1851, 2718] and enigmatic cell culture data that is open to other interpretations [4601, 4603]. N does not bind Scabrous [2460, 4601, 4823] but associates with it in a complex [3456], so a third protein must serve as a bridge to attach them.

numb (30B, sensory defect). **Pathway:** US(−) of *N* [1651, 3579], *Su(H)* [3027, 4542], and *ttk* [1650]; US(+) of *pros* [3550]. **PHENOTYPES: Null:** 4 sockets [3579]. *N.B.:* The *numb* allele used here [3579] was considered null [4417] but may not be [552]. **LOF:** switches IIb to IIa, shaft to socket, and neuron to sheath [1307, 3579, 4542]. **GOF:** switches IIa to IIb, socket to shaft, and sheath to neuron (all = opposite of LOF) [1307, 3027, 3579, 4542, 4789]. **Extreme GOF:** 4 neurons [4542]. **PROTEIN: Function:** determinant of bristle cell fates (zygotic form: 556 a.a.) that is unequally partitioned in SOP mitoses: inherited by IIb, shaft cell, glial cell, and neuron [1447, 3579, 4417, 4542]. **Location:** cell cortex (i.e., inner face of cell

membrane) [1447, 3579, 4542]. **Domains:** PTB, PEST, zinc finger?, basic, binding sites for EH and SH3 domains [2530, 3725, 4417, 4473]. **Partners:** binds Partner of Numb [2609] via its PTB domain, N [1307, 1651], Nak [765, 4900], Miranda [3892, 3893], and Amos [1928]. Also binds H [4542] but not Su(H) [4542]. (Its homolog mNumb binds Seven-in-absentia [4473].)

numb-associated kinase (*nak*, 37B4–7). **Pathway:** same pathway as *numb*. GOF effects are sensitive to dose of *numb* [765]. **PHENOTYPES: LOF:** unknown. **GOF** (≈ $numb^{LOF}$): switches IIb to IIa, shaft to socket, and maybe neuron to sheath [765]. **Extreme GOF** (≈ $numb^{null}$): 4 sockets [765]. **PROTEIN: Function:** putative serine/threonine kinase (1490 a.a.) [765]. **Location:** cell cortex when overexpressed [765]. **Domains:** putative S/T kinase domain near N terminus, and Numb-binding 11-mer peptide near C terminus [765]. **Partners:** binds Numb's PTB domain [765].

prospero (*pros*, 86E1–2, named for the magician who sets others' fates in Shakespeare's *The Tempest*). **Pathway:** DS(+) of *numb* and DS(−) of *N* (and *Su(H)* and *ttk*?) in IIa: $numb^{GOF}$ or N^{LOF} turns *pros* ON in IIa daughters, while N^{GOF} turns *pros* OFF in IIIb daughters [3550]. **PHENOTYPES: Null:** switches IIb to IIa (≈ $numb^{LOF}$, partial expressivity); stunting of axon and dendrite [2680, 3550]. **GOF:** switches IIa to IIb (≈ $numb^{GOF}$, ~100% expressivity); rarely switches neuron to sheath [2680, 3550]. **PROTEIN: Function:** transcription factor [1753] that helps specify IIb cell identity [2680]. Prospero has two alternately spliced isoforms (1374 and 1403 a.a.) [785, 4053]. **Location:** IIb (but not IIa) nucleus, then basal cortex during IIb and IIIb mitoses, then nuclear in sheath cell while fading from neuron [1447, 2680, 3550]. **Domains:** homeo (partial), opa (2x), PEST (3x), masked NLS [1020], stretches of Ala, Ser, Thr, and Asn, plus a sequence (≤32 a.a.) that causes asymmetric localization in mitotic neuroblasts [1855, 4436]. **Partners:** binds Miranda [2749, 3892, 3893].

scratch (64A, scarred eye facets). **Pathway:** interacts with *deadpan* (another pan-neural gene), although its immediate US regulators are unknown [3714]. **PHENOTYPES: Null:** wild-type (i.e., no apparent defects) [3611]. **GOF:** 2 neurons (and maybe 2 sheath cells). **PROTEIN: Function:** pan-neural transcription factor (664 a.a.) [3611]. **Location:** neurons. **Domains:** zinc finger (5x).

Serrate (*Ser*, 97F, serrated wing margin). **Pathway:** Notch pathway [4859]. **PHENOTYPES: Null:** wild-type in bristles. When *Delta* function is also eliminated, Ser^{null} switches IIa to IIb, socket to shaft, and (less often) sheath to neuron ($\approx N^{\text{LOF}}$) [4859]. **GOF:** switches shaft to socket [2008]. **PROTEIN: Function:** ligand for Notch receptor (1404 a.a.) [1252, 4302]. **Location:** transmembrane (single-pass) [1252, 4302]. **Domains:** signal peptide, EGF-like (14x), Cys-rich, transmembrane [1252, 1943, 1944, 4302]. **Partners:** binds N [2263, 3544, 4601].

shibire (*shi*, 14A, means "paralyzed" in Japanese). **Pathway:** presumed to act in the Notch pathway based on bristle-pattern effects (App. 5). **PHENOTYPES: Null:** unknown. Null alleles exist [1595] but have not been analyzed in mosaics for effects on bristles. **LOF:** switches socket to shaft [1803, 3425]. **GOF:** unknown. **PROTEIN: Function:** endocytosis of Notch receptor [3271, 3863]. Homolog of mammalian dynamin. Two alternately spliced isoforms: 836 and 883 a.a. [740, 4443]. **Location:** cytoplasmic (free) or submembrane (clathrin-coated pits) [4549]. **Domains:** PH [2458, 4437], tripartite GTPase motif, and Pro/Arg-rich C terminus (ratio of ~4 basic/acidic residues) with 5 potential binding sites for SH3 domains [740, 1095, 4396, 4549, 4802]. **Partners:** binds various SH3-domain proteins [1588, 3913], including Amphiphysin-2 [4655, 4656] whose 3-D structure is known [3218]. Binds itself during assembly of "collars" [954, 1852, 2184, 4240, 4549], leading to new juxtapositions of domains that may explain GTPase activation [4802].

Suppressor of Hairless (*Su(H)*, 35B9–10). **Pathway:** same pathway as *numb*. $Su(H)^{\text{LOF}}$ synergistically suppresses $numb^{\text{LOF}}$ [4542]. DS(−) of *numb*: double-null $\approx$ $Su(H)^{\text{null}}$ [4542], and double-GOF $\approx Su(H)^{\text{GOF}}$ [3027]. Antagonistic to *H*, but epistatic relationship (US vs. DS) is unclear: double mutant of H^{LOF} and $Su(H)^{\text{null}}$ has mixture of H^{LOF} and $Su(H)^{\text{null}}$ traits [3827]. **PHENOTYPES: Null:** apparently switches IIa to IIb [999, 3820], but little–if any–switching of sheath-to-neuron identity [3027]. **LOF** (near time of SOP mitosis): switches socket to shaft (but not sheath to neuron) [3827, 4542]. **GOF:** switches shaft to socket (but not neuron to sheath) [200, 3027, 3827, 4542]. Rarely switches IIb to IIa to yield a 4-socket cluster ($\approx H^{\text{null}}$) [3027]. **PROTEIN: Function:** transcription factor (594 a.a.) [1131, 3826]. Enhances transcription by using N as co-activator [1329, 2211] but probably blocks transcription in absence of N by recruiting a histone deacetylase complex [1913, 2129, 2941]. In adults, it acts physiologically in the socket cell to permit deflections of the shaft to be transduced into electrical impulses [218]. **Location:** nuclear in cultured cells, but

moves to cyt. if N is added; returns to nuc. if N binds Dl [1269]; returns to cyt. if Numb is added [1307]. Also shows N-dependent movement from cyt. to nuc. in socket cells *in vivo* [1448]. Detectable in epidermal cells and SOP lineage (equal amounts in IIa and IIb); strong expression in socket cell after IIa mitosis [1448, 3027]. **Domains:** novel (2-part?) DNA-binding domain, nonfunctional integrase motif, NLS (2x) [788, 1331, 3825, 3826]. **Partners:** binds N [1269, 2747, 4244] and H [171, 492, 1332, 2657] but not Numb [1307, 4542]. Binds as monomer to DNA at GTGGGAA (high affinity) and GTGAGAA (low affinity) [788, 1332, 4408]. These sites are paired in promoters of a subset of E(spl)-C genes [171, 1131, 2453].

tramtrack (*ttk*, 100D, railroad pattern of gene expression in embryonic germ band). **Pathway:** DS(+) of *N*: ttk^{LOF} prevents N^{GOF} from changing neuron to sheath [1651]. DS(−) of *numb*: double-LOF $\approx ttk^{\text{LOF}}$ [1650]; ttk^{LOF} does not alter Numb asymmetry [1650]; Ttk appears after Numb [3502]. **PHENOTYPES: Null:** 4 neurons ($\approx$ extreme $numb^{\text{GOF}}$) [1650]. **GOF:** 2-shafts-2 sockets (switches IIb to IIa) [1650, 3502]. Excess blocks mitosis [3502]. **Extreme GOF** (rare): 4 sockets ($\approx numb^{\text{null}}$) [1650]. **PROTEIN: Function:** transcription factor, which has two alternately spliced isoforms: 69 kD (641 a.a.) and 88 kD (811 a.a.) [1650, 1731, 3537]. **Location:** nucleus [3502]. The 69 kD isoform is detected in IIa (not IIb), sheath (not neuron), socket and shaft cells, and slightly in epidermal cells [1650, 1651, 3502]; while 88 kD is also detected in 3 bristle cells (probably sheath, socket, shaft) in eyes but acts differently [2391, 2529]. **Domains:** zinc finger (C2H2 type, 2x), BTB [211, 1516, 4596], PEST [1731]. **Partners:** binds dCtBP (via BTB domain) [4596] and binds itself as homo-oligomer (via BTB) [211, 1516]. Isoforms bind different DNA sequences [1187, 2123, 3537]: GGTCCTGC (69 kD) or AGGG^{C}/$_{\text{T}}$GG (88 kD).

twins (85F, mirror-image duplications of the wing disc). **PHENOTYPES: LOF** ($\approx numb^{\text{LOF}}$ and nak^{GOF}): 2-shafts-2 sockets (switches IIb to IIa) [3904]. **GOF:** unknown. **PROTEIN: Function:** regulatory subunit of serine/threonine protein phosphatase PP2A, which has two alternately spliced isoforms: 443 and 499 a.a. [2761, 4418]. PP2A dephosphorylates activated kinases in various signaling pathways [2861]. **Subcellular location:** unknown. Uniform in discs [2761, 4418]. **Partners:** binds the PP2A catalytic subunit [2861], which in turn binds Axin [1919].

Genes That Can Transform One Type of Bristle Into Another or Into a Different Type of Sense Organ

In parentheses after each gene name are the salivary gland map location and the origin of the name. The **PHENOTYPES** section only lists bristle-related phenotypes (see Fig. 2.8). The **PROTEIN** section presents available data on function, length, subcellular location, domains, and binding partners. For further information, see *FlyBase* and *The Interactive Fly*.

Abbreviations: CS (chemosensory), DS (downstream in causation; "+" activated or "−" inhibited by US gene), GOF (gain of function), LOF (partial loss of function), MS (mechanosensory), US (upstream). Evidence for the hierarchical order of genes in a pathway (US or DS) is given from a DS perspective only. Protein domains are defined in Appendix 1. Numbers of repeats (e.g., "6x") are in parentheses. Other genes whose embryonic mutant phenotypes suggest that they may belong to this group are *BarH1* and *BarH2* [1456, 1843]. Certain sensilla campaniformia on the wing are transformed to bristles by *ash2* (a member of the Trithorax Group of regulators; cf. Ch. 8) [15].

absent solo-MD neurons and olfactory sensilla (*amos*, 36F; "MD" stands for multiple dendritic). **Pathway:** DS(+) of *lozenge* [1587]; interacts positively (dose-dependent) with *daughterless* [1587]. **PHENOTYPES: Null:** unknown. **LOF:** fewer sensilla basiconica and trichodea on the antenna [1587]. **GOF:** extra sensilla basiconica, trichodea, and coeloconica on the antenna, ectopic olfactory sensilla elsewhere, and conversions of bristles into olfactory sensilla [1587]. **PROTEIN: Function:** transcription factor (198 a.a.) [1587, 1928]. **Location:** unknown (presumably nuclear). **Domains:** bHLH (C-terminal) [1587, 1928]. **Partners:** binds Daughterless [1928], and these het-erodimers bind E boxes [1928]. Also binds Numb [1928], but the function of this interaction is unknown.

atonal (*ato*, 84F, missing chordotonal organs). **Pathway:** US(−) of *cut* [2038]. **PHENOTYPES: Null:** missing chordotonal organs (due to failure of SOP initiation) [2040, 2042]. **GOF:** transforms bristles into chordotonal organs [2038] and causes ectopic chordotonal organs [764, 2040]. **PROTEIN: Function:** transcription factor (312 a.a.) [2040]. **Location:** chordotonal SOPs [2040]. **Domains:** bHLH (C-terminal), acidic (transcriptional activation?) section [2040]. Atonal's specificity for evoking chordotonal (vs. bristle) organs resides in its bHLH domain [764]. **Partners:** binds Daughterless, and these heterodimers bind E boxes [2040].

cut (7B1-2, scalloped margin of the wing). **Pathway:** DS(−) of *ato*. Expressing *ato* in bristle SOPs abolishes *cut* expression there, but *ato* is not activated in transformed chordotonal SOPs of *cut*[null] embryos [2038]. Was thought to be DS(+) of *poxn* because (1) ubiquitous *poxn* activates *cut* in SOPs where *cut* is not normally expressed, and (2) reporter gene driven by a piece of the *cut* promoter is expressed in CS (but not MS) bristles [4481]. However, other evidence (see *poxn*) indicates independence. **PHENOTYPES: Null:** partial or complete transformation of MS bristles into chordotonal organs [378]. **GOF:** unknown for adult bristles. In the embryo, transforms chordotonal organs into external sense organs [372]. **PROTEIN: Function:** transcription factor (2175 a.a.) [370], which acts as a repressor by recruiting histone deacetylases [2528]. **Location:** nucleus of SOP and all descendants in embryonic external sense organs [370, 371] and adult bristles (both MS and CS) [149, 2000]. **Domains:**

homeo, cut repeats (3x), highly acidic stretches of Glu and Asp (4x), regions rich in Gln (with His or Ala), Pro and Ala, Ala alone, or Asn alone [370]. **Partners:** binds DNA [87] and chromatin-modifying proteins [2528].

paired box-neuro (*poxn*, a.k.a. *pox neuro*, 52D1, has "paired box" DNA-binding motif and is transcribed in the CNS). **Pathway:** independent of *cut*. Surprisingly, *poxn* is expressed normally in SOPs of *cut*[null] embryos [4481], and *cut* is expressed normally in SOPs of *poxn*[null] wings and legs [149]. **PHENOTYPES: Null:** transforms CS (poly-innervated) into MS (mono-innervated) bristles on the dorsal tibia and wing margin [149], and suppresses some CS bristle SOPs (suggesting a proneural role). **LOF:** milder version of null phenotype [149], plus differentiation defects within the bristle organ. **GOF:** transforms MS into CS bristles on the dorsal tibia [149, 3141, 3142]. **PROTEIN: Function:** putative transcription factor (425 a.a.) [401,950]. **Location:** nuclei of CS bristle SOPs and most or all SOP descendents [149, 950, 3142]. **Domains:** "paired" domain, poly-Ala stretches, a Pro-rich segment (15/45 a.a.), a highly charged region (14/25 a.a.), and an acidic region (18/45 a.a.) [401, 950]. **Partners:** binds DNA, including the *cut* A3 enhancer [4481].

Genes That Can Alter Bristle Number by Directly Affecting SOP Equivalence Groups or Inhibitory Fields

Some of the genes catalogued below are also listed in Appendix 3 because their mutations affect both bristle development (App. 3) and patterning (App. 5). Redundancy is minimized by stating only those **PHENOTYPES** here that pertain to patterns. For genes that are listed in both appendices, the reader is referred to Appendix 3 for protein domains and binding partners. See Table 6.2 for upstream "prepattern" genes that influence bristle patterning indirectly.

In parentheses after each gene name are the salivary gland map location and the origin of the name, except where obvious. The **PROTEIN** section presents available data on function, length, subcellular location, domains, and binding partners. For further information, see *FlyBase* and *The Interactive Fly*.

Abbreviations: "≈" (resembles), a.k.a. (also known as), cyt. (cytoplasm), DS (downstream in causation; "+" activated or "−" inhibited by US gene), GOF (gain of function by mutant allele or transgene construct), LOF (partial loss of function), MC (macrochaete), mC (microchaete), nuc. (nucleus), PNC (proneural cluster), SOP (sensory organ precursor), US (upstream), UTR (untranslated region). To avoid redundancy, data about hierarchical rank within a pathway are given at the DS gene only. Thus, the link "$a \rightarrow b$" would be entered under gene b as "DS(+) of a" followed by epistasis evidence, but under gene a as just "US(+) of b" without the data. Likewise, "$c \dashv d$" would be listed for d as "DS(−) of c" with evidence, but for c as "US(−) of d" without the data. Protein domains are described in Appendix 1. Numbers of repeats (e.g., "6x") are in parentheses.

Many of the tabulated genes perform similar duties in the embryonic CNS and PNS [2018, 2019, 2062]. Some genes whose mutant alleles add or delete bristles were omitted because they act redundantly with a listed gene (e.g., $m\alpha$ and $m4$ [92]) or because their effects are too slight or too poorly studied: *aristaless* [3798], *Dense* [2500], *Domina* [4139], *echinus* [4185], *l(3)ecdysoneless* [3990], *puckered* [2713], *Suppressor of deltex* [1276], *Tufted* [102, 1802], etc. [951, 1514, 4333]. *Scutoid*^{GOF} (not listed) [126] involves misexpression of Snail (a zinc-finger protein) in the scutellum, where it blocks AS-C function (i.e., suppresses SOPs) by competing for the same E boxes [1334]. The role of the pair-rule gene *odd Oz* is uncertain: it is expressed in SOPs [2497], but no mutational effects on bristles have been reported. The only known maternal effect on bristle number is from isoalleles of *tumorous head 1* [3220]. Additional pattern-affecting genes were recovered in a recent GOF screen [6].

Shaggy (Sgg) was once thought to be a downstream agent of Notch in lateral inhibition [2018, 3681, 3963], but subsequent evidence argues that it is an upstream regulator of PNCs (Table 6.2) [2890]. The old view was based on three main facts, each of which is open to other interpretations. **Fact #1:** Notal *sgg*^{null} clones only cause extra bristles in proneural areas [1797, 3681, 3963], and this confinement to PNCs suggests that *sgg* functions in lateral inhibition. **Rebuttal:** Although bristle density does increase in *sgg*^{null} clones [886, 3373], the density is more similar to *ac*^{GOF} [1577] than to *N*^{LOF} [459]. The lack of *N*^{LOF}-like "tufting" argues that the Notch pathway still mediates lateral inhibition despite *sgg*^{null}. Moreover, *sgg*^{null} clones put bristles on the wing [351, 353, 3349, 3958], so *sgg* must act upstream of AS-C. **Fact #2:** Extra doses of the AS-C do not add bristles to *sgg*^{null} clones [3956], so *sgg* must not be affecting the proneural stage. **Rebuttal:** If the Notch pathway is still functioning in sgg^{null} clones (as argued above), then increasing the dose of proneural agents cannot force bristles any closer than a minimum distance dictated by the inhibitory fields. The absence of an overt phenotype does not preclude a genetic interaction. **Fact #3:** Alleles of *sgg* can interact with *Abruptex*-class *Notch* (N^{Ax}) alleles [1797, 3681, 3963]. **Rebuttal:** N^{Ax} mutations involve Fringe [990] and Wg signaling [459, 886], so this interaction could be due to Wg (and only indirectly to Notch). Wg is the likely culprit in *sgg*-mediated hyperplasia because the macrochaete sites most affected by sgg^{LOF} are those that rely on *wg* [3373, 3374, 3956, 4368]. **Conclusion:** Sgg appears to affect SOP initiation solely via the Wg pathway [353, 886, 3374]. In contrast, Dsh affects SOPs by both the proneural and lateral-inhibitory routes by short-circuiting the Wg and Notch pathways [151, 459, 1054]: "Wg → Dsh ⊣ {Notch ⊣ AS-C}."

achaete (*ac*, 1B1–7, AS-C, missing bristle or "chaete" in Latin). **Pathway:** same proneural pathway as *da* [949]. DS(−) of *N*, *Su(H)*, and E(spl)-C (see text): compound of ac^{null} sc^{null} with N^{LOF} [997, 1797, 1802], Dl^{LOF} [997], or E(spl)-C deletion [1794] ≈ ac^{null} sc^{null}. DS of *H*: H^{GOF} fails to rescue ac^{null} sc^{null} [197]. DS(−) of *h*: *ac* is upregulated in h^{LOF} [3982]. Paradoxically, it is also US(+) of *h* (see *h*). DS(−) of *emc*: ac^{null} sc^{null} is epistatic to emc^{LOF} [997]. US(+) of *Dl* [1672, 2359] (but see *Dl*), *Brd*, *sca*, E(spl)-C genes *m4*, *m7*, and *m8* [2885, 3974], and *klu* [2250]. DS(−) of *tam* (a.k.a. *pyd*): more Ac in tam^{LOF} [4237]. **PHENOTYPES: Null:** unknown except when nearby enhancers are also deleted, in which case ac^{null} removes many mCs and a few specific MCs [1538]. In combination with sc^{null}, ac^{null} eliminates nearly all bristles [1379, 1462, 1802, 3812] by suppressing SOP initiation [912]. **LOF:** fewer mCs and loss of certain MCs [1802, 4185]. **GOF** ("*Hairy wing*" alleles [635]): extra bristles (including mCs and MCs) on the thorax, wing blade, and legs [191, 1348, 1577, 1802]. **PROTEIN: Function:** proneural transcription factor (bHLH type; 201 a.a.) [4485]. **Location:** nucleus in PNC cells, more in SOPs [912, 3763]. **Domains:** bHLH, C-terminal acidic region [4485], which could activate transcription [2878]. **Partners:** binds certain "E box" DNA sequences as an Ac/Da heterodimer (rather than binding itself or as Ac/Sc) [588, 589, 918, 2359, 4452, 4453]. Also binds Emc but not Hairy or M8 [589, 4452]. Ac binds Pannier and Chip (as an Ac/Da dimer) in a multimeric complex that may bridge ~30-kb gaps between E boxes and regional *cis*-enhancers at the AS-C [3504]. How Ac activates transcription is unknown; some other bHLH proteins do so by recruiting histone acetyltransferases [2737].

asense (*ase*, 1B1–7, AS-C, sensory defect). **Pathway:** DS(+) of *ac* and *sc* temporally [438]. **PHENOTYPES: Null:** defects in stout bristles along anterior wing margin [1079, 2039] and loss of tergital mCs [2690]. **GOF:** extra bristles, but these are probably artifacts of Ase mimicking Ac or Sc because *l'sc*^{GOF} also causes extra notal bristles despite *l'sc* not being transcribed in wild-type wing discs [438, 1079]. **PROTEIN: Function:** pan-neural transcription factor (bHLH type; 486 a.a.) [1540]. **Location:** nucleus in SOPs only [438, 1079]. **Domains/partners:** see Appendix 3.

Bearded (*Brd*, 71A1–2, extra bristles) – one of 6 related genes in the Brd Complex, which act alike [2382, 2383, 4842]. **Pathway:** same pathway as *N*. The extra-bristle phenotype of *Brd*^{GOF} is partly suppressed by an extra dose of N^+ [2500]. DS(+) of AS-C: expression in wing disc PNCs is abolished in ac^{null} sc^{null} [3974]. US of *H* [2500]. **PHENOTYPES: Null:** wild-type (i.e., no apparent defects) [2500], probably due to redundancy [2382]. **GOF:** tufts at MC and mC sites (but bare areas persist between mC sites; cf. *neu*^{LOF}) or mild bristle loss, depending on allele and body region [2382, 2385, 2499, 2500]. **PROTEIN: Function:** unknown (81 a.a.) [2499]. Brd resembles the E(spl)-C proteins Mα, M2, M4, and M6 [92, 3075, 4767], and the 3′ UTRs of their mRNAs also share motifs ("Brd," "GY," and "K" boxes) that suggest a common function [2499, 2500]. That function probably involves regulation of proneural genes whose mRNAs have motifs ("proneural boxes") complementary to GY boxes in their 3′ UTRs [2386]. **Location:** unknown; Brd is transcribed in PNCs [2499]. **Domains:** no obvious motifs except for one putative amphipathic α-helix [2383, 2499].

big brain (30F). **Pathway:** Notch pathway (see GOF data). **PHENOTYPES: Null:** 2–4x increase in bristle number (milder than N^{LOF}) [3519]. **GOF:** essentially wild-type, except when overexpressed concomitantly with *N* or *Dl*, with which it interacts synergistically [1075]. **PROTEIN: Function:** may form channels in cell membrane for exchange of small molecules (700 a.a.) [573, 3520] to aid reception of (or response to) Delta signal [1075]. **Location:** possibly transmembrane (6-pass) [3520], although antibodies detect it on cytoplasmic face of plasma membrane, on adherens junctions, and in vesicles [1075]. Colocalizes (intracellularly) with Dl and N in PNCs [1075]. **Domains:** hydrophobic stretches (6x), Gln-rich in C-terminal half [573, 3520]. **Partners:** binds N [4601, 4823].

canoe (*cno*, 82E4–F2, dorsal surface of embryo remains open). **Pathway:** Notch pathway based on synergistic interaction with N^{LOF}, but probably also used in Ras signaling because Cno binds Ras [2363]. Also, *cno*^{LOF}

interacts with *sca* [2882]. **PHENOTYPES: LOF:** extra MCs [2882]. **GOF:** unknown. **PROTEIN: Function:** intercellular signaling (1893 a.a.) [652, 2882]? **Location:** intercellular (adherens) junctions [4236]. **Domains:** PDZ, Ras-binding, kinesin-like, myosin (class V)-like [2882, 3417, 3418, 4236]. **Partners:** binds Pyd [4236] and Ras [2363].

C-terminal Binding Protein (*CtBP*; 87D7–9). **Pathway:** dose-sensitive interactions with *h* [3430]. **PHENOTYPES: LOF:** extra bristles (MCs) on the scutellum, other parts of the notum, and head [3430]. **GOF:** unknown. **PROTEIN: Function:** regulates Hairy's repression of AS-C (386 a.a.) [3430], probably as a co-repressor like Groucho (cf. Fig. 3.12e) [2129, 3112, 4213]. However, dCtBP can also behave as a co-activator in other contexts [3377]. **Location:** not known for discs. **Domains:** (unmapped) dimerization and peptide-binding domains [3430]. **Partners:** binds itself, <u>PLSLV</u> in Hairy, and <u>PVNLA</u> in E(spl)-C's Mδ (but not M8, Emc, or Gro) [3430]; binds <u>PxDLS</u> in Knirps, Krüppel, and Snail [3112, 3113].

Cyclin A (*CycA*, 68E). **PHENOTYPES: LOF** (due to P-element insertion in first intron): removal of specific MCs (DC, SC, etc.), socketless shafts, shaftless sockets, and other defects indicative of an insufficient number of mitoses [4415]. **GOF:** induces cells to enter S phase [4047], but effects on bristle patterning (aside from ommatidial defects [1087]) have not been reported. **PROTEIN: Function:** regulates G_2/M transition [2270]. Degradation of Cyclin A at metaphase allows cells to enter mitosis [2482, 4415, 4633] (491 a.a.) [2481]. **Location:** PNCs, less in background epidermal cells [2225]. In embryos, CycA associates with spindle poles and chromatin during mitosis (up to metaphase) but is cytoplasmic in interphase [2481, 2664]. **Domains:** Cdc2-binding "cyclin-box" domain, "destruction box" motif (needed for ubiquitination and degradation) [975, 1512, 3009, 3926]. **Partners:** binds Roughex [143], Cdc2, and other cyclin-dependent kinases [1793].

daughterless (*da*, 31E). **Pathway:** same proneural pathway as *ac* and *sc* [949]. DS(−) of E(spl)-C: *m8* and *m5* reduce Da activation of reporter gene [3171]. US(+) of *Dl* [2359]. **PHENOTYPES: Null:** unknown for discs. In embryos *da*^{null} blocks all PNS development [688] after SOP emergence [4435] by preventing mitosis [1755]. **LOF:** in presence of heterozygous AS-C deletion, haploid dose of *da*⁺ removes certain MCs [949]. **GOF:** unknown. **PROTEIN: Function:** proneural transcription factor (bHLH type; 710 a.a.) [689, 906]. Da is a ubiquitous partner for AS-C proteins (cf. Tango [4022]). **Location:** nucleus; ubiquitous

in discs [905, 4435]. **Domains:** bHLH, leucine zipper, PRD (2x), opa (2x), poly-G (2x), poly-A, PEST [689, 906]. **Partners:** binds certain "E box" DNA motifs as a heterodimer with bHLH AS-C proteins (Ac, Sc, L'Sc, Ase) or, less so, as a homodimer [588, 589, 918, 2039, 2359, 3171, 4452, 4453]. Also binds Emc [589, 4452], Amos [1928], and – under some [68] but not other [4452] conditions – various E(spl)-C proteins. Binds Chip (as a Da/Ac or Da/Sc dimer) in a complex with Pannier (notum only) [3504]. May also bind Net [2741].

Delta (*Dl*, 92A2, delta-shaped tips of wing veins). **Pathway:** US(+) of *N* [2542], but *Dl*^{LOF} suppresses *Abruptex* alleles (see text) [992, 1798, 4780], and *Dl* is downregulated by *N*^{GOF} [3270]. Equivocal with respect to AS-C: DS(+) of AS-C (ectopic expression of *sc* activates *Dl* transcription in congruent areas of wing disc [1854]) but also US(−) of AS-C (*Dl*^{LOF} upregulates *ac* [3270], and *ac*^{null} *sc*^{null} is epistatic to *Dl*^{LOF} [997]). Interacts with *sca* (see *sca*). **PHENOTYPES: Null** (assayed in mosaics): extra bristles (tufts at MC sites and high-density mCs) [2690, 4859]. **LOF:** t.s. LOF alleles cause extra bristles when mutants are heat-pulsed at a certain time but bristle loss (due to fate switches) when pulsed later [1797, 3272, 3273, 3277]. Haplo-insufficiency causes mild increase in bristle density [997, 4466]. **GOF:** unknown. **PROTEIN: Function:** ligand for Notch receptor (832 a.a.) [3022]. Serrate is not a redundant ligand in PNCs [4859]. See Figures 2.2, 2.3, and 3.6. **Location:** transmembrane (single-pass) [2300, 4465] but also secreted under some conditions [2264, 3479]. Detected on surfaces and in cytoplasmic vesicles of cells in PNCs [2296] and notal stripes [3270]. **Domains/partners:** see Appendix 3.

deltex (*dx*, 6A, *Delta*-like gene on the X). **Pathway:** US(+) of *N* [2746]. **PHENOTYPES: Null:** unknown [580, 1053]. **LOF:** extra bristles and disrupted patterns [1570]. **GOF:** no effects on bristle number reported. **PROTEIN: Function:** unknown (737 a.a.) [580]. **Location:** cytoplasm [580, 1053]. **Domains/partners:** see Appendix 3.

Dichaete (*D*, 70D1–2, a.k.a. *fish-hook*, missing MCs). **Pathway:** interacts with AS-C: *D*^{GOF} is partly suppressed by certain *sc*^{LOF} alleles [2561]. **PHENOTYPES:** effects of *D* inversions on bristles are probably not actually due to *D* but rather to breakpoints in *mirror* (Iro-C; cf. Table 6.2) [3079]. **LOF:** unknown (with regard to bristles) because mutant cells become necrotic [2977]. **GOF:** removes certain MCs [3405, 3707, 4178]. **PROTEIN: Function:** activates transcription directly [2635] and may sterically facilitate DNA binding by other transcription factors (382 a.a.) [3045, 3707]. **Location:** nucleus [2635]. **Domains:** Sox (a subgroup of HMG), NLS (within Sox), short stretches of Ala, Gln, or Ser, plus 11 repeats of a 5-a.a. consensus [2977, 3045, 3707]. **Partners:** binds <u>AACAAT</u> and <u>AACAAAG</u> [2635].

Enhancer of split (*E(spl)*, a.k.a. *m8*, 96F11–14, enhances *split*, a N^{LOF} allele) – one of 7 bHLH genes in the E(spl)-C that act alike in many contexts [3075] (but see [871, 2548, 2549]). **Pathway:** DS(+) of *N*: M8 levels are lower in N^{LOF} and higher in N^{GOF} [2052]. DS(+) of *kuz*: $m8^{GOF}$ rescues kuz^{LOF} [4025]. DS(+) of *Su(H)*: Su(H)-binding sites in *m8* promoter are needed for expression of *m8* in PNCs (but not in SOPs) [2453]. US(−) of *da* [3171], *ac*, and *sc* [918, 991, 1794, 2063]. Paradoxically, it is also DS(+) of AS-C: expression in PNCs is abolished in ac^{null} sc^{null} [3974] and augmented in sc^{GOF} wing discs [871]. Also regulated by other pathways [3075, 4568]. **PHENOTYPES: Null:** wild-type, due to redundancy of other E(spl)-C genes [1018, 1794, 3806]. Deletion of all 7 bHLH E(spl)-C genes causes extra bristles (tufts at MC sites and high-density mCs) [999, 1794]. **GOF:** missing bristles [2384, 2548, 3037], due to stifling of SOP initiation [3037, 4256]. **PROTEIN: Function:** transcription factor (bHLH type; 179 a.a.) [2246, 2274, 3171]. **Location:** nucleus of non-SOP PNC cells [871, 2052, 2053]. **Domains/partners:** see Appendix 3.

Epidermal growth factor receptor homolog (*Egfr*, 57F, a.k.a. *DER*, *torpedo*). **Pathway:** EGFR pathway, but also interacts with Notch pathway [3465, 3634]. **PHENOTYPES: LOF:** loss of MCs from ocellar, supra-alar, humeral, and posterior postalar sites (the scutellum is unaffected) but extra bristles at postvertical and anterior postalar sites [814]. Whether this spatial heterogeneity stems from Egfr expression or ligand access is unknown. Also causes slight (∼20%) increases in mC density [1042]. **GOF:** unknown. **PROTEIN: Function:** receptor tyrosine kinase (1459 a.a. vs. 1410 a.a. for the Type-1 vs. -2 splicing variants) [813]. **Location:** transmembrane (single-pass) [813]. During the larval period, *Egfr* is transcribed in imaginal discs [2125, 3779, 3928] and in nondividing (polytene) larval tissues [2152]. Egfr is expressed in eye discs [4843]. **Domains:** signal peptide, Cys-rich extracellular (ligand-binding) region, transmembrane, tyrosine kinase [813, 2573].

extramacrochaetae (*emc*, 61D1–2). **Pathway:** interacts with AS-C (see text and Fig. 3.12) and with genes of the Notch pathway [997, 1349, 2690, 3982, 4453]. DS(+) of *tam* (a.k.a. *pyd*): less *emc* RNA in tam^{LOF}, but *tam* RNA is normal in emc^{LOF} [4237]; synergistic effect on bristles in emc^{LOF} tam^{LOF} double mutant. **PHENOTYPES: Null:** extreme LOF defects [1349, 2690]. **LOF:** extra bristles (mainly MCs) at normal and ectopic sites on thorax and wings [997, 1349, 2959] and conversion of mCs to MCs on ter-

gites [2690]. **GOF:** absence of specific MCs and reduced number of mCs [997, 1349, 2690, 4453]. **PROTEIN: Function:** antagonist of proneural genes (199 a.a.) [1156, 1388]. Also involved in cell proliferation [206]. **Location:** expressed unevenly throughout the wing disc (cf. Fig. 6.2) [913, 1156, 4453]. **Domains:** HLH (but no adjoining basic domain), Gln-rich C-terminal region [1156, 1388]. **Partners:** binds Ac, Sc, and Da [68, 3430, 4453].

fasciclin II (*fasII*, 4B1–2, glycoprotein on certain axon fascicles in embryonic CNS). **Pathway:** $fasII^{LOF}$ phenotypes are enhanced by LOF mutations in the Ableson tyrosine kinase gene (even when the latter are heterozygous) [1351], although the latter mutations cause no such phenotypes on their own. **PHENOTYPES: Null** (assayed in mosaics): missing MCs on the head (verticals, postverticals, ocellars, but not orbitals) and notum (and reduced density of mCs) due to failure of AS-C activation needed for SOP initiation [1351]. Also missing ocelli (due to absence of *atonal* expression). **LOF:** similar to null phenotype but less extreme. **GOF:** extra bristles on the head (near the postverticals) and other regions (including scutellum) [1351]. **PROTEIN: Function:** neural cell adhesion molecule (two splicing isoforms: 811 vs. 873 a.a.) [1620, 2553, 4757, 4829]. **Location:** expressed throughout the eye disc, most intensely in the morphogenetic furrow and in the prospective ocellar region [1351]. **Domains:** signal peptide, Ig-like (C2-type; 5x), fibronectin (type III [1956]; 2x), transmembrane (873-a.a. isoform only) [1620, 1725]. **Partners:** Binds itself to mediate homophilic adhesion [1619, 1620].

groucho (*gro*, 96F11–14, extra bristles like bushy eyebrows of comedian Groucho Marx). **Pathway:** DS of *H*: tufting trait of gro^{null} is epistatic to balding trait of H^{null} in double-mutant clones [197]. Interacts synergistically with *h* [1794]. **PHENOTYPES: Null:** extra bristles (tufts at MC sites and high-density mCs) with virtually no intervening epidermal cells [997, 1794, 2690]; ectopic bristles on scutellum and wing (due to interaction with *h*) [1794]. **LOF:** extra bristles near the eyes [3265]. **GOF:** unknown. **PROTEIN: Function:** co-repressor (719 a.a.) [1746]. That is, it inhibits transcription via a DNA-binding protein that tethers it to DNA [1243]. Can silence transcription at long (≥1 kb) range [4865], apparently by multimerizing [738] and recruiting histone H3 [3232]. **Location:** nuclear [692, 4255]. **Domains:** WD (7x), leucine-zipper-like sequences (2x in Pro-rich N-terminal region) that mediate the tetramerization needed for Gro to repress transcription [738, 1243, 1746, 3392]. **Partners:** binds itself [738], *Drosophila* Histone H1, and histone deacetylase Rpd3 to form a complex that apparently self-polymerizes and spreads to silence chromatin [737]. Gro is recruited to DNA-binding proteins by various motifs, including (1) <u>WRPW</u> at C-termini of Hairy, Deadpan, and E(spl)-C bHLH proteins [3278, 3430]; (2) <u>WRPY</u> at C-termini of proteins that have a Runt Homology domain [107, 1243, 4616]; (3) <u>FRPW</u> near the N-terminus of Huckebein [1526]; and (4) <u>LFTIDSILG</u> in dGoosecoid [2066] and a related motif in Engrailed (cf. Fig. 3.12e) [1243, 2064, 4351]. Gro also binds Dorsal [1110, 1243] and dTcf [692, 2496].

hairy (*h*, 66D15, extra bristles). **Pathway:** US(−) of *ac* (see *ac*) but also DS(+) of *ac* and/or *sc*: downregulated in double-null for *ac* and *sc* [3982] (see text and Fig. 3.12). **PHENOTYPES: Null:** extra mCs and misalignments [3193]. **LOF:** extra mCs in various regions [1802, 1981, 2561, 2959]. **GOF:** ubiquitous expression of *hs-hairy*[+] transgene erases ectopic bristles caused by *h*[LOF] or *ac*[GOF], but (surprisingly) does not alter the normal pattern [3697]. Paradoxically, *hs-hairy*[+] induces extra bristles in the eye [507]. **PROTEIN: Function:** transcription factor antagonist for proneural genes (bHLH type; 337 a.a.) [3697]. **Location:** nucleus [663, 3697]. **Domains:** bHLH, <u>WRPW</u> (Groucho-binding) motif, "Orange" (Sc-repressing?) domain, NLS (3x), opa, Ala stretch, Pro- and Ser-rich carboxy half [976, 3430, 3697]. **Partners:** binds "N box" (but not "E box") DNA sequences [4452], Gro [1242, 3430], and CtBP [3430] but not Ac or Sc [589, 3430, 4452]. However, its ability to bind CtBP may be nonfunctional [4865].

Hairless (*H*, 92E12–14, bristle loss). **Pathway:** antagonistic to *Su(H)*, but epistatic relationship is unclear: double mutant of *H*[LOF] and *Su(H)*[null] has mixture of *H*[LOF] and *Su(H)*[null] traits [3827]. Similar antagonism (but equivocal epistasis) to *N* and E(spl)-C [171, 197, 437, 2627]. DS of *Brd*: compound of *H*[LOF] and *Brd*[GOF] looks like *H*[LOF] [2500]. US of *gro*, *ac*, and *sc* [197]. **PHENOTYPES: Null** (≈ *Su(H)*[GOF]): balding due to failure of SOP initiation [198]. **GOF** (≈ *Su(H)*[LOF]): extra bristles (milder than *N*[LOF]) [2657] or balding (SOP status unknown), depending on stage when transgene is overexpressed [200]. **PROTEIN: Function:** antagonist for Su(H). Thought to act by blocking Su(H)'s DNA-binding domain (1059 a.a.) [200, 492, 2659], H also recruits the co-repressor dCtBP [1329] ([2129] sequel). **Location:** nucleus [2657, 2658]; *H* mRNA is ubiquitous in discs [200] as is H protein [2658]. **Domains/partners:** see Appendix 3.

klumpfuss (*klu*, 68A1–2, means "clubfoot" in German, refers to leg defect). **Pathway:** DS(+) of AS-C: *klu*[LOF] interacts synergistically with *ac*[LOF] for bristle loss and is epistatic to *ac*[GOF]. **PHENOTYPES: LOF:** partial loss of MCs

(head and notum) and other bristles (wing margin and legs). **PROTEIN: Function:** putative transcription factor (750 a.a.) [2250]. **Location:** PNCs, but only transiently expressed in SOPs. **Domains:** zinc finger (C2H2 type, 4x), putative NLS (3x), stretches of poly-A (3x) and poly-N (3x), stretches rich in Q, H, or P near N-terminus, and tandem collagen-like repeats (4x) of G(R or K)E.

kuzbanian (*kuz*, 34C4–5, named for a popular puppet with tufts of hair). **Pathway:** US(+) of *N* [3239, 4025] and US(+) of *m8* [4025]. **PHENOTYPES: Null** (assayed in mosaics): absence of bristles within a clone but dense tufts of bristles when a clone border skirts a MC or mC site [3640]. **LOF:** extra bristles (tufts at MC sites and high-density mCs) but also patchy bristle loss [4025]. **GOF:** unknown. **PROTEIN: Function:** protease (1239 a.a.) for Delta [3479, 3640] and probably also for Notch [2298, 3153, 3239] (see [3479] sequel for debate). **Location:** transmembrane (single-pass) [3640]. Expressed ubiquitously in imaginal discs [4025]. **Domains:** ADAM, signal peptide, prodomain, Cys-rich, transmembrane [3640].

mastermind (*mam*, 50C20–23, enlarged CNS). **Pathway:** Notch pathway [1000, 1514, 2394, 4780], possibly US of Delta [1816]. **PHENOTYPES: LOF** (as deduced from constructs that act in a dominant-negative manner): extra MCs (due to more SOPs) on the notum (≈ *N*[LOF]) [1816]. **GOF:** unknown. **PROTEIN: Function:** possibly a transcription factor (1596 a.a.) [1816, 4007] or co-activator for Notch [3355, 4812]. **Location:** nuclear; ubiquitous in leg and wing discs [305, 3787, 4007]. **Domains/partners:** see Appendix 3.

neuralized (*neu*, cf. its enhancer trap "A101," 85C, neural hyperplasia in the embryo). **Pathway:** Notch pathway, but not for wing veins or margin [2387]. US(+) of *N* [1000, 2542]. **PHENOTYPES: Null:** balding [1058] due to 4-neuron phenotype [2387, 4814]. **LOF:** extra bristles (tufts at MC and mC sites ≈ *Brd*[GOF]) [2387, 4814]. Initially, it was thought that the extra SOPs arise autonomously via a failure of lateral inhibition [2387, 4814]. However, recent evidence demonstrates nonautonomy ([2387] sequels). **GOF:** balding due to suppression of SOPs [2387]. **Extreme GOF:** extra bristles (tufts at MC and mC sites) apparently due to a dominant-negative effect of Neu at high doses [2387]. **PROTEIN: Function:** ubiquitin ligase (754 a.a.; see [2387] sequels) [417, 3463]. Expressed in SOPs, but neither *neu* transcripts [2387] nor the A101 enhancer trap are detectable in other PNC cells [417, 1925, 2387], so Neu may only be needed in SOPs to inhibit other PNC cells. **Location:** cell membrane [2387, 4814]. **Domains:** see Appendix 3.

Notch (*N*, 3C7, notched wing tip). **Pathway:** DS(+) of *Dl*: double mutant of N^{GOF} and $Dl^{LOF} \approx N^{GOF}$ [2542] (but see *Dl*). DS(+) of *dx*: N^{GOF} rescues dx^{LOF} phenotype [2746]. DS(+) of *kuz* and *Psn*: expressing N^{GOF} in kuz^{LOF} or Psn^{LOF} background gives N^{GOF} phenotype [4025, 4810]. DS(+) of *neu*: N^{GOF} rescues neu^{LOF} [1000, 2542]. US(+) of *E(spl)* [2052, 2542]. US(−) of AS-C [918, 1797, 1802, 3270, 3983]. Interacts with *sca* (see *sca*). **PHENOTYPES: Null** (assayed in mosaics): balding [996, 1058, 1797, 2690] (presumably due to 4-neuron phenotype [1742]), plus extra SOPs [3689] (presumably due to lack of SOP inhibition). **LOF:** extra bristles (by blocking SOP inhibition) in heterozygotes, although one LOF allele removes bristles (by stifling PNCs?) [460]. Heat-sensitive LOF alleles cause extra bristles when mutants are heat-pulsed at a certain time but bristle loss (due to fate switches) when pulsed later [603, 1742, 1802, 3689, 3891]. Haplo-insufficiency causes mild increase in bristle density [460, 997]. **GOF:** fewer bristles [2690] or balding [996, 997] due to failure of SOP initiation [197, 460, 3863]. **PROTEIN: Function:** receptor for Delta and signal transducer (2703 a.a.) [2210, 3022, 4611]. (Serrate does not participate in lateral inhibition [4859].) See Figures 2.2, 2.3, and 3.6. **Location:** transmembrane (single-pass) [2210, 4611]; ubiquitous in discs [2070, 2296]. **Domains/partners:** see Appendix 3.

Notchless (*Nle*, 21C7–8). **PHENOTYPES: Null:** phenotypically wild-type, although Nle^{null}/+ enhances bristle-loss phenotype of a N^{GOF} (*Abruptex*) allele [3662]. **GOF:** slight reduction in bristle number. **PROTEIN: Function:** possible adaptor for multiprotein complex that includes Notch [3662]. **Location:** unknown. **Domains:** WD (9x) [3662]. **Partners:** binds Notch [3662].

polychaetoid (*pyd*, 85B6, extra MCs, a.k.a. *tam* [4236, 4237]). **Pathway:** US(+) of *emc*; US(−) of *ac* [4237]. LOF alleles interact synergistically with ac^{GOF} and h^{LOF} [1802, 3068, 3069], as well as (in dose-dependent manner) with LOF alleles of *N*, *Dl*, and *emc* [733]. Because pyd^{LOF} mainly affects PNCs, *pyd* likely acts in or with the Notch pathway. **PHENOTYPES: Null:** extra MCs and mCs (due to more SOPs) on the notum [733]. **LOF:** extra bristles on notum [3066, 3067, 4237] and legs [1802]. **GOF:** unknown. **PROTEIN: Function:** probably intercellular signaling (two isoforms: major = 1367 a.a.; minor = 1289 a.a.) [533, 1061, 4237, 4574], possibly via Ras modulation [4792] of gap junctions [1476] or direct action in the nucleus [1181, 1576, 1920]. **Location:** intercellular (adherens and septate) junctions [2148, 4236, 4237] and possibly the nucleus [1576]. Localization to junction vs. membrane is controlled by RNA splicing [4574]. **Domains:** PDZ (3x), SH3, GUK, and

a Pro-rich C-terminal region. Homologous to mammalian ZO-1 [4237] and member of the MAGUK family of scaffolding proteins [1062]. **Partners:** binds Canoe [4236] and dCortactin (which binds actin filaments) [2148]. Its homolog (ZO-1) binds a gap junction protein (Connexin43) [1476].

Presenilin (*Psn*, 77A–C, refers to senility in humans afflicted with Alzheimer's disease). **Pathway:** US(+) of *N* [712, 4810], although it can also affect Wingless signaling [891, 3124]. **PHENOTYPES: Null:** expanded PNCs where ~all PNC cells become SOPs, apparently due to failure of lateral inhibition [4810], similar to $Su(H)^{null}$. **GOF:** unknown. **PROTEIN: Function:** needed for ligand-dependent proteolytic processing of Notch [4156, 4810] and for its subcellular localization [1652], although whether Psn acts enzymatically or indirectly is unclear [1723, 2602, 3533, 4162, 4710] (541 a.a.). **Location:** vertebrate homolog localizes to adherens junctions [1432], as well as to endoplasmic reticulum and Golgi membranes [4710]. **Domains:** transmembrane (8x) [4710]. **Partners:** binds Notch [3533, 3534] and β-catenin (Armadillo ortholog) [86, 3003, 4060].

scabrous (*sca*, 49C2–D4, rough eyes). **Pathway:** interacts with *N* and *Dl* [179, 437, 1921, 2885, 3490] at the level of the receptor [2462] but may not be in the Notch pathway per se [186, 2461]. DS(+) of AS-C: expression in wing disc PNCs is abolished in ac^{null} sc^{null} [2885, 3974] (see text). **PHENOTYPES: Null** (same as LOF): extra bristles (usually doublets at normal sites) [1921, 2885]. **GOF:** unknown for bristles, but overexpression in the eye disrupts spacing of R8 cells, which are patterned like bristle SOPs [1158, 2461]. **PROTEIN: Function:** probably a diffusible SOP-inhibitor (774 a.a.) [1921, 2885], but may act redundantly [1158, 2885]. Sca receptor is unknown [2461]. **Location:** extracellular (secreted) [2461] and in intracellular vesicles [186, 1921]. **Domains:** signal peptide, ~200-a.a. "FReD" (Fibrinogen-Related) domain (seen in many secreted proteins) near C-terminus [1921, 2463] linked to dimerization domain by Pro-rich region. Sca is secreted as a soluble glycoprotein dimer (like many growth factors) held together by disulfide bonds [2461]. **Partners:** binds heparin [2461]. Does not bind N [2460, 4601, 4823] but associates with it in a complex [3456], so a third protein must serve as a bridge to attach them.

scute (*sc*, 1B1–7, AS-C, loss of MCs from scutum – i.e., dorsal thorax). **Pathway:** same proneural pathway as *da* [949]. DS(−) of *emc*: *sc* is upregulated in emc^{LOF} [912, 3982] (see *emc*). DS(−) of *H*, *N*, and E(spl)-C (see *ac*). US(+) of *Dl* [1672, 1854] (but see *Dl*), *Brd*, *sca*, and E(spl)-C genes *m4*, *m7*, and *m8* [2885, 3974]. **PROTEIN: Function:** proneural transcription factor (bHLH type; 345 a.a.) [4485]. **Location:** nucleus in PNC cells, more in SOPs [912]. **PHENOTYPES: Null:** removes certain MCs [1538]. Compound with *ac*[null] removes nearly all bristles (see *ac*). **LOF:** removes allele-specific subsets of MCs [1453, 1538] and chemosensory bristles on legs [1802]. **GOF:** extra MCs [912, 918, 2038]. **Domains:** bHLH, C-terminal acidic region [4485], which could activate transcription [2878]. **Partners:** binds certain "E box" DNA sequences as a Sc/Da heterodimer (rather than binding itself or as Ac/Sc) [588, 589, 918, 2039, 4452, 4453]. Also binds Emc and several E(spl)-C bHLH proteins but not M8 or Hairy [589, 3430, 4452]. Sc binds Pannier and Chip (as a Sc/Da dimer) in a multimeric complex that may bridge ~30-kb gaps between E boxes and regional *cis*-enhancers at the AS-C [3504].

senseless (*sens*, 85D, loss of sensory neurons). **Pathway:** same proneural pathway as *da* and the AS-C [3127], although its US vs. DS rank is ambiguous. **PHENOTYPES:** a.k.a. *Lyra*, whose elimination of wing margin [2, 4] is attributable to a GOF (ectopic) activation of *sens* [3127]. **LOF:** balding [3127] due to interference with SOP survival or differentiation (or both). **GOF:** extra bristles within and outside regions that normally contain bristles [3127]. **PROTEIN: Function:** activator of proneural genes and mediator of cross-talk with the Notch pathway (541 a.a.) [3127]. **Location:** nuclear in SOPs of the PNS but wanes in SOP descendants [3127]. Sens is manifest after proneural gene products accumulate but before the pan-neural marker *A101* (*neu*) is detectable. **Domains:** zinc finger (C2H2 type, 4x), NLS [3127]. **Partners:** binds the nucleotide sequence <u>TAAATCAC</u>, which is found upstream of three of Sens's supposed target genes (*ac*, *sc*, and *m8*) [3127].

shibire (*shi*, 14A, means "paralyzed" in Japanese). **Pathway:** Notch pathway: phenotypes and sensitive periods of $shi^{LOF} \approx N^{LOF}$ [1803, 3425, 3863, 3891], and shi^{LOF} N^{LOF} double mutant is more affected than single mutants [3863], but unclear whether *shi* is US or DS of *N* [3863]. **PHENOTYPES: LOF:** extra bristles (tufts at MC sites and high-density mCs) [1802, 1803, 3425]. **GOF:** unknown [2561]. **PROTEIN: Function:** endocytosis of Notch receptor [3271, 3863], but it is unclear is why endocytosis is needed for Notch signaling [2320]. Two alternately spliced isoforms: 836 and 883 a.a. [740, 4443]. **Location:** cytoplasmic (free) or submembrane (clathrin-coated pits) [4549]. **Domains/partners:** see Appendix 3.

Suppressor of Hairless (*Su(H)*, 35B9–10). **Pathway:** Notch pathway. Chimeric construct of Su(H) plus VP16 activation domain mimics N's effects [871]. Antagonistic to *H*, but epistatic ranking (US vs. DS) is unclear: double mutant of H^{LOF} and $Su(H)^{null}$ has mixture of H^{LOF} and $Su(H)^{null}$ traits [3827]. **PHENOTYPES: Null:** extra MCs [3820] due to extra SOPs per PNC [3826]. **GOF:** balding due to failure of SOP initiation [3827]. **PROTEIN: Function:** transcription factor (594 a.a.) [1131, 3826]: co-activator with N [1329, 2211] but repressor sans N [1330, 2129, 2941]. **Location:** nucleus in PNCs, less in background epidermal cells [1448]. Equal amounts in SOP vs. non-SOP PNC cells [1448]. **Domains/partners:** see Appendix 3.

Signal Transduction Pathways: Hedgehog, Decapentaplegic, and Wingless

Three of the 5 chief transduction pathways used by discs are outlined below and diagramed in Figure 5.6. The other two cardinal pathways – Notch and EGFR – are discussed in Chapters 2 (Fig. 2.2) and 6 (Fig. 6.12), respectively. Abbreviations: "MF" (morphogenetic furrow), "vh" (vertebrate homolog).

1. Hedgehog signaling pathway [1554, 1974, 2075, 2791, 3685] (see [2832] for overview). Although the agents below constitute the standard version of this pathway, notable deviations have been found [1554, 3077, 3176, 3739, 4282]. However, the heretical proposal that Hh controls target genes in Bolwig's organ without employing Cubitus interruptus [4216] has been refuted [2834]. Aside from the components below, the zinc-finger protein Combgap regulates the levels of Cubitus interruptus in leg, wing, and eye discs [621, 4217] but acts in the Wingless pathway in optic lobes [4017]. Another possible player is *oroshigane* (unknown role), which acts upstream of *patched* [1171]. Other genes have been implicated [937, 1683, 2086] (e.g., *shifted* [2858]), but how their products fit into the chain remains to be determined.

1.A. Hedgehog (Hh, named for the lawn of spiky denticles in LOF embryos [1974, 2893, 3151]) is a 471-a.a. polypeptide (nascent form) [2467, 2895, 4254] that cleaves itself (between residues 257 and 258) into C- and N-terminal moieties [2466, 3338, 3433, 4246]: "Hh_C" bears the catalytic active site [2466], whereas "Hh_N" is the signaling fragment [1231, 3433]. Hh_C is soluble. So is Hh_N, but to a lesser extent because the cholesterol tail that it acquires during the autoproteolysis reaction [255, 2705, 3432, 3434] confers an affinity for membranes, Patched, or both [255, 572, 1973, 4275]. Hh_N also has to be acylated on an N-terminal cysteine

to be active [2464]. Diffusion of Hh_N is impeded by extracellular heparin [2466], and Hh_N requires a specific subset of proteoglycans (processed via Tout-velu and Sulfateless [277, 4275, 4375]) in order to traverse cells. Without its cholesterol tail, Hh_N escapes this need and travels 5 times farther but is less able to turn ON target genes [572, 4275]. When Hh is artificially tethered to cell membranes, it stills convey signals but only to adjacent cells [572, 4136]. Unlike vertebrates, the fly has no *hh* paralogs [3674]. In thoracic discs, *hh* is expressed exclusively in the P compartment [2467, 2832, 4227, 4228, 4254].

1.B. Patched (Ptc, name based on denticle pattern) is a 1286-a.a. transmembrane protein [644, 1895, 3036] that serves as the canonical Hh receptor [754, 755, 2691], although other Hh receptors may exist [3506]. Hh's cholesterol tail is evidently bound by 5 of Ptc's 12 transmembrane spans that comprise its sterol-sensing domain (SSD) [255, 572]. This domain also seen in Dispatched, a protein that interacts with Hh's tail to enable Hh to detach from the cells that make it [572]. Ptc's binding and signaling functions are likely mediated by separate domains [255] because mutations can block signaling without affecting binding [754], and this effect can also be achieved by C-terminal truncation [2073]. This uncoupling is shown by mutations in Ptc's SSD that derepress Smo without altering Ptc's ability to sequester Hh [4174]. Such mutations interfere with the "Ptc ⊣ Smo" link (see below) by altering the endocytic trafficking of Ptc [2711]. Ptc is an unusual receptor insofar as (1) it actively represses the pathway in the absence of ligand [1982, 3487] and (2) it negatively controls its own expression [1828, 1973, 2072, 4136]. Transcription of *ptc* is confined to the A compartment

[3372]. In P cells, *en* ⊣ *ci* so that Ci cannot activate *ptc* [65, 1647, 1828, 4229]. Ptc is upregulated in border cells that receive Hh [642, 754, 1841, 1982, 2074]. The purpose of this upregulation was thought to be the boosting of responses to Hh, and responses are indeed sensitive to relative levels of Hh vs. Ptc [2072]. However, the "Hh → Ptc" link serves mainly to sequester Hh and prevent its long-range diffusion [277, 572, 754, 755, 2074]. In the absence of Hh, Ptc inhibits Smo in a stable receptor complex [4130] (but see [2073]), although Ptc has a high turnover rate [2072]. Whether the "Ptc ⊣ Smo" link is direct or indirect remains unclear [1022]. When Hh binds Ptc, the Ptc-Smo complex changes shape [1973, 1980] so that Smo is no longer repressed [755, 2074]. This derepression is correlated with (1) an increase in Smo phosphorylation, and (2) a dissociation of the Ptc-Smo complex, whereby Ptc and Hh enter the endocytic pathway (for lysosomal destruction?) and Smo stays at the surface [1022, 3506] (but see [4881]).

1.C. Smoothened (Smo, named after embryo defect) is a 1028-a.a. transmembrane (7-span) protein [61, 4441] that functions downstream of Ptc [61, 754, 1264, 1894, 2993, 3487]. Because of its resemblance to G protein-coupled receptors, Smo was initially thought to be the Hh receptor [61, 4441], but subsequent work showed that Hh physically interacts with Ptc instead [754, 755]. The evidence implies that Ptc and Smo interact as receptor (Ptc) and co-receptor (Smo), although Smo can activate the pathway by itself [1022, 1974]. Ptc appears to regulate Smo catalytically (vs. stoichiometrically) [1980, 2117], but how Smo relays the signal is unclear [2116, 2117]. Smo may be a Frizzled-type receptor that has evolved the ability to activate itself without a ligand [755]. No *bona fide* G proteins have yet been implicated [2074] (but see [1893]). Smo is expressed in both A and P compartments of wing discs [2832] along basolateral cell surfaces [1022]. Smo appears to be degraded by a PKA-dependent route that is blocked in the presence of Hh [62], and in this way Hh evidently amplifies the perceived volume of its signal [2117].

1.D. Costal2 (Cos2, named for anterior part of the wing that duplicates [642, 1596, 3959, 4642]) is a 1201-a.a. protein that has conserved motifs of the kinesin heavy-chain (KHC) motor proteins [3612, 3976] – including domains for microtubule-binding, ATP-binding, and homodimer formation (heptad repeats). Like KHC, Cos2 binds microtubules but, unlike KHC, binding is aborted not by ATP but by receipt of a Hh signal [3612, 3976]. Genetically, *cos2* acts upstream of *ci* [1264, 3976, 4539]. Cos2 directly binds Ci [4068, 4536] and persists in binding Ci even after microtubules dissolve [3976], so its function may be to simply

restrain Ci from entering the nucleus by tethering it to the cytoskeleton [3612, 3685, 3976] in the same way that Numb is supposed to tether Notch to the membrane (cf. Fig. 2.2). Consistent with this "**Handcuff Scenario,**" Cos2 interacts with Ci in a saturable, stoichiometric way [4540], and cytoplasmic retention is abolished by deleting the C-terminal domain of Ci that binds Cos2 (a.a. #941–1065) [4536]. In wing discs, *cos2* mRNA is uniform, but Cos2 accumulates (in a Hh-independent way) in the A compartment [3976]. Cos2 is phosphorylated when cells receive a Hh signal, but whether this event is causally related to the release of the complex from microtubules is not known [3612]. Also unknown is the kinase [2075], which (based on mutational dissection) cannot be Fused – thus precluding one obvious scenario (viz., Fused phosphorylates Cos2 so that it cannot bind microtubules [1974, 4536]). Releasing a transducer from microtubules may also be a step in the Dpp pathway [1084].

1.E. Fused (Fu, inaccurately named for fused wing veins [1198]) is an 805-a.a. protein with the sequence hallmarks of a serine-threonine kinase [365, 3460, 3920, 4280, 4281]. Its catalytic domain is needed for signal relay [3461, 4280], but its substrates are unknown [2906, 4536]. Fu homodimerizes [116] and affiliates with the Cos2-Ci complex [4068]. Like Cos2, it is phosphorylated [3506, 4283] when the complex detaches from microtubules in response to Hh [3612, 4283]. Fu and Cos2 can enter the nucleus, but their entry is not regulated by Hh [2833]. Transcription of *fu* appears uniform in leg and wing discs [4281], but Fu is more intense in the A compartment [78]. Genetically, *fu* acts upstream of *ci* as a positive regulator [78, 1264, 2967, 3341, 3991], but the effect of Fu on Ci may occur by a separate route from either Cos2 [1264, 3461, 3739] or PKA [1554, 1701, 1974, 2074, 3739]. Paradoxically, *fu*^LOF inhibits Hh signaling while increasing the amount of detectable Ci-155, although reports differ as to whether the excess Ci-155 arises at the expense of Ci-75 [3177, 4539]. Fu might be converting Ci-155 to a form that is highly active but labile and hence less detectable [3177, 4536, 4809], although the increase could be a trivial side effect of less Fu → lower response to Hh → less Ptc in border zone → greater diffusion range of Hh → more cells expressing Ci-155 [4539]. Alternatively, Fu may just be the key that unlocks the handcuffs in the Handcuff Scenario [2479, 2833]. The Fu-Ci stoichiometry is consistent with the latter idea [4540], as is the otherwise odd fact that *fu*'s LOF and GOF phenotypes are similar [116].

1.F. Suppressor of fused (Su(fu)) is a 468-a.a. protein whose only obvious motif is a PEST sequence [3368], although even that motif may be dispensable for the protein's function [1015]. Su(fu) was thought to serve as a bridge between Fu and Ci [2906], but Fu and Ci directly bind each other [4068]. The rigid Fu:Su(fu) stoichiometry explains the dose sensitivity of *fu* relative to *Su(fu)* alleles [3368, 3459, 4280] as well as similar titrations with *cos2* [3461]. Like Fu, Su(fu) exerts effects on Ci-155 levels that make no sense unless an unstable "matured" form of Ci-155 exists [3177]. Some sort of maturation must occur because an uncleavable Ci-155 construct cannot enter the nucleus [2833, 4536] or upregulate *ptc* [2832] without the Hh pathway being active. Dose-sensitive interactions of *Su(fu)* with *ci* suggest that Su(fu) physically blocks Ci maturation [3177]. Amazingly, Fu and Su(fu) can both be removed from the transduction machinery without altering the wild-type phenotype [78, 3459], so Ci-155 cannot require catalytic alteration by Fu [2832]. Ci's maturation might just involve (1) avoiding PKA-mediated phosphorylation [749, 4538] or (2) allowing phosphorylation but then escaping proteolysis by rapidly entering the nucleus [4540]. In either case, maturation appears to require that all Su(fu) be "cleaned off" from Ci before Ci enters the nucleus [2833, 4068, 4536]. Otherwise, the Su(fu)-bound Ci complex cannot activate transcription at high levels. Oddly, Cos2 (once thought to only antagonize Hh signaling) assists this cleansing (hence abetting Hh) [4536] in some way that does not entail competitive binding (because Cos2 and Su(fu) bind different parts of the Ci protein).

1.G. PKA is cAMP-dependent Protein kinase A [3920], and DC0 (0 = zero, not the letter O) is its catalytic subunit [2118, 2403, 2822]. LOF alleles of *DC0* mimic *hh*[GOF] phenotypes in discs [2058, 2059, 2491, 2533, 3238], plus Hh's effect on Ci [2072] (but see [2214]). Double mutants show that *DC0* acts downstream of *smo* [754, 755, 4441], but no direct Smo-PKA link is known [751, 1554, 1701, 2214]. PKA inhibits Hh signaling by phosphorylating Ci [731, 749, 751, 3466, 4539] (in addition to other kinases [731]), although, ironically, PKA may also be needed for Ci to mature into its active form [4540] (but see [4538]). Phosphorylation marks Ci for proteolytic cleavage (and inactivation) by the ubiquitin pathway as gated by Slimb [2668, 4538]. Hh signaling reduces Ci phosphorylation [731], but this effect could be mediated by a phosphatase acting in parallel with PKA [731, 4538]. A PKA-anchoring protein has been identified in discs [1704, 4470] that could (1) put PKA near Smo [3685] or (2) attach PKA to the same microtubule sites as the Cos2-Fu-Su(fu)-Ci complex [3685]. The latter idea seems to obviate the need to postulate ≥2 active Ci isoforms and would explain why Cos2 and PKA do not fit a linear pathway

[3176, 4538]. To wit, *cos2*[LOF] mutations might block Hh signaling by undocking the Fu-Su(fu)-Ci remnant from a separately anchored PKA. Similar reasoning might also explain why deleting both Su(fu) and Fu yields a wild-type phenotype [1974]. The ability of excess Ci to bypass inhibition by PKA [65] is also consistent with the notion that PKA cannot access Ci freely via 2nd-order kinetics. The basic idea can be visualized as a "**Cyclops Scenario**," wherein a Cyclops (PKA) cannot eat (phosphorylate and hence proteolyze) Odysseus (Ci) unless he is a captive (held by Cos2-Fu-Su(fu) handcuffs) in a cave (microtubule network) [2833, 4537]. (PKA's phosphorylation of Ci per se has no impact on nuclear import [4540].) This scenario could be true even if ≥2 active Ci isoforms exist. Scaffold-dependent phosphorylation, as seen here, may be used elsewhere to insulate signaling pathways from one another [3454]. *N.B.:* One snag in trying to analyze this system is that each cog probably binds several other cogs [2906, 4068], thus confounding any attempt to chart a linear chain of epistases via genetics [2479]. A similar mess has been encountered with a protein complex (Eyes absent, Dachshund, Sine oculis) that switches ON eye development (cf. Fig. 7.3 and Ch. 8) [553, 743, 3387].

1.H. Slimb ("**S**upernumerary <u>limb</u>") is a 510-a.a. protein [2060, 4279] whose F-domain and WD40 motifs allow it to form a bridge between ubiquitin ligases and proteins that have been tagged (e.g., by phosphorylation) for ubiquitination [791, 2668, 4031]. Ci appears to be such a protein [4538], although its proteasome-dependent cleavage [731] appears to occur sans ubiquitination [731] or further degradation [1974]. Epistasis experiments place *slimb* downstream of, or parallel to, *DC0*, and upstream of *ci* [4538]. Slimb is also used by the Wg pathway (see below) [2060, 2856, 4279]. The rationale for Slimb's rapid turnover of activators like Ci and Arm is probably that it allows quick responses [4294]. Such speed is undoubtedly useful in the eye where ommatidia are spawned by a "forest fire" that flashes across the disc, although it may be useless in leg and wing discs where cell states change at a more languid pace.

1.I. Cubitus interruptus is named for a gap in the cubital (4th) wing vein [4092, 4094, 4106, 4107]. Formerly called Ci[D] [3818], Ci is a 1396-a.a. protein [2832] in the Gli family of transcription factors [3686] that bind the core consensus <u>TGGGTGGTC</u> [2980, 4503] (but see [154]). In thoracic discs, *ci* mRNA is found only in the A compartment, where it is uniform [1135, 2967, 3177, 3818, 3991]. Full-length Ci (155 kD) predominates near the A/P boundary [65, 2967, 3177, 3991] in a band ~7–9 cells wide [1828, 4539, 4540]. More anteriorly (beyond the range of Hh), Ci is cleaved (somewhere in the a.a. #650–700 interval [155]; numbering as per [2832]) through Slimb-mediated [4538], proteasome-dependent [4539] proteolysis to yield a 75-kD N-terminal fragment [155] (Fig. 5.7). Ci-155 and Ci-75 function respectively as transcriptional activator vs. repressor [155, 731, 2832]. Ci-155 is mainly cytoplasmic due to C-terminal sections (including a.a. #703–836) that counteract an NLS (a.a. #596–614) in Ci-75 [4539]. Because Ci's DNA-binding domain of 5 tandem zinc fingers (Cys_2His_2 class; a.a. #453–603) [65, 3194, 3292] is N-terminal to the cleavage site, both Ci-155 and Ci-75 can bind DNA and hence compete for target promoters [154, 751, 2980, 3685]. However, Ci-155 and Ci-75 have distinct (overlapping) sets of target genes [2832], so other factors must edit their specificities [2980] (e.g., see [1686, 1827]). Domains that are C-terminal to the cleavage site (and hence cropped from Ci-75) include (1) a nuclear export signal (a.a. #675–860) [731, 4539] initially thought to be a cytoplasmic tether site [155], (2) 5 PKA phosphorylation sites [3466, 4539], (3) an acidic section that activates transcription [65, 1828], and (4) a binding domain (a.a. #1020–1160) that recruits the co-activator dCBP [54, 752]. Unlike Ci-155, Ci-75 is a repressor because it lacks the latter two domains, while retaining an alanine-rich "repressor" section (a.a. #1–453) [65]. Until recently, the genetics of *ci* was muddled by the neomorphic ci^D allele [154, 2832, 3191, 3818]. That mutation turns out to involve a quirky fusion of *ci* with its neighboring gene *pan* [3831, 4502]. In what may be the ultimate example of crossed-wire "short circuiting" (Inv^D is a contender [3934]), the chimeric Ci^D protein (Ci-Pan) regulates Hh target genes (via Ci's zinc fingers) in response to Wg (via Pan's Arm-binding domain; cf. Wg pathway). Why on earth the effector genes for the Hh and Wg pathways (*ci* and *pan*) should abut one other (~10 kb apart) in the wild-type fly [519, 1088, 4439] is baffling [3831]. It is hard to believe that this juxtaposition is an evolutionary accident given the proximity of Ci and Pan binding sites at jointly regulated promoters [3386] (cf. another peculiar case [671]). From a developmental standpoint, the riddle is why ci^D's effects are so mild in heterozygous condition. Ci may not be the only transcriptional effector for Hh [2494] (but see [4540]).

1.J. _Drosophila_ _C_REB-_binding_ _p_rotein (dCBP, where the "CREB" bZip transcription factor is the _c_AMP _r_esponse _e_lement _b_inding protein [88, 3879, 4393, 4817]) is a gigantic 3190-a.a. (332-kD) protein (a.k.a. Nejire) [54] that binds Ci (as a co-activator) and increases (by ~60-fold) Ci's activation of target gene transcription [54]. The ability of dCBP to also bind phosphorylated dCREB2 [54]

means that PKA might have an indirect effect on Ci (besides its direct effect [3466]). Namely, by phosphorylating dCREB2, PKA could cause it to "steal" dCBP from Ci if the amount of dCBP is limiting [54, 100, 2023]. CBP proteins stimulate transcription [3279] by acetylating histones [202, 1553, 2726, 3174, 3334] (as do their associated factors [2489, 2911, 3998, 4518, 4718]), and this is presumably the role of dCBP as a co-activator in the Hh pathway. In the Wg pathway, however, dCBP plays an inhibitory role (see below) by acetylating a transcription factor (Pan) so that it can no longer bind a co-activator (Arm) [4533]. This involvement of dCBP in multiple pathways is consistent with CBP's ability to integrate diverse inputs [705, 1506, 1553, 2023, 2124] by dint of its huge size [2024].

1.K. Target genes include the following. Some of these links have been confirmed in cell culture [65, 731, 4503], although in most cases it is unclear whether the effect is direct or indirect. A likely target in the wing pouch (turned OFF by Hh) is *Egfr* [1643, 4604], and an indirect target (via *en*) is *polyhomeotic* [2728, 2729].

1. Hh ⊣ *hh* itself, which is kept OFF in A cells of leg and wing discs by Ci-75 (when no Hh signal is received), although *hh* is unresponsive to Ci-155 [2832]$^\Delta$. An autoregulatory loop of this sort also operates in the developing eye [1786, 3238].

2. Hh → *ptc*, whose low-level transcription in A cells is upregulated by Hh near the A/P boundary [2834]$^\Delta$ via binding of Ci-155 to the *ptc* promoter [731]$^\Delta$ but whose lower level in remaining A cells is not due to Ci-75 [2832]$^\Delta$.

3. Hh → *en*, which is turned ON by Ci-155 in A cells near the A/P boundary [2834]$^\Delta$ during late 3rd instar [78]$^\Delta$ (cf. similar effects in ptc^{LOF} A clones [3177]$^\Delta$). This incursion of the *en*-ON state into the A compartment is stopped by fu^{LOF} [1537, 3739].

4. Hh → *dpp*, which is turned ON by Ci-155 along the A/P border in wing discs [65, 754]$^\Delta$ by binding of Ci-155 to a *dpp* enhancer [2980]$^\Delta$, whereas Ci-75 reduces *dpp*'s basal (default) transcription level to zero beyond the range of Hh [2834]$^\Delta$. Likewise, *dpp* is turned ON dorsally in leg discs [3739]$^\Delta$, turned ON ventrally in antennal anlagen [697]$^\Delta$, and turned ON in the MF of the eye disc before [406]$^\Delta$ and during [4387]$^\Delta$ retinal differentiation except at the MF tips [406]$^\Delta$.

5. Hh ⊣ *tkv* near the A/P border in wing discs [4251] via *master of thickveins* (*mtv*): Hh → *mtv* ⊣ *tkv* [1327].

6. Hh → *vein* (as a direct target) in the wing pouch [4604], as well as in the dorsal head and MF [80].

7. Hh ➡ *knot* (a.k.a. *collier*) in the wing pouch [2894]$^\Delta$ (cf. *roadkill* [4502]).

8. Hh ➡ *blistered* (via Knot [4479]) in the wing pouch [3145].

9. Hh ➡ {*araucan* AND *caupolican*} in the wing pouch [2993] IF a high level of Dpp is also present [1537].

10. Hh ➡ *wg* ventrally in leg discs [231, 2466], dorsally in antennal anlagen [697]$^\Delta$, and anteriorly in the eye rudiment [1781]$^\Delta$, but Hh ⊣ *wg* in the dorsal head [3664].

11. Hh ➡ *hairy* anterior to the A/P boundary in leg discs [1778] and anterior to the MF in eye discs [4387]$^\Delta$, but Hh ⊣ *hairy* within the MF itself (a threshold effect?) [1616].

12. Hh ➡ *atonal* in the MF during its progression [725]$^\Delta$ and at its inception [406], but high levels of Hh ⊣ *atonal* at the MF's rear [1077]. A "Hh ➡ *atonal*" link also operates in the developing antenna [2055] and larval eye organ [4216], but in the latter case neither Fu nor Ci is needed for Hh transduction.

13. Hh ➡ *orthodenticle* in the eye disc [3664].

14. Hh ➡ *glass* in the developing eye [2632].

15. Hh ➡ *rough* in the developing eye [1077].

16. Hh ➡ *scabrous* in the developing eye [3238]$^\Delta$.

17. Hh ➡ *shortsighted* ahead of the MF in the eye disc [4388].

1.L. Summary of the basic chain: Hh ⊣ Ptc ⊣ Smo ➡ Fu ➡ Ci ➡ {target genes}. In the absence of Hh, Ptc keeps the pathway OFF by repressing Smo. Binding of Hh to Ptc alters Ptc so that Smo is freed to (1) stimulate Fu, which somehow enables Ci to enter the nucleus and activate target genes; (2) liberate the Ci-Cos2-Fu-Su(fu) complex from its microtubule tether, which also helps release Ci for nuclear transit; and (3) prevent PKA from inactivating Ci by phosphorylation and subsequent cleavage. How Hh represses target genes is unknown.

2. Decapentaplegic signaling pathway [2733] (see [3498] for a primer.) Aside from the components below, other candidates include *crossveinless 2* [854], *lilliputian* [3097, 4195], *Merlin* [2776], *expanded* [2776], *shortsighted* [4388], *vrille* [1434], p38 mitogen-activated protein kinase [13], and unidentified genes at 54F–55A and 66B–C [3114] and elsewhere [1939, 3034]. Based on what is known about the vertebrate TGF-β pathway, (1) Protein Phosphatase 2A is likely downstream of (and phosphorylated by) Tkv [1627], and (2) microtubules may sequester Mad [1084]. Also, vertebrate Mad is escorted to the receptor complex (≈ Punt-Tkv) by SARA (Smad Anchor for Receptor Activation) [138, 3221, 3897, 4403], and flies may use the same trick because they have an ortholog (dSara) [288, 1172]. Certain components of the Dpp pathway are replaced by others during embryogenesis (e.g., see [88, 2990, 3586]), oogenesis (e.g., see [4703]), and patterning of the larval midgut (e.g., see [4203, 4226]).

2.A. Decapentaplegic (Dpp) is so named because at least 15 ("decapenta-") of the 19 imaginal discs are incapacitated (-plegic) by *dpp* mutations [4033]. Despite the presence of 5 differently spliced *dpp* transcripts [4056], there is only one Dpp protein [3096]. Like other members of the TGF-β family, Dpp (BMP4 subfamily [2733, 2735]) is translated as a precursor (588 a.a.) [3224] and cleaved into (1) an N-terminal piece (Dpp$_N$, a.a. #1–456) that fosters dimerization and secretion and (2) a C-terminal ligand (Dpp$_C$, a.a. #457–588) for intercellular signaling [1427, 3244]. The ligand has a consensus N-glycosylation site [1427, 3585] and 7 conserved cysteines [3096]. Six of the cysteines join pairwise in disulfide bridges to weave the TGF-β "knot" [2734], and the 7th links the Dpp monomers [4608]. The protease that cleaves the precursor was thought to be a BMP1 relative [2696] but is probably a subtilisin cousin [1232]. Aside from Dpp, 5 other TGF-β family members have been characterized: dActivin [2369, 3222], Glass bottom boat (Gbb, a.k.a. 60A) [1068, 4614], Maverick [3110], Myoglianin [2578], and Screw [109] (cf. EST DS07149 [3498] and [3674]). Among these paralogs, only Gbb is known to affect Dpp signaling in discs, but its influence is slight and limited mainly to wing veins [1674, 2203, 4610]. Historically, *dpp* genetics was complicated because most *dpp* mutations disjoin *cis*-enhancers from the coding region [1426, 4056, 4608] (cf. the *achaete-scute* Complex, Ch. 3). The locus spans ~55 kb [3098, 4056] and is organized into (1) an ~8 kb "Hin" (Haplo-insufficient) coding section [1428, 1987] with enhancers for embryonic expression [1872, 2003, 3833]; (2) a 5′ "Shortvein" region where mutations cause defects in wing veins, larval midgut, and other structures [1952, 2738, 3850]; and (3) a 3′ "Disk" region where enhancers drive expression in various parts of discs [347, 2739, 4033].

2.B. Punt (a.k.a. Activin Receptor Type II; name based on embryo phenotype) is a transmembrane (1-pass) receptor (516 a.a.) with a cysteine-rich extracellular domain and an intracellular kinase domain [772, 2495, 3668]. Punt appears to be the only "type II" receptor in *D. melanogaster* [3932], although there are 3 "type I" receptors (Thick veins, Saxophone, and Baboon) [511, 2495, 4754]. Type-I and -II proteins form hetero-oligomeric receptor complexes in flies and vertebrates [4267, 4472, 4585, 4752] – a mix-and-match strategy that accommodates diverse ligands [2732]. Both kinds of proteins are serine-threonine kinases (type II are constitutive; type I are ligand dependent) [4267, 4753], although their catalytic domains resemble receptor tyrosine kinases (RTKs) more than S-T kinases [1711, 1953], so TGF-β receptors may have evolved from RTKs or vice versa [3221]. Indeed, type-II receptors were recently found to naturally autophosphorylate on tyrosines [2418] as well as serines [2619], and like RTKs they assemble into a complex via ligand binding [505, 2731, 4753]. Dpp dimers bind type-I (Tkv) and type-II (Punt) proteins cooperatively (vs. sequentially) [2495, 2733, 3328, 4609], and there are ≥2 Punt monomers in the complex [3932].

2.C. Thick veins (Tkv, named for its LOF effect on wing veins [980, 995, 3328, 4189, 4271]) is a type-I transmembrane (1-pass) receptor (563 a.a.) [512, 3073, 3328] with a cysteine-rich extracellular domain, an intracellular kinase domain, and a juxtamembrane GS (Glycine-Serine-rich) domain that is presumably phosphorylated by Punt in the presence of Dpp [1874]. After it is activated (phosphorylated) by Punt, Tkv puts phosphates on serines in Mad's C-terminal SSxS tail [1983], and Mad then relays the signal to the nucleus. Saxophone (Sax, 570 a.a.) is another type-I receptor [512, 3073, 3328, 4771], but its LOF and GOF effects (mainly on wing veins) are meager [3972]. Sax does not normally serve as a Dpp receptor in discs [1674, 3109, 3498], despite initial suspicions that it transduces Dpp like Tkv [512, 3073, 3668, 3972, 4609]. Sax's ability to boost the Tkv signal [3972] must therefore occur indirectly by downstream cross-talk between Dpp and a TGF-β ligand other than Dpp [3406] – possibly Gbb [3498].

2.D. Dally (Division abnormally delayed) is a putative cell-surface proteoglycan (626 a.a.) [3038] with motifs diagnostic of the "GRIP" subfamily (Glypican-Related Integral-membrane Proteoglycan): (1) an N-terminal signal peptide, (2) a hydrophobic C-terminus needed for glypican-type anchoring to the outer leaflet of the membrane [966], (3) consensus sequences for glycosaminoglycan links, and (4) 14 conserved cysteines of vertebrate GRIPs (cf. type-III "receptors" in the TGF-β family [139]). LOF alleles of *dally* have mild *dpp*-like effects on eye size and wing venation [2556, 3038] (cf. similar vein defects in *sugarless*LOF [1673]), and they enhance *dpp*LOF phenotypes in double mutants [2004]. The ability of *dally*LOF to reduce expression of *dpp* target genes (*omb* and *spalt*) without affecting *dpp* expression itself suggests that *dally* acts downstream of *dpp*, but the ability of extra doses of *dpp*+ to rescue *dally*LOF defects argues that *dally* acts upstream of *dpp* [2004]. The latter is more consistent with Dally's supposed role (based on molecular features) as

a co-receptor for Dpp and Wg on receiving cells [2004, 2556, 3345].

2.E. Mothers against dpp is a 455-a.a. transcription factor [2217, 3852]. The gene name is capitalized because of a dominant maternal (hence "Mothers") effect that *Mad*LOF mutations have on the lethality of *dpp*LOF embryos [3499, 3852]. The name is a pun on the American association "Mothers Against Drunk Driving," but "*against* Dpp" is misleading because Mad actually assists (rather than hinders) signaling (*Mad*LOF enhances *dpp*LOF). Its abbreviation "Mad" is also unfortunate because it can be confused with (1) vertebrate Mad (an unrelated bHLH-Zip transcription factor [1594]) or its fly ortholog dMad [2596, 4820], (2) *mad* (the unrelated fly gene *many abnormal discs*) [2561], or (3) the MADS box (an unrelated DNA-binding domain). Genetically, *Mad* functions downstream of *dpp* [3094] and *tkv* [3095]. Mad is the founder of the Smad family of regulators [137, 138, 174, 4638], which contains 3 subgroups [1815, 2332, 2736]: (1) pathway-specific Smads (e.g., Mad and dSmad2 [511]), (2) common mediators (e.g., Medea), and (3) inhibitory Smads (e.g., Dad). Smad proteins share one or two homology domains with Mad (separated by a proline-rich linker). The MH1 domain binds DNA [2217, 3898], whereas MH2 binds the receptor [756, 2579]. MH2 also mediates oligomerization [2380] and activates transcription [2568] (autonomously and by recruiting CBP [138, 2332, 4534]). Oddly, MH1 inhibits the CBP-binding ability of MH2 [4534], while MH2 inhibits the DNA-binding ability of MH1 [2217]. Evidently, phosphorylation cures Mad's self-defeatism by reshaping the molecule [1815, 2332, 3498, 4751]. Pathway-specific Smads end C-terminally in SSxS (where "x" is variable) and cannot transmit signals unless the serines bear phosphates (but see [2380]). Phosphorylation of Mad's SSxS tail by Tkv [1983, 3095] triggers (1) formation of Mad homo-oligomers, (2) formation of Mad-Medea hetero-oligomers, and (3) entry of the latter complexes into the nucleus [959, 1983, 4703], although Mad can enter sans Medea [959, 4703]. Mad binds the consensus nucleotide sequence GCCGnCGC (where "n" is variable) [2217]. Mad is expressed ubiquitously in wing discs (and presumably other discs) [3095].

2.F. Medea is a 745-a.a. protein [959, 1938, 1983, 4703] (but see [4782]) that belongs to the "common mediator" subgroup of the Smad family [1815]. Like other subgroup members, Medea has MH1 and MH2 (Mad homology) domains and lacks a C-terminal SSxS [1983]. It also has an opa motif that is lacking in its human ortholog Smad4 [1938]. *Medea* was found in the same screen as *Mad* [3499] and was likewise named for its maternal enhancement of *dpp*LOF embryos' lethality. (In Greek mythology, Medea was a sorceress who killed her children.) Genetically, *Medea* acts downstream of *dpp* [4703] and *tkv* [1938] (but see [959]). Medea binds Mad (via Medea's MH2 domain) only when Mad's <u>SSxS</u> serines are phosphorylated [959, 1983, 4703], and this binding is needed for Medea to enter the nucleus [4703]. In the nucleus, Medea is a partner for Mad in DNA binding [4703, 4782], although the binding complex's constitution and configuration are uncertain [1027, 4638]. In contrast to Mad, Dpp signaling does not involve phosphorylation of Medea, and endogenously high levels of Dpp can activate target genes (*omb* in wings) without Medea [4703].

2.G. Daughters against dpp (Dad) is a 568-a.a. Smad protein [4404] whose name parodies Mad. Like other members of the inhibitor subgroup, Dad has MH2 but lacks MH1 and <u>SSxS</u> [3498, 4637]. Genetically, *Dad* acts downstream of *dpp* but upstream of *omb*, whose expression it represses [4404]. This repression is due to Dad's ability to prevent Mad phosphorylation by Tkv [1983] – a blockage that seems steric and stoichiometric because (1) Dad competes with Mad for binding to Tkv (Dad binds more tightly) [1983] and (2) overgrowth caused by excess Mad can be rescued to wild-type by overexpressing Dad coincidently [4404]. Dpp signaling activates *Dad* transcription, which represses Dpp signaling [4404] – ergo, a negative feedback loop. This loop should make cells in the Dad zone (near the Dpp stripe; Fig. 6.2) perceive the Dpp gradient as flatter than it really is [4637, 4638]. How much the perceived and actual slopes differ is hard to guess without knowing affinities and amounts of all the relevant players. The only certainty is that Dad can't be buffering Dpp's signal too strongly because in that case Dpp could not act as a morphogen (cf. Fig. 6.3).

2.H. Brinker (Brk) is a 704-a.a. transcription factor with weak homeodomain homology, 3 opa repeats, 3 possible NLSs, several copies of a motif that recruits the dCtBP co-repressor [619, 2049, 2867], and an <u>FKPY</u> motif that recruits the Groucho co-repressor [4866]. Brk binds (and needs) Groucho more strongly than dCtBP [4866]. Its N-terminal helix-turn-helix (homeo-like) domain binds a specific DNA sequence at *omb* [3978] and other loci [2416, 3696, 4866]. Because *brk*[null] somatic clones express Dpp target genes (*Dad*, *omb*, *spalt*) in areas where they are normally OFF [2049, 2867], Brk's normal duty may be to keep these genes OFF. The ability of *brk*[null] to turn ON *omb* and *spalt* in the absence of Dpp signaling (in *brk*[null] clones that are also *Mad*[null] or *tkv*[null]) argues that Dpp acts via Brk – i.e., Dpp ⊣ *brk* ⊣ {*omb* AND *spalt*} [619, 2049, 2050].

However, *spalt* upregulation does not attain *in situ* levels (nor do structures attain a "Dpp peak" fate as fully as in *tkv*[GOF] clones) [619, 2049], so Brk may act partly in parallel with Dpp [2049]. Additional evidence for downstream (in-series) placement of Brk is the ability of *brk*[LOF] to restore tiny *dpp*[LOF] discs to normal size [619], although here too *spalt* expression is abnormal. Dpp represses *brk* transcription [619, 2049, 2867], so Brk forms a gradient that overlaps and opposes the Dpp gradient. (Its expression complements Dad's [2867]; Fig. 6.2.) Indeed, Brk may help cells distinguish Dpp levels at the shallow end of the Dpp gradient [619, 2049]: if Mad overcomes Brk at a target promoter, then the gene should switch ON, whereas if Brk wins, then the gene should turn OFF [619, 2049, 2233]. Thresholds may thus be sharpened [3347, 3406]. Brk cannot be regulating itself because *brk*[null] clones express a *brk* enhancer-trapped reporter in a normal pattern [2867]. Rather, *brk* is regulated by Schnurri (see below) [1749, 2727]. Brk may also participate in other pathways [2397].

2.I. *Drosophila* CREB-binding protein (dCBP) is a huge 3190-a.a. protein (cf. Hh pathway) [54] that activates transcription when recruited to DNA by DNA-binding proteins [202, 2726, 3174]. Ci is one such recruiter, and so is Mad [127, 4534]. This co-activator role for dCBP in Dpp signaling agrees with the genetic placement of *dCBP* downstream of *sax* [4534].

2.J. Schnurri (Shn) is a 2528-a.a. protein with 7 (or 8 [4059]) zinc fingers, an acidic (activator?) domain, and multiple opa motifs [108]. Until recently, it was unclear where Shn fit with other components of the pathway [3223, 3498], especially because some of its interactions are allele specific [753]. We now know that Shn binds DNA and physically interacts with Mad as a cofactor to affect transcription of target genes [944]. Its chief target appears to be *brinker* (Shn ⊣ *brk*) [1749, 2727] (but see [4372]). Shn is needed for Dpp signaling in wing discs (for A-P patterning and vein formation) [4373] as well as during embryogenesis and midgut formation [569, 1622, 4059]. Shn (with Punt) also transduces signals from a ligand other than Dpp during spermatogenesis [2751].

2.K. Target genes include:

1. Dpp ⊣ *dpp* itself in an analog manner by a Tkv-mediated autocrine loop until a set point is reached (≈ thermostat) in wing discs [1674]. However, Dpp → *dpp* at the tips of the MF in the developing eye [4387]△ and at the anterior margin [3388]. Given Dpp's ability to diffuse, the latter loop should incite the *dpp*-ON state to spread like wildfire. Spreading is indeed

seen wherever the link works [724], but it is normally damped by "Wg ⊣ *dpp*" [4387]$^\Delta$.

2. Dpp ⊣ *tkv* in prospective veins of pupal wing discs [980] but not earlier (in the pouch), except at unnaturally high doses (cf. Ch. 6) [1674, 2457].

3. Dpp ⊣ *brk* within ~20 cells from the Dpp stripe in wing discs [619, 2049, 2867] and leg discs [2049]. Repression of *brk* (in the wing at least) is mediated by Shn [2727]$^\Delta$: Dpp → *shn* ⊣ *brk*.

4. Dpp → *Dad* in the same zone as *brk* [4404].

5. Dpp → *optomotor-blind* (*omb*) in the leg disc [2049]$^\Delta$, in the eye disc [3978], and in the same zone of the wing disc as *Dad* except for the notum [619]$^\Delta$. The leg and eye *cis*-enhancers reside in *omb*'s second intron, whereas the wing enhancer is in a 284-b.p. piece that lies 27 kb upstream of the transcription start site [3978].

6. Dpp → *spalt* in the wing pouch in a narrower band than *omb* [619]$^\Delta$.

7. Dpp → *vestigial* in the wing pouch [619]$^\Delta$.

8. Dpp → {*araucan* AND *caupolican*} in the wing pouch [1537] IF Hh is also present (and transduced to Ci-155).

9. Dpp ⊣ *homothorax* in the wing pouch [157] via *vestigial* (at least in part).

10. Dpp ⊣ *wg* in leg discs [487, 2059, 2082, 2954, 4277]$^\Delta$ and eye discs [1780, 4387]$^\Delta$, but Dpp → *wg* in the notal portion of the wing disc [3763, 3764, 4369] (but see [4368]).

11. Dpp → *pannier* in the notal portion of the wing disc [3764, 4369].

12. Dpp → *u-shaped* in the notal portion of the wing disc [3764, 4369].

13. Dpp → {*achaete* AND *scute*} in the notal portion of the wing disc [4368] via *spalt* (at least in part) [987].

14. Dpp ⊣ *BarH1* in the notal portion of the wing disc [3763].

15. Dpp → *aristaless* in leg and wing discs (at a certain threshold) IF Wg is also present [643]$^\Delta$.

16. Dpp → *dachshund* in leg discs (at a certain threshold) IF Wg is also present [2456], although this same link works without Wg in the eye disc (before initiation of the MF) [930].

17. Dpp → *Distal-less* in leg discs (at a certain threshold) IF Wg is also present [2456]$^\Delta$, and the same is true for antennal anlagen [1037].

18. Dpp → *hairy* in leg discs IF Hh is also present [1778]. The same link operates in eye discs but does not require Hh [1616].

19. Dpp → *eyes absent* in eye discs before initiation of the MF [930].

20. Dpp → *sine oculis* in eye discs before initiation of the MF [930].

21. Dpp → *hh* when *dpp* is misexpressed at the anterior edge of the eye rudiment [406].

22. Dpp ⊣ *eyeless* in eye discs IF Wg is also present [2465].

2.L. Summary of the basic chain: Dpp → Punt/Tkv → Mad → {target genes}. Binding of Dpp to Punt/Tkv causes (1) phosphorylation of Mad by Tkv, (2) entry of activated Mad (complexed with Medea) into the nucleus, and (3) stimulation of target gene transcription by dCBP that is recruited by Mad. How Dpp represses target genes (aside from *brk*) is unknown.

3. Wingless signaling pathway [598, 2720, 4708] (see [3919] for a synopsis and [3411] for mechanics). Aside from the components listed below, *fringe connection* (*frc*) [3855], *gammy legs* (*gam*) [599], *Lobe* [758], *naked* [1603], *nemo* [304, 2872, 4475], *skinhead* [3371], *split ends* (*spen*) [2554], *tartaruga* (*trt*) [2290], and at least 6 other unidentified genes have been implicated [891, 2984], although some of them may affect the pathway indirectly via Arm's secondary (nonsignaling) role in cell junctions [3873]. For example, *arc* encodes a PDZ-domain protein that co-localizes with Arm in junctions [2571], but its LOF phenotype (curved wings) does not reveal any obvious involvement with Wg transduction per se. The Bright-family protein Eyelid (a.k.a. Osa [4468]) inhibits Wg signaling [4389], but whether it belongs in the pathway is unclear. The zinc-finger protein Combgap acts downstream of dAxin to transduce Wg in optic lobes [4017], but it transduces Hh signals in leg, wing, and eye discs [621, 4217]. "*Notum*" antagonizes Wg (at high Wg doses) in a negative feedback loop (Wg → *Notum* ⊣ Wg signaling) [1494] that resembles *Dad*'s effect in the Dpp pathway (Dpp → *Dad* ⊣ Dpp signaling) [4404]. One heretical result in need of further investigation is the apparent ability of Notch to respond to Wg without being cleaved from the membrane [4601, 4603], as well as the attendant notion that Notch can serve as a receptor for Wg [726].

3.A. Wingless (Wg) is a secreted protein [920, 1541, 4442, 4455], which, in nascent form (468 a.a.), contains a signal peptide, 23 conserved Cys residues, and one N-linked glycosylation site [175, 590, 3599]. It is the best characterized of the fly's four analyzed Wnt genes [1150, 1277, 1592, 2315, 3700] – one of which (*DWnt-4*) antagonizes Wg signaling [557, 1482]. Three more Wnts exist in the genome, equaling a total of seven [3674]. Wg can be secreted with or without glycosylation [3561], and both forms are active as ligands, as are nonsecretable Wnt constructs that are tethered to the cell surface [3269]. The degree of glycosylation affects Wg's ability to bind glycosaminoglycans of the extracellular matrix [3561]. That binding is apparently needed for signaling *in vivo* because removal of Porcupine (Porc) – which helps glycosylate Wg before secretion [2110] – causes wg^{LOF}-like phenotypes.

3.B. *Drosophila* Frizzled2 (Dfz2, named for disturbances in bristle polarity) is a 694-a.a. (7-pass) transmembrane receptor for Wg [310, 595, 4867]. Dfz2 is one of 4 Frizzled-family proteins identified in *D. melanogaster* [422, 1971, 3674, 3762, 3977]. The only other well-studied member is Frizzled (Fz) [26, 2088, 2325, 3264], which functions in a "planar cell polarity" (PCP) pathway [2326, 2884, 3912, 4702] that controls the orientation of bristles [1449, 1640, 1810], hairs [22, 846, 4430, 4739, 4839], and photoreceptor clusters [1639, 3912, 4170, 4365, 4873]. The unknown ligand for the PCP pathway is probably another Wnt [310, 421, 2781, 3912, 4573]. The Wg and PCP pathways were originally thought to use distinct receptors (i.e., Dfz2 vs. Fz) [422, 3912, 4867], despite (1) their shared use of Dsh (see 3.E below) [3316], (2) pathway interactions in GOF studies [4365], and (3) redundancy of Fz and Dfz2 under some LOF conditions [311, 312, 2187, 2984] (but see [595, 4867]). However, this idea seemed untenable when a $Dfz2^{null}$ allele became available in 1999, and Fz and Dfz2 were shown to be interchangeable – thus affirming their redundancy [734] (but see [421, 2899]). Nevertheless, in 2000, Wg was indeed found to have a 10-fold greater affinity for Dfz2 than for Fz [3691]. Conceivably, the elusive ligand for planar polarity prefers Fz because accessory factors form a co-receptor complex with it but not with Dfz2 [421, 734]. Transduction of Wg stops in the absence of both Fz and Dfz2 (arguing against other receptors), except in the notum where a third Wnt may preside [678]. Like all Frizzled family members, Dfz2 has a conserved ~120-a.a. extracellular domain (with 10 cysteines at fixed sites) and an idiosyncratic cytoplasmic domain [3203, 4544, 4784], wherein lies part of its specificity [421, 3691]. It is unclear how Dfz2 physically transduces the Wg signal [310, 421, 598, 4708] and relays it to Dsh [422, 2883, 3130, 3919, 4795]. Receptor dimerization is probably involved given what is known about fibroblast growth factor (FGF) [3322, 3403, 3404], which resembles Wg in its use of glycosaminoglycans (GAGs) as co-receptors [3561, 3784]. Indeed, FGF [2555] and Wg receptors [2556, 4400] both rely on Dally [3038], apparently for this purpose. Relay of signal from Dfz2 to Dsh may involve Wg-dependent recruitment of Dsh to Dfz2 (via binding of Dfz2's C-terminus by Dsh's PDZ domain) and subsequent activation of Dsh by a membrane-linked kinase (Casein kinase 2?) [3344].

3.C. Dally (Division abnormally delayed) is a putative cell-surface proteoglycan (626 a.a.) [3038, 3345] (see Dpp pathway above). Dally's GAG moiety is synthesized via Sugarless (Sgl, a UDP-glucose dehydrogenase) [338, 921, 1664, 1673] and Sulfateless (Sfl, a heparan sulfate deacetylase/sulphotransferase) [2556, 4400]. Indeed, LOF mutations in all three genes (*dally*, *sgl*, and *sfl*) cause wg^{LOF}-like phenotypes. Dally is probably a co-receptor for Wg (i.e., part of a complex with Wg and Dfz2) but could instead just be helping Wg reach Dfz2 by holding Wg at the membrane [2556]. In embryos, a protein resembling Dally is expressed in D-V ectodermal stripes adjacent to *dally*-ON stripes [2204]. The gene *dally-like* (*dlp*) acts differently from *dally* because its GOF phenotype mimics *dally*'s LOF phenotype in wings [166]. Like Dally, Dlp is a glypican (765 a.a.), but instead of helping Wg reach its receptor, Dlp sequesters Wg and reduces its potency as a ligand [166].

3.D. Arrow (Arr, named for segment polarity defects [3152]) is a one-pass transmembrane protein (1678 a.a.) with a signal peptide, 4 EGF-like repeats, and 3 "LDLR" (low-density lipoprotein receptor) repeats [4570]. It acts downstream of Wg and upstream of Dsh [3607, 4570, 4573]. Unlike *dally*, *sgl*, and *sfl*, the LOF effects of *arr* cannot be suppressed by excess Wg, so Arr is probably not functioning in presenting the Wg signal [4570]. The identity of arr^{null} and wg^{null} phenotypes argues that *arr* is squarely in the transduction chain (not in a parallel or redundant pathway) [4570], but exactly where and how Arr might interact with Dfz2 is unclear [4571]. The best guess at present is that Arr and Dfz2 cooperate to form a receptor complex [270, 3241, 4242, 4570].

3.E. Dishevelled (Dsh, named for disturbances in bristle polarity) is a 623-a.a. cytoplasmic protein [4795] with three conserved domains: PDZ (a scaffolding motif; cf. App. 1), DIX (Dishevelled-Axin), and DEP (Dishevelled-EGL-10-Pleckstrin) [422, 2262, 4278]. The DIX and PDZ domains are crucial for Wg signaling (and subcellular localization [152]), whereas the DEP and PDZ domains are involved in the PCP pathway [422, 423, 2238, 2883, 4795] (but see [152]). Downstream of the DEP-PDZ branchpoint, the PCP signal is transduced by dRac1 (GTPase) [1132, 1136, 1195], RhoA (GTPase) [4170], Misshapen (kinase) [3261], JNK (Jun N-terminal Kinase) [423, 2781], and dJun (leucine-zipper transcription factor) [4566]. Other nodes (of uncertain linkage [2883, 4173, 4566]) in the PCP chain include Dachsous [21], Expanded [367], Prickle (a.k.a. Spiny Legs [1641, 1795, 2884]), Starry Night (a.k.a. Flamingo) [704, 4430], and Van Gogh (a.k.a. Strabismus) [25, 4263, 4716]. (Apical-basal polarity uses an entirely separate circuit [2983].) Dsh also affects the Notch pathway by binding Notch directly [151, 356]. Dsh is hyperphosphorylated by Casein kinase 2 in response to Wg [3149, 3683, 4365, 4676, 4795], although reports differ as to whether phosphorylation alone is sufficient to relay the signal [422, 3683, 4676]. Genetically, *dsh* functions upstream of *sgg* as a negative regulator [882, 1833, 3683, 3924]. Dsh lacks an obvious catalytic domain and seems to act rather as an Axin-binding adaptor that recruits Sgg-inhibiting agents to the Axin-Sgg-Arm complex [3316, 3994]. Overexpression of Dsh (or Wg or Dfz2) inactivates Sgg by causing it to be serine-phosphorylated [3683], apparently via Protein kinase C [859, 3889]. Dsh and Arm both help pattern the V side of the leg, but only Arm is essential for cell viability there [2262, 4278].

3.F. Shaggy (Sgg, named for extra bristles, a.k.a. Zeste-white 3) is a serine-threonine kinase (514 a.a. for the most prevalent of the 5 isoforms) [419, 3682, 3920, 3922] homologous to vertebrate GSK-3β (glycogen synthase kinase-3β) [3921, 4741]. In keeping with its role in other systems, Sgg keeps the pathway tonically OFF in the absence of Wg [4741]. Genetically *sgg* acts upstream of *arm* [3228, 3315, 3318, 3924, 4796], and its net effect is to reduce the pool size of free Arm by causing Arm to be phosphorylated, ubiquitinated, and degraded [3204, 3228, 3315]. Sgg can phosphorylate Arm [3683], but Sgg may not be Arm's natural kinase [3228, 3344, 4062, 4678, 4679]. Because Sgg can phosphorylate dAxin [3683] and probably also APC, one or both of these agents might relay the Wg signal from Sgg to Arm (cf. dAxin). Indeed, Dsh-VH inhibits both of these reactions when it binds Axin [2238]. This inhibition is attributable to reduced enzymatic activity of Sgg and ultimately to serine phosphorylation of Sgg by an unidentified kinase [3683].

3.G. E-APC (Epithelial Adenomatous Polyposis Coli; a.k.a. dAPC2) is the product of a *Drosophila* homolog of vertebrate *APC* [2775, 3468, 4832, 4833]. Whereas *E-APC* is expressed ubiquitously, the other known homolog – *dAPC* (*Drosophila APC*) – appears to function only in photoreceptor cells [39, 1766]. In vertebrates, APC binds Sgg-VH, Arm-VH, and Axin [1593, 2239, 2996, 3675, 4040] and helps target Arm-VH for degradation [2239, 2914, 2997, 3304, 3307]. E-APC likewise binds Arm [2775, 4832, 4833] and seems to play a similar role [2775]. Its association with "adherens" junctions (see below) [2608] suggests that E-APC ferries Arm to Axin [322, 4374], and it may actually be the bridge (instead of dAxin) between Dsh and Sgg [3344, 4833]. Consistent with a ferrying role, E-APC has nuclear export sequences that must be intact for Arm to exit the nucleus [3650].

3.H. Slimb ("Supernumerary limb") is a 511-a.a. protein [2060, 4279] whose F-domain and WD40 motifs enable it to link ubiquitin ligases with specific ligase substrates [791, 2668, 4031], including Arm [2856]. The effects of *slimb*[null] are not spatially congruent with those of *sgg*[null] [2060, 4279], probably because Slimb also participates in other pathways (e.g., Hedgehog signaling above; cf. IκB degradation [4476]).

3.I. dAxin (739 a.a.) is the *Drosophila* homolog of vertebrate Axin [1698, 4677, 4861]. It uses a central conserved sequence to bind Arm and an N-terminal "RGS" (regulator of G-protein signaling) domain [2282] to bind dAPC (but see [4040]). Notwithstanding an initial report to the contrary [1698], dAxin also binds Sgg [3683, 4677] near dAxin's Arm-binding site. The 80-a.a. DIX domain shared by Axin and Dsh (at their C- and N-termini, respectively) can also mediate protein binding [1919, 2238]. Indeed, vertebrate Axin is thought to be the scaffold on which a protein complex is built [598, 1182, 3316], consisting of APC, PP2A, the homologs of Dsh, Sgg, Arm, and Slimb [1919, 2238, 2523, 3309, 4495], and Casein kinase [3724]. Genetically, *dAxin* behaves as a negative regulator of Wg signaling upstream of *arm* [1698, 2465, 4677]. Given that (1) Sgg phosphorylates dAxin in Wg-stimulated cells [3683], (2) phospho-Axin has a greater affinity for Arm-VH [4679] (but see [1959]), and (3) phospho-Axin keeps Arm-VH attached to the ubiquitination machinery [2241], it follows that dAxin may be relaying the Wg signal by releasing Arm when Dsh blocks Sgg [1182, 4679]. When Axin is dephosphorylated by PP2A (serine-threonine Protein Phosphatase 2A, another Axin-binding partner [1919, 3849, 4679]), Sgg dissociates from the complex and Sgg-dependent phosphorylation of Arm subsides [3683]. Thus, Arm's release is also likely due to Slimb's low affinity for unphosphorylated Arm [1736, 2241, 4701].

3.J. Armadillo (Arm, named for embryo segmentation defect) is an 843-a.a. protein (a 721-a.a. isoform exists in the CNS [2604]) with 12 protein-binding "arm" repeats (cf. App. 1) [892, 3320, 3598] and an acidic C-terminal region that activates transcription [892, 3146, 4439]. Arm is homologous to vertebrate β-catenin [3320, 4622]. Like its counterpart, Arm is detectable in the cytosol but enriched at cell-cell adherens junctions [3306, 3314, 3320, 3597, 4745]. Adherens junctions are crucial for epithelial integrity [890, 1947, 2273, 2985, 3167, 4295] and possibly some types of signaling (e.g., Wg [2608], EGFR [1001, 1905, 4081], or Notch [184]). Within these junctions, Arm acts as a linchpin between transmembranous dE-cadherin [796, 893, 3168, 3749, 4796] (a.k.a. Shotgun [4268, 4416]) and cortical dα-catenin [3166, 3169, 3409], which in turn is anchored to the actin cytoskeleton [956, 1937, 2185, 3412, 4807] via dRac1 [1133], etc. [1648, 2368, 3679]. Transport of Arm to the junction depends on Presenilin [3124]. Retention of Arm at the membrane depends on PP2A [1578]. Arm's junctional role is separable from its signaling function in the nucleus [893, 2860, 3204, 3749, 4880], although various cadherins can reduce signaling by diverting Arm into junctions [1603]. Wg signaling affects *dE-cadherin* transcription [4796] but not *arm* transcription, which is uniform in discs [3598]. Rather, Wg dictates the amount of Arm post-translationally [273, 1345, 3318, 3320, 3597] by blocking its rapid turnover [7, 4701]. In the nucleus, Arm is a co-activator for Pan [519, 4439] and possibly for other transcription factors as well [3873]. The deployment of Arm to the cytoplasm vs. nucleus appears to reflect Arm's relative affinities for dAxin vs. Pan [4352].

3.K. Pangolin (Pan, named for an armadillo-like embryo segmentation defect, a.k.a. dTcf = *Drosophila* T-cell factor) is a 751-a.a. DNA-binding protein [519, 1088, 4439] that can bend and unwind DNA by virtue of its HMG domain [266, 691, 810, 1481] (cf. App. 1). Genetically, *pan* acts downstream of *arm* [519, 3591, 4439]. Pan binds Arm [519, 4439] when the cytoplasmic pool of Arm enlarges sufficiently after Wg-dependent Arm stabilization [2996, 3228, 3318, 3749]. Pan may help Arm enter the nucleus [266, 1917, 2337], although Arm can do so independently [691, 4439], apparently by using its arm motifs instead of a NLS (cf. App. 1) [1180]. Within the nucleus, the Arm-Pan complex activates target genes via (1) Pan's binding to <u>CCTTTGATCTT</u> [3591, 4439] and (2) Arm's stimulating of transcription via its C-terminus [892, 1917, 4439] (but see [2697]). In the absence of Arm, Pan uses a clever trick to keep Wg-responsive genes completely OFF [3147]: it binds and escorts the co-repressor Groucho (Gro) into the nucleus [692, 2496], and Gro then elicits remodeling of the chromatin [737, 849]. Pan's binding of Arm or Gro may be mutually exclusive [3147], and Arm's binding of Pan or dE-cadherin is competitive [1593, 3202, 3204, 3749], so dE-cadherin and Gro apparently set thresholds that the level of Arm must exceed before Arm can help Pan turn ON genes [849, 893, 2860]. Pan might also bind Gammy legs, which acts downstream of Arm (as a cofactor?) and behaves stoichiometrically as if it is part of a multiprotein complex [599]. A milder antagonist of Arm is dCBP (cf. Hh and Dpp pathways): dCBP binds Pan and acetylates a lysine in Pan's Arm-binding domain, thus decreasing Pan's affinity for Arm [4533]. Still another means of influencing Pan in mammals and nematodes (and maybe flies) is by phosphorylation of Pan, which blocks Wg signaling [1996]. The ability of Pan to bind Arm may also be inhibited by a small (81-a.a.) Arm-binding protein "Icat" (<u>I</u>nhibitor of β-<u>cat</u>enin and TCF-4) found in *Xenopus* [4232]. In fly embryos, the zinc-finger protein Teashirt is an auxiliary Arm-binding partner and modulator of Wg signaling [1344, 1345], and the nuclear protein Lines may be playing a similar role [1759]. Pan can also regulate target genes negatively (in response to Wg signaling) by competing with activators at overlapping binding sites in their promoters [3386] – the phenomenon of "quenching" [1600].

3.L. Naked cuticle (Nkd, named for embryo defect) is a 928-a.a. hydrophilic protein with a putative Ca^{2+}-binding (EF hand) motif, implying ion fluxes in Wg signaling [4863]. The *nkd* gene is activated wherever Wg signal is received, but *nkd*^{GOF} transgenes cause *wg*^{LOF} phenotypes. Thus, Nkd seems to be part of a negative feedback loop (cf. Slimb [4038]) – analogous, perhaps, to Dad's role in the Dpp pathway. However, disabling *nkd* in somatic clones has little effect, so it may act redundantly with another gene [4863]. Nkd binds Dsh and appears to act at the level of Dsh (or immediately downstream) in the Wg pathway [3659].

3.M. Target genes include the following, plus possibly *dE2F* (turned OFF at the wing margin) [1126, 2079, 3979].

1. Wg ⊣ *wg* itself in a paracrine [3690] but not autocrine [595] manner at the wing margin. This link apparently does not operate within the blade proper [353]. In the notal part of the wing disc, Wg → *wg*; however (for some unknown reason), this feedback loop only pushes the *wg*-ON domain a few cells beyond its *pannier*-dependent realm [4369], instead of expanding to fill the entire disc.
2. Wg ⊣ *Dfz2* (partially) in wing discs [595].
3. Wg → *Dfz3* in wing, leg, and eye discs [3977].
4. Wg ⊣ *arr* in leg discs [4570], which suggests that *wg*-ON cells should be relatively deaf to their own signal.
5. Wg → *nkd* wherever Wg is received [4863].
6. Wg ⊣ *dpp* in leg discs [487, 2059, 2082, 2954, 4277]$^\Delta$ and eye discs [4387]$^\Delta$ except at the anterior margin [1780]. In contrast, Wg → *dpp* in the optic lobe of the brain [2130].
7. Wg → *aristaless* in leg and wing discs (at a certain threshold) IF Dpp is also present [643]$^\Delta$.
8. Wg → *dachshund* in leg discs (at a certain threshold) IF Dpp is also present [2456].
9. Wg → *Distal-less* in leg discs (at a certain threshold) IF Dpp is also present [2456]$^\Delta$, and the same is true for antennal anlagen [1037]. In wing discs, however, Wg → *Distal-less* independently of Dpp [595]$^\Delta$.

10. Wg ➞ *hairy* in leg discs IF Hh is also present [1778].
11. Wg ➞ *H15* in leg discs [3762]$^{\Delta}$.
12. Wg ➞ *omb* in wing discs [1626], although the control appears to be indirect [3978].
13. Wg ➞ {*achaete* AND *scute*} at the wing margin [3691]$^{\Delta}$ and prospective notum [3373]$^{\Delta}$, but Wg ⊣ {*achaete* AND *scute*} in the developing eye [597].
14. Wg ⊣ {*araucan* AND *caupolican*} at the wing margin [1537].
15. Wg ⊣ *string* (an indirect target that is turned OFF by *achaete* and *scute*) at the wing margin [2079, 3979].
16. Wg ➞ *Delta* at the wing margin [595]$^{\Delta}$.
17. Wg ➞ *Serrate* at the wing margin [595]$^{\Delta}$.
18. Wg ➞ *vestigial* in the wing pouch [734]$^{\Delta}$. The same link can be artificially triggered in leg discs [2753], where it requires the cooperation of Dpp [2754]$^{\Delta}$.
19. Wg ➞ *scalloped* in the wing pouch [353]$^{\Delta}$.
20. Wg ⊣ *homothorax* in the wing pouch [157] via *vestigial* (at least in part), but Wg ➞ *homothorax* in the wing hinge [678] and head capsule [3380].
21. Wg ⊣ *ventral veinless* (a.k.a. *drifter*) in the wing pouch [703, 990]$^{\Delta}$.
22. Wg ⊣ *vein* in the wing pouch [4604] and dorsal head [80].

23. Wg ⊣ *apterous* in the wing pouch in 2nd instar but not thereafter [4683].
24. Wg ⊣ *spalt* in the notal portion of the wing disc [987].
25. Wg ➞ *BarH1* in the notal portion of the wing disc [3763].
26. Wg ➞ *orthodenticle* in the dorsal head [3665].
27. Wg ➞ *mirror* in the developing eye [1781]$^{\Delta}$.
28. Wg ⊣ *WR122-lacZ* (an enhancer trap expressed at the equator) in the developing eye [1781].
29. Wg ⊣ *eyeless* in eye discs IF Dpp is also present [2465].
30. Wg ➞ *atonal* in the developing antenna [2055].

3.N. Summary of the basic chain: Wg ➞ Dfz2 ➞ Arr ➞ Dsh ⊣ Sgg ⊣ Arm ➞ Pan ➞ {target genes}. Binding of Wg to Dfz2/Arr alters Dsh so that it can disrupt the complex of Sgg, dAxin, E-APC, PP2A, Slimb, Arm, and ubiquitinating enzymes. Dsh causes an unknown kinase to phosphorylate Sgg, thus lowering Sgg's own kinase activity. PP2A removes the phosphate that Sgg had put on a protein (dAxin? E-APC? Arm?) that keeps Arm in the complex. Arm is freed and escapes proteolysis. Its concentration rises. Arm displaces Gro from Gro-Pan dimers, and the Arm-Pan dimers activate Wg-target genes [269]. How Wg represses target genes is unclear [1759].

Commentaries on the Pithier Figures

The following musings concern certain figures that warrant further scrutiny. Through these distillations, I attempt to draw some general conclusions about how the fly's control system operates and why it evolved this way. Some of the annotations also offer historical perspective.

Figure 2.1 One corollary issue (symbolized by the hourglass) is: how do cells measure time over periods longer than a mitotic cycle? Cells presumably need to do so in order to know when to stop dividing and start differentiating. Possible timekeeping devices include the "**POU Hourglass**," which limits the number of mitoses in certain neuroblasts in the fly CNS [313, 314, 3713, 4804, 4816]. This clock gauges the declining amount of the POU-domain proteins Pdm-1 and Pdm-2. Mammalian oligodendrocytes use an "**HLH Hourglass**" that triggers differentiation when the amount of specific HLH-domain proteins drops below a critical threshold [2293, 2294]. An oscillator based on this sort of mechanism may be involved in vertebrate somitogenesis [945]. Other protein clocks appear to count mitoses and meioses leading to sperm and egg differentiation [1762, 4631]. Another strategy involves using a cascade of transcription factors to trigger different events in different phases of the cascade [485]. For RNA clocks, see [223, 1206], and for general discourses on (noncircadian) timekeeping, see [867, 1227, 2762, 3767, 4011].

A deeper issue concerns how structures are represented abstractly in the genome [462, 2724]. Given that we know most of the genes involved in bristle development (cf. App. 3), we can begin to ask how "bristle" is "written" in "gene language." The answer is not obvious. Clearly, a bristle is the outcome of a chain of cellular events, but each event uses genes that are also used elsewhere. No "master gene" exists here, nor is there a hierarchy of dedicated subordinate genes [1114, 2410]. If we could interrogate the shaft cell, for example, it would be unable to define its identity without reciting its history. In this sense, the genes act as a program [98], not as a blueprint [136, 1094]. How the "Make a bristle!" program is encoded remains enigmatic.

Figure 2.2 The Nuclear Notch Model is considered proven [443, 711, 1329, 2514], although Su(H) can affect transcription without Notch [1329, 2129, 2255, 2540, 2941]. Catalytic activation of Su(H) by Notch (or a binding partner) is still possible but hard to reconcile with the pathway's sensitivity to the dosage of Notch and Delta [118, 1797]. The ability of Notch's intracellular domain to act alone quashed the idea [1204, 1742, 1899, 1900] that Notch's primary role is in cell adhesion [112, 3022, 4161]. Nevertheless, Notch is involved in epithelial-mesenchymal transitions [1737, 1744, 4269], and its removal may foster neuroblast delamination [4129]. Another old idea was that Notch's role is to keep cells in an undifferentiated state [112, 113, 1308]. That idea is contradicted by the ability of Notch-activated cells to differentiate as socket or sheath (cf. Fig. 2.1). This figure was adapted from [1251] and *op. cit.*

Figure 2.3 In other signaling pathways, the input-output complexity is distributed among many proteins (cf. Figs. 5.6 and 6.12), not concentrated in one big molecule as here (although the cytoplasmic domain of Egfr is also intricate [2007, 3184]). The variety of domains suggests that Notch was cobbled together evolutionarily from parts of other proteins [3289], and the multiplicity of each type of domain implies repeated episodes of tandem duplication [1824].

Our grasp of this gadgetry is hampered by our ignorance of "cellomics" [1389] – the 3-D protein shapes of the cogs and their gearing stoichiometry [58, 136, 1249]. See [75, 459, 3271, 4206, 4207] for Notch's nuances and [1758, 2310, 4041, 4499, 4837, 4838] for the larger issue of input-output kinetics. John Dexter, a student of Morgan's [2283], found the first *Notch* mutant in 1913, and Morgan himself discovered an allele (later lost) in 1915 [470].

Figure 2.4 Despite knowing the DNA sequence and binding sites for the promoter in **g**, we remain ignorant about how it works [871]. Indeed, we are pretty clueless about "promoter logic" in general [393, 2836, 4165, 4755]. Yes, we know that *trans*-acting factors along the DNA compute a certain output of mRNA [1230, 4464, 4836], but we still need to figure out how they interact *physically* with one another and with the polymerase machinery (cf. Ch. 8).

The gene complexes in **a** and **b** probably evolved from single genes that duplicated repeatedly [1877, 3075, 3178, 3616, 3982, 4617], as apparently happened in the Hox complexes [146]. Unequal crossing over was first deduced at the *Bar* locus by A. H. Sturtevant in 1925 [4179], and tandem duplications were conjectured for the Bithorax Complex by E. B. Lewis in 1951 [2505]. Evolutionarily, such events may have been facilitated by transposons [1401, 1416, 1602]. After a duplication, the paralogs would act redundantly [1517, 1824, 3384, 4584] until *cis*-regulatory regions have time to diverge [1114, 1419, 2630, 4260, 4519]. Divergence could occur by (1) asymmetric loss of old *cis*-control sites [694, 1265, 2316, 2629, 4266] or (2) asymmetric creation of new ones that capture foreign *trans*-acting regulators [1114, 1151, 1347, 3379, 4259, 4704]. Point mutations could lead to new sites that are small ($\leq\sim$10 b.p.) [662, 1477, 2597, 2616, 2763, 3477], but larger sites probably arose via cut-and-paste transpositions [1046, 2231, 2236, 2480, 3075, 3816, 3840] or inversions [593, 666, 2583, 3615]. To the extent that nearby genes share *cis*-enhancers, each complex will tend to remain intact [699, 871, 1112, 1579, 2508, 2672]. Remaining mysteries that are worth pondering include

1. How do tandem duplicates avoid cosuppression [339, 340, 1823, 1825, 3231, 3396]?
2. Why did colinearity evolve in Hox complexes but not elsewhere [1113, 2508, 2672, 3494]?
3. How on earth did *groucho* get into a complex of unrelated (bHLH) genes whose products appear to mainly interact with Groucho at a protein (vs. *cis*-genetic) level?

See [146, 2538, 4786] for other case studies, [2034] for the idea of "gene sharing," and [662, 866, 968, 1440, 3144, 3496, 4298] for overview.

Figure 3.1 No sketch or photo can do justice to the splendor of this fly. In 1683 (the earliest known account of this species), Christian Mentzel (1622–1701) waxed poetic [3361]:

Its eyes are purple. The back is yellow and gibbous. The tail also is yellow, signed by six dark wasp-like stripes or lines. The small wings are a little longer than the body and very transparent. In bright daylight they are splendidly iridescent like a rainbow. . . . Back and head are full of certain bristles on both sides.

In 1764, W. F. von Gleichen offered another ode [2982]:

The head, which is somewhat broader than the thorax, is shaped towards the rear like that of other flies, but by far more beautiful in color and the red eyes on both sides of it are very ornamental. The latter are separated by a broad band which is set with fine hairs on both sides. . . . The brown chest-plate looks like a tile, arched and covered all over with small hairs, especially towards the narrower end. The six long hairy legs are attached to the thorax. . . . Two small bubble-like balancers can be seen between the middle and rear thigh and on both sides under the chest-plate. . . . The body is very hairy and divided into five large rings. It reflects through the two long net-like and iridescent wings which are interwoven by many nerves.

The ultimate paean, however, is Curt Stern's 1954 classic, "Two or three bristles" [4096]:

I marvel at the clear-cut form of the head with giant red eyes, the antennae, and elaborate mouth parts, at the arch of the sturdy thorax bearing a pair of beautifully iridescent, transparent wings and three pairs of legs, at the design of the simple abdomen composed of a series of ringlike segments. A shining, waxed armor of chitin entirely covers the body of the insect. In some regions this armor is bare, but in other regions there arise short or long outgrowths – the bristles – strong and wide at the base and gently tapering off to a fine point. Narrow grooves, as in fluted columns with a slightly baroque twist, extend along their lengths. A short stalk fits each bristle into a round socket within the body armor so that the bristle can be moved within this articulation. There is a regular arrangement of these bristles. . . . With surrealistic clarity the dark colored bristles and hairs project from the light brownish surface of the animal, delicate but stiff, in rigid symmetry.

In fact, bristle grooves do not twist helically around the shaft, although it is easy to see how Stern could have thought so based on side views. Rather, they converge like chevrons at a seam along the top [4639, 4641].

Newcomers may wonder: Why did T. H. Morgan choose this inconspicuous species as his experimental organism? In his 1925 opus with coauthors Calvin Bridges and Alfred Sturtevant, Morgan explains [2951]:

The rapid rate of multiplication of the vinegar fly, owing to the large number of offspring produced, and to the brevity of the life cycle, makes it excellent material for genetic purposes. At room temperature or a little above (24 to 25°C), it takes only about 10 days to pass through a complete generation from imago to imago. Under favorable conditions twenty-five generations may be bred in a year.

Figure 3.4 The AS-C's ~8 *cis*-enhancer elements (**a**) are using an "OR" logic: if any of them is activated by a *trans*-acting factor (cf. Fig. 6.14), then the cell will become proneural via expression of *ac* and/or *sc*. It is easy to see, therefore, how evolution could insert or delete such modules without harming the function of the locus [1094] – a system property called "flexible robustness" [1440]. Indeed, this complex is one of the clearest examples of *cis*-regulatory modularity in any metazoan [968].

Figure 3.6 The feedback circuit in **a** blurs the distinction between "competence" (mediated by the AS-C) and "determination" (mediated by the Delta-Notch duality) [1455, 1654]. Its bistable quality ensures discrete cell types [2162, 2163], but the final pattern of those states is somewhat stochastic. Hence, fly genes do not dictate bristle sites any more than human genes dictate the swirls of ridges on our fingertips [923, 1159, 1606]. This notion of "emergent properties" in control systems was a favorite theme for the great embryologist Paul Weiss [4593, 4594].

Figure 3.10 These patterns are obviously much richer than the emblematic French flag of Wolpert's Positional Information Hypothesis [4723], and it is likely that a variety of tricks (aside from gradients) help organize the bristles (cf. Fig. 3.11) [1805]. For example, SOPs might align into transverse rows via filopodia [3051] or via homophilic binding proteins on the sides of each cell [3048]. Similar strategies may be used in the pedicel of the haltere, where sensilla campaniformia align into strikingly similar transverse rows [1866, 3623].

Figure 3.12 The importance of temporal factors in gene expression (especially "rate genes"; cf. **i**) was a favorite theme for the controversial geneticist Richard Goldschmidt (1878–1958) [1521, 1524, 3398]. His mental playground was "physiological genetics" [1520]. Upon that foundation the edifice of "developmental genetics" was built [3494] by Curt Stern (his protégé) [4095, 4100, 4104], Ernst Hadorn [1667], C. H. Waddington [4510, 4513, 4515], and others outside the Fly World [2309, 2692]. Indeed, "heterochronic" mutations that tinker with the timing of events are thought to have been instrumental in evolution [978, 1114,

1581, 2788, 2789]. Two intriguing mutations of this kind have been documented in butterfly wings, where a time window appears to govern scale pigmentation [2278].

Figure 4.1 Histoblast nests are replaced by ectopic discs when the Bithorax Complex is disabled [828, 1285, 2670, 2745, 3930, 4434] (cf. Fig. 4.2), a default state that has intriguing evolutionary implications [263, 664, 682, 1429, 2503, 4550, 4552]. For example, $Ubx^{bxd\,LOF}$ adults have four pairs of legs [2506], a condition that predates the order Insecta [4266]. *N.B.:* In wild-type embryos, the leg discs are only partly within the neuroectoderm [389] (not totally immersed as depicted here).

Figure 4.3 Despite its demise in 1969 and its formal "burial" in 1978 [4346], Stern's Prepattern Hypothesis was resurrected in the 1980s as a way of explaining embryonic segmentation (see text), where odd sets of signals along the anterior-posterior axis of fly embryos demarcate body segments. It lives on today in derivative models that explain bristle pattern formation in terms of upstream "prepattern" genes (cf. Fig. 6.14c and Table 6.2). Indeed, both models have proven useful in different contexts (cf. Fig. 8.3).

Quirks of antenna-leg topology (**a**) include (1) unlike walking legs, antennal "legs" face up as shown [2933, 3441, 3775, 4145, 4212] because eye discs rotate ~180° after arising, thereby inverting their dorsal-ventral axis [4146]; (2) in more extreme genotypes, the sternopleura (a sclerite proximal to Co) replaces anterior head vibrissae [1461]; and (3) joint membranes that arise at the arista base [1260] resemble tarsal joints but may differ developmentally. Given the homology between the arista and the claws [620, 1085], it seems odd that the A/P lineage boundary bisects the claws [2443, 2449, 4076] but skirts the arista entirely [2931, 4146, 4148]. Nevertheless, tarsal and aristal primordia are similarly bisected by the *engrailed* ON/OFF line [1697]. The arista may be using a recent gene circuit that evolution inserted into the tarsal algorithm [1085, 1119]. Antenna-leg homologies are further explored in Chapter 8.

Figure 5.3 Both models assume that cells adopt fates due to "social" signals that, ultimately, are properties of the whole population (cf. Ch. 1). One benefit of this strategy is robustness [1839]: no cell is ever given so much authority that its malfunction is catastrophic [858, 1440]. Thus, the system withstands sporadic cell death [1776, 3440, 3449], deviant mitotic rates [2935, 3947–3949, 3961, 3962, 3965], and even massive tissue loss [525, 1299, 2698]. For the same reason, discs have tolerated allometric changes during evolution [4114]. This tenacity suggests that the system is *goal*-driven [2237] (cf. Driesch's entelechy [2989, 3743]), but in fact it must be *rule*-driven [1558, 2160, 4605] because the abnormal limbs that it produces when perturbed (**c–e, g–i**) obey Bateson's Rule [240, 539]. Generative rules of this kind constrain the paths that evolution can take [56, 622, 1583, 1584, 1690, 2767] – an insight that was famously formalized in Waddington's "epigenetic landscape" [1114, 1488, 4513].

Figure 5.6 Despite their differences, these pathways have essentially the same input-output capability (cf. Figs. 2.2 and 6.12) [3287, 3288], so the evolutionary decision to use one or another in specific patterning events was probably arbitrary and accidental [2301].

The ability of so many proteins to bind one another makes these networks highly nonlinear [2220]. The transient aggregates of transducer proteins have been called "transducisomes" [556, 4405, 4406] in analogy with conglomerate machines like the ribosome [687], replisome [187, 2490], spliceosome [563, 3996, 4061], enhanceosome [649, 1317, 2109, 2826], and transcriptosome [1703, 2304, 2488, 3279], not to mention the mammoth Mediator [412, 2665] and the zoo of chromatin-remodeling contraptions [1, 2227, 2228] (e.g., Brahma [849, 908, 2114], MSL [1633, 2183, 3760], NuRD [41, 2271], Pc-G [2622, 3869], SWI/SNF [252, 1752, 4199], and even one that is specific to the fly's 4th chromosome [2412]).

In general, the benefits of multiprotein consolidation include (1) greater speed by confining reagents [1106, 3651]; (2) greater efficiency by managing the order, orientation, and stoichiometry of reactions [1062, 3297, 3299, 3834, 4406]; (3) greater specificity by increasing the requisite number of docking domains [3874, 4252, 4253]; (4) reduced interpathway cross-talk by forming insulated corridors [1062, 1389, 2007, 3297, 4896]; and (5) sharper output by fostering allosteric cooperativity among the partners [829, 1555, 3473, 4597]. Hence, the old view of the cell as a "soup" of freely diffusing molecules [565, 1146, 2875, 3302, 4064] is giving way to the idea that cells rely more on "solid-state" chemistry [36, 58, 441, 442] and jigsaw-puzzle computations [440], although the interactions may still be fairly sloppy [2236].

Another design feature of these (and other) pathways is a reliance on phosphorylation. Adding or removing phosphates is an easy way of (1) altering docking domains [2136, 3874, 4200], (2) amplifying signals [441, 1950, 2311], (3) creating feedback loops [1292, 1675, 2943, 3010, 3347], (4) recycling transducers [59, 4606], and (5) modulating transcription factors [3874, 4176]. Moreover, proteins that have multiple phosphorylation sites can integrate the various positive and negative inputs combinatorially [440, 3184] in

the same way that a neuron "computes" an output from its multiple inputs [2275, 2276].

Figure 5.11 Despite recovering so many genes with quasisegmental expression patterns, we still do not know what makes one leg segment different from another. All leg segments are thought to use the same kind of gradient to specify cell positions along their length [386, 2425, 4143], but if that is true, then different segments should need different identity codes so that they can interpret the iterated gradient in different ways. In fact, though, few homeotic mutations have been found that transform one type of leg segment into another (see text). We also do not know what genes create the membranes between segments [3421]. For inventories of segmentation defects, see [1810, 4512, 4515]. For evolutionary context, see [10, 4230]. Similar annuli arise in the legs of grasshoppers [718, 3971] and cockroaches [3132] but in an odd sequence [234]. The tarsus of ancestral arthropods was unsegmented [415], and it may have subdivided by heterotopic deployment of *spineless* from the antenna to the leg (cf. Ch. 8) [1119].

Mapping of unassigned rings (from photos) is based on layouts in [834] (his Fig. 11d), [1311] (their Fig. 14d), [2287], and [4484]. Circuitry links are mainly from [344, 1573, 2287, 3523, 3525]. The "*dac* ⊣ *bab*" and "*al* ⊣ *bab*" links, which help delimit the *bab*-ON domain, are from J. P. Couso (pers. comm.), who also has found that *al*'s influence on *bab* from a distance is mediated by Egfr. The question mark (lower left) concerns how Ser and Dl act. GOF studies indicate that either Ser or Dl alone is sufficient to induce expression of downstream genes [3525], implying "OR" logic. However, each gene has a LOF phenotype (absence of joints), ruling out a simple redundancy and implying "AND" logic [344].

Although *fringe* is crucial for joint formation [344, 3525], it is excluded here because its expression data contradict its LOF defects [999, 1988, 3525]. Also omitted: *18 wheeler* and *klumpfuss* (LOF alleles suppress joints but rings are unmapped) [344]$^\Delta$, *Arrowhead* (spiral stripe vs. ring) [928], *BarH2* (same as *BarH1*) [2287], *enabled* ($\approx$ *dpn* but not ON in claws) [3505], *extradenticle* (Exd is nuclear in *hth*-ON domain and cytoplasmic elsewhere) [9]$^\Delta$, *grain* (Fe and Ti spots, T5 ring) [509], *l(3)1215* (probable wide-zone type) [539, 792, 1810], *mβ* ($\approx$ *m8* but stronger and detectable proximally) [871]$^\Delta$, *orthodenticle* (very proximal crescent) [4654], *pipsqueak* ($\approx$ *Dll*) [4567], *prickle* (uniform but absent from tarsal joints) [1641], *Toll-like receptor* ($\approx$ HZ76) [760], Iro-C enhancer-trap lines *iro^{Sc2}* and *iro$^{B6.8}$* ($\approx$ *ss*) [3079], and enhancer-trap lines c803 ($\approx$ *Dll*) [2684], 05271,

07022, and k10209 ($\approx ap$) [1397]. See *FlyView* (*flyview.uni-muenster.de*) for more enhancer-trap expression patterns [2027] and [4048] for enzyme patterns. A similar diagram of some of these domains is given in [1341, 3523] with further annotations.

Figure 6.2 The kaleidoscope of patterns collected here reveals how balkanized the disc is, despite the cellular homogeneity seen in ordinary histological preparations [981]. Until the 1980s, such differences could only be inferred from indirect approaches such as fate mapping. Now we know that cell fates reflect identities that are cobbled together from subsets of genes. In short, circuitry dictates destiny. The mythical "prepattern" prophesied by Stern [4095] turned out to be a rich "landscape of transcriptional regulators" [981] (cf. Fig. 6.14).

Genes omitted (relation to depicted patterns is in parentheses where "NOT-" means "complementary to"): *extradenticle* (Exd-nuclear $\approx$ NOT-*Dll* but patchy in notum) [130]$^\Delta$, *four jointed* ($\approx vg$) [483]$^\Delta$, *invected* ($= en$) [842], *master of thick veins* ($\approx$ NOT-*tkv* in the pouch) [1327], *mβ* ($\approx emc$ but also ON in P intervins) [871], *ptc* ($\approx dpp$ but weakly ON in A region; cf. Fig. 6.3) [4188], *scalloped* ($= vg$) [1686]$^\Delta$, *scute* ($= ac$ [3689]$^\Delta$, *Serrate* ($= ap$ in 2nd and early-3rd instar) [163]$^\Delta$, *spalt-related* ($= spalt$) [984], *Toll-like receptor* (hinge, part of wing margin, and 2 large notal spots) [760]. Aldehyde oxidase and other enzymes (not shown) have unique patterns [2342, 4049]. Nubbin is expressed and required in the hinge as well as the pouch (not shown) [157, 3103], notwithstanding reports to the contrary [793] – a fact that has caused some confusion in circuit analysis (compare [2252] vs. [3104]). For galleries of expression pattern images, see [353, 834, 2092, 2252, 3089, 4683], and for enhancer-trap patterns, see [4210] and *FlyView* (*flyview.uni-muenster.de*) [2027]. The link "Knot ➔ *bs*" [4479] was omitted due to space limitations, and *bs* is also regulated independently by Dpp [3145].

Figure 6.3 These gradients bear a striking resemblance to Wolpert's scheme for positional information (cf. Fig. 4.3b). However, there is no evidence that (1) cells "record" any morphogen levels as lasting memories or that (2) cells "interpret" any levels aside from the few that delimit target gene expression zones [224, 3087]. Some of the A-P zones are later subdivided to create wing veins (cf. Fig. 6.11), but even during that "read out" phase it is not clear that any cell actually knows where it is. Rather, cells seem to adopt progressively more restricted *qualitative* states. Such states surely depend on position, but not on PI per se [2448]. In general, the patterning mechanisms discovered in discs show features

of both PI (the gradient stage) and prepatterns (local cues $\approx$ "singularities").

As for why the A-P axis needs two gradients, one idea is that cells cannot discriminate Dpp levels near the source and hence must rely on the higher-resolution Hh gradient [224, 865]. However, the same argument could be made for the D-V axis, which manages very nicely with only one gradient.

To the extent that these morphogens dictate cell fates over large distances, their source zones are acting as "organizers" [1040, 1699, 3745, 4510]. In 1936, the renowned scholar Joseph Needham reasoned that organizers probably use sterols as morphogens [3063], thus anticipating the 1996 discovery of Hh's cholesterol adduct [3434].

While Wg controls *vg* in most of the pouch [2254] (notwithstanding suspicions to the contrary [835]), *vg* is activated by Notch at the D/V boundary [2216, 3091, 3490, 4849] via a separate *cis*-enhancer [2219, 2254, 3089, 4684]. Cells that receive both Wg and Vg adopt a "blade" fate, whereas those that get only Wg (from a *wg*-ON ring not shown) adopt a "hinge" fate [2252, 2570] via *homothorax* [678] and respond by proliferating [3088].

Details omitted: (1) A vs. P halves of the pouch differ in size and shape [3972], (2) A vs. P Dpp gradients differ in shape and slope [1327]$^\Delta$, (3) all gradients are probably exponential (e.g., see [1169, 2832, 4251, 4265]), and (4) the *dpp*-ON stripe "varies in width from ~2 cells at the center of the wing pouch to ~8 cells at the periphery" and even vanishes at one point where it crosses the wing margin [3497] (but see [4251]). This sector shape recalls leg disc's wedges (cf. Fig. 5.8 and 2nd-instar *wg*-ON sector in Fig. 6.2) and suggests a common geometry for leg and wing patterning. Registration of the *dpp*-ON zone with vein 3 is based on ref. [350], but in other specimens this zone seems to fill the entire area between veins 3 and 4 [4188].

N.B.: The "Hh ➔ *dpp*" and "Dpp ➔ *omb*" links work similarly in wing and leg discs (cf. Fig. 5.8), but the leg's "Hh ➔ *wg*" link is inoperative in the wing. Other wiring differences include (1) *vg* is not turned ON in leg discs [4681], (2) *spalt* is only ON in a few proximal cells there [220], and (3) *Dll* is solely regulated by Wg in the wing [3089] but is under dual control ({Wg AND Dpp} ➔ *Dll*) in the leg (cf. Ch. 5) [1037]. The "En ⊣ *dpp*" link may not actually function anterior to the A/P line as shown here. Seth Blair (pers. comm.) does find evidence for such repression when he monitors *dpp* expression with *lacZ* driven by the lone *dpp* enhancer "BS3.0" (the usual method [347]), but not with more reliable inserts of *lacZ* at the *dpp* locus (i.e., enhancer traps). If the anterior

"En $\dashv$ *dpp*" effect is spurious, then the Registration Riddle becomes moot (see text).

Figure 6.7 The creation of a third type of cellular state (here a fringe of bristles) at the interface between two others is a common theme in development [825, 3119, 4075, 4114, 4397] (e.g., the vertebrate neural crest [3853, 4135, 4868]; cf. Figs. 5.12c, 6.6b, and 6.11). This **"Stratification Device"** is the epitome of an emergent property [2989, 4098]. It was used as a bootstrapping trick in Meinhardt's Boundary Model II in 1983 (cf. Fig. 5.4) [2808], and its power as a tool for tissue elaboration was recognized much earlier by Paul Weiss in 1961 [4592] and by Curt Stern in 1936 [4088]: "...any physiological activity of a cell or a cell area, if liable to be influenced by the neighboring areas, will lead to new differences. The part of a homogeneous area A bordering on another area B will become different from that part of A which borders on C." A vivid incarnation of this phenomenon was documented by Whiting in 1934: a novel zone of pigmentation arises in genetically mosaic wasp eyes when pigment precursor molecules diffuse across the border between territories [4634].

N.B.: Partial elimination of the margin in *Lyra* mutants [2, 4] turns out to be due to a neomorphic mutation at the *senseless* locus [3128], which normally has nothing to do with margin formation [3127].

Figure 6.9 These dualities (**a.i** and **b.i**) underscore the progressive nature of disc development (cf. Fig. 5.5 for the leg disc). After establishing two cardinal regions (wing vs. notum, or eye vs. antenna), each disc is further subdivided into parts (cf. Figs. 5.8, 6.3, 6.14, 7.3, and 8.3) that are then patterned at a finer scale (cf. Figs. 3.3, 3.8, 3.9, 6.11, 7.7, and 7.9). The notion that embryos are built stepwise by this kind of elaboration and embellishment [825] was asserted by Hans Driesch in 1894 [1101] (as translated by T. H. Morgan [2945]; cf. a different translation by Curt Stern [4096]):

Development proceeds from a few prearranged conditions, that are given in the structure of the egg, and these conditions, by reacting on each other, produce new conditions, and these may in turn react on the first ones, etc. With every effect there is at the same time a new cause, and the possibility of a new specific action, i.e. the development of a specific receiving station for stimuli. In this way there develops from the simple conditions existing in the egg the complicated form of the embryo.

N.B.: The Iroquois Complex (not shown) acts downstream of Vn (Vn $\rightarrow$ Iro-C $\rightarrow$ notum state) [4543]: *ara*^{GOF} transforms wing to notum [4543], and Iro-C^{LOF} clones in

the notum switch cells to "hinge" fate [1060]. Transdetermination frequencies can deviate considerably from the trends depicted here, depending on genotype, culture conditions, and other factors [3883].

Figure 6.10 Other examples of this Venn logic include (1) *wingless* in the notum, which is turned OFF wherever Pannier AND U-shaped overlap (cf. Fig. 6.14) [1380, 1671, 3764]; (2) *spalt* in the antenna, which is only turned ON where Homothorax AND Distal-less overlap (cf. Figs. 7.3 and 8.3) [1085]; and (3) Extradenticle, which appears throughout the leg and antenna but only goes to the nucleus where Homothorax is also present [8, 9, 130, 677, 2673, 3589]. In the Vg/Sd case, a third state emerges at the overlap between two cardinal states. Evolution has undoubtedly "invented" a wide spectrum of genetic circuits by playing with this sort of genetic "**Kaleidoscope Toy**" (see [2353]). The ultimate example may be the eye, where ≥ 4 "early eye" genes must be coexpressed in order to specify eye identity [2353]. A variation on this theme is the use of "NOT" conditions. For example, Hairy stripes suppress *achaete* on the tarsus (Hairy $\dashv$ *achaete*) to leave a complementary pattern of Achaete stripes (cf. Fig. 3.9), and Extramacrochaetae performs a similar masking function on the notum (cf. Ch. 3).

Like feedback circuits (cf. Fig. 3.6), combinatorial control mechanisms defy reductionism because, in this case, the whole is literally more than the sum of the parts [2407, 4644]. When two genes can only function in combination (as here), each can be turned ON separately in other places without harming development [694, 1046]. This freedom helps explain why so many genes are expressed in cells where they have no apparent function [1253, 1654, 2845] (e.g., *period* [1693, 1694], *sevenless* [194, 4363], and *string* [2480]). To wit, evolution is evidently "tinkering" harmlessly [1114, 2005, 3494], and some of these playful links may lead to opportunistic pleiotropy [815, 1046, 1865, 2845, 4110]. Examples of such tinkering are apparent in many "housekeeping" enzymes (e.g., aldehyde oxidase [2339, 2340, 4050, 4051], glucose-6-phosphate dehydrogenase [924], isocitrate dehydrogenase [2341], NADP$^+$-dependent malic enzyme [1235, 2344], and 6-phosphogluconate dehydrogenase [924]). The functional irrelevance of these patterns is shown by their liberal evolution in the leg, wing, and eye discs of various fly species [4048, 4049]. Similar frivolity is familiar in the pigment patterns of dogs, cats, and other mammals [3011–3013, 4819].

How Sd interferes with gene expression is unclear [3936]. One possibility (shown here) is that Sd recruits a co-repressor ("?"). This notion finds support in (1) the ability of excess Sd to turn OFF target genes even when Vg is present, which suggests a Sd partner aside from Vg [1686, 4463]; (2) the identification of a candidate co-repressor for Sd's human ortholog TEF-1 [727, 728]; and (3) the existence of a dominant *vg* allele (vg^{79d5}) whose Vg protein binds Sd but may lack an activation domain [3291], thus mimicking a co-repressor. This figure was adapted from [445, 1686].

N.B.: The exclusive need for *vg* in the wing is shown by vg^{null} flies, which lack wings and halteres but otherwise look surprisingly normal [3291]. The Vg circle in **c** should more aptly be a subset of the Sd circle (not an overlapping set) because there is no known region where *vg* is ON without *sd* also being ON [3936], although one may yet be found. Vg's subcellular location sans Sd (**d**) was assessed in cell culture [1686] and with a Vg construct lacking its Sd-interaction domain [3936]. It is not known whether Sd uses Vg-like partners in regions where it is ON but *vg* is OFF. Vg alters Sd's DNA-binding specificity [445, 3936], and Vg:Sd stoichiometry is critical for all known functions of both genes [2570]. The above reasoning has been tested by making a chimeric Vg-Sd construct [4054]. Expression of that fusion protein in various tissues successfully mimics the effects of Vg/Sd heterodimers.

Figure 6.11 The flow of this circuit (from broad to narrow districts) resembles the embryo's segmentation hierarchy (cf. Fig. 4.2) [981, 984, 4188], and it nicely shows how deep a regulatory network can be [968]. It is surprising that evolution opted for such complexity to allocate only two cell states, given that the same periodicity could theoretically have been created much more simply [1805, 2806].

Genes omitted: *abrupt* (ΔL5) [320]$^\Delta$, *caupolican* ($\approx$ *ara* in its expression and regulation) [984]$^\Delta$, *emc* ($\approx$ *vein* at Ev, but $\approx$ $m\beta$ at 30 h AP) [205], *nemo* ($\approx$ *argos* at 30 h AP) [4475], *scute* ($\approx$ *ac*) [320], *short gastrulation* ($\approx$ *tkv*) [4831], *spitz* ($\approx$ *Star*) [1643], *ventral veinless* ($\approx$ *Dl* at 24–28 h AP but OFF along margin) [984]$^\Delta$. It is unclear whether *Dl* is actually expressed in ~3-cell-wide "vein" stripes as probed by anti-Dl antibodies [205]$^\Delta$ or in ~8-cell-wide provein stripes as probed with antisense RNA [989].

The "*kek1* ⊣ *Egfr*" link is inferred from (1) *Egfr* down-regulation where *kek1* is expressed, (2) $kek1^{LOF}$ effects elsewhere [1445], and (3) $kek1^{GOF}$ effects on veins [4604]. The lack of a $kek1^{LOF}$ vein phenotype may be due to redundant paralogs [1445]. The antagonism between *rho* and *net* [466] is shown in **c** but omitted from **e**. Evidence exists for a "*net* → *bs*" link [320], which could be indirect (i.e., "*net* ⊣ *rho* ⊣ *bs*"). For further interactions, see [980, 995, 1041, 1368, 4189]. Regulation of *caup* appears identical to *ara*, and both genes require high levels of Dpp (not shown) via a route that somehow bypasses *omb* and *spalt* [1537].

Aside from ptc^{LOF} (cf. Fig. 6.6e, f), other A-region clones that induce L3 outside their perimeter (not shown) include DCO^{null} [3238], DCO^{null} dpp^{null} (but not singly dpp^{null}) [2058], and en^{GOF} [4848]. An ectopic L2 is induced at the front edge of *Distal-less*^{null} clones that reside anterior to L2 [1561]. $Egfr^{LOF}$ clones can elicit an ectopic vein outside their perimeter in the vicinity of any extant vein when they overlap it [1042], and $knot^{LOF}$ clones (marked here as "i") can extend their ectopic veins a few cell diameters into surrounding wild-type territory [2894].

Figure 6.12 More RTK molecules can be deployed to improve sensitivity at minimal cost in energy because subsequent steps are only executed if receptors dimerize. This "assembly on demand" strategy is a clever engineering trick [441]. Another trick is the use of a 2-D surface (the membrane) to increase the likelihood of productive collisions among recruited components during the assembly process [441].

The intricacy of this network recalls Dearden and Akam's lament that such diagrams resemble "explosions in a spaghetti factory" [1010]. Their remedy is to go "*in silico*" because computers can pick up where our intuitions leave off [1758, 4499]. Time will tell whether the cell's full "instruction manual" is so sophisticated that only math wizards can hope to understand it. For now at least, any literate car mechanic or house electrician should be able to figure out the basics.

Why is this pathway so complex when its input-output relationship seems to be a trivial relay (devoid of true computation) like a one-step JAK-STAT [958] or Notch [446] pathway (cf. Fig. 2.2) [1846]? Proposed answers include (1) the multitude of nodes allows for interpathway cross-talk [215, 375, 1663, 2892, 3622], (2) the negative loops offer flexibility for operating in either digital (solenoid) or analog (rheostat) mode [1222, 1292, 2654], and (3) the added "frills" are useful for modulating the duration and amplitude of the signal [2701, 3783] (cf. the ~6 h needed for Boss-Sev to have an effect [2991]). In short, the pathway performs like a finely tuned (and easily tunable) musical instrument.

Dos is also thought to bind Drk, dShc, PI3 kinase, and PLCγ *in vivo* [2639], but these liaisons are dispensable [243, 1830]. Dos is phosphorylated by Sev [1829] (Csw is not [71]), but whether Dos is likewise modified by Egfr is unclear. Apparently, Dos's duty is to recruit Csw so that Csw can dephosphorylate Dos [1829] or a non-Dos target [1830]. In Torso signaling, Csw dephosphorylates Torso [808, 1403] so as to rebuff the SH2 domain of dRasGap (a Gap1-like protein) [808]. In Egfr and Sev signaling, Csw's targets are unknown [808, 1403]: dephosphorylation may rebuff other dampers (e.g., PLCγ), and Csw's chief role may be steric because it partially transmits a Sev signal even when catalytically inert [71]. Also murky is how Dos and dShc affect Ras1 or agents downstream. Whether dShc affects the pathway at Ras1 or farther downstream is unclear [392, 2623]. Other targets of dMAPK (links not shown) may include Sos and dMEK [2007]. Other phosphatases for dMAPK must exist aside from PTP-ER [2133] but are as yet unknown [2355] (cf. MKP-1 [4205] and MKP-3 [634, 1675, 2972], etc. [799, 2202, 2205, 3065]). Additional components remain to be identified (see text) [1237, 2532]. For further nuances, see [148, 381, 2957, 3899].

Figure 7.5 If Fz can localize asymmetrically within each cell in the eye as it does in the wing [4166], then the Vector Model would be favored. The identity of factor X is one of the great unsolved mysteries of fly genetics [447, 2883, 4173, 4366]. Once it is found, we may be able to solve the general problem of how cells adopt particular orientations within epithelia: the "**Planar Polarity Puzzle.**" The robustness of this polarity mechanism in the eye (100% correct "read out" of the signal over a span of >100 cells) [2883, 4366] may stem from the superposition of multiple gradients with opposing slopes [4173, 4573].

A related mystery (albeit a morphogenetic one) is the "**Gearing Riddle**" [3563]. Namely, how do adjacent clusters rotate relative to one another? Some corollary issues are (1) how do outer cells make and break contacts with neighbors during the pivoting process?, (2) do the outer cells push or pull themselves along?, and (3) do they get traction by gripping their neighbors or by gaining purchase on a noncellular substratum? One clue is that solitary ommatidia can still rotate [181], so gearing per se may not be used.

N.B.: Notch-dependent "symmetry breaking" is also thought to occur during selection of bristle SOPs (cf. Fig. 3.6) [3954]. Neither of the initial biasing events (Type 1 or 2) requires transcription, so they could happen fast. A precedent exists for Type-2 biasing in planar polarity: Starry night (a 7-pass cadherin; a.k.a. Flamingo) localizes to specific faces of wing cells at certain times in a Fz-dependent manner [447, 704, 4430].

Figure 7.6 By adding a long-range inhibitor (Argos) to its short-range activator (Spitz), the vocal cell reduces the risk that the activator will mistakenly affect cells beyond its intended range (**c, d**). This trick for reducing the noise in sloppy signals is known as "lateral inhibition" [1479, 2806, 2815, 2818, 3207]. Argos apparently plays this same role in other tissues [1528, 3346, 3772, 4117] as well as possibly setting the intervals between ommatidial clusters (cf. Fig. 7.7) [4035].

In this regard, the nests of sensilla on the legs are intriguing. Those nests have constant numbers of sensilla trichodea (e.g., St8 on the coxa) or campaniformia (e.g., Sc11 on the femur) [526, 3705, 3807], and the number per nest can be changed by manipulating the Ras-MAPK pathway (L. Held, unpub. obs.). If Argos were to enable Spitz to behave like a membrane-bound ligand, then stepwise inductions via Spitz could "count" cells centrifugally in concentric rings around a founder cell – in effect computing a Fibonacci series based on the tiling geometries of the competent cells.

The idea that induction merely "evokes" pre-existing abilities in responding cells is implicit in the Stop the Clock! Model. This same idea was endorsed long ago by Holtfreter [1885] and Waddington [4508].

N.B.: The peripodial membrane (omitted here) might play a role in photoreceptor patterning, but at present the evidence is only suggestive [773, 1473, 3508]. It may also be directing the odd vectors of regeneration that characterize this disc [3705]. The Stop the Clock! Model resembles the classic **Clock and Wavefront Model,** which was designed to explain the periodicity of somites in vertebrates [868, 4010]. In the latter model, the "Stop!" signal is conveyed by a wavefront instead of by juxtacrine signals [1805]. Surprisingly, vertebrate somitogenesis uses homologs of some of the fly's MF circuitry genes, including *hairy* [2224, 2484, 2988, 3236], *Notch* [856, 2051, 2061, 2095, 3451], and *Delta* [1878]. See [2786, 4111] for reviews, [3799, 4850] for related models, and [1181] for juxtacrine signaling in general.

Figure 7.7 The nuclei of disc cells migrate along the cell's apical-basal axis during mitosis [20, 2650, 2744, 3422, 3426, 3539]. Curiously, R cells undergo similarly choreographed shifts during their differentiation [1742, 4355, 4364, 4715]. Such shifts were thought to be instrumental in cell signaling [533] or gene expression [2392, 3672], but they cannot be critical because differentiation proceeds fairly normally when the migrations are blocked [1239, 2483]. Perhaps, what matters here is the quiescence [604, 4387]. Cell cycle arrest is a general feature of proneural fields [2079, 4427, 4898], and the eye's *atonal* stripe (Fig. 7.9) resembles AS-C stripes on the thorax and legs (cf. Figs. 3.8 and 3.9)

[2020]. (Why evolution chose *atonal* for R8s and the AS-C for bristles is unknown [4897].) Quiescence might ensure precision in cell signaling (by homogenizing cell shapes? [287]) without affecting signals per se [1003, 2971, 3655]. Mitotic domains can be decisive in the embryo [110, 616, 1258, 3652, 4014], and wavefronts of various kinds seem to assign fates in other systems as well [1560, 1662, 1805].

This basic model was originally proposed not in flies but in birds. In 1972, Donald Ede invented it to explain feather lattices in chicken skin [1140]. (Amazingly, bird feathers and fly ommatidia may be patterned by similar gene cassettes [790, 1784, 2100, 3131, 3200, 3282, 3382, 4647].) Versions of this model were later advocated for the fly eye [3574] using Sca [179, 2461, 2885], Argos [4035], or other molecules [204] as the inhibitor [1804, 2632].

N.B.: One quirk of the actual process (not shown) is that ommatidia arise first in the equatorial portion of the MF (the "firing center" [1638, 2631]), with later ones appearing at progressively more peripheral sites every ~15 min. [182, 1076, 4712, 4873]. This sequential strategy avoids the stochastic errors ("stacking flaws" [1639]) inherent in synchronous patterning (cf. Fig. 3.9c) and may thus ensure a greater uniformity in R8 intervals [3563], which are ~7 cell diameters [4035, 4364]. Nevertheless, a problem arises at the MF's outer tips, where new nodes do not have two prior nodes to delimit their "crevices" as the columns are getting taller [2632]. Some other device must help in this regard.

Figure 7.9 More genes undoubtedly remain to be incorporated into the circuit in **b**. Until they are found and subjected to LOF-GOF analysis, this puzzle will be unsolvable [2962]. Fortunately, future screens should be easier with clever new techniques that bypass the lethality of mutated vital genes [4132, 4292].

Genes omitted: *eyelid* (≈ *hairy*) [4389], *m8* (≈ *mδ*) [229], *orthodenticle* (posterior to column 9) [4461], *shortsighted* (≈ *hairy*) [4388], *Toll-like receptor* (two stripes that "sandwich" the MF) [760], *vein* (weak expression in MF) [80]. Other genes are modulated in their expression along the A-P axis but show no sharp boundaries: *Delta* [186] and *Notch* [186].

Links not shown due to ambiguities: "*hh* ➔ *hairy*" anterior to the MF [1786, 3238, 4387] but "*hh* ⊣ *hairy*" within

the MF (a threshold effect?) [1616], regulation of *ato* by Egfr (which is positive but has negative aftereffects) [4035], regulation of *ato* by Hh (which is positive or negative depending on the level of Hh) [1077], and regulation of *ato* by *groucho* and *Hairless* (which is also complex) [725]. Also not shown (to avoid clutter): "*ato* ➔ *boss*" [1076], "*dpp* ➔ *hairy*" [1616], "*hh* ➔ *glass*" [2632], "*hh* ➔ *rough*" [1077], and "*hh* ➔ *sca*" [2632, 3238]. Early (IC) and late (IG) expression of *ato* are controlled by separate *cis*-enhancers [4208]. For further details, see [182, 183, 720, 1076].

Figure 8.1 The BX-C is unusual in the vast amount of DNA that it devotes to *cis*-regulation (**a**) [1990, 2509]: only 1.4% of the locus actually encodes protein [2703] (cf. AS-C [1538], *dpp* [347, 2739, 4056], and *homothorax* [2360] loci). Note the similarity of the stages in **b** to the phases of information processing proposed in Wolpert's Positional Information Hypothesis (cf. Fig. 4.3b). The 1987 "**Open for Business Model**" of Mark Peifer, François Karch, and Welcome Bender [3313] has been quite influential in this field. It assumes that successively larger portions of the BX-C are available to *trans*-acting regulators in successively more posterior parasegments, but availability per se does not assure transcription. The relevant factors might be unevenly distributed within a parasegment, in which case *Ubx*, say, would be turned ON nonuniformly (cf. Fig. 8.2). See [1770, 2304, 2305, 4165, 4719] for "nucleosome logic."

Figure 8.2 In essence, Ubx enables certain groups of cells in T3 discs to become distinct from the corresponding groups in T2 (i.e., it endows them with the competence to respond differently to the same ensemble of high-level "prepattern" regulators). Based on this one bit of difference (≈ "I am not my brother"), evolution has made various changes in the low-level "realizator" genes that control growth rates, pattern elements, etc. (≈ "I therefore do things differently") [4563, 4564]. Interestingly, an analogous process occurred in the evolution of male vs. female sexual behavior. In that case, certain areas of the male brain were evolutionarily "painted" with the protein Fruitless, and they subseqently acquired circuits for the courtship of females [173]. (Pax6 in the eyes of arthropods vs. vertebrates is yet another example [3382].) What a difference one bit can make!

References

1. Aalfs, J.D. and Kingston, R.E. (2000). What does 'chromatin remodeling' mean? *Trends Biochem.* **25**, 548–555.

2. Abbott, L.A. (1986). Restoration of wing margins in *Lyra* mutants of *Drosophila melanogaster*. *Dev. Biol.* **115**, 233–248.

3. Abbott, L.A. and Lindenmayer, A. (1981). Models for growth of clones in hexagonal cell arrangements: applications in *Drosophila* wing disc epithelia and plant epidermal tissues. *J. Theor. Biol.* **90**, 495–544.

4. Abbott, L.A. and Sprey, T.E. (1990). Components of positional information in the developing wing margin of the *Lyra* mutant of *Drosophila*. *Roux's Arch. Dev. Biol.* **198**, 448–459.

5. Abbott, L.C., Karpen, G.H., and Schubiger, G. (1981). Compartmental restrictions and blastema formation during pattern regulation in *Drosophila* imaginal leg discs. *Dev. Biol.* **87**, 64–75.

6. Abdelilah-Seyfried, S., Chan, Y.-M., Zeng, C., Justice, N.J., Younger-Shepherd, S., Sharp, L.E., Barbel, S., Meadows, S.A., Jan, L.Y., and Jan, Y.N. (2000). A gain-of-function screen for genes that affect the development of the *Drosophila* adult external sensory organ. *Genetics* **155**, 733–752.

7. Aberle, H., Bauer, A., Stappert, J., Kispert, A., and Kemler, R. (1997). β-catenin is a target for the ubiquitin-proteasome pathway. *EMBO J.* **16**, 3797–3804.

8. Abu-Shaar, M. and Mann, R.S. (1998). Generation of multiple antagonistic domains along the proximodistal axis during *Drosophila* leg development. *Development* **125**, 3821–3830.

9. Abu-Shaar, M., Ryoo, H.D., and Mann, R.S. (1999). Control of the nuclear localization of Extradenticle by competing nuclear import and export signals. *Genes Dev.* **13**, 935–945.

10. Abzhanov, A. and Kaufman, T.C. (2000). Homologs of *Drosophila* appendage genes in the patterning of arthropod limbs. *Dev. Biol.* **227**, 673–689.

11. Adachi, M., Fukuda, M., and Nishida, E. (1999). Two co-existing mechanisms for nuclear import of MAP kinase: passive diffusion of a monomer and active transport of a dimer. *EMBO J.* **18**, 5347–5358.

12. Adachi-Yamada, T., Fujimura-Kamada, K., Nishida, Y., and Matsumoto, K. (1999). Distortion of proximodistal information causes JNK-dependent apoptosis in *Drosophila* wing. *Nature* **400**, 166–169.

13. Adachi-Yamada, T., Nakamura, M., Irie, K., Tomoyasu, Y., Sano, Y., Mori, E., Goto, S., Ueno, N., Nishida, Y., and Matsumoto, K. (1999). p38 mitogen-activated protein kinase can be involved in Transforming Growth Factor β superfamily signal transduction in *Drosophila* wing morphogenesis. *Mol. Cell. Biol.* **19**, 2322–2329.

14. Adams, M.D., Celniker, S.E., Holt, R.A., Evans, C.A., Gocayne, J.D., Amanatides, P.G., Scherer, S.E., Li, P.W., Hoskins, R.A., Galle, R.F., George, R.A., Lewis, S.E., Richards, S., Ashburner, M., Henderson, S.N., Sutton, G.G., Wortman, J.R., Yandell, M.D., Zhang, Q., Chen, L.X., Brandon, R.C., Rogers, Y.-H.C., Blazej, R.G., Champe, M., Pfeiffer, B.D., Wan, K.H., Doyle, C., Baxter, E.G., Helt, G., Nelson, C.R., Gabor Miklos, G.L., Abril, J.F., Agbayani, A., An, H.-J., Andrews-Pfannkoch, C.A., Baldwin, D., Ballew, R.M., Basu, A., Baxendale, J., Bayraktaroglu, L., Beasley, E.M., Beeson, K.Y., Benos, P.V., Berman, B.P., Bhandari, D., Bolshakov, S., Borkova, D., Botchan, M.R., Bouck, J., Brokstein, P., Brottier, P., Burtis, K.C., Busam, D.A., Butler, H., Cadieu, E., Center, A., Chandra, I., Cherry, J.M., Cawley, S., Dahlke, C., Davenport, L.B., Davies, P., de Pablos, B., Delcher, A., Deng, Z., Deslattes Mays, A., Dew, I., Dietz, S.M., Dodson, K., Doup, L.E., Downes, M., Dugan-Rocha, S., Dunkov, B.C., Dunn, P., Durbin, K.J., Evangelista, C.C., Ferraz, C., Ferriera, S., Fleischmann, W., Fosler, C., Gabrielian, A.E., Garg, N.S., Gelbart, W.M., Glasser, K., Glodek, A., Gong, F., Gorrell, J.H., Gu, Z., Guan, P., Harris, M., Harris, N.L., Harvey, D., Heiman, T.J., Hernandez, J.R., Houck, J., Hostin, D., Houston, K.A., Howland, T.J., Wei, M.-H., Ibegwam, C., Jalali, M., Kalush, F., Karpen, G.H., Ke, Z., Kennison, J.A., Ketchum, K.A., Kimmel, B.E., Kodira, C.D., Kraft, C., Kravitz, S., Kulp, D., Lai, Z., Lasko, P., Lei, Y., Levitsky, A.A., Li, J., Li, Z., Liang, Y., Lin, X., Liu, X., Mattei, B., McIntosh, T.C., McLeod, M.P., McPherson, D., Merkulov, G., Milshina, N.V., Mobarry, C., Morris, J., Moshrefi, A., Mount, S.M., Moy, M., Murphy, B., Murphy, L., Muzny, D.M., Nelson, D.L., Nelson, D.R., Nelson, K.A., Nixon, K., Nusskern, D.R., Pacleb, J.M., Palazzolo, M., Pittman, G.S., Pan, S., Pollard, J., Puri, V., Reese, M.G., Reinert, K., Remington, K., Saunders, R.D.C., Scheeler,

F., Shen, H., Shue, B.C., Sidén-Kiamos, I., Simpson, M., Skupski, M.P., Smith, T., Spier, E., Spradling, A.C., Stapleton, M., Strong, R., Sun, E., Svirskas, R., Tector, C., Turner, R., Venter, E., Wang, A.H., Wang, X., Wang, Z.-Y., Wassarman, D.A., Weinstock, G.M., Weissenbach, J., Williams, S.M., Woodage, T., Worley, K.C., Wu, D., Yang, S., Yao, Q.A., Ye, J., Yeh, R.-F., Zaveri, J.S., Zhan, M., Zhang, G., Zhao, Q., Zheng, L., Zheng, X.H., Zhong, F.N., Zhong, W., Zhou, X., Zhu, S., Zhu, X., Smith, H.O., Gibbs, R.A., Myers, E.W., Rubin, G.M. and Venter, J.C. (2000). The genome sequence of *Drosophila melanogaster*. *Science* **287**, 2185–2195. [See also 2001 *Nature* **411**, 259–260.]

15. Adamson, A.L. and Shearn, A. (1996). Molecular genetic analysis of *Drosophila ash2*, a member of the trithorax group required for imaginal disc pattern formation. *Genetics* **144**, 621–633.

16. Adler, P. (1978). Mutants of the Bithorax Complex and determinative states in the thorax of *Drosophila melanogaster*. *Dev. Biol.* **65**, 447–461.

17. Adler, P.N. (1978). Positional information in imaginal discs transformed by homoeotic mutations. Fate map and regulative behavior of fragments of haltere discs transformed by *bithorax³* and *postbithorax*. *W. Roux's Arch.* **185**, 271–292.

18. Adler, P.N. (1979). Position-specific interaction between cells of the imaginal wing and haltere discs of *Drosophila melanogaster*. *Dev. Biol.* **70**, 262–267.

19. Adler, P.N. (1981). Growth during pattern regulation in imaginal discs. *Dev. Biol.* **87**, 356–373.

20. Adler, P.N. (1987). Growth and pattern regulation in insects. *In* "Invertebrate Models: Cell Receptors and Cell Communication," (A.H. Greenberg, ed.). Karger: Basel, pp. 172–189.

21. Adler, P.N., Charlton, J., and Liu, J. (1998). Mutations in the cadherin superfamily member gene *dachsous* cause a tissue polarity phenotype by altering *frizzled* signaling. *Development* **125**, 959–968.

22. Adler, P.N., Krasnow, R.E., and Liu, J. (1997). Tissue polarity points from cells that have higher Frizzled levels towards cells that have lower Frizzled levels. *Curr. Biol.* **7**, 940–949.

23. Adler, P.N. and MacQueen, M. (1981). Partial coupling of the cell cycles of neighboring imaginal disc cells. *Exp. Cell Res.* **133**, 452–456.

24. Adler, P.N. and Taylor, J. (2001). Asymmetric cell division: plane but not simple. *Curr. Biol.* **11**, R233–R236.

25. Adler, P.N., Taylor, J., and Charlton, J. (2000). The domineering non-autonomy of *frizzled* and *Van Gogh* clones in the *Drosophila* wing is a consequence of a disruption in local signaling. *Mechs. Dev.* **96**, 197–207.

26. Adler, P.N., Vinson, C., Park, W.J., Conover, S., and Klein, L. (1990). Molecular structure of *frizzled*, a *Drosophila* tissue polarity gene. *Genetics* **126**, 401–416.

27. Adler, R. and Hatlee, M. (1989). Plasticity and differentiation of embryonic retinal cells after terminal mitosis. *Science* **243**, 391–393.

28. Affolter, M., Marty, T., and Vigano, M.A. (1999). Balancing import and export in development. *Genes Dev.* **13**, 913–915.

29. Affolter, M., Montagne, J., Walldorf, U., Groppe, J., Kloter, U., LaRosa, M., and Gehring, W.J. (1994). The *Drosophila* SRF homolog is expressed in a subset of tracheal cells and maps within a genomic region required for tracheal development. *Development* **120**, 743–753.

30. Affolter, M., Nellen, D., Nussbaumer, U., and Basler, K. (1994). Multiple requirements for the receptor serine/threonine kinase *thick veins* reveal novel functions of TGFβ homologs during *Drosophila* embryogenesis. *Development* **120**, 3105–3117.

31. Affolter, M., Percival-Smith, A., Müller, M., Billeter, M., Qian, Y.Q., Otting, G., Wüthrich, K., and Gehring, W.J. (1991). Similarities between the homeodomain and the Hin recombinase DNA-binding domain. *Cell* **64**, 879–880.

32. Agarwal, P. (1995). The *Cell Programming Language*. *Artif. Life* **2**, 37–77.

33. Agnel, M., Röder, L., Griffin-Shea, R., and Vola, C. (1992). The spatial expression of *Drosophila rotund* gene reveals that the imaginal discs are organized in domains along the proximal-distal axis. *Roux's Arch. Dev. Biol.* **201**, 284–295.

34. Agnès, F., Suzanne, M., and Noselli, S. (1999). The *Drosophila* JNK pathway controls the morphogenesis of imaginal discs during metamorphosis. *Development* **126**, 5453–5462.

35. Agrawal, N., Kango, M., Mishra, A., and Sinha, P. (1995). Neoplastic transformation and aberrant cell-cell interactions in genetic mosaics of *lethal(2)giant larvae (lgl)*, a tumor suppressor gene of *Drosophila*. *Dev. Biol.* **172**, 218–229.

36. Agutter, P.S. and Wheatley, D.N. (2000). Random walks and cell size. *BioEssays* **22**, 1018–1023.

37. Ahmad, K. and Henikoff, S. (2001). Modulation of transcription factor counteracts heterochromatic gene silencing in *Drosophila*. *Cell* **104**, 839–847.

38. Ahmad, S.M. and Baker, B.S. (2001). Sex-specific deployment of FGF-signalling in *Drosophila* recruits mesodermal cells into the male genital imaginal disc. *Proc. 42nd Ann. Drosophila Res. Conf.* **Abstracts Vol.**, a49.

39. Ahmed, Y., Hayashi, S., Levine, A., and Wieschaus, E. (1998). Regulation of Armadillo by a *Drosophila* APC inhibits neuronal apoptosis during retinal development. *Cell* **93**, 1171–1182.

40. Ahrendt, W.R. and Savant, C.J., Jr. (1960). "Servomechanism Practice." McGraw-Hill, New York.

41. Ahringer, J. (2000). NuRD and SIN3: histone deacetylase complexes in development. *Trends Genet.* **16**, 351–356.

42. Aitken, A. (1995). 14-3-3 proteins on the MAP. *Trends Biochem. Sci.* **20**, 95–97.

43. Aitken, A. (1996). 14-3-3 and its possible role in coordinating multiple signalling pathways. *Trends Cell Biol.* **6**, 341–347.

44. Aitken, A., Collinge, D.B., van Heusden, B.P.H., Isobe, T., Roseboom, P.H., Rosenfeld, G., and Soll, J. (1992). 14-3-3 proteins: a highly conserved, widespread family of eukaryotic proteins. *Trends Biochem. Sci.* **17**, 498–501. [See also 2001 *BioEssays* **23**, 936–946.]

45. Akam, M. (1984). A common segment in genes for segments of *Drosophila*. *Nature* **308**, 402–403.

46. Akam, M. (1984). A molecular theme for homoeotic genes. *BioEssays* **1**, 78–79.

47. Akam, M. (1987). The molecular basis for metameric pattern in the *Drosophila* embryo. *Development* **101**, 1–22.

48. Akam, M. (1989). *Hox* and HOM: homologous gene clusters in insects and vertebrates. *Cell* **57**, 347–349.

49. Akam, M. (1989). Making stripes inelegantly. *Nature* **341**, 282–283.

50. Akam, M. (1991). Wondrous transformation. *Nature* **349**, 282.

51. Akam, M. (1998). *Hox* genes: From master genes to micromanagers. *Curr. Biol.* **8**, R676–R678.

52. Akam, M. (1998). *Hox* genes, homeosis and the evolution of segment identity: no need for hopeless monsters. *Int. J. Dev. Biol.* **42**, 445–451.

53. Akam, M.E. and Martinez-Arias, A. (1985). The distribution of *Ultrabithorax* transcripts in *Drosophila* embryos. *EMBO J.* **4**, 1689–1700.

54. Akimaru, H., Chen, Y., Dai, P., Hou, D.-X., Nonaka, M., Smolik, S.M., Armstrong, S., Goodman, R.H., and Ishii, S. (1997). *Drosophila* CBP is a co-activator of *cubitus interruptus* in *hedgehog* signalling. *Nature* **386**, 735–738.

55. Alber, T. (1993). How GCN4 binds DNA. *Curr. Biol.* **3**, 182–184.

56. Alberch, P. (1982). Developmental constraints in evolutionary processes. *In* "Evolution and Development," (J.T. Bonner, ed.). Springer-Verlag: Berlin, pp. 313–332.

57. Alberga, A., Boulay, J.-L., Kempe, E., Dennefeld, C., and Haenlin, M. (1991). The *snail* gene required for mesoderm formation in *Drosophila* is expressed dynamically in derivatives of all three germ layers. *Development* **111**, 983–992.

58. Alberts, B. (1998). The cell as a collection of protein machines: preparing the next generation of molecular biologists. *Cell* **92**, 291–294.

59. Alberts, B., Bray, D., Lewis, J., Raff, M., Roberts, K., and Watson, J.D. (1994). "Molecular Biology of the Cell." 3rd ed. Garland, New York.

60. Albright, T.D., Jessell, T.M., Kandel, E.R., and Posner, M.I. (2000). Neural science: a century of progress and the mysteries that remain. *Cell* **100 (Suppl.)**, S1–S55.

61. Alcedo, J., Ayzenzon, M., Von Ohlen, T., Noll, M., and Hooper, J.E. (1996). The *Drosophila smoothened* gene encodes a seven-pass membrane protein, a putative receptor for the Hedgehog signal. *Cell* **86**, 221–232.

62. Alcedo, J., Zou, Y., and Noll, M. (2000). Posttranscriptional regulation of Smoothened is part of a self-correcting mechanism in the Hedgehog signaling system. *Molec. Cell* **6**, 457–465.

63. Alevizopoulos, A. and Mermod, N. (1997). Transforming growth factor-β: the breaking open of a black box. *BioEssays* **19**, 581–591.

64. Alexander, R.M. (1995). Big flies have bigger cells. *Nature* **375**, 20.

65. Alexandre, C., Jacinto, A., and Ingham, P.W. (1996). Transcriptional activation of *hedgehog* target genes in *Drosophila* is mediated directly by the Cubitus interruptus protein, a member of the GLI family of zinc finger DNA-binding proteins. *Genes Dev.* **10**, 2003–2013.

66. Alexandre, E., Graba, Y., Fasano, L., Gallet, A., L., P., De Zulueta, P., Pradel, J., Kerridge, S., and Jacq, B. (1996). The *Drosophila* Teashirt homeotic protein is a DNA-binding protein and *modulo*, a HOM-C regulated modifier of variegation, is a likely candidate for being a direct target gene. *Mechs. Dev.* **59**, 191–204.

67. Ali, Z., Drees, B., Coleman, K.G., Gustavson, E., Karr, T.L., Kauvar, L.M., Poole, S.J., Soeller, W., Weir, M.P., and Kornberg, T. (1985). The engrailed locus of *Drosophila melanogaster*: genetic, developmental, and molecular studies. *Cold Spr. Harb. Symp. Quant. Biol.* **50**, 229–233.

68. Alifragis, P., Poortinga, G., Parkhurst, S.M., and Delidakis, C. (1997). A network of interacting transcriptional regulators involved in *Drosophila* neural fate specification revealed by the yeast two-hybrid system. *Proc. Natl. Acad. Sci. USA* **94**, 13099–13104.

69. Allada, R., White, N.E., So, W.V., Hall, J.C., and Rosbash, M. (1998). A mutant *Drosophila* homolog of mammalian *Clock* disrupts circadian rhythms and transcription of *period* and *timeless*. *Cell* **93**, 791–804.

70. Allard, J.D., Chang, H.C., Herbst, R., McNeill, H., and Simon, M.A. (1996). The SH2-containing tyrosine phosphatase corkscrew is required during signaling by sevenless, Ras1 and Raf. *Development* **122**, 1137–1146.

71. Allard, J.D., Herbst, R., Carroll, P.M., and Simon, M.A. (1998). Mutational analysis of the SRC homology 2 domain protein-tyrosine phosphatase Corkscrew. *J. Biol. Chem.* **273** #21, 13129–13135.

72. Allen, G.E. (1978). "Thomas Hunt Morgan: The Man and His Science." Princeton Univ. Pr., Princeton, N. J.

73. Alonso, M.C. and Cabrera, C.V. (1988). The *achaete-scute* gene complex of *Drosophila melanogaster* comprises four homologous genes. *EMBO J.* **7**, 2585–2591.

74. Alpert, M. (1999). Not just fun and games. *Sci. Am.* **280** #4, 40–42.

75. Alton, A.K., Fechtel, K., Kopczynski, C.C., Shepard, S.B., Kooh, P.J., and Muskavitch, M.A.T. (1989). Molecular genetics of *Delta*, a locus required for ectodermal differentiation in *Drosophila*. *Dev. Genet.* **10**, 261–272.

76. Alton, A.K., Fechtel, K., Terry, A.L., Meikle, S.B., and Muskavitch, M.A.T. (1988). Cytogenetic definition and morphogenetic analysis of *Delta*, a gene affecting neurogenesis in *Drosophila melanogaster*. *Genetics* **118**, 235–245.

77. Alvarado, D. and Duffy, J.B. (2001). Functional characterization of KEKKON1 during *Drosophila* oogenesis. *Proc. 42nd Ann. Drosophila Res. Conf.* **Abstracts Vol.**, a167.

78. Alves, G., Limbourg-Bouchon, B., Tricoire, H., Brissard-Zahraoui, J., Lamour-Isnard, C., and Busson, D. (1998). Modulation of Hedgehog target gene expression by the Fused serine-threonine kinase in wing imaginal discs. *Mechs. Dev.* **78**, 17–31.

79. Ambros, V. (1999). Cell cycle-dependent sequencing of cell fate decisions in *Caenorhabditis elegans* vulva precursor cells. *Development* **126**, 1947–1956.

80. Amin, A., Li, Y., and Finkelstein, R. (1999). Hedgehog activates the EGF receptor pathway during *Drosophila* head development. *Development* **126**, 2623–2630.

81. Anderson, D.T. (1963). The embryology of *Dacus tryoni*. 2. Development of imaginal discs in the embryo. *J. Embryol. Exp. Morph.* **11**, 339–351.

82. Anderson, D.T. (1972). The development of holometabolous insects. *In* "Developmental Systems: Insects," Vol. 1 (S.J. Counce and C.H. Waddington, eds.). Acad. Pr.: New York, pp. 165–242. [See also 2001 *Evol. Dev.* **3**, 59–72.]

83. Anderson, J.M. (1996). Cell signalling: MAGUK magic. *Curr. Biol.* **6**, 382–384.

84. Anderson, M.G., Certel, S.J., Certel, K., Lee, T., Montell, D.J., and Johnson, W.A. (1996). Function of the *Drosophila* POU domain transcription factor Drifter as an upstream regulator of Breathless receptor tyrosine kinase expression in developing trachea. *Development* **122**, 4169–4178.

85. Anderson, M.G., Perkins, G.L., Chittick, P., Shrigley, R.J., and Johnson, W.A. (1995). *drifter*, a *Drosophila* POU-domain transcription factor, is required for correct differentiation and migration of tracheal cells and midline glia. *Genes Dev.* **9**, 123–137.

86. Anderton, B.H. (1999). Alzheimer's disease: clues from flies and worms. *Curr. Biol.* **9**, R106–R109.

87. Andrés, V., Chiara, M.D., and Mahdavi, V. (1994). A new bipartite DNA-binding domain: cooperative interaction

between the *cut* repeat and homeo domain of the *cut* homeo proteins. *Genes Dev.* **8**, 245–257.

88. Andrew, D.J., Baig, A., Bhanot, P., Smolik, S.M., and Henderson, K.D. (1997). The *Drosophila dCREB-A* gene is required for dorsal/ventral patterning of the larval cuticle. *Development* **124**, 181–193.

89. Andrew, D.J., Horner, M.A., Petitt, M.G., Smolik, S.M., and Scott, M.P. (1994). Setting limits on homeotic gene function: restraint of *Sex combs reduced* activitiy by *teashirt* and other homeotic genes. *EMBO J.* **13**, 1132–1144.

90. Andrews, D.W. (2000). Transport across membranes: a question of navigation. *Cell* **102**, 139–144.

91. Angerer, L.M. and Angerer, R.C. (1999). Regulative development of the sea urchin embryo: Signalling cascades and morphogen gradients. *Sems. Cell Dev. Biol.* **10**, 327–334.

92. Apidianakis, Y., Nagel, A.C., Chalkiadaki, A., Preiss, A., and Delidakis, C. (1999). Overexpression of the *m4* and *mα* genes of the *E(spl)-Complex* antagonizes Notch mediated lateral inhibition. *Mechs. Dev.* **86**, 39–50.

93. Aplin, A.C. and Kaufman, T.C. (1997). Homeotic transformation of legs to mouthparts by *proboscipedia* expression in *Drosophila* imaginal discs. *Mechs. Dev.* **62**, 51–60.

94. Appel, L.F., Prout, M., Abu-Shumays, R., Hammonds, A., Garbe, J.C., Fristrom, D., and Fristrom, J. (1993). The *Drosophila* Stubble-stubbloid gene encodes an apparent transmembrane serine protease required for epithelial morphogenesis. *Proc. Natl. Acad. Sci. USA* **90**, 4937–4941.

95. Apter, M.J. and Wolpert, L. (1965). Cybernetics and development. I. Information theory. *J. Theor. Biol.* **8**, 244–257.

96. Araujo, H. and Bier, E. (2000). *sog* and *dpp* exert opposing maternal functions to modify Toll signaling and pattern the dorsoventral axis of the *Drosophila* embryo. *Development* **127**, 3631–3644.

97. Arber, S. and Caroni, P. (1996). Specificity of single LIM motifs in targeting and LIM/LIM interactions in situ. *Genes Dev.* **10**, 289–300.

98. Arbib, M.A. (1972). Automata theory in the context of theoretical embryology. *In* "Foundations of Mathematical Biology," Vol. 2 (R. Rosen, ed.). Acad. Pr.: New York, pp. 141–215.

99. Argyris, T.S. (1972). Chalones and the control of normal, regenerative, and neoplastic growth of the skin. *Amer. Zool.* **12**, 137–149.

100. Arias, J., Alberts, A.S., Brindle, P., Claret, F.X., Smeal, T., Karin, M., Feramisco, J., and Montminy, M. (1994). Activation of cAMP and mitogen responsive genes relies on a common nuclear factor. *Nature* **370**, 226–229.

101. Arking, R. (1975). Temperature-sensitive cell-lethal mutants of *Drosophila*: isolation and characterization. *Genetics* **80**, 519–537.

102. Arnheim, N., Jr. (1967). The regional effects of two mutants in *Drosophila* analyzed by means of mosaics. *Genetics* **56**, 253–263.

103. Arnone, M.I. and Davidson, E.H. (1997). The hardwiring of development: organization and function of genomic regulatory systems. *Development* **124**, 1851–1864.

104. Arnosti, D.N., Barolo, S., Levine, M., and Small, S. (1996). The eve stripe 2 enhancer employs multiple modes of transcriptional synergy. *Development* **122**, 205–214.

105. Arnosti, D.N., Gray, S., Barolo, S., Zhou, J., and Levine, M. (1996). The gap protein knirps mediates both quenching and direct repression in the *Drosophila* embryo. *EMBO J.* **15**, 3659–3666.

106. Aronheim, A., Engelberg, D., Li, N., Al-Alawi, N., Schlessinger, J., and Karin, M. (1994). Membrane targeting of the nucleotide exchange factor Sos is sufficient for activating the Ras signaling pathway. *Cell* **78**, 949–961.

107. Aronson, B.D., Fisher, A.L., Blechman, K., Caudy, M., and Gergen, J.P. (1997). Groucho-dependent and -independent repression activities of Runt domain proteins. *Mol. Cell. Biol.* **17**, 5581–5587.

108. Arora, K., Dai, H., Kazuko, S.G., Jamal, J., O'Connor, M.B., Letsou, A., and Warrior, R. (1995). The *Drosophila schnurri* gene acts in the Dpp/TGFβ signaling pathway and encodes a transcription factor homologous to the human MBP family. *Cell* **81**, 781–790.

109. Arora, K., Levine, M.S., and O'Connor, M.B. (1994). The *screw* gene encodes a ubiquitously expressed member of the TGF-β family required for specification of dorsal cell fates in the *Drosophila* embryo. *Genes Dev.* **8**, 2588–2601.

110. Arora, K. and Nüsslein-Volhard, C. (1992). Altered mitotic domains reveal fate map changes in *Drosophila* embryos mutant for zygotic dorsoventral patterning genes. *Development* **114**, 1003–1024.

111. Arora, K., Rodrigues, V., Joshi, S., Shanbhag, S., and Siddiqi, O. (1987). A gene affecting the specificity of the chemosensory neurons of *Drosophila*. *Nature* **330**, 62–63.

112. Artavanis-Tsakonas, S., Matsuno, K., and Fortini, M.E. (1995). Notch signaling. *Science* **268**, 225–232.

113. Artavanis-Tsakonas, S., Rand, M.D., and Lake, R.J. (1999). Notch signaling: cell fate control and signal integration in development. *Science* **284**, 770–776.

114. Artavanis-Tsakonas, S. and Simpson, P. (1991). Choosing a cell fate: a view from the *Notch* locus. *Trends Genet.* **7**, 403–408.

115. Artero, R.D., Akam, M., and Pérez-Alonso, M. (1992). Oligonucleotide probes detect splicing variants *in situ* in *Drosophila* embryos. *Nucl. Acids Res.* **20**, 5687–5690.

116. Ascano, M., Nybakken, K.E., Vallance, J.E., and Robbins, D.J. (2001). Disruption of a Hedgehog signaling complex. *Proc. 42nd Ann. Drosophila Res. Conf.* **Abstracts Vol.**, a160.

117. Ashburner, M. (1982). The genetics of a small autosomal region of *Drosophila melanogaster* containing the structural gene for alcohol dehydrogenase. III. Hypomorphic and hypermorphic mutations affecting the expression of *Hairless*. *Genetics* **101**, 447–459.

118. Ashburner, M. (1989). "*Drosophila*: A Laboratory Handbook." CSH Pr., Cold Spring Harbor, N.Y.

119. Ashburner, M. (1993). Epilogue. *In* "The Development of *Drosophila melanogaster*," Vol. 2 (M. Bate and A. Martinez Arias, eds.). Cold Spring Harbor Lab. Pr.: Plainview, N. Y., pp. 1493–1506.

120. Ashburner, M. (1995). Maps and legends. *Trends Genet.* **11**, 33–34.

121. Ashburner, M. and Berendes, H.D. (1978). Puffing of polytene chromosomes. *In* "The Genetics and Biology of *Drosophila*," Vol. 2b (M. Ashburner and T.R.F. Wright, eds.). Acad. Pr.: New York, pp. 315–395.

122. Ashburner, M., Carson, H.L., Novitski, E., Thompson, J.N., and Wright, T.R.F., eds. (1976–1986). "The Genetics and Biology of *Drosophila*." Vol. 1 (a-c), 2 (a-d), 3 (a-e). Acad. Pr., London.

123. Ashburner, M., Chihara, C., Meltzer, P., and Richards, G. (1974). Temporal control of puffing activity in polytene chromosomes. *Cold Spr. Harb. Symp. Quant. Biol.* **38**, 655–662.

124. Ashburner, M. and Drysdale, R. (1994). FlyBase–The

Drosophila genetic database. *Development* **120**, 2077–2079.

125. Ashburner, M. and Lawrence, P. (1988). Fruitful fruitflies in Crete. *Nature* **336**, 620–621.

126. Ashburner, M., Tsubota, S., and Woodruff, R.C. (1982). The genetics of a small chromosome region of *Drosophila melanogaster* containing the structural gene for alcohol dehydrogenase. IV. Scutoid, an antimorphic mutation. *Genetics* **102**, 401–420.

127. Ashe, H.L., Mannervik, M., and Levine, M. (2000). Dpp signaling thresholds in the dorsal ectoderm of the *Drosophila* embryo. *Development* **127**, 3305–3312.

128. Ashley, J.A. and Katz, F.N. (1994). Competition and position-dependent targeting in the development of the *Drosophila* R7 visual projections. *Development* **120**, 1537–1547.

129. Aslam, H. and Baron, M. (2001). Suppressor of deltex requires a ubiquitin ligase domain to downregulate the Notch signal. *Proc. 42nd Ann. Drosophila Res. Conf.* **Abstracts Vol.**, a168.

130. Aspland, S.E. and White, R.A.H. (1997). Nucleocytoplasmic localisation of *extradenticle* protein is spatially regulated throughout development in *Drosophila*. *Development* **124**, 741–747.

131. Assa-Munt, N., Mortishire-Smith, R.J., Aurora, R., Herr, W., and Wright, P.E. (1993). The solution structure of the Oct-1 POU-specific domain reveals a striking similarity to the bacteriophage l repressor DNA-binding domain. *Cell* **73**, 193–205.

132. Aster, J.C., Robertson, E.S., Hasserjian, R.P., Turner, J.R., Kieff, E., and Sklar, J. (1997). Oncogenic forms of NOTCH1 lacking either the primary binding site for RBP-Jκ or nuclear localization sequences retain the ability to associate with RBP-Jκ and activate transcription. *J. Biol. Chem.* **272** #17, 11336–11343.

133. Astor, J.C. and Adami, C. (2001). A developmental model for the evolution of artificial neural networks. *Artif. Life* **6**, 189–218.

134. Asturias, F.J., Jiang, Y.W., Myers, L.C., Gustafsson, C.M., and Kornberg, R.D. (1999). Conserved structures of mediator and RNA polymerase II holoenzyme. *Science* **283**, 985–987.

135. Atchley, W.R. and Fitch, W.M. (1997). A natural classification of the basic helix-loop-helix class of transcription factors. *Proc. Natl. Acad. Sci. USA* **94**, 5172–5176.

136. Atlan, H. and Koppel, M. (1990). The cellular computer DNA: program or data. *Bull. Math. Biol.* **52**, 335–348.

137. Attisano, L. and Wrana, J.L. (1998). Mads and Smads in TGFβ signalling. *Curr. Opin. Cell Biol.* **10**, 188–194.

138. Attisano, L. and Wrana, J.L. (2000). Smads as transcriptional co-modulators. *Curr. Opin. Cell Biol.* **12**, 235–243.

139. Attisano, L., Wrana, J.L., López-Casillas, F., and Massagué, J. (1994). TGF-β receptors and actions. *Bioch. Biophys. Acta* **1222**, 71–80.

140. Audibert, A., Juge, F., and Simonelig, M. (1998). The suppressor of forked protein of *Drosophila*, a homologue of the human 77K protein required for mRNA 3′-end formation, accumulates in mitotically-active cells. *Mechs. Dev.* **72**, 53–63.

141. Audibert, A. and Simonelig, M. (1999). The *suppressor of forked* gene of *Drosophila*, which encodes a homologue of human CstF-77K involved in mRNA 3′-end processing, is required for progression through mitosis. *Mechs. Dev.* **82**, 41–50.

142. Auerbach, C. (1936). The development of the legs, wings, and halteres in wild type and some mutant strains of *Drosophila melanogaster*. *Trans. Roy. Soc. Edin.* **58**, 787–815.

143. Avedisov, S.N., Krasnoselskaya, I., Mortin, M., and Thomas, B.J. (2000). Roughex mediates G_1 arrest through a physical association with Cyclin A. *Mol. Cell. Biol.* **20**, 8220–8229.

144. Averof, M. (1997). Arthropod evolution: Same *Hox* genes, different body plans. *Curr. Biol.* **7**, R634–R636.

145. Averof, M. and Cohen, S.M. (1997). Evolutionary origin of insect wings from ancestral gills. *Nature* **385**, 627–630.

146. Averof, M., Dawes, R., and Ferrier, D. (1996). Diversification of arthropod Hox genes as a paradigm for the evolution of gene functions. *Sems. Cell Dev. Biol.* **7**, 539–551.

147. Avery, L. and Wasserman, S. (1992). Ordering gene function: the interpretation of epistasis in regulatory hierarchies. *Trends Genet.* **8**, 312–316.

148. Avruch, J., Zhang, X.-f., and Kyriakis, J.M. (1994). Raf meets Ras: completing the framework of a signal transduction pathway. *Trends Biochem. Sci.* **19**, 279–284.

149. Awasaki, T. and Kimura, K.-i. (1997). *pox-neuro* is required for development of chemosensory bristles in *Drosophila*. *J. Neurobiol.* **32**, 707–721.

150. Awasaki, T. and Kimura, K.-i. (2001). Multiple function of *poxn* gene in larval PNS development and in adult appendage formation of *Drosophila*. *Dev. Genes Evol.* **211**, 20–29.

151. Axelrod, J.D., Matsuno, K., Artavanis-Tsakonas, S., and Perrimon, N. (1996). Interaction between Wingless and Notch signaling pathways mediated by Dishevelled. *Science* **271**, 1826–1832. [See also 2001 *Curr. Biol.* **11**, 1729–1738.]

152. Axelrod, J.D., Miller, J.R., Shulman, J.M., Moon, R.T., and Perrimon, N. (1998). Differential recruitment of Dishevelled provides signaling specificity in the planar cell polarity and Wingless signaling pathways. *Genes Dev.* **12**, 2610–2622.

153. Ayyar, S. and White, R.A.H. (2001). *achintya* and *vismay*: two novel, tandem-repeated TALE-class homeobox genes. *Proc. 42nd Ann. Drosophila Res. Conf.* **Abstracts Vol.**, a177.

154. Aza-Blanc, P. and Kornberg, T.B. (1999). Ci: a complex transducer of the Hedgehog signal. *Trends Genet.* **15**, 458–462.

155. Aza-Blanc, P., Ramírez-Weber, F.-A., Laget, M.-P., Schwartz, C., and Kornberg, T.B. (1997). Proteolysis that is inhibited by Hedgehog targets Cubitus interruptus protein to the nucleus and converts it to a repressor. *Cell* **89**, 1043–1053.

156. Azpiazu, N. and Morata, G. (1998). Functional and regulatory interactions between *Hox* and *extradenticle* genes. *Genes Dev.* **12**, 261–273.

157. Azpiazu, N. and Morata, G. (2000). Function and regulation of *homothorax* in the wing imaginal disc of *Drosophila*. *Development* **127**, 2685–2693.

158. Babu, P. (1977). Early developmental subdivisions of the wing disk in *Drosophila*. *Mol. Gen. Genet.* **151**, 289–294.

159. Babu, P. and Bhat, S.G. (1986). Autonomy of the *wingless* mutation in *Drosophila melanogaster*. *Mol. Gen. Genet.* **205**, 483–486.

160. Bach, I. (2000). The LIM domain: regulation by association. *Mechs. Dev.* **91**, 5–17.

161. Bach, I., Rodriguez-Esteban, C., Carriére, C., Bhushan, A., Krones, A., Rose, D.W., Glass, C.K., Andersen, B., Izpisúa

Belmonte, J.C., and Rosenfeld, M.G. (1999). RLIM inhibits functional activity of LIM homeodomain transcription factors via recruitment of the histone deacetylase complex. *Nature Genet.* **22**, 394–399.

162. Bachiller, D., Macías, A., Duboule, D., and Morata, G. (1994). Conservation of a functional hierarchy between mammalian and insect Hox/HOM genes. *EMBO J.* **13**, 1930–1941.

163. Bachmann, A. and Knust, E. (1998). Dissection of *cis*-regulatory elements of the *Drosophila* gene *Serrate*. *Dev. Genes Evol.* **208**, 346–351.

164. Bachmann, A. and Knust, E. (1998). Positive and negative control of *Serrate* expression during early development of the *Drosophila* wing. *Mechs. Dev.* **76**, 67–78.

165. Bae, K., Lee, C., Sidote, D., Chuang, K.-Y., and Edery, I. (1998). Circadian regulation of a *Drosophila* homolog of the mammalian *Clock* gene: PER and TIM function as positive regulators. *Mol. Cell. Biol.* **18**, 6142–6151.

166. Baeg, G.-H., Lin, X., Khare, N., Baumgartner, S., and Perrimon, N. (2001). Heparan sulfate proteoglycans are critical for the organization of the extracellular distribution of Wingless. *Development* **128**, 87–94.

167. Baek, K.-H., Fabian, J.R., Sprenger, F., Morrison, D.K., and Ambrosio, L. (1996). The activity of D-raf in Torso signal transduction is altered by serine substitution, N-terminal deletion and membrane targeting. *Dev. Biol.* **175**, 191–204.

168. Baeuerle, P.A. (1998). IκB-NF-κB structures: At the interface of inflammation control. *Cell* **95**, 729–731.

169. Bai, C., Sen, P., Hofmann, K., Ma, L., Goebl, M., Harper, J.W., and Elledge, S.J. (1996). *SKP1* connects cell cycle regulators to the ubiquitin proteolysis machinery through a novel motif, the F-box. *Cell* **86**, 263–274.

170. Bai, J.-M., Chiu, W.-H., Wang, J.-C., Tzeng, T.-H., Perrimon, N., and Hsu, J.-C. (2001). The cell adhesion molecule Echinoid defines a new pathway that antagonizes the *Drosophila* EGF receptor signaling pathway. *Development* **128**, 591–601.

171. Bailey, A.M. and Posakony, J.W. (1995). Suppressor of Hairless directly activates transcription of *Enhancer of split* Complex genes in response to Notch receptor activity. *Genes Dev.* **9**, 2609–2622.

172. Baker, A., Kaplan, C.P., and Pool, M.R. (1996). Protein targeting and translocation; a comparative survey. *Biol. Rev.* **71**, 637–702.

173. Baker, B.S., Taylor, B.J., and Hall, J.C. (2001). Are complex behaviors specified by dedicated regulatory genes? Reasoning from *Drosophila*. *Cell* **105**, 13–24.

174. Baker, J.C. and Harland, R.M. (1997). From receptor to nucleus: the Smad pathway. *Curr. Opin. Gen. Dev.* **7**, 467–473.

175. Baker, N.E. (1987). Molecular cloning of sequences from *wingless*, a segment polarity gene in *Drosophila*: the spatial distribution of a transcript in embryos. *EMBO J.* **6**, 1765–1773.

176. Baker, N.E. (1988). Localization of transcripts from the *wingless* gene in whole *Drosophila* embryos. *Development* **103**, 289–298.

177. Baker, N.E. (1988). Transcription of the segment-polarity gene *wingless* in the imaginal discs of *Drosophila*, and the phenotype of a pupal-lethal *wg* mutation. *Development* **102**, 489–497.

178. Baker, N.E. (2000). Notch signaling in the nervous system. Pieces still missing from the puzzle. *BioEssays* **22**, 264–273.

179. Baker, N.E., Mlodzik, M., and Rubin, G.M. (1990). Spacing differentiation in the developing *Drosophila* eye: a fibrinogen-related lateral inhibitor encoded by *scabrous*. *Science* **250**, 1370–1377.

180. Baker, N.E. and Rubin, G.M. (1989). Effect on eye development of dominant mutations in *Drosophila* homologue of the EGF receptor. *Nature* **340**, 150–153.

181. Baker, N.E. and Rubin, G.M. (1989). *Ellipse* mutations in the *Drosophila* homologue of the EGF receptor affect pattern formation, cell division, and cell death in eye imaginal discs. *Dev. Biol.* **150**, 381–396.

182. Baker, N.E., Yu, S., and Han, D. (1996). Evolution of proneural atonal expression during distinct regulatory phases in the developing *Drosophila* eye. *Curr. Biol.* **6**, 1290–1301.

183. Baker, N.E. and Yu, S.-Y. (1997). Proneural function of neurogenic genes in the developing *Drosophila* eye. *Curr. Biol.* **7**, 122–132. [See also 2001 *Development* **128**, 3889–3898.]

184. Baker, N.E. and Yu, S.-Y. (1998). The R8-photoreceptor equivalence group in *Drosophila*: fate choice precedes regulated *Delta* transcription and is independent of *Notch* gene dose. *Mechs. Dev.* **74**, 3–14.

185. Baker, N.E. and Yu, S.-Y. (2001). The EGF receptor defines domains of cell cycle progression and survival to regulate cell number in the developing *Drosophila* eye. *Cell* **104**, 699–708. [See also 2001 *Sems. Cell Dev. Biol.* **12**, 499–507.]

186. Baker, N.E. and Zitron, A.E. (1995). *Drosophila* eye development: *Notch* and *Delta* amplify a neurogenic pattern conferred on the morphogenetic furrow by *scabrous*. *Mechs. Dev.* **49**, 173–189.

187. Baker, T.A. and Bell, S.P. (1998). Polymerases and the replisome: machines within machines. *Cell* **92**, 295–305.

188. Baker, W.K. (1967). A clonal system of differential gene activity in *Drosophila*. *Dev. Biol.* **16**, 1–17.

189. Baker, W.K. (1978). A clonal analysis reveals early developmental restrictions in the *Drosophila* head. *Dev. Biol.* **62**, 447–463.

190. Baker, W.K., Marcey, D.J., and McElwain, M.C. (1985). On the development of ectopic eyes in *Drosophila melanogaster* produced by the mutation extra eye (*ee*). *Genetics* **111**, 67–88.

191. Balcells, L., Modolell, J., and Ruiz-Gómez, M. (1988). A unitary basis for different *Hairy-wing* mutations of *Drosophila melanogaster*. *EMBO J.* **7**, 3899–3906.

192. Bally-Cuif, L., Dubois, L., and Vincent, A. (1998). Molecular cloning of *Zcoe2*, the zebrafish homolog of *Xenopus Xcoe2* and mouse *EBF-2*, and its expression during primary neurogenesis. *Mechs. Dev.* **77**, 85–90.

193. Bandziulis, R.J., Swanson, M.S., and Dreyfuss, G. (1989). RNA-binding proteins as developmental regulators. *Genes Dev.* **3**, 431–437.

194. Banerjee, U., Renfranz, P.J., Hinton, D.R., Rabin, B.A., and Benzer, S. (1987). The *sevenless*+ protein is expressed apically in cell membranes of developing *Drosophila* retina; it is not restricted to cell R7. *Cell* **51**, 151–158.

195. Banerjee, U., Renfranz, P.J., Pollock, J.A., and Benzer, S. (1987). Molecular characterization and expression of *sevenless*, a gene involved in neuronal pattern formation in the *Drosophila* eye. *Cell* **49**, 281–291.

196. Banerjee, U. and Zipursky, S.L. (1990). The role of cell-cell interaction in the development of the *Drosophila* visual system. *Neuron* **4**, 177–187.

197. Bang, A.G., Bailey, A.M., and Posakony, J.W. (1995). *Hairless* promotes stable commitment to the sensory organ precursor cell fate by negatively regulating the activity of the *Notch* signaling pathway. *Dev. Biol.* **172**, 479–494.

198. Bang, A.G., Hartenstein, V., and Posakony, J.W. (1991). *Hairless* is required for the development of adult sensory organ precursor cells in *Drosophila. Development* **111**, 89–104.

199. Bang, A.G. and Kintner, C. (2000). Rhomboid and Star facilitate presentation and processing of the *Drosophila* TGF-α homolog Spitz. *Genes Dev.* **14**, 177–186. [See also 2001 references: *Cell* **107**, 161–171 and 173–182.]

200. Bang, A.G. and Posakony, J.W. (1992). The *Drosophila* gene *Hairless* encodes a novel basic protein that controls alternative cell fates in adult sensory organ development. *Genes Dev.* **6**, 1752–1769.

201. Bangs, P. and White, K. (2000). Regulation and execution of apoptosis during *Drosophila* development. *Dev. Dynamics* **218**, 68–79.

202. Bannister, A.J. and Kouzarides, T. (1996). The CBP coactivator is a histone acetyltransferase. *Nature* **384**, 641–643.

203. Bantignies, F., Goodman, R.H., and Smolik, S.M. (2000). Functional interaction between the coactivator *Drosophila* CREB-binding protein and ASH1, a member of the trithorax group of chromatin modifiers. *Mol. Cell. Biol.* **20**, 9317–9330. [See also 2001 *Science* **294**, 1331–1334.]

204. Baonza, A., Casci, T., and Freeman, M. (2001). A primary role for the epidermal growth factor receptor in ommatidial spacing in the *Drosophila* eye. *Curr. Biol.* **11**, 396–404.

205. Baonza, A., de Celis, J.F., and García-Bellido, A. (2000). Relationships between *extramacrochaetae* and *Notch* signalling in *Drosophila* wing development. *Development* **127**, 2383–2393.

206. Baonza, A. and García-Bellido, A. (1999). Dual role of *extramacrochaetae* in cell proliferation and cell differentiation during wing morphogenesis in *Drosophila. Mechs. Dev.* **80**, 133–146.

207. Baonza, A., Roch, F., and Martín-Blanco, E. (2000). DER signaling restricts the boundaries of the wing field during *Drosophila* development. *Proc. Natl. Acad. Sci. USA* **97**, 7331–7335.

208. Bar-Sagi, D., Rotin, D., Batzer, A., Mandiyan, V., and Schlessinger, J. (1993). SH3 domains direct cellular localization of signaling molecules. *Cell* **74**, 83–91.

209. Barbash, D.A. and Cline, T.W. (1995). Genetic and molecular analysis of the autosomal component of the primary sex determination signal of *Drosophila melanogaster. Genetics* **141**, 1451–1471.

210. Bard, J.B.L. (1984). Butterfly wing patterns: how good a determining mechanism is the simple diffusion of a single morphogen? *J. Embryol. Exp. Morph.* **84**, 255–274.

211. Bardwell, V.J. and Treisman, R. (1994). The POZ domain: A conserved protein-protein interaction motif. *Genes Dev.* **8**, 1664–1677.

212. Barges, S., Mihaly, J., Galloni, M., Hagstrom, K., Müller, M., Shanower, G., Schedl, P., Gyurkovics, H., and Karch, F. (2000). The *Fab-8* boundary defines the distal limit of the bithorax complex *iab-7* domain and insulates *iab-7* from initiation elements and a PRE in the adjacent *iab-8* domain. *Development* **127**, 779–790.

213. Barinaga, M. (1991). Dimers direct development. *Science* **251**, 1176–1177.

214. Barinaga, M. (1995). Focusing on the *eyeless* gene. *Science* **267**, 1766–1767.

215. Barinaga, M. (1995). Two major signaling pathways meet at MAP kinase. *Science* **269**, 1673.

216. Barlow, P.W. (1978). Endopolyploidy: towards an understanding of its biological significance. *Acta Biotheor.* **27**, 1–18.

217. Barnes, J., ed. (1984). "The Complete Works of Aristotle." Vol. 1. Princeton Univ. Pr., Princeton, N. J. [*N.B.*: Aristotle's comment about vinegar "gnats" – which Peyer (1947) and Sturtevant (1965, p. 43) conclude must be *D. melanogaster* – is in *History of Animals*, Book 5, part 19, line 552b5; *cf.* Book 4:8:535a4, and *Generation of Animals*, Book 1:16:721a5 *ff.*]

218. Barolo, S., Walker, R.G., Polyanovsky, A.D., Freschi, G., Keil, T., and Posakony, J.W. (2000). A Notch-independent activity of Suppressor of Hairless is required for normal mechanoreceptor physiology. *Cell* **103**, 957–969. [See also 2001 *Curr. Biol.* **11**, 789–792.]

219. Barrington, E.J.W., ed. (1979). "Hormones and Evolution." Vol. 1. Acad. Pr., New York.

220. Barrio, R., de Celis, J.F., Bolshakov, S., and Kafatos, F.C. (1999). Identification of regulatory regions driving the expression of the *Drosophila spalt* complex at different developmental stages. *Dev. Biol.* **215**, 33–47.

221. Barrio, R., Shea, M.J., Carulli, J., Lipkow, K., Gaul, U., Frommer, G., Schuh, R., Jäckle, H., and Kafatos, F.C. (1996). The *spalt-related* gene of *Drosophila melanogaster* is a member of an ancient gene family, defined by the adjacent, region-specific homeotic gene *spalt. Dev. Genes Evol.* **206**, 315–325.

222. Bartel, P.L., Chien, C.-T., Sternglanz, R., and Fields, S. (1993). Using the two-hybrid system to detect protein-protein interactions. *In* "Cellular Interactions in Development: A Practical Approach," (D.A. Hartley, ed.). Oxford Univ. Pr.: New York, pp. 153–179.

223. Bashirullah, A., Halsell, S.R., Cooperstock, R.L., Kloc, M., Karaiskakis, A., Fisher, W.W., Fu, W., Hamilton, J.K., Etkin, L.D., and Lipshitz, H.D. (1999). Joint action of two RNA degradation pathways controls the timing of maternal transcript elimination at the midblastula transition in *Drosophila melanogaster. EMBO J.* **18**, 2610–2620.

224. Basler, K. (2000). Waiting periods, instructive signals and positional information. *EMBO J.* **19**, 1169–1175.

225. Basler, K., Christen, B., and Hafen, E. (1991). Ligand-independent activation of the Sevenless receptor tyrosine kinase changes the fate of cells in the developing *Drosophila* eye. *Cell* **64**, 1069–1081.

226. Basler, K. and Hafen, E. (1988). Control of photoreceptor cell fate by the *sevenless* protein requires a functional tyrosine kinase domain. *Cell* **54**, 299–311.

227. Basler, K. and Hafen, E. (1988). *Sevenless* and *Drosophila* eye development: a tyrosine kinase controls cell fate. *Trends Genet.* **4**, 74–79.

228. Basler, K. and Hafen, E. (1989). Dynamics of *Drosophila* eye development and temporal requirements of *sevenless* expression. *Development* **107**, 723–731.

229. Basler, K. and Hafen, E. (1991). Specification of cell fate in the developing eye of *Drosophila. BioEssays* **13**, 621–631.

230. Basler, K., Siegrist, P., and Hafen, E. (1989). The spatial and temporal expression pattern of *sevenless* is exclusively controlled by gene-internal elements. *EMBO J.* **8**, 2381–2386.

231. Basler, K. and Struhl, G. (1994). Compartment boundaries and the control of *Drosophila* limb pattern by *hedgehog* protein. *Nature* **368**, 208–214.

232. Basler, K., Yen, D., Tomlinson, A., and Hafen, E. (1990). Reprogramming cell fate in the developing *Drosophila*

retina: transformation of R7 cells by ectopic expression of *rough*. *Genes Dev.* **4**, 728–739.

233. Bass, B.L. (2000). Double-stranded RNA as a template for gene silencing. *Cell* **101**, 235–238.

234. Bastiani, M.J., De Couet, H.G., Quinn, J.M.A., Karlstrom, R.O., Kotrla, K., Goodman, C.S., and Ball, E.E. (1992). Position-specific expression of the annulin protein during grasshopper embryogenesis. *Dev. Biol.* **154**, 129–142.

235. Batchelor, A.H., Piper, D.E., de la Brousse, F.C., McKnight, S.L., and Wolberger, C. (1998). The structure of GABPα/β: an ETS domain-ankyrin repeat heterodimer bound to DNA. *Science* **279**, 1037–1041.

236. Bate, M. (1993). The mesoderm and its derivatives. *In* "The Development of *Drosophila melanogaster*," Vol. 2 (M. Bate and A. Martinez Arias, eds.). Cold Spring Harbor Lab. Pr.: Plainview, N. Y., pp. 1013–1090.

237. Bate, M. and Martinez Arias, A. (1991). The embryonic origin of imaginal discs in *Drosophila*. *Development* **112**, 755–761.

238. Bate, M. and Martinez Arias, A., eds. (1993). "The Development of *Drosophila melanogaster*." Cold Spring Harbor Lab. Pr., Plainview, N. Y.

239. Bate, M., Rushton, E., and Currie, D.A. (1991). Cells with persistent *twist* expression are the embryonic precursors of adult muscles in *Drosophila*. *Development* **113**, 79–89.

240. Bateson, W. (1894). "Materials for the Study of Variation Treated with Especial Regard to Discontinuity in the Origin of Species." MacMillan, London. (*N.B.*: Quotes were excerpted from pages 16, 68, and 85.)

241. Batterham, P., Crew, J.R., Sokac, A.M., Andrews, J.R., Pasquini, G.M.F., Davies, A.G., Stocker, R.F., and Pollock, J.A. (1996). Genetic analysis of the *lozenge* gene complex in *Drosophila melanogaster*: adult visual system phenotypes. *J. Neurogenetics* **10**, 193–220.

242. Baumgartner, S. and Noll, M. (1991). Network of interactions among pair-rule genes regulating *paired* expression during primordial segmentation of *Drosophila*. *Mechs. Dev.* **33**, 1–18.

243. Bausenwein, B.S., Schmidt, M., Mielke, B., and Raabe, T. (2000). In vivo functional analysis of the Daughter of Sevenless protein in receptor tyrosine kinase signaling. *Mechs. Dev.* **90**, 205–215.

244. Bax, B. and Jhoti, H. (1995). Putting the pieces together. *Curr. Biol.* **5**, 1119–1121.

245. Baxevanis, A.D. and Vinson, C.R. (1993). Interactions of coiled coils in transcription factors: where is the specificity? *Curr. Op. Gen. Dev.* **3**, 278–285.

246. Bayer, C., von Kalm, L., and Fristrom, J.W. (1996). Gene regulation in imaginal disc and salivary gland development during *Drosophila* metamorphosis. *In* "Metamorphosis: Postembryonic Reprogramming of Gene Expression in Amphibian and Insect Cells," (L.I. Gilbert, J.R. Tata, and B.G. Atkinson, eds.). Acad. Pr.: New York, pp. 321–361.

247. Bayer, C.A., Holley, B., and Fristrom, J.W. (1996). A switch in *Broad-Complex* zinc-finger isoform expression is regulated posttranscriptionally during the metamorphosis of *Drosophila* imaginal discs. *Dev. Biol.* **177**, 1–14. [See also 2001 *Development* **128**, 3729–3737.]

248. Bayer, C.A., von Kalm, L., and Fristrom, J.W. (1997). Relationships between protein isoforms and genetic functions demonstrate functional redundancy at the *Broad-Complex* during *Drosophila* metamorphosis. *Dev. Biol.* **187**, 267–282.

249. Baylies, M.K. and Bate, M. (1996). *twist*: a myogenic switch in *Drosophila*. *Science* **272**, 1481–1484.

250. Baylies, M.K., Bate, M., and Ruiz Gomez, M. (1997). The specification of muscle in *Drosophila*. *Cold Spr. Harb. Symp. Quant. Biol.* **62**, 385–393.

251. Baylies, M.K., Bate, M., and Ruiz Gomez, M. (1998). Myogenesis: a view from *Drosophila*. *Cell* **93**, 921–927.

252. Bazett-Jones, D.P., Côte, J., Landel, C.C., Peterson, C.L., and Workman, J.L. (1999). The SWI/SNF complex creates loop domains in DNA and polynucleosome arrays and can disrupt DNA-histone contacts within these domains. *Mol. Cell. Biol.* **19**, 1470–1478.

253. Bazett-Jones, D.P., Leblanc, B., Herfort, M., and Moss, T. (1994). Short-range DNA looping by the *Xenopus* HMG-box transcription factor, xUBF. *Science* **264**, 1134–1137.

254. Beachy, P.A. (1990). A molecular view of the *Ultrabithorax* homeotic gene of *Drosophila*. *Trends Genet.* **6**, 46–51.

255. Beachy, P.A., Cooper, M.K., Young, K.E., von Kessler, D.P., Park, W.-J., Tanaka Hall, T.M., Leahy, D.J., and Porter, J.A. (1997). Multiple roles of cholesterol in Hedgehog protein biogenesis and signaling. *Cold Spr. Harb. Symp. Quant. Biol.* **62**, 191–204.

256. Beachy, P.A., Helfand, S.L., and Hogness, D.S. (1985). Segmental distribution of bithorax complex proteins during *Drosophila* development. *Nature* **313**, 545–551.

257. Beadle, G.W. (1970). Alfred Henry Sturtevant (1891–1970). *Amer. Philos. Soc. Yearbook* **1970**, 166–171.

258. Beamonte, D. and Modolell, J. (1989). Search for *Drosophila* genes encoding a conserved domain present in the *achaete-scute* complex and *myc* proteins. *Mol. Gen. Genet.* **215**, 281–285.

259. Beardsley, T. (1991). Smart genes. *Sci. Am.* **265** #2, 86–95.

260. Becker, H.J. (1957). Über Röntgenmosaikflecken und Defektmutationen am Auge von *Drosophila* und die Entwicklungsphysiologie des Auges. *Z. Indukt. Abstamm. Vererbungsl.* **88**, 333–373.

261. Becker, H.J. (1966). Genetic and variegation mosaics in the eye of *Drosophila*. *Curr. Top. Dev. Biol.* **1**, 155–171.

262. Becskei, A. and Serrano, L. (2000). Engineering stability in gene networks by autoregulation. *Nature* **405**, 590–593.

263. Beeman, R.W., Stuart, J.J., Brown, S.J., and Denell, R.E. (1993). Structure and function of the homeotic gene complex (HOM-C) in the beetle, *Tribolium castaneum*. *BioEssays* **15**, 439–444.

264. Begemann, G., Michon, A.-M., van der Voorn, L., Wepf, R., and Mlodzik, M. (1995). The *Drosophila* orphan nuclear receptor Seven-up requires the Ras pathway for its function in photoreceptor determination. *Development* **121**, 225–235.

265. Begemann, G., Paricio, N., Artero, R., Kiss, I., Pérez-Alonso, M., and Mlodzik, M. (1997). *muscleblind*, a gene required for photoreceptor differentiation in *Drosophila*, encodes novel nuclear Cys₃His-type zinc-finger-containing proteins. *Development* **124**, 4321–4331.

266. Behrens, J., von Kries, J.P., Kühl, M., Bruhn, L., Wedlich, D., Grosschedl, R., and Birchmeier, W. (1996). Functional interaction of β-catenin with the transcription factor LEF-1. *Nature* **382**, 638–642.

267. Beiman, M., Shilo, B.-Z., and Volk, T. (1996). Heartless, a *Drosophila* FGF receptor homolog, is essential for cell migration and establishment of several mesodermal lineages. *Genes Dev.* **10**, 2993–3002.

268. Beitel, G.J. and Krasnow, M.A. (2000). Genetic control of

epithelial tube size in the *Drosophila* tracheal system. *Development* **127**, 3271–3282.

269. Bejsovec, A. (1999). Signal transduction: Wnt signalling shows its versatility. *Curr. Biol.* **9**, R684–R687.

270. Bejsovec, A. (2000). Wnt signaling: an embarrassment of receptors. *Curr. Biol.* **10**, R919–R922.

271. Bejsovec, A. and Martinez Arias, A. (1991). Roles of *wingless* in patterning the larval epidermis of *Drosophila*. *Development* **113**, 471–485.

272. Bejsovec, A. and Wieschaus, E. (1993). Segment polarity gene interactions modulate epidermal patterning in *Drosophila* embryos. *Development* **119**, 501–517.

273. Bejsovec, A. and Wieschaus, E. (1995). Signaling activities of the *Drosophila wingless* gene are separately mutable and appear to be transduced at the cell surface. *Genetics* **139**, 309–320.

274. Bell, A.C. and Felsenfeld, G. (1999). Stopped at the border: boundaries and insulators. *Curr. Opin. Gen. Dev.* **9**, 191–198.

275. Bell, A.C., West, A.G., and Felsenfeld, G. (2001). Insulators and boundaries: versatile regulatory elements in the eukaryotic genome. *Science* **291**, 447–450.

276. Bellaïche, Y., Gho, M., Kaltschmidt, J.A., Brand, A.H., and Schweisguth, F. (2001). Frizzled regulates localization of cell-fate determinants and mitotic spindle rotation during asymmetric cell division. *Nature Cell Biol.* **3**, 50–57.

277. Bellaiche, Y., The, I., and Perrimon, N. (1998). *Tout-velu* is a *Drosophila* homologue of the putative tumour suppressor *EXT-1* and is needed for Hh diffusion. *Nature* **394**, 85–88.

278. Bellen, H.J., O'Kane, C.J., Wilson, C., Grossniklaus, U., Pearson, R.K., and Gehring, W.J. (1989). P-element-mediated enhancer detection: a versatile method to study development in *Drosophila*. *Genes Dev.* **3**, 1288–1300.

279. Bellen, H.J. and Smith, R.F. (1995). FlyBase: a virtual *Drosophila* cornucopia. *Trends Genet.* **11**, 456–457.

280. Bellen, H.J., Wilson, C., and Gehring, W.J. (1990). Dissecting the complexity of the nervous system by enhancer detection. *BioEssays* **12**, 199–204.

281. Belote, J.M. and Baker, B.S. (1982). Sex determination in *Drosophila melanogaster*: analysis of transformer-2, a sex-transforming locus. *Proc. Natl. Acad. Sci. USA* **79**, 1568–1572.

282. Bender, W. (1985). Homeotic gene products as growth factors. *Cell* **43**, 559–560.

283. Bender, W., Akam, M., Karch, F., Beachy, P.A., Peifer, M., Spierer, P., Lewis, E.B., and Hogness, D.S. (1983). Molecular genetics of the bithorax complex in *Drosophila melanogaster*. *Science* **221**, 23–29.

284. Bender, W. and Hudson, A. (2000). P element homing to the *Drosophila* bithorax complex. *Development* **127**, 3981–3992.

285. Benedyk, M.J., Mullen, J.R., and DiNardo, S. (1994). *odd-paired*: a zinc finger pair-rule protein required for the timely activation of *engrailed* and *wingless* in *Drosophila* embryos. *Genes Dev.* **8**, 105–117.

286. Benezra, R. (1994). An intermolecular disulfide bond stabilizes E2A homodimers and is required for DNA binding at physiological temperatures. *Cell* **79**, 1057–1067.

287. Benlali, A., Draskovic, I., Hazelett, D.J., and Treisman, J.E. (2000). *act up* controls actin polymerization to alter cell shape and restrict Hedgehog signaling in the *Drosophila* eye disc. *Cell* **101**, 271–281.

288. Bennett, D.H. and Alphey, L.S. (2001). Role of PP1 in dpp/TGFβ signalling. *Proc. 42nd Ann. Drosophila Res. Conf.* **Abstracts Vol.**, a139.

289. Bennett, R.L. and Hoffmann, F.M. (1992). Increased levels of the *Drosophila* Abelson tyrosine kinase in nerves and muscles: subcellular localization and mutant phenotypes imply a role in cell-cell interactions. *Development* **116**, 953–966.

290. Benveniste, R.J., Thor, S., Thomas, J.B., and Taghert, P.H. (1998). Cell type-specific regulation of the *Drosophila FMRF-NH2* neuropeptide gene by Apterous, a LIM homeodomain transcription factor. *Development* **125**, 4757–4765.

291. Benzer, S. (1971). From the gene to behavior. *JAMA (J. Amer. Med. Assoc.)* **218**, 1015–1022.

292. Benzer, S. (1973). Genetic dissection of behavior. *Sci. Am.* **229** #6, 24–37.

293. Benzer, S. (1991). The fly and eye. *In* "Development of the Visual System," *Proc. Retina Res. Found. Symp.*, Vol. 3 (D.M.-K. Lam and C.J. Shatz, eds.). MIT Press: Cambridge, Mass., pp. 9–34.

294. Berardi, M.J., Sun, C., Zehr, M., Abildgaard, F., Peng, J., Speck, N.A., and Bushweller, J.H. (1999). The Ig fold of the core binding factor a Runt domain is a member of a family of structurally and functionally related Ig-fold DNA-binding domains. *Structure* **7**, 1247–1256.

295. Berdnik, D., Török, T., Bulgheresi, S., Petronczki, M., and Knoblich, J.A. (2000). Genetic control of asymmetric cell division. *Proc. 41st Ann. Drosophila Res. Conf.* **Abstracts Vol.**, a253.

296. Berg, J.M. and Shi, Y. (1996). The galvanization of biology: a growing appreciation for the roles of zinc. *Science* **271**, 1081–1085.

297. Berkowitz, A. (1996). Networks of neurons, networks of genes. *Neuron* **17**, 199–202.

298. Bernard, F. (1937). Recherches sur la morphogénèse des yeux composés d'arthropodes. Développement, croissance, réduction. *Bull. Biol. Fr. Belg.* **Suppl. 23**, 1–162 (+ 6 plates).

299. Bernfield, M., Götte, M., Park, P.W., Reizes, O., Fitzgerald, M.L., Lincecum, J., and Zako, M. (2000). Functions of cell surface heparan sulfate proteoglycans. *Annu. Rev. Biochem.* **68**, 729–777.

300. Berry, M. and Gehring, W. (2000). Phosphorylation status of the SCR homeodomain determines its functional activity: essential role for protein phosphatase 2A,B'. *EMBO J.* **19**, 2946–2957.

301. Bershadsky, A., Chausovsky, A., Becker, E., Lyubimova, A., and Geiger, B. (1996). Involvement of microtubules in the control of adhesion-dependent signal transduction. *Curr. Biol.* **6**, 1279–1289.

302. Berthelsen, J., Kilstrup-Nielsen, C., Blasi, F., Mavilio, F., and Zappavigna, V. (1999). The subcellular localization of PBX1 and EXD proteins depends on nuclear import and export signals and is modulated by association with PREP1 and HTH. *Genes Dev.* **13**, 946–953.

303. Bertuccioli, C., Fasano, L., Jun, S., Wang, S., Sheng, G., and Desplan, C. (1996). In vivo requirement for the paired domain and homeodomain of the *paired* segmentation gene product. *Development* **122**, 2673–2685.

304. Bessette, D.C. and Verheyen, E.M. (2001). Genetic characterization of *nemo* function during development in *Drosophila melanogaster*. *Proc. 42nd Ann. Drosophila Res. Conf.* **Abstracts Vol.**, a205.

305. Bettler, D., Pearson, S., and Yedvobnick, B. (1996). The

nuclear protein encoded by the *Drosophila* neurogenic gene *mastermind* is widely expressed and associates with specific chromosomal regions. *Genetics* **143**, 859–875.

306. Beuchle, D., Struhl, G., and Müller, J. (2001). Polycomb group proteins and heritable silencing of *Drosophila* Hox genes. *Development* **128**, 993–1004.

307. Bhadra, U., Pal-Bhadra, M., and Birchler, J.A. (1999). Role of the *male specific lethal* (*msl*) genes in modifying the effects of sex chromosomal dosage in *Drosophila*. *Genetics* **152**, 249–268.

308. Bhadra, U., Pal-Bhadra, M., and Birchler, J.A. (2001). Evidence that the MSL complex blocks a response of genes to histone acetylation. *Proc. 42nd Ann. Drosophila Res. Conf.* **Abstracts Vol.**, a101.

309. Bhalla, U.S. and Iyengar, R. (1999). Emergent properties of networks of biological signaling pathways. *Science* **283**, 381–387.

310. Bhanot, P., Brink, M., Samos, C.H., Hsieh, J.-C., Wang, Y., Macke, J.P., Andrew, D., Nathans, J., and Nusse, R. (1996). A new member of the *frizzled* family from *Drosophila* functions as a Wingless receptor. *Nature* **382**, 225–230.

311. Bhanot, P., Fish, M., Jemison, J.A., Nusse, R., Nathans, J., and Cadigan, K.M. (1999). Frizzled and DFrizzled-2 function as redundant receptors for Wingless during *Drosophila* embryonic development. *Development* **126**, 4175–4186.

312. Bhat, K.M. (1998). *frizzled* and *frizzled 2* play a partially redundant role in Wingless signaling and have similar requirements to Wingless in Neurogenesis. *Cell* **95**, 1027–1036.

313. Bhat, K.M., Poole, S.J., and Schedl, P. (1995). The *miti-mere* and *pdm1* genes collaborate during specification of the RP2/sib lineage in *Drosophila* neurogenesis. *Mol. Cell. Biol.* **15**, 4052–4063.

314. Bhat, K.M. and Schedl, P. (1994). The *Drosophila miti-mere* gene, a member of the POU family, is required for the specification of the RP2/sibling lineage during neurogenesis. *Development* **120**, 1483–1501.

315. Bhojwani, J., Singh, A., Misquitta, L., Mishra, A., and Sinha, P. (1995). Search for *Drosophila* genes based on patterned expression of mini-*white* reporter gene of a P *lacW* vector in adult eyes. *Roux's Arch. Dev. Biol.* **205**, 114–121.

316. Bianchi, M.E., Falciola, L., Ferrari, S., and Lilley, D.M.J. (1992). The DNA binding site of HMG1 protein is composed of two similar segments (HMG boxes), both of which have counterparts in other eukaryotic regulatory proteins. *EMBO J.* **11**, 1055–1063.

317. Bianchi, M.E. and Lilley, D.M.J. (1995). Applying a genetic cantilever. *Nature* **375**, 532.

318. Biddle, R.L. (1932). The bristles of hybrids between *Drosophila melanogaster* and *Drosophila simulans*. *Genetics* **17**, 153–174.

319. Biehs, B., François, V., and Bier, E. (1996). The *Drosophila short gastrulation* gene prevents Dpp from autoactivating and suppressing neurogenesis in the neuroectoderm. *Genes Dev.* **10**, 2922–2934.

320. Biehs, B., Sturtevant, M.A., and Bier, E. (1998). Boundaries in the *Drosophila* wing imaginal disc organize vein-specific genetic programs. *Development* **125**, 4245–4257.

321. Bienz, M. (1997). Endoderm induction in *Drosophila*: the nuclear targets of the inducing signals. *Curr. Opin. Gen. Dev.* **7**, 683–688.

322. Bienz, M. (1999). APC: the plot thickens. *Curr. Opin. Gen. Dev.* **9**, 595–603.

323. Bienz, M. and Müller, J. (1995). Transcriptional silencing of homeotic genes in *Drosophila*. *BioEssays* **17**, 775–784.

324. Bienz, M., Saari, G., Tremml, G., Müller, J., Züst, B., and Lawrence, P.A. (1988). Differential regulation of *Ultrabithorax* in two germ layers of *Drosophila*. *Cell* **53**, 567–576.

325. Bienz, M. and Tremml, G. (1988). Domain of *Ultrabithorax* expression in *Drosophila* visceral mesoderm from autoregulation and exclusion. *Nature* **333**, 576–578.

326. Bier, E. (1998). Localized activation of RTK/MAPK pathways during *Drosophila* development. *BioEssays* **20**, 189–194.

327. Bier, E. (2000). "The Coiled Spring: How Life Begins." Cold Spr. Harb. Lab. Pr., Cold Spring Harbor, N. Y.

328. Bier, E. (2000). Drawing lines in the *Drosophila* wing: initiation of wing vein development. *Curr. Opin. Gen. Dev.* **10**, 393–398.

329. Bier, E., H., V., Shepherd, S., Lee, K., McCall, K., Barbel, S., Ackerman, L., Carretto, R., Uemura, T., Grell, E., Jan, L.Y., and Jan, Y.N. (1989). Searching for pattern and mutation in the *Drosophila* genome with a P-*lacZ* vector. *Genes Dev.* **3**, 1273–1287.

330. Bier, E., Jan, L.Y., and Jan, Y.N. (1990). *rhomboid*, a gene required for dorsoventral axis establishment and peripheral nervous system development in *Drosophila melanogaster*. *Genes Dev.* **4**, 190–203.

331. Bier, E., Vaessin, H., Younger-Shepherd, S., Jan, L.Y., and Jan, Y.N. (1992). *deadpan*, an essential pan-neural gene in *Drosophila*, encodes a helix-loop-helix protein similar to the *hairy* gene product. *Genes Dev.* **6**, 2137–2151.

332. Biggs, W.H., III, Zavitz, K.H., Dickson, B., van der Straten, A., Brunner, D., Hafen, E., and Zipursky, S.L. (1994). The *Drosophila rolled* locus encodes a MAP kinase required in the sevenless signal transduction pathway. *EMBO J.* **13**, 1628–1635.

333. Bilder, D., Li, M., and Perrimon, N. (2000). Cooperative regulation of cell polarity and growth by *Drosophila* tumor suppressors. *Science* **289**, 113–116.

334. Bilder, D. and Scott, M.P. (1998). *Hedgehog* and *Wingless* induce metameric pattern in the *Drosophila* visceral mesoderm. *Dev. Biol.* **201**, 43–56.

335. Billeter, M., Güntert, P., Luginbühl, P., and Wüthrich, K. (1996). Hydration and DNA recognition by homeodomains. *Cell* **85**, 1057–1065.

336. Billin, A.N., Cockerill, K.A., and Poole, S.J. (1991). Isolation of a family of *Drosophila* POU domain genes expressed in early development. *Mechs. Dev.* **34**, 75–84.

337. Billin, A.N. and Poole, S.J. (1995). Expression domains of the Cf1a POU domain protein during *Drosophila* development. *Roux's Arch. Dev. Biol.* **204**, 502–508.

338. Binari, R.C., Staveley, B.E., Johnson, W.A., Godavarti, R., Sasisekharan, R., and Manoukian, A.S. (1997). Genetic evidence that heparin-like glycosaminoglycans are involved in *wingless* signaling. *Development* **124**, 2623–2632.

339. Bingham, P.M. (1997). Cosuppression comes to the animals. *Cell* **90**, 385–387.

340. Birchler, J.A., Pal Bhadra, M., and Bhadra, U. (2000). Making noise about silence: repression of repeated genes in animals. *Curr. Opin. Gen. Dev.* **10**, 211–216.

341. Birdsall, K., Zimmerman, E., Teeter, K., and Gibson, G. (2000). Genetic variation for the positioning of wing veins in *Drosophila melanogaster*. *Evol. Dev.* **2**, 16–24.

342. Birge, R.B. and Hanafusa, H. (1993). Closing in on SH2 specificity. *Science* **262**, 1522–1524.

343. Birmingham, L. (1942). Boundaries of differentiation of cephalic imaginal discs in *Drosophila. J. Exp. Zool.* **91**, 345–363.

344. Bishop, S.A., Klein, T., Martinez Arias, A., and Couso, J.P. (1999). Composite signalling from *Serrate* and *Delta* establishes leg segments in *Drosophila* through *Notch. Development* **126**, 2993–3003.

345. Black, D.L. (2000). Protein diversity from alternative splicing: a challenge for bioinformatics and post-genome biology. *Cell* **103**, 367–370.

346. Black, R.A. and White, J.M. (1998). ADAMs: focus on the protease domain. *Curr. Opin. Cell Biol.* **10**, 654–659.

347. Blackman, R.K., Sanicola, M., Raftery, L.A., Gillevet, T., and Gelbart, W.M. (1991). An extensive 3′ *cis*-regulatory region directs the imaginal disk expression of *decapentaplegic*, a member of the TGF-β family in *Drosophila. Development* **111**, 657–665.

348. Blackwell, T.K. and Weintraub, H. (1990). Differences and similarities in DNA-binding preferences of MyoD and E2A protein complexes revealed by binding site selection. *Science* **250**, 1104–1110.

349. Blackwood, E.M. and Kadonaga, J.T. (1998). Going the distance: a current view of enhancer action. *Science* **281**, 60–63.

350. Blair, S.S. (1992). *engrailed* expression in the anterior lineage compartment of the developing wing blade of *Drosophila. Development* **115**, 21–33.

351. Blair, S.S. (1992). *shaggy* (*zeste-white 3*) and the formation of supernumerary bristle precursors in the developing wing blade of *Drosophila. Dev. Biol.* **152**, 263–278.

352. Blair, S.S. (1993). Mechanisms of compartment formation: evidence that non-proliferating cells do not play a critical role in defining the D/V lineage restriction in the developing wing of *Drosophila. Development* **119**, 339–351.

353. Blair, S.S. (1994). A role for the segment polarity gene *shaggy-zeste white 3* in the specification of regional identity in the developing wing of *Drosophila. Dev. Biol.* **162**, 229–244.

354. Blair, S.S. (1995). Compartments and appendage development in *Drosophila. BioEssays* **17**, 299–309.

355. Blair, S.S. (1995). Hedgehog digs up an old friend. *Nature* **373**, 656–657.

356. Blair, S.S. (1996). Notch and Wingless signals collide. *Science* **271**, 1822–1823.

357. Blair, S.S. (1997). Limb development: Marginal Fringe benefits. *Curr. Biol.* **7**, R686–R690.

358. Blair, S.S. (1999). *Drosophila* imaginal disc development: patterning the adult fly. *In* "Development: Genetics, Epigenetics and Environmental Regulation," (V.E.A. Russo, D.J. Cove, L.G. Edgar, R. Jaenisch, and F. Salamini, eds.). Springer: N. Y., pp. 347–370.

359. Blair, S.S. (1999). Eye development: Notch lends a handedness. *Curr. Biol.* **9**, R356–R360.

360. Blair, S.S. (2000). Notch signaling: Fringe really is a glycosyltransferase. *Curr. Biol.* **10**, R608–R612.

361. Blair, S.S., Brower, D.L., Thomas, J.B., and Zavortink, M. (1994). The role of *apterous* in the control of dorsoventral compartmentalization and PS integrin gene expression in the developing wing of *Drosophila. Development* **120**, 1805–1815.

362. Blair, S.S., Giangrande, A., Skeath, J.B., and Palka, J. (1992). The development of normal and ectopic sensilla in the wings of *hairy* and *Hairy wing* mutants of *Drosophila. Mechs. Dev.* **38**, 3–16.

363. Blair, S.S. and Palka, J. (1985). Axon guidance in the wing of *Drosophila. Trends Neurosci.* **8**, 284–288.

364. Blair, S.S. and Ralston, A. (1997). Smoothened-mediated Hedgehog signalling is required for the maintenance of the anterior-posterior lineage restriction in the developing wing of *Drosophila. Development* **124**, 4053–4063.

365. Blanchet-Tournier, M.-F., Tricoire, H., Busson, D., and Lamour-Isnard, C. (1995). The segment-polarity gene *fused* is highly conserved in *Drosophila. Gene* **161**, 157–162.

366. Blank, V., Kourilsky, P., and Israël, A. (1992). NF-κB and related proteins: Rel/dorsal homologies meet ankyrin-like repeats. *Trends Biochem. Sci.* **17**, 135–140.

367. Blaumueller, C.M. and Mlodzik, M. (2000). The *Drosophila* tumor suppressor *expanded* regulates growth, apoptosis, and patterning during development. *Mechs. Dev.* **92**, 251–262.

368. Blaumueller, C.M., Qi, H., Zagouras, P., and Artavanis-Tsakonas, S. (1997). Intracellular cleavage of Notch leads to a heterodimeric receptor on the plasma membrane. *Cell* **90**, 281–291.

369. Blobel, C.P. (1997). Metalloprotease-disintegrins: links to cell adhesion and cleavage of TNFα and Notch. *Cell* **90**, 589–592.

370. Blochlinger, K., Bodmer, R., Jack, J., Jan, L.Y., and Jan, Y.N. (1988). Primary structure and expression of a product from *cut*, a locus involved in specifying sensory organ identity in *Drosophila. Nature* **333**, 629–635.

371. Blochlinger, K., Bodmer, R., Jan, L.Y., and Jan, Y.N. (1990). Patterns of expression of Cut, a protein required for external sensory organ development in wild-type and *cut* mutant *Drosophila* embryos. *Genes Dev.* **4**, 1322–1331.

372. Blochlinger, K., Jan, L.Y., and Jan, Y.N. (1991). Transformation of sensory organ identity by ectopic expression of Cut in *Drosophila. Genes Dev.* **5**, 1124–1135.

373. Blochlinger, K., Jan, L.Y., and Jan, Y.N. (1993). Postembryonic patterns of expression of *cut*, a locus regulating sensory organ identity in *Drosophila. Development* **117**, 441–450.

374. Bloomquist, B.T., Shortridge, R.D., Schneuwly, S., Perdew, M., Montell, C., Steller, H., Rubin, G., and Pak, W.L. (1988). Isolation of a putative phospholipase C gene of *Drosophila, norpA*, and its role in phototransduction. *Cell* **54**, 723–733.

375. Blumer, K.J. and Johnson, G.L. (1994). Diversity in function and regulation of MAP kinase pathways. *Trends Biochem. Sci.* **19**, 236–240.

376. Bock, I.R. and Wheeler, M.R. (1972). The *Drosophila melanogaster* species group. *In* "Studies in Genetics," Vol. 7 (Pub. #7213) (M.R. Wheeler, ed.). Univ. of Texas Pr.: Austin, Texas, pp. 1–102.

377. Bodenstein, D. (1950). The postembryonic development of *Drosophila. In* "Biology of *Drosophila*," (M. Demerec, ed.). Hafner: New York, pp. 275–367.

378. Bodmer, R., Barbel, S., Sheperd, S., Jack, J.W., Jan, L.Y., and Jan, Y.N. (1987). Transformation of sensory organs by mutations of the *cut* locus of *D. melanogaster. Cell* **51**, 293–307. [See also 2001 *Mechs. Dev.* **105**, 57–68.]

379. Bodmer, R., Jan, L.Y., and Jan, Y.-N. (1993). A late role for a subset of neurogenic genes to limit sensory precursor recruitments in *Drosophila* embryos. *Roux's Arch. Dev. Biol.* **202**, 371–381.

380. Bodnar, J.W. (1997). Programming the *Drosophila* embryo. *J. Theor. Biol.* **188**, 391–445.

381. Boguski, M.S. and McCormick, F. (1993). Proteins regulating Ras and its relatives. *Nature* **366**, 643–654.

382. Bohley, P. (1996). Surface hydrophobicity and intracellular degradation of proteins. *Biol. Chem.* **377**, 425–435.

383. Bohmann, D., Ellis, M.C., Staszewski, L.M., and Mlodzik, M. (1994). *Drosophila* Jun mediates Ras-dependent photoreceptor determination. *Cell* **78**, 973–986.

384. Bohn, H. (1965). Analyse der Regenerationfähigkeit der Insektenextremität durch Amputations- und Transplantationsversuche an Larven der Afrikanischen Schabe (*Leucophaea maderae* Fabr.). II. Achsendetermination. *Roux'Arch. Entw.-mech.* **156**, 449–503.

385. Bohn, H. (1976). Regeneration of proximal tissues from a more distal amputation level in the insect leg (*Blaberus craniifer, Blattaria*). *Dev. Biol.* **53**, 285–293.

386. Bohn, H. (1976). Tissue interactions in the regenerating cockroach leg. *In* "Insect Development," *Symp. Roy. Entomol. Soc. Lond.*, Vol. 8 (P.A. Lawrence, ed.). Wiley: New York, pp. 170–185.

387. Boissonade, J. (1994). Long-range inhibition. *Nature* **369**, 188–189.

388. Boivin, A. and Dura, J.-M. (1998). *In vivo* chromatin accessibility correlates with gene silencing in *Drosophila*. *Genetics* **150**, 1539–1549.

389. Bolinger, R.A. and Panganiban, G. (2001). Specification of neural and imaginal identities in the embryonic limb primordium of *Drosophila melanogaster*. *Proc. 42nd Ann. Drosophila Res. Conf.* **Abstracts Vol.**, a195.

390. Bonabeau, E. (1997). From classical models of morphogenesis to agent-based models of pattern formation. *Artif. Life* **3**, 191–211.

391. Bonfini, L., Karlovich, C.A., Dasgupta, C., and Banerjee, U. (1992). The *Son of sevenless* gene product: a putative activator of Ras. *Science* **255**, 603–606.

392. Bonfini, L., Migliaccio, E., Pelicci, G., Lanfrancone, L., and Pelicci, P. (1996). Not all Shc's roads lead to Ras. *Trends Biochem. Sci.* **21**, 257–261.

393. Bonifer, C. (2000). Developmental regulation of eukaryotic gene loci. *Trends Genet.* **16**, 310–315.

394. Bonini, N.M., Bui, Q.T., Gray-Board, G.L., and Warrick, J.M. (1997). The *Drosophila eyes absent* gene directs ectopic eye formation in a pathway conserved between flies and vertebrates. *Development* **124**, 4819–4826.

395. Bonini, N.M. and Choi, K.-W. (1995). Early decisions in *Drosophila* eye morphogenesis. *Curr. Opin. Gen. Dev.* **5**, 507–515.

396. Bonini, N.M. and Fortini, M.E. (1999). Surviving *Drosophila* eye development: integrating cell death with differentiation during formation of a neural structure. *BioEssays* **21**, 991–1003.

397. Bonini, N.M., Leiserson, W.M., and Benzer, S. (1993). The *eyes absent* gene: genetic control of cell survival and differentiation in the developing *Drosophila* eye. *Cell* **72**, 379–395.

398. Bonini, N.M., Leiserson, W.M., and Benzer, S. (1998). Multiple roles of the *eyes absent* gene in *Drosophila*. *Dev. Biol.* **196**, 42–57.

399. Boole, G. (1854). "An Investigation of the Laws of Thought, on which are founded the Mathematical Theories of Logic and Probabilities." Walton and Maberley, London.

400. Bopp, D., Burri, M., Baumgartner, S., Frigerio, G., and Noll, M. (1986). Conservation of a large protein domain in the segmentation gene *paired* and in functionally related genes of *Drosophila*. *Cell* **47**, 1033–1040.

401. Bopp, D., Jamet, E., Baumgartner, S., Burri, M., and Noll, M. (1989). Isolation of two tissue-specific *Drosophila* paired box genes, *Pox meso* and *Pox neuro*. *EMBO J.* **8**, 3447–3457.

402. Boriack-Sjodin, P.A., Margarit, S.M., Bar-Sagi, D., and Kuriyan, J. (1998). The structural basis of the activation of Ras by Sos. *Nature* **394**, 337–343.

403. Bork, P. (1993). Hundreds of ankyrin-like repeats in functionally diverse proteins: mobile modules that cross phyla horizontally? *Proteins: Struct., Funct., Genet.* **17**, 363–374.

404. Bork, P., Downing, A.K., Kieffer, B., and Campbell, I.D. (1996). Structure and distribution of modules in extracellular proteins. *Quart. Rev. Biophys.* **29**, 119–167.

405. Bornemann, D., Miller, E., and Simon, J. (1998). Expression and properties of wild-type and mutant forms of the *Drosophila* Sex comb on Midleg (SCM) repressor protein. *Genetics* **150**, 675–686.

406. Borod, E.R. and Heberlein, U. (1998). Mutual regulation of *decapentaplegic* and *hedgehog* during the initiation of differentiation in the *Drosophila* retina. *Dev. Biol.* **197**, 187–197.

407. Botas, J. and Auwers, L. (1996). Chromosomal binding sites of Ultrabithorax homeotic proteins. *Mechs. Dev.* **56**, 129–138.

408. Botas, J., Moscoso del Prado, J., and García-Bellido, A. (1982). Gene-dose titration analysis in the search of trans-regulatory genes in *Drosophila*. *EMBO J.* **1**, 307–310.

409. Botquin, V., Hess, H., Fuhrmann, G., Anastassiadis, C., Gross, M.K., Vriend, G., and Schöler, H.R. (1998). New POU dimer configuration mediates antagonistic control of an osteopontin preimplantation enhancer by Oct-4 and Sox-2. *Genes Dev.* **12**, 2073–2090.

410. Botstein, D. (1993). Genetics in the post-sequence era. *Trends Genet.* **9**, 101.

411. Boube, M., Benassayag, C., Seroude, L., and Cribbs, D.L. (1997). Ras1-mediated modulation of *Drosophila* homeotic function in cell and segment identity. *Genetics* **146**, 619–628.

412. Boube, M., Faucher, C., Joulia, L., Cribbs, D.L., and Bourbon, H.-M. (2000). *Drosophila* homologs of transcriptional mediator complex subunits are required for adult cell and segment identity specification. *Genes Dev.* **14**, 2906–2917.

413. Boube, M., Seroude, L., and Cribbs, D.L. (1998). Homeotic *proboscipedia* cell identity functions respond to cell signaling pathways along the proximo-distal axis. *Int. J. Dev. Biol.* **42**, 431–436.

414. Bouchard, M., St-Amand, J., and Côté, S. (2000). Combinatorial activity of pair-rule proteins on the *Drosophila gooseberry* early enhancer. *Dev. Biol.* **222**, 135–146.

415. Boudreaux, H.B. (1979). Significance of intersegmental tendon system in Arthropod phylogeny and a monophyletic classification of Arthropoda. *In* "Arthropod Phylogeny," (A.P. Gupta, ed.). Van Nostrand Reinhold: New York, pp. 551–586.

416. Boulet, A.M., Lloyd, A., and Sakonju, S. (1991). Molecular definition of the morphogenetic and regulatory functions and the *cis*-regulatory elements of the *Drosophila Abd-B* homeotic gene. *Development* **111**, 393–405.

417. Boulianne, G.L., de la Concha, A., Campos-Ortega, J.A., Jan, L.Y., and Jan, Y.N. (1991). The *Drosophila* neurogenic gene *neuralized* encodes a novel protein and is expressed in precursors of larval and adult neurons. *EMBO J.* **10**, 2975–2983.

418. Bourgouin, C., Lundgren, S.E., and Thomas, J.B. (1992).

apterous is a *Drosophila* LIM domain gene required for the development of a subset of embryonic muscles. *Neuron* **9**, 549–561.

419. Bourouis, M., Moore, P., Ruel, L., Grau, Y., Heitzler, P., and Simpson, P. (1990). An early embryonic product of the gene *shaggy* encodes a serine/threonine protein kinase related to the CDC28/cdc2+ subfamily. *EMBO J.* **9**, 2877–2884.

420. Boussy, I.A., Yu, H., Leone, M.A., Ogura, K., and Itoh, M. (2001). *P* element repression may involve "homology-dependent gene silencing." *Proc. 42nd Ann. Drosophila Res. Conf.* **Abstracts Vol.**, a38.

421. Boutros, M., Mihaly, J., Bouwmeester, T., and Mlodzik, M. (2000). Signaling specificity by Frizzled receptors in *Drosophila*. *Science* **288**, 1825–1828. [See also 2001 *Development* **128**, 4829–4835.]

422. Boutros, M. and Mlodzik, M. (1999). Dishevelled: at the crossroads of divergent intracellular signaling pathways. *Mechs. Dev.* **83**, 27–37.

423. Boutros, M., Paricio, N., Strutt, D.I., and Mlodzik, M. (1998). Dishevelled activates JNK and discriminates between JNK pathways in planar polarity and *wingless* signaling. *Cell* **94**, 109–118.

424. Bowles, J., Schepers, G., and Koopman, P. (2000). Phylogeny of the SOX family of developmental transcription factors based on sequence and structural indicators. *Dev. Biol.* **227**, 239–255.

425. Bownes, M. (1975). Adult deficiencies and duplications of head and thoracic structures resulting from microcautery of blastoderm stage *Drosophila* embryos. *J. Embryol. Exp. Morph.* **34**, 33–54.

426. Bownes, M., Bournias-Vardiabasis, N., and Spare, W.J. (1979). Genetic analysis of the *spineless-aristapedia* homoeotic mutants of *Drosophila melanogaster*. *Mol. Gen. Genet.* **174**, 67–74.

427. Bownes, M. and Seiler, M. (1977). Developmental effects of exposing *Drosophila* embryos to ether vapour. *J. Exp. Zool.* **199**, 9–23.

428. Bowtell, D.D.L., Simon, M.A., and Rubin, G.M. (1988). Nucleotide sequence and structure of the *sevenless* gene of *Drosophila melanogaster*. *Genes Dev.* **2**, 620–634.

429. Boyle, M., Bonini, N., and DiNardo, S. (1997). Expression and function of *clift* in the development of somatic gonadal precursors within the *Drosophila* mesoderm. *Development* **124**, 971–982.

430. Brabant, M.C. and Brower, D.L. (1993). PS2 integrin requirements in *Drosophila* embryo and wing morphogenesis. *Dev. Biol.* **157**, 49–59.

431. Brabant, M.C., Fristrom, D., Bunch, T.A., and Brower, D.L. (1996). Distinct spatial and temporal functions for PS integrins during *Drosophila* wing morphogenesis. *Development* **122**, 3307–3317.

432. Bradley, B.P. (1980). Developmental stability of *Drosophila melanogaster* under artificial and natural selection in constant and fluctuating environments. *Genetics* **95**, 1033–1042.

433. Braitenberg, V. (1967). Patterns of projection in the visual system of the fly. I. Retina-lamina projections. *Exp. Brain Res.* **3**, 271–298.

434. Brakefield, P.M. and French, V. (1999). Butterfly wings: the evolution of development of colour patterns. *BioEssays* **21**, 391–401.

435. Brand, A.H. and Perrimon, N. (1993). Targeted gene expression as a means of altering cell fates and generating dominant phenotypes. *Development* **118**, 401–415.

436. Brand, M. and Campos-Ortega, J.A. (1988). Two groups of interrelated genes regulate early neurogenesis in *Drosophila melanogaster*. *Roux's Arch. Dev. Biol.* **197**, 457–470.

437. Brand, M. and Campos-Ortega, J.A. (1990). Second-site modifiers of the *split* mutation of *Notch* define genes involved in neurogenesis in *Drosophila melanogaster*. *Roux's Arch. Dev. Biol.* **198**, 275–285.

438. Brand, M., Jarman, A.P., Jan, L.Y., and Jan, Y.N. (1993). *asense* is a *Drosophila* neural precursor gene and is capable of initiating sense organ formation. *Development* **119**, 1–17.

439. Brasted, A. (1941). An analysis of the expression of the mutant "engrailed" in *Drosophila melanogaster*. *Genetics* **26**, 347–373.

440. Bray, D. (1995). Protein molecules as computational elements in living cells. *Nature* **376**, 307–312.

441. Bray, D. (1998). Signaling complexes: biophysical constraints on intracellular communication. *Annu. Rev. Biophys. Biomol. Struct.* **27**, 59–75.

442. Bray, D. and Lay, S. (1997). Computer-based analysis of the binding steps in protein complex formation. *Proc. Natl. Acad. Sci. USA* **94**, 13493–13498.

443. Bray, S. (1998). A Notch affair. *Cell* **93**, 499–503.

444. Bray, S. (1998). Notch signalling in *Drosophila*: three ways to use a pathway. *Sems. Cell Dev. Biol.* **9**, 591–597.

445. Bray, S. (1999). *Drosophila* development: Scalloped and Vestigial take wing. *Curr. Biol.* **9**, R245–R247.

446. Bray, S. (2000). Notch. *Curr. Biol.* **10**, R433–R435.

447. Bray, S. (2000). Planar polarity: out of joint? *Curr. Biol.* **10**, R155–R158.

448. Bray, S. (2000). Specificity and promiscuity among proneural proteins. *Neuron* **25**, 1–5.

449. Bray, S. and Furriols, M. (2001). Notch pathway: making sense of Suppressor of Hairless. *Curr. Biol.* **11**, R217–R221.

450. Bredt, D.S. (2000). Reeling CASK into the nucleus. *Nature* **404**, 241–242.

451. Breen, T.R. (1999). Mutant alleles of the *Drosophila trithorax* gene produce common and unusual homeotic and other developmental phenotypes. *Genetics* **152**, 319–344.

452. Breen, T.R., Chinwalla, V., and Harte, P.J. (1995). *Trithorax* is required to maintain *engrailed* expression in a subset of *engrailed*-expressing cells. *Mechs. Dev.* **52**, 89–98.

453. Breen, T.R. and Duncan, I.M. (1986). Maternal expression of genes that regulate the bithorax complex of *Drosophila melanogaster*. *Dev. Biol.* **118**, 442–456.

454. Breen, T.R. and Harte, P.J. (1993). *trithorax* regulates multiple homeotic genes in the bithorax and Antennapedia complexes and exerts different tissue-specific, parasegment-specific and promoter-specific effects on each. *Development* **117**, 119–134.

455. Breiling, A., Bonte, E., Ferrari, S., Becker, P.B., and Paro, R. (1999). The *Drosophila* Polycomb protein interacts with nucleosomal core particles in vitro via its repression domain. *Mol. Cell. Biol.* **19**, 8451–8460. [See also 2001 *Nature* **412**, 651–655.]

456. Breitling, R. and Gerber, J.-K. (2000). Origin of the paired domain. *Dev. Genes Evol.* **210**, 644–650.

457. Brennan, C.A., Ashburner, M., and Moses, K. (1998). Ecdysone pathway is required for furrow progression in the developing *Drosophila* eye. *Development* **125**, 2653–2664.

458. Brennan, C.A., Li, T.-R., Bender, M., Hsiung, F., and Moses, K. (2001). *Broad-complex*, but not *Ecdysone receptor*, is

required for progression of the morphogenetic furrow in the *Drosophila* eye. *Development* **128**, 1–11.

459. Brennan, K., Klein, T., Wilder, E., and Martinez Arias, A. (1999). Wingless modulates the effects of dominant negative Notch molecules in the developing wing of *Drosophila*. *Dev. Biol.* **216**, 210–229.

460. Brennan, K., Tateson, R., Lewis, K., and Martinez Arias, A. (1997). A functional analysis of *Notch* mutations in *Drosophila*. *Genetics* **147**, 177–188.

461. Brennan, K., Tateson, R., Lieber, T., Couso, J.P., Zecchini, V., and Martinez Arias, A. (1999). The *Abruptex* mutations of *Notch* disrupt the establishment of proneural clusters in *Drosophila*. *Dev. Biol.* **216**, 230–242.

462. Brenner, S. (1981). Genes and development. *In* "Cellular Controls in Differentiation," (C.W. Lloyd and D.A. Rees, eds.). Acad. Pr.: New York, pp. 3–7.

463. Brenner, S. (1993). Thoughts on genetics at the *fin de siècle*. *Trends Genet.* **9**, 104.

464. Brenner, S., Dove, W., Herskowitz, I., and Thomas, R. (1990). Genes and development: molecular and logical themes. *Genetics* **126**, 479–486.

465. Brent, R. (2000). Genomic biology. *Cell* **100**, 169–183.

466. Brentrup, D., Lerch, H.-P., Jäckle, H., and Noll, M. (2000). Regulation of *Drosophila* wing vein patterning: *net* encodes a bHLH protein repressing *rhomboid* and is repressed by rhomboid-dependent Egfr signalling. *Development* **127**, 4729–4741.

467. Brentrup, D., Lerch, H.-P., and Noll, M. (1996). The *Drosophila net* gene, required for intervein fate in wings, encodes a putative bHLH transcription factor. *Experientia* **52**, A16.

468. Bresnick, A.R. (1999). Molecular mechanisms of nonmuscle myosin-II regulation. *Curr. Opin. Cell Biol.* **11**, 26–33.

469. Bridges, C.B. (1920). The mutant *crossveinless* in *Drosophila melanogaster*. *Proc. Natl. Acad. Sci. USA* **6**, 660–663.

470. Bridges, C.B. and Brehme, K.S. (1944). "The Mutants of *Drosophila melanogaster*." Carnegie Inst. (Publication #552), Washington, D. C.

471. Bridges, C.B. and Morgan, T.H. (1923). The third-chromosome group of mutant characters of *Drosophila melanogaster*. *Carnegie Inst. Wash. Publ.* **327**, 1–251.

472. Briscoe, J. and Ericson, J. (1999). The specification of neuronal identity by graded sonic hedgehog signalling. *Sems. Cell Dev. Biol.* **10**, 353–362.

473. Britten, R.J. and Davidson, E.H. (1969). Gene regulation for higher cells: a theory. *Science* **165**, 349–357.

474. Britton, J.S. and Edgar, B.A. (1998). Environmental control of the cell cycle in *Drosophila*: nutrition activates mitotic and endoreplicative cells by distinct mechanisms. *Development* **125**, 2149–2158.

475. Brizuela, B.J. and Kennison, J.A. (1997). The *Drosophila* homeotic gene *moira* regulates expression of *engrailed* and HOM genes in imaginal tissues. *Mechs. Dev.* **65**, 209–220.

476. Broadie, K.S. and Bate, M. (1991). The development of adult muscles in *Drosophila*: albation of identified muscle precursor cells. *Development* **113**, 103–118.

477. Broadus, J. and Doe, C.Q. (1997). Extrinsic cues, intrinsic cues and microfilaments regulate asymmetric protein localization in *Drosophila* neuroblasts. *Curr. Biol.* **7**, 827–835.

478. Broadus, J., Fuerstenberg, S., and Doe, C.Q. (1998). Staufen-dependent localization of *prospero* mRNA contributes to neuroblast daughter-cell fate. *Nature* **391**, 792–795.

479. Broadus, J., McCabe, J.R., Endrizzi, B., Thummel, C.S., and Woodard, C.T. (1999). The *Drosophila* bFTZ-F1 orphan nuclear receptor provides competence for stage-specific responses to the steroid hormone ecdysone. *Molec. Cell* **3**, 143–149.

480. Brock, D.A. and Gomer, R.H. (1999). A cell-counting factor regulating structure size in *Dictyostelium*. *Genes Dev.* **13**, 1960–1969.

481. Brock, H.W. (1987). Sequence and genomic structure of *ras* homologues Dm*ras*85D and Dm*ras*64B of *Drosophila melanogaster*. *Gene* **51**, 129–137.

482. Brock, H.W. and van Lohuizen, M. (2001). The Polycomb group–no longer an exclusive club? *Curr. Opin. Gen. Dev.* **11**, 175–181.

483. Brodsky, M.H. and Steller, H. (1996). Positional information along the dorsal-ventral axis of the *Drosophila* eye: graded expression of the *four-jointed* gene. *Dev. Biol.* **173**, 428–446.

484. Brody, T. (1999). The Interactive Fly: gene networks, development and the Internet. *Trends Genet.* **15**, 333–334.

485. Brody, T. and Odenwald, W.F. (2000). Programmed transformations in neuroblast gene expression during *Drosophila* CNS lineage development. *Dev. Biol.* **226**, 34–44. [See also 2001 references: *Cell* **106**, 511–521; *Dev. Cell* **1**, 313–314; *Nature* **413**, 471–473.]

486. Brogiolo, W., Stocker, H., Ikeya, T., Rintelen, F., Fernandez, R., and Hafen, E. (2001). An evolutionarily conserved function of the *Drosophila* insulin receptor and insulin-like peptides in growth control. *Curr. Biol.* **11**, 213–221.

487. Brook, W.J. and Cohen, S.M. (1996). Antagonistic interactions between Wingless and Decapentaplegic responsible for dorsal-ventral pattern in the *Drosophila* leg. *Science* **273**, 1373–1377.

488. Brook, W.J., Diaz-Benjumea, F.J., and Cohen, S.M. (1996). Organizing spatial pattern in limb development. *Annu. Rev. Cell Dev. Biol.* **12**, 161–180.

489. Brook, W.J., Ostafichuk, L.M., Piorecky, J., Wilkinson, M.D., Hodgetts, D.J., and Russell, M.A. (1993). Gene expression during imaginal disc regeneration detected using enhancer-sensitive P-elements. *Development* **177**, 1287–1297.

490. Brooke, N.M., Garcia-Fernàndez, J., and Holland, P.W.H. (1998). The ParaHox gene cluster is an evolutionary sister of the Hox gene cluster. *Nature* **392**, 920–922. [See also 2001 *J. Anat.* **199**, 13–23.]

491. Brou, C., Logeat, F., Gupta, N., Bessia, C., LeBail, O., Doedens, J.R., Cumano, A., Roux, P., Black, R.A., and Israël, A. (2000). A novel proteolytic cleavage involved in Notch signaling: the role of the disintegrin-metalloprotease TACE. *Molec. Cell* **5**, 207–216.

492. Brou, C., Logeat, F., Lecourtois, M., Vandekerckhove, J., Kourilsky, P., Schweisguth, F., and Israël, A. (1994). Inhibition of the DNA-binding activity of *Drosophila* Suppressor of Hairless and of its human homolog, KBF2/RBP-Jκ, by direct protein-protein interaction with *Drosophila* Hairless. *Genes Dev.* **8**, 2491–2503. [See also 2001 *Curr. Biol.* **11**, 789–792.]

493. Brower, D.L. (1984). Posterior-to-anterior transformation in *engrailed* wing imaginal disks of *Drosophila*. *Nature* **310**, 496–497.

494. Brower, D.L. (1985). The sequential compartmentalization of *Drosophila* segments revisited. *Cell* **41**, 361–364.

495. Brower, D.L. (1986). *engrailed* gene expression in *Drosophila* imaginal discs. *EMBO J.* **5**, 2649–2656.

496. Brower, D.L. (1987). *Ultrabithorax* gene expression in *Drosophila* imaginal discs and larval nervous system. *Development* **101**, 83–92.

497. Brower, D.L., Lawrence, P.A., and Wilcox, M. (1981). Clonal analysis of the undifferentiated wing disk of *Drosophila*. *Dev. Biol.* **86**, 448–455.

498. Brower, D.L., Piovant, M., and Reger, L.A. (1985). Developmental analysis of *Drosophila* position-specific antigens. *Dev. Biol.* **108**, 120–130.

499. Brower, D.L., Smith, R.J., and Wilcox, M. (1980). A monoclonal antibody specific for diploid epithelial cells in *Drosophila*. *Nature* **285**, 403–405.

500. Brower, D.L., Smith, R.J., and Wilcox, M. (1982). Cell shapes on the surface of the *Drosophila* wing imaginal disc. *J. Embryol. Exp. Morph.* **67**, 137–151.

501. Brown, C.E., Lechner, T., Howe, L., and Workman, J.L. (2000). The many HATs of transcriptional coactivators. *Trends Biochem. Sci.* **25**, 15–19.

502. Brown, J.L., Sonoda, S., Ueda, H., Scott, M.P., and Wu, C. (1991). Repression of the *Drosophila fushi tarazu* (*ftz*) segmentation gene. *EMBO J.* **10**, 665–674.

503. Brown, J.L. and Wu, C. (1993). Repression of *Drosophila* pair-rule segmentation genes by ectopic expression of *tramtrack*. *Development* **117**, 45–58.

504. Brown, N.H., Gregory, S.L., and Martin-Bermudo, M.D. (2000). Integrins as mediators of morphogenesis in *Drosophila*. *Dev. Biol.* **223**, 1–16.

505. Brown, N.H. and Hartley, D.A. (1994). Exploring signalling pathways. *Nature* **370**, 414–415.

506. Brown, N.L., Paddock, S.W., Sattler, C.A., Cronmiller, C., Thomas, B.J., and Carroll, S.B. (1996). *daughterless* is required for *Drosophila* photoreceptor cell determination, eye morphogenesis, and cell cycle progression. *Dev. Biol.* **179**, 65–78.

507. Brown, N.L., Sattler, C.A., Markey, D.R., and Carroll, S.B. (1991). *hairy* gene function in the *Drosophila* eye: normal expression is dispensable but ectopic expression alters cell fates. *Development* **113**, 1245–1256.

508. Brown, N.L., Sattler, C.A., Paddock, S.W., and Carroll, S.B. (1995). Hairy and Emc negatively regulate morphogenetic furrow progression in the *Drosophila* eye. *Cell* **80**, 879–887.

509. Brown, S. and Hombría, J.C.-G. (2000). *Drosophila grain* encodes a GATA transcription factor required for cell rearrangement during morphogenesis. *Development* **127**, 4867–4876.

510. Brückner, K., Perez, L., Clausen, H., and Cohen, S. (2000). Glycosyltransferase activity of Fringe modulates Notch-Delta interactions. *Nature* **406**, 411–415.

511. Brummel, T., Abdollah, S., Haerry, T.E., Shimell, M.J., Merriam, J., Raftery, L., Wrana, J.L., and O'Connor, M.B. (1999). The *Drosophila* Activin receptor Baboon signals through dSmad2 and controls cell proliferation but not patterning during larval development. *Genes Dev.* **13**, 98–111.

512. Brummel, T.J., Twombly, V., Marqués, G., Wrana, J.L., Newfeld, S.J., Attisano, L., Massagué, J., O'Connor, M.B., and Gelbart, W.M. (1994). Characterization and relationship of Dpp receptors encoded by the *saxophone* and *thick veins* genes in *Drosophila*. *Cell* **78**, 251–261.

513. Brunk, B.P. and Adler, P.N. (1990). *Aristapedioid*: a gain of function, homeotic mutation in *Drosophila melanogaster*. *Genetics* **124**, 145–156.

514. Brunk, B.P., Martin, E.C., and Adler, P.N. (1991). Molecular genetics of the *Posterior Sex Combs/Suppressor 2 of zeste* region of *Drosophila*: aberrant expression of the *Suppressor 2 of zeste* gene results in abnormal bristle development. *Genetics* **128**, 119–132.

515. Brunner, A., Twardzik, T., and Schneuwly, S. (1994). The *Drosophila giant lens* gene plays a dual role in eye and optic lobe development: inhibition of differentiation of ommatidial cells and interference in photoreceptor axon guidance. *Mechs. Dev.* **48**, 175–185.

516. Brunner, D., Dücker, K., Oellers, N., Hafen, E., Scholz, H., and Klämbt, C. (1994). The ETS domain protein Pointed-P2 is a target of MAP kinase in the Sevenless signal transduction pathway. *Nature* **370**, 386–389.

517. Brunner, D., Oellers, N., Szabad, J., Biggs, W.H., III, Zipursky, S.L., and Hafen, E. (1994). A gain-of-function mutation in *Drosophila* MAP kinase activates multiple receptor tyrosine kinase signaling pathways. *Cell* **76**, 875–888.

518. Brunner, E., Brunner, D., Fu, W., Hafen, E., and Basler, K. (1999). The dominant mutation *Glazed* is a gain-of-function allele of *wingless* that, similar to loss of APC, interferes with normal eye development. *Dev. Biol.* **206**, 178–188.

519. Brunner, E., Peter, O., Schweizer, L., and Basler, K. (1997). *pangolin* encodes a Lef-1 homologue that acts downstream of Armadillo to transduce the Wingless signal in *Drosophila*. *Nature* **385**, 829–833.

520. Bryant, P.J. (1969). Mosaicism. [Book review of "Genetic Mosaics and Other Essays" by C. Stern]. *BioScience* **19**, 1126.

521. Bryant, P.J. (1970). Cell lineage relationships in the imaginal wing disc of *Drosophila melanogaster*. *Dev. Biol.* **22**, 389–411.

522. Bryant, P.J. (1971). Regeneration and duplication following operations *in situ* on the imaginal discs of *Drosophila melanogaster*. *Dev. Biol.* **26**, 637–651.

523. Bryant, P.J. (1974). Determination and pattern formation in the imaginal discs of *Drosophila*. *Curr. Top. Dev. Biol.* **8**, 41–80.

524. Bryant, P.J. (1975). Pattern formation in the imaginal wing disc of *Drosophila melanogaster*: fate map, regeneration and duplication. *J. Exp. Zool.* **193**, 49–78.

525. Bryant, P.J. (1975). Regeneration and duplication in imaginal discs. *In* "Cell Patterning," (R. Porter and J. Rivers, eds.). Elsevier: Amsterdam, pp. 71–93.

526. Bryant, P.J. (1978). Pattern formation in imaginal discs. *In* "The Genetics and Biology of Drosophila," Vol. 2c (M. Ashburner and T.R.F. Wright, eds.). Acad. Pr.: New York, pp. 229–335.

527. Bryant, P.J. (1979). Pattern formation, growth control and cell interactions in *Drosophila* imaginal discs. *In* "Determinants of Spatial Organization," *Symp. Soc. Dev. Biol.*, Vol. 37 (S. Subtelny and I.R. Konigsberg, eds.). Acad. Pr.: New York, pp. 295–316.

528. Bryant, P.J. (1987). Experimental and genetic analysis of growth and cell proliferation in *Drosophila* imaginal discs. *In* "Genetic Regulation of Development," (W.F. Loomis, ed.). Alan R. Liss: New York, pp. 339–372.

529. Bryant, P.J. (1988). Localized cell death caused by mutations in a *Drosophila* gene coding for a transforming growth factor-β homolog. *Dev. Biol.* **128**, 386–395.

530. Bryant, P.J. (1991). In memoriam: Howard A. Schneiderman (1927–1990). *Dev. Biol.* **146**, 1–3.

531. Bryant, P.J. (1993). The Polar Coordinate Model goes molecular. *Science* **259**, 471–472.

532. Bryant, P.J. (1996). Cell proliferation control in *Drosophila*: flies are not worms. *BioEssays* **18**, 781–784.

533. Bryant, P.J. (1997). Junction genetics. *Dev. Genet.* **20**, 75–90.

534. Bryant, P.J. (1999). Fickle fingers of cell fate? *Curr. Biol.* **9**, R655–R657.

535. Bryant, P.J., Adler, P.N., Duranceau, C., Fain, M.J., Glenn, S., Hsei, B., James, A.A., Littlefield, C.L., Reinhardt, C.A., Strub, S., and Schneiderman, H.A. (1978). Regulative interactions between cells from different imaginal disks of *Drosophila melanogaster. Science* **201**, 928–930.

536. Bryant, P.J., Bryant, S.V., and French, V. (1977). Biological regeneration and pattern formation. *Sci. Am.* **237** #1, 66–81.

537. Bryant, P.J. and Fraser, S.E. (1988). Wound healing, cell communication, and DNA synthesis during imaginal disc regeneration in *Drosophila. Dev. Biol.* **127**, 197–208.

538. Bryant, P.J. and Girton, J.R. (1980). Genetics of pattern formation. *In* "Development and Neurobiology of *Drosophila*," (O. Siddiqi, P. Babu, L.M. Hall, and J.C. Hall, eds.). Plenum: New York, pp. 109–127.

539. Bryant, P.J., Girton, J.R., and Martin, P. (1980). Physical and pattern continuity in the insect epidermis. *In* "Insect Biology in the Future," (M. Locke and D.S. Smith, eds.). Acad. Pr.: New York, pp. 517–542.

540. Bryant, P.J. and Hsei, B.W. (1977). Pattern formation in asymmetrical and symmetrical imaginal discs of *Drosophila melanogaster. Amer. Zool.* **17**, 595–611.

541. Bryant, P.J., Huettner, B., Held, L.I., Jr., Ryerse, J., and Szidonya, J. (1988). Mutations at the *fat* locus interfere with cell proliferation control and epithelial morphogenesis in *Drosophila. Dev. Biol.* **129**, 541–554.

542. Bryant, P.J. and Levinson, P. (1985). Intrinsic growth control in the imaginal primordia of *Drosophila*, and the autonomous action of a lethal mutation causing overgrowth. *Dev. Biol.* **107**, 355–363.

543. Bryant, P.J. and Schmidt, O. (1990). The genetic control of cell proliferation in *Drosophila* imaginal discs. *In* "Growth Factors in Cell and Developmental Biology," *J. Cell Sci. Suppl.*, Vol. 13 (M.D. Waterfield, ed.). Company of Biologists, Ltd.: Cambridge, pp. 169–189.

544. Bryant, P.J. and Schneiderman, H.A. (1969). Cell lineage, growth, and determination in the imaginal leg discs of *Drosophila melanogaster. Dev. Biol.* **20**, 263–290.

545. Bryant, P.J. and Schubiger, G. (1971). Giant and duplicated imaginal discs in a new lethal mutant of *Drosophila melanogaster. Dev. Biol.* **24**, 233–263.

546. Bryant, P.J. and Simpson, P. (1984). Intrinsic and extrinsic control of growth in developing organs. *Q. Rev. Biol.* **59**, 387–415.

547. Bryant, S.V., Bryant, P.J., and French, V. (1981). Distal regeneration and symmetry. *Science* **212**, 993–1002.

548. Bryant, S.V. and Muneoka, K. (1986). Views of limb development and regeneration. *Trends Genet.* **2**, 153–159.

549. Buchenau, P., Hodgson, J., Strutt, H., and Arndt-Jovin, D.J. (1998). The distribution of Polycomb-group proteins during cell division and development in *Drosophila* embryos: impact on models for silencing. *J. Cell Biol.* **141**, 469–481.

550. Buckles, G.R., Smith, Z.D.J., and Katz, F.N. (1992). *mip* causes hyperinnervation of a retinotopic map in *Drosophila* by excessive recruitment of R7 photoreceptor cells. *Neuron* **8**, 1015–1029.

551. Budd, G.E. (1999). Does evolution in body patterning genes drive morphological change–or vice versa? *BioEssays* **21**, 326–332.

552. Buescher, M., Yeo, S.L., Udolph, G., Zavortink, M., Yang, X., Tear, G., and Chia, W. (1998). Binary sibling neuronal cell fate decisions in the *Drosophila* embryonic central nervous system are nonstochastic and require *inscuteable*-mediated asymmetry of ganglion mother cells. *Genes Dev.* **12**, 1858–1870.

553. Bui, Q.T., Zimmerman, J.E., Liu, H., and Bonini, N.M. (2000). Molecular analysis of *Drosophila eyes absent* mutants reveals features of the conserved Eya domain. *Genetics* **155**, 709–720.

554. Bui, Q.T., Zimmerman, J.E., Liu, H., Gray-Board, G.L., and Bonini, N.M. (2000). Functional analysis of an eye enhancer of the *Drosophila eyes absent* gene: differential regulation by eye specification genes. *Dev. Biol.* **221**, 355–364.

555. Bulinski, S.C., Qiu, X., Durtschi, R.B., and Zeng, C. (2001). Dissecting external sensory organ formation through gain-of-function genetics. *Proc. 42nd Ann. Drosophila Res. Conf.* **Abstracts Vol.**, a268.

556. Burak, W.R. and Shaw, A.S. (2000). Signal transduction: hanging on a scaffold. *Curr. Opin. Cell Biol.* **12**, 211–216.

557. Buratovich, M.A., Anderson, S., Gieseler, K., Pradel, J., and Wilder, E.L. (2000). *DWnt-4* and *Wingless* have distinct activities in the *Drosophila* dorsal epidermis. *Dev. Genes Evol.* **210**, 111–119.

558. Buratovich, M.A. and Bryant, P.J. (1995). Duplication of *l(2)gd* imaginal discs in *Drosophila* is mediated by ectopic expression of *wg* and *dpp*. *Dev. Biol.* **168**, 452–463.

559. Buratovich, M.A. and Bryant, P.J. (1997). Enhancement of overgrowth by gene interactions in *lethal(2)giant discs* imaginal discs from *Drosophila melanogaster. Genetics* **147**, 657–670.

560. Buratovich, M.A., Phillips, R.G., and Whittle, J.R.S. (1997). Genetic relationships between the mutations *spade* and *Sternopleural* and the *wingless* gene in *Drosophila* development. *Dev. Biol.* **185**, 244–260.

561. Burbelo, P.D. and Hall, A. (1995). Hot numbers in signal transduction. *Curr. Biol.* **5**, 95–96.

562. Burden, S.J. (2000). Wnts as retrograde signals for axon and growth cone differentiation. *Cell* **100**, 495–497.

563. Burge, C.B., Tuschl, T., and Sharp, P.A. (1998). Splicing of precursors to mRNAs by the spliceosomes. *In* "The RNA World," (R.F. Gesteland, T.R. Cech, and J.F. Atkins, eds.). Cold Spr. Harb. Lab. Pr.: Cold Spring Harbor, pp. 525–560.

564. Burgess, E.A. and Duncan, I. (1990). Direct control of antennal identity by the *spineless-aristapedia* gene of *Drosophila. Mol. Gen. Genet.* **221**, 347–352.

565. Burgess, S.M. and Kleckner, N. (1999). Collisions between yeast chromosomal loci in vivo are governed by three layers of organization. *Genes Dev.* **13**, 1871–1883.

566. Bürglin, T.R. (1991). The TEA domain: a novel, highly conserved DNA-binding motif. *Cell* **66**, 11–12.

567. Bürglin, T.R. (1997). Analysis of TALE superclass homeobox genes (MEIS, PBC, KNOX, Iroquois, TGIF) reveals a novel domain conserved between plants and animals. *Nucl. Acids Res.* **25**, 4173–4180.

568. Bürglin, T.R. (1998). The PBC domain contains a MEINOX domain: coevolution of Hox and TALE homeobox genes? *Dev. Genes Evol.* **208**, 113–116.

569. Burke, R. and Basler, K. (1996). Dpp receptors are autonomously required for cell proliferation in the entire developing *Drosophila* wing. *Development* **122**, 2261–2269.

570. Burke, R. and Basler, K. (1996). Hedgehog-dependent

patterning in the *Drosophila* eye can occur in the absence of Dpp signaling. *Dev. Biol.* **179**, 360–368.

571. Burke, R. and Basler, K. (1997). Hedgehog signaling in *Drosophila* eye and limb development–conserved machinery, divergent roles? *Curr. Opin. Neurobiol.* **7**, 55–61.

572. Burke, R., Nellen, D., Bellotto, M., Hafen, E., Senti, K.-A., Dickson, B.J., and Basler, K. (1999). Dispatched, a novel sterol-sensing domain protein dedicated to the release of cholesterol-modified Hedgehog from signaling cells. *Cell* **99**, 803–815. [See also 2001 references: *Curr. Biol.* **11**, 1147–1152; *Development* **128**, 5119–5127; *Science* **293**, 2080–2084.]

573. Burris, P.A., Zhang, Y., Rusconi, J.C., and Corbin, V. (1998). The pore-forming and cytoplasmic domains of the neurogenic gene product, BIG BRAIN, are conserved between *Drosophila virilis* and *Drosophila melanogaster*. *Gene* **206**, 69–76.

574. Burrus, L.W. (1994). Wnt-1 as a short-range signaling molecule. *BioEssays* **16**, 155–157.

575. Burz, D.S., Rivera-Pomar, R., Jäckle, H., and Hanes, S.D. (1998). Cooperative DNA-binding by Bicoid provides a mechanism for threshold-dependent gene activation in the *Drosophila* embryo. *EMBO J.* **17**, 5998–6009.

576. Bush, A., Hiromi, Y., and Cole, M. (1996). *biparous*: A novel bHLH gene expressed in neuronal and glial precursors in *Drosophila*. *Dev. Biol.* **180**, 759–772 (cf. erratum: Dev. Biol. 205, 332).

577. Bush, A., Hiromi, Y., and Cole, M. (1999). Erratum. *Dev. Biol.* **205**, 332.

578. Bush, G., diSibio, G., Miyamoto, A., Denault, J.-B., Leduc, R., and Weinmaster, G. (2001). Ligand-induced signaling in the absence of furin processing of Notch1. *Dev. Biol.* **229**, 494–502.

579. Bushweller, J.H. (2000). CBF–a biophysical perspective. *Sems. Cell Dev. Biol.* **11**, 377–382.

580. Busseau, I., Diederich, R.J., Xu, T., and Artavanis-Tsakonas, S. (1994). A member of the Notch group of interacting loci, *deltex* encodes a cytoplasmic basic protein. *Genetics* **136**, 585–596.

581. Busturia, A. and Lawrence, P.A. (1994). Regulation of cell number in *Drosophila*. *Nature* **370**, 561–563.

582. Busturia, A. and Morata, G. (1988). Ectopic expression of homeotic genes caused by the elimination of the *Polycomb* gene in *Drosophila* imaginal epidermis. *Development* **104**, 713–720.

583. Busturia, A., Vernos, I., Macias, A., Casanova, J., and Morata, G. (1990). Different forms of *Ultrabithorax* proteins generated by alternative splicing are functionally equivalent. *EMBO J.* **9**, 3551–3555.

584. Busturia, A., Wightman, C.D., and Sakonju, S. (1997). A silencer is required for maintenance of transcriptional repression throughout *Drosophila* development. *Development* **124**, 4343–4350. [See also 2001 *Genetics* **158**, 291–307.]

585. Butler, S.J., Ray, S., and Hiromi, Y. (1997). *klingon*, a novel member of the *Drosophila* immunoglobulin superfamily, is required for the development of the R7 photoreceptor neuron. *Development* **124**, 781–792.

586. Cabral, J.H.M., Petosa, C., Sutcliffe, M.J., Raza, S., Byron, O., Poy, F., Marfatia, S.M., Chishti, A.H., and Liddington, R.C. (1996). Crystal structure of a PDZ domain. *Nature* **382**, 649–652.

587. Cabrera, C.V. (1992). The generation of cell diversity during early neurogenesis in *Drosophila*. *Development* **115**, 893–901.

588. Cabrera, C.V. and Alonso, M.C. (1991). Transcriptional activation by heterodimers of the *achaete-scute* and *daughterless* gene products of *Drosophila*. *EMBO J.* **10**, 2965–2973.

589. Cabrera, C.V., Alonso, M.C., and Huikeshoven, H. (1994). Regulation of scute function by extramacrochaetae in vitro and in vivo. *Development* **120**, 3595–3603.

590. Cabrera, C.V., Alonso, M.C., Johnston, P., Phillips, R.G., and Lawrence, P.A. (1987). Phenocopies induced with antisense RNA identify the *wingless* gene. *Cell* **50**, 659–663.

591. Cabrera, C.V., Botas, J., and Garcia-Bellido, A. (1985). Distribution of *Ultrabithorax* proteins in mutants of *Drosophila* bithorax complex and its transregulatory genes. *Nature* **318**, 569–571.

592. Cacace, A.M., Michaud, N.R., Therrien, M., Mathes, K., Copeland, T., Rubin, G.M., and Morrison, D.K. (1999). Identification of constitutive and Ras-inducible phosphorylation sites of KSR: implications for 14-3-3 binding, mitogen-activated protein kinase binding, and KSR overexpression. *Mol. Cell. Biol.* **19**, 229–240.

593. Cáceres, M., Barbadilla, A., and Ruiz, A. (1999). Recombination rate predicts inversion size in diptera. *Genetics* **153**, 251–259.

594. Cadavid, A.L.M., Ginzel, A., and Fischer, J.A. (2000). The function of the *Drosophila* Fat facets deubiquitinating enzyme in limiting photoreceptor cell number is intimately associated with endocytosis. *Development* **127**, 1727–1736.

595. Cadigan, K.M., Fish, M.P., Rulifson, E.J., and Nusse, R. (1998). Wingless repression of *Drosophila frizzled 2* expression shapes the Wingless morphogen gradient in the wing. *Cell* **93**, 767–777.

596. Cadigan, K.M., Grossniklaus, U., and Gehring, W.J. (1994). Localized expression of *sloppy paired* protein maintains the polarity of *Drosophila* parasegments. *Genes Dev.* **8**, 899–913.

597. Cadigan, K.M. and Nusse, R. (1996). *wingless* signaling in the *Drosophila* eye and embryonic epidermis. *Development* **122**, 2801–2812.

598. Cadigan, K.M. and Nusse, R. (1997). Wnt signaling: a common theme in animal development. *Genes Dev.* **11**, 3286–3305.

599. Cadigan, K.M., Parker, D.S., Jemison, J., Klinedienst, S., and Kohen, R. (2001). The role of the *gammy legs* gene in Wingless signaling. *Proc. 42nd Ann. Drosophila Res. Conf.* **Abstracts Vol.**, a55.

600. Cagan, R. (1993). Cell fate specification in the developing *Drosophila* retina. *Development* **1993 Suppl.**, 19–28.

601. Cagan, R.L., Krämer, H., Hart, A.C., and Zipursky, S.L. (1992). The bride of sevenless and sevenless interaction: internalization of a transmembrane ligand. *Cell* **69**, 393–399.

602. Cagan, R.L. and Ready, D.F. (1989). The emergence of order in the *Drosophila* pupal retina. *Dev. Biol.* **136**, 346–362.

603. Cagan, R.L. and Ready, D.F. (1989). *Notch* is required for successive cell decisions in the developing *Drosophila* retina. *Genes Dev.* **3**, 1099–1112.

604. Cagan, R.L., Thomas, B.J., and Zipursky, S.L. (1993). The role of induction in cell choice and cell cycle in the developing *Drosophila* retina. *In* "Molecular Basis of Morphogenesis," *Symp. Soc. Dev. Biol.*, Vol. 51 (M. Bernfield, ed.). Wiley-Liss: New York, pp. 109–133.

605. Cagan, R.L. and Zipursky, S.L. (1992). Cell choice and patterning in the *Drosophila* retina. *In* "Determinants of Neuronal Identity," (M. Shankland and E.R. Macagno, eds.). Acad. Pr.: New York, pp. 189–224.

606. Cai, H., N. and Shen, P. (2001). Effects of *cis* arrangement of chromatin insulators on enhancer-blocking activity. *Science* **291**, 493–495.

607. Cai, H.N. and Levine, M. (1997). The gypsy insulator can function as a promoter-specific silencer in the *Drosophila* embryo. *EMBO J.* **16**, 1732–1741.

608. Cai, J., Lan, Y., Appel, L.F., and Weir, M. (1994). Dissection of the *Drosophila* Paired protein: functional requirements for conserved motifs. *Mechs. Dev.* **47**, 139–150.

609. Cai, Y., Chia, W., and Yang, X. (2001). A family of Snail-related zinc finger proteins regulates two distinct and parallel mechanisms that mediate *Drosophila* neuroblast asymmetric divisions. *EMBO J.* **20**, 1704–1714.

610. Cairns, J., Stent, G.S., and Watson, J.D., eds. (1966). "Phage and the Origins of Molecular Biology." Cold Spring Harbor Lab. Quant. Biol., Cold Spring Harbor, N. Y.

611. Callaerts, P., Halder, G., and Gehring, W.J. (1997). *Pax-6* in development and evolution. *Annu. Rev. Neurosci.* **20**, 483–532.

612. Callahan, C.A., Bonkovsky, J.L., Scully, A.L., and Thomas, J.B. (1996). *derailed* is required for muscle attachment site selection in *Drosophila*. *Development* **122**, 2761–2767.

613. Callahan, P.S. (1979). Evolution of antennae, their sensilla and the mechanism of scent detection in Arthropoda. *In* "Arthropod Phylogeny," (A.P. Gupta, ed.). Van Nostrand Reinhold: New York, pp. 259–298.

614. Calleja, M., Herranz, H., Estella, C., Casal, J., Lawrence, P., Simpson, P., and Morata, G. (2000). Generation of medial and lateral dorsal body domains by the *pannier* gene of *Drosophila*. *Development* **127**, 3971–3980.

615. Calleja, M., Moreno, E., Pelaz, S., and Morata, G. (1996). Visualization of gene expression in living adult *Drosophila*. *Science* **274**, 252–255.

616. Cambridge, S.B., Davis, R.L., and Minden, J.S. (1997). *Drosophila* mitotic domain boundaries as cell fate boundaries. *Science* **277**, 825–828.

617. Campbell, G. and Tomlinson, A. (1995). Initiation of the proximodistal axis in insect legs. *Development* **121**, 619–628.

618. Campbell, G. and Tomlinson, A. (1998). The roles of the homeobox genes *aristaless* and *Distal-less* in patterning the legs and wings of *Drosophila*. *Development* **125**, 4483–4493.

619. Campbell, G. and Tomlinson, A. (1999). Transducing the Dpp morphogen gradient in the wing of *Drosophila*: regulation of Dpp targets by *brinker*. *Cell* **96**, 553–562.

620. Campbell, G., Weaver, T., and Tomlinson, A. (1993). Axis specification in the developing *Drosophila* appendage: the role of *wingless*, *decapentaplegic*, and the homeobox gene *aristaless*. *Cell* **74**, 1113–1123.

621. Campbell, G.L. and Tomlinson, A. (2000). Transcriptional regulation of the Hedgehog effector Ci by the zinc-finger gene *combgap*. *Development* **127**, 4095–4103.

622. Campbell, J.H. (1985). An organizational interpretation of evolution. *In* "Evolution at a Crossroads: The New Biology and the New Philosophy of Science," (D.J. Depew and B.H. Weber, eds.). MIT Pr.: Cambridge, Mass., pp. 133–167.

623. Campbell, S., Inamdar, M., Rodrigues, V., Raghavan, V., Palazzolo, M., and Chovnick, A. (1992). The *scalloped* gene encodes a novel, evolutionarily conserved transcrip-

tion factor required for sensory organ differentiation in *Drosophila*. *Genes Dev.* **6**, 367–379.

624. Campos, A.R., Fischbach, K.-F., and Steller, H. (1992). Survival of photoreceptor neurons in the compound eye of *Drosophila* depends on connections with the optic ganglia. *Development* **114**, 355–366.

625. Campos-Ortega, J.A. (1980). On compound eye development in *Drosophila melanogaster*. *Curr. Top. Dev. Biol.* **15**, 347–371.

626. Campos-Ortega, J.A. (1988). Cellular interactions during early neurogenesis of *Drosophila melanogaster*. *Trends Neurosci.* **11**, 400–405.

627. Campos-Ortega, J.A. (1993). Early neurogenesis in *Drosophila melanogaster*. *In* "The Development of *Drosophila melanogaster*," Vol. 2 (M. Bate and A. Martinez Arias, eds.). Cold Spring Harbor Laboratory Pr.: New York,

628. Campos-Ortega, J.A. (1998). The genetics of the *Drosophila achaete-scute* gene complex: a historical appraisal. *Int. J. Dev. Biol.* **42**, 291–297.

629. Campos-Ortega, J.A. and Gateff, E.A. (1976). The development of ommatidial patterning in metamorphosed eye imaginal discs implants of *Drosophila melanogaster*. *W. Roux's Arch.* **179**, 373–392.

630. Campos-Ortega, J.A. and Haenlin, M. (1992). Regulatory signals and signal molecules in early neurogenesis of *Drosophila melanogaster*. *Roux's Arch. Dev. Biol.* **201**, 1–11.

631. Campos-Ortega, J.A. and Hartenstein, V. (1997). "The Embryonic Development of *Drosophila melanogaster*." 2nd ed. Springer, Berlin.

632. Campos-Ortega, J.A., Jürgens, G., and Hofbauer, A. (1979). Cell clones and pattern formation: studies on *sevenless*, a mutant of *Drosophila melanogaster*. *W. Roux's Arch.* **186**, 27–50.

633. Campos-Ortega, J.A. and Waitz, M. (1978). Cell clones and pattern formation: developmental restrictions in the compound eye of *Drosophila*. *W. Roux's Arch.* **184**, 155–170.

634. Camps, M., Nichols, A., Gillieron, C., Antonsson, B., Muda, M., Chabert, C., Boschert, U., and Arkinstall, S. (1998). Catalytic activation of the phosphatase MKP-3 by ERK2 mitogen-activated protein kinase. *Science* **280**, 1262–1265.

635. Campuzano, S., Balcells, L., Villares, R., Carramolino, L., García-Alonso, L., and Modolell, J. (1986). Excess function of *Hairy-wing* mutations caused by *gypsy* and *copia* insertions within structural genes of the *achaete-scute* locus of *Drosophila*. *Cell* **44**, 303–312.

636. Campuzano, S., Carramolino, L., Cabrera, C.V., Ruíz-Gómez, M., Villares, R., Boronat, A., and Modolell, J. (1985). Molecular genetics of the *achaete-scute* gene complex of *D. melanogaster*. *Cell* **40**, 327–338.

637. Campuzano, S. and Modolell, J. (1992). Patterning of the *Drosophila* nervous system: the *achaete-scute* gene complex. *Trends Genet.* **8**, 202–208.

638. Canagarajah, B.J., Khokhlatchev, A., Cobb, M.H., and Goldsmith, E.J. (1997). Activation mechanism of the MAP kinase ERK2 by dual phosphorylation. *Cell* **90**, 859–869.

639. Canon, J. and Banerjee, U. (2000). Runt and Lozenge function in *Drosophila* development. *Sems. Cell Dev. Biol.* **11**, 327–336.

640. Cant, K., Knowles, B.A., Mooseker, M.S., and Cooley, L. (1994). *Drosophila* Singed, a fascin homolog, is required for actin bundle formation during oogenesis and bristle extension. *J. Cell Biol.* **125**, 369–380.

641. Capdevila, J. and Belmonte, J.C.I. (1999). Extracellular modulation of the Hedgehog, Wnt and TGF-β signalling pathways during embryonic development. *Curr. Opin. Gen. Dev.* **9**, 427–433.

642. Capdevila, J., Estrada, M.P., Sánchez-Herrero, E., and Guerrero, I. (1994). The *Drosophila* segment polarity gene *patched* interacts with *decapentaplegic* in wing development. *EMBO J.* **13**, 71–82.

643. Capdevila, J. and Guerrero, I. (1994). Targeted expression of the signaling molecule decapentaplegic induces pattern duplications and growth alterations in *Drosophila* wings. *EMBO J.* **13**, 4459–4468.

644. Capdevila, J., Pariente, F., Sampedro, J., Alonso, J.L., and Guerrero, I. (1994). Subcellular localization of the segment polarity protein patched suggests an interaction with the wingless reception complex in *Drosophila* embryos. *Development* **120**, 987–998.

645. Capdevila, M.P., Botas, J., and García-Bellido, A. (1986). Genetic interactions between the *Polycomb* locus and the *Antennapedia* and *Bithorax* complexes of *Drosophila*. *Roux's Arch. Dev. Biol.* **195**, 417–432.

646. Capdevila, M.P. and Garcia-Bellido, A. (1974). Development and genetic analysis of *bithorax* phenocopies in *Drosophila*. *Nature* **250**, 500–502.

647. Capdevila, M.P. and García-Bellido, A. (1978). Phenocopies of *bithorax* mutants. *W. Roux's Arch.* **185**, 105–126.

648. Capovilla, M. and Botas, J. (1998). Functional dominance among Hox genes: repression dominates activation in the regulation of *dpp*. *Development* **125**, 4949–4957.

649. Carey, M. (1998). The enhanceosome and transcriptional synergy. *Cell* **92**, 5–8.

650. Carlson, E.A. (1966). "The Gene: A Critical History." W. B. Saunders, Philadephia.

651. Carlson, E.A. (1981). "Genes, Radiation, and Society: The Life and Work of H. J. Muller." Cornell Univ. Pr., Ithaca, N. Y.

652. Carmena, A. and Baylies, M. (2001). Canoe: a possible link between Ras and Notch signaling pathways during *Drosophila* mesoderm differentiation. *Proc. 42nd Ann. Drosophila Res. Conf.* **Abstracts Vol.**, a145.

653. Carmena, A., Murugasu-Oei, B., Menon, D., Jiménez, F., and Chia, W. (1998). *inscuteable* and *numb* mediate asymmetric muscle progenitor cell divisions during *Drosophila* myogenesis. *Genes Dev.* **12**, 304–315.

654. Carpenter, G. (2000). The EGF receptor: a nexus for trafficking and signaling. *BioEssays* **22**, 697–707.

655. Carr, A. and Biggin, M.D. (1999). A comparison of *in vivo* and *in vitro* DNA-binding specificities suggests a new model for homeoprotein DNA binding in *Drosophila* embryos. *EMBO J.* **18**, 1598–1608.

656. Carramolino, L., Ruiz-Gomez, M., del Carmen Guerrero, M., Campuzano, S., and Modolell, J. (1982). DNA map of mutations at the *scute* locus of *Drosophila melanogaster*. *EMBO J.* **1**, 1185–1191.

657. Carroll, S.B. (1990). Zebra patterns in fly embryos: activation of stripes or repression of interstripes? *Cell* **60**, 9–16.

658. Carroll, S.B. (1995). Homeotic genes and the evolution of arthropods and chordates. *Nature* **376**, 479–485.

659. Carroll, S.B. (1998). From pattern to gene, from gene to pattern. *Int. J. Dev. Biol.* **42**, 305–309.

660. Carroll, S.B. (2000). Endless forms: the evolution of gene regulation and morphological diversity. *Cell* **101**, 577–580.

661. Carroll, S.B., DiNardo, S., O'Farrell, P.H., White, R.A.H., and Scott, M.P. (1988). Temporal and spatial relationships between segmentation and homeotic gene expression in *Drosophila* embryos: distributions of the *fushi tarazu*, *engrailed*, *Sex combs reduced*, *Antennapedia*, and *Ultrabithorax* proteins. *Genes Dev.* **2**, 350–360.

662. Carroll, S.B., Grenier, J.K., and Weatherbee, S.D. (2001). "From DNA to Diversity: Molecular Genetics and the Evolution of Animal Design." Blackwell Science, Malden, Mass.

663. Carroll, S.B., Laughon, A., and Thalley, B.S. (1988). Expression, function, and regulation of the *hairy* segmentation protein in the *Drosophila* embryo. *Genes Dev.* **2**, 883–890.

664. Carroll, S.B., Weatherbee, S.D., and Langeland, J.A. (1995). Homeotic genes and the regulation and evolution of insect wing number. *Nature* **375**, 58–61.

665. Carroll, S.B. and Whyte, J.S. (1989). The role of the *hairy* gene during *Drosophila* morphogenesis: stripes in imaginal discs. *Genes Dev.* **3**, 905–916.

666. Carson, H.L. (1986). Patterns of inheritance. *Amer. Zool.* **26**, 797–809.

667. Carson, H.L. and Bryant, P.J. (1979). Change in a secondary sexual character as evidence of incipient speciation in *Drosophila silvestris*. *Proc. Natl. Acad. Sci. USA* **76**, 1929–1932.

668. Carson, H.L., Hardy, D.E., Spieth, H.T., and Stone, W.S. (1970). The evolutionary biology of the Hawaiian Drosophilidae. *In* "Essays in Evolution and Genetics," (M.K. Hecht and W.C. Steere, eds.). Appleton-Century-Crofts: New York, pp. 437–543.

669. Carson, H.L. and Kaneshiro, K.Y. (1976). *Drosophila* of Hawaii: systematics and ecological genetics. *Annu. Rev. Ecol. Syst.* **7**, 311–345.

670. Carthew, R.W., Kauffmann, R.C., Kladny, S., Li, S., and Zhang, J. (1999). Cell determination in the *Drosophila* eye. *In* "Cell Lineage and Fate Determination," (S.A. Moody, ed.). Acad. Pr.: New York, pp. 235–248.

671. Carthew, R.W. and Rubin, G.M. (1990). *seven in absentia*, a gene required for specification of R7 cell fate in the *Drosophila* eye. *Cell* **63**, 561–577.

672. Casanova, J. (1989). Mutations in the *spalt* gene of *Drosophila* cause ectopic expression of *Ultrabithorax* and *Sex combs reduced*. *Roux's Arch. Dev. Biol.* **198**, 137–140.

673. Casanova, J., Sánchez-Herrero, E., and Morata, G. (1985). *Contrabithorax* and the control of spatial expression of the bithorax complex genes of *Drosophila*. *J. Embryol. Exp. Morph.* **90**, 179–196.

674. Casanova, J. and Struhl, G. (1993). The *torso* receptor localizes as well as transduces the spatial signal specifying terminal body pattern in *Drosophila*. *Nature* **362**, 152–155.

675. Casares, F., Bender, W., Merriam, J., and Sánchez-Herrero, E. (1997). Interactions of *Drosophila Ultrabithorax* regulatory regions with native and foreign promoters. *Genetics* **145**, 123–137.

676. Casares, F., Calleja, M., and Sánchez-Herrero, E. (1996). Functional similarity in appendage specification by the *Ultrabithorax* and *abdominal-A Drosophila* HOX genes. *EMBO J.* **15**, 3934–3942.

677. Casares, F. and Mann, R.S. (1998). Control of antennal versus leg development in *Drosophila*. *Nature* **392**, 723–726. [See also 2001 references: *Curr. Biol.* **11**, R1025–R1027; *Science* **293**, 1477–1480.]

678. Casares, F. and Mann, R.S. (2000). A dual role for *homothorax* in inhibiting wing blade development and specifying proximal wing identities in *Drosophila*. *Development* **127**, 1499–1508.

679. Casares, F., Sánchez, L., Guerrero, I., and Sánchez-Herrero,

E. (1997). The genital disc of *Drosophila melanogaster*. I. Segmental and compartmental organization. *Dev. Genes Evol.* **207**, 216–228.

680. Casares, F. and Sánchez-Herrero, E. (1995). Regulation of the *infraabdominal* regions of the bithorax complex of *Drosophila* by gap genes. *Development* **121**, 1855–1866.

681. Casci, T., Vinós, J., and Freeman, M. (1999). Sprouty, an intracellular inhibitor of Ras signaling. *Cell* **96**, 655–665.

682. Castelli-Gair, J. (1998). Implications of the spatial and temporal regulation of *Hox* genes on development and evolution. *Int. J. Dev. Biol.* **42**, 437–444.

683. Castelli-Gair, J. (1998). The *lines* gene of *Drosophila* is required for specific functions of the Abdominal-B HOX protein. *Development* **125**, 1269–1274.

684. Castelli-Gair, J. and Akam, M. (1995). How the Hox gene *Ultrabithorax* specifies two different segments: the significance of spatial and temporal regulation within metameres. *Development* **121**, 2973–2982.

685. Castelli-Gair, J., Greig, S., Micklem, G., and Akam, M. (1994). Dissecting the temporal requirements for homeotic gene function. *Development* **120**, 1983–1995.

686. Castelli-Gair, J., Müller, J., and Bienz, M. (1992). Function of an *Ultrabithorax* minigene in imaginal cells. *Development* **114**, 877–886.

687. Cate, J.H., Yusupov, M.M., Yusupova, G.Z., Earnest, T.N., and Noller, H.F. (1999). X-ray crystal structures of 70*S* ribosome functional complexes. *Science* **285**, 2095–2104.

688. Caudy, M., Grell, E.H., Dambly-Chaudière, C., Ghysen, A., Jan, L.Y., and Jan, Y.N. (1988). The maternal sex determination gene *daughterless* has zygotic activity necessary for the formation of peripheral neurons in *Drosophila*. *Genes Dev.* **2**, 843–852.

689. Caudy, M., Vässin, H., Brand, M., Tuma, R., Jan, L.Y., and Jan, Y.N. (1988). *daughterless*, a *Drosophila* gene essential for both neurogenesis and sex determination, has sequence similarities to *myc* and the *achaete-scute* complex. *Cell* **55**, 1061–1067.

690. Cavalli, G. and Paro, R. (1999). Epigenetic inheritance of active chromatin after removal of the main transactivator. *Science* **286**, 955–958.

691. Cavallo, R., Rubenstein, D., and Peifer, M. (1997). Armadillo and dTCF: a marriage made in the nucleus. *Curr. Opin. Gen. Dev.* **7**, 459–466.

692. Cavallo, R.A., Cox, R.T., Moline, M.M., Roose, J., Polevoy, G.A., Clevers, H., Peifer, M., and Bejsovec, A. (1998). *Drosophila* Tcf and Groucho interact to repress Wingless signalling activity. *Nature* **395**, 604–608.

693. Cavener, D.R. (1987). Combinatorial control of structural genes in *Drosophila*: solutions that work for the animal. *BioEssays* **7**, 103–107.

694. Cavener, D.R. (1992). Transgenic animal studies on the evolution of genetic regulatory circuitries. *BioEssays* **14**, 237–244.

695. Caveney, S. (1985). Intercellular communication. *In* "Comprehensive Insect Physiology, Biochemistry, and Pharmacology," Vol. 2 (G.A. Kerkut and L.I. Gilbert, eds.). Pergamon Pr.: Oxford, pp. 319–370. [See also 2001 *Nature* **411**, 759–762.]

696. Cavodeassi, F., del Corral, R.D., Campuzano, S., and Domínguez, M. (1999). Compartments and organising boundaries in the *Drosophila* eye: the role of the homeodomain Iroquois proteins. *Development* **126**, 4933–4942.

697. Cavodeassi, F., Modolell, J., and Campuzano, S. (2000). The Iroquois homeobox genes function as dorsal selectors in the *Drosophila* head. *Development* **127**, 1921–1929.

698. Celniker, S.E. (2000). The *Drosophila* genome. *Curr. Opin. Gen. Dev.* **10**, 612–616.

699. Celniker, S.E., Sharma, S., Keelan, D.J., and Lewis, E.B. (1990). The molecular genetics of the bithorax complex of *Drosophila*: cis-regulation in the *Abdominal-B* domain. *EMBO J.* **9**, 4277–4286.

700. Celotto, A.M. and Graveley, B.R. (2001). Alternative splicing of the *Drosophila Dscam* pre-mRNA is both temporally and spatially regulated. *Genetics* **159**, 599–608.

701. Cerdá-Olmedo, E. (1998). The development of concepts on development: a dialogue with Antonio García-Bellido. *Int. J. Dev. Biol.* **42**, 233–236.

702. Ceresa, B.P. and Schmid, S.L. (2000). Regulation of signal transduction by endocytosis. *Curr. Opin. Cell Biol.* **12**, 204–210.

703. Certel, K., Hudson, A., Carroll, S.B., and Johnson, W.A. (2000). Restricted patterning of *vestigial* expression in *Drosophila* wing imaginal discs requires synergistic activation by both Mad and the Drifter POU domain transcription factor. *Development* **127**, 3173–3183.

704. Chae, J., Kim, M.-J., Goo, J.H., Collier, S., Gubb, D., Charlton, J., Adler, P.N., and Park, W.J. (1999). The *Drosophila* tissue polarity gene *starry night* encodes a member of the protocadherin family. *Development* **126**, 5421–5429.

705. Chakravarti, D., LaMorte, V.J., Nelson, M.C., Nakajima, T., Schulman, I.G., Juguilon, H., Montminy, M., and Evans, R.M. (1996). Role of CBP/P300 in nuclear receptor signalling. *Nature* **383**, 99–103.

706. Chan, S.-K., Jaffe, L., Capovilla, M., Botas, J., and Mann, R.S. (1994). The DNA binding specificity of Ultrabithorax is modulated by cooperative interactions with Extradenticle, another homeoprotein. *Cell* **78**, 603–615.

707. Chan, S.-K. and Mann, R.S. (1993). The segment identity functions of Ultrabithorax are contained within its homeo domain and carboxy-terminal sequences. *Genes Dev.* **7**, 796–811.

708. Chan, S.-K., Pöpperl, H., Krumlauf, R., and Mann, R.S. (1996). An extradenticle-induced conformational change in a HOX protein overcomes an inhibitory function of the conserved hexapeptide motif. *EMBO J.* **15**, 2476–2487.

709. Chan, S.-K., Ryoo, H.-D., Gould, A., Krumlauf, R., and Mann, R.S. (1997). Switching the in vivo specificity of a minimal Hox-responsive element. *Development* **124**, 2007–2014.

710. Chan, S.-K. and Struhl, G. (1997). Sequence-specific RNA binding by Bicoid. *Nature* **388**, 634.

711. Chan, Y.-M. and Jan, Y.N. (1998). Roles for proteolysis and trafficking in Notch maturation and signal transduction. *Cell* **94**, 423–426.

712. Chan, Y.-M. and Jan, Y.N. (1999). Presenilins, processing of β-amyloid precursor protein, and Notch signaling. *Neuron* **23**, 201–204.

713. Chandebois, R. (1977). Cell sociology and the problem of position effect: pattern formation, origin and role of gradients. *Acta Biotheor.* **26**, 203–238.

714. Chandebois, R. and Faber, J. (1983). "Automation in Animal Development." *Monographs in Developmental Biology*, Vol. 16. Karger, Basel.

715. Chang, H.-Y. and Ready, D.F. (2000). Rescue of photoreceptor degeneration in rhodopsin-null *Drosophila* mutants by activated Rac1. *Science* **290**, 1978–1980.

716. Chang, H.C. and Rubin, G.M. (1997). 14-3-3ε positively regulates Ras-mediated signaling in *Drosophila*. *Genes Dev.* **11**, 1132–1139.

717. Chang, H.C., Solomon, N.M., Wassarman, D.A., Karim, F.D., Therrien, M., Rubin, G.M., and Wolff, T. (1995). *phyllopod* functions in the fate determination of a subset of photoreceptors in *Drosophila*. *Cell* **80**, 463–472.

718. Chang, W.S., Zachow, K.R., and Bentley, D. (1993). Expression of epithelial alkaline phosphatase in segmentally iterated bands during grasshopper limb morphogenesis. *Development* **118**, 651–663.

719. Chang, Y.-L., King, B.O., O'Connor, M., Mazo, A., and Huang, D.-H. (1995). Functional reconstruction of *trans* regulation of the *Ultrabithorax* promoter by the products of two antagonistic genes, *trithorax* and *Polycomb*. *Mol. Cell. Biol.* **15**, 6601–6612.

720. Chang, Y.W. and Chien, C.T. (2001). Eye formation genes *eyeless* and *dachshund* are negatively regulated by proneural genes *atonal* and *daughterless* during *Drosophila* eye development. *Proc. 42nd Ann. Drosophila Res. Conf.* **Abstracts Vol.**, a255.

721. Chang, Z., Price, B.D., Bockheim, S., Boedigheimer, M.J., Smith, R., and Laughon, A. (1993). Molecular and genetic characterization of the *Drosophila tartan* gene. *Dev. Biol.* **160**, 315–332.

722. Chanut, F. and Heberlein, U. (1995). Role of the morphogenetic furrow in establishing polarity in the *Drosophila* eye. *Development* **121**, 4085–4094.

723. Chanut, F. and Heberlein, U. (1997). Retinal morphogenesis in *Drosophila*: hints from an eye-specific *decapentaplegic* allele. *Dev. Genet.* **20**, 197–207.

724. Chanut, F. and Heberlein, U. (1997). Role of *decapentaplegic* in initiation and progression of the morphogenetic furrow in the developing *Drosophila* retina. *Development* **124**, 559–567.

725. Chanut, F., Luk, A., and Heberlein, U. (2000). A screen for dominant modifiers of *roDom*, a mutation that disrupts morphogenetic furrow progression in *Drosophila*, identifies Groucho and Hairless as regulators of *atonal* expression. *Genetics* **156**, 1203–1217.

726. Chao, J.L. and Sun, Y.H. (2001). *eyg* is a downstream effector of *Notch* signaling on cell proliferation in *Drosophila* eye. *Proc. 42nd Ann. Drosophila Res. Conf.* **Abstracts Vol.**, a195.

727. Chaudhary, S., Brou, C., Valentin, M.-E., Burton, N., Tora, L., Chambon, P., and Davidson, I. (1994). A cell-specific factor represses stimulation of transcription in vitro by Transcriptional Enhancer Factor 1. *Mol. Cell. Biol.* **14**, 5290–5299.

728. Chaudhary, S., Tora, L., and Davidson, I. (1995). Characterization of a HeLa cell factor which negatively regulates transcriptional activation *in vitro* by Transcriptional Enhancer Factor-1 (TEF-1). *J. Biol. Chem.* **270** #8, 3631–3637.

729. Cheah, P.Y., Chia, W., and Yang, X. (2000). Jumeaux, a novel *Drosophila* winged-helix family protein, is required for generating asymmetric sibling neuronal cell fates. *Development* **127**, 3325–3335.

730. Chelsky, D., Ralph, R., and Jonak, G. (1989). Sequence requirements for synthetic peptide-mediated translocation to the nucleus. *Mol. Cell. Biol.* **9**, 2487–2492.

731. Chen, C.-H., von Kessler, D.P., Park, W., Wang, B., Ma, Y., and Beachy, P.A. (1999). Nuclear trafficking of Cubitus interruptus in the transcriptional regulation of Hedgehog target gene expression. *Cell* **98**, 305–316.

732. Chen, C.-K., Kühnlein, R.P., Eulenberg, K.G., Vincent, S., Affolter, M., and Schuh, R. (1998). The transcription factors KNIRPS and KNIRPS RELATED control cell migration and branch morphogenesis during *Drosophila* tracheal development. *Development* **125**, 4959–4968.

733. Chen, C.-M., Freedman, J.A., Bettler, D.R., Jr., Manning, S.D., Giep, S.N., Steiner, J., and Ellis, H.M. (1996). *polychaetoid* is required to restrict segregation of sensory organ precursors from proneural clusters in *Drosophila*. *Mechs. Dev.* **57**, 215–227.

734. Chen, C.-m. and Struhl, G. (1999). Wingless transduction by the Frizzled and Frizzled2 proteins of *Drosophila*. *Development* **126**, 5441–5452.

735. Chen, E.H. and Baker, B.S. (1997). Compartmental organization of the *Drosophila* genital imaginal discs. *Development* **124**, 205–218.

736. Chen, F. and Rebay, I. (2000). *split ends*, a new component of the *Drosophila* EGF receptor pathway, regulates development of midline glial cells. *Curr. Biol.* **10**, 943–946.

737. Chen, G., Fernandez, J., Mische, S., and Courey, A.J. (1999). A functional interaction between the histone deacetylase Rpd3 and the corepressor Groucho in *Drosophila* development. *Genes Dev.* **13**, 2218–2230.

738. Chen, G., Nguyen, P.H., and Courey, A.J. (1998). A role for Groucho tetramerization in transcriptional repression. *Mol. Cell. Biol.* **18**, 7259–7268.

739. Chen, H., Fre, S., Slepnev, V.I., Capua, M.R., Takei, K., Butler, M.H., Di Fiore, P.P., and De Camilli, P. (1998). Epsin is an EH-domain-binding protein implicated in clathrin-mediated endocytosis. *Nature* **394**, 793–797.

740. Chen, M.S., Obar, R.A., Schroeder, C.C., Austin, T.W., Poodry, C.A., Wadsworth, S.C., and Vallee, R.B. (1991). Multiple forms of dynamin are encoded by *shibire*, a *Drosophila* gene involved in endocytosis. *Nature* **351**, 583–586.

741. Chen, P., Nordstrom, W., Gish, B., and Abrams, J.M. (1996). *grim*, a novel cell death gene in *Drosophila*. *Genes Dev.* **10**, 1773–1782.

742. Chen, P., Rodriguez, A., Erskine, R., Thach, T., and Abrams, J.M. (1998). *Dredd*, a novel effector of the apoptosis activators *reaper*, *grim*, and *hid* in *Drosophila*. *Dev. Biol.* **201**, 202–216.

743. Chen, R., Amoui, M., Zhang, Z., and Mardon, G. (1997). Dachshund and Eyes absent proteins form a complex and function synergistically to induce ectopic eye development in *Drosophila*. *Cell* **91**, 893–903.

744. Chen, R., Halder, G., Zhang, Z., and Mardon, G. (1999). Signaling by the TGF-β homolog *decapentaplegic* functions reiteratively within the network of genes controlling retinal cell fate determination in *Drosophila*. *Development* **126**, 935–943.

745. Chen, T., Bunting, M., Karim, F.D., and Thummel, C.S. (1992). Isolation and characterization of five *Drosophila* genes that encode an *ets*-related DNA binding domain. *Dev. Biol.* **151**, 176–191.

746. Chen, X. and Fischer, J.A. (2000). *In vivo* structure/function analysis of the *Drosophila fat facets* deubiquitinating enzyme gene. *Genetics* **156**, 1829–1836.

747. Chen, X., Li, Q., and Fischer, J.A. (2000). Genetic analysis of the *Drosophila DNAprim* gene: the function of the 60-kD primase subunit of DNA polymerase opposes the *fat facets* signaling pathway in the developing eye. *Genetics* **156**, 1787–1795.

748. Chen, X., Overstreet, E., Wood, S.A., and Fischer, J.A. (2000). On the conservation of function of the *Drosophila* Fat facets deubiquitinating enzyme and Fam, its mouse homolog. *Dev. Genes Evol.* **210**, 603–610.

749. Chen, Y., Cardinaux, J.-R., Goodman, R.H., and Smolik, S.M. (1999). Mutants of *cubitus interruptus* that are independent of PKA regulation are independent of *hedgehog* signaling. *Development* **126**, 3607–3616.

750. Chen, Y., Fischer, W.H., and Gill, G.N. (1997). Regulation of the *ERBB-2* promoter by RBPJκ and Notch. *J. Biol. Chem.* **272** #22, 14110–14114.

751. Chen, Y., Gallaher, N., Goodman, R.H., and Smolik, S.M. (1998). Protein kinase A directly regulates the activity and proteolysis of *cubitus interruptus. Proc. Natl. Acad. Sci. USA* **95**, 2349–2354. [See also 2001 *Genetics* **158**, 1157–1166.]

752. Chen, Y., Goodman, R.H., and Smolik, S.M. (2000). Cubitus interruptus requires *Drosophila* CREB-binding protein to activate *wingless* expression in the *Drosophila* embryo. *Mol. Cell. Biol.* **20**, 1616–1625.

753. Chen, Y., Riese, M.J., Killinger, M.A., and Hoffmann, F.M. (1998). A genetic screen for modifiers of *Drosophila decapentaplegic* signaling identifies mutations in *punt, Mothers against dpp* and the BMP-7 homologue, *60A. Development* **125**, 1759–1768.

754. Chen, Y. and Struhl, G. (1996). Dual roles for Patched in sequestering and transducing Hedgehog. *Cell* **87**, 553–563.

755. Chen, Y. and Struhl, G. (1998). In vivo evidence that Patched and Smoothened constitute distinct binding and transducing components of a Hedgehog receptor complex. *Development* **125**, 4943–4948.

756. Chen, Y.-G., Hata, A., Lo, R.S., Wotton, D., Shi, Y., Pavletich, N., and Massagué, J. (1998). Determinants of specificity in TGF-β signal transduction. *Genes Dev.* **12**, 2144–2152.

757. Cheng, M.K. and Shearn, A. (2001). Molecular and genetic interactions between *ash2,* a trithorax group gene, and *skittles,* a gene encoding 1-phosphatidylinositol-4-phosphate kinase. *Proc. 42nd Ann. Drosophila Res. Conf.* **Abstracts Vol.**, a94.

758. Chern, J.J. and Choi, K.W. (2001). Cloning and characterizing of Lobe mutation. *Proc. 42nd Ann. Drosophila Res. Conf.* **Abstracts Vol.**, a185.

759. Cheyette, B.N.R., Green, P.J., Martin, K., Garren, H., Hartenstein, V., and Zipursky, S.L. (1994). The *Drosophila sine oculis* locus encodes a homeodomain-containing protein required for the development of the entire visual system. *Neuron* **12**, 977–996.

760. Chiang, C. and Beachy, P.A. (1994). Expression of a novel *Toll*-like gene spans the parasegment boundary and contributes to *hedgehog* function in the adult eye of *Drosophila. Mechs. Dev.* **47**, 225–239.

761. Chicurel, M.E., Chen, C.S., and Ingber, D.E. (1998). Cellular control lies in the balance of forces. *Curr. Opin. Cell Biol.* **10**, 232–239.

762. Chicurel, M.E., Singer, R.H., Meyer, C.J., and Ingber, D.E. (1998). Integrin binding and mechanical tension induce movement of mRNA and ribosomes to focal adhesions. *Nature* **392**, 730–733.

763. Chien, C.-T., Bartel, P.L., Sternglanz, R., and Fields, S. (1991). The two-hybrid system: A method to identify and clone genes for proteins that interact with a protein of interest. *Proc. Natl. Acad. Sci. USA* **88**, 9578–9582.

764. Chien, C.-T., Hsiao, C.-D., Jan, L.Y., and Jan, Y.N. (1996). Neuronal type information encoded in the basic-helix-loop-helix domain of proneural genes. *Proc. Natl. Acad. Sci. USA* **93**, 13239–13244.

765. Chien, C.-T., Wang, S., Rothenberg, M., Jan, L.Y., and Jan, Y.N. (1998). Numb-associated kinase interacts with the phosphotyrosine binding domain of Numb and antagonizes the function of Numb *in vivo. Mol. Cell. Biol.* **18**, 598–607.

766. Chihara, T. and Hayashi, S. (2000). Control of tracheal tubulogenesis by Wingless signaling. *Development* **127**, 4433–4442.

767. Child, G. (1935). Phenogenetic studies on scute-1 of *Drosophila melanogaster*. I. The associations between the bristles and the effects of genetic modifiers and temperature. *Genetics* **20**, 109–126.

768. Child, G. (1935). Phenogenetic studies on scute-1 of *Drosophila melanogaster*. II. The temperature-effective period. *Genetics* **20**, 127–155.

769. Child, G. (1936). Phenogenetic studies in scute of *D. melanogaster*. III. The effect of temperature on scute 5. *Genetics* **21**, 808–816.

770. Child, G.P., Blanc, R., and Plough, H.H. (1940). Somatic effects of temperature on development in *Drosophila melanogaster*. I. Phenocopies and reversal of dominance. *Physiol. Zool.* **13**, 56–64.

771. Childs, S.R. and O'Connor, M.B. (1994). Two domains of the *tolloid* protein contribute to its unusual genetic interaction with *decapentaplegic. Dev. Biol.* **162**, 209–220.

772. Childs, S.R., Wrana, J.L., Arora, K., Attisano, L., O'Connor, M.B., and Massagué, J. (1993). Identification of a *Drosophila* activin receptor. *Proc. Natl. Acad. Sci. USA* **90**, 9475–9479.

773. Cho, K.-O., Chern, J., Izaddoost, S., and Choi, K.-W. (2000). Novel signaling from the peripodial membrane is essential for eye disc patterning in *Drosophila. Cell* **103**, 331–342.

774. Cho, K.-O. and Choi, K.-W. (1998). Fringe is essential for mirror symmetry and morphogenesis in the *Drosophila* eye. *Nature* **396**, 272–276.

775. Cho, K.W.Y. and Blitz, I.L. (1998). BMPs, Smads and metalloproteases: extracellular and intracellular modes of negative regulation. *Curr. Opin. Gen. Dev.* **8**, 443–449.

776. Choi, C.Y., Lee, Y.M., Kim, Y.H., Park, T., Jeon, B.H., Schulz, R.A., and Kim, Y. (1999). The homeodomain transcription factor NK-4 acts as either a transcriptional activator or repressor and interacts with the p300 coactivator and the Groucho corepressor. *J. Biol. Chem.* **274** #44, 31543–31552.

777. Choi, K.-W. and Benzer, S. (1994). Rotation of photoreceptor clusters in the developing *Drosophila* eye requires the *nemo* gene. *Cell* **78**, 125–136.

778. Choi, K.-W., Mozer, B., and Benzer, S. (1996). Independent determination of symmetry and polarity in the *Drosophila* eye. *Proc. Natl. Acad. Sci. USA* **93**, 5737–5741.

779. Chou, W.-H., Hall, K.J., Wilson, D.B., Wideman, C.L., Townson, S.M., Chadwell, L.V., and Britt, S.G. (1996). Identification of a novel *Drosophila* opsin reveals specific patterning of the R7 and R8 photoreceptor cells. *Neuron* **17**, 1101–1115.

780. Chou, W.-H., Huber, A., Bentrop, J., Schulz, S., Schwab, K., Chadwell, L.V., Paulsen, R., and Britt, S.G. (1999). Patterning of the R7 and R8 photoreceptor cells of *Drosophila*: evidence for induced and default cell-fate specification. *Development* **126**, 607–616.

781. Choy, E., Chiu, V.K., Silletti, J., Feoktistov, M., Morimoto, T., Michaelson, D., Ivanov, I.E., and Philips, M.R. (1999). Endomembrane trafficking of Ras: the CAAX motif targets proteins to the ER and Golgi. *Cell* **98**, 69–80.

782. Christen, B. and Bienz, M. (1994). Imaginal disc silencers from *Ultrabithorax*: evidence for *Polycomb* response elements. *Mechs. Dev.* **48**, 255–266.

783. Christian, J.L. (2000). BMP, Wnt and Hedgehog signals: how far can they go? *Curr. Opin. Cell Biol.* **12**, 244–249.

784. Chu, J. and Panganiban, G. (2001). Distal-less specifies antennal identity by interacting with Extradenticle and Homothorax. *Proc. 42nd Ann. Drosophila Res. Conf.* **Abstracts Vol.**, a118.

785. Chu-LaGraff, Q., Wright, D.M., McNeil, L.K., and Doe, C.Q. (1991). The *prospero* gene encodes a divergent homeodomain protein that controls neuronal identity in *Drosophila*. *Development* **Suppl. 2**, 79–85.

786. Chuang, P.-T. and Kornberg, T.B. (2000). On the range of Hedgehog signaling. *Curr. Opin. Gen. Dev.* **10**, 515–522.

787. Chun, J. (1999). Developmental neurobiology: a genetic Cheshire cat? *Curr. Biol.* **9**, R651–R654.

788. Chung, C.-N., Hamaguchi, Y., Honjo, T., and Kawaichi, M. (1994). Site-directed mutagenesis study on DNA binding regions of the mouse homologue of Suppressor of Hairless, RBP-Jκ. *Nucleic Acids Res.* **22**, 2938–2944.

789. Chung, Y.D., Zhu, J., Han, Y.-G., and Kernan, M.J. (2001). *nompA* encodes a PNS-specific, ZP domain protein required to connect mechanosensory dendrites to sensory structures. *Neuron* **29**, 415–428.

790. Chuong, C.-M., Chodankar, R., Widelitz, R.B., and Jiang, T.-X. (2000). *Evo-Devo* of feathers and scales: building complex epithelial appendages. *Curr. Opin. Gen. Dev.* **10**, 449–456.

791. Ciechanover, A., Orian, A., and Schwartz, A.L. (2000). Ubiquitin-mediated proteolysis: biological regulation via destruction. *BioEssays* **22**, 442–451.

792. Ciechanska, E. and Brook, W. (2001). The role of l(3)1215 in joint formation in the *Drosophila* leg. *Proc. 42nd Ann. Drosophila Res. Conf.* **Abstracts Vol.**, a196.

793. Cifuentes, F.J. and García-Bellido, A. (1997). Proximodistal specification in the wing disc of *Drosophila* by the nubbin gene. *Proc. Natl. Acad. Sci. USA* **94**, 11405–11410.

794. Citri, Y., Colot, H.V., Jacquier, A.C., Yu, Q., Hall, J.C., Baltimore, D., and Rosbash, M. (1987). A family of unusually spliced biologically active transcripts encoded by a *Drosophila* clock gene. *Nature* **326**, 42–47.

795. Clandinin, T.R. and Zipursky, S.L. (2000). Afferent growth cone interactions control synaptic specificity in the *Drosophila* visual system. *Neuron* **28**, 427–436.

796. Clark, H.F., Brentrup, D., Schneitz, K., Bieber, A., Goodman, C., and Noll, M. (1995). *Dachsous* encodes a member of the cadherin superfamily that controls imaginal disc morphogenesis in *Drosophila*. *Genes Dev.* **9**, 1530–1542.

797. Clark, W.C. and Russell, M.A. (1977). The correlation of lysosomal activity and adult phenotype in a cell-lethal mutant of *Drosophila*. *Dev. Biol.* **57**, 160–173.

798. Clarke, P.G.H. (1981). Chance, repetition, and error in the development of normal nervous systems. *Persp. Biol. Med.* **25**, 2–19.

799. Clarke, P.R. (1994). Switching off MAP kinases. *Curr. Biol.* **4**, 647–650.

800. Claxton, J.H. (1964). The determination of patterns with special reference to that of the central primary skin follicles in sheep. *J. Theor. Biol.* **7**, 302–317.

801. Claxton, J.H. (1967). Patterns of abdominal tergite bristles in wild-type and scute *Drosophila melanogaster*. *Genetics* **55**, 525–545.

802. Claxton, J.H. (1969). Mosaic analysis of bristle displacement in *Drosophila*. *Genetics* **63**, 883–896.

803. Claxton, J.H. (1971). Cell numbers and acrostichal row numbers in hairy and non-hairy flies. *Dros. Info. Serv.* **46**, 133–134.

804. Claxton, J.H. (1974). Some quantitative features of *Drosophila* sternite bristle patterns. *Aust. J. Biol. Sci.* **27**, 533–543.

805. Claxton, J.H. (1976). Developmental origin of even spacing between the microchaetes of *Drosophila melanogaster*. *Aust. J. Biol. Sci.* **29**, 131–135.

806. Claxton, J.H. (1982). Temporal pattern of bristle development on *Drosophila melanogaster* sternites. *Aust. J. Biol. Sci.* **35**, 653–660.

807. Claxton, J.H. and Kongsuwan, K. (1976). Non-autonomy in achaete mosaics of *Drosophila*. *Genet. Res., Camb.* **27**, 11–22.

808. Cleghon, V., Feldmann, P., Ghiglione, C., Copeland, T.D., Perrimon, N., Hughes, D.A., and Morrison, D.K. (1998). Opposing actions of CSW and RasGAP modulate the strength of Torso RTK signaling in the *Drosophila* terminal pathway. *Molec. Cell* **2**, 719–727.

809. Cleghon, V., Gayko, U., Copeland, T.D., Perkins, L.A., Perrimon, N., and Morrison, D.K. (1996). *Drosophila* terminal structure development is regulated by the compensatory activities of positive and negative phosphotyrosine signaling sites on the Torso RTK. *Genes Dev.* **10**, 566–577.

810. Clevers, H. and van de Wetering, M. (1997). TCF/LEF factors earn their wings. *Trends Genet.* **13**, 485–489.

811. Cleves, A.E. (1997). Protein transport: the nonclassical ins and outs. *Curr. Biol.* **7**, R318–R320.

812. Clifford, R. and Schüpbach, T. (1992). The torpedo (DER) receptor tyrosine kinase is required at multiple times during *Drosophila* embryogenesis. *Development* **115**, 853–872.

813. Clifford, R. and Schüpbach, T. (1994). Molecular analysis of the *Drosophila* EGF receptor homolog reveals that several genetically defined classes of alleles cluster in subdomains of the receptor protein. *Genetics* **137**, 531–550.

814. Clifford, R.J. and Schüpbach, T. (1989). Coordinately and differentially mutable activities of *torpedo*, the *Drosophila melanogaster* homolog of the vertebrate EGF receptor gene. *Genetics* **123**, 771–787.

815. Cline, T.W. (1989). The affairs of *daughterless* and the promiscuity of developmental regulators. *Cell* **59**, 231–234.

816. Cline, T.W. (1993). The *Drosophila* sex determination signal: how do flies count to two? *Trends Genet.* **9**, 385–390.

817. Cline, T.W. and Meyer, B.J. (1996). *Vive la différence:* Males *vs.* females in flies *vs.* worms. *Annu. Rev. Genet.* **30**, 637–702.

818. Clyne, P.J., Certel, S.J., de Bruyne, M., Zaslavsky, L., Johnson, W.A., and Carlson, J.R. (1999). The odor specificities of a subset of olfactory receptor neurons are governed by Acj6, a POU-domain transcription factor. *Neuron* **22**, 339–347.

819. Clyne, P.J., Warr, C.G., Freeman, M.R., Lessing, D., Kim, J., and Carlson, J.R. (1999). A novel family of divergent seven-transmembrane proteins: candidate odorant receptors in *Drosophila*. *Neuron* **22**, 327–338.

820. Cobb, M. (1995). The fly of the lords. *Evolution* **49**, 581–583.

821. Cobb, M. (1999). What and how do maggots smell? *Biol. Rev.* **74**, 425–459.

822. Cobb, M.H. and Goldsmith, E.J. (1995). How MAP kinases are regulated. *J. Biol. Chem.* **270** #25, 14843–14846.

823. Cobb, M.H. and Goldsmith, E.J. (2000). Dimerization in MAP-kinase signaling. *Trends Biochem. Sci.* **25**, 7–9.

824. Codd, E.F. (1968). "Cellular Automata." Acad. Pr., New York.

825. Coen, E. (1999). "The Art of Genes: How Organisms Make Themselves." Oxford Univ. Pr., New York.

826. Cohen, B., McGuffin, M.E., Pfeifle, C., Segal, D., and Cohen, S.M. (1992). *apterous*, a gene required for imaginal disc development in *Drosophila* encodes a member of the LIM family of developmental regulatory proteins. *Genes Dev.* **6**, 715–729.

827. Cohen, B., Simcox, A.A., and Cohen, S.M. (1993). Allocation of the thoracic imaginal primordia in the *Drosophila* embryo. *Development* **117**, 597–608.

828. Cohen, B., Wimmer, E.A., and Cohen, S.M. (1991). Early development of leg and wing primordia in the *Drosophila* embryo. *Mechs. Dev.* **33**, 229–240.

829. Cohen, G.B., Ren, R., and Baltimore, D. (1995). Modular binding domains in signal transduction proteins. *Cell* **80**, 237–248.

830. Cohen, P.T.W., Brewis, N.D., Hughes, V., and Mann, D.J. (1990). Protein serine/threonine phosphatases: an expanding family. *FEBS Letters* **268**, 355–359.

831. Cohen, S. and Hyman, A.A. (1994). When is a determinant a determinant? *Curr. Biol.* **5**, 420–422.

832. Cohen, S. and Jürgens, G. (1991). *Drosophila* headlines. *Trends Genet.* **7**, 267–272.

833. Cohen, S.M. (1990). Specification of limb development in the *Drosophila* embryo by positional cues from segmentation genes. *Nature* **343**, 173–177.

834. Cohen, S.M. (1993). Imaginal disc development. *In* "The Development of *Drosophila melanogaster*," Vol. 2 (M. Bate and A. Martinez Arias, eds.). Cold Spring Harbor Lab. Pr.: Plainview, N. Y., pp. 747–841.

835. Cohen, S.M. (1996). Controlling growth of the wing: Vestigial integrates signals from the compartment boundaries. *BioEssays* **18**, 855–858.

836. Cohen, S.M., Brönner, G., Küttner, F., Jürgens, G., and Jäckle, H. (1989). *Distal-less* encodes a homoeodomain protein required for limb development in *Drosophila*. *Nature* **338**, 432–434.

837. Cohen, S.M. and Di Nardo, S. (1993). wingless: from embryo to adult. *Trends Genet.* **9**, 189–192.

838. Cohen, S.M. and Jürgens, G. (1989). Proximal-distal pattern formation in *Drosophila*: cell autonomous requirement for *Distal-less* gene activity in limb development. *EMBO J.* **8**, 2045–2055.

839. Cohen, S.M. and Jürgens, G. (1989). Proximal-distal pattern formation in *Drosophila*: graded requirement for *Distal-less* gene activity during limb development. *Roux's Arch. Dev. Biol.* **198**, 157–169.

840. Cole, E.S. and Palka, J. (1982). The pattern of campaniform sensilla on the wing and haltere of *Drosophila melanogaster* and several of its homeotic mutants. *J. Embryol. Exp. Morph.* **71**, 41–61.

841. Coleman, J.E. (1992). Zinc proteins: Enzymes, storage proteins, transcription factors, and replication proteins. *Annu. Rev. Biochem.* **61**, 897–946.

842. Coleman, K.G., Poole, S.J., Weir, M.P., Soeller, W.C., and Kornberg, T. (1987). The *invected* gene of *Drosophila*: sequence analysis and expression studies reveal a close kinship to the *engrailed* gene. *Genes Dev.* **1**, 19–28.

843. Colledge, M., Dean, R.A., Scott, G.K., Langeberg, L.K., Huganir, R.L., and Scott, J.D. (2000). Targeting of PKA to glutamate receptors through a MAGUK-AKAP complex. *Neuron* **27**, 107–119.

844. Colley, N.J. (2000). Actin' up with Rac1. *Science* **290**, 1902–1903.

845. Collier, C.P., Wong, E.W., Belohradsky, M., Raymo, F.M., Stoddart, J.F., Kuekes, P.J., Williams, R.S., and Heath, J.R. (1999). Electrically configurable molecular-based logic gates. *Science* **285**, 391–394.

846. Collier, S. and Gubb, D. (1997). *Drosophila* tissue polarity requires the cell-autonomous activity of the *fuzzy* gene, which encodes a novel transmembrane protein. *Development* **124**, 4029–4037.

847. Collins, M.K.L., Perkins, G.R., Rodriguez-Tarduchy, G., Nieto, M.A., and López-Rivas, A. (1994). Growth factors as survival factors: regulation of apoptosis. *BioEssays* **16**, 133–138.

848. Collins, R.T., Furukawa, T., Tanese, N., and Treisman, J. (1999). Osa associates with the Brahma chromatin remodeling complex and promotes the activation of some target genes. *EMBO J.* **18**, 7029–7040.

849. Collins, R.T. and Treisman, J.E. (2000). Osa-containing Brahma chromatin remodeling complexes are required for the repression of Wingless target genes. *Genes Dev.* **14**, 3140–3152.

850. Combs, J.D. (1937). Genetic and environmental factors affecting the development of the sex-combs of *Drosophila melanogaster*. *Genetics* **22**, 427–433.

851. Condic, M.L., Fristrom, D., and Fristrom, J.W. (1991). Apical cell shape changes during *Drosophila* imaginal leg disc elongation: a novel morphogenetic mechanism. *Development* **111**, 23–33.

852. Condie, J.M. and Brower, D.L. (1989). Allelic interactions at the *engrailed* locus of *Drosophila*: *engrailed* protein expression in imaginal discs. *Dev. Biol.* **135**, 31–42.

853. Condron, B.G., Patel, N.H., and Zinn, K. (1994). *engrailed* controls glial/neuronal cell fate decisions at the midline of the central nervous system. *Neuron* **13**, 541–554.

854. Conley, C.A., Silburn, R., Singer, M.A., Ralston, A., Rohwer-Nutter, D., Olson, D.J., Gelbart, W., and Blair, S.S. (2000). Crossveinless 2 contains cysteine-rich domains and is required for high levels of BMP-like activity during the formation of the cross veins in *Drosophila*. *Development* **127**, 3947–3959.

855. Conlon, I. and Raff, M. (1999). Size control in animal development. *Cell* **96**, 235–244.

856. Conlon, R.A., Reaume, A.G., and Rossant, J. (1995). *Notch1* is required for the coordinate segmentation of somites. *Development* **121**, 1533–1545.

857. Connolly, J.P., Augustine, J.G., and Francklyn, C. (1999). Mutational analysis of the *engrailed* homeodomain recognition helix by phage display. *Nucl. Acids Res.* **27**, 1182–1189.

858. Conrad, M. (1990). The geometry of evolution. *BioSystems* **24**, 61–81.

859. Cook, D., Fry, M.J., Hughes, K., Sumathipala, R., Woodgett, J.R., and Dale, T.C. (1996). Wingless inactivates glycogen synthase kinase-3 via an intracellular signalling pathway which involves a protein kinase C. *EMBO J.* **15**, 4526–4536.

860. Cook, T. and Desplan, C. (2001). Identification of the homeodomain protein Prospero as a regulator of rhodopsin gene expression in the adult *Drosophila* eye. *Proc. 42nd Ann. Drosophila Res. Conf.* **Abstracts Vol.**, a122. [See also 2001 *Sems. Cell Dev. Biol.* **12**, 509–518.]

861. Cook, T.A. (1914). "The Curves of Life." Constable, London.

862. Cooke, J. (1975). The emergence and regulation of spatial organization in early animal development. *Annu. Rev. Biophys. Bioeng.* **4**, 185–217.

863. Cooke, J. (1978). Embryonic and regenerating patterns. *Nature* **271**, 705–706.

864. Cooke, J. (1988). The early embryo and the formation of body pattern. *Am. Sci.* **76**, 35–41.

865. Cooke, J. (1995). Morphogens in vertebrate development: how do they work? *BioEssays* **17**, 93–96. [See also 2001 references: *Curr. Biol.* **11**, R851–R854; *Nature Rev. Genet.* **2**, 620–630.]

866. Cooke, J., Nowak, M.A., Boerlijst, M., and Maynard-Smith, J. (1997). Evolutionary origins and maintenance of redundant gene expression during metazoan development. *Trends Genet.* **13**, 360–363.

867. Cooke, J. and Smith, J.C. (1990). Measurement of developmental time by cells of early embryos. *Cell* **60**, 891–894.

868. Cooke, J. and Zeeman, E.C. (1976). A clock and wavefront model for control of the number of repeated structures during animal morphogenesis. *J. Theor. Biol.* **58**, 455–476. [See also 2001 *Nature* **412**, 780–781.]

869. Cooper, M.T.D. and Bray, S.J. (1999). Frizzled regulation of Notch signalling polarizes cell fate in the *Drosophila* eye. *Nature* **397**, 526–530.

870. Cooper, M.T.D. and Bray, S.J. (2000). R7 photoreceptor specification requires Notch activity. *Curr. Biol.* **10**, 1507–1510.

871. Cooper, M.T.D., Tyler, D.M., Furriols, M., Chalkiadaki, A., Delidakis, C., and Bray, S. (2000). Spatially restricted factors cooperate with Notch in the regulation of *Enhancer of split* genes. *Dev. Biol.* **221**, 390–403.

872. Copeland, B.J. and Proudfoot, D. (1999). Alan Turing's forgotten ideas in computer science. *Sci. Am.* **280** #4, 98–103.

873. Copeland, J.W.R., Nasiadka, A., Dietrich, B.H., and Krause, H.M. (1996). Patterning of the *Drosophila* embryo by a homeodomain-deleted Ftz polypeptide. *Nature* **379**, 162–165. [See also 2001 *Genes Dev.* **15**, 1716–1723.]

874. Corfas, G. and Dudai, Y. (1990). Adaptation and fatigue of a mechanosensory neuron in wild-type *Drosophila* and in memory mutants. *J. Neurosci.* **10**, 491–499.

875. Cornell, M., Evans, D.A.P., Mann, R., Fostier, M., Flasza, M., Monthatong, M., Artavanis-Tsakonas, S., and Baron, M. (1999). The *Drosophila melanogaster Suppressor of deltex* gene, a regulator of the Notch receptor signaling pathway, is an E3 Class ubiquitin ligase. *Genetics* **152**, 567–576.

876. Cornell, R.A. and Kimelman, D. (1994). Combinatorial signaling in development. *BioEssays* **16**, 577–581.

877. Coulter, D.E., Swaykus, E.A., Beran-Koehn, M.A., Goldberg, D., Wieschaus, E., and Schedl, P. (1990). Molecular analysis of *odd-skipped*, a zinc finger encoding segmentation gene with a novel pair-rule expression pattern. *EMBO J.* **8**, 3795–3804.

878. Coulter, D.E. and Wieschaus, E. (1988). Gene activities and segmental patterning in *Drosophila*: analysis of *odd-skipped* and pair-rule double mutants. *Genes Dev.* **2**, 1812–1823.

879. Courey, A.J. (2001). Cooperativity in transcriptional control. *Curr. Biol.* **10**, R250–R252.

880. Couso, J.P., Bate, M., and Martínez-Arias, A. (1993). A *wingless*-dependent polar coordinate system in *Drosophila* imaginal discs. *Science* **259**, 484–489.

881. Couso, J.P. and Bishop, S.A. (1998). Proximo-distal development in the legs of *Drosophila*. *Int. J. Dev. Biol.* **42**, 345–352.

882. Couso, J.P., Bishop, S.A., and Martinez Arias, A. (1994). The wingless signalling pathway and the patterning of the wing margin in *Drosophila*. *Development* **120**, 621–636. [See also 2001 *Dev. Biol.* **234**, 13–23.]

883. Couso, J.P., Galindo, M.I., and Greig, S. (2001). Intercalation of proximal-distal fates independently of *wg* and *dpp*. *Proc. 42nd Ann. Drosophila Res. Conf.* **Abstracts Vol.**, a196.

884. Couso, J.P. and González-Gaitán, M. (1993). Embryonic limb development in *Drosophila*. *Trends Genet.* **9**, 371–373.

885. Couso, J.P., Knust, E., and Martinez Arias, A. (1995). *Serrate* and *wingless* cooperate to induce *vestigial* gene expression and wing formation in *Drosophila*. *Curr. Biol.* **5**, 1437–1448.

886. Couso, J.P. and Martinez Arias, A. (1994). *Notch* is required for *wingless* signaling in the epidermis of *Drosophila*. *Cell* **79**, 259–272.

887. Cowan, W.M. (1973). Neuronal death as a regulative mechanism in the control of cell number in the nervous system. *In* "Development and Aging in the Nervous System," (M. Rockstein and M.L. Sussman, eds.). Acad. Pr.: New York, pp. 19–41.

888. Cowan, W.M., Fawcett, J.W., O'Leary, D.D.M., and Stanfield, B.B. (1984). Regressive events in neurogenesis. *Science* **225**, 1258–1265.

889. Cowman, A.F., Zuker, C.S., and Rubin, G.M. (1986). An opsin gene expressed in only one photoreceptor cell type of the *Drosophila* eye. *Cell* **44**, 705–710.

890. Cox, R.T., Kirkpatrick, C., and Peifer, M. (1996). Armadillo is required for adherens junction assembly, cell polarity, and morphogenesis during *Drosophila* embryogenesis. *J. Cell Biol.* **134**, 133–148.

891. Cox, R.T., McEwen, D.G., Myster, D.L., Duronio, R.J., Loureiro, J., and Peifer, M. (2000). A screen for mutations that suppress the phenotype of *Drosophila armadillo*, the β-catenin homolog. *Genetics* **155**, 1725–1740.

892. Cox, R.T., Pai, L.-M., Kirkpatrick, C., Stein, J., and Peifer, M. (1999). Roles of the C terminus of Armadillo in Wingless signaling in *Drosophila*. *Genetics* **153**, 319–332.

893. Cox, R.T., Pai, L.-M., Miller, J.R., Orsulic, S., Stein, J., McCormick, C.A., Audeh, Y., Wang, W., Moon, R.T., and Peifer, M. (1999). Membrane-tethered *Drosophila* Armadillo cannot transduce Wingless signal on its own. *Development* **126**, 1327–1335.

894. Cox, R.T. and Peifer, M. (1998). Wingless signaling: The inconvenient complexities of life. *Curr. Biol.* **8**, R140–R144.

895. Coyle-Thompson, C.A. and Banerjee, U. (1993). The *strawberry notch* gene functions with *Notch* in common developmental pathways. *Development* **119**, 377–395. [See also 2001 *Mechs. Dev.* **109**, 241–251.]

896. Craven, S.E. and Bredt, D.S. (1998). PDZ proteins organize synaptic signaling pathways. *Cell* **93**, 495–498.

897. Crew, J.R., Batterham, P., and Pollock, J.A. (1997). Developing compound eye in *lozenge* mutants of *Drosophila*: lozenge expression in the R7 equivalence group. *Dev. Genes Evol.* **206**, 481–493.

898. Crews, C.M. and Erikson, R.L. (1993). Extracellular signals and reversible protein phosphorylation: what to Mek of it all. *Cell* **74**, 215–217.

899. Crews, S.T. (1998). Control of cell lineage-specific development and transcription by bHLH-PAS proteins. *Genes Dev.* **12**, 607–620.

900. Crews, S.T. and Fan, C.-M. (1999). Remembrance of things

PAS: regulation of development by bHLH-PAS proteins. *Curr. Opin. Gen. Dev.* **9**, 580–587.

901. Crews, S.T., Thomas, J.B., and Goodman, C.S. (1988). The *Drosophila single-minded* gene encodes a nuclear protein with sequence similarity to the *per* gene product. *Cell* **52**, 143–151.

902. Crick, F. (1970). Diffusion in embryogenesis. *Nature* **225**, 420–422.

903. Crick, F.H.C. (1971). The scale of pattern formation. *Symp. Soc. Exp. Biol.* **25**, 429–438.

904. Crick, F.H.C. and Lawrence, P.A. (1975). Compartments and polyclones in insect development. *Science* **189**, 340–347.

905. Cronmiller, C. and Cummings, C.A. (1993). The *daughterless* gene product in *Drosophila* is a nuclear protein that is broadly expressed throughout the organism during development. *Mechs. Dev.* **42**, 159–169. [See also 2001 *Development* **128**, 4705–4714.]

906. Cronmiller, C., Schedl, P., and Cline, T.W. (1988). Molecular characterization of *daughterless*, a *Drosophila* sex determination gene with multiple roles in development. *Genes Dev.* **2**, 1666–1676.

907. Crosby, M.A., Lundquist, E.A., Tautvydas, R.M., and Johnson, J.J. (1993). The 3′ regulatory region of the *Abdominal-B* gene: genetic analysis supports a model of reiterated and interchangeable regulatory elements. *Genetics* **134**, 809–824.

908. Crosby, M.A., Miller, C., Alon, T., Watson, K.L., Verrijzer, C.P., Goldman-Levi, R., and Zak, N.B. (1999). The *trithorax* group gene *moira* encodes a Brahma-associated putative chromatin-remodeling factor in *Drosophila melanogaster*. *Mol. Cell. Biol.* **19**, 1159–1170.

909. Crozatier, M., Valle, D., Dubois, L., Ibnsouda, S., and Vincent, A. (1996). *collier*, a novel regulator of *Drosophila* head development, is expressed in a single mitotic domain. *Curr. Biol.* **6**, 707–718.

910. Crozatier, M. and Vincent, A. (1999). Requirement for the *Drosophila* COE transcription factor Collier in formation of an embryonic muscle: transcriptional response to Notch signalling. *Development* **126**, 1495–1504. [See also 2001 *Mechs. Dev.* **108**, 3–12.]

911. Cubadda, Y., Heitzler, P., Ray, R.P., Bourouis, M., Ramain, P., Gelbart, W., Simpson, P., and Haenlin, M. (1997). *u-shaped* encodes a zinc-finger protein that regulates the proneural genes *achaete* and *scute* during the formation of bristles in *Drosophila*. *Genes Dev.* **11**, 3083–3095.

912. Cubas, P., de Celis, J.-F., Campuzano, S., and Modolell, J. (1991). Proneural clusters of *achaete-scute* expression and the generation of sensory organs in the *Drosophila* imaginal wing disc. *Genes Dev.* **5**, 996–1008.

913. Cubas, P. and Modolell, J. (1992). The *extramacrochaetae* gene provides information for sensory organ patterning. *EMBO J.* **11**, 3385–3393.

914. Cubas, P., Modolell, J., and Ruiz-Gómez, M. (1994). The helix-loop-helix extramacrochaetae protein is required for proper specification of many cell types in the *Drosophila* embryo. *Development* **120**, 2555–2565.

915. Cuenoud, B. and Schepartz, A. (1993). Altered specificity of DNA-binding proteins with transition metal dimerization domains. *Science* **259**, 510–513.

916. Cui, X. and Doe, C.Q. (1995). The role of the cell cycle and cytokinesis in regulating neuroblast sublineage gene expression in the *Drosophila* CNS. *Development* **121**, 3233–3243.

917. Culí, J., Martín-Blanco, E., and Modolell, J. (2001). The EGF receptor and N signalling pathways act antagonistically in *Drosophila* mesothorax bristle patterning. *Development* **128**, 299–308.

918. Culí, J. and Modolell, J. (1998). Proneural gene self-stimulation in neural precursors: an essential mechanism for sense organ development that is regulated by *Notch* signaling. *Genes Dev.* **12**, 2036–2047.

919. Cullen, C.F. and Milner, M.J. (1991). Parameters of growth in primary cultures and cell lines established from *Drosophila* imaginal discs. *Tissue Cell* **23**, 29–39.

920. Cumberledge, S. and Krasnow, M.A. (1993). Intercellular signalling in *Drosophila* segment formation reconstructed *in vitro*. *Nature* **363**, 549–552.

921. Cumberledge, S. and Reichsman, F. (1997). Glycosaminoglycans and WNTs: just a spoonful of sugar helps the signal do down. *Trends Genet.* **13**, 421–423.

922. Cummings, F.W. and Prothero, J.W. (1978). A model of pattern formation in multicellular organisms. *Collective Phenom.* **3**, 41–53.

923. Cummins, H. and Midlo, C. (1943). "Finger Prints, Palms and Soles. An Introduction to Dermatoglyphics." Dover, New York.

924. Cunningham, G.N., Smith, N.M., Makowski, M.K., and Kuhn, D.T. (1983). Enzyme patterns in *D. melanogaster* imaginal discs: Distribution of glucose-6-phosphate and 6-phospho-gluconate dehydrogenase. *Mol. Gen. Genet.* **191**, 238–243.

925. Curcio, C.A., Sloan, K.R., Jr., Packer, O., Hendrickson, A.E., and Kalina, R.E. (1987). Distribution of cones in human and monkey retina: individual variability and radial asymmetry. *Science* **236**, 579–582.

926. Curtin, K., D., Huang, Z.J., and Rosbash, M. (1995). Temporally regulated nuclear entry of the *Drosophila period* protein contributes to the circadian clock. *Neuron* **14**, 365–372.

927. Curtiss, J. and Heilig, J.S. (1995). Establishment of *Drosophila* imaginal precursor cells is controlled by the *Arrowhead* gene. *Development* **121**, 3819–3825.

928. Curtiss, J. and Heilig, J.S. (1997). *Arrowhead* encodes a LIM homeodomain protein that distinguishes subsets of *Drosophila* imaginal cells. *Dev. Biol.* **190**, 129–141.

929. Curtiss, J. and Heilig, J.S. (1998). DeLIMiting development. *BioEssays* **20**, 58–69.

930. Curtiss, J. and Mlodzik, M. (2000). Morphogenetic furrow initiation and progression during eye development in *Drosophila*: the roles of *decapentaplegic*, *hedgehog*, and *eyes absent*. *Development* **127**, 1325–1336.

931. Cutforth, T. and Gaul, U. (1997). The genetics of visual system development in *Drosophila*: specification, connectivity and asymmetry. *Curr. Opin. Neurobiol.* **7**, 48–54.

932. Cutforth, T. and Rubin, G.M. (1994). Mutations in *Hsp83* and *cdc37* impair signaling by the Sevenless receptor tyrosine kinase in *Drosophila*. *Cell* **77**, 1027–1036.

933. Czerny, T., Halder, G., Kloter, U., Souabni, A., Gehring, W.J., and Busslinger, M. (1999). *twin of eyeless*, a second *Pax-6* gene of *Drosophila*, acts upstream of *eyeless* in the control of eye development. *Molec. Cell* **3**, 297–307.

934. Czerny, T., Schaffner, G., and Busslinger, M. (1993). DNA sequence recognition by Pax proteins: bipartite structure of the paired domain and its binding site. *Genes Dev.* **7**, 2048–2061.

935. D'Avino, P.P. and Thummel, C.S. (1998). *crooked legs* encodes a family of zinc finger proteins required for leg morphogenesis and ecdysone-regulated gene expression

during *Drosophila* metamorphosis. *Development* **125**, 1733–1745.

936. D'Avino, P.P. and Thummel, C.S. (2000). The ecdysone regulatory pathway controls wing morphogenesis and integrin expression during *Drosophila* metamorphosis. *Dev. Biol.* **220**, 211–224.

937. D'Costa, A.R., Reifegerste, R., Siera, S., and Moses, K. (2001). Characterisation of a dominant modifier of hedgehog. *Proc. 42nd Ann. Drosophila Res. Conf.* **Abstracts Vol.**, a271.

938. Daga, A., Karlovich, C.A., Dumstrei, K., and Banerjee, U. (1996). Patterning of cells in the *Drosophila* eye by Lozenge, which shares homologous domains with AML1. *Genes Dev.* **10**, 1194–1205.

939. Dahanukar, A. and Wharton, R.P. (1996). The Nanos gradient in *Drosophila* embryos is generated by translational regulation. *Genes Dev.* **10**, 2610–2620.

940. Dahl, E., Koseki, H., and Balling, R. (1997). *Pax* genes and organogenesis. *BioEssays* **19**, 755–765.

941. Dahlsveen, I.K. and McNeill, H. (2001). Identification and characterisation of proteins that interact wtih the transcription factor Mirror. *Proc. 42nd Ann. Drosophila Res. Conf.* **Abstracts Vol.**, a123.

942. Dahmann, C. and Basler, K. (1999). Compartment boundaries at the edge of development. *Trends Genet.* **15**, 320–326.

943. Dahmann, C. and Basler, K. (2000). Opposing transcriptional outputs of Hedgehog signaling and Engrailed control compartmental cell sorting at the *Drosophila* A/P boundary. *Cell* **100**, 411–422.

944. Dai, H., Hogan, C., Gopalakrishnan, B., Torres-Vazquez, J., Nguyen, M., Park, S., Raftery, L.A., Warrior, R., and Arora, K. (2000). The zinc finger protein Schnurri acts as a Smad partner in mediating the transcriptional response to Decapentaplegic. *Dev. Biol.* **227**, 373–387.

945. Dale, K.J. and Pourquié, O. (2000). A clock-work somite. *BioEssays* **22**, 72–83. [See also 2001 *J. Anat.* **199**, 169–175.]

946. Dale, L. and Bownes, M. (1980). Is regeneration in *Drosophila* the result of epimorphic regulation? *W. Roux's Arch.* **189**, 91–96.

947. Dale, L. and Bownes, M. (1985). Pattern regulation in fragments of *Drosophila* wing discs which show variable wound healing. *J. Embryol. Exp. Morph.* **85**, 95–109.

948. Dambly-Chaudière, C. and Ghysen, A. (1987). Independent subpatterns of sense organs require independent genes of the *achaete-scute* complex in *Drosophila* larvae. *Genes Dev.* **1**, 297–306.

949. Dambly-Chaudière, C., Ghysen, A., Jan, L.Y., and Jan, Y.N. (1988). The determination of sense organs in *Drosophila*: interaction of *scute* with *daughterless. Roux's Arch. Dev. Biol.* **197**, 419–423.

950. Dambly-Chaudière, C., Jamet, E., Burri, M., Bopp, D., Basler, K., Hafen, E., Dumont, N., Spielmann, P., Ghysen, A., and Noll, M. (1992). The paired box gene *pox neuro*: a determinant of poly-innervated sense organs in *Drosophila. Cell* **69**, 159–172.

951. Dambly-Chaudière, C. and Leyns, L. (1992). The determination of sense organs in *Drosophila*: a search for interacting genes. *Int. J. Dev. Biol.* **36**, 85–91.

952. Dambly-Chaudière, C. and Vervoort, M. (1998). The bHLH genes in neural development. *Int. J. Dev. Biol.* **42**, 269–273.

953. Damelin, M. and Silver, P.A. (2000). Mapping interactions between nuclear transport factors in living cells reveals pathways through the nuclear pore complex. *Molec. Cell* **5**, 133–140.

954. Damke, H. (1996). Dynamin and receptor-mediated endocytosis. *FEBS Letters* **389**, 48–51.

955. Daniel, A., Dumstrei, K., Lengyel, J.A., and Hartenstein, V. (1999). The control of cell fate in the embryonic visual system by *atonal, tailless* and EGFR signaling. *Development* **126**, 2945–2954.

956. Daniel, J.M. and Reynolds, A.B. (1997). Tyrosine phosphorylation and cadherin/catenin function. *BioEssays* **19**, 883–891.

957. Danpure, C.J. (1995). How can the products of a single gene be localized to more than one intracellular compartment? *Trends Cell Biol.* **5**, 230–238.

958. Darnell, J.E., Jr. (1997). STATs and gene regulation. *Science* **277**, 1630–1635.

959. Das, P., Maduzia, L.L., Wang, H., Finelli, A.L., Cho, S.-H., Smith, M.M., and Padgett, R.W. (1998). The *Drosophila* gene *Medea* demonstrates the requirement for different classes of Smads in *dpp* signaling. *Development* **125**, 1519–1528.

960. Dasen, J.S. and Rosenfeld, M.G. (1999). Combinatorial codes in signaling and synergy: lessons from pituitary development. *Curr. Opin. Gen. Dev.* **9**, 566–574.

961. Datar, S.A., Jacobs, H.W., de la Cruz, A.F.A., Lehner, C.F., and Edgar, B.A. (2000). The *Drosophila* Cyclin D-Cdk4 complex promotes cellular growth. *EMBO J.* **19**, 4543–4554.

962. Datta, R.K. and Mukherjee, A.S. (1971). Developmental genetics of the mutant combgap in *Drosophila melanogaster*. I. Effect on the morphology and chaetotaxy of the prothoracic leg. *Genetics* **68**, 269–286.

963. Datta, S., Stark, K., and Kankel, D.R. (1993). Enhancer detector analysis of the extent of genomic involvement in nervous system development in *Drosophila melanogaster. J. Neurobiol.* **24**, 824–841.

964. Daubresse, G., Deuring, R., Moore, L., Papoulas, O., Zakrajsek, I., Waldrip, W.R., Scott, M.P., Kennison, J.A., and Tamkun, J.W. (1999). The *Drosophila kismet* gene is related to chromatin-remodeling factors and is required for both segmentation and segment identity. *Development* **126**, 1175–1187.

965. Daum, G., Eisenmann-Tappe, I., Fries, H.-W., Troppmair, J., and Rapp, U.R. (1994). The ins and outs of Raf kinases. *Trends Biochem. Sci.* **19**, 474–480.

966. David, G. (1993). Integral membrane heparan sulfate proteoglycans. *FASEB J.* **7**, 1023–1030.

967. Davidson, E.H. (1993). Later embryogenesis: regulatory circuitry in morphogenetic fields. *Development* **118**, 665–690.

968. Davidson, E.H. (2001). "Genomic Regulatory Systems: Development and Evolution." Acad. Pr., New York.

969. Davies, E., Hsiao, F., Williams, A., and Rebay, I. (2001). Eyes absent mediates cross-talk between a network of retinal determination genes and the RTK/Ras/MAPK signaling pathway. *Proc. 42nd Ann. Drosophila Res. Conf.* **Abstracts Vol.**, a55. [See also 2001 *Dev. Cell* **1**, 51–61.]

970. Davis, C.G. (1990). The many faces of epidermal growth factor repeats. *New Biol.* **2**, 410–419.

971. Davis, R.J. (2000). Signal transduction by the JNK group of MAP kinases. *Cell* **103**, 239–252.

972. Dawid, I.B. (1998). LIM protein interactions: *Drosophila* enters the stage. *Trends Genet.* **14**, 480–482.

973. Dawid, I.B., Breen, J.J., and Toyama, R. (1998). LIM domains: multiple roles as adapters and functional modifiers in protein interactions. *Trends Genet.* **14**, 156–162.

974. Dawkins, R. (1996). "Climbing Mount Improbable." Norton, N.Y.

975. Dawson, I.A., Roth, S., and Artavanis-Tsakonas, S. (1995). The *Drosophila* cell cycle gene *fizzy* is required for normal degradation of cyclins A and B during mitosis and has homology to the *CDC20* gene of *Saccharomyces cerevisiae*. *J. Cell Biol.* **129**, 725–737.

976. Dawson, S.R., Turner, D.L., Weintraub, H., and Parkhurst, S.M. (1995). Specificity for the Hairy/Enhancer of split basic helix-loop-helix (bHLH) proteins maps outside the bHLH domain and suggests two separable modes of transcriptional repression. *Mol. Cell. Biol.* **15**, 6923–6931.

977. Day, S.J. and Lawrence, P.A. (2000). Measuring dimensions: the regulation of size and shape. *Development* **127**, 2977–2987.

978. de Beer, G. (1958). "Embryos and Ancestors." 3rd ed. Clarendon Pr., Oxford.

979. de Beer, T., Carter, R.E., Lobel-Rice, K.E., Sorkin, A., and Overduin, M. (1998). Structure and Asn-Pro-Phe binding pocket of the Eps15 homology domain. *Science* **281**, 1357–1360.

980. de Celis, J.F. (1997). Expression and function of *decapentaplegic* and *thick veins* during the differentiation of the veins in the *Drosophila* wing. *Development* **124**, 1007–1018.

981. de Celis, J.F. (1998). Positioning and differentiation of veins in the *Drosophila* wing. *Int. J. Dev. Biol.* **42**, 335–343.

982. de Celis, J.F. (1999). The function of *vestigial* in *Drosophila* wing development: how are tissue-specific responses to signalling pathways specified? *BioEssays* **21**, 542–545.

983. de Celis, J.F., Baonza, A., and García-Bellido, A. (1995). Behavior of *extramacrochaetae* mutant cells in the morphogenesis of the *Drosophila* wing. *Mechs. Dev.* **53**, 209–221.

984. de Celis, J.F. and Barrio, R. (2000). Function of the *spalt/spalt-related* gene complex in positioning the veins in the *Drosophila* wing. *Mechs. Dev.* **91**, 31–41.

985. de Celis, J.F., Barrio, R., del Arco, A., and García-Bellido, A. (1993). Genetic and molecular characterization of a Notch mutation in its Delta- and Serrate-binding domain in *Drosophila*. *Proc. Natl. Acad. Sci. USA* **90**, 4037–4041.

986. de Celis, J.F., Barrio, R., and Kafatos, F.C. (1996). A gene complex acting downstream of *dpp* in *Drosophila* wing morphogenesis. *Nature* **381**, 421–424.

987. de Celis, J.F., Barrio, R., and Kafatos, F.C. (1999). Regulation of the *spalt/spalt-related* gene complex and its function during sensory organ development in the *Drosophila* thorax. *Development* **126**, 2653–2662.

988. de Celis, J.F. and Bray, S. (1997). Feed-back mechanisms affecting Notch activation at the dorsoventral boundary in the *Drosophila* wing. *Development* **124**, 3241–3251.

989. de Celis, J.F., Bray, S., and Garcia-Bellido, A. (1997). Notch signalling regulates *veinlet* expression and establishes boundaries between veins and interveins in the *Drosophila* wing. *Development* **124**, 1919–1928.

990. de Celis, J.F. and Bray, S.J. (2000). The *Abruptex* domain of Notch regulates negative interactions between Notch, its ligands and Fringe. *Development* **127**, 1291–1302.

991. de Celis, J.F., de Celis, J., Ligoxygakis, P., Preiss, A., Delidakis, C., and Bray, S. (1996). Functional relationships between *Notch*, *Su(H)* and the bHLH genes of the *E(spl)* complex: the *E(spl)* genes mediate only a subset of *Notch* activities during imaginal development. *Development* **122**, 2719–2728.

992. de Celis, J.F. and Garcia-Bellido, A. (1994). Modifications of the Notch function by *Abruptex* mutations in *Drosophila melanogaster*. *Genetics* **136**, 183–194.

993. de Celis, J.F. and García-Bellido, A. (1994). Roles of the *Notch* gene in *Drosophila* wing morphogenesis. *Mechs. Dev.* **46**, 109–122.

994. de Celis, J.F., Garcia-Bellido, A., and Bray, S.J. (1996). Activation and function of *Notch* at the dorsal-ventral boundary of the wing imaginal disc. *Development* **122**, 359–369.

995. de Celis, J.F., Llimargas, M., and Casanova, J. (1995). *ventral veinless*, the gene encoding the Cf1a transcription factor, links positional information and cell differentiation during embryonic and imaginal development in *Drosophila melanogaster*. *Development* **121**, 3405–3416.

996. de Celis, J.F., Marí-Beffa, M., and García-Bellido, A. (1991). Cell-autonomous role of Notch, an epidermal growth factor homologue, in sensory organ differentiation in *Drosophila*. *Proc. Natl. Acad. Sci. USA* **88**, 632–636.

997. de Celis, J.F., Marí-Beffa, M., and García-Bellido, A. (1991). Function of trans-acting genes of the *achaete-scute* complex in sensory organ patterning in the mesonotum of *Drosophila*. *Roux's Arch. Dev. Biol.* **200**, 64–76.

998. de Celis, J.F. and Ruiz-Gómez, M. (1995). *groucho* and *hedgehog* regulate *engrailed* expression in the anterior compartment of the *Drosophila* wing. *Development* **121**, 3467–3476.

999. de Celis, J.F., Tyler, D.M., de Celis, J., and Bray, S.J. (1998). Notch signalling mediates segmentation of the *Drosophila* leg. *Development* **125**, 4617–4626. [See also 2001 *Mechs. Dev.* **105**, 115–127.]

1000. de la Concha, A., Dietrich, U., Weigel, D., and Campos-Ortega, J.A. (1988). Functional interactions of neurogenic genes of *Drosophila melanogaster*. *Genetics* **118**, 499–508.

1001. De Lorenzo, C.M., Huwe, A.W., Spillane, M., and Bryant, P.J. (2001). The Dlg multimeric complex and its function in cell proliferation control. *Proc. 42nd Ann. Drosophila Res. Conf.* **Abstracts Vol.**, a156.

1002. de Nooij, J.C., Graber, K.H., and Hariharan, I.K. (2000). Expression of the cyclin-dependent kinase inhibitor Dacapo is regulated by Cyclin E. *Mechs. Dev.* **97**, 73–83.

1003. de Nooij, J.C. and Hariharan, I.K. (1995). Uncoupling cell fate determination from patterned cell division in the *Drosophila* eye. *Science* **270**, 983–985.

1004. de Nooij, J.C., Letendre, M.A., and Hariharan, I.K. (1996). A cyclin-dependent kinase inhibitor, Dacapo, is necessary for timely exit from the cell cycle during *Drosophila* embryogenesis. *Cell* **87**, 1237–1247.

1005. De Robertis, E.M., Oliver, G., and Wright, C.V.E. (1989). Determination of axial polarity in the vertebrate embryo: homeodomain proteins and homeogenetic induction. *Cell* **57**, 189–191.

1006. de Vetten, N., Quattrocchio, F., Mol, J., and Koes, R. (1997). The *an11* locus controlling flower pigmentation in petunia encodes a novel WD-repeat protein conserved in yeast, plants, and animals. *Genes Dev.* **11**, 1422–1434.

1007. de Zulueta, P., Alexandre, E., Jacq, B., and Kerridge, S. (1994). Homeotic complex and *teashirt* genes co-operate to establish trunk segmental identities in *Drosophila*. *Development* **120**, 2278–2296.

1008. Deak, I.I. (1980). A model linking segmentation, compartmentalization and regeneration in *Drosophila* development. *J. Theor. Biol.* **84**, 477–504.

1009. Dearden, P. and Akam, M. (2000). A role for *Fringe* in segment morphogenesis but not segment formation in the grasshopper, *Shistocerca gregaria*. *Dev. Genes Evol.* **210**, 329–336.

1010. Dearden, P. and Akam, M. (2000). Segmentation *in silico*. *Nature* **406**, 131–132.

1011. Dearolf, C.R., Topol, J., and Parker, C.S. (1989). The *caudal* gene product is a direct activator of *fushi tarazu* transcription during *Drosophila* embryogenesis. *Nature* **341**, 340–343.

1012. DeCamillis, M., Cheng, N., Pierre, D., and Brock, H.W. (1992). The *polyhomeotic* gene of *Drosophila* encodes a chromatin protein that shares polytene chromosome-binding sites with *Polycomb*. *Genes Dev.* **6**, 223–232.

1013. Decoville, M., Giacomello, E., Leng, M., and Locker, D. (2001). DSP1, an HMG-like protein, is involved in the regulation of homeotic genes. *Genetics* **157**, 237–244.

1014. Dekker, N., Cox, M., Boelens, R., Verrijzer, C.P., van der Vliet, P.C., and Kaptein, R. (1993). Solution structure of the POU-specific DNA-binding domain of Oct-1. *Nature* **362**, 852–855.

1015. Delattre, M., Briand, S., Paces-Fessy, M., and Blanchet-Tournier, M.-F. (1999). The *Suppressor of fused* gene, involved in Hedgehog signal transduction in *Drosophila*, is conserved in mammals. *Dev. Genes Evol.* **209**, 294–300.

1016. Delgado, R., Maureira, C., Oliva, C., Kidokoro, Y., and Labarca, P. (2000). Size of vesicle pools, rates of mobilization, and recycling at neuromuscular synapses of a *Drosophila* mutant, *shibire*. *Neuron* **28**, 941–953.

1017. Delidakis, C. and Artavanis-Tsakonas, S. (1992). The Enhancer of split [E(spl)] locus of *Drosophila* encodes seven independent helix-loop-helix proteins. *Proc. Natl. Acad. Sci. USA* **89**, 8731–8735.

1018. Delidakis, C., Preiss, A., Hartley, D.A., and Artavanis-Tsakonas, S. (1991). Two genetically and molecularly distinct functions involved in early neurogenesis reside within the *Enhancer of split* locus of *Drosophila melanogaster*. *Genetics* **129**, 803–823.

1019. Demerec, M., ed. (1950). "Biology of *Drosophila*." Hafner, New York.

1020. Demidenko, Z., Badenhorst, P., Jones, T., Bi, X., and Mortin, M.A. (2001). Regulated nuclear export of the homeodomain transcription factor Prospero. *Development* **128**, 1359–1367.

1021. DeMille, M.M.C., Kimmel, B.E., and Rubin, G.M. (1996). A *Drosophila* gene regulated by *rough* and *glass* shows similarity to *ena* and *VASP*. *Gene* **183**, 103–108.

1022. Denef, N., Neubüser, D., Perez, L., and Cohen, S.M. (2000). Hedgehog induces opposite changes in turnover and subcellular localization of Patched and Smoothened. *Development* **102**, 521–531.

1023. Denell, R. (1994). Discovery and genetic definition of the *Drosophila* Antennapedia Complex. *Genetics* **138**, 549–552.

1024. Denell, R.E. (1978). Homoeosis in *Drosophila*: II. A genetic analysis of Polycomb. *Genetics* **90**, 277–289.

1025. Denell, R.E., Brown, S.J., and Beeman, R.W. (1996). Evolution of the organization and function of insect homeotic complexes. *Sems. Cell Dev. Biol.* **7**, 527–538.

1026. Denell, R.E. and Frederick, R.D. (1983). Homoeosis in *Drosophila*: a description of the Polycomb lethal syndrome. *Dev. Biol.* **97**, 34–47.

1027. Derynck, R., Zhang, Y., and Feng, X.-H. (1998). Smads: transcriptional activators of TGF-β responses. *Cell* **95**, 737–740.

1028. Deshpande, G., Stukey, J., and Schedl, P. (1995). *scute* (*sis-*

b) function in *Drosophila* sex determination. *Mol. Cell. Biol.* **15**, 4430–4440.

1029. Deshpande, N., Chopra, A., Rangarajan, A., Shashidhara, L.S., Rodrigues, V., and Krishna, S. (1997). The human Transcription Enhancer Factor-1, TEF-1, can substitute for *Drosophila scalloped* during wingblade development. *J. Biol. Chem.* **272** #16, 10664–10668.

1030. Desplan, C. (1997). Eye development: governed by a dictator or a junta? *Cell* **91**, 861–864.

1031. Desplan, C., Theis, J., and O'Farrell, P.H. (1985). The *Drosophila* developmental gene, *engrailed*, encodes a sequence-specific DNA binding activity. *Nature* **318**, 630–635.

1032. Devoe, R.D. (1985). The eye: electrical activity. *In* "Comprehensive Insect Physiology, Biochemistry, and Pharmacology," Vol. 6 (G.A. Kerkut and L.I. Gilbert, eds.). Pergamon Pr.: Oxford, pp. 277–354.

1033. Dewdney, A.K. (1988). The hodgepodge machine makes waves. *Sci. Am.* **259** #2, 104–107.

1034. Dewdney, A.K. (1989). "The Turing Omnibus: 61 Excursions in Computer Science." Computer Sci. Pr., Rockville, MD.

1035. Dhalluin, C., Yan, K.S., Plotnikova, O., Lee, K.W., Zeng, L., Kuti, M., Mujtaba, S., Goldfarb, M.P., and Zhou, M.-M. (2000). Structural basis of SNT PTB domain interactions with distinct neurotrophic receptors. *Molec. Cell* **6**, 921–929.

1036. Dhordain, P., Albagli, O., Lin, R.J., Ansieau, S., Quief, S., Leutz, A., Kerckaert, J.-P., Evans, R.M., and Leprince, D. (1997). Corepressor SMRT binds the BTB/POZ repressing domain of the LAZ3/BCL6 oncoprotein. *Proc. Natl. Acad. Sci. USA* **94**, 10762–10767.

1037. Diaz-Benjumea, F.J., Cohen, B., and Cohen, S.M. (1994). Cell interaction between compartments establishes the proximal-distal axis of *Drosophila* legs. *Nature* **372**, 175–179.

1038. Diaz-Benjumea, F.J. and Cohen, S.M. (1993). Interaction between dorsal and ventral cells in the imaginal disc directs wing development in *Drosophila*. *Cell* **75**, 741–752.

1039. Diaz-Benjumea, F.J. and Cohen, S.M. (1994). *wingless* acts through the *shaggy/zeste-white 3* kinase to direct dorsal-ventral axis formation in the *Drosophila* leg. *Development* **120**, 1661–1670.

1040. Diaz-Benjumea, F.J. and Cohen, S.M. (1995). Serrate signals through Notch to establish a Wingless-dependent organizer at the dorsal/ventral compartment boundary of the *Drosophila* wing. *Development* **121**, 4215–4225.

1041. Diaz-Benjumea, F.J. and García-Bellido, A. (1990). Genetic analysis of the wing vein pattern of *Drosophila*. *Roux's Arch. Dev. Biol.* **198**, 336–354.

1042. Díaz-Benjumea, F.J. and García-Bellido, A. (1990). Behaviour of cells mutant for an EGF *receptor* homologue of *Drosophila* in genetic mosaics. *Proc. Roy. Soc. Lond. B* **242**, 36–44.

1043. Diaz-Benjumea, F.J. and Hafen, E. (1994). The *sevenless* signalling cassette mediates *Drosophila* EGF receptor function during epidermal development. *Development* **120**, 569–578.

1044. DiBello, P.R., Withers, D.A., Bayer, C.A., Fristrom, J.W., and Guild, G.M. (1991). The *Drosophila Broad-Complex* encodes a family of related proteins containing zinc fingers. *Genetics* **129**, 385–397.

1045. Dickinson, M.H., Lehmann, F.-O., and Sane, S.P. (1999). Wing rotation and the aerodynamic basis of insect flight. *Science* **284**, 1954–1960.

1046. Dickinson, W.J. (1988). On the architecture of regulatory systems: evolutionary insights and implications. *BioEssays* **8**, 204–208.

1047. Dickson, B. (1995). Nuclear factors in sevenless signalling. *Trends Genet.* **11**, 106–111.

1048. Dickson, B. and Hafen, E. (1993). Genetic dissection of eye development in *Drosophila*. In "The Development of *Drosophila melanogaster*," Vol. 2 (M. Bate and A. Martinez Arias, eds.). Cold Spring Harbor Lab. Pr.: Plainview, N. Y., pp. 1327–1362.

1049. Dickson, B., Sprenger, F., and Hafen, E. (1992). Prepattern in the developing *Drosophila* eye revealed by an activated torso-sevenless chimeric receptor. *Genes Dev.* **6**, 2327–2339.

1050. Dickson, B., Sprenger, F., Morrison, D., and Hafen, E. (1992). Raf functions downstream of Ras1 in the Sevenless signal transduction pathway. *Nature* **360**, 600–603.

1051. Dickson, B.J. (1998). Photoreceptor development: breaking down the barriers. *Curr. Biol.* **8**, R90–R92.

1052. Dickson, B.J., Domínguez, M., van der Straten, A., and Hafen, E. (1995). Control of *Drosophila* photoreceptor cell fates by Phyllopod, a novel nuclear protein acting downstream of the Raf kinase. *Cell* **80**, 453–462.

1053. Diederich, R.J., Matsuno, K., Hing, H., and Artavanis-Tsakonas, S. (1994). Cytosolic interaction between deltex and Notch ankyrin repeats implicates deltex in the Notch signaling pathway. *Development* **120**, 473–481.

1054. Dierick, H. and Bejsovec, A. (1999). Cellular mechanisms of Wingless/Wnt signal transduction. *Curr. Topics Dev. Biol.* **43**, 153–190.

1055. Dierick, H.A. and Bejsovec, A. (1998). Functional analysis of Wingless reveals a link between intercellular ligand transport and dorsal-cell-specific signaling. *Development* **125**, 4729–4738.

1056. Dietrich, M.R. (2000). From hopeful monsters to homeotic effects: Richard Goldschmidt's integration of development, evolution, and genetics. *Amer. Zool.* **40**, 738–747.

1057. Dietrich, U. and Campos-Ortega, J.A. (1980). The effect of temperature on *shibire*^ts^ cell clones in the compound eye of *Drosophila melanogaster*. *W. Roux's Arch.* **188**, 55–63.

1058. Dietrich, U. and Campos-Ortega, J.A. (1984). The expression of neurogenic loci in imaginal epidermal cells of *Drosophila melanogaster*. *J. Neurogen.* **1**, 315–332.

1059. Dietrich, W. (1909). Die Facettenaugen der Dipteren. *Z. Wiss. Zool.* **92**, 465–539.

1060. Diez del Corral, R., Aroca, P., Gómez-Skarmeta, J.L., Cavodeassi, F., and Modolell, J. (1999). The Iroquois homeodomain proteins are required to specify body wall identity in *Drosophila*. *Genes Dev.* **13**, 1754–1761.

1061. Dimitratos, S.D., Woods, D.F., and Bryant, P.J. (1997). Camguk, Lin-2, and CASK: novel membrane-associated guanylate kinase homologs that also contain CaM kinase domains. *Mechs. Dev.* **63**, 127–130.

1062. Dimitratos, S.D., Woods, D.F., Stathakis, D.G., and Bryant, P.J. (1999). Signaling pathways are focused at specialized regions of the plasma membrane by scaffolding proteins of the MAGUK family. *BioEssays* **21**, 912–921. [See also 2001 *Trends Genet.* **17**, 511–519.]

1063. DiNardo, S., Heemskerk, J., Dougan, S., and O'Farrell, P.H. (1994). The making of a maggot: patterning the *Drosophila* embryonic epidermis. *Curr. Opinion Gen. Dev.* **4**, 529–534.

1064. DiNardo, S., Kuner, J.M., Theis, J., and O'Farrell, P.H. (1985). Development of embryonic pattern in *D. melanogaster* as revealed by accumulation of the nuclear *engrailed* protein. *Cell* **43**, 59–69.

1065. DiNardo, S., Sher, E., Heemskerk-Jongens, J., Kassis, J.A., and O'Farrell, P.H. (1988). Two-tiered regulation of spatially patterned *engrailed* gene expression during *Drosophila* embryogenesis. *Nature* **332**, 604–609.

1066. Dingwall, C. and Laskey, R.A. (1998). Nuclear import: a tale of two sites. *Curr. Biol.* **8**, R922–R924.

1067. Djian, P. (1998). Evolution of simple repeats in DNA and their relation to human disease. *Cell* **94**, 155–160.

1068. Doctor, J.S., Jackson, P.D., Rashka, K.E., Visalli, M., and Hoffmann, F.M. (1992). Sequence, biochemical characterization, and developmental expression of a new member of the TGF-β superfamily in *Drosophila melanogaster*. *Dev. Biol.* **151**, 491–505.

1069. Doe, C.Q. (1992). Molecular markers for identified neuroblasts and ganglion mother cells in the *Drosophila* central nervous system. *Development* **116**, 855–863.

1070. Doe, C.Q., Chu-LaGraff, Q., Wright, D.M., and Scott, M.P. (1991). The *prospero* gene specifies cell fates in the *Drosophila* central nervous system. *Cell* **65**, 451–464.

1071. Doe, C.Q. and Goodman, C.S. (1985). Early events in insect neurogenesis. I. Development and segmental differences in the pattern of neuronal precursor cells. *Dev. Biol.* **111**, 193–205.

1072. Doe, C.Q. and Goodman, C.S. (1985). Early events in insect neurogenesis. II. The role of cell interactions and cell lineage in the determination of neuronal precursor cells. *Dev. Biol.* **111**, 206–219.

1073. Doe, C.Q., Kuwada, J.Y., and Goodman, C.S. (1985). From epithelium to neuroblasts to neurons: the role of cell interactions and cell lineage during insect neurogenesis. *Phil. Trans. Roy. Soc. Lond. B* **312**, 67–81.

1074. Doherty, D., Feger, G., Younger-Shepherd, S., Jan, L.Y., and Jan, Y.N. (1996). Delta is a ventral to dorsal signal complementary to Serrate, another Notch ligand, in *Drosophila* wing formation. *Genes Dev.* **10**, 421–434.

1075. Doherty, D., Jan, L.Y., and Jan, Y.N. (1997). The *Drosophila* neurogenic gene *big brain*, which encodes a membrane-associated protein, acts cell autonomously and can act synergistically with *Notch* and *Delta*. *Development* **124**, 3881–3893.

1076. Dokucu, M.E., Zipursky, S.L., and Cagan, R.L. (1996). Atonal, Rough and the resolution of proneural clusters in the developing *Drosophila* retina. *Development* **122**, 4139–4147.

1077. Domínguez, M. (1999). Dual role for Hedgehog in the regulation of the proneural gene *atonal* during ommatidia development. *Development* **126**, 2345–2353.

1078. Domínguez, M., Brunner, M., Hafen, E., and Basler, K. (1996). Sending and receiving the Hedgehog signal: control by the *Drosophila* Gli protein Cubitus interruptus. *Science* **272**, 1621–1625.

1079. Domínguez, M. and Campuzano, S. (1993). *asense*, a member of the *Drosophila achaete-scute* complex, is a proneural and neural differentiation gene. *EMBO J.* **12**, 2049–2060.

1080. Domínguez, M. and de Celis, J.F. (1998). A dorsal/ventral boundary established by Notch controls growth and polarity in the *Drosophila* eye. *Nature* **396**, 276–278.

1081. Domínguez, M. and Hafen, E. (1996). Genetic dissection of cell fate specification in the developing eye of *Drosophila*. *Sems. Cell Dev. Biol.* **7**, 219–226.

1082. Domínguez, M. and Hafen, E. (1997). Hedgehog directly controls initiation and propagation of retinal differentiation in the *Drosophila* eye. *Genes Dev.* **11**, 3254–3264.

1083. Domínguez, M., Wasserman, J.D., and Freeman, M. (1998). Multiple functions of the EGF receptor in *Drosophila* eye development. *Curr. Biol.* **8**, 1039–1048.

1084. Dong, C., Li, Z., Alvarez, R., Jr., Feng, X.-H., and Goldschmidt-Clermont, P.J. (2000). Microtubule binding to Smads may regulate TGFβ activity. *Molec. Cell* **5**, 27–34.

1085. Dong, P.D.S., Chu, J., and Panganiban, G. (2000). Coexpression of the homeobox genes *Distal-less* and *homothorax* determines *Drosophila* antennal identity. *Development* **127**, 209–216. [See also 2001 *Development* **128**, 2365–2372.]

1086. Dong, X., Tsuda, L., Zavitz, K.H., Lin, M., Li, S., Carthew, R.W., and Zipursky, S.L. (1999). *ebi* regulates epidermal growth factor receptor signaling pathways in *Drosophila*. *Genes Dev.* **13**, 954–965.

1087. Dong, X., Zavitz, K.H., Thomas, B.J., Lin, M., Campbell, S., and Zipursky, S.L. (1997). Control of G_1 in the developing *Drosophila* eye: *rca1* regulates Cyclin A. *Genes Dev.* **11**, 94–105.

1088. Dooijes, D., van Beest, M., van de Wetering, M., Boulanger, G., Jones, T., Clevers, H., and Mortin, M.A. (1998). Genomic organization of the segment polarity gene *pan* in *Drosophila melanogaster*. *Mol. Gen. Genet.* **258**, 45–52.

1089. Doolittle, R.F. (1995). The multiplicity of domains in proteins. *Annu. Rev. Biochem.* **64**, 287–314.

1090. Doolittle, W.F. (1988). Hierarchical approaches to genome evolution. *In* "Philosophy and Biology," (M. Matthen and B. Linsky, eds.). Univ. Calgary Pr.: Calgary, Alberta, Canada, pp. 101–133.

1091. Dorfman, R. and Shilo, B.-Z. (2001). Biphasic activation of the BMP pathway patterns the *Drosophila* embryonic dorsal region. *Development* **128**, 965–972.

1092. Dorsett, D. (1999). Distant liaisons: long-range enhancer-promoter interactions in *Drosophila*. *Curr. Opin. Gen. Dev.* **9**, 505–514.

1093. Dougan, S. and DiNardo, S. (1992). *Drosophila wingless* generates cell type diversity among *engrailed* expressing cells. *Nature* **360**, 347–350.

1094. Dover, G. (2000). How genomic and developmental dynamics affect evolutionary processes. *BioEssays* **22**, 1153–1159.

1095. Downing, A.K., Driscoll, P.C., Gout, I., Salim, K., Zvelebil, M.J., and Waterfield, M.D. (1994). Three-dimensional solution structure of the pleckstrin homology domain from dynamin. *Curr. Biol.* **4**, 884–891.

1096. Downward, J. (1995). KSR: a novel player in the RAS pathway. *Cell* **83**, 831–834.

1097. Doyle, D.A., Lee, A., Lewis, J., Kim, E., Sheng, M., and MacKinnon, R. (1996). Crystal structures of a complexed and peptide-free membrane protein-binding domain: molecular basis of peptide recognition by PDZ. *Cell* **85**, 1067–1076.

1098. Doyle, H.J. and Bishop, J.M. (1993). Torso, a receptor tyrosine kinase required for embryonic pattern formation, shares substrates with the Sevenless and EGF-R pathways in *Drosophila*. *Genes Dev.* **7**, 633–646.

1099. Dreyfuss, G., Matunis, M.J., Piñol-Roma, S., and Burd, C.G. (1993). hnRNP proteins and the biogenesis of mRNA. *Annu. Rev. Biochem.* **62**, 289–321.

1100. Dreyfuss, G., Swanson, M.S., and Piñol-Roma, S. (1998). Heterogeneous nuclear ribonucleoprotein particles and the pathway of mRNA formation. *Trends Biochem. Sci.* **13**, 86–91.

1101. Driesch, H. (1894). "Analytische Theorie der organischen Entwicklung." Engelmann, Leipzig.

1102. Driever, W. and Nüsslein-Volhard, C. (1988). The *bicoid* protein determines position in the *Drosophila* embryo in a concentration-dependent manner. *Cell* **54**, 95–104.

1103. Driever, W. and Nüsslein-Volhard, C. (1988). A gradient of *bicoid* protein in *Drosophila* embryos. *Cell* **54**, 83–93.

1104. Driever, W., Siegel, V., and Nüsslein-Volhard, C. (1990). Autonomous determination of anterior structures in the early *Drosophila* embryo by the *bicoid* morphogen. *Development* **109**, 811–820.

1105. Driever, W., Thoma, G., and Nüsslein-Volhard, C. (1989). Determination of spatial domains of zygotic gene expression in the *Drosophila* embryo by the affinity of binding sites for the bicoid morphogen. *Nature* **340**, 363–367.

1106. Dröge, P. and Müller-Hill, B. (2001). High local protein concentrations at promoters: strategies in prokaryotic and eukaryotic cells. *BioEssays* **23**, 179–183.

1107. Dübendorfer, K. and Nöthiger, R. (1982). A clonal analysis of cell lineage and growth in the male and female genital disc of *Drosophila melanogaster*. *W. Roux's Arch.* **191**, 42–55.

1108. Dubinin, N.P. (1932). Step-allelomorphism and the theory of centres of the gene, achaete-scute. *J. Genet.* **26**, 37–58.

1109. Dubinin, N.P. (1933). Step-allelomorphism in *Drosophila melanogaster*. *J. Genetics* **27**, 443–464.

1110. Dubnicoff, T., Valentine, S.A., Chen, G., Shi, T., Lengyel, J.A., Paroush, Z., and Courey, A.J. (1997). Conversion of Dorsal from an activator to a repressor by the global corepressor Groucho. *Genes Dev.* **11**, 2952–2957.

1111. Duboule, D. (1994). Temporal colinearity and the phylotypic progression: a basis for the stability of a vertebrate Bauplan and the evolution of morphologies through heterochrony. *Development* **1994 Suppl.**, 135–142.

1112. Duboule, D. (1998). Vertebrate *Hox* gene regulation: clustering and/or colinearity? *Curr. Opin. Gen. Dev.* **8**, 514–518.

1113. Duboule, D. and Morata, G. (1994). Colinearity and functional hierarchy among genes of the homeotic complexes. *Trends Genet.* **10**, 358–364.

1114. Duboule, D. and Wilkins, A.S. (1998). The evolution of "bricolage". *Trends Genet.* **14**, 54–59.

1115. Dudley, R. (1999). Unsteady aerodynamics. *Science* **284**, 1937–1939.

1116. Dudley, R. (2000). "The Biomechanics of Insect Flight: Form, Function, Evolution." Princeton Univ. Pr., Princeton, N. J.

1117. Duffy, J.B. and Perrimon, N. (1994). The Torso pathway in *Drosophila*: lessons on receptor tyrosine kinase signaling and pattern formation. *Dev. Biol.* **166**, 380–395.

1118. Duggan, A., García-Añoveros, J., and Corey, D.P. (2000). Insect mechanoreception: What a long, strange TRP it's been. *Curr. Biol.* **10**, R384–R387.

1119. Duncan, D.M., Burgess, E.A., and Duncan, I. (1998). Control of distal antennal identity and tarsal development in *Drosophila* by *spineless-aristapedia*, a homolog of the mammalian dioxin receptor. *Genes Dev.* **12**, 1290–1303.

1120. Duncan, I. (1987). The bithorax complex. *Annu. Rev. Genet.* **21**, 285–319.

1121. Duncan, I. (1996). How do single homeotic genes control multiple segment identities? *BioEssays* **18**, 91–94.

1122. Duncan, I.M. (1982). Polycomblike: a gene that appears to be required for the normal expression of the bithorax and antennapedia complexes of *Drosophila melanogaster*. *Genetics* **102**, 49–70.

1123. Dunne, J.F. (1981). Growth dynamics in the regeneration of a fragment of the wing imaginal disc of *Drosophila melanogaster*. *Dev. Biol.* **87**, 379–382.

1124. Dura, J.-M. and Ingham, P. (1988). Tissue- and stage-specific control of homeotic and segmentation gene expression in *Drosophila* embryos by the *polyhomeotic* gene. *Development* **103**, 733–741.

1125. Duranceau, C., Glenn, S.L., and Schneiderman, H.A. (1980). Positional information as a regulator of growth in the imaginal wing disc of *Drosophila melanogaster*. In "Insect Biology in the Future," (M. Locke and D. Smith, eds.). Acad. Pr.: New York, pp. 479–516.

1126. Duronio, R.J. (1999). Establishing links between developmental signaling pathways and cell-cycle regulation in *Drosophila*. *Curr. Opin. Gen. Dev.* **9**, 81–88.

1127. Dvir, A., Conaway, J.W., and Conaway, R.C. (2001). Mechanism of transcription initiation and promoter escape by RNA polymerase II. *Curr. Opin. Gen. Dev.* **11**, 209–214.

1128. Dye, C.A., Lee, J.-K., Atkinson, R.C., Brewster, R., Han, P.-L., and Bellen, H.J. (1998). The *Drosophila sanpodo* gene controls sibling cell fate and encodes a tropomodulin homolog, an actin/tropomyosin-associated protein. *Development* **125**, 1845–1856.

1129. Dyson, S. and Gurdon, J.B. (1998). The interpretation of position in a morphogen gradient as revealed by occupancy of activin receptors. *Cell* **93**, 557–568.

1130. Easter, S.S., Jr. (2000). Let there be sight. *Neuron* **27**, 193–195.

1131. Eastman, D.S., Slee, R., Skoufos, E., Bangalore, L., Bray, S., and Delidakis, C. (1997). Synergy between Suppressor of Hairless and Notch in regulation of *Enhancer of split mγ* and *mδ* expression. *Mol. Cell. Biol.* **17**, 5620–5628.

1132. Eaton, S. (1997). Planar polarization of *Drosophila* and vertebrate epithelia. *Curr. Opin. Cell Biol.* **9**, 860–866.

1133. Eaton, S., Auvinen, P., Luo, L., Jan, Y.N., and Simons, K. (1995). CDC42 and Rac1 control different actin-dependent processes in the *Drosophila* wing disc epithelium. *J. Cell Biol.* **131**, 151–164.

1134. Eaton, S. and Cohen, S. (1996). Wnt signal transduction: more than one way to skin a (β-)cat? *Trends Cell Biol.* **6**, 287–289.

1135. Eaton, S. and Kornberg, T.B. (1990). Repression of *ci-D* in posterior compartments of *Drosophila* by *engrailed*. *Genes Dev.* **4**, 1068–1077.

1136. Eaton, S., Wepf, R., and Simons, K. (1996). Roles for Rac1 and Cdc42 in planar polarization and hair outgrowth in the wing of *Drosophila*. *J. Cell Biol.* **135**, 1277–1289.

1137. Eberhard, W.G. (2001). Species-specific genitalic copulatory courtship in sepsid flies (Diptera, Sepsidae, *Microsepsis*) and theories of genitalic evolution. *Evolution* **55**, 93–102.

1138. Eberlein, S. and Russell, M.A. (1983). Effects of deficiencies in the *engrailed* region of *Drosophila melanogaster*. *Dev. Biol.* **100**, 227–237.

1139. Eck, M.J., Dhe-Paganon, S., Trüb, T., Nolte, R.T., and Shoelson, S.E. (1996). Structure of the IRS-1 PTB domain bound to the juxtamembrane region of the insulin receptor. *Cell* **85**, 695–705.

1140. Ede, D.A. (1972). Cell behaviour and embryonic development. *Internat. J. Neurosci.* **3**, 165–174.

1141. Edelman, G.M. (1993). Neural Darwinism: selection and reentrant signaling in higher brain function. *Neuron* **10**, 115–125.

1142. Edgar, B.A. and Lehner, C.F. (1996). Developmental control of cell cycle regulators: a fly's perspective. *Science* **274**, 1646–1652.

1143. Edgar, B.A., Odell, G.M., and Schubiger, G. (1989). A genetic switch, based on negative regulation, sharpens stripes in *Drosophila* embryos. *Dev. Genet.* **10**, 124–142.

1144. Edlund, T. and Jessell, T.M. (1999). Progression from extrinsic to intrinsic signaling in cell fate specification: a view from the nervous system. *Cell* **96**, 211–224.

1145. Edwards, J.S. (1994). In memoriam. Sir Vincent Brian Wigglesworth (1899–1994). *Dev. Biol.* **166**, 361–362.

1146. Egan, S.E. and Weinberg, R.A. (1993). The pathway to signal achievement. *Nature* **365**, 781–783.

1147. Eggert, T., Hauck, B., Hildebrandt, N., Gehring, W.J., and Walldorf, U. (1998). Isolation of a *Drosophila* homolog of the vertebrate homeobox gene Rx and its possible role in brain and eye development. *Proc. Natl. Acad. Sci. USA* **95**, 2343–2348.

1148. Eigen, M. (1971). Selforganization of matter and the evolution of biological macromolecules. *Naturwiss.* **58**, 465–523.

1149. Eisen, J.S. and Youssef, N.N. (1980). Fine structural aspects of the developing compound eye of the honey bee, *Apis mellifera* L. *J. Ultrastr. Res.* **71**, 79–94.

1150. Eisenberg, L.M., Ingham, P.W., and Brown, A.M.C. (1992). Cloning and characterization of a novel *Drosophila Wnt* gene, *Dwnt-5*, a putative downstream target of the homeobox gene *Distal-less*. *Dev. Biol.* **154**, 73–83.

1151. Eizinger, A., Jungblut, B., and Sommer, R.J. (1999). Evolutionary change in the functional specificity of genes. *Trends Genet.* **15**, 197–202.

1152. Ellenberger, T., Fass, D., Arnaud, M., and Harrison, S.C. (1994). Crystal structure of transcription factor E47: E-box recognition by a basic region helix-loop-helix dimer. *Genes Dev.* **8**, 970–980.

1153. Ellenberger, T.E., Brandl, C.J., Struhl, K., and Harrison, S.C. (1992). The GCN4 basic region leucine zipper binds DNA as a dimer of uninterrupted α helices: crystal structure of the protein-DNA complex. *Cell* **71**, 1223–1237.

1154. Elliott, G. and O'Hare, P. (1997). Intercellular trafficking and protein delivery by a herpesvirus structural protein. *Cell* **88**, 223–233.

1155. Ellis, H.M. (1994). Embryonic expression and function of the *Drosophila* helix-loop-helix gene, *extramacrochaetae*. *Mechs. Dev.* **47**, 65–72.

1156. Ellis, H.M., Spann, D.R., and Posakony, J.W. (1990). *extramacrochaetae*, a negative regulator of sensory organ development in *Drosophila*, defines a new class of helix-loop-helix proteins. *Cell* **61**, 27–38.

1157. Ellis, M.C., O'Neill, E.M., and Rubin, G.M. (1993). Expression of *Drosophila glass* protein and evidence for negative regulation of its activity in non-neuronal cells by another DNA-binding protein. *Development* **119**, 855–865.

1158. Ellis, M.C., Weber, U., Wiersdorff, V., and Mlodzik, M. (1994). Confrontation of *scabrous* expressing and non-expressing cells is essential for normal ommatidial spacing in the *Drosophila* eye. *Development* **120**, 1959–1969. [Cf. 2001 *Dev. Biol.* **240**, 361–376.]

1159. Elsdale, T. and Wasoff, F. (1976). Fibroblast cultures and dermatoglyphics: the topology of two planar patterns. *W. Roux's Arch.* **180**, 121–147.

1160. Elsholtz, H.P., Albert, V.R., Treacy, M.N., and Rosenfeld, M.G. (1990). A two-base change in a POU factor-binding site switches pituitary-specific to lymphoid-specific gene expression. *Genes Dev.* **4**, 43–51.

1161. Elstob, P.R., Brodu, V., and Gould, A.P. (2001). *spalt*-dependent switching between two cell fates that are induced by the *Drosophila* EGF receptor. *Development* **128**, 723–732.

1162. Emerald, B.S. and Roy, J.K. (1997). Homeotic transformation in *Drosophila*. *Nature* **389**, 684.

1163. Emerald, B.S. and Roy, J.K. (1998). Organising activities of *engrailed, hedgehog, wingless* and *decapentaplegic* in the genital discs of *Drosophila melanogaster*. *Dev. Genes Evol.* **208**, 504–516.

1164. Emery, J.F. and Bier, E. (1995). Specificity of CNS and PNS regulatory subelements comprising pan-neural enhancers of the *deadpan* and *scratch* genes is achieved by repression. *Development* **121**, 3549–3560.

1165. Emlen, D.J. and Nijhout, H.F. (2000). The development and evolution of exaggerated morphologies in insects. *Annu. Rev. Entomol.* **45**, 661–708.

1166. Emmons, R.B., Duncan, D., Estes, P.A., Kiefel, P., Mosher, J.T., Sonnenfeld, M., Ward, M.P., Duncan, I., and Crews, S.T. (1999). The Spineless-Aristapedia and Tango bHLH-PAS proteins interact to control antennal and tarsal development in *Drosophila*. *Development* **126**, 3937–3945.

1167. Engstrom, L., Noll, E., and Perrimon, N. (1997). Paradigms to study signal transduction pathways in *Drosophila*. *Curr. Topics Dev. Biol.* **35**, 229–261.

1168. Engström, Y., Schneuwly, S., and Gehring, W.J. (1992). Spatial and temporal expression of an *Antennapedia/lac Z* gene construct integrated into the endogenous *Antennapedia* gene of *Drosophila melanogaster*. *Roux's Arch. Dev. Biol.* **201**, 65–80.

1169. Entchev, E.V., Schwabedissen, A., and González-Gaitán, M. (2000). Gradient formation of the TGF-β homolog Dpp. *Cell* **103**, 981–991.

1170. Epper, F. and Nöthiger, R. (1982). Genetic and developmental evidence for a repressed genital primordium in *Drosophila melanogaster*. *Dev. Biol.* **94**, 163–175.

1171. Epps, J.L., Jones, J.B., and Tanda, S. (1997). *oroshigane*, a new segment polarity gene of *Drosophila melanogaster*, functions in Hedgehog signal transduction. *Genetics* **145**, 1041–1052.

1172. Epstein, E.A., Martin, J.L., Haerry, T., Lincecum, J., and O'Connor, M.B. (2001). Genetic investigation of the *sara* and *syndecan* region. *Proc. 42nd Ann. Drosophila Res. Conf.* **Abstracts Vol.**, a145.

1173. Erickson, J.W. and Cline, T.W. (1993). A bZIP protein, Sisterless-a, collaborates with bHLH transcription factors early in *Drosophila* development to determine sex. *Genes Dev.* **7**, 1688–1702.

1174. Erickson, J.W. and Cline, T.W. (1998). Key aspects of the primary sex determination mechanism are conserved across the genus *Drosophila*. *Development* **125**, 3259–3268.

1175. Erkner, A., Gallet, A., Angelats, C., Fasano, L., and Kerridge, S. (1999). The role of Teashirt in proximal leg development in *Drosophila*: ectopic *teashirt* expression reveals different cell behaviours in ventral and dorsal domains. *Dev. Biol.* **215**, 221–232.

1176. Erneux, T., Hiernaux, J., and Nicolis, G. (1978). Turing's theory in morphogenesis. *Bull. Math. Biol.* **40**, 771–789.

1177. Erwin, D.H. and Valentine, J.W. (1984). "Hopeful monsters," transposons, and metazoan radiation. *Proc. Natl. Acad. Sci. USA* **81**, 5482–5483.

1178. Eskens, A.A.C., Sprey, T.E., and Westra, A. (1981). Morphological indications for compartment boundaries in the imaginal wing disc of *Drosophila melanogaster*. *Neth. J. Zool.* **31**, 773–776.

1179. Estrada, B. and Sánchez-Herrero, E. (2001). The Hox gene *Abdominal-B* antagonizes appendage development in the genital disc of *Drosophila*. *Development* **128**, 331–339.

1180. Fagotto, F., Glück, U., and Gumbiner, B.M. (1998). Nuclear localization signal-independent and importin/karyopherin-independent nuclear import of β-catenin. *Curr. Biol.* **8**, 181–190.

1181. Fagotto, F. and Gumbiner, B.M. (1996). Cell contact-dependent signaling. *Dev. Biol.* **180**, 445–454.

1182. Fagotto, F., Jho, E.-h., Zeng, L., Kurth, T., Joos, T., Kaufmann, C., and Costantini, F. (1999). Domains of Axin involved in protein-protein interactions, Wnt pathway inhibition, and intracellular localization. *J. Cell Biol.* **145**, 741–756.

1183. Fain, M.J. and Schneiderman, H.A. (1979). Regulative interaction of mature imaginal disc fragments with embryonic and immature disc tissues in *Drosophila*. *W. Roux's Arch.* **187**, 1–11.

1184. Fain, M.J. and Schneiderman, H.A. (1979). Wound healing and regenerative response of fragments of the *Drosophila* wing imaginal disc cultured *in vitro*. *J. Insect Physiol.* **25**, 913–924.

1185. Fain, M.J. and Stevens, B. (1982). Alterations in the cell cycle of *Drosophila* imaginal disc cells precede metamorphosis. *Dev. Biol.* **92**, 247–258.

1186. Fairall, L., Harrison, S.D., Travers, A.A., and Rhodes, D. (1992). Sequence-specific DNA binding by a two zinc-finger peptide from the *Drosophila melanogaster* Tramtrack protein. *J. Mol. Biol.* **226**, 349–366.

1187. Fairall, L., Schwabe, J.W.R., Chapman, L., Finch, J.T., and Rhodes, D. (1993). The crystal structure of a two zinc-finger peptide reveals an extension to the rules for zinc-finger/DNA recognition. *Nature* **366**, 483–487.

1188. Falvo, J.W., Thanos, D., and Maniatis, T. (1995). Reversal of intrinsic DNA bends in the IFNb gene enhancer by transcription factors and the architectural protein HMG I(Y). *Cell* **83**, 1101–1111.

1189. Fan, X., Brass, L.F., Poncz, M., Spitz, F., Maire, P., and Manning, D.R. (2000). The α subunits of Gz and Gi interact with the *eyes absent* transcription cofactor Eya2, preventing its interaction with the Six class of homeodomain-containing proteins. *J. Biol. Chem.* **275** #41, 32129–32134.

1190. Fankhauser, G. (1945). The effects of changes in chromosome number on amphibian development. *Q. Rev. Biol.* **20**, 20–78.

1191. Fankhauser, G. (1955). The role of nucleus and cytoplasm. *In* "Analysis of Development," (B.H. Willier, P.A. Weiss, and V. Hamburger, eds.). Hafner: New York, pp. 126–150.

1192. Fanning, A.S. and Anderson, J.M. (1996). Protein-protein interactions: PDZ domain networks. *Curr. Biol.* **6**, 1385–1388.

1193. Fanto, M., Mayes, C.A., and Mlodzik, M. (1998). Linking cell-fate specification to planar polarity: determination of the R3/R4 photoreceptors is a prerequisite for the interpretation of the Frizzled mediated polarity signal. *Mechs. Dev.* **74**, 51–58.

1194. Fanto, M. and Mlodzik, M. (1999). Asymmetric Notch activation specifies photoreceptors R3 and R4 and planar polarity in the *Drosophila* eye. *Nature* **397**, 523–526.

1195. Fanto, M., Weber, U., Strutt, D.I., and Mlodzik, M. (2000). Nuclear signaling by Rac and Rho GTPases is required in the establishment of epithelial planar polarity in the *Drosophila* eye. *Curr. Biol.* **10**, 979–988.

1196. Farrar, M.A., Alberola-Ila, J., and Perlmutter, R.M. (1996). Activation of the Raf-1 kinase cascade by coumermycin-induced dimerization. *Nature* **383**, 178–181.

1197. Fasano, L., Röder, L., Coré, N., Alexandre, E., Vola, C., Jacq, B., and Kerridge, S. (1991). The gene *teashirt* is required for the development of *Drosophila* embryonic trunk segments and encodes a protein with widely spaced zinc finger motifs. *Cell* **64**, 63–79.

1198. Fausto-Sterling, A. and Hsieh, L. (1978). Pattern formation in the wing veins of the *fused* mutant (*Drosophila melanogaster*). *Dev. Biol.* **63**, 358–369.

1199. Fauvarque, M.-O. and Dura, J.-M. (1993). *polyhomeotic* regulatory sequences induce developmental regulator-dependent variegation and targeted P-element insertions in *Drosophila*. *Genes Dev.* **7**, 1508–1520.

1200. Featherstone, C. (1997). Src structure crystallizes 20 years of oncogene research. *Science* **275**, 1066.

1201. Feger, G., Vaessin, H., Su, T.T., Wolff, E., Jan, L.Y., and Jan, Y.N. (1995). *dpa*, a member of the MCM family, is required for mitotic DNA replication but not endoreplication in *Drosophila*. *EMBO J.* **14**, 5387–5398.

1202. Fehon, R.G., Gauger, A., and Schubiger, G. (1987). Cellular recognition and adhesion in embryos and imaginal discs of *Drosophila melanogaster*. *In* "Genetic Regulation of Development," *Symp. Soc. Dev. Biol.*, Vol. 45 (W.F. Loomis, ed.). Alan R. Liss: New York, pp. 141–170.

1203. Fehon, R.G., Johansen, K., Rebay, I., and Artavanis-Tsakonas, S. (1991). Complex cellular and subcellular regulation of Notch expression during embryonic and imaginal development of *Drosophila*: Implications for Notch function. *J. Cell Biol.* **113**, 657–669.

1204. Fehon, R.G., Kooh, P.J., Rebay, I., Regan, C.L., Xu, T., Muskavitch, M.A.T., and Artavanis-Tsakonas, S. (1990). Molecular interactions between the protein products of the neurogenic loci *Notch* and *Delta*, two EGF-homologous genes in *Drosophila*. *Cell* **61**, 523–534.

1205. Fehon, R.G. and Schubiger, G. (1985). Dissociation and sorting out of *Drosophila* imaginal disc cells. *Dev. Biol.* **108**, 465–473.

1206. Feinbaum, R. and Ambros, V. (1999). The timing of *lin-4* RNA accumulation controls the timing of postembryonic developmental events in *Caenorhabditis elegans*. *Dev. Biol.* **210**, 87–95.

1207. Feldman, P., Eicher, E.N., Leevers, S.J., Hafen, E., and Hughes, D.A. (1999). Control of growth and differentiation by *Drosophila* RasGAP, a homolog of p120 Ras-GTPase-activating protein. *Mol. Cell. Biol.* **19**, 1928–1937.

1208. Felsenfeld, A.L. and Kennison, J.A. (1995). Positional signaling by *hedgehog* in *Drosophila* imaginal disc development. *Development* **121**, 1–10.

1209. Feng, G.-S. and Pawson, T. (1994). Phosphotyrosine phosphatases with SH2 domains: regulators of signal transduction. *Trends Genet.* **10**, 54–58.

1210. Feng, S., Chen, J.K., Yu, H., Simon, J.A., and Schreiber, S.L. (1994). Two binding orientations for peptides to the Src SH3 domain: development of a general model for SH3-ligand interactions. *Science* **266**, 1241–1247.

1211. Ferguson, E.L. and Anderson, K.V. (1992). *decapentaplegic* acts as a morphogen to organize dorsal-ventral pattern in the *Drosophila* embryo. *Cell* **71**, 451–461.

1212. Ferguson, E.L. and Anderson, K.V. (1992). Localized enhancement and repression of the activity of the TGF-β

family member, *decapentaplegic*, is necessary for dorsal-ventral pattern formation in the *Drosophila* embryo. *Development* **114**, 583–597.

1213. Ferguson, K.M., Kavran, J.M., Sankaran, V.G., Fournier, E., Isakoff, S.J., Skolnik, E.Y., and Lemmon, M.A. (2000). Structural basis for discrimination of 3-phosphoinositides by pleckstrin homology domains. *Molec. Cell* **6**, 373–384.

1214. Fernandes, J., Celniker, S.E., Lewis, E.B., and VijayRaghavan, K. (1994). Muscle development in the four-winged *Drosophila* and the role of the *Ultrabithorax* gene. *Curr. Biol.* **4**, 957–964.

1215. Fernandez, R., Tabarini, D., Azpiazu, N., Frasch, M., and Schlessinger, J. (1995). The *Drosophila* insulin receptor homolog: a gene essential for embryonic development encodes two receptor isoforms with different signaling potential. *EMBO J.* **14**, 3373–3384.

1216. Fernandez-Almonacid, R. and Rosen, O.M. (1987). Structure and ligand specificity of the *Drosophila melanogaster* insulin receptor. *Mol. Cell. Biol.* **7**, 2718–2727.

1217. Fernández-Fúnez, P., Lu, C.-H., Rincón-Limas, D.E., García-Bellido, A., and Botas, J. (1998). The relative expression amounts of *apterous* and its co-factor *dLdb/Chip* are critical for dorso-ventral compartmentalization in the *Drosophila* wing. *EMBO J.* **17**, 6846–6853.

1218. Fernandez-Funez, P., Rickhof, G., Morata, G., and Botas, J. (2001). dlim1 plays an essential role in the specification of ventral appendages. *Proc. 42nd Ann. Drosophila Res. Conf.* **Abstracts Vol.**, a198.

1219. Ferrandon, D., Koch, I., Westhof, E., and Nüsslein-Volhard, C. (1997). RNA-RNA interaction is required for the formation of specific *bicoid* mRNA 3′ UTR-STAUFEN ribonucleoprotein particles. *EMBO J.* **16**, 1751–1758.

1220. Ferrari, S., Harley, V.R., Pontiggia, A., Goodfellow, P.N., Lovell-Badge, R., and Bianchi, M.E. (1992). SRY, like HMG1, recognizes sharp angles in DNA. *EMBO J.* **11**, 4497–4506.

1221. Ferré-D'Amaré, A.R., Pognonec, P., Roeder, R.G., and Burley, S.K. (1994). Structure and function of the b/HLH/Z domain of USF. *EMBO J.* **13**, 180–189.

1222. Ferrell, J.E., Jr. and Machleder, E.M. (1998). The biochemical basis of an all-or-none cell fate switch in *Xenopus* oocytes. *Science* **280**, 895–898.

1223. Ferrier, D.E.K. and Holland, P.W.H. (2001). Ancient origin of the Hox gene cluster. *Nature Rev. Gen.* **2**, 33–38.

1224. Ferris, G.F. (1950). External morphology of the adult. *In* "Biology of *Drosophila*," (M. Demerec, ed.). Hafner: New York, pp. 368–419.

1225. Ferrus, A. (1979). Cell functions in morphogenesis: clonal analysis of new morphogenetic mutations in *Drosophila melanogaster*. *Dev. Biol.* **68**, 16–28.

1226. Ferrús, A. and Garcia-Bellido, A. (1976). Morphogenetic mutants detected in mitotic recombination clones. *Nature* **260**, 425–426.

1227. ffrench-Constant, C. (1994). How do embryonic cells measure time? *Curr. Biol.* **4**, 415–419.

1228. Fields, S. and Song, O.-k. (1989). A novel genetic system to detect protein-protein interactions. *Nature* **340**, 245–246.

1229. Fields, S. and Sternglanz, R. (1994). The two-hybrid system: an assay for protein-protein interactions. *Trends Genet.* **10**, 286–292.

1230. Fiering, S., Whitelaw, E., and Martin, D.I.K. (2000). To be or not to be active: the stochastic nature of enhancer action. *BioEssays* **22**, 381–387.

1231. Fietz, M.J., Jacinto, A., Taylor, A.M., Alexandre, C., and

Ingham, P.W. (1995). Secretion of the amino-terminal fragment of the Hedgehog protein is necessary and sufficient for *hedgehog* signalling in *Drosophila. Curr. Biol.* **6**, 643–650.

1232. Finelli, A.L., Bossie, C.A., Xie, T., and Padgett, R.W. (1994). Mutational analysis of the *Drosophila tolloid* gene, a human BMP-1 homologue. *Development* **120**, 861–870.

1233. Finelli, A.L., Xie, T., Bossie, C.A., Blackman, R.K., and Padgett, R.W. (1995). The *tolkin* gene is a *tolloid*/BMP-1 homologue that is essential for *Drosophila* development. *Genetics* **141**, 271–281.

1234. Fini, M.E., Strissel, K.J., and West-Mays, J.A. (1997). Perspectives on eye development. *Dev. Genet.* **20**, 175–185.

1235. Finkbohner, J.D., Cunningham, G.N., and Kuhn, D.T. (1985). Tissue distribution of malic enzyme-NADP$^+$ in *D. melanogaster* imaginal discs. *Roux's Arch. Dev. Biol.* **194**, 217–223.

1236. Finnerty, J.R. and Martindale, M.Q. (1998). The evolution of the Hox cluster: insights from outgroups. *Curr. Opin. Gen. Dev.* **8**, 681–687. [See also 2001 *Curr. Biol.* **11**, 759–763.]

1237. Firth, L., Manchester, J., Lorenzen, J.A., Baron, M., and Perkins, L.A. (2000). Identification of genomic regions that interact with a viable allele of the *Drosophila* protein tyrosine phosphatase Corkscrew. *Genetics* **156**, 733–748.

1238. Fischer, J.A., Leavell, S.K., and Li, Q. (1997). Mutagenesis screens for interacting genes reveal three roles for *fat facets* during *Drosophila* eye development. *Dev. Genet.* **21**, 167–174.

1239. Fischer-Vize, J.A. and Mosley, K.L. (1994). *marbles* mutants: uncoupling cell determination and nuclear migration in the developing *Drosophila* eye. *Development* **120**, 2609–2618.

1240. Fischer-Vize, J.A., Rubin, G.M., and Lehmann, R. (1992). The *fat facets* gene is required for *Drosophila* eye and embryo development. *Development* **116**, 985–1000.

1241. Fischer-Vize, J.A., Vize, P.D., and Rubin, G.M. (1992). A unique mutation in the *Enhancer of split* gene complex affects the fates of the mystery cells in the developing *Drosophila* eye. *Development* **115**, 89–101.

1242. Fisher, A. and Caudy, M. (1998). The function of hairy-related bHLH repressor proteins in cell fate decisions. *BioEssays* **20**, 298–306.

1243. Fisher, A.L. and Caudy, M. (1998). Groucho proteins: transcriptional corepressors for specific subsets of DNA-binding transcription factors in vertebrates and invertebrates. *Genes Dev.* **12**, 1931–1940.

1244. Fisher, A.L., Ohsako, S., and Caudy, M. (1996). The WRPW motif of the Hairy-related basic helix-loop-helix repressor proteins acts as a 4-amino-acid transcription repression and protein-protein interaction domain. *Mol. Cell. Biol.* **16**, 2670–2677.

1245. Fisher, F. and Goding, C.R. (1992). Single amino acid substitutions alter helix-loop-helix protein specificity for bases flanking the core CANNTG motif. *EMBO J.* **11**, 4103–4109.

1246. Fitzpatrick, C.A., Sharkov, N., and Katzen, A.L. (2001). *Drosophila myb* may play a role in regulating differentiation as well as proliferation during development. *Proc. 42nd Ann. Drosophila Res. Conf.* **Abstracts Vol.**, a64.

1247. Fjose, A., McGinnis, W.J., and Gehring, W.J. (1985). Isolation of a homoeo box-containing gene from the *engrailed* region of *Drosophila* and the spatial distribution of its transcripts. *Nature* **313**, 284–289.

1248. Flegel, W.A., Singson, A.W., Margolis, J.S., Bang, A.G., Posakony, J.W., and Murre, C. (1993). *Dpbx*, a new homeobox gene closely related to the human proto-oncogene *pbx1*: molecular structure and developmental expression. *Mechs. Dev.* **41**, 155–161.

1249. Fleming, R.J. (1998). Structural conservation of Notch receptors and ligands. *Sems. Cell Dev. Biol.* **9**, 599–607.

1250. Fleming, R.J., Gu, Y., and Hukriede, N.A. (1997). *Serrate*-mediated activation of *Notch* is specifically blocked by the product of the gene *fringe* in the dorsal compartment of the *Drosophila* wing imaginal disc. *Development* **124**, 2973–2981.

1251. Fleming, R.J., Purcell, K., and Artavanis-Tsakonas, S. (1997). The NOTCH receptor and its ligands. *Trends Cell Biol.* **7**, 437–441.

1252. Fleming, R.J., Scottgale, T.N., Diederich, R.J., and Artavanis-Tsakonas, S. (1990). The gene *Serrate* encodes a putative EGF-like transmembrane protein essential for proper ectodermal development in *Drosophila melanogaster. Genes Dev.* **4**, 2188–2201.

1253. Flenniken, A.M., Gale, N.W., Yancopoulos, G.D., and Wilkinson, D.G. (1996). Distinct and overlapping expression patterns of ligands for Eph-related receptor tyrosine kinases during mouse embryogenesis. *Dev. Biol.* **179**, 382–401.

1254. Fletcher, J.C. and Thummel, C.S. (1995). The ecdysone-inducible *Broad-Complex* and *E74* early genes interact to regulate target gene transcription and *Drosophila* metamorphosis. *Genetics* **141**, 1025–1035.

1255. Flores, G.V., Daga, A., Kalhor, H.R., and Banerjee, U. (1998). Lozenge is expressed in pluripotent precursor cells and patterns multiple cell types in the *Drosophila* eye through the control of cell-specific transcription factors. *Development* **125**, 3681–3687.

1256. Flores, G.V., Duan, H., Yan, H., Nagaraj, R., Fu, W., Zou, Y., Noll, M., and Banerjee, U. (2000). Combinatorial signaling in the specification of unique cell fates. *Cell* **103**, 75–85.

1257. Flores-Saaib, R.D., Jia, S., and Courey, A.J. (2001). Activation and repression by the C-terminal domain of Dorsal. *Development* **128**, 1869–1879.

1258. Foe, V.E. (1989). Mitotic domains reveal early commitment of cells in *Drosophila* embryos. *Development* **107**, 1–22.

1259. Foe, V.E., Odell, G.M., and Edgar, B.A. (1993). Mitosis and morphogenesis in the *Drosophila* embryo: point and counterpoint. *In* "The Development of *Drosophila melanogaster*," Vol. 1 (M. Bate and A. Martinez Arias, eds.). Cold Spring Harbor Lab. Pr.: Plainview, N. Y., pp. 149–300.

1260. Foelix, R.F., Stocker, R.F., and Steinbrecht, R.A. (1989). Fine structure of a sensory organ in the arista of *Drosophila melanogaster* and some other dipterans. *Cell Tissue Res.* **258**, 277–287.

1261. Follette, P.J., Duronio, R.J., and O'Farrell, P.H. (1998). Fluctuations in Cyclin E levels are required for multiple rounds of endocycle S phase in *Drosophila. Curr. Biol.* **8**, 235–238.

1262. Follette, P.J. and O'Farrell, P.H. (1997). Cdks and the *Drosophila* cell cycle. *Curr. Opin. Gen. Dev.* **7**, 17–22.

1263. Follette, P.J. and O'Farrell, P.H. (1997). Connecting cell behavior to patterning: lessons from the cell cycle. *Cell* **88**, 309–314.

1264. Forbes, A.J., Nakano, Y., Taylor, A.M., and Ingham, P.W. (1993). Genetic analysis of *hedgehog* signalling in the *Drosophila* embryo. *Development* **1993 Suppl.**, 115–124.

1265. Force, A., Lynch, M., Pickett, F.B., Amores, A., Yan, Y.-l., and Postlethwait, J. (1999). Preservation of duplicate genes by complementary, degenerate mutations. *Genetics* **151**, 1531–1545.

1266. Ford, B.J. (1998). The earliest views. *Sci. Am.* **278** #4, 50–53.

1267. Fornerod, M., Ohno, M., Yoshida, M., and Mattaj, I.W. (1997). CRM1 is an export receptor for leucine-rich nuclear export signals. *Cell* **90**, 1051–1060.

1268. Fortini, M. (2000). Fringe benefits to carbohydrates. *Nature* **406**, 357–358.

1269. Fortini, M.E. and Artavanis-Tsakonas, S. (1994). The Suppressor of Hairless protein participates in Notch receptor signaling. *Cell* **79**, 273–282.

1270. Fortini, M.E. and Bonini, N.M. (2000). Modeling human neurodegenerative diseases in *Drosophila*. *Trends Genet.* **16**, 161–167.

1271. Fortini, M.E., Rebay, I., Caron, L.A., and Artavanis-Tsakonas, S. (1993). An activated Notch receptor blocks cell-fate commitment in the developing *Drosophila* eye. *Nature* **365**, 555–557.

1272. Fortini, M.E. and Rubin, G.M. (1990). Analysis of *cis*-acting requirements of the *Rh3* and *Rh4* genes reveals a bipartite organization to the rhodopsin promoters in *Drosophila melanogaster*. *Genes Dev.* **4**, 444–463.

1273. Fortini, M.E., Simon, M.A., and Rubin, G.M. (1992). Signalling by the *sevenless* protein tyrosine kinase is mimicked by Ras1 activation. *Nature* **355**, 559–561.

1274. Foster, G.G. (1975). Negative complementation at the Notch locus of *Drosophila melanogaster*. *Genetics* **81**, 99–120.

1275. Fostier, M. and Baron, M. (2001). Suppressor of deltex needs deltex to down-regulate the Notch pathway during wing development. *Proc. 42nd Ann. Drosophila Res. Conf.* **Abstracts Vol.**, a34.

1276. Fostier, M., Evans, D.A.P., Artavanis-Tsakonas, S., and Baron, M. (1998). Genetic characterization of the *Drosophila melanogaster Suppressor of deltex* gene: a regulator of Notch signaling. *Genetics* **150**, 1477–1485.

1277. Fradkin, L.G., Noordermeer, J.N., and Nusse, R. (1995). The *Drosophila* Wnt protein DWnt-3 is a secreted glycoprotein localized on the axon tracts of the embryonic CNS. *Dev. Biol.* **168**, 202–213.

1278. François, V., Solloway, M., O'Neill, J.W., Emery, J., and Bier, E. (1994). Dorsal-ventral patterning of the *Drosophila* embryo depends on a putative negative growth factor encoded by the *short gastrulation* gene. *Genes Dev.* **8**, 2602–2616.

1279. Frankfort, B.J., Nolo, R., Zhang, Z., Bellen, H.J., and Mardon, G. (2001). *senseless* repression of *rough* is required for R8 photoreceptor differentiation in the developing *Drosophila* eye. *Neuron* **32**, 403–414.

1280. Franks, R.G. and Crews, S.T. (1994). Transcriptional activation domains of the single-minded bHLH protein are required for CNS midline cell development. *Mechs. Dev.* **45**, 269–277.

1281. Frasch, M. (1995). Induction of visceral and cardiac mesoderm by ectodermal Dpp in the early *Drosophila* embryo. *Nature* **374**, 464–467.

1282. Fraser, A. (1963). Variation of scutellar bristles in *Drosophila*. I. Genetic leakage. *Genetics* **48**, 497–514.

1283. Fraser, A.G., McCarthy, N.J., and Evan, G.I. (1997). drICE is an essential caspase required for apoptotic activity in *Drosophila* cells. *EMBO J.* **16**, 6192–6199.

1284. Fraser, S.E. and Harland, R.M. (2000). The molecular metamorphosis of experimental embryology. *Cell* **100**, 41–55.

1285. Frayne, E.G. and Sato, T. (1991). The *Ultrabithorax* gene of *Drosophila* and the specification of abdominal histoblasts. *Dev. Biol.* **146**, 265–277.

1286. Freeman, M. (1991). First, trap your enhancer. *Curr. Biol.* **1**, 378–381.

1287. Freeman, M. (1994). Misexpression of the *Drosophila argos* gene, a secreted regulator of cell determination. *Development* **120**, 2297–2304.

1288. Freeman, M. (1994). The *spitz* gene is required for photoreceptor determination in the *Drosophila* eye where it interacts with the EGF receptor. *Mechs. Dev.* **48**, 25–33.

1289. Freeman, M. (1996). Reiterative use of the EGF receptor triggers differentiation of all cell types in the *Drosophila* eye. *Cell* **87**, 651–660.

1290. Freeman, M. (1997). Cell determination strategies in the *Drosophila* eye. *Development* **124**, 261–270. [See also 2001 *Mechs. Dev.* **108**, 13–27.]

1291. Freeman, M. (1998). Complexity of EGF receptor signalling revealed in *Drosophila*. *Curr. Opin. Gen. Dev.* **8**, 407–411.

1292. Freeman, M. (2000). Feedback control of intercellular signalling in development. *Nature* **408**, 313–319. [See also 2001 *EMBO J.* **20**, 2528–2535.]

1293. Freeman, M., Kimmel, B.E., and Rubin, G.M. (1992). Identifying targets of the *rough* homeobox gene of *Drosophila*: evidence that *rhomboid* functions in eye development. *Development* **116**, 335–346.

1294. Freeman, M., Klämbt, C., Goodman, C.S., and Rubin, G.M. (1992). The *argos* gene encodes a diffusible factor that regulates cell fate decisions in the *Drosophila* eye. *Cell* **69**, 963–975.

1295. Freemont, P.S. (1993). The RING finger: a novel protein sequence motif related to the zinc finger. *Annals N. Y. Acad. Sci.* **684**, 174–192.

1296. Freigang, J., Proba, K., Leder, L., Diederichs, K., Sonderegger, P., and Welte, W. (2000). The crystal structure of the ligand binding module of Axonin-1/TAG-1 suggests a zipper mechanism for neural cell adhesion. *Cell* **101**, 425–433.

1297. French, V. (1980). Positional information around the segments of the cockroach leg. *J. Embryol. Exp. Morph.* **59**, 281–313.

1298. French, V. (1981). Pattern regulation and regeneration. *Phil. Trans. Roy. Soc. Lond. B* **295**, 601–617.

1299. French, V. (1982). Leg regeneration in insects: cell interactions and lineage. *Amer. Zool.* **22**, 79–90.

1300. French, V. (1983). Development and evolution of the insect segment. *In* "Development and Evolution," *Symp. Brit. Soc. Dev. Biol.*, Vol. 6 (B.C. Goodwin, N. Holder, and C.C. Wylie, eds.). Cambridge Univ. Pr.: Cambridge, pp. 161–193.

1301. French, V. (1984). A model of insect limb regeneration. *In* "Pattern Formation: A Primer in Developmental Biology," (G.M. Malacinski and S.V. Bryant, eds.). Macmillan: New York, pp. 339–364.

1302. French, V. (1984). The structure of supernumerary leg regenerates in the cricket. *J. Embryol. Exp. Morph.* **81**, 185–209.

1303. French, V., Bryant, P.J., and Bryant, S.V. (1976). Pattern regulation in epimorphic fields. *Science* **193**, 969–981.

1304. French, V. and Daniels, G. (1994). The beginning and the end of insect limbs. *Curr. Biol.* **4**, 34–37.

1305. Friedrich, M., Rambold, I., and Melzer, R.R. (1996). The early stages of ommatidial development in the flour beetle *Tribolium castaneum* (Coleoptera; Tenebrionidae). *Dev. Genes Evol.* **206**, 136–146.

1306. Frigerio, G., Burri, M., Bopp, D., Baumgartner, S., and Noll, M. (1986). Structure of the segmentation gene *paired* and the *Drosophila* PRD gene set as part of a gene network. *Cell* **47**, 735–746.

1307. Frise, E., Knoblich, J.A., Younger-Shepherd, S., Jan, L.Y., and Jan, Y.N. (1996). The *Drosophila* Numb protein inhibits signaling of the Notch receptor during cell-cell interaction in sensory organ lineage. *Proc. Natl. Acad. Sci. USA* **93**, 11925–11932.

1308. Frisén, J. and Lendahl, U. (2001). Oh no, Notch again! *BioEssays* **23**, 3–7.

1309. Fristrom, D. (1969). Cellular degeneration in the production of some mutant phenotypes in *Drosophila melanogaster. Mol. Gen. Genet.* **103**, 363–379.

1310. Fristrom, D. (1972). Chemical modification of cell death in the *Bar* eye of *Drosophila. Molec. Gen. Genet.* **115**, 10–18.

1311. Fristrom, D. and Fristrom, J.W. (1993). The metamorphic development of the adult epidermis. *In* "The Development of *Drosophila melanogaster*," Vol. 2 (M. Bate and A. Martinez Arias, eds.). Cold Spring Harbor Lab. Pr.: Plainview, N. Y., pp. 843–897.

1312. Fristrom, D., Gotwals, P., Eaton, S., Kornberg, T.B., Sturtevant, M., Bier, E., and Fristrom, J. (1994). *blistered*: a gene required for vein/intervein formation in wings of *Drosophila. Development* **120**, 2661–2671.

1313. Fristrom, D., Wilcox, M., and Fristrom, J. (1993). The distribution of PS integrins, laminin A and F-actin during key stages in *Drosophila* wing development. *Development* **117**, 509–523.

1314. Fristrom, J.W. (1970). The developmental biology of *Drosophila. Annu. Rev. Genet.* **4**, 325–346.

1315. Fristrom, J.W. (1972). The biochemistry of imaginal disk development. *In* "The Biology of Imaginal Disks," *Results and Problems in Cell Differentiation, Vol. 5,* (H. Ursprung and R. Nöthiger, eds.). Springer-Verlag: Berlin, pp. 109–154.

1316. Fritz, C.C. and Green, M.R. (1992). Fishing for partners. *Curr. Biol.* **2**, 403–405.

1317. Fry, C.J. and Peterson, C.L. (2001). Chromatin remodeling enzymes: who's on first? *Curr. Biol.* **11**, R185–R197.

1318. Fryxell, K.J. and Wood, C.P. (1995). Genetic mosaic analysis of the *equatorial-less* mutation in *Drosophila melanogaster. Dev. Genet.* **16**, 264–272.

1319. Fu, W., Duan, H., Frei, E., and Noll, M. (1998). *shaven* and *sparkling* are mutations in separate enhancers of the *Drosophila Pax2* homolog. *Development* **125**, 2943–2950.

1320. Fu, W. and Noll, M. (1997). The *Pax2* homolog *sparkling* is required for development of cone and pigment cells in the *Drosophila* eye. *Genes Dev.* **11**, 2066–2078.

1321. Fuerstenberg, S., Broadus, J., and Doe, C.Q. (1998). Asymmetry and cell fate in the *Drosophila* embryonic CNS. *Int. J. Dev. Biol.* **42**, 379–383.

1322. Fuerstenberg, S., Peng, C.-Y., Alvarez-Ortiz, P., Hor, T., and Doe, C.Q. (1998). Identification of Miranda protein domains regulating asymmetric cortical localization, cargo binding, and cortical release. *Molec. Cell. Neurosci.* **12**, 325–339.

1323. Fujioka, M., Emi-Sarker, Y., Yusibova, G.L., Goto, T., and Jaynes, J.B. (1999). Analysis of an *even-skipped* rescue transgene reveals both composite and discrete neuronal and early blastoderm enhancers, and multi-stripe positioning by gap gene repressor gradients. *Development* **126**, 2527–2538.

1324. Fujioka, M., Jaynes, J.B., and Goto, T. (1995). Early *even-skipped* stripes act as morphogenetic gradients at the sin-

gle cell level to establish *engrailed* expression. *Development* **121**, 4371–4382.

1325. Fujioka, M., Miskiewicz, P., Raj, L., Gulledge, A.A., Weir, M., and Goto, T. (1996). *Drosophila* Paired regulates late *even-skipped* expression through a composite binding site for the paired domain and the homeodomain. *Development* **122**, 2697–2707.

1326. Fuks, F., Burgers, W.A., Brehm, A., Hughes-Davies, L., and Kouzarides, T. (2000). DNA methyltransferase Dnmt1 associates with histone deacetylase activity. *Nature Genet.* **24**, 88–91.

1327. Funakoshi, Y., Minami, M., and Tabata, T. (2001). *mtv* shapes the activity gradient of the Dpp morphogen through regulation of *thickveins. Development* **128**, 67–74.

1328. Furman, D.P., Rodin, S.N., and Ratner, V.A. (1979). Structural and functional analysis of the scute locus in *Drosophila melanogaster. Theor. Appl. Genet.* **55**, 231–238.

1329. Furriols, M. and Bray, S. (2000). Dissecting the mechanisms of Suppressor of Hairless function. *Dev. Biol.* **227**, 520–532.

1330. Furriols, M. and Bray, S. (2001). A model Notch response element detects Suppressor of Hairless-dependent molecular switch. *Curr. Biol.* **11**, 60–64.

1331. Furukawa, T., Kawaichi, M., Matsunami, N., Ryo, H., Nishida, Y., and Honjo, T. (1991). The *Drosophila* RBP-Jκ gene encodes the binding protein for the immunoglobulin Jκ recombination signal sequence. *J. Biol. Chem.* **266** #34, 23334–23340.

1332. Furukawa, T., Kobayakawa, Y., Tamura, K., Kimura, K.-I., Kawaichi, M., Tanimura, T., and Honjo, T. (1995). Suppressor of Hairless, the *Drosophila* homologue of RBP-Jκ, transactivates the neurogenic gene *E(spl)m8. Jpn. J. Genet.* **70**, 505–524.

1333. Fuse, N., Hirose, S., and Hayashi, S. (1996). Determination of wing cell fate by the *escargot* and *snail* genes in *Drosophila. Development* **122**, 1059–1067.

1334. Fuse, N., Matakatsu, H., Taniguchi, M., and Hayashi, S. (1999). Snail-type zinc finger proteins prevent neurogenesis in *Scutoid* and transgenic animals of *Drosophila. Dev. Genes Evol.* **209**, 573–580.

1335. Fuss, B. and Hoch, M. (1998). *Drosophila* endoderm development requires a novel homeobox gene which is a target of Wingless and Dpp signalling. *Mechs. Dev.* **79**, 83–97.

1336. Fuss, B., Meissner, T., Bauer, R., Lehmann, C., Eckardt, F., and Hoch, M. (2001). Control of endoreduplication domains in the *Drosophila* gut by the *knirps* and *knirps-related* genes. *Mechs. Dev.* **100**, 15–23.

1337. Gabay, L., Seger, R., and Shilo, B.-Z. (1997). In situ activation pattern of *Drosophila* EGF receptor pathway during development. *Science* **277**, 1103–1106.

1338. Gabay, L., Seger, R., and Shilo, B.-Z. (1997). MAP kinase in situ activation atlas during *Drosophila* embryogenesis. *Development* **124**, 3535–3541.

1339. Gajewski, K., Choi, C.Y., Kim, Y., and Schulz, R.A. (2000). Genetically distinct cardial cells within the *Drosophila* heart. *Genesis* **28**, 36–43.

1340. Galant, R., Skeath, J.B., Paddock, S., Lewis, D.L., and Carroll, S.B. (1998). Expression pattern of a butterfly *achaete-scute* homolog reveals the homology of butterfly wing scales and insect sensory bristles. *Curr. Biol.* **8**, 807–813.

1341. Galindo, M.I. and Couso, J.P. (2000). Intercalation of cell fates during tarsal development in *Drosophila. BioEssays* **22**, 777–780.

1342. Gallant, P., Li, C., Montero, L., Johnston, L.A., and Eisenman, R.N. (2001). Genetic dissection of Myc function in *Drosophila. Proc. 42nd Ann. Drosophila Res. Conf.* **Abstracts Vol.**, a82.

1343. Gallant, P., Shiio, Y., Cheng, P.F., Parkhurst, S.M., and Eisenman, R.N. (1996). Myc and Max homologs in *Drosophila. Science* **274**, 1523–1527.

1344. Gallet, A., Angelats, C., Erkner, A., Charroux, B., Fasano, L., and Kerridge, S. (1999). The C-terminal domain of Armadillo binds to hypophosphorylated Teashirt to modulate Wingless signalling in *Drosophila. EMBO J.* **18**, 2208–2217.

1345. Gallet, A., Erkner, A., Charroux, B., Fasano, L., and Kerridge, S. (1998). Trunk-specific modulation of Wingless signalling in *Drosophila* by Teashirt binding to Armadillo. *Curr. Biol.* **8**, 893–902.

1346. Galloni, M. and Edgar, B.A. (1999). Cell-autonomous and non-autonomous growth-defective mutants of *Drosophila melanogaster. Development* **126**, 2365–2375.

1347. Ganfornina, M.D. and Sánchez, D. (1999). Generation of evolutionary novelty by functional shift. *BioEssays* **21**, 432–439.

1348. García Alonso, L. and García-Bellido, A. (1986). Genetic analysis of *Hairy-wing* mutations. *Roux's Arch. Dev. Biol.* **195**, 259–264.

1349. García Alonso, L.A. and García-Bellido, A. (1988). *Extramacrochaetae*, a *trans*-acting gene of the *achaete-scute* complex of *Drosophila* involved in cell communication. *Roux's Arch. Dev. Biol.* **197**, 328–338.

1350. García-Alonso, L., Fetter, R.D., and Goodman, C.S. (1996). Genetic analysis of *Laminin A* in *Drosophila*: extracellular matrix containing laminin A is required for ocellar axon pathfinding. *Development* **122**, 2611–2621.

1351. García-Alonso, L., VanBerkum, M.F.A., Grenningloh, G., Schuster, C., and Goodman, C.S. (1995). Fasciclin II controls proneural gene expression in *Drosophila. Proc. Natl. Acad. Sci. USA* **92**, 10501–10505.

1352. García-Añoveros, J. and Corey, D.P. (1997). The molecules of mechanosensation. *Annu. Rev. Neurosci.* **20**, 567–594.

1353. Garcia-Bellido, A. (1966). Changes in selective affinity following transdetermination in imaginal disc cells of *Drosophila melanogaster. Exp. Cell Res.* **44**, 382–392.

1354. Garcia-Bellido, A. (1981). From the gene to the pattern: chaeta differentiation. *In* "Cellular Controls in Differentiation," (C.W. Lloyd and D.A. Rees, eds.). Acad. Pr.: New York, pp. 281–304.

1355. Garcia-Bellido, A. (1998). The engrailed story. *Genetics* **148**, 539–544.

1356. García-Bellido, A. (1966). Pattern reconstruction by dissociated imaginal disk cells of *Drosophila melanogaster. Dev. Biol.* **14**, 278–306.

1357. García-Bellido, A. (1972). Pattern formation in imaginal disks. *In* "The Biology of Imaginal Disks," *Results and Problems in Cell Differentiation, Vol. 5*, (H. Ursprung and R. Nöthiger, eds.). Springer-Verlag: Berlin, pp. 59–91.

1358. García-Bellido, A. (1975). Genetic control of wing disc development in *Drosophila. In* "Cell Patterning," *Ciba Found. Symp.*, Vol. 29 (R. Porter and J. Rivers, eds.). Elsevier: Amsterdam, pp. 161–182.

1359. García-Bellido, A. (1977). Inductive mechanisms in the process of wing vein formation in *Drosophila. W. Roux's Arch.* **182**, 93–106.

1360. García-Bellido, A. (1979). Genetic analysis of the achaete-scute system of *Drosophila melanogaster. Genetics* **91**, 491–520.

1361. García-Bellido, A. (1983). Comparative anatomy of cuticular patterns in the genus *Drosophila. In* "Development and Evolution," *Symp. Brit. Soc. Dev. Biol.*, Vol. 6 (B.C. Goodwin, N. Holder, and C.C. Wylie, eds.). Cambridge Univ. Pr.: Cambridge, pp. 227–255.

1362. García-Bellido, A. (1985). Cell lineages and genes. *Phil. Trans. Roy. Soc. Lond. B* **312**, 101–128.

1363. García-Bellido, A. (1986). Genetic analysis of morphogenesis. *In* "Genetics, Development, and Evolution," (J.P. Gustafson, G.L. Stebbins, and F.J. Ayala, eds.). Plenum Pr.: New York, pp. 187–209.

1364. García-Bellido, A. (1993). Coming of age. *Trends Genet.* **9**, 102–103.

1365. García-Bellido, A. and Capdevila, M.P. (1978). The initiation and maintenance of gene activity in a developmental pathway of *Drosophila. In* "The Clonal Basis of Development," *Symp. Soc. Dev. Biol.*, Vol. 36 (S. Subtelny and I.M. Sussex, eds.). Acad. Pr.: New York, pp. 3–21.

1366. García-Bellido, A., Cortés, F., and Milán, M. (1994). Cell interactions in the control of size in *Drosophila* wings. *Proc. Natl. Acad. Sci. USA* **91**, 10222–10226.

1367. Garcia-Bellido, A. and Dapena, J. (1974). Induction, detection and characterization of cell differentiation mutants in *Drosophila. Molec. Gen. Genet.* **128**, 117–130.

1368. García-Bellido, A. and de Celis, J.F. (1992). Developmental genetics of the venation pattern of *Drosophila. Annu. Rev. Genet.* **26**, 277–304.

1369. García-Bellido, A., Lawrence, P.A., and Morata, G. (1979). Compartments in animal development. *Sci. Am.* **241** #1, 102–110.

1370. Garcia-Bellido, A. and Merriam, J.R. (1969). Cell lineage of the imaginal discs in *Drosophila* gynandromorphs. *J. Exp. Zool.* **170**, 61–76.

1371. Garcia-Bellido, A. and Merriam, J.R. (1971). Clonal parameters of tergite development in *Drosophila. Dev. Biol.* **26**, 264–276.

1372. Garcia-Bellido, A. and Merriam, J.R. (1971). Genetic analysis of cell heredity in imaginal discs of *Drosophila melanogaster. Proc. Natl. Acad. Sci. USA* **68**, 2222–2226.

1373. Garcia-Bellido, A. and Merriam, J.R. (1971). Parameters of the wing imaginal disc development of *Drosophila melanogaster. Dev. Biol.* **24**, 61–87.

1374. Garcia-Bellido, A. and Nöthiger, R. (1976). Maintenance of determination by cells of imaginal discs of *Drosophila* after dissociation and culture in vivo. *W. Roux's Arch.* **180**, 189–206.

1375. Garcia-Bellido, A. and Ripoll, P. (1978). Cell lineage and differentiation in *Drosophila. In* "Genetic Mosaics and Cell Differentiation," *Results and Problems in Cell Differentiation, Vol. 9*, (W.J. Gehring, ed.). Springer-Verlag: Berlin, pp. 119–156.

1376. Garcia-Bellido, A., Ripoll, P., and Morata, G. (1973). Developmental compartmentalisation of the wing disk of *Drosophila. Nature New Biol.* **245**, 251–253.

1377. Garcia-Bellido, A., Ripoll, P., and Morata, G. (1976). Developmental compartmentalization in the dorsal mesothoracic disc of *Drosophila. Dev. Biol.* **48**, 132–147.

1378. Garcia-Bellido, A. and Santamaria, P. (1972). Developmental analysis of the wing disc in the mutant *engrailed* of *Drosophila melanogaster. Genetics* **72**, 87–104.

1379. García-Bellido, A. and Santamaria, P. (1978). Developmental analysis of the achaete-scute system of *Drosophila melanogaster*. *Genetics* **88**, 469–486.

1380. García-García, M.J., Ramain, P., Simpson, P., and Modolell, J. (1999). Different contributions of *pannier* and *wingless* to the patterning of the dorsal mesothorax of *Drosophila*. *Development* **126**, 3523–3532.

1381. Gardner, M. (1971). On cellular automata, self-reproduction, the Garden of Eden and the game "life". *Sci. Am.* **224** #2, 112–117.

1382. Gardner, M. (1983). "Wheels, Life and Other Mathematical Amusements." W. H. Freeman, New York.

1383. Gardner, T.S. and Collins, J.J. (2000). Neutralizing noise in gene networks. *Nature* **405**, 520–521.

1384. Garen, A. (1992). Looking for the homunculus in *Drosophila*. *Genetics* **131**, 5–7.

1385. Garner, C.C., Nash, J., and Huganir, R.L. (2000). PDZ domains in synapse assembly and signalling. *Trends Cell Biol.* **10**, 274–280.

1386. Garoia, F., Guerra, D., Pezzoli, M.C., López-Varea, A., Cavicchi, S., and García-Bellido, A. (2000). Cell behaviour of *Drosophila fat* cadherin mutations in wing development. *Mechs. Dev.* **94**, 95–109.

1387. Garrell, J. and Campuzano, S. (1991). The helix-loop-helix domain: a common motif for bristles, muscles, and sex. *BioEssays* **13**, 493–498.

1388. Garrell, J. and Modolell, J. (1990). The *Drosophila extramacrochaetae* locus, an antagonist of proneural genes that, like these genes, encodes a helix-loop-helix protein. *Cell* **61**, 39–48.

1389. Garrington, T.P. and Johnson, G.L. (1999). Organization and regulation of mitogen-activated protein kinase signaling pathways. *Curr. Opin. Cell Biol.* **11**, 211–218.

1390. Garrity, P.A., Lee, C.-H., Salecker, I., Robertson, H.C., Desai, C.J., Zinn, K., and Zipursky, S.L. (1999). Retinal axon target selection in *Drosophila* is regulated by a receptor protein tyrosine phosphatase. *Neuron* **22**, 707–717.

1391. Garrity, P.A., Rao, Y., Salecker, I., McGlade, J., Pawson, T., and Zipursky, S.L. (1996). *Drosophila* photoreceptor axon guidance and targeting requires the Dreadlocks SH2/SH3 adapter protein. *Cell* **85**, 639–650.

1392. Garrity, P.A. and Zipursky, S.L. (1995). Neuronal target recognition. *Cell* **83**, 177–185.

1393. Gasser, S.M. (2001). Positions of potential: nuclear organization and gene expression. *Cell* **104**, 639–642.

1394. Gateff, E. (1978). Malignant and benign neoplasms of *Drosophila melanogaster*. *In* "The Genetics and Biology of *Drosophila*," Vol. 2b (M. Ashburner and T.R.F. Wright, eds.). Acad. Pr.: New York, pp. 181–275.

1395. Gateff, E. (1978). Malignant neoplasms of genetic origin in *Drosophila melanogaster*. *Science* **200**, 1448–1459.

1396. Gateff, E. and Schneiderman, H.A. (1974). Developmental capacities of benign and malignant neoplasms of *Drosophila*. *W. Roux's Arch.* **176**, 23–65.

1397. Gates, J. and Thummel, C.S. (2000). An enhancer trap screen for ecdysone-inducible genes required for *Drosophila* adult leg morphogenesis. *Genetics* **156**, 1765–1776.

1398. Gaul, U. and Jäckle, H. (1990). Role of gap genes in early *Drosophila* development. *Adv. Genet.* **27**, 239–275.

1399. Gaul, U., Mardon, G., and Rubin, G.M. (1992). A putative Ras GTPase activating protein acts as a negative regulator of signaling by the Sevenless receptor tyrosine kinase. *Cell* **68**, 1007–1019.

1400. Gaunt, S.J. and Singh, P.B. (1990). Homeogene expression patterns and chromosomal imprinting. *Trends Genet.* **6**, 208–212.

1401. Gause, M., Hovhannisyan, H., Kan, T., Kuhfittig, S., Mogila, V., and Georgiev, P. (1998). *hobo* induced rearrangements in the *yellow* locus influence the insulation effect of the *gypsy* su(Hw)-binding region in *Drosophila melanogaster*. *Genetics* **149**, 1393–1405.

1402. Gautier, P., Ledent, V., Massaer, M., Dambly-Chaudière, C., and Ghysen, A. (1997). *tap*, a *Drosophila* bHLH gene expressed in chemosensory organs. *Gene* **191**, 15–21.

1403. Gayko, U., Cleghon, V., Copeland, T., Morrison, D.K., and Perrimon, N. (1999). Synergistic activities of multiple phosphotyrosine residues mediate full signaling from the *Drosophila* Torso receptor tyrosine kinase. *Proc. Natl. Acad. Sci. USA* **96**, 523–528.

1404. Gehring, W. (1966). Übertragung und Änderung der Determinationsqualitäten in Antennensheiben-Kulturen von *Drosophila melanogaster*. *J. Embryol. Exp. Morph.* **15**, 77–111.

1405. Gehring, W. (1967). Clonal analysis of determination dynamics in cultures of imaginal disks in *Drosophila melanogaster*. *Dev. Biol.* **16**, 438–456.

1406. Gehring, W. (1972). The stability of the determined state in cultures of imaginal disks in *Drosophila*. *In* "The Biology of Imaginal Disks," *Results and Problems in Cell Differentiation, Vol. 5*, (H. Ursprung and R. Nöthiger, eds.). Springer-Verlag: Berlin, pp. 35–58.

1407. Gehring, W. (2000). Reply to Meyer-Rochow. *Trends Genet.* **16**, 245.

1408. Gehring, W. and Seippel, S. (1967). Die Imaginalzellen des Clypeo-Labrums und die Bildung des Rüssels von *Drosophila melanogaster*. *Rev. Suisse Zool.* **74**, 589–596.

1409. Gehring, W.J. (1976). Developmental genetics of *Drosophila*. *Annu. Rev. Genet.* **10**, 209–252.

1410. Gehring, W.J., ed. (1978). "Genetic Mosaics and Cell Differentiation." *Results and Problems in Cell Differentiation, Vol. 9*, Springer-Verlag, Berlin.

1411. Gehring, W.J. (1978). Imaginal discs: determination. *In* "The Genetics and Biology of *Drosophila*," Vol. 2c (M. Ashburner and T.R.F. Wright, eds.). Acad. Pr.: New York, pp. 511–554.

1412. Gehring, W.J. (1985). Homeotic genes, the homeo box, and the genetic control of development. *Cold Spr. Harb. Symp. Quant. Biol.* **50**, 243–251.

1413. Gehring, W.J. (1985). The molecular basis of development. *Sci. Am.* **253** #4, 153–162.

1414. Gehring, W.J. (1987). Homeo boxes in the study of development. *Science* **236**, 1245–1252.

1415. Gehring, W.J. (1996). The master control gene for morphogenesis and evolution of the eye. *Genes to Cells* **1**, 11–15.

1416. Gehring, W.J. (1998). "Master Control Genes in Development and Evolution: The Homeobox Story." Yale Univ. Pr., New Haven. [See also 2001 *BioEssays* **23**, 763–766.]

1417. Gehring, W.J., Affolter, M., and Bürglin, T. (1994). Homeodomain proteins. *Annu. Rev. Biochem.* **63**, 487–526.

1418. Gehring, W.J. and Hiromi, Y. (1986). Homeotic genes and the homeobox. *Annu. Rev. Genet.* **20**, 147–173.

1419. Gehring, W.J. and Ikeo, K. (1999). *Pax6*: mastering eye morphogenesis and eye evolution. *Trends Genet.* **15**, 371–377.

1420. Gehring, W.J., Müller, M., Affolter, M., Percival-Smith, A., Billeter, M., Qian, Y.Q., Otting, G., and Wüthrich, K. (1990). The structure of the homeodomain and its functional implications. *Trends Genet.* **6**, 323–329.

1421. Gehring, W.J. and Nöthiger, R. (1973). The imaginal discs of *Drosophila*. *In* "Developmental Systems: Insects," Vol. 2 (S.J. Counce and C.H. Waddington, eds.). Acad. Pr.: New York, pp. 211–290.

1422. Gehring, W.J., Qian, Y.Q., Billeter, M., Furukubo-Tokunaga, K., Schier, A.F., Resendez-Perez, D., Affolter, M., Otting, G., and Wüthrich, K. (1994). Homeodomain-DNA recognition. *Cell* **78**, 211–223.

1423. Gehring, W.J. and Schubiger, G. (1975). Expression of homeotic mutations in duplicated and regenerated antennae of *Drosophila melanogaster*. *J. Embryol. Exp. Morph.* **33**, 459–469.

1424. Geigy, R. (1931). Erzeugung rein imaginaler Defekte durch ultraviolette Eibestrahlung bei *Drosophila melanogaster*. *Roux' Arch. Entw. Mech.* **125**, 406–447.

1425. Gekakis, N., Saez, L., Delahaye-Brown, A.-M., Myers, M.P., Sehgal, A., Young, M.W., and Weitz, C.J. (1995). Isolation of *timeless* by PER protein interaction: defective interaction between *timeless* protein and long-period mutant PER. *Science* **270**, 811–815.

1426. Gelbart, W.M. (1982). Synapsis-dependent allelic complementation at the decapentaplegic gene complex in *Drosophila melanogaster*. *Proc. Natl. Acad. Sci. USA* **79**, 2636–2640.

1427. Gelbart, W.M. (1989). The *decapentaplegic* gene: a TGF-β homologue controlling pattern formation in *Drosophila*. *Development* **107 (Suppl.)**, 65–74.

1428. Gelbart, W.M., Irish, V.F., St. Johnston, R.D., Hoffmann, F.M., Blackman, R.K., Segal, D., Posakony, L.M., and Grimaila, R. (1985). The Decapentaplegic Gene Complex in *Drosophila melanogaster*. *Cold Spr. Harb. Symp. Quant. Biol.* **50**, 119–125.

1429. Gellon, G. and McGinnis, W. (1998). Shaping animal body plans in development and evolution by modulation of *Hox* expression patterns. *BioEssays* **20**, 116–125.

1430. Geng, W., He, B., Wang, M., and Adler, P.N. (2000). The *tricornered* gene, which is required for the integrity of epidermal cell extensions, encodes the *Drosophila* nuclear DBF2-related kinase. *Genetics* **156**, 1817–1828.

1431. Genova, J.L., Jong, S., Camp, J.T., and Fehon, R.G. (2000). Functional analysis of *Cdc42* in actin filament assembly, epithelial morphogenesis, and cell signaling during *Drosophila* development. *Dev. Biol.* **221**, 181–194.

1432. Georgakopoulos, A., Marambaud, P., Efthimiopoulos, S., Shioi, J., Cui, W., Li, H.-C., Schütte, M., Gordon, R., Holstein, G.R., Martinelli, G., Mehta, P., Friedrich, V.L., Jr., and Robakis, N.K. (1999). Presenilin-1 forms complexes with the cadherin/catenin cell-cell adhesion system and is recruited to intercellular and synaptic contacts. *Molec. Cell* **4**, 893–902.

1433. George, F.H. (1965). "Cybernetics and Biology." Oliver and Boyd, London.

1434. George, H. and Terracol, R. (1997). The *vrille* gene of *Drosophila* is a maternal enhancer of *decapentaplegic* and encodes a new member of the bZIP family of transcription factors. *Genetics* **146**, 1345–1363.

1435. Georgiev, P., Tikhomirova, T., Yelagin, V., Belenkaya, T., Gracheva, E., Parshikov, A., Evgen'ev, M.B., Samarina, O.P., and Corces, V.G. (1997). Insertions of hybrid *P* elements in the *yellow* gene of *Drosophila* cause a large variety of mutant phenotypes. *Genetics* **146**, 583–594.

1436. Gerasimova, T.I. and Corces, V.G. (1998). Polycomb and trithorax group proteins mediate the function of a chromatin insulator. *Cell* **92**, 511–521.

1437. Gerhardt, M., Schuster, H., and Tyson, J.J. (1990). A cellular automaton model of excitable media including curvature and dispersion. *Science* **247**, 1563–1566.

1438. Gerhart, J. (1989). The primacy of cell interactions in development. *Trends Genet.* **5**, 233–236.

1439. Gerhart, J. (2000). Inversion of the chordate body axis: are there alternatives? *Proc. Natl. Acad. Sci. USA* **97**, 4445–4448.

1440. Gerhart, J. and Kirschner, M. (1997). "Cells, Embryos, and Evolution." Blackwell Science, Malden, Mass.

1441. Geyer, P.K. (1997). The role of insulator elements in defining domains of gene expression. *Curr. Opin. Gen. Dev.* **7**, 242–248.

1442. Geyer, P.K. and Corces, V.G. (1987). Separate regulatory elements are responsible for the complex pattern of tissue-specific and developmental transcription of the *yellow* locus in *Drosophila melanogaster*. *Genes Dev.* **1**, 996–1004.

1443. Geyer, P.K. and Corces, V.G. (1992). DNA position-specific repression of transcription by a *Drosophila* zinc finger protein. *Genes Dev.* **6**, 1865–1873.

1444. Ghazi, A. and VijayRaghavan, K. (2000). Control by combinatorial codes. *Nature* **408**, 419–420.

1445. Ghiglione, C., Carraway, K.L., III, Amundadottir, L.T., Boswell, R.E., Perrimon, N., and Duffy, J.B. (1999). The transmembrane molecule Kekkon 1 acts in a feedback loop to negatively regulate the activity of the *Drosophila* EGF receptor during oogenesis. *Cell* **96**, 847–856.

1446. Ghiglione, C., Perrimon, N., and Perkins, L.A. (1999). Quantitative variations in the level of MAPK activity control patterning of the embryonic termini in *Drosophila*. *Dev. Biol.* **205**, 181–193.

1447. Gho, M., Bellaïche, Y., and Schweisguth, F. (1999). Revisiting the *Drosophila* microchaete lineage: a novel intrinsically asymmetric cell division generates a glial cell. *Development* **126**, 3573–3584.

1448. Gho, M., Lecourtois, M., Géraud, G., Posakony, J.W., and Schweisguth, F. (1996). Subcellular localization of Suppressor of Hairless in *Drosophila* sense organ cells during Notch signaling. *Development* **122**, 1673–1682.

1449. Gho, M. and Schweisguth, F. (1998). Frizzled signalling controls orientation of asymmetric sense organ precursor cell divisions in *Drosophila*. *Nature* **393**, 178–181.

1450. Ghosh, R.N., Chen, Y.-T., DeBiasio, R., DeBiasio, R.L., Conway, B.R., Minor, L.K., and Demarest, K.T. (2000). Cell-based, high-content screen for receptor internalization, recycling and intracellular trafficking. *BioTechniques* **29**, 170–175.

1451. Ghysen, A. (1980). The projection of sensory neurons in the central nervous system of *Drosophila*: choice of the appropriate pathway. *Dev. Biol.* **78**, 521–541.

1452. Ghysen, A., ed. (1998). "Developmental Genetics of *Drosophila*." *Int. J. Dev. Biol.* **42** #3, Univ. Basque Country Pr., Bilbao, Spain.

1453. Ghysen, A. and Dambly-Chaudière, C. (1988). From DNA to form: the *achaete-scute* complex. *Genes Dev.* **2**, 495–501.

1454. Ghysen, A. and Dambly-Chaudière, C. (1989). Genesis of the *Drosophila* peripheral nervous system. *Trends Genet.* **5**, 251–255.

1455. Ghysen, A. and Dambly-Chaudière, C. (1992). Development of the peripheral nervous system in *Drosophila*. *In* "Determinants of Neuronal Identity," (M. Shankland and E.R. Macagno, eds.). Acad. Pr.: New York, pp. 225–292.

1456. Ghysen, A. and Dambly-Chaudière, C. (1993). The specification of sensory neuron identity in *Drosophila*. *BioEssays* **15**, 293–298.

1457. Ghysen, A. and Dambly-Chaudière, C. (2000). A genetic programme for neuronal connectivity. *Trends Genet.* **16**, 221–226.

1458. Ghysen, A., Dambly-Chaudière, C., Jan, L.Y., and Jan, Y.-N. (1993). Cell interactions and gene interactions in peripheral neurogenesis. *Genes Dev.* **7**, 723–733.

1459. Ghysen, A., Dambly-Chaudière, C., Jan, L.Y., and Jan, Y.N. (1982). Segmental differences in the protein content of *Drosophila* imaginal discs. *EMBO J.* **1**, 1373–1379.

1460. Ghysen, A. and Janson, R. (1980). Formation of central patterns by receptor cell axons in *Drosophila*. In "Development and Neurobiology of *Drosophila*," (O. Siddiqi, P. Babu, L.M. Hall, and J.C. Hall, eds.). Plenum: New York, pp. 247–265.

1461. Ghysen, A., Janson, R., and Santamaria, P. (1983). Segmental determination of sensory neurons in *Drosophila*. *Dev. Biol.* **99**, 7–26.

1462. Ghysen, A. and Richelle, J. (1979). Determination of sensory bristles and pattern formation in *Drosophila*. II. The achaete-scute locus. *Dev. Biol.* **70**, 438–452.

1463. Giangrande, A. (1994). Glia in the fly wing are clonally related to epithelial cells and use the nerve as a pathway for migration. *Development* **120**, 523–534.

1464. Giangrande, A. (1995). Proneural genes influence gliogenesis in *Drosophila*. *Development* **121**, 429–438.

1465. Giangrande, A., Murray, M.A., and Palka, J. (1993). Development and organization of glial cells in the peripheral nervous system of *Drosophila melanogaster*. *Development* **117**, 895–904.

1466. Gibson, G. (1999). Redesigning the fruitfly. *Curr. Biol.* **9**, R86–R89.

1467. Gibson, G. (2000). Evolution: *Hox* genes and the cellared wine principle. *Curr. Biol.* **10**, R452–R455.

1468. Gibson, G. and Gehring, W.J. (1988). Head and thoracic transformations caused by ectopic expression of *Antennapedia* during *Drosophila* development. *Development* **102**, 657–675.

1469. Gibson, G. and Hogness, D.S. (1996). Effect of polymorphism in the *Drosophila* regulatory gene *Ultrabithorax* on homeotic stability. *Science* **271**, 200–203.

1470. Gibson, G. and Wagner, G. (2000). Canalization in evolutionary genetics: a stablilizing theory? *BioEssays* **22**, 372–380.

1471. Gibson, J.B., Parsons, P.A., and Spickett, S.G. (1961). Correlations between chaeta number and fly size in *Drosophila melanogaster*. *Heredity* **16**, 349–354.

1472. Gibson, M.C. and Schubiger, G. (1999). Hedgehog is required for activation of *engrailed* during regeneration of fragmented *Drosophila* imaginal discs. *Development* **126**, 1591–1599.

1473. Gibson, M.C. and Schubiger, G. (2000). Peripodial cells regulate proliferation and patterning of *Drosophila* imaginal discs. *Cell* **103**, 343–350. [See also 2001 *BioEssays* **23**, 691–697.]

1474. Gibson, T.J., Hyvönen, M., Musacchio, A., Saraste, M., and Birney, E. (1994). PH domain: the first anniversary. *Trends Biochem. Sci.* **19**, 349–353.

1475. Giebel, B. and Campos-Ortega, J.A. (1997). Functional dissection of the *Drosophila* Enhancer of split protein, a suppressor of neurogenesis. *Proc. Natl. Acad. Sci. USA* **94**, 6250–6254.

1476. Giepmans, B.N.G. and Moolenaar, W.H. (1998). The gap junction protein connexin43 interacts with the second PDZ domain of the zona occludens-1 protein. *Curr. Biol.* **8**, 931–934.

1477. Gierer, A. (1974). Molecular models and combinatorial principles in cell differentiation and morphogenesis. *Cold Spr. Harb. Symp. Quant. Biol.* **38**, 951–961.

1478. Gierer, A. and Meinhardt, H. (1972). A theory of biological pattern formation. *Kybernetik* **12**, 30–39.

1479. Gierer, A. and Meinhardt, H. (1974). Biological pattern formation involving lateral inhibition. In "Lectures on Mathematics in the Life Sciences," Vol. 7 Am. Math. Soc.: Providence, Rhode Island, pp. 163–183.

1480. Giese, K., Cox, J., and Grosschedl, R. (1992). The HMG domain of Lymphoid Enhancer Factor 1 bends DNA and facilitates assembly of functional nucleoprotein structures. *Cell* **69**, 185–195.

1481. Giese, K., Pagel, J., and Grosschedl, R. (1997). Functional analysis of DNA bending and unwinding by the high mobility group domain of LEF-1. *Proc. Natl. Acad. Sci. USA* **94**, 12845–12850.

1482. Gieseler, K., Graba, Y., Mariol, M.-C., Wilder, E.L., Martinez-Arias, A., Lamaire, P., and Pradel, J. (1999). Antagonist activity of *DWnt-4* and *wingless* in the *Drosophila* embryonic ventral ectoderm and in heterologous *Xenopus* assays. *Mechs. Dev.* **85**, 123–131. [See also 2001 *Dev. Biol.* **232**, 339–350.]

1483. Giesen, K., Hummel, T., Stollewerk, A., Harrison, S., Travers, A., and Klämbt, C. (1997). Glial development in the *Drosophila* CNS requires concomitant activation of glial and repression of neuronal differentiation genes. *Development* **124**, 2307–2311.

1484. Gigliani, F., Longo, F., Gaddini, L., and Battaglia, P.A. (1996). Interactions among the bHLH domains of the proteins encoded by the *Enhancer of split* and *achaete-scute* gene complexes of *Drosophila*. *Mol. Gen. Genet.* **251**, 628–634.

1485. Gilbert, L.I. (1994). Howard A. Schneiderman (February 9, 1927–December 5, 1990). *Biogr. Memoirs Natl. Acad. Sci. U. S. A.* **63**, 481–502.

1486. Gilbert, S.F. (1991). Induction and the origins of developmental genetics. In "A Conceptual History of Modern Embryology," *Developmental Biology: A Comprehensive Synthesis*, Vol. 7 (S.F. Gilbert, ed.). Plenum: New York, pp. 181–206.

1487. Gilbert, S.F. (2000). "Developmental Biology". 6th ed. Sinauer, Sunderland, Mass.

1488. Gilbert, S.F. (2000). Diachronic biology meets evo-devo: C. H. Waddington's approach to evolutionary developmental biology. *Amer. Zool.* **40**, 729–737.

1489. Gilbert, S.F. and Sarkar, S. (2000). Embracing complexity: organicism for the 21st century. *Dev. Dynamics* **219**, 1–9.

1490. Gildea, J.J., Lopez, R., and Shearn, A. (2000). A screen for new trithorax group genes identified *little imaginal discs*, the *Drosophila melanogaster* homologue of human retinoblastoma binding protein 2. *Genetics* **156**, 645–663.

1491. Gindhart, J.G., Jr. and Kaufman, T.C. (1995). Identification of *Polycomb* and *trithorax* group responsive elements in the regulatory region of the *Drosophila* homeotic gene *Sex combs reduced*. *Genetics* **139**, 797–814.

1492. Ginger, R.S., Drury, L., Baader, C., Zhukovskaya, N.V., and Williams, J.G. (1998). A novel *Dictyostelium* cell surface protein important for both cell adhesion and cell sorting. *Development* **125**, 3343–3352.

1493. Giniger, E. (1998). A role for Abl in Notch signaling. *Neuron* **20**, 667–681.

1494. Giraldez, A.J. and Cohen, S. (2001). Notum antagonizes wingless signaling during development. *Proc. 42nd Ann. Drosophila Res. Conf.* **Abstracts Vol.**, a206.

1495. Girton, J.R. (1981). Pattern triplications produced by a cell-lethal mutation in *Drosophila*. *Dev. Biol.* **84**, 164–172.

1496. Girton, J.R. (1982). Genetically induced abnormalities in *Drosophila*: two or three patterns? *Amer. Zool.* **22**, 65–77.

1497. Girton, J.R. (1983). Morphological and somatic clonal analyses of pattern triplications. *Dev. Biol.* **99**, 202–209.

1498. Girton, J.R. and Berns, M.W. (1982). Pattern abnormalities induced in *Drosophila* imaginal discs by an ultraviolet laser microbeam. *Dev. Biol.* **91**, 73–77.

1499. Girton, J.R. and Bryant, P.J. (1980). The use of cell lethal mutations in the study of *Drosophila* development. *Dev. Biol.* **77**, 233–243.

1500. Girton, J.R. and Jeon, S.H. (1994). Novel embryonic and adult homeotic phenotypes are produced by *pleiohomeotic* mutations in *Drosophila*. *Dev. Biol.* **161**, 393–407.

1501. Girton, J.R. and Kumor, A.L. (1985). The role of cell death in the induction of pattern abnormalities in a cell-lethal mutation of *Drosophila*. *Dev. Genet.* **5**, 93–102.

1502. Girton, J.R., Langner, K., and Cejka, N. (1986). Developmental genetics of a P element induced allele of *suppressor-of-forked* in *Drosophila melanogaster*. *Roux's Arch. Dev. Biol.* **195**, 334–337.

1503. Girton, J.R. and Russell, M.A. (1980). A clonal analysis of pattern duplication in a temperature-sensitive cell-lethal mutant of *Drosophila melanogaster*. *Dev. Biol.* **77**, 1–21.

1504. Girton, J.R. and Russell, M.A. (1981). An analysis of compartmentalization in pattern duplications induced by a cell-lethal mutation in *Drosophila*. *Dev. Biol.* **85**, 55–64.

1505. Gisselbrecht, S., Skeath, J.B., Doe, C.Q., and Michelson, A.M. (1996). *heartless* encodes a fibroblast growth factor receptor (DFR1/DFGF-R2) involved in the directional migration of early mesodermal cells in the *Drosophila* embryo. *Genes Dev.* **10**, 3003–3017.

1506. Glass, C.K., Rose, D.W., and Rosenfeld, M.G. (1997). Nuclear receptor coactivators. *Curr. Opin. Cell Biol.* **9**, 222–232.

1507. Glazer, L. and Shilo, B.-Z. (1991). The *Drosophila* FGF-R homolog is expressed in the embryonic tracheal system and appears to be required for directed tracheal cell extension. *Genes Dev.* **5**, 697–705.

1508. Glicksman, M.A. and Brower, D.L. (1988). Misregulation of homeotic gene expression in *Drosophila* larvae resulting from mutations at the *extra sex combs* locus. *Dev. Biol.* **126**, 219–227.

1509. Glicksman, M.A. and Brower, D.L. (1990). Persistent ectopic expression of *Drosophila* homeotic genes resulting from maternal deficiency of the *extra sex combs* gene product. *Dev. Biol.* **142**, 422–431.

1510. Glise, B. and Noselli, S. (1997). Coupling of Jun amino-terminal kinase and Decapentaplegic signaling pathways in *Drosophila* morphogenesis. *Genes Dev.* **11**, 1738–1747.

1511. Gloor, H. (1947). Phänokopie-Versuche mit Äther an *Drosophila*. *Rev. Suisse Zool.* **54**, 637–712.

1512. Glotzer, M., Murray, A.W., and Kirschner, M.W. (1991). Cyclin is degraded by the ubiquitin pathway. *Nature* **349**, 132–138.

1513. Glover, J.N.M. and Harrison, S.C. (1995). Crystal structure of the heterodimeric bZIP transcription factor c-Fos-c-Jun bound to DNA. *Nature* **373**, 257–261.

1514. Go, M.J. and Artavanis-Tsakonas, S. (1998). A genetic screen for novel components of the Notch signaling pathway during *Drosophila* bristle development. *Genetics* **150**, 211–220.

1515. Goberdhan, D.C.I. and Wilson, C. (1998). JNK, cytoskeletal regulator and stress response kinase? A *Drosophila* perspective. *BioEssays* **20**, 1009–1019.

1516. Godt, D., Couderc, J.-L., Cramton, S.E., and Laski, F.A. (1993). Pattern formation in the limbs of *Drosophila*: bric à brac is expressed in both a gradient and a wave-like pattern and is required for specification and proper segmentation of the tarsus. *Development* **119**, 799–812.

1517. Gogarten, J.P. and Olendzenski, L. (1999). Orthologs, paralogs and genome comparisons. *Curr. Opin. Gen. Dev.* **9**, 630–636.

1518. Goger, M., Gupta, V., Kim, W.-Y., Shigesada, K., Ito, Y., and Werner, M.H. (1999). Molecular insights into PEBP2/CBFβ-SMMHC associated acute leukemia revealed from the structure of PEBP2/CBFβ. *Nature Struct. Biol.* **6**, 620–623.

1519. Gogos, J.A., Hsu, T., Bolton, J., and Kafatos, F.C. (1992). Sequence discrimination by alternatively spliced isoforms of a DNA binding zinc finger domain. *Science* **257**, 1951–1955.

1520. Goldschmidt, R. (1938). "Physiological Genetics." McGraw-Hill, New York.

1521. Goldschmidt, R. (1940). "The Material Basis of Evolution." Yale Univ. Pr., New Haven.

1522. Goldschmidt, R.B. (1949). Phenocopies. *Sci. Am.* **181** #10, 46–49.

1523. Goldschmidt, R.B. (1952). Homoeotic mutants and evolution. *Acta Biotheor.* **10**, 87–104.

1524. Goldschmidt, R.B. (1955). "Theoretical Genetics." Univ. Calif. Pr., Berkeley.

1525. Goldstein, B. (2000). When cells tell their neighbors which direction to divide. *Dev. Dynamics* **218**, 23–29.

1526. Goldstein, R.E., Jiménez, G., Cook, O., Gur, D., and Paroush, Z. (1999). Huckebein repressor activity in *Drosophila* terminal patterning is mediated by Groucho. *Development* **126**, 3747–3755.

1527. Golembo, M., Raz, E., and Shilo, B.-Z. (1996). The *Drosophila* embryonic midline is the site of Spitz processing, and induces activation of the EGF receptor in the ventral ectoderm. *Development* **122**, 3363–3370.

1528. Golembo, M., Schweitzer, R., Freeman, M., and Shilo, B.-Z. (1996). *argos* transcription is induced by the *Drosophila* EGF receptor pathway to form an inhibitory feedback loop. *Development* **122**, 223–230.

1529. Golembo, M., Yarnitzky, T., Volk, T., and Shilo, B.-Z. (1999). Vein expression is induced by the EGF receptor pathway to provide a positive feedback loop in patterning the *Drosophila* embryonic ventral ectoderm. *Genes Dev.* **13**, 158–162.

1530. Golic, K.G. (1991). Site-specific recombination between homologous chromosomes in *Drosophila*. *Science* **252**, 958–961.

1531. Golic, K.G. (1993). Generating mosaics by site-specific recombination. *In* "Cellular Interactions in Development: A Practical Approach," (D.A. Hartley, ed.). Oxford Univ. Pr.: New York, pp. 1–31.

1532. Golic, K.G. and Lindquist, S. (1989). The FLP recombinase of yeast catalyzes site-specific recombination in the *Drosophila* genome. *Cell* **59**, 499–509.

1533. Golling, G., Li, L.-H., Pepling, M., Stebbins, M., and Gergen, J.P. (1996). *Drosophila* homologs of the proto-oncogene product PEBP2/CBFβ regulate the DNA-binding properties of Runt. *Mol. Cell. Biol.* **16**, 932–942.

1534. Golovnin, A., Gause, M., Georgieva, S., Gracheva, E., and

Georgiev, P. (1999). The su(Hw) insulator can disrupt enhancer-promoter interactions when located more than 20 kilobases away from the *Drosophila* achaete-scute complex. *Mol. Cell. Biol.* **19**, 3443–3456.

1535. Gomes, R., Karess, R.E., Ohkura, H., Glover, D.M., and Sunkel, C.E. (1993). Abnormal anaphase resolution (*aar*): a locus required for progression through mitosis in *Drosophila. J. Cell Sci.* **104**, 583–593.

1536. Gómez-Skarmeta, J.-L., Diez del Corral, R., de la Calle-Mustienes, E., Ferrés-Marcó, D., and Modolell, J. (1996). *araucan* and *caupolican*, two members of the novel Iroquois Complex, encode homeoproteins that control proneural and vein-forming genes. *Cell* **85**, 95–105. [See also 2001 *Development* **128**, 2847–2855.]

1537. Gómez-Skarmeta, J.L. and Modolell, J. (1996). *araucan* and *caupolican* provide a link between compartment subdivisions and patterning of sensory organs and veins in the *Drosophila* wing. *Genes Dev.* **10**, 2935–2945.

1538. Gómez-Skarmeta, J.L., Rodríguez, I., Martínez, C., Culí, J., Ferrés-Marcó, D., Beamonte, D., and Modolell, J. (1995). *Cis*-regulation of *achaete* and *scute*: shared enhancer-like elements drive their coexpression in proneural clusters of the imaginal discs. *Genes Dev.* **9**, 1869–1882.

1539. Gong, Q., Rangarajan, R., Seeger, M., and Gaul, U. (1999). The Netrin receptor Frazzled is required in the target for establishment of retinal projections in the *Drosophila* visual system. *Development* **126**, 1451–1456.

1540. González, F., Romani, S., Cubas, P., Modolell, J., and Campuzano, S. (1989). Molecular analysis of the *asense* gene, a member of the *achaete-scute* complex of *Drosophila melanogaster*, and its novel role in optic lobe development. *EMBO J.* **8**, 3553–3562.

1541. González, F., Swales, L., Bejsovec, A., Skaer, H., and Martinez Arias, A. (1991). Secretion and movement of *wingless* protein in the epidermis of the *Drosophila* embryo. *Mechs. Dev.* **35**, 43–54.

1542. González-Crespo, S., Abu-Shaar, M., Torres, M., Martínez-A., C., Mann, R.S., and Morata, G. (1998). Antagonism between *extradenticle* function and Hedgehog signalling in the developing limb. *Nature* **394**, 196–200.

1543. González-Crespo, S. and Morata, G. (1995). Control of *Drosophila* adult pattern by *extradenticle*. *Development* **121**, 2117–2125.

1544. González-Crespo, S. and Morata, G. (1996). Genetic evidence for the subdivision of the arthropod limb into coxopodite and telopodite. *Development* **122**, 3921–3928.

1545. González-Gaitán, M., Capdevila, M.P., and García-Bellido, A. (1994). Cell proliferation patterns in the wing imaginal disc of *Drosophila*. *Mechs. Dev.* **40**, 183–200.

1546. González-Gaitán, M. and Jäckle, H. (1999). The range of *spalt*-activating Dpp signalling is reduced in endocytosis-defective *Drosophila* wing discs. *Mechs. Dev.* **87**, 143–151.

1547. González-Gaitán, M.A., Micol, J.-L., and García-Bellido, A. (1990). Developmental genetic analysis of *Contrabithorax* mutations in *Drosophila melanogaster*. *Genetics* **126**, 139–155.

1548. González-Hoyuela, M., Barbas, J.A., and Rodríguez-Tébar, A. (2001). The autoregulation of retinal ganglion cell number. *Development* **128**, 117–124.

1549. González-Reyes, A. and Morata, G. (1990). The developmental effect of overexpressing a *Ubx* product in *Drosophila* embryos is dependent on its interactions with other homeotic products. *Cell* **61**, 515–522.

1550. González-Reyes, A., Urquia, N., Gehring, W.J., Struhl, G., and Morata, G. (1990). Are cross-regulatory interactions between homoeotic genes functionally significant? *Nature* **344**, 78–80.

1551. Goode, S. and Perrimon, N. (1997). Brainiac and Fringe are similar pioneer proteins that impart specificity to Notch signaling during *Drosophila* development. *Cold Spr. Harb. Symp. Quant. Biol.* **62**, 177–184.

1552. Goodman, C.S. and Coughlin, B.C. (2000). The evolution of evo-devo biology. *Proc. Natl. Acad. Sci. USA* **97**, 4424–4425.

1553. Goodman, R.H. and Smolik, S. (2000). CBP/p300 in cell growth, transformation, and development. *Genes Dev.* **14**, 1553–1577.

1554. Goodrich, L.V. and Scott, M.P. (1998). Hedgehog and Patched in neural development and disease. *Neuron* **21**, 1243–1257.

1555. Goodsell, D.S. and Olson, A.J. (2000). Structural symmetry and protein function. *Annu. Rev. Biophys. Biomol. Struct.* **29**, 105–153.

1556. Goodstein, J.R. (1991). "Millikan's School." W. W. Norton, New York.

1557. Goodwin, B. (1994). "How the Leopard Changed Its Spots." Charles Scribner's Sons, New York.

1558. Goodwin, B.C. (1982). Development and evolution. *J. Theor. Biol.* **97**, 43–55.

1559. Gopal, S., Schroeder, M., Pieper, U., Sczyrba, A., Aytekin-Kurban, G., Bekiranov, S., Fajardo, J.E., Eswar, N., Sanchez, R., Sali, A., and Gaasterland, T. (2001). Homology-based annotation yields 1,042 new candidate genes in the *Drosophila melanogaster* genome. *Nature Genet.* **27**, 337–340.

1560. Gordon, R. (1999). "The Hierarchical Genome and Differentiation Waves: Novel Unification of Development, Genetics and Evolution." Vols. 1 and 2. World Scientific, Singapore.

1561. Gorfinkiel, N., Morata, G., and Guerrero, I. (1997). The homeobox gene *Distal-less* induces ventral appendage development in *Drosophila*. *Genes Dev.* **11**, 2259–2271.

1562. Gorfinkiel, N., Sánchez, L., and Guerrero, I. (1999). *Drosophila* terminalia as an appendage-like structure. *Mechs. Dev.* **86**, 113–123.

1563. Goriely, A., Dumont, N., Dambly-Chaudière, C., and Ghysen, A. (1991). The determination of sense organs in *Drosophila*: effect of the neurogenic mutations in the embryo. *Development* **113**, 1395–1404.

1564. Gorina, S. and Pavletich, N.P. (1996). Structure of the p53 tumor suppressor bound to the ankyrin and SH3 domains of 53BP2. *Science* **274**, 1001–1005.

1565. Görlich, D. (1997). Nuclear protein import. *Curr. Op. Cell Biol.* **9**, 412–419.

1566. Görlich, D. (1998). Transport into and out of the cell nucleus. *EMBO J.* **17**, 2721–2727.

1567. Görlich, D. and Kutay, U. (1999). Transport between the cell nucleus and the cytoplasm. *Annu. Rev. Cell Dev. Biol.* **15**, 607–660.

1568. Görlich, D. and Mattaj, I.W. (1996). Nucleocytoplasmic transport. *Science* **271**, 1513–1518.

1569. Görlich, D., Prehn, S., Laskey, R.A., and Hartmann, E. (1994). Isolation of a protein that is essential for the first step of nuclear protein import. *Cell* **79**, 767–778.

1570. Gorman, M.J. and Girton, J.R. (1992). A genetic analysis of *deltex* and its interaction with the *Notch* locus in *Drosophila melanogaster*. *Genetics* **131**, 99–112.

1571. Goto, S. and Hayashi, S. (1997). Cell migration within the embryonic limb primordium of *Drosophila* as revealed by a novel fluorescence method to visualize mRNA and protein. *Dev. Genes Evol.* **207**, 194–198.

1572. Goto, S. and Hayashi, S. (1997). Specification of the embryonic limb primordium by graded activity of Decapentaplegic. *Development* **124**, 125–132.

1573. Goto, S. and Hayashi, S. (1999). Proximal to distal cell communication in the *Drosophila* leg provides a basis for an intercalary mechanism of limb patterning. *Development* **126**, 3407–3413.

1574. Goto, S., Tanimura, T., and Hotta, Y. (1995). Enhancer-trap detection of expression patterns corresponding to the polar coordinate system in the imaginal discs of *Drosophila melanogaster*. *Roux's Arch. Dev. Biol.* **204**, 378–391.

1575. Gotoh, N., Toyoda, M., and Shibuya, M. (1997). Tyrosine phosphorylation sites at amino acids 239 and 240 of Shc are involved in Epidermal growth factor-induced mitogenic signaling that is distinct from Ras/mitogen-activated protein kinase activation. *Mol. Cell. Biol.* **17**, 1824–1831.

1576. Gottardi, C.J., Arpin, M., Fanning, A.S., and Louvard, D. (1996). The junction-associated protein, zonula occludens-1, localizes to the nucleus before the maturation and during the remodeling of cell-cell contacts. *Proc. Natl. Acad. Sci. USA* **93**, 10779–10784.

1577. Gottlieb, F.J. (1964). Genetic control of pattern determination in *Drosophila*. The action of Hairy-wing. *Genetics* **49**, 739–760.

1578. Götz, J., Probst, A., Mistl, C., Nitsch, R.M., and Ehler, E. (2000). Distinct role of protein phosphatase 2A subunit Cα in the regulation of E-cadherin and β-catenin during development. *Mechs. Dev.* **93**, 83–93.

1579. Gould, A., Morrison, A., Sproat, G., White, R.A.H., and Krumlauf, R. (1997). Positive cross-regulation and enhancer sharing: two mechanisms for specifying overlapping *Hox* expression patterns. *Genes Dev.* **11**, 900–913.

1580. Gould, A.P., Brookman, J.J., Strutt, D.I., and White, R.A.H. (1990). Targets of homeotic gene control in *Drosophila*. *Nature* **348**, 308–312.

1581. Gould, S.J. (1977). "Ontogeny and Phylogeny." Harvard Univ. Pr., Cambridge, Mass.

1582. Gould, S.J. (1977). The return of hopeful monsters. *Nat. Hist.* **86** #6, 22–30. [Cf. 2001 *Genetics* **159**, 1383–1392.]

1583. Gould, S.J. (1980). The evolutionary biology of constraint. *Proc. Amer. Acad. Arts Sci.* **109** #2, 39–52.

1584. Gould, S.J. (1980). Is a new and general theory of evolution emerging? *Paleobiol.* **6**, 119–130.

1585. Gould, S.J. (1981). What color is a zebra? *Nat. Hist.* **90** #8, 16–22.

1586. Goulding, S.E., White, N.M., and Jarman, A.P. (2000). *cato* encodes a basic helix-loop-helix transcription factor implicated in the correct differentiation of *Drosophila* sense organs. *Dev. Biol.* **221**, 120–131.

1587. Goulding, S.E., zur Lage, P., and Jarman, A.P. (2000). *amos*, a proneural gene for *Drosophila* olfactory sense organs that is regulated by *lozenge*. *Neuron* **25**, 69–78.

1588. Gout, I., Dhand, R., Hiles, I.D., Fry, M.J., Panayotou, G., Das, P., Truong, O., Totty, N.F., Hsuan, J., Booker, G.W., Campbell, I.D., and Waterfield, M.D. (1993). The GTPase dynamin binds to and is activated by a subset of SH3 domains. *Cell* **75**, 25–36.

1589. Gowher, H., Leismann, O., and Jeltsch, A. (2000). DNA of *Drosophila melanogaster* contains 5-methylcytosine. *EMBO J.* **19**, 6918–6923.

1590. Goyal, L., McCall, K., Agapite, J., Hartwieg, E., and Steller, H. (2000). Induction of apoptosis by *Drosophila reaper*, *hid* and *grim* through inhibition of IAP function. *EMBO J.* **19**, 589–597.

1591. Graba, Y., Aragnol, D., and Pradel, J. (1997). *Drosophila* Hox complex downstream targets and the function of homeotic genes. *BioEssays* **19**, 379–388.

1592. Graba, Y., Gieseler, K., Aragnol, D., Laurenti, P., Mariol, M.-C., Berenger, H., Sagnier, T., and Pradel, J. (1995). *DWnt-4*, a novel *Drosophila Wnt* gene acts downstream of homeotic complex genes in the visceral mesoderm. *Development* **121**, 209–218.

1593. Graham, T.A., Weaver, C., Mao, F., Kimelman, D., and Xu, W. (2000). Crystal structure of a β-catenin/Tcf complex. *Cell* **103**, 885–896.

1594. Grandori, C., Cowley, S.M., James, L.P., and Eisenman, R.N. (2000). The Myc/Max/Mad network and the transcriptional control of cell behavior. *Annu. Rev. Cell Dev. Biol.* **16**, 653–699.

1595. Grant, D., Unadkat, S., Katzen, A., Krishnan, K.S., and Ramaswami, M. (1998). Probable mechanisms underlying interallelic complementation and temperature-sensitivity of mutations at the *shibire* locus of *Drosophila melanogaster*. *Genetics* **149**, 1019–1030.

1596. Grau, Y. and Simpson, P. (1987). The segment polarity gene *costal-2* in *Drosophila*. I. The organization of both primary and secondary embryonic fields may be affected. *Dev. Biol.* **122**, 186–200.

1597. Graveley, B.R. (2001). Alternative splicing: increasing diversity in the proteomic world. *Trends Genet.* **17**, 100–107.

1598. Graves, B. and Schubiger, G. (1981). Regional differences in the developing foreleg of *Drosophila melanogaster*. *Dev. Biol.* **85**, 334–343.

1599. Graves, B.J. and Schubiger, G. (1982). Cell cycle changes during growth and differentiation of imaginal leg discs in *Drosophila melanogaster*. *Dev. Biol.* **93**, 104–110.

1600. Gray, S. and Levine, M. (1996). Short-range transcriptional repressors mediate both quenching and direct repression within complex loci in *Drosophila*. *Genes Dev.* **10**, 700–710.

1601. Gray, S., Szymanski, P., and Levine, M. (1994). Short-range repression permits multiple enhancers to function autonomously within a complex promoter. *Genes Dev.* **8**, 1829–1838.

1602. Gray, Y.H.M. (2000). It takes two transposons to tango: transposable-element-mediated chromosomal rearrangements. *Trends Genet.* **16**, 461–468.

1603. Greaves, S., Sanson, B., White, P., and Vincent, J.-P. (1999). A screen for identifying genes interacting with armadillo, the *Drosophila* homolog of β-catenin. *Genetics* **153**, 1753–1766.

1604. Greco, V., Hannus, M., and Eaton, S. (2001). Argosomes: a potential vehicle for the spread of morphogens through epithelia. *Cell* **106**, 633–645.

1605. Green, D.R. (2000). Apoptotic pathways: paper wraps stone blunts scissors. *Cell* **102**, 1–4.

1606. Green, H. and Thomas, J. (1978). Pattern formation by cultured human epidermal cells: development of curved ridges resembling dermatoglyphics. *Science* **200**, 1385–1388.

1607. Green, J.B.A. and Cooke, J. (1991). Induction, gradient models and the role of negative feedback in body pattern formation in the amphibian embryo. *Sems. Dev. Biol.* **2**, 95–106.

1608. Green, J.B.A. and Smith, J.C. (1991). Growth factors as morphogens: do gradients and thresholds establish body plan? *Trends Genet.* **7**, 245–250.

1609. Green, J.B.A., Smith, J.C., and Gerhart, J.C. (1994). Slow emergence of a multithreshold response to activin requires cell-contact-dependent sharpening but not prepattern. *Development* **120**, 2271–2278.

1610. Green, P.B. (1980). Organogenesis–a biophysical view. *Annu. Rev. Plant Physiol.* **31**, 51–82.

1611. Greenberg, R.M. and Adler, P.N. (1982). Protein synthesis and accumulation in *Drosophila melanogaster* imaginal discs: identification of a protein with a nonrandom spatial distribution. *Dev. Biol.* **89**, 273–286.

1612. Greenwald, I. (1994). Structure/function studies of lin-12/Notch proteins. *Curr. Op. Gen. Dev.* **4**, 556–562.

1613. Greenwald, I. (1998). LIN-12/Notch signaling: lessons from worms and flies. *Genes Dev.* **12**, 1751–1762.

1614. Greenwald, I. and Rubin, G.M. (1992). Making a difference: the role of cell-cell interactions in establishing separate identities for equivalent cells. *Cell* **68**, 271–281.

1615. Greenwood, S. and Struhl, G. (1997). Different levels of Ras activity can specify distinct transcriptional and morphological consequences in early *Drosophila* embryos. *Development* **124**, 4879–4886.

1616. Greenwood, S. and Struhl, G. (1999). Progression of the morphogenetic furrow in the *Drosophila* eye: the roles of Hedgehog, Decapentaplegic and the Raf pathway. *Development* **126**, 5795–5808.

1617. Grenier, J.K. and Carroll, S.B. (2000). Functional evolution of the Ultrabithorax protein. *Proc. Natl. Acad. Sci. USA* **97**, 704–709.

1618. Grenier, J.K., Garber, T.L., Warren, R., Whitington, P.M., and Carroll, S. (1997). Evolution of the entire arthropod Hox gene set predated the origin and radiation of the onychophoran/arthropod clade. *Curr. Biol.* **7**, 547–553.

1619. Grenningloh, G., Bieber, A.J., Rehm, E.J., Snow, P.M., Traquina, Z.R., Hortsch, M., Patel, N.H., and Goodman, C.S. (1990). Molecular genetics of neuronal recognition in *Drosophila*: Evolution and function of immunoglobulin superfamily cell adhesion molecules. *Cold Spr. Harb. Symp. Quant. Biol.* **55**, 327–340.

1620. Grenningloh, G., Rehm, E.J., and Goodman, C.S. (1991). Genetic analysis of growth cone guidance in *Drosophila*: Fasciclin II functions as a neuronal recognition molecule. *Cell* **67**, 45–57.

1621. Grether, M.E., Abrams, J.M., Agapite, J., White, K., and Steller, H. (1995). The *head involution defective* gene of *Drosophila melanogaster* functions in programmed cell death. *Genes Dev.* **9**, 1694–1708.

1622. Grieder, N.C., Nellen, D., Burke, R., Basler, K., and Affolter, M. (1995). *schnurri* is required for Drosophila Dpp signaling and encodes a zinc finger protein similar to the mammalian transcription factor PRDII-BF1. *Cell* **81**, 791–800.

1623. Grigliatti, T. and Suzuki, D.T. (1971). Temperature-sensitive mutations in *Drosophila melanogaster*, VIII. The homeotic mutant, ss^{a40a}. *Proc. Natl. Acad. Sci. USA* **68**, 1307–1311.

1624. Grillenzoni, N., van Helden, J., Dambly-Chaudière, C., and Ghysen, A. (1998). The *iroquois* complex controls the somatotopy of *Drosophila* notum mechanosensory projections. *Development* **125**, 3563–3569.

1625. Grimaldi, D.A. (1987). Amber fossil Drosophilidae (Diptera), with particular reference to the Hispaniolan taxa. *Amer. Mus. Novitates* **2880**, 1–23.

1626. Grimm, S. and Pflugfelder, G.O. (1996). Control of the gene *optomotor-blind* in *Drosophila* wing development by *decapentaplegic* and *wingless*. *Science* **271**, 1601–1604.

1627. Griswold-Prenner, I., Kamibayashi, C., Maruoka, E.M., Mumby, M.C., and Derynck, R. (1998). Physical and functional interactions between type I transforming growth factor β receptors and Bα, a WD-40 repeat subunit of phosphatase 2A. *Mol. Cell. Biol.* **18**, 6595–6604.

1628. Gritzan, U., Hatini, V., and DiNardo, S. (1999). Mutual antagonism between signals secreted by adjacent Wingless and Engrailed cells leads to specification of complementary regions of the *Drosophila* parasegment. *Development* **126**, 4107–4115.

1629. Grosschedl, R., Giese, K., and Pagel, J. (1994). HMG domain proteins: architectural elements in the assembly of nucleoprotein structures. *Trends Genet.* **10**, 94–100.

1630. Grossman, J., ed. (1993). "The Chicago Manual of Style." 14th ed. Univ. Chicago Pr., Chicago.

1631. Grossniklaus, U., Cadigan, K.M., and Gehring, W.J. (1994). Three maternal coordinate systems cooperate in the patterning of the *Drosophila* head. *Development* **120**, 3155–3171.

1632. Gruss, P. and Walther, C. (1992). Pax in development. *Cell* **69**, 719–722. [Cf. 2002 *Trends Genet.* **18**, 41–47.]

1633. Gu, W., Wei, X., Pannuti, A., and Lucchesi, J.C. (2000). Targeting the chromatin-remodeling MSL complex of *Drosophila* to its sites of action on the X chromosome requires both acetyl transferase and ATPase activities. *EMBO J.* **19**, 5202–5211. [See also 2001 *Dev. Biol.* **234**, 275–288.]

1634. Gu, Y., Hukriede, N.A., and Fleming, R.J. (1995). *Serrate* expression can functionally replace *Delta* activity during neuroblast segregation in the *Drosophila* embryo. *Development* **121**, 855–865.

1635. Gubb, D. (1985). Domains, compartments and determinative switches in *Drosophila* development. *BioEssays* **2**, 27–31.

1636. Gubb, D. (1985). Further studies on *engrailed* mutants in *Drosophila melanogaster*. *Roux's Arch. Dev. Biol.* **194**, 236–246.

1637. Gubb, D. (1986). Turing's fly. *Nature* **323**, 675.

1638. Gubb, D. (1993). Genes controlling cellular polarity in *Drosophila*. *Development* **1993 Suppl.**, 269–277.

1639. Gubb, D. (1998). Cellular polarity, mitotic synchrony and axes of symmetry during growth. Where does the information come from? *Int. J. Dev. Biol.* **42**, 369–377.

1640. Gubb, D. and García-Bellido, A. (1982). A genetic analysis of the determination of cuticular polarity during development in *Drosophila melanogaster*. *J. Embryol. Exp. Morph.* **68**, 37–57.

1641. Gubb, D., Green, C., Huen, D., Coulson, D., Johnson, G., Tree, D., Collier, S., and Roote, J. (1999). The balance between isoforms of the Prickle LIM domain protein is critical for planar polarity in *Drosophila* imaginal discs. *Genes Dev.* **13**, 2315–2327.

1642. Guichard, A., Bergeret, E., and Griffin-Shea, R. (1997). Overexpression of RnRacGAP in *Drosophila melanogaster* deregulates cytoskeletal organisation in cellularising embryos and induces discrete imaginal phenotypes. *Mechs. Dev.* **61**, 49–62.

1643. Guichard, A., Biehs, B., Sturtevant, M.A., Wickline, L., Chacko, J., Howard, K., and Bier, E. (1999). *rhomboid* and *Star* interact synergistically to promote EGFR/MAPK signaling during *Drosophila* wing vein development. *Development* **126**, 2663–2676.

1644. Guichard, A., Roark, M., Ronshaugen, M., and Bier, E. (2000). *brother of rhomboid*, a *rhomboid*-related gene expressed during early *Drosophila* oogenesis, promotes EGF-R/MAPK signaling. *Dev. Biol.* **226**, 255–266.

1645. Guild, G.M., Connelly, P.S., Vranich, K.A., Shaw, M.K., and Tilney, L.G. (2001). Actin filament turnover removes bundles from *Drosophila* bristle cells. *Proc. 42nd Ann. Drosophila Res. Conf.* **Abstracts Vol.**, a75.

1646. Guillemin, K., Groppe, J., Dücker, K., Treisman, R., Hafen, E., Affolter, M., and Krasnow, M.A. (1996). The *pruned* gene encodes the *Drosophila* serum response factor and regulates cytoplasmic outgrowth during terminal branching of the tracheal system. *Development* **122**, 1353–1362.

1647. Guillén, I., Mullor, J.L., Capdevila, J., Sánchez-Herrero, E., Morata, G., and Guerrero, I. (1995). The function of *engrailed* and the specification of *Drosophila* wing pattern. *Development* **121**, 3447–3456.

1648. Gumbiner, B.M. (1993). Proteins associated with the cytoplasmic surface of adhesion molecules. *Neuron* **11**, 551–564.

1649. Gumbiner, B.M. (1998). Propagation and localization of Wnt signaling. *Curr. Opin. Gen. Dev.* **8**, 430–435.

1650. Guo, M., Bier, E., Jan, L.Y., and Jan, Y.N. (1995). *tramtrack* acts downstream of *numb* to specify distinct daughter cell fates during asymmetric cell divisions in the *Drosophila* PNS. *Neuron* **14**, 913–925.

1651. Guo, M., Jan, L.Y., and Jan, Y.N. (1996). Control of daughter cell fates during asymmetric division: Interaction of Numb and Notch. *Neuron* **17**, 27–41.

1652. Guo, Y., Livne-Bar, I., Zhou, L., and Boulianne, G.L. (1999). *Drosophila presenilin* is required for neuronal differentiation and affects Notch subcellular localization and signaling. *J. Neurosci.* **19**, 8435–8442.

1653. Gupta, B.P., Flores, G.V., Banerjee, U., and Rodrigues, V. (1998). Patterning an epidermal field: *Drosophila* Lozenge, a member of the AML-1/Runt family of transcription factors, specifies olfactory sense organ type in a dose-dependent manner. *Dev. Biol.* **203**, 400–411.

1654. Gurdon, J.B. (1992). The generation of diversity and pattern in animal development. *Cell* **68**, 185–199.

1655. Gurdon, J.B., Dyson, S., and St. Johnston, D. (1998). Cells' perception of position in a concentration gradient. *Cell* **95**, 159–162. [See also 2001 references: *Curr. Biol.* **11**, 1578–1585; *Nature* **413**, 797–803.]

1656. Gurdon, J.B., Harger, P., Mitchell, A., and Lemaire, P. (1994). Activin signalling and response to a morphogen gradient. *Nature* **371**, 487–492.

1657. Gurdon, J.B., Mitchell, A., and Mahony, D. (1995). Direct and continuous assessment by cells of their position in a morphogen gradient. *Nature* **376**, 520–521.

1658. Gurganus, M.C., Fry, J.D., Nuzhdin, S.V., Pasyukova, E.G., Lyman, R.F., and Mackay, T.F.C. (1998). Genotype-environment interaction at quantitative trait loci affecting sensory bristle number in *Drosophila melanogaster*. *Genetics* **149**, 1883–1898.

1659. Gustavson, E., Goldsborough, A.S., Ali, Z., and Kornberg, T.B. (1996). The *Drosophila engrailed* and *invected* genes: partners in regulation, expression and function. *Genetics* **142**, 893–906.

1660. Gutjahr, T., Frei, E., Spicer, C., Baumgartner, S., White, R.A.H., and Noll, M. (1995). The Polycomb-group gene, *extra sex combs*, encodes a nuclear member of the WD-40 repeat family. *EMBO J.* **14**, 4296–4306.

1661. Gutowitz, H., ed. (1991). "Cellular Automata: Theory and Experiment." M.I.T. Pr., Cambridge, Mass.

1662. Haas, H.J. (1968). On the epigenetic mechanisms of patterns in the insect integument. (A reappraisal of older concepts). *Int. Rev. Gen. Exp. Zool.* **3**, 1–51.

1663. Hackel, P.O., Zwick, E., Prenzel, N., and Ullrich, A. (1999). Epidermal growth factor receptors: critical mediators of multiple receptor pathways. *Curr. Opin. Cell Biol.* **11**, 184–189.

1664. Häcker, U., Lin, X., and Perrimon, N. (1997). The *Drosophila sugarless* gene modulates Wingless signaling and encodes an enzyme involved in polysaccharide biosynthesis. *Development* **124**, 3565–3573.

1665. Hacohen, N., Kramer, S., Sutherland, D., Hiromi, Y., and Krasnow, M.A. (1998). *sprouty* encodes a novel antagonist of FGF signaling that patterns apical branching of the *Drosophila* airways. *Cell* **92**, 253–263.

1666. Hadley, N.F. (1986). The arthropod cuticle. *Sci. Am.* **255** #1, 104–112.

1667. Hadorn, E. (1961). "Developmental Genetics and Lethal Factors." Methuen, London (transl. fr. 1955 German original, Thieme Verlag, Stuttgart, by U. Mittwoch).

1668. Hadorn, E. (1965). Problems of determination and transdetermination. *Brookhaven Symp. Biol.* **18**, 148–161.

1669. Hadorn, E. (1968). Transdetermination in cells. *Sci. Am.* **219** #11, 110–120.

1670. Hadorn, E. (1978). Transdetermination. *In* "The Genetics and Biology of *Drosophila*," Vol. 2c (M. Ashburner and T.R.F. Wright, eds.). Acad. Pr.: New York, pp. 555–617.

1671. Haenlin, M., Cubadda, Y., Blondeau, F., Heitzler, P., Lutz, Y., Simpson, P., and Ramain, P. (1997). Transcriptional activity of Pannier is regulated negatively by heterodimerization of the GATA DNA-binding domain with a cofactor encoded by the *u-shaped* gene of *Drosophila*. *Genes Dev.* **11**, 3096–3108.

1672. Haenlin, M., Kunisch, M., Kramatschek, B., and Campos-Ortega, J.A. (1994). Genomic regions regulating early embryonic expression of the *Drosophila* neurogenic gene *Delta*. *Mechs. Dev.* **47**, 99–110.

1673. Haerry, T.E., Heslip, T.R., Marsh, J.L., and O'Connor, M.B. (1997). Defects in glucuronate biosynthesis disrupt Wingless signaling in *Drosophila*. *Development* **124**, 3055–3064.

1674. Haerry, T.E., Khalsa, O., O'Connor, M.B., and Wharton, K.A. (1998). Synergistic signaling by two BMP ligands through the SAX and TKV receptors controls wing growth and patterning in *Drosophila*. *Development* **125**, 3977–3987.

1675. Hafen, E. (1998). Kinases and phosphatases – a marriage is consummated. *Science* **280**, 1212–1213.

1676. Hafen, E. and Basler, K. (1990). Mechanisms of positional signalling in the developing eye of *Drosophila* studied by ectopic expression of *sevenless* and *rough*. *In* "Growth Factors in Cell and Developmental Biology," *J. Cell Sci. Suppl.*, Vol. 13 (M.D. Waterfield, ed.). Company of Biologists, Ltd.: Cambridge, pp. 157–168.

1677. Hafen, E., Basler, K., Edstroem, J.-E., and Rubin, G.M. (1987). *Sevenless*, a cell-specific homeotic gene of *Drosophila*, encodes a putative transmembrane receptor with a tyrosine kinase domain. *Science* **236**, 55–63.

1678. Hafen, E., Dickson, B., Raabe, T., Brunner, D., Oellers, N., and van der Straten, A. (1993). Genetic analysis of the sevenless signal transduction pathway of *Drosophila*. *Development* **1993 Suppl.**, 41–46.

1679. Hagman, J., Gutch, M.J., Lin, H., and Grosschedl, R. (1995). EBF contains a novel zinc coordination motif and multiple dimerization and transcriptional activation domains. *EMBO J.* **14**, 2907–2916.

1680. Hagstrom, K., Muller, M., and Schedl, P. (1996). *Fab-7* functions as a chromatin domain boundary to ensure proper segment specification by the *Drosophila* bithorax complex. *Genes Dev.* **10**, 3202–3215.

1681. Hagstrom, K., Muller, M., and Schedl, P. (1997). A *Polycomb* and GAGA dependent silencer adjoins the *Fab-7* boundary in the *Drosophila* bithorax complex. *Genetics* **146**, 1365–1380.

1682. Hagting, A., Jackman, M., Simpson, K., and Pines, J. (1999). Translocation of cyclin B1 to the nucleus at prophase requires a phosphorylation-dependent nuclear import signal. *Curr. Biol.* **9**, 680–689.

1683. Haines, N. and van den Heuvel, M. (2000). A directed mutagenesis screen in *Drosophila melanogaster* reveals new mutants that influence *hedgehog* signaling. *Genetics* **156**, 1777–1785.

1684. Halder, G., Callaerts, P., Flister, S., Walldorf, U., Kloter, U., and Gehring, W.J. (1998). *Eyeless* initiates the expression of both *sine oculis* and *eyes absent* during *Drosophila* compound eye development. *Development* **125**, 2181–2191.

1685. Halder, G., Callaerts, P., and Gehring, W.J. (1995). Induction of ectopic eyes by targeted expression of the *eyeless* gene in *Drosophila*. *Science* **267**, 1788–1792.

1686. Halder, G., Polaczyk, P., Kraus, M.E., Hudson, A., Kim, J., Laughon, A., and Carroll, S. (1998). The Vestigial and Scalloped proteins act together to directly regulate wing-specific gene expression in *Drosophila*. *Genes Dev.* **12**, 3900–3909. [See also 2001 references: *Development* **128**, 3295–3305; *Science* **292**, 1080–1081; *Science* **292**, 1164–1167.]

1687. Halfar, K., Rommel, C., Stocker, H., and Hafen, E. (2001). Ras controls growth, survival and differentiation in the *Drosophila* eye by different thresholds of MAP kinase activity. *Development* **128**, 1687–1696.

1688. Halfon, M.S., Carmena, A., Gisselbrecht, S., Sackerson, C.M., Jiménez, F., Baylies, M.K., and Michelson, A.M. (2000). Ras pathway specificity is determined by the integration of multiple signal-activated and tissue-restricted transcription factors. *Cell* **103**, 63–74.

1689. Hall, A. (1994). A biochemical function for Ras–at last. *Science* **264**, 1413–1414.

1690. Hall, B.K. (1983). Epigenetic control in development and evolution. *In* "Development and Evolution," (B.C. Goodwin, N. Holder, and C.G. Wylie, eds.). Cambridge Univ. Pr.: Cambridge, pp. 353–379.

1691. Hall, B.K. (2000). Evo-devo or devo-evo–does it matter? *Evol. Dev.* **2**, 177–178.

1692. Hall, J.C. (1994). The mating of a fly. *Science* **264**, 1702–1714.

1693. Hall, J.C. (1995). Tripping along the trail to the molecular mechanisms of biological clocks. *Trends Neurosci.* **18**, 230–240.

1694. Hall, J.C. (1996). Are cycling gene products as internal zeitgebers no longer the zeitgeist of chronobiology? *Neuron* **17**, 799–802.

1695. Hall, J.C., Gelbart, W.M., and Kankel, D.R. (1976). Mosaic systems. *In* "The Genetics and Biology of *Drosophila*," Vol. 1a (M. Ashburner and E. Novitski, eds.). Acad. Pr.: New York, N. Y., pp. 265–314.

1696. Halle, J.-P. and Meisterernst, M. (1996). Gene expression: increasing evidence for a transcriptosome. *Trends Genet.* **12**, 161–163.

1697. Hama, C., Ali, Z., and Kornberg, T.B. (1990). Region-specific recombination and expression are directed by portions of the *Drosophila engrailed* promoter. *Genes Dev.* **4**, 1079–1093.

1698. Hamada, F., Tomoyasu, Y., Takatsu, Y., Nakamura, M., Nagai, S.-i., Suzuki, A., Fujita, F., Shibuya, H., Toyoshima, K., Ueno, N., and Akiyama, T. (1999). Negative regulation of Wingless signaling by D-Axin, a *Drosophila* homolog of Axin. *Science* **283**, 1739–1742.

1699. Hamburger, V. (1988). "The Heritage of Experimental Embryology: Hans Spemann and the Organizer." Oxford Univ. Pr., New York.

1700. Hamiche, A., Sandaltzopoulos, R., Gdula, D.A., and Wu, C. (1999). ATP-dependent histone octamer sliding mediated by the chromatin-remodeling complex NURF. *Cell* **97**, 833–842.

1701. Hammerschmidt, M., Brook, A., and McMahon, A.P. (1997). The world according to *hedgehog*. *Trends Genet.* **13**, 14–21.

1702. Hammond, S.M., Caudy, A.A., and Hannon, G.J. (2001). Post-transcriptional gene silencing by double-stranded RNA. *Nature Rev. Gen.* **2**, 110–119.

1703. Hampsey, M. and Reinberg, D. (1999). RNA polymerase II as a control panel for multiple coactivator complexes. *Curr. Opin. Gen. Dev.* **9**, 132–139.

1704. Han, J.-D., Baker, N.E., and Rubin, C.S. (1997). Molecular characterization of a novel A kinase anchor protein from *Drosophila melanogaster*. *J. Biol. Chem.* **272** #42, 26611–26619.

1705. 1705. Han, K., Levine, M.S., and Manley, J.L. (1989). Synergistic activation and repression of transcription by *Drosophila* homeobox proteins. *Cell* **56**, 573–583.

1706. Han, K. and Manley, J.L. (1993). Functional domains of the *Drosophila* Engrailed protein. *EMBO J.* **12**, 2723–2733.

1707. Handford, P.A., Mayhew, M., Baron, M., Winship, P.R., Campbell, I.D., and Brownlee, G.G. (1991). Key residues involved in calcium-binding motifs in EGF-like domains. *Nature* **351**, 164–167.

1708. Hanes, S.D. and Burz, D.S. (2001). Isolation of mutations that disrupt cooperative DNA binding by the *Drosophila* Bicoid protein. *Proc. 42nd Ann. Drosophila Res. Conf.* **Abstracts Vol.**, a35. [See also 2001 *J. Mol. Biol.* **305**, 219–230.]

1709. Hanes, S.D., Riddihough, G., Ish-Horowicz, D., and Brent, R. (1994). Specific DNA recognition and intersite spacing are critical for action of the Bicoid morphogen. *Mol. Cell. Biol.* **14**, 3364–3375.

1710. Hankinson, O. (1995). The aryl hydrocarbon receptor complex. *Annu. Rev. Pharmacol. Toxicol.* **35**, 307–340.

1711. Hanks, S.K. and Hunter, T. (1995). The eukaryotic protein kinase superfamily: kinase (catalytic) domain structure and classification. *FASEB J.* **9**, 576–596.

1712. Hanna-Rose, W. and Hansen, U. (1996). Active repression mechanisms of eukaryotic transcription repressors. *Trends Genet.* **12**, 229–234.

1713. Hannah-Alava, A. (1958). Developmental genetics of the posterior legs in *Drosophila melanogaster*. *Genetics* **43**, 878–905.

1714. Hannah-Alava, A. (1958). Morphology and chaetotaxy of the legs of *Drosophila melanogaster*. *J. Morph.* **103**, 281–310.

1715. Hannah-Alava, A. (1960). Genetic mosaics. *Sci. Am.* **202** #5, 118–130.

1716. Hannah-Alava, A. and Stern, C. (1957). The sexcombs in males and intersexes of *Drosophila melanogaster. J. Exp. Zool.* **134**, 533–556.

1717. Harbecke, R., Meise, M., Holz, A., Klapper, R., Naffin, E., Nordhoff, V., and Janning, W. (1996). Larval and imaginal pathways in early development of *Drosophila. Int. J. Dev. Biol.* **40**, 197–204.

1718. Harbury, P.B., Kim, P.S., and Alber, T. (1994). Crystal structure of an isoleucine-zipper trimer. *Nature* **371**, 80–83.

1719. Harbury, P.B., Zhang, T., Kim, P.S., and Alber, T. (1993). A switch between two-, three-, and four-stranded coiled coils in GCN4 leucine zipper mutants. *Science* **262**, 1401–1407.

1720. Hardie, R.C. (1985). Functional organization of the fly retina. *In* "Progress in Sensory Physiology," Vol. 5 (D. Ottoson, ed.). Springer-Verlag: Berlin, pp. 1–79.

1721. Hardie, R.C. (1986). The photoreceptor array of the dipteran retina. *Trends Neurosci.* **9**, 419–423.

1722. Harding, K., Rushlow, C., Doyle, H.J., Hoey, T., and Levine, M. (1986). Cross-regulatory interactions among pair-rule genes in *Drosophila. Science* **233**, 953–959.

1723. Hardy, J. and Israël, A. (1999). In search of γ-secretase. *Nature* **398**, 466–467.

1724. Hariharan, I.K., Carthew, R.W., and Rubin, G.M. (1991). The *Drosophila Roughened* mutation: activation of a *rap* homolog disrupts eye development and interferes with cell determination. *Cell* **67**, 717–722.

1725. Harrelson, A.L. and Goodman, C.S. (1988). Growth cone guidance in insects: fasciclin II is a member of the immunoglobulin superfamily. *Science* **242**, 700–708.

1726. Harris, W.A., Stark, W.S., and Walker, J.A. (1976). Genetic dissection of the photoreceptor system in the compound eye of *Drosophila melanogaster. J. Physiol.* **256**, 415–439.

1727. Harrison, L.G. (1993). "Kinetic Theory of Living Pattern." *Developmental and Cell Biology Series*, Vol. 28. Cambridge Univ. Pr., Cambridge.

1728. Harrison, L.G. and Tan, K.Y. (1988). Where may reaction-diffusion mechanisms be operating in metameric patterning of *Drosophila* embryos? *BioEssays* **8**, 118–124.

1729. Harrison, S.C. (1991). A structural taxonomy of DNA-binding domains. *Nature* **353**, 715–719.

1730. Harrison, S.C. (1996). Peptide-surface association: the case of PDZ and PTB domains. *Cell* **86**, 341–343.

1731. Harrison, S.D. and Travers, A.A. (1990). The *tramtrack* gene encodes a *Drosophila* finger protein that interacts with the *ftz* transcriptional regulatory region and shows a novel embryonic expression pattern. *EMBO J.* **9**, 207–216.

1732. Hart, A.C., Krämer, H., Van Vactor, D.L., Jr., Paidhungat, M., and Zipursky, S.L. (1990). Induction of cell fate in the *Drosophila* retina: the bride of sevenless protein is predicted to contain a large extracellular domain and seven transmembrane segments. *Genes Dev.* **4**, 1835–1847.

1733. Hart, A.C., Krämer, H., and Zipursky, S.L. (1993). Extracellular domain of the boss transmembrane ligand acts as an antagonist of the sev receptor. *Nature* **361**, 732–736.

1734. Hart, C.M., Ketel, C.S., and Simon, J.A. (2001). Characterization of the ESC-E(Z) complex and functional partnership. *Proc. 42nd Ann. Drosophila Res. Conf.* **Abstracts Vol.**, a95.

1735. Hart, K. and Bienz, M. (1996). A test for cell autonomy, based on di-cistronic messenger translation. *Development* **122**, 747–751.

1736. Hart, M., Concordet, J.-P., Lassot, I., Albert, I., del los Santos, R., Durand, H., Perret, C., Rubinfeld, B., Margottin,

E., Benarous, R., and Polakis, P. (1999). The F-box protein β-TrCP associates with phosphorylated β-catenin and regulates its activity in the cell. *Curr. Biol.* **9**, 207–210.

1737. Hartenstein, A.Y., Rugendorff, A., Tepass, U., and Hartenstein, V. (1992). The function of the neurogenic genes during epithelial development in the *Drosophila* embryo. *Development* **116**, 1203–1220.

1738. Hartenstein, K., Sinha, P., Mishra, A., Schenkel, H., Török, I., and Mechler, B.M. (1997). The *congested-like tracheae* gene of *Drosophila melanogaster* encodes a member of the mitochondrial carrier family required for gas-filling of the tracheal system and expansion of the wings after eclosion. *Genetics* **147**, 1755–1768.

1739. Hartenstein, V. (1993). "Atlas of *Drosophila* Development." Cold Spring Harbor, N.Y.

1740. Hartenstein, V. and Campos-Ortega, J.A. (1984). Early neurogenesis in wild-type *Drosophila melanogaster. Roux's Arch. Dev. Biol.* **193**, 308–325.

1741. Hartenstein, V. and Posakony, J.W. (1989). Development of adult sensilla on the wing and notum of *Drosophila melanogaster. Development* **107**, 389–405.

1742. Hartenstein, V. and Posakony, J.W. (1990). A dual function of the *Notch* gene in *Drosophila* sensillum development. *Dev. Biol.* **142**, 13–30.

1743. Hartenstein, V. and Posakony, J.W. (1990). Sensillum development in the absence of cell division: the sensillum phenotype of the *Drosophila* mutant string. *Dev. Biol.* **138**, 147–158.

1744. Hartenstein, V., Tepass, U., and Gruszynski-deFeo, E. (1996). Proneural and neurogenic genes control specification and morphogenesis of stomatogastric nerve cell precursors in *Drosophila. Dev. Biol.* **173**, 213–227.

1745. Hartenstein, V., Younossi-Hartenstein, A., and Lekven, A. (1994). Delamination and division in the *Drosophila* neurectoderm: spatiotemporal pattern, cytoskeletal dynamics, and common control by neurogenic and segment polarity genes. *Dev. Biol.* **165**, 480–499.

1746. Hartley, D.A., Preiss, A., and Artavanis-Tsakonas, S. (1988). A deduced gene product from the *Drosophila* neurogenic locus, *Enhancer of Split*, shows homology to mammalian G-protein β subunit. *Cell* **55**, 785–795.

1747. Hartley, D.A., Xu, T., and Artavanis-Tsakonas, S. (1987). The embryonic expression of the *Notch* locus of *Drosophila melanogaster* and the implications of point mutations in the extracellular EGF-like domain of the predicted protein. *EMBO J.* **6**, 3407–3417.

1748. Hartman, J.L., IV, Garvik, B., and Hartwell, L. (2001). Principles for the buffering of genetic variation. *Science* **291**, 1001–1004.

1749. Hartmann, B., Marty, T., and Affolter, M. (2001). Functional analysis of the *Drosophila* Schnurri protein *in vivo. Proc. 42nd Ann. Drosophila Res. Conf.* **Abstracts Vol.**, a162.

1750. Hartmann, C., Taubert, H., Jäckle, H., and Pankratz, M.J. (1994). A two-step mode of stripe formation in the *Drosophila* blastoderm requires interactions among primary pair rule genes. *Mechs. Dev.* **45**, 3–13.

1751. Haskell, R.E. (1993). "Introduction to Computer Engineering: Logic Design and the 8086 Microprocessor." Prentice-Hall, Englewood Cliffs, N. J.

1752. Hassan, A.H., Neely, K.E., and Workman, J.L. (2001). Histone acetyltransferase complexes stabilize SWI/SNF binding to promoter nucleosomes. *Cell* **104**, 817–827.

1753. Hassan, B., Li, L., Bremer, K.A., Chang, W., Pinsonneault, J., and Vaessin, H. (1997). Prospero is a panneural transcription factor that modulates homeodomain protein activity. *Proc. Natl. Acad. Sci. USA* **94**, 10991–10996.

1754. Hassan, B. and Vaessin, H. (1996). Regulatory interactions during early neurogenesis in *Drosophila. Dev. Genet.* **18**, 18–27.

1755. Hassan, B. and Vaessin, H. (1997). Daughterless is required for the expression of cell cycle genes in peripheral nervous system precursors of *Drosophila* embryos. *Dev. Genet.* **21**, 117–122.

1756. Hassan, B.A. and Bellen, H.J. (2000). Doing the MATH: is the mouse a good model for fly development? *Genes Dev.* **14**, 1852–1865.

1757. Hastie, N.D. (1991). *Pax* in our time. *Curr. Biol.* **1**, 342–344.

1758. Hasty, J., McMillen, D., Isaacs, F., and Collins, J.J. (2001). Computational studies of gene regulatory networks: *in numero* molecular biology. *Nature Rev. Genet.* **2**, 268–279.

1759. Hatini, V., Bokor, P., Goto-Mandeville, R., and DiNardo, S. (2000). Tissue and stage-specific modulation of Wingless signaling by the segment polarity gene *lines. Genes Dev.* **14**, 1364–1376.

1760. Hauck, B., Gehring, W.J., and Walldorf, U. (1999). Functional analysis of an eye specific enhancer of the *eyeless* gene in *Drosophila. Proc. Natl. Acad. Sci. USA* **96**, 564–569.

1761. Hawkins, N. and Garriga, G. (1998). Asymmetric cell division: from A to Z. *Genes Dev.* **12**, 3625–3638.

1762. Hawkins, N.C., Thorpe, J., and Schüpbach, T. (1996). *encore*, a gene required for the regulation of germ line mitosis and oocyte differentiation during *Drosophila* oogenesis. *Development* **122**, 281–290.

1763. Hay, B.A., Wolff, T., and Rubin, G.M. (1994). Expression of baculovirus P35 prevents cell death in *Drosophila. Development* **120**, 2121–2129.

1764. Hayashi, S. (1996). A Cdc2 dependent checkpoint maintains diploidy in *Drosophila. Development* **122**, 1051–1058.

1765. Hayashi, S., Hirose, S., Metcalfe, T., and Shirras, A.D. (1993). Control of imaginal cell development by the *escargot* gene of *Drosophila. Development* **118**, 105–115.

1766. Hayashi, S., Rubinfeld, B., Souza, B., Polakis, P., and Wieschaus, E. (1997). A *Drosophila* homolog of the tumor suppressor gene adenomatous polyposis coli down-regulates β-catenin but its zygotic expression is not essential for the regulation of Armadillo. *Proc. Natl. Acad. Sci. USA* **94**, 242–247.

1767. Hayashi, S. and Scott, M.P. (1990). What determines the specificity of action of *Drosophila* homeodomain proteins? *Cell* **63**, 883–894.

1768. Hayashi, T., Kojima, T., and Saigo, K. (1998). Specification of primary pigment cell and outer photoreceptor fates by *BarH1* homeobox gene in the developing *Drosophila* eye. *Dev. Biol.* **200**, 131–145.

1769. Hayes, B. (1999). *E pluribus unum. Am. Sci.* **87**, 10–14.

1770. Hayes, J.J. and Hansen, J.C. (2001). Nucleosomes and the chromosome fiber. *Curr. Opin. Gen. Dev.* **11**, 124–129.

1771. Hayes, P.H. (1982). "Mutant Analysis of Determination in *Drosophila*." Ph.D. dissertation, Dept. of Genetics, Univ. of Alberta, Edmonton, Alberta, Canada.

1772. Hayes, P.H., Sato, T., and Denell, R.E. (1984). Homoeosis in *Drosophila*: the Ultrabithorax larval syndrome. *Proc. Natl. Acad. Sci. USA* **81**, 545–549.

1773. Haynie, J. and Schubiger, G. (1979). Absence of distal to proximal intercalary regeneration in imaginal wing discs of *Drosophila melanogaster. Dev. Biol.* **68**, 151–161.

1774. Haynie, J.L. (1982). Homologies of positional information in thoracic imaginal discs of *Drosophila melanogaster. W. Roux's Archiv.* **191**, 293–300.

1775. Haynie, J.L. and Bryant, P.J. (1976). Intercalary regeneration in imaginal wing disk of *Drosophila melanogaster. Nature* **259**, 659–662.

1776. Haynie, J.L. and Bryant, P.J. (1977). The effects of X-rays on the proliferation dynamics of cells in the imaginal wing disc of *Drosophila melanogaster. W. Roux's Arch.* **183**, 85–100.

1777. Haynie, J.L. and Bryant, P.J. (1986). Development of the eye-antenna imaginal disc and morphogenesis of the adult head in *Drosophila melanogaster. J. Exp. Zool.* **237**, 293–308.

1778. Hays, R., Buchanan, K.T., Neff, C., and Orenic, T.V. (1999). Patterning of *Drosophila* leg sensory organs through combinatorial signaling by Hedgehog, Decapentaplegic and Wingless. *Development* **126**, 2891–2899.

1779. Hays, R., Gibori, G.B., and Bejsovec, A. (1997). Wingless signaling generates pattern through two distinct mechanisms. *Development* **124**, 3727–3736.

1780. Hazelett, D.J., Bourouis, M., Walldorf, U., and Treisman, J.E. (1998). *decapentaplegic* and *wingless* are regulated by *eyes absent* and *eyegone* and interact to direct the pattern of retinal differentiation in the eye disc. *Development* **125**, 3741–3751.

1781. Heberlein, U., Borod, E.R., and Chanut, F.A. (1998). Dorsoventral patterning in the *Drosophila* retina by *wingless. Development* **125**, 567–577.

1782. Heberlein, U., Hariharan, I.K., and Rubin, G.M. (1993). *Star* is required for neuronal differentiation in the *Drosophila* retina and displays dosage-sensitive interactions with *Ras1. Dev. Biol.* **160**, 51–63.

1783. Heberlein, U., Mlodzik, M., and Rubin, G.M. (1991). Cell-fate determination in the developing *Drosophila* eye: role of the *rough* gene. *Development* **112**, 703–712.

1784. Heberlein, U. and Moses, K. (1995). Mechanisms of *Drosophila* retinal morphogenesis: the virtues of being progressive. *Cell* **81**, 987–990.

1785. Heberlein, U. and Rubin, G.M. (1991). *Star* is required in a subset of photoreceptor cells in the developing *Drosophila* retina and displays dosage sensitive interactions with *rough. Dev. Biol.* **144**, 353–361.

1786. Heberlein, U., Singh, C.M., Luk, A.Y., and Donohoe, T.J. (1995). Growth and differentiation in the *Drosophila* eye coordinated by *hedgehog. Nature* **373**, 709–711.

1787. Heberlein, U. and Treisman, J.E. (2000). Early retinal development in *Drosophila. In* "Vertebrate Eye Development," *Results and Problems in Cell Differentiation*, Vol. 31 (M.E. Fini, ed.). Springer: Berlin, pp. 37–50.

1788. Heberlein, U., Wolff, T., and Rubin, G.M. (1993). The TGFβ homolog *dpp* and the segment polarity gene *hedgehog* are required for propagation of a morphogenetic wave in the *Drosophila* retina. *Cell* **75**, 913–926.

1789. Heemskerk, J. and DiNardo, S. (1994). *Drosophila* hedgehog acts as a morphogen in cellular patterning. *Cell* **76**, 449–460.

1790. Heemskerk, J., DiNardo, S., Kostriken, R., and O'Farrell, P.H. (1991). Multiple modes of *engrailed* regulation in the progression towards cell fate determination. *Nature* **352**, 404–410.

1791. Heilig, J.S., Freeman, M., Laverty, T., Lee, K.J., Campos,

A.R., Rubin, G.M., and Steller, H. (1991). Isolation and characterization of the *disconnected* gene of *Drosophila melanogaster. EMBO J.* **10**, 809–815.

1792. Heintzen, C., Loros, J.J., and Dunlap, J.C. (2001). The PAS protein VIVID defines a clock-associated feedback loop that represses light input, modulates gating, and regulates clock resetting. *Cell* **104**, 453–464.

1793. Heitz, F., Morris, M.C., Fesquet, D., Cavadore, J.-C., Dorée, M., and Divita, G. (1997). Interactions of cyclins with cyclin-dependent kinases: a common interactive mechanism. *Biochemistry* **36**, 4995–5003.

1794. Heitzler, P., Bourouis, M., Ruel, L., Carteret, C., and Simpson, P. (1996). Genes of the *Enhancer of split* and *achaete-scute* complexes are required for a regulatory loop between *Notch* and *Delta* during lateral signalling in *Drosophila. Development* **122**, 161–171.

1795. Heitzler, P., Coulson, D., Saenz-Robles, M.-T., Ashburner, M., Roote, J., Simpson, P., and Gubb, D. (1993). Genetic and cytogenetic analysis of the 43A-E region containing the segment polarity gene *costa* and the cellular polarity genes *prickle* and *spiny-legs* in *Drosophila melanogaster. Genetics* **135**, 105–115.

1796. Heitzler, P., Haenlin, M., Ramain, P., Calleja, M., and Simpson, P. (1996). A genetic analysis of *pannier*, a gene necessary for viability of dorsal tissues and bristle positioning in *Drosophila. Genetics* **143**, 1271–1286.

1797. Heitzler, P. and Simpson, P. (1991). The choice of cell fate in the epidermis of *Drosophila. Cell* **64**, 1083–1092.

1798. Heitzler, P. and Simpson, P. (1993). Altered epidermal growth factor-like sequences provide evidence for a role of *Notch* as a receptor in cell fate decisions. *Development* **117**, 1113–1123.

1799. Held, L.I., Jr. (1977). "Analysis of Bristle-Pattern Formation in *Drosophila*." Ph.D. dissertation, Dept. of Molecular Biology, Univ. of California, Berkeley.

1800. Held, L.I., Jr. (1979). A high-resolution morphogenetic map of the second-leg basitarsus in *Drosophila melanogaster. Wilhelm Roux's Arch.* **187**, 129–150.

1801. Held, L.I., Jr. (1979). Pattern as a function of cell number and cell size on the second-leg basitarsus of *Drosophila. Wilhelm Roux's Arch.* **187**, 105–127.

1802. Held, L.I., Jr. (1990). Arrangement of bristles as a function of bristle number on a leg segment in *Drosophila melanogaster. Roux's Arch. Dev. Biol.* **199**, 48–62.

1803. Held, L.I., Jr. (1990). Sensitive periods for abnormal patterning on a leg segment in *Drosophila melanogaster. Roux's Arch. Dev. Biol.* **199**, 31–47.

1804. Held, L.I., Jr. (1991). Bristle patterning in *Drosophila. BioEssays* **13**, 633–640.

1805. Held, L.I., Jr. (1992). "Models for Embryonic Periodicity." *Monographs in Developmental Biology*, Vol. 24. Karger, Basel.

1806. Held, L.I., Jr. (1993). Segment-polarity mutations cause stripes of defects along a leg segment in *Drosophila. Dev. Biol.* **157**, 240–250.

1807. Held, L.I., Jr. (1995). Axes, boundaries and coordinates: the ABCs of fly leg development. *BioEssays* **17**, 721–732.

1808. Held, L.I., Jr. and Bryant, P.J. (1984). Cell interactions controlling the formation of bristle patterns in *Drosophila. In* "Pattern Formation: A Primer in Developmental Biology," (G.M. Malacinski and S.V. Bryant, eds.). Macmillan: New York, pp. 291–322.

1809. Held, L.I., Jr., Duarte, C.M., and Derakhshanian, K. (1986). Extra joints and misoriented bristles on *Drosophila* legs. *In*

"Progress in Developmental Biology (Part A)," (H. Slavkin, ed.). Alan R. Liss: N. Y., pp. 293–296.

1810. Held, L.I., Jr., Duarte, C.M., and Derakhshanian, K. (1986). Extra tarsal joints and abnormal cuticular polarities in various mutants of *Drosophila melanogaster. Roux's Arch. Dev. Biol.* **195**, 145–157.

1811. Held, L.I., Jr. and Heup, M. (1996). Genetic mosaic analysis of *decapentaplegic* and *wingless* gene function in the *Drosophila* leg. *Dev. Genes Evol.* **206**, 180–194.

1812. Held, L.I., Jr., Heup, M.A., Sappington, J.M., and Peters, S.D. (1994). Interactions of *decapentaplegic, wingless*, and *Distal-less* in the *Drosophila* leg. *Roux's Arch. Dev. Biol.* **203**, 310–319.

1813. Held, L.I., Jr. and Pham, T.T. (1983). Accuracy of bristle placement on a leg segment in *Drosophila melanogaster. J. Morph.* **178**, 105–110.

1814. Heldin, C.-H. (1995). Dimerization of cell surface receptors in signal transduction. *Cell* **80**, 213–223.

1815. Heldin, C.-H., Miyazono, K., and ten Dijke, P. (1997). TGF-β signalling from cell membrane to nucleus through SMAD proteins. *Nature* **390**, 465–471.

1816. Helms, W., Lee, H., Ammerman, M., Parks, A.L., Muskavitch, M.A.T., and Yedvobnick, B. (1999). Engineered truncations in the *Drosophila* Mastermind protein disrupt Notch pathway function. *Dev. Biol.* **215**, 358–374.

1817. Helps, N.R., Cohen, P.T.W., Bahri, S.M., Chia, W., and Babu, K. (2001). Interaction with protein phosphatase 1 is essential for *bifocal* function during the morphogenesis of the *Drosophila* compound eye. *Mol. Cell. Biol.* **21**, 2154–2164.

1818. Hemmati-Brivanlou, A. and Melton, D. (1997). Vertebrate embryonic cells will become nerve cells unless told otherwise. *Cell* **88**, 13–17.

1819. Hemmings, B.A. (1997). PH domains–a universal membrane adapter. *Science* **275**, 1899.

1820. Henderson, K.D., Isaac, D.D., and Andrew, D.J. (1999). Cell fate specification in the *Drosophila* salivary gland: the integration of homeotic gene function with the DPP signaling cascade. *Dev. Biol.* **205**, 10–21.

1821. Hendrickson, J.E. and Sakonju, S. (1995). *Cis* and *trans* interactions between the *iab* regulatory regions and *abdominal-A* and *Abdominal-B* in *Drosophila melanogaster. Genetics* **139**, 835–848.

1822. Henery, C.C., Bard, J.B.L., and Kaufman, M.H. (1992). Tetraploidy in mice, embryonic cell number, and the grain of the developmental map. *Dev. Biol.* **152**, 233–241.

1823. Henikoff, S. (1998). Conspiracy of silence among repeated transgenes. *BioEssays* **20**, 532–535.

1824. Henikoff, S., Greene, E.A., Pietrokovski, S., Bork, P., Attwood, T.K., and Hood, L. (1997). Gene families: the taxonomy of protein paralogs and chimeras. *Science* **278**, 609–614.

1825. Henikoff, S. and Matzke, M.A. (1997). Exploring and explaining epigenetic effects. *Trends Genet.* **13**, 293–295.

1826. Henke, K. and Maas, H. (1946). Über sensible Perioden der allgemeinen Körpergliederung von *Drosophila. Nachr. Akad. Wiss. Göttingen. Math.-Phys. Kl. IIb, Biol.-Physiol.-Chem. Abt.* **1**, 3–4.

1827. Hepker, J., Blackman, R.K., and Holmgren, R. (1999). Cubitus interruptus is necessary but not sufficient for direct activation of a wing-specific *decapentaplegic* enhancer. *Development* **126**, 3669–3677.

1828. Hepker, J., Wang, Q.-T., Motzny, C.K., Holmgren, R., and Orenic, T.V. (1997). *Drosophila cubitus interruptus* forms a negative feedback loop with *patched* and regulates expression of Hedgehog target genes. *Development* **124**, 549–558.

1829. Herbst, R., Carroll, P.M., Allard, J.D., Schilling, J., Raabe, T., and Simon, M.A. (1996). Daughter of sevenless is a substrate of the phosphotyrosine phosphatase Corkscrew and functions during Sevenless signaling. *Cell* **85**, 899–909.

1830. Herbst, R., Zhang, X., Qin, J., and Simon, M.A. (1999). Recruitment of the protein tyrosine phosphatase CSW by DOS is an essential step during signaling by the Sevenless receptor tyrosine kinase. *EMBO J.* **18**, 6950–6961.

1831. Hériché, J.-K. and O'Farrell, P.H. (2001). Genetic analysis of Cul-1 in *Drosophila. Proc. 42nd Ann. Drosophila Res. Conf.* **Abstracts Vol.**, a136.

1832. Herr, W. and Cleary, M.A. (1995). The POU domain: versatility in transcriptional regulation by a flexible two-in-one DNA-binding domain. *Genes Dev.* **9**, 1679–1693.

1833. Heslip, T.R., Theisen, H., Walker, H., and Marsh, J.L. (1997). SHAGGY and DISHEVELLED exert opposite effects on *wingless* and *decapentaplegic* expression and on positional identity in imaginal discs. *Development* **124**, 1069–1078.

1834. Heuer, J.G. and Kaufman, T.C. (1992). Homeotic genes have specific functional roles in the establishment of the *Drosophila* embryonic peripheral nervous system. *Development* **115**, 35–47.

1835. Hewitt, G.F., Strunk, B.S., Margulies, C., Priputin, T., Wang, X.-D., Amey, R., Pabst, B.A., Kosman, D., Reinitz, J., and Arnosti, D.N. (1999). Transcriptional repression by the *Drosophila* Giant protein: *cis* element positioning provides an alternative means of interpreting an effector gradient. *Development* **126**, 1201–1210.

1836. Hidalgo, A. (1991). Interactions between segment polarity genes and the generation of the segmental pattern in *Drosophila. Mechs. Dev.* **35**, 77–87.

1837. Hidalgo, A. (1994). Three distinct roles for the *engrailed* gene in *Drosophila* wing development. *Curr. Biol.* **4**, 1087–1098.

1838. Hidalgo, A. (1996). The roles of *engrailed. Trends Genet.* **12**, 1–4.

1839. Hidalgo, A. (1998). Growth and patterning from the *engrailed* interface. *Int. J. Dev. Biol.* **42**, 317–324.

1840. Hidalgo, A. (1998). Wing patterning *knot* untangled. *BioEssays* **20**, 449–452.

1841. Hidalgo, A. and Ingham, P. (1990). Cell patterning in the *Drosophila* segment: spatial regulation of the segment polarity gene *patched. Development* **110**, 291–301.

1842. Higashijima, S.-i., Kojima, T., Michiue, T., Ishimaru, S., Emori, Y., and Saigo, K. (1992). Dual *Bar* homeo box genes of *Drosophila* required in two photoreceptor cells, R1 and R6, and primary pigment cells for normal eye development. *Genes Dev.* **6**, 50–60.

1843. Higashijima, S.-i., Michiue, T., Emori, Y., and Saigo, K. (1992). Subtype determination of *Drosophila* embryonic external sensory organs by redundant homeo box genes *BarH1* and *BarH2. Genes Dev.* **6**, 1005–1018.

1844. High, S. and Dobberstein, B. (1992). Mechanisms that determine the transmembrane disposition of proteins. *Curr. Opin. Cell Biol.* **4**, 581–586.

1845. Hildreth, P.E. (1965). Doublesex, a recessive gene that transforms both males and females of *Drosophila* into intersexes. *Genetics* **51**, 659–678.

1846. Hill, C.S. and Treisman, R. (1995). Transcriptional regulation by extracellular signals: mechanisms and specificity. *Cell* **80**, 199–211.

1847. Hill, R. and Van Heyningen, V. (1992). Mouse mutations and human disorders are Paired. *Trends Genet.* **8**, 119–120.

1848. Hill, R.E. and Davidson, D.R. (1994). Seeing eye to eye. *Curr. Biol.* **4**, 1155–1157.

1849. Hillier, B.J., Christopherson, K.S., Prehoda, K.E., Bredt, D.S., and Lim, W.A. (1999). Unexpected modes of PDZ domain scaffolding revealed by structure of nNOS-syntrophin complex. *Science* **284**, 812–815.

1850. Hing, H., Xiao, J., Harden, N., Lim, L., and Zipursky, S.L. (1999). Pak functions downstream of Dock to regulate photoreceptor axon guidance in *Drosophila. Cell* **97**, 853–863.

1851. Hing, H.K., Sun, X., and Artavanis-Tsakonas, S. (1994). Modulation of wingless signaling by Notch in *Drosophila. Mechs. Dev.* **47**, 261–268.

1852. Hinshaw, J.E. and Schmid, S.L. (1995). Dynamin self-assembles into rings suggesting a mechanism for coated vesicle budding. *Nature* **374**, 190–192.

1853. Hinton, H.E. (1970). Some little known surface structures. *In* "Insect Ultrastructure," *5th Symp. Roy. Entomol. Soc. Lond.*, (A.C. Neville, ed.). Blackwell: Oxford, pp. 41–58.

1854. Hinz, U., Giebel, B., and Campos-Ortega, J.A. (1994). The basic-helix-loop-helix domain of *Drosophila* lethal of scute protein is sufficient for proneural function and activates neurogenic genes. *Cell* **76**, 77–87.

1855. Hirata, J., Nakagoshi, H., Nabeshima, Y.-i., and Matsuzaki, F. (1995). Asymmetric segregation of the homeodomain protein Prospero during *Drosophila* development. *Nature* **377**, 627–630.

1856. Hiromi, Y., Mlodzik, M., West, S.R., Rubin, G.M., and Goodman, C.S. (1993). Ectopic expression of *seven-up* causes cell fate changes during ommatidial assembly. *Development* **118**, 1123–1135.

1857. Hirota, Y., Okabe, M., Imai, T., Kurusu, M., Yamamoto, A., Miyao, S., Nakamura, M., Sawamoto, K., and Okano, H. (1999). Musashi and Seven in absentia downregulate Tramtrack through distinct mechanisms in *Drosophila* eye development. *Mechs. Dev.* **87**, 93–101.

1858. Hirsch, J.A. and Aggarwal, A.K. (1995). Structure of the Even-skipped homeodomain complexed to AT-rich DNA: new perspectives on homeodomain specificity. *EMBO J.* **14**, 6280–6291.

1859. Ho, M.-W., Bolton, E., and Saunders, P.T. (1983). Bithorax phenocopy and pattern formation. I. Spatiotemporal characteristics of the phenocopy response. *Exp. Cell Biol.* **51**, 282–290.

1860. Ho, M.-W., Saunders, P.T., and Bolton, E. (1983). Bithorax phenocopy and pattern formation. II. A model of prepattern formation. *Exp. Cell Biol.* **51**, 291–299.

1861. Hobert, O. and Ruvkun, G. (1999). *Pax* genes in *Caenorhabditis elegans*: a new twist. *Trends Genet.* **15**, 214–216.

1862. Hobert, O. and Westphal, H. (2000). Functions of LIM-homeobox genes. *Trends Genet.* **16**, 75–83.

1863. Hodges, A. (1983). "Alan Turing: The Enigma." Simon & Schuster, New York.

1864. Hodges, R.S. (1992). Unzipping the secrets of coiled-coils. *Curr. Biol.* **2**, 122–124.

1865. Hodgkin, J. (1998). Seven types of pleiotropy. *Int. J. Dev. Biol.* **42**, 501–505.

1866. Hodgkin, N.M. and Bryant, P.J. (1978). Scanning electron microscopy of the adult of *Drosophila melanogaster. In* "The Genetics and Biology of *Drosophila*," Vol. 2c (M. Ashburner and T.R.F. Wright, eds.). Acad. Pr.: New York, pp. 337–358.

1867. Hodgson, J.W., Cheng, N.N., Sinclair, D.A.R., Kyba, M., Randsholt, N.B., and Brock, H.W. (1997). The *polyhomeotic* locus of *Drosophila melanogaster* is transcriptionally and post-transcriptionally regulated during embryogenesis. *Mechs. Dev.* **66**, 69–81.

1868. Hoey, T. and Levine, M. (1988). Divergent homeo box proteins recognize similar DNA sequences in *Drosophila*. *Nature* **332**, 858–861.

1869. Hofbauer, A. and Campos-Ortega, J.A. (1976). Cell clones and pattern formation: genetic eye mosaics in *Drosophila melanogaster*. *W. Roux's Arch.* **179**, 275–289.

1870. Hoffman, M. (1991). New role found for a common protein "motif". *Science* **253**, 742.

1871. Hoffmann, F.M. (1990). Developmental functions of *decapentaplegic*, a member of the TGF-β family in *Drosophila*. *In* "Genetics of Pattern Formation and Growth Control," *Symp. Soc. Dev. Biol.*, Vol. 48 (A.P. Mahowald, ed.). Wiley-Liss: New York, pp. 103–124.

1872. Hoffmann, F.M. and Goodman, W. (1987). Identification in transgenic animals of the *Drosophila decapentaplegic* sequences required for embryonic dorsal pattern formation. *Genes Dev.* **1**, 615–625.

1873. Hofstadter, D.R. (1979). "Gödel, Escher, Bach: An Eternal Golden Braid." Random House, New York.

1874. Hogan, B.L.M. (1994). Sorting out the signals. *Curr. Biol.* **4**, 1122–1124.

1875. Hoge, M. (1915). The influence of temperature on the development of a Mendelian character. *J. Exp. Zool.* **18**, 241–297.

1876. Holland, P.W.H. (1999). The future of evolutionary developmental biology. *Nature* **402 Suppl.**, C41–C44.

1877. Holland, P.W.H. (1999). Gene duplication: past, present and future. *Sems. Cell Dev. Biol.* **10**, 541–547.

1878. Holley, S.A., Geisler, R., and Nüsslein-Volhard, C. (2000). Control of *her1* expression during zebrafish somitogenesis by a *Delta*-dependent oscillator and an independent wave-front activity. *Genes Dev.* **14**, 1678–1690.

1879. Holliday, R. (1988). Successes and limitations of molecular biology. *J. Theor. Biol.* **132**, 253–262.

1880. Holliday, R. (1990). Mechanisms for the control of gene activity during development. *Biol. Rev.* **65**, 431–471.

1881. Holliday, R. (1990). Paradoxes between genetics and development. *J. Cell Sci.* **97**, 395–398.

1882. Hollingsworth, M.J. (1960). The morphology of intersexes in *Drosophila subobscura*. *J. Exp. Zool.* **143**, 123–151.

1883. Hollingsworth, M.J. (1964). Sex-combs of intersexes and the arrangement of the chaetae on the legs of *Drosophila*. *J. Morph.* **115**, 35–51.

1884. Holmes, F.L. (2000). Seymour Benzer and the definition of the gene. *In* "The Concept of the Gene in Development and Evolution: Historical and Epistemological Perspectives," (P.J. Beurton, R. Falk, and H.-J. Rheinberger, eds.). Cambridge Univ. Pr.: Cambridge, pp. 115–155.

1885. Holtfreter, J. (1951). Some aspects of embryonic induction. *Growth* **15 (Suppl.)**, 117–152.

1886. Holz, A., Meise, M., and Janning, W. (1997). Adepithelial cells in *Drosophila melanogaster*: origin and cell lineage. *Mechs. Dev.* **62**, 93–101.

1887. Homyk, T., Jr., Isono, K., and Pak, W.L. (1985). Developmental and physiological analysis of a conditional mutation affecting photoreceptor and optic lobe development in *Drosophila melanogaster*. *J. Neurogenet.* **2**, 309–324.

1888. Honda, H., Kodama, R., Takeuchi, T., Yamanaka, H., Watanabe, K., and Eguchi, G. (1984). Cell behaviour in a polygonal cell sheet. *J. Embryol. Exp. Morph.* **83 (Suppl.)**, 313–327.

1889. Honda, H., Tanemura, M., and Yoshida, A. (1990). Estimation of neuroblast numbers in insect neurogenesis using the lateral inhibition hypothesis of cell differentiation. *Development* **110**, 1349–1352.

1890. Honda, H. and Yamanaka, H. (1984). A computer simulation of geometrical configurations during cell division. *J. Theor. Biol.* **106**, 423–435.

1891. Honjo, T. (1996). The shortest path from the surface to the nucleus: RBP-Jκ/Su(H) transcription factor. *Genes to Cells* **1**, 1–9.

1892. Honma, T. and Goto, K. (2001). Complexes of MADS-box proteins are sufficient to convert leaves into floral organs. *Nature* **409**, 525–529.

1893. Hooper, J. (2001). Modulation of Hedgehog signaling by heterotrimeric G proteins. *Proc. 42nd Ann. Drosophila Res. Conf.* **Abstracts Vol.**, a334.

1894. Hooper, J.E. (1994). Distinct pathways for autocrine and paracrine Wingless signalling in *Drosophila* embryos. *Nature* **372**, 461–464.

1895. Hooper, J.E. and Scott, M.P. (1989). The *Drosophila patched* gene encodes a putative membrane protein required for segmental patterning. *Cell* **59**, 751–765.

1896. Hooper, N.M. (1998). Membrane biology: Do glycolipid microdomains really exist? *Curr. Biol.* **8**, R114–R116.

1897. Hopcroft, J.E. (1984). Turing machines. *Sci. Am.* **250 #5**, 86–98.

1898. Hopmann, R. and Miller, K. (2001). Bristle growth in the absence of actin bundles. *Proc. 42nd Ann. Drosophila Res. Conf.* **Abstracts Vol.**, a76. [Cf. 2002 *Cell* **108**, 233–246.]

1899. Hoppe, P.E. and Greenspan, R.J. (1986). Local function of the Notch gene for embryonic ectodermal pathway choice in *Drosophila*. *Cell* **46**, 773–783.

1900. Hoppe, P.E. and Greenspan, R.J. (1990). The *Notch* locus of *Drosophila* is required in epidermal cells for epidermal development. *Development* **109**, 875–885.

1901. Hoppler, S. and Bienz, M. (1994). Specification of a single cell type by a *Drosophila* homeotic gene. *Cell* **76**, 689–702.

1902. Horard, B., Tatout, C., Poux, S., and Pirrotta, V. (2000). Structure of a Polycomb response element and in vitro binding of Polycomb group complexes containing GAGA factor. *Mol. Cell. Biol.* **20**, 3187–3197.

1903. Horsfield, J., Penton, A., Secombe, J., Hoffmann, F.M., and Richardson, H. (1998). *decapentaplegic* is required for arrest in G1 phase during *Drosophila* eye development. *Development* **125**, 5069–5078.

1904. Horvitz, H.R. and Herskowitz, I. (1992). Mechanisms of asymmetric cell division: Two Bs or not two Bs, that is the question. *Cell* **68**, 237–255.

1905. Hoschuetzky, H., Aberle, H., and Kemler, R. (1994). β-catenin mediates the interaction of the cadherin-catenin complex with Epidermal Growth Factor Receptor. *J. Cell Biol.* **127**, 1375–1380.

1906. Hou, X.S. and Perrimon, N. (1997). The JAK-STAT pathway in *Drosophila*. *Trends Genet.* **13**, 105–110.

1907. Hough, C.D., Woods, D.F., Park, S., and Bryant, P.J. (1997). Organizing a functional junctional complex requires specific domains of the *Drosophila* MAGUK Discs large. *Genes Dev.* **11**, 3242–3253.

1908. Howard, K.R. and Struhl, G. (1990). Decoding positional information: regulation of the pair-rule gene *hairy*. *Development* **110**, 1223–1231.

1909. Howells, A.J. (1972). Levels of RNA and DNA in *Drosophila melanogaster* at different stages of development: a comparison between one bobbed and two phenotypically non-bobbed stocks. *Biochem. Genet.* **6**, 217–230.

1910. Howes, R. and Bray, S. (2000). Pattern formation: Wingless on the move. *Curr. Biol.* **10**, R222–R226.

1911. Howes, R., Wasserman, J.D., and Freeman, M. (1998). *In vivo* analysis of Argos structure-function. *J. Biol. Chem.* **273** #7, 4275–4281.

1912. Hoyle, H.D., Turner, F.R., and Raff, E.C. (2000). A transient specialization of the microtubule cytoskeleton is required for differentiation of the *Drosophila* visual system. *Dev. Biol.* **221**, 375–389.

1913. Hsieh, J.J.-D., Zhou, S., Chen, L., Young, D.B., and Hayward, S.D. (1999). CIR, a corepressor linking the DNA binding factor CBF1 to the histone deacetylase complex. *Proc. Natl. Acad. Sci. USA* **96**, 23–28.

1914. Hsieh, J.J.D., Henkel, T., Salmon, P., Robey, E., Peterson, M.G., and Hayward, S.D. (1996). Truncated mammalian Notch1 activates CBF1/RBPJκ-repressed genes by a mechanism resembling that of Epstein-Barr virus EBNA2. *Mol. Cell. Biol.* **16**, 952–959.

1915. Hsiung, F. and Moses, K. (2001). Star mediates Spi maturation via direct physical interaction. *Proc. 42nd Ann. Drosophila Res. Conf.* **Abstracts Vol.**, a170. [See also 2001 *Mechs. Dev.* **107**, 13–23.]

1916. Hsu, J.-C. and Perrimon, N. (1994). A temperature-sensitive MEK mutation demonstrates the conservation of the signaling pathways activated by receptor tyrosine kinases. *Genes Dev.* **8**, 2176–2187.

1917. Hsu, S.-C., Galceran, J., and Grosschedl, R. (1998). Modulation of transcriptional regulation by LEF-1 in response to Wnt-1 signaling and association with β-catenin. *Mol. Cell. Biol.* **18**, 4807–4818.

1918. Hsu, T., Gogos, J.A., Kirsh, S.A., and Kafatos, F.C. (1992). Multiple zinc finger forms resulting from developmentally regulated alternative splicing of a transcription factor gene. *Science* **257**, 1946–1950.

1919. Hsu, W., Zeng, L., and Costantini, F. (1999). Identification of a domain of Axin that binds to the serine/threonine protein phosphatase 2A and a self-binding domain. *J. Biol. Chem.* **274** #6, 3439–3445.

1920. Hsueh, Y.-P., Wang, T.-F., Yang, F.-C., and Sheng, M. (2000). Nuclear translocation and transcription regulation by the membrane-associated guanylate kinase CASK/LIN-2. *Nature* **404**, 298–302.

1921. Hu, X., Lee, E.-C., and Baker, N.E. (1995). Molecular analysis of *scabrous* mutant alleles from *Drosophila melanogaster* indicates a secreted protein with two functional domains. *Genetics* **141**, 607–617.

1922. Huang, A.M. and Rubin, G.M. (2000). A misexpression screen identifies genes that can modulate RAS1 pathway signaling in *Drosophila melanogaster. Genetics* **156**, 1219–1230.

1923. Huang, A.M., Rusch, J., and Levine, M. (1997). An anteroposterior Dorsal gradient in the *Drosophila* embryo. *Genes Dev.* **11**, 1963–1973.

1924. Huang, F. (1998). Syntagms in development and evolution. *Int. J. Dev. Biol.* **42**, 487–494.

1925. Huang, F., Dambly-Chaudière, C., and Ghysen, A. (1991). The emergence of sense organs in the wing disc of *Drosophila. Development* **111**, 1087–1095.

1926. Huang, F., van Helden, J., Dambly-Chaudière, C., and Ghysen, A. (1995). Contribution of the gene *extra-macrochaetae* to the precise positioning of bristles in *Drosophila. Roux's Arch. Dev. Biol.* **204**, 336–343.

1927. Huang, J.-D., Schwyter, D.H., Shirokawa, J.M., and Courey, A.J. (1993). The interplay between multiple enhancer and silencer elements defines the pattern of *decapentaplegic* expression. *Genes Dev.* **7**, 694–704.

1928. Huang, M.-L., Hsu, C.-H., and Chien, C.-T. (2000). The proneural gene *amos* promotes multiple dendritic neuron formation in the *Drosophila* peripheral nervous system. *Neuron* **25**, 57–67.

1929. Huang, X., Crute, B.E., Sun, C., Tang, Y.-Y., Kelley, J.J., III, Lewis, A.F., Hartman, K.L., Laue, T.M., Speck, N.A., and Bushweller, J.H. (1998). Overexpression, purification, and biophysical characterization of the heterodimerization domain of the core-binding factor β subunit. *J. Biol. Chem.* **273** #4, 2480–2487.

1930. Huang, X., Peng, J.W., Speck, N.A., and Bushweller, J.H. (1999). Solution structure of core binding factor β and map of the CBFα binding site. *Nature Struct. Biol.* **6**, 624–627.

1931. Huang, Y., Baker, R.T., and Fischer-Vize, J.A. (1995). Control of cell fate by a deubiquitinating enzyme encoded by the *fat facets* gene. *Science* **270**, 1828–1831.

1932. Huang, Y. and Fischer-Vize, J.A. (1996). Undifferentiated cells in the developing *Drosophila* eye influence facet assembly and require the Fat facets ubiquitin-specific protease. *Development* **122**, 3207–3216.

1933. Huang, Z. and Kunes, S. (1996). Hedgehog, transmitted along retinal axons, triggers neurogenesis in the developing visual centers of the *Drosophila* brain. *Cell* **86**, 411–422.

1934. Huang, Z. and Kunes, S. (1998). Signals transmitted along retinal axons in *Drosophila*: Hedgehog signal reception and the cell circuitry of lamina cartridge assembly. *Development* **125**, 3753–3764.

1935. Huang, Z.J., Edery, I., and Rosbash, M. (1993). PAS is a dimerization domain common to *Drosophila* Period and several transcription factors. *Nature* **364**, 259–262.

1936. Huber, A.H., Nelson, W.J., and Weis, W.I. (1997). Three-dimensional structure of the armadillo repeat region of β-catenin. *Cell* **90**, 871–882.

1937. Huber, O., Bierkamp, C., and Kemler, R. (1996). Cadherins and catenins in development. *Curr. Opin. Cell Biol.* **8**, 685–691.

1938. Hudson, J.B., Podos, S.D., Keith, K., Simpson, S.L., and Ferguson, E.L. (1998). The *Drosophila Medea* gene is required downstream of *dpp* and encodes a functional homolog of human Smad4. *Development* **125**, 1407–1420.

1939. Huggins, L.G., Cho, S.-H., Harrison, D., and Padgett, R.W. (2001). A dominant modifier screen identifies potential new components of the Dpp signal transduction pathway. *Proc. 42nd Ann. Drosophila Res. Conf.* **Abstracts Vol.**, a162.

1940. Hughes, S.C. and Krause, H.M. (2001). Establishment and maintenance of parasegmental compartments. *Development* **128**, 1109–1118.

1941. Huh, J.R., Yoo, S.J., and Hay, B.A. (2001). Visualizing activated caspases in *Drosophila. Proc. 42nd Ann. Drosophila Res. Conf.* **Abstracts Vol.**, a315.

1942. Huiskes, R., Ruimerman, R., van Lenthe, G.H., and Janssen, J.D. (2000). Effects of mechanical forces on maintenance and adaptation of form in trabecular bone. *Nature* **405**, 704–706.

1943. Hukriede, N.A. and Fleming, R.J. (1997). *Beaded of Goldschmidt*, an antimorphic allele of *Serrate*, encodes a protein lacking transmembrane and intracellular domains. *Genetics* **145**, 359–374.

1944. Hukriede, N.A., Gu, Y., and Fleming, R.J. (1997). A dominant-negative form of Serrate acts as a general antagonist of Notch activation. Development 124, 3427–3437.

1945. Hülskamp, M., Schröder, C., Pfeifle, C., Jäckle, H., and Tautz, D. (1989). Posterior segmentation of the Drosophila embryo in the absence of a maternal posterior organizer gene. Nature 338, 629–632.

1946. Hülskamp, M. and Tautz, D. (1991). Gap genes and gradients–the logic behind the gaps. BioEssays 13, 261–268.

1947. Hülsken, J., Behrens, J., and Birchmeier, W. (1994). Tumor-suppressor gene products in cell contacts: the cadherin-APC-armadillo connection. Curr. Opin. Cell Biol. 6, 711–716.

1948. Hummel, T., Krukkert, K., Roos, J., Davis, G., and Klämbt, C. (2000). Drosophila Futsch/22C10 is a MAP1B-like protein required for dendritic and axonal development. Neuron 26, 357–370.

1949. Hunt, T. (1989). Cytoplasmic anchoring proteins and the control of nuclear localization. Cell 59, 949–951.

1950. Hunter, T. (2000). Signaling–2000 and beyond. Cell 100, 113–127.

1951. Huppert, S.S., Jacobsen, T.L., and Muskavitch, M.A.T. (1997). Feedback regulation is central to Delta-Notch signalling required for Drosophila wing vein morphogenesis. Development 124, 3283–3291.

1952. Hursh, D.A., Padgett, R.W., and Gelbart, W.M. (1993). Cross regulation of decapentaplegic and Ultrabithorax transcription in the embryonic visceral mesoderm of Drosophila. Development 117, 1211–1222.

1953. Huse, M., Chen, Y.-G., Massagué, J., and Kuriyan, J. (1999). Crystal structure of the cytoplasmic domain of the Type I TGFβ receptor in complex with FKBP12. Cell 96, 425–436.

1954. Huwe, A.W., Spillane, M.A., and Bryant, P.J. (2001). The nuts and bolts of the cell: PDZ-containing proteins and their potential binding partners. Proc. 42nd Ann. Drosophila Res. Conf. Abstracts Vol., a329. [See also 2001 Annu. Rev. Neurosci. 24, 1–29.]

1955. Huxley, J.S. and de Beer, G.R. (1934). "The Elements of Experimental Embryology." Cambridge Univ. Pr., Cambridge.

1956. Hynes, R. (1985). Molecular biology of fibronectin. Annu. Rev. Cell Biol. 1, 67–90.

1957. Hynes, R.O. (1999). Cell adhesion: old and new questions. Trends Genet. 15, M33–M37.

1958. Ichimura, T., Isobe, T., Okuyama, T., Takahashi, N., Araki, K., Kuwano, R., and Takahashi, Y. (1988). Molecular cloning of cDNA coding for brain-specific 14-3-3 protein, a protein kinase-dependent activator of tyrosine and tryptophan hydroxylases. Proc. Natl. Acad. Sci. USA 85, 7084–7088.

1959. Ikeda, S., Kishida, S., Yamamoto, H., Murai, H., Koyama, S., and Kikuchi, A. (1998). Axin, a negative regulator of the Wnt signaling pathway, forms a complex with GSK-3β and β-catenin and promotes GSK-3β-dependent phosphorylation of β-catenin. EMBO J. 17, 1371–1384.

1960. Incardona, J.P. and Eaton, S. (2000). Cholesterol in signal transduction. Curr. Opin. Cell Biol. 12, 193–203.

1961. Indrasamy, H., McKenzie, J.A., Woods, R., and Batterham, P. (2001). The contribution of Notch signalling pathway to bristle asymmetry. Proc. 42nd Ann. Drosophila Res. Conf. Abstracts Vol., a269.

1962. Ingham, P. (1988). The molecular genetics of embryonic pattern formation in Drosophila. Nature 335, 25–34.

1963. Ingham, P., Martinez-Arias, A., Lawrence, P.A., and Howard, K. (1985). Expression of engrailed in the parasegment of Drosophila. Nature 317, 634–636.

1964. Ingham, P. and Smith, J. (1992). Crossing the threshold. Curr. Biol. 2, 465–467.

1965. Ingham, P.W. (1984). A gene that regulates the bithorax complex differentially in larval and adult cells of Drosophila. Cell 37, 815–823.

1966. Ingham, P.W. (1985). A clonal analysis of the requirement for the trithorax gene in the diversification of segments in Drosophila. J. Embryol. Exp. Morph. 89, 349–365.

1967. Ingham, P.W. (1985). Genetic control of the spatial pattern of selector gene expression in Drosophila. Cold Spr. Harb. Symp. Quant. Biol. 50, 201–208.

1968. Ingham, P.W. (1991). Segment polarity genes and cell patterning within the Drosophila body segment. Curr. Op. Gen. Dev. 1, 261–267.

1969. Ingham, P.W. (1993). Localized hedgehog activity controls spatial limits of wingless transcription in the Drosophila embryo. Nature 366, 560–562.

1970. Ingham, P.W. (1994). Hedgehog points the way. Curr. Biol. 4, 347–350.

1971. Ingham, P.W. (1996). Has the quest for a Wnt receptor finally frizzled out? Trends Genet. 12, 382–384.

1972. Ingham, P.W. (1998). Boning up on Hedgehog's movements. Nature 394, 16–17.

1973. Ingham, P.W. (1998). The patched gene in development and cancer. Curr. Opin. Gen. Dev. 8, 88–94.

1974. Ingham, P.W. (1998). Transducing Hedgehog: the story so far. EMBO J. 17, 3505–3511. [See also 2001 Genes Dev. 15, 3059–3087.]

1975. Ingham, P.W. (2000). Hedgehog signalling: how cholesterol modulates the signal. Curr. Biol. 10, R180–R183.

1976. Ingham, P.W. and Fietz, M.J. (1995). Quantitative effects of hedgehog and decapentaplegic activity on the patterning of the Drosophila wing. Curr. Biol. 5, 432–440.

1977. Ingham, P.W. and Hidalgo, A. (1993). Regulation of wingless transcription in the Drosophila embryo. Development 117, 283–291.

1978. Ingham, P.W. and Martinez Arias, A. (1992). Boundaries and fields in early embryos. Cell 68, 221–235.

1979. Ingham, P.W. and Nakano, Y. (1990). Cell patterning and segment polarity genes in Drosophila. Dev. Growth Differ. 32, 563–574.

1980. Ingham, P.W., Nystedt, S., Nakano, Y., Brown, W., Stark, D., van den Heuvel, M., and Taylor, A.M. (2000). Patched represses the Hedgehog signalling pathway by promoting modification of the Smoothened protein. Curr. Biol. 10, 1315–1318.

1981. Ingham, P.W., Pinchin, S.M., Howard, K.R., and Ish-Horowicz, D. (1985). Genetic analysis of the hairy locus in Drosophila melanogaster. Genetics 111, 463–486.

1982. Ingham, P.W., Taylor, A.M., and Nakano, Y. (1991). Role of the Drosophila patched gene in positional signalling. Nature 353, 184–187.

1983. Inoue, H., Imamura, T., Ishidou, Y., Takase, M., Udagawa, Y., Oka, Y., Tsuneizumi, K., Tabata, T., Miyazono, K., and Kawabata, M. (1998). Interplay of signal mediators of Decapentaplegic (Dpp): molecular characterization of Mothers against dpp, Medea, and Daughters against dpp. Mol. Biol. Cell 9, 2145–2156.

1984. Ip, Y.T. and Hemavathy, K. (1997). Drosophila development: Delimiting patterns by repression. Curr. Biol. 7, R216–R218.

1985. Ip, Y.T., Park, R.E., Kosman, D., Yazdanbakhsh, K., and Levine, M. (1992). *dorsal-twist* interactions establish *snail* expression in the presumptive mesoderm of the *Drosophila* embryo. *Genes Dev.* **6**, 1518–1530.

1986. Irish, V., Lehmann, R., and Akam, M. (1989). The *Drosophila* posterior-group gene *nanos* functions by repressing *hunchback* activity. *Nature* **338**, 646–648.

1987. Irish, V.F. and Gelbart, W.M. (1987). The decapentaplegic gene is required for dorsal-ventral patterning of the *Drosophila* embryo. *Genes Dev.* **1**, 868–879.

1988. Irvine, K.D. (1999). Fringe, Notch, and making developmental boundaries. *Curr. Opin. Gen. Dev.* **9**, 434–441.

1989. Irvine, K.D., Botas, J., Jha, S., Mann, R.S., and Hogness, D.S. (1993). Negative autoregulation by *Ultrabithorax* controls the level and pattern of its expression. *Development* **117**, 387–399.

1990. Irvine, K.D., Helfand, S.L., and Hogness, D.S. (1991). The large upstream control region of the *Drosophila* homeotic gene *Ultrabithorax*. *Development* **111**, 407–424.

1991. Irvine, K.D. and Vogt, T.F. (1997). Dorsal-ventral signaling in limb development. *Curr. Opin. Cell Biol.* **9**, 867–876.

1992. Irvine, K.D. and Wieschaus, E. (1994). *fringe*, a boundary-specific signaling molecule, mediates interactions between dorsal and ventral cells during *Drosophila* wing development. *Cell* **79**, 595–606.

1993. Irwin, T. (1988). "Aristotle's First Principles." Clarendon Pr., Oxford.

1994. Isaksson, A., Peverali, F.A., Kockel, L., Mlodzik, M., and Bohmann, D. (1997). The deubiquitination enzyme Fat facets negatively regulates RTK/Ras/MAPK signalling during *Drosophila* eye development. *Mechs. Dev.* **68**, 59–67.

1995. Ishimaru, S., Williams, R., Clark, E., Hanafusa, H., and Gaul, U. (1999). Activation of the *Drosophila* C3G leads to cell fate changes and overproliferation during development, mediated by the RAS-MAPK pathway and RAP1. *EMBO J.* **18**, 145–155.

1996. Ishitani, T., Ninomiya-Tsuji, J., Nagai, S.-i., Nishita, M., Meneghini, M., Barker, N., Waterman, M., Bowerman, B., Clevers, H., Shibuya, H., and Matsumoto, K. (1999). The TAK1-NLK-MAPK-related pathway antagonizes signalling between β-catenin and transcription factor TCF. *Nature* **399**, 798–802.

1997. Ives, P.T. (1935). The temperature-effective period of the scute-1 phenotype. *Proc. Natl. Acad. Sci. USA* **21**, 646–650.

1998. Ives, P.T. (1939). The effects of high temperature on bristle frequencies in scute and wild-type males of *Drosophila melanogaster*. *Genetics* **24**, 315–331.

1999. Iwanami, M. and Hiromi, Y. (2001). Regulating the neuronal induction by Argos and Sprouty in the eye development. *Proc. 42nd Ann. Drosophila Res. Conf.* **Abstracts Vol.**, a256.

2000. Jack, J. and DeLotto, Y. (1995). Structure and regulation of a complex locus: the *cut* gene of *Drosophila*. *Genetics* **139**, 1689–1700.

2001. Jack, J., Dorsett, D., Delotto, Y., and Liu, S. (1991). Expression of the *cut* locus in the *Drosophila* wing margin is required for cell type specification and is regulated by a distant enhancer. *Development* **113**, 735–747.

2002. Jack, T. and McGinnis, W. (1990). Establishment of the *Deformed* expression stripe requires the combinatorial action of coordinate, gap and pair-rule proteins. *EMBO J.* **9**, 1187–1198.

2003. Jackson, P.D. and Hoffmann, F.M. (1994). Embryonic expression patterns of the *Drosophila decapentaplegic* gene:

2004. Jackson, S.M., Nakato, H., Sugiura, M., Jannuzi, A., Oakes, R., Kaluza, V., Golden, C., and Selleck, S.B. (1997). *dally*, a *Drosophila* glypican, controls cellular responses to the TGF-β-related morphogen, Dpp. *Development* **124**, 4113–4120.

2005. Jacob, F. (1977). Evolution and tinkering. *Science* **196**, 1161–1166.

2006. Jacob, L., Opper, M., Metzroth, B., Phannavong, B., and Mechler, B.M. (1987). Structure of the *l(2)gl* gene of *Drosophila* and delimitation of its tumor suppressor domain. *Cell* **50**, 215–225.

2007. Jacobs, D., Glossip, D., Xing, H., Muslin, A.J., and Kornfeld, K. (1999). Multiple docking sites on substrate proteins form a modular system that mediates recognition by ERK MAP kinase. *Genes Dev.* **13**, 163–175.

2008. Jacobsen, T.L., Brennan, K., Martinez Arias, A., and Muskavitch, M.A.T. (1998). *Cis*-interactions between Delta and Notch modulate neurogenic signalling in *Drosophila*. *Development* **125**, 4531–4540. [Cf. 2002 *Dev. Biol.* **241**, 313–326.]

2009. Jacobson, M.D., Weil, M., and Raff, M.C. (1997). Programmed cell death in animal development. *Cell* **88**, 347–354.

2010. Jacquemin, P., Hwang, J.-J., Martial, J.A., Dollé, P., and Davidson, I. (1996). A novel family of developmentally regulated mammalian transcription factors containing the TEA/ATTS DNA binding domain. *J. Biol. Chem.* **271** #36, 21775–21785.

2011. Jaffe, L., Ryoo, H.-D., and Mann, R.S. (1997). A role for phosphorylation by casein kinase II in modulating Antennapedia activity in *Drosophila*. *Genes Dev.* **11**, 1327–1340.

2012. Jagla, K., Bellard, M., and Frasch, M. (2001). A cluster of *Drosophila* homeobox genes involved in mesoderm differentiation programs. *BioEssays* **23**, 125–133.

2013. Jagla, K., Jagla, T., Heitzler, P., Dretzen, G., Bellard, F., and Bellard, M. (1997). *ladybird*, a tandem of homeobox genes that maintain late *wingless* expression in terminal and dorsal epidermis of the *Drosophila* embryo. *Development* **124**, 91–100.

2014. Jakobs, R., de Lorenzo, C., Spiess, E., Strand, D., and Mechler, B.M. (1996). Homo-oligomerization domains in the *lethal(2)giant larvae* tumor suppressor protein, p127 of *Drosophila*. *J. Mol. Biol.* **264**, 484–496.

2015. James, A.A. and Bryant, P.J. (1981). Mutations causing pattern deficiencies and duplications in the imaginal wing disk of *Drosophila melanogaster*. *Dev. Biol.* **85**, 39–54.

2016. James, R.M., Klerkx, A.H.E.M., Keighren, M., Flockhart, J.H., and West, J.D. (1995). Restricted distribution of tetraploid cells in mouse tetraploid-diploid chimaeras. *Dev. Biol.* **167**, 213–226.

2017. Jan, Y.-N. and Jan, L.Y. (2000). Polarity in cell division: what frames thy fearful symmetry? *Cell* **100**, 599–602.

2018. Jan, Y.N. and Jan, L.Y. (1993). The peripheral nervous system. *In* "The Development of *Drosophila melanogaster*," Vol. 2 (M. Bate and A. Martinez Arias, eds.). Cold Spring Harbor Lab. Pr.: Plainview, N. Y., pp. 1207–1244.

2019. Jan, Y.N. and Jan, L.Y. (1994). Genetic control of cell fate specification in *Drosophila* peripheral nervous system. *Annu. Rev. Genet.* **28**, 373–393.

2020. Jan, Y.N. and Jan, L.Y. (1995). Maggot's hair and bug's eye: role of cell interactions and intrinsic factors in cell fate specification. *Neuron* **14**, 1–5.

separate regulatory elements control blastoderm expression and lateral ectodermal expression. *Dev. Dynamics* **199**, 28–44.

2021. Jan, Y.N. and Jan, L.Y. (1998). Asymmetric cell division. *Nature* **392**, 775–778.

2022. Jang, C.C. and Sun, Y.H. (2001). Molecular analysis of the *eyg-toe* gene complex in *Drosophila*. *Proc. 42nd Ann. Drosophila Res. Conf.* **Abstracts Vol.**, a198.

2023. Janknecht, R. and Hunter, T. (1996). A growing coactivator network. *Nature* **383**, 22–23.

2024. Janknecht, R. and Hunter, T. (1996). Transcriptional control: versatile molecular glue. *Curr. Biol.* **6**, 951–954.

2025. Janknecht, R., Sander, C., and Pongs, O. (1991). (HX)$_n$ repeats: a pH-controlled protein-protein interaction motif of eukaryotic transcription factors? *FEBS Letters* **295**, 1–2.

2026. Janning, W. (1978). Gynandromorph fate maps in *Drosophila*. *In* "Genetic Mosaics and Cell Differentiation," *Results and Problems in Cell Differentiation, Vol. 9*, (W.J. Gehring, ed.). Springer-Verlag: Berlin, pp. 1–28.

2027. Janning, W. (1997). *FlyView*, a *Drosophila* image database, and other *Drosophila* databases. *Sems. Cell Dev. Biol.* **8**, 469–475.

2028. Janning, W., Labhart, C., and Nöthiger, R. (1983). Cell lineage restrictions in the genital disc of *Drosophila* revealed by *Minute* gynandromorphs. *Roux's Arch. Dev. Biol.* **192**, 337–346.

2029. Janning, W., Pfreundt, J., and Tiemann, R. (1979). The distribution of anlagen in the early embryo of *Drosophila*. *In* "Cell Lineage, Stem Cells and Cell Determination," (N. Le Douarin, ed.). North-Holland Pub. Co.: Amsterdam, pp. 83–98.

2030. Jans, D.A. (1994). Nuclear signaling pathways for polypeptide ligands and their membrane receptors? *FASEB J.* **8**, 841–847. [See also 2001 *Trends Genet.* **17**, 625–626.]

2031. Jans, D.A. and Hassan, G. (1998). Nuclear targeting by growth factors, cytokines, and their receptors: a role in signaling? *BioEssays* **20**, 400–411. [See also 2001 *Cell* **106**, 1–4.]

2032. Jans, D.A. and Hübner, S. (1996). Regulation of protein transport to the nucleus: central role of phosphorylation. *Physiol. Rev.* **76**, 651–685.

2033. Jans, D.A., Xiao, C.-Y., and Lam, M.H.C. (2000). Nuclear targeting signal recognition: a key control point in nuclear transport? *BioEssays* **22**, 532–544.

2034. Janssens, H. and Gehring, W.J. (1999). Isolation and characterization of *drosocrystallin*, a lens crystallin gene of *Drosophila melanogaster*. *Dev. Biol.* **207**, 204–214.

2035. Jantzen, H.-M., Admon, A., Bell, S.P., and Tjian, R. (1990). Nucleolar transcription factor hUBF contains a DNA-binding motif with homology to HMG proteins. *Nature* **344**, 830–836.

2036. Jarman, A.P. (1996). Epithelial polarity in the *Drosophila* compound eye: eyes left or right? *Trends Genet.* **12**, 121–123.

2037. Jarman, A.P. (2000). Developmental genetics: Vertebrates and insects see eye to eye. *Curr. Biol.* **10**, R857–R859.

2038. Jarman, A.P. and Ahmed, I. (1998). The specificity of proneural genes in determining *Drosophila* sense organ identity. *Mechs. Dev.* **76**, 117–125.

2039. Jarman, A.P., Brand, M., Jan, L.Y., and Jan, Y.N. (1993). The regulation and function of the helix-loop-helix gene, *asense*, in *Drosophila* neural precursors. *Development* **119**, 19–29.

2040. Jarman, A.P., Grau, Y., Jan, L.Y., and Jan, Y.N. (1993). *atonal* is a proneural gene that directs chordotonal organ formation in the *Drosophila* peripheral nervous system. *Cell* **73**, 1307–1321.

2041. Jarman, A.P., Grell, E.H., Ackerman, L., Jan, L.Y., and Jan, Y.N. (1994). *atonal* is the proneural gene for *Drosophila* photoreceptors. *Nature* **369**, 398–400.

2042. Jarman, A.P., Sun, Y., Jan, L.Y., and Jan, Y.N. (1995). Role of the proneural gene, *atonal*, in formation of *Drosophila* chordotonal organs and photoreceptors. *Development* **121**, 2019–2030.

2043. Jarriault, S., Brou, C., Logeat, F., Schroeter, E.H., Kopan, R., and Israel, A. (1995). Signalling downstream of activated mammalian Notch. *Nature* **377**, 355–358.

2044. Järvilehto, M. (1985). The eye: vision and perception. *In* "Comprehensive Insect Physiology, Biochemistry, and Pharmacology," Vol. 6 (G.A. Kerkut and L.I. Gilbert, eds.). Pergamon Pr.: New York, pp. 355–429.

2045. Jaw, T.J., You, L.-R., Knoepfler, P.S., Yao, L.-C., Pai, C.-Y., Tang, C.-Y., Chang, L.-P., Berthelsen, J., Blasi, F., Kamps, M.P., and Sun, Y.H. (2000). Direct interaction of two homeoproteins, Homothorax and Extradenticle, is essential for EXD nuclear localization and function. *Mechs. Dev.* **91**, 279–291.

2046. Jayaraman, P.-S., Hirst, K., and Goding, C.R. (1994). The activation domain of a basic helix-loop-helix protein is masked by repressor interaction with domains distinct from that required for transcription regulation. *EMBO J.* **13**, 2192–2199.

2047. Jaynes, J.B. and O'Farrell, P.H. (1988). Activation and repression of transcription by homoeodomain-containing proteins that bind a common site. *Nature* **336**, 744–749.

2048. Jaynes, J.B. and O'Farrell, P.H. (1991). Active repression of transcription by the Engrailed homeodomain protein. *EMBO J.* **10**, 1427–1433.

2049. Jazwinska, A., Kirov, N., Wieschaus, E., Roth, S., and Rushlow, C. (1999). The *Drosophila* gene *brinker* reveals a novel mechanism of Dpp target gene regulation. *Cell* **96**, 563–573.

2050. Jazwinska, A., Rushlow, C., and Roth, S. (1999). The role of *brinker* in mediating the graded response to Dpp in early *Drosophila* embryos. *Development* **126**, 3323–3334.

2051. Jen, W.-C., Gawantka, V., Pollet, N., Niehrs, C., and Kintner, C. (1999). Periodic repression of Notch pathway genes governs the segmentation of *Xenopus* embryos. *Genes Dev.* **13**, 1486–1499.

2052. Jennings, B., de Celis, J., Delidakis, C., Preiss, A., and Bray, S. (1995). Role of *Notch* and *achaete-scute* complex in the expression of *Enhancer of split* bHLH proteins. *Development* **121**, 3745–3752.

2053. Jennings, B., Preiss, A., Delidakis, C., and Bray, S. (1994). The Notch signalling pathway is required for *Enhancer of split* bHLH protein expression during neurogenesis in the *Drosophila* embryo. *Development* **120**, 3537–3548.

2054. Jennings, B.H., Tyler, D.M., and Bray, S.J. (1999). Target specificities of *Drosophila* Enhancer of split basic helix-loop-helix proteins. *Mol. Cell. Biol.* **19**, 4600–4610.

2055. Jhaveri, D., Sen, A., Reddy, V., and Rodrigues, V. (2000). Sense organ identity in the *Drosophila* antenna is specified by the expression of the proneural gene *atonal*. *Mechs. Dev.* **99**, 101–111.

2056. Jhaveri, D., Sen, A., and Rodrigues, V. (2000). Mechanisms underlying olfactory neuronal connectivity in *Drosophila* – the Atonal lineage organizes the periphery while sensory neurons and glia pattern the olfactory lobe. *Dev. Biol.* **226**, 73–87.

2057. Jiang, J., Hoey, T., and Levine, M. (1991). Autoregulation of a segmentation gene in *Drosophila*: combinatorial interaction of the *even-skipped* homeo box protein with a distal enhancer element. *Genes Dev.* **5**, 265–277.

2058. Jiang, J. and Struhl, G. (1995). Protein kinase A and Hedgehog signaling in *Drosophila* limb development. *Cell* **80**, 563–572.

2059. Jiang, J. and Struhl, G. (1996). Complementary and mutually exclusive activities of Decapentaplegic and Wingless organize axial patterning during *Drosophila* leg development. *Cell* **86**, 401–409.

2060. Jiang, J. and Struhl, G. (1998). Regulation of the Hedgehog and Wingless signalling pathways by the F-box/WD40-repeat protein Slimb. *Nature* **391**, 493–496.

2061. Jiang, Y.-J., Aerne, B.L., Smithers, L., Haddon, C., Ish-Horowicz, D., and Lewis, J. (2000). Notch signalling and the synchronization of the somite segmentation clock. *Nature* **408**, 475–479.

2062. Jiménez, F. and Modolell, J. (1993). Neural fate specification in *Drosophila*. *Curr. Op. Gen. Dev.* **3**, 626–632.

2063. Jiménez, G. and Ish-Horowicz, D. (1997). A chimeric Enhancer-of-split transcriptional activator drives neural development and *achaete-scute* expression. *Mol. Cell. Biol.* **17**, 4355–4362.

2064. Jiménez, G., Paroush, Z., and Ish-Horowicz, D. (1997). Groucho acts as a corepressor for a subset of negative regulators, including Hairy and Engrailed. *Genes Dev.* **11**, 3072–3082.

2065. Jiménez, G., Pinchin, S.M., and Ish-Horowicz, D. (1996). *In vivo* interactions of the *Drosophila* Hairy and Runt transcriptional repressors with target promoters. *EMBO J.* **15**, 7088–7098.

2066. Jiménez, G., Verrijzer, C.P., and Ish-Horowicz, D. (1999). A conserved motif in Goosecoid mediates Groucho-dependent repression in *Drosophila* embryos. *Mol. Cell. Biol.* **19**, 2080–2087.

2067. Jin, M.-H., Sawamoto, K., Ito, M., and Okano, H. (2000). The interaction between the *Drosophila* secreted protein Argos and the Epidermal Growth Factor Receptor inhibits dimerization of the receptor and binding of secreted Spitz to the receptor. *Mol. Cell. Biol.* **20**, 2098–2107.

2068. Joazeiro, C.A.P., Wing, S.S., Huang, H.-k., Leverson, J.D., Hunter, T., and Liu, Y.-C. (1999). The tyrosine kinase negative regulator c-Cbl as a RING-type, E2-dependent ubiquitin-protein ligase. *Science* **286**, 309–312.

2069. Jockusch, E.L., Nulsen, C., Newfeld, S.J., and Nagy, L.M. (2000). Leg development in flies versus grasshoppers: differences in *dpp* expression do not lead to differences in the expression of downstream components of the leg patterning pathway. *Development* **127**, 1617–1626.

2070. Johansen, K.M., Fehon, R.G., and Artavanis-Tsakonas, S. (1989). The *Notch* gene product is a glycoprotein expressed on the cell surface of both epidermal and neuronal precursor cells during *Drosophila* development. *J. Cell Biol.* **109**, 2427–2440.

2071. John, A., Smith, S.T., and Jaynes, J.B. (1995). Inserting the Ftz homeodomain into Engrailed creates a dominant transcriptional repressor that specifically turns off Ftz target genes in vivo. *Development* **121**, 1801–1813.

2072. Johnson, R.L., Grenier, J.K., and Scott, M.P. (1995). *patched* overexpression alters wing disc size and pattern: transcriptional and post-transcriptional effects on *hedgehog* targets. *Development* **121**, 4161–4170.

2073. Johnson, R.L., Milenkovic, L., and Scott, M.P. (2000). In vivo functions of the Patched protein: requirement of the C terminus for target gene inactivation but not Hedgehog sequestration. *Molec. Cell* **6**, 467–478.

2074. Johnson, R.L. and Scott, M.P. (1997). Control of cell growth and fate by *patched* genes. *Cold Spr. Harb. Symp. Quant. Biol.* **62**, 205–215.

2075. Johnson, R.L. and Scott, M.P. (1998). New players and puzzles in the Hedgehog signaling pathway. *Curr. Opin. Gen. Dev.* **8**, 450–456.

2076. Johnson, R.L. and Tabin, C. (1995). The long and short of *hedgehog* signaling. *Cell* **81**, 313–316.

2077. Johnson, S.A. and Milner, M.J. (1987). The final stages of wing development in *Drosophila melanogaster*. *Tissue Cell* **19**, 505–513.

2078. Johnston, L.A. (1998). Uncoupling growth from the cell cycle. *BioEssays* **20**, 283–286.

2079. Johnston, L.A. and Edgar, B.A. (1998). Wingless and Notch regulate cell-cycle arrest in the developing *Drosophila* wing. *Nature* **394**, 82–84.

2080. Johnston, L.A., Ostrow, B.D., Jasoni, C., and Blochlinger, K. (1998). The homeobox gene *cut* interacts genetically with the homeotic genes *proboscipedia* and *Antennapedia*. *Genetics* **149**, 131–142.

2081. Johnston, L.A., Prober, D.A., Edgar, B.A., Eisenman, R.N., and Gallant, P. (1999). *Drosophila myc* regulates cellular growth during development. *Cell* **98**, 779–790.

2082. Johnston, L.A. and Schubiger, G. (1996). Ectopic expression of *wingless* in imaginal discs interferes with *decapentaplegic* expression and alters cell determination. *Development* **122**, 3519–3529.

2083. Johnston, N.V. (1966). The pattern of abdominal microchaetae in *Drosophila*. *Aust. J. Biol. Sci.* **19**, 155–166.

2084. Johnston, S.H., Rauskolb, C., Wilson, R., Prabhakaran, B., Irvine, K.D., and Vogt, T.F. (1997). A family of mammalian *Fringe* genes implicated in boundary determination and the *Notch* pathway. *Development* **124**, 2245–2254.

2085. Joliot, A., Maizel, A., Rosenberg, D., Trembleau, A., Dupas, S., Volovitch, M., and Prochiantz, A. (1998). Identification of a signal sequence necessary for the unconventional secretion of Engrailed homeoprotein. *Curr. Biol.* **8**, 856–863.

2086. Jones, C., Reifegerste, R., and Moses, K. (2001). Molecular and functional characterization of a dominant modifier of hedgehog. *Proc. 42nd Ann. Drosophila Res. Conf.* **Abstracts Vol.**, a199.

2087. Jones, D.H., Ley, S., and Aitken, A. (1995). Isoforms of 14-3-3 protein can form homo- and heterodimers in vivo and in vitro: implications for function as adapter proteins. *FEBS Letters* **368**, 55–58.

2088. Jones, K.H., Liu, J., and Adler, P.N. (1996). Molecular analysis of EMS-induced *frizzled* mutations in *Drosophila melanogaster*. *Genetics* **142**, 205–215.

2089. Jones, N. (1990). Transcriptional regulation by dimerization: two sides to an incestuous relationship. *Cell* **61**, 9–11.

2090. Jones, N.A., Kuo, Y.M., Sun, Y.H., and Beckendorf, S.K. (1998). The *Drosophila Pax* gene *eye gone* is required for embryonic salivary gland duct development. *Development* **125**, 4163–4174.

2091. Jones, S. and Thornton, J.M. (1996). Principles of protein-protein interactions. *Proc. Natl. Acad. Sci. USA* **93**, 13–20.

2092. Jönsson, F. and Knust, E. (1996). Distinct functions of the *Drosophila* genes *Serrate* and *Delta* revealed by ectopic expression during wing development. *Dev. Genes Evol.* **206**, 91–101.

2093. Jordan, J.D., Landau, E.M., and Iyengar, R. (2000). Signaling networks: the origins of cellular multitasking. *Cell* **103**, 193–200.

2094. Jorgensen, E.M. and Garber, R.L. (1987). Function and misfunction of the two promoters of the *Drosophila Antennapedia* gene. *Genes Dev.* **1**, 544–555.

2095. Jouve, C., Palmeirim, I., Henrique, D., Beckers, J., Gossler, A., Ish-Horowicz, D., and Pourquié, O. (2000). Notch signalling is required for cyclic expression of the hairy-like gene *HES1* in the presomitic mesoderm. *Development* **127**, 1421–1429. [See also 2001 *Genes Dev.* **15**, 2642–2647.]

2096. Ju, B.-G., Jeong, S., Bae, E., Hyun, S., Carroll, S.B., Yim, J., and Kim, J. (2000). Fringe forms a complex with Notch. *Nature* **405**, 191–195.

2097. Judd, B.H. (1998). Genes and chromomeres: a puzzle in three dimensions. *Genetics* **150**, 1–9.

2098. Jun, S. and Desplan, C. (1996). Cooperative interactions between paired domain and homeodomain. *Development* **122**, 2639–2650.

2099. Jun, S., Wallen, R.V., Goriely, A., Kalionis, B., and Desplan, C. (1998). Lune/eye gone, a Pax-like protein, uses a partial paired domain and a homeodomain for DNA recognition. *Proc. Natl. Acad. Sci. USA* **95**, 13720–13725.

2100. Jung, H.-S., Francis-West, P.H., Widelitz, R.B., Jiang, T.-X., Ting-Berreth, S., Tickle, C., Wolpert, L., and Chuong, C.-M. (1998). Local inhibitory action of BMPs and their relationships with activators in feather formation: implications for periodic patterning. *Dev. Biol.* **196**, 11–23.

2101. Jürgens, G. (1988). Head and tail development of the *Drosophila* embryo involves *spalt*, a novel homeotic gene. *EMBO J.* **7**, 189–196.

2102. Jürgens, G. and Gateff, E. (1979). Pattern specification in imaginal discs of *Drosophila melanogaster*. Developmental analysis of a temperature-sensitive mutant producing duplicated legs. *W. Roux's Arch.* **186**, 1–25.

2103. Jürgens, G. and Hartenstein, V. (1993). The terminal regions of the body pattern. *In* "The Development of *Drosophila melanogaster*," Vol. 1 (M. Bate and A. Martinez Arias, eds.). Cold Spring Harbor Lab. Pr.: Plainview, N. Y., pp. 687–746.

2104. Jürgens, G., Lehmann, R., Schardin, M., and Nüsslein-Volhard, C. (1986). Segmental organisation of the head in the embryo of *Drosophila melanogaster*. *Roux's Arch. Dev. Biol.* **195**, 359–377.

2105. Jürgens, G., Wieschaus, E., Nüsslein-Volhard, C., and Kluding, H. (1984). Mutations affecting the pattern of the larval cuticle in *Drosophila melanogaster*. II. Zygotic loci on the third chromosome. *Roux's Arch. Dev. Biol.* **193**, 283–295.

2106. Jursnich, V.A. and Burtis, K.C. (1993). A positive role in differentiation for the male doublesex protein of *Drosophila*. *Dev. Biol.* **155**, 235–249.

2107. Jursnich, V.A., Fraser, S.E., Held, L.I., Jr., Ryerse, J., and Bryant, P.J. (1990). Defective gap-junctional communication associated with imaginal disc overgrowth and degeneration caused by mutations of the *dco* gene in *Drosophila*. *Dev. Biol.* **140**, 413–429.

2108. Kadlec, L., Ghabrial, A., and Schüpbach, T. (2001). The *gritz* gene encodes a novel putative ligand of the *Drosophila* Egf receptor. *Proc. 42nd Ann. Drosophila Res. Conf.* **Abstracts Vol.**, a152.

2109. Kadonaga, J.T. (1998). Eukaryotic transcription: an interlaced network of transcription factors and chromatin-modifying machines. *Cell* **92**, 307–313.

2110. Kadowaki, T., Wilder, E., Klingensmith, J., Zachary, K., and Perrimon, N. (1996). The segment polarity gene *porcupine* encodes a putative multitransmembrane protein involved in Wingless processing. *Genes Dev.* **10**, 3116–3128.

2111. Kafatos, F.C. (1976). Sequential cell polymorphism: a fundamental concept in developmental biology. *Adv. Insect Physiol.* **12**, 1–15.

2112. Kagoshima, H., Akamatsu, Y., Ito, Y., and Shigesada, K. (1996). Functional dissection of the α and β subunits of transcription factor PEBP2 and the redox susceptibility of its DNA binding activity. *J. Biol. Chem.* **271** #51, 33074–33082.

2113. Kagoshima, H., Shigesada, K., Satake, M., Ito, Y., Miyoshi, H., Ohki, M., Pepling, M., and Gergen, P. (1993). The Runt domain identifies a new family of heteromeric transcriptional regulators. *Trends Genet.* **9**, 338–341.

2114. Kal, A.J., Mahmoudi, T., Zak, N.B., and Verrijzer, C.P. (2000). The *Drosophila* Brahma complex is an essential coactivator for the *trithorax* group protein Zeste. *Genes Dev.* **14**, 1058–1071.

2115. Kalderon, D. (1995). Responses to Hedgehog. *Curr. Biol.* **5**, 580–582.

2116. Kalderon, D. (1997). Hedgehog signalling: Ci complex cuts and clasps. *Curr. Biol.* **7**, R759–R762.

2117. Kalderon, D. (2000). Transducing the Hedgehog signal. *Cell* **103**, 371–374.

2118. Kalderon, D. and Rubin, G.M. (1988). Isolation and characterization of *Drosophila* cAMP-dependent protein kinase genes. *Genes Dev.* **2**, 1539–1556.

2119. Kalionis, B. and O'Farrell, P.H. (1993). A universal target sequence is bound in vitro by diverse homeodomains. *Mechs. Dev.* **43**, 57–70.

2120. Kalmes, A., Merdes, G., Neumann, B., Strand, D., and Mechler, B.M. (1996). A serine-kinase associated with the p127-*l(2)gl* tumour suppressor of *Drosophila* may regulate the binding of p127 to nonmuscle myosin II heavy chain and the attachment of p127 to the plasma membrane. *J. Cell Sci.* **109**, 1359–1368.

2121. Kalthoff, K. (2001). "Analysis of Biological Development." McGraw-Hill, New York.

2122. Kamachi, Y., Uchikawa, M., and Kondoh, H. (2000). Pairing SOX off with partners in the regulation of embryonic development. *Trends Genet.* **16**, 182–187.

2123. Kamashev, D.E., Balandina, A.V., and Karpov, V.L. (2000). Tramtrack protein-DNA interactions. *J. Biol. Chem.* **275** #46, 36056–36061.

2124. Kamei, Y., Xu, L., Heinzel, T., Torchia, J., Kurokawa, R., Gloss, B., Lin, S.-C., Heyman, R.A., Rose, D.W., Glass, C.K., and Rosenfeld, M.G. (1996). A CBP integrator complex mediates transcriptional activation and AP-1 inhibition by nuclear receptors. *Cell* **85**, 403–414.

2125. Kammermeyer, K.L. and Wadsworth, S.C. (1987). Expression of *Drosophila* epidermal growth factor receptor homologue in mitotic cell populations. *Development* **100**, 201–210.

2126. Kango-Singh, M., Singh, A., and Sun, Y.H. (2001). Requirement of *hedgehog* (*hh*) for the induction of ectopic eyes by the *Drosophila* PAX6 gene–*eyeless* (*ey*). *Proc. 42nd Ann. Drosophila Res. Conf.* **Abstracts Vol.**, a47.

2127. Kania, M.A., Bonner, A.S., Duffy, J.B., and Gergen, J.P. (1990). The *Drosophila* segmentation gene *runt* encodes a novel nuclear regulatory protein that is also expressed in the developing nervous system. *Genes Dev.* **4**, 1701–1713.

2128. Kankel, D.R., Ferrús, A., Garen, S.H., Harte, P.J., and Lewis,

P.E. (1980). The structure and development of the nervous system. *In* "The Genetics and Biology of *Drosophila*," Vol. 2d (M. Ashburner and T.R.F. Wright, eds.). Acad. Pr.: New York, pp. 295–368.

2129. Kao, H.-Y., Ordentlich, P., Koyano-Nakagawa, N., Tang, Z., Downes, M., Kintner, C.R., Evans, R.M., and Kadesch, T. (1998). A histone deacetylase corepressor complex regulates the Notch signal transduction pathway. *Genes Dev.* **12**, 2269–2277. [See also 2001 *Curr. Biol.* **11**, 789–792.]

2130. Kaphingst, K. and Kunes, S. (1994). Pattern formation in the visual centers of the *Drosophila* brain: *wingless* acts via *decapentaplegic* to specify the dorsoventral axis. *Cell* **78**, 437–448.

2131. Karim, F.D., Chang, H.C., Therrien, M., Wassarman, D.A., Laverty, T., and Rubin, G.M. (1996). A screen for genes that function downstream of Ras1 during *Drosophila* eye development. *Genetics* **143**, 315–329.

2132. Karim, F.D. and Rubin, G.M. (1998). Ectopic expression of activated Ras1 induces hyperplastic growth and increased cell death in *Drosophila* imaginal tissues. *Development* **125**, 1–9.

2133. Karim, F.D. and Rubin, G.M. (1999). PTP-ER, a novel tyrosine phosphatase, functions downstream of Ras1 to downregulate MAP kinase during *Drosophila* eye development. *Molec. Cell* **3**, 741–750.

2134. Karim, F.D. and Thummel, C.S. (1991). Ecdysone coordinates the timing and amounts of *E74A* and *E74B* transcription in *Drosophila*. *Genes Dev.* **5**, 1067–1079.

2135. Karim, F.D., Urness, L.D., Thummel, C.S., Klemsz, M.J., McKercher, S.R., Celada, A., Van Beveren, C., Maki, R.A., Gunther, C.V., Nye, J.A., and Graves, B.J. (1990). The ETS-domain: a new DNA-binding motif that recognizes a purine-rich core DNA sequence. *Genes Dev.* **4**, 1451–1453.

2136. Karin, M. and Hunter, T. (1995). Transcriptional control by protein phosphorylation: signal transmission from the cell surface to the nucleus. *Curr. Biol.* **5**, 747–757.

2137. Karlovich, C.A., Bonfini, L., McCollam, L., Rogge, R.D., Daga, A., Czech, M.P., and Banerjee, U. (1995). In vivo functional analysis of the Ras exchange factor Son of sevenless. *Science* **268**, 576–579.

2138. Karlsson, J. (1979). A major difference between transdetermination and homeosis. *Nature* **279**, 426–428.

2139. Karlsson, J. (1980). Distal regeneration in proximal fragments of the wing disc of *Drosophila*. *J. Embryol. Exp. Morph.* **59**, 315–323.

2140. Karlsson, J. (1981). The distribution of regenerative potential in the wing disc of *Drosophila*. *J. Embryol. Exp. Morph.* **61**, 303–316.

2141. Karlsson, J. (1981). Sequence of regeneration in the *Drosophila* wing disc. *J. Embryol. Exp. Morph.* **65 (Suppl.)**, 37–47.

2142. Karlsson, J. (1984). Morphogenesis and compartments in *Drosophila*. *In* "Pattern Formation: A Primer in Developmental Biology," (G.M. Malacinski and S.V. Bryant, eds.). Macmillan: New York, pp. 323–337.

2143. Karlsson, J. and Smith, R.J. (1981). Regeneration from duplicating fragments of the *Drosophila* wing disc. *J. Embryol. Exp. Morph.* **66**, 117–126.

2144. Karpen, G.H. and Schubiger, G. (1981). Extensive regulatory capabilities of a *Drosophila* imaginal disk blastema. *Nature* **294**, 744–747.

2145. Karpilow, J., Kolodkin, A., Bork, T., and Venkatesh, T. (1989). Neuronal development in the *Drosophila* compound eye: *rap* gene function is required in photorecep-

tor cell R8 for ommatidial pattern formation. *Genes Dev.* **3**, 1834–1844.

2146. Karr, T.L., Weir, M.P., Ali, Z., and Kornberg, T. (1989). Patterns of engrailed protein in early *Drosophila* embryos. *Development* **105**, 605–612.

2147. Katsani, K.R., Hajibagheri, M.A.N., and Verrijzer, C.P. (1999). Co-operative DNA binding by GAGA transcription factor requires the conserved BTB/POZ domain and reorganizes promoter topology. *EMBO J.* **18**, 698–708.

2148. Katsube, T., Takahisa, M., Ueda, R., Hashimoto, N., Kobayashi, M., and Togashi, S. (1998). Cortactin associates with the cell-cell junction protein ZO-1 in both *Drosophila* and mouse. *J. Biol. Chem.* **273** #45, 29672–29677.

2149. Katz, M.J. and Grenander, U. (1982). Developmental matching and the numerical matching hypothesis for neuronal cell death. *J. Theor. Biol.* **98**, 501–517.

2150. Katz, M.J. and Lasek, R.J. (1978). Evolution of the nervous system: role of ontogenetic mechanisms in the evolution of matching populations. *Proc. Natl. Acad. Sci. USA* **75**, 1349–1352.

2151. Katzen, A.L., Jackson, J., Harmon, B.P., Fung, S.-M., Ramsay, G., and Bishop, J.M. (1998). *Drosophila myb* is required for the G_2/M transition and maintenance of diploidy. *Genes Dev.* **12**, 831–843.

2152. Katzen, A.L., Kornberg, T., and Bishop, J.M. (1991). Expression during *Drosophila* development of DER, a gene related to *erbB*-1 and *neu*: correlations with mutant phenotypes. *Dev. Biol.* **145**, 287–301.

2153. Kauffman, S. (1975). Control circuits for determination and transdetermination: interpreting positional information in a binary epigenetic code. *In* "Cell Patterning," *Ciba Found. Symp.*, Vol. 29 (R. Porter and J. Rivers, eds.). Elsevier: Amsterdam, pp. 201–221.

2154. Kauffman, S. (1996). Evolving evolvability. *Nature* **382**, 309–310.

2155. Kauffman, S.A. (1971). Gene regulation networks: a theory for their global structure and behaviors. *Curr. Top. Dev. Biol.* **6**, 145–182.

2156. Kauffman, S.A. (1973). Control circuits for determination and transdetermination. *Science* **181**, 310–318.

2157. Kauffman, S.A. (1981). Bifurcations in insect morphogenesis. I. *In* "Nonlinear Phenomena in Physics and Biology," (R.H. Enns, B.L. Jones, R.M. Miura, and S.S. Rangnekar, eds.). Plenum: New York, pp. 401–450.

2158. Kauffman, S.A. (1981). Bifurcations in insect morphogenesis. II. *In* "Nonlinear Phenomena in Physics and Biology," (R.H. Enns, B.L. Jones, R.M. Miura, and S.S. Rangnekar, eds.). Plenum: New York, pp. 451–484.

2159. Kauffman, S.A. (1981). Pattern formation in the *Drosophila* embryo. *Phil. Trans. Roy. Soc. Lond. B* **295**, 567–594.

2160. Kauffman, S.A. (1983). Developmental constraints: internal factors in evolution. *In* "Development and Evolution," *Symp. Brit. Soc. Dev. Biol.*, Vol. 6 (B.C. Goodwin, N. Holder, and C.C. Wylie, eds.). Cambridge Univ. Pr.: Cambridge, pp. 195–225.

2161. Kauffman, S.A. (1984). Pattern generation and regeneration. *In* "Pattern Formation: A Primer in Developmental Biology," (G.M. Malacinski and S.V. Bryant, eds.). Macmillan: New York, pp. 73–102.

2162. Kauffman, S.A. (1987). Developmental logic and its evolution. *BioEssays* **6**, 82–87.

2163. Kauffman, S.A. (1993). "The Origins of Order: Self-organization and Selection in Evolution." Oxford Univ. Pr., Oxford.

2164. Kauffman, S.A., Shymko, R.M., and Trabert, K. (1978). Control of sequential compartment formation in *Drosophila*. *Science* **199**, 259–270.

2165. Kauffmann, R.C., Li, S., Gallagher, P.A., Zhang, J., and Carthew, R.W. (1996). Ras1 signaling and transcriptional competence in the R7 cell of *Drosophila*. *Genes Dev.* **10**, 2167–2178.

2166. Kaufman, T.C. (1983). The genetic regulation of segmentation in *Drosophila melanogaster*. *In* "Time, Space, and Pattern in Embryonic Development," (W.R. Jeffery and R.A. Raff, eds.). Alan R. Liss: New York, pp. 365–383.

2167. Kaufman, T.C., Seeger, M.A., and Olsen, G. (1990). Molecular and genetic organization of the Antennapedia gene complex of *Drosophila melanogaster*. *Adv. Genet.* **27**, 309–362.

2168. Kavaler, J., Fu, W., Duan, H., Noll, M., and Posakony, J.W. (1999). An essential role for the *Drosophila Pax2* homolog in the differentiation of adult sensory organs. *Development* **126**, 2261–2272.

2169. Kazemi-Esfarjani, P. and Benzer, S. (2000). Genetic suppression of polyglutamine toxicity in *Drosophila*. *Science* **287**, 1837–1840.

2170. Kehl, B.T., Cho, K.-O., and Choi, K.-W. (1998). *mirror*, a *Drosophila* homeobox gene in the *iroquois* complex, is required for sensory organ and alula formation. *Development* **125**, 1217–1227.

2171. Kehle, J., Beuchle, D., Treuheit, S., Christen, B., Kennison, J.A., Bienz, M., and Müller, J. (1998). dMi-2, a Hunchback-interacting protein that functions in *Polycomb* repression. *Science* **282**, 1897–1900.

2172. Keightley, P.D. (1995). Loci with large effects. *Curr. Biol.* **5**, 485–487.

2173. Keil, T.A. (1997). Functional morphology of insect mechanoreceptors. *Microsc. Res. Tech.* **39**, 506–531.

2174. Keil, T.A. and Steinbrecht, R.A. (1984). Mechanosensitive and olfactory sensilla of insects. *In* "Insect Ultrastructure," Vol. 2 (R.C. King and H. Akai, eds.). Plenum: New York, pp. 477–516.

2175. Keil, T.A. and Steiner, C. (1990). Morphogenesis of the antenna of the male silkmoth, *Antheraea polyphemus*. II. Differential mitoses of "dark" precursor cells create the anlagen of sensilla. *Tissue & Cell* **22**, 705–720.

2176. Keisman, E.L. and Baker, B.S. (2001). The *Drosophila* sex determination hierarchy modulates *wingless* and *decapentaplegic* signaling to deploy *dachshund* sex-specifically in the genital imaginal disc. *Development* **128**, 1643–1656.

2177. Keller, C.A., Erickson, M.S., and Abmayr, S.M. (1997). Misexpression of *nautilus* induces myogenesis in cardioblasts and alters the pattern of somatic muscle fibers. *Dev. Biol.* **181**, 197–212.

2178. Keller, E.F. (1996). *Drosophila* embryos as transitional objects: the work of Donald Poulson and Christiane Nüsslein-Volhard. *Hist. Studies Phys. Biol. Sci.* **26**, 313–346.

2179. Keller, E.F. (2000). Decoding the genetic program ... or, some circular logic in the logic of circularity. *In* "The Concept of the Gene in Development and Evolution: Historical and Epistemological Perspectives," (P.J. Beurton, R. Falk, and H.-J. Rheinberger, eds.). Cambridge Univ. Pr.: Cambridge, pp. 159–177.

2180. Keller, S.A., Mao, Y., Struffi, P., Margulies, C., Yurk, C.E., Anderson, A.R., Amey, R.L., Moore, S., Ebels, J.M., Foley, K., Corado, M., and Arnosti, D.N. (2000). dCtBP-dependent and -independent repression activities of the *Drosophila* Knirps protein. *Mol. Cell. Biol.* **20**, 7247–7258.

2181. Kellerman, K.A., Mattson, D.M., and Duncan, I. (1990). Mutations affecting the stability of the *fushi tarazu* protein of *Drosophila*. *Genes Dev.* **4**, 1936–1950.

2182. Kelley, M.R., Kidd, S., Deutsch, W.A., and Young, M.W. (1987). Mutations altering the structure of epidermal growth factor-like coding sequences at the *Drosophila Notch* locus. *Cell* **51**, 539–548.

2183. Kelley, R.L. and Kuroda, M.I. (2001). The MSL dosage compensation complex modifies chromatin when redirected to autosomes. *Proc. 42nd Ann. Drosophila Res. Conf.* **Abstracts Vol.**, a102. [Cf. 2001 *EMBO J.* **20**, 2236–2245.]

2184. Kelly, R.B. (1995). Ringing necks with dynamin. *Nature* **374**, 116–117.

2185. Kemler, R. (1993). From cadherins to catenins: cytoplasmic protein interactions and regulation of cell adhesion. *Trends Genet.* **9**, 317–321.

2186. Kenan, D.J., Query, C.C., and Keene, J.D. (1991). RNA recognition: towards identifying determinants of specificity. *Trends Biochem. Sci.* **16**, 214–220.

2187. Kennerdell, J.R. and Carthew, R.W. (1998). Use of dsRNA-mediated genetic interference to demonstrate that *frizzled* and *frizzled 2* act in the Wingless pathway. *Cell* **95**, 1017–1026.

2188. Kennison, J.A. (1993). Transcriptional activation of *Drosophila* homeotic genes from distant regulatory elements. *Trends Genet.* **9**, 75–79.

2189. Kennison, J.A. (1995). The Polycomb and trithorax group proteins of *Drosophila*: trans-regulators of homeotic gene function. *Annu. Rev. Genet.* **29**, 289–303.

2190. Kennison, J.A. and Russell, M.A. (1987). Dosage-dependent modifiers of homoeotic mutations in *Drosophila melanogaster*. *Genetics* **116**, 75–86.

2191. Kenyon, C. (1994). If birds can fly, why can't we? Homeotic genes and evolution. *Cell* **78**, 175–180.

2192. Kenyon, C. (1995). A perfect vulva every time: gradients and signaling cascades in *C. elegans*. *Cell* **82**, 171–174.

2193. Kenyon, K.L., Clouser, C.R., and Pignoni, F. (2001). Modulation of Sine oculis function by direct interaction with different co-factors. *Proc. 42nd Ann. Drosophila Res. Conf.* **Abstracts Vol.**, a194.

2194. Keplinger, B.L., Guo, X., Quine, J., Feng, Y., and Cavener, D.R. (2001). Complex organization of promoter and enhancer elements regulate the tissue- and developmental stage-specific expression of the *Drosophila melanogaster Gld* gene. *Genetics* **157**, 699–716.

2195. Kerber, B., Monge, I., Mueller, M., Mitchell, P.J., and Cohen, S.M. (2001). The AP-2 transcription factor is required for joint formation and cell survival in *Drosophila* leg development. *Development* **128**, 1231–1238.

2196. Kernan, M., Cowan, D., and Zuker, C. (1994). Genetic dissection of mechanosensory transduction: mechanoreception-defective mutations of *Drosophila*. *Neuron* **12**, 1195–1206.

2197. Kerppola, T. and Curran, T. (1995). Zen and the art of Fos and Jun. *Nature* **373**, 199–200.

2198. Kerridge, S. and Morata, G. (1982). Developmental effects of some newly induced *Ultrabithorax* alleles of *Drosophila*. *J. Embryol. Exp. Morph.* **68**, 211–234.

2199. Kerszberg, M. (1999). Morphogen propagation and action: Towards molecular models. *Sems. Cell Dev. Biol.* **10**, 297–302.

2200. Kerszberg, M. and Changeux, J.-P. (1994). A model for

reading morphogenetic gradients: autocatalysis and competition at the gene level. *Proc. Natl. Acad. Sci. USA* **91**, 5823–5827.

2201. Kerszberg, M. and Changeux, J.-P. (1994). Partners make patterns in morphogenesis. *Curr. Biol.* **4**, 1046–1047.

2202. Keyse, S.M. (2000). Protein phosphatases and the regulation of mitogen-activated protein kinase signalling. *Curr. Opin. Cell Biol.* **12**, 186–192.

2203. Khalsa, O., Yoon, J.-w., Torres-Schumann, S., and Wharton, K.A. (1998). TGF-β/BMP superfamily members, Gbb-60A and Dpp, cooperate to provide pattern information and establish cell identity in the *Drosophila* wing. *Development* **125**, 2723–2734. [See also 2001 *Development* **128**, 3913–3925.]

2204. Khare, N. and Baumgartner, S. (2000). Dally-like protein, a new *Drosophila* glypican with expression overlapping with *wingless*. *Mechs. Dev.* **99**, 199–202.

2205. Kharitonenkov, A., Chen, Z., Sures, I., Wang, H., Schilling, J., and Ullrich, A. (1997). A family of proteins that inhibit signalling through tyrosine kinase receptors. *Nature* **386**, 181–186.

2206. Khokhlatchev, A.V., Canagarajah, B., Wilsbacher, J., Robinson, M., Atkinson, M., Goldsmith, E., and Cobb, M.H. (1998). Phosphorylation of the MAP kinase ERK2 promotes its homodimerization and nuclear translocation. *Cell* **93**, 605–615.

2207. Khorasanizadeh, S. and Rastinejad, F. (1999). Transcription factors: the right combination for the DNA lock. *Curr. Biol.* **9**, R456–R458.

2208. Kidd, S. (1992). Characterization of the *Drosophila cactus* locus and analysis of interactions between cactus and dorsal proteins. *Cell* **71**, 623–635.

2209. Kidd, S., Baylies, M.K., Gasic, G.P., and Young, M.W. (1989). Structure and distribution of the Notch protein in developing *Drosophila*. *Genes Dev.* **3**, 1113–1129.

2210. Kidd, S., Kelley, M.R., and Young, M.W. (1986). Sequence of the Notch locus of *Drosophila melanogaster*: relationship of the encoded protein to mammalian clotting and growth factors. *Mol. Cell. Biol.* **6**, 3094–3108.

2211. Kidd, S., Lieber, T., and Young, M.W. (1998). Ligand-induced cleavage and regulation of nuclear entry of Notch in *Drosophila melanogaster* embryos. *Genes Dev.* **12**, 3728–3740.

2212. Kidwell, M.G. and Lisch, D.R. (2001). Transposable elements, parasitic DNA, and genome evolution. *Evolution* **55**, 1–24. [See also 2001 *Nature* **411**, 146–149.]

2213. Kiehle, C.P. and Schubiger, G. (1985). Cell proliferation changes during pattern regulation in imaginal leg discs of *Drosophila melanogaster*. *Dev. Biol.* **109**, 336–346.

2214. Kiger, J.A., Jr., Eklund, J.L., Younger, S.H., and O'Kane, C.J. (1999). Transgenic inhibitors identify two roles for protein kinase A in *Drosophila* development. *Genetics* **152**, 281–290.

2215. Kilbey, B.J. (1995). Charlotte Auerbach (1899–1994). *Genetics* **141**, 1–5.

2216. Kim, J., Irvine, K.D., and Carroll, S.B. (1995). Cell recognition, signal induction, and symmetrical gene activation at the dorsal-ventral boundary of the developing *Drosophila* wing. *Cell* **82**, 795–802.

2217. Kim, J., Johnson, K., Chen, H.J., Carroll, S., and Laughon, A. (1997). *Drosophila* Mad binds to DNA and directly mediates activation of vestigial by Decapentaplegic. *Nature* **388**, 304–308.

2218. Kim, J., Magee, J., and Carroll, S.B. (1997). Intercompartmental signaling and the regulation of *vestigial* expression at the dorsoventral boundary of the developing *Drosophila* wing. *Cold Spr. Harb. Symp. Quant. Biol.* **62**, 283–291.

2219. Kim, J., Sebring, A., Esch, J.J., Kraus, M.E., Vorwerk, K., Magee, J., and Carroll, S.B. (1996). Integration of positional signals and regulation of wing formation and identity by *Drosophila vestigial* gene. *Nature* **382**, 133–138.

2220. Kim, L. and Kimmel, A.R. (2000). GSK3, a master switch regulating cell-fate specification and tumorigenesis. *Curr. Opin. Gen. Dev.* **10**, 508–514.

2221. Kim, M.-S., Repp, A., and Smith, D.P. (1998). LUSH odorant-binding protein mediates chemosensory responses to alcohols in *Drosophila melanogaster*. *Genetics* **150**, 711–721.

2222. Kimble, J. and Simpson, P. (1997). The LIN-12/Notch signaling pathway and its regulation. *Annu. Rev. Cell Dev. Biol.* **13**, 333–361.

2223. Kimmel, B.E., Heberlein, U., and Rubin, G.M. (1990). The homeo domain protein *rough* is expressed in a subset of cells in the developing *Drosophila* eye where it can specify photoreceptor cell subtype. *Genes Dev.* **4**, 712–727.

2224. Kimmel, C.B. (1996). Was *Urbilateria* segmented? *Trends Genet.* **12**, 329–331.

2225. Kimura, K.-i., Usui-Ishihara, A., and Usui, K. (1997). G2 arrest of cell cycle ensures a determination process of sensory mother cell formation in *Drosophila*. *Dev. Genes Evol.* **207**, 199–202.

2226. Kingsley, D.M. (1994). The TGF-β superfamily: new members, new receptors, and new genetic tests of function in different organisms. *Genes Dev.* **8**, 133–146.

2227. Kingston, R.E., Bunker, C.A., and Imbalzano, A.N. (1996). Repression and activation by multiprotein complexes that alter chromatin structure. *Genes Dev.* **10**, 905–920.

2228. Kingston, R.E. and Narlikar, G.J. (1999). ATP-dependent remodeling and acetylation as regulators of chromatin fluidity. *Genes Dev.* **13**, 2339–2352.

2229. Kirby, B.S. and Bryant, P.J. (1982). Growth is required for cell competition in the imaginal discs of *Drosophila melanogaster*. *W. Roux's Arch.* **191**, 289–291.

2230. Kirby, B.S., Bryant, P.J., and Schneiderman, H.A. (1982). Regeneration following duplication in imaginal wing disc fragments of *Drosophila melanogaster*. *Dev. Biol.* **90**, 259–271.

2231. Kirchhamer, C.V., Yuh, C.-H., and Davidson, E.H. (1996). Modular cis-regulatory organization of developmentally expressed genes: two genes transcribed territorially in the sea urchin embryo, and additional examples. *Proc. Natl. Acad. Sci. USA* **93**, 9322–9328.

2232. Kirchhausen, T. (1998). Vesicle formation: Dynamic dynamin lives up to its name. *Curr. Biol.* **8**, R792–R794.

2233. Kirkpatrick, H., Johnson, K., and Laughon, A. (2001). Repression of Dpp targets by binding of Brinker to mad sites. *J. Biol. Chem.* **276** #21, 18216–18222.

2234. Kirschfeld, K. (1967). Die Projektion der optischen Umwelt auf das Raster der Rhabdomere im Komplexauge von MUSCA. *Exp. Brain Res.* **3**, 248–270.

2235. Kirschfeld, K., Feiler, R., and Franceschini, N. (1978). A photostable pigment within the rhabdomere of fly photoreceptors no. 7. *J. Comp. Physiol.* **125**, 275–284.

2236. Kirschner, M. (1992). Evolution of the cell. *In* "Molds, Molecules, and Metazoa: Growing Points in Evolutionary

Biology," (P.R. Grant and H.S. Horn, eds.). Princeton Univ. Pr.: Princeton, pp. 99–126.

2237. Kirschner, M., Gerhart, J., and Mitchison, T. (2000). Molecular "vitalism". *Cell* **100**, 79–88.

2238. Kishida, S., Yamamoto, H., Hino, S.-i., Ikeda, S., Kishida, M., and Kikuchi, A. (1999). DIX domains of Dvl and Axin are necessary for protein interactions and their ability to regulate β-catenin stability. *Mol. Cell. Biol.* **19**, 4414–4422.

2239. Kishida, S., Yamamoto, H., Ikeda, S., Kishida, M., Sakamoto, I., Koyama, S., and Kikuchi, A. (1998). Axin, a negative regulator of the Wnt signaling pathway, directly interacts with adenomatous polyposis coli and regulates the stabilization of β-catenin. *J. Biol. Chem.* **273** #18, 10823–10826.

2240. Kissinger, C.R., Liu, B., Martin-Blanco, E., Kornberg, T.B., and Pabo, C.O. (1990). Crystal structure of an engrailed homeodomain-DNA complex at 2.8 Å resolution: a framework for understanding homeodomain-DNA interactions. *Cell* **63**, 579–590.

2241. Kitagawa, M., Hatakeyama, S., Shirane, M., Matsumoto, M., Ishida, N., Hattori, K., Nakamichi, I., Kikuchi, A., Nakayama, K.-i., and Nakayama, K. (1999). An F-box protein, FWD1, mediates ubiquitin-dependent proteolysis of β-catenin. *EMBO J.* **18**, 2401–2410.

2242. Klämbt, C. (1993). The *Drosophila* gene *pointed* encodes two ETS-like proteins which are involved in the development of the midline glial cells. *Development* **117**, 163–176.

2243. Klämbt, C. (1997). *Drosophila* development: a receptor for ommatidial recruitment. *Curr. Biol.* **7**, R132–R135.

2244. Klämbt, C. (2000). EGF receptor signalling: The importance of presentation. *Curr. Biol.* **10**, R388–R391.

2245. Klämbt, C., Glazer, L., and Shilo, B.-Z. (1992). *breathless*, a *Drosophila* FGF receptor homolog, is essential for migration of tracheal and specific midline glial cells. *Genes Dev.* **6**, 1668–1678.

2246. Klämbt, C., Knust, E., Tietze, K., and Campos-Ortega, J.A. (1989). Closely related transcripts encoded by the neurogenic gene complex Enhancer of split of *Drosophila melanogaster*. *EMBO J.* **8**, 203–210.

2247. Klämbt, C., Müller, S., Lützelschwab, R., Rossa, R., Totzke, F., and Schmidt, O. (1989). The *Drosophila melanogaster l(2)gl* gene encodes a protein homologous to the cadherin cell-adhesion molecule family. *Dev. Biol.* **133**, 425–436.

2248. Klein, T., Brennan, K., and Martinez Arias, A. (1997). An intrinsic dominant negative activity of Serrate that is modulated during wing development in *Drosophila*. *Dev. Biol.* **189**, 123–134.

2249. Klein, T. and Campos-Ortega, J.A. (1992). Second-site modifiers of the *Delta* wing phenotype in *Drosophila melanogaster*. *Roux's Arch. Dev. Biol.* **202**, 49–60.

2250. Klein, T. and Campos-Ortega, J.A. (1997). *klumpfuss*, a *Drosophila* gene encoding a member of the EGR family of transcription factors, is involved in bristle and leg development. *Development* **124**, 3123–3134.

2251. Klein, T., Couso, J.P., and Martinez Arias, A. (1998). Wing development and specification of dorsal cell fates in the absence of *apterous* in *Drosophila*. *Curr. Biol.* **8**, 417–420.

2252. Klein, T. and Martinez Arias, A. (1998). Different spatial and temporal interactions between *Notch*, *wingless*, and *vestigial* specify proximal and distal pattern elements of the wing in *Drosophila*. *Dev. Biol.* **194**, 196–212.

2253. Klein, T. and Martinez Arias, A. (1998). Interactions among Delta, Serrate and Fringe modulate Notch activity during *Drosophila* wing development. *Development* **125**, 2951–2962.

2254. Klein, T. and Martinez Arias, A. (1999). The Vestigial gene product provides a molecular context for the interpretation of signals during the development of the wing in *Drosophila*. *Development* **126**, 913–925.

2255. Klein, T., Seugnet, L., Haenlin, M., and Martinez Arias, A. (2000). Two different activities of *Suppressor of Hairless* during wing development in *Drosophila*. *Development* **127**, 3553–3566.

2256. Klemm, J.D., Beals, C.R., and Crabtree, G.R. (1997). Rapid targeting of nuclear proteins to the cytoplasm. *Curr. Biol.* **7**, 638–644.

2257. Klemm, J.D. and Pabo, C.O. (1996). Oct-1 POU domain-DNA interactions: cooperative binding of isolated subdomains and effects of covalent linkage. *Genes Dev.* **10**, 27–36.

2258. Klemm, J.D., Rould, M.A., Aurora, R., Herr, W., and Pabo, C.O. (1994). Crystal structure of the Oct-1 POU domain bound to an octamer site: DNA recognition with tethered DNA-binding modules. *Cell* **77**, 21–32.

2259. Klingenberg, C.P. and Nijhout, H.F. (1999). Genetics of fluctuating asymmetry: a developmental model of developmental instability. *Evolution* **53**, 358–375.

2260. Klingenberg, C.P. and Zaklan, S.D. (2000). Morphological integration between developmental compartments in the *Drosophila* wing. *Evolution* **54**, 1273–1285.

2261. Klingensmith, J. and Nusse, R. (1994). Signaling by *wingless* in *Drosophila*. *Dev. Biol.* **166**, 396–414.

2262. Klingensmith, J., Nusse, R., and Perrimon, N. (1994). The *Drosophila* segment polarity gene *dishevelled* encodes a novel protein required for response to the *wingless* signal. *Genes Dev.* **8**, 118–130.

2263. Klueg, K.M. and Muskavitch, M.A.T. (1999). Ligand-receptor interactions and trans-endocytosis of Delta, Serrate and Notch: members of the Notch signalling pathway in *Drosophila*. *J. Cell Sci.* **112**, 3289–3297.

2264. Klueg, K.M., Parody, T.R., and Muskavitch, M.A.T. (1998). Complex proteolytic processing acts on Delta, a transmembrane ligand for Notch, during *Drosophila* development. *Mol. Biol. Cell* **9**, 1709–1723.

2265. Klug, A. and Schwabe, J.W.R. (1995). Zinc fingers. *FASEB J.* **9**, 597–604. [See also 2001 *Annu. Rev. Biochem* **70**, 313–340.]

2266. Kmita, M., van der Hoeven, F., Zákány, J., Krumlauf, R., and Duboule, D. (2000). Mechanisms of *Hox* gene colinearity: transposition of the anterior *Hoxb1* gene into the posterior *HoxD* complex. *Genes Dev.* **14**, 198–211.

2267. Knoblich, J.A., Jan, L.Y., and Jan, Y.N. (1995). Asymmetric segregation of Numb and Prospero during cell division. *Nature* **377**, 624–627.

2268. Knoblich, J.A., Jan, L.Y., and Jan, Y.N. (1997). Asymmetric segregation of the *Drosophila* Numb protein during mitosis: facts and speculations. *Cold Spr. Harb. Symp. Quant. Biol.* **62**, 71–77.

2269. Knoblich, J.A., Jan, L.Y., and Jan, Y.N. (1997). The N terminus of the *Drosophila* Numb protein directs membrane association and actin-dependent asymmetric localization. *Proc. Natl. Acad. Sci. USA* **94**, 13005–13010.

2270. Knoblich, J.A. and Lehner, C.F. (1993). Synergistic action of *Drosophila* cyclins A and B during the G_2-M transition. *EMBO J.* **12**, 65–74.

2271. Knoepfler, P.S. and Eisenman, R.N. (1999). Sin meets NuRD and other tails of repression. *Cell* **99**, 447–450.

2272. Knoll, A.H. and Carroll, S.B. (1999). Early animal evolution: emerging views from comparative biology and geology. *Science* **284**, 2129–2137.

2273. Knust, E. and Leptin, M. (1996). Adherens junctions in the *Drosophila* embryo: the role of E-cadherin in their establishment and morphogenetic function. *BioEssays* **18**, 609–612.

2274. Knust, E., Schrons, H., Grawe, F., and Campos-Ortega, J.A. (1992). Seven genes of the *Enhancer of split* Complex of *Drosophila melanogaster* encode helix-loop-helix proteins. *Genetics* **132**, 505–518.

2275. Koch, C. (1997). Computation and the single neuron. *Nature* **385**, 207–210.

2276. Koch, C. (1999). "Biophysics of Computation: Information Processing in Single Neurons." Oxford Univ. Pr., New York.

2277. Koch, C. and Laurent, G. (1999). Complexity and the nervous system. *Science* **284**, 96–98.

2278. Koch, P.B., Lorenz, U., Brakefield, P.M., and ffrench-Constant, R.H. (2000). Butterfly wing pattern mutants: developmental heterochrony and co-ordinately regulated phenotypes. *Dev. Genes Evol.* **210**, 536–544.

2279. Kockel, L., Vorbrüggen, G., Jäckle, H., Mlodzik, M., and Bohmann, D. (1997). Requirement for *Drosophila* 14-3-3ζ in Raf-dependent photoreceptor development. *Genes Dev.* **11**, 1140–1147.

2280. Kockel, L., Zeitlinger, J., Staszewski, L.M., Mlodzik, M., and Bohmann, D. (1997). Jun in *Drosophila* development: redundant and nonredundant functions and regulation by two MAPK signal transduction pathways. *Genes Dev.* **11**, 1748–1758.

2281. Kockel, L., Zeitlinger, J., Staszewski, L.M., Mlodzik, M., and Bohmann, D. (1998). Erratum. [Kockel *et al.*'s 1997 finding of a synergistic interaction between *dJun* and *dMAPK* was not reproducible.]. *Genes Dev.* **12**, 447.

2282. Koelle, M.R. (1997). A new family of G-protein regulators–the RGS proteins. *Curr. Opin. Cell Biol.* **9**, 143–147.

2283. Kohler, R.E. (1994). "Lords of the Fly: *Drosophila* Genetics and the Experimental Life." Univ. Chicago Pr., Chicago.

2284. Kohn, K.W. (1999). Molecular interaction map of the mammalian cell cycle control and DNA repair systems. *Mol. Biol. Cell* **10**, 2703–2734.

2285. Koizumi, K., Stivers, C., Brody, T., Zangeneh, S., Mozer, B., and Odenwald, W.F. (2001). A search for *Drosophila* neural precursor genes identifies *ran*. *Dev. Genes Evol.* **211**, 67–75.

2286. Kojima, T., Ishimaru, S., Higashijima, S.-i., Takayama, E., Akimaru, H., Sone, M., Emori, Y., and Saigo, K. (1991). Identification of a different-type homeobox gene, *BarH1*, possibly causing Bar (*B*) and *Om(1D)* mutations in *Drosophila*. *Proc. Natl. Acad. Sci. USA* **88**, 4343–4347.

2287. Kojima, T., Sato, M., and Saigo, K. (2000). Formation and specification of distal leg segments in *Drosophila* by dual *Bar* homeobox genes, *BarH1* and *BarH2*. *Development* **127**, 769–778.

2288. Kolodkin, A.L., Pickup, A.T., Lin, D.M., Goodman, C.S., and Banerjee, U. (1994). Characterization of *Star* and its interactions with *sevenless* and *EGF receptor* during photoreceptor cell development in *Drosophila*. *Development* **120**, 1731–1745.

2289. Komachi, K., Redd, M.J., and Johnson, A.D. (1994). The WD repeats of Tup1 interact with the homeo domain protein α2. *Genes Dev.* **8**, 2857–2867.

2290. Kon, C.Y., Lopes da Silva, S., Cadigan, K., and Nusse, R. (2001). Tartaruga, a histone deacetylase complex member which may play a role in wingless signalling. *Proc. 42nd Ann. Drosophila Res. Conf.* **Abstracts Vol.**, a87.

2291. Kondo, S. (1992). A mechanistic model for morphogenesis and regeneration of limbs and imaginal discs. *Mechs. Dev.* **39**, 161–170.

2292. Kondo, T. and Duboule, D. (1999). Breaking colinearity in the mouse *HoxD* complex. *Cell* **97**, 407–417.

2293. Kondo, T. and Raff, M. (2000). Basic helix-loop-helix proteins and the timing of oligodendrocyte differentiation. *Development* **127**, 2989–2998.

2294. Kondo, T. and Raff, M. (2000). The Id4 HLH protein and the timing of oligodendrocyte differentiation. *EMBO J.* **19**, 1998–2007.

2295. Kondo, T., Zákány, J., and Duboule, D. (1998). Control of colinearity in *AbdB* genes of the mouse *HoxD* complex. *Molec. Cell* **1**, 289–300.

2296. Kooh, P.J., Fehon, R.G., and Muskavitch, M.A.T. (1993). Implications of dynamic patterns of Delta and Notch expression for cellular interactions during *Drosophila* development. *Development* **117**, 493–507.

2297. Kooyman, D.L., Byrne, G.W., McClellan, S., Nielsen, D., Tone, M., Waldmann, H., Coffman, T.M., McCurry, K.R., Platt, J.L., and Logan, J.S. (1995). In vivo transfer of GPI-linked complement restriction factors from erythrocytes to the endothelium. *Science* **269**, 89–92.

2298. Kopan, R. and Cagan, R. (1997). Notch on the cutting edge. *Trends Genet.* **13**, 465–467.

2299. Kopan, R., Schroeter, E.H., Weintraub, H., and Nye, J.S. (1996). Signal transduction by activated mNotch: importance of proteolytic processing and its regulation by the extracellular domain. *Proc. Natl. Acad. Sci. USA* **93**, 1683–1688.

2300. Kopczynski, C.C., Alton, A.K., Fechtel, K., Kooh, P.J., and Muskavitch, M.A.T. (1988). *Delta*, a *Drosophila* neurogenic gene, is transcriptionally complex and encodes a protein related to blood coagulation factors and epidermal growth factor of vertebrates. *Genes Dev.* **2**, 1723–1735.

2301. Kopp, A., Blackman, R.K., and Duncan, I. (1999). Wingless, Decapentaplegic and EGF receptor signaling pathways interact to specify dorso-ventral pattern in the adult abdomen of *Drosophila*. *Development* **126**, 3495–3507.

2302. Kopp, A. and Duncan, I. (1997). Control of cell fate and polarity in the adult abdominal segments of *Drosophila* by *optomotor-blind*. *Development* **124**, 3715–3726.

2303. Kopp, A., Muskavitch, M.A.T., and Duncan, I. (1997). The roles of *hedgehog* and *engrailed* in patterning adult abdominal segments of *Drosophila*. *Development* **124**, 3703–3714.

2304. Kornberg, R.D. (1999). Eukaryotic transcriptional control. *Trends Genet.* **15**, M46–M49.

2305. Kornberg, R.D. and Lorch, Y. (1999). Twenty-five years of the nucleosome, fundamental particle of the eukaryote chromosome. *Cell* **98**, 285–294.

2306. Kornberg, T. (1981). *engrailed*: A gene controlling compartment and segment formation in *Drosophila*. *Proc. Natl. Acad. Sci. USA* **78**, 1095–1099.

2307. Kornberg, T., Sidén, I., O'Farrell, P., and Simon, M. (1985). The *engrailed* locus of *Drosophila*: *in situ* localization of transcripts reveals compartment-specific expression. *Cell* **40**, 45–53.

2308. Kornfeld, K., Saint, R.B., Beachy, P.A., Harte, P.J., Peattie, D.A., and Hogness, D.S. (1989). Structure and expression of a family of *Ultrabithorax* mRNAs generated by alternative splicing and polyadenylation in *Drosophila*. *Genes Dev.* **3**, 243–258.

2309. Korzh, V. and Grunwald, D. (2001). Nadine Dobrovol-skaïa-Zavadskaïa and the dawn of developmental genetics. *BioEssays* **23**, 365–371.

2310. Koshland, D.E., Jr. (1998). The era of pathway quantification. *Science* **280**, 852–853.

2311. Koshland, D.E., Jr., Goldbeter, A., and Stock, J.B. (1982). Amplification and adaptation in regulatory and sensory systems. *Science* **217**, 220–225.

2312. Koushika, S.P., Lisbin, M.J., and White, K. (1996). ELAV, a *Drosophila* neuron-specific protein, mediates the generation of an alternatively spliced neural protein isoform. *Curr. Biol.* **12**, 1634–1641. [See also 2001 *Genes Dev.* **15**, 2546–2561.]

2313. Koushika, S.P., Soller, M., and White, K. (2000). The neuron-enriched splicing pattern of *Drosophila erect wing* is dependent on the presence of ELAV protein. *Mol. Cell. Biol.* **20**, 1836–1845.

2314. Kozmik, Z., Czerny, T., and Busslinger, M. (1997). Alternatively spliced insertions in the paired domain restrict the DNA sequence specificity of Pax6 and Pax8. *EMBO J.* **16**, 6793–6803.

2315. Kozopas, K.M., Samos, C.H., and Nusse, R. (1998). *DWnt-2*, a *Drosophila* Wnt gene required for the development of the male reproductive tract, specifies a sexually dimorphic cell fate. *Genes Dev.* **12**, 1155–1165.

2316. Krakauer, D.C. and Nowak, M.A. (1999). Evolutionary preservation of redundant duplicated genes. *Sems. Cell Dev. Biol.* **10**, 555–559.

2317. Kramatschek, B. and Campos-Ortega, J.A. (1994). Neuroectodermal transcription of the *Drosophila* neurogenic genes *E(spl)* and *HLH-m5* is regulated by proneural genes. *Development* **120**, 815–826.

2318. Krämer, H. (1993). Patrilocal cell-cell interactions: *sevenless* captures its bride. *Trends Cell Biol.* **3**, 103–105.

2319. Krämer, H., Cagan, R.L., and Zipursky, S.L. (1991). Interaction of *bride of sevenless* membrane-bound ligand and the *sevenless* tyrosine-kinase receptor. *Nature* **352**, 207–212.

2320. Krämer, H. and Phistry, M. (1999). Genetic analysis of *hook*, a gene required for endocytic trafficking in *Drosophila*. *Genetics* **151**, 675–684.

2321. Kramer, S., Okabe, M., Hacohen, N., Krasnow, M.A., and Hiromi, Y. (1999). Sprouty: a common antagonist of FGF and EGF signaling pathways in *Drosophila*. *Development* **126**, 2515–2525.

2322. Kramer, S., West, S.R., and Hiromi, Y. (1995). Cell fate control in the *Drosophila* retina by the orphan receptor seven-up: its role in the decisions mediated by the ras signaling pathway. *Development* **121**, 1361–1372.

2323. Krantz, D.E. and Zipursky, S.L. (1990). *Drosophila* chaoptin, a member of the leucine-rich repeat family, is a photoreceptor cell-specific adhesion molecule. *EMBO J.* **9**, 1969–1977.

2324. Krasnow, M.A., Saffman, E.E., Kornfeld, K., and Hogness, D.S. (1989). Transcriptional activation and repression by *Ultrabithorax* proteins in cultured *Drosophila* cells. *Cell* **57**, 1031–1043.

2325. Krasnow, R.E. and Adler, P.N. (1994). A single *frizzled* protein has a dual function in tissue polarity. *Development* **120**, 1883–1893.

2326. Krasnow, R.E., Wong, L.L., and Adler, P.N. (1995). *dishevelled* is a component of the *frizzled* signaling pathway in *Drosophila*. *Development* **121**, 4095–4102.

2327. Kraut, R. and Campos-Ortega, J.A. (1996). *inscuteable*, a neural precursor gene of *Drosophila*, encodes a candidate for a cytoskeleton adaptor protein. *Dev. Biol.* **174**, 65–81.

2328. Kraut, R., Chia, W., Jan, L.Y., Jan, Y.N., and Knoblich, J.A. (1996). Role of *inscuteable* in orienting asymmetric cell divisions in *Drosophila*. *Nature* **383**, 50–55.

2329. Kraut, R. and Levine, M. (1991). Mutually repressive interactions between the gap genes *giant* and *Krüppel* define middle body regions of the *Drosophila* embryo. *Development* **111**, 611–621.

2330. Krebs, J.E. and Dunaway, M. (1998). The scs and scs' insulator elements impart a *cis* requirement on enhancer-promoter interactions. *Molec. Cell* **1**, 301–308.

2331. Kretzschmar, D., Brunner, A., Wiersdorff, V., Pflugfelder, G.O., Heisenberg, M., and Schneuwly, S. (1992). *giant lens*, a gene involved in cell determination and axon guidance in the visual system of *Drosophila melanogaster*. *EMBO J.* **11**, 2531–2539.

2332. Kretzschmar, M. and Massagué, J. (1998). SMADs: mediators and regulators of TGF-β signaling. *Curr. Opin. Gen. Dev.* **8**, 103–111.

2333. Krishnan, B., Dryer, S.E., and Hardin, P.E. (1999). Circadian rhythms in olfactory responses of *Drosophila melanogaster*. *Nature* **400**, 375–378.

2334. Krylov, D., Mikhailenko, I., and Vinson, C. (1994). A thermodynamic scale for leucine zipper stability and dimerization specificity: e and g interhelical interactions. *EMBO J.* **13**, 2849–2861.

2335. Kubota, K., Goto, S., Eto, K., and Hayashi, S. (2000). EGF receptor attenuates Dpp signaling and helps to distinguish the wing and leg cell fates in *Drosophila*. *Development* **127**, 3769–3776.

2336. Kucharczuk, K.L., Love, C.M., Dougherty, N.M., and Goldhamer, D.J. (1999). Fine-scale transgenic mapping of the *MyoD* core enhancer: *MyoD* is regulated by distinct but overlapping mechanisms in myotomal and non-myotomal muscle lineages. *Development* **126**, 1957–1965.

2337. Kühl, M. and Wedlich, D. (1997). Wnt signalling goes nuclear. *BioEssays* **19**, 101–104.

2338. Kühn, A. (1971). "Lectures on Developmental Physiology." 2nd ed. Springer-Verlag, Berlin.

2339. Kuhn, D.T. and Cunningham, G.N. (1977). Aldehyde oxidase compartmentalization in *Drosophila melanogaster* wing imaginal disks. *Science* **196**, 875–877.

2340. Kuhn, D.T. and Cunningham, G.N. (1978). Aldehyde oxidase distribution in *Drosophila melanogaster* mature imaginal discs, histoblasts and rings of imaginal cells. *J. Exp. Zool.* **204**, 1–9.

2341. Kuhn, D.T. and Cunningham, G.N. (1986). Isocitrate dehydrogenase in *D. melanogaster* imaginal discs: Pattern development and alteration by homoeotic mutant genes. *Dev. Genet.* **7**, 21–34.

2342. Kuhn, D.T., Fogerty, S.C., Eskens, A.A.C., and Sprey, T.E. (1983). Developmental compartments in the *Drosophila melanogaster* wing disc. *Dev. Biol.* **95**, 399–413.

2343. Kuhn, D.T. and Freeland, D.E. (1996). Expression patterns of developmental genes reveal segment and parasegment organization of *D. melanogaster* genital discs. *Mechs. Dev.* **56**, 61–72.

2344. Kuhn, D.T. and Sprey, T.E. (1987). Regulation of NADP-malic enzyme in the eye-antennal imaginal disc of *D. melanogaster/D. simulans* hybrids: Evidence for *cis*- and *trans*- regulation. *Genetics* **115**, 277–281.

2345. Kühnlein, R.P., Brönner, G., Taubert, H., and Schuh, R.

(1997). Regulation of *Drosophila spalt* gene expression. *Mechs. Dev.* **66**, 107–118.

2346. Kühnlein, R.P., Frommer, G., Friedrich, M., Gonzalez-Gaitan, M., Weber, A., Wagner-Bernholz, J.F., Gehring, W.J., Jäckle, H., and Schuh, R. (1994). *spalt* encodes an evolutionarily conserved zinc finger protein of novel structure which provides homeotic gene function in the head and tail region of the *Drosophila* embryo. *EMBO J.* **13**, 168–179.

2347. Kühnlein, R.P. and Schuh, R. (1996). Dual function of the region-specific homeotic gene *spalt* during *Drosophila* tracheal system development. *Development* **122**, 2215–2223.

2348. Kukalová-Peck, J. (1983). Origin of the insect wing and wing articulation from the arthropodan leg. *Can. J. Zool.* **61**, 1618–1669.

2349. Kukalová-Peck, J. (1985). Ephemeroid wing venation based upon new gigantic Carboniferous mayflies and basic morphology, phylogeny, and metamorphosis of pterygote insects (Insecta, Ephemerida). *Can. J. Zool.* **63**, 933–955.

2350. Kulkarni, M.M. and Arnosti, D.N. (2001). In vivo characterization of short-range transcriptional repression. *Proc. 42nd Ann. Drosophila Res. Conf.* **Abstracts Vol.**, a124.

2351. Kumar, J. and Moses, K. (1997). Transcription factors in eye development: a gorgeous mosaic? *Genes Dev.* **11**, 2023–2028.

2352. Kumar, J.P. and Moses, K. (2000). Cell fate specification in the *Drosophila* retina. *In* "Vertebrate Eye Development," *Results and Problems in Cell Differentiation*, Vol. 31 (M.E. Fini, ed.). Springer: Berlin, pp. 93–114.

2353. Kumar, J.P. and Moses, K. (2001). EGF receptor and Notch signaling act upstream of Eyeless/Pax6 to control eye specification. *Cell* **104**, 687–697. [See also 2001 references: *Development* **128**, 2689–2697; *Dev. Genes Evol.* **211**, 406–414; *Nature Rev. Genet.* **2**, 846–857; *Sems. Cell Dev. Biol.* **12**, 469–474.]

2354. Kumar, J.P. and Ready, D.F. (1995). Rhodopsin plays an essential structural role in *Drosophila* photoreceptor development. *Development* **121**, 4359–4370.

2355. Kumar, J.P., Tio, M., Hsiung, F., Akopyan, S., Gabay, L., Seger, R., Shilo, B.-Z., and Moses, K. (1998). Dissecting the roles of the *Drosophila* EGF receptor in eye development and MAP kinase activation. *Development* **125**, 3875–3885.

2356. Kuner, J.M., Nakanishi, M., Ali, Z., Drees, B., Gustavson, E., Theis, J., Kauvar, L., Kornberg, T., and O'Farrell, P.H. (1985). Molecular cloning of *engrailed*: a gene involved in the development of pattern in *Drosophila melanogaster*. *Cell* **42**, 309–316.

2357. Kunes, S. (1999). Stop or Go in the target zone. *Neuron* **22**, 639–640.

2358. Kunes, S. and Steller, H. (1991). Ablation of *Drosophila* photoreceptor cells by conditional expression of a toxin gene. *Genes Dev.* **5**, 970–983.

2359. Kunisch, M., Haenlin, M., and Campos-Ortega, J.A. (1994). Lateral inhibition mediated by the *Drosophila* neurogenic gene Delta is enhanced by proneural proteins. *Proc. Natl. Acad. Sci. USA* **91**, 10139–10143.

2360. Kurant, E., Eytan, D., and Salzberg, A. (2001). Mutational analysis of the *Drosophila homothorax* gene. *Genetics* **157**, 689–698.

2361. Kurant, E., Pai, C.-y., Sharf, R., Halachmi, N., Sun, Y.H., and Salzberg, A. (1998). *dorsotonals/homothorax*, the *Drosophila* homologue of *meis1*, interacts with *extradenticle* in patterning of the embryonic PNS. *Development* **125**, 1037–1048.

2362. Kurata, S., Go, M.J., Artavanis-Tsakonas, S., and Gehring, W.J. (2000). *Notch* signaling and the determination of appendage identity. *Proc. Natl. Acad. Sci. USA* **97**, 2117–2122.

2363. Kuriyama, M., Harada, N., Kuroda, S., Yamamoto, T., Nakafuku, M., Iwamatsu, A., Yamamoto, D., Prasad, R., Croce, C., Canaani, E., and Kaibuchi, K. (1996). Identification of AF-6 and Canoe as putative targets for Ras. *J. Biol. Chem.* **271** #2, 607–610.

2364. Kuriyan, J. and Cowburn, D. (1997). Modular peptide recognition domains in eukaryotic signaling. *Annu. Rev. Biophys. Biomol. Struct.* **26**, 259–288.

2365. Kuriyan, J. and Darnell, J.E., Jr. (1999). An SH2 domain in disguise. *Nature* **398**, 22–25.

2366. Kurokawa, M., Tanaka, T., Tanaka, K., Hirano, N., Ogawa, S., Mitani, K., Yazaki, Y., and Hirai, H. (1996). A conserved cysteine residue in the *runt* homology domain of AML1 is required for the DNA binding ability and the transforming activity on fibroblasts. *J. Biol. Chem.* **271** #28, 16870–16876.

2367. Kurusu, M., Nagao, T., Walldorf, U., Flister, S., Gehring, W.J., and Furukubo-Tokunaga, K. (2000). Genetic control of development of the mushroom bodies, the associative learning centers in the *Drosophila* brain, by the *eyeless*, *twin of eyeless*, and *dachshund* genes. *Proc. Natl. Acad. Sci. USA* **97**, 2140–2144.

2368. Küssel-Andermann, P., El-Amraoui, A., Safieddine, S., Nouaille, S., Perfettini, I., Lecuit, M., Cossart, P., Wolfrum, U., and Petit, C. (2000). Vezatin, a novel transmembrane protein, bridges myosin VIIA to the cadherin-catenins complex. *EMBO J.* **19**, 6020–6029.

2369. Kutty, G., Kutty, R.K., Samuel, W., Duncan, T., Jaworski, C., and Wiggert, B. (1998). Identification of a new member of Transforming Growth Factor-Beta superfamily in *Drosophila*: the first invertebrate activin gene. *Biochem. Biophys. Res. Comm.* **246**, 644–649.

2370. Kuzin, B., Roberts, I., Peunova, N., and Enikolopov, G. (1996). Nitric oxide regulates cell proliferation during *Drosophila* development. *Cell* **87**, 639–649.

2371. Kuziora, M.A. (1993). *Abdominal-B* protein isoforms exhibit distinct cuticular transformations and regulatory activities when ectopically expressed in *Drosophila* embryos. *Mechs. Dev.* **42**, 125–137.

2372. Kuziora, M.A. and McGinnis, W. (1988). Autoregulation of a *Drosophila* homeotic selector gene. *Cell* **55**, 477–485.

2373. Kyba, M. and Brock, H.W. (1998). The *Drosophila* Polycomb group protein Psc contacts ph and Pc through specific conserved domains. *Mol. Cell. Biol.* **18**, 2712–2720.

2374. Kyba, M. and Brock, H.W. (1998). The SAM domain of Polyhomeotic, RAE28, and Scm mediates specific interactions through conserved residues. *Dev. Genet.* **22**, 74–84.

2375. Kylsten, P. and Saint, R. (1997). Imaginal tissues of *Drosophila melanogaster* exhibit different modes of cell proliferation control. *Dev. Biol.* **192**, 509–522.

2376. La Rosée, A., Häder, T., Taubert, H., Rivera-Pomar, R., and Jäckle, H. (1997). Mechanism and Bicoid-dependent control of *hairy* stripe 7 expression in the posterior region of the *Drosophila* embryo. *EMBO J.* **16**, 4403–4411.

2377. Lacalli, T.C. (1990). Modeling the *Drosophila* pair-rule pattern by reaction-diffusion: gap input and pattern control in a 4-morphogen system. *J. Theor. Biol.* **144**, 171–194.

2378. Lacalli, T.C. and Harrison, L.G. (1991). From gradient to segments: models for pattern formation in early *Drosophila* embryogenesis. *Sems. Dev. Biol.* **2**, 107–117.

2379. Lacalli, T.C., Wilkinson, D.A., and Harrison, L.G. (1988).

Theoretical aspects of stripe formation in relation to *Drosophila* segmentation. *Development* **104**, 105–113.

2380. Lagna, G. and Hemmati-Brivanlou, A. (1999). A molecular basis for Smad specificity. *Dev. Dynamics* **214**, 269–277.

2381. Lai, C., Lyman, R.F., Long, A.D., Langley, C.H., and Mackay, T.F.C. (1994). Naturally occurring variation in bristle number and DNA polymorphisms at the *scabrous* locus of *Drosophila melanogaster*. *Science* **266**, 1697–1702.

2382. Lai, E.C., Bodner, R., Kavaler, J., Freschi, G., and Posakony, J.W. (2000). Antagonism of Notch signaling activity by members of a novel protein family encoded by the *Bearded* and *Enhancer of split* gene complexes. *Development* **127**, 291–306.

2383. Lai, E.C., Bodner, R., and Posakony, J.W. (2000). The *Enhancer of split* Complex of *Drosophila* includes four Notch-regulated members of the Bearded gene family. *Development* **127**, 3441–3455.

2384. Lai, E.C., Burks, C., and Posakony, J.W. (1998). The K box, a conserved 3′ UTR sequence motif, negatively regulates accumulation of *Enhancer of split* Complex transcripts. *Development* **125**, 4077–4088.

2385. Lai, E.C. and Posakony, J.W. (1997). The Bearded box, a novel 3′ UTR sequence motif, mediates negative post-transcriptional regulation of *Bearded* and *Enhancer of split* Complex gene expression. *Development* **124**, 4847–4856.

2386. Lai, E.C. and Posakony, J.W. (1998). Regulation of *Drosophila* neurogenesis by RNA: RNA duplexes? *Cell* **93**, 1103–1104.

2387. Lai, E.C. and Rubin, G.M. (2001). *neuralized* functions cell-autonomously to regulate a subset of Notch-dependent processes during adult *Drosophila* development. *Dev. Biol.* **231**, 217–233. [See also 2001 references: *Dev. Cell* **1**, 725–731, 783–794, and 807–816; *Proc. Natl. Acad. Sci. USA* **98**, 5637–5642.]

2388. Lai, E.p. (2000). Bristling all over. *Curr. Biol.* **10**, R431.

2389. Lai, K.-M.V., Olivier, J.P., Gish, G.D., Henkemeyer, M., McGlade, J., and Pawson, T. (1995). A *Drosophila shc* gene product is implicated in signaling by the DER receptor tyrosine kinase. *Mol. Cell. Biol.* **15**, 4810–4818.

2390. Lai, Z.-C., Fetchko, M., and Li, Y. (1997). Repression of *Drosophila* photoreceptor cell fate through cooperative action of two transcriptional repressors Yan and Tramtrack. *Genetics* **147**, 1131–1137.

2391. Lai, Z.-C. and Li, Y. (1999). Tramtrack69 is positively and autonomously required for *Drosophila* photoreceptor development. *Genetics* **152**, 299–305.

2392. Lai, Z.-C. and Rubin, G.M. (1992). Negative control of photoreceptor development in *Drosophila* by the product of the *yan* gene, an ETS domain protein. *Cell* **70**, 609–620.

2393. LaJeunesse, D. and Shearn, A. (1995). Trans-regulation of thoracic homeotic selector genes of the Antennapedia and bithorax complexes by the trithorax group genes: *absent, small, and homeotic discs 1* and *2*. *Mechs. Dev.* **53**, 123–139.

2394. Lake, R.J., Wakabayashi-Ito, N., and Artavanis-Tsakonas, S. (2001). Elucidating the role of Mastermind in Notch signaling. *Proc. 42nd Ann. Drosophila Res. Conf.* **Abstracts Vol.**, a147.

2395. Lakin-Thomas, P.L. (2000). Circadian rhythms: new functions for old clock genes? *Trends Genet.* **16**, 135–142.

2396. Lamka, M.L., Boulet, A.M., and Sakonju, S. (1992). Ectopic expression of UBX and ABD-B proteins during *Drosophila* embryogenesis: competition, not a functional hierarchy, explains phenotypic suppression. *Development* **116**, 841–854.

2397. Lammel, U., Meadows, L., and Saumweber, H. (2000).

Analysis of *Drosophila* salivary gland, epidermis and CNS development suggests an additional function of *brinker* in anterior-posterior cell fate specification. *Mechs. Dev.* **92**, 179–191.

2398. Landauer, W. (1958). On phenocopies, their developmental physiology and genetic meaning. *Am. Nat.* **92**, 201–213.

2399. Landauer, W. (1959). The phenocopy concept: illusion or reality? *Experientia* **15**, 409–412.

2400. Landecker, H.L., Sinclair, D.A.R., and Brock, H.W. (1994). Screen for enhancers of *Polycomb* and *Polycomblike* in *Drosophila melanogaster*. *Dev. Genet.* **15**, 425–434.

2401. Landschulz, W.H., Johnson, P.F., and McKnight, S.L. (1988). The leucine zipper: a hypothetical structure common to a new class of DNA binding proteins. *Science* **240**, 1759–1764.

2402. Landsman, D. and Bustin, M. (1993). A signature for the HMG-1 Box DNA-binding proteins. *BioEssays* **15**, 539–546.

2403. Lane, M.E. and Kalderon, D. (1993). Genetic investigation of cAMP-dependent protein kinase function in *Drosophila* development. *Genes Dev.* **7**, 1229–1243.

2404. Lane, M.E., Sauer, K., Wallace, K., Jan, Y.N., Lehner, C.F., and Vaessin, H. (1996). Dacapo, a cyclin-dependent kinase inhibitor, stops cell proliferation during *Drosophila* development. *Cell* **87**, 1225–1235.

2405. Lanoue, B.R. and Jacobs, J.R. (1999). *rhomboid* function in the midline of the *Drosophila* CNS. *Dev. Genet.* **25**, 321–330.

2406. Larsen, E., Lee, T., and Glickman, N. (1996). Antenna to leg transformation: dynamics of developmental competence. *Dev. Genet.* **19**, 333–339.

2407. Larsen, E. and McLaughlin, H.M.G. (1987). The morphogenetic alphabet: lessons for simple-minded genes. *BioEssays* **7**, 130–132.

2408. Larsen, E. and Zorn, A. (1989). Cell patterns associated with normal and mutant morphogenesis in silver-impregnated imaginal discs of *Drosophila*. *Trans. Am. Microsc. Soc.* **108**, 51–57.

2409. Larsen, E.W. (1992). Tissue strategies as developmental constraints: implications for animal evolution. *Trends Ecol. Evol.* **7**, 414–417.

2410. Larsen, E.W. (1997). Evolution of development: The shuffling of ancient modules by ubiquitous bureaucracies. *In* "Physical Theory in Biology: Foundations and Explorations," (C.J. Lumsden, W.A. Brandts, and L.E.H. Trainor, eds.). World Scientific: Singapore, pp. 431–441.

2411. Larsen-Rapport, E.W. (1986). Imaginal disc determination: molecular and cellular correlates. *Annu. Rev. Entomol.* **31**, 145–175.

2412. Larsson, J., Chen, J.D., Rasheva, V., Rasmuson-Lestander, A., and Pirrotta, V. (2001). Painting of fourth, a chromosome-specific protein in *Drosophila*. *Proc. Natl. Acad. Sci. USA* **98**, 6273–6278.

2413. Latter, B.D.H. (1970). Selection for a threshold character in *Drosophila*. III. Genetic control of variability in plateaued populations. *Genet. Res., Camb.* **15**, 285–300.

2414. Laubichler, M.D. (1999). Seeing is believing, but what do we see? *Science* **284**, 58.

2415. Laudet, V., Stehelin, D., and Clevers, H. (1993). Ancestry and diversity of the HMG box superfamily. *Nucleic Acids Res.* **21**, 2493–2501.

2416. Laughon, A.S., Kirkpatrick, H., and Johnson, K. (2001). Repression of Dpp targets by Brinker. *Proc. 42nd Ann. Drosophila Res. Conf.* **Abstracts Vol.**, a125.

2417. Lavrov, S.A. and Cavalli, G. (2001). Chromatin organisation of the cellular memory module (CMM). *Proc. 42nd Ann. Drosophila Res. Conf.* **Abstracts Vol.**, a96.

2418. Lawler, S., Feng, X.-H., Chen, R.-H., Maruoka, E.M., Turck, C.W., Griswold-Prenner, I., and Derynck, R. (1997). The type II transforming growth factor-β receptor autophosphorylates not only on serine and threonine but also on tyrosine residues. *J. Biol. Chem.* **272** #23, 14850–14859.

2419. Lawrence, N., Klein, T., Brennan, K., and Martinez Arias, A. (2000). Structural requirements for Notch signalling with Delta and Serrate during the development and patterning of the wing disc of *Drosophila*. *Development* **127**, 3185–3195.

2420. Lawrence, P. (1979). Squaring the circle. *Nature* **280**, 722–723.

2421. Lawrence, P. and Morata, G. (1994). Homeobox genes: their function in *Drosophila* segmentation and pattern formation. *Cell* **78**, 181–189.

2422. Lawrence, P.A. (1966). Development and determination of hairs and bristles in the milkweed bug *Oncopeltus fasciatus* (Lygaeidae, Hemiptera). *J. Cell Sci.* **1**, 475–498.

2423. Lawrence, P.A. (1966). Gradients in the insect segment: the orientation of hairs in the milkweed bug *Oncopeltus fasciatus*. *J. Exp. Biol.* **44**, 607–620.

2424. Lawrence, P.A. (1967). The insect epidermal cell–"A simple model of the embryo". *In* "Insects and Physiology," (J.W.L. Beament and J.E. Treherne, eds.). Oliver & Boyd: London, pp. 53–68.

2425. Lawrence, P.A. (1970). Polarity and patterns in the postembryonic development of insects. *Adv. Insect Physiol.* **7**, 197–266.

2426. Lawrence, P.A. (1971). The organization of the insect segment. *Symp. Soc. Exp. Biol.* **25**, 379–392.

2427. Lawrence, P.A. (1973). The development of spatial patterns in the integument of insects. *In* "Developmental Systems: Insects," Vol. 2 (S.J. Counce and C.H. Waddington, eds.). Acad. Pr.: New York, pp. 157–209.

2428. Lawrence, P.A. (1975). The cell cycle and cellular differentiation in insects. *In* "Cell Cycle and Cell Differentiation," (J. Reinert and H. Holtzer, eds.). Springer: Berlin, pp. 111–121.

2429. Lawrence, P.A., ed. (1976). "Insect Development." *Symp. Roy. Entomol. Soc. Lond.*, Vol. 8. Wiley, New York.

2430. Lawrence, P.A. (1981). The cellular basis of segmentation in insects. *Cell* **26**, 3–10.

2431. Lawrence, P.A. (1984). Homoeotic selector genes–a working definition. *BioEssays* **1**, 227–229.

2432. Lawrence, P.A. (1985). Compartment genes in hand. *Nature* **313**, 268–269.

2433. Lawrence, P.A. (1987). Pair-rule genes: do they paint stripes or draw lines? *Cell* **51**, 879–880.

2434. Lawrence, P.A. (1992). "The Making of a Fly: The Genetics of Animal Design." Blackwell Sci. Pub., Oxford.

2435. Lawrence, P.A. (1997). Straight and wiggly affinities. *Nature* **389**, 546–547.

2436. Lawrence, P.A., Casal, J., and Struhl, G. (1999). *hedgehog* and *engrailed*: pattern formation and polarity in the *Drosophila* abdomen. *Development* **126**, 2431–2439.

2437. Lawrence, P.A., Casal, J., and Struhl, G. (1999). The Hedgehog morphogen and gradients of cell affinity in the abdomen of *Drosophila*. *Development* **126**, 2441–2449.

2438. Lawrence, P.A., Johnston, P., Macdonald, P., and Struhl, G. (1987). Borders of parasegments in *Drosophila* embryos are delimited by the *fushi tarazu* and *even-skipped* genes. *Nature* **328**, 440–442.

2439. Lawrence, P.A. and Locke, M. (1997). A man for our season. *Nature* **386**, 757–758.

2440. Lawrence, P.A. and Morata, G. (1976). The compartment hypothesis. *In* "Insect Development," *Symp. Roy. Entomol. Soc. Lond.*, Vol. 8 (P.A. Lawrence, ed.). Wiley: New York, pp. 132–149.

2441. Lawrence, P.A. and Morata, G. (1976). Compartments in the wing of *Drosophila*: A study of the *engrailed* gene. *Dev. Biol.* **50**, 321–337.

2442. Lawrence, P.A. and Morata, G. (1977). The early development of mesothoracic compartments in *Drosophila*: An analysis of cell lineage and fate mapping and an assessment of methods. *Dev. Biol.* **56**, 40–51.

2443. Lawrence, P.A. and Morata, G. (1979). Pattern formation and compartments in the tarsus of *Drosophila*. *In* "Determinants of Spatial Organization," *Symp. Soc. Dev. Biol.*, Vol. 37 (S. Subtelny and I.R. Konigsberg, eds.). Acad. Pr.: New York, pp. 317–323.

2444. Lawrence, P.A. and Morata, G. (1992). Lighting up *Drosophila*. *Nature* **356**, 107–108.

2445. Lawrence, P.A. and Morata, G. (1993). A no-wing situation. *Nature* **366**, 305–306.

2446. Lawrence, P.A. and Sampedro, J. (1993). *Drosophila* segmentation: after the first three hours. *Development* **119**, 971–976.

2447. Lawrence, P.A. and Struhl, G. (1982). Further studies of the *engrailed* phenotype in *Drosophila*. *EMBO J.* **1**, 827–833.

2448. Lawrence, P.A. and Struhl, G. (1996). Morphogens, compartments, and pattern: lessons from *Drosophila*? *Cell* **85**, 951–961. [See also 2001 *Cell* **105**, 559–562.]

2449. Lawrence, P.A., Struhl, G., and Morata, G. (1979). Bristle patterns and compartment boundaries in the tarsi of *Drosophila*. *J. Embryol. Exp. Morphol.* **51**, 195–208.

2450. Lawrence, P.A. and Tomlinson, A. (1991). A marriage is consummated. *Nature* **352**, 193.

2451. Lear, B.C., Skeath, J.B., and Patel, N.H. (1999). Neural cell fate in *rca1* and *cycA* mutants: the roles of intrinsic and extrinsic factors in asymmetric division in the *Drosophila* central nervous system. *Mechs. Dev.* **88**, 207–219.

2452. Lebovitz, R.M. and Ready, D.F. (1986). Ommatidial development in *Drosophila* eye disc fragments. *Dev. Biol.* **117**, 663–671.

2453. Lecourtois, M. and Schweisguth, F. (1995). The neurogenic Suppressor of Hairless DNA-binding protein mediates the transcriptional activation of the *Enhancer of split* Complex genes triggered by Notch signaling. *Genes Dev.* **9**, 2598–2608.

2454. Lecourtois, M. and Schweisguth, F. (1998). Indirect evidence for *Delta*-dependent intracellular processing of Notch in *Drosophila* embryos. *Curr. Biol.* **8**, 771–774.

2455. Lecuit, T., Brook, W.J., Ng, M., Calleja, M., Sun, H., and Cohen, S.M. (1996). Two distinct mechanisms for long-range patterning by Decapentaplegic in the *Drosophila* wing. *Nature* **381**, 387–393.

2456. Lecuit, T. and Cohen, S.M. (1997). Proximal-distal axis formation in the *Drosophila* leg. *Nature* **388**, 139–145.

2457. Lecuit, T. and Cohen, S.M. (1998). Dpp receptor levels contribute to shaping the Dpp morphogen gradient in the *Drosophila* wing imaginal disc. *Development* **125**, 4901–4907.

2458. Lee, A., Frank, D.W., Marks, M.S., and Lemmon, M.A. (1999). Dominant-negative inhibition of receptor-mediated endocytosis by a dynamin-1 mutant with a defective pleckstrin homology domain. *Curr. Biol.* **9**, 261–264.

2459. Lee, C., Parikh, V., Itsukaichi, T., Bae, K., and Edery, I. (1996). Resetting the *Drosophila* clock by photic regulation of PER and a PER-TIM complex. *Science* **271**, 1740–1744.

2460. Lee, E.-C. and Baker, N.E. (1996). Gp300Sca is not a high affinity Notch ligand. *Biochem. Biophys. Res. Comm.* **225**, 720–725.

2461. Lee, E.-C., Hu, X., Yu, S.-Y., and Baker, N.E. (1996). The *scabrous* gene encodes a secreted glycoprotein dimer and regulates proneural development in *Drosophila* eyes. *Molec. Cell Biol.* **16**, 1179–1188.

2462. Lee, E.-C., Yu, S.-Y., and Baker, N.E. (2000). The Scabrous protein can act as an extracellular antagonist of Notch signaling in the *Drosophila* wing. *Curr. Biol.* **10**, 931–934.

2463. Lee, E.-C., Yu, S.-Y., Hu, X., Mlodzik, M., and Baker, N.E. (1998). Functional analysis of the Fibrinogen-related *scabrous* gene from *Drosophila melanogaster* identifies potential effector and stimulatory protein domains. *Genetics* **150**, 663–673.

2464. Lee, J.D., Kraus, P., Gaiano, N., Nery, S., Kohtz, J., Fishell, G., Loomis, C.A., and Treisman, J.E. (2001). An acylatable residue of Hedgehog is differentially required in *Drosophila* and mouse limb development. *Dev. Biol.* **233**, 122–136.

2465. Lee, J.D. and Treisman, J.E. (2001). The role of Wingless signaling in establishing the anteroposterior and dorsoventral axes of the eye disc. *Development* **128**, 1519–1529.

2466. Lee, J.J., Ekker, S.C., von Kessler, D.P., Porter, J.A., Sun, B.I., and Beachy, P.A. (1994). Autoproteolysis in *hedgehog* protein biogenesis. *Science* **266**, 1528–1537.

2467. Lee, J.J., von Kessler, D.P., Parks, S., and Beachy, P.A. (1992). Secretion and localized transcription suggest a role in positional signaling for products of the segmentation gene *hedgehog*. *Cell* **71**, 33–50.

2468. Lee, K.J., Freeman, M., and Steller, H. (1991). Expression of the *disconnected* gene during development of *Drosophila melanogaster*. *EMBO J.* **10**, 817–826.

2469. Lee, T., Hacohen, N., Krasnow, M., and Montell, D.J. (1996). Regulated Breathless receptor tyrosine kinase activity required to pattern cell migration and branching in the *Drosophila* tracheal system. *Genes Dev.* **10**, 2912–2921.

2470. Lee, T. and Montell, D.J. (1997). Multiple Ras signals pattern the *Drosophila* ovarian follicle cells. *Dev. Biol.* **185**, 25–33.

2471. Lee, T.I., Wyrick, J.J., Koh, S.S., Jennings, E.G., Gadbois, E.L., and Young, R.A. (1998). Interplay of positive and negative regulators in transcription initiation by RNA polymerase II holoenzyme. *Mol. Cell. Biol.* **18**, 4455–4462.

2472. Lees, A.D. (1941). Operations on the pupal wing of *Drosophila melanogaster*. *J. Genet.* **42**, 115–142.

2473. Lees, A.D. (1942). Homology of the campaniform organs on the wing of *Drosophila melanogaster*. *Nature* **150**, 375.

2474. Lees, A.D. and Picken, L.E.R. (1945). Shape in relation to fine structure in the bristles of *Drosophila melanogaster*. *Proc. Roy. Soc. Lond. (Ser. B.)* **132**, 396–423. [See also 2001 *Development* **128**, 2793–2802.]

2475. Lees, A.D. and Waddington, C.H. (1942). The development of the bristles in normal and some mutant types of *Drosophila melanogaster*. *Proc. Roy. Soc. (London), Ser. B* **131**, 87–101.

2476. Leevers, S.J. (2001). Growth control: invertebrate insulin surprises! *Curr. Biol.* **11**, R209–R212.

2477. Leevers, S.J., Paterson, H.F., and Marshall, C.J. (1994). Requirement for Ras in Raf activation is overcome by targeting Raf to the plasma membrane. *Nature* **369**, 411–414.

2478. Leevers, S.J., Vanhaesebroeck, B., and Waterfield, M.D. (1999). Signalling through phosphoinositide 3-kinases: the lipids take center stage. *Curr. Opin. Cell Biol.* **11**, 219–225.

2479. Lefers, M.A., Wang, Q.T., and Holmgren, R.A. (2001). Genetic dissection of the *Drosophila* Cubitus interruptus signaling complex. *Dev. Biol.* **236**, 411–420.

2480. Lehman, D.A., Patterson, B., Johnston, L.A., Balzer, T., Britton, J.S., Saint, R., and Edgar, B.A. (1999). *Cis*-regulatory elements of the mitotic regulator, *string/Cdc25*. *Development* **126**, 1793–1803.

2481. Lehner, C.F. and O'Farrell, P.H. (1989). Expression and function of *Drosophila* Cyclin A during embryonic cell cycle progression. *Cell* **56**, 957–968.

2482. Lehner, C.F. and O'Farrell, P.H. (1990). The roles of *Drosophila* cyclins A and B in mitotic control. *Cell* **61**, 535–547.

2483. Lei, Y. and Warrior, R. (2000). The *Drosophila Lissencephaly1* (*DLis1*) gene is required for nuclear migration. *Dev. Biol.* **226**, 57–72.

2484. Leimeister, C., Dale, K., Fischer, A., Klamt, B., Hrabe de Angelis, M., Radtke, F., McGrew, M.J., Pourquié, O., and Gessler, M. (2000). Oscillating expression of *c-Hey2* in the presomitic mesoderm suggests that the segmentation clock may use combinatorial signaling through multiple interacting bHLH factors. *Dev. Biol.* **227**, 91–103.

2485. Leiserson, W.M., Benzer, S., and Bonini, N.M. (1998). Dual functions of the *Drosophila eyes absent* gene in the eye and embryo. *Mechs. Dev.* **73**, 193–202.

2486. Leloup, J.-C. and Goldbeter, A. (2000). Modeling the molecular regulatory mechanism of circadian rhythms in *Drosophila*. *BioEssays* **22**, 84–93.

2487. Lemmon, M.A., Ferguson, K.M., and Schlessinger, J. (1996). PH domains: Diverse sequences with a common fold recruit signaling molecules to the cell surface. *Cell* **85**, 621–624.

2488. Lemon, B. and Tjian, R. (2000). Orchestrated response: a symphony of transcription factors for gene control. *Genes Dev.* **14**, 2551–2569.

2489. Lemon, B.D. and Freedman, L.P. (1999). Nuclear receptor cofactors as chromatin remodelers. *Curr. Opin. Gen. Dev.* **9**, 499–504.

2490. Lemon, K.P. and Grossman, A.D. (2000). Movement of replicating DNA through a stationary replisome. *Molec. Cell* **6**, 1321–1330.

2491. Lepage, T., Cohen, S.M., Diaz-Benjumea, F.J., and Parkhurst, S.M. (1995). Signal transduction by cAMP-dependent protein kinase A in *Drosophila* limb patterning. *Nature* **373**, 711–715.

2492. Leshko-Lindsay, L. and Corces, V.G. (1997). The role of selectins in *Drosophila* eye and bristle development. *Development* **124**, 169–180.

2493. Lesokhin, A.M., Yu, S.-Y., Katz, J., and Baker, N.E. (1999). Several levels of EGR receptor signaling during photoreceptor specification in wild-type, *Ellipse*, and null mutant *Drosophila*. *Dev. Biol.* **205**, 129–144.

2494. Lessing, D. and Nusse, R. (1998). Expression of *wingless* in the *Drosophila* embryo: a conserved *cis*-acting element lacking conserved Ci-binding sites is required for *patched*-mediated repression. *Development* **125**, 1469–1476.

2495. Letsou, A., Arora, K., Wrana, J.L., Simin, K., Twombly, V.,

Jamal, J., Staehling-Hampton, K., Hoffmann, F.M., Gelbart, W.M., Massagué, J., and O'Connor, M.B. (1995). *Drosophila* Dpp signaling is mediated by the *punt* gene product: a dual ligand-binding type II receptor of the TGF-β receptor family. *Cell* **80**, 899–908.

2496. Levanon, D., Goldstein, R.E., Bernstein, Y., Tang, H., Goldenberg, D., Stifani, S., Paroush, Z., and Groner, Y. (1998). Transcriptional repression by AML1 and LEF-1 is mediated by the TLE/Groucho corepressors. *Proc. Natl. Acad. Sci. USA* **95**, 11590–11595.

2497. Levine, A., Weiss, C., and Wides, R. (1997). Expression of the pair-rule gene *odd Oz* (*odz*) in imaginal tissues. *Dev. Dynamics* **209**, 1–14.

2498. Levine, M. and Hoey, T. (1988). Homeobox proteins as sequence-specific transcription factors. *Cell* **55**, 537–540.

2499. Leviten, M.W., Lai, E.C., and Posakony, J.W. (1997). The *Drosophila* gene *Bearded* encodes a novel small protein and shares 3′ UTR sequence motifs with multiple *Enhancer of split* Complex genes. *Development* **124**, 4039–4051.

2500. Leviten, M.W. and Posakony, J.W. (1996). Gain-of-function alleles of *Bearded* interfere with alternative cell fate decisions in *Drosophila* adult sensory organ development. *Dev. Biol.* **176**, 264–283.

2501. Lewin, B. (1998). The mystique of epigenetics. *Cell* **93**, 301–303.

2502. Lewis, C.T. (1970). Structure and function in some external receptors. *In* "Insect Ultrastructure," *5th Symp. Roy. Entomol. Soc. Lond.*, (A.C. Neville, ed.). Blackwell: Oxford, pp. 59–76.

2503. Lewis, D.L., DeCamillis, M., and Bennett, R.L. (2000). Distinct roles of the homeotic genes Ubx and abd-A in beetle embryonic abdominal appendage development. *Proc. Natl. Acad. Sci. USA* **97**, 4504–4509.

2504. Lewis, E. (1995). Remembering Sturtevant. *Genetics* **141**, 1227–1230.

2505. Lewis, E.B. (1951). Pseudoallelism and gene evolution. *Cold Spr. Harb. Symp. Quant. Biol.* **16**, 159–174.

2506. Lewis, E.B. (1963). Genes and developmental pathways. *Amer. Zool.* **3**, 33–56.

2507. Lewis, E.B. (1978). A gene complex controlling segmentation in *Drosophila*. *Nature* **276**, 565–570.

2508. Lewis, E.B. (1992). Clusters of master control genes regulate the development of higher organisms. *JAMA* **267**, 1524–1531.

2509. Lewis, E.B. (1994). Homeosis: the first 100 years. *Trends Genet.* **10**, 341–343.

2510. Lewis, E.B. (1998). The *bithorax complex*: the first fifty years. *Int. J. Dev. Biol.* **42**, 403–415.

2511. Lewis, E.B., Knafels, J.D., Mathog, D.R., and Celniker, S.E. (1995). Sequence analysis of the cis-regulatory regions of the bithorax complex of *Drosophila*. *Proc. Natl. Acad. Sci. USA* **92**, 8403–8407.

2512. Lewis, J. (1981). Simpler rules for epimorphic regeneration: the polar-coordinate model without polar coordinates. *J. Theor. Biol.* **88**, 371–392.

2513. Lewis, J. (1982). Continuity and discontinuity in pattern formation. *In* "Developmental Order: Its Origin and Regulation," *Symp. Soc. Dev. Biol.*, Vol. 40 (S. Subtelny and P.B. Green, eds.). Liss: New York, pp. 511–531.

2514. Lewis, J. (1998). A short cut to the nucleus. *Nature* **393**, 304–305.

2515. Lewis, J., Slack, J.M.W., and Wolpert, L. (1977). Thresholds in development. *J. Theor. Biol.* **65**, 579–590.

2516. Lewis, J.H. and Wolpert, L. (1976). The principle of nonequivalence in development. *J. Theor. Biol.* **62**, 479–490.

2517. Lewis, P.H. (1947). Pc: Polycomb. *Dros. Info. Serv.* **21**, 69.

2518. Lewontin, R. (2000). Computing the organism. *Nat. Hist.* **109** #3, 94.

2519. Leyns, L., Dambly-Chaudière, C., and Ghysen, A. (1989). Two different sets of *cis* elements regulate *scute* to establish two different sensory patterns. *Roux's Arch. Dev. Biol.* **198**, 227–232.

2520. Leyns, L., Gómez-Skarmeta, J.-L., and Dambly-Chaudière, C. (1996). *iroquois*: a prepattern gene that controls the formation of bristles on the thorax of *Drosophila*. *Mechs. Dev.* **59**, 63–72.

2521. Li, E. (1999). The mojo of methylation. *Nature Genet.* **23**, 5–6.

2522. Li, L. and Vaessin, H. (2000). Pan-neural Prospero terminates cell proliferation during *Drosophila* neurogenesis. *Genes Dev.* **14**, 147–151.

2523. Li, L., Yuan, H., Weaver, C.D., Mao, J., Farr III, G.H., Sussman, D.J., Jonkers, J., Kimelman, D., and Wu, D. (1999). Axin and Frat1 interact with Dvl and GSK, bridging Dvl to GSK in Wnt-mediated regulation of LEF-1. *EMBO J.* **18**, 4233–4240.

2524. Li, L.-H. and Gergen, J.P. (1999). Differential interactions between Brother proteins and Runt domain proteins in the *Drosophila* embryo and eye. *Development* **126**, 3313–3322. [See also 2001 *Development* **128**, 2639–2648.]

2525. Li, P., He, X., Gerrero, M.R., Mok, M., Aggarwal, A., and Rosenfeld, M.G. (1993). Spacing and orientation of bipartite DNA-binding motifs as potential functional determinants for POU domain factors. *Genes Dev.* **7**, 2483–2496.

2526. Li, P., Yang, X., Wasser, M., Cai, Y., and Chia, W. (1997). Inscuteable and Staufen mediate asymmetric localization and segregation of *prospero* RNA during *Drosophila* neuroblast cell divisions. *Cell* **90**, 437–447.

2527. Li, Q.-J., Pazdera, T.M., and Minden, J.S. (1999). *Drosophila* embryonic pattern repair: how embryos respond to cyclin E-induced ectopic division. *Development* **126**, 2299–2307.

2528. Li, S., Aufiero, B., Schiltz, R.L., and Walsh, M.J. (2000). Regulation of the homeodomain CCAAT displacement/*cut* protein function by histone acetyltransferases p300/CREB-binding protein (CBP)-associated factor and CBP. *Proc. Natl. Acad. Sci. USA* **97**, 7166–7171.

2529. Li, S., Li, Y., Carthew, R.W., and Lai, Z.-C. (1997). Photoreceptor cell differentiation requires regulated proteolysis of the transcriptional repressor Tramtrack. *Cell* **90**, 469–478.

2530. Li, S.-C., Songyang, Z., Vincent, S.J.F., Zwahlen, C., Wiley, S., Cantley, L., Kay, L.E., Forman-Kay, J., and Pawson, T. (1997). High-affinity binding of the *Drosophila* Numb phosphotyrosine-binding domain to peptides containing a Gly-Pro-(p)Tyr motif. *Proc. Natl. Acad. Sci. USA* **94**, 7204–7209.

2531. Li, W., Melnick, M., and Perrimon, N. (1998). Dual function of Ras in Raf activation. *Development* **125**, 4999–5008.

2532. Li, W., Noll, E., and Perrimon, N. (2000). Identification of autosomal regions involved in *Drosophila* Raf function. *Genetics* **156**, 763–774.

2533. Li, W., Ohlmeyer, J.T., Lane, M.E., and Kalderon, D. (1995). Function of protein kinase A in hedgehog signal transduction and *Drosophila* imaginal disc development. *Cell* **80**, 553–562.

2534. Li, X., Gutjahr, T., and Noll, M. (1993). Separable regulatory elements mediate the establishment and maintenance of cell states by the *Drosophila* segment-polarity gene *gooseberry*. *EMBO J.* **12**, 1427–1436.

2535. Li, X., Murre, C., and McGinnis, W. (1999). Activity regulation of a Hox protein and a role for the homeodomain in inhibiting transcriptional activation. *EMBO J.* **18**, 198–211.

2536. Li, X. and Noll, M. (1993). Role of the *gooseberry* gene in *Drosophila* embryos: maintenance of *wingless* expression by a *wingless-gooseberry* autoregulatory loop. *EMBO J.* **12**, 4499–4509.

2537. Li, X. and Noll, M. (1994). Compatibility between enhancers and promoters determines the transcriptional specificity of *gooseberry* and *gooseberry neuro* in the *Drosophila* embryo. *EMBO J.* **13**, 400–406.

2538. Li, X. and Noll, M. (1994). Evolution of distinct developmental functions of three *Drosophila* genes by acquisition of different *cis*-regulatory regions. *Nature* **367**, 83–87.

2539. Li, X., Veraksa, A., and McGinnis, W. (1999). A sequence motif distinct from Hox binding sites controls the specificity of a Hox response element. *Development* **126**, 5581–5589.

2540. Li, Y. and Baker, N.E. (2001). Proneural enhancement by Notch overcomes Suppressor-of-Hairless repressor function in the developing *Drosophila* eye. *Curr. Biol.* **11**, 330–338.

2541. Liang, Z. and Biggin, M.D. (1998). Eve and ftz regulate a wide array of genes in blastoderm embryos: the selector homeoproteins directly or indirectly regulate most genes in *Drosophila*. *Development* **125**, 4471–4482.

2542. Lieber, T., Kidd, S., Alcamo, E., Corbin, V., and Young, M.W. (1993). Antineurogenic phenotypes induced by truncated Notch proteins indicate a role in signal transduction and may point to a novel function for Notch in nuclei. *Genes Dev.* **7**, 1949–1965.

2543. Lieber, T., Wesley, C.S., Alcamo, E., Hassel, B., Krane, J.F., Campos-Ortega, J.A., and Young, M.W. (1992). Single amino acid substitutions in EGF-like elements of Notch and Delta modify *Drosophila* development and affect cell adhesion in vitro. *Neuron* **9**, 847–859.

2544. Lienhard, M.C. and Stocker, R.F. (1987). Sensory projection patterns of supernumerary legs and aristae in *D. melanogaster*. *J. Exp. Zool.* **244**, 187–201.

2545. Lienhard, M.C. and Stocker, R.F. (1991). The development of the sensory neuron pattern in the antennal disc of wild-type and mutant (lz^3, ss^a) *Drosophila melanogaster*. *Development* **112**, 1063–1075.

2546. Lietzke, S.E., Bose, S., Cronin, T., Klarlund, J., Chawla, A., Czech, M.P., and Lambright, D.G. (2000). Structural basis of 3-phosphoinositide recognition by pleckstrin homology domains. *Molec. Cell* **6**, 385–394.

2547. Lifson, S. (1994). What is information for molecular biology? *BioEssays* **16**, 373–375.

2548. Ligoxygakis, P., Bray, S.J., Apidianakis, Y., and Delidakis, C. (1999). Ectopic expression of individual *E(spl)* genes has differential effects on different cell fate decisions and underscores the biphasic requirement for Notch activity in wing margin establishment in *Drosophila*. *Development* **126**, 2205–2214.

2549. Ligoxygakis, P., Yu, S.-Y., Delidakis, C., and Baker, N.E. (1998). A subset of *Notch* functions during *Drosophila* eye development require *Su(H)* and the *E(spl)* gene complex. *Development* **125**, 2893–2900.

2550. Lill, N.L., Douillard, P., Awwad, R.A., Ota, S., Lupher, M.L., Jr., Miyake, S., Meissner-Lula, N., Hsu, V.W., and Band, H. (2000). The evolutionarily conserved N-terminal region of Cbl is sufficient to enhance down-regulation of the epidermal growth factor receptor. *J. Biol. Chem.* **275**#1, 367–377.

2551. Lilly, M.A. and Spradling, A.C. (1996). The *Drosophila* endocycle is controlled by Cyclin E and lacks a checkpoint ensuring S-phase completion. *Genes Dev.* **10**, 2514–2526.

2552. Lim, W.A., Richards, F.M., and Fox, R.O. (1994). Structural determinants of peptide-binding orientation and of sequence specificity in SH3 domains. *Nature* **372**, 375–379.

2553. Lin, D.M., Fetter, R.D., Kopczynski, C., Grenningloh, G., and Goodman, C.S. (1994). Genetic analysis of Fasciclin II in *Drosophila*: defasciculation, refasciculation, and altered fasciculation. *Neuron* **13**, 1055–1069.

2554. Lin, H., Cho, S., and Cadigan, K.M. (2001). *Split ends*, a putative tissue-specific factor in Wingless signaling. *Proc. 42nd Ann. Drosophila Res. Conf.* **Abstracts Vol.**, a163.

2555. Lin, X., Buff, E.M., Perrimon, N., and Michelson, A.M. (1999). Heparan sulfate proteoglycans are essential for FGF receptor signaling during *Drosophila* embryonic development. *Development* **126**, 3715–3723.

2556. Lin, X. and Perrimon, N. (1999). Dally cooperates with *Drosophila* Frizzled 2 to transduce Wingless signalling. *Nature* **400**, 281–284.

2557. Lindenmayer, A. (1975). Developmental algorithms for multicellular organisms: a survey of L-systems. *J. Theor. Biol.* **54**, 3–22.

2558. Lindenmayer, A. (1982). Developmental algorithms: lineage versus interactive control mechanisms. *In* "Developmental Order: Its Origin and Regulation," *Symp. Soc. Dev. Biol.*, Vol. 40 (S. Subtelny and P.B. Green, eds.). Alan R. Liss: New York, pp. 219–245.

2559. Lindenmayer, A. and Prusinkiewicz, P. (1989). Developmental models of multicellular organisms: a computer graphics perspective. *In* "Artificial Life," (C.G. Langton, ed.). Addison-Wesley: New York, pp. 221–249.

2560. Lindsley, D.L. and Grell, E.H. (1968). "Genetic Variations of *Drosophila melanogaster*." *Carnegie Inst. Wash. Publ.*, Vol. 627. Washington, D. C.

2561. Lindsley, D.L. and Zimm, G.G. (1992). "The Genome of *Drosophila melanogaster*." Acad. Pr., New York.

2562. Lipshitz, H.D. and Kankel, D.R. (1985). Developmental interactions between the peripheral and central nervous system in *Drosophila melanogaster*: analysis of the mutant, *two-faced*. *Dev. Biol.* **107**, 1–12.

2563. Lisbin, M.J., Gordon, M., Yannoni, Y.M., and White, K. (2000). Function of RRM domains of *Drosophila melanogaster* ELAV: RNP1 mutations and RRM domain replacements with ELAV family proteins and SXL. *Genetics* **155**, 1789–1798.

2564. Little, J.W., Byrd, C.A., and Brower, D.L. (1990). Effect of *abx*, *bx* and *pbx* mutations on expression of homeotic genes in *Drosophila* larvae. *Genetics* **124**, 899–908.

2565. Little, J.W., Shepley, D.P., and Wert, D.W. (1999). Robustness of a gene regulatory circuit. *EMBO J.* **18**, 4299–4307.

2566. Littlewood, T.D. and Evan, G.I. (1998). "Helix-loop-helix Transcription Factors." 3rd ed. *Protein Profile Series*, Oxford Univ. Pr., Oxford.

2567. Liu, D., Bienkowska, J., Petosa, C., Collier, R.J., Fu, H., and Liddington, R. (1995). Crystal structure of the zeta isoform of the 14-3-3 protein. *Nature* **376**, 191–194.

2568. Liu, F., Hata, A., Baker, J.C., Doody, J., Cárcamo, J., Harland, R.M., and Massagué, J. (1996). A human Mad protein acting as a BMP-regulated transcriptional activator. *Nature* **381**, 620–623.

2569. Liu, H., Ma, C., and Moses, K. (1996). Identification

and functional characterization of conserved promoter elements from *glass*: a retinal development gene of *Drosophila. Mechs. Dev.* **56**, 73–82.

2570. Liu, X., Grammont, M., and Irvine, K.D. (2000). Roles for *scalloped* and *vestigial* in regulating cell affinity and interactions between the wing blade and the wing hinge. *Dev. Biol.* **228**, 287–303.

2571. Liu, X. and Lengyel, J.A. (2000). *Drosophila arc* encodes a novel adherens junction-associated PDZ domain protein required for wing and eye development. *Dev. Biol.* **221**, 419–434.

2572. Liu, Z.-P., Galindo, R.L., and Wasserman, S.A. (1997). A role for CKII phosphorylation of the Cactus PEST domain in dorsoventral patterning of the *Drosophila* embryo. *Genes Dev.* **11**, 3413–3422.

2573. Livneh, E., Glazer, L., Segal, D., Schlessinger, J., and Shilo, B.-Z. (1985). The *Drosophila* EGF receptor gene homolog: conservation of both hormone binding and kinase domains. *Cell* **40**, 599–607.

2574. Llimargas, M. (1999). The *Notch* pathway helps to pattern the tips of the *Drosophila* tracheal branches by selecting cell fates. *Development* **126**, 2355–2364.

2575. Llimargas, M. and Casanova, J. (1997). *ventral veinless*, a POU domain transcription factor, regulates different transduction pathways required for tracheal branching in *Drosophila. Development* **124**, 3273–3281.

2576. Llimargas, M. and Casanova, J. (1999). EGF signalling regulates cell invagination as well as cell migration during formation of tracheal system in *Drosophila. Dev. Genes Evol.* **209**, 174–179.

2577. Lloyd, C.W. and Rees, D.A., eds. (1981). "Cellular Controls in Differentiation." Acad. Pr., New York.

2578. Lo, P.C.H. and Frasch, M. (1999). Sequence and expression of *myoglianin*, a novel *Drosophila* gene of the TGF-β superfamily. *Mechs. Dev.* **86**, 171–175.

2579. Lo, R.S., Chen, Y.-G., Shi, Y., Pavletich, N.P., and Massagué, J. (1998). The L3 loop: a structural motif determining specific interactions between SMAD proteins and TGF-β receptors. *EMBO J.* **17**, 996–1005.

2580. Lo, S.H., Weisberg, E., and Chen, L.B. (1994). Tensin: a potential link between the cytoskeleton and signal transduction. *BioEssays* **16**, 817–823.

2581. Locke, J. and Hanna, S. (1996). *engrailed* gene dosage determines whether certain recessive *cubitus interruptus* alleles exhibit dominance of the adult wing phenotype in *Drosophila. Dev. Genet.* **19**, 340–349.

2582. Locke, J., Kotarski, M.A., and Tartof, K.D. (1988). Dosage-dependent modifiers of position effect variegation in *Drosophila* and a mass action model that explains their effect. *Genetics* **120**, 181–198.

2583. Locke, J., Podemski, L., and Ferrer, C. (2001). Inversion of the whole arm of chromosome 4 in *Drosophila* species. *Proc. 42nd Ann. Drosophila Res. Conf.* **Abstracts Vol.**, a103. [See also 2001 *Chromosoma* **110**, 305–312.]

2584. Locke, M. (1984). Epidermal cells [Arthropoda]. *In* "Biology of the Integument," Vol. 1 (J. Bereiter-Hahn, A.G. Matoltsy, and K.S. Richards, eds.). Springer-Verlag: Berlin, pp. 502–522.

2585. Locke, M. (1985). The structure of epidermal feet during their development. *Tissue & Cell* **17**, 901–921.

2586. Locke, M. (1987). The very rapid induction of filopodia in insect cells. *Tissue & Cell* **19**, 301–318.

2587. Locke, M. and Huie, P. (1981). Epidermal feet in insect morphogenesis. *Nature* **293**, 733–735.

2588. Locke, M. and Huie, P. (1981). Epidermal feet in pupal segment morphogenesis. *Tissue & Cell* **13**, 787–803.

2589. Logeat, F., Bessia, C., Brou, C., LeBail, O., Jarriault, S., Seidah, N.G., and Israël, A. (1998). The Notch1 receptor is cleaved constitutively by a furin-like convertase. *Proc. Natl. Acad. Sci. USA* **95**, 8108–8112.

2590. Löhr, U., Yussa, M., and Pick, L. (2001). *Drosophila fushi tarazu*: a gene on the border of homeotic function. *Curr. Biol.* **11**, 1403–1412. [See also 2001 *Curr. Biol.* **11**, 1473–1478.]

2591. Lohs-Schardin, M., Sander, K., Cremer, C., Cremer, T., and Zorn, C. (1979). Localized ultraviolet laser microbeam irradiation of early *Drosophila* embryos: fate maps based on location and frequency of adult defects. *Dev. Biol.* **68**, 533–545.

2592. Long, A.D., Lyman, R.F., Langley, C.H., and Mackay, T.F.C. (1998). Two sites in the *Delta* gene region contribute to naturally occurring variation in bristle number in *Drosophila melanogaster. Genetics* **149**, 999–1017.

2593. Long, A.D., Lyman, R.F., Morgan, A.H., Langley, C.H., and Mackay, T.F.C. (2000). Both naturally occurring insertions of transposable elements and intermediate frequency polymorphisms at the *achaete-scute* complex are associated with variation in bristle number in *Drosophila melanogaster. Genetics* **154**, 1255–1269.

2594. Longley, R.L., Jr. and Ready, D.F. (1995). Integrins and the development of three-dimensional structure in the *Drosophila* compound eye. *Dev. Biol.* **171**, 415–433.

2595. Lonie, A., D'Andrea, R., Paro, R., and Saint, R. (1994). Molecular characterization of the *Polycomblike* gene of *Drosophila melanogaster*, a *trans*-acting negative regulator of homeotic gene expression. *Development* **120**, 2629–2636.

2596. Loo, L.W., Yost, C., and Eisenman, R.N. (2001). Functional characterization of dMad, a transcriptional repressor, during development. *Proc. 42nd Ann. Drosophila Res. Conf.* **Abstracts Vol.**, a125.

2597. Loomis, W.F. and Kuspa, A. (1992). Spontaneous generation of enhancers by point mutations. *Trends Genet.* **8**, 229.

2598. Lopez, A., Salvaing, J., Deutsch, J., and Peronnet, F. (2001). Involvement of CORTO in Polycomb and trithorax functions. *Proc. 42nd Ann. Drosophila Res. Conf.* **Abstracts Vol.**, a96.

2599. López, A.J. (1995). Developmental role of transcription factor isoforms generated by alternative splicing. *Dev. Biol.* **172**, 396–411.

2600. Lopez, A.J. and Hogness, D.S. (1991). Immunochemical dissection of the Ultrabithorax homeoprotein family in *Drosophila melanogaster. Proc. Natl. Acad. Sci. USA* **88**, 9924–9928.

2601. Lopez, M., Oettgen, P., Akbarali, Y., Dendorfer, U., and Libermann, T.A. (1994). ERP, a new member of the *ets* transcription factor/oncoprotein family: cloning, characterization, and differential expression during B-lymphocyte development. *Mol. Cell. Biol.* **14**, 3292–3309.

2602. Lopez-Schier, H. and St. Johnston, D. (2001). *agoraphobic* modulates Notch receptor cleavage and signalling in *Drosophila. Proc. 42nd Ann. Drosophila Res. Conf.* **Abstracts Vol.**, a158. [Cf. 2002 *Dev. Cell* **2**, 79–89.]

2603. Lorick, K.L., Jensen, J.P., Fang, S., Ong, A.M., Hatakeyama, S., and Weissman, A.M. (1999). RING fingers mediate ubiquitin-conjugating enzyme (E2)-dependent ubiquitination. *Proc. Natl. Acad. Sci. USA* **96**, 11364–11369.

2604. Loureiro, J. and Peifer, M. (1998). Roles of Armadillo, a *Drosophila* catenin, during central nervous system development. *Curr. Biol.* **8**, 622–632.

2605. Love, J.J., Li, X., Case, D.A., Giese, K., Grosschedl, R., and Wright, P.E. (1995). Structural basis for DNA bending by the architectural transcription factor LEF-1. *Nature* **376**, 791–795.

2606. Lu, B., Ackerman, L., Jan, L.Y., and Jan, Y.-N. (1999). Modes of protein movement that lead to the asymmetric localization of Partner of Numb during *Drosophila* neuroblast division. *Molec. Cell* **4**, 883–891.

2607. Lu, B., Jan, L.Y., and Jan, Y.-N. (1998). Asymmetric cell division: lessons from flies and worms. *Curr. Opin. Genet. Dev.* **8**, 392–399.

2608. Lu, B., Roegiers, F., Jan, L.Y., and Jan, Y.N. (2001). Adherens junctions inhibit asymmetric division in the *Drosophila* epithelium. *Nature* **409**, 522–525.

2609. Lu, B., Rothenberg, M., Jan, L.Y., and Jan, Y.N. (1998). Partner of Numb colocalizes with Numb during mitosis and directs Numb asymmetric localization in *Drosophila* neural and muscle progenitors. *Cell* **95**, 225–235.

2610. Lu, C.-H., Rincón-Limas, D.E., and Botas, J. (2000). Conserved overlapping and reciprocal expression of *msh/Msx1* and *apterous/Lhx2* in *Drosophila* and mice. *Mechs. Dev.* **99**, 177–181.

2611. Lu, F.M. and Lux, S.E. (1996). Constitutively active human Notch1 binds to the transcription factor CBF1 and stimulates transcription through a promoter containing a CBF1-responsive element. *Proc. Natl. Acad. Sci. USA* **93**, 5663–5667.

2612. Lu, X. and Li, Y. (1999). *Drosophila Src42A* is a negative regulator of RTK signaling. *Dev. Biol.* **208**, 233–243.

2613. Lu, X., Melnick, M.B., Hsu, J.-C., and Perrimon, N. (1994). Genetic and molecular analyses of mutations involved in *Drosophila raf* signal transduction. *EMBO J.* **13**, 2592–2599.

2614. Lucchesi, J.C. (1983). Curt Stern: 1902–1981. *Genetics* **103**, 1–4.

2615. Lucchesi, J.C. (1994). Sturtevant's mantle and the (lost?) art of chromosome mechanics. *Genetics* **136**, 707–708.

2616. Ludwig, M.Z., Patel, N.H., and Kreitman, M. (1998). Functional analysis of eve stripe 2 enhancer evolution in *Drosophila*: rules governing conservation and change. *Development* **125**, 949–958.

2617. Lunde, K., Biehs, B., Nauber, U., and Bier, E. (1998). The *knirps* and *knirps-related* genes organize development of the second wing vein in *Drosophila*. *Development* **125**, 4145–4154.

2618. Luo, H., Asha, H., Kockel, L., Parke, T., Mlodzik, M., and Dearolf, C.R. (1999). The *Drosophila* Jak kinase Hopscotch is required for multiple developmental processes in the eye. *Dev. Biol.* **213**, 432–441. [See also 2001 *BioEssays* **23**, 1138–1147.]

2619. Luo, K. and Lodish, H.F. (1997). Positive and negative regulation of type II TGF-β receptor signal transduction by autophosphorylation on multiple serine residues. *EMBO J.* **16**, 1970–1981.

2620. Luo, Z., Tzivion, G., Belshaw, P.J., Vavvas, D., Marshall, M., and Avruch, J. (1996). Oligomerization activates c-Raf-1 through a Ras-dependent mechanism. *Nature* **383**, 181–185.

2621. Lupas, A. (1996). Coiled coils: new structures and new functions. *Trends Biochem. Sci.* **21**, 375–382.

2622. Lupo, R., Breiling, A., Bianchi, M.E., and Orlando, V. (2001). *Drosophila* chromosome condensation proteins Topoisomerase II and Barren colocalize with Polycomb and maintain *Fab-7* PRE silencing. *Molec. Cell* **7**, 127–136.

2623. Luschnig, S., Krauss, J., Bohmann, K., Desjeux, I., and Nüsslein-Volhard, C. (2000). The *Drosophila* SHC adaptor protein is required for signaling by a subset of receptor tyrosine kinases. *Molec. Cell* **5**, 231–241.

2624. Lux, S.E., John, K.M., and Bennett, V. (1990). Analysis of cDNA for human erythrocyte ankyrin indicates a repeated structure with homology to tissue-differentiation and cell-cycle control proteins. *Nature* **344**, 36–42.

2625. Lyko, F. (2001). DNA methylation learns to fly. *Trends Genet.* **17**, 169–172.

2626. Lyman, D. and Young, M.W. (1993). Further evidence for function of the *Drosophila* Notch protein as a transmembrane receptor. *Proc. Natl. Acad. Sci. USA* **90**, 10395–10399.

2627. Lyman, D.F. and Yedvobnick, B. (1995). *Drosophila Notch* receptor activity suppresses *Hairless* function during adult external sensory organ development. *Genetics* **141**, 1491–1505.

2628. Lyman, R.F. and Mackay, T.F.C. (1998). Candidate quantitative trait loci and naturally occurring phenotypic variation for bristle number in *Drosophila melanogaster*: the *Delta-Hairless* gene region. *Genetics* **149**, 983–998.

2629. Lynch, M. and Conery, J.S. (2000). The evolutionary fate and consequences of duplicate genes. *Science* **290**, 1151–1155.

2630. Lynch, M. and Force, A. (2000). The probability of duplicate gene preservation by subfunctionalization. *Genetics* **154**, 459–473.

2631. Ma, C. and Moses, K. (1995). *wingless* and *patched* are negative regulators of the morphogenetic furrow and can affect tissue polarity in the developing *Drosophila* compound eye. *Development* **121**, 2279–2289.

2632. Ma, C., Zhou, Y., Beachy, P.A., and Moses, K. (1993). The segment polarity gene *hedgehog* is required for progression of the morphogenetic furrow in the developing *Drosophila* eye. *Cell* **75**, 927–938.

2633. Ma, J. and Karplus, M. (1997). Molecular switch in signal transduction: reaction paths of the conformational changes in *ras* p21. *Proc. Natl. Acad. Sci. USA* **94**, 11905–11910.

2634. Ma, P.C.M., Rould, M.A., Weintraub, H., and Pabo, C.O. (1994). Crystal structure of MyoD bHLH domain-DNA complex: perspectives on DNA recognition and implications for transcriptional activation. *Cell* **77**, 451–459.

2635. Ma, Y., Niemitz, E.L., Nambu, P.A., Shan, X., Sackerson, C., Fujioka, M., Goto, T., and Nambu, J.R. (1998). Gene regulatory functions of *Drosophila* Fish-hook, a high mobility group domain Sox protein. *Mechs. Dev.* **73**, 169–182.

2636. Maas, A.-H. (1948). Über die Auslösbarkeit von Temperatur-Modifikationen während der Embryonal-Entwicklung von *Drosophila melanogaster* Meigen. *W. Roux' Arch. Entw.-Mech. Org.* **143**, 515–572.

2637. Macara, I.G. (1999). Nuclear transport: randy couples. *Curr. Biol.* **9**, R436–R439.

2638. MacBean, I.T., McKenzie, J.A., and Parsons, P.A. (1971). A pair of closely linked genes controlling high scutellar chaeta number in *Drosophila*. *Theor. Appl. Genet.* **41**, 227–235.

2639. MacDougall, L.K. and Waterfield, M.D. (1996). Receptor signalling: to Sevenless, a daughter. *Curr. Biol.* **6**, 1250–1253.

2640. MacDowell, E.C. (1915). Bristle inheritance in *Drosophila*. I. Extra bristles. *J. Exp. Zool.* **19**, 61–97.

2641. Macías, A. and Morata, G. (1996). Functional hierarchy and phenotypic suppression among *Drosophila* homeotic genes: the *labial* and *empty spiracles* genes. *EMBO J.* **15**, 334–343.

2642. Macías, A., Pelaz, S., and Morata, G. (1994). Genetic factors controlling the expression of the *abdominal-A* gene of *Drosophila* within its domain. *Mechs. Dev.* **46**, 15–25.

2643. MacKay, T.F.C. (1995). The genetic basis of quantitative variation: numbers of sensory bristles of *Drosophila melanogaster* as a model system. *Trends Genet.* **11**, 464–470.

2644. Mackay, T.F.C. (1995). The nature of quantitative genetic variation revisited: lessons from *Drosophila* bristles. *BioEssays* **18**, 113–121.

2645. Mackay, T.F.C. and Langley, C.H. (1990). Molecular and phenotypic variation in the *achaete-scute* region of *Drosophila melanogaster*. *Nature* **348**, 64–65.

2646. MacWilliams, H.K. (1978). A model of gradient interpretation based on morphogen binding. *J. Theor. Biol.* **72**, 385–411.

2647. Madhani, H.D. and Fink, G.R. (1998). The riddle of MAP kinase signaling specificity. *Trends Genet.* **14**, 151–155.

2648. Madhavan, M.M. and Madhavan, K. (1980). Morphogenesis of the epidermis of adult abdomen of *Drosophila*. *J. Embryol. Exp. Morph.* **60**, 1–31.

2649. Madhavan, M.M. and Madhavan, K. (1982). Pattern regulation in tergite of *Drosophila*: a model. *J. Theor. Biol.* **95**, 731–748.

2650. Madhavan, M.M. and Schneiderman, H.A. (1977). Histological analysis of the dynamics of growth of imaginal discs and histoblast nests during the larval development of *Drosophila melanogaster*. *W. Roux's Arch.* **183**, 269–305.

2651. Madore, B.F. and Freedman, W.L. (1987). Self-organizing structures. *Am. Sci.* **75**, 252–259.

2652. Maegley, K.A., Admiraal, S.J., and Herschlag, D. (1996). Ras-catalyzed hydrolysis of GTP: a new perspective from model studies. *Proc. Natl. Acad. Sci. USA* **93**, 8160–8166.

2653. Magee, T. and Marshall, C. (1999). New insights into the interaction of Ras with the plasma membrane. *Cell* **98**, 9–12.

2654. Mahadevan, L.C. (1991). Growth factors, intracellular signals and developmental decisions: can growth factors be morphogens? *Sems. Dev. Biol.* **2**, 339–343.

2655. Mahaffey, J.W. and Kaufman, T.C. (1987). Distribution of *Sex combs reduced* gene products in *Drosophila melanogaster*. *Genetics* **117**, 51–60.

2656. Mahaffey, J.W. and Kaufman, T.C. (1987). The homeotic genes of the Antennapedia Complex and the Bithorax Complex of *Drosophila*. *In* "Developmental Genetics of Higher Organisms," (G.M. Malacinski, ed.). Macmillan: New York, pp. 329–359.

2657. Maier, D., Marquart, J., Thompson-Fontaine, A., Beck, I., Wurmbach, E., and Preiss, A. (1997). In vivo structure-function analysis of *Drosophila* HAIRLESS. *Mechs. Dev.* **67**, 97–106.

2658. Maier, D., Nagel, A.C., Johannes, B., and Preiss, A. (1999). Subcellular localization of Hairless protein shows a major focus of activity within the nucleus. *Mechs. Dev.* **89**, 195–199.

2659. Maier, D., Stumm, G., Kuhn, K., and Preiss, A. (1992). *Hairless*, a *Drosophila* gene involved in neural development, encodes a novel, serine rich protein. *Mechs. Dev.* **38**, 143–156.

2660. Maixner, A., Hecker, T.P., Phan, Q.N., and Wassarman, D.A. (1998). A screen for mutations that prevent lethality caused by expression of activated Sevenless and Ras1 in the *Drosophila* embryo. *Dev. Genet.* **23**, 347–361.

2661. Maizel, A., Bensaude, O., Prochiantz, A., and Joliot, A. (1999). A short region of its homeodomain is necessary for Engrailed nuclear export and secretion. *Development* **126**, 3183–3190.

2662. Majumdar, A., Nagaraj, R., and Banerjee, U. (1997). *strawberry notch* encodes a conserved nuclear protein that functions downstream of *Notch* and regulates gene expression along the developing wing margin of *Drosophila*. *Genes Dev.* **11**, 1341–1353.

2663. Makkerh, J.P.S., Dingwall, C., and Laskey, R.A. (1996). Comparative mutagenesis of nuclear localization signals reveals the importance of neutral and acidic amino acids. *Curr. Biol.* **6**, 1025–1027.

2664. Maldonado-Codina, G. and Glover, D.M. (1992). Cyclins A and B associate with chromatin and the polar regions of spindles, respectively, and do not undergo complete degradation at anaphase in syncytial *Drosophila* embryos. *J. Cell Biol.* **116**, 967–976.

2665. Malik, S. and Roeder, R.G. (2000). Transcriptional regulation through Mediator-like coactivators in yeast and metazoan cells. *Trends Biochem. Sci.* **25**, 277–283. [See also 2001 references: *Molec. Cell* **8**, 9–19; *Development* **128**, 3095–3104.]

2666. Mallin, D.R., Myung, J.S., Patton, J.S., and Geyer, P.K. (1998). Polycomb group repression is blocked by the *Drosophila suppressor of Hairy-wing* [*su(Hw)*] insulator. *Genetics* **148**, 331–339.

2667. Manfruelli, P., Arquier, N., Hanratty, W.P., and Sémériva, M. (1996). The tumor suppressor gene, *lethal(2)giant larvae (l(2)gl)*, is required for cell shape change of epithelial cells during *Drosophila* development. *Development* **122**, 2283–2294.

2668. Maniatis, T. (1999). A ubiquitin ligase complex essential for the NF-κB, Wnt/Wingless, and Hedgehog signaling pathways. *Genes Dev.* **13**, 505–510.

2669. Mankidy, R., Abbeyquaye, T., and Thackeray, J.R. (2001). Genetic and molecular analysis of PLC-γ function. *Proc. 42nd Ann. Drosophila Res. Conf.* **Abstracts Vol.**, a170.

2670. Mann, R.S. (1994). *engrailed*-mediated repression of *Ultrabithorax* is necessary for the parasegment 6 identity in *Drosophila*. *Development* **120**, 3205–3212.

2671. Mann, R.S. (1995). The specificity of homeotic gene function. *BioEssays* **17**, 855–863.

2672. Mann, R.S. (1997). Why are *Hox* genes clustered? *BioEssays* **19**, 661–664.

2673. Mann, R.S. and Abu-Shaar, M. (1996). Nuclear import of the homeodomain protein Extradenticle in response to Wg and Dpp signalling. *Nature* **383**, 630–633.

2674. Mann, R.S. and Affolter, M. (1998). Hox proteins meet more partners. *Curr. Opin. Gen. Dev.* **8**, 423–429.

2675. Mann, R.S. and Chan, S.-K. (1996). Extra specificity from *extradenticle*: the partnership between HOX and PBX/EXD homeodomain proteins. *Trends Genet.* **12**, 258–262.

2676. Mann, R.S. and Hogness, D.S. (1990). Functional dissection of Ultrabithorax proteins in *D. melanogaster*. *Cell* **60**, 597–610.

2677. Mann, R.S. and Morata, G. (2000). The developmental and molecular biology of genes that subdivide the body of *Drosophila. Annu. Rev. Cell Dev. Biol.* **16**, 243–271.

2678. Mannervik, M., Nibu, Y., Zhang, H., and Levine, M. (1999). Transcriptional coregulators in development. *Science* **284**, 606–609.

2679. Manning, G. and Krasnow, M.A. (1993). Development of the *Drosophila* tracheal system. *In* "The Development of *Drosophila melanogaster*," Vol. 1 (M. Bate and A. Martinez Arias, eds.). Cold Spring Harbor Lab. Pr.: Plainview, N. Y., pp. 609–685.

2680. Manning, L. and Doe, C.Q. (1999). Prospero distinguishes sibling cell fate without asymmetric localization in the *Drosophila* adult external sense organ lineage. *Development* **126**, 2063–2071.

2681. Manoukian, A.S. and Krause, H.M. (1992). Concentration-dependent activities of the *even-skipped* protein in *Drosophila* embryos. *Genes Dev.* **6**, 1740–1751.

2682. Manoukian, A.S. and Krause, H.M. (1993). Control of segmental asymmetry in *Drosophila* embryos. *Development* **118**, 785–796.

2683. Manoukian, A.S., Yoffe, K.B., Wilder, E.L., and Perrimon, N. (1995). The *porcupine* gene is required for *wingless* autoregulation in *Drosophila*. *Development* **121**, 4037–4044.

2684. Manseau, L., Baradaran, A., Brower, D., Budhu, A., Elefant, F., Phan, H., Philp, A.V., Yang, M., Glover, D., Kaiser, K., Palter, K., and Selleck, S. (1997). GAL4 enhancer traps expressed in the embryo, larval brain, imaginal discs, and ovary of *Drosophila*. *Dev. Dynamics* **209**, 310–322.

2685. Mansfield, E., Hersperger, E., Biggs, J., and Shearn, A. (1994). Genetic and molecular analysis of *hyperplastic discs*, a gene whose product is required for regulation of cell proliferation in *Drosophila melanogaster* imaginal discs and germ cells. *Dev. Biol.* **165**, 507–526.

2686. Maquat, L.E. and Carmichael, G.G. (2001). Quality control of mRNA function. *Cell* **104**, 173–176.

2687. Marcu, O., Cros, N., and Marsh, J.L. (2001). Identification of *wg* autoregulatory elements. *Proc. 42nd Ann. Drosophila Res. Conf.* **Abstracts Vol.**, a115.

2688. Marcus, J.M. (2001). The development and evolution of crossveins in insect wings. *J. Anat.* **199**, 211–216.

2689. Mardon, G., Solomon, N.M., and Rubin, G.M. (1994). *dachshund* encodes a nuclear protein required for normal eye and leg development in *Drosophila*. *Development* **120**, 3473–3486.

2690. Marí-Beffa, M., de Celis, J.F., and García-Bellido, A. (1991). Genetic and developmental analyses of chaetae pattern formation in *Drosophila* tergites. *Roux's Arch. Dev. Biol.* **200**, 132–142.

2691. Marigo, V., Davey, R.A., Zuo, Y., Cunningham, J.M., and Tabin, C.J. (1996). Biochemical evidence that Patched is the Hedgehog receptor. *Nature* **384**, 176–179.

2692. Markert, C.L. and Ursprung, H. (1971). "Developmental Genetics." Prentice-Hall, Englewood Cliffs, N. J.

2693. Markow, T.A., ed. (1994). "Developmental Instability: Its Origins and Evolutionary Implications." Klüwer, London.

2694. Markus, M. and Hess, B. (1990). Isotropic cellular automaton for modelling excitable media. *Nature* **347**, 56–58.

2695. Marquart, J., Alexief-Damianof, C., Preiss, A., and Maier, D. (1999). Rapid divergence in the course of *Drosophila* evolution reveals structural important domains of the Notch antagonist Hairless. *Dev. Genes Evol.* **209**, 155–164.

2696. Marqués, G., Musacchio, M., Shimell, M.J., Wünnenberg-Stapleton, K., Cho, K.W.Y., and O'Connor, M.B. (1997). Production of a DPP activity gradient in the early *Drosophila*

embryo through the opposing actions of the SOG and TLD proteins. *Cell* **91**, 417–426.

2697. Marsh, J.L., Syed, A., Theisen, H., and Sanchez, S. (2001). A novel function for Arm and dTcf: negative regulation of gene expression in response to Wg/Wnt signaling. *Proc. 42nd Ann. Drosophila Res. Conf.* **Abstracts Vol.**, a206.

2698. Marsh, J.L. and Theisen, H. (1999). Regeneration in insects. *Sems. Cell Dev. Biol.* **10**, 365–375.

2699. Marsh, M. and McMahon, H.T. (1999). The structural era of endocytosis. *Science* **285**, 215–220.

2700. Marshall, C.J. (1994). MAP kinase kinase kinase, MAP kinase kinase and MAP kinase. *Curr. Opin. Gen. Dev.* **4**, 82–89.

2701. Marshall, C.J. (1995). Specificity of receptor tyrosine kinase signaling: transient versus sustained extracellular signal-regulated kinase activation. *Cell* **80**, 179–185.

2702. Marshall, C.J. (1996). Raf gets it together. *Nature* **383**, 127–128.

2703. Martin, C.H., Mayeda, C.A., Davis, C.A., Ericsson, C.L., Knafels, J.D., Mathog, D.R., Celniker, S.E., Lewis, E.B., and Palazzolo, M.J. (1995). Complete sequence of the bithorax complex of *Drosophila*. *Proc. Natl. Acad. Sci. USA* **92**, 8398–8402.

2704. Martin, E.C. and Adler, P.N. (1993). The *Polycomb* group gene *Posterior Sex Combs* encodes a chromosomal protein. *Development* **117**, 641–655.

2705. Martin, G. (1996). Pass the butter . . . *Science* **274**, 203–204.

2706. Martin, G.R. (1995). Why thumbs are up. *Nature* **374**, 410–411.

2707. Martin, J.F., Hersperger, E., Simcox, A., and Shearn, A. (2000). *minidiscs* encodes a putative amino acid transporter subunit required non-autonomously for imaginal cell proliferation. *Mechs. Dev.* **92**, 155–167.

2708. Martin, K.A., Poeck, B., Roth, H., Ebens, A.J., Ballard, L.C., and Zipursky, S.L. (1995). Mutations disrupting neuronal connectivity in the *Drosophila* visual system. *Neuron* **14**, 229–240.

2709. Martin, P., Martin, A., and Shearn, A. (1977). Studies of *l(3)c43^{hs1}* a polyphasic, temperature-sensitive mutant of *Drosophila melanogaster* with a variety of imaginal disc defects. *Dev. Biol.* **55**, 213–232.

2710. Martin, P.F. (1982). Direct determination of the growth rate of *Drosophila* imaginal discs. *J. Exp. Zool.* **222**, 97–102.

2711. Martín, V., Carrillo, G., Torroja, C., and Guerrero, I. (2001). The sterol-sensing domain of Patched protein seems to control Smoothened activity through Patched vesicular trafficking. *Curr. Biol.* **11**, 601–607.

2712. Martín-Bermudo, M.D., Carmena, A., and Jiménez, F. (1995). Neurogenic genes control gene expression at the transcriptional level in early neurogenesis and in mesectoderm specification. *Development* **121**, 219–224.

2713. Martín-Blanco, E. (1998). Regulatory control of signal transduction during morphogenesis in *Drosophila*. *Int. J. Dev. Biol.* **42**, 363–368.

2714. Martín-Blanco, E., Pastor-Pareja, J.C., and García-Bellido, A. (2000). JNK and *decapentaplegic* signaling control adhesiveness and cytoskeletal dynamics during thorax closure in *Drosophila*. *Proc. Natl. Acad. Sci. USA* **97**, 7888–7893.

2715. Martín-Blanco, E., Roch, F., Noll, E., Baonza, A., Duffy, J.B., and Perrimon, N. (1999). A temporal switch in DER signaling controls the specification and differentiation of veins and interveins in the *Drosophila* wing. *Development* **126**, 5739–5747.

2716. Martinez Arias, A. (1989). A cellular basis for pattern formation in the insect epidermis. *Trends Genet.* **5**, 262–267. [See also 2001 *Trends Genet.* **17**, 574–579.]

2717. Martinez Arias, A. (1993). Development and patterning of the larval epidermis of *Drosophila*. *In* "The Development of *Drosophila melanogaster*," Vol. 1 (M. Bate and A. Martinez-Arias, eds.). Cold Spring Harbor Laboratory Pr.: Plainview, N. Y., pp. 517–608.

2718. Martinez Arias, A. (1998). Interactions between Wingless and Notch during the assignation of cell fates in *Drosophila*. *Int. J. Dev. Biol.* **42**, 325–333.

2719. Martinez Arias, A., Baker, N.E., and Ingham, P.W. (1988). Role of segment polarity genes in the definition and maintenance of cell states in the *Drosophila* embryo. *Development* **103**, 157–170.

2720. Martinez Arias, A., Brown, A.M.C., and Brennan, K. (1999). Wnt signalling: pathway or network? *Curr. Opin. Gen. Dev.* **9**, 447–454.

2721. Martínez, C. and Modolell, J. (1991). Cross-regulatory interactions between the proneural *achaete* and *scute* genes of *Drosophila*. *Science* **251**, 1485–1487.

2722. Martínez, C., Modolell, J., and Garrell, J. (1993). Regulation of the proneural gene *achaete* by helix-loop-helix proteins. *Mol. Cell. Biol.* **13**, 3514–3521.

2723. Martinez-Arias, A., Ingham, P.W., Scott, M.P., and Akam, M.E. (1987). The spatial and temporal deployment of *Dfd* and *Scr* transcripts throughout development of *Drosophila*. *Development* **100**, 673–683.

2724. Martinez-Arias, A. and Lawrence, P.A. (1985). Parasegments and compartments in the *Drosophila* embryo. *Nature* **313**, 639–642.

2725. Martinez-Arias, A. and White, R.A.H. (1988). *Ultrabithorax* and *engrailed* expression in *Drosophila* embryos mutant for segmentation genes of the pair-rule class. *Development* **102**, 325–338.

2726. Martínez-Balbás, M.A., Bannister, A.J., Martin, K., Haus-Seuffert, P., Meisterernst, M., and Kouzarides, T. (1998). The acetyltransferase activity of CBP stimulates transcription. *EMBO J.* **17**, 2886–2893.

2727. Marty, T., Müller, B., Basler, K., and Affolter, M. (2000). Schnurri mediates Dpp-dependent repression of *brinker* transcription. *Nature Cell Biol.* **2**, 745–749. [See also 2001 *EMBO J.* **20**, 3298–3305.]

2728. Maschat, F., Serrano, N., Randsholt, N.B., and Géraud, G. (1998). *engrailed* and *polyhomeotic* interactions are required to maintain the A/P boundary of the *Drosophila* developing wing. *Development* **125**, 2771–2780.

2729. Maschat, F., Toulza, E., and Montagne, J. (2001). Engrailed and D-SRF are acting as cofactors on *polyhomeotic* activation during wing morphogenesis. *Proc. 42nd Ann. Drosophila Res. Conf.* **Abstracts Vol.**, a187.

2730. Mason, E.D., Williams, S., Grotendorst, G.R., and Marsh, J.L. (1997). Combinatorial signaling by Twisted Gastrulation and Decapentaplegic. *Mechs. Dev.* **64**, 61–75.

2731. Massagué, J. (1992). Receptors for the TGF-β family. *Cell* **69**, 1067–1070.

2732. Massagué, J. (1996). TGFβ signaling: receptors, transducers, and Mad proteins. *Cell* **85**, 947–950.

2733. Massagué, J. (1998). TGF-β signal transduction. *Annu. Rev. Biochem.* **67**, 753–791.

2734. Massagué, J., Attisano, L., and Wrana, J.L. (1994). The TGF-β family and its composite receptors. *Trends Cell Biol.* **4**, 172–178.

2735. Massagué, J., Blain, S.W., and Lo, R.S. (2000). TGFβ signaling in growth control, cancer, and heritable disorders. *Cell* **103**, 295–309.

2736. Massagué, J. and Wotton, D. (2000). Transcriptional control by the TGF-β/Smad signaling system. *EMBO J.* **19**, 1745–1754.

2737. Massari, M.E., Grant, P.A., Pray-Grant, M.G., Berger, S.L., Workman, J.L., and Murre, C. (1999). A conserved motif present in a class of helix-loop-helix proteins activates transcription by direct recruitment of the SAGA complex. *Molec. Cell* **4**, 63–73.

2738. Masucci, J.D. and Hoffmann, F.M. (1993). Identification of two regions from the *Drosophila decapentaplegic* gene required for embryonic midgut development and larval viability. *Dev. Biol.* **159**, 276–287.

2739. Masucci, J.D., Miltenberger, R.J., and Hoffmann, F.M. (1990). Pattern-specific expression of the *Drosophila decapentaplegic* gene in imaginal discs is regulated by 3′ cis-regulatory elements. *Genes Dev.* **4**, 2011–2023.

2740. Mata, J., Curado, S., Ephrussi, A., and Rørth, P. (2000). Tribbles coordinates mitosis and morphogenesis in *Drosophila* by regulating String/CDC25 proteolysis. *Cell* **101**, 511–522.

2741. Matakatsu, H., Brentrup, D., and Hayashi, S. (2001). Repression of wing vein differentiation by interaction between Plexus and two bHLH proteins. *Proc. 42nd Ann. Drosophila Res. Conf.* **Abstracts Vol.**, a200.

2742. Matakatsu, H., Tadokoro, R., Gamo, S., and Hayashi, S. (1999). Repression of the wing vein development in *Drosophila* by the nuclear matrix protein Plexus. *Development* **126**, 5207–5216.

2743. Máthé, E., Bates, H., Huikeshoven, H., Deák, P., Glover, D.M., and Cotterill, S. (2000). Importin-α3 is required at multiple stages of *Drosophila* development and has a role in the completion of oogenesis. *Dev. Biol.* **223**, 307–322.

2744. Mathi, S.K. and Larsen, E. (1988). Patterns of cell division in imaginal discs of *Drosophila*. *Tissue & Cell* **20**, 461–472.

2745. Mathog, D.R. (1991). Suppression of abdominal legs in *Drosophila melanogaster*. *Roux's Arch. Dev. Biol.* **199**, 449–457.

2746. Matsuno, K., Diederich, R.J., Go, M.J., Blaumueller, C.M., and Artavanis-Tsakonas, S. (1995). Deltex acts as a positive regulator of Notch signaling through interactions with the Notch ankyrin repeats. *Development* **121**, 2633–2644.

2747. Matsuno, K., Go, M.J., Sun, X., Eastman, D.S., and Artavanis-Tsakonas, S. (1997). Suppressor of Hairless-independent events in Notch signaling imply novel pathway elements. *Development* **124**, 4265–4273.

2748. Matsuo, T., Takahashi, K., Kondo, S., Kaibuchi, K., and Yamamoto, D. (1997). Regulation of cone cell formation by Canoe and Ras in the developing *Drosophila* eye. *Development* **124**, 2671–2680.

2749. Matsuzaki, F., Ohshiro, T., Ikeshima-Kataoka, H., and Izumi, H. (1998). miranda localizes staufen and prospero asymmetrically in mitotic neuroblasts and epithelial cells in early *Drosophila* embryogenesis. *Development* **125**, 4089–4098.

2750. Mattaj, I.W. and Conti, E. (1999). Snail mail to the nucleus. *Nature* **399**, 208–210.

2751. Matunis, E., Tran, J., Gönczy, P., Caldwell, K., and DiNardo, S. (1997). *punt* and *schnurri* regulate a somatically derived signal that restricts proliferation of committed progenitors in the germline. *Development* **124**, 4383–4391.

2752. Maurel-Zaffran, C. and Treisman, J.E. (2000). *pannier* acts upstream of *wingless* to direct dorsal eye disc development in *Drosophila*. *Development* **127**, 1007–1016.

2753. Maves, L. and Schubiger, G. (1995). *wingless* induces transdetermination in developing *Drosophila* imaginal discs. *Development* **121**, 1263–1272.

2754. Maves, L. and Schubiger, G. (1998). A molecular basis for transdetermination in *Drosophila* imaginal discs: interactions between *wingless* and *decapentaplegic* signaling. *Development* **125**, 115–124.

2755. Maves, L. and Schubiger, G. (1999). Cell determination and transdetermination in *Drosophila* imaginal discs. *Curr. Topics Dev. Biol.* **43**, 115–151.

2756. Mayer, B.J. (1999). Endocytosis: EH domains lend a hand. *Curr. Biol.* **9**, R70–R73.

2757. Mayer, B.J. and Eck, M.J. (1995). Minding your p's and q's. *Curr. Biol.* **5**, 364–367.

2758. Mayer, B.J., Hirai, H., and Sakai, R. (1995). Evidence that SH2 domains promote processive phosphorylation by protein-tyrosine kinases. *Curr. Biol.* **5**, 296–305.

2759. Mayer, B.J., Ren, R., Clark, K.L., and Baltimore, D. (1993). A putative modular domain present in diverse signaling proteins. *Cell* **73**, 629–630.

2760. Mayer, M.P. and Bukau, B. (1999). Molecular chaperones: the busy life of Hsp90. *Curr. Biol.* **9**, R322–R325.

2761. Mayer-Jaekel, R.E., Ohkura, H., Gomes, R., Sunkel, C.E., Baumgartner, S., Hemmings, B.A., and Glover, D.M. (1993). The 55 kd regulatory subunit of *Drosophila* protein phosphatase 2A is required for anaphase. *Cell* **72**, 621–633.

2762. Maynard Smith, J. (1968). The counting problem. *In* "Towards a Theoretical Biology. I. Prolegomena," (C.H. Waddington, ed.). Aldine Pub. Co.: Chicago, pp. 120–124.

2763. Maynard Smith, J. (1970). Natural selection and the concept of a protein space. *Nature* **225**, 563–564.

2764. Maynard Smith, J. (1983). Evolution and development. *In* "Development and Evolution," *Symp. Brit. Soc. Dev. Biol.*, Vol. 6 (B.C. Goodwin, N. Holder, and C.C. Wylie, eds.). Cambridge Univ. Pr.: Cambridge, pp. 33–45.

2765. Maynard Smith, J. (1998). "Shaping Life: Genes, Embryos and Evolution." Yale Univ. Pr., New Haven.

2766. Maynard Smith, J. (1999). The idea of information in biology. *Quart. Rev. Biol.* **74**, 395–400.

2767. Maynard Smith, J., Burian, R., Kauffman, S., Alberch, P., Campbell, J., Goodwin, B., Lande, R., Raup, D., and Wolpert, L. (1985). Developmental constraints and evolution. *Q. Rev. Biol.* **60**, 265–287.

2768. Maynard Smith, J. and Sondhi, K.C. (1960). The genetics of a pattern. *Genetics* **45**, 1039–1050.

2769. Maynard Smith, J. and Sondhi, K.C. (1961). The arrangement of bristles in *Drosophila*. *J. Embryol. Exp. Morph.* **9**, 661–672.

2770. McAdams, H.H. and Shapiro, L. (1995). Circuit simulation of genetic networks. *Science* **269**, 650–656.

2771. McCabe, J., French, V., and Partridge, L. (1997). Joint regulation of cell size and cell number in the wing blade of *Drosophila melanogaster*. *Genet. Res.* **69**, 61–68.

2772. McCall, K. and Bender, W. (1996). Probes for chromatin accessibility in the *Drosophila* bithorax complex respond differently to *Polycomb*-mediated repression. *EMBO J.* **15**, 569–580.

2773. McCall, K., O'Connor, M.B., and Bender, W. (1994). Enhancer traps in the *Drosophila* bithorax complex mark parasegmental domains. *Genetics* **138**, 387–399.

2774. McCall, K. and Steller, H. (1997). Facing death in the fly: genetic analysis of apoptosis in *Drosophila*. *Trends Genet.* **13**, 222–226.

2775. McCartney, B.M., Dierick, H.A., Kirkpatrick, C., Moline, M.M., Baas, A., Peifer, M., and Bejsovec, A. (1999). *Drosophila* APC2 is a cytoskeletally-associated protein that regulates Wingless signaling in the embryonic epidermis. *J. Cell Biol.* **146**, 1303–1318.

2776. McCartney, B.M., Kulikauskas, R.M., LaJeunesse, D.R., and Fehon, R.G. (2000). The *Neurofibromatosis-2* homologue, *Merlin*, and the tumor suppressor *expanded* function together in *Drosophila* to regulate cell proliferation and differentiation. *Development* **127**, 1315–1324.

2777. McCarty, J.H. (1998). The Nck SH2/SH3 adaptor protein: a regulator of multiple intracellular signal transduction events. *BioEssays* **20**, 913–921.

2778. McCormick, A., Coré, N., Kerridge, S., and Scott, M.P. (1995). Homeotic response elements are tightly linked to tissue-specific elements in a transcriptional enhancer of the *teashirt* gene. *Development* **121**, 2799–2812.

2779. McDowell, N. and Gurdon, J.B. (1999). Activin as a morphogen in *Xenopus* mesoderm induction. *Sems. Cell Dev. Biol.* **10**, 311–317.

2780. McDowell, N., Zorn, A.M., Crease, D.J., and Gurdon, J.B. (1997). Activin has direct long-range signalling activity and can form a concentration gradient by diffusion. *Curr. Biol.* **7**, 671–681.

2781. McEwen, D.G. and Peifer, M. (2000). Wnt signaling: Moving in a new direction. *Curr. Biol.* **10**, R562–R564.

2782. McGinnis, W. (1994). A century of homeosis, a decade of homeoboxes. *Genetics* **137**, 607–611.

2783. McGinnis, W., Garber, R.L., Wirz, J., Kuroiwa, A., and Gehring, W.J. (1984). A homologous protein-coding sequence in *Drosophila* homeotic genes and its conservation in other metazoans. *Cell* **37**, 403–408.

2784. McGinnis, W. and Krumlauf, R. (1992). Homeobox genes and axial patterning. *Cell* **68**, 283–302.

2785. McGinnis, W., Levine, M.S., Hafen, E., Kuroiwa, A., and Gehring, W.J. (1984). A conserved DNA sequence in homoeotic genes of the *Drosophila* Antennapedia and bithorax complexes. *Nature* **308**, 428–433.

2786. McGrew, M.J. and Pourquié, O. (1998). Somitogenesis: segmenting a vertebrate. *Curr. Opin. Gen. Dev.* **8**, 487–493.

2787. McIver, S.B. (1985). Mechanoreception. *In* "Comprehensive Insect Physiology, Biochemistry, and Pharmacology," Vol. 6 (G.A. Kerkut and L.I. Gilbert, eds.). Pergamon: New York, pp. 71–132.

2788. McKinney, M.L., ed. (1988). "Heterochrony in Evolution: A Multidisciplinary Approach." Plenum Pr., New York.

2789. McKinney, M.L. and McNamara, K.J. (1991). "Heterochrony: The Evolution of Ontogeny." Plenum Pr., New York.

2790. McKnight, S.L. (1991). Molecular zippers in gene regulation. *Sci. Am.* **264** #4, 54–64.

2791. McMahon, A.P. (2000). More surprises in the Hedgehog signaling pathway. *Cell* **100**, 185–188.

2792. McNeill, H. (2000). Sticking together and sorting things out: adhesion as a force in development. *Nature Rev. Gen.* **1**, 100–108.

2793. McNeill, H. and Downward, J. (1999). Apoptosis: Ras to the rescue in the fly eye. *Curr. Biol.* **9**, R176–R179.

2794. McNeill, H., Yang, C.-H., Brodsky, M., Ungos, J., and Simon, M.A. (1997). *mirror* encodes a novel PBX-class homeoprotein that functions in the definition of the dorsal-ventral border in the *Drosophila* eye. *Genes Dev.* **11**, 1073–1082.

2795. McNiven, M.A., Cao, H., Pitts, K.R., and Yoon, Y. (2000). The dynamin family of mechanoenzymes: pinching in new places. *Trends Bioch. Sci.* **25**, 115–120. [See also 2001 *Curr. Biol.* **11**, R850.]

2796. Mechler, B.M., McGinnis, W., and Gehring, W. (1985). Molecular cloning of *lethal(2)giant larvae*, a recessive oncogene of *Drosophila melanogaster*. *EMBO J.* **4**, 1551–1557.

2797. Meier, P. and Evan, G. (1998). Dying like flies. *Cell* **95**, 295–298.

2798. Meinertzhagen, I.A. (1975). The development of neuronal connection patterns in the visual systems of insects. *In* "Cell Patterning," *Ciba Found. Symp.*, Vol. 29 (R. Porter and J. Rivers, eds.). Elsevier: Amsterdam, pp. 265–288. [Cf. 2001 *Neuron* **32**, 225–235, 237–248, 381–384.]

2799. Meinertzhagen, I.A. (1993). Sleeping neuroblasts: the proliferation of optic lobe stem cells in *Drosophila* is regulated by extrinsic factors that include short-range interactions with ingrowing retinal axons and glial cells. *Curr. Biol.* **3**, 904–906.

2800. Meinertzhagen, I.A. (2000). Wiring the fly's eye. *Neuron* **28**, 310–313.

2801. Meinertzhagen, I.A. and Hanson, T.E. (1993). The development of the optic lobe. *In* "The Development of *Drosophila melanogaster*," Vol. 2 (M. Bate and A. Martinez Arias, eds.). Cold Spring Harbor Lab. Pr.: Plainview, N. Y., pp. 1363–1491.

2802. Meinhardt, H. (1978). Models for the ontogenetic development of higher organisms. *Rev. Physiol. Biochem. Pharmacol.* **80**, 47–104.

2803. Meinhardt, H. (1978). Space-dependent cell determination under the control of a morphogen gradient. *J. Theor. Biol.* **74**, 307–321.

2804. Meinhardt, H. (1980). Cooperation of compartments for the generation of positional information. *Z. Naturforsch.* **35c**, 1086–1091.

2805. Meinhardt, H. (1982). Generation of structures in a developing organism. *In* "Developmental Order: Its Origin and Regulation," (S. Subtelny and P.B. Green, eds.). Alan R. Liss: New York, pp. 439–461.

2806. Meinhardt, H. (1982). "Models of Biological Pattern Formation." Acad. Pr., New York.

2807. Meinhardt, H. (1982). The role of compartmentalization in the activation of particular control genes and in the generation of proximo-distal positional information in appendages. *Amer. Zool.* **22**, 209–220.

2808. Meinhardt, H. (1983). Cell determination boundaries as organizing regions for secondary embryonic fields. *Dev. Biol.* **96**, 375–385.

2809. Meinhardt, H. (1984). Models for pattern formation during development of higher organisms. *In* "Pattern Formation: A Primer in Developmental Biology," (G.M. Malacinski and S.V. Bryant, eds.). Macmillan: New York, pp. 47–72.

2810. Meinhardt, H. (1984). Models for positional signaling, the threefold subdivision of segments and the pigmentation pattern of molluscs. *J. Embryol. Exp. Morph.* **83 (Suppl.)**, 289–311.

2811. Meinhardt, H. (1986). Hierarchical inductions of cell states: a model for segmentation in *Drosophila*. *J. Cell Sci. Suppl.* **4**, 357–381.

2812. Meinhardt, H. (1986). The threefold subdivision of segments and the initiation of legs and wings in insects. *Trends Genet.* **2**, 36–41.

2813. Meinhardt, H. (1991). Determination borders as organizing regions in the generation of secondary embryonic fields: the initiation of legs and wings. *Sems. Dev. Biol.* **2**, 129–138.

2814. Meinhardt, H. (1994). Biological pattern formation: new observations provide support for theoretical predictions. *BioEssays* **16**, 627–632.

2815. Meinhardt, H. and Gierer, A. (1974). Applications of a theory of biological pattern formation based on lateral inhibition. *J. Cell Sci.* **15**, 321–346.

2816. Meinhardt, H. and Gierer, A. (1980). Generation and regeneration of sequence of structures during morphogenesis. *J. Theor. Biol.* **85**, 429–450.

2817. Meinhardt, H. and Gierer, A. (1981). Generation of spatial sequences of structures during development of higher organisms. *In* "Lectures on Mathematics in the Life Sciences," Vol. 14 Am. Math. Soc.: Providence, Rhode Island, pp. 1–20.

2818. Meinhardt, H. and Gierer, A. (2000). Pattern formation by local self-activation and lateral inhibition. *BioEssays* **22**, 753–760.

2819. Meise, M. and Janning, W. (1993). Cell lineage of larval and imaginal thoracic anlagen cells of *Drosophila melanogaster*, as revealed by single-cell transplantations. *Development* **118**, 1107–1121.

2820. Meise, M. and Janning, W. (1994). Localization of thoracic imaginal-disc precursor cells in the early embryo of *Drosophila melanogaster*. *Mechs. Dev.* **48**, 109–117.

2821. Meisner, H., Daga, A., Buxton, J., Fernández, B., Chawla, A., Banerjee, U., and Czech, M.P. (1997). Interactions of *Drosophila* Cbl with epidermal growth factor receptors and role of Cbl in R7 photoreceptor cell development. *Mol. Cell. Biol.* **17**, 2217–2225.

2822. Meléndez, A., Li, W., and Kalderon, D. (1995). Activity, expression and function of a second *Drosophila* protein kinase A catalytic subunit gene. *Genetics* **141**, 1507–1520.

2823. Melnick, M.B., Perkins, L.A., Lee, M., Ambrosio, L., and Perrimon, N. (1993). Developmental and molecular characterization of mutations in the *Drosophila-raf* serine/threonine protein kinase. *Development* **118**, 127–138.

2824. Melzer, R.R., Sprenger, J., Nicastro, D., and Smola, U. (1999). Larva-adult relationships in an ancestral dipteran: a re-examination of sensillar pathways across the antenna and leg anlagen of *Chaoborus crystallinus* (DeGeer, 1776; Chaoboridae). *Dev. Genes Evol.* **209**, 103–112.

2825. Mercader, N., Leonardo, E., Azpiazu, N., Serrano, A., Morata, G., Martínez-A, C., and Torres, M. (1999). Conserved regulation of proximodistal limb axis development by Meis1/Hth. *Nature* **402**, 425–429.

2826. Merika, M. and Thanos, D. (2001). Enhanceosomes. *Curr. Opin. Gen. Dev.* **11**, 205–208.

2827. Merli, C., Bergstrom, D.E., Cygan, J.A., and Blackman, R.K. (1996). Promoter specificity mediates the independent regulation of neighboring genes. *Genes Dev.* **10**, 1260–1270. [See also 2001 *Genes Dev.* **15**, 2515–2519.]

2828. Merriam, J.R. (1978). Estimating primordial cell numbers in *Drosophila* imaginal discs and histoblasts. *In* "Genetic Mosaics and Cell Differentiation," *Results and Problems in Cell Differentiation, Vol. 9*, (W.J. Gehring, ed.). Springer-Verlag: Berlin, pp. 71–96.

2829. Merrill, V.K.L., Diederich, R.J., Turner, F.R., and Kaufman, T.C. (1989). A genetic and developmental analysis of mutations in *labial*, a gene necessary for proper head formation in *Drosophila melanogaster*. *Dev. Biol.* **135**, 376–391.

2830. Merrill, V.K.L., Turner, F.R., and Kaufman, T.C. (1987). A genetic and developmental analysis of mutations in the *Deformed* locus in *Drosophila melanogaster*. *Dev. Biol.* **122**, 379–395.

2831. Merritt, D.J., Hawken, A., and Whitington, P.M. (1993). The role of the *cut* gene in the specification of central projections by sensory axons in *Drosophila*. *Neuron* **10**, 741–752.

2832. Méthot, N. and Basler, K. (1999). Hedgehog controls limb development by regulating the activities of distinct transcriptional activator and repressor forms of Cubitus interruptus. *Cell* **96**, 819–831. [See also 2001 *Development* **128**, 4361–4370.]

2833. Méthot, N. and Basler, K. (2000). Suppressor of fused opposes Hedgehog signal transduction by impeding nuclear accumulation of the activator form of Cubitus interruptus. *Development* **127**, 4001–4010.

2834. Méthot, N. and Basler, K. (2001). An absolute requirement for Cubitus interruptus in Hedgehog signaling. *Development* **128**, 733–742.

2835. Meyer-Rochow, V.B. (2000). The eye: monophyletic, polyphyletic or perhaps biphyletic? *Trends Genet.* **16**, 244–245 (*cf.* Gehring's response).

2836. Meyerowitz, E.M. (1997). Plants and the logic of development. *Genetics* **145**, 5–9.

2837. Mglinets, V.A. and Kostina, I.V. (1978). Genetic control of bristle length in *Drosophila*. *Genetika* **14**, 285–293.

2838. Micchelli, C.A. and Blair, S.S. (1999). Dorsoventral lineage restriction in wing imaginal discs requires Notch. *Nature* **401**, 473–476. [See also 2001 references: *Cell* **106**, 785–794; *Curr. Biol.* **11**, R1017–R1021.]

2839. Micchelli, C.A., Rulifson, E.J., and Blair, S.S. (1997). The function and regulation of *cut* expression on the wing margin of *Drosophila*: Notch, Wingless and a dominant negative role for Delta and Serrate. *Development* **124**, 1485–1495.

2840. Michaelson, J. (1987). Cell selection in development. *Biol. Rev.* **62**, 115–139.

2841. Michaely, P. and Bennett, V. (1992). The ANK repeat: a ubiquitous motif involved in macromolecular recognition. *Trends Cell Biol.* **2**, 127–129.

2842. Michelson, A.M. (1996). A new turn (or two) for Twist. *Science* **272**, 1449–1450.

2843. Middleton, J., Neil, S., Wintle, J., Clark-Lewis, I., Moore, H., Lam, C., Auer, M., Hub, E., and Rot, A. (1997). Transcytosis and surface presentation of IL-8 by venular endothelial cells. *Cell* **91**, 385–395.

2844. Mihaly, J., Hogga, I., Gausz, J., Gyurkovics, H., and Karch, F. (1997). In situ dissection of the *Fab-7* region of the bithorax complex into a chromatin domain boundary and a *Polycomb*-response element. *Development* **124**, 1809–1820.

2845. Miklos, G.L.G. and Rubin, G.M. (1996). The role of the Genome Project in determining gene function: insights from model organisms. *Cell* **86**, 521–529.

2846. Milán, M. (1998). Cell cycle control in the *Drosophila* wing. *BioEssays* **20**, 969–971.

2847. Milán, M., Baonza, A., and García-Bellido, A. (1997). Wing surface interactions in venation patterning in *Drosophila*. *Mechs. Dev.* **67**, 203–213.

2848. Milán, M., Campuzano, S., and García-Bellido, A. (1996). Cell cycling and patterned cell proliferation in the wing primordium of *Drosophila*. *Proc. Natl. Acad. Sci. USA* **93**, 640–645.

2849. Milán, M., Campuzano, S., and García-Bellido, A. (1997). Developmental parameters of cell death in the wing disc of *Drosophila*. *Proc. Natl. Acad. Sci. USA* **94**, 5691–5696.

2850. Milán, M. and Cohen, S.M. (1999). Notch signaling is not sufficient to define the affinity boundary between dorsal and ventral compartments. *Molec. Cell* **4**, 1073–1078.

2851. Milán, M. and Cohen, S.M. (1999). Regulation of LIM homeodomain activity in vivo: a tetramer of dLDB and Apterous confers activity and capacity for regulation by dLMO. *Molec. Cell* **4**, 267–273.

2852. Milán, M. and Cohen, S.M. (2000). Subdividing cell populations in the developing limbs of *Drosophila*: Do wing veins and leg segments define units of growth control? *Dev. Biol.* **217**, 1–9.

2853. Milán, M. and Cohen, S.M. (2000). Temporal regulation of Apterous activity during development of the *Drosophila* wing. *Development* **127**, 3069–3078.

2854. Milán, M., Diaz-Benjumea, F.J., and Cohen, S.M. (1998). *Beadex* encodes an LMO protein that regulates Apterous LIM-homeodomain activity in *Drosophila* wing development: a model for *LMO* oncogene function. *Genes Dev.* **12**, 2912–2920.

2855. Milburn, M.V., Tong, L., deVos, A.M., Brünger, A., Yamaizumi, Z., Nishimura, S., and Kim, S.-H. (1990). Molecular switch for signal transduction: structural differences between active and inactive forms of protooncogenic *ras* proteins. *Science* **247**, 939–945.

2856. Milétich, I. and Limbourg-Bouchon, B. (2000). *Drosophila* null *slimb* clones transiently deregulate Hedgehog-independent transcription of *wingless* in all limb discs, and induce *decapentaplegic* transcription linked to imaginal disc regeneration. *Mechs. Dev.* **93**, 15–26.

2857. Miller, A. (1950). The internal anatomy and histology of the imago of *Drosophila melanogaster*. *In* "Biology of *Drosophila*," (M. Demerec, ed.). Hafner: New York, pp. 420–534.

2858. Miller, C. and Blair, S. (2001). Genetic interactions between *shifted* and members of the Hedgehog signaling pathway. *Proc. 42nd Ann. Drosophila Res. Conf.* **Abstracts Vol.**, a188.

2859. Miller, D.T. and Cagan, R.L. (1998). Local induction of patterning and programmed cell death in the developing *Drosophila* retina. *Development* **125**, 2327–2335.

2860. Miller, J.R. and Moon, R.T. (1996). Signal transduction through β-catenin and specification of cell fate during embryogenesis. *Genes Dev.* **10**, 2527–2539.

2861. Millward, T.A., Zolnierowicz, S., and Hemmings, B.A. (1999). Regulation of protein kinase cascades by protein phosphatase 2A. *Trends Biochem. Sci.* **24**, 186–191.

2862. Milner, M.J., Bleasby, A.J., and Kelly, S.L. (1984). The role of the peripodial membrane of leg and wing imaginal discs of *Drosophila melanogaster* during evagination and differentiation in vitro. *Roux's Arch. Dev. Biol.* **193**, 180–186.

2863. Milner, M.J., Bleasby, A.J., and Pyott, A. (1983). The role of the peripodial membrane in the morphogenesis of the eye-antennal disc of *Drosophila melanogaster*. *Roux's Arch. Dev. Biol.* **192**, 164–170.

2864. Milner, M.J., Bleasby, A.J., and Pyott, A. (1984). Cell interactions during the fusion *in vitro* of *Drosophila* eye-antennal imaginal discs. *Roux's Arch. Dev. Biol.* **193**, 406–413.

2865. Milner, M.J. and Haynie, J.L. (1979). Fusion of *Drosophila* eye-antennal imaginal discs during differentiation in vitro. *W. Roux's Arch.* **185**, 363–370.

2866. Milton, C.C., McKenzie, J.A., Woods, R.E., and Batterham, P. (2001). Effect of chaperone genes on asymmetry. *Proc. 42nd Ann. Drosophila Res. Conf.* **Abstracts Vol.**, a269.

2867. Minami, M., Kinoshita, N., Kamoshida, Y., Tanimoto, H., and Tabata, T. (1999). *brinker* is a target of Dpp in *Drosophila* that negatively regulates Dpp-dependent genes. *Nature* **398**, 242–246.

2868. Minelli, A. (2000). Limbs and tail as evolutionarily diverging duplicates of the main body axis. *Evol. Dev.* **2**, 157–165.

2869. Minelli, A. and Bortoletto, S. (1988). Myriapod metamerism and arthropod segmentation. *Biol. J. Linnean Soc.* **33**, 323–343. [Cf. 2001 *Dev. Genes Evol.* **211**, 509–521.]

2870. Minsky, M., ed. (1968). "Semantic Information Processing." M.I.T. Pr., Cambridge, Mass. [See also 2001 *Am. Sci.* **89**, 204–208.]

2871. Minsky, M. (1985). "The Society of Mind." Simon & Schuster, New York.

2872. Mirkovic, I., Smith, J.L., Gorski, S.M., and Verheyen, E.M. (2001). Nemo functions at multiple stages during *Drosophila* epidermal development. *Proc. 42nd Ann. Drosophila Res. Conf.* **Abstracts Vol.**, a174. [See also 2001 *Mechs. Dev.* **101**, 119–132.]

2873. Miskiewicz, P., Morrissey, D., Lan, Y., Raj, L., Kessler, S., Fujioka, M., Goto, T., and Weir, M. (1996). Both the paired domain and homeodomain are required for in vivo function of *Drosophila* Paired. *Development* **122**, 2709–2718.

2874. Missler, M. and Südhof, T.C. (1998). Neurexins: three genes and 1001 products. *Trends Genet.* **14**, 20–26.

2875. Misteli, T. (2001). Protein dynamics: implications for nuclear architecture and gene expression. *Science* **291**, 843–847.

2876. Mitchell, H.K., Edens, J., and Petersen, N.S. (1990). Stages of cell hair construction in *Drosophila*. *Dev. Genet.* **11**, 133–140.

2877. Mitchell, H.K. and Petersen, N.S. (1989). Epithelial differentiation in *Drosophila* pupae. *Dev. Genet.* **10**, 42–52.

2878. Mitchell, P.J. and Tjian, R. (1989). Transcriptional regulation in mammalian cells by sequence-specific DNA binding proteins. *Science* **245**, 371–378.

2879. Mitchelson, A., Simonelig, M., Williams, C., and O'Hare, K. (1993). Homology with *Saccharomyces cerevisiae RNA14* suggests that phenotypic suppression in *Drosophila melanogaster* by *suppressor of forked* occurs at the level of RNA stability. *Genes Dev.* **7**, 241–249.

2880. Mittal, R., Ahmadian, M.R., Goody, R.S., and Wittinghofer, A. (1996). Formation of a transition-state analog of the Ras GTPase reaction by RAS.GDP, tetrafluoroaluminate, and GTPase-activating proteins. *Science* **273**, 115–117.

2881. Mittenthal, J.E. (1981). The rule of normal neighbors: A hypothesis for morphogenetic pattern regulation. *Dev. Biol.* **88**, 15–26.

2882. Miyamoto, H., Nihonmatsu, I., Kondo, S., Ueda, R., Togashi, S., Hirata, K., Ikegami, Y., and Yamamoto, D. (1995). *canoe* encodes a novel protein containing a GLGF/DHR motif and functions with *Notch* and *scabrous* in common developmental pathways in *Drosophila*. *Genes Dev.* **9**, 612–625.

2883. Mlodzik, M. (1999). Planar polarity in the *Drosophila* eye: a multifaceted view of signaling specificity and cross-talk. *EMBO J.* **18**, 6873–6879.

2884. Mlodzik, M. (2000). Spiny legs and prickled bodies: new insights and complexities in planar polarity establishment. *BioEssays* **22**, 311–315.

2885. Mlodzik, M., Baker, N.E., and Rubin, G.M. (1990). Isolation and expression of *scabrous*, a gene regulating neurogenesis in *Drosophila*. *Genes Dev.* **4**, 1848–1861.

2886. Mlodzik, M., Gibson, G., and Gehring, W.J. (1990). Effects of ectopic expression of *caudal* during *Drosophila* development. *Development* **109**, 271–277.

2887. Mlodzik, M., Hiromi, Y., Goodman, C.S., and Rubin, G.M. (1992). The presumptive R7 cell of the developing *Drosophila* eye receives positional information independent of *sevenless*, *boss* and *sina*. *Mechs. Dev.* **37**, 37–42.

2888. Mlodzik, M., Hiromi, Y., Weber, U., Goodman, C.S., and Rubin, G.M. (1990). The *Drosophila seven-up* gene, a member of the steroid receptor gene superfamily, controls photoreceptor cell fate. *Cell* **60**, 211–224.

2889. Moazed, D. and O'Farrell, P.H. (1992). Maintenance of the *engrailed* expression pattern by *Polycomb* group genes in *Drosophila*. *Development* **116**, 805–810.

2890. Modolell, J. (1997). Patterning of the adult peripheral nervous system of *Drosophila*. *Persp. Dev. Neurobiol.* **4**, 285–296.

2891. Modolell, J. and Campuzano, S. (1998). The *achaete-scute* complex as an integrating device. *Int. J. Dev. Biol.* **42**, 275–282.

2892. Moghal, N. and Sternberg, P.W. (1999). Multiple positive and negative regulators of signaling by the EGF receptor. *Curr. Opin. Cell Biol.* **11**, 190–196.

2893. Mohler, J. (1988). Requirements for *hedgehog*, a segmental polarity gene, in patterning larval and adult cuticle of *Drosophila*. *Genetics* **120**, 1061–1072.

2894. Mohler, J., Seecoomar, M., Agarwal, S., Bier, E., and Hsai, J. (2000). Activation of *knot* (*kn*) specifies the 3-4 intervein region in the *Drosophila* wing. *Development* **127**, 55–63.

2895. Mohler, J. and Vani, K. (1992). Molecular organization and embryonic expression of the *hedgehog* gene involved in cell-cell communication in segmental patterning of *Drosophila*. *Development* **115**, 957–971.

2896. Mohler, J.D. (1965). The influence of some crossveinless-like genes on the crossveinless phenocopy sensitivity in *Drosophila melanogaster*. *Genetics* **51**, 329–340.

2897. Mohler, J.D. (1965). Preliminary genetic analysis of crossveinless-like strains of *Drosophila melanogaster*. *Genetics* **51**, 641–651.

2898. Mohler, J.D. and Swedberg, G.S. (1964). Wing vein development in crossveinless-like strains of *Drosophila melanogaster*. *Genetics* **50**, 1403–1419.

2899. Moline, M.M., Dierick, H.A., Southern, C., and Bejsovec, A. (2000). Non-equivalent roles of *Drosophila* Frizzled and Dfrizzled2 in embryonic Wingless signal transduction. *Curr. Biol.* **10**, 1127–1130.

2900. Moline, M.M., Southern, C., and Bejsovec, A. (1999). Directionality of Wingless protein transport influences epidermal patterning in the *Drosophila* embryo. *Development* **126**, 4375–4384.

2901. Møller, A.P. and Swaddle, J.P. (1997). "Asymmetry, Developmental Stability, and Evolution." Oxford Univ. Pr., Oxford.

2902. Mollereau, B., Dominguez, M., Webel, R., Colley, N.J., Keung, B., De Celis, J.F., and Desplan, C. (2001). Two-step process for photoreceptor formation in *Drosophila*. *Nature* **412**, 911–913.

2903. Mollereau, B., Wernet, M.F., Beaufils, P., Killian, D., Pichaud, F., Kühnlein, R., and Desplan, C. (2000). A green fluorescent protein enhancer trap screen in *Drosophila* photoreceptor cells. *Mechs. Dev.* **93**, 151–160.

2904. Moloney, D.J., Panin, V.M., Johnston, S.H., Chen, J., Shao, L., Wilson, R., Wang, Y., Stanley, P., Irvine, K.D., Haltiwanger, R.S., and Vogt, T.F. (2000). Fringe is a glycosyltransferase that modifies Notch. *Nature* **406**, 369–375.

2905. Monge, I. and Mitchell, P.J. (1998). DAP-2, the *Drosophila* homolog of transcription factor AP-2. *Mechs. Dev.* **76**, 191–195.

2906. Monnier, V., Dussillol, F., Alves, G., Lamour-Isnard, C., and Plessis, A. (1998). Suppressor of fused links Fused and

Cubitus interruptus on the Hedgehog signalling pathway. *Curr. Biol.* **8**, 583–586.

2907. Montagne, J., Groppe, J., Guillemin, K., Krasnow, M.A., Gehring, W.J., and Affolter, M. (1996). The *Drosophila* Serum Response Factor gene is required for the formation of intervein tissue of the wing and is allelic to *blistered*. *Development* **122**, 2589–2597.

2908. Montell, C. (1989). Molecular genetics of *Drosophila* vision. *BioEssays* **11**, 43–48.

2909. Montell, C. (1999). Visual transduction in *Drosophila*. *Annu. Rev. Cell Dev. Biol.* **15**, 231–268.

2910. Montminy, M. (1993). Trying on a new pair of SH2s. *Science* **261**, 1694–1695.

2911. Montminy, M. (1997). Something new to hang your HAT on. *Nature* **387**, 654–655.

2912. Moodie, S.A., Willumsen, B.M., Weber, M.J., and Wolfman, A. (1993). Complexes of Ras.GTP with Raf-1 and mitogen-activated protein kinase kinase. *Science* **260**, 1658–1661.

2913. Moodie, S.A. and Wolfman, A. (1994). The 3 Rs of life: Ras, Raf and growth regulation. *Trends Genet.* **10**, 44–48.

2914. Moon, R.T. and Miller, J.R. (1997). The APC tumor suppressor protein in development and cancer. *Trends Genet.* **13**, 256–258.

2915. Moore, B.W. and McGregor, D. (1965). Chromatographic and electrophoretic fractionation of soluble proteins of brain and liver. *J. Biol. Chem.* **240** #4, 1647–1653.

2916. Moore, B.W. and Perez, V.J. (1968). Specific acidic proteins of the nervous system. *In* "Physiological and Biochemical Aspects of Nervous Integration," (F.D. Carlson, ed.). Prentice-Hall: Englewood Cliffs, N. J., pp. 343–359.

2917. Moore, J.A. (1986). Science as a way of knowing–genetics. *Amer. Zool.* **26**, 583–747.

2918. Moore, J.A. (1987). Science as a way of knowing–developmental biology. *Amer. Zool.* **27**, 415–573.

2919. Moore, J.A. (1993). "Science as a Way of Knowing. The Foundations of Modern Biology." Harvard Univ. Pr., Cambridge, Mass.

2920. Moore, P.B. (1999). Structural motifs in RNA. *Annu. Rev. Biochem.* **67**, 287–300.

2921. Morata, G. (1975). Analysis of gene expression during development in the homeotic mutant *Contrabithorax* of *Drosophila melanogaster*. *J. Embryol. Exp. Morph.* **34**, 19–31.

2922. Morata, G. (1982). The mode of action of the bithorax genes of *Drosophila melanogaster*. *Amer. Zool.* **22**, 57–64.

2923. Morata, G. (1993). Homeotic genes of *Drosophila*. *Curr. Opin. Gen. Dev.* **3**, 606–614.

2924. Morata, G. and Garcia-Bellido, A. (1976). Developmental analysis of some mutants of the bithorax system of *Drosophila*. *W. Roux's Arch.* **179**, 125–143.

2925. Morata, G. and Kerridge, S. (1981). Sequential functions of the bithorax complex of *Drosophila*. *Nature* **290**, 778–781.

2926. Morata, G. and Kerridge, S. (1982). The role of position in determining homoeotic gene function in *Drosophila*. *Nature* **300**, 191–192.

2927. Morata, G., Kornberg, T., and Lawrence, P.A. (1983). The phenotype of *engrailed* mutations in the antenna of *Drosophila*. *Dev. Biol.* **99**, 27–33.

2928. Morata, G. and Lawrence, P.A. (1975). Control of compartment development by the *engrailed* gene in *Drosophila*. *Nature* **255**, 614–617.

2929. Morata, G. and Lawrence, P.A. (1977). The development of *wingless*, a homeotic mutation of *Drosophila*. *Dev. Biol.* **56**, 227–240.

2930. Morata, G. and Lawrence, P.A. (1977). Homeotic genes, compartments and cell determination in *Drosophila*. *Nature* **265**, 211–216.

2931. Morata, G. and Lawrence, P.A. (1978). Anterior and posterior compartments in the head of *Drosophila*. *Nature* **274**, 473–474.

2932. Morata, G. and Lawrence, P.A. (1978). Cell lineage and homeotic mutants in the development of imaginal discs of *Drosophila*. *In* "The Clonal Basis of Development," *Symp. Soc. Dev. Biol.*, Vol. 36 (S. Subtelny and I.M. Sussex, eds.). Acad. Pr.: New York, pp. 45–60.

2933. Morata, G. and Lawrence, P.A. (1979). Development of the eye-antenna imaginal disc of *Drosophila*. *Dev. Biol.* **70**, 355–371.

2934. Morata, G., Macías, A., Urquía, N., and González-Reyes, A. (1990). Homoeotic genes. *Sems. Cell Biol.* **1**, 219–227.

2935. Morata, G. and Ripoll, P. (1975). *Minutes*: mutants of *Drosophila* autonomously affecting cell division rate. *Dev. Biol.* **42**, 211–221.

2936. Morata, G. and Sánchez-Herrero, E. (1998). Pulling the fly's leg. *Nature* **392**, 657–658.

2937. Morata, G. and Sánchez-Herrero, E. (1999). Patterning mechanisms in the body trunk and the appendages of *Drosophila*. *Development* **126**, 2823–2828.

2938. Morata, G. and Struhl, G. (1990). Fly fishing downstream. *Nature* **348**, 587–588.

2939. Morcillo, P., Rosen, C., Baylies, M.K., and Dorsett, D. (1997). Chip, a widely expressed chromosomal protein required for segmentation and activity of a remote wing margin enhancer in *Drosophila*. *Genes Dev.* **11**, 2729–2740.

2940. Morcillo, P., Rosen, C., and Dorsett, D. (1996). Genes regulating the remote wing margin enhancer in the *Drosophila cut* locus. *Genetics* **144**, 1143–1154.

2941. Morel, V. and Schweisguth, F. (2000). Repression by Suppressor of Hairless and activation by Notch are required to define a single row of *single-minded* expressing cells in the *Drosophila* embryo. *Genes Dev.* **14**, 377–388.

2942. Moreno, E. and Morata, G. (1999). *Caudal* is the Hox gene that specifies the most posterior *Drosophila* segment. *Nature* **400**, 873–877.

2943. Morgan, D.O. (1997). Cyclin-dependent kinases: engines, clocks, and microprocessors. *Annu. Rev. Cell Dev. Biol.* **13**, 261–291.

2944. Morgan, R., In der Rieden, P., Hooiveld, M.H.W., and Durston, A.J. (2000). Identifying HOX paralog groups by the PBX-binding region. *Trends Genet.* **16**, 66–67.

2945. Morgan, T.H. (1901). "Regeneration." MacMillan, New York.

2946. Morgan, T.H. (1910). Chance or purpose in the origin and evolution of adaptation. *Science* **31**, 201–210.

2947. Morgan, T.H. (1910). Chromosomes and heredity. *Am. Nat.* **44**, 449–496.

2948. Morgan, T.H. (1910). Sex limited inheritance in *Drosophila*. *Science* **32**, 120–122.

2949. Morgan, T.H. (1934). "Embryology and Genetics." Greenwood Pr., Westport, Conn.

2950. Morgan, T.H. and Bridges, C.B. (1919). The origin of gynandromorphs. *In* "Contributions to the Genetics of *Drosophila melanogaster*," Vol. Pub. No. 278 Carnegie Inst. Wash.: Wash. D. C., pp. 1–122.

2951. Morgan, T.H., Bridges, C.B., and Sturtevant, A.H. (1925). The Genetics of *Drosophila*. *Bibliogr. Genet.* **2**, 1–262.

2952. Morimoto, A.M., Jordan, K.C., Tietze, K., Britton, J.S., O'Neill, E.M., and Ruohola-Baker, H. (1996). Pointed, an

ETS domain transcription factor, negatively regulates the EGF receptor pathway in *Drosophila* oogenesis. *Development* **122**, 3745–3754.

2953. Morimoto, R.I. (1998). Regulation of the heat shock transcriptional response: cross talk between a family of heat shock factors, molecular chaperones, and negative regulators. *Genes Dev.* **12**, 3788–3796.

2954. Morimura, S., Maves, L., and Hoffmann, F.M. (1996). *decapentaplegic* overexpression affects *Drosophila* wing and leg imaginal disc development and *wingless* expression. *Dev. Biol.* **177**, 136–151.

2955. Morita, H. and Shiraishi, A. (1985). Chemoreception physiology. *In* "Comprehensive Insect Physiology, Biochemistry, and Pharmacology," Vol. 6 (G.A. Kerkut and L.I. Gilbert, eds.). Pergamon: New York, pp. 133–170.

2956. Moritz, K.B. and Sauer, H.W. (1996). Boveri's contributions to developmental biology–a challenge for today. *Int. J. Dev. Biol.* **40**, 27–47.

2957. Morrison, D.K. and Cutler, R.E., Jr. (1997). The complexity of Raf-1 regulation. *Curr. Opin. Cell Biol.* **9**, 174–179.

2958. Morrow, E.M., Furukawa, T., Lee, J.E., and Cepko, C.L. (1999). NeuroD regulates multiple functions in the developing neural retina in rodent. *Development* **126**, 23–36.

2959. Moscoso del Prado, J. and Garcia-Bellido, A. (1984). Cell interactions in the generation of chaetae pattern in *Drosophila*. *Roux's Arch. Dev. Biol.* **193**, 246–251.

2960. Moscoso del Prado, J. and Garcia-Bellido, A. (1984). Genetic regulation of the *Achaete-scute* complex of *Drosophila melanogaster*. *Roux's Arch. Dev. Biol.* **193**, 242–245.

2961. Moses, K. (1991). The role of transcription factors in the developing *Drosophila* eye. *Trends Genet.* **7**, 250–255.

2962. Moses, K., ed. (2002). "*Drosophila* Eye Development." *Results and Problems in Cell Differentiation*, Springer, Berlin.

2963. Moses, K., Ellis, M.C., and Rubin, G.M. (1989). The *glass* gene encodes a zinc-finger protein required by *Drosophila* photoreceptor cells. *Nature* **340**, 531–536.

2964. Moses, K. and Rubin, G.M. (1991). *glass* encodes a site-specific DNA-binding protein that is regulated in response to positional signals in the developing *Drosophila* eye. *Genes Dev.* **5**, 583–593.

2965. Moss, E.G. (2000). Non-coding RNAs: lightning strikes twice. *Curr. Biol.* **10**, R436–R439.

2966. Mottus, R., Sobel, R.E., and Grigliatti, T.A. (2000). Mutational analysis of a histone deacetylase in *Drosophila melanogaster*: missense mutations suppress gene silencing associated with position effect variegation. *Genetics* **154**, 657–668.

2967. Motzny, C.K. and Holmgren, R. (1995). The *Drosophila* cubitus interruptus protein and its role in the *wingless* and *hedgehog* signal transduction pathways. *Mechs. Dev.* **52**, 137–150.

2968. Moyer, S.E. and Stepak, S.P. (1971). Inheritance of trident and its role in detecting ebony heterozygotes in *D. melanogaster*. *Dros. Info. Serv.* **46**, 133.

2969. Moyle, W.R., Campbell, R.K., Myers, R.V., Bernard, M.P., Han, Y., and Wang, X. (1994). Co-evolution of ligand-receptor pairs. *Nature* **368**, 251–255.

2970. Mozer, B.A. and Benzer, S. (1994). Ingrowth of photoreceptor axons induces transcription of a retrotransposon in the developing *Drosophila* brain. *Development* **120**, 1049–1058.

2971. Mozer, B.A. and Easwarachandran, K. (1999). Pattern formation in the absence of cell proliferation: tissue-specific regulation of cell cycle progression by *string* (*stg*) during *Drosophila* eye development. *Dev. Biol.* **213**, 54–69.

2972. Muda, M., Boschert, U., Dickinson, R., Martinou, J.-C., Martinou, I., M., C., Schlegel, W., and Arkinstall, S. (1996). MKP-3, a novel cytosolic protein-tyrosine phosphatase that exemplifies a new class of mitogen-activated protein kinase phosphatase. *J. Biol. Chem.* **271**, 4319–4326.

2973. Mueller, B.K. (1999). Growth cone guidance: first steps towards a deeper understanding. *Annu. Rev. Neurosci.* **22**, 351–388.

2974. Mugat, B., Brodu, V., Kejzlarova-Lepesant, J., Antoniewski, C., Bayer, C.A., Fristrom, J.W., and Lepesant, J.-A. (2000). Dynamic expression of *Broad-Complex* isoforms mediates temporal control of an ecdysteroid target gene at the onset of *Drosophila* metamorphosis. *Dev. Biol.* **227**, 104–117.

2975. Mukherjee, A., Lakhotia, S.C., and Roy, J.K. (1995). *l(2)gl* gene regulates late expression of segment polarity genes in *Drosophila*. *Mechs. Dev.* **51**, 227–234.

2976. Mukherjee, A. and Roy, J.K. (1995). Mutations in tumour suppressor genes, *l(2)gl* and *l(2)gd*, alter the expression of *wingless* in *Drosophila*. *J. Biosci.* **20**, 333–339.

2977. Mukherjee, A., Shan, X., Mutsuddi, M., Ma, Y., and Nambu, J.R. (2000). The *Drosophila* Sox gene, *fish-hook*, is required for postembryonic development. *Development* **217**, 91–106.

2978. Mukherjee, A.S. (1965). The effects of sexcombless on the forelegs of *Drosophila melanogaster*. *Genetics* **51**, 285–304.

2979. Mukherjee, A.S. and Hildreth, P.E. (1971). Interaction of the mutants *sx*, *tra* and *dsx* in *Drosophila melanogaster*. Chaetotaxy of the basitarsus of the forelegs in males and females. *Genetica* **42**, 338–352.

2980. Müller, B. and Basler, K. (2000). The repressor and activator forms of Cubitus interruptus control Hedgehog target genes through common generic Gli-binding sites. *Development* **127**, 2999–3007.

2981. Müller, C. and Leutz, A. (2001). Chromatin remodeling in development and differentiation. *Curr. Opin. Gen. Dev.* **11**, 167–174.

2982. Müller, G.H. (1976). *Drosophila*: a contribution to its morphology and development by W. F. von Gleichen in 1764. *J. Nat. Hist.* **10**, 581–597.

2983. Müller, H.-A.J. (2000). Genetic control of epithelial cell polarity: lessons from *Drosophila*. *Dev. Dynamics* **218**, 52–67.

2984. Müller, H.-A.J., Samanta, R., and Wieschaus, E. (1999). Wingless signaling in the *Drosophila* embryo: zygotic requirements and the role of the *frizzled* genes. *Development* **126**, 577–586.

2985. Müller, H.-A.J. and Wieschaus, E. (1996). *armadillo*, *bazooka*, and *stardust* are critical for early stages in formation of the zonula adherens and maintenance of the polarized blastoderm epithelium in *Drosophila*. *J. Cell Biol.* **134**, 149–163.

2986. Muller, H.J. (1932). Further studies on the nature and causes of gene mutations. *Proc. 6th Internat. Congr. Genet.* **1**, 213–255.

2987. Müller, J. and Bienz, M. (1991). Long range repression conferring boundaries of *Ultrabithorax* expression in the *Drosophila* embryo. *EMBO J.* **10**, 3147–3155.

2988. Müller, M., v. Weizsäcker, E., and Campos-Ortega, J.A. (1996). Expression domains of a zebrafish homologue of the *Drosophila* pair-rule gene *hairy* correspond to primordia of alternating somites. *Development* **122**, 2071–2078.

2989. Müller, W.A. (1996). From the Aristotelian soul to genetic and epigenetic information: the evolution of the modern concepts in developmental biology at the turn of the century. *Int. J. Dev. Biol.* **40**, 21–26.

2990. Mullins, M.C. (1998). Holy Tolloido: Tolloid cleaves SOG/Chordin to free DPP/BMPs. *Trends Genet.* **14**, 127–129.

2991. Mullins, M.C. and Rubin, G.M. (1991). Isolation of temperature-sensitive mutations of the tyrosine kinase receptor sevenless (*sev*) in *Drosophila* and their use in determining its time of action. *Proc. Natl. Acad. Sci. USA* **88**, 9387–9391.

2992. Mullor, J.L., Calleja, M., Capdevila, J., and Guerrero, I. (1997). Hedgehog activity, independent of Decapentaplegic, participates in wing disc patterning. *Development* **124**, 1227–1237.

2993. Mullor, J.L. and Guerrero, I. (2000). A gain-of-function mutant of *patched* dissects different responses to the Hedgehog gradient. *Dev. Biol.* **228**, 211–224.

2994. Mumm, J.S. and Kopan, R. (2000). Notch signaling: from the outside in. *Dev. Biol.* **228**, 151–165.

2995. Mumm, J.S., Schroeter, E.H., Saxena, M.T., Griesemer, A., Tian, X., Pan, D.J., Ray, W.J., and Kopan, R. (2000). A ligand-induced extracellular cleavage regulates γ-secretase-like proteolytic activation of Notch1. *Molec. Cell* **5**, 197–206.

2996. Munemitsu, S., Albert, I., Rubinfeld, B., and Polakis, P. (1996). Deletion of an amino-terminal sequence stabilizes β-catenin in vivo and promotes hyperphosphorylation of the adenomatous polyposis coli tumor suppressor protein. *Mol. Cell. Biol.* **16**, 4088–4094.

2997. Munemitsu, S., Albert, I., Souza, B., Rubinfeld, B., and Polakis, P. (1995). Regulation of intracellular β-catenin levels by the adenomatous polyposis coli (APC) tumor-suppressor protein. *Proc. Natl. Acad. Sci. USA* **92**, 3046–3050.

2998. Muneoka, K. and Bryant, S. (1984). Regeneration and development of vertebrate appendages. *Symp. Zool. Soc. Lond.* **52**, 177–196.

2999. Muneoka, K. and Bryant, S.V. (1982). Evidence that patterning mechanisms in developing and regenerating limbs are the same. *Nature* **298**, 369–371.

3000. Munro, S. and Freeman, M. (2000). The Notch signalling regulator Fringe acts in the Golgi apparatus and requires the glycosyltransferase signature motif DxD. *Curr. Biol.* **10**, 813–820.

3001. Muravyova, E., Golovnin, A., Gracheva, E., Parshikov, A., Belenkaya, T., Pirrotta, V., and Georgiev, P. (2001). Loss of insulator activity by paired Su(Hw) chromatin insulators. *Science* **291**, 495–498.

3002. Muravyova, E., Golovnin, A., Gracheva, E., Pirrota, V., and Georgiev, P. (2001). Paired su(Hw) insulators lose enhancer blocking activity and may instead facilitate enhancer-promoter interactions. *Proc. 42nd Ann. Drosophila Res. Conf.* **Abstracts Vol.**, a92.

3003. Murayama, M., Tanaka, S., Palacino, J., Murayama, O., Honda, T., Sun, X., Yasutake, K., Nihonmatsu, N., Wolozin, B., and Takashima, A. (1998). Direct association of presenilin-1 with β-catenin. *FEBS Letters* **433**, 73–77.

3004. Murphey, R.K., Possidente, D., Pollack, G., and Merritt, D.J. (1989). Modality-specific axonal projections in the CNS of the flies *Phormia* and *Drosophila*. *J. Comp. Neurol.* **290**, 185–200.

3005. Murphey, R.K., Possidente, D.R., Vandervorst, P., and Ghysen, A. (1989). Compartments and the topography of leg afferent projections in *Drosophila*. *J. Neurosci.* **9**, 3209–3217.

3006. Murphy, C. (1967). Determination of the dorsal mesothoracic disc in *Drosophila*. *Dev. Biol.* **15**, 368–394.

3007. Murphy, C. and Tokunaga, C. (1970). Cell lineage in the dorsal mesothoracic disc of *Drosophila*. *J. Exp. Zool.* **175**, 197–219.

3008. Murphy, F.V., IV, Sweet, R.M., and Churchill, M.E.A. (1999). The structure of a chromosomal high mobility group protein-DNA complex reveals sequence-neutral mechanisms important for non-sequence-specific DNA recognition. *EMBO J.* **18**, 6610–6618.

3009. Murray, A. (1995). Cyclin ubiquitination: the destructive end of mitosis. *Cell* **81**, 149–152.

3010. Murray, A.W. and Kirschner, M.W. (1989). Dominoes and clocks: the union of two views of the cell cycle. *Science* **246**, 614–621.

3011. Murray, J.D. (1981). On pattern formation mechanisms for lepidopteran wing patterns and mammalian coat markings. *Phil. Trans. Roy. Soc. Lond.* B **295**, 473–496.

3012. Murray, J.D. (1981). A pre-pattern formation mechanism for animal coat markings. *J. Theor. Biol.* **88**, 161–199.

3013. Murray, J.D. (1989). "Mathematical Biology." Springer-Verlag, Berlin.

3014. Murray, J.D. (1990). Turing's theory of morphogenesis–its influence on modelling biological pattern and form. *Bull. Math. Biol.* **52**, 119–152.

3015. Murray, M.A., Fessler, L.I., and Palka, J. (1995). Changing distributions of extracellular matrix components during early wing morphogenesis in *Drosophila*. *Dev. Biol.* **168**, 150–165.

3016. Murre, C., McCaw, P.S., and Baltimore, D. (1989). A new DNA binding and dimerization motif in immunoglobulin enhancer binding, *daughterless, MyoD,* and *myc* proteins. *Cell* **56**, 777–783.

3017. Murre, C., McCaw, P.S., Vaessin, H., Caudy, M., Jan, L.Y., Jan, Y.N., Cabrera, C.V., Buskin, J.N., Hauschka, S.D., Lassar, A.B., Weintraub, H., and Baltimore, D. (1989). Interactions between heterologous helix-loop-helix proteins generate complexes that bind specifically to a common DNA sequence. *Cell* **58**, 537–544.

3018. Murzin, A.G. (1992). Structural principles for the propeller assembly of β-sheets: the preference for seven-fold symmetry. *Proteins* **14**, 191–201.

3019. Musacchio, A., Wilmanns, M., and Saraste, M. (1994). Structure and function of the SH3 domain. *Prog. Biophys. Molec. Biol.* **61**, 283–297.

3020. Musacchio, M. and Perrimon, N. (1996). The *Drosophila kekkon* genes: novel members of both the leucine-rich repeat and immunoglobulin superfamilies expressed in the CNS. *Dev. Biol.* **178**, 63–76.

3021. Mushegian, A.R. and Koonin, E.V. (1996). Sequence analysis of eukaryotic developmental proteins: ancient and novel domains. *Genetics* **144**, 817–828.

3022. Muskavitch, M.A.T. (1994). Delta-Notch signaling and *Drosophila* cell fate choice. *Dev. Biol.* **166**, 415–430.

3023. Muslin, A.J., Tanner, J.W., Allen, P.M., and Shaw, A.S. (1996). Interaction of 14-3-3 with signaling proteins is mediated by the recognition of phosphoserine. *Cell* **84**, 889–897.

3024. Myers, M.P., Wager-Smith, K., Wesley, C.S., Young, M.W., and Sehgal, A. (1995). Positional cloning and sequence analysis of the *Drosophila* clock gene, *timeless. Science* **270**, 805–808.

3025. Nagaraj, R., Pickup, A.T., Howes, R., Moses, K., Freeman, M., and Banerjee, U. (1999). Role of the EGF receptor pathway in growth and patterning of the *Drosophila* wing through the regulation of *vestigial*. *Development* **126**, 975–985.

3026. Nagata, T., Gupta, V., Sorce, D., Kim, W.-Y., Sali, A., Chait, B.T., Shigesada, K., Ito, Y., and Werner, M.H. (1999). Immunoglobulin motif DNA recognition and heterodimerization of the PEBP2/CBF Runt domain. *Nature Struct. Biol.* **6**, 615–619.

3027. Nagel, A.C., Maier, D., and Preiss, A. (2000). Su(H)-independent activity of Hairless during mechano-sensory organ formation in *Drosophila*. *Mechs. Dev.* **94**, 3–12.

3028. Nagel, A.C. and Preiss, A. (1999). *Notch*^spl is deficient for inductive processes in the eye, and *E(spl)*^D enhances *split* by interfering with proneural activity. *Dev. Biol.* **208**, 406–415.

3029. Nagel, A.C., Yu, Y., and Preiss, A. (1999). *Enhancer of split* [*E(spl)*^D] is a Gro-independent, hypermorphic mutation in *Drosophila*. *Dev. Genet.* **25**, 168–179.

3030. Nagl, W. (1978). "Endopolyploidy and Polyteny in Differentiation and Evolution." North-Holland, New York.

3031. Nagorcka, B.N. (1988). A pattern formation mechanism to control spatial organization in the embryo of *Drosophila melanogaster*. *J. Theor. Biol.* **132**, 277–306.

3032. Nagy, L. (1998). Changing patterns of gene regulation in the evolution of arthropod morphology. *Amer. Zool.* **38**, 818–828.

3033. Nakagoshi, H., Hoshi, M., Nabeshima, Y.-i., and Matsuzaki, F. (1998). A novel homeobox gene mediates the Dpp signal to establish functional specificity within target cells. *Genes Dev.* **12**, 2724–2734.

3034. Nakamura, M., Nishida, Y., and Ueno, N. (2001). Genetic screening to isolate novel Dpp signal downstream components. *Proc. 42nd Ann. Drosophila Res. Conf.* **Abstracts Vol.**, a163.

3035. Nakamura, M., Okano, H., Blendy, J.A., and Montell, C. (1994). Musashi, a neural RNA-binding protein required for *Drosophila* adult external sensory organ development. *Neuron* **13**, 67–81. [See also 2001 *Nature* **411**, 94–98.]

3036. Nakano, Y., Guerrero, I., Hidalgo, A., Taylor, A., Whittle, J.R.S., and Ingham, P.W. (1989). A protein with several possible membrane-spanning domains encoded by the *Drosophila* segment polarity gene *patched*. *Nature* **341**, 508–513.

3037. Nakao, K. and Campos-Ortega, J.A. (1996). Persistent expression of genes of the *Enhancer of Split* Complex suppresses neural development in *Drosophila*. *Neuron* **16**, 275–286.

3038. Nakato, H., Futch, T.A., and Selleck, S.B. (1995). The *division abnormally delayed* (*dally*) gene: a putative integral membrane proteoglycan required for cell division patterning during postembryonic development of the nervous system in *Drosophila*. *Development* **121**, 3687–3702.

3039. Nakielny, S. and Dreyfuss, G. (1997). Import and export of the nuclear protein import receptor transportin by a mechanism independent of GTP hydrolysis. *Curr. Biol.* **8**, 89–95.

3040. Nakielny, S. and Dreyfuss, G. (1999). Transport of proteins and RNAs in and out of the nucleus. *Cell* **99**, 677–690.

3041. Namba, R. and Minden, J.S. (1999). Fate mapping of *Drosophila* embryonic mitotic domain 20 reveals that the larval visual system is derived from a subdomain of a few cells. *Dev. Biol.* **212**, 465–476.

3042. Namba, R., Pazdera, T.M., Cerrone, R.L., and Minden, J.S. (1997). *Drosophila* embryonic pattern repair: how embryos respond to *bicoid* dosage alteration. *Development* **124**, 1393–1403.

3043. Nambu, J.R., Franks, R.G., Hu, S., and Crews, S.T. (1990). The *single-minded* gene of *Drosophila* is required for the expression of genes important for the development of CNS midline cells. *Cell* **63**, 63–75.

3044. Nambu, J.R., Lewis, J.O., Wharton, K.A., Jr., and Crews, S.T. (1991). The *Drosophila single-minded* gene encodes a helix-loop-helix protein that acts as a master regulator of CNS midline development. *Cell* **67**, 1157–1167.

3045. Nambu, P.A. and Nambu, J.R. (1996). The *Drosophila fish-hook* gene encodes a HMG domain protein essential for segmentation and CNS development. *Development* **122**, 3467–3475.

3046. Nardi, J.B. (1981). Epithelial invagination: adhesive properties of cells can govern position and directionality of epithelial folding. *Differentiation* **20**, 97–103.

3047. Nardi, J.B. (1981). Induction of invagination in insect epithelium: paradigm for embryonic invagination. *Science* **214**, 564–566.

3048. Nardi, J.B. (1994). Rearrangement of epithelial cell types in an insect wing monolayer is accompanied by differential expression of a cell surface protein. *Dev. Dynamics* **199**, 315–325.

3049. Nardi, J.B. and Kafatos, F.C. (1976). Polarity and gradients in lepidopteran wing epidermis. I. Changes in graft polarity, form, and cell density accompanying transpositions and reorientations. *J. Embryol. Exp. Morph.* **36**, 469–487.

3050. Nardi, J.B. and Kafatos, F.C. (1976). Polarity and gradients in lepidopteran wing epidermis. II. The differential adhesiveness model: gradient of a non-diffusible cell surface parameter. *J. Embryol. Exp. Morph.* **36**, 489–512.

3051. Nardi, J.B. and Magee-Adams, S.M. (1986). Formation of scale spacing patterns in a moth wing. I. Epithelial feet may mediate cell rearrangement. *Dev. Biol.* **116**, 265–277.

3052. Nardi, J.B. and Norby, S.W. (1987). Evocation of venation pattern in the wing of a moth, *Manduca sexta*. *J. Morphol.* **193**, 53–62.

3053. Nash, D. (1965). The expression of "Hairless" in *Drosophila* and the role of two closely linked modifiers of opposite effect. *Genet. Res., Camb.* **6**, 175–189.

3054. Nash, D. (1970). The mutational basis for the "allelic" modifier mutants, *Enhancer* and *Suppressor of Hairless*, of *Drosophila melanogaster*. *Genetics* **64**, 471–479.

3055. Nash, W.G. (1976). Patterns of pigmentation color states regulated by the *y* locus in *Drosophila melanogaster*. *Dev. Biol.* **48**, 336–343.

3056. Nash, W.G. and Yarkin, R.J. (1974). Genetic regulation and pattern formation: a study of the *yellow* locus in *Drosophila melanogaster*. *Genet. Res., Camb.* **24**, 19–26.

3057. Nassar, N., Horn, G., Herrmann, C., Scherer, A., McCormick, F., and Wittinghofer, A. (1995). The 2.2 Å crystal structure of the Ras-binding domain of the serine/threonine kinase c-Raf1 in complex with Rap1A and a GTP analogue. *Nature* **375**, 554–560.

3058. Nassif, C., Daniel, A., Lengyel, J.A., and Hartenstein, V. (1998). The role of morphogenetic cell death during *Drosophila* embryonic head development. *Dev. Biol.* **197**, 170–186.

3059. Natzle, J.E. (1993). Temporal regulation of *Drosophila* imaginal disc morphogenesis: a hierarchy of primary and secondary 20-hydroxyecdysone-responsive loci. *Dev. Biol.* **155**, 516–532.

3060. Nauber, U., Pankratz, M.J., Kienlin, A., Seifert, E., Klemm, U., and Jäckle, H. (1988). Abdominal segmentation of the

Drosophila embryo requires a hormone receptor-like protein encoded by the gap gene *knirps*. *Nature* **336**, 489–492.

3061. Nayak, S.V. and Singh, R.N. (1983). Sensilla on the tarsal segments and mouthparts of adult *Drosophila melanogaster* Meigen (Diptera: Drosophilidae). *Int. J. Insect Morphol. Embryol.* **12**, 273–291.

3062. Needham, J. (1933). On the dissociability of the fundamental processes in ontogenesis. *Biol. Rev.* **8**, 180–223.

3063. Needham, J. (1936). "Order and Life." Yale Univ. Pr., New Haven. (*N.B.* : Needham's Fig. 11 predates Waddington's 1940 idea of the "epigenetic landscape"; *cf.* Yoxen's paper).

3064. Needham, J. (1959). "A History of Embryology." 2nd ed. Cambridge Univ. Pr., Cambridge.

3065. Neel, B.G. and Tonks, N.K. (1997). Protein tyrosine phosphatases in signal transduction. *Curr. Opin. Cell Biol.* **9**, 193–204.

3066. Neel, J.V. (1940). The interrelations of temperature, body size, and character expression in *Drosophila melanogaster*. *Genetics* **25**, 225–250.

3067. Neel, J.V. (1940). The pattern of supernumerary macrochaetae in certain *Drosophila* mutants. *Genetics* **25**, 251–277.

3068. Neel, J.V. (1941). Studies on the interaction of mutations affecting the chaetae of *Drosophila melanogaster*. I. The interaction of hairy, polychaetoid, and Hairy wing. *Genetics* **26**, 52–68.

3069. Neel, J.V. (1943). Studies on the interaction of mutations affecting the chaetae of *Drosophila melanogaster*. II. The relation of character expression to size in flies homozygous for polychaetoid, hairy, Hairy wing, and the combinations of these factors. *Genetics* **28**, 49–68.

3070. Neel, J.V. (1983). Curt Stern: 1902–1981. *Annu. Rev. Genet.* **17**, 1–10.

3071. Neel, J.V. (1987). Curt Stern: August 30, 1902–October 23, 1981. *Biogr. Memoirs Natl. Acad. Sci. U. S.* **56**, 442–473.

3072. Neer, E.J., Schmidt, C.J., Nambudripad, R., and Smith, T.F. (1994). The ancient regulatory-protein family of WD-repeat proteins. *Nature* **371**, 297–300.

3073. Nellen, D., Affolter, M., and Basler, K. (1994). Receptor serine/threonine kinases implicated in the control of *Drosophila* body pattern by *decapentaplegic*. *Cell* **78**, 225–237.

3074. Nellen, D., Burke, R., Struhl, G., and Basler, K. (1996). Direct and long-range action of a DPP morphogen gradient. *Cell* **85**, 357–368.

3075. Nellesen, D.T., Lai, E.C., and Posakony, J.W. (1999). Discrete enhancer elements mediate selective responsiveness of *Enhancer of split* Complex genes to common transcriptional activators. *Dev. Biol.* **213**, 33–53.

3076. Nelson, H.B., Heiman, R.G., Bolduc, C., Kovalick, G.E., Whitley, P., Stern, M., and Beckingham, K. (1997). Calmodulin point mutations affect *Drosophila* development and behavior. *Genetics* **147**, 1783–1798.

3077. Nestoras, K., Lee, H., and Mohler, J. (1997). Role of *knot* (*kn*) in wing patterning in *Drosophila*. *Genetics* **147**, 1203–1212.

3078. Netter, S., Faucheux, M., Roignant, J.Y., Antoniewski, C., and Théodore, L. (2001). Developmental dynamics of Polycomb group proteins in vivo. *Proc. 42nd Ann. Drosophila Res. Conf.* **Abstracts Vol.**, a97.

3079. Netter, S., Fauvarque, M.-O., del Corral, R.D., Dura, J.-M., and Coen, D. (1998). *white*⁺ transgene insertions presenting a dorsal/ventral pattern define a single cluster of homeobox genes that is silenced by the *Polycomb*-group proteins in *Drosophila melanogaster*. *Genetics* **149**, 257–275.

3080. Neubig, R.R. and Thomsen, W.J. (1989). How does a key fit a flexible lock? Structure and dynamics in receptor function. *BioEssays* **11**, 136–141.

3081. Neufeld, T.P., de la Cruz, A.F.A., Johnston, L.A., and Edgar, B.A. (1998). Coordination of growth and cell division in the *Drosophila* wing. *Cell* **93**, 1183–1193.

3082. Neufeld, T.P., Tang, A.H., and Rubin, G.M. (1998). A genetic screen to identify components of the *sina* signaling pathway in *Drosophila* eye development. *Genetics* **148**, 277–286.

3083. Neuhold, L.A. and Wold, B. (1993). HLH forced dimers: Tethering MyoD to E47 generates a dominant positive myogenic factor insulated from negative regulation by Id. *Cell* **74**, 1033–1042.

3084. Neuman-Silberberg, F.S., Schejter, E., Hoffmann, F.M., and Shilo, B.-Z. (1984). The *Drosophila ras* oncogenes: structure and nucleotide sequence. *Cell* **37**, 1027–1033.

3085. Neuman-Silberberg, F.S. and Schupbach, T. (1994). Dorsoventral axis formation in *Drosophila* depends on the correct dosage of the gene *gurken*. *Development* **120**, 2457–2463.

3086. Neuman-Silberberg, F.S. and Schüpbach, T. (1993). The *Drosophila* dorsoventral patterning gene *gurken* produces a dorsally localized RNA and encodes a TGFα-like protein. *Cell* **75**, 165–174.

3087. Neumann, C. and Cohen, S. (1997). Morphogens and pattern formation. *BioEssays* **19**, 721–729.

3088. Neumann, C.J. and Cohen, S.M. (1996). Distinct mitogenic and cell fate specification functions of *wingless* in different regions of the wing. *Development* **122**, 1781–1789.

3089. Neumann, C.J. and Cohen, S.M. (1996). A hierarchy of cross-regulation involving *Notch*, *wingless*, *vestigial* and *cut* organizes the dorsal/ventral axis of the *Drosophila* wing. *Development* **122**, 3477–3485.

3090. Neumann, C.J. and Cohen, S.M. (1996). *Sternopleural* is a regulatory mutation of *wingless* with both dominant and recessive effects on larval development of *Drosophila melanogaster*. *Genetics* **142**, 1147–1155.

3091. Neumann, C.J. and Cohen, S.M. (1997). Long-range action of Wingless organizes the dorsal-ventral axis of the *Drosophila* wing. *Development* **124**, 871–880.

3092. Neumann, C.J. and Cohen, S.M. (1998). Boundary formation in *Drosophila* wing: Notch activity attenuated by the POU protein Nubbin. *Science* **281**, 409–413.

3093. Neumann, C.J. and Nuesslein-Volhard, C. (2000). Patterning of the zebrafish retina by a wave of sonic hedgehog activity. *Science* **289**, 2137–2139.

3094. Newfeld, S.J., Chartoff, E.H., Graff, J.M., Melton, D.A., and Gelbart, W.M. (1996). *Mothers against dpp* encodes a conserved cytoplasmic protein required in DPP/TGF-β responsive cells. *Development* **122**, 2099–2108.

3095. Newfeld, S.J., Mehra, A., Singer, M.A., Wrana, J.L., Attisano, L., and Gelbart, W.M. (1997). *Mothers against dpp* participates in a DPP/TGF-β responsive serine-threonine kinase signal transduction cascade. *Development* **124**, 3167–3176.

3096. Newfeld, S.J., Padgett, R.W., Findley, S.D., Richter, B.G., Sanicola, M., de Cuevas, M., and Gelbart, W.M. (1997). Molecular evolution at the *decapentaplegic* locus in *Drosophila*. *Genetics* **145**, 297–309.

3097. Newfeld, S.J., Su, M.A., and Wisotzkey, R.G. (2001). A screen for modifiers of decapentaplegic mutant phenotypes identifies lilliputian, the only member of the Fragile-X/Burkitt's lymphoma family of transcription factors in *Drosophila melanogaster. Proc. 42nd Ann. Drosophila Res. Conf.* **Abstracts Vol.**, a164. [See also 2001 *Genetics* **157**, 717–725.]

3098. Newfeld, S.J. and Takaesu, N.T. (1999). Local transposition of a *hobo* element within the *decapentaplegic* locus of *Drosophila. Genetics* **151**, 177–187.

3099. Newfeld, S.J., Wisotzkey, R.G., and Kumar, S. (1999). Molecular evolution of a developmental pathway: phylogenetic analyses of Transforming Growth Factor-β family ligands, receptors and Smad signal transducers. *Genetics* **152**, 783–795.

3100. Ng, H.H. and Bird, A. (2000). Histone deacetylases: silencers for hire. *Trends Biochem. Sci.* **25**, 121–126.

3101. Ng, J., Hart, C.M., Morgan, K., and Simon, J.A. (2000). A *Drosophila* ESC-E(Z) protein complex is distinct from other Polycomb group complexes and contains covalently modified ESC. *Mol. Cell. Biol.* **20**, 3069–3078.

3102. Ng, J., Li, R., Morgan, K., and Simon, J. (1997). Evolutionary conservation and predicted structure of the *Drosophila* extra sex combs repressor protein. *Mol. Cell. Biol.* **17**, 6663–6672.

3103. Ng, M., Diaz-Benjumea, F.J., and Cohen, S.M. (1995). *nubbin* encodes a POU-domain protein required for proximal-distal patterning in the *Drosophila* wing. *Development* **121**, 589–599.

3104. Ng, M., Diaz-Benjumea, F.J., Vincent, J.P., Wu, J., and Cohen, S.M. (1996). Specification of the wing by localized expression of *wingless* protein. *Nature* **381**, 316–318.

3105. Ng, M. and Yanofsky, M.F. (2001). Function and evolution of the plant MADS-box gene family. *Nature Rev. Genet.* **2**, 186–195.

3106. Nguyen, D.N.T., Liu, Y., Litsky, M.L., and Reinke, R. (1997). The *sidekick* gene, a member of the immunoglobulin superfamily, is required for pattern formation in the *Drosophila* eye. *Development* **124**, 3303–3312.

3107. Nguyen, D.N.T., Rohrbaugh, M., and Lai, Z.-C. (2000). The *Drosophila* homolog of Onecut homeodomain proteins is a neural-specific transcriptional activator with a potential role in regulating neural differentiation. *Mechs. Dev.* **97**, 57–72.

3108. Nguyen, J.T., Turck, C.W., Cohen, F.E., Zuckermann, R.N., and Lim, W.A. (1998). Exploiting the basis of proline recognition by SH3 and WW domains: design of N-substituted inhibitors. *Science* **282**, 2088–2092.

3109. Nguyen, M., Park, S., Marqués, G., and Arora, K. (1998). Interpretation of a BMP activity gradient in *Drosophila* embryos depends on synergistic signaling by two type I receptors, SAX and TKV. *Cell* **95**, 495–506.

3110. Nguyen, M., Parker, L., and Arora, K. (2000). Identification of *maverick*, a novel member of the TGF-β superfamily in *Drosophila. Mechs. Dev.* **95**, 201–206.

3111. Nguyen, T., Jamal, J., Shimell, M.J., Arora, K., and O'Connor, M.B. (1994). Characterization of *tolloid-related-1*: a BMP-1-like product that is required during larval and pupal stages of *Drosophila* development. *Dev. Biol.* **166**, 569–586.

3112. Nibu, Y., Zhang, H., Bajor, E., Barolo, S., Small, S., and Levine, M. (1998). dCtBP mediates transcriptional repression by Knirps, Krüppel and Snail in the *Drosophila* embryo. *EMBO J.* **17**, 7009–7020. [See also 2001 *EMBO J.* **20**, 2246–2253.]

3113. Nibu, Y., Zhang, H., and Levine, M. (1998). Interaction of short-range repressors with *Drosophila* CtBP in the embryo. *Science* **280**, 101–104. [See also 2001 *BioEssays* **23**, 683–690.]

3114. Nicholls, R.E. and Gelbart, W.M. (1998). Identification of chromosomal regions involved in *decapentaplegic* function in *Drosophila. Genetics* **149**, 203–215.

3115. Niehrs, C. and Pollet, N. (1999). Synexpression groups in eukaryotes. *Nature* **402**, 483–487.

3116. Niessing, D., Dostatni, N., Jäckle, H., and Rivera-Pomar, R. (1999). Sequence interval within the PEST motif of Bicoid is important for translational repression of *caudal* mRNA in the anterior region of the *Drosophila* embryo. *EMBO J.* **18**, 1966–1973.

3117. Nigg, E.A. (1997). Nucleocytoplasmic transport: signals, mechanisms and regulation. *Nature* **386**, 779–787.

3118. Niimi, T., Seimiya, M., Kloter, U., Flister, S., and Gehring, W.J. (1999). Direct regulatory interaction of the *eyeless* protein with an eye-specific enhancer in the *sine oculis* gene during eye induction in *Drosophila. Development* **126**, 2253–2260.

3119. Nijhout, H.F. (1991). "The Development and Evolution of Butterfly Wing Patterns." Smithsonian Pr., Washington.

3120. Nilson, L.A. and Schüpbach, T. (1999). EGF receptor signaling in *Drosophila* oogenesis. *Curr. Topics Dev. Biol.* **44**, 203–243.

3121. Nilsson, D.-E. (1996). Eye ancestry: old genes for new eyes. *Curr. Biol.* **6**, 39–42.

3122. Niwa, N., Inoue, Y., Nozawa, A., Saito, M., Misumi, Y., Ohuchi, H., Yoshioka, H., and Noji, S. (2000). Correlation of diversity of leg morphology in *Gryllus bimaculatus* (cricket) with divergence in *dpp* expression pattern during leg development. *Development* **127**, 4373–4381.

3123. Nolan, G.P. and Baltimore, D. (1992). The inhibitory ankyrin and activator Rel proteins. *Curr. Op. Gen. Dev.* **2**, 211–220.

3124. Noll, E., Medina, M., Hartley, D., Zhou, J., Perrimon, N., and Kosik, K.S. (2000). Presenilin affects Arm/β-catenin localization and function in *Drosophila. Dev. Biol.* **227**, 450–464.

3125. Noll, M. (1993). Evolution and role of *Pax* genes. *Curr. Opin. Gen. Dev.* **3**, 595–605.

3126. Noll, R., Sturtevant, M.A., Gollapudi, R.R., and Bier, E. (1994). New functions of the *Drosophila rhomboid* gene during embryonic and adult development are revealed by a novel genetic method, enhancer piracy. *Development* **120**, 2329–2338.

3127. Nolo, R., Abbott, L.A., and Bellen, H.J. (2000). Senseless, a Zn finger transcription factor, is necessary and sufficient for sensory organ development in *Drosophila. Cell* **102**, 349–362.

3128. Nolo, R., Abbott, L.A., and Bellen, H.J. (2001). *Drosophila Lyra* mutations are gain-of-function mutations of *senseless. Genetics* **157**, 307–315.

3129. Noordermeer, J., Klingensmith, J., and Nusse, R. (1995). Differential requirements for segment polarity genes in *wingless* signaling. *Mechs. Dev.* **51**, 145–155.

3130. Noordermeer, J., Klingensmith, J., Perrimon, N., and Nusse, R. (1994). *dishevelled* and *armadillo* act in the Wingless signalling pathway in *Drosophila. Nature* **367**, 80–83.

3131. Noramly, S. and Morgan, B.A. (1998). BMPs mediate lateral inhibition at successive stages in feather tract development. *Development* **125**, 3775–3787.

3132. Norbeck, B.A. and Denburg, J.L. (1991). Pattern formation during insect leg segmentation: studies with a prepattern of a cell surface antigen. *Roux's Arch. Dev. Biol.* **199**, 476–491.

3133. Nordenskiöld, E. (1936). "The History of Biology." Tudor, New York.

3134. Nordstrom, W., Chen, P., Steller, H., and Abrams, J.M. (1996). Activation of the *reaper* gene during ectopic cell killing in *Drosophila. Dev. Biol.* **180**, 213–226.

3135. North, G. (1986). Descartes and the fruitfly. *Nature* **322**, 404–405.

3136. Norton, P.A., Hynes, R.O., and Rees, D.J.G. (1990). *sevenless*: seven found? *Cell* **61**, 15–16.

3137. Nosseli, S. and Agnès, F. (1999). Roles of the JNK signaling pathway in *Drosophila* morphogenesis. *Curr. Opin. Gen. Dev.* **9**, 466–472.

3138. Nöthiger, R. (1964). Differenzierungsleistungen in Kombinaten, hergestellt aus Imaginalscheiben verschiedener Arten, Geschlechter und Körpersegmente von *Drosophila. Roux' Arch. Entw.-Mech.* **155**, 269–301.

3139. Nöthiger, R. (1972). The larval development of imaginal disks. *In* "The Biology of Imaginal Disks," *Results and Problems in Cell Differentiation, Vol. 5*, (H. Ursprung and R. Nöthiger, eds.). Springer-Verlag: Berlin, pp. 1–34.

3140. Nöthiger, R. and Schubiger, G. (1966). Developmental behaviour of fragments of symmetrical and asymmetrical imaginal discs of *Drosophila melanogaster* (Diptera). *J. Embryol. Exp. Morph.* **16**, 355–368.

3141. Nottebohm, E., Dambly-Chaudière, C., and Ghysen, A. (1992). Connectivity of chemosensory neurons is controlled by the gene *poxn* in *Drosophila. Nature* **359**, 829–832.

3142. Nottebohm, E., Usui, A., Therianos, S., Kimura, K.-i., Dambly-Chaudière, C., and Ghysen, A. (1994). The gene *poxn* controls different steps of the formation of chemosensory organs in *Drosophila. Neuron* **12**, 25–34.

3143. Noujdin, N.I. (1936). Genetic analysis of certain problems of the physiology of development of *Drosophila melanogaster. Biol. Zh. (Russia)* **5**, 571–624.

3144. Nowak, M.A., Boerlijst, M.C., Cooke, J., and Maynard Smith, J. (1997). Evolution of genetic redundancy. *Nature* **388**, 167–171.

3145. Nussbaumer, U., Halder, G., Groppe, J., Affolter, M., and Montagne, J. (2000). Expression of the *blistered/DSRF* gene is controlled by different morphogens during *Drosophila* trachea and wing development. *Mechs. Dev.* **96**, 27–36.

3146. Nusse, R. (1997). A versatile transcriptional effector of Wingless signaling. *Cell* **89**, 321–323.

3147. Nusse, R. (1999). Wnt targets: repression and activation. *Trends Genet.* **15**, 1–3.

3148. Nusse, R., Brown, A., Papkoff, J., Scambler, P., Shackleford, G., McMahon, A., Moon, R., and Varums, H. (1991). A new nomenclature for *int*-1 and related genes: the *Wnt* gene family. *Cell* **64**, 231.

3149. Nusse, R., Samos, C.H., Brink, M., Willert, K., Cadigan, K.M., Wodarz, A., Fish, M., and Rulifson, E. (1997). Cell culture and whole animal approaches to understanding signaling by Wnt proteins in *Drosophila. Cold Spr. Harb. Symp. Quant. Biol.* **62**, 185–190.

3150. Nusse, R. and Varmus, H.E. (1992). *Wnt* genes. *Cell* **69**, 1073–1087.

3151. Nüsslein-Volhard, C. and Wieschaus, E. (1980). Mutations affecting segment number and polarity in *Drosophila. Nature* **287**, 795–801.

3152. Nüsslein-Volhard, C., Wieschaus, E., and Kluding, H. (1984). Mutations affecting the pattern of the larval cuticle in *Drosophila melanogaster*. I. Zygotic loci on the second chromosome. *Roux's Arch. Dev. Biol.* **193**, 267–282.

3153. Nye, J.S. (1997). Developmental signaling: Notch signals Kuz it's cleaved. *Curr. Biol.* **7**, R716–R720.

3154. O'Brochta, D.A. and Bryant, P.J. (1985). A zone of non-proliferating cells at a lineage restriction boundary in *Drosophila. Nature* **313**, 138–141.

3155. O'Brochta, D.A. and Bryant, P.J. (1987). Distribution of S-phase cells during the regeneration of *Drosophila* imaginal wing discs. *Dev. Biol.* **119**, 137–142.

3156. O'Connor, M.B., Binari, R., Perkins, L.A., and Bender, W. (1988). Alternative RNA products from the *Ultrabithorax* domain of the bithorax complex. *EMBO J.* **7**, 435–445.

3157. O'Farrell, P.H., Desplan, C., DiNardo, S., Kassis, J.A., Kuner, J.M., Sher, E., Theis, J., and Wright, D. (1985). Embryonic pattern in *Drosophila*: the spatial distribution and sequence-specific DNA binding of engrailed protein. *Cold Spr. Harb. Symp. Quant. Biol.* **50**, 235–242.

3158. O'Keefe, D.D. and Thomas, J.B. (2001). *Drosophila* wing development in the absence of dorsal identity. *Development* **128**, 703–710.

3159. O'Keefe, D.D., Thor, S., and Thomas, J.B. (1998). Function and specificity of LIM domains in *Drosophila* nervous system and wing development. *Development* **125**, 3915–3923.

3160. O'Neil, K.T., Hoess, R.H., and DeGrado, W.F. (1990). Design of DNA-binding peptides based on the leucine zipper motif. *Science* **249**, 774–778.

3161. O'Neill, E.M., Ellis, M.C., Rubin, G.M., and Tjian, R. (1995). Functional domain analysis of glass, a zinc-finger-containing transcription factor in *Drosophila. Proc. Natl. Acad. Sci. USA* **92**, 6557–6561.

3162. O'Neill, E.M., Rebay, I., Tjian, R., and Rubin, G.M. (1994). The activities of two Ets-related transcription factors required for *Drosophila* eye development are modulated by the Ras/MAPK pathway. *Cell* **78**, 137–147.

3163. O'Shea, E.K., Klemm, J.D., Kim, P.S., and Alber, T. (1991). X-ray structure of the GCN4 leucine zipper, a two-stranded, parallel coiled coil. *Science* **254**, 539–544.

3164. O'Tousa, J.E., Baehr, W., Martin, R.L., Hirsh, J., Pak, W.L., and Applebury, M.L. (1985). The *Drosophila ninaE* gene encodes an opsin. *Cell* **40**, 839–850.

3165. Oberlander, H. (1985). The imaginal discs. *In* "Comprehensive Insect Physiology, Biochemistry, and Pharmacology," Vol. 2 (G.A. Kerkut and L.I. Gilbert, eds.). Pergamon Pr.: Oxford, pp. 151–182.

3166. Oda, H. and Tsukita, S. (1999). Dynamic features of adherens junctions during *Drosophila* embryonic epithelial morphogenesis revealed by a Dα-catenin-GFP fusion protein. *Dev. Genes Evol.* **209**, 218–225.

3167. Oda, H., Tsukita, S., and Takeichi, M. (1998). Dynamic behavior of the cadherin-based cell-cell adhesion system during *Drosophila* gastrulation. *Dev. Biol.* **203**, 435–450.

3168. Oda, H., Uemura, T., Harada, Y., Iwai, Y., and Takeichi, M. (1994). A *Drosophila* homolog of cadherin associated with Armadillo and essential for embryonic cell-cell adhesion. *Dev. Biol.* **165**, 716–726.

3169. Oda, H., Uemura, T., Shiomi, K., Nagafuchi, A., Tsukita, S., and Takeichi, M. (1993). Identification of a *Drosophila* homologue of α-catenin and its association with the *armadillo* protein. *J. Cell Biol.* **121**, 1133–1140.

3170. Oeda, E., Oka, Y., Miyazono, K., and Kawabata, M. (1998). Interaction of *Drosophila* inhibitors of apoptosis with

Thick veins, a type I serine/threonine kinase receptor for Decapentaplegic. *J. Biol. Chem.* **273**#16, 9353–9356.

3171. Oellers, N., Dehio, M., and Knust, E. (1994). bHLH proteins encoded by the *Enhancer of split* complex of *Drosophila* negatively interfere with transcriptional activation mediated by proneural genes. *Mol. Gen. Genet.* **244**, 465–473.

3172. Ogawa, E., Inuzuka, M., Maruyama, M., Satake, M., Naito-Fujimoto, M., Ito, Y., and Shigesada, K. (1993). Molecular cloning and characterization of PEBP2β, the heterodimeric partner of a novel *Drosophila* runt-related DNA binding protein PEBP2α. *Virology* **194**, 314–331.

3173. Ogawa, E., Maruyama, M., Kagoshima, H., Inuzuka, M., Lu, J., Satake, M., Shigesada, K., and Ito, Y. (1993). PEBP2/PEA2 represents a family of transcription factors homologous to the products of the *Drosophila* runt gene and the human *AML1* gene. *Proc. Natl. Acad. Sci. USA* **90**, 6859–6863.

3174. Ogryzko, V.V., Schiltz, R.L., Russanova, V., Howard, B.H., and Nakatani, Y. (1996). The transcriptional coactivators p300 and CBP are histone acetyltransferases. *Cell* **87**, 953–959.

3175. Ohkuma, Y., Horikoshi, M., Roeder, R.G., and Desplan, C. (1990). Binding site-dependent direct activation and repression of in vitro transcription by *Drosophila* homeodomain proteins. *Cell* **61**, 475–484.

3176. Ohlmeyer, J.T. and Kalderon, D. (1997). Dual pathways for induction of *wingless* expression by protein kinase A and Hedgehog in *Drosophila* embryos. *Genes Dev.* **11**, 2250–2258.

3177. Ohlmeyer, J.T. and Kalderon, D. (1998). Hedgehog stimulates maturation of Cubitus interruptus into a labile transcriptional activator. *Nature* **396**, 749–753.

3178. Ohno, S. (1970). "Evolution by Gene Duplication." Springer-Verlag, Berlin.

3179. Ohsako, S., Hyer, J., Panganiban, G., Oliver, I., and Caudy, M. (1994). hairy function as a DNA-binding helix-loop-helix repressor of *Drosophila* sensory organ formation. *Genes Dev.* **8**, 2743–2755.

3180. Ohtsuki, S., Levine, M., and Cai, H.N. (1998). Different core promoters possess distinct regulatory activities in the *Drosophila* embryo. *Genes Dev.* **12**, 547–556.

3181. Okabe, M. and Okano, H. (1997). Two-step induction of chordotonal organ precursors in *Drosophila* embryogenesis. *Development* **124**, 1045–1053.

3182. Okano, H. (1995). Two major mechanisms regulating cell-fate decisions in the developing nervous system. *Develop. Growth Differ.* **37**, 619–629.

3183. Okano, H., Hayashi, S., Tanimura, T., Sawamoto, K., Yoshikawa, S., Watanabe, J., Iwasaki, M., Hirose, S., Mikoshiba, K., and Montell, C. (1992). Regulation of *Drosophila* neural development by a putative secreted protein. *Differentiation* **52**, 1–11.

3184. Olayioye, M.A., Neve, R.M., Lane, H.A., and Hynes, N.E. (2000). The ErbB signaling network: receptor heterodimerization in development and cancer. *EMBO J.* **19**, 3159–3167.

3185. Olivier, J.P., Raabe, T., Henkemeyer, M., Dickson, B., Mbamalu, G., Margolis, B., Schlessinger, J., Hafen, E., and Pawson, T. (1993). A *Drosophila* SH2-SH3 adaptor protein implicated in coupling the Sevenless tyrosine kinase to an activator of Ras Guanine Nucleotide Exchange, Sos. *Cell* **73**, 179–191.

3186. Olson, E.N. (1990). MyoD family: a paradigm for development? *Genes Dev.* **4**, 1454–1461.

3187. Oppenheim, R.W. (1981). Neuronal cell death and some related regressive phenomena during neurogenesis: a selective historical review and progress report. *In* "Studies in Developmental Neurobiology. Essays in Honor of Viktor Hamburger," (W.M. Cowan, ed.). Oxford Univ. Pr.: New York, pp. 74–133.

3188. Oppenheim, R.W. (1989). The neurotrophic theory and naturally occurring motoneuron death. *Trends Neurosci.* **12**, 252–255.

3189. Oppenheim, R.W. (1991). Cell death during development of the nervous system. *Annu. Rev. Neurosci.* **14**, 453–501.

3190. Oppenheimer, J.M. (1955). Problems, concepts and their history. *In* "Analysis of Development," (B.H. Willier, P.A. Weiss, and V. Hamburger, eds.). W. B. Saunders: Philadephia, pp. 1–24.

3191. Orenic, T., Chidsey, J., and Holmgren, R. (1987). *Cell* and *cubitus interruptus Dominant*: two segment polarity genes on the fourth chromosome in *Drosophila*. *Dev. Biol.* **124**, 50–56.

3192. Orenic, T.V. and Carroll, S.B. (1992). The cell biology of pattern formation during *Drosophila* development. *Int. Rev. Cytol.* **139**, 121–155.

3193. Orenic, T.V., Held, L.I., Jr., Paddock, S.W., and Carroll, S.B. (1993). The spatial organization of epidermal structures: *hairy* establishes the geometrical pattern of *Drosophila* leg bristles by delimiting the domains of *achaete* expression. *Development* **118**, 9–20.

3194. Orenic, T.V., Slusarski, D.C., Kroll, K.L., and Holmgren, R.A. (1990). Cloning and characterization of the segment polarity gene *cubitus interruptus Dominant* of *Drosophila*. *Genes Dev.* **4**, 1053–1067.

3195. Orgogozo, V., Schweisguth, F., and Bellaïche, Y. (2001). Lineage, cell polarity and *inscuteable* function in the peripheral nervous system of the *Drosophila* embryo. *Development* **128**, 631–643.

3196. Orlando, V., Jane, E.P., Chinwalla, V., Harte, P.J., and Paro, R. (1998). Binding of Trithorax and Polycomb proteins to the bithorax complex: dynamic changes during early *Drosophila* embryogenesis. *EMBO J.* **17**, 5141–5150.

3197. Orlando, V. and Paro, R. (1995). Chromatin multiprotein complexes involved in the maintenance of transcription patterns. *Curr. Opin. Gen. Dev.* **5**, 174–179.

3198. Oro, A.E., McKeown, M., and Evans, R.M. (1992). The *Drosophila* retinoid X receptor homolog *ultraspiracle* functions in both female reproduction and eye morphogenesis. *Development* **115**, 449–462.

3199. Oro, A.E., Ong, E.S., Margolis, J.S., Posakony, J.W., McKeown, M., and Evans, R.M. (1988). The *Drosophila* gene *knirps-related* is a member of the steriod-receptor gene superfamily. *Nature* **336**, 493–496.

3200. Oro, A.E. and Scott, M.P. (1998). Splitting hairs: dissecting roles of signaling systems in epidermal development. *Cell* **95**, 575–578.

3201. Orr, H.T. (2001). Beyond the Qs in the polyglutamine diseases. *Genes Dev.* **15**, 925–932. [See also 2001 *Nature* **413**, 739–743.]

3202. Orsulic, S., Huber, O., Aberle, H., Arnold, S., and Kemler, R. (1999). E-cadherin binding prevents β-catenin nuclear localization and β-catenin/LEF-1-mediated transactivation. *J. Cell Sci.* **112**, 1237–1245.

3203. Orsulic, S. and Peifer, M. (1996). Cell-cell signalling: Wingless lands at last. *Curr. Biol.* **6**, 1363–1367.

3204. Orsulic, S. and Peifer, M. (1996). An in vivo structure-function study of Armadillo, the β-catenin homologue, reveals both separate and overlapping regions of the protein required for cell adhesion and for Wingless signaling. *J. Cell Biol.* **134**, 1283–1300.

3205. Oshiro, T., Yagami, T., Zhang, C., and Matsuzaki, F. (2000). Role of cortical tumour-suppressor proteins in asymmetric division of *Drosophila* neuroblast. *Nature* **408**, 593–596.

3206. Osten-Sacken, C.R. (1884). An essay of comparative chaetotaxy, or the arrangement of characteristic bristles of Diptera. *Trans. Entomol. Soc. Lond.* **1884** #4, 497–517.

3207. Oster, G. (1988). Lateral inhibition models of developmental processes. *Math. Biosci.* **90**, 265–286.

3208. Oster, G. and Weliky, M. (1991). Pattern and morphogenesis. *Sems. Dev. Biol.* **2**, 139–150.

3209. Ota, K., Sakaguchi, M., von Heijne, G., Hamasaki, N., and Mihara, K. (1998). Forced transmembrane orientation of hydrophilic polypeptide segments in multispanning membrane proteins. *Molec. Cell* **2**, 495–503.

3210. Ouweneel, W.J. (1969). Influence of environmental factors on the homoeotic effect of *loboid-ophthalmoptera* in *Drosophila melanogaster. W. Roux' Arch.* **164**, 15–36.

3211. Ouweneel, W.J. (1969). Morphology and development of *loboid-ophthalmoptera*, a homoeotic strain in *Drosophila melanogaster. W. Roux' Arch.* **164**, 1–14.

3212. Ouweneel, W.J. (1970). Developmental capacities of young and mature, wild-type and *opht* eye imaginal discs in *Drosophila melanogaster. W. Roux' Arch.* **166**, 76–88.

3213. Ouweneel, W.J. (1972). Determination, regulation, and positional information in insect development. *Acta Biotheor.* **21**, 115–131.

3214. Ouweneel, W.J. (1976). Developmental genetics of homoeosis. *Adv. Genet.* **18**, 179–248.

3215. Overduin, M., Rios, C.B., Mayer, B.J., Baltimore, D., and Cowburn, D. (1992). Three-dimensional solution structure of the src homology 2 domain of c-abl. *Cell* **70**, 697–704.

3216. Overton, J. (1967). The fine structure of developing bristles in wild type and mutant *Drosophila melanogaster. J. Morph.* **122**, 367–379.

3217. Owen, D.J. and Evans, P.R. (1998). A structural explanation for the recognition of tyrosine-based endocytotic signals. *Science* **282**, 1327–1332.

3218. Owen, D.J., Wigge, P., Vallis, Y., Moore, J.D.A., Evans, P.R., and McMahon, H.T. (1998). Crystal structure of the amphiphysin-2 SH3 domain and its role in the prevention of dynamin ring formation. *EMBO J.* **17**, 5273–5285.

3219. Pabo, C.O. and Sauer, R.T. (1992). Transcription factors: structural families and principles of DNA recognition. *Annu. Rev. Biochem.* **61**, 1053–1095.

3220. Packert, G. and Kuhn, D.T. (1998). The *tumorous-head-1* locus affects bristle number of the *Drosophila melanogaster* cuticle. *Genetics* **148**, 743–752.

3221. Padgett, R.W. (1999). Intracellular signaling: fleshing out the TGFβ pathway. *Curr. Biol.* **9**, R408–R411.

3222. Padgett, R.W., Das, P., and Huang, H. (2001). Characterizing the *Drosophila* activin pathway. *Proc. 42nd Ann. Drosophila Res. Conf.* **Abstracts Vol.**, a272.

3223. Padgett, R.W., Das, P., and Krishna, S. (1998). TGF-β signaling, Smads, and tumor suppressors. *BioEssays* **20**, 382–390.

3224. Padgett, R.W., St. Johnston, R.D., and Gelbart, W.M. (1987). A transcript from a *Drosophila* pattern gene predicts a protein homologous to the transforming growth factor-β family. *Nature* **325**, 81–84.

3225. Pagès, F. and Kerridge, S. (2000). Morphogen gradients: a question of time or concentration? *Trends Genet.* **16**, 40–44.

3226. Pai, C.-Y., Kuo, T.-S., Jaw, T.J., Kurant, E., Chen, C.-T., Bessarab, D.A., Salzberg, A., and Sun, Y.H. (1998). The Homothorax homeoprotein activates the nuclear localization of another homeoprotein, Extradenticle, and suppresses eye development in *Drosophila. Genes Dev.* **12**, 435–446.

3227. Pai, L.-M., Barcelo, G., and Schüpbach, T. (2000). *D-cbl*, a negative regulator of the Egfr pathway, is required for dorsoventral patterning in *Drosophila* oogenesis. *Cell* **103**, 51–61.

3228. Pai, L.-M., Orsulic, S., Bejsovec, A., and Peifer, M. (1997). Negative regulation of Armadillo, a Wingless effector in *Drosophila. Development* **124**, 2255–2266.

3229. Pak, W.L., Conrad, S.K., Kremer, N.E., Larrivee, D.C., Schinz, R.H., and Wong, F. (1980). Photoreceptor function. *In* "Development and Neurobiology of *Drosophila*," (O. Siddiqi, P. Babu, L.M. Hall, and J.C. Hall, eds.). Plenum: New York, pp. 331–346.

3230. Pak, W.L. and Grabowski, S.R. (1978). Physiology of the *visual* and *flight* systems. *In* "The Genetics and Biology of *Drosophila*," Vol. 2a (M. Ashburner and T.R.F. Wright, eds.). Acad. Pr.: New York, pp. 553–604.

3231. Pal-Bhadra, M., Bhadra, U., and Birchler, J.A. (1997). Cosuppression in *Drosophila*: gene silencing of *Alcohol dehydrogenase* by *white-Adh* transgenes is *Polycomb* dependent. *Cell* **90**, 479–490.

3232. Palaparti, A., Baratz, A., and Stifani, S. (1997). The Groucho/Transducin-like Enhancer of split transcriptional repressors interact with the genetically defined amino-terminal silencing domain of histone H3. *J. Biol. Chem.* **272** #42, 26604–26610.

3233. Palka, J. (1979). Mutants and mosaics: tools in insect developmental neurobiology. *In* "Aspects of Developmental Neurobiology," *Soc. Neurosci. Symp.*, Vol. 4 (J.A. Ferrendelli, ed.). Society for Neuroscience: Bethesda, MD, pp. 209–227.

3234. Palka, J., Schubiger, M., and Hart, H.S. (1981). The path of axons in *Drosophila* wings in relation to compartment boundaries. *Nature* **294**, 447–449.

3235. Palka, J., Schubiger, M., and Schwaninger, H. (1990). Neurogenic and antineurogenic effects from modifications at the *Notch* locus. *Development* **109**, 167–175.

3236. Palmeirim, I., Henrique, D., Ish-Horowicz, D., and Pourquié, O. (1997). Avian *hairy* gene expression identifies a molecular clock linked to vertebrate segmentation and somitogenesis. *Cell* **91**, 639–648.

3237. Palsson, A. and Gibson, G. (2000). Quantitative developmental genetic analysis reveals that the ancestral dipteran wing vein prepattern is conserved in *Drosophila melanogaster. Dev. Genes Evol.* **210**, 617–622.

3238. Pan, D. and Rubin, G.M. (1995). cAMP-dependent protein kinase and *hedgehog* act antagonistically in regulating *decapentaplegic* transcription in *Drosophila* imaginal discs. *Cell* **80**, 543–552.

3239. Pan, D. and Rubin, G.M. (1997). Kuzbanian controls proteolytic processing of Notch and mediates lateral inhibition during *Drosophila* and vertebrate neurogenesis. *Cell* **90**, 271–280.

3240. Pan, D. and Rubin, G.M. (1998). Targeted expression of *teashirt* induces ectopic eyes in *Drosophila. Proc. Natl. Acad. Sci. USA* **95**, 15508–15512.

3241. Pandur, P. and Kühl, M. (2001). An arrow for wingless to take-off. *BioEssays* **23**, 207–210.

3242. Panganiban, G. (2000). *Distal-less* function during *Drosophila* appendage and sense organ development. *Dev. Dynamics* **218**, 554–562.

3243. Panganiban, G., Nagy, L., and Carroll, S.B. (1994). The role of the *Distal-less* gene in the development and evolution of insect limbs. *Curr. Biol.* **4**, 671–675.

3244. Panganiban, G.E.F., Rashka, K.E., Neitzel, M.D., and Hoffman, F.M. (1990). Biochemical characterization of the *Drosophila dpp* protein, a member of the Transforming Growth Factor β family of growth factors. *Molec. Cell Biol.* **10**, 2669–2677. [See also 2001 *Development* **128**, 2209–2220.]

3245. Panin, V.M., Lei, L., Moloney, D.J., Shao, L., Haltiwanger, R.S., and Irvine, K.D. (2001). How does glycosylation by Fringe modulate Notch signaling? *Proc. 42nd Ann. Drosophila Res. Conf.* **Abstracts Vol.**, a153.

3246. Panin, V.M., Papayannopoulos, V., Wilson, R., and Irvine, K.D. (1997). Fringe modulates Notch-ligand interactions. *Nature* **387**, 908–912.

3247. Pankratz, M.J. and Jäckle, H. (1990). Making stripes in the *Drosophila* embryo. *Trends Genet.* **6**, 287–292.

3248. Pankratz, M.J. and Jäckle, H. (1993). Blastoderm segmentation. *In* "The Development of *Drosophila melanogaster*," Vol. 1 (M. Bate and A. Martinez Arias, eds.). Cold Spring Harbor Lab. Pr.: Plainview, N. Y., pp. 467–516.

3249. Pannuti, A. and Lucchesi, J.C. (2000). Recycling to remodel: evolution of dosage-compensation complexes. *Curr. Opin. Gen. Dev.* **10**, 644–650.

3250. Pantelouris, E.M. and Waddington, C.H. (1955). Regulation capacities of the wing- and haltere-discs of wild type and bithorax *Drosophila*. *Roux' Arch. Entw.-Mech.* **147**, 539–546.

3251. Paolella, D.N., Palmer, C.R., and Schepartz, A. (1994). DNA targets for certain bZIP proteins distinguished by an intrinsic bend. *Science* **264**, 1130–1133.

3252. Papageorgiou, S. (1989). Cartesian or polar co-ordinates in pattern formation? *J. Theor. Biol.* **141**, 281–283.

3253. Papageorgiou, S. (1996). Distalization in insects and amphibians. *BioEssays* **17**, 1089–1090.

3254. Papaioannou, V.E. (1997). T-box family reunion. *Trends Genet.* **13**, 212–213.

3255. Papaioannou, V.E. and Silver, L.M. (1998). The T-box gene family. *BioEssays* **20**, 9–19.

3256. Papatsenko, D., Nazina, A., and Desplan, C. (2001). A conserved regulatory element present in all *Drosophila rhodopsin* genes mediates Pax6 functions and participates in the fine-tuning of cell-specific expression. *Mechs. Dev.* **101**, 143–153.

3257. Papatsenko, D., Sheng, G., and Desplan, C. (1997). A new rhodopsin in R8 photoreceptors of *Drosophila*: evidence for coordinate expression with Rh3 in R7 cells. *Development* **124**, 1665–1673.

3258. Papayannopoulos, V., Tomlinson, A., Panin, V.M., Rauskolb, C., and Irvine, K.D. (1998). Dorsal-ventral signaling in the *Drosophila* eye. *Science* **281**, 2031–2034.

3259. Papert, S. (1980). "Mindstorms: Children, Computers, and Powerful Ideas." Basic Bks., New York.

3260. Papoulas, O., Beek, S.J., Moseley, S.L., McCallum, C.M., Sarte, M., Shearn, A., and Tamkun, J.W. (1998). The *Drosophila* trithorax group proteins BRM, ASH1 and ASH2 are subunits of distinct protein complexes. *Development* **125**, 3955–3966.

3261. Paricio, N., Feiguin, F., Boutros, M., Eaton, S., and Mlodzik, M. (1999). The *Drosophila* STE20-like kinase Misshapen is required downstream of the Frizzled receptor in planar polarity signaling. *EMBO J.* **18**, 4669–4678.

3262. Park, J.M., Gim, B.S., Kim, J.M., Yoon, J.H., Kim, H.-S., Kang, J.-G., and Kim, Y.-J. (2001). *Drosophila* Mediator complex is broadly utilized by diverse gene-specific transcription factors at different types of core promoters. *Molec. Cell. Biol.* **21**, 2312–2323.

3263. Park, M., Yaich, L.E., and Bodmer, R. (1998). Mesodermal cell fate decisions in *Drosophila* are under the control of the lineage genes *numb*, *Notch*, and *sanpodo*. *Mechs. Dev.* **75**, 117–126.

3264. Park, W.-J., Liu, J., and Adler, P.N. (1994). The *frizzled* gene of *Drosophila* encodes a membrane protein with an odd number of transmembrane domains. *Mechs. Dev.* **45**, 127–137.

3265. Parkhurst, S.M. (1998). Groucho: making its Marx as a transcriptional co-repressor. *Trends Genet.* **14**, 130–132.

3266. Parkhurst, S.M., Bopp, D., and Ish-Horowicz, D. (1990). X:A ratio, the primary sex-determining signal in *Drosophila*, is transduced by helix-loop-helix proteins. *Cell* **63**, 1179–1191.

3267. Parkhurst, S.M. and Ish-Horowicz, D. (1991). Misregulating segmentation gene expression in *Drosophila*. *Development* **111**, 1121–1135.

3268. Parkhurst, S.M. and Ish-Horowicz, D. (1992). Common denominators for sex. *Curr. Biol.* **2**, 629–631.

3269. Parkin, N.T., Kitajewski, J., and Varmus, H.E. (1993). Activity of Wnt-1 as a transmembrane protein. *Genes Dev.* **7**, 2181–2193.

3270. Parks, A.L., Huppert, S.S., and Muskavitch, M.A.T. (1997). The dynamics of neurogenic signalling underlying bristle development in *Drosophila melanogaster*. *Mechs. Dev.* **63**, 61–74.

3271. Parks, A.L., Klueg, K.M., Stout, J.R., and Muskavitch, M.A.T. (2000). Ligand endocytosis drives receptor dissociation and activation in the Notch pathway. *Development* **127**, 1373–1385.

3272. Parks, A.L. and Muskavitch, M.A.T. (1993). *Delta* function is required for bristle organ determination and morphogenesis in *Drosophila*. *Dev. Biol.* **157**, 484–496.

3273. Parks, A.L., Turner, R., and Muskavitch, M.A.T. (1995). Relationships between complex Delta expression and the specification of retinal cell fates during *Drosophila* eye development. *Mechs. Dev.* **50**, 201–216.

3274. Parks, H.B. (1936). Cleavage patterns in *Drosophila* and mosaic formation. *Ann. Ent. Soc. Am.* **29**, 350–392.

3275. Paro, R., Strutt, H., and Cavalli, G. (1998). Heritable chromatin states induced by the Polycomb and trithorax group genes. *In* "Epigenetics," *Novartis Foundation Symposium*, Vol. 214 (D.J. Chadwick and G. Cardew, eds.). Wiley & Sons: New York, pp. 51–66. [See also 2001 *BioEssays* **23**, 561–562.]

3276. Paro, R. and Zink, B. (1992). The *Polycomb* gene is differentially regulated during oogenesis and embryogenesis of *Drosophila melanogaster*. *Mechs. Dev.* **40**, 37–46.

3277. Parody, T.R. and Muskavitch, M.A.T. (1993). The pleiotropic function of *Delta* during postembryonic development of *Drosophila melanogaster*. *Genetics* **135**, 527–539.

3278. Paroush, Z., Finley, R.L., Jr., Kidd, T., Wainwright, S.M., Ingham, P.W., Brent, R., and Ish-Horowicz, D. (1994). Groucho is required for *Drosophila* neurogenesis, segmentation, and sex determination and interacts directly with Hairy-related bHLH proteins. *Cell* **79**, 805–815.

3279. Parvin, J.D. and Young, R.A. (1998). Regulatory targets in the RNA polymerase II holoenzyme. *Curr. Opin. Gen. Dev.* **8**, 565–570.

3280. Pascal, S.M., Singer, A.U., Gish, G., Yamazaki, T., Shoelson, S.E., Pawson, T., Kay, L.E., and Forman-Kay, J.D. (1994). Nuclear magnetic resonance structure of an SH2 domain of phospholipase C-γ1 complexed with a high affinity binding peptide. *Cell* **77**, 461–472.

3281. Passner, J.M., Ryoo, H.D., Shen, L., Mann, R.S., and Aggarwal, A.K. (1999). Structure of a DNA-bound Ultrabithorax-Extradenticle homeodomain complex. *Nature* **397**, 714–719.

3282. Patel, K., Makarenkova, H., and Jung, H.-S. (1999). The role of long range, local and direct signalling molecules during chick feather bud development involving the BMPs, Follistatin and the Eph receptor tyrosine kinase Eph-A4. *Mechs. Dev.* **86**, 51–62.

3283. Paterson, J. and O'Hare, K. (1991). Structure and transcription of the *singed* locus of *Drosophila melanogaster*. *Genetics* **129**, 1073–1084.

3284. Patikoglou, G. and Burley, S.K. (1997). Eukaryotic transcription factor-DNA complexes. *Annu. Rev. Biophys. Biomol. Struct.* **26**, 289–325. [See also 2001 *Molec. Cell* **8**, 937–946.]

3285. Pattatucci, A.M. and Kaufman, T.C. (1991). The homeotic gene *Sex combs reduced* of *Drosophila melanogaster* is differentially regulated in the embryonic and imaginal stages of development. *Genetics* **129**, 443–461.

3286. Pattatucci, A.M., Otteson, D.C., and Kaufman, T.C. (1991). A functional and structural analysis of the *Sex combs reduced* locus of *Drosophila melanogaster*. *Genetics* **129**, 423–441.

3287. Pattee, H.H. (1969). How does a molecule become a message? *In* "Communication in Development," *Symp. Soc. Dev. Biol.*, Vol. 28 (A. Lang, ed.). Acad. Pr.: New York, pp. 1–16.

3288. Pattee, H.H. (1972). The nature of hierarchical controls in living matter. *In* "Foundations of Mathematical Biology," Vol. 1 (R. Rosen, ed.). Acad. Pr.: New York, pp. 1–22.

3289. Patthy, L. (1991). Modular exchange principles in proteins. *Curr. Opinion Struct. Biol.* **1**, 351–361.

3290. Paulus, H.F. (1979). Eye structure and the monophyly of the Arthropoda. *In* "Arthropod Phylogeny," (A.P. Gupta, ed.). Van Nostrand Reinhold: New York, pp. 299–383.

3291. Paumard-Rigal, S., Zider, A., Vaudin, P., and Silber, J. (1998). Specific interactions between *vestigial* and *scalloped* are required to promote wing tissue proliferation in *Drosophila melanogaster*. *Dev. Genes Evol.* **208**, 440–446.

3292. Pavletich, N.P. and Pabo, C.O. (1993). Crystal structure of a five-finger GLI-DNA complex: new perspectives on zinc fingers. *Science* **261**, 1701–1707.

3293. Pawley, J. (2000). The 39 steps: a cautionary tale of quantitative 3-D fluorescence microscopy. *BioTechniques* **28**, 884–888.

3294. Pawson, T. (1993). Signal transduction–a conserved pathway from the membrane to the nucleus. *Dev. Genet.* **14**, 333–338.

3295. Pawson, T. (1995). Protein modules and signalling networks. *Nature* **373**, 573–580.

3296. Pawson, T. and Gish, G.D. (1992). SH2 and SH3 domains: from structure to function. *Cell* **71**, 359–362.

3297. Pawson, T. and Nash, P. (2000). Protein-protein interactions define specificity in signal transduction. *Genes Dev.* **14**, 1027–1047.

3298. Pawson, T. and Schlessinger, J. (1993). SH2 and SH3 domains. *Curr. Biol.* **3**, 434–442.

3299. Pawson, T. and Scott, J.D. (1997). Signaling through scaffold, anchoring, and adaptor proteins. *Science* **278**, 2075–2080.

3300. Pazman, C., Mayes, C.A., Fanto, M., Haynes, S.R., and Mlodzik, M. (2000). Rasputin, the *Drosophila* homologue of the RasGAP SH3 binding protein, functions in Ras- and Rho-mediated signaling. *Development* **127**, 1715–1725.

3301. Pearson, M.J. (1974). Polyteny and the functional significance of the polytene cell cycle. *J. Cell Sci.* **15**, 457–479.

3302. Pederson, T. (2001). Protein mobility within the nucleus– what are the right moves? *Cell* **104**, 635–638.

3303. Peel, D.J., Johnson, S.A., and Milner, M.J. (1990). The ultrastructure of imaginal disc cells in primary cultures and during cell aggregation in continuous cell lines. *Tissue & Cell* **22**, 749–758.

3304. Peifer, M. (1993). Cancer, catenins, and cuticle pattern: a complex connection. *Science* **262**, 1667–1668.

3305. Peifer, M. (1994). The two faces of Hedgehog. *Science* **266**, 1492–1493.

3306. Peifer, M. (1995). Cell adhesion and signal transduction: the Armadillo connection. *Trends Cell Biol.* **5**, 224–229.

3307. Peifer, M. (1996). Regulating cell proliferation: as easy as APC. *Science* **272**, 974–975.

3308. Peifer, M. (1998). Birds of a feather flock together. *Nature* **395**, 324–325.

3309. Peifer, M. (1999). Neither straight nor narrow. *Nature* **400**, 213–215.

3310. Peifer, M. and Bejsovec, A. (1992). Knowing your neighbors: cell interactions determine intrasegmental patterning in *Drosophila*. *Trends Genet.* **8**, 243–249.

3311. Peifer, M. and Bender, W. (1986). The anterobithorax and bithorax mutations of the bithorax complex. *EMBO J.* **5**, 2293–2303.

3312. Peifer, M., Berg, S., and Reynolds, A.B. (1994). A repeating amino acid motif shared by proteins with diverse cellular roles. *Cell* **76**, 789–791.

3313. Peifer, M., Karch, F., and Bender, W. (1987). The bithorax complex: control of segmental identity. *Genes Dev.* **1**, 891–898.

3314. Peifer, M., Orsulic, S., Sweeton, D., and Wieschaus, E. (1993). A role for the *Drosophila* segment polarity gene *armadillo* in cell adhesion and cytoskeletal integrity during oogenesis. *Development* **118**, 1191–1207.

3315. Peifer, M., Pai, L.-M., and Casey, M. (1994). Phosphorylation of the *Drosophila* adherens junction protein armadillo: roles for wingless signal and zeste-white 3 kinase. *Dev. Biol.* **166**, 543–556.

3316. Peifer, M. and Polakis, P. (2000). Wnt signaling in oncogenesis and embryogenesis–a look outside the nucleus. *Science* **287**, 1606–1609.

3317. Peifer, M., Rauskolb, C., Williams, M., Riggleman, B., and Wieschaus, E. (1991). The segment polarity gene *armadillo* interacts with the *wingless* signalling pathway in both embryonic and adult pattern formation. *Development* **111**, 1029–1043.

3318. Peifer, M., Sweeton, D., Casey, M., and Wieschaus, E. (1994). *wingless* signal and Zeste-white 3 kinase trigger opposing changes in the intracellular distribution of Armadillo. *Development* **120**, 369–380.

3319. Peifer, M. and Wieschaus, E. (1990). Mutations in the *Drosophila* gene *extradenticle* affect the way specific homeo domain proteins regulate segmental identity. *Genes Dev.* **4**, 1209–1223.

3320. Peifer, M. and Wieschaus, E. (1990). The segment polarity gene *armadillo* encodes a functionally modular protein that is the *Drosophila* homolog of human plakoglobin. *Cell* **63**, 1167–1178.

3321. Pelaz, S., Urquía, N., and Morata, G. (1993). Normal and ectopic domains of the homeotic gene *Sex combs reduced* of *Drosophila*. *Development* **117**, 917–923.

3322. Pellegrini, L., Burke, D.F., von Delft, F., Mulloy, B., and Blundell, T.L. (2000). Crystal structure of fibroblast growth factor receptor ectodomain bound to ligand and heparin. *Nature* **407**, 1029–1034.

3323. Pellegrini, L., Tan, S., and Richmond, T.J. (1995). Structure of serum response factor core bound to DNA. *Nature* **376**, 490–498.

3324. Peltenburg, L.T.C. and Murre, C. (1996). Engrailed and Hox homeodomain proteins contain a related Pbx interaction motif that recognizes a common structure present in Pbx. *EMBO J.* **15**, 3385–3393.

3325. Peltenburg, L.T.C. and Murre, C. (1997). Specific residues in the Pbx homeodomain differentially modulate the DNA-binding activity of Hox and Engrailed proteins. *Development* **124**, 1089–1098.

3326. Peng, C.-Y., Manning, L., Albertson, R., and Doe, C.Q. (2000). The tumour-suppressor genes *lgl* and *dlg* regulate basal protein targeting in *Drosophila* neuroblasts. *Nature* **408**, 596–600.

3327. Pennisi, E. (1997). Multiple clocks keep time in fruit fly tissues. *Science* **278**, 1560–1561.

3328. Penton, A., Chen, Y., Staehling-Hampton, K., Wrana, J.L., Attisano, L., Szidonya, J., Cassill, J.A., Massagué, J., and Hoffmann, F.M. (1994). Identification of two bone morphogenetic protein type 1 receptors in Drosophila and evidence that Brk25D is a *decapentaplegic* receptor. *Cell* **78**, 239–250.

3329. Penton, A. and Hoffmann, M. (1996). *Decapentaplegic* restricts the domain of *wingless* during *Drosophila* limb patterning. *Nature* **382**, 162–165.

3330. Penton, A., Selleck, S.B., and Hoffmann, F.M. (1997). Regulation of cell cycle synchronization by *decapentaplegic* during *Drosophila* eye development. *Science* **275**, 203–206.

3331. Pepling, M.E. and Gergen, J.P. (1995). Conservation and function of the transcriptional regulatory protein Runt. *Proc. Natl. Acad. Sci. USA* **92**, 9087–9091.

3332. Percival-Smith, A. and Hayden, D.J. (1998). Analysis in *Drosophila melanogaster* of the interaction between Sex combs reduced and Extradenticle activity in the determination of tarsus and arista identity. *Genetics* **150**, 189–198.

3333. Percival-Smith, A., Weber, J., Gilfoyle, E., and Wilson, P. (1997). Genetic characterization of the role of the two HOX proteins, Proboscipedia and Sex Combs Reduced, in determination of adult antennal, tarsal, maxillary palp and proboscis identities in *Drosophila melanogaster*. *Development* **124**, 5049–5062.

3334. Perissi, V., Dasen, J.S., Kurokawa, R., Wang, Z., Korzus, E., Rose, D.W., Glass, C.K., and Rosenfeld, M.G. (1999). Factor-specific modulation of CREB-binding protein acetyltransferase activity. *Proc. Natl. Acad. Sci. USA* **96**, 3652–3657.

3335. Periz, G. and Fortini, M.E. (1999). Ca^{2+}-ATPase function is required for intracellular trafficking of the Notch receptor in *Drosophila*. *EMBO J.* **18**, 5983–5993.

3336. Perkins, L.A., Johnson, M.R., Melnick, M.B., and Perrimon, N. (1996). The nonreceptor protein tyrosine phosphatase Corkscrew functions in multiple receptor tyrosine kinase pathways in *Drosophila*. *Dev. Biol.* **180**, 63–81.

3337. Perkins, L.A., Larsen, I., and Perrimon, N. (1992). *corkscrew* encodes a putative protein tyrosine phosphatase that functions to transduce the terminal signal from the receptor tyrosine kinase torso. *Cell* **70**, 225–236.

3338. Perler, F.B. (1998). Protein splicing of inteins and Hedgehog autoproteolysis: structure, function, and evolution. *Cell* **92**, 1–4.

3339. Perlman, D. and Halvorson, H.O. (1983). A putative signal peptidase recognition site and sequence in eukaryotic and prokaryotic signal peptides. *J. Mol. Biol.* **167**, 391–409.

3340. Perrimon, N. (1993). The torso receptor protein-tyrosine kinase signaling pathway: an endless story. *Cell* **74**, 219–222.

3341. Perrimon, N. (1994). The genetic basis of patterned baldness in *Drosophila*. *Cell* **76**, 781–784.

3342. Perrimon, N. (1994). Signalling pathways initiated by receptor protein tyrosine kinases in *Drosophila*. *Curr. Opin. Cell Biol.* **6**, 260–266.

3343. Perrimon, N. (1995). Hedgehog and beyond. *Cell* **80**, 517–520.

3344. Perrimon, N. (1996). Serpentine proteins slither into the Wingless and Hedgehog fields. *Cell* **86**, 513–516.

3345. Perrimon, N. and Bernfield, M. (2000). Specificities of heparan sulphate proteoglycans in developmental processes. *Nature* **404**, 725–728.

3346. Perrimon, N. and Duffy, J.B. (1998). Sending all the right signals. *Nature* **396**, 18–19.

3347. Perrimon, N. and McMahon, A.P. (1999). Negative feedback mechanisms and their roles during pattern formation. *Cell* **97**, 13–16.

3348. Perrimon, N. and Perkins, L.A. (1997). There must be 50 ways to rule the signal: the case of the *Drosophila* EGF receptor. *Cell* **89**, 13–16.

3349. Perrimon, N. and Smouse, D. (1989). Multiple functions of a *Drosophila* homeotic gene, *zeste-white 3*, during segmentation and neurogenesis. *Dev. Biol.* **135**, 287–305.

3350. Perry, M.M. (1968). Further studies on the development of the eye of *Drosophila melanogaster*. I. The ommatidia. *J. Morph.* **124**, 227–248.

3351. Perry, M.M. (1968). Further studies on the development of the eye of *Drosophila melanogaster*. II. The interommatidial bristles. *J. Morph.* **124**, 249–262.

3352. Peruski, L.F., Jr., Wadzinski, B.E., and Johnson, G.L. (1993). Analysis of the multiplicity, structure, and function of protein serine/threonine phosphatases. *Adv. Prot. Phosphatases* **7**, 9–30.

3353. Perutz, M.F., Johnson, T., Suzuki, M., and Finch, J.T. (1994). Glutamine repeats as polar zippers: Their possible role in inherited neurodegenerative diseases. *Proc. Natl. Acad. Sci. USA* **91**, 5355–5358.

3354. Perutz, M.F., Staden, R., Moens, L., and De Baere, I. (1993). Polar zippers. *Curr. Biol.* **3**, 249–253.

3355. Petcherski, A.G. and Kimble, J. (2000). Mastermind is a putative activator for Notch. *Curr. Biol.* **10**, R471–R473.

3356. Petersen, N.S., Lankenau, D.-H., Mitchell, H.K., Young, P., and Corces, V.G. (1994). *forked* proteins are components of fiber bundles present in developing bristles of *Drosophila melanogaster*. *Genetics* **136**, 173–182.

3357. Pettigrew, J.B. (1908). "Design in Nature." Vol. 1. Longmans, Green, & Co., London.

3358. Pettmann, B. and Henderson, C.E. (1998). Neuronal cell death. *Neuron* **20**, 633–647.

3359. Peverali, F.A., Isaksson, A., Papavassiliou, A.G., Plastina, P., Staszewski, L.M., Mlodzik, M., and Bohmann, D. (1996). Phosphorylation of *Drosophila* Jun by the MAP kinase Rolled regulates photoreceptor differentiation. *EMBO J.* **15**, 3943–3950.

3360. Pevny, L.H. and Lovell-Badge, R. (1997). *Sox* genes find their feet. *Curr. Opin. Genet. Dev.* **7**, 338–344.

3361. Peyer, B. (1947). An early description of *Drosophila*. *J. Heredity* **38**, 194–199.

3362. Peyer, B. and Hadorn, E. (1965). Zum Manifestationsmuster der Mutante "*multiple wing hairs*" (*mwh*) von *Drosophila melanogaster*. *Arch. Klaus-Stift. Vererb.-Forsch.* **40**, 19–26.

3363. Pfeiffer, S., Alexandre, C., Calleja, M., and Vincent, J.-P. (2000). The progeny of *wingless*-expressing cells deliver the signal at a distance in *Drosophila* embryos. *Curr. Biol.* **10**, 321–324.

3364. Pfeiffer, S. and Vincent, J.-P. (1999). Signalling at a distance: Transport of Wingless in the embryonic epidermis of *Drosophila*. *Sems. Cell Dev. Biol.* **10**, 303–309.

3365. Pflugfelder, G.O., Roth, H., and Poeck, B. (1992). A homology domain shared between *Drosophila optomotor-blind* and mouse *Brachyury* is involved in DNA binding. *Biochem. Biophys. Res. Comm.* **186**, 918–925.

3366. Pflugfelder, G.O., Roth, H., Poeck, B., Kerscher, S., Schwarz, H., Jonschker, B., and Heisenberg, M. (1992). The lethal(1)optomotor-blind gene of *Drosophila melanogaster* is a major organizer of optic lobe development: isolation and characterization of the gene. *Proc. Natl. Acad. Sci. USA* **89**, 1199–1203.

3367. Pflugfelder, G.O., Schwarz, H., Roth, H., Poeck, B., Sigl, A., Kerscher, S., Jonschker, B., Pak, W.L., and Heisenberg, M. (1990). Genetic and molecular characterization of the *optomotor-blind* gene locus in *Drosophila melanogaster*. *Genetics* **126**, 91–104.

3368. Pham, A., Therond, P., Alves, G., Tournier, F.B., Busson, D., Lamour-Isnard, C., Bouchon, B.L., Préat, T., and Tricoire, H. (1995). The *Suppressor of fused* gene encodes a novel PEST protein involved in *Drosophila* segment polarity establishment. *Genetics* **140**, 587–598.

3369. Phillips, J.P. and Forrest, H.S. (1980). Ommochromes and pteridines. *In* "The Genetics and Biology of *Drosophila*," Vol. 2d (M. Ashburner and T.R.F. Wright, eds.). Acad. Pr.: New York, pp. 541–623.

3370. Phillips, M.D. and Shearn, A. (1990). Mutations in *polycombeotic*, a *Drosophila* Polycomb-group gene, cause a wide range of maternal and zygotic phenotypes. *Genetics* **125**, 91–101.

3371. Phillips, R.G. (2001). A novel gene, *skinhead*, regulates *vestigial* expression and may be involved in Wingless signal transduction. *Proc. 42nd Ann. Drosophila Res. Conf.* **Abstracts Vol.**, a200.

3372. Phillips, R.G., Roberts, I.J.H., Ingham, P.W., and Whittle, J.R.S. (1990). The *Drosophila* segment polarity gene *patched* is involved in a position-signalling mechanism in imaginal discs. *Development* **110**, 105–114.

3373. Phillips, R.G., Warner, N.L., and Whittle, J.R.S. (1999). Wingless signaling leads to an asymmetric response to Decapentaplegic-dependent signaling during sense organ patterning on the notum of *Drosophila melanogaster*. *Dev. Biol.* **207**, 150–162.

3374. Phillips, R.G. and Whittle, J.R.S. (1993). *wingless* expression mediates determination of peripheral nervous system elements in late stages of *Drosophila* wing disc development. *Development* **118**, 427–438.

3375. Phillips, S.E.V. (1994). Built by association: structure and function of helix-loop-helix DNA-binding proteins. *Curr. Biol.* **2**, 1–4.

3376. Phillis, R.W., Bramlage, A.T., Wotus, C., Whittaker, A., Gramates, L.S., Seppala, D., Farahanchi, F., Caruccio, P., and Murphey, R.K. (1993). Isolation of mutations affecting neural circuitry required for grooming behavior in *Drosophila melanogaster*. *Genetics* **133**, 581–592.

3377. Phippen, T.M., Sweigart, A.L., Moniwa, M., Krumm, A., Davie, J.R., and Parkhurst, S.M. (2000). *Drosophila* C-terminal binding protein functions as a context-dependent transcriptional co-factor and interferes with both Mad and Groucho transcriptional repression. *J. Biol. Chem.* **275** #48, 37628–37637.

3378. Pi, H., Wu, H.-J., and Chien, C.-T. (2001). A dual function of *phyllopod* in *Drosophila* external sensory organ development: cell fate specification of sensory organ precursor and its progeny. *Development* **128**, 2699–2710.

3379. Piatigorsky, J. and Wistow, G.J. (1989). Enzyme/crystallins: gene sharing as an evolutionary strategy. *Cell* **57**, 197–199.

3380. Pichaud, F. and Casares, F. (2000). *homothorax* and *iroquois-C* genes are required for the establishment of territories within the developing eye disc. *Mechs. Dev.* **96**, 15–25.

3381. Pichaud, F. and Desplan, C. (2001). A new visualization approach for identifying mutations that affect differentiation and organization of the *Drosophila* ommatidia. *Development* **128**, 815–826.

3382. Pichaud, F., Treisman, J., and Desplan, C. (2001). Reinventing a common strategy for patterning the eye. *Cell* **105**, 9–12.

3383. Pick, L. (1998). Segmentation: painting stripes from flies to vertebrates. *Dev. Genet.* **23**, 1–10.

3384. Pickett, F.B. and Meeks-Wagner, D.R. (1995). Seeing double: appreciating genetic redundancy. *The Plant Cell* **7**, 1347–1356.

3385. Pickup, A.T. and Banerjee, U. (1999). The role of Star in the production of an activated ligand for the EGF receptor signaling pathway. *Development* **205**, 254–259. [See also 2001 *Curr. Biol.* **12**, R21–R23, 2002 *Genes Dev.* **16**, 222–234, and 2001 *Mechs. Dev.* **107**, 13–23.]

3386. Piepenburg, O., Vorbrüggen, G., and Jäckle, H. (2000). *Drosophila* segment borders result from unilateral repression of Hedgehog activity by Wingless signaling. *Molec. Cell* **6**, 203–209.

3387. Pignoni, F., Hu, B., Zavitz, K.H., Xiao, J., Garrity, P.A., and Zipursky, S.L. (1997). The eye-specification proteins So and Eya form a complex and regulate multiple steps in *Drosophila* eye development. *Cell* **91**, 881–891.

3388. Pignoni, F. and Zipursky, S.L. (1997). Induction of *Drosophila* eye development by Decapentaplegic. *Development* **124**, 271–278.

3389. Pilpel, Y. and Lancet, D. (1999). Good reception in fruitfly antennae. *Nature* **398**, 285–287.

3390. Pimentel, A.C. and Venkatesh, T.R. (2001). The *retina aberrant in pattern (rap)* gene encodes the *Drosophila* Fizzy related protein (Fzr) and regulates cell cycle and pattern formation in the developing eye. *Proc. 42nd Ann. Drosophila Res. Conf.* **Abstracts Vol.**, a65.

3391. Pinsonneault, J., Florence, B., Vaessin, H., and McGinnis, W. (1997). A model for *extradenticle* function as a switch that changes HOX proteins from repressors to activators. *EMBO J.* **16**, 2032–2042.

3392. Pinto, M. and Lobe, C.G. (1996). Products of the *grg* (*groucho-related gene*) family can dimerize through the amino-terminal Q domain. *J. Biol. Chem.* **271** #51, 33026–33031.

3393. Piper, D.E., Batchelor, A.H., Chang, C.-P., Cleary, M.L., and Wolberger, C. (1999). Structure of a HoxB1-Pbx1 heterodimer bound to DNA: role of the hexapeptide and a fourth homeodomain helix in complex formation. *Cell* **96**, 587–597.

3394. Pirrotta, V. (1995). Chromatin complexes regulating gene expression in *Drosophila*. *Curr. Opin. Gen. Dev.* **5**, 466–472.

3395. Pirrotta, V. (1997). Chromatin-silencing mechanisms in *Drosophila* maintain patterns of gene expression. *Trends Genet.* **13**, 314–318.

3396. Pirrotta, V. (1998). Polycombing the genome: PcG, trxG, and chromatin silencing. *Cell* **93**, 333–336.

3397. Pirrotta, V., Chan, C.S., McCabe, D., and Qian, S. (1995). Distinct parasegmental and imaginal enhancers and the establishment of the expression pattern of the *Ubx* gene. *Genetics* **141**, 1439–1450.

3398. Piternick, L.K., ed. (1980). "Richard Goldschmidt: Controversial Geneticist and Creative Biologist. A Critical Review of His Contributions." Birkhäuser, Basel.

3399. Placzek, M. and Skaer, H. (1999). Airway patterning: a paradigm for restricted signaling. *Curr. Biol.* **9**, R506–R510.

3400. Plasterk, R.H.A. and Ketting, R.F. (2000). The silence of the genes. *Curr. Opin. Gen. Dev.* **10**, 562–567.

3401. Plautz, J.D., Kaneko, M., Hall, J.C., and Kay, S.A. (1997). Independent photoreceptive circadian clocks throughout *Drosophila*. *Science* **278**, 1632–1635.

3402. Plaza, S., Prince, F., Jaeger, J., Kloter, U., Flister, S., Benassayag, C., Cribbs, D., and Gehring, W.J. (2001). Molecular basis for the inhibition of *Drosophila* eye development by *Antennapedia*. *EMBO J.* **20**, 802–811. [See also 2001 *Development* **128**, 3307–3319.]

3403. Plotnikov, A.N., Hubbard, S.R., Schlessinger, J., and Mohammadi, M. (2000). Crystal structures of two FGF-FGFR complexes reveal the determinants of ligand-receptor specificity. *Cell* **101**, 413–424.

3404. Plotnikov, A.N., Schlessinger, J., Hubbard, S.R., and Mohammadi, M. (1999). Structural basis for FGF receptor dimerization and activation. *Cell* **98**, 641–650.

3405. Plunkett, C.R. (1926). The interaction of genetic and environmental factors in development. *J. Exp. Zool.* **46**, 181–244.

3406. Podos, S.D. and Ferguson, E.L. (1999). Morphogen gradients: new insights from DPP. *Trends Genet.* **15**, 396–402.

3407. Poeck, B., Fischer, S., Gunning, D., Zipursky, S.L., and Salecker, I. (2001). Glial cells mediate target layer selection of retinal axons in the developing visual system of *Drosophila*. *Neuron* **29**, 99–113. [See also 2001 *Neuron* **30**, 437–450.]

3408. Poeck, B., Hofbauer, A., and Pflugfelder, G.O. (1993). Expression of the *Drosophila optomotor-blind* gene transcript in neuronal and glial cells of the developing nervous system. *Development* **117**, 1017–1029.

3409. Pokutta, S. and Weis, W.I. (2000). Structure of the dimerization and β-catenin-binding region of α-catenin. *Molec. Cell* **5**, 533–543.

3410. Polaczyk, P.J., Gasperini, R., and Gibson, G. (1998). Naturally occurring genetic variation affects *Drosophila* photoreceptor determination. *Dev. Genes Evol.* **207**, 462–470.

3411. Polakis, P. (2000). Wnt signaling and cancer. *Genes Dev.* **14**, 1837–1851.

3412. Pollard, T.D., Blanchoin, L., and Mullins, R.D. (2000). Molecular mechanisms controlling actin filament dynamics in nonmuscle cells. *Annu. Rev. Biophys. Biomol. Struct.* **29**, 545–576.

3413. Pollard, V.W., Michael, W.M., Nakielny, S., Siomi, M.C., Wang, F., and Dreyfuss, G. (1996). A novel receptor-mediated nuclear protein import pathway. *Cell* **86**, 985–994.

3414. Pollock, J.A. and Benzer, S. (1988). Transcript localization of four opsin genes in the three visual organs of *Drosophila*: RH2 is ocellus specific. *Nature* **333**, 779–782.

3415. Polyak, S. (1957). "The Vertebrate Visual System." Univ. Chicago Pr., Chicago.

3416. Polymenis, M. and Schmidt, E.V. (1999). Coordination of cell growth with cell division. *Curr. Opin. Gen. Dev.* **9**, 76–80.

3417. Ponting, C.P. (1995). AF-6/cno: neither a kinesin nor a myosin, but a bit of both. *Trends Biochem. Sci.* **20**, 265–266.

3418. Ponting, C.P. and Benjamin, D.R. (1996). A novel family of Ras-binding domains. *Trends Biochem. Sci.* **21**, 422–425.

3419. Ponting, C.P., Phillips, C., Davies, K.E., and Blake, D.J. (1997). PDZ domains: targeting signalling molecules to sub-membranous sites. *BioEssays* **19**, 469–479.

3420. Poodry, C.A. (1975). A temporal pattern in the development of sensory bristles in *Drosophila*. *W. Roux's Arch.* **178**, 203–213.

3421. Poodry, C.A. (1980). Epidermis: morphology and development. *In* "The Genetics and Biology of *Drosophila*," Vol. 2d (M. Ashburner and T.R.F. Wright, eds.). Acad. Pr.: New York, pp. 443–497.

3422. Poodry, C.A. (1980). Imaginal discs: morphology and development. *In* "The Genetics and Biology of *Drosophila*," Vol. 2d (M. Ashburner and T.R.F. Wright, eds.). Acad. Pr.: New York, pp. 407–441.

3423. Poodry, C.A. (1990). *shibire*, a neurogenic mutant of *Drosophila*. *Dev. Biol.* **138**, 464–472.

3424. Poodry, C.A., Bryant, P.J., and Schneiderman, H.A. (1971). The mechanism of pattern reconstruction by dissociated imaginal discs of *Drosophila melanogaster*. *Dev. Biol.* **26**, 464–477.

3425. Poodry, C.A., Hall, L., and Suzuki, D.T. (1973). Developmental properties of *shibire*[ts]: a pleiotropic mutation affecting larval and adult locomotion and development. *Dev. Biol.* **32**, 373–386.

3426. Poodry, C.A. and Schneiderman, H.A. (1970). The ultrastructure of the developing leg of *Drosophila melanogaster*. *W. Roux' Arch.* **166**, 1–44.

3427. Poodry, C.A. and Schneiderman, H.A. (1976). Pattern formation in *Drosophila melanogaster*: the effects of mutations on polarity in the developing leg. *W. Roux's Arch.* **180**, 175–188.

3428. Poodry, C.A. and Woods, D.F. (1990). Control of the developmental timer for *Drosophila* pupariation. *Roux's Arch. Dev. Biol.* **199**, 219–227.

3429. Poole, S.J., Kauvar, L.M., Drees, B., and Kornberg, T. (1985). The *engrailed* locus of *Drosophila*: structural analysis of an embryonic transcript. *Cell* **40**, 37–43.

3430. Poortinga, G., Watanabe, M., and Parkhurst, S.M. (1998). *Drosophila* CtBP: a Hairy-interacting protein required for embryonic segmentation and Hairy-mediated transcriptional repression. *EMBO J.* **17**, 2067–2078.

3431. Popadic, A., Abzhanov, A., Rusch, D., and Kaufman, T.C. (1998). Understanding the genetic basis of morphological evolution: the role of homeotic genes in the diversification of the arthropod bauplan. *Int. J. Dev. Biol.* **42**, 453–461.

3432. Porter, J.A., Ekker, S.C., Park, W.-J., von Kessler, D.P., Young, K.E., Chen, C.-H., Ma, Y., Woods, A.S., Cotter, R.J., Koonin, E.V., and Beachy, P.A. (1996). Hedgehog patterning activity: role of a lipophilic modification mediated by the carboxy-terminal autoprocessing domain. *Cell* **86**, 21–34.

3433. Porter, J.A., von Kessler, D.P., Ekker, S.C., Young, K.E., Lee, J.J., Moses, K., and Beachy, P.A. (1995). The product of *hedgehog* autoproteolytic cleavage active in local and long-range signalling. *Nature* **374**, 363–366.

3434. Porter, J.A., Young, K.E., and Beachy, P.A. (1996). Cholesterol modification of Hedgehog signaling proteins in animal development. *Science* **274**, 255–259.

3435. Portin, P. (1975). Allelic negative complementation at the Abruptex locus of *Drosophila melanogaster*. *Genetics* **81**, 121–133.

3436. Portin, P. (1981). The antimorphic mode of action of lethal Abruptex alleles of the Notch locus in *Drosophila melanogaster*. *Hereditas* **95**, 247–251.

3437. Posakony, J.W. (1994). Nature versus nurture: asymmetric cell divisions in *Drosophila* bristle development. *Cell* **76**, 415–418.

3438. Posakony, L.G., Raftery, L.A., and Gelbart, W.M. (1991). Wing formation in *Drosophila melanogaster* requires *decapentaplegic* gene function along the anterior-posterior compartment boundary. *Mechs. Dev.* **33**, 69–82.

3439. Postlethwait, J.H. (1974). Development of the temperature-sensitive homoeotic mutant *ophthalmoptera* of *Drosophila melanogaster*. *Dev. Biol.* **36**, 212–217.

3440. Postlethwait, J.H. (1975). Pattern formation in the wing and haltere imaginal discs after irradiation of *Drosophila melanogaster* first instar larvae. *W. Roux' Arch.* **178**, 29–50.

3441. Postlethwait, J.H. (1978). Clonal analysis of *Drosophila* cuticular patterns. In "The Genetics and Biology of *Drosophila*," Vol. 2c (M. Ashburner and T.R.F. Wright, eds.). Acad. Pr.: New York, pp. 359–441.

3442. Postlethwait, J.H. (1978). Development of cuticular patterns in the legs of a cell lethal mutant of *Drosophila melanogaster*. *W. Roux's Arch.* **185**, 37–57.

3443. Postlethwait, J.H. and Girton, J. (1974). Development of antennal-leg homoeotic mutants in *Drosophila melanogaster*. *Genetics* **76**, 767–774.

3444. Postlethwait, J.H. and Girton, J.R. (1974). Development in genetic mosaics of aristapedia, a homoeotic mutant of *Drosophila melanogaster*. *Genetics* **76**, 767–774.

3445. Postlethwait, J.H. and Schneiderman, H.A. (1969). A clonal analysis of determination in *Antennapedia*, a homoeotic mutant of *Drosophila melanogaster*. *Proc. Natl. Acad. Sci. USA* **64**, 176–183.

3446. Postlethwait, J.H. and Schneiderman, H.A. (1971). A clonal analysis of development in *Drosophila melanogaster*: morphogenesis, determination, and growth in the wild-type antenna. *Dev. Biol.* **24**, 477–519.

3447. Postlethwait, J.H. and Schneiderman, H.A. (1971). Pattern formation and determination in the antenna of the homoeotic mutant *Antennapedia* of *Drosophila melanogaster*. *Dev. Biol.* **25**, 606–640.

3448. Postlethwait, J.H. and Schneiderman, H.A. (1973). Developmental genetics of *Drosophila* imaginal discs. *Annu. Rev. Genetics* **7**, 381–433.

3449. Postlethwait, J.H. and Schneiderman, H.A. (1973). Pattern formation in imaginal discs of *Drosophila melanogaster* after irradiation of embryos and young larvae. *Dev. Biol.* **32**, 345–360.

3450. Potter, C.J., Huang, H., and Xu, T. (2001). *Drosophila Tsc1* functions with *dTsc2* to antagonize insulin signaling in regulating cell growth, cell proliferation, and organ size. *Cell* **105**, 357–368. [See also 2002 *BioEssays* **24**, 54–64; 2001 *Cell* **105**, 345–355; 2001 *Genes Dev.* **15**, 1383–1392.]

3451. Pourquié, O. (1999). Notch around the clock. *Curr. Opin. Gen. Dev.* **9**, 559–565. [See also 2001 *J. Anat.* **199**, 169–175.]

3452. Poux, S., Kostic, C., and Pirrotta, V. (1996). Hunchback-independent silencing of late *Ubx* enhancers by a Polycomb Group Response Element. *EMBO J.* **15**, 4713–4722.

3453. Poux, S., McCabe, D., and Pirrotta, V. (2001). Recruitment of components of Polycomb Group chromatin complexes in *Drosophila*. *Development* **128**, 75–85. [See also 2001 *Nature* **412**, 655–660.]

3454. Pouysségur, J. (2000). An arresting start for MAPK. *Science* **290**, 1515–1518.

3455. Powe, A.C., Jr., Strathdee, D., Cutforth, T., D'Souza-Correia, T., Gaines, P., Thackeray, J., Carlson, J., and Gaul, U. (1999). In vivo functional analysis of *Drosophila* Gap1: involvement of Ca^{2+} and IP_4 regulation. *Mechs. Dev.* **81**, 89–101.

3456. Powell, P.A., Wesley, C., Spencer, S., and Cagan, R.L. (2001). Scabrous complexes with Notch to mediate boundary formation. *Nature* **409**, 626–630.

3457. Pradel, J. and White, R.A.H. (1998). From selectors to realizators. *Int. J. Dev. Biol.* **42**, 417–421.

3458. Prado, A., Canal, I., and Ferrús, A. (1999). The haplolethal region at the 16F gene cluster of *Drosophila melanogaster*: structure and function. *Genetics* **151**, 163–175.

3459. Préat, T. (1992). Characterization of *Suppressor of fused*, a complete suppressor of the *fused* segment gene of *Drosophila melanogaster*. *Genetics* **132**, 725–736.

3460. Préat, T., Thérond, P., Lamour-Isnard, C., Limbourg-Bouchon, B., Tricoire, H., Erk, I., Mariol, M.-C., and Busson, D. (1990). A putative serine/threonine protein kinase encoded by the segment-polarity *fused* gene of *Drosophila*. *Nature* **347**, 87–89.

3461. Préat, T., Thérond, P., Limbourg-Bouchon, B., Pham, A., Tricoire, H., Busson, D., and Lamour-Isnard, C. (1993). Segmental polarity in *Drosophila melanogaster*: genetic dissection of *fused* in a *Suppressor of fused* background reveals interaction with *costal-2*. *Genetics* **135**, 1047–1062.

3462. Prelich, G. (1999). Suppression mechanisms: themes from variations. *Trends Genet.* **15**, 261–266.

3463. Price, B.D., Chang, Z., Smith, R., Bockheim, S., and Laughon, A. (1993). The *Drosophila neuralized* gene encodes a C_3HC_4 zinc finger. *EMBO J.* **12**, 2411–2418.

3464. Price, J.V., Clifford, R.J., and Schüpbach, T. (1989). The maternal ventralizing locus *torpedo* is allelic to *faint little ball*, an embryonic lethal, and encodes the *Drosophila* EGF receptor homolog. *Cell* **56**, 1085–1092.

3465. Price, J.V., Savenye, E.D., Lum, D., and Breitkreutz, A. (1997). Dominant enhancers of *Egfr* in *Drosophila melanogaster*: genetic links between the *Notch* and *Egfr* signaling pathways. *Genetics* **147**, 1139–1153.

3466. Price, M.A. and Kalderon, D. (1999). Proteolysis of Cubitus interruptus in *Drosophila* requires phosphorylation by Protein Kinase A. *Development* **126**, 4331–4339.

3467. Price, M.D. and Lai, Z.-C. (1999). The *yan* gene is highly conserved in *Drosophila* and its expression suggests a complex role throughout development. *Dev. Genes Evol.* **209**, 207–217.

3468. Price, M.H., McCartney, B.M., Rayner, J., Bejsovec, A., and Peifer, M. (2001). Characterization of new alleles of *Drosophila APC2*, a cytoskeletally associated protein which negatively regulates the Wingless pathway. *Proc. 42nd Ann. Drosophila Res. Conf.* **Abstracts Vol.**, a172.

3469. Primakoff, P. and Myles, D.G. (2000). The ADAM gene family. *Trends Genet.* **16**, 83–87.

3470. Prober, D.A. and Edgar, B.A. (2000). Ras1 promotes cellular growth in the *Drosophila* wing. *Cell* **100**, 435–446.

3471. Prokopenko, S.N., He, Y., Lu, Y., and Bellen, H.J. (2000). Mutations affecting the development of the peripheral nervous system in *Drosophila*: a molecular screen for novel proteins. *Genetics* **156**, 1691–1715.

3472. Prout, M., Damania, Z., Soong, J., Fristrom, D., and Fristrom, J. (1997). Autosomal mutations affecting adhesion between wing surfaces in *Drosophila melanogaster*. *Genetics* **146**, 275–285.

3473. Ptashne, M. and Gann, A. (1998). Imposing specificity by localization: mechanism and evolvabilty. *Curr. Biol.* **8**, R812–R822.

3474. Pueyo, J.I., Galindo, M.I., Bishop, S.A., and Couso, J.P. (2000). Proximal-distal leg development in *Drosophila* requires the *apterous* gene and the *Lim1* homologue *dlim1*. *Development* **127**, 5391–5402.

3475. Pulido, D., Campuzano, S., Koda, T., Modolell, J., and Barbacid, M. (1992). D*trk*, a *Drosophila* gene related to the *trk* family of neurotrophin receptors, encodes a novel class of neural cell adhesion molecule. *EMBO J.* **11**, 391–404.

3476. Punzo, C., Seimiya, M., Schnupf, P., Gehring, W.J., and Plaza, S. (2001). The sine oculis enhancer is regulated by two *Drosophila* Pax-6 genes. *Proc. 42nd Ann. Drosophila Res. Conf.* **Abstracts Vol.**, a116.

3477. Purugganan, M.D. (1998). The molecular evolution of development. *BioEssays* **20**, 700–711.

3478. Purves, D. (1988). "Body and Brain: A Trophic Theory of Neural Connections." Harvard, Cambridge, Mass.

3479. Qi, H., Rand, M.D., Wu, X., Sestan, N., Wang, W., Rakic, P., Xu, T., and Artavanis-Tsakonas, S. (1999). Processing of the Notch ligand Delta by the metalloprotease Kuzbanian. *Science* **283**, 91–94. [But cf. 2002 *Genes Dev.* **16**, 209–221.]

3480. Qian, S., Capovilla, M., and Pirrotta, V. (1991). The *bx* region enhancer, a distant *cis*-control element of the *Drosophila Ubx* gene and its regulation by *hunchback* and other segmentation genes. *EMBO J.* **10**, 1415–1425.

3481. Qian, S., Capovilla, M., and Pirrotta, V. (1993). Molecular mechanisms of pattern formation by the BRE enhancer of the *Ubx* gene. *EMBO J.* **12**, 3865–3877.

3482. Qian, Y.Q., Billeter, M., Otting, G., Müller, M., Gehring, W.J., and Wüthrich, K. (1989). The structure of the *Antennapedia* homeodomain determined by NMR spectroscopy in solution: comparison with prokaryotic repressors. *Cell* **59**, 573–580.

3483. Qin, Y., Luo, Z.-Q., Smyth, A.J., Gao, P., von Bodman, S.B., and Farrand, S.K. (2000). Quorum-sensing signal binding results in dimerization of TraR and its release from membranes into the cytoplasm. *EMBO J.* **19**, 5212–5221.

3484. Queenan, A.M., Ghabrial, A., and Schüpbach, T. (1997). Ectopic activation of *torpedo/Egfr*, a *Drosophila* receptor tyrosine kinase, dorsalizes both the eggshell and the embryo. *Development* **124**, 3871–3880.

3485. Quilliam, L.A., Khosravi-Far, R., Huff, S.Y., and Der, C.J. (1995). Guanine nucleotide exchange factors: activators of the Ras superfamily of proteins. *BioEssays* **17**, 395–404.

3486. Quiring, R., Walldorf, U., Kloter, U., and Gehring, W.J. (1994). Homology of the *eyeless* gene of *Drosophila* to the *Small eye* gene in mice and *Aniridia* in humans. *Science* **265**, 785–789.

3487. Quirk, J., van den Heuvel, M., Henrique, D., Marigo, V., Jones, T.A., Tabin, C., and Ingham, P.W. (1997). The *smoothened* gene and Hedgehog signal transduction in *Drosophila* and vertebrate development. *Cold Spr. Harb. Symp. Quant. Biol.* **62**, 217–226.

3488. Raabe, T., Olivier, J.P., Dickson, B., Liu, X., Gish, G.D., Pawson, T., and Hafen, E. (1995). Biochemical and genetic analysis of the Drk SH2/SH3 adaptor protein of *Drosophila*. *EMBO J.* **14**, 2509–2518.

3489. Raabe, T., Riesgo-Escovar, J., Liu, X., Bausenwein, B.S., Deak, P., Maröy, P., and Hafen, E. (1996). DOS, a novel pleckstrin homology domain-containing protein required for signal transduction between Sevenless and Ras1 in *Drosophila*. *Cell* **85**, 911–920.

3490. Rabinow, L. and Birchler, J.A. (1990). Interactions of *vestigial* and *scabrous* with the *Notch* locus of *Drosophila melanogaster*. *Genetics* **125**, 41–50.

3491. Radke, K., Baek, K.-H., and Ambrosio, L. (1997). Characterization of maternal and zygotic D-raf proteins: dominant negative effects on Torso signal transduction. *Genetics* **145**, 163–171.

3492. Raff, M.C. (1996). Size control: the regulation of cell numbers in animal development. *Cell* **86**, 173–175.

3493. Raff, M.C., Barres, B.A., Burne, J.F., Coles, H.S., Ishizaki, Y., and Jacobson, M.D. (1993). Programmed cell death and the control of cell survival: lessons from the nervous system. *Science* **262**, 695–700.

3494. Raff, R.A. (1996). "The Shape of Life: Genes, Development, and the Evolution of Animal Form." Univ. Chicago Pr., Chicago.

3495. Raff, R.A. (2000). Evo-devo: the evolution of a new discipline. *Nature Rev. Gen.* **1**, 74–79.

3496. Raff, R.A. and Sly, B.J. (2000). Modularity and dissociation in the evolution of gene expression territories in development. *Evol. Dev.* **2**, 102–113.

3497. Raftery, L.A., Sanicola, M., Blackman, R.K., and Gelbart, W.M. (1991). The relationship of *decapentaplegic* and *engrailed* expression in *Drosophila* imaginal disks: do these genes mark the anterior-posterior compartment boundary? *Development* **113**, 27–33.

3498. Raftery, L.A. and Sutherland, D.J. (1999). TGF-β family signal transduction in *Drosophila* development: from *Mad* to Smads. *Dev. Biol.* **210**, 251–268.

3499. Raftery, L.A., Twombly, V., Wharton, K., and Gelbart, W.M. (1995). Genetic screens to identify elements of the *decapentaplegic* signaling pathway in *Drosophila*. *Genetics* **139**, 241–254.

3500. Raj, L., Vivekanand, P., Das, T.K., Badam, E., Fernandes, M., Finley, R.L., Jr., Brent, R., Appel, L.F., Hanes, S.D., and Weir, M. (2000). Targeted localized degradation of Paired protein in *Drosophila* development. *Curr. Biol.* **10**, 1265–1272.

3501. Ralston, A. and Blair, S.S. (2001). BMP-like signaling and the crossveinless 2-dependent pathway of crossvein development. *Proc. 42nd Ann. Drosophila Res. Conf.* **Abstracts Vol.**, a201.

3502. Ramaekers, G., Usui, K., Usui-Ishihara, A., Ramaekers, A., Ledent, V., Ghysen, A., and Dambly-Chaudière, C. (1997). Lineage and fate in *Drosophila*: role of the gene *tramtrack* in sense organ development. *Dev. Genes Evol.* **207**, 97–106.

3503. Ramain, P., Heitzler, P., Haenlin, M., and Simpson, P. (1993).

pannier, a negative regulator of *achaete* and *scute* in *Drosophila*, encodes a zinc finger protein with homology to the vertebrate transcription factor GATA-1. *Development* 119, 1277–1291.

3504. Ramain, P., Khechumian, R., Khechumian, K., Arbogast, N., Ackermann, C., and Heitzler, P. (2000). Interactions between Chip and the Achaete/Scute-Daughterless heterodimers are required for Pannier-driven proneural patterning. *Molec. Cell* 6, 781–790.

3505. Ramel, M.-C. and Katz, F.N. (2001). A novel role for *enabled* in leg segmentation: expression and genetic analysis. *Proc. 42nd Ann. Drosophila Res. Conf.* **Abstracts Vol.**, a201.

3506. Ramírez-Weber, F.-A., Casso, D.J., Aza-Blanc, P., Tabata, T., and Kornberg, T.B. (2000). Hedgehog signal transduction in the posterior compartment of the *Drosophila* wing imaginal disc. *Molec. Cell* 6, 479–485.

3507. Ramírez-Weber, F.-A. and Kornberg, T.B. (1999). Cytonemes: cellular processes that project to the principal signaling center in *Drosophila* imaginal discs. *Cell* 97, 599–607.

3508. Ramírez-Weber, F.-A. and Kornberg, T.B. (2000). Signaling reaches to new dimensions in *Drosophila* imaginal discs. *Cell* 103, 189–192.

3509. Ramos, R.G.P., Igloi, G.L., Lichte, B., Baumann, U., Maier, D., Schneider, T., Brandstätter, J.H., Fröhlich, A., and Fischbach, K.-F. (1993). The *irregular chiasm C-roughest* locus of *Drosophila*, which affects axonal projections and programmed cell death, encodes a novel immunoglobulin-like protein. *Genes Dev.* 7, 2533–2547.

3510. Randall, L.L. and Hardy, S.J.S. (1989). Unity of function in the absence of consensus in sequence: role of leader peptides in export. *Science* 243, 1156–1159.

3511. Randazzo, F.M., Seeger, M.A., Huss, C.A., Sweeney, M.A., Cecil, J.K., and Kaufman, T.C. (1993). Structural changes in the Antennapedia complex of *Drosophila pseudoobscura*. *Genetics* 134, 319–330.

3512. Randsholt, N.B., Maschat, F., and Santamaria, P. (2000). *polyhomeotic* controls *engrailed* expression and the *hedgehog* signaling pathway in imaginal discs. *Mechs. Dev.* 95, 89–99.

3513. Ranganathan, R. and Ross, E.M. (1997). PDZ domain proteins: scaffolds for signaling complexes. *Curr. Biol.* 7, R770–R773.

3514. Rangarajan, R., Gong, Q., and Gaul, U. (1999). Migration and function of glia in the developing *Drosophila* eye. *Development* 126, 3285–3292.

3515. Ransom, R. (1975). Computer analysis of division patterns in the *Drosophila* head disc. *J. Theor. Biol.* 53, 445–462.

3516. Ransom, R. (1982). Eye and head development. *In* "A Handbook of *Drosophila* Development," (R. Ransom, ed.). Elsevier: Amsterdam, pp. 123–152.

3517. Ransom, R. (1982). "A Handbook of *Drosophila* Development." Elsevier, Amsterdam.

3518. Ransom, R. and Matela, R.J. (1984). Computer modelling of cell division during development using a topological approach. *J. Embryol. Exp. Morph.* 83 (**Suppl.**), 233–259.

3519. Rao, Y., Bodmer, R., Jan, L.Y., and Jan, Y.N. (1992). The *big brain* gene of *Drosophila* functions to control the number of neuronal precursors in the peripheral nervous system. *Development* 116, 31–40.

3520. Rao, Y., Jan, L.Y., and Jan, Y.N. (1990). Similarity of the product of the *Drosophila* neurogenic gene *big brain* to transmembrane channel proteins. *Nature* 345, 163–167.

3521. Rao, Z., Handford, P., Mayhew, M., Knott, V., Brownlee, G.G., and Stuart, D. (1995). The structure of a Ca^{2+}-binding epidermal growth factor-like domain: its role in protein-protein interactions. *Cell* 82, 131–141.

3522. Rasmuson, M. (1952). Variation in bristle number of *Drosophila melanogaster*. *Acta Zool.* 33, 277–307.

3523. Rauskolb, C. (2001). The establishment of segmentation in the *Drosophila* leg. *Development* 128, 4511–4521.

3524. Rauskolb, C., Correia, T., and Irvine, K.D. (1999). Fringe-dependent separation of dorsal and ventral cells in the *Drosophila* wing. *Nature* 401, 476–480.

3525. Rauskolb, C. and Irvine, K.D. (1999). Notch-mediated segmentation and growth control of the *Drosophila* leg. *Dev. Biol.* 210, 339–350.

3526. Rauskolb, C., Peifer, M., and Wieschaus, E. (1993). *extradenticle*, a regulator of homeotic gene activity, is a homolog of the homeobox-containing human proto-oncogene *pbx1*. *Cell* 74, 1101–1112.

3527. Rauskolb, C., Smith, K.M., Peifer, M., and Wieschaus, E. (1995). *extradenticle* determines segmental identities throughout *Drosophila* development. *Development* 121, 3663–3673.

3528. Rauskolb, C. and Wieschaus, E. (1994). Coordinate regulation of downstream genes by *extradenticle* and the homeotic selector proteins. *EMBO J.* 13, 3561–3569.

3529. Ray, K., Hartenstein, V., and Rodrigues, V. (1993). Development of the taste bristles on the labellum of *Drosophila melanogaster*. *Dev. Biol.* 155, 26–37.

3530. Ray, K. and Rodrigues, V. (1994). The function of the proneural genes *achaete* and *scute* in the spatio-temporal patterning of the adult labellar bristles of *Drosophila melanogaster*. *Roux's Arch. Dev. Biol.* 203, 340–350.

3531. Ray, K. and Rodrigues, V. (1995). Cellular events during development of the olfactory sense organs in *Drosophila melanogaster*. *Dev. Biol.* 167, 426–438.

3532. Ray, R.P., Arora, K., Nüsslein-Volhard, C., and Gelbart, W.M. (1991). The control of cell fate along the dorsal-ventral axis of the *Drosophila* embryo. *Development* 113, 35–54.

3533. Ray, W.J., Yao, M., Mumm, J., Schroeter, E.H., Saftig, P., Wolfe, M., Selkoe, D.J., Kopan, R., and Goate, A.M. (1999). Cell surface Presenilin-1 participates in the γ-secretase-like proteolysis of Notch. *J. Biol. Chem.* 274 #51, 36801–36807.

3534. Ray, W.J., Yao, M., Nowotny, P., Mumm, J., Zhang, W., Wu, J.Y., Kopan, R., and Goate, A.M. (1999). Evidence for a physical interaction between presenilin and Notch. *Proc. Natl. Acad. Sci. USA* 96, 3263–3268.

3535. Raz, E. and Shilo, B.-Z. (1992). Dissection of the *faint little ball* (*flb*) phenotype: determination of the development of the *Drosophila* central nervous system by early interactions in the ectoderm. *Development* 114, 113–123.

3536. Raz, E. and Shilo, B.-Z. (1993). Establishment of ventral cell fates in the *Drosophila* embryonic ectoderm requires DER, the EGF receptor homolog. *Genes Dev.* 7, 1937–1948.

3537. Read, D. and Manley, J.L. (1992). Alternatively spliced transcripts of the *Drosophila tramtrack* gene encode zinc finger proteins with distinct DNA binding specificities. *EMBO J.* 11, 1035–1044. [See also 2001 *Proc. Natl. Acad. Sci. USA* 98, 9724–9729.]

3538. Ready, D.F. (1989). A multifaceted approach to neural development. *Trends Neurosci.* 12, 102–110.

3539. Ready, D.F., Hanson, T.E., and Benzer, S. (1976). Development of the *Drosophila* retina, a neurocrystalline lattice. *Dev. Biol.* 53, 217–240.

3540. Ready, D.F., Tomlinson, A., and Lebovitz, R.M. (1986). Building an ommatidium: geometry and genes. *In* "Development of Order in the Visual System," (S.R. Hilfer and J.B. Sheffield, eds.). Springer-Verlag: Berlin, pp. 97–125.

3541. Rebagliati, M. (1989). An RNA recognition motif in the *bicoid* protein. *Cell* **58**, 231–232.

3542. Rebay, I., Chen, F., Hsiao, F., Kolodziej, P.A., Kuang, B.H., Laverty, T., Suh, C., Voas, M., Williams, A., and Rubin, G.M. (2000). A genetic screen for novel components of the Ras/Mitogen-activated protein kinase signaling pathway that interact with the *yan* gene of *Drosophila* identifies *split ends*, a new RNA recognition motif-containing protein. *Genetics* **154**, 695–712.

3543. Rebay, I., Fehon, R.G., and Artavanis-Tsakonas, S. (1993). Specific truncations of *Drosophila* Notch define dominant activated and dominant negative forms of the receptor. *Cell* **74**, 319–329.

3544. Rebay, I., Fleming, R.J., Fehon, R.G., Cherbas, L., Cherbas, P., and Artavanis-Tsakonas, S. (1991). Specific EGF repeats of Notch mediate interactions with Delta and Serrate: implications for Notch as a multifunctional receptor. *Cell* **67**, 687–699.

3545. Rebay, I. and Rubin, G.M. (1995). Yan functions as a general inhibitor of differentiation and is negatively regulated by activation of the Ras1/MAPK pathway. *Cell* **81**, 857–866.

3546. Rebecchi, M.J. and Scarlata, S. (1998). Pleckstrin homology domains: a common fold with diverse functions. *Annu. Rev. Biophys. Biomol. Struct.* **27**, 503–528.

3547. Rechsteiner, M. and Rogers, S.W. (1996). PEST sequences and regulation by proteolysis. *Trends Biochem. Sci.* **21**, 267–271.

3548. Reddy, G.V., Gupta, B., Ray, K., and Rodrigues, V. (1997). Development of the *Drosophila* olfactory sense organs utilizes cell-cell interactions as well as lineage. *Development* **124**, 703–712.

3549. Reddy, G.V. and Rodrigues, V. (1999). A glial cell arises from an additional division within the mechanosensory lineage during development of the microchaete on the *Drosophila* notum. *Development* **126**, 4617–4622.

3550. Reddy, G.V. and Rodrigues, V. (1999). Sibling cell fate in the *Drosophila* adult external sense organ lineage is specified by Prospero function, which is regulated by Numb and Notch. *Development* **126**, 2083–2092.

3551. Reed, B.H. and Orr-Weaver, T.L. (1997). The *Drosophila* gene *morula* inhibits mitotic functions in the endo cell cycle and the mitotic cell cycle. *Development* **124**, 3543–3553.

3552. Reed, C.T., Murphy, C., and Fristrom, D. (1975). The ultrastructure of the differentiating pupal leg of *Drosophila melanogaster*. *W. Roux's Arch.* **178**, 285–302.

3553. Reeve, E.C.R. (1960). Some genetic tests on asymmetry of sternopleural chaeta number in *Drosophila*. *Genet. Res. Camb.* **1**, 151–172.

3554. Reeve, E.C.R. (1961). Modifying the sternopleural hair pattern in *Drosophila* by selection. *Genet. Res. Camb.* **2**, 158–160.

3555. Reeve, E.C.R. and Robertson, F.W. (1954). Studies in quantitative inheritance. VI. Sternite chaeta number in *Drosophila*: A metameric quantitative character. *Z. Vererbungslehre* **86**, 269–288.

3556. Regulski, M., Harding, K., Kostriken, R., Karch, F., Levine, M., and McGinnis, W. (1985). Homeo box genes of the Antennapedia and Bithorax complexes of *Drosophila*. *Cell* **43**, 71–80.

3557. Reh, T.A. and Cagan, R.L. (1994). Intrinsic and extrinsic signals in the developing vertebrate and fly eyes: viewing vertebrate and invertebrate eyes in the same light. *Persp. Dev. Neurobiol.* **2**, 183–190.

3558. Reich, A., Sapir, A., and Shilo, B.-Z. (1999). Sprouty is a general inhibitor of receptor tyrosine kinase signaling. *Development* **126**, 4139–4147.

3559. Reichman-Fried, M., Dickson, B., Hafen, E., and Shilo, B.-Z. (1994). Elucidation of the role of breathless, a *Drosophila* FGF receptor homolog, in tracheal cell migration. *Genes Dev.* **8**, 428–439.

3560. Reichman-Fried, M. and Shilo, B.-Z. (1995). Breathless, a *Drosophila* FGF receptor homolog, is required for the onset of tracheal cell migration and tracheole formation. *Mechs. Dev.* **52**, 265–273.

3561. Reichsman, F., Smith, L., and Cumberledge, S. (1996). Glycosaminoglycans can modulate extracellular localization of the *wingless* protein and promote signal transduction. *J. Cell Biol.* **135**, 819–827.

3562. Reifegerste, R., Ma, C., and Moses, K. (1997). A polarity field is established early in the development of the *Drosophila* compound eye. *Mechs. Dev.* **68**, 69–79.

3563. Reifegerste, R. and Moses, K. (1999). Genetics of epithelial polarity and pattern in the *Drosophila* retina. *BioEssays* **21**, 275–285.

3564. Reinhardt, C.A. and Bryant, P.J. (1981). Wound healing in the imaginal discs of *Drosophila*. II. Transmission electron microscopy of normal and healing wing discs. *J. Exp. Zool.* **216**, 45–61.

3565. Reinhardt, C.A., Hodgkin, N.M., and Bryant, P.J. (1977). Wound healing in the imaginal discs of *Drosophila*. I. Scanning electron microscopy of normal and healing wing discs. *Dev. Biol.* **60**, 238–257.

3566. Reinhardt, C.A. and Sanchez, L. (1982). Effects of the "*sexcombless*" chromosome (In(1)sx) on males and females of *Drosophila melanogaster*. *W. Roux's Arch.* **191**, 264–269.

3567. Reinitz, J., Kosman, D., Vanario-Alonso, C.E., and Sharp, D.H. (1998). Stripe forming architecture of the gap gene system. *Dev. Genet.* **23**, 11–27. [See also 2001 *Nature* **411**, 151–152.]

3568. Reinke, R., Krantz, D.E., Yen, D., and Zipursky, S.L. (1988). Chaoptin, a cell surface glycoprotein required for *Drosophila* photoreceptor cell morphogenesis, contains a repeat motif found in yeast and human. *Cell* **52**, 291–301.

3569. Reinke, R. and Zipursky, S.L. (1988). Cell-cell interaction in the *Drosophila* retina: the *bride of sevenless* gene is required in photoreceptor cell R8 for R7 cell development. *Cell* **55**, 321–330.

3570. Reiter, C., Schimansky, T., Nie, Z., and Fischbach, K.-F. (1996). Reorganization of membrane contacts prior to apoptosis in the *Drosophila* retina: the role of the IrreC-rst protein. *Development* **122**, 1931–1940.

3571. Rendel, J.M. (1959). Canalization of the scute phenotype of *Drosophila*. *Evolution* **13**, 425–439.

3572. Rendel, J.M. (1962). The relationship between gene and phenotype. *J. Theor. Biol.* **2**, 296–308.

3573. Rendel, J.M. (1967). "Canalisation and Gene Control." Logos Pr./Elek Books, London.

3574. Renfranz, P.J. and Benzer, S. (1989). Monoclonal antibody probes discriminate early and late mutant defects in development of the *Drosophila* retina. *Dev. Biol.* **136**, 411–429.

3575. Reppert, S.M. and Sauman, I. (1995). *period* and *timeless* tango: a dance of two clock genes. *Neuron* **15**, 983–986.

3576. Reuter, R. and Leptin, M. (1994). Interacting functions of *snail, twist* and *huckebein* during the early development of germ layers in *Drosophila. Development* **120**, 1137–1150.

3577. Révet, B., von Wilcken-Bergmann, B., Bessert, H., Barker, A., and Müller-Hill, B. (1998). Four dimers of λ repressor bound to suitably spaced pairs of λ operators form octamers and DNA loops over large distances. *Curr. Biol.* **9**, 151–154.

3578. Rhodes, D. and Klug, A. (1993). Zinc fingers. *Sci. Am.* **268** #2, 56–65.

3579. Rhyu, M.S., Jan, L.Y., and Jan, Y.N. (1994). Asymmetric distribution of numb protein during division of the sensory organ precursor cell confers distinct fates to daughter cells. *Cell* **76**, 477–491.

3580. Rhyu, M.S. and Knoblich, J.A. (1995). Spindle orientation and asymmetric cell fate. *Cell* **82**, 523–526.

3581. Ribbeck, K., Kutay, U., Paraskeva, E., and Görlich, D. (1998). The translocation of transportin-cargo complexes through nuclear pores is independent of both Ran and energy. *Curr. Biol.* **9**, 47–50.

3582. Richards, A.G. (1951). "The Integument of Arthropods." Oxford Univ. Pr., London.

3583. Richards, A.G. and Richards, P.A. (1979). The cuticular protuberances of insects. *Int. J. Insect Morphol. & Embryol.* **8**, 143–157.

3584. Richelle, J. and Ghysen, A. (1979). Determination of sensory bristles and pattern formation in *Drosophila*. I. A model. *Dev. Biol.* **70**, 418–437.

3585. Richter, B., Long, M., Lewontin, R.C., and Nitasaka, E. (1997). Nucleotide variation and conservation at the *dpp* locus, a gene controlling early development in *Drosophila. Genetics* **145**, 311–323.

3586. Ricos, M.G., Harden, N., Sem, K.P., Lim, L., and Chia, W. (1999). *Dcdc42* acts in TGF-β signaling during *Drosophila* morphogenesis: distinct roles for the Drac1/JNK and Dcdc42/TGF-β cascades in cytoskeletal regulation. *J. Cell Sci.* **112**, 1225–1235.

3587. Riddihough, G. and Ish-Horowicz, D. (1991). Individual stripe regulatory elements in the *Drosophila hairy* promoter respond to maternal, gap, and pair-rule genes. *Genes Dev.* **5**, 840–854.

3588. Ridley, M. (1999). "Genome: The Autobiography of a Species in 23 Chapters." Harper-Collins, New York.

3589. Rieckhof, G.E., Casares, F., Ryoo, H.D., Abu-Shaar, M., and Mann, R.S. (1997). Nuclear translocation of Extradenticle requires *homothorax*, which encodes an Extradenticle-related homeodomain protein. *Cell* **91**, 171–183.

3590. Riese, D.J., II and Stern, D.F. (1998). Specificity within the EGF family/ErbB receptor family signaling network. *BioEssays* **20**, 41–48.

3591. Riese, J., Yu, X., Munnerlyn, A., Eresh, S., Hsu, S.-C., Grosschedl, R., and Bienz, M. (1997). LEF-1, a nuclear factor coordinating signaling inputs from *wingless* and *decapentaplegic. Cell* **88**, 777–787.

3592. Riesgo-Escovar, J.R. and Hafen, E. (1997). *Drosophila* Jun kinase regulates expression of *decapentaplegic* via the ETS-domain protein Aop and the AP-1 transcription factor DJun during dorsal closure. *Genes Dev.* **11**, 1717–1727.

3593. Riesgo-Escovar, J.R., Jenni, M., Fritz, A., and Hafen, E. (1996). The *Drosophila* Jun-N-terminal kinase is required for cell morphogenesis but not for DJun-dependent cell fate specification in the eye. *Genes Dev.* **10**, 2759–2768.

3594. Rietdorf, J., Siegert, F., Dharmawardhane, S., Firtel, R.A., and Weijer, C.J. (1997). Analysis of cell movement and signalling during ring formation in an activated Gα1 mutant of *Dictyostelium discoideum* that is defective in prestalk zone formation. *Dev. Biol.* **181**, 79–90.

3595. Rietveld, A., Neutz, S., Simons, K., and Eaton, S. (1999). Association of sterol- and glycosylphosphatidylinositol-linked proteins with *Drosophila* raft lipid microdomains. *J. Biol. Chem.* **274** #17, 12049–12054.

3596. Riezman, H., Woodman, P.G., van Meer, G., and Marsh, M. (1997). Molecular mechanisms of endocytosis. *Cell* **91**, 731–738.

3597. Riggleman, B., Schedl, P., and Wieschaus, E. (1990). Spatial expression of the *Drosophila* segment polarity gene *armadillo* is posttranscriptionally regulated by *wingless. Cell* **63**, 549–560.

3598. Riggleman, B., Wieschaus, E., and Schedl, P. (1989). Molecular analysis of the *armadillo* locus: uniformly distributed transcripts and a protein with novel internal repeats are associated with a *Drosophila* segment polarity gene. *Genes Dev.* **3**, 96–113.

3599. Rijsewijk, F., Schuermann, M., Wagenaar, E., Parren, P., Weigel, D., and Nusse, R. (1987). The *Drosophila* homolog of the mouse mammary oncogene *int-1* is identical to the segment polarity gene *wingless. Cell* **50**, 649–657.

3600. Rincón-Limas, D.E., Lu, C.-H., Canal, I., and Botas, J. (2000). The level of DLDB/CHIP controls the activity of the LIM homeodomain protein Apterous: evidence for a functional tetramer complex *in vivo. EMBO J.* **19**, 2602–2614.

3601. Rincón-Limas, D.E., Lu, C.-H., Canal, I., Calleja, M., Rodríguez-Esteban, C., Izpisúa-Belmonte, J.C., and Botas, J. (1999). Conservation of the expression and function of *apterous* orthologs in *Drosophila* and mammals. *Proc. Natl. Acad. Sci. USA* **96**, 2165–2170.

3602. Ripoll, P. (1972). The embryonic organization of the imaginal wing disc of *Drosophila melanogaster. W. Roux' Arch.* **169**, 200–215.

3603. Ripoll, P., El Messal, M., Laran, E., and Simpson, P. (1988). A gradient of affinities for sensory bristles across the wing blade of *Drosophila melanogaster. Development* **103**, 757–767.

3604. Ritossa, F. (1976). The bobbed locus. *In* "The Genetics and Biology of *Drosophila*," Vol. 1b (M. Ashburner and E. Novitski, eds.). Acad. Pr.: New York, pp. 801–846.

3605. Rivera-Pomar, R. and Jäckle, H. (1996). From gradients to stripes in *Drosophila* embryogenesis: filling in the gaps. *Trends Genet.* **12**, 478–483.

3606. Rivera-Pomar, R., Niessing, D., Schmidt-Ott, U., Gehring, W.J., and Jäckle, H. (1996). RNA binding and translational suppression by bicoid. *Nature* **379**, 746–749.

3607. Rives, A.F., Wehrli, M., and DiNardo, S. (2001). The role of *arrow* in the *wingless* signaling pathway. *Proc. 42nd Ann. Drosophila Res. Conf.* **Abstracts Vol.**, a207.

3608. Rivlin, P.K., Gong, A., Schneiderman, A.M., and Booker, R. (2001). The role of *Ultrabithorax* in the patterning of adult thoracic muscles in *Drosophila melanogaster. Dev. Genes Evol.* **211**, 55–66.

3609. Rivlin, P.K., Schneiderman, A.M., and Booker, R. (2000). Imaginal pioneers prefigure the formation of adult thoracic muscles in *Drosophila melanogaster. Dev. Biol.* **222**, 450–459.

3610. Rivlin, R. (1986). "The Algorithmic Image: Graphic Visions of the Computer Age." Microsoft Pr., Richmond, Wash.

3611. Roark, M., Sturtevant, M.A., Emery, J., Vaessin, H., Grell, E., and Bier, E. (1995). *scratch*, a pan-neural gene encoding a zinc-finger protein related to *snail*, promotes neuronal development. *Genes Dev.* **9**, 2384–2398.

3612. Robbins, D.J., Nybakken, K.E., Kobayashi, R., Sisson, J.C., Bishop, J.M., and Thérond, P.P. (1997). Hedgehog elicits signal transduction by means of a large complex containing the kinesin-related protein Costal2. *Cell* **90**, 225–234.

3613. Roberts, P. (1961). Bristle formation controlled by the achaete locus in genetic mosaics of *Drosophila melanogaster*. *Genetics* **46**, 1241–1243.

3614. Roberts, P. (1964). Mosaics involving aristapedia, a homeotic mutant of *Drosophila melanogaster*. *Genetics* **49**, 593–598.

3615. Roberts, P.A. (1976). The genetics of chromosome aberration. *In* "The Genetics and Biology of *Drosophila*," Vol. 1a (M. Ashburner and E. Novitski, eds.). Acad. Pr.: New York, pp. 67–184.

3616. Roberts, P.A. and Broderick, D.J. (1982). Properties and evolutionary potential of newly induced tandem duplications in *Drosophila melanogaster*. *Genetics* **102**, 75–89.

3617. Robertson, F.W. (1959). Studies in quantitative inheritance. XII. Cell size and number in relation to genetic and environmental variation of body size in *Drosophila*. *Genetics* **44**, 869–896.

3618. Robertson, H.M. (2001). Molecular evolution of the insect chemoreceptor superfamily and underannotation of the *Drosophila* genome. *Proc. 42nd Ann. Drosophila Res. Conf.* **Abstracts Vol.**, a327.

3619. Robertson, M. (1981). Gene families, hopeful monsters and the selfish genetics of DNA. *Nature* **293**, 333–334.

3620. Robinow, S., Campos, A.R., Yao, K.-M., and White, K. (1988). The *elav* gene product of *Drosophila*, required in neurons, has three RNP consensus motifs. *Science* **242**, 1570–1572.

3621. Robinow, S., Draizen, T.A., and Truman, J.W. (1997). Genes that induce apoptosis: transcriptional regulation in identified, doomed neurons of the *Drosophila* CNS. *Dev. Biol.* **190**, 206–213.

3622. Robinson, M.J. and Cobb, M.H. (1997). Mitogen-activated protein kinase pathways. *Curr. Opin. Cell Biol.* **9**, 180–186.

3623. Roch, F. and Akam, M. (2000). *Ultrabithorax* and the control of cell morphology in *Drosophila* halteres. *Development* **127**, 97–107.

3624. Roch, F., Baonza, A., Martín-Blanco, E., and García-Bellido, A. (1998). Genetic interactions and cell behaviour in *blistered* mutants during proliferation and differentiation of the *Drosophila* wing. *Development* **125**, 1823–1832.

3625. Rodgers, M.E. and Shearn, A. (1977). Patterns of protein synthesis in imaginal discs of *Drosophila melanogaster*. *Cell* **12**, 915–921.

3626. Rodrigues, V. (1988). Spatial coding of olfactory information in the antennal lobe of *Drosophila melanogaster*. *Brain Res.* **453**, 299–307.

3627. Rodriguez, I. and Basler, K. (1997). Control of compartmental affinity boundaries by Hedgehog. *Nature* **389**, 614–618.

3628. Rodríguez, I., Hernández, R., Modolell, J., and Ruiz-Gómez, M. (1990). Competence to develop sensory organs is temporally and spatially regulated in *Drosophila* epidermal primordia. *EMBO J.* **9**, 3583–3592.

3629. Roegiers, F., Younger-Shepherd, S., Jan, L.Y., and Jan, Y.N. (2001). Genetic analysis of two types of asymmetric divisions of sensory organ precursor cell lineage. *Proc. 42nd Ann. Drosophila Res. Conf.* **Abstracts Vol.**, a50.

3630. Roegiers, F., Younger-Shepherd, S., Jan, L.Y., and Jan, Y.N. (2001). Two types of asymmetric divisions in the *Drosophila* sensory organ precursor cell lineage. *Nature Cell Biol.* **3**, 58–67. [Cf. 2001 *Dev. Biol.* **240**, 361–376.]

3631. Rogers, B.T. and Kaufman, T.C. (1997). Structure of the insect head in ontogeny and phylogeny: a view from *Drosophila*. *Int. Rev. Cytol.* **174**, 1–84.

3632. Rogers, S., Wells, R., and Rechsteiner, M. (1986). Amino acid sequences common to rapidly degraded proteins: the PEST Hypothesis. *Science* **234**, 364–368.

3633. Rogge, R., Cagan, R., Majumdar, A., Dulaney, T., and Banerjee, U. (1992). Neuronal development in the *Drosophila* retina: the sextra gene defines an inhibitory component in the developmental pathway of R7 photoreceptor cells. *Proc. Natl. Acad. Sci. USA* **89**, 5271–5275.

3634. Rogge, R., Green, P.J., Urano, J., Horn-Saban, S., Mlodzik, M., Shilo, B.-Z., Hartenstein, V., and Banerjee, U. (1995). The role of *yan* in mediating the choice between cell division and differentiation. *Development* **121**, 3947–3958.

3635. Rogge, R.D., Karlovich, C.A., and Banerjee, U. (1991). Genetic dissection of a neurodevelopmental pathway: *Son of sevenless* functions downstream of the *sevenless* and EGF receptor tyrosine kinases. *Cell* **64**, 39–48.

3636. Roignant, J.Y., Faucheux, M., Lepesant, J.-A., Théodore, L., and Antoniewski, C. (2001). Identification of Batman as a putative partner of the Broad proteins. *Proc. 42nd Ann. Drosophila Res. Conf.* **Abstracts Vol.**, a97.

3637. Romani, S., Campuzano, S., Macagno, E.R., and Modolell, J. (1989). Expression of *achaete* and *scute* genes in *Drosophila* imaginal discs and their function in sensory organ development. *Genes Dev.* **3**, 997–1007.

3638. Rommel, C. and Hafen, E. (1998). Ras–a versatile cellular switch. *Curr. Opin. Gen. Dev.* **8**, 412–418.

3639. Rommel, C., Radziwill, G., Moelling, K., and Hafen, E. (1997). Negative regulation of Raf activity by binding of 14-3-3 to the amino terminus of Raf in vivo. *Mechs. Dev.* **64**, 95–104.

3640. Rooke, J., Pan, D., Xu, T., and Rubin, G.M. (1996). KUZ, a conserved metalloprotease-disintegrin protein with two roles in *Drosophila* neurogenesis. *Science* **273**, 1227–1231.

3641. Rooke, J.E. and Xu, T. (1998). Positive and negative signals between interacting cells for establishing neural fate. *BioEssays* **20**, 209–214.

3642. Rørth, P. and Montell, D.J. (1992). *Drosophila* C/EBP: a tissue-specific DNA-binding protein required for embryonic development. *Genes Dev.* **6**, 2299–2311.

3643. Rørth, P., Szabo, K., Bailey, A., Laverty, T., Rehm, J., Rubin, G.M., Weigmann, K., Milán, M., Benes, V., Ansorge, W., and Cohen, S.M. (1998). Systematic gain-of-function genetics in *Drosophila*. *Development* **125**, 1049–1057.

3644. Rose, S.M. (1952). A hierarchy of self-limiting reactions as the basis of cellular differentiation and growth control. *Am. Nat.* **86**, 337–354.

3645. Rose, S.M. (1957). Cellular interaction during differentiation. *Biol. Rev.* **32**, 351–382.

3646. Rose, S.M. (1970). Differentiation during regeneration caused by migration of repressors in bioelectric fields. *Amer. Zool.* **10**, 91–99.

3647. Roseland, C.R. and Schneiderman, H.A. (1979). Regulation and metamorphosis of the abdominal histoblasts of *Drosophila melanogaster*. *W. Roux's Arch.* **186**, 235–265.

3648. Rosenberg, M.I., Phippen, T.M., Secombe, J., Sweigart, A.L., and Parkhurst, S.M. (2001). Cofactors for Hairy-mediated repression. *Proc. 42nd Ann. Drosophila Res. Conf.* **Abstracts Vol.**, a125.

3649. Rosenfeld, M.G. (1991). POU-domain transcription factors: pou-er-ful developmental regulators. *Genes Dev.* **5**, 897–907.

3650. Rosin-Arbesfeld, R., Townsley, F., and Bienz, M. (2000). The APC tumor suppressor has a nuclear export function. *Nature* **406**, 1009–1012.

3651. Rossky, P.J. (2001). Molecules at the edge. *Nature* **410**, 645–648.

3652. Rothe, M., Pehl, M., Taubert, H., and Jäckle, H. (1992). Loss of gene function through rapid mitotic cycles in the *Drosophila* embryo. *Nature* **359**, 156–159.

3653. Rothe, M., Wimmer, E.A., Pankratz, M.J., González-Gaitán, M., and Jäckle, H. (1994). Identical transacting factor requirement for *knirps* and *knirps-related* gene expression in the anterior but not in the posterior region of the *Drosophila* embryo. *Mechs. Dev.* **46**, 169–181.

3654. Rothenfluh, A., Young, M.W., and Saez, L. (2000). A TIMELESS-independent function for PERIOD proteins in the *Drosophila* clock. *Neuron* **26**, 505–514.

3655. Roush, W. (1995). Sifting mitosis, cell fate in fly eyes. *Science* **270**, 916–917.

3656. Roush, W. (1996). Receptor for vital protein finally found. *Science* **273**, 309.

3657. Roush, W. (1996). "Smart" genes use many cues to set cell fate. *Science* **272**, 652–653.

3658. Roush, W. (1997). A "master control" gene for fly eyes shares its power. *Science* **275**, 618–619.

3659. Rousset, R., Mack, J.A., Wharton, K.A., Jr., Axelrod, J.D., Cadigan, K.M., Fish, M.P., Nusse, R., and Scott, M.P. (2001). *naked cuticle* targets *dishevelled* to antagonize Wnt signal transduction. *Genes Dev.* **15**, 658–671.

3660. Roy, S. and VijayRaghavan, K. (1997). Homeotic genes and the regulation of myoblast migration, fusion, and fibre-specific gene expression during adult myogenesis in *Drosophila*. *Development* **124**, 3333–3341.

3661. Roy, S. and VijayRaghavan, K. (1999). Muscle pattern diversification in *Drosophila*: the story of imaginal myogenesis. *BioEssays* **21**, 486–498.

3662. Royet, J., Bouwmeester, T., and Cohen, S.M. (1998). *Notchless* encodes a novel WD40-repeat-containing protein that modulates Notch signaling activity. *EMBO J.* **17**, 7351–7360.

3663. Royet, J. and Finkelstein, R. (1995). Pattern formation in *Drosophila* head development: the role of the *orthodenticle* homeobox gene. *Development* **121**, 3561–3572.

3664. Royet, J. and Finkelstein, R. (1996). *hedgehog, wingless* and *orthodenticle* specify adult head development in *Drosophila*. *Development* **122**, 1849–1858.

3665. Royet, J. and Finkelstein, R. (1997). Establishing primordia in the *Drosophila* eye-antennal imaginal disc: the roles of *decapentaplegic, wingless*, and *hedgehog*. *Development* **124**, 4793–4800.

3666. Rozovskaia, T., Rozenblatt-Rosen, O., Sedkov, Y., Burakov, D., Yano, T., Nakamura, T., Petruk, S., Ben-Simchon, L., Croce, C.M., Mazo, A., and Canaani, E. (2000). Self-association of the SET domains of human ALL-1 and of *Drosophila* TRITHORAX and ASH1 proteins. *Oncogene* **19**, 351–357.

3667. Rozovskaia, T., Tillib, S., Smith, S., Sedkov, Y., Rozenblatt-Rosen, O., Petruk, S., Yano, T., Nakamura, T., Ben-Simchon, L., Gildea, J., Croce, C.M., Shearn, A., Canaani, E., and Mazo, A. (1999). Trithorax and ASH1 interact directly and associate with the trithorax group-responsive *bxd* region of the *Ultrabithorax* promoter. *Mol. Cell. Biol.* **19**, 6441–6447.

3668. Ruberte, E., Marty, T., Nellen, D., Affolter, M., and Basler, K. (1995). An absolute requirement for both the type II and type I receptors, punt and thick veins, for Dpp signaling in vivo. *Cell* **80**, 889–897.

3669. Rubin, G.M. (1988). *Drosophila melanogaster* as an experimental organism. *Science* **240**, 1453–1459.

3670. Rubin, G.M. (1989). Development of the *Drosophila* retina: inductive events studied at single cell resolution. *Cell* **57**, 519–520.

3671. Rubin, G.M. (1991). Signal transduction and the fate of the R7 photoreceptor in *Drosophila*. *Trends Genet.* **7**, 372–377.

3672. Rubin, G.M., Chang, H.C., Karim, F., Laverty, T., Michaud, N.R., Morrison, D.K., Rebay, I., Tang, A., Therrien, M., and Wassarman, D.A. (1997). Signal transduction downstream from RAS in *Drosophila*. *Cold Spr. Harb. Symp. Quant. Biol.* **62**, 347–352.

3673. Rubin, G.M. and Lewis, E.B. (2000). A brief history of *Drosophila*'s contributions to genome research. *Science* **287**, 2216–2218.

3674. Rubin, G.M., Yandell, M.D., Wortman, J.R., Gabor Miklos, G.L., Nelson, C.R., Hariharan, I.K., Fortini, M.E., Li, P.W., Apweiler, R., Fleischmann, W., Cherry, J.M., Henikoff, S., Skupski, M.P., Misra, S., Ashburner, M., Birney, E., Boguski, M.S., Brody, T., Brokstein, P., Celniker, S.E., Chervitz, S.A., Coates, D., Cravchik, A., Gabrielian, A., Galle, R.F., Gelbart, W.M., George, R.A., Goldstein, L.S.B., Gong, F., Guan, P., Harris, N.L., Hay, B.A., Hoskins, R.A., Li, J., Li, Z., Hynes, R.O., Jones, S.J.M., Kuehl, P.M., Lemaitre, B., Littleton, J.T., Morrison, D.K., Mungall, C., O'Farrell, P.H., Pickeral, O.K., Shue, C., Vosshall, L.B., Zhang, J., Zhao, Q., Zheng, X.H., Zhong, F., Zhong, W., Gibbs, R., Venter, J.C., Adams, M.D., and Lewis, S. (2000). Comparative genomics of the eukaryotes. *Science* **287**, 2204–2215.

3675. Rubinfeld, B., Albert, I., Porfiri, E., Fiol, C., Munemitsu, S., and Polakis, P. (1996). Binding of GSK3β to the APC-β-catenin complex and regulation of complex assembly. *Science* **272**, 1023–1026.

3676. Rudd, C.E. (1999). Adaptors and molecular scaffolds in immune cell signaling. *Cell* **96**, 5–8.

3677. Ruddle, F.H., Bartels, J.L., Bentley, K.L., Kappen, C., Murtha, M.T., and Pendleton, J.W. (1994). Evolution of *Hox* genes. *Annu. Rev. Genet.* **28**, 423–442.

3678. Ruden, D.M., Wang, X., Cui, W., Mori, D., and Alterman, M. (1999). A novel follicle-cell-dependent dominant female sterile allele, *StarKojak*, alters receptor tyrosine kinase signaling in *Drosophila*. *Dev. Biol.* **207**, 393–407.

3679. Rüdiger, M. (1998). Vinculin and α-catenin: shared and unique functions in adherens junctions. *BioEssays* **20**, 733–740.

3680. Rudolph, K.M., Liaw, G.-J., Daniel, A., Green, P., Courey, A.J., Hartenstein, V., and Lengyel, J.A. (1997). Complex regulatory region mediating *tailless* expression in early embryonic patterning and brain development. *Development* **124**, 4297–4308.

3681. Ruel, L., Bourouis, M., Heitzler, P., Pantesco, V., and Simpson, P. (1993). *Drosophila shaggy* kinase and rat glycogen synthase kinase-3 have conserved activities and act downstream of *Notch*. *Nature* **362**, 557–560.

3682. Ruel, L., Pantesco, V., Lutz, Y., Simpson, P., and Bourouis, M. (1993). Functional significance of a family of protein kinases encoded at the *shaggy* locus in *Drosophila*. *EMBO J.* **12**, 1657–1669.

3683. Ruel, L., Stambolic, V., Ali, A., Manoukian, A.S., and Woodgett, J.R. (1999). Regulation of the protein kinase activity of Shaggy$^{Zeste-white3}$ by components of the Wingless pathway in *Drosophila* cells and embryos. *J. Biol. Chem.* **274** #31, 21790–21796. [See also 2001 *Cell* **105**, 721–732.]

3684. Ruiz Gómez, M. and Bate, M. (1997). Segregation of myogenic lineages in *Drosophila* requires Numb. *Development* **124**, 4857–4866.

3685. Ruiz i Altaba, A. (1997). Catching a Gli-mpse of Hedgehog. *Cell* **90**, 193–196.

3686. Ruiz i Altaba, A. (1999). Gli proteins and Hedgehog signaling. *Trends Genet.* **15**, 418–425.

3687. Ruiz-Gómez, M. (1998). Muscle patterning and specification in *Drosophila*. *Int. J. Dev. Biol.* **42**, 283–290.

3688. Ruiz-Gómez, M. and Modolell, J. (1987). Deletion analysis of the *achaete-scute* locus of *Drosophila melanogaster*. *Genes Dev.* **1**, 1238–1246.

3689. Rulifson, E.J. and Blair, S.S. (1995). *Notch* regulates *wingless* expression and is not required for reception of the paracrine *wingless* signal during wing margin neurogenesis in *Drosophila*. *Development* **121**, 2813–2824.

3690. Rulifson, E.J., Micchelli, C.A., Axelrod, J.D., Perrimon, N., and Blair, S.S. (1996). *wingless* refines its own expression domain on the *Drosophila* wing margin. *Nature* **384**, 72–74 (erratum: 1996 *Nature* **384**: 597).

3691. Rulifson, E.J., Wu, C.-H., and Nusse, R. (2000). Pathway specificity by the bifunctional receptor Frizzled is determined by affinity for Wingless. *Molec. Cell* **6**, 117–126.

3692. Ruohola-Baker, H., Grell, E., Chou, T.-B., Baker, D., Jan, L.Y., and Jan, Y.N. (1993). Spatially localized Rhomboid is required for establishment of the dorsal-ventral axis in *Drosophila* oogenesis. *Cell* **73**, 953–965.

3693. Rusch, J. and Levine, M. (1997). Regulation of a *dpp* target gene in the *Drosophila* embryo. *Development* **124**, 303–311.

3694. Rusconi, J.C. and Corbin, V. (1998). Evidence for a novel Notch pathway required for muscle precursor selection in *Drosophila*. *Mechs. Dev.* **79**, 39–50.

3695. Rusconi, J.C. and Corbin, V. (1999). A widespread and early requirement for a novel Notch function during *Drosophila* embryogenesis. *Dev. Biol.* **215**, 388–398.

3696. Rushlow, C., Colosimo, P.F., Lin, M.-c., Xu, M., and Kirov, N. (2001). Transcriptional regulation of the *Drosophila* gene *zen* by competing Smad and Brinker inputs. *Genes Dev.* **15**, 340–351.

3697. Rushlow, C.A., Hogan, A., Pinchin, S.M., Howe, K.M., Lardelli, M., and Ish-Horowicz, D. (1989). The *Drosophila hairy* protein acts in both segmentation and bristle patterning and shows homology to N-*myc*. *EMBO J.* **8**, 3095–3103.

3698. Rushton, E., Drysdale, R., Abmayr, S.M., Michelson, A.M., and Bate, M. (1995). Mutations in a novel gene, *myoblast city*, provide evidence in support of the founder cell hypothesis for *Drosophila* muscle development. *Development* **121**, 1979–1988.

3699. Russell, E.S. (1917). "Form and Function: A Contribution to the History of Animal Morphology." E. P. Dutton, New York.

3700. Russell, J., Gennissen, A., and Nusse, R. (1992). Isolation and expression of two novel *Wnt/wingless* gene homologues in *Drosophila*. *Development* **115**, 475–485.

3701. Russell, M. (1982). Imaginal discs. *In* "A Handbook of *Drosophila* Development," (R. Ransom, ed.). Elsevier: Amsterdam, pp. 95–121.

3702. Russell, M.A. (1974). Pattern formation in the imaginal discs of a temperature-sensitive cell-lethal mutant of *Drosophila melanogaster*. *Dev. Biol.* **40**, 24–39.

3703. Russell, M.A. (1985). Positional information in imaginal

discs: A Cartesian coordinate model. *In* "Mathematical Essays on Growth and the Emergence of Form," (P.L. Antonelli, ed.). Univ. of Alberta Pr.: Edmonton, pp. 169–183.

3704. Russell, M.A. (1985). Positional information in insect segments. *Dev. Biol.* **108**, 269–283.

3705. Russell, M.A., Girton, J.R., and Morgan, K. (1977). Pattern formation in a ts-cell-lethal mutant of *Drosophila*: the range of phenotypes induced by larval heat treatments. *W. Roux's Arch.* **183**, 41–59.

3706. Russell, S. and Ashburner, M. (1996). Ecdysone-regulated chromosome puffing in *Drosophila melanogaster*. *In* "Metamorphosis: Postembryonic Reprogramming of Gene Expression in Amphibian and Insect Cells," (L.I. Gilbert, J.R. Tata, and B.G. Atkinson, ed.). Acad. Pr.: New York, pp. 109–144.

3707. Russell, S.R.H., Sanchez-Soriano, N., Wright, C.R., and Ashburner, M. (1996). The *Dichaete* gene of *Drosophila melanogaster* encodes a SOX-domain protein required for embryonic segmentation. *Development* **122**, 3669–3676.

3708. Rusten, T.E., Cantera, R., Urban, J., Technau, G., Kafatos, F.C., and Barrio, R. (2001). Spalt modifies EGFR-mediated induction of chordotonal precursors in the embryonic PNS of *Drosophila* promoting the development of oenocytes. *Development* **128**, 711–722.

3709. Rutherford, S.L. (2000). From genotype to phenotype: buffering mechanisms and the storage of genetic information. *BioEssays* **22**, 1095–1105.

3710. Rutila, J.E., Suri, V., Le, M., So, W.V., Rosbash, M., and Hall, J.C. (1998). CYCLE is a second bHLH-PAS clock protein essential for circadian rhythmicity and transcription of *Drosophila period* and *timeless*. *Cell* **93**, 805–814.

3711. Rutledge, B.J., Zhang, K., Bier, E., Jan, Y.N., and Perrimon, N. (1992). The *Drosophila spitz* gene encodes a putative EGF-like growth factor involved in dorsal-ventral axis formation and neurogenesis. *Genes Dev.* **6**, 1503–1517.

3712. Ruvkin, G. and Finney, M. (1991). Regulation of transcription and cell identity by POU domain proteins. *Cell* **64**, 475–478.

3713. Ryan, A.K. and Rosenfeld, M.G. (1997). POU domain family values: flexibility, partnerships, and developmental codes. *Genes Dev.* **11**, 1207–1225.

3714. Rybtsova, N., Waltzer, L., and Haenlin, M. (2001). Searching for new proneural genes which might regulate the activity of the pan-neural genes deadpan, scratch and snail in *Drosophila*. *Proc. 42nd Ann. Drosophila Res. Conf.* **Abstracts Vol.**, a267.

3715. Ryoo, H.D. and Mann, R.S. (1999). The control of trunk Hox specificity and activity by Extradenticle. *Genes Dev.* **13**, 1704–1716.

3716. Ryoo, H.D., Marty, T., Casares, F., Affolter, M., and Mann, R.S. (1999). Regulation of Hox target genes by a DNA bound Homothorax/Hox/Extradenticle complex. *Development* **126**, 5137–5148. [See also 2001 *Development* **128**, 3405–3413.]

3717. Sackerson, C., Fujioka, M., and Goto, T. (1999). The *evenskipped* locus is contained in a 16-kb chromatin domain. *Dev. Biol.* **211**, 39–52.

3718. Saebøe-Larssen, S., Lyamouri, M., Merriam, J., Oksvold, M.P., and Lambertsson, A. (1998). Ribosomal protein insufficiency and the Minute syndrome in *Drosophila*: a dose-response relationship. *Genetics* **148**, 1215–1224.

3719. Saenz-Robles, M.T., Maschat, F., Tabata, T., Scott, M.P., and Kornberg, T.B. (1995). Selection and characterization of sequences with high affinity for the engrailed protein of *Drosophila*. *Mechs. Dev.* **53**, 185–195.

3720. Saez, L. and Young, M.W. (1996). Regulation of nuclear entry of the *Drosophila* clock proteins Period and Timeless. *Neuron* **17**, 911–920.

3721. Saget, O., Forquignon, F., Santamaria, P., and Randsholt, N.B. (1998). Needs and targets for the *multi sex combs* gene product in *Drosophila melanogaster*. *Genetics* **149**, 1823–1838.

3722. Sahut-Barnola, I., Godt, D., Laski, F.A., and Couderc, J.-L. (1995). *Drosophila* ovary morphogenesis: analysis of terminal filament formation and identification of a gene required for this process. *Dev. Biol.* **170**, 127–135.

3723. Saint, R., Kalionis, B., Lockett, T.J., and Elizur, A. (1988). Pattern formation in the developing eye of *Drosophila melanogaster* is regulated by the homeo-box gene, *rough*. *Nature* **334**, 151–154.

3724. Sakanaka, C., Leong, P., Xu, L., Harrison, S.D., and Williams, L.T. (1999). Casein kinase 1ε in the Wnt pathway: regulation of β-catenin function. *Proc. Natl. Acad. Sci. USA* **96**, 12548–12552.

3725. Salcini, A.E., Confalonieri, S., Doria, M., Santolini, E., Tassi, E., Minenkova, O., Cesareni, G., Pelicci, P.G., and Di Fiore, P.P. (1997). Binding specificity and in vivo targets of the EH domain, a novel protein-protein interaction module. *Genes Dev.* **11**, 2239–2249.

3726. Salecker, I., Clandinin, T.R., and Zipursky, S.L. (1998). Hedgehog and Spitz: making a match between photoreceptor axons and their targets. *Cell* **95**, 587–590.

3727. Salser, S.J. and Kenyon, C. (1996). A *C. elegans* Hox gene switches on, off, on and off again to regulate proliferation, differentiation and morphogenesis. *Development* **122**, 1651–1661.

3728. Salzberg, A., D'Evelyn, D., Schulze, K.L., Lee, J.-K., Strumpf, D., Tsai, L., and Bellen, H.J. (1994). Mutations affecting the pattern of the PNS in *Drosophila* reveal novel aspects of neuronal development. *Neuron* **13**, 269–287.

3729. Samakovlis, C., Hacohen, N., Manning, G., Sutherland, D.C., Guillemin, K., and Krasnow, M.A. (1996). Development of the *Drosophila* tracheal system occurs by a series of morphologically distinct but genetically coupled branching events. *Development* **122**, 1395–1407.

3730. Sampedro, J., Johnston, P., and Lawrence, P.A. (1993). A role for *wingless* in the segmental gradient of *Drosophila*? *Development* **117**, 677–687.

3731. Samson, M.-L. (1998). Evidence for 3' untranslated region-dependent autoregulation of the *Drosophila* gene encoding the neuronal nuclear RNA-binding protein ELAV. *Genetics* **150**, 723–733.

3732. Sánchez, L., Casares, F., Gorfinkiel, N., and Guerrero, I. (1997). The genital disc of *Drosophila melanogaster*. II. Role of the genes *hedgehog*, *decapentaplegic* and *wingless*. *Dev. Genes Evol.* **207**, 229–241. [See also 2001 *BioEssays* **23**, 698–707.]

3733. Sánchez, L., Gorfinkiel, N., and Guerrero, I. (2001). Sex determination genes control the development of the *Drosophila* genital disc, modulating the response to Hedgehog, Wingless and Decapentaplegic signals. *Development* **128**, 1033–1043.

3734. Sánchez, L. and Nöthiger, R. (1983). Sex determination and dosage compensation in *Drosophila melanogaster*: production of male clones in XX females. *EMBO J.* **2**, 485–491.

3735. Sánchez-García, I., Osada, H., Forster, A., and Rabbitts, T.H. (1993). The cysteine-rich LIM domains inhibit DNA binding by the associated homeodomain in Isl-1. *EMBO J.* **12**, 4243–4250.

3736. Sánchez-García, I. and Rabbitts, T.H. (1994). The LIM domain: a new structural motif found in zinc-finger-like proteins. *Trends Genet.* **10**, 315–320.

3737. Sánchez-Herrero, E. (1988). Heads or tails?: a homeotic gene for both. *Trends Genet.* **4**, 119–120.

3738. Sánchez-Herrero, E. (1991). Control of the expression of the bithorax complex genes abdominal-A and Abdominal-B by *cis*-regulatory regions in *Drosophila* embryos. *Development* **111**, 437–449.

3739. Sánchez-Herrero, E., Couso, J.P., Capdevila, J., and Guerrero, I. (1996). The *fu* gene discriminates between pathways to control *dpp* expression in *Drosophila* imaginal discs. *Mechs. Dev.* **55**, 159–170.

3740. Sánchez-Herrero, E., Vernós, I., Marco, R., and Morata, G. (1985). Genetic organization of *Drosophila* bithorax complex. *Nature* **313**, 108–113.

3741. Sander, K. (1991). Wilhelm Roux and the rest: developmental theories 1885–1895. *Roux's Arch. Dev. Biol.* **200**, 297–299.

3742. Sander, K. (1992). Hans Driesch the critical mechanist: "Analytische Theorie der organischen Entwicklung". *Roux's Arch. Dev. Biol.* **201**, 331–333.

3743. Sander, K. (1993). Entelechy and the ontogenetic machine–work and views of Hans Driesch from 1895 to 1910. *Roux's Arch. Dev. Biol.* **202**, 67–69.

3744. Sander, K. (1994). Of gradients and genes: developmental concepts of Theodor Boveri and his students. *Roux's Arch. Dev. Biol.* **203**, 295–297.

3745. Sander, K. (1996). Hilde Mangold (1898–1924) and Spemann's organizer: achievement and tragedy. *Roux's Arch. Dev. Biol.* **205**, 323–332.

3746. Sang, J.H. (1961). Environmental control of mutant expression. *In* "Insect Polymorphism," *Symp. Roy. Entomol. Soc. Lond.*, Vol. 1 (J.S. Kennedy, ed.). E. W. Classey: Farington, Oxon, U.K., pp. 91–102.

3747. Sanicola, M., Sekelsky, J., Elson, S., and Gelbart, W.M. (1995). Drawing a stripe in *Drosophila* imaginal disks: negative regulation of *decapentaplegic* and *patched* expression by *engrailed*. *Genetics* **139**, 745–756.

3748. Sanson, B., Alexandre, C., Fascetti, N., and Vincent, J.-P. (1999). Engrailed and Hedgehog make the range of Wingless asymmetric in *Drosophila* embryos. *Cell* **98**, 207–216.

3749. Sanson, B., White, P., and Vincent, J.-P. (1996). Uncoupling cadherin-based adhesion from *wingless* signalling in *Drosophila*. *Nature* **383**, 627–630.

3750. Santamaria, P. (1983). Analysis of haploid mosaics in *Drosophila*. *Dev. Biol.* **96**, 285–295.

3751. Santamaría, P. (1979). Heat shock induced phenocopies of dominant mutants of the Bithorax Complex in *Drosophila melanogaster*. *Mol. Gen. Genet.* **172**, 161–163.

3752. Santamaría, P. (1998). Genesis versus epigenesis: the odd jobs of the *Polycomb* group of genes. *Int. J. Dev. Biol.* **42**, 463–469.

3753. Santamaría, P., Deatrick, J., and Randsholt, N.B. (1989). Pattern triplications following genetic ablation on the wing of *Drosophila*: Effect of eliminating the *polyhomeotic* gene. *Roux's Arch. Dev. Biol.* **198**, 65–77.

3754. Santamaría, P. and Gans, M. (1980). Chimaeras of *Drosophila melanogaster* obtained by injection of haploid nuclei. *Nature* **287**, 143–144.

3755. Santamaría, P. and Randsholt, N.B. (1995). Characterization of a region of the X chromosome of *Drosophila* including *multi sex combs* (*mxc*), a *Polycomb* group gene which also functions as a tumour suppressor. *Mol. Gen. Genet.* **246**, 282–290.

3756. Santarén, J.F., Assiego, R., and García-Bellido, A. (1993). Patterns of protein synthesis in the imaginal discs of *Drosophila melanogaster*: a comparison between different discs and stages. *Roux's Arch. Dev. Biol.* **203**, 131–139.

3757. Santolini, E., Puri, C., Salcini, A.E., Gagliani, M.C., Pelicci, P.G., Tacchetti, C., and Di Fiore, P.P. (2000). Numb is an endocytic protein. *J. Cell Biol.* **151**, 1345–1351.

3758. Sapir, A., Schweitzer, R., and Shilo, B.-Z. (1998). Sequential activation of the EGF receptor pathway during *Drosophila* oogenesis establishes the dorsoventral axis. *Development* **125**, 191–200.

3759. Sasai, Y., Kageyama, R., Tagawa, Y., Shigemoto, R., and Nakanishi, S. (1992). Two mammalian helix-loop-helix factors structurally related to *Drosophila hairy* and *Enhancer of split*. *Genes Dev.* **6**, 2620–2634.

3760. Sass, G., Fehr, K., and Lucchesi, J.C. (2001). The chromatin-remodeling MSL complex of *Drosophila* spreads in cis to activated loci on the X chromosome. *Proc. 42nd Ann. Drosophila Res. Conf.* **Abstracts Vol.**, a89.

3761. Sathe, S.S. and Harte, P.J. (1995). The *Drosophila extra sex combs* protein contains WD motifs essential for its function as a repressor of homeotic genes. *Mechs. Dev.* **52**, 77–87.

3762. Sato, A., Kojima, T., Ui-Tei, K., Miyata, Y., and Saigo, K. (1999). Dfrizzled-3, a new *Drosophila* Wnt receptor, acting as an attenuator of Wingless signaling in *wingless* hypomorphic mutants. *Development* **126**, 4421–4430.

3763. Sato, M., Kojima, T., Michiue, T., and Saigo, K. (1999). *Bar* homeobox genes are latitudinal prepattern genes in the developing *Drosophila* notum whose expression is regulated by the concerted functions of *decapentaplegic* and *wingless*. *Development* **126**, 1457–1466.

3764. Sato, M. and Saigo, K. (2000). Involvement of *pannier* and *u-shaped* in regulation of Decapentaplegic-dependent *wingless* expression in developing *Drosophila* notum. *Mechs. Dev.* **93**, 127–138.

3765. Sato, T. (1984). A new homoeotic mutation affecting antennae and legs. *Dros. Info. Serv.* **60**, 180–182.

3766. Sato, T. and Denell, R.E. (1985). Homoeosis in *Drosophila*: anterior and posterior transformations of Polycomb lethal embryos. *Dev. Biol.* **110**, 53–64.

3767. Satoh, N. (1982). Timing mechanisms in early embryonic development. *Differentiation* **22**, 156–163.

3768. Sauer, F. and Jäckle, H. (1995). Heterodimeric *Drosophila* gap gene protein complexes acting as transcriptional repressors. *EMBO J.* **14**, 4773–4780.

3769. Sauer, R.T. (1990). Scissors and helical forks. *Nature* **347**, 514–515.

3770. Saulier-Le Dréan, B., Nasiadka, A., Dong, J., and Krause, H.M. (1998). Dynamic changes in the functions of Odd-skipped during early *Drosophila* embryogenesis. *Development* **125**, 4851–4861.

3771. Sawamoto, K., Okabe, M., Tanimura, T., Mikoshiba, K., Nishida, Y., and Okano, H. (1996). The *Drosophila* secreted protein Argos regulates signal transduction in the Ras/MAPK pathway. *Dev. Biol.* **178**, 13–22.

3772. Sawamoto, K., Okano, H., Kobayakawa, Y., Hayashi, S., Mikoshiba, K., and Tanimura, T. (1994). The function of *argos* in regulating cell fate decisions during *Drosophila* eye and wing vein development. *Dev. Biol.* **164**, 267–276.

3773. Scandalios, J.G. and Wright, T.R.F., eds. (1990). "Genetic Regulatory Hierarchies in Development." *Advances in Genetics*, Vol. 27. Acad. Pr., New York.

3774. Scanga, S., Gupta, R., Nohyenk, D., and Manoukian, A. (2001). Regulation of the Wingless morphogen gradient by heparin-like glycosaminoglycans. *Proc. 42nd Ann. Drosophila Res. Conf.* **Abstracts Vol.**, a149.

3775. Scanga, S., Manoukian, A., and Larsen, E. (1995). Time- and concentration-dependent response of the *Drosophila* antenna imaginal disc to *Antennapedia*. *Dev. Biol.* **169**, 673–682.

3776. Schaefer, M., Petronczki, M., Dorner, D., Forte, M., and Knoblich, J.A. (2001). Heterotrimeric G proteins direct two modes of asymmetric cell division in the *Drosophila* nervous system. *Cell* **107**, 183–194. [See also 2001 *Cell* **107**, 125–128.]

3777. Schaefer, M., Shevchenko, A., Shevchenko, A., and Knoblich, J.A. (2000). A protein complex containing Inscuteable and the Gα-binding protein Pins orients asymmetric cell divisions in *Drosophila*. *Curr. Biol.* **10**, 353–362.

3778. Scheffzek, K., Ahmadian, M.R., Kabsch, W., Wiesmüller, L., Lautwein, A., Schmitz, F., and Wittinghofer, A. (1997). The Ras-RasGAP complex: structural basis for GTPase activation and its loss in oncogenic Ras mutants. *Science* **277**, 333–338.

3779. Schejter, E.D., Segal, D., Glazer, L., and Shilo, B.-Z. (1986). Alternative 5′ exons and tissue-specific expression of the *Drosophila* EGF receptor homolog transcripts. *Cell* **46**, 1091–1101.

3780. Schejter, E.D. and Shilo, B.-Z. (1989). The *Drosophila* EGF receptor homolog (DER) gene is allelic to *faint little ball*, a locus essential for embryonic development. *Cell* **56**, 1093–1104.

3781. Schlake, T., Schorpp, M., and Boehm, T. (2000). Formation of regulator/target gene relationships during evolution. *Gene* **256**, 29–34.

3782. Schlessinger, J. (1993). How receptor tyrosine kinases activate Ras. *Trends Biochem. Sci.* **18**, 273–275.

3783. Schlessinger, J. (2000). Cell signaling by receptor tyrosine kinases. *Cell* **103**, 211–225.

3784. Schlessinger, J., Plotnikov, A.N., Ibrahimi, O.A., Eliseenkova, A.V., Yeh, B.K., Yayon, A., Linhardt, R.J., and Mohammadi, M. (2000). Crystal structure of a ternary FGF-FGFR-heparin complex reveals a dual role for heparin in FGFR binding and dimerization. *Molec. Cell* **6**, 743–750.

3785. Schlessinger, J. and Ullrich, A. (1992). Growth factor signaling by receptor tyrosine kinases. *Neuron* **9**, 383–391.

3786. Schmeichel, K.L. and Beckerle, M.C. (1994). The LIM domain is a modular protein-binding interface. *Cell* **79**, 211–219.

3787. Schmid, A.T., Tinley, T.L., and Yedvobnick, B. (1996). Transcription of the neurogenic gene *mastermind* during *Drosophila* development. *J. Exp. Zool.* **274**, 207–220.

3788. Schmid, H. (1985). Transdetermination in the homeotic eye-antenna imaginal disc of *Drosophila melanogaster*. *Dev. Biol.* **107**, 28–37.

3789. Schmid, H., Gendre, N., and Stocker, R.F. (1986). Surgical generation of supernumerary appendages for studying neuronal specificity in *Drosophila melanogaster*. *Dev. Biol.* **113**, 160–173.

3790. Schmidt, K. (1994). A puzzle: how similar signals yield different effects. *Science* **266**, 566–567.

3791. Schmidt-Ott, U., González-Gaitán, M., and Technau, G.M. (1995). Analysis of neural elements in head-mutant

Drosophila embryos suggests segmental origin of the optic lobes. *Roux's Arch. Dev. Biol.* **205**, 31–44.

3792. Schmidt-Ott, U. and Technau, G.M. (1994). Fate-mapping in the procephalic region of the embryonic *Drosophila* head. *Roux's Arch. Dev. Biol.* **203**, 367–373.

3793. Schmucker, D., Clemens, J.C., Shu, H., Worby, C.A., Xiao, J., Muda, M., Dixon, J.E., and Zipursky, S.L. (2000). *Drosophila* Dscam is an axon guidance receptor exhibiting extraordinary molecular diversity. *Cell* **101**, 671–684.

3794. Schneiderman, H.A. (1969). Control systems in insect development. *In* "Biology and the Physical Sciences," (S. Devons, ed.). Columbia Univ. Pr.: New York, pp. 186–208.

3795. Schneiderman, H.A. (1976). New ways to probe pattern formation and determination in insects. *In* "Insect Development," *Symp. Roy. Entomol. Soc. Lond.*, Vol. 8 (P.A. Lawrence, ed.). Wiley: New York, pp. 3–34.

3796. Schneiderman, H.A. (1979). Pattern formation and determination in insects. *In* "Mechanisms of Cell Change," (J. Ebert and T. Okada, eds.). Wiley: New York, pp. 243–272.

3797. Schneiderman, H.A. and Bryant, P.J. (1971). Genetic analysis of developmental mechanisms in *Drosophila*. *Nature* **234**, 187–194.

3798. Schneitz, K., Spielmann, P., and Noll, M. (1993). Molecular genetics of *aristaless*, a *prd*-type homeo box gene involved in the morphogenesis of proximal and distal pattern elements in a subset of appendages in *Drosophila*. *Genes Dev.* **7**, 114–129.

3799. Schnell, S. and Maini, P.K. (2000). Clock and induction model for somitogenesis. *Dev. Dynamics* **217**, 415–420.

3800. Schnepp, B., Donaldson, T., Grumbling, G., Ostrowski, S., Schweitzer, R., Shilo, B.-Z., and Simcox, A. (1998). EGF domain swap converts a *Drosophila* EGF receptor activator into an inhibitor. *Genes Dev.* **12**, 908–913.

3801. Schnepp, B., Grumbling, G., Donaldson, T., and Simcox, A. (1996). Vein is a novel component in the *Drosophila* epidermal growth factor receptor pathway with similarity to the neuregulins. *Genes Dev.* **10**, 2302–2313.

3802. Schneuwly, S., Klemenz, R., and Gehring, W.J. (1987). Redesigning the body plan of *Drosophila* by ectopic expression of the homoeotic gene *Antennapedia*. *Nature* **325**, 816–818.

3803. Schober, M., Schaefer, M., and Knoblich, J.A. (1999). Bazooka recruits Inscuteable to orient asymmetric cell divisions in *Drosophila* neuroblasts. *Nature* **402**, 548–551.

3804. Scholz, H., Deatrick, J., Klaes, A., and Klämbt, C. (1993). Genetic dissection of *pointed*, a *Drosophila* gene encoding two ETS-related proteins. *Genetics* **135**, 455–468.

3805. Schroeter, E.H., Kisslinger, J.A., and Kopan, R. (1998). Notch-1 signalling requires ligand-induced proteolytic release of intracellular domain. *Nature* **393**, 382–386.

3806. Schrons, H., Knust, E., and Campos-Ortega, J.A. (1992). The *Enhancer of split* Complex and adjacent genes in the 96F region of *Drosophila melanogaster* are required for segregation of neural and epidermal progenitor cells. *Genetics* **132**, 481–503.

3807. Schubiger, G. (1968). Anlageplan, Determinationszustand und Transdeterminationsleistungen der männlichen Vorderbeinscheibe von *Drosophila melanogaster*. *W. Roux' Arch. Entw.-Mech. Org.* **160**, 9–40.

3808. Schubiger, G. (1971). Regeneration, duplication and transdetermination in fragments of the leg disc of *Drosophila melanogaster*. *Dev. Biol.* **26**, 277–295.

3809. Schubiger, G. (1974). Acquisition of differentiative com-

petence in the imaginal leg discs of *Drosophila*. *W. Roux' Arch.* **174**, 303–311.

3810. Schubiger, G. and Alpert, G.D. (1975). Regeneration and duplication in a temperature sensitive homeotic mutant of *Drosophila melanogaster*. *Dev. Biol.* **42**, 292–304.

3811. Schubiger, G. and Schubiger, M. (1978). Distal transformation in *Drosophila* leg imaginal disc fragments. *Dev. Biol.* **67**, 286–295.

3812. Schubiger, M. and Palka, J. (1985). Genetic suppression of putative guidepost cells: effect on establishment of nerve pathways in *Drosophila* wings. *Dev. Biol.* **108**, 399–410.

3813. Schubiger, M. and Palka, J. (1987). Changing spatial patterns of DNA replication in the developing wing of *Drosophila*. *Dev. Biol.* **123**, 145–153.

3814. Schubiger, M. and Truman, J.W. (2000). The RXR ortholog USP suppresses early metamorphic processes in *Drosophila* in the absence of ecdysteroids. *Development* **127**, 1151–1159.

3815. Schulz, C. and Tautz, D. (1995). Zygotic *caudal* regulation by *hunchback* and its role in abdominal segment formation of the *Drosophila* embryo. *Development* **121**, 1023–1028.

3816. Schüpbach, T. and Wieschaus, E. (1998). Probing for gene specificity in epithelial development. *Int. J. Dev. Biol.* **42**, 249–255.

3817. Schüpbach, T., Wieschaus, E., and Nöthiger, R. (1978). The embryonic organization of the genital disc studied in genetic mosaics of *Drosophila melanogaster*. *W. Roux's Arch.* **185**, 249–270.

3818. Schwartz, C., Locke, J., Nishida, C., and Kornberg, T.B. (1995). Analysis of *cubitus interruptus* regulation in *Drosophila* embryos and imaginal disks. *Development* **121**, 1625–1635.

3819. Schwarz-Sommer, Z., Huijser, P., Nacken, W., Saedler, H., and Sommer, H. (1990). Genetic control of flower development by homeotic genes in *Antirrhinum majus*. *Science* **250**, 931–936.

3820. Schweisguth, F. (1995). *Suppressor of Hairless* is required for signal reception during lateral inhibition in the *Drosophila* pupal notum. *Development* **121**, 1875–1884.

3821. Schweisguth, F. (1999). Dominant-negative mutation in the β2 and β6 proteasome subunit genes affect alternative cell fate decisions in the *Drosophila* sense organ lineage. *Proc. Natl. Acad. Sci. USA* **96**, 11382–11386.

3822. Schweisguth, F. (2000). Cell polarity: Fixing cell polarity with Pins. *Curr. Biol.* **10**, R265–R267.

3823. Schweisguth, F., Gho, M., and Lecourtois, M. (1996). Control of cell fate choices by lateral signaling in the adult peripheral nervous system of *Drosophila melanogaster*. *Dev. Genet.* **18**, 28–39.

3824. Schweisguth, F. and Lecourtois, M. (1998). The activity of *Drosophila Hairless* is required in pupae but not in embryos to inhibit Notch signal transduction. *Dev. Genes Evol.* **208**, 19–27.

3825. Schweisguth, F., Nero, P., and Posakony, J.W. (1994). The sequence similarity of the *Drosophila* Suppressor of Hairless protein to the integrase domain has no functional significance *in vivo*. *Dev. Biol.* **166**, 812–814.

3826. Schweisguth, F. and Posakony, J.W. (1992). *Suppressor of Hairless*, the *Drosophila* homolog of the mouse recombination signal-binding protein gene, controls sensory organ cell fates. *Cell* **69**, 1199–1212.

3827. Schweisguth, F. and Posakony, J.W. (1994). Antagonistic activities of *Suppressor of Hairless* and *Hairless* control alternative cell fates in the *Drosophila* adult epidermis. *Development* **120**, 1433–1441.

3828. Schweitzer, R., Howes, R., Smith, R., Shilo, B.-Z., and Freeman, M. (1995). Inhibition of *Drosophila* EGF receptor activation by the secreted protein Argos. *Nature* **376**, 699–702.

3829. Schweitzer, R., Shaharabany, M., Seger, R., and Shilo, B.-Z. (1995). Secreted Spitz triggers the DER signaling pathway and is a limiting component in embryonic ventral ectoderm determination. *Genes Dev.* **9**, 1518–1529.

3830. Schweitzer, R. and Shilo, B.-Z. (1997). A thousand and one roles for the *Drosophila* EGF receptor. *Trends Genet.* **13**, 191–196.

3831. Schweizer, L. and Basler, K. (1998). *Drosophila ci^D* encodes a hybrid Pangolin/Cubitus interruptus protein that diverts the Wingless into the Hedgehog signaling pathway. *Mechs. Dev.* **78**, 141–151.

3832. Schwenk, H. (1947). Untersuchungen über die Entwicklung der Borsten bei *Drosophila*. *Nachr. Akad. Wiss. Göttingen, Math-physik. Kl. IIb, Biol.-Physiol.-Chem. Abt.* **(1947)**, 14–16.

3833. Schwyter, D.H., Huang, J.-D., Dubnicoff, T., and Courey, A.J. (1995). The *decapentaplegic* core promoter region plays an integral role in the spatial control of transcription. *Mol. Cell. Biol.* **15**, 3960–3968.

3834. Scott, J.D. and Pawson, T. (2000). Cell communication: the inside story. *Sci. Am.* **282** #6, 72–79.

3835. Scott, K., Brady, R., Jr., Cravchik, A., Morozov, P., Rzhetsky, A., Zuker, C., and Axel, R. (2001). A chemosensory gene family encoding candidate gustatory and olfactory receptors in *Drosophila*. *Cell* **104**, 661–673.

3836. Scott, M.P. (1985). Molecules and puzzles from the antennapedia homoeotic gene complex of *Drosophila*. *Trends Genet.* **1**, 74–80.

3837. Scott, M.P. (1986). More on the homeobox. *BioEssays* **5**, 88–89.

3838. Scott, M.P. (1987). Complex loci of *Drosophila*. *Annu. Rev. Biochem.* **56**, 195–227.

3839. Scott, M.P. (1992). Vertebrate homeobox gene nomenclature. *Cell* **71**, 551–553.

3840. Scott, M.P. (1994). Intimations of a creature. *Cell* **79**, 1121–1124.

3841. Scott, M.P. (1997). A common language. *Cold Spr. Harb. Symp. Quant. Biol.* **62**, 555–562.

3842. Scott, M.P. (1999). Hox proteins reach out round DNA. *Nature* **397**, 649–651.

3843. Scott, M.P. and Carroll, S.B. (1987). The segmentation and homeotic gene network in early *Drosophila* development. *Cell* **51**, 689–698.

3844. Scott, M.P. and Weiner, A.J. (1984). Structural relationships among genes that control development: sequence homology between the Antennapedia, Ultrabithorax, and fushi tarazu loci of *Drosophila*. *Proc. Natl. Acad. Sci. USA* **81**, 4115–4119.

3845. Scowcroft, W.R., Green, M.M., and Latter, B.D.H. (1968). Dosage at the *scute* locus, and canalisation of anterior and posterior scutellar bristles in *Drosophila melanogaster*. *Genetics* **60**, 373–388.

3846. Sears, R., Leone, G., DeGregori, J., and Nevins, J.R. (1999). Ras enhances Myc protein stability. *Molec. Cell* **3**, 169–179.

3847. Sedkov, Y., Tillib, S., Mizrokhi, L., and Mazo, A. (1994). The bithorax complex is regulated by *trithorax* earlier during *Drosophila* embryogenesis than is the Antennapedia complex, correlating with a bithorax-like expression pattern of distinct early *trithorax* transcripts. *Development* **120**, 1907–1917.

3848. Sedlak, B.J., Manzo, R., and Stevens, M. (1984). Localized cell death in *Drosophila* imaginal wing disc epithelium caused by the mutation *apterous-blot*. *Dev. Biol.* **104**, 489–496.

3849. Seeling, J.M., Miller, J.R., Gil, R., Moon, R.T., White, R., and Virshup, D.M. (1999). Regulation of β-catenin signaling by the B56 subunit of protein phosphatase 2A. *Science* **283**, 2089–2091.

3850. Segal, D. and Gelbert, W.M. (1985). Shortvein, a new component of the decapentaplegic gene complex in *Drosophila melanogaster*. *Genetics* **109**, 119–143.

3851. Seimiya, M. and Gehring, W.J. (2000). The *Drosophila* homeobox gene *optix* is capable of inducing ectopic eyes by an *eyeless*-independent mechanism. *Development* **127**, 1879–1886.

3852. Sekelsky, J.J., Newfeld, S.J., Raftery, L.A., Chartoff, E.H., and Gelbart, W.M. (1995). Genetic characterization and cloning of *Mothers against dpp*, a gene required for *decapentaplegic* function in *Drosophila melanogaster*. *Genetics* **139**, 1347–1358.

3853. Selleck, M.A.J. and Bronner-Fraser, M. (1995). Origins of the avian neural crest: the role of neural plate-epidermal interactions. *Development* **121**, 525–538.

3854. Selleck, S.B. (2000). Proteoglycans and pattern formation: sugar biochemistry meets developmental genetics. *Trends Genet.* **16**, 206–212.

3855. Selva, E.M., Hong, K., Baeg, G.H., Beverley, S.M., Turco, S.J., Perrimon, N., and Hacker, U. (2001). Dual role of the fringe connection gene in both heparin sulfate and fringe-dependent signalling events. *Nature Cell Biol.* **3**, 809–815.

3856. Sengar, A.S., Wang, W., Bishay, J., Cohen, S., and Egan, S.E. (1999). The EH and SH3 domain Ese proteins regulate endocytosis by linking to dynamin and Eps15. *EMBO J.* **18**, 1159–1171.

3857. Sepp, K.J. and Auld, V.J. (1999). Conversion of *lacZ* enhancer trap lines to *GAL4* lines using targeted transposition in *Drosophila melanogaster*. *Genetics* **151**, 1093–1101.

3858. Serafini, T. (1999). Finding a partner in a crowd: neuronal diversity and synaptogenesis. *Cell* **98**, 133–136.

3859. Serikaku, M.A. and O'Tousa, J.E. (1994). *sine oculis* is a homeobox gene required for *Drosophila* visual system development. *Genetics* **138**, 1137–1150.

3860. Serrano, N., Brock, H.W., and Maschat, F. (1997). *β3-tubulin* is directly repressed by the Engrailed protein in *Drosophila*. *Development* **124**, 2527–2536.

3861. Serrano, N. and Maschat, F. (1998). Molecular mechanism of *polyhomeotic* activation by Engrailed. *EMBO J.* **17**, 3704–3713.

3862. Serrano, N. and O'Farrell, P.H. (1997). Limb morphogenesis: connections between patterning and growth. *Curr. Biol.* **7**, R186–R195.

3863. Seugnet, L., Simpson, P., and Haenlin, M. (1997). Requirement for dynamin during Notch signaling in *Drosophila* neurogenesis. *Dev. Biol.* **192**, 585–598.

3864. Sewalt, R.G.A.B., Gunster, M.J., van der Vlag, J., Satijn, D.P.E., and Otte, A.P. (1999). C-terminal binding protein is a transcriptional repressor that interacts with a specific class of vertebrate Polycomb proteins. *Mol. Cell. Biol.* **19**, 777–787.

3865. Seybold, W.D. and Sullivan, D.T. (1978). Protein synthetic patterns during differentiation of imaginal discs *in vitro*. *Dev. Biol.* **65**, 69–80.

3866. Seydoux, G. and Greenwald, I. (1989). Cell autonomy of *lin-12* function in a cell fate decision in *C. elegans. Cell* **57**, 1237–1245.

3867. Shanbhag, S.R., Singh, K., and Singh, R.N. (1992). Ultrastructure of the femoral chordotonal organs and their novel synaptic organization in the legs of *Drosophila melanogaster* Meigen (Diptera: Drosophilidae). *Int. J. Insect Morphol. & Embryol.* **21**, 311–322.

3868. Shanower, G., Muller, M., Blanton, J., Gyurkovics, H., and Schedl, P. (2001). *grappa*: a new *trithorax/Polycomb*-group gene involved in telomere silencing. *Proc. 42nd Ann. Drosophila Res. Conf.* **Abstracts Vol.**, a98.

3869. Shao, Z., Raible, F., Mollaaghababa, R., Guyon, J.R., Wu, C.-t., Bender, W., and Kingston, R.E. (1999). Stabilization of chromatin structure by PRC1, a Polycomb complex. *Cell* **98**, 37–46.

3870. Shapiro, L. and Colman, D.R. (1999). The diversity of cadherins and implications for a synaptic adhesive code in the CNS. *Neuron* **23**, 427–430.

3871. Sharkey, M., Graba, Y., and Scott, M.P. (1997). *Hox* genes in evolution: protein surfaces and paralog groups. *Trends Genet.* **13**, 145–151.

3872. Sharp, E.J., Martin, E.C., and Adler, P.N. (1994). Directed overexpression of *Suppressor 2 of zeste* and *Posterior Sex Combs* results in bristle abnormalities in *Drosophila melanogaster. Dev. Biol.* **161**, 379–392.

3873. Sharpe, C., Lawrence, N., and Martinez Arias, A. (2001). Wnt signalling: a theme with nuclear variations. *BioEssays* **23**, 311–318.

3874. Sharrocks, A.D., Yang, S.-H., and Galanis, A. (2000). Docking domains and substrate specificity determination for MAP kinases. *Trends Biochem.* **25**, 448–453.

3875. Shashidhara, L.S., Agrawal, N., Bajpai, R., Bharathi, V., and Sinha, P. (1999). Negative regulation of dorsoventral signaling by the homeotic gene *Ultrabithorax* during haltere development in *Drosophila. Dev. Biol.* **212**, 491–502.

3876. Shaw, G. (1996). The pleckstrin homology domain: an intriguing multifunctional protein module. *BioEssays* **18**, 35–46.

3877. Shaw, S.R. and Meinertzhagen, I.A. (1986). Evolutionary progression at synaptic connections made by identified homologous neurones. *Proc. Natl. Acad. Sci. USA* **83**, 7961–7965.

3878. Shaw, V.K. and Bryant, P.J. (1975). Intercalary leg regeneration in the large milkweed bug *Oncopeltus fasciatus. Dev. Biol.* **45**, 187–191.

3879. Shaywitz, A.J. and Greenberg, M.E. (1999). CREB: a stimulus-induced transcription factor activated by a diverse array of extracellular signals. *Annu. Rev. Biochem.* **68**, 821–861.

3880. Shearn, A. (1977). Mutational dissection of imaginal disc development in *Drosophila melanogaster. Amer. Zool.* **17**, 585–594.

3881. Shearn, A. (1978). Mutational dissection of imaginal disc development. *In* "The Genetics and Biology of *Drosophila*," Vol. 2c (M. Ashburner and T.R.F. Wright, eds.). Acad. Pr.: New York, pp. 443–510.

3882. Shearn, A. (1980). What is the normal function of genes which give rise to homeotic mutations? *In* "Development and Neurobiology of *Drosophila*," (O. Siddiqi, P. Babu, L.M. Hall, and J.C. Hall, eds.). Plenum: New York, pp. 155–162.

3883. Shearn, A. (1985). Analysis of transdetermination. *In* "Comprehensive Insect Physiology, Biochemistry, and Pharmacology," Vol. 2 (G.A. Kerkut and L.I. Gilbert, eds.). Pergamon Pr.: Oxford, pp. 183–199.

3884. Shearn, A. (1989). The *ash-1, ash-2* and *trithorax* genes of *Drosophila melanogaster* are functionally related. *Genetics* **121**, 517–525.

3885. Shearn, A., Davis, K.T., and Hersperger, E. (1978). Transdetermination of *Drosophila* imaginal discs cultured *in vitro. Dev. Biol.* **65**, 536–540.

3886. Shearn, A. and Garen, A. (1974). Genetic control of imaginal disc development in *Drosophila. Proc. Natl. Acad. Sci. USA* **71**, 1393–1397.

3887. Shearn, A., Martin, A., Davis, K., and Hersperger, E. (1984). Genetic analysis of transdetermination in *Drosophila*. I. The effects of varying growth parameters using a temperature-sensitive mutation. *Dev. Biol.* **106**, 135–146.

3888. Shearn, A., Rice, T., Garen, A., and Gehring, W. (1971). Imaginal disc abnormalities in lethal mutants of *Drosophila. Proc. Natl. Acad. Sci. USA* **68**, 2594–2598.

3889. Sheldahl, L.C., Park, M., Malbon, C.C., and Moon, R.T. (1999). Protein kinase C is differentially stimulated by Wnt and Frizzled homologs in a G-protein-dependent manner. *Curr. Biol.* **9**, 695–698.

3890. Sheldon, B.L. (1968). Studies on the scutellar bristles of *Drosophila melanogaster*. I. Basic variability, some temperature and culture effects, and responses to short-term selection in the Oregon-RC strain. *Aust. J. Biol. Sci.* **21**, 721–740.

3891. Shellenbarger, D.L. and Mohler, J.D. (1978). Temperature-sensitive periods and autonomy of pleiotropic effects of $l(1)N^{ts1}$, a conditional Notch lethal in *Drosophila. Dev. Biol.* **62**, 432–446.

3892. Shen, C.-P., Jan, L.Y., and Jan, Y.N. (1997). Miranda is required for the asymmetric localization of Prospero during mitosis in *Drosophila. Cell* **90**, 449–458.

3893. Shen, C.-P., Knoblich, J.A., Chan, Y.-M., Jiang, M.-M., Jan, L.Y., and Jan, Y.N. (1998). Miranda as a multidomain adapter linking apically localized Inscuteable and basally localized Staufen and Prospero during asymmetric cell division in *Drosophila. Genes Dev.* **12**, 1837–1846.

3894. Shen, W. and Mardon, G. (1997). Ectopic eye development in *Drosophila* induced by directed *dachshund* expression. *Development* **124**, 45–52.

3895. Sheng, G., Thouvenot, E., Schmucker, D., Wilson, D.S., and Desplan, C. (1997). Direct regulation of *rhodopsin 1* by *Pax-6/eyeless* in *Drosophila*: evidence for a conserved function in photoreceptors. *Genes Dev.* **11**, 1122–1131.

3896. Shepard, S.B., Broverman, S.A., and Muskavitch, M.A.T. (1989). A tripartite interaction among alleles of *Notch, Delta*, and *Enhancer of split* during imaginal development of *Drosophila melanogaster. Genetics* **122**, 429–438.

3897. Shi, Y. (2001). Structural insights on Smad function in TGFβ signaling. *BioEssays* **23**, 223–232.

3898. Shi, Y., Wang, Y.-F., Jayaraman, L., Yang, H., Massagué, J., and Pavletich, N.P. (1998). Crystal structure of a Smad MH1 domain bound to DNA: insights on DNA binding in TGF-β signaling. *Cell* **94**, 585–594.

3899. Shields, J.M., Pruitt, K., McFall, A., Shaub, A., and Der, C.J. (2000). Understanding Ras: "It ain't over 'til it's over". *Trends Cell Biol.* **10**, 147–154.

3900. Shilo, B.-Z. and Raz, E. (1991). Developmental control by the *Drosophila* EGF receptor homolog DER. *Trends Genet.* **7**, 388–392.

3901. Shimell, M.J., Ferguson, E.L., Childs, S.R., and O'Connor, M.B. (1991). The *Drosophila* dorsal-ventral patterning gene *tolloid* is related to human bone morphogenetic protein 1. *Cell* **67**, 469–481.

3902. Shimell, M.J., Simon, J., Bender, W., and O'Connor, M.B. (1994). Enhancer point mutation results in a homeotic transformation in *Drosophila*. *Science* **264**, 968–971.

3903. Shine, I. and Wrobel, S. (1976). "Thomas Hunt Morgan, Pioneer of Genetics." Kentucky Univ. Pr., Lexington, Kentucky.

3904. Shiomi, K., Takeichi, M., Nishida, Y., Nishi, Y., and Uemura, T. (1994). Alternative cell fate choice induced by low-level expression of a regulator of protein phosphatase 2A in the *Drosophila* peripheral nervous system. *Development* **120**, 1591–1599.

3905. Shirras, A.D. and Couso, J.P. (1996). Cell fates in the adult abdomen of *Drosophila* are determined by *wingless* during pupal development. *Dev. Biol.* **175**, 24–36.

3906. Shishido, E., Higashijima, S.-i., Emori, Y., and Saigo, K. (1993). Two FGF-receptor homologues of *Drosophila*: one is expressed in mesodermal primordium in early embryos. *Development* **117**, 751–761.

3907. Shore, P. and Sharrocks, A.D. (1995). The MADS-box family of transcription factors. *Eur. J. Biochem.* **229**, 1–13.

3908. Shoresh, M., Orgad, S., Shmueli, O., Werczberger, R., Gelbaum, D., Abiri, S., and Segal, D. (1998). Overexpression *Beadex* mutations and loss-of-function *heldup-a* mutations in *Drosophila* affect the 3′ regulatory and coding components, respectively, of the *Dlmo* Gene. *Genetics* **150**, 283–299.

3909. Shostak, S. (1998). "Death of Life: The Legacy of Molecular Biology." MacMillan, London.

3910. Shtutman, M., Zhurinsky, J., Simcha, I., Albanese, C., D'Amico, M., Pestell, R., and Ben-Ze'ev, A. (1999). The cyclin D1 gene is a target of the β-catenin/LEF-1 pathway. *Proc. Natl. Acad. Sci. USA* **96**, 5522–5527.

3911. Shubin, N., Tabin, C., and Carroll, S. (1997). Fossils, genes and the evolution of animal limbs. *Nature* **388**, 639–648.

3912. Shulman, J.M., Perrimon, N., and Axelrod, J.D. (1998). Frizzled signaling and the developmental control of cell polarity. *Trends Genet.* **14**, 452–458. [See also 2001 references: *Cell* **106**, 355–366; *Curr. Biol.* **11**, R506–R509; *Development* **128**, 3209–3220; *Genes Dev.* **15**, 1182–1187.]

3913. Shupliakov, O., Löw, P., Grabs, D., Gad, H., Chen, H., David, C., Takei, K., De Camilli, P., and Brodin, L. (1997). Synaptic vesicle endocytosis impaired by disruption of dynamin-SH3 domain interactions. *Science* **276**, 259–263.

3914. Sibatani, A. (1977). Possible interaction of positional information and its interpretation in development: a model. *Australian Biochem. Soc.* **10**, 74.

3915. Sibatani, A. (1981). The Polar Co-ordinate Model for pattern regulation in epimorphic fields: a critical appraisal. *J. Theor. Biol.* **93**, 433–489.

3916. Sibatani, A. (1983). A plausible molecular interpretation of the polar coordinate model with a centripetal degeneracy of distinctive field values. *J. Theor. Biol.* **103**, 421–428.

3917. Sibbons, J.P., Shandala, T., and Saint, R. (2001). Characterisation of the *Drosophila dead ringer* gene in eye development. *Proc. 42nd Ann. Drosophila Res. Conf.* **Abstracts Vol.**, a272.

3918. Siddiqi, O. and Rodrigues, V. (1980). Genetic analysis of a complex chemoreceptor. *In* "Development and Neurobiology of *Drosophila*," (O. Siddiqi, P. Babu, L.M. Hall, and J.C. Hall, eds.). Plenum: New York, pp. 347–359.

3919. Siegfried, E. (1999). Role of *Drosophila* Wingless signaling in cell fate determination. *In* "Cell Lineage and Fate Determination," (S.A. Moody, ed.). Acad. Pr.: New York, pp. 249–271. [See also 2001 references concerning Wingless secretion (*BioEssays* **23**, 869–872; *Cell* **105**, 197–207, 209–219, and 613–624; *Curr. Biol.* **11**, R638–R639); binding (*Dev. Biol.* **235**, 467–475; *Nature* **412**, 86–90); transduction (*Cell* **105**, 391–402 and 563–566; *Curr. Biol.* **11**, R524–R526; *Dev. Biol.* **235**, 303–313 and 433–448); and other Wnts (*Dev. Biol.* **232**, 339–350; *Mechs. Dev.* **103**, 117–120).]

3920. Siegfried, E., Ambrosio, L., and Perrimon, N. (1990). Serine/threonine protein kinases in *Drosophila*. *Trends Genet.* **6**, 357–362.

3921. Siegfried, E., Chou, T.-B., and Perrimon, N. (1992). *wingless* signaling acts through *zeste-white 3*, the *Drosophila* homolog of *glycogen synthase kinase-3*, to regulate *engrailed* and establish cell fate. *Cell* **71**, 1167–1179.

3922. Siegfried, E., Perkins, L.A., Capaci, T.M., and Perrimon, N. (1990). Putative protein kinase product of the *Drosophila* segment-polarity gene *zeste-white3*. *Nature* **345**, 825–829.

3923. Siegfried, E. and Perrimon, N. (1994). *Drosophila* wingless: a paradigm for the function and mechanism of Wnt signaling. *BioEssays* **16**, 395–404.

3924. Siegfried, E., Wilder, E.L., and Perrimon, N. (1994). Components of *wingless* signalling in *Drosophila*. *Nature* **367**, 76–80.

3925. Siegler, M.V.S. and Jia, X.X. (1999). Engrailed negatively regulates the expression of cell adhesion molecules Connectin and Neuroglian in embryonic *Drosophila* nervous system. *Neuron* **22**, 265–276.

3926. Sigrist, S., Jacobs, H., Stratmann, R., and Lehner, C.F. (1995). Exit from mitosis is regulated by *Drosophila fizzy* and the sequential destruction of cyclins A, B and B3. *EMBO J.* **14**, 4827–4838.

3927. Sigrist, S.J. and Lehner, C.F. (1997). *Drosophila fizzy-related* down-regulates mitotic cyclins and is required for cell proliferation arrest and entry into endocycles. *Cell* **90**, 671–681.

3928. Simcox, A. (1997). Differential requirement for EGF-like ligands in *Drosophila* wing development. *Mechs. Dev.* **62**, 41–50.

3929. Simcox, A.A., Grumbling, G., Schnepp, B., Bennington-Mathias, C., Hersperger, E., and Shearn, A. (1996). Molecular, phenotypic, and expression analysis of *vein*, a gene required for growth of the *Drosophila* wing disc. *Dev. Biol.* **177**, 475–489.

3930. Simcox, A.A., Hersperger, E., Shearn, A., Whittle, J.R.S., and Cohen, S.M. (1991). Establishment of imaginal discs and histoblast nests in *Drosophila*. *Mechs. Dev.* **34**, 11–20.

3931. Simcox, A.A., Wurst, G., Hersperger, E., and Shearn, A. (1987). The *defective dorsal discs* gene of *Drosophila* is required for the growth of specific imaginal discs. *Dev. Biol.* **122**, 559–567. (*N.B.*: The "*ddd*" gene studied here was renamed "*vein*".)

3932. Simin, K., Bates, E.A., Horner, M.A., and Letsou, A. (1998). Genetic analysis of Punt, a type II Dpp receptor that functions throughout the *Drosophila melanogaster* life cycle. *Genetics* **148**, 801–813.

3933. Simmonds, A., Hughes, S., Tse, J., Cocquyt, S., and Bell, J. (1997). The effect of dominant *vestigial* alleles upon *vestigial*-mediated wing patterning during development of *Drosophila melanogaster*. *Mechs. Dev.* **67**, 17–33.

3934. Simmonds, A.J. and Bell, J.B. (1998). A genetic and molecular analysis of an *invected*[Dominant] mutation in *Drosophila melanogaster*. *Genome* **41**, 381–390.

3935. Simmonds, A.J., Brook, W.J., Cohen, S.M., and Bell, J.B.

(1995). Distinguishable functions for *engrailed* and *invected* in anterior-posterior patterning in the *Drosophila* wing. *Nature* **376**, 424–427.

3936. Simmonds, A.J., Liu, X., Soanes, K.H., Krause, H.M., Irvine, K.D., and Bell, J.B. (1998). Molecular interactions between Vestigial and Scalloped promote wing formation in *Drosophila. Genes Dev.* **12**, 3815–3820. [See also 2001 *Development* **128**, 3295–3305.]

3937. Simmons, D.L. (1993). Dissecting the modes of interactions amongst cell adhesion molecules. *Development* **1993 Suppl.**, 193–203.

3938. Simon, J. (1995). Locking in stable states of gene expression: transcriptional control during *Drosophila* development. *Curr. Opin. Cell Biol.* **7**, 376–385.

3939. Simon, J., Bornemann, D., Lunde, K., and Schwartz, C. (1995). The *extra sex combs* product contains WD40 repeats and its time of action implies a role distinct from other *Polycomb* group products. *Mechs. Dev.* **53**, 197–208. [See also 2001 *Genes Dev.* **15**, 2509–2514.]

3940. Simon, J., Peifer, M., Bender, W., and O'Connor, M. (1990). Regulatory elements of the bithorax complex that control expression along the anterior-posterior axis. *EMBO J.* **9**, 3945–3956.

3941. Simon, M.A. (1994). Signal transduction during the development of the *Drosophila* R7 photoreceptor. *Dev. Biol.* **166**, 431–442.

3942. Simon, M.A. (2000). Receptor tyrosine kinases: specific outcomes from general signals. *Cell* **103**, 13–15.

3943. Simon, M.A., Bowtell, D.D.L., Dodson, G.S., Laverty, T.R., and Rubin, G.M. (1991). Ras1 and a putative guanine nucleotide exchange factor perform crucial steps in signaling by the Sevenless protein tyrosine kinase. *Cell* **67**, 701–716.

3944. Simon, M.A., Dodson, G.S., and Rubin, G.M. (1993). An SH3-SH2-SH3 protein is required for p21^{Ras1} activation and binds to Sevenless and Sos proteins in vitro. *Cell* **73**, 169–177.

3945. Simonelig, M., Elliott, K., Mitchelson, A., and O'Hare, K. (1996). Interallelic complementation at the *suppressor of forked* locus of *Drosophila* reveals complementation between Suppressor of forked proteins mutated in different regions. *Genetics* **142**, 1225–1235.

3946. Simons, K. and Ikonen, E. (1997). Functional rafts in cell membranes. *Nature* **387**, 569–572.

3947. Simpson, P. (1976). Analysis of the compartments of the wing of *Drosophila melanogaster* mosaic for a temperature-sensitive mutation that reduces mitotic rate. *Dev. Biol.* **54**, 100–115.

3948. Simpson, P. (1979). Parameters of cell competition in the compartments of the wing disc of *Drosophila. Dev. Biol.* **69**, 182–193.

3949. Simpson, P. (1981). Growth and cell competition in *Drosophila. J. Embryol. Exp. Morph.* **65 (Suppl.)**, 77–88.

3950. Simpson, P. (1990). Lateral inhibition and the development of the sensory bristles of the adult peripheral nervous system of *Drosophila. Development* **109**, 509–519.

3951. Simpson, P. (1990). *Notch* and the choice of cell fate in *Drosophila* neuroepithelium. *Trends Genet.* **6**, 343–345.

3952. Simpson, P. (1993). Flipping fruit-flies: a powerful new technique for generating *Drosophila* mosaics. *Trends Genet.* **9**, 227–228.

3953. Simpson, P. (1996). *Drosophila* development: A prepattern for sensory organs. *Curr. Biol.* **6**, 948–950.

3954. Simpson, P. (1997). Notch signalling in development: on

equivalence groups and asymmetric developmental potential. *Curr. Op. Gen. Dev.* **7**, 537–542.

3955. Simpson, P., Berreur, P., and Berreur-Bonnenfant, J. (1980). The initiation of pupariation in *Drosophila*: dependence on growth of the imaginal discs. *J. Embryol. Exp. Morph.* **57**, 155–165.

3956. Simpson, P. and Carteret, C. (1989). A study of *shaggy* reveals spatial domains of expression of *achaete-scute* alleles on the thorax of *Drosophila. Development* **106**, 57–66.

3957. Simpson, P. and Carteret, C. (1990). Proneural clusters: equivalence groups in the epithelium of *Drosophila. Development* **110**, 927–932.

3958. Simpson, P., El Messal, M., Moscoso del Prado, J., and Ripoll, P. (1988). Stripes of positional homologies across the wing blade of *Drosophila melanogaster. Development* **103**, 391–401.

3959. Simpson, P. and Grau, Y. (1987). The segment polarity gene *costal-2* in *Drosophila*. II. The origin of imaginal pattern duplications. *Dev. Biol.* **122**, 201–209.

3960. Simpson, P., Lawrence, P.A., and Maschat, F. (1981). Clonal analysis of two wing-scalloping mutants of *Drosophila. Dev. Biol.* **84**, 206–211.

3961. Simpson, P. and Morata, G. (1980). The control of growth in the imaginal discs of *Drosophila. In* "Development and Neurobiology of *Drosophila*," (O. Siddiqi, P. Babu, L.M. Hall, and J.C. Hall, eds.). Plenum: New York, pp. 129–139.

3962. Simpson, P. and Morata, G. (1981). Differential mitotic rates and patterns of growth in compartments in the *Drosophila* wing. *Dev. Biol.* **85**, 299–308.

3963. Simpson, P., Ruel, L., Heitzler, P., and Bourouis, M. (1993). A dual role for the protein kinase *shaggy* in the repression of *achaete-scute. Development* **1993 Suppl.**, 29–39.

3964. Simpson, P. and Schneiderman, H.A. (1975). Isolation of temperature sensitive mutations blocking clone development in *Drosophila melanogaster*, and the effects of a temperature sensitive cell lethal mutation on pattern formation in imaginal discs. *Wilhelm Roux's Arch.* **178**, 247–275.

3965. Simpson, P. and Schneiderman, H.A. (1976). A temperature sensitive mutation that reduces mitotic rate in *Drosophila melanogaster. W. Roux's Arch.* **179**, 215–236.

3966. Simpson, P., Woehl, R., and Usui, K. (1999). The development and evolution of bristle patterns in Diptera. *Development* **126**, 1349–1364.

3967. Sinclair, A.H., Berta, P., Palmer, M.S., Hawkins, J.R., Griffiths, B.L., Smith, M.J., Foster, J.W., Frischauf, A.-M., Lovell-Badge, R., and Goodfellow, P.N. (1990). A gene from the human sex-determining region encodes a protein with homology to a conserved DNA-binding motif. *Nature* **346**, 240–244.

3968. Sinclair, D.A.R., Clegg, N.J., Antonchuk, J., Milne, T.A., Stankunas, K., Ruse, C., Grigliatti, T.A., Kassis, J.A., and Brock, H.W. (1998). *Enhancer of Polycomb* is a suppressor of position-effect variegation in *Drosophila melanogaster. Genetics* **148**, 211–220.

3969. Sinclair, D.A.R., Grigliatti, T.A., and Kaufman, T.C. (1984). Effects of a temperature-sensitive *Minute* mutation on gene expression in *Drosophila melanogaster. Genet. Res., Camb.* **43**, 257–275.

3970. Sinclair, D.A.R., Milne, T.A., Hodgson, J.W., Shellard, J., Salinas, C.A., Kyba, M., Randazzo, F., and Brock, H.W. (1998). The *Additional sex combs* gene of *Drosophila* encodes a chromatin protein that binds to shared and unique Polycomb group sites on polytene chromosomes. *Development* **125**, 1207–1216.

3971. Singer, M.A., Hortsch, M., Goodman, C.S., and Bentley, M. (1992). Annulin, a protein expressed at limb segment boundaries in the grasshopper embryo, is homologous to protein cross-linking transglutaminases. *Dev. Biol.* **154**, 143–159.

3972. Singer, M.A., Penton, A., Twombly, V., Hoffmann, F.M., and Gelbart, W.M. (1997). Signaling through both type I DPP receptors is required for anterior-posterior patterning of the entire *Drosophila* wing. *Development* **124**, 79–89.

3973. Singh, A., Kango-Singh, M., and Sun, Y.H. (2001). *teashirt* collaborates with Wg to suppress ventral eye development. *Proc. 42nd Ann. Drosophila Res. Conf.* **Abstracts Vol.**, a202.

3974. Singson, A., Leviten, M.W., Bang, A.G., Hua, X.H., and Posakony, J.W. (1994). Direct downstream targets of proneural activators in the imaginal disc include genes involved in lateral inhibitory signaling. *Genes Dev.* **8**, 2058–2071.

3975. Sipos, L., Mihály, J., Karch, F., Schedl, P., Gausz, J., and Gyurkovics, H. (1998). Transvection in the *Drosophila Abd-B* domain: extensive upstream sequences are involved in anchoring distant *cis*-regulatory regions to the promoter. *Genetics* **149**, 1031–1050.

3976. Sisson, J.C., Ho, K.S., Suyama, K., and Scott, M.P. (1997). Costal2, a novel kinesin-related protein in the Hedgehog signaling pathway. *Cell* **90**, 235–245.

3977. Sivasankaran, R., Calleja, M., Morata, G., and Basler, K. (2000). The Wingless target gene *Dfz3* encodes a new member of the *Drosophila* Frizzled family. *Mechs. Dev.* **91**, 427–431.

3978. Sivasankaran, R., Vigano, M.A., Müller, B., Affolter, M., and Basler, K. (2000). Direct transcriptional control of the Dpp target *omb* by the DNA binding protein Brinker. *EMBO J.* **19**, 6162–6172.

3979. Skaer, H. (1998). Who pulls the string to pattern cell division in *Drosophila? Trends Genet.* **14**, 337–339.

3980. Skaer, N. and Simpson, P. (2000). Genetic analysis of bristle loss in hybrids between *Drosophila melanogaster* and *D. simulans* provides evidence for divergence of *cis*-regulatory sequences in the *achaete-scute* gene complex. *Dev. Biol.* **221**, 148–167.

3981. Skeath, J.B. (1999). At the nexus between pattern formation and cell-type specification: the generation of individual neuroblast fates in the *Drosophila* embryonic central nervous system. *BioEssays* **21**, 922–931.

3982. Skeath, J.B. and Carroll, S.B. (1991). Regulation of *achaete-scute* gene expression and sensory organ pattern formation in the *Drosophila* wing. *Genes Dev.* **5**, 984–995.

3983. Skeath, J.B. and Carroll, S.B. (1992). Regulation of proneural gene expression and cell fate during neuroblast segregation in the *Drosophila* embryo. *Development* **114**, 939–946.

3984. Skeath, J.B. and Carroll, S.B. (1994). The *achaete-scute* complex: generation of cellular pattern and fate within the *Drosophila* nervous system. *FASEB J.* **8**, 714–721.

3985. Skeath, J.B. and Doe, C.Q. (1998). Sanpodo and Notch act in opposition to Numb to distinguish sibling neuron fates in the *Drosophila* CNS. *Development* **125**, 1857–1865.

3986. Skeath, J.B., Panganiban, G., Selegue, J., and Carroll, S.B. (1992). Gene regulation in two dimensions: the proneural *achaete* and *scute* genes are controlled by combinations of axis-patterning genes through a common intergenic control region. *Genes Dev.* **6**, 2606–2619.

3987. Skripsky, T. and Lucchesi, J.C. (1982). Intersexuality resulting from the interaction of sex-specific lethal mutations in *Drosophila melanogaster. Dev. Biol.* **94**, 153–162.

3988. Slack, J.M.W. (1987). Morphogenetic gradients–past and present. *Trends Bioch. Sci.* **12**, 200–204.

3989. Slack, J.M.W. (1994). How to make the gradient. *Nature* **371**, 477–478.

3990. Sliter, T.J. (1989). Imaginal disc-autonomous expression of a defect in sensory bristle patterning caused by the *lethal(3)ecdysoneless¹ (l(3)ecd¹)* mutation of *Drosophila melanogaster. Development* **106**, 347–354.

3991. Slusarski, D.C., Motzny, C.K., and Holmgren, R. (1995). Mutations that alter the timing and pattern of *cubitus interruptus* gene expression in *Drosophila melanogaster. Genetics* **139**, 229–240.

3992. Small, S., Arnosti, D.N., and Levine, M. (1993). Spacing ensures autonomous expression of different stripe enhancers in the *even-skipped* promoter. *Development* **119**, 767–772.

3993. Small, S., Blair, A., and Levine, M. (1996). Regulation of two pair-rule stripes by a single enhancer in the *Drosophila* embryo. *Dev. Biol.* **175**, 314–324.

3994. Smalley, M.J., Sara, E., Paterson, H., Naylor, S., Cook, D., Jayatilake, H., Fryer, L.G., Hutchinson, L., Fry, M.J., and Dale, T.C. (1999). Interaction of Axin and Dvl-2 proteins regulates Dvl-2-stimulated TCF-dependent transcription. *EMBO J.* **18**, 2823–2835.

3995. Smith, A.V. and Orr-Weaver, T.L. (1991). The regulation of the cell cycle during *Drosophila* embryogenesis: the transition to polyteny. *Development* **112**, 997–1008.

3996. Smith, C.W.J. and Valcárcel, J. (2000). Alternative pre-mRNA splicing: the logic of combinatorial control. *Trends Biochem. Sci.* **25**, 381–388.

3997. Smith, E., Pannuti, A., Allis, C.D., and Lucchesi, J.C. (2001). Mapping of MSL proteins and H4 acetylated at lysine 16 on dosage compensation genes. *Proc. 42nd Ann. Drosophila Res. Conf.* **Abstracts Vol.**, a90.

3998. Smith, E.R., Belote, J.M., Schiltz, R.L., Yang, X.-J., Moore, P.A., Berger, S.L., Nakatani, Y., and Allis, C.D. (1998). Cloning of *Drosophila* GCN5: conserved features among metazoan GCN5 family members. *Nucl. Acids Res.* **26**, 2948–2954.

3999. Smith, J. (1996). How to tell a cell where it is. *Nature* **381**, 367–368.

4000. Smith, J. (1997). *Brachyury* and the T-box genes. *Curr. Opin. Gen. Dev.* **7**, 474–480.

4001. Smith, J. (1999). T-box genes: what they do and how they do it. *Trends Genet.* **15**, 154–158.

4002. Smith, R.K., Carroll, P.M., Allard, J.D., and Simon, M.A. (2002). MASK, a large ankyrin repeat and KH domain-containing protein involved in *Drosophila* receptor tyrosine kinase signaling. *Development* **129**, 71–82.

4003. Smith, S.T. and Jaynes, J.B. (1996). A conserved region of engrailed, shared among all en-, gsc-, Nk1-, Nk2- and msh-class homeoproteins, mediates active transcriptional repression in vivo. *Development* **122**, 3141–3150.

4004. Smith, T.F., Gaitatzes, C., Saxena, K., and Neer, E.J. (1999). The WD repeat: a common architecture for diverse functions. *Trends Biochem. Sci.* **24**, 181–185.

4005. Smolen, P., Baxter, D.A., and Byrne, J.H. (2000). Mathematical modeling of gene networks. *Neuron* **26**, 567–580.

4006. Smolik-Utlaut, S.M. (1990). Dosage requirements of *Ultrabithorax* and *bithoraxoid* in the determination of segment identity in *Drosophila melanogaster. Genetics* **124**, 357–366.

4007. Smoller, D., Friedel, C., Schmid, A., Bettler, D., Lam, L., and Yedvobnick, B. (1990). The *Drosophila* neurogenic locus *mastermind* encodes a nuclear protein unusually rich in amino acid homopolymers. *Genes Dev.* **4**, 1688–1700.

4008. Snodgrass, R.E. (1935). "Principles of Insect Morphology." McGraw-Hill, New York.

4009. Snodgrass, R.E. (1958). Evolution of arthropod mechanisms. *Smithsonian Misc. Coll.* **138** #2, i-77.

4010. Snow, M.H.L. and Tam, P.P.L. (1980). Timing in embryological development. *Nature* **286**, 107. [See also 2001 *Cell* **106**, 133–136.]

4011. Soll, D.R. (1983). A new method for examining the complexity and relationships of "timers" in developing systems. *Dev. Biol.* **95**, 73–91.

4012. Sondek, J., Bohm, A., Lambright, D.G., Hamm, H.E., and Sigler, P.B. (1996). Crystal structure of a G$_A$ protein βγ dimer at 2.1 Å resolution. *Nature* **379**, 369–374.

4013. Sondhi, K.C. (1962). The evolution of a pattern. *Evolution* **16**, 186–191.

4014. Sondhi, K.C. (1963). The biological foundations of animal patterns. *Quart. Rev. Biol.* **38**, 289–327.

4015. Sondhi, K.C. (1965). Genetic control of an anteroposterior gradient and its bearing on structural orientation in *Drosophila*. *Genetics* **51**, 653–657.

4016. Sondhi, K.C. (1970). Dynamics of pattern formation. *Genetica* **41**, 111–118.

4017. Song, Y., Chung, S., and Kunes, S. (2000). Combgap relays Wingless signal reception to the determination of cortical cell fate in the *Drosophila* visual system. *Molec. Cell* **6**, 1143–1154.

4018. Song, Z., McCall, K., and Steller, H. (1997). DCP-1, a *Drosophila* cell death protease essential for development. *Science* **275**, 536–540.

4019. Songyang, Z., Fanning, A.S., Fu, C., Xu, J., Marfatia, S.M., Chishti, A.H., Crompton, A., Chan, A.C., Anderson, J.M., and Cantley, L.C. (1997). Recognition of unique carboxyl-terminal motifs by distinct PDZ domains. *Science* **275**, 73–77.

4020. Songyang, Z., Shoelson, S.E., Chaudhuri, M., Gish, G., Pawson, T., Haser, W.G., King, F., Roberts, T., Ratnofsky, S., Lechleider, R.J., Neel, B.G., Birge, R.B., Fajardo, J.E., Chou, M.M., Hanafusa, H., Schaffhausen, B., and Cantley, L.C. (1993). SH2 domains recognize specific phosphopeptide sequences. *Cell* **72**, 767–778.

4021. Sonnenblick, B.P. (1950). The early embryology of *Drosophila melanogaster*. *In* "Biology of *Drosophila*," (M. Demerec, ed.). Hafner: New York, pp. 62–167.

4022. Sonnenfeld, M., Ward, M., Nystrom, G., Mosher, J., Stahl, S., and Crews, S. (1997). The *Drosophila tango* gene encodes a bHLH-PAS protein that is orthologous to mammalian Arnt and controls CNS midline and tracheal development. *Development* **124**, 4571–4582.

4023. Sonnenfeld, M.J. and Jacobs, J.R. (1994). Mesectodermal cell fate analysis in *Drosophila* midline mutants. *Mechs. Dev.* **46**, 3–13.

4024. Sotillos, S. and Campuzano, S. (2000). *DRacGAP*, a novel *Drosophila* gene, inhibits EGFR/Ras signalling in the developing imaginal wing disc. *Development* **127**, 5427–5438.

4025. Sotillos, S., Roch, F., and Campuzano, S. (1997). The metalloprotease-disintegrin Kuzbanian participates in *Notch* activation during growth and patterning of *Drosophila* imaginal discs. *Development* **124**, 4769–4779.

4026. Soto, M.C., Chou, T.-B., and Bender, W. (1995). Comparison of germline mosaics of genes in the Polycomb Group of *Drosophila melanogaster*. *Genetics* **140**, 231–243.

4027. Spana, E.P. and Doe, C.Q. (1996). Numb antagonizes Notch signaling to specify sibling neuron cell fates. *Neuron* **17**, 21–26.

4028. Spana, E.P., Kopczynski, C., Goodman, C.S., and Doe, C.Q. (1995). Asymmetric localization of Numb autonomously determines sibling neuron identity in the *Drosophila* CNS. *Development* **121**, 3489–3494.

4029. Speicher, S.A., Thomas, U., Hinz, U., and Knust, E. (1994). The *Serrate* locus of *Drosophila* and its role in morphogenesis of the wing imaginal discs: control of cell proliferation. *Development* **120**, 535–544.

4030. Spemann, H. (1938). "Embryonic Development and Induction." Yale Univ. Pr., New Haven.

4031. Spencer, E., Jiang, J., and Chen, Z.J. (1999). Signal-induced ubiquitination of IκBα by the F-box protein Slimb/β-TrCP. *Genes Dev.* **13**, 284–294.

4032. Spencer, F.A. (1984). "The *Decapentaplegic* Gene Complex and Adult Pattern Formation in *Drosophila*." Ph.D. Dissertation, Harvard U.,

4033. Spencer, F.A., Hoffmann, F.M., and Gelbart, W.M. (1982). Decapentaplegic: A gene complex affecting morphogenesis in *Drosophila melanogaster*. *Cell* **28**, 451–461.

4034. Spencer, S.A. and Cagan, R. (2001). *E(Elp)24D*, an EGF-receptor inhibitor, is essential for R8 patterning. *Proc. 42nd Ann. Drosophila Res. Conf.* **Abstracts Vol.**, a194.

4035. Spencer, S.A., Powell, P.A., Miller, D.T., and Cagan, R.L. (1998). Regulation of EGF receptor signaling establishes pattern across the developing *Drosophila* retina. *Development* **125**, 4777–4790.

4036. Spickett, S.G. (1963). Genetic and developmental studies of a quantitative character. *Nature* **199**, 870–873.

4037. Spiegelman, S. (1945). Physiological competition as a regulatory mechanism in morphogenesis. *Q. Rev. Biol.* **20**, 121–146.

4038. Spiegelman, V.S., Slaga, T.J., Pagano, M., Minamoto, T., Ronai, Z., and Fuchs, S.Y. (2000). Wnt/β-catenin signaling induces the expression and activity of βTrCP ubiquitin ligase receptor. *Molec. Cell* **5**, 877–882.

4039. Spieth, H.T. (1974). Courtship behavior in *Drosophila*. *Annu. Rev. Entomol.* **19**, 385–405.

4040. Spink, K.E., Polakis, P., and Weis, W.I. (2000). Structural basis of the Axin-adenomatous polyposis coli interaction. *EMBO J.* **19**, 2270–2279.

4041. Spitzer, N.C. and Sejnowski, T.J. (1997). Biological information processing: bits of progress. *Science* **277**, 1060–1061.

4042. Sprague, G.F., Jr. (1990). Combinatorial associations of regulatory proteins and the control of cell type in yeast. *Adv. Genet.* **27**, 33–62.

4043. Sprang, S.R. (1997). G protein mechanisms: insights from structural analysis. *Annu. Rev. Biochem.* **66**, 639–678.

4044. Sprang, S.R. (1997). GAP into the breach. *Science* **277**, 329–330.

4045. Sprenger, F. and Nüsslein-Volhard, C. (1992). Torso receptor activity is regulated by a diffusible ligand produced at the extracellular terminal regions of the *Drosophila* egg. *Cell* **71**, 987–1001.

4046. Sprenger, F., Stevens, L.M., and Nüsslein-Volhard, C. (1989). The *Drosophila* gene *torso* encodes a putative receptor tyrosine kinase. *Nature* **338**, 478–483.

4047. Sprenger, F., Yakubovich, N., and O'Farrell, P.H. (1997).

S-phase function of *Drosophila* cyclin A and its downregulation in G1 phase. *Curr. Biol.* **7**, 488–499.

4048. Sprey, T.E. (1977). Aldehyde oxidase distribution in the imaginal discs of some diptera. *W. Roux's Arch.* **183**, 1–15.

4049. Sprey, T.E., Eskens, A.A.C., and Kuhn, D.T. (1982). Enzyme distribution patterns in the imaginal wing disc of *Drosophila melanogaster* and other diptera: A subdivision of compartments into territories. *W. Roux's Arch.* **191**, 301–308.

4050. Sprey, T.E. and Kuhn, D.T. (1987). The regulation of aldehyde oxidase in imaginal wing discs of *Drosophila* hybrids: Evidence for *cis*- and *trans*-acting control elements. *Genetics* **115**, 283–294.

4051. Sprey, T.E., Segal, D., Sprey-Pieters, H.E., and Kuhn, D.T. (1981). Influence of homoeotic genes on the aldehyde oxidase pattern in imaginal discs of *Drosophila melanogaster*. *Dev. Genet.* **2**, 75–87.

4052. Spudich, J.L., Yang, C.-S., Jung, K.-H., and Spudich, E.N. (2000). Retinylidene proteins: structures and functions from archaea to humans. *Annu. Rev. Cell Dev. Biol.* **16**, 365–392.

4053. Srinivasan, S., Peng, C.-Y., Nair, S., Skeath, J.B., Spana, E.P., and Doe, C.Q. (1998). Biochemical analysis of Prospero protein during asymmetric cell division: cortical Prospero is highly phosphorylated relative to nuclear Prospero. *Dev. Biol.* **204**, 478–487.

4054. Srivastava, A., Mackay, J.O., and Bell, J.B. (2001). Rescue of a *scalloped* mutation using a Vestigial, Scalloped TEA domain chimera. *Proc. 42nd Ann. Drosophila Res. Conf.* **Abstracts Vol.**, a203.

4055. St. Johnston, D. and Nüsslein-Volhard, C. (1992). The origin of pattern and polarity in the *Drosophila* embryo. *Cell* **68**, 201–219.

4056. St. Johnston, R.D., Hoffmann, F.M., Blackman, R.K., Segal, D., Grimaila, R., Padgett, R.W., Irick, H.A., and Gelbart, W.M. (1990). Molecular organization of the *decapentaplegic* gene in *Drosophila melanogaster*. *Genes Dev.* **4**, 1114–1127.

4057. Stade, K., Ford, C.S., Guthrie, C., and Weis, K. (1997). Exportin 1 (Crm1p) is an essential nuclear export factor. *Cell* **90**, 1041–1050.

4058. Staehling-Hampton, K., Hoffmann, F.M., Baylies, M.K., Rushton, E., and Bate, M. (1994). *dpp* induces mesodermal gene expression in *Drosophila*. *Nature* **372**, 783–786.

4059. Staehling-Hampton, K., Laughon, A.S., and Hoffmann, F.M. (1995). A *Drosophila* protein related to the human zinc finger transcription factor PRDII/MBPI/HIV-EP1 is required for *dpp* signaling. *Development* **121**, 3393–3403.

4060. Stahl, B., Diehlmann, A., and Südhof, T.C. (1999). Direct interaction of Alzheimer's disease-related Presenilin 1 with Armadillo protein p0071. *J. Biol. Chem.* **274** #14, 9141–9148.

4061. Staley, J.P. and Guthrie, C. (1998). Mechanical devices of the spliceosome: motors, clocks, springs, and things. *Cell* **92**, 315–326.

4062. Stambolic, V., Ruel, L., and Woodgett, J.R. (1996). Lithium inhibits glycogen synthase kinase-3 activity and mimics Wingless signalling in intact cells. *Curr. Biol.* **6**, 1664–1668.

4063. Standley, H.J., Zorn, A.M., and Gurdon, J.B. (2001). eFGF and its mode of action in the community effect during *Xenopus* myogenesis. *Development* **128**, 1347–1357.

4064. Stanford, N.P., Szczelkun, M.D., Marko, J.F., and Halford, S.E. (2000). One- and three-dimensional pathways for proteins to reach specific DNA sites. *EMBO J.* **19**, 6546–6557.

4065. Stark, J., Bonacum, J., Remsen, J., and DeSalle, R. (1999). The evolution and development of dipteran wing veins: a systematic approach. *Annu. Rev. Entomol.* **44**, 97–129.

4066. Stassen, M.J., Bailey, D., Nelson, S., Chinwalla, V., and Harte, P.J. (1995). The *Drosophila trithorax* proteins contain a novel variant of the nuclear receptor type DNA binding domain and an ancient conserved motif found in other chromosomal proteins. *Mechs. Dev.* **52**, 209–223.

4067. Stauber, M., Prell, A., and Schmidt-Ott, U. (2001). Evolution of early developmental genes in the insect order Diptera. *Proc. 42nd Ann. Drosophila Res. Conf.* **Abstracts Vol.**, a12. [Cf. 2002 *Proc. Natl. Acad. Sci. USA* **99**, 274–279.]

4068. Stegman, M.A., Vallance, J.E., Elangovan, G., Sosinski, J., Cheng, Y., and Robbins, D.J. (2000). Identification of a tetrameric Hedgehog signaling complex. *J. Biol. Chem.* **275** #29, 21809–21812.

4069. Steinberg, J. and Haglund, K. (1994). A screen door to understanding *Drosophila* genetics. (Landmark Interviews Series). *J. NIH Res.* **6**, 68–77.

4070. Steinberg, M.S. (1963). Reconstruction of tissues by dissociated cells. *Science* **141**, 401–408.

4071. Steinberg, M.S. (1970). Does differential adhesion govern self-assembly processes in histogenesis? Equilibrium configurations and the emergence of a hierarchy among populations of embryonic cells. *J. Exp. Zool.* **173**, 395–434.

4072. Steinberg, M.S. (1981). The adhesive specification of tissue self-organization. *In* "Morphogenesis and Pattern Formation," (T.G. Connelly, L.L. Brinkley, and B.M. Carlson, eds.). Raven: New York, pp. 179–203.

4073. Steinberg, M.S. and Poole, T.J. (1981). Strategies for specifying form and pattern: adhesion-guided multicellular assembly. *Phil. Trans. Roy. Soc. Lond.* **B 295**, 451–460.

4074. Steinberg, M.S. and Poole, T.J. (1982). Cellular adhesive differentials as determinants of morphogenetic movements and organ segregation. *In* "Developmental Order: Its Origin and Regulation," *Symp. Soc. Dev. Biol.*, Vol. 40 (S. Subtelny and P.B. Green, eds.). Alan R. Liss: New York, pp. 351–378.

4075. Steindler, D.A. (1993). Glial boundaries in the developing nervous system. *Annu. Rev. Neurosci.* **16**, 445–470.

4076. Steiner, E. (1976). Establishment of compartments in the developing leg imaginal discs of *Drosophila melanogaster*. *W. Roux's Arch.* **180**, 9–30.

4077. Steinmann-Zwicky, M. (1993). Sex determination in *Drosophila*: *sis-b*, a major numerator element of the X:A ratio in the soma, does not contribute to the X:A ratio in the germ line. *Development* **117**, 763–767.

4078. Steller, H. (1995). Mechanisms and genes of cellular suicide. *Science* **267**, 1445–1449.

4079. Steller, H., Fischbach, K.-F., and Rubin, G.M. (1987). *disconnected*: a locus required for neuronal pathway formation in the visual system of *Drosophila*. *Cell* **50**, 1139–1153.

4080. Steller, H. and Grether, M.E. (1994). Programmed cell death in *Drosophila*. *Neuron* **13**, 1269–1274.

4081. Stemerdink, C. and Jacobs, J.R. (1997). Argos and Spitz group genes function to regulate midline glial cell number in *Drosophila* embryos. *Development* **124**, 3787–3796.

4082. Steneberg, P., Hemphälä, J., and Samakovlis, C. (1999). Dpp and Notch specify the fusion cell fate in dorsal branches of the *Drosophila* trachea. *Mechs. Dev.* **87**, 153–163.

4083. Stent, G.S. (1971). "Molecular Genetics: An Introductory Narrative." W. H. Freeman, San Francisco.

4084. Stent, G.S. (1977). Explicit and implicit semantic content of the genetic information. *In* "Foundational Problems in the Special Sciences," (R. Butts and J. Hintikka, eds.). D. Reidel: Dordrecht, Holland, pp. 131–149.

4085. Stent, G.S. (1981). Strength and weakness of the genetic approach to the development of the nervous system. *Annu. Rev. Neurosci.* **4**, 163–194.

4086. Stent, G.S. (1985). The role of cell lineage in development. *Phil. Trans. Roy. Soc. Lond. B* **312**, 3–19.

4087. Stent, G.S. (1998). Developmental cell lineage. *Int. J. Dev. Biol.* **42**, 237–241.

4088. Stern, C. (1936). Genetics and ontogeny. *Am. Nat.* **70**, 29–35.

4089. Stern, C. (1936). Somatic crossing over and segregation in *Drosophila melanogaster. Genetics* **21**, 625–730.

4090. Stern, C. (1940). The prospective significance of imaginal discs in *Drosophila. J. Morphol.* **67**, 107–122.

4091. Stern, C. (1940). Recent work on the relation between genes and developmental processes. *Growth Suppl.* **1**, 19–36.

4092. Stern, C. (1943). Genic action as studied by means of the effects of different doses and combinations of alleles. *Genetics* **28**, 441–475.

4093. Stern, C. (1944). The journey, not the goal. *Scientific Monthly* **58**, 96–100.

4094. Stern, C. (1948). The effects of changes in quantity, combination, and position of genes. *Science* **108**, 615–621.

4095. Stern, C. (1954). Genes and developmental patterns. *Proc. 9th Internat. Congr. Genetics* **Part I**, 355–369.

4096. Stern, C. (1954). Two or three bristles. *Am. Sci.* **42**, 213–247.

4097. Stern, C. (1956). The genetic control of developmental competence and morphogenetic tissue interactions in genetic mosaics. *Roux's Arch. Dev. Biol.* **149**, 1–25.

4098. Stern, C. (1956). Genetic mechanisms in the localized initiation of differentiation. *Cold Spr. Harb. Symp. Quant. Biol.* **21**, 375–382.

4099. Stern, C. (1963). The cell lineage of the sternopleura in *Drosophila melanogaster. Dev. Biol.* **7**, 365–378.

4100. Stern, C. (1968). "Genetic Mosaics and Other Essays." Harvard Univ. Pr., Cambridge, Mass.

4101. Stern, C. (1969). Gene expression in genetic mosaics. *Genetics* **61 (Suppl.)**, 199–211.

4102. Stern, C. (1969). Richard Benedict Goldschmidt: April 12, 1878–April 24, 1958. *Persp. Biol. Med.* **12**, 178–203.

4103. Stern, C. (1971). From crossing-over to developmental genetics. *In* "Stadler Genetics Symposia," Vol. Vols. 1-2 (G. Kimbar and G. Redei, eds.). Univ. of Missouri Pr.: Columbia, pp. 21–28.

4104. Stern, C. (1974). A geneticist's journey. *In* "Chromosomes and Cancer," (J. German, ed.). Wiley & Sons: New York, pp. xii–xxv.

4105. Stern, C. and Hannah, A.M. (1950). The sex combs in gynanders of *Drosophila melanogaster. Portug. Acta. Biol., Ser. A* **R. B. Goldschmidt Vol.**, 798–812.

4106. Stern, C. and Schaeffer, E.W. (1943). On primary attributes of alleles in *Drosophila melanogaster. Proc. Natl. Acad. Sci. USA* **29**, 351–361.

4107. Stern, C. and Schaeffer, E.W. (1943). On wild-type isoalleles in *Drosophila melanogaster. Proc. Natl. Acad. Sci. USA* **29**, 361–367.

4108. Stern, C. and Swanson, D.L. (1957). The control of the ocellar bristle by the scute locus in *Drosophila melanogaster. J. Faculty Sci. (Zool.), Hokkaido Univ.* **Ser. 6, Vol. 13**, 303–307.

4109. Stern, C. and Tokunaga, C. (1967). Nonautonomy in differentiation of pattern-determining genes in *Drosophila*. I. The sex comb of eyeless-Dominant. *Proc. Natl. Acad. Sci. USA* **57**, 658–664.

4110. Stern, C. and Tokunaga, C. (1968). Autonomous pleiotropy in *Drosophila. Proc. Natl. Acad. Sci. USA* **60**, 1252–1259.

4111. Stern, C.D. and Vasiliauskas, D. (1998). Clocked gene expression in somite formation. *BioEssays* **20**, 528–531.

4112. Stern, D.L. (1998). A role of *Ultrabithorax* in morphological differences between *Drosophila* species. *Nature* **396**, 463–466.

4113. Stern, D.L. (2000). Evolutionary developmental biology and the problem of variation. *Evolution* **54**, 1079–1091.

4114. Stern, D.L. and Emlen, D.J. (1999). The developmental basis for allometry in insects. *Development* **126**, 1091–1101.

4115. Sternberg, P.W. (1993). Falling off the knife edge. *Curr. Biol.* **3**, 763–765.

4116. Sternberg, P.W. and Alberola-Ila, J. (1998). Conspiracy theory: RAS and RAF do not act alone. *Cell* **95**, 447–450.

4117. Stevens, L. (1998). Twin peaks: Spitz and Argos star in patterning of the *Drosophila* egg. *Cell* **95**, 291–294.

4118. Stevens, M.E. and Brower, D.L. (1986). Disruption of positional fields in *apterous* imaginal discs of *Drosophila. Dev. Biol.* **117**, 326–330.

4119. Stevens, M.E. and Bryant, P.J. (1986). Temperature-dependent expression of the *apterous* phenotype in *Drosophila melanogaster. Genetics* **112**, 217–228.

4120. Stevenson, R.D., Hill, M.F., and Bryant, P.J. (1995). Organ and cell allometry in Hawaiian *Drosophila*: how to make a big fly. *Proc. Roy. Soc. Lond. B* **259**, 105–110.

4121. Stewart, M., Murphy, C., and Fristrom, J.W. (1972). The recovery and preliminary characterization of X chromosome mutants affecting imaginal discs of *Drosophila melanogaster. Dev. Biol.* **27**, 71–83.

4122. Stewart, S., Sundaram, M., Zhang, Y., Lee, J., Han, M., and Guan, K.-L. (1999). Kinase suppressor of Ras forms a multiprotein signaling complex and modulates MEK localization. *Mol. Cell. Biol.* **19**, 5523–5534.

4123. Stochaj, U. and Rother, K.L. (1999). Nucleocytoplasmic trafficking of proteins: with or without Ran? *BioEssays* **21**, 579–589.

4124. Stocker, H. and Hafen, E. (2000). Genetic control of cell size. *Curr. Opin. Gen. Dev.* **10**, 529–535.

4125. Stocker, R.F. (1994). The organization of the chemosensory system in *Drosophila melanogaster*: a review. *Cell Tiss. Res.* **275**, 3–26.

4126. Stocker, R.F. and Gendre, N. (1988). Peripheral and central nervous effects of *Lozenge³*: a *Drosophila* mutant lacking basiconic antennal sensilla. *Dev. Biol.* **127**, 12–24.

4127. Stocker, R.F., Gendre, N., and Batterham, P. (1993). Analysis of the antennal phenotype in the *Drosophila* mutant *Lozenge. J. Neurogen.* **9**, 29–53.

4128. Stokoe, D., Macdonald, S.G., Cadwallader, K., Symons, M., and Hancock, J.F. (1994). Activation of Raf as a result of recruitment to the plasma membrane. *Science* **264**, 1463–1467.

4129. Stollewerk, A. (2000). Changes in cell shape in the ventral neuroectoderm of *Drosophila melanogaster* depend on the activity of the *achaete-scute* complex genes. *Dev. Genes Evol.* **210**, 190–199.

4130. Stone, D.M., Hynes, M., Armanini, M., Swanson, T.A., Gu, Q., Johnson, R.L., Scott, M.P., Pennica, D., Goddard, A., Phillips, H., Noll, M., Hooper, J.E., de Sauvage, F., and Rosenthal, A. (1996). The tumor-suppressor gene *patched* encodes a candidate receptor for Sonic hedgehog. *Nature* **384**, 129–134.

4131. Stowers, R.S., Russell, S., and Garza, D. (1999). The 82F late puff contains the *L82* gene, an essential member of a novel gene family. *Dev. Biol.* **213**, 116–130.

4132. Stowers, R.S. and Schwarz, T.L. (1999). A genetic method for generating *Drosophila* eyes composed exclusively of mitotic clones of a single genotype. *Genetics* **152**, 1631–1639.

4133. Strand, D., Jakobs, R., Merdes, G., Neumann, B., Kalmes, A., Heid, H.W., Husmann, I., and Mechler, B.M. (1994). The *Drosophila lethal(2)giant larvae* tumor suppressor protein forms homo-oligomers and is associated with nonmuscle myosin II heavy chain. *J. Cell Biol.* **127**, 1361–1373.

4134. Strand, D., Raska, I., and Mechler, B.M. (1994). The *Drosophila lethal(2)giant larvae* tumor suppressor protein is a component of the cytoskeleton. *J. Cell Biol.* **127**, 1345–1360.

4135. Streit, A. and Stern, C.D. (1999). Establishment and maintenance of the border of the neural plate in the chick: involvement of FGF and BMP activity. *Mechs. Dev.* **82**, 51–66.

4136. Strigini, M. and Cohen, S.M. (1997). A Hedgehog activity gradient contributes to AP axial patterning of the *Drosophila* wing. *Development* **124**, 4697–4705.

4137. Strigini, M. and Cohen, S.M. (1999). Formation of morphogen gradients in the *Drosophila* wing. *Sems. Cell Dev. Biol.* **10**, 335–344.

4138. Strigini, M. and Cohen, S.M. (2000). Wingless gradient formation in the *Drosophila* wing. *Curr. Biol.* **10**, 293–300.

4139. Strödicke, M., Karberg, S., and Korge, G. (2000). *Domina (Dom)*, a new *Drosophila* member of the FKH/WH gene family, affects morphogenesis and is a suppressor of position-effect variegation. *Mechs. Dev.* **96**, 67–78.

4140. Strub, S. (1977). Developmental potentials of the cells of the male foreleg disc of *Drosophila*. I. Pattern regulation in intact fragments. *W. Roux's Arch.* **181**, 309–320.

4141. Strub, S. (1977). Developmental potentials of the cells of the male foreleg disc of *Drosophila*. II. Regulative behaviour of dissociated fragments. *W. Roux's Arch.* **182**, 75–92.

4142. Strub, S. (1979). Heteromorphic regeneration in the developing imaginal primordia of *Drosophila*. *In* "Cell Lineage, Stem Cells and Cell Determination," (N. Le Douarin, ed.). North-Holland Pub. Co.: Amsterdam, pp. 311–324.

4143. Strub, S. (1979). Leg regeneration in insects: an experimental analysis in *Drosophila* and a new interpretation. *Dev. Biol.* **69**, 31–45.

4144. Struhl, G. (1977). Developmental compartments in the proboscis of *Drosophila*. *Nature* **270**, 723–725.

4145. Struhl, G. (1981). Anterior and posterior compartments in the proboscis of *Drosophila*. *Dev. Biol.* **84**, 372–385.

4146. Struhl, G. (1981). A blastoderm fate map of compartments and segments of the *Drosophila* head. *Dev. Biol.* **84**, 386–396.

4147. Struhl, G. (1981). A gene product required for correct initiation of segmental determination in *Drosophila*. *Nature* **293**, 36–41.

4148. Struhl, G. (1981). A homoeotic mutation transforming leg to antenna in *Drosophila*. *Nature* **292**, 635–638.

4149. Struhl, G. (1982). Genes controlling segmental specification in the *Drosophila* thorax. *Proc. Natl. Acad. Sci. USA* **79**, 7380–7384.

4150. Struhl, G. (1982). Spineless-aristapedia: a homeotic gene that does not control the development of specific compartments in *Drosophila*. *Genetics* **102**, 737–749.

4151. Struhl, G. (1983). Role of the *esc*$^+$ gene product in ensuring the selective expression of segment-specific homeotic genes in *Drosophila*. *J. Embryol. Exp. Morph.* **76**, 297–331.

4152. Struhl, G. (1984). Splitting the bithorax complex of *Drosophila*. *Nature* **308**, 454–457.

4153. Struhl, G. (1984). A universal genetic key to body plan? *Nature* **310**, 10–11.

4154. Struhl, G. (1989). Differing strategies for organizing anterior and posterior body pattern in *Drosophila* embryos. *Nature* **338**, 741–744.

4155. Struhl, G. and Adachi, A. (1998). Nuclear access and action of Notch *in vivo*. *Cell* **93**, 649–660.

4156. Struhl, G. and Adachi, A. (2000). Requirements for Presenilin-dependent cleavage of Notch and other transmembrane proteins. *Molec. Cell* **6**, 625–636. [See also 2002 *Dev. Cell* **2**, 69–78 and 79–89.]

4157. Struhl, G., Barbash, D.A., and Lawrence, P.A. (1997). Hedgehog acts by distinct gradient and signal relay mechanisms to organise cell type and cell polarity in the *Drosophila* abdomen. *Development* **124**, 2155–2165.

4158. Struhl, G., Barbash, D.A., and Lawrence, P.A. (1997). Hedgehog organises the pattern and polarity of epidermal cells in the *Drosophila* abdomen. *Development* **124**, 2143–2154.

4159. Struhl, G. and Basler, K. (1993). Organizing activity of wingless protein in *Drosophila*. *Cell* **72**, 527–540.

4160. Struhl, G. and Brower, D. (1982). Early role of the *esc*$^+$ gene product in the determination of segments in *Drosophila*. *Cell* **31**, 285–292.

4161. Struhl, G., Fitzgerald, K., and Greenwald, I. (1993). Intrinsic activity of the Lin-12 and Notch intracellular domains *in vivo*. *Cell* **74**, 331–345.

4162. Struhl, G. and Greenwald, I. (1999). Presenilin is required for activity and nuclear access of Notch in *Drosophila*. *Nature* **398**, 522–525. [See also 2001 *Proc. Natl. Acad. Sci. USA* **98**, 229–234.]

4163. Struhl, G., Struhl, K., and Macdonald, P.M. (1989). The gradient morphogen *bicoid* is a concentration-dependent transcriptional activator. *Cell* **57**, 1259–1273.

4164. Struhl, K. (1991). Mechanisms for diversity in gene expression patterns. *Neuron* **7**, 177–181.

4165. Struhl, K. (1999). Fundamentally different logic of gene regulation in eukaryotes and prokaryotes. *Cell* **98**, 1–4.

4166. Strutt, D.I. (2001). Asymmetric localization of Frizzled and the establishment of cell polarity in the *Drosophila* wing. *Molec. Cell* **7**, 367–375.

4167. Strutt, D.I. and Mlodzik, M. (1995). Ommatidial polarity in the *Drosophila* eye is determined by the direction of furrow progression and local interactions. *Development* **121**, 4247–4256.

4168. Strutt, D.I. and Mlodzik, M. (1996). The regulation of *hedgehog* and *decapentaplegic* during *Drosophila* eye imaginal disc development. *Mechs. Dev.* **58**, 39–50.

4169. Strutt, D.I. and Mlodzik, M. (1997). Hedgehog is an indirect regulator of morphogenetic furrow progression in the *Drosophila* eye disc. *Development* **124**, 3233–3240.

4170. Strutt, D.I., Weber, U., and Mlodzik, M. (1997). The role of RhoA in tissue polarity and Frizzled signalling. *Nature* **387**, 292–295.

4171. Strutt, D.I., Wiersdorff, V., and Mlodzik, M. (1995). Regulation of furrow progression in the *Drosophila* eye by cAMP-dependent protein kinase A. *Nature* **373**, 705–709.

4172. Strutt, H., Cavalli, G., and Paro, R. (1997). Co-localization of Polycomb protein and GAGA factor on regulatory elements responsible for the maintenance of homeotic gene expression. *EMBO J.* **16**, 3621–3632.

4173. Strutt, H. and Strutt, D. (1999). Polarity determination in the *Drosophila* eye. *Curr. Opin. Gen. Dev.* **9**, 442–446.

4174. Strutt, H., Thomas, C., Nakano, Y., Stark, D., Neave, B., Taylor, A.M., and Ingham, P.W. (2001). Mutations in the sterol-sensing domain of Patched suggest a role for vesicular trafficking in Smoothened regulation. *Curr. Biol.* **11**, 608–613.

4175. Stuart, E.T., Kioussi, C., and Gruss, P. (1993). Mammalian Pax genes. *Annu. Rev. Genet.* **27**, 219–236.

4176. Stukenberg, P.T., Lustig, K.D., McGarry, T.J., King, R.W., Kuang, J., and Kirschner, M.W. (1997). Systematic identification of mitotic phosphoproteins. *Curr. Biol.* **7**, 338–348.

4177. Sturm, R.A. and Herr, W. (1988). The POU domain is a bipartite DNA-binding structure. *Nature* **336**, 601–604.

4178. Sturtevant, A.H. (1918). An analysis of the effects of selection. *Carnegie Inst. Wash. Publ.* **264**, 1–68.

4179. Sturtevant, A.H. (1925). The effects of unequal crossing-over at the *Bar* locus in *Drosophila*. *Genetics* **10**, 117–147.

4180. Sturtevant, A.H. (1929). The claret mutant type of *Drosophila simulans*: a study of chromosome elimination and of cell-lineage. *Z. wiss. Zool.* **135**, 323–356.

4181. Sturtevant, A.H. (1932). The use of mosaics in the study of the developmental effects of genes. *Proc. 6th Internat. Congr. Genet.* **1**, 304–307.

4182. Sturtevant, A.H. (1961). "Genetics and Evolution. Selected papers of A. H. Sturtevant." W. H. Freeman, San Francisco. (*N. B.*: Sturtevant's retort to Stern's argument is on p. 145.)

4183. Sturtevant, A.H. (1965). The "fly room." *Am. Sci.* **53**, 303–307. [See also 2001 *Genetics* **159**, 1–5.]

4184. Sturtevant, A.H. (1965). "A History of Genetics." Harper & Row, New York.

4185. Sturtevant, A.H. (1970). Studies on the bristle pattern of *Drosophila*. *Dev. Biol.* **21**, 48–61.

4186. Sturtevant, A.H. and Morgan, T.H. (1923). Reverse mutation of the *Bar* gene correlated with crossing over. *Science* **57**, 746–747.

4187. Sturtevant, A.H. and Schultz, J. (1931). The inadequacy of the sub-gene hypothesis of the nature of the scute allelomorphs of *Drosophila*. *Proc. Natl. Acad. Sci. U. S.* **17**, 265–270.

4188. Sturtevant, M.A., Biehs, B., Marin, E., and Bier, E. (1997). The *spalt* gene links the A/P compartment boundary to a linear adult structure in the *Drosophila* wing. *Development* **124**, 21–32.

4189. Sturtevant, M.A. and Bier, E. (1995). Analysis of the genetic hierarchy guiding wing vein development in *Drosophila*. *Development* **121**, 785–801.

4190. Sturtevant, M.A., O'Neill, J.W., and Bier, E. (1994). Down-regulation of *Drosophila Egf-r* mRNA levels following hyperactivated receptor signaling. *Development* **120**, 2593–2600.

4191. Sturtevant, M.A., Roark, M., and Bier, E. (1993). The *Drosophila rhomboid* gene mediates the localized formation of wing veins and interacts genetically with components of the EGF-R signaling pathway. *Genes Dev.* **7**, 961–973.

4192. Sturtevant, M.A., Roark, M., O'Neill, J.W., Biehs, B., Colley, N., and Bier, E. (1996). The *Drosophila* Rhomboid protein is concentrated in patches at the apical cell surface. *Dev. Biol.* **174**, 298–309.

4193. Stüttem, I. and Campos-Ortega, J.A. (1997). Autonomous and non-autonomous phenotypic expression of *Notch* mutant cells in *Drosophila* embryogenesis. *Dev. Genes Evol.* **207**, 82–89.

4194. Su, M.-T., Fujioka, M., Goto, T., and Bodmer, R. (1999). The *Drosophila* homeobox genes *zfh-1* and *even-skipped* are required for cardiac-specific differentiation of a *numb*-dependent lineage decision. *Development* **126**, 3241–3251.

4195. Su, M.A., Wisotzkey, R.G., and Newfeld, S.J. (2001). A screen for modifiers of *decapentaplegic* mutant phenotypes identifies *lilliputian*, the only member of the Fragile-X/Burkitt's Lymphoma family of transcription factors in *Drosophila melanogaster*. *Genetics* **157**, 717–725.

4196. Su, T.T. and O'Farrell, P.H. (1998). Size control: cell proliferation does not equal growth. *Curr. Biol.* **8**, R687–R689.

4197. Subramaniam, V., Bomze, H.M., and López, A.J. (1994). Functional differences between Ultrabithorax protein isoforms in *Drosophila melanogaster*: evidence from elimination, substitution and ectopic expression of specific isoforms. *Genetics* **136**, 979–991.

4198. Subramaniam, V., Jovin, T.M., and Rivera-Pomar, R.V. (2001). Aromatic amino acids are critical for stability of the Bicoid homeodomain. *J. Biol. Chem.* **276** #24, 21506–21511.

4199. Sudarsanam, P. and Winston, F. (2000). The Swi/Snf family: nucleosome-remodeling complexes and transcriptional control. *Trends Genet.* **16**, 345–351.

4200. Sudol, M. and Hunter, T. (2000). New wrinkles for an old domain. *Cell* **103**, 1001–1004.

4201. Sulston, J.E., Albertson, D.G., and Thomson, J.N. (1980). The *Caenorhabditis elegans* male: postembryonic development of nongonadal structures. *Dev. Biol.* **78**, 542–576.

4202. Sulston, J.E., Schierenberg, E., White, J.G., and Thomson, J.N. (1983). The embryonic cell lineage of the nematode *Caenorhabditis elegans*. *Dev. Biol.* **100**, 64–119.

4203. Sun, B., Hursh, D.A., Jackson, D., and Beachy, P.A. (1995). Ultrabithorax protein is necessary but not sufficient for full activation of *decapentaplegic* expression in the visceral mesoderm. *EMBO J.* **14**, 520–535.

4204. Sun, F.-L. and Elgin, S.C.R. (1999). Putting boundaries on silence. *Cell* **99**, 459–462.

4205. Sun, H., Charles, C.H., Lau, L.F., and Tonks, N.K. (1993). MKP-1 (3CH134), an immediate early gene product, is a dual specificity phosphatase that dephosphorylates MAP kinase in vivo. *Cell* **75**, 487–493.

4206. Sun, X. and Artavanis-Tsakonas, S. (1996). The intracellular deletions of DELTA and SERRATE define dominant negative forms of the *Drosophila* Notch ligands. *Development* **122**, 2465–2474.

4207. Sun, X. and Artavanis-Tsakonas, S. (1997). Secreted forms of DELTA and SERRATE define antagonists of Notch signaling in *Drosophila*. *Development* **124**, 3439–3448.

4208. Sun, Y., Jan, L.Y., and Jan, Y.N. (1998). Transcriptional regulation of *atonal* during development of the *Drosophila* peripheral nervous system. *Development* **125**, 3731–3740.

4209. Sun, Y., Jan, L.Y., and Jan, Y.N. (2000). Ectopic *scute* induces *Drosophila* ommatidia development without R8 founder photoreceptors. *Proc. Natl. Acad. Sci. USA* **97**, 6815–6819.

4210. Sun, Y.H., Tsai, C.-J., Green, M.M., Chao, J.-L., Yu, C.-T., Jaw, T.J., Yeh, J.-Y., and Bolshakov, V.N. (1995). *white* as a reporter gene to detect transcriptional silencers specifying position-specific gene expression during *Drosophila melanogaster* eye development. *Genetics* **141**, 1075–1086.

4211. Sunio, A., Metcalf, A.B., and Krämer, H. (1999). Genetic dissection of endocytic trafficking in *Drosophila* using a horseradish peroxidase-Bride of sevenless chimera: *hook* is required for normal maturation of multivesicular endosomes. *Mol. Biol. Cell* **10**, 847–859. [Cf. 2002 *Cell* **108**, 261–269.]

4212. Sunkel, C.E. and Whittle, J.R.S. (1987). *Brista*: a gene involved in the specification and differentiation of distal cephalic and thoracic structures in *Drosophila melanogaster. Roux's Arch. Dev. Biol.* **196**, 124–132.

4213. Sutrias-Grau, M. and Arnosti, D.N. (2001). Functional characterization of the dCtBP corepressor in the *Drosophila* embryo. *Proc. 42nd Ann. Drosophila Res. Conf.* **Abstracts Vol.**, a126. [Cf. 2001 *Dev. Biol.* **239**, 229–240.]

4214. Suzuki, D.T. (1970). Temperature-sensitive mutations in *Drosophila melanogaster. Science* **170**, 695–706.

4215. Suzuki, T., Maurel-Zaffran, C., Gahmon, G., and Dickson, B.J. (2001). The receptor tyrosine phosphatase DLAR controls R7 axon targeting in the *Drosophila* visual system. *Proc. 42nd Ann. Drosophila Res. Conf.* **Abstracts Vol.**, a253. [Cf. 2001 *Neuron* **32**, 225–235.]

4216. Suzuki, T. and Saigo, K. (2000). Transcriptional regulation of *atonal* required for *Drosophila* larval eye development by concerted action of Eyes absent, Sine oculis and Hedgehog signaling independent of Fused kinase and Cubitus interruptus. *Development* **127**, 1531–1540.

4217. Svendsen, P.C., Marshall, S.D.G., Kyba, M., and Brook, W.J. (2000). The *combgap* locus encodes a zinc-finger protein that regulates *cubitus interruptus* during limb development in *Drosophila melanogaster. Development* **127**, 4083–4093.

4218. Swammerdam, J. (1758). "The Book of Nature." (T. Flloyd, translator), C. G. Seyffert, London. (Reprint Ed., 1978, Arno Press, N. Y.).

4219. Swindells, M.B., Orengo, C.A., Jones, D.T., Hutchinson, E.G., and Thornton, J.M. (1998). Contemporary approaches to protein structure classification. *BioEssays* **20**, 884–891.

4220. Symes, K., Yordán, C., and Mercola, M. (1994). Morphological differences in *Xenopus* embryonic mesodermal cells are specified as an early response to distinct threshold concentrations of activin. *Development* **120**, 2339–2346.

4221. Szabad, J. and Bryant, P.J. (1982). The mode of action of "discless" mutations in *Drosophila melanogaster. Dev. Biol.* **93**, 240–256.

4222. Szabad, J., Schpbach, T., and Wieschaus, E. (1979). Cell lineage and development in the larval epidermis of *Drosophila melanogaster. Dev. Biol.* **73**, 256–271.

4223. Szabad, J., Simpson, P., and Nöthiger, R. (1979). Regeneration and compartments in *Drosophila. J. Embryol. Exp. Morph.* **49**, 229–241.

4224. Szabó, K., Jékely, G., and Rørth, P. (2001). Cloning and expression of *sprint*, a *Drosophila* homologue of RIN1. *Mechs. Dev.* **101**, 259–262.

4225. Szebenyi, A.L. (1969). Cleaning behaviour in *Drosophila melanogaster. Anim. Behav.* **17**, 641–651.

4226. Szüts, D., Eresh, S., and Bienz, M. (1998). Functional intertwining of Dpp and EGFR signaling during *Drosophila* endoderm induction. *Genes Dev.* **12**, 2022–2035.

4227. Tabata, T., Eaton, S., and Kornberg, T.B. (1992). The *Drosophila hedgehog* gene is expressed specifically in posterior compartment cells and is a target of *engrailed* regulation. *Genes Dev.* **6**, 2635–2645.

4228. Tabata, T. and Kornberg, T.B. (1994). *Hedgehog* is a signaling protein with a key role in patterning *Drosophila* imaginal discs. *Cell* **76**, 89–102.

4229. Tabata, T., Schwartz, C., Gustavson, E., Ali, Z., and Kornberg, T.B. (1995). Creating a *Drosophila* wing de novo, the role of *engrailed*, and the compartment border hypothesis. *Development* **121**, 3359–3369.

4230. Tabin, C.J., Carroll, S.B., and Panganiban, G. (1999). Out on a limb: parallels in vertebrate and invertebrate limb patterning and the origin of appendages. *Amer. Zool.* **39**, 650–663. [See also 2001 *Evol. Dev.* **3**, 343–354.]

4231. Taghert, P.H., Doe, C.Q., and Goodman, C.S. (1984). Cell determination and regulation during development of neuroblasts and neurones in grasshopper embryo. *Nature* **307**, 163–165.

4232. Tago, K.-i., Nakamura, T., Nishita, M., Hyodo, J., Nagai, S.-i., Murata, Y., Adachi, S., Ohwada, S., Morishita, Y., Shibuya, H., and Akiyama, T. (2000). Inhibition of Wnt signaling by ICAT, a novel β-catenin-interacting protein. *Genes Dev.* **14**, 1741–1749.

4233. Taguchi, A., Sawamoto, K., and Okano, H. (2000). Mutations modulating the argos-regulated signaling pathway in *Drosophila* eye development. *Genetics* **154**, 1639–1648.

4234. Tahirov, T.H., Inoue-Bungo, T., Morii, H., Fujikawa, A., Sasaki, M., Kimura, K., Shiina, M., Sato, K., Kumasaka, T., Yamamoto, M., Ishii, S., and Ogata, K. (2001). Structural analyses of DNA recognition by the AML1/Runx-1 Runt domain and its allosteric control by CBFβ. *Cell* **104**, 755–767.

4235. Takagaki, Y. and Manley, J.L. (1994). A polyadenlyation factor subunit is the human homologue of the *Drosophila suppressor of forked* protein. *Nature* **372**, 471–474.

4236. Takahashi, K., Matsuo, T., Katsube, T., Ueda, R., and Yamamoto, D. (1998). Direct binding between two PDZ domain proteins Canoe and ZO-1 and their roles in regulation of the Jun N-terminal kinase pathway in *Drosophila* morphogenesis. *Mechs. Dev.* **78**, 97–111.

4237. Takahisa, M., Togashi, S., Suzuki, T., Kobayashi, M., Murayama, A., Kondo, K., Miyake, T., and Ueda, R. (1996). The *Drosophila tamou* gene, a component of the activating pathway of *extramacrochaetae* expression, encodes a protein homologous to mammalian cell-cell junction-associated protein ZO-1. *Genes Dev.* **10**, 1783–1795.

4238. Takano, T.S. (1998). Loss of notum macrochaetae as an interspecific hybrid anomaly between *Drosophila melanogaster* and *D. simulans. Genetics* **149**, 1435–1450.

4239. Takano-Shimizu, T. (2000). Genetic screens for factors involved in the notum bristle loss of interspecific hybrids between *Drosophila melanogaster* and *D. simulans. Genetics* **156**, 269–282.

4240. Takei, K., McPherson, P.S., Schmid, S.L., and De Camilli, P. (1995). Tubular membrane invaginations coated by dynamin rings are induced by GTP-γS in nerve terminals. *Nature* **374**, 186–190.

4241. Talbert, P.B. and Garber, R.L. (1994). The *Drosophila* homeotic mutation *Nasobemia (Antp^{Ns})* and its revertants: an analysis of mutational reversion. *Genetics* **138**, 709–720.

4242. Tamai, K., Semenov, M., Kato, Y., Spokony, R., Liu, C., Katsuyama, Y., Hess, F., Saint-Jeannet, J.-P., and He, X. (2000). LDL-receptor-related proteins in Wnt signal transduction. *Nature* **407**, 530–535.

4243. Tamkun, J.W., Deuring, R., Scott, M.P., Kissinger, M., Pattatucci, A.M., Kaufman, T.C., and Kennison, J.A. (1992). brahma: a regulator of *Drosophila* homeotic genes structurally related to the yeast transcriptional activator SNF2/SWI2. *Cell* **68**, 561–572.

4244. Tamura, K., Taniguchi, Y., Minoguchi, S., Sakai, T., Tun, T., Furukawa, T., and Honjo, T. (1995). Physical interaction between a novel domain of the receptor Notch and the transcription factor RBP-Jκ/Su(H). *Curr. Biol.* **5**, 1416–1423.

4245. Tan, P.B.O. and Kim, S.K. (1999). Signaling specificity: the RTK/RAS/MAP kinase pathway in metazoans. *Trends Genet.* **15**, 145–149.

4246. Tanaka Hall, T.M., Porter, J.A., Young, K.E., Koonin, E.V., Beachy, P.A., and Leahy, D.J. (1997). Crystal structure of a Hedgehog autoprocessing domain: homology between Hedgehog and self-splicing proteins. *Cell* **91**, 85–97.

4247. Tanaka-Matakatsu, M. and Thomas, B. (2001). Shattered is a new rough eye mutant which interacts with roughex, string and fizzy-related. *Proc. 42nd Ann. Drosophila Res. Conf.* **Abstracts Vol.**, a66.

4248. Tanenbaum, S.B., Gorski, S.M., Rusconi, J.C., and Cagan, R.L. (2000). A screen for dominant modifiers of the *irreC-rst* cell death phenotype in the developing *Drosophila* retina. *Genetics* **156**, 205–217.

4249. Tang, A.H., Neufeld, T.P., Kwan, E., and Rubin, G.M. (1997). PHYL acts to down-regulate TTK88, a transcriptional repressor of neuronal cell fates, by a SINA-dependent mechanism. *Cell* **90**, 459–467.

4250. Tang, A.H., Neufeld, T.P., Rubin, G.M., and Müller, H.-A.J. (2001). Transcriptional regulation of cytoskeletal functions and segmentation by a novel maternal pair-rule gene, *lilliputian*. *Development* **128**, 801–813.

4251. Tanimoto, H., Itoh, S., ten Dijke, P., and Tabata, T. (2000). Hedgehog creates a gradient of DPP activity in *Drosophila* wing imaginal discs. *Molec. Cell* **5**, 59–71.

4252. Tanoue, T., Adachi, M., Moriguchi, T., and Nishida, E. (2000). A conserved docking motif in MAP kinases common to substrates, activators and regulators. *Nature Cell Biol.* **2**, 110–116.

4253. Tanoue, T., Maeda, R., Adachi, M., and Nishida, E. (2001). Identification of a docking groove on ERK and p38 MAP kinases that regulates the specificity of docking interactions. *EMBO J.* **20**, 466–479.

4254. Tashiro, S., Michiue, T., Higashijima, S.-i., Zenno, S., Ishimaru, S., Takahashi, F., Orihara, M., Kojima, T., and Saigo, K. (1993). Structure and expression of *hedgehog*, a *Drosophila* segment-polarity gene required for cell-cell communication. *Gene* **124**, 183–189.

4255. Tata, F. and Hartley, D.A. (1993). The role of the *Enhancer of split* complex during cell fate determination in *Drosophila*. *Development* **1993 Suppl.**, 139–148.

4256. Tata, F. and Hartley, D.A. (1995). Inhibition of cell fate in *Drosophila* by *Enhancer of split* genes. *Mechs. Dev.* **51**, 305–315.

4257. Taubert, H. and Szabad, J. (1987). Genetic control of cell proliferation in female germ line cells of *Drosophila*: Mosaic analysis of five discless mutations. *Mol. Gen. Genet.* **209**, 545–551.

4258. Tautz, D. (1996). Selector genes, polymorphisms, and evolution. *Science* **271**, 160–161.

4259. Tautz, D. (2000). Evolution of transcriptional regulation. *Curr. Opin. Gen. Dev.* **10**, 575–579. [See also 2001 *Evol. Dev.* **3**, 109–119.]

4260. Tautz, D. (2000). A genetic uncertainty problem. *Trends Genet.* **16**, 475–477.

4261. Tayler, T.D. and Garrity, P.A. (2001). Regulation of R1-R6 photoreceptor axon targeting. *Proc. 42nd Ann. Drosophila Res. Conf.* **Abstracts Vol.**, a253.

4262. Taylor, A.M., Nakano, Y., Mohler, J., and Ingham, P.W. (1993). Contrasting distributions of patched and hedgehog proteins in the *Drosophila* embryo. *Mechs. Dev.* **42**, 89–96.

4263. Taylor, J., Abramova, N., Charlton, J., and Adler, P.N. (1998). *Van Gogh*: a new *Drosophila* tissue polarity gene. *Genetics* **150**, 199–210.

4264. Tear, G. (1999). Neuronal guidance: a genetic perspective. *Trends Genet.* **15**, 113–118.

4265. Teleman, A.A. and Cohen, S.M. (2000). Dpp gradient formation in the *Drosophila* wing imaginal disc. *Cell* **103**, 971–980.

4266. Telford, M.J. and Thomas, R.H. (1998). Of mites and *zen*: expression studies in a chelicerate arthropod confirm *zen* is a divergent Hox gene. *Dev. Genes Evol.* **208**, 591–594.

4267. ten Dijke, P., Miyazono, K., and Heldin, C.-H. (1996). Signalling via hetero-oligomeric complexes of type I and type II serine/threonine kinase receptors. *Curr. Opin. Cell Biol.* **8**, 139–145.

4268. Tepass, U., Gruszynski-DeFeo, E., Haag, T.A., Omatyar, L., Török, T., and Hartenstein, V. (1996). *shotgun* encodes *Drosophila* E-cadherin and is preferentially required during cell rearrangement in the neurectoderm and other morphogenetically active epithelia. *Genes Dev.* **10**, 672–685.

4269. Tepass, U. and Hartenstein, V. (1995). Neurogenic and proneural genes control cell fate specification in the *Drosophila* endoderm. *Development* **121**, 393–405.

4270. Tepass, U., Pellikka, M., Tanentzapf, G., Pinto, M., Hong, H., Smith, C., McGlade, J., and Ready, D. (2001). Crumbs, the *Drosophila* homolog of human CRB1 (retinitis pigmentosa 12/RP12), is required for photoreceptor cell morphogenesis. *Proc. 42nd Ann. Drosophila Res. Conf.* **Abstracts Vol.**, a45.

4271. Terracol, R. and Lengyel, J.A. (1994). The *thick veins* gene of *Drosophila* is required for dorsoventral polarity of the embryo. *Genetics* **138**, 165–178.

4272. Thackeray, J.R., Gaines, P.C.W., Ebert, P., and Carlson, J.R. (1998). *small wing* encodes a phospholipase C-γ that acts as a negative regulator of R7 development in *Drosophila*. *Development* **125**, 5033–5042.

4273. Thaker, H.M. and Kankel, D.R. (1992). Mosaic analysis gives an estimate of the extent of genomic involvement in the development of the visual system in *Drosophila melanogaster*. *Genetics* **131**, 883–894.

4274. Thanos, D. and Maniatis, T. (1995). Virus induction of human IFNβ gene expression requires the assembly of an enhanceosome. *Cell* **83**, 1091–1100.

4275. The, I., Bellaiche, Y., and Perrimon, N. (1999). Hedgehog movement is regulated through *tout velu*-dependent synthesis of a heparan sulfate proteoglycan. *Molec. Cell* **4**, 633–639.

4276. The, I. and Perrimon, N. (2000). Morphogen diffusion: the case of the Wingless protein. *Nature Cell Biol.* **2**, E79–E82.

4277. Theisen, H., Haerry, T.E., O'Connor, M.B., and Marsh, J.L. (1996). Developmental territories created by mutual antagonism between Wingless and Decapentaplegic. *Development* **122**, 3939–3948.

4278. Theisen, H., Purcell, J., Bennett, M., Kansagara, D., Syed, A., and Marsh, J.L. (1994). *dishevelled* is required during *wingless* signaling to establish both cell polarity and cell identity. *Development* **120**, 347–360.

4279. Theodosiou, N.A., Zhang, S., Wang, W.-Y., and Xu, T. (1998). *slimb* coordinates *wg* and *dpp* expression in the dorsal-ventral and anterior-posterior axes during limb development. *Development* **125**, 3411–3416.

4280. Thérond, P., Alves, G., Limbourg-Bouchon, B., Tricoire, H., Guillemet, E., Brissard-Zahraoui, J., Lamour-Isnard, C., and Busson, D. (1996). Functional domains of Fused, a serine-threonine kinase required for signaling in *Drosophila. Genetics* **142**, 1181–1198.

4281. Therond, P., Busson, D., Guillemet, E., Limbourg-Bouchon, B., Preat, T., Terracol, R., Tricoire, H., and Lamour-Isnard, C. (1993). Molecular organisation and expression pattern of the segment polarity gene *fused* of *Drosophila melanogaster. Mechs. Dev.* **44**, 65–80.

4282. Thérond, P.P., Bouchon, B.L., Gallet, A., Dussilol, F., Pietri, T., van den Heuvel, M., and Tricoire, H. (1999). Differential requirements of the Fused kinase for Hedgehog signalling in the *Drosophila* embryo. *Development* **126**, 4039–4051.

4283. Thérond, P.P., Knight, J.D., Kornberg, T.B., and Bishop, J.M. (1996). Phosphorylation of the fused protein kinase in response to signaling from hedgehog. *Proc. Natl. Acad. Sci. USA* **93**, 4224–4228.

4284. Therrien, M., Chang, H.C., Solomon, N.M., Karim, F.D., Wassarman, D.A., and Rubin, G.M. (1995). KSR, a novel protein kinase required for RAS signal transduction. *Cell* **83**, 879–888.

4285. Therrien, M., Michaud, N.R., Rubin, G.M., and Morrison, D.K. (1996). KSR modulates signal propagation within the MAPK cascade. *Genes Dev.* **10**, 2684–2695.

4286. Therrien, M., Morrison, D.K., Wong, A.M., and Rubin, G.M. (2000). A genetic screen for modifiers of a kinase suppressor of Ras-dependent rough eye phenotype in *Drosophila. Genetics* **156**, 1231–1242.

4287. Therrien, M., Wong, A.M., Kwan, E., and Rubin, G.M. (1999). Functional analysis of CNK in RAS signaling. *Proc. Natl. Acad. Sci. USA* **96**, 13259–13263.

4288. Therrien, M., Wong, A.M., and Rubin, G.M. (1998). CNK, a RAF-binding multidomain protein required for RAS signaling. *Cell* **95**, 343–353.

4289. Thieffry, D., Huerta, A.M., Pérez-Rueda, E., and Collado-Vides, J. (1998). From specific gene regulation to genomic networks: a global analysis of transcriptional regulation in *Escherichia coli. BioEssays* **20**, 433–440.

4290. Thisse, C., Perrin-Schmitt, F., Stoetzel, C., and Thisse, B. (1991). Sequence-specific transactivation of the *Drosophila twist* gene by the *dorsal* gene product. *Cell* **65**, 1191–1201.

4291. Thomas, B.J., Gunning, D.A., Cho, J., and Zipursky, S.L. (1994). Cell cycle progression in the developing *Drosophila* eye: *roughex* encodes a novel protein required for the establishment of G_1. *Cell* **77**, 1003–1014.

4292. Thomas, B.J. and Wassarman, D.A. (1999). A fly's eye view of biology. *Trends Genet.* **15**, 184–190.

4293. Thomas, B.J., Zavitz, K.H., Dong, X., Lane, M.E., Weigmann, K., Finley, R.L., Jr., Brent, R., Lehner, C.F., and Zipursky, S.L. (1997). *roughex* down-regulates G_2 cyclins in G_1. *Genes Dev.* **11**, 1289–1298.

4294. Thomas, D. and Tyers, M. (2000). Transcriptional regulation: kamikazi activators. *Curr. Biol.* **10**, R341–R343.

4295. Thomas, G.H. (2001). Spectrin: the ghost in the machine. *BioEssays* **23**, 152–160.

4296. Thomas, G.H., Zarnescu, D.C., Juedes, A.E., Bales, M.A., Londergan, A., Korte, C.C., and Kiehart, D.P. (1998). *Drosophila* β$_{Heavy}$-spectrin is essential for development and contributes to specific cell fates in the eye. *Development* **125**, 2125–2134.

4297. Thomas, J.B., Crews, S.T., and Goodman, C.S. (1988). Molecular genetics of the *single-minded* locus: a gene involved in the development of the *Drosophila* nervous system. *Cell* **52**, 133–141.

4298. Thomas, J.H. (1993). Thinking about genetic redundancy. *Trends Genet.* **9**, 395–399.

4299. Thomas, J.O. and Travers, A.A. (2001). HMG1 and 2, and related "architectural" DNA-binding proteins. *Trends Biochem. Sci.* **26**, 167–174.

4300. Thomas, U., Ebitsch, S., Gorczyca, M., Koh, Y.H., Hough, C.D., Woods, D., Gundelfinger, E.D., and Budnik, V. (2000). Synaptic targeting and localization of Discs-large is a stepwise process controlled by different domains of the protein. *Curr. Biol.* **10**, 1108–1117.

4301. Thomas, U., Jönsson, F., Speicher, S.A., and Knust, E. (1995). Phenotypic and molecular characterization of *SerD*, a dominant allele of the *Drosophila* gene *Serrate. Genetics* **139**, 203–213.

4302. Thomas, U., Speicher, S.A., and Knust, E. (1991). The *Drosophila* gene *Serrate* encodes an EGF-like transmembrane protein with a complex expression pattern in embryos and wing discs. *Development* **111**, 749–761.

4303. Thompson, A. and Kavaler, J. (2001). The role of Pax2 in glial cell differentiation during *Drosophila* sense organ development. *Proc. 42nd Ann. Drosophila Res. Conf.* **Abstracts Vol.**, a262.

4304. Thompson, C.C. and McKnight, S.L. (1992). Anatomy of an enhancer. *Trends Genet.* **8**, 232–236.

4305. Thompson, D.W. (1913). "On Aristotle as a Biologist." Clarendon Pr., Oxford.

4306. Thompson, D.W. (1917). "On Growth and Form." Cambridge Univ. Pr., Cambridge.

4307. Thompson, J.N., Jr. (1974). Studies on the nature and function of polygenic loci in *Drosophila*. II. The subthreshold wing vein pattern revealed in selection experiments. *Heredity* **33**, 389–401.

4308. Thompson, J.N., Jr., Hellack, J.J., and Kennedy, J.S. (1982). Polygenic analysis of pattern formation in *Drosophila*: specification of campaniform sensilla positions on the wing. *Dev. Genet.* **3**, 115–128.

4309. Thompson, J.N., Jr., Spivey, W.E., and Duncan, D.K. (1985). Temperature-dependent responses to a developmental gradient in the *Drosophila* wing. *Experientia* **41**, 1346–1347.

4310. Thor, S., Andersson, S.G.E., Tomlinson, A., and Thomas, J.B. (1999). A LIM-homeodomain combinatorial code for motor-neuron pathway selection. *Nature* **397**, 76–80.

4311. Thummel, C.S. (1996). Flies on steroids–*Drosophila* metamorphosis and the mechanisms of steriod hormone action. *Trends Genet.* **12**, 306–310. [See also 2001 *Dev. Cell.* **1**, 453–465.]

4312. Thüringer, F., Cohen, S.M., and Bienz, M. (1993). Dissection of an indirect autoregulatory response of a homeotic *Drosophila* gene. *EMBO J.* **12**, 2419–2430.

4313. Tie, F., Furuyama, T., and Harte, P.J. (1998). The *Drosophila* Polycomb Group proteins ESC and E(Z) bind directly to each other and co-localize at multiple chromosome sites. *Development* **125**, 3483–3496.

4314. Tie, F., Furuyama, T., Prasad-Sinha, J., Jane, E., and Harte, P. (2001). Purification and characterization of components of the ESC complex. *Proc. 42nd Ann. Drosophila Res. Conf.* **Abstracts Vol.**, a90.

4315. Tie, F., Furuyama, T., Prasad-Sinha, J., Jane, E., and Harte, P.J. (2001). The *Drosophila* Polycomb Group proteins ESC and E(Z) are present in a complex containing the histone-binding protein p55 and the histone deacetylase RPD3. *Development* **128**, 275–286. [See also 2001 *Proc. Natl. Acad. Sci. USA* **98**, 9730–9735.]

4316. Tie, F. and Harte, P. (2001). Characterization of ESC self-association and phosphorylation. *Proc. 42nd Ann. Drosophila Res. Conf.* **Abstracts Vol.**, a98.

4317. Tien, A.-C., Hsei, H.-Y., and Chien, C.-T. (1999). Dynamic expression and cellular localization of the *Drosophila* 14-3-3ε during embryonic development. *Mechs. Dev.* **81**, 209–212.

4318. Tietze, K., Oellers, N., and Knust, E. (1992). Enhancer of split^D, a dominant mutation of *Drosophila*, and its use in the study of functional domains of a helix-loop-helix protein. *Proc. Natl. Acad. Sci. USA* **89**, 6152–6156.

4319. Tillib, S., Petruk, S., Sedkov, Y., Kuzin, A., Fujioka, M., Goto, T., and Mazo, A. (1999). Trithorax- and Polycomb-group response elements within an *Ultrabithorax* transcription maintenance unit consist of closely situated but separable sequences. *Mol. Cell. Biol.* **19**, 5189–5202.

4320. Timms, J.F., Swanson, K.D., Marie-Cardine, A., Raab, M., Rudd, C.E., Schraven, B., and Neel, B.G. (1999). SHPS-1 is a scaffold for assembling distinct adhesion-regulated multi-protein complexes in macrophages. *Curr. Biol.* **9**, 927–930.

4321. Tio, M., Ma, C., and Moses, K. (1994). *spitz*, a *Drosophila* homolog of transforming growth factor-α, is required in the founding photoreceptor cells of the compound eye facets. *Mechs. Dev.* **48**, 13–23.

4322. Tio, M. and Moses, K. (1997). The *Drosophila* TGFα homolog Spitz acts in photoreceptor recruitment in the developing retina. *Development* **124**, 343–351.

4323. Tio, M., Zavortink, M., Yang, X., and Chia, W. (1999). A functional analysis of *inscuteable* and its roles during *Drosophila* asymmetric cell divisions. *J. Cell Sci.* **112**, 1541–1551.

4324. Tiong, S.Y., Girton, J.R., Hayes, P.H., and Russell, M.A. (1977). Effect of regeneration on compartment specificity of the bithorax mutant of *Drosophila melanogaster*. *Nature* **268**, 435–437.

4325. Tiong, S.Y.K., Nash, D., and Bender, W. (1995). *Dorsal wing*, a locus that affects dorsoventral wing patterning in *Drosophila*. *Development* **121**, 1649–1656. [See also 2001 *Development* **128**, 3263–3268.]

4326. Tiong, S.Y.K. and Russell, M.A. (1986). Effect of the *bithorax* mutation on determination in duplicating *Drosophila* imaginal discs. *Dev. Biol.* **113**, 271–281.

4327. Tiong, S.Y.K. and Russell, M.A. (1990). Clonal analysis of segmental and compartmental homoeotic transformations in Polycomb mutants of *Drosophila melanogaster*. *Dev. Biol.* **141**, 306–318.

4328. Tiong, S.Y.K., Whittle, J.R.S., and Gribbin, M.C. (1987). Chromosomal continuity in the abdominal region of the bithorax complex of *Drosophila* is not essential for its contribution to metameric identity. *Development* **101**, 135–142.

4329. Tissot, M. and Stocker, R.F. (2000). Metamorphosis in *Drosophila* and other insects: the fate of neurons throughout the stages. *Progr. Neurobiol.* **62**, 89–111.

4330. Tix, S., Bate, M., and Technau, G.M. (1989). Pre-existing neuronal pathways in the leg imaginal discs of *Drosophila*. *Development* **107**, 855–862.

4331. Tix, S., Minden, J.S., and Technau, G.M. (1989). Pre-existing neuronal pathways in the developing optic lobes of *Drosophila*. *Development* **105**, 739–746.

4332. Tjian, R. and Maniatis, T. (1994). Transcriptional activation: a complex puzzle with few easy pieces. *Cell* **77**, 5–8. [See also 2001 references: *Curr. Biol.* **11**, R510–R513; *Science* **293**, 1054–1055.]

4333. Toba, G., Ohsako, T., Miyata, N., Ohtsuka, T., Seong, K.-H., and Aigaki, T. (1999). The gene search system: a method for efficient detection and rapid molecular identification of genes in *Drosophila melanogaster*. *Genetics* **151**, 725–737.

4334. Tobler, H. (1966). Zellspezifische Determination und Beziehung zwischen Proliferation und Transdetermination in Bein- und Flügelprimordien von *Drosophila melanogaster*. *J. Embryol. Exp. Morph.* **16**, 609–633.

4335. Tobler, H. (1969). Beeinflussung der Borstendifferenzierung und Musterbildung durch Mitomycin bei *Drosophila melanogaster*. *Experientia* **25**, 213–214.

4336. Tobler, H. and Huber, S. (1972). Effect of nitrogen mustard (HN-2) on the development of the male foreleg of *Drosophila melanogaster*. *Dros. Info. Serv.* **49**, 99–100.

4337. Tobler, H. and Maier, V. (1970). Zur Wirkung von Senfgaslösungen auf die Differenzierung des Borstenorganes und auf die Transdeterminationsfrequenz bei *Drosophila melanogaster*. *W. Roux' Arch.* **164**, 303–312.

4338. Tobler, H., Rothenbuhler, V., and Nothiger, R. (1973). A study of the differentiation of bracts in *Drosophila melanogaster* using two mutations, H^2 and sv^{de}. *Experientia* **29**, 370–371.

4339. Toby, G., Law, S.F., and Golemis, E.A. (1998). Vectors to target protein domains to different cellular compartments. *BioTechniques* **24**, 637–640.

4340. Toering, S.J. (2001). Cellular localization of *Drosophila* Sprouty protein. *Proc. 42nd Ann. Drosophila Res. Conf.* **Abstracts Vol.**, a165.

4341. Toffoli, T. and Margolus, N. (1987). "Cellular Automata Machines: A New Environment for Modeling." M.I.T. Pr., Cambridge, Mass.

4342. Toh, Y. (1985). Structure of campaniform sensilla on the haltere of *Drosophila* prepared by cryofixation. *J. Ultrastruct. Res.* **93**, 92–100.

4343. Tokunaga, C. (1961). The differentiation of a secondary sex comb under the influence of the gene engrailed in *Drosophila melanogaster*. *Genetics* **46**, 157–176.

4344. Tokunaga, C. (1962). Cell lineage and differentiation on the male foreleg of *Drosophila melanogaster*. *Dev. Biol.* **4**, 489–516.

4345. Tokunaga, C. (1966). *Msc: Multiple sex comb. Dros. Info. Serv.* **41**, 57.

4346. Tokunaga, C. (1978). Genetic mosaic studies of pattern formation in *Drosophila melanogaster*, with special reference to the prepattern hypothesis. *In* "Genetic Mosaics and Cell Differentiation," *Results and Problems in Cell Differentiation, Vol. 9*, (W.J. Gehring, ed.). Springer-Verlag: Berlin, pp. 157–204.

4347. Tokunaga, C. (1982). Curt Stern, 1902–1981, *in memoriam*. *Jpn. J. Genet.* **57**, 459–466.

4348. Tokunaga, C. and Gerhart, J.C. (1976). The effect of growth and joint formation on bristle pattern in *D. melanogaster*. *J. Exp. Zool.* **198**, 79–96.

4349. Tokunaga, C. and Stern, C. (1965). The developmental autonomy of extra sex combs in *Drosophila melanogaster*. *Dev. Biol.* **11**, 50–81.

4350. Tokunaga, C. and Stern, C. (1969). Determination of bristle direction in *Drosophila*. *Dev. Biol.* **20**, 411–425.

4351. Tolkunova, E.N., Fujioka, M., Kobayashi, M., Deka, D., and Jaynes, J.B. (1998). Two distinct types of repression domain in Engrailed: one interacts with the Groucho corepressor and is preferentially active on integrated target genes. *Mol. Cell. Biol.* **18**, 2804–2814.

4352. Tolwinski, N.S. and Wieschaus, E. (2001). Armadillo nuclear import is regulated by cytoplasmic anchor Axin and nuclear anchor dTCF/Pan. *Development* **128**, 2107–2117.

4353. Tomilin, A., Reményi, A., Lins, K., Bak, H., Leidel, S., Vriend, G., Wilmanns, M., and Schöler, H.R. (2000). Synergism with the coactivator OBF-1 (OCA-B, BOB-1) is mediated by a specific POU dimer configuration. *Cell* **103**, 853–864.

4354. Tomita, K., Ishibashi, M., Nakahara, K., Ang, S.-L., Nakanishi, S., Guillemot, F., and Kageyama, R. (1996). Mammalian *hairy* and *Enhancer of Split* homolog 1 regulates differentiation of retinal neurons and is essential for eye morphogenesis. *Neuron* **16**, 723–734.

4355. Tomlinson, A. (1985). The cellular dynamics of pattern formation in the eye of *Drosophila*. *J. Embryol. Exp. Morph.* **89**, 313–331.

4356. Tomlinson, A. (1988). Cellular interactions in the developing *Drosophila* eye. *Development* **104**, 183–193.

4357. Tomlinson, A. (1989). An eye to the main chance. *Nature* **340**, 510–511.

4358. Tomlinson, A. (1989). Short-range positional signals in the developing *Drosophila* eye. *Development* **1989 Suppl.**, 59–63.

4359. Tomlinson, A. (1991). *sevenless* and its bride: a marriage of molecules? *Curr. Biol.* **1**, 132–134.

4360. Tomlinson, A., Bowtell, D.D.L., Hafen, E., and Rubin, G.M. (1987). Localization of the *sevenless* protein, a putative receptor for positional information, in the eye imaginal disc of *Drosophila*. *Cell* **51**, 143–150.

4361. Tomlinson, A., Kimmel, B.E., and Rubin, G.M. (1988). *rough*, a *Drosophila* homeobox gene required in photoreceptors R2 and R5 for inductive interactions in the developing eye. *Cell* **55**, 771–784.

4362. Tomlinson, A. and Ready, D.F. (1986). *Sevenless*: a cell-specific homeotic mutation of the *Drosophila* eye. *Science* **231**, 400–402.

4363. Tomlinson, A. and Ready, D.F. (1987). Cell fate in the *Drosophila* ommatidium. *Dev. Biol.* **123**, 264–275.

4364. Tomlinson, A. and Ready, D.F. (1987). Neuronal differentiation in the *Drosophila* ommatidium. *Dev. Biol.* **120**, 366–376.

4365. Tomlinson, A., Strapps, W.R., and Heemskerk, J. (1997). Linking Frizzled and Wnt signaling in *Drosophila* development. *Development* **124**, 4515–4521.

4366. Tomlinson, A. and Struhl, G. (1999). Decoding vectorial information from a gradient: sequential roles of the receptors Frizzled and Notch in establishing planar polarity in the *Drosophila* eye. *Development* **126**, 5725–5738.

4367. Tomlinson, A. and Struhl, G. (2001). Delta/Notch and Boss/Sevenless signals act combinatorially to specify the *Drosophila* R7 photoreceptor. *Molec. Cell* **7**, 487–495.

4368. Tomoyasu, Y., Nakamura, M., and Ueno, N. (1998). Role of Dpp signalling in prepattern formation of the dorsocentral mechanosensory organ in *Drosophila melanogaster*. *Development* **125**, 4215–4224.

4369. Tomoyasu, Y., Ueno, N., and Nakamura, M. (2000). The Decapentaplegic morphogen gradient regulates the notal *wingless* expression through induction of *pannier* and *u-shaped* in *Drosophila*. *Mechs. Dev.* **96**, 37–49.

4370. Torres, M. and Sánchez, L. (1991). The sisterless-b function of the *Drosophila* gene *scute* is restricted to the stage when the X:A ratio determines the activity of *Sex-lethal*. *Development* **113**, 715–722.

4371. Torres, R., Firestein, B.L., Dong, H., Staudinger, J., Olson, E.N., Huganir, R.L., Bredt, D.S., Gale, N.W., and Yancopoulos, G.D. (1998). PDZ proteins bind, cluster, and synaptically colocalize with Eph receptors and their ephrin ligands. *Neuron* **21**, 1453–1463.

4372. Torres-Vazquez, J., Park, S., Warrior, R., and Arora, K. (2001). The transcription factor Schnurri plays a dual role in mediating Dpp signaling during embryogenesis. *Development* **128**, 1657–1670.

4373. Torres-Vazquez, J., Warrior, R., and Arora, K. (2000). *schnurri* is required for *dpp*-dependent patterning of the *Drosophila* wing. *Dev. Biol.* **227**, 388–402.

4374. Townsley, F.M. and Bienz, M. (2000). Actin-dependent membrane association of a *Drosophila* epithelial APC protein and its effect on junctional Armadillo. *Curr. Biol.* **10**, 1339–1348.

4375. Toyoda, H., Kinoshita-Toyoda, A., and Selleck, S.B. (2000). Structural analysis of glycosaminoglycans in *Drosophila* and *Caenorhabditis elegans* and demonstration that *tout-velu*, a *Drosophila* gene related to EXT tumor suppressors, affects heparan sulfate *in vivo*. *J. Biol. Chem.* **275** #4, 2269–2275.

4376. Traas, J., Hülskamp, M., Gendreau, E., and Höfte, H. (1998). Endoreduplication and development: rule without dividing? *Curr. Opin. Plant Biol.* **1**, 498–503. [See also 2001 *Cell* **105**, 297–306.]

4377. Travers, A. (1999). An engine for nucleosome remodeling. *Cell* **96**, 311–314.

4378. Travers, A.A. (1992). The reprogramming of transcriptional competence. *Cell* **69**, 573–575.

4379. Travis, J. (1993). Cell biologists explore "tiny caves". *Science* **262**, 1208–1209.

4380. Treacy, M.N., Neilson, L.I., Turner, E.E., He, X., and Rosenfeld, M.G. (1992). Twin of I-POU: a two amino acid difference in the I-POU homeodomain distinguishes an activator from an inhibitor of transcription. *Cell* **68**, 491–505.

4381. Treier, M., Bohmann, D., and Mlodzik, M. (1995). JUN cooperates with the ETS domain protein Pointed to induce photoreceptor R7 fate in the *Drosophila* eye. *Cell* **83**, 753–760.

4382. Treier, M., Seufert, W., and Jentsch, S. (1992). *Drosophila UbcD1* encodes a highly conserved ubiquitin-conjugating enzyme involved in selective protein degradation. *EMBO J.* **11**, 367–372.

4383. Treisman, J., Harris, E., and Desplan, C. (1991). The paired box encodes a second DNA-binding domain in the Paired homeo domain protein. *Genes Dev.* **5**, 594–604.

4384. Treisman, J., Harris, E., Wilson, D., and Desplan, C. (1992). The homeodomain: a new face for the helix-turn-helix? *BioEssays* **14**, 145–150.

4385. Treisman, J.E. (1999). A conserved blueprint for the eye? *BioEssays* **21**, 843–850.

4386. Treisman, J.E. (2001). *Drosophila* homologues of the transcriptional coactivation complex subunits TRAP240 and TRAP230 are required for identical processes in eye-antennal disc development. *Development* **128**, 603–615.

4387. Treisman, J.E. and Heberlein, U. (1998). Eye development in *Drosophila*: formation of the eye field and control of differentiation. *Curr. Top. Dev. Biol.* **39**, 119–158. [See also 2001 *Mechs. Dev.* **108**, 13–27.]

4388. Treisman, J.E., Lai, Z.-C., and Rubin, G.M. (1995). *short-sighted* acts in the *decapentaplegic* pathway in *Drosophila* eye development and has homology to a mouse TGF-β-responsive gene. *Development* **121**, 2835–2845.

4389. Treisman, J.E., Luk, A., Rubin, G.M., and Heberlein, U. (1997). *eyelid* antagonizes *wingless* signaling during *Drosophila* development and has homology to the Bright family of DNA-binding proteins. *Genes Dev.* **11**, 1949–1962.

4390. Treisman, J.E. and Rubin, G.M. (1995). *wingless* inhibits morphogenetic furrow movement in the *Drosophila* eye disc. *Development* **121**, 3519–3527.

4391. Treisman, J.E. and Rubin, G.M. (1996). Targets of *glass* regulation in the *Drosophila* eye disc. *Mechs. Dev.* **56**, 17–24.

4392. Treisman, R. (1995). Inside the MADS box. *Nature* **376**, 468–469.

4393. Treisman, R. (1996). Regulation of transcription by MAP kinase cascades. *Curr. Opin. Cell Biol.* **8**, 205–215. [See also 2001 *Nature* **411**, 330–334.]

4394. Triezenberg, S.J., Kingsbury, R.C., and McKnight, S.L. (1988). Functional dissection of VP16, the *trans*-activator of herpes simplex virus immediate early gene expression. *Genes Dev.* **2**, 718–729.

4395. Trott, R.L., Kalive, M., Paroush, Z., and Bidwai, A.P. (2001). *Drosophila melanogaster* casein kinase II interacts with and phosphorylates the basic helix-loop-helix proteins m5, m7, and m8 derived from the *Enhancer of split* Complex. *J. Biol. Chem.* **276** #3, 2159–2167.

4396. Trowbridge, I.S. (1993). Dynamin, SH3 domains and endocytosis. *Curr. Biol.* **3**, 773–775.

4397. True, J.R., Edwards, K.A., Yamamoto, D., and Carroll, S.B. (1999). *Drosophila* wing melanin patterns form by vein-dependent elaboration of enzymatic prepatterns. *Curr. Biol.* **9**, 1382–1391.

4398. Trujillo-Cenóz, O. (1985). The eye: development, structure and neural connections. *In* "Comprehensive Insect Physiology, Biochemistry, and Pharmacology," Vol. 6 (G.A. Kerkut and L.I. Gilbert, eds.). Pergamon Pr.: Oxford, pp. 171–223.

4399. Tsuda, L., Inoue, Y.H., Yoo, M.-A., Mizuno, M., Hata, M., Lim, Y.-M., Adachi-Yamada, T., Ryo, H., Masamune, Y., and Nishida, Y. (1993). A protein kinase similar to MAP kinase activator acts downstream of the Raf kinase in *Drosophila*. *Cell* **72**, 407–414.

4400. Tsuda, M., Kamimura, K., Nakato, H., Archer, M., Staatz, W., Fox, B., Humphrey, M., Olson, S., Futch, T., Kaluza, V., Siegfried, E., Stam, L., and Selleck, S.B. (1999). The cell-surface proteoglycan Dally regulates Wingless signalling in *Drosophila*. *Nature* **400**, 276–280.

4401. Tsuji, T., Sato, A., Hiratani, I., Taira, M., Saigo, K., and Kojima, T. (2000). Requirements of *Lim1*, a *Drosophila* LIM-homeobox gene, for normal leg and antennal development. *Development* **127**, 4315–4323.

4402. Tsujimura, H. (2001). Neuronal programmed cell death is controlled by cell death genes, hid and reaper, in the reorganization of the CNS during *Drosophila* metamorphosis. *Proc. 42nd Ann. Drosophila Res. Conf.* **Abstracts Vol.**, a260.

4403. Tsukazaki, T., Chiang, T.A., Davison, A.F., Attisano, L., and Wrana, J.L. (1998). SARA, a FYVE domain protein that recruits Smad2 to the TGFβ receptor. *Cell* **95**, 779–791.

4404. Tsuneizumi, K., Nakayama, T., Kamoshida, Y., Kornberg, T.B., Christian, J.L., and Tabata, T. (1997). *Daughters against dpp* modulates *dpp* organizing activity in *Drosophila* wing development. *Nature* **389**, 627–631.

4405. Tsunoda, S., Sierralta, J., Sun, Y., Bodner, R., Suzuki, E., Becker, A., Socolich, M., and Zuker, C.S. (1997). A multivalent PDZ-domain protein assembles signalling complexes in a G-protein-coupled cascade. *Nature* **388**, 243–249.

4406. Tsunoda, S., Sierralta, J., and Zuker, C.S. (1998). Specificity in signaling pathways: assembly into multimolecular signaling complexes. *Curr. Opin. Gen. Dev.* **8**, 419–422.

4407. Tucker, J.B., Milner, M.J., Currie, D.A., Muir, J.W., Forrest, D.A., and Spencer, M.-J. (1986). Centrosomal microtubule-organizing centres and a switch in the control of protofilament number for cell surface-associated microtubules during *Drosophila* wing morphogenesis. *Eur. J. Cell Biol.* **41**, 279–289.

4408. Tun, T., Hamaguchi, Y., Matsunami, N., Furukawa, T., Honjo, T., and Kawaichi, M. (1994). Recognition sequence of a highly conserved DNA binding protein RBP-Jκ. *Nucleic Acids Res.* **22**, 965–971.

4409. Turing, A.M. (1937). On computable numbers, with an application to the Entscheidungs problem. *Proc. Lond. Math Soc., Ser. 2* **42**, 230–265.

4410. Turing, A.M. (1952). The chemical basis of morphogenesis. *Phil. Trans. Roy. Soc. London, Ser. B* **237**, 37–72.

4411. Tweedie, S., Ng, H.-H., Barlow, A.L., Turner, B.M., Hendrich, B., and Bird, A. (1999). Vestiges of a DNA methylation system in *Drosophila melanogaster*? *Nature Genet.* **23**, 389–390.

4412. Tyler, A. (1947). An auto-antibody concept of cell structure, growth and differentiation. *Growth* **10 (Suppl.)**, 7–19.

4413. Tyler, J.K. and Kadonaga, J.T. (1999). The "dark side" of chromatin remodeling: repressive effects on transcription. *Cell* **99**, 443–446.

4414. Udvardy, A. (1999). Dividing the empire: boundary chromatin elements delimit the territory of enhancers. *EMBO J.* **18**, 1–8.

4415. Ueda, R., Togashi, S., Takahisa, M., Tsurumura, S., Mikuni, M., Kondo, K., and Miyake, T. (1992). Sensory mother cell division is specifically affected in a *Cyclin-A* mutant of *Drosophila melanogaster*. *EMBO J.* **11**, 2935–2939.

4416. Uemura, T., Oda, H., Kraut, R., Hayashi, S., Kataoka, Y., and Takeichi, M. (1996). Zygotic *Drosophila* E-cadherin expression is required for processes of dynamic epithelial cell rearrangement in the *Drosophila* embryo. *Genes Dev.* **10**, 659–671.

4417. Uemura, T., Shepherd, S., Ackerman, L., Jan, L.Y., and Jan, Y.N. (1989). *numb*, a gene required in determination of cell fate during sensory organ formation in *Drosophila* embryos. *Cell* **58**, 349–360.

4418. Uemura, T., Shiomi, K., Togashi, S., and Takeichi, M. (1993). Mutation of *twins* encoding a regulator of protein phosphatase 2A leads to pattern duplication in *Drosophila* imaginal discs. *Genes Dev.* **7**, 429–440.

4419. Ullman, K.S., Powers, M.A., and Forbes, D.J. (1997). Nuclear export receptors: from importin to exportin. *Cell* **90**, 967–970.

4420. Ullrich, B., Ushkaryov, Y.A., and Südhof, T.C. (1995). Cartography of neurexins: more than 1000 isoforms generated by alternative splicing and expressed in distinct subsets of neurons. *Neuron* **14**, 497–507.

4421. Underwood, E.M., Turner, F.R., and Mahowald, A.P. (1980). Analysis of cell movements and fate mapping during early embryogenesis in *Drosophila melanogaster*. *Dev. Biol.* **74**, 286–301.

4422. Urness, L.D. and Thummel, C.S. (1995). Molecular analysis of a steroid-induced regulatory hierarchy: the *Drosophila* E74A protein directly regulates *L71-6* transcription. *EMBO J.* **14**, 6239–6246.

4423. Ursprung, H. (1963). Development and genetics of patterns. *Amer. Zool.* **3**, 71–86.

4424. Ursprung, H. (1972). The fine structure of imaginal disks.

In "The Biology of Imaginal Disks," *Results and Problems in Cell Differentiation, Vol. 5*, (H. Ursprung and R. Nöthiger, eds.). Springer-Verlag: Berlin, pp. 93–107.

4425. Ursprung, H. and Hadorn, E. (1962). Weitere Untersuchungen über Musterbildung in Kombinaten aus teilweise dissoziierten Flügel-Imaginalscheiben von *Drosophila melanogaster. Dev. Biol.* **4**, 40–66.

4426. Ursprung, H. and Nöthiger, R., eds. (1972). "The Biology of Imaginal Disks." *Results and Problems in Cell Differentiation, Vol. 5*, Springer-Verlag, Berlin.

4427. Usui, K. and Kimura, K.-i. (1992). Sensory mother cells are selected from among mitotically quiescent cluster of cells in the wing disc of *Drosophila. Development* **116**, 601–610.

4428. Usui, K. and Kimura, K.-i. (1993). Sequential emergence of the evenly spaced microchaetes on the notum of *Drosophila. Roux's Arch. Dev. Biol.* **203**, 151–158.

4429. Usui, K. and Simpson, P. (2000). Cellular basis of the dynamic behavior of the imaginal thoracic discs during *Drosophila* metamorphosis. *Dev. Biol.* **225**, 13–25.

4430. Usui, T., Shima, Y., Shimada, Y., Hirano, S., Burgess, R.W., Schwarz, T.L., Takeichi, M., and Uemura, T. (1999). Flamingo, a seven-pass transmembrane cadherin, regulates planar cell polarity under the control of Frizzled. *Cell* **98**, 585–595. [See also 2001 *Curr. Biol.* **11**, 859–863.]

4431. Usui-Ishihara, A., Ghysen, A., and Kimura, K.-I. (1995). Peripheral axonal pathway and cleaning behavior are correlated in *Drosophila* microchaetes. *Dev. Biol.* **167**, 398–401.

4432. Usui-Ishihara, A., Simpson, P., and Usui, K. (2000). Larval multidendrite neurons survive metamorphosis and participate in the formation of imaginal sensory axonal pathways in the notum of *Drosophila. Dev. Biol.* **225**, 357–369.

4433. Utley, R.T., Ikeda, K., Grant, P.A., Côté, J., Steger, D.J., Eberharter, A., John, S., and Workman, J.L. (1998). Transcriptional activators direct histone acetyltransferase complexes to nucleosomes. *Nature* **394**, 498–502.

4434. Vachon, G., Cohen, B., Pfeifle, C., McGuffin, M.E., Botas, J., and Cohen, S.M. (1992). Homeotic genes of the Bithorax Complex repress limb development in the abdomen of the *Drosophila* embryo through the target gene *Distalless. Cell* **71**, 437–450.

4435. Vaessin, H., Brand, M., Jan, L.Y., and Jan, Y.N. (1994). *daughterless* is essential for neuronal precursor differentiation but not for initiation of neuronal precursor formation in *Drosophila* embryo. *Development* **120**, 935–945.

4436. Vaessin, H., Grell, E., Wolff, E., Bier, E., Jan, L.Y., and Jan, Y.N. (1991). *prospero* is expressed in neuronal precursors and encodes a nuclear protein that is involved in the control of axonal outgrowth in *Drosophila. Cell* **67**, 941–953.

4437. Vallis, Y., Wigge, P., Marks, B., Evans, P.R., and McMahon, H.T. (1999). Importance of the pleckstrin homology domain of dynamin in clathrin-mediated endocytosis. *Curr. Biol.* **9**, 257–260.

4438. Van de Bor, V., Walther, R., and Giangrande, A. (2000). Some fly sensory organs are gliogenic and require *glide/gcm* in a precursor that divides symmetrically and produces glial cells. *Development* **127**, 3735–3743.

4439. van de Wetering, M., Cavallo, R., Dooijes, D., van Beest, M., van Es, J., Loureiro, J., Ypma, A., Hursh, D., Jones, T., Bejsovec, A., Peifer, M., Mortin, M., and Clevers, H. (1997). Armadillo coactivates transcription driven by the product of the *Drosophila* segment polarity gene *dTCF. Cell* **88**, 789–799.

4440. van den Heuvel, M., Harryman-Samos, C., Klingensmith, J., Perrimon, N., and Nusse, R. (1993). Mutations in the segment polarity genes *wingless* and *porcupine* impair secretion of the wingless protein. *EMBO J.* **12**, 5293–5302.

4441. van den Heuvel, M. and Ingham, P.W. (1996). *smoothened* encodes a receptor-like serpentine protein required for *hedgehog* signalling. *Nature* **382**, 547–551.

4442. van den Heuvel, M., Nusse, R., Johnston, P., and Lawrence, P.A. (1989). Distribution of the *wingless* gene product in *Drosophila* embryos: A protein involved in cell-cell communication. *Cell* **59**, 739–749.

4443. van der Bliek, A.M. and Meyerowitz, E.M. (1991). Dynamin-like protein encoded by the *Drosophila shibire* gene associated with vesicular traffic. *Nature* **351**, 411–414.

4444. van der Geer, P. and Pawson, T. (1995). The PTB domain: a new protein module implicated in signal transduction. *Trends Biochem. Sci.* **20**, 277–280.

4445. van der Geer, P., Wiley, S., Lai, V.K.-M., Olivier, J.P., Gish, G.D., Stephens, R., Kaplan, D., Shoelson, S., and Pawson, T. (1995). A conserved amino-terminal Shc domain binds to phosphotyrosine motifs in activated receptors and phosphopeptides. *Curr. Biol.* **5**, 404–412.

4446. van der Hoeven, F., Zákány, J., and Duboule, D. (1996). Gene transpositions in the *HoxD* complex reveal a hierarchy of regulatory controls. *Cell* **85**, 1025–1035.

4447. van der Straten, A., Rommel, C., Dickson, B., and Hafen, E. (1997). The heat shock protein 83 (Hsp83) is required for Raf-mediated signalling in *Drosophila. EMBO J.* **16**, 1961–1969.

4448. van der Vlag, J., den Blaauwen, J.L., Sewalt, R.G.A.B., van Driel, R., and Otte, A.P. (2000). Transcriptional repression mediated by Polycomb group proteins and other chromatin-associated repressors is selectively blocked by insulators. *J. Biol. Chem.* **275** #1, 697–704.

4449. van der Voorn, L. and Ploegh, H.L. (1992). The WD-40 repeat. *FEBS Letters* **307**, 131–134.

4450. van Dijk, M.A. and Murre, C. (1994). *extradenticle* raises the DNA binding specificity of homeotic selector gene products. *Cell* **78**, 617–624.

4451. Van Doren, M., Bailey, A.M., Esnayra, J., Ede, K., and Posakony, J.W. (1994). Negative regulation of proneural gene activity: hairy is a direct transcriptional repressor of *achaete. Genes Dev.* **8**, 2729–2742.

4452. Van Doren, M., Ellis, H.M., and Posakony, J.W. (1991). The *Drosophila extramacrochaetae* protein antagonizes sequence-specific DNA binding by *daughterless/achaete-scute* protein complexes. *Development* **113**, 245–255.

4453. Van Doren, M., Powell, P.A., Pasternak, D., Singson, A., and Posakony, J.W. (1992). Spatial regulation of proneural gene activity: auto- and cross-activation of *achaete* is antagonized by *extramacrochaetae. Genes Dev.* **6**, 2592–2605.

4454. van Holde, K. and Zlatanova, J. (1996). Chromatin architectural proteins and transcription factors: a structural connection. *BioEssays* **18**, 697–700.

4455. van Leeuwen, F., Harryman Samos, C., and Nusse, R. (1994). Biological activity of soluble *wingless* protein in cultured *Drosophila* imaginal disc cells. *Nature* **368**, 342–344.

4456. van Meyel, D.J., O'Keefe, D.D., Jurata, L.W., Thor, S., Gill, G.N., and Thomas, J.B. (1999). Chip and Apterous physically interact to form a functional complex during *Drosophila* development. *Molec. Cell* **4**, 259–265.

4457. van Meyel, D.J., O'Keefe, D.D., Thor, S., Jurata, L.W., Gill, G.N., and Thomas, J.B. (2000). Chip is an essential cofactor for Apterous in the regulation of axon guidance in *Drosophila*. *Development* **127**, 1823–1831.

4458. Van Vactor, D., Jr., Krantz, D.E., Reinke, R., and Zipursky, S.L. (1988). Analysis of mutants in Chaoptin, a photoreceptor cell-specific glycoprotein in *Drosophila*, reveals its role in cellular morphogenesis. *Cell* **52**, 281–290.

4459. Van Vactor, D.L., Jr., Cagan, R.L., Krämer, H., and Zipursky, S.L. (1991). Induction in the developing compound eye of *Drosophila*: multiple mechanisms restrict R7 induction to a single retinal precursor cell. *Cell* **67**, 1145–1155.

4460. Van Valen, L. (1962). A study of fluctuating asymmetry. *Evolution* **16**, 125–142.

4461. Vandendries, E.R., Johnson, D., and Reinke, R. (1996). *orthodenticle* is required for photoreceptor cell development in the *Drosophila* eye. *Dev. Biol.* **173**, 243–255.

4462. Vandervorst, P. and Ghysen, A. (1980). Genetic control of sensory connections in *Drosophila*. *Nature* **286**, 65–67.

4463. Varadarajan, S. and VijayRaghavan, K. (1999). *scalloped* functions in a regulatory loop with *vestigial* and *wingless* to pattern the *Drosophila* wing. *Dev. Genes Evol.* **209**, 10–17.

4464. Vashee, S., Melcher, K., Ding, W.V., Johnston, S.A., and Kodadek, T. (1998). Evidence for two modes of cooperative DNA binding *in vivo* that do not involve direct protein-protein interactions. *Curr. Biol.* **8**, 452–458.

4465. Vässin, H., Bremer, K.A., Knust, E., and Campos-Ortega, J.A. (1987). The neurogenic gene Delta of *Drosophila melanogaster* is expressed in neurogenic territories and encodes a putative transmembrane protein with EGF-like repeats. *EMBO J.* **6**, 3431–3440.

4466. Vässin, H. and Campos-Ortega, J.A. (1987). Genetic analysis of *Delta*, a neurogenic gene of *Drosophila melanogaster*. *Genetics* **116**, 433–445.

4467. Vässin, H., Vielmetter, J., and Campos-Ortega, J.A. (1985). Genetic interactions in early neurogenesis of *Drosophila melanogaster*. *J. Neurogenetics* **2**, 291–308.

4468. Vázquez, M., Moore, L., and Kennison, J.A. (1999). The trithorax group gene *osa* encodes an ARID-domain protein that genetically interacts with the Brahma chromatin-remodeling factor to regulate transcription. *Development* **126**, 733–742.

4469. Venkatesh, S. and Singh, R.N. (1984). Sensilla on the third antennal segment of *Drosophila melanogaster* Meigen (Diptera: Drosophilidae). *Int. J. Insect Morphol. Embryol.* **13**, 51–63.

4470. Venkatesh, T.R., Hyatt, V., Korayan, A., Brindzei, N., Fermin, H., Mbogho, M., Shamloula, H., Lightowlers, Z.M.A., and Pimentel, A. (2001). *rugose* (*rg*) encodes a *Drosophila* A kinase anchor protein (DAKAP550) and functions in multiple signaling pathways. *Proc. 42nd Ann. Drosophila Res. Conf.* **Abstracts Vol.**, a166.

4471. Venkatesh, T.R., Zipursky, S.L., and Benzer, S. (1985). Molecular analysis of the development of the compound eye in *Drosophila*. *Trends Neurosci.* **8**, 251–257.

4472. Ventura, F., Doody, J., Liu, F., Wrana, J.L., and Massagué, J. (1994). Reconstitution and transphosphorylation of TGF-β receptor complexes. *EMBO J.* **13**, 5581–5589.

4473. Verdi, J.M., Schmandt, R., Bashirullah, A., Jacob, S., Salvino, R., Craig, C.G., Amgen EST Program, Lipshitz, H.D., and McGlade, C.J. (1996). Mammalian NUMB is an evolutionarily conserved signaling adapter protein that specifies cell fate. *Curr. Biol.* **6**, 1134–1145.

4474. Verheyen, E.M. and Cooley, L. (1994). Profilin mutations disrupt multiple actin-dependent processes dur-

ing *Drosophila* development. *Development* **120**, 717–728.

4475. Verheyen, E.M., Mirkovic, I., MacLean, S.J., Langmann, C., Andrews, B.C., and MacKinnon, C. (2001). The tissue polarity gene *nemo* carries out multiple roles in patterning during *Drosophila* development. *Mechs. Dev.* **101**, 119–132.

4476. Verma, I.M., Stevenson, J.K., Schwarz, E.M., Antwerp, D.V., and Miyamoto, S. (1995). Rel/NF-κB/IκB family: intimate tales of association and dissociation. *Genes Dev.* **9**, 2723–2735.

4477. Verrijzer, C.P., van Oosterhout, J.A.W.M., van Weperen, W.W., and van der Vliet, P.C. (1991). POU proteins bend DNA via the POU-specific domain. *EMBO J.* **10**, 3007–3014.

4478. Vervoort, M. (2000). *hedgehog* and wing development in *Drosophila*: a morphogen at work? *BioEssays* **22**, 460–468.

4479. Vervoort, M., Crozatier, M., Valle, D., and Vincent, A. (1999). The COE transcription factor Collier is a mediator of short-range Hedgehog-induced patterning of the *Drosophila* wing. *Curr. Biol.* **9**, 632–639.

4480. Vervoort, M., Dambly-Chaudière, C., and Ghysen, A. (1997). Cell fate determination in *Drosophila*. *Curr. Opin. Neurobiol.* **7**, 21–28.

4481. Vervoort, M., Zink, D., Pujol, N., Victoir, K., Dumont, N., Ghysen, A., and Dambly-Chaudière, C. (1995). Genetic determinants of sense organ identity in *Drosophila*: regulatory interactions between *cut* and *poxn*. *Development* **121**, 3111–3120.

4482. Vieira, A.V., Lamaze, C., and Schmid, S.L. (1996). Control of EGF receptor signaling by clathrin-mediated endocytosis. *Science* **274**, 2086–2089.

4483. Vignali, M., Steger, D.J., Neely, K.E., and Workman, J.L. (2000). Distribution of acetylated histones resulting from Gal4-VP16 recruitment of SAGA and NuA4 complexes. *EMBO J.* **19**, 2629–2640.

4484. Villano, J.L. and Katz, F.N. (1995). *four-jointed* is required for intermediate growth in the proximal-distal axis in *Drosophila*. *Development* **121**, 2767–2777.

4485. Villares, R. and Cabrera, C.V. (1987). The *achaete-scute* gene complex of *D. melanogaster*: conserved domains in a subset of genes required for neurogenesis and their homology to *myc*. *Cell* **50**, 415–424.

4486. Villee, C.A. (1942). The phenomenon of homeosis. *Am. Nat.* **76**, 494–506.

4487. Vincent, J.-P. (1994). Morphogens dropping like flies? *Trends Genet.* **10**, 383–385.

4488. Vincent, J.-P. (1998). Compartment boundaries: where, why and how? *Int. J. Dev. Biol.* **42**, 311–315.

4489. Vincent, J.-P. and Lawrence, P.A. (1994). Drosophila *wingless* sustains *engrailed* expression only in adjoining cells: evidence from mosaic embryos. *Cell* **77**, 909–915.

4490. Vincent, J.-P. and Lawrence, P.A. (1994). It takes three to distalize. *Nature* **372**, 132–133.

4491. Vincent, J.-P. and O'Farrell, P.H. (1992). The state of *engrailed* expression is not clonally transmitted during early *Drosophila* development. *Cell* **68**, 923–931.

4492. Vincent, S., Ruberte, E., Grieder, N.C., Chen, C.-K., Haerry, T., Schuh, R., and Affolter, M. (1997). DPP controls tracheal cell migration along the dorsoventral body axis of the *Drosophila* embryo. *Development* **124**, 2741–2750.

4493. Vinson, C.R., Hai, T., and Boyd, S.M. (1993). Dimerization specificity of the leucine zipper-containing bZIP motif on

DNA binding: prediction and rational design. *Genes Dev.* **7**, 1047–1058.

4494. Vinson, C.R., Sigler, P.B., and McKnight, S.L. (1989). Scissors-grip model for DNA recognition by a family of leucine zipper proteins. *Science* **246**, 911–916.

4495. Virshup, D.M. (2000). Protein phosphatase 2A: a panoply of enzymes. *Curr. Opin. Cell Biol.* **12**, 180–185.

4496. Vogel, G. (1999). Many modes of transport for an embryo's signals. *Science* **285**, 1003–1005.

4497. Vogt, T.F. and Duboule, D. (1999). Antagonists go out on a limb. *Cell* **99**, 563–566.

4498. Von Allmen, G., Hogga, I., Spierer, A., Karch, F., Bender, W., Gyurkovics, H., and Lewis, E. (1996). Splits in the fruitfly Hox gene complexes. *Nature* **380**, 116.

4499. von Dassow, G., Meir, E., Munro, E.M., and Odell, G.M. (2000). The segment polarity network is a robust developmental module. *Nature* **406**, 188–192.

4500. von Heijne, G. (1986). A new method for predicting signal sequence cleavage sites. *Nuc. Acids Res.* **14**, 4683–4690.

4501. von Heijne, G. (1995). Membrane protein assembly: rules of the game. *BioEssays* **17**, 25–30.

4502. Von Ohlen, T. and Hooper, J.E. (1999). The ci^D mutation encodes a chimeric protein whose activity is regulated by Wingless signaling. *Dev. Biol.* **208**, 147–156.

4503. Von Ohlen, T., Lessing, D., Nusse, R., and Hooper, J.E. (1997). Hedgehog signaling regulates transcription through cubitus interruptus, a sequence-specific DNA binding protein. *Proc. Natl. Acad. Sci. USA* **94**, 2404–2409.

4504. Vosshall, L.B., Amrein, H., Morozov, P.S., Rzhetsky, A., and Axel, R. (1999). A spatial map of olfactory receptor expression in the *Drosophila* antenna. *Cell* **96**, 725–736.

4505. Vosshall, L.B., Price, J.L., Sehgal, A., Saez, L., and Young, M.W. (1994). Block in nuclear localization of *period* protein by a second clock mutation, *timeless*. *Science* **263**, 1606–1609.

4506. Vosshall, L.B., Wong, A.M., and Axel, R. (2000). An olfactory sensory map in the fly brain. *Cell* **102**, 147–159. [See also 2001 *Nature* **414**, 204–208.]

4507. Vreezen, W.J. and Veldkamp, J.F. (1969). Selection and temperature effects on extra dorsocentral bristles in *Drosophila melanogaster*. *Genetica* **40**, 19–39.

4508. Waddington, C.H. (1940). Genes as evocators in development. *Growth* **1 (Suppl.)**, 37–44.

4509. Waddington, C.H. (1940). The genetic control of wing development in *Drosophila*. *J. Genet.* **41**, 75–139.

4510. Waddington, C.H. (1940). "Organizers and Genes." Cambridge Univ. Pr., Cambridge.

4511. Waddington, C.H. (1941). Body-colour genes in *Drosophila*. *Proc. Zool. Soc. Lond. Ser. A* **111**, 173–180.

4512. Waddington, C.H. (1943). The development of some "leg genes" in *Drosophila*. *J. Genet.* **45**, 29–43.

4513. Waddington, C.H. (1957). "The Strategy of the Genes: A Discussion of Some Aspects of Theoretical Biology." George Allen & Unwin, London.

4514. Waddington, C.H. (1960). Experiments on canalizing selection. *Genet. Res. Camb.* **1**, 140–150.

4515. Waddington, C.H. (1962). "New Patterns in Genetics and Development." Columbia Univ. Pr., New York.

4516. Waddington, C.H. (1966). Fields and gradients. *In* "Major Problems in Developmental Biology," *Symp. Soc. Dev. Biol.*, Vol. 25 (M. Locke, ed.). Acad. Pr.: New York, pp. 105–124.

4517. Waddington, C.H. (1973). The morphogenesis of patterns in *Drosophila*. *In* "Developmental Systems: Insects," Vol.

2 (S.J. Counce and C.H. Waddington, eds.). Acad. Pr.: New York, pp. 499–535.

4518. Wade, P.A. and Wolffe, A.P. (1997). Chromatin: Histone acetyltransferases in control. *Curr. Biol.* **7**, R82–R84. [See also 2001 *BioEssays* **23**, 820–830.]

4519. Wagner, A. (1998). The fate of duplicated genes: loss or new function? *BioEssays* **20**, 785–788. [See also 2001 references: *BioEssays* **23**, 873–876; *Trends Genet.* **17**, 237–239.]

4520. Wagner, G.P. and Altenberg, L. (1996). Complex adaptations and the evolution of evolvability. *Evolution* **50**, 967–976.

4521. Wagner, G.P., Chiu, C.-H., and Laubichler, M. (2000). Developmental evolution as a mechanistic science: the inference from developmental mechanisms to evolutionary processes. *Amer. Zool.* **40**, 819–831.

4522. Wagner-Bernholz, J.T., Wilson, C., Gibson, G., Schuh, R., and Gehring, W.J. (1991). Identification of target genes of the homeotic gene *Antennapedia* by enhancer detection. *Genes Dev.* **5**, 2467–2480.

4523. Wai, P., Truong, B., and Bhat, K.M. (1999). Cell division genes promote asymmetric interaction between Numb and Notch in the *Drosophila* CNS. *Development* **126**, 2759–2770.

4524. Wainwright, S.M. and Ish-Horowicz, D. (1992). Point mutations in the *Drosophila* hairy gene demonstrate in vivo requirements for basic, helix-loop-helix, and WRPW domains. *Mol. Cell. Biol.* **12**, 2475–2483.

4525. Wakimoto, B.T. and Kaufman, T.C. (1981). Analysis of larval segmentation in lethal genotypes associated with the Antennapedia gene complex in *Drosophila melanogaster*. *Dev. Biol.* **81**, 51–64.

4526. Waksman, G., Shoelson, S.E., Pant, N., Cowburn, D., and Kuriyan, J. (1993). Binding of a high affinity phosphotyrosyl peptide to the Src SH2 domain: crystal structures of the complexed and peptide-free forms. *Cell* **72**, 779–790.

4527. Walker, R.G., Willingham, A.T., and Zuker, C.S. (2000). A *Drosophila* mechanosensory transduction channel. *Science* **287**, 2229–2234.

4528. Wall, M.A., Coleman, D.E., Lee, E., Iñiguez-Lluhi, J.A., Posner, B.A., Gilman, A.G., and Sprang, S.R. (1995). The structure of the G protein heterotrimer $G_{i\alpha1}\beta_1\gamma_2$. *Cell* **83**, 1047–1058.

4529. Wallace, B. (1985). Reflections on the still-"hopeful monster." *Quart. Rev. Biol.* **60**, 31–42.

4530. Wallis, M. (1975). The molecular evolution of pituitary hormones. *Biol. Rev.* **50**, 35–98.

4531. Walt, H. and Tobler, H. (1978). Ultrastructural analysis of differentiating bristle organs in wild-type, *shaven-depilate* and Mitomycin C-treated larvae of *Drosophila melanogaster*. *Biol. Cellulaire* **32**, 291–298.

4532. Walter, J., Dever, C.A., and Biggin, M.D. (1994). Two homeo domain proteins bind with similar specificity to a wide range of DNA sites in *Drosophila* embryos. *Genes Dev.* **8**, 1678–1692.

4533. Waltzer, L. and Bienz, M. (1998). *Drosophila* CBP represses the transcription factor TCF to antagonize Wingless signalling. *Nature* **395**, 521–525.

4534. Waltzer, L. and Bienz, M. (1999). A function of CBP as a transcriptional co-activator during Dpp signaling. *EMBO J.* **18**, 1630–1641.

4535. Wandless, T.J. (1996). SH2 domains: a question of independence. *Curr. Biol.* **6**, 125–127.

4536. Wang, G., Amanai, K., Wang, B., and Jiang, J. (2000). Interactions with Costal 2 and Suppressor of fused regulate nuclear translocation and activity of Cubitus interruptus. *Genes Dev.* **14**, 2893–2905.

4537. Wang, G., Amanai, K., Wang, B., and Jiang, J. (2001). Interactions with Costal 2 and suppressor of fused regulate nuclear translocation and activity of Cubitus interruptus. *Proc. 42nd Ann. Drosophila Res. Conf.* **Abstracts Vol. Addendum**, 7.

4538. Wang, G., Wang, B., and Jiang, J. (1999). Protein kinase A antagonizes Hedgehog signaling by regulating both the activator and repressor forms of Cubitus interruptus. *Genes Dev.* **13**, 2828–2837.

4539. Wang, Q.T. and Holmgren, R.A. (1999). The subcellular localization and activity of *Drosophila* Cubitus interruptus are regulated at multiple levels. *Development* **126**, 5097–5106.

4540. Wang, Q.T. and Holmgren, R.A. (2000). Nuclear import of Cubitus interruptus is regulated by Hedgehog via a mechanism distinct from Ci stabilization and Ci activation. *Development* **127**, 3131–3139.

4541. Wang, S., Wang, Q., Crute, B.E., Melnikova, I.N., Keller, S.R., and Speck, N.A. (1993). Cloning and characterization of subunits of the T-cell receptor and murine leukemia virus enhancer core-binding factor. *Mol. Cell. Biol.* **13**, 3324–3339.

4542. Wang, S., Younger-Shepherd, S., Jan, L.Y., and Jan, Y.N. (1997). Only a subset of the binary cell fate decisions mediated by Numb/Notch signaling in *Drosophila* sensory organ lineage requires *Suppressor of Hairless. Development* **124**, 4435–4446.

4543. Wang, S.-H., Simcox, A., and Campbell, G. (2000). Dual role for *Drosophila* epidermal growth factor receptor signaling in early wing disc development. *Genes Dev.* **14**, 2271–2276. [See also 2001 *Genes Dev.* **11**, 470–475.]

4544. Wang, Y., Macke, J.P., Abella, B.S., Andreasson, K., Worley, P., Gilbert, D.J., Copeland, N.G., Jenkins, N.A., and Nathans, J. (1996). A large family of putative transmembrane receptors homologous to the product of the *Drosophila* tissue polarity gene *frizzled. J. Biol. Chem.* **271** #8, 4468–4476.

4545. Wang, Z. and Moran, M.F. (1996). Requirement for the adapter protein GRB2 in EGF receptor endocytosis. *Science* **272**, 1935–1939.

4546. Wappner, P., Gabay, L., and Shilo, B.-Z. (1997). Interactions between the EGF receptor and DPP pathways establish distinct cell fates in the tracheal placodes. *Development* **124**, 4707–4716.

4547. Ward, E.J. and Skeath, J.B. (2000). Characterization of a novel subset of cardiac cells and their progenitors in the *Drosophila* embryo. *Development* **127**, 4959–4969.

4548. Ward, M.P., Mosher, J.T., and Crews, S.T. (1998). Regulation of bHLH-PAS protein subcellular localization during *Drosophila* embryogenesis. *Development* **125**, 1599–1608.

4549. Warnock, D.E. and Schmid, S.L. (1996). Dynamin GTPase, a force-generating molecular switch. *BioEssays* **18**, 885–893.

4550. Warren, R. and Carroll, S. (1995). Homeotic genes and diversification of the insect body plan. *Curr. Opin. Gen. Dev.* **5**, 459–465.

4551. Warren, R.W. (1993). Defining a neural "ground state" and photoreceptor cell identities in the *Drosophila* eye. *BioEssays* **15**, 827–829.

4552. Warren, R.W., Nagy, L., Selegue, J., Gates, J., and Carroll, S. (1994). Evolution of homeotic gene regulation and function in flies and butterflies. *Nature* **372**, 458–461.

4553. Warrior, R. and Levine, M. (1990). Dose-dependent regulation of pair-rule stripes by gap proteins and the initiation of segment polarity. *Development* **110**, 759–767.

4554. Wassarman, D.A., Solomon, N.M., Chang, H.C., Karim, F.D., Therrien, M., and Rubin, G.M. (1996). Protein phosphatase 2A positively and negatively regulates Ras1-mediated photoreceptor development in *Drosophila. Genes Dev.* **10**, 272–278.

4555. Wassarman, D.A., Therrien, M., and Rubin, G.M. (1995). The Ras signaling pathway in *Drosophila. Curr. Opin. Gen. Dev.* **5**, 44–50.

4556. Wasserman, J.D. and Freeman, M. (1997). Control of EGF receptor activation in *Drosophila. Trends Cell Biol.* **7**, 431–436.

4557. Wasserman, J.D. and Freeman, M. (1998). An autoregulatory cascade of EGF receptor signaling patterns the *Drosophila* egg. *Cell* **95**, 355–364.

4558. Wasserman, J.D., Urban, S., and Freeman, M. (2000). A family of *rhomboid*-like genes: *Drosophila rhomboid-1* and *roughoid/rhomboid-3* cooperate to activate EGF receptor signaling. *Genes Dev.* **14**, 1651–1663.

4559. Wasylyk, B., Hahn, S.L., and Giovane, A. (1993). The Ets family of transcription factors. *Eur. J. Biochem.* **211**, 7–18.

4560. Watson, J.D. (1970). "Molecular Biology of the Gene." 2nd ed. W. A. Benjamin, New York.

4561. Watson, K.L., Justice, R.W., and Bryant, P.J. (1994). *Drosophila* in cancer research: the first fifty tumor suppressor genes. *J. Cell Sci.* **18 (suppl.)**, 19–33.

4562. Watts, C. (1997). Inside the gearbox of the dendritic cell. *Nature* **388**, 724–725.

4563. Weatherbee, S.D. and Carroll, S.B. (1999). Selector genes and limb identity in arthropods and vertebrates. *Cell* **97**, 283–286.

4564. Weatherbee, S.D., Halder, G., Kim, J., Hudson, A., and Carroll, S. (1998). Ultrabithorax regulates genes at several levels of the wing-patterning hierarchy to shape the development of the *Drosophila* haltere. *Genes Dev.* **12**, 1474–1482.

4565. Weaver, T.A. and White, R.A.H. (1995). *headcase*, an imaginal specific gene required for adult morphogenesis in *Drosophila melanogaster. Development* **121**, 4149–4160.

4566. Weber, U., Paricio, N., and Mlodzik, M. (2000). Jun mediates Frizzled-induced R3/R4 cell fate distinction and planar polarity determination in the *Drosophila* eye. *Development* **127**, 3619–3629.

4567. Weber, U., Siegel, V., and Mlodzik, M. (1995). *pipsqueak* encodes a novel nuclear protein required downstream of *seven-up* for the development of photoreceptors R3 and R4. *EMBO J.* **14**, 6247–6257.

4568. Wech, I., Bray, S., Delidakis, C., and Preiss, A. (1999). Distinct expression patterns of different *Enhancer of split* bHLH genes during embryogenesis of *Drosophila melanogaster. Dev. Genes Evol.* **209**, 370–375.

4569. Wegner, M., Drolet, D.W., and Rosenfeld, M.G. (1993). POU-domain proteins: structure and function of developmental regulators. *Curr. Opin. Cell Biol.* **5**, 488–498.

4570. Wehrli, M., Dougan, S.T., Caldwell, K., O'Keefe, L., Schwartz, S., Vaizel-Ohayon, D., Schejter, E., Tomlinson, A., and DiNardo, S. (2000). *arrow* encodes an LDL-receptor-related protein essential for Wingless signalling. *Nature* **407**, 527–530.

4571. Wehrli, M., Rives, A., and DiNardo, S. (2001). How is the Wg signal transduced across the cell membrane? *Proc. 42nd Ann. Drosophila Res. Conf.* **Abstracts Vol.**, a155.

4572. Wehrli, M. and Tomlinson, A. (1995). Epithelial planar polarity in the developing *Drosophila* eye. *Development* **121**, 2451–2459.

4573. Wehrli, M. and Tomlinson, A. (1998). Independent regulation of anterior/posterior and equatorial/polar polarity in the *Drosophila* eye; evidence for the involvement of Wnt signaling in the equatorial/polar axis. *Development* **125**, 1421–1432.

4574. Wei, X. and Ellis, H.M. (2001). Localization of the *Drosophila* MAGUK protein Polychaetoid is controlled by alternative splicing. *Mechs. Dev.* **100**, 217–231.

4575. Weigmann, K. and Cohen, S.M. (1999). Lineage-tracing cells born in different domains along the PD axis of the developing *Drosophila* leg. *Development* **126**, 3823–3830.

4576. Weigmann, K., Cohen, S.M., and Lehner, C.F. (1997). Cell cycle progression, growth and patterning in imaginal discs despite inhibition of cell division after inactivation of *Drosophila* Cdc2 kinase. *Development* **124**, 3555–3563.

4577. Weihe, U., Milan, M., and Cohen, S.M. (2001). Posttranscriptional regulation of Apterous activity during wing development. *Proc. 42nd Ann. Drosophila Res. Conf.* **Abstracts Vol.**, a188. [See also 2001 *Development* **128**, 4615–4622.]

4578. Weiner, J. (1999). "Time, Love, Memory." Random House, New York.

4579. Weinkove, D. and Leevers, S.J. (2000). The genetic control of organ growth: insights from *Drosophila*. *Curr. Opin. Gen. Dev.* **10**, 75–80.

4580. Weinmaster, G. (1997). The ins and outs of Notch signaling. *Molec. Cell. Neurosci.* **9**, 91–102.

4581. Weinmaster, G. (1998). Notch signaling: direct or what? *Curr. Opin. Gen. Dev.* **8**, 436–442.

4582. Weinmaster, G. (2000). Notch signal transduction: a real Rip and more. *Curr. Opin. Gen. Dev.* **10**, 363–369.

4583. Weinstein, A. (1920). Homologous genes and linear linkage in *Drosophila virilis*. *Proc. Natl. Acad. Sci. USA* **6**, 625–639.

4584. Weintraub, H. (1993). The MyoD family and myogenesis: redundancy, networks, and thresholds. *Cell* **75**, 1241–1244.

4585. Weis-Garcia, F. and Massagué, J. (1996). Complementation between kinase-defective and activation-defective TGF-β receptors reveals a novel form of receptor cooperativity essential for signaling. *EMBO J.* **15**, 276–289.

4586. Weismann, A. (1864). Die nachembryonale Entwicklung der Musciden nach Beobachtungen an *Musca vomitoria* und *Sarcophaga carnaria*. *Z. wiss. Zool.* **14**, 187–336.

4587. Weiss, A., Herzig, A., Jacobs, H., and Lehner, C.F. (1998). Continuous *Cyclin E* expression inhibits progression through endoreduplication cycles in *Drosophila*. *Curr. Biol.* **8**, 239–242.

4588. Weiss, A. and Schlessinger, J. (1998). Switching signals on or off by receptor dimerization. *Cell* **94**, 277–280.

4589. Weiss, P. (1939). "Principles of Development." Henry Holt & Co., New York.

4590. Weiss, P. (1947). The problem of specificity in growth and development. *Yale J. Biol. Med.* **19**, 235–278 (reprinted in Weiss's 1968 book).

4591. Weiss, P. (1950). Perspectives in the field of morphogenesis. *Quart. Rev. Biol.* **25**, 177–198.

4592. Weiss, P. (1961). From cell to molecule. *In* "The Molecular Control of Cellular Activity," (J.M. Allen, ed.). McGraw-Hill: New York, pp. 1–72 (reprinted in Weiss's 1968 book).

4593. Weiss, P. (1969). The living system: determinism stratified. *In* "Beyond Reductionism: New Perspectives in the Life Sciences," (A. Koestler and J.R. Smythies, eds.). MacMillan: New York, pp. 3–55.

4594. Weiss, P.A. (1968). "Dynamics of Development: Experiments and Inferences." Acad. Pr., New York.

4595. Welte, M.A., Duncan, I., and Lindquist, S. (1995). The basis for a heat-induced developmental defect: defining crucial lesions. *Genes Dev.* **9**, 2240–2250.

4596. Wen, Y., Nguyen, D., Li, Y., and Lai, Z.-C. (2000). The N-terminal BTB/POZ domain and C-terminal sequences are essential for Tramtrack69 to specify cell fate in the developing *Drosophila* eye. *Genetics* **156**, 195–203.

4597. Weng, G., Bhalla, U.S., and Iyengar, R. (1999). Complexity in biological signaling systems. *Science* **284**, 92–96.

4598. Wente, S.R. (2000). Gatekeepers of the nucleus. *Science* **288**, 1374–1377.

4599. Werb, Z. and Yan, Y. (1998). A cellular striptease act. *Science* **282**, 1279–1280. [Cf. 2002 *BioEssays* **24**, 8–12.]

4600. Wernet, M.F., Mollereau, B., and Desplan, C. (2001). Identifying genes involved in opsin regulation and photoreceptor cell type specification using a Gal4 enhancer trap screen. *Proc. 42nd Ann. Drosophila Res. Conf.* **Abstracts Vol.**, a270.

4601. Wesley, C.S. (1999). Notch and Wingless regulate expression of cuticle patterning genes. *Mol. Cell. Biol.* **19**, 5743–5758.

4602. Wesley, C.S. and Saez, L. (2000). Analysis of Notch lacking the carboxyl terminus identified in *Drosophila* embryos. *J. Cell Biol.* **149**, 683–696.

4603. Wesley, C.S. and Saez, L. (2000). Notch responds differently to Delta and Wingless in cultured *Drosophila* cells. *J. Biol. Chem.* **275** #13, 9099–9101.

4604. Wessells, R.J., Grumbling, G., Donaldson, T., Wang, S.-H., and Simcox, A. (1999). Tissue-specific regulation of Vein/EGF receptor signaling in *Drosophila*. *Dev. Biol.* **216**, 243–259.

4605. West-Eberhard, M.J. (1989). Phenotypic plasticity and the origins of diversity. *Annu. Rev. Ecol. Syst.* **20**, 249–278.

4606. Westheimer, F.H. (1987). Why nature chose phosphates. *Science* **235**, 1173–1178.

4607. Whalen, J.H. and Simon, M.A. (2001). Genetic analysis of *fat-head*, a novel homeotic gene involved in eye and wing morphogenesis. *Proc. 42nd Ann. Drosophila Res. Conf.* **Abstracts Vol.**, a189.

4608. Wharton, K., Ray, R.P., Findley, S.D., Duncan, H.E., and Gelbart, W.M. (1996). Molecular lesions associated with alleles of *decapentaplegic* identify residues necessary for TGF-β/BMP cell signaling in *Drosophila melanogaster*. *Genetics* **142**, 493–505.

4609. Wharton, K.A. (1995). How many receptors does it take? *BioEssays* **17**, 13–16.

4610. Wharton, K.A., Cook, J.M., Torres-Schumann, S., de Castro, K., Borod, E., and Phillips, D.A. (1999). Genetic analysis of the bone morphogenetic protein-related gene, *gbb*, identifies multiple requirements during *Drosophila* development. *Genetics* **152**, 629–640.

4611. Wharton, K.A., Johansen, K.M., Xu, T., and Artavanis-Tsakonas, S. (1985). Nucleotide sequence from the neurogenic locus Notch implies a gene product that shares homology with proteins containing EGF-like repeats. *Cell* **43**, 567–581.

4612. Wharton, K.A., Jr., Franks, R.G., Kasai, Y., and Crews, S.T. (1994). Control of CNS midline transcription by asymmetric E-box-like elements: similarity to xenobiotic responsive regulation. *Development* **120**, 3563–3569.

4613. Wharton, K.A., Ray, R.P., and Gelbart, W.M. (1993). An activity gradient of *decapentaplegic* is necessary for the specification of dorsal pattern elements in the *Drosophila* embryo. *Development* **117**, 807–822.

4614. Wharton, K.A., Thomsen, G.H., and Gelbart, W.M. (1991). *Drosophila* 60A gene, another transforming growth factor β family member, is closely related to human bone morphogenetic proteins. *Proc. Natl. Acad. Sci. USA* **88**, 9214–9218.

4615. Wharton, K.A., Yedvobnick, B., Finnerty, V.G., and Artavanis-Tsakonas, S. (1985). *opa:* A novel family of transcribed repeats shared by the Notch locus and other developmentally regulated loci in *D. melanogaster. Cell* **40**, 55–62.

4616. Wheeler, J.C., Shigesada, K., Gergen, J.P., and Ito, Y. (2000). Mechanisms of transcriptional regulation by Runt domain proteins. *Sems. Cell Dev. Biol.* **11**, 369–375.

4617. Wheeler, S.R., Brown, S.J., Panganiban, G., and Skeath, J.B. (2001). The evolution of neural patterning and the *achaete-scute* complex. *Proc. 42nd Ann. Drosophila Res. Conf.* **Abstracts Vol.**, a259.

4618. White, K. (2000). Cell death: *Drosophila* Apaf-1–no longer in the (d)Ark. *Curr. Biol.* **10**, R167–R169.

4619. White, K., Grether, M.E., Abrams, J.M., Young, L., Farrell, K., and Steller, H. (1994). Genetic control of programmed cell death in *Drosophila. Science* **264**, 677–683.

4620. White, K., Tahaoglu, E., and Steller, H. (1996). Cell killing by the *Drosophila* gene *reaper. Science* **271**, 805–807.

4621. White, N.M. and Jarman, A.P. (2000). *Drosophila* Atonal controls photoreceptor R8-specific properties and modulates both Receptor Tyrosine Kinase and Hedgehog signalling. *Development* **127**, 1681–1689.

4622. White, P., Aberle, H., and Vincent, J.-P. (1998). Signaling and adhesion activities of mammalian β-catenin and plakoglobin in *Drosophila. J. Cell Biol.* **140**, 183–195.

4623. White, R. (1994). Homeotic genes seek partners. *Curr. Biol.* **4**, 48–50.

4624. White, R.A.H. and Akam, M.E. (1985). *Contrabithorax* mutations cause inappropriate expression of *Ultrabithorax* products in *Drosophila. Nature* **318**, 567–569.

4625. White, R.A.H. and Wilcox, M. (1984). Protein products of the bithorax complex in *Drosophila. Cell* **39**, 163–171.

4626. White, R.A.H. and Wilcox, M. (1985). Distribution of *Ultrabithorax* proteins in *Drosophila. EMBO J.* **4**, 2035–2043.

4627. White, R.A.H. and Wilcox, M. (1985). Regulation of the distribution of *Ultrabithorax* proteins in *Drosophila. Nature* **318**, 563–567.

4628. White, R.H. (1961). Analysis of the development of the compound eye in the mosquito, *Aedes aegypti. J. Exp. Zool.* **148**, 223–239.

4629. White, R.H. (1963). Evidence for the existence of a differentiation center in the developing eye of the mosquito. *J. Exp. Zool.* **152**, 139–147.

4630. White, R.J. (1990). Cell type-specific enhancement in the *Drosophila* embryo by consensus homeodomain binding sites. *BioEssays* **12**, 537–539.

4631. White-Cooper, H., Schäfer, M.A., Alphey, L.S., and Fuller, M.T. (1998). Transcriptional and post-transcriptional control mechanisms coordinate the onset of spermatid differentiation with meiosis I in *Drosophila. Development* **125**, 125–134.

4632. Whitelaw, E. and Martin, D.I.K. (2001). Retrotransposons as epigenetic mediators of phenotypic variation in mammals. *Nature Genet.* **27**, 361–365.

4633. Whitfield, W.G.F., Gonzalez, C., Maldonado-Codina, G., and Glover, D.M. (1990). The A- and B-type cyclins of *Drosophila* are accumulated and destroyed in temporally distinct events that define separable phases of the G2-M transition. *EMBO J.* **9**, 2563–2572.

4634. Whiting, A.R. (1934). Eye colours in the parasitic wasp *Habrobracon* and their behaviour in multiple recessives and in mosaics. *J. Genetics* **29**, 99–107.

4635. Whiting, M.F. and Wheeler, W.C. (1994). Insect homeotic transformation. *Nature* **368**, 696.

4636. Whitlock, K.E. (1993). Development of *Drosophila* wing sensory neurons in mutants with missing or modified cell surface molecules. *Development* **117**, 1251–1260.

4637. Whitman, M. (1997). Feedback from inhibitory SMADs. *Nature* **389**, 549–551.

4638. Whitman, M. (1998). Smads and early developmental signaling by the TGFβ superfamily. *Genes Dev.* **12**, 2445–2462.

4639. Whitten, J.M. (1969). Coordinated development in the fly foot: sequential cuticle secretion. *J. Morph.* **127**, 73–104.

4640. Whitten, J.M. (1972). Comparative anatomy of the tracheal system. *Annu. Rev. Ent.* **17**, 373–402.

4641. Whitten, J.M. (1976). Some observations on cellular organization and pattern in flies. *In* "The Insect Integument," (H.R. Hepburn, ed.). Elsevier: New York, pp. 277–297.

4642. Whittle, J.R.S. (1976). Clonal analysis of a genetically caused duplication of the anterior wing in *Drosophila melanogaster. Dev. Biol.* **51**, 257–268.

4643. Whittle, J.R.S. (1976). Mutations affecting the development of the wing. *In* "Insect Development," *Symp. Roy. Entomol. Soc. Lond.,* Vol. 8 (P.A. Lawrence, ed.). Wiley: New York, pp. 118–131.

4644. Whittle, J.R.S. (1983). Litany and creed in the genetic analysis of development. *In* "Development and Evolution," *Symp. Brit. Soc. Dev. Biol.,* Vol. 6 (B.C. Goodwin, N. Holder, and C.C. Wylie, eds.). Cambridge Univ. Pr.: Cambridge, pp. 59–74.

4645. Whittle, J.R.S. (1990). Pattern formation in imaginal discs. *Seminars Cell Biol.* **1**, 241–252.

4646. Whittle, J.R.S. (1998). How is developmental stability sustained in the face of genetic variation? *Int. J. Dev. Biol.* **42**, 495–499.

4647. Widelitz, R.B., Jiang, T.-X., Chen, C.-W.J., Stott, N.S., and Chuong, C.-M. (1999). Wnt-7a in feather morphogenesis: involvement of anterior-posterior asymmetry and proximal-distal elongation demonstrated with an in vitro reconstitution model. *Development* **126**, 2577–2587.

4648. Wiersdorff, V., Lecuit, T., Cohen, S.M., and Mlodzik, M. (1996). *Mad* acts downstream of Dpp receptors, revealing a differential requirement for *dpp* signaling in initiation and propagation of morphogenesis in the *Drosophia* eye. *Development* **122**, 2153–2162.

4649. Wieschaus, E. (1978). Cell lineage relationships in the *Drosophila* embryo. *In* "Genetic Mosaics and Cell Differentiation," *Results and Problems in Cell Differentiation, Vol. 9,* (W.J. Gehring, ed.). Springer-Verlag: Berlin, pp. 97–118.

4650. Wieschaus, E. (1996). Embryonic transcription and the control of developmental pathways. *Genetics* **142**, 5–10.

4651. Wieschaus, E. and Gehring, W. (1976). Clonal analysis of primordial disc cells in the early embryo of *Drosophila melanogaster. Dev. Biol.* **50**, 249–263.

4652. Wieschaus, E. and Gehring, W. (1976). Gynandromorph

analysis of the thoracic disc primordia in *Drosophila melanogaster. W. Roux's Arch.* **180**, 31–46.

4653. Wieschaus, E., Nüsslein-Volhard, C., and Jürgens, G. (1984). Mutations affecting the pattern of the larval cuticle in *Drosophila melanogaster*. III. Zygotic loci on the X-chromosome and fourth chromosome. *Roux's Arch. Dev. Biol.* **193**, 296–307.

4654. Wieschaus, E., Perrimon, N., and Finkelstein, R. (1992). *orthodenticle* activity is required for the development of medial structures in the larval and adult epidermis of *Drosophila. Development* **115**, 801–811.

4655. Wigge, P. and McMahon, H.T. (1998). The amphiphysin family of proteins and their role in endocytosis at the synapse. *Trends Neur. Sci.* **21**, 339–344.

4656. Wigge, P., Vallis, Y., and McMahon, H.T. (1997). Inhibition of receptor-mediated endocytosis by the amphiphysin SH3 domain. *Curr. Biol.* **7**, 554–560.

4657. Wigglesworth, V.B. (1940). Local and general factors in the development of "pattern" in *Rhodnius prolixus* (Hemiptera). *J. Exp. Zool.* **17**, 180–200.

4658. Wigglesworth, V.B. (1961). The epidermal cell. *In* "The Cell and the Organism," (J.A. Ramsay and V.B. Wigglesworth, eds.). Cambridge Univ. Pr.: Cambridge, pp. 127–143.

4659. Wigglesworth, V.B. (1961). Insect polymorphism–a tentative synthesis. *In* "Insect Polymorphism," *Symp. Roy. Entomol. Soc. Lond.*, Vol. 1 (J.S. Kennedy, ed.). E. W. Classey: Faringdon, Oxon, U.K., pp. 103–113.

4660. Wigglesworth, V.B. (1964). Homeostasis in insect growth. *Symp. Soc. Exp. Biol.* **18**, 265–281.

4661. Wigglesworth, V.B. (1972). "The Principles of Insect Physiology." 7th ed. Chapman and Hall, London.

4662. Wigglesworth, V.B. (1973). Evolution of insect wings and flight. *Nature* **246**, 127–129.

4663. Wigglesworth, V.B. (1988). The control of pattern as seen in the integument of an insect. *BioEssays* **9**, 23–27.

4664. Wilcox, M., DiAntonio, A., and Leptin, M. (1989). The function of PS integrins in *Drosophila* wing morphogenesis. *Development* **107**, 891–897.

4665. Wilcox, M. and Smith, R.J. (1977). Regenerative interactions between *Drosophila* imaginal discs of different types. *Dev. Biol.* **60**, 287–297.

4666. Wilder, E.L. and Perrimon, N. (1995). Dual functions of *wingless* in the *Drosophila* leg imaginal disc. *Development* **121**, 477–488.

4667. Wilder, E.L. and Perrimon, N. (1996). Genes involved in postembryonic cell proliferation in *Drosophila*. *In* "Metamorphosis: Postembryonic Reprogramming of Gene Expression in Amphibian and Insect Cells," (L.I. Gilbert, J.R. Tata, and B.G. Atkinson, eds.). Acad. Pr.: New York, pp. 363–400.

4668. Wilkin, M.B., Becker, M.N., Mulvey, D., Phan, I., Chao, A., Cooper, K., Chung, H.-J., Campbell, I.D., Baron, M., and MacIntyre, R. (2000). *Drosophila* Dumpy is a gigantic extracellular protein required to maintain tension at epidermal-cuticle attachment sites. *Curr. Biol.* **10**, 559–567.

4669. Wilkins, A.S. (1985). The limits of molecular biology. *BioEssays* **3**, 3.

4670. Wilkins, A.S. (1989). Organizing the *Drosophila* posterior pattern: why has the fruit fly made life so complicated for itself? *BioEssays* **11**, 67–69.

4671. Wilkins, A.S. (1993). "Genetic Analysis of Animal Development." 2nd ed. Wiley-Liss, New York.

4672. Wilkins, A.S. (1995). Singling out the tip cell of the Malpighian tubules–lessons from neurogenesis. *BioEssays* **17**, 199–202.

4673. Wilkins, A.S. (1995). What's in a (biological) term?... Frequently, a great deal of ambiguity. *BioEssays* **17**, 375–377.

4674. Wilkins, A.S. (1997). Canalization: a molecular genetic perspective. *BioEssays* **19**, 257–262.

4675. Wilkins, A.S. and Gubb, D. (1991). Pattern formation in the embryo and imaginal discs of *Drosophila*: what are the links? *Dev. Biol.* **145**, 1–12.

4676. Willert, K., Brink, M., Wodarz, A., Varmus, H., and Nusse, R. (1997). Casein kinase 2 associates with and phosphorylates Dishevelled. *EMBO J.* **16**, 3089–3096.

4677. Willert, K., Logan, C.Y., Arora, A., Fish, M., and Nusse, R. (1999). A *Drosophila Axin* homolog, *Daxin*, inhibits Wnt signaling. *Development* **126**, 4165–4173.

4678. Willert, K. and Nusse, R. (1998). β-catenin: a key mediator of Wnt signaling. *Curr. Opin. Gen. Dev.* **8**, 95–102.

4679. Willert, K., Shibamoto, S., and Nusse, R. (1999). Wnt-induced dephosphorylation of Axin releases β-catenin from the Axin complex. *Genes Dev.* **13**, 1768–1773.

4680. Williams, A.F. and Barclay, A.N. (1988). The immunoglobulin superfamily–domains for cell surface recognition. *Annu. Rev. Immunol.* **6**, 381–405.

4681. Williams, J.A., Bell, J.B., and Carroll, S.B. (1991). Control of *Drosophila* wing and haltere development by the nuclear *vestigial* gene product. *Genes Dev.* **5**, 2481–2495.

4682. Williams, J.A. and Carroll, S.B. (1993). The origin, patterning and evolution of insect appendages. *BioEssays* **15**, 567–577.

4683. Williams, J.A., Paddock, S.W., and Carroll, S.B. (1993). Pattern formation in a secondary field: a hierarchy of regulatory genes subdivides the developing *Drosophila* wing disc into discrete subregions. *Development* **117**, 571–584.

4684. Williams, J.A., Paddock, S.W., Vorwerk, K., and Carroll, S.B. (1994). Organization of wing formation and induction of a wing-patterning gene at the dorsal/ventral compartment boundary. *Nature* **368**, 299–305.

4685. Williams, R.W. and Herrup, K. (1988). The control of neuron number. *Annu. Rev. Neurosci.* **11**, 423–453.

4686. Willis, J.H. (1986). The paradigm of stage-specific gene sets in insect metamorphosis: time for revision! *Arch. Insect Biochem. Physiol.* **Suppl. 1**, 47–57.

4687. Wilson, C., Pearson, R.K., Bellen, H.J., O'Kane, C.J., Grossniklaus, U., and Gehring, W. (1989). P-element-mediated enhancer detection: an efficient method for isolating and characterizing developmentally regulated genes in *Drosophila. Genes Dev.* **3**, 1301–1313.

4688. Wilson, D., Sheng, G., Lecuit, T., Dostatni, N., and Desplan, C. (1993). Cooperative dimerization of Paired class homeo domains on DNA. *Genes Dev.* **7**, 2120–2134.

4689. Wilson, D.S., Guenther, B., Desplan, C., and Kuriyan, J. (1995). High resolution crystal structure of a Paired (Pax) class cooperative homeodomain dimer on DNA. *Cell* **82**, 709–719.

4690. Wilson, K.L., Zastrow, M.S., and Lee, K.K. (2001). Lamins and disease: insights into nuclear infrastructure. *Cell* **104**, 647–650.

4691. Wilson, P.A. and Melton, D.A. (1994). Mesodermal patterning by an inducer gradient depends on secondary cell-cell communication. *Curr. Biol.* **4**, 676–686.

4692. Wilson, T.G. (1980). Correlation of phenotypes of the apterous mutation in *Drosophila melanogaster. Dev. Genet.* **1**, 195–204.

4693. Wilson, T.G. (1981). Expression of phenotypes in a temperature-sensitive allele of the apterous mutation in *Drosophila melanogaster*. *Dev. Biol.* **85**, 425–433.

4694. Wilton, J.C. and Matthews, G.M. (1996). Polarised membrane traffic in hepatocytes. *BioEssays* **18**, 229–236.

4695. Wimmer, E.A., Cohen, S.M., Jäckle, H., and Desplan, C. (1997). *buttonhead* does not contribute to a combinatorial code proposed for *Drosophila* head development. *Development* **124**, 1509–1517.

4696. Winchester, G. (1996). The Morgan lineage. *Curr. Biol.* **6**, 100–101.

4697. Winfree, A.T. (1980). "The Geometry of Biological Time." Springer-Verlag, Berlin.

4698. Winfree, A.T. (1984). A continuity principle for regeneration. *In* "Pattern Formation: A Primer in Developmental Biology," (G.M. Malacinski and S.V. Bryant, eds.). Macmillan: New York, pp. 103–124.

4699. Winfree, A.T., Winfree, E.M., and Seifert, H. (1985). Organizing centers in a cellular excitable medium. *Physica* **17D**, 109–115.

4700. Winick, J., Abel, T., Leonard, M.W., Michelson, A.M., Chardon-Loriaux, I., Holmgren, R.A., Maniatis, T., and Engel, J.D. (1993). A GATA family transcription factor is expressed along the embryonic dorsoventral axis in *Drosophila melanogaster*. *Development* **119**, 1055–1065.

4701. Winston, J.T., Strack, P., Beer-Romero, P., Chu, C.Y., Elledge, S.J., and Harper, J.W. (1999). The SCF$^{\beta-TRCP}$-ubiquitin ligase complex associates specifically with phosphorylated destruction motifs in IκBα and β-catenin and stimulates IκBα ubiquitination in vitro. *Genes Dev.* **13**, 270–283.

4702. Winter, C.G., Wang, B., Ballew, A., Royou, A., Karess, R., Axelrod, J.D., and Luo, L. (2001). *Drosophila* Rho-associated kinase (Drok) links Frizzled-mediated planar cell polarity signaling to the actin cytoskeleton. *Cell* **105**, 81–91.

4703. Wisotzkey, R.G., Mehra, A., Sutherland, D.J., Dobens, L.L., Liu, X., Dohrmann, C., Attisano, L., and Raftery, L.A. (1998). *Medea* is a *Drosophila* Smad4 homolog that is differentially required to potentiate DPP responses. *Development* **125**, 1433–1445.

4704. Wistow, G. (1993). Lens crystallins: gene recruitment and evolutionary dynamism. *Trends Biochem. Sci.* **18**, 301–306.

4705. Wittinghofer, F. (1998). Caught in the act of the switch-on. *Nature* **394**, 317–320.

4706. Wittwer, F., van der Straten, A., Keleman, K., Dickson, B.J., and Hafen, E. (2001). Lilliputian: an AF4/FMR2-related protein that controls cell identity and cell growth. *Development* **128**, 791–800.

4707. Wodarz, A. (2000). Tumor suppressors: linking cell polarity and growth control. *Curr. Biol.* **10**, R624–R626.

4708. Wodarz, A. and Nusse, R. (1998). Mechanisms of Wnt signaling in development. *Annu. Rev. Cell Dev. Biol.* **14**, 59–88.

4709. Wodarz, A., Ramrath, A., Kuchinke, U., and Knust, E. (1999). Bazooka provides an apical cue for Inscuteable localization in *Drosophila* neuroblasts. *Nature* **402**, 544–547.

4710. Wolfe, M.S., Xia, W., Ostaszewski, B.L., Diehl, T.S., Kimberly, W.T., and Selkoe, D.J. (1999). Two transmembrane aspartates in presenilin-1 required for presenilin endoproteolysis and g-secretase activity. *Nature* **398**, 513–517.

4711. Wolfe, S.A., Nekludova, L., and Pabo, C.O. (1999). DNA recognition by Cys$_2$His$_2$ zinc finger proteins. *Annu. Rev. Biophys. Biomol. Struct.* **3**, 183–212.

4712. Wolff, T. and Ready, D.F. (1991). The beginning of pattern formation in the *Drosophila* compound eye: the morphogenetic furrow and the second mitotic wave. *Development* **113**, 841–850.

4713. Wolff, T. and Ready, D.F. (1991). Cell death in normal and rough eye mutants of *Drosophila*. *Development* **113**, 825–839.

4714. Wolff, T. and Ready, D.F. (1991). In search of a role for growth factors in *Drosophila* eye development. *Sems. Dev. Biol.* **2**, 305–316.

4715. Wolff, T. and Ready, D.F. (1993). Pattern formation in the *Drosophila* retina. *In* "The Development of *Drosophila melanogaster*," Vol. 2 (M. Bate and A. Martinez Arias, eds.). Cold Spring Harbor Lab. Pr.: Plainview, N. Y., pp. 1277–1325.

4716. Wolff, T. and Rubin, G.M. (1998). *strabismus*, a novel gene that regulates tissue polarity and cell fate decisions in *Drosophila*. *Development* **125**, 1149–1159.

4717. Wolffe, A.P. (1994). Architectural transcription factors. *Science* **264**, 1100–1101.

4718. Wolffe, A.P. (1997). Sinful repression. *Nature* **387**, 16–17.

4719. Wolffe, A.P., Khochbin, S., and Dimitrov, S. (1997). What do linker histones do in chromatin? *BioEssays* **19**, 249–255.

4720. Wolfram, S. (1984). Cellular automata as models of complexity. *Nature* **311**, 419–424.

4721. Wolfsberg, T.G., Primakoff, P., Myles, D.G., and White, J.M. (1995). ADAM, a novel family of membrane proteins containing A̲ D̲isintegrin A̲nd M̲etalloprotease domain: multipotential functions in cell-cell and cell-matrix interactions. *J. Cell Biol.* **131**, 275–278.

4722. Wolfsberg, T.G. and White, J.M. (1996). ADAMs in fertilization and development. *Dev. Biol.* **180**, 389–401.

4723. Wolpert, L. (1968). The French Flag Problem: a contribution to the discussion on pattern development and regulation. *In* "Towards a Theoretical Biology. I. Prolegomena," (C.H. Waddington, ed.). Aldine Pub. Co.: Chicago, pp. 125–133.

4724. Wolpert, L. (1969). Positional information and the spatial pattern of cellular differentiation. *J. Theor. Biol.* **25**, 1–47.

4725. Wolpert, L. (1971). Positional information and pattern formation. *Curr. Top. Dev. Biol.* **6**, 183–224.

4726. Wolpert, L. (1974). Positional information and the development of pattern and form. *In* "Lectures on Mathematics in the Life Sciences," Vol. 6 Am. Math. Soc.: Providence, Rhode Island, pp. 27–41.

4727. Wolpert, L. (1978). Cell position and cell lineage in pattern formation and regulation. *In* "Stem Cells and Tissue Homeostasis," *Brit. Soc. Cell Biol. Sympos.*, Vol. 2 (B. Lord, C.S. Potten, and R. Cole, eds.). Cambridge Univ. Pr.: Cambridge, pp. 29–47.

4728. Wolpert, L. (1978). Pattern formation in biological development. *Sci. Am.* **239** #4, 154–164.

4729. Wolpert, L. (1981). Positional information and pattern formation. *Phil. Trans. Roy. Soc. Lond.* B **295**, 441–450.

4730. Wolpert, L. (1984). Molecular problems of positional information. *BioEssays* **1**, 175–177.

4731. Wolpert, L. (1985). Gradients, position and pattern: a history. *In* "A History of Embryology," *Symp. Brit. Soc. Dev. Biol.*, Vol. 8 (T.J. Horder, J.A. Witkowski, and C.C. Wylie, eds.). Cambridge U. Pr.: New York, pp. 347–362.

4732. Wolpert, L. (1989). Positional information revisited. *Development* **1989 Suppl.**, 3–12.

4733. Wolpert, L. (1994). Positional information and pattern formation in development. *Dev. Genet.* **15**, 485–490.

4734. Wolpert, L. (1996). One hundred years of positional information. *Trends Genet.* **12**, 359–364.

4735. Wolpert, L. and Lewis, J.H. (1975). Towards a theory of development. *Fed. Proc.* **34**, 14–20.

4736. Wolpert, L. and Stein, W.D. (1984). Positional information and pattern formation. *In* "Pattern Formation: A Primer in Developmental Biology," (G.M. Malacinski and S.V. Bryant, eds.). Macmillan: New York, pp. 3–21.

4737. Wolsky, A. and Wolsky, M.I. (1990). Constraints in the development and evolution of the arthropodan compound eye. *In* "Organizational Constraints on the Dynamics of Evolution," (J. Maynard Smith and G. Vida, eds.). Manchester Univ. Pr.: Manchester, pp. 133–140.

4738. Wong, E.S.M., Lim, J., Low, B.C., Chen, Q., and Guy, G.R. (2001). Evidence for direct interaction between Sprouty and Cbl. *J. Biol. Chem.* **276** #8, 5866–5875.

4739. Wong, L.L. and Adler, P.N. (1993). Tissue polarity genes of *Drosophila* regulate the subcellular location for prehair initiation in pupal wing cells. *J. Cell Biol.* **123**, 209–221.

4740. Wong, W.T., Schumacher, C., Salcini, A.E., Romano, A., Castagnino, P., Pelicci, P.G., and Di Fiore, P.P. (1995). A protein-binding domain, EH, identified in the receptor tyrosine kinase substrate Eps15 and conserved in evolution. *Proc. Natl. Acad. Sci. USA* **92**, 9530–9534.

4741. Woodgett, J.R. (1991). A common denominator linking glycogen metabolism, nuclear oncogenes and development. *Trends Biochem. Sci.* **16**, 177–181.

4742. Woodhouse, E., Hersperger, E., and Shearn, A. (1998). Growth, metastasis, and invasiveness of *Drosophila* tumors caused by mutations in specific tumor suppressor genes. *Dev. Genes Evol.* **207**, 542–550.

4743. Woods, D.F. and Bryant, P.J. (1992). Genetic control of cell interactions in developing *Drosophila* epithelia. *Annu. Rev. Genet.* **26**, 305–350.

4744. Woods, D.F. and Bryant, P.J. (1993). ZO-1, DlgA and PSD-95/SAP90: homologous proteins in tight, septate and synaptic cell junctions. *Mechs. Dev.* **44**, 85–89.

4745. Woods, D.F., Wu, J.-W., and Bryant, P.J. (1997). Localization of proteins to the apico-lateral junctions of *Drosophila* epithelia. *Dev. Genet.* **20**, 111–118.

4746. Woods, R.E., Hercus, M.J., and Hoffmann, A.A. (1998). Estimating the heritability of fluctuating asymmetry in field *Drosophila*. *Evolution* **52**, 816–824.

4747. Woods, R.E., Sgrò, C.M., Hercus, M.J., and Hoffmann, A.A. (1999). The association between fluctuating asymmetry, trait variability, trait heritability, and stress: a multiply replicated experiment on combined stresses in *Drosophila melanogaster*. *Evolution* **53**, 493–505.

4748. Wootton, R. (1999). How flies fly. *Nature* **400**, 112–113.

4749. Wootton, R.J. (1992). Functional morphology of insect wings. *Annu. Rev. Entomol.* **37**, 113–140.

4750. Wootton, R.J. and Kukalová-Peck, J. (2000). Flight adaptations in Palaeozoic Palaeoptera (Insecta). *Biol. Rev.* **75**, 129–167.

4751. Wrana, J. and Pawson, T. (1997). Mad about SMADs. *Nature* **388**, 28–29.

4752. Wrana, J.L., Attisano, L., Cárcamo, J., Zentella, A., Doody, J., Laiho, M., Wang, X.-F., and Massagué, J. (1992). TGFβ signals through a heteromeric protein kinase receptor complex. *Cell* **71**, 1003–1014.

4753. Wrana, J.L., Attisano, L., Wieser, R., Ventura, F., and Massagué, J. (1994). Mechanism of activation of the TGF-β receptor. *Nature* **370**, 341–347.

4754. Wrana, J.L., Tran, H., Attisano, L., Arora, K., Childs, S.R., Massagué, J., and O'Connor, M.B. (1994). Two distinct transmembrane serine/threonine kinases from *Drosophila melanogaster* form an activin receptor complex. *Mol. Cell. Biol.* **14**, 944–950.

4755. Wray, G.A. (1998). Promoter logic. *Science* **279**, 1871–1872. [See also 2001 *Am. Sci.* **89**, 204–208.]

4756. Wreden, C., Verrotti, A.C., Schisa, J.A., Lieberfarb, M.E., and Strickland, S. (1997). *Nanos* and *pumilio* establish embryonic polarity in *Drosophila* by promoting posterior deadenylation of *hunchback* mRNA. *Development* **124**, 3015–3023.

4757. Wright, J.W. and Copenhaver, P.F. (2000). Different isoforms of Fasciclin II play distinct roles in the guidance of neuronal migration during insect embryogenesis. *Dev. Biol.* **225**, 59–78.

4758. Wright, J.W., Snyder, M.A., Schwinof, K.M., Combes, S., and Copenhaver, P.F. (1999). A role for fasciclin II in the guidance of neuronal migration. *Development* **126**, 3217–3228.

4759. Wu, C.-F., Berneking, J.M., and Barker, D.L. (1983). Acetylcholine synthesis and accumulation in the CNS of *Drosophila* larvae: Analysis of *shibire^{ts}*, a mutant with a temperature-sensitive block in synaptic transmission. *J. Neurochem.* **40**, 1386–1396.

4760. Wu, J. and Cohen, S.M. (1999). Proximodistal axis formation in the *Drosophila* leg: subdivision into proximal and distal domains by Homothorax and Distal-less. *Development* **126**, 109–117.

4761. Wu, J. and Cohen, S.M. (2000). Proximal distal axis formation in the *Drosophila* leg: distinct functions of Teashirt and Homothorax in the proximal leg. *Mechs. Dev.* **94**, 47–56.

4762. Wu, J.Y. and Rao, Y. (1999). Fringe: defining borders by regulating the Notch pathway. *Curr. Opin. Neurobiol.* **9**, 537–543.

4763. Wu, Q. and Maniatis, T. (1999). A striking organization of a large family of human cadherin-like cell adhesion genes. *Cell* **97**, 779–790.

4764. Wu, X., Vakani, R., and Small, S. (1998). Two distinct mechanisms for differential positioning of gene expression borders involving the *Drosophila* gap protein giant. *Development* **125**, 3765–3774.

4765. Wu, Z., Li, Q., Fortini, M.E., and Fischer, J.A. (1999). Genetic analysis of the role of the *Drosophila fat facets* gene in the ubiquitin pathway. *Dev. Genet.* **25**, 312–320.

4766. Wülbeck, C. and Simpson, P. (2000). Expression of *achaete-scute* homologues in discrete proneural clusters on the developing notum of the medfly *Ceratitis capitata* suggests a common origin for the stereotyped bristle patterns of higher Diptera. *Development* **127**, 1411–1420.

4767. Wurmbach, E., Wech, I., and Preiss, A. (1999). The *Enhancer of split* complex of *Drosophila melanogaster* harbors three classes of Notch responsive genes. *Mechs. Dev.* **80**, 171–180.

4768. Wurst, G., Hersperger, E., and Shearn, A. (1984). Genetic analysis of transdetermination in *Drosophila*. II. Transdetermination to wing of leg discs from a mutant which lacks wing discs. *Dev. Biol.* **106**, 147–155. (*N.B.*: The "L6" gene studied here was renamed "*vein*.")

4769. Wyman, R.J. (1986). Sequential induction and a homeotic switch of cell fate. *Trends Neurosci.* **9**, 339–340.

4770. Xiao, B., Smerdon, S.J., Jones, D.H., Dodson, G.G., Soneji, Y., Aitken, A., and Gamblin, S.J. (1995). Structure of a 14-3-3 protein and implications for coordination of multiple signalling pathways. *Nature* **376**, 188–191.

4771. Xie, T., Finelli, A.L., and Padgett, R.W. (1994). The

Drosophila saxophone gene: a serine-threonine kinase receptor of the TGF-β superfamily. *Science* **263**, 1756–1759.

4772. Xing, H., Kornfeld, K., and Muslin, A.J. (1997). The protein kinase KSR interacts with 14-3-3 protein and Raf. *Curr. Biol.* **7**, 294–300.

4773. Xiong, W.-C. and Montell, C. (1993). *tramtrack* is a transcriptional repressor required for cell fate determination in the *Drosophila* eye. *Genes Dev.* **7**, 1085–1096.

4774. Xiong, W.-C. and Montell, C. (1995). Defective glia induce neuronal apoptosis in the *repo* visual system of *Drosophila. Neuron* **14**, 581–590.

4775. Xu, C., Kauffmann, R.C., Zhang, J., Kladny, S., and Carthew, R.W. (2000). Overlapping activators and repressors delimit transcriptional response to receptor tyrosine kinase signals in the *Drosophila* eye. *Cell* **103**, 87–97.

4776. Xu, H.E., Rould, M.A., Xu, W., Epstein, J.A., Maas, R.L., and Pabo, C.O. (1999). Crystal structure of the human Pax6 paired domain-DNA complex reveals specific roles for the linker region and carboxy-terminal subdomain in DNA binding. *Genes Dev.* **13**, 1263–1275.

4777. Xu, P.-X., Zhang, X., Heaney, S., Yoon, A., Michelson, A.M., and Maas, R.L. (1999). Regulation of *Pax6* expression is conserved between mice and flies. *Development* **126**, 383–395.

4778. Xu, T. and Artavanis-Tsakonas, S. (1990). *deltex*, a locus interacting with the neurogenic genes, *Notch, Delta* and *mastermind* in *Drosophila melanogaster. Genetics* **126**, 665–677.

4779. Xu, T., Caron, L.A., Fehon, R.G., and Artavanis-Tsakonas, S. (1992). The involvement of the *Notch* locus in *Drosophila* oogenesis. *Development* **115**, 913–922.

4780. Xu, T., Rebay, I., Fleming, R.J., Scottgale, T.N., and Artavanis-Tsakonas, S. (1990). The *Notch* locus and the genetic circuitry involved in early *Drosophila* neurogenesis. *Genes Dev.* **4**, 464–475.

4781. Xu, T. and Rubin, G.M. (1993). Analysis of genetic mosaics in developing and adult *Drosophila* tissues. *Development* **117**, 1223–1237.

4782. Xu, X., Yin, Z., Hudson, J.B., Ferguson, E.L., and Frasch, M. (1998). Smad proteins act in combination with synergistic and antagonistic regulators to target Dpp responses to the *Drosophila* mesoderm. *Genes Dev.* **12**, 2354–2370.

4783. Xu, X.S., Kuspa, A., Fuller, D., Loomis, W.F., and Knecht, D.A. (1996). Cell-cell adhesion prevents mutant cells lacking myosin II from penetrating aggregation streams of *Dictyostelium. Dev. Biol.* **175**, 218–226.

4784. Xu, Y.K. and Nusse, R. (1998). The Frizzled CRD domain is conserved in diverse proteins including several receptor tyrosine kinases. *Curr. Biol.* **8**, R405–R406.

4785. Xue, L., Li, X., and Noll, M. (2001). Multiple protein functions of Paired in *Drosophila* development and their conservation in the Gooseberry and Pax3 homologs. *Development* **128**, 395–405.

4786. Xue, L. and Noll, M. (1996). The functional conservation of proteins in evolutionary alleles and the dominant role of enhancers in evolution. *EMBO J.* **15**, 3722–3731.

4787. Yaffe, M.B., Rittinger, K., Volinia, S., Caron, P.R., Aitken, A., Leffers, H., Gamblin, S.J., Smerdon, S.J., and Cantley, L.C. (1997). The structural basis for 14-3-3:phosphopeptide binding specificity. *Cell* **91**, 961–971.

4788. Yagi, Y., Suzuki, T., and Hayashi, S. (1998). Interaction between *Drosophila* EGF receptor and *vnd* determines three dorsoventral domains of the neuroectoderm. *Development* **125**, 3625–3633.

4789. Yaich, L., Ooi, J., Park, M., Borg, J.-P., Landry, C., Bodmer, R., and Margolis, B. (1998). Functional analysis of the Numb phosphotyrosine-binding domain using site-directed mutagenesis. *J. Biol. Chem.* **273** #17, 10381–10388.

4790. Yamamoto, D. (1994). Signaling mechanisms in induction of the R7 photoreceptor in the developing *Drosophila* retina. *BioEssays* **16**, 237–244.

4791. Yamamoto, D., Nihonmatsu, I., Matsuo, T., Miyamoto, H., Kondo, S., Hirata, K., and Ikegami, Y. (1996). Genetic interactions of *pokkuri* with *seven in absentia, tramtrack* and downstream components of the *sevenless* pathway in R7 photoreceptor induction in *Drosophila melanogaster. Roux's Arch. Dev. Biol.* **205**, 215–224.

4792. Yamamoto, T., Harada, N., Kano, K., Taya, S.-i., Canaani, E., Matsuura, Y., Mizoguchi, A., Ide, C., and Kaibuchi, K. (1997). The Ras target AF-6 interacts with ZO-1 and serves as a peripheral component of tight junctions in epithelial cells. *J. Cell Biol.* **139**, 785–795.

4793. Yamamoto, Y., Girard, F., Bello, B., Affolter, M., and Gehring, W.J. (1997). The *cramped* gene of *Drosophila* is a member of the *Polycomb*-group, and interacts with *mus209*, the gene encoding Proliferating Cell Nuclear Antigen. *Development* **124**, 3385–3394.

4794. Yan, R., Luo, H., Darnell, J.E., Jr., and Dearolf, C.R. (1996). A JAK-STAT pathway regulates wing vein formation in *Drosophila. Proc. Natl. Acad. Sci. USA* **93**, 5842–5847.

4795. Yanagawa, S., van Leeuwen, F., Wodarz, A., Klingensmith, J., and Nusse, R. (1995). The Dishevelled protein is modified by Wingless signaling in *Drosophila. Genes Dev.* **9**, 1087–1097.

4796. Yanagawa, S.-i., Lee, J.-S., Haruna, T., Oda, H., Uemura, T., Takeichi, M., and Ishimoto, A. (1997). Accumulation of Armadillo induced by Wingless, Dishevelled, and dominant-negative Zeste-white 3 leads to elevated DE-cadherin in *Drosophila* clone 8 wing disc cells. *J. Biol. Chem.* **272** #40, 25243–25251.

4797. Yang, C.-H., Simon, M.A., and McNeill, H. (1999). *mirror* controls planar polarity and equator formation through repression of *fringe* expression and through control of cell affinities. *Development* **126**, 5857–5866.

4798. Yang, H.-P., Tanikawa, A.Y., and Kondrashov, A.S. (2001). Molecular nature of 11 spontaneous *de novo* mutations in *Drosophila melanogaster. Genetics* **157**, 1285–1292.

4799. Yang, L. and Baker, N.E. (2001). Role of the EGFR/Ras/Raf pathway in specification of photoreceptor cells in the *Drosophila* retina. *Development* **128**, 1183–1191.

4800. Yang, S.-H., Whitmarsh, A.J., Davis, R.J., and Sharrocks, A.D. (1998). Differential targeting of MAP kinases to the ETS-domain transcription factor Elk-1. *EMBO J.* **17**, 1740–1749.

4801. Yang, S.-H., Yates, P.R., Whitmarsh, A.J., Davis, R.J., and Sharrocks, A.D. (1998). The Elk-1 ETS-domain transcription factor contains a mitogen-activated protein kinase targeting motif. *Mol. Cell. Biol.* **18**, 710–720.

4802. Yang, W. and Cerione, R.A. (1999). Endocytosis: Is dynamin a "blue collar" or "white collar" worker? *Curr. Biol.* **9**, R511–R514.

4803. Yang, X., Bahri, S., Klein, T., and Chia, W. (1997). Klumpfuss, a putative *Drosophila* zinc finger transcription factor, acts to differentiate between the identities of two secondary precursor cells within one neuroblast lineage. *Genes Dev.* **11**, 1396–1408.

4804. Yang, X., Yeo, S., Dick, T., and Chia, W. (1993). The role of a *Drosophila* POU homeo domain gene in the specification of neural precursor cell identity in the developing embryonic central nervous system. *Genes Dev.* **7**, 504–516.

4805. Yao, K.-M., Samson, M.-L., Reeves, R., and White, K. (1993). Gene *elav* of *Drosophila melanogaster*: a prototype for neuronal-specific RNA binding protein gene family that is conserved in flies and humans. *J. Neurobiol.* **24**, 723–739.

4806. Yao, L.-C., Liaw, G.-J., Pai, C.-Y., and Sun, Y.H. (1999). A common mechanism for antenna-to-leg transformation in *Drosophila*: suppression of *homothorax* transcription by four HOM-C genes. *Dev. Biol.* **211**, 268–276.

4807. Yap, A.S., Brieher, W.M., and Gumbiner, B.M. (1997). Molecular and functional analysis of cadherin-based adherens junctions. *Annu. Rev. Cell Dev. Biol.* **13**, 119–146.

4808. Yarnitzky, T., Min, L., and Volk, T. (1997). The *Drosophila* neuregulin homolog Vein mediates inductive interactions between myotubes and their epidermal attachment cells. *Genes Dev.* **11**, 2691–2700.

4809. Ye, Y. and Fortini, M.E. (2000). Proteolysis and developmental signal transduction. *Sems. Cell Dev. Biol.* **11**, 211–221.

4810. Ye, Y., Lukinova, N., and Fortini, M.E. (1999). Neurogenic phenotypes and altered Notch processing in *Drosophila Presenilin* mutants. *Nature* **398**, 525–529.

4811. Yeaman, C., Grindstaff, K.K., Hansen, M.D.H., and Nelson, W.J. (1999). Cell polarity: versatile scaffolds keep things in place. *Curr. Biol.* **9**, R515–R517.

4812. Yedvobnick, B. and Kumar, A. (2001). Mastermind enhances the activation of a Notch pathway target in cell culture. *Proc. 42nd Ann. Drosophila Res. Conf.* **Abstracts Vol.**, a260. [See also 2001 *Genesis* **30**, 250–258.]

4813. Yedvobnick, B., Smoller, D., Young, P., and Mills, D. (1988). Molecular analysis of the neurogenic locus *mastermind* of *Drosophila melanogaster*. *Genetics* **118**, 483–497.

4814. Yeh, E., Zhou, L., Rudzik, N., and Boulianne, G.L. (2000). Neuralized functions cell autonomously to regulate *Drosophila* sense organ development. *EMBO J.* **19**, 4827–4837. [See also 2001 *Curr. Biol.*, **11**, 1675–1679.]

4815. Yenush, L., Fernandez, R., Myers, M.G., Jr., Grammer, T.C., Sun, X.J., Blenis, J., Pierce, J.H., Schlessinger, J., and White, M.F. (1996). The *Drosophila* insulin receptor activates multiple signaling pathways but requires insulin receptor substrate proteins for DNA synthesis. *Mol. Cell. Biol.* **16**, 2509–2517.

4816. Yeo, S.L., Lloyd, A., Kozak, K., Dinh, A., Dick, T., Yang, X., Sakonju, S., and Chia, W. (1995). On the functional overlap between two *Drosophila* POU homeo domain genes and the cell fate specification of a CNS neural precursor. *Genes Dev.* **9**, 1223–1236.

4817. Yin, J.C.P., Wallach, J.S., Wilder, E.L., Klingensmith, J., Dang, D., Perrimon, N., Zhou, H., Tully, T., and Quinn, W.G. (1995). A *Drosophila* CREB/CREM homolog encodes multiple isoforms, including a cyclic AMP-dependent protein kinase-responsive transcriptional activator and antagonist. *Mol. Cell. Biol.* **15**, 5123–5130.

4818. Yin, Z., Xu, X.-L., and Frasch, M. (1997). Regulation of the Twist target gene *tinman* by modular *cis*-regulatory elements during early mesoderm development. *Development* **124**, 4971–4982.

4819. Yoshida, H., Kunisada, T., Kusakabe, M., Nishikawa, S., and Nishikawa, S.-I. (1996). Distinct stages of melanocyte differentiation revealed by analysis of nonuniform pigmentation patterns. *Development* **122**, 1207–1214.

4820. Yost, C., Britton, J., Loo, L.W., Edgar, B., and Eisenman, R. (2001). The role of dMyc and dMad in endoreplication. *Proc. 42nd Ann. Drosophila Res. Conf.* **Abstracts Vol.**, a86.

4821. Young, M.W. (1998). The molecular control of circadian behavioral rhythms and their entrainment in *Drosophila*. *Annu. Rev. Biochem.* **67**, 135–152.

4822. Young, M.W. (2000). The tick-tock of the biological clock. *Sci. Am.* **282** #3, 64–71.

4823. Young, M.W. and Wesley, C.S. (1997). Diverse roles for the Notch receptor in the development of *D. melanogaster*. *Persp. Dev. Neurobiol.* **4**, 345–355.

4824. Younger-Shepherd, S., Vaessin, H., Bier, E., Jan, L.Y., and Jan, Y.N. (1992). *deadpan*, an essential pan-neural gene encoding an HLH protein, acts as a denominator in *Drosophila* sex determination. *Cell* **70**, 911–922.

4825. Younossi-Hartenstein, A., Tepass, U., and Hartenstein, V. (1993). Embryonic origin of the imaginal discs of the head of *Drosophila melanogaster*. *Roux's Arch. Dev. Biol.* **203**, 60–73.

4826. Yoxen, E. (1985). Form and strategy in biology: reflections on the career of C. H. Waddington. *In* "A History of Embryology," *8th Symp. Brit. Soc. Dev. Biol.*, (T.J. Horder, J.A. Witkowski, and C.C. Wylie, eds.). Cambridge Univ. Pr.: Cambridge, pp. 309–329.

4827. Yu, F., Morin, X., Cai, Y., Yang, X., and Chia, W. (2000). Analysis of *partner of inscuteable*, a novel player of *Drosophila* asymmetric divisions, reveals two distinct steps in Inscuteable apical localization. *Cell* **100**, 399–409.

4828. Yu, H., Rosen, M.K., Shin, T.B., Seidel-Dugan, C., Brugge, J.S., and Schreiber, S.L. (1992). Solution structure of the SH3 domain of Src and identification of its ligand-binding site. *Science* **258**, 1665–1668.

4829. Yu, H.-H., Huang, A.S., and Kolodkin, A.L. (2000). Semaphorin-1a acts in concert with the cell adhesion molecules fasciclin II and connectin to regulate axon fasciculation in *Drosophila*. *Genetics* **156**, 723–731.

4830. Yu, K., Srinivasan, S., Shimmi, O., Biehs, B., Rashka, K., E., Kimelman, D., O'Connor, M.B., and Bier, E. (2000). Processing of the *Drosophila* Sog protein creates a novel BMP inhibitory activity. *Development* **127**, 2143–2154.

4831. Yu, K., Sturtevant, M.A., Biehs, B., François, V., Padgett, R.W., Blackman, R.K., and Bier, E. (1996). The *Drosophila decapentaplegic* and *short gastrulation* genes function antagonistically during adult wing vein development. *Development* **122**, 4033–4044.

4832. Yu, X. and Bienz, M. (1999). Ubiquitous expression of a *Drosophila* adenomatous polyposis coli homolog and its localization in cortical actin caps. *Mechs. Dev.* **84**, 69–73.

4833. Yu, X., Waltzer, L., and Bienz, M. (1999). A new *Drosophila* APC homologue associated with adhesive zones of epithelial cells. *Nature Cell Biol.* **1**, 144–151.

4834. Yu, Y. and Pick, L. (1995). Non-periodic cues generate seven *ftz* stripes in the *Drosophila* embryo. *Mechs. Dev.* **50**, 163–175.

4835. Yuan, Y.P., Schultz, J., Mlodzik, M., and Bork, P. (1997). Secreted Fringe-like signaling molecules may be glycosyltransferases. *Cell* **88**, 9–11.

4836. Yuh, C.-H., Bolouri, H., and Davidson, E.H. (1998). Genomic *cis*-regulatory logic: experimental and computational analysis of a sea urchin gene. *Science* **279**, 1896–1902.

4837. Yuh, C.-H., Bolouri, H., and Davidson, E.H. (2001).

Cis-regulatory logic in the *endo16* gene: switching from a specification to a differentiation mode of control. *Development* **128**, 617–629.

4838. Yuh, C.-H., Moore, J.G., and Davidson, E.H. (1996). Quantitative functional interrelations within the *cis*-regulatory system of the *S. purpuratus Endo16* gene. *Development* **122**, 4045–4056.

4839. Yun, U.J., Kim, S.Y., Liu, J., Adler, P.N., Bae, E., Kim, J., and Park, W.J. (1999). The Inturned protein of *Drosophila melanogaster* is a cytoplasmic protein located at the cell periphery in wing cells. *Dev. Genet.* **25**, 297–305.

4840. Zacharuk, R.Y. (1980). Ultrastructure and function of insect chemosensilla. *Annu. Rev. Entomol.* **25**, 27–47.

4841. Zacharuk, R.Y. (1985). Antennae and sensilla. *In* "Comprehensive Insect Physiology, Biochemistry, and Pharmacology," Vol. 6 (G.A. Kerkut and L.I. Gilbert, eds.). Pergamon: New York, pp. 1–69.

4842. Zaffran, S. and Frasch, M. (2000). *Barbu*: an E(spl) *m4/mα*-related gene that antagonizes Notch signaling and is required for the establishment of ommatidial polarity. *Development* **127**, 1115–1130.

4843. Zak, N.B. and Shilo, B.-Z. (1992). Localization of DER and the pattern of cell divisions in wild-type and *Ellipse* eye imaginal discs. *Dev. Biol.* **149**, 448–456.

4844. Zákány, J., Gérard, M., Favier, B., and Duboule, D. (1997). Deletion of a *HoxD* enhancer induces transcriptional heterochrony leading to transposition of the sacrum. *EMBO J.* **16**, 4393–4402.

4845. Zalokar, M., Erk, I., and Santamaria, P. (1980). Distribution of ring-X chromosomes in the blastoderm of gynandromorphic *D. melanogaster*. *Cell* **19**, 133–141.

4846. Zappavigna, V., Sartori, D., and Mavilio, F. (1994). Specificity of HOX protein function depends on DNA-protein and protein-protein interactions, both mediated by the homeo domain. *Genes Dev.* **8**, 732–744.

4847. Zaret, K. and Wolffe, A.P. (2001). Chromosomes and expression mechanisms: The post-genomic era of gene control. *Curr. Opin. Gen. Dev.* **11**, 121–123.

4848. Zecca, M., Basler, K., and Struhl, G. (1995). Sequential organizing activities of engrailed, hedgehog, and decapentaplegic in the *Drosophila* wing. *Development* **121**, 2265–2278.

4849. Zecca, M., Basler, K., and Struhl, G. (1996). Direct and long-range action of a Wingless morphogen gradient. *Cell* **87**, 833–844.

4850. Zeeman, E.C. (1974). Primary and secondary waves in developmental biology. *In* "Lectures on Mathematics in the Life Sciences," Vol. 7 Am. Math. Soc.: Providence, Rhode Island, pp. 69–161.

4851. Zeidler, M.P., Perrimon, N., and Strutt, D.I. (1999). The *four-jointed* gene is required in the *Drosophila* eye for ommatidial polarity specification. *Curr. Biol.* **9**, 1363–1372.

4852. Zeidler, M.P., Perrimon, N., and Strutt, D.I. (2000). Multiple roles for *four-jointed* in planar polarity and limb patterning. *Dev. Biol.* **228**, 181–196. [See also 2001 *Development* **128**, 3533–3542.]

4853. Zeitlinger, J. and Bohmann, D. (1999). Thorax closure in *Drosophila*: involvement of Fos and the JNK pathway. *Development* **126**, 3947–3956.

4854. Zeitlinger, J., Kockel, L., Peverali, F.A., Jackson, D.B., Mlodzik, M., and Bohmann, D. (1997). Defective dorsal closure and loss of epidermal *decapentaplegic* expression in *Drosophila fos* mutants. *EMBO J.* **16**, 7393–7401.

4855. Zelhof, A.C., Ghbeish, N., Tsai, C., Evans, R.M., and McKeown, M. (1997). A role for Ultraspiracle, the *Drosophila* RXR, in morphogenetic furrow movement and photoreceptor cluster formation. *Development* **124**, 2499–2506.

4856. Zelhof, A.C., Yao, T.-P., Chen, J.D., Evans, R.M., and McKeown, M. (1995). Seven-up inhibits Ultraspiracle-based signaling pathways in vitro and in vivo. *Mol. Cell. Biol.* **15**, 6736–6745.

4857. Zelzer, E., Wappner, P., and Shilo, B.-Z. (1997). The PAS domain confers target gene specificity of *Drosophila* bHLH/PAS proteins. *Genes Dev.* **11**, 2079–2089.

4858. Zeng, C., Justice, N.J., Abdelilah, S., Chan, Y.-M., Jan, L.Y., and Jan, Y.N. (1998). The *Drosophila* LIM-only gene, *dLMO*, is mutated in *Beadex* alleles and might represent an evolutionarily conserved function in appendage development. *Proc. Natl. Acad. Sci. USA* **95**, 10637–10642.

4859. Zeng, C., Younger-Shepherd, S., Jan, L.Y., and Jan, Y.N. (1998). Delta and Serrate are redundant Notch ligands required for asymmetric cell divisions within the *Drosophila* sensory organ lineage. *Genes Dev.* **12**, 1086–1091.

4860. Zeng, H., Qian, Z., Myers, M.P., and Rosbash, M. (1996). A light-entrainment mechanism for the *Drosophila* circadian clock. *Nature* **380**, 129–135.

4861. Zeng, L., Fagotto, F., Zhang, T., Hsu, W., Vasicek, T.J., Perry III, W.L., Lee, J.J., Tilghman, S.M., Gumbiner, B.M., and Costantini, F. (1997). The mouse *Fused* locus encodes Axin, an inhibitor of the Wnt signaling pathway that regulates embryonic axis formation. *Cell* **90**, 181–192.

4862. Zeng, W., Andrew, D.J., Mathies, L.D., Horner, M.A., and Scott, M.P. (1993). Ectopic expression and function of the *Antp* and *Scr* homeotic genes: the N terminus of the homeodomain is critical to functional specificity. *Development* **118**, 339–352.

4863. Zeng, W., Wharton, K.A., Jr., Mack, J.A., Wang, K., Gadbaw, M., Suyama, K., Klein, P.S., and Scott, M.P. (2000). *naked cuticle* encodes an inducible antagonist of Wnt signalling. *Nature* **403**, 789–794.

4864. Zhang, C.-C., Müller, J., Hoch, M., Jäckle, H., and Bienz, M. (1991). Target sequences for *hunchback* in a control region conferring *Ultrabithorax* expression boundaries. *Development* **113**, 1171–1179.

4865. Zhang, H. and Levine, M. (1999). Groucho and dCtBP mediate separate pathways of transcriptional repression in the *Drosophila* embryo. *Proc. Natl. Acad. Sci. USA* **96**, 535–540.

4866. Zhang, H., Levine, M., and Ashe, H.L. (2001). Brinker is a sequence-specific transcriptional repressor in the *Drosophila* embryo. *Genes Dev.* **15**, 261–266. [See also 2001 *EMBO J.* **20**, 5725–5736.]

4867. Zhang, J. and Carthew, R.W. (1998). Interactions between Wingless and DFz2 during *Drosophila* wing development. *Development* **125**, 3075–3085.

4868. Zhang, J. and Jacobson, A.G. (1993). Evidence that the border of the neural plate may be positioned by the interaction between signals that induce ventral and dorsal mesoderm. *Dev. Dynamics* **196**, 79–90.

4869. Zhang, Q. and Lu, X. (2000). *semang* affects the development of a subset of cells in the *Drosophila* compound eye. *Mechs. Dev.* **95**, 113–122.

4870. Zhang, Q., Zheng, Q., and Lu, X. (1999). A genetic screen for modifiers of *Drosophila Src42A* identifies mutations in *Egfr*, *rolled* and a novel signaling gene. *Genetics* **151**, 697–711.

4871. Zhang, Z., Murphy, A., Hu, J.C., and Kodadek, T. (1999).

Genetic selection of short peptides that support protein oligomerization *in vivo*. *Curr. Biol.* **9**, 417–420.

4872. Zhao, J.J., Lazzarini, R.A., and Pick, L. (1993). The mouse *Hox-1.3* gene is functionally equivalent to the *Drosophila Sex combs reduced* gene. *Genes Dev.* **7**, 343–354.

4873. Zheng, L., Zhang, J., and Carthew, R.W. (1995). *frizzled* regulates mirror-symmetric pattern formation in the *Drosophila* eye. *Development* **121**, 3045–3055.

4874. Zheng, N. and Gierasch, L.M. (1996). Signal sequences: the same yet different. *Cell* **86**, 849–852.

4875. Zhong, W., Feder, J.N., Jiang, M.-M., Jan, L.Y., and Jan, Y.N. (1996). Asymmetric localization of a mammalian numb homolog during mouse cortical neurogenesis. *Neuron* **17**, 43–53.

4876. Zhong, W., Jiang, M.-M., Weinmaster, G., Jan, L.Y., and Jan, Y.N. (1997). Differential expression of mammalian Numb, Numblike and Notch1 suggests distinct roles during mouse cortical neurogenesis. *Development* **124**, 1887–1897.

4877. Zhou, J., Barolo, S., Szymanski, P., and Levine, M. (1996). The Fab-7 element of the bithorax complex attenuates enhancer-promoter interactions in the *Drosophila* embryo. *Genes Dev.* **10**, 3195–3201.

4878. Zhou, M.-M., Ravichandran, K.S., Olejniczak, E.T., Petros, A.M., Meadows, R.P., Sattler, M., Harlan, J.E., Wade, W.S., Burakoff, S.J., and Fesik, S.W. (1995). Structure and ligand recognition of the phosphotyrosine binding domain of Shc. *Nature* **378**, 584–592.

4879. Zhu, A. and Kuziora, M.A. (1996). Functional domains in the Deformed protein. *Development* **122**, 1577–1587.

4880. Zhu, A.J. and Watt, F.M. (1999). β-catenin signalling modulates proliferative potential of human epidermal keratinocytes independently of intercellular adhesion. *Development* **126**, 2285–2298.

4881. Zhu, A.J., Zheng, L., Suyama, K., Ho, K.S., and Scott, M.P. (2001). Hedgehog signaling moves Smoothened protein to the cell surface. *Proc. 42nd Ann. Drosophila Res. Conf.* **Abstracts Vol.**, a155.

4882. Zhu, L. and Skoultchi, A.I. (2001). Coordinating cell proliferation and differentiation. *Curr. Opin. Gen. Dev.* **10**, 91–97.

4883. Zhu, L., Wilken, J., Phillips, N.B., Narendra, U., Chan, G., Stratton, S.M., Kent, S.B., and Weiss, M.A. (2000). Sexual dimorphism in diverse metazoans is regulated by a novel class of intertwined zinc fingers. *Genes Dev.* **14**, 1750–1764.

4884. Zhuo, N., Tyler, D.M., Joglekar, S., Sultan, R., and Baker, N.E. (2001). Mutations affecting cell competition and growth. *Proc. 42nd Ann. Drosophila Res. Conf.* **Abstracts Vol.**, a86.

4885. Ziemer, A., Tietze, K., Knust, E., and Campos-Ortega, J.A. (1988). Genetic analysis of *Enhancer of split*, a locus involved in neurogenesis in *Drosophila melanogaster*. *Genetics* **119**, 63–74.

4886. Zilian, O., Frei, E., Burke, R., Brentrup, D., Gutjahr, T., Bryant, P.J., and Noll, M. (1999). *double-time* is identical to *discs overgrown*, which is required for cell survival, proliferation and growth arrest in *Drosophila* imaginal discs. *Development* **126**, 5409–5420.

4887. Zill, S.N. and Seyfarth, E.-A. (1996). Exoskeletal sensors for walking. *Sci. Am.* **275** #1, 86–90.

4888. Zimmerman, J.E., Bui, Q.T., Liu, H., and Bonini, N.M. (2000). Molecular genetic analysis of *Drosophila eyes absent* mutants reveals an eye enhancer element. *Genetics* **154**, 237–246.

4889. Zink, D. and Paro, R. (1995). *Drosophila* Polycomb-group regulated chromatin inhibits the accessibility of a *trans*-activator to its target DNA. *EMBO J.* **14**, 5660–5671.

4890. Zipursky, S.L. and Rubin, G.M. (1994). Determination of neuronal cell fate: lessons from the R7 neuron of *Drosophila*. *Annu. Rev. Neurosci.* **17**, 373–397.

4891. Zollman, S., Godt, D., Privé, G.G., Couderc, J.-L., and Laski, F.A. (1994). The BTB domain, found primarily in zinc finger proteins, defines an evolutionarily conserved family that includes several developmentally regulated genes in *Drosophila*. *Proc. Natl. Acad. Sci. USA* **91**, 10717–10721.

4892. Zorin, I.D., Gerasimova, T.I., and Corces, V.G. (1999). The *lawc* gene is a new member of the *trithorax*-group that affects the function of the *gypsy* insulator of *Drosophila*. *Genetics* **152**, 1045–1055.

4893. Zrzavy, J. and Stys, P. (1995). Evolution of metamerism in arthropoda: developmental and morphological perspectives. *Q. Rev. Biol.* **70**, 279–295. [See also 2001 *Evol. Dev.* **3**, 332–342.]

4894. Zuker, C.S. (1994). On the evolution of eyes: would you like it simple or compound? *Science* **265**, 742–743.

4895. Zuker, C.S., Cowman, A.F., and Rubin, G.M. (1985). Isolation and structure of a rhodopsin gene from *D. melanogaster*. *Cell* **40**, 851–858.

4896. Zuker, C.S. and Ranganathan, R. (1999). The path to specificity. *Science* **283**, 650–651.

4897. zur Lage, P., Jan, Y.N., and Jarman, A.P. (1997). Requirement for EGF receptor signalling in neural recruitment during formation of *Drosophila* chordotonal sense organ clusters. *Curr. Biol.* **7**, 166–175.

4898. zur Lage, P. and Jarman, A.P. (1999). Antagonism of EGFR and Notch signalling in the reiterative recruitment of *Drosophila* adult chordotonal sense organ precursors. *Development* **126**, 3149–3157.

4899. Zusman, S.B. and Wieschaus, E. (1987). A cell marker system and mosaic patterns during early embryonic development in *Drosophila melanogaster*. *Genetics* **115**, 725–736.

4900. Zwahlen, C., Li, S.-C., Kay, L.E., Pawson, T., and Forman-Kay, J.D. (2000). Multiple modes of peptide recognition by the PTB domain of the cell fate determinant Numb. *EMBO J.* **19**, 1505–1515.

Index

abbreviations, xiii, xiv. *See also* genes,
  particular
abdomen, xi, 49, 84, 141, 239
  sternites, 2, 30, 31, 56
  tergites, 2, 30–32, 56, 62, 122, 151
Akam, Michael, 79, 84, 246, 247, 303
anatomy, 2, 32
  antenna, 83, 199, 251
  bilateral symmetry, 191, 193, 209
  bristles, 5, 7, 30, 63
  encoding, xi, 37, 41, 45, 83, 101, 129, 191,
    213, 239, 246, 254, 255, 297
  geometry, 31, 32, 99, 105, 128, 137, 189,
    201, 211
  gradients, 30, 65, 99
  head, 41, 96, 199
  leg, 65, 99, 127, 131
  metameres, 76, 79, 84, 239, 243–247, 254
  mirror-symmetry planes, 62, 97, 99, 136,
    167, 188, 198–201, 209
  periodicity, 77, 84, 136, 225, 303, 304
  sclerites, 2, 86, 89, 245, 299
  sexual dimorphism, 1, 56, 62–65, 72, 80
  thorax, 41, 193
  wing, 139, 153, 155, 159, 177
antenna
  anatomy, 83, 199, 251
  axes. *See under* axes
  bipolar duality (vs. eye), 114, 169, 235,
    302
  circuitry, genetic, 205, 249–252, 300
  compartments, 4, 199
  duplication vs. regeneration, 96, 169, 242
  evolution, 252
  fate map, 199
  gene expression patterns, 91, 193, 205,
    251, 288–296, 302
  homeosis to leg, 80–85, 129, 249–252
  homology to leg, 83, 249, 251, 299
  Hox gene irrelevance, 249
  Hox gene misexpression in, 249, 251
  identity (vs. eye), 169
  identity (vs. leg), 83, 246, 249–254
  part of eye disc, xiii, 96, 169, 199, 302
  sensilla, 29, 191, 199, 246, 276
  topology (vs. leg), 83, 199, 299
  wiring of axons, 191

apoptosis, 63, 77, 80, 92, 99, 100, 105, 118,
  119, 123, 128, 135, 139, 157, 160,
  227–229
Aristotle, xi, xiii
arthropods, 93, 99, 246, 300, 305
Ashburner, Michael, xii
asymmetry
  bilateral, 56, 67, 87
  bristle patterns, 31
  cellular, 7, 11, 24, 274, 304
  circuitry, 89, 107
  eye D/V, 203–211
  fluctuating, 21
  growth, 154
  growth potential, 118–119
  mitotic, 7, 11, 24
  ommatidia, 209–211
  symmetry-breaking, 47, 209–211, 304
  wing A/P, 188, 301
  wing D/V, 160, 164–167
Auerbach, Charlotte, 76
axes
  antenna D-V, 199
  antenna proximal-distal, 83, 251
  as reference lines, 84, 87, 89, 97, 101–105,
    109, 151, 188, 203, 205
  body and limb, 30, 31, 155, 239
  bristle development, 30
  cell apical-basal, 24, 87, 124, 146, 273, 293,
    304
  chordotonal, 27
  diffusion along, 65
  embryo A-P, 76, 79, 84, 87–90, 299
  embryo D-V, 87–91, 170, 239
  embryo left-right, 87
  eye A-P, 201, 208–211, 215, 227, 233–236,
    305
  eye centrifugal, 229
  eye D-V, 202–211, 227, 233, 236, 299
  eye rhabdomeres, 199
  interdependence, 115
  larva, 205
  leg A-P, 89, 112
  leg A-P vs. wing A-P, 137
  leg disc centrifugal, 129, 131
  leg disc medial-lateral, 92
  leg disc upper(stalk)-lower, 92

leg D-V, 61, 65, 67, 91, 97–99, 103, 109, 112,
  115, 118, 119, 124–127, 132
leg proximal-distal, 29, 30, 61, 67, 80, 83,
  91, 115–118, 127–135, 148, 166, 201, 251,
  300
morphogen usage, 158, 249
notum A-P, 30, 31, 37, 68
optic lobe medial-lateral, 196
positional information, 81, 87
sternite A-P, 30
tarsus D-V, 30, 62
tergite A-P, 30
tergite D-V, 30
thorax A-P, 107, 132
thorax D-V, 205
wing A-P, 89, 137–156, 165, 167, 177, 301
wing A-P vs. D-V, 158–161, 165
wing centripetal, 125
wing crossveins, 189
wing D-V, 96, 136, 137, 146, 156–167, 170,
  301
wing proximal-distal, 132, 151, 159, 171

Baker, Nicholas, 225
Baker, William, 202
Bang, Anne, 23
Basler, Konrad, 151, 157
Bateson, William, 99, 237
Becker, Hans, 202
Bender, Welcome, 305
Benzer, Seymour, 202, 225
Bernard, Francis, 201
Blair, Seth, xii, 173, 301
Bodenstein, Dietrich, xii, 76
boundaries. *See also* compartment
  boundaries
  absence, 93, 95, 105, 123, 155, 203
  as axes, 84, 87, 89, 97, 101–105, 109, 188,
    203, 205
  as chains of adhering cells, 151, 173
  as guidelines, 61–65, 163, 164
  cis-enhancers for, 171, 245
  clonal, 67, 201, 202
  convergence/divergence (leg outgrowths),
    100
  creation of, 78, 159–161, 173, 203–207,
    236, 302

boundaries (*contd.*)
　crossing of, 132, 136, 141, 148–154, 160, 173, 185, 201–204
　deformations, 151
　denoted by slash mark (vs. hyphen), xiii
　diffusion barriers, 125, 126, 143
　discontinuities, 93
　effects, 105, 115
　eye equator, 198–211, 296
　fuzzy, 128, 167, 193, 305
　gene dosage, 52, 69
　gene expression (ON/OFF), 89, 121, 131, 136, 153, 157–160, 165, 173, 177, 186, 191, 203, 207, 209, 234, 235, 299, 303
　gene expression at, 167, 171, 207, 209, 296
　gene expression, complementary, 78, 129, 131, 134, 135
　Hairy/Achaete, 63
　idiosyncrasies, 166
　insulating, 190
　intersections, 46, 90, 101, 137, 157, 234, 301
　kinks (natural), 112, 148
　L$_{2\&7}$, 124
　leg dAC/vAC, 100, 109, 111, 117, 119, 132
　leg segment, 136
　maintenance, 148–153, 166, 173, 273
　male/female, 1–4, 52, 67
　metamere, 79
　midline, 59, 62, 63, 72, 77, 89, 99, 109, 113, 118, 131, 151, 164, 177, 262
　models, 100–111, 115, 122, 130, 163, 165, 302
　moving, 137, 229–234
　mutant/wild-type, 47, 52
　neural/non-neural, 151
　offsets, 97, 99, 109, 148, 224, 225
　parasegment, 78, 84, 90
　peripodial membrane, 139
　proneural clusters, 191
　rewiring of control, 171
　segment, 84
　sharpening, 78, 81, 84, 114, 184
　sharpness, 45, 167
　shifts, 89, 111, 119–123, 145, 188–190, 288
　smooth vs. ragged, 160, 173
　straddling of, 153, 164
　straightening of, 135, 148–153, 160, 173–174
　wing margin, 142, 147, 157, 159, 173, 245, 301
　Wingless/Engrailed, 78, 89, 90
　yellow/brown, 2
　zones, 111, 139, 302
bracts, 5, 7, 28–29, 61, 63, 67, 99
Bray, Sarah, 224
Brehme, Katherine, xii
Bridges, Calvin, xii, 37, 299
bristle patterns. *See also* bristles
　alignment, 67, 68
　ancestral, 62, 63, 135
　ancient, 31
　antineural gradients, 69
　antineural mask, 69
　antineural RNAs, 73
　antineural stripes, 59, 61, 62, 69, 75
　asymmetry, 56, 67
　basitarsus, 56, 62
　bristle density, 31, 49, 53–56, 73, 163, 187, 230

bristle displacements, 37, 39, 43, 47, 49, 53, 68, 69, 163, 191, 229, 230
brushes, 62
cell-size dependence, 56, 68
confluent lawns, 163, 229
constant vs. variable, 31, 56, 67
constellations, 32, 68
CS vs. MS bristles, 67
evolution, 31, 36, 62, 65, 69, 255
fine-tuning, 50, 62–69, 75, 228
functions, 31, 62, 65
furry stripes, 59
general problem, 31
genetic control, 31–75, 190–194, 230, 278–284
geometry, 31, 56, 59
gradients, 62
growth-dependence, 39, 50, 57, 63, 67
heterochronic superposition, 56, 57, 67
indeterminacy, 56, 59, 67, 68
inhibitory fields, 50, 55–67, 71–75, 194, 195, 279
isotropic, 32
lattice, 32, 229, 230
leg, 56–62, 65, 67
macrochaetes, 190–194, 255
macrochaetes vs. microchaetes, 31, 32, 56, 67, 68
misalignment, 63
modules, 62, 63
mutant phenotypes, 278–284
natural variation, 31, 37, 55, 56
notum, 37, 41, 55–61, 65, 68, 190–194, 255
other species, 67, 69, 255
periodic, 67
precision, 31, 56, 59–62, 65, 67, 69
proneural clusters, 39, 41–75, 191–195, 225, 229
proneural competence, 49, 59, 69, 72, 75
proneural fields, 62, 68, 71, 72, 304
proneural gradients, 164, 191
proneural landscape, 69
proneural machinery, 190
proneural plateaus, 62
proneural potential, 67, 68, 69
proneural RNA motifs, 73
proneural spots, 59, 62, 71
proneural stripes, 59–63, 67–71, 75, 163, 280, 304
proneural subclusters, 43, 46, 47, 50, 53, 69
reconstitution, 67
rotation, 63, 65, 67, 106
rows, 31, 32, 56–62, 67, 68, 99
rows, alignment of, 61, 65, 67
rows, double (wing), 159, 189
rows, extra, 61, 68, 80, 187
rows, transverse (leg), 31, 62, 63, 65, 299
rows, triple (wing), 148, 159, 164, 189
scutellum, 49, 61
sex comb, 32, 36, 62–67, 106, 246
sex combs, extra, 80, 134, 135, 148, 248
sex dimorphisms, 56, 62, 63, 65
size-independence, 55, 62, 67
spacing, 31, 50, 55–68, 99, 159, 187, 224, 225, 279
sternites, 56
sternopleura, 56, 86
stochastic factors, 51, 72, 75, 201, 203, 299
symmetry, 62
tandem, 32, 65, 74
tergites, 31, 56

tufts, 23, 49, 52, 53, 59
variety, 32
vs. vein patterns, 175
wing margin, 65, 67, 159
bristles. *See also* bristle patterns; sensory organ precursor (SOP)
　abdominal, 55
　anatomy, 5, 30, 63
　bracted vs. bractless, 7, 65, 67, 99, 134
　cell fate transformations, 7–28, 52, 271–275
　cell lineage, 4, 5, 21, 27, 30, 44
　central, 63
　chemosensory, 5, 27, 45, 57, 62, 67, 99
　CNS projections, 191
　deformed, 23, 48
　dendrite, 5, 23, 27
　determination of type, 27, 63, 72
　differentiation, 5, 11, 29, 36, 48
　elongation, 29
　evolution, 28, 29, 30
　extra, xiv, 20, 31, 45, 49–55, 59–62, 68–75, 80, 124, 164, 166, 194, 229, 278–284
　eye, 30, 57, 199, 209
　function, 5
　genetic control, 5–28, 271–277
　grooves, 23, 29, 298
　head, 31, 61
　homologies, 63
　inducers, 39, 50
　inhibitors, 53, 56
　leg, 5, 21, 27–31, 36, 56, 57, 61–67
　lengths, 29–31, 39, 41, 56, 62, 65, 73
　macrochaetes, xiii, 23, 31–62, 67–71, 75, 99, 190–195, 278–284
　macrochaetes vs. microchaetes, 20, 29–30, 57, 68, 191
　mechanosensory, 5, 27, 45, 57, 67, 99
　microchaetes, xiii, 31–62, 67–71, 191, 278–284
　migration, 63
　misoriented, 65
　missing, xiv, 20, 23, 31, 37, 45, 48–52, 71, 194, 278–284
　mutant phenotypes, 271–277
　notum, 29–31, 36, 39, 53–57, 62–68, 75
　number, 31, 67
　origin of glial cell, 5
　orphan, 63
　photosensitivity, 5
　pigmentation, 5, 23, 30, 63, 65, 99
　polarity, 29, 65, 67, 80, 205, 293
　sex comb, 7, 30, 36, 72, 149
　shaft-to-socket spectrum, 23
　shapes, 99
　size, 27, 29–30, 68
　sternites, 30
　sternopleura, 55, 56
　tergites, 30, 48, 65, 68
　thoracic, 4, 30
　timing of determination, 27, 30, 63, 68, 72
　timing of differentiation, 29, 39, 57, 67, 92
　timing of radioinsensitivity, 39
　transformation to sensilla, 28
　types (MS vs. CS), 5, 27, 65, 67, 99, 159
　vibrissae (head), 299
　vs. photoreceptors, 225
　vs. scales (lepidopteran), 29, 47, 61, 65, 67
　vs. sensilla, 27–28

wing blade (ectopic), 49, 61
wing margin, 5, 18, 27, 31, 48, 56, 143, 159–167, 189, 272
Britten, Roy, 244
Brody, Thomas, xiii
Bryant, Peter, xii, 33, 80, 83, 93, 94, 139, 155
Bryant, Susan, 93

Cadigan, Kenneth, 158
Cagan, Ross, 212, 217, 227
Campbell, Gerard, 115
Carroll, Sean, 41
Castelli-Gair, James, 246, 247
Cavodeassi, Florencia, 203
cell adhesion, 65, 81, 87, 90, 92, 107, 132–136, 148–153, 160, 161, 173–177, 193, 204, 223, 257–261, 274, 281, 297
cell affinities, 87, 90, 107, 132, 148–153, 160, 173, 174, 193, 204
cell behaviors, 63, 85, 93, 96, 107, 122, 123, 132, 153, 190, 203, 235, 252
cell competition, 105, 153
cell components
    cytoskeleton, 11, 257, 295
    lysosomes, 27
    microfilaments, 11, 29, 125
    microtubules, 15, 109, 125
    microvilli, 9, 87, 109, 179, 199
    nuclear matrix, 249
    nuts and bolts, 263
    recycling, 300
    rhabdomeres, 197, 199, 223, 224
    ribosomes, 29
cell cortex, 7, 9, 11, 15, 24, 179, 273, 274
cell cycle. See under mitosis
cell death, 63, 80, 87, 89, 95–97, 100, 113, 119, 135, 156, 174, 227–228, 299
cell instructions, 4, 5, 9, 18, 20, 93, 99, 101, 115, 173, 205, 215, 216, 217, 228, 254, 255. See also pattern formation: rules
cell jostling, 90, 91, 111, 149, 151, 173
cell junctions, 20, 72, 92
    adherens, 11, 75, 179, 180, 263, 294
    making and breaking, 201, 304
cell lineage, 1–4. See also compartment boundaries
    adult vs. larva, 86
    bracts, 5, 7, 28, 61, 63
    bristles, 4, 5, 27, 30, 44
    clonal analysis, 1–4, 86, 91, 99, 123, 201, 202
    clone fragmentation, 91
    clone fusions, 153
    clone outlines, 3, 39, 67, 91, 201
    clone overlaps, 91
    clone roundness vs. raggedness, 132, 149–153, 160, 190, 201, 203
    clone shapes, 133, 151, 153
    compartments, 4, 44, 63, 85, 89, 99, 103, 104, 146, 150, 243, 245
    compartments, function of, 87, 107, 148
    compartments, regeneration of, 121
    fluid vs. stereotyped, 3, 5, 7, 90, 91, 111, 149, 151, 173
    heart, 4, 9
    indeterminate, 3, 4, 85, 86, 91, 111, 123, 136, 149, 202, 212, 252
    Minute technique, 4, 91, 202
    mixing, 105
    muscle, 4, 9
    nervous system, 4, 11

Proximity-vs.-Pedigree Rule, 4
    sensilla, 1, 4
    sex comb, 63
    sibling rivalry, 9, 10
    sternopleura, 86
    strategies, 1
    tracing, 3, 90
cell migration, 86, 91, 99, 258
cell movements, 61–67, 86-91, 131, 148–153, 177, 211, 230
cell packing, 53, 55, 122, 134, 149, 174, 208, 233, 304
cell polarity, 24, 29, 65, 67, 79, 80, 87, 131–136, 205, 211, 293, 304
cell psychology
    amnesia, 85, 249
    antisocial behavior, 153
    delirium, 107
    how cells think, xii, 11, 83, 99, 111, 114, 124, 134, 136, 138, 143, 149, 154, 189, 190
    memory, 10, 65, 81–84, 143, 158, 211, 239, 248, 251
    myopia, 93, 94, 173
    not goal-oriented, 255, 300
    obedience to rules, 300. See also rules
    schizophrenia, 65, 107
    sibling rivalry, 9, 10
    sociability vs. introspection, 1–4, 85, 299
    suicidal tendencies, 228
    what cells know, xii, 83, 85, 158, 244, 249, 254, 255
cell rearrangements, 63–67, 89–91, 139, 174, 193, 209, 211, 304
cell recruitment, 29, 30, 87, 208, 212–220, 224, 227–229
cell shapes, 87, 99, 122, 124, 135, 151, 215, 233, 272, 273, 305
cell signaling, 1–4, 50, 72. See also circuitry; signaling pathways
    amplification, 13, 47, 188, 209, 211, 286, 300
    amplitude, 53, 74, 103, 215
    amplitude effects, 113, 154
    amplitude modulation, 145, 303
    analog vs. digital responses, 182
    anisotropic, 37, 59, 65, 75
    apical, 44
    as evocation, 304
    attenuation, 27, 134, 189, 264
    attenuation, signal-dependent, 147, 182
    autocrine, 52, 151, 167, 188
    community effect, 85
    competence to respond, 29, 35, 39, 43, 45, 109, 111, 145, 148, 153, 155, 164, 175, 215, 304, 305
    contact-limited, 139, 145, 153, 157, 185, 285, 293
    contact-mediated, 47, 53, 72, 93, 103, 135, 164, 165, 212, 215
    context-dependent, 217
    deafness, enforced, 35, 107, 150–153, 173, 229, 231, 235
    deafness, natural, 55, 61, 107, 113, 117, 121, 145, 148, 150, 163–167, 183, 187, 188, 207, 209, 227, 295
    default states, 11, 59, 85, 169, 216, 227
    diffusion-mediated, 39, 46–49, 52–55, 72, 75, 80–85, 98, 101, 105, 118, 123–126, 133, 137–141, 167, 169, 182, 183, 203–209

direct vs. signal relay, 139, 141, 156, 167, 186, 215
    duration, 182, 303
    ecdysone, 11, 217, 229
    endocrine, 92, 217
    endocytosis, 53
    gating into pathways, 10
    hijacking by heritable determinant, 10
    information content, 216, 217
    intercellular negotiation, 94, 95
    juxtacrine, 180, 212, 224, 304
    ligand-independent, 182
    mitosis-dependent, 217
    mufflers, 134
    muteness, enforced, 107, 153, 209
    muteness, natural, 103, 107, 145, 148, 163–166, 183, 187, 207
    need for mitotic quiescence, 44, 234, 304
    one-bit (Stop! or Go!) signals, 215–218, 304
    paracrine, 92, 125, 133, 166, 167, 231, 295
    perception modulation, 125, 134, 143, 147, 158, 188–190, 286, 291
    PI coordinates, 93, 99, 103
    potentiation, 166, 207
    primers vs. boosters, 216
    qualitative vs. quantitative, 35, 83, 179
    rate, 9, 181, 287
    rectification, 147
    resetting, 179, 182
    rheostat vs. solenoid mode, 182, 303
    second messengers, 141
    sensitivity of reception, 230, 303
    signal-to-noise ratio, 81, 124, 211, 228, 304
    signal-to-receiver ratio, 60, 167, 286, 303
    specificity, 179
    strategies, xii, 1, 9
    subthreshold, 36, 47, 68, 69, 117, 134, 174
    transcytosis, 27, 124, 146, 167
    trans-ingestion of receptor, 175
cell size, 29–30, 55–57, 65, 67, 92
cell sorting, 90, 134, 148–151
cell states. See also circuitry
    affinities, 132, 148–153. See also cell affinities
    as singularities, 158
    automatic sequence of, 215, 229
    axon projections, 55, 191
    Boolean, 7–12, 18–23, 84–85
    border vs. non-border, 149–153, 173, 205
    cellular automata, 235
    compartmental, 85, 89, 107, 145, 148–153, 158–163, 167, 173, 188, 202–208, 248
    competence, 148, 217
    created at interfaces, 87, 89, 107, 131–134, 142, 145, 153–155, 159, 160, 164, 177, 208–211, 234, 302, 303
    default. See under circuitry
    dependent vs. independent, 89, 90, 117
    determined vs. specified, 81, 90, 173, 247–249
    differentiated, 81, 87, 224
    eye vs. antenna, 169
    firm vs. transitory biases, 90, 173, 218, 248, 252
    heritable vs. not heritable, 89, 90, 132, 143, 246–252
    implementation by gene hierarchies, 149, 174, 177, 303
    inherited by discs from embryo, 107, 164
    intrinsic vs. extrinsic, 1–4, 9, 90–91, 124, 143, 149, 217

cell states (*contd.*)
  leg dAC vs. vAC, 111, 113
  leg proximal vs. distal, 134
  proneural, 75. *See also under* bristle
    patterns
  quadrants in wing pouch, 158
  qualitative vs. quantitative, 29–30, 139,
    150, 246, 301
  quantitative shifts in, 134, 135, 143, 147,
    154–156, 189, 190
  Ras-dependent diversity, 179
  specified vs. determined, 90, 111
  spectrum within eye MF, 205
  switching, 7–12, 18–23, 52, 84–85, 101,
    123, 132, 150, 182, 216–219, 223, 229,
    242, 254, 261
  switching, fast vs. slow, 287
  vein identities, various, 154–157
  vein vs. intervein, 141, 182, 187, 188, 259,
    261, 303
  wing vs. hinge, 139, 158, 172, 301, 302
  wing vs. notum, 169, 171–173
cell types
  discrete nature of, 72, 246, 299
  encoding of, 10, 213, 243, 246, 254
  immiscible, 90, 107, 132–134, 148, 150
  iteration, 217
  laser ablation, 217
  neural vs. non-neural, 20, 216, 218, 224
  ommatidium, 208–209, 223
  photoreceptors, 196, 197, 223, 253. *See
    also* photoreceptors
  sequential emergence, 216
  signaler vs. receiver, 107, 145, 148,
    163–166, 183, 187, 207
  squamous, 87, 122
  vein vs. intervein, 174–175
cells, cultured, 50, 61, 92, 217, 288, 303
Child, George, 39
circuit diagrams
  for bristle cell fates, 7, 25, 27
  for bristle patterning, 35, 41, 43, 47, 71, 75,
    193
  for compartments, 89, 111, 142, 145, 163,
    205, 207
  for disc identity, 169, 239, 251
  for disc initiation, 89
  for embryo segmentation, 78
  for gene regulation, 17, 19, 41, 43, 47, 71,
    75, 78, 111, 112, 117, 121, 127, 131, 141,
    142, 145, 163, 169, 172, 177, 193, 205,
    207, 211, 233, 239, 251
  for leg segmentation, 127, 131
  for photoreceptor cell fates, 233
  for protein networks, 15, 23, 71, 109, 179
  for regeneration, 104, 121
  for signal transduction, 9, 109, 179
  for vein patterning, 177
  symbols, 25
circuitry. *See also* cell signaling; cell states;
    circuit diagrams; circuits; codes;
    computation; computer metaphor;
    gene regulation; links; logic;
    mechanisms; uncoupling
  amplification, 13, 47, 122, 188, 209, 211,
    286, 300
  analog to digital, 72, 81, 158, 209, 251
  analog vs. digital, 11, 21, 79, 145, 160, 189,
    208, 246, 247, 303
  antagonism, 18, 21, 24, 25, 48–51, 61,
    71–75, 114–121, 129, 134, 157, 161, 167,

    169, 175, 179–182, 188, 234, 273, 275,
    281–284, 292, 295, 303
antagonism vs. cooperation, 117
antagonism, self-evoked, 161
antenna, 205, 249
antenna vs. leg, 249–252, 300
asymmetric, 107, 118–119, 164–167,
  203–211
auto-activation, 47–51, 72, 91, 107, 171,
  219
autocatalysis, 47, 50, 51, 69, 72
auto-repression, 18, 107, 132
biasing, 10, 45, 47, 52, 72, 90, 109, 132, 164,
  174, 191, 207–211, 216, 218, 227, 229,
  248–252, 304
biasing by Fringe, 164–167, 203–208
bipolar duality, 114, 167–173, 302
bistable seesaws, 25, 47, 106, 114–118,
  170, 209, 299, 302
branched control, xiv, 51, 161, 165
buffering, 21, 71, 73, 291
canalization, 73, 174
cascades, xiv, 11, 20, 92, 139, 144, 174, 179,
  199, 211–218, 224, 244, 249, 265, 297
cell shaping, 23
circuit breakers, 137, 145
clocks, circadian, 5, 263
competition, 9, 10, 21, 23, 46–53, 59, 68,
  161, 209, 211, 227, 229
damping, 21, 47, 69, 72, 134, 163–167, 182,
  235, 292, 304
default states, 7, 11, 47, 59, 65, 85, 148,
  167–173, 197, 216, 227, 228, 251, 299
design flaws, 228
design principles, cell level, 73, 270, 304
design principles, gene level, 19, 20, 73,
  79, 91, 129, 142, 145, 148, 161, 173, 182,
  207, 248, 270, 299
design principles, protein level, 71–73, 79,
  161, 163, 260, 263, 270, 300, 303
design principles, tissue level, 57, 59,
  62–67, 73, 87, 89, 107, 131–134, 142, 145,
  153, 158–164, 177, 199, 228, 234, 235,
  270, 302–305
devices, various, 270
discs vs. embryo, 86, 109, 113
dorsal discs vs. ventral discs, 171
driving factors, 25
dual control, 79, 92, 124, 183, 234, 254,
  301
effective range, 69, 72, 73, 107, 119, 161,
  209, 218, 304
evolution, xi, 10, 31, 36, 63, 69, 72, 77, 136,
  158, 173, 186, 223, 239, 244–247, 255,
  262, 288, 299, 302, 305
evolutionary relics, 36, 174
feedback loops, xiv, 248, 252, 299, 300, 302
feedback loops, negative, 164, 248,
  291–295, 303
feedback loops, positive, 47–53, 72, 73, 79,
  89, 121, 209, 211, 234, 248
fidelity, 10, 21, 72, 136, 209, 304
fine-tuning, 47, 50, 61–69, 75, 151, 161,
  174–177, 228, 305
flip-flop, 25, 114
general problem, xii
glitches, 138
ground states, 47, 53, 134, 135, 170, 171,
  216, 243, 249
impedance of ligand diffusion, 125, 143,
  146, 285

in parallel vs. in series, 9, 44, 148, 161, 164,
  181, 183, 291
indeterminacy, 56
installation, 161
limiting factors, xiii, 11, 12, 21–25, 29, 38,
  60, 69, 72, 92, 113, 160, 182
modules, 72, 179, 205, 208, 209, 305
mutual activation, 44, 48, 90, 91, 133, 171
mutual exclusivity, 114
mutual repression, 50, 56, 77, 134
neural networks, xiii, 255
noise, 17, 21, 51, 228
on-then-off switching, 48, 91, 184
oogenesis, 170, 179, 181, 289
optimization, 79
orchestration, xii, 79, 133
overreaction, 63
overrides, 10, 91, 107, 119, 128, 133, 137,
  145, 164, 171, 190, 193, 246, 248
physiological range, 50, 125, 132, 144, 151,
  253
plasticity, 91
positive vs. negative control, 69, 79
precision, 45, 46, 73, 81, 211, 304, 305
priming, 48
protein-level, 72, 161, 263
quantitative-to-qualitative, 81
quirks, 85, 123, 132, 133, 161, 182, 235
rate control, 30, 68, 72, 73, 208, 229, 234
reconfiguration, 255
rectifiers, 147
rewiring, 63, 171, 245, 255
rheostats vs. switches, 30, 156, 182, 303
robustness, 21, 73, 79, 81, 91, 98, 106, 136,
  145, 174, 299, 304
safety switches, 145
saturation, 53, 72, 160, 167
scalar vs. vector, 72, 81, 209, 211
segmentation genes, 78, 79
sensing of amplitude, 147
sensing of cell number, 145
sensing of concentration, 81, 83, 158, 209,
  211
sensing of polarity, 81, 131, 136, 209–211,
  293
sensing of position, 81, 83, 142, 158, 177,
  211, 212, 251
sensing of size, 81, 124
sensing of slope, 81, 92, 158, 205, 227
short-circuiting, xiv, 15, 157, 167, 195, 288
simulations, 79
slack, 73
solid-state, 300
SOP selection, 73, 75
stability vs. instability, 47, 51, 170, 174,
  211, 228, 237
starting conditions, 236
steady-state, 90
stoichiometry, 13, 21, 61, 73, 163, 165, 171,
  175, 286, 287, 291, 295, 298, 300, 303
superimposing, 10, 69
symbols, xiii, 11, 25
symmetry-breaking, 51, 148
synergy, 52, 71, 75, 157, 182, 185, 191,
  252–254
system properties, 57, 73, 79, 90, 299
temporal control, 112, 121, 123, 129,
  133–137, 141, 145, 161–167, 177, 196,
  205, 208–209, 218, 233, 247
thresholds, 21, 23, 39, 43–51, 69–73, 81,
  119, 128, 133, 134, 160, 167, 169, 183,

193, 218, 228, 243, 247, 295, 297. *See also under* positional information
time constraints, 10, 29, 41, 77, 154, 156
toggling of activator-repressor modes, 19, 61, 107, 139, 255, 288
toggling of Delta-Serrate modes, 163, 165, 207
tracking, 79, 91, 124, 183, 190
transduction, 20, 27, 35, 72, 81, 134, 285–296
triggers, 43, 47, 51, 72, 115–119, 155, 157, 169, 207, 217, 297
triggers, scalar vs. temporal, 11, 133, 215
virtuosity, 72, 79
wing veins vs. tracheal branching, 174
wing vs. notum, 87, 167–173, 190–193
wiring, *cis-trans*, 136
circuits, versatile
Dpp-Wg antagonism, eye disc, 234
Dpp-Wg antagonism, leg disc, 104, 106, 112–115
Dpp-Wg cooperation, leg disc, 115–118
Dpp-Wg cooperation, notum, 190–193
Dpp-Wg cooperation, wing disc, 157–158, 189–190
Hh-Dpp, antenna, 205, 249
Hh-Dpp, leg disc, 105–111, 128–129, 249
Hh-Dpp, wing disc, 111, 137–145
Hh-Dpp-Wg, eye MF initiation, 234–236
Hh-Dpp-Wg, eye MF movement, 229–234
Hh-Wg, antenna, 205, 249
Hh-Wg, embryo, 87–91
Hh-Wg, leg disc, 105–111, 128–129, 249
Notch, bristle differentiation, 9–15, 25
Notch, bristle spacing, 47–53, 59, 75, 229
Notch, eye equator, 203–209
Notch, leg segmentation, 127, 135–136
Notch, ommatidial chirality, 209–211
Notch, photoreceptor identity, 216
Notch, R8p spacing, 227
Notch, wing margin, 161–167
Notch, wing veins, 175–177
Notch-EGFR, chordotonal organs, 74
Notch-EGFR, eye-antenna bipolar duality, 169, 235, 302
Notch-Wg, leg bipolar duality, 114
Notch-Wg, wing margin, 161–167
PCP (Wnt?), cell polarity (in general), 293
PCP (Wnt?), chirality of ommatidia, 209–211
PCP (Wnt?), polarity of leg joints, 131, 136
codes. *See also* circuitry
abstract, 89, 243
amino acid, 257
area, 81, 85, 158, 191
binary, 7, 9, 27, 84, 85, 143, 145, 243, 246, 249
birthplace, 191
bitmap, 83
bristles, 37, 41, 44
capacity, 83
combinatorial, 10, 35, 77, 79, 83, 115, 128, 143, 179, 212, 219, 233, 240, 246, 247, 251, 254, 259, 302
combinatorial vs. hierarchical, 78
decoding, 209
disc identity, 84, 85, 240, 251, 254, 302
Enigma, 33
general problem, xi, 37, 224, 255, 297
genetic, 255
histotypes, 149, 169, 213, 243, 246, 254, 262

Hox, 28, 191, 239, 240, 246–249
leg segments, 300
models, 7, 10, 212
Morse, 9
nonsense words, 7
Numb, 7, 9
patterns, 101
photoreceptors, 219, 224, 233
positions, 84, 101, 104, 124, 158, 213
transcription factors, 218, 219
transdetermination, 85, 169
zinc fingers, 19
Cohen, Stephen, xii, 129, 137, 143, 157, 173
compartment boundaries, 4, 91, 105, 273. *See also* boundaries; cell lineage
blastoderm A/P, 87, 89, 97
border zones, 107, 111, 125, 137–139, 142, 145, 150–159, 163, 173, 207
border zones, widths, 112, 137, 142–151, 167, 173, 287, 301
eye A/P, 91, 199, 201, 299
eye D/V, 165, 202–211
in growth control, 92, 118–119, 144, 148
kinks, 112, 148
lack of D/V in leg, 100, 111, 159, 170, 203
leg A/P, 61, 63, 89, 97, 100, 104, 105, 109, 114, 121, 125, 129, 201, 203, 289, 299
notum A/P, 193
peripodial membrane, 139
reestablishment of, 105, 123
role in disc initiation, 87, 91
thorax A/P, 89, 96, 97, 103, 111, 128, 148, 156, 287, 288
wing A/P, 87, 89, 92, 94, 105, 125, 137–154, 173, 177, 186, 189
wing D/V, 142, 146, 153, 159, 160, 173
compartments. *See under* cell lineage
competence, 91, 227, 228, 235. *See also under* bristle patterns; cell signaling; prepattern
based on transcription factors, 179, 217, 218
landscape, 84, 134
need for mitotic quiescence, 69, 304
region-specific, 109, 111, 117, 148, 153, 164, 166
states, 148, 217
suppression of, 135
vs. determination, 299
window, 28, 29, 72
computation. *See also* circuitry; computer metaphor
absolute vs. relative, 47, 147, 167, 209
addition, 45, 72
arithmetic, 72, 73, 188
bristles, 71–74, 161
comparison, 47, 71, 72, 91, 167
critical mass, 133
distances, 173
division, 72
exponential, 72, 81, 83, 158, 301
Fibonacci series, 304
growth rates, 124
integration, 77, 124, 205, 288, 300
multiplication, 72
net force vectors, 137
patterns, xii
ratio, 68, 72, 89, 124
shortest path, 93, 123
step function, 147
subtraction, 59, 61, 72, 128, 302

titration, 20–23, 71–73, 287
with RNA, 73
computer metaphor. *See also* circuitry; computation
abstract symbolism, 9, 29, 72, 81, 89, 165, 243, 297
algorithms, 23, 31, 32, 72
binary digit (bit), 9, 84, 305
bitmaps and pixels, 83
cellular automata, 69, 235
cybernetics, 173
gating, 10
hardware vs. software, 255
infinite loops, 137, 145
information processing, 84, 124, 205, 300
information theory, 81
input/output, 13, 17, 18, 23, 68, 72, 77, 123, 129, 131, 145, 163, 164, 209, 217, 246, 248, 254, 297–303
memory registers, 7, 10
modular subroutines, 7, 169, 299
program for building a bristle, 36, 297
program for building an eye, 255
program vs. blueprint, 297
resetting of variables, 7, 179, 182
servomechanisms, 153–155
Conway, John, 235
Cooper, Michael, 224
Couso, Juan Pablo, 300
Crick, Francis, 101
cuticle, xi, 5, 28, 87
pigmentation, 5, 28, 299
secretion, 5, 29, 40, 41, 73, 87, 199
thickness, 174
trichomes (hairs), 28, 65, 99, 159, 199, 247
two-dimensionality, 31

Dahmann, Christian, 151
Davidson, Eric, xiii, 244
Dearden, Peter, 303
Demerec, Milislav, 76
determinants
asymmetric segregation, 7, 11, 24, 274
cytoplasmic, 9
heritable, 5
determination. *See also* cell states
adult vs. larva, 9, 86, 87
all-or-none, 72
appendage tips, 72
binary decision trees, 27
bristle. *See* bristles
endoderm, 72
gender. *See* genetics, sex determination
glial cell fates, 7, 29, 72, 271
imaginal discs. *See under* imaginal discs
leg vs. wing, 91
Malpighian tubule, 72
mesectoderm, 72, 177
muscle, 4, 9, 72, 259
neural. *See* neurogenesis
potency, 91
role of HLH genes, 72
salivary ducts, 72, 91
stability, 86
states, 86, 90
states, maintenance, 90, 247–249
stepwise, 133, 302
tracheae, 72, 174
vs. competence, 299
vs. differentiation, 175, 185
wing veins, 177

development, stages of
  embryo. *See under* embryo
  larval instars, 49, 85, 87, 92, 121, 133, 148,
      161, 208, 224
  pupal period, 1, 92, 189, 208, 233
  pupariation, xiii, 1, 85
Dexter, John, 298
Dietrich, Wilhelm, 202
differentiation, 5
  bristle vs. sensillum, 28
  bristle vs. vein, 175
  general problem, xi
  macrochaete vs. microchaete, 29, 57
  shaft vs. socket, 23
  vs. determination, 175, 185
  vs. growth, 196
  waves, 208, 212, 215, 227, 229, 233, 305
DNA
  bending, 45, 259–265, 295
  binding, 68, 72, 191, 219, 224, 248, 254.
      See also under protein domains,
      particular
  binding affinities, 17, 45, 61, 84, 139, 259,
      260, 261, 264, 275
  binding screens, 61
  binding sites, 15, 17, 19, 49–52, 61, 171,
      172, 190, 248, 253, 259–265, 275, 281,
      284, 287, 290, 295. *See also* DNA motifs
  binding sites, overlapping, 239, 295
  binding sites, swapping, 171
  binding specificity, 91, 259–263, 303
  binding, competitive, 17, 18, 79, 84, 107,
      182, 247, 278, 288, 291, 295
  binding, cooperative, 79, 84, 263, 265
  binding, nonspecific, 260
  bookmarking, 249
  chromatin, 248, 249
  chromatin, spreading, 249, 281
  cloning, 38, 40, 148
  coding vs. non-coding, 41, 305
  endoreplication, 29, 30, 41, 68, 86
  euchromatin vs. heterochromatin, 43, 248
  footprinting, 239
  homology screens, 27
  inverted repeats, 17
  looping, 41, 45, 161, 191, 249, 261
  methylation, 249
  open vs. closed states, 249, 305
  ploidy, 29, 86
  replication, 249
DNA motifs. *See also* protein domains,
      particular
  E box, 17, 48–55, 68–72, 258, 263
  homeobox, 28, 29, 85, 90, 115, 132–136,
      149, 260–261
  N box, 17, 50, 71, 258
  T box, 112, 141
Dong, Si, 251
Driesch, Hans, 81, 93, 300, 302
Drosophila
  Aristotle's "gnat", xi, 256
  artificial selection, 193
  genus, 62, 63, 72, 135, 174, 237
  giant, xi, 92
  Hawaiian, xi, 67, 92, 193
  hybrids, 71
  melanogaster, xi, 202, 212, 298
  other species, 31, 36, 63, 67
  polymorphisms, 37
Dubinin, N. P., 37
Duncan, Ian, 251

Ede, Donald, 224, 225, 305
embryo
  axes. *See under* axes
  blastoderm, 35, 77, 79, 89, 90, 97, 237
  blastoderm cells, 77, 86, 87
  blastoderm clones, 90, 199
  cell transplantation, 77, 86
  cleavage, 4
  dorsal closure, 151
  ectoderm, 77, 86–90, 109, 191
  endoderm, 72
  epidermis, 125
  fate maps, 76, 77, 91, 245
  gastrulation, 177
  gliogenesis, 72
  maternal gene products, 87
  mesectoderm, 72, 177
  mesoderm, 239
  metameres, 76, 79, 84, 239, 246, 247, 254
  myogenesis, 4, 9, 28, 72, 259
  neuroectoderm, 50, 299
  neurogenesis. *See* neurogenesis
  pair-rule stripes, 35, 45, 77–79, 84, 177
  quirks of patterning, 86
  salivary ducts, 72, 91
  segmental gradients, 84
  segmental identities, 79
  segmentation, 76–80, 239, 299
  segments vs. parasegments, 78, 84,
      237–246, 305
  stages, 85, 86, 90
  tracheal system, 72, 151, 174
  Wingless stripes, 78, 109
embryonic fields, 37, 43, 81, 86, 87, 190
emergent properties, 87, 89, 101, 107,
      131–134, 142, 145, 153, 159, 160, 164,
      177, 208–209, 213, 234, 236, 299, 302, 303
endocytosis, 26, 53, 180, 259
engineering, mechanical, 177, 227, 303
epidermis, 31
epiphanies, 270
equivalence groups, 37, 39, 43, 47, 49
evolution
  accidents, 31, 288
  allometry, 299
  arbitrariness, 158, 300
  artificial selection, 31, 55, 193
  Bateson's Rule, 237, 300
  by atavism, 246
  by cobbling, xiv, 62, 297
  by co-option, 10, 31, 72, 109, 136, 239, 246,
      261
  by crosslinking, xi, 36, 246, 255, 298, 305
  by elaboration, 63
  by gene duplication, 298
  by gene sharing, 298
  by genetic drift, 68, 71
  by heterochrony, 69, 299
  by heterotopy, 300
  by hopeful monsters, 237, 246, 267
  by tinkering, 63, 148, 255, 299, 302
  by transposon jumping, 298
  constraints, 92, 158, 300
  default states, 299
  entelechy, 300
  epigenetic landscape, 300
  evo-devo biology, 189
  evolvability, 255, 264, 265
  fine-tuning, 228
  frivolity, 31, 302
  genetic workload, 247

improvisation, xiv
inelegance, xiv, 35
legacies, 31, 132
loss through disuse, 68
missing links, 63
  of anatomy, 255
  of antennae vs. legs, 252, 300
  of appendages, 243
  of binding affinities, 264
  of body vs. leg segmentation, 136
  of bract-bristle adhesion, 67
  of bristle patterns, 31, 36, 62, 65, 69, 255
  of bristles, 28, 29, 30
  of circuitry, xi, 10, 31, 36, 63, 69, 72, 77,
      136, 158, 173, 186, 223, 239, 244–247,
      255, 262, 288, 299, 302, 305
  of competence, 36, 246
  of endocrine systems, 246
  of eyes, 197, 223
  of gene complexes, 36, 239, 243, 255, 298
  of gene hierarchies, 303, 305
  of genitalia, xii, 252
  of halteres, 243, 246
  of HLH genes, 68, 72, 305
  of insects, 299
  of leg segments, 300
  of legs, 132
  of metameres, 243
  of modules, 63, 255, 264, 265, 297, 299
  of opsins, 223
  of organ shapes, 148
  of protein domains, 257, 265
  of proteins, 68
  of receptor proteins, 286, 290, 297
  of sex combs, 36, 63, 135
  of sex determination, 72
  of sexual behavior, 305
  of tissue polarity, 235
  of toggle switches, 255
  of vein patterns, 174
  of wings, 86
  opportunism, xiv, 302
  optimization, 79
  reprograming, 10, 63, 72, 246
  strategies, 165, 303, 305
  vs. engineering, 158
eye
  anatomy, 197, 199
  axes. *See under* axes
  ectopic, 252–254
  equator, 198, 202–211, 296
  equators, extra, 203, 207
  fovea, 199
  identity ("eyeness"), 240, 247, 302
  missing, 169, 228, 231, 252
  ommatidia. *See* ommatidia
  other insects, 212, 302
  perimeter, 230
  pigmentation, 197, 302
  scar phenotype, 229
  small, 169, 207, 228, 231
  split phenotype, 17, 50
eye disc, 96
  bipolar duality, 114, 169, 235, 302
  cell death, 87
  compartments, A vs. P, 91
  fate map, 87, 199
  gene expression patterns, 169, 203–209,
      216–220, 233
  head capsule, 204, 296
  initiation, 77, 207

invagination, 91
lattice tightening, 87, 227–228
maxillary palp, 191, 199, 240
MF (morphogenetic furrow), 44, 208–211, 215, 227, 233
MF engine, 229–234
MF initiation vs. progression, 231, 234–236, 253
MF vs. compartment boundaries, 229–234
MF, adhesive molecules, 281
MF, speed of, 229, 234
MF, straightening of, 151
MFs, collision of opposing, 236
MFs, extra, 212, 231–236
MFs, hot spot for, 231
mitotic band, 208, 211, 233
nomenclature, xiii, 199
ommatidia. *See* ommatidia
peculiarities, 91
photoreceptors. *See* photoreceptors
regional markers, 207
role of EGFR pathway, 179
role of peripodial membrane, 304
transcription factors, 182, 218
eye field
as a cellular automaton, 235
initiation, 253
margins, 231, 234–236, 253
MF initiation site, 235
mosquito, 212

fate maps, xiii, 92
embryo, 77, 91, 245
eye disc, 87, 199
leg disc, 96, 99, 105, 106, 111, 128
peripodial membrane, 87, 122, 139
wing disc, 94, 139, 156, 193, 301
Fernández-Fúnez, Pedro, 160
Freeman, Matthew, 216, 217
French, Vernon, 93

García-Bellido, Antonio, 4, 32, 52, 85, 148, 155, 243
Gehring, Walter, xii
gene complexes
ANT-C, 237–243, 248, 260
AS-C, 17, 28, 30–75, 191, 193, 245, 272, 299, 305
Bar-C, 28, 134, 169, 194, 219, 220, 228, 233, 298
Brd-C, 272, 279
Broad-C, 20, 217, 223, 259
BX-C, 38, 45, 237–243, 246, 248, 260, 298, 299, 305
E(spl)-C, 10, 15, 17, 44, 48–51, 61, 68, 73, 218, 258
engrailed-invected, 141, 149, 243
homeobox, 243
homunculi, 45, 239
Hox, 239, 298
Iro-C, 96, 142, 166, 186, 187, 190–194, 203–208, 235, 236, 242, 243, 300, 302
Kni-C, 186
Spalt-C, 186, 187, 224
gene expression. *See also under* eye disc; leg disc; wing disc
antenna, 91, 193, 205, 251, 288–296, 302
antenna vs. leg, 159, 251
arcs, 115, 127
at boundaries, 167, 171, 207, 209, 296

at interfaces, 87, 89, 107, 131–134, 142, 145, 153–155, 159, 160, 164, 177, 234, 302, 303
bands, 77, 127, 131, 136, 137, 141, 186, 205, 251
basal transcription level, 107
biscuits vs. bands, 151, 157
biscuits vs. doughnuts, 137, 138, 151
cages, 122, 123, 125, 137
cell-size independence, 92
circles, 104, 115, 117, 127, 129, 131, 137, 205, 251
coexpression, forced, 143, 157, 253, 265, 291
coexpression, natural, 59, 131, 135, 164–166, 171, 172, 179, 183, 208, 302
coexpression, paradoxical, 47, 91, 107, 114, 128, 139, 142, 219
coexpression, spatial but not temporal, 71
compartment-specific, 89, 104, 111, 112, 117, 121, 141, 146, 148, 153, 158–164, 203–207, 245, 285, 287
complementarity, 41–44, 59, 69, 78, 125, 129–135, 141, 143, 158, 169, 174, 175, 184, 188, 193, 301, 302
constitutive, 107
drivers, artificial, xiii, 7, 13, 30, 50, 51, 107, 113, 182, 216, 253
drivers, strong, 144, 157
drivers, weak, 138, 145, 154, 157
dynamics, 69, 78, 79, 127, 131, 134, 164, 167, 171, 188–191, 208, 218, 224, 234, 246, 247
enhancer traps, 112, 114, 127, 193, 205, 223, 300, 301
enzyme patterns, 301, 302
fuzzy zones, 128, 169, 193, 305
gradients, 127, 129, 135, 143, 177, 235
halos, 143, 156
hiatuses, 117
maintenance, 90, 107, 131, 160, 164, 166, 171, 247–249
nonfunctional, 172, 189, 218
nonuniform, 45, 69, 78, 84, 154, 243, 247, 249, 281, 305
ON/OFF boundaries, 89, 121, 131, 136, 153, 157–160, 165, 173, 177, 186, 191, 203, 207, 209, 234, 235, 299, 303
ON-then-OFF, 48, 91, 151, 184
overlaps, 77, 79, 128, 142, 145, 157, 167, 169, 172, 193, 251, 302, 303
overlaps, transient, 127, 129, 131, 134
parabolic regions, 104, 112
perdurance, xv, 125, 139, 147, 166, 167
periodic, 127, 136, 300
punctate, 157, 233
response to trauma, 98, 100, 119, 123
rings, 105, 112, 126–136, 141, 151, 205, 218
sectors, 104–106, 109, 112, 117, 121–128, 133, 141, 146, 205, 301
segmental vs. parasegmental, 246
single files of cells, 127, 135, 139, 145, 175, 177, 217
spreading, 119–122, 127, 145, 166, 171, 242, 288, 291
stratified (rainbow), 78, 131, 138–142, 157, 164, 188–193, 234
stripes, antenna, 251
stripes, antineural, 59–62, 69
stripes, ectopic, 68, 115, 117, 157, 171
stripes, embryo, 35, 77, 78, 89–91, 109, 293

stripes, eye D/V boundary, 207, 209
stripes, eye D-V axis, 233
stripes, eye furrow, 215, 233, 234, 304, 305
stripes, eye margins, 234
stripes, interveins, 177, 183
stripes, leg A/P boundary, 109, 111, 125, 129
stripes, leg disc, 112, 127, 128, 251
stripes, leg proximal-distal axis, 61, 112, 302, 304
stripes, notum, 104, 190–193, 304
stripes, pigment, 193
stripes, proneural, 59, 61, 62, 71, 163, 280
stripes, tarsus, 61, 302
stripes, widths of, 47, 59, 61, 89, 112, 127, 128, 137, 140–151, 163, 167, 173–177, 233, 287, 303
stripes, wing A/P boundary, 111, 128, 137, 142–147, 301
stripes, wing D/V boundary, 141, 146, 157, 159, 163, 164
stripes, wing disc, 141, 142, 157
stripes, wing veins, 47, 175, 177, 183–186, 303
gene families *See also* protein domains, particular
Bearded, 17, 271, 272
Bright, 223, 292
Frizzled, 293
Gli, 287
GRIP, 290
Hedgehog, 105
Hox, 239, 248
IκB, 258
LIM-HD, 71, 260
MAGUK, 260, 283
Notch, 258
olfactory receptor, 191
POU-HD, 71, 260
Pox-Pax, 242, 262
Rel, 258
RTK, 179
Shc, 264
Smad, 290
steroid receptor, 217
Tcf, 258
TGF-α, 170
TGF-β, 105, 115, 189, 289, 290
Wnt, 105, 115, 292
gene groups
Minute, 21, 29, 52, 56
Pc-G, 65, 111, 135, 207, 239, 242, 243, 247–251, 264, 300
Trx-G, 111, 239, 242, 243, 247–251, 276
gene regulation. *See also* circuitry
activator-repressor switching, 19, 61, 107, 139, 255, 288
activators, 9, 13, 17, 51, 72, 73, 107, 133, 193, 239, 248, 249, 254, 261, 262, 290
activators, unknown, 107
auto-activation, 47–51, 72, 91, 107, 171, 193, 219, 233, 234, 248, 291, 295
auto-regulation, 72, 147, 171, 248
auto-repression, 17, 18, 107, 132, 147, 153, 166, 182, 225, 248, 263, 285, 288, 291, 295
basal transcription apparatus, 45, 298
batteries, 87, 217, 246
bipolar regulators, 19, 107, 255, 288
Britten-Davidson Model, 244
by cell-surface receptors (direct), 12, 175

gene regulation (*contd.*)
  by chromatin-remodeling, 17, 71, 111, 207, 239, 247–249, 264, 282, 295, 300
  by micromanagers, 245–249, 253, 254
  by nuclear import, 131, 133, 172, 181, 242, 251, 287, 288, 294
  by proteolysis of regulators, 19, 20, 105, 107, 133, 182, 224, 263, 271, 295
  cis-enhancers, 44–45, 79, 84, 92, 107, 128, 136
  cis-enhancers, arrays of, 18, 21, 35–44, 52, 53, 72, 129, 131, 141, 164, 174, 188, 239, 289, 299
  cis-enhancers, boundary vs. quadrant, 157, 164, 171, 245, 301
  cis-enhancers, cell-type specific, 11, 48, 51, 219, 224, 254
  cis-enhancers, colinear vs. scrambled, 41, 45, 239
  cis-enhancers, constitutive, 171
  cis-enhancers, disc-specific, 292
  cis-enhancers, distant, 161
  cis-enhancers, embryonic vs. imaginal, 245, 249
  cis-enhancers, evolution of, 255
  cis-enhancers, genomic repertoire of, 255
  cis-enhancers, parasegment-specific, 243
  cis-enhancers, region-specific, 28, 35, 36, 41–48, 191, 219, 239, 245, 249, 253, 289, 299
  cis-enhancers, shared, 44, 298
  cis-enhancers, stage-specific, 45, 86, 92, 128, 164, 174, 188, 227, 249, 289, 305
  cis-regulatory region, 17, 71, 87, 239
  cis-silencers, 62, 239
  co-activators, 12–15, 91, 223, 264, 265, 288, 291, 295
  combinatorial, 77, 136, 219, 245, 253, 254, 301
  co-repressors, 12, 17, 49, 51, 61, 71, 207, 259, 264, 265, 281, 295, 302
  default states, 173. *See also under* circuitry
  direct vs. indirect, xiii, 51, 107, 111, 128, 139, 141, 156, 164, 172, 187, 191, 193, 205, 219, 245, 248, 271, 288
  enhanceosomes, 254, 255, 260, 300
  enhancer-promoter bridging, 41, 45, 160, 161, 191, 279, 284
  enhancer-promoter specificity, 45
  hierarchies, 45, 69, 75–79, 83, 103, 133, 136, 141, 149, 182, 186, 217, 239, 244–248, 253, 303
  histone acetylation, 248, 279, 288
  histone deacetylation, 17, 248, 249, 275, 276, 281
  insulators, 161
  introns, 49, 171, 292
  leg vs. wing disc, 137, 157, 164–170
  licensing agents, 233, 254
  locked-OFF vs. passive-OFF, 172, 173, 285, 294, 295
  Mediator complex, 45, 135, 254, 300
  networks, 79, 205, 246, 253, 303
  open-for-business idea, 218
  post-transcriptional, 48, 69, 72, 78, 79, 133, 234
  post-translational, 68
  promoters, 13, 15, 18, 45, 48, 50, 52, 55, 61, 72, 81, 253
  puffing cascade, 20, 217

qualitative vs. quantitative, 44, 72, 81, 138, 139, 158
quenching vs. squelching, 17, 71, 79, 249, 295
Ras-dependent, 179
relays, 112, 141, 145, 177, 205, 215, 233
repressors, 13, 17, 18, 47, 48, 68, 72, 107, 135, 187, 190, 239, 248, 249, 261, 264
repressors, unknown, 171
stage-specific, 166, 246, 248, 249
synergy, 12, 52, 71, 75, 157, 181–185, 191, 252–254, 273, 275, 279, 281–283
trans-acting factors, 36, 43–45, 69, 72, 79, 245, 249, 298, 299
transcription cofactors, 190
transcription factors, 35, 51, 107, 181, 186, 190, 218, 260
transcription factors, landscape of, 301
triggers, 100, 133, 234
triggers, interface, 87, 89, 107, 131–134, 142, 145, 153–155, 159, 160, 164, 167, 171, 177, 207–209, 234, 296, 303
triggers, temporal, 91, 121, 123, 129, 137, 141, 145, 164, 299
genes. *See also* gene regulation; genes, particular; genetics; mutations; phenotypes; protein domains, particular
  abbreviations, xv, 37
  adult vs. larval, 87
  annulus, 127
  antineural, xiii, 48, 53, 59, 62, 71–75, 163, 164, 180, 281, 282
  autonomy in clones, 1, 33, 35, 39, 48, 50, 72, 80–85, 89, 124, 133, 141, 143, 148, 160, 165, 177, 183, 186–190, 209, 219, 245, 251, 282, 290
  axis, 76, 147
  boundary, 165
  cascades, xiv, 20, 217
  cassettes, xi, 47, 72, 305
  cell cycle, 49, 258
  cell type, 243
  co-adapted, 71
  co-expressed (naturally), 59, 135, 164–166, 171, 179, 208
  colinearity, 45, 298
  competence, 35, 45
  cooperativity, 172
  default states, 11, 59, 169, 171, 251, 299
  disc-specific, xi, 159
  dispensable, xiii, 24, 48, 86, 117, 157–159, 175, 182, 216, 218, 235
  dispensable pairs, 79, 287
  duplications, 36
  early eye, 169, 235, 242, 252–254, 302
  event-counting, 11
  executive, 135
  field-specific, 242, 243, 254
  gap, 77, 79, 84, 136, 147, 158, 186, 239
  genomic repertoire of, 255
  homeotic, 28, 65, 78, 83–86, 237–240, 247–249, 260
  homologs, xii, 11, 13, 23, 26, 75, 109, 170, 179, 181, 239, 257, 274, 275, 283, 294, 304
  housekeeping, 29, 85, 100, 302
  Hox, 79, 80, 245–249, 254, 260
  human, 13, 171, 254, 258, 259, 265, 290, 299, 303
  instructive vs. permissive, xiv, 9, 36, 130–134, 189, 191, 215, 234, 249

interchangeability, 44
intronless, 17, 41, 73, 265
limiting factors, 21, 29, 69
mammals, 13, 15, 23, 26, 75, 225, 249, 258–265, 274, 275, 283, 295
map locations, xii
master, 85, 171, 242, 252–254, 260, 297
maternal-effect, 76
memory, 10, 65, 81, 83, 84, 129, 158, 239, 248, 251
metamere identity, 239, 254
misexpression, xiii, 86
modifiers, 12
necessary vs. sufficient, xiii, 37, 118, 132, 136, 245
nematodes, 261, 263, 295
neural precursor, 271
neurogenic, xiii
nomenclature, xii
nonautonomy, 39, 47, 48, 80, 133, 141, 177, 183, 186, 203, 209, 228, 231, 246, 303
ON/OFF states, 1, 28, 79, 83, 84, 89, 107, 111, 143, 148–153, 158–160, 179, 239, 243–249
orthologs, xiii, 171, 258, 265, 283, 289, 290, 303
overexpression, xiii, 49
pair-rule, 18, 35, 77, 79, 84, 136, 158, 262, 278
pan-neural, 72, 182, 272, 274, 279, 284
paralogs, xii, 10, 141, 149, 181, 186, 219, 223, 253, 285, 289, 298, 303
pathways, xiii, xiv
periodic-zone, 136
pinwheel, 97
plant, 239, 260, 261
prepattern, 27, 33, 35, 39, 45, 80, 148, 164, 193, 219, 278, 299
proneural, xiii, xiv, 37, 39, 41, 45, 49–53, 71–75, 161, 225, 229, 271, 279, 280, 283, 284
rate, 299
realizator, 134, 149, 174, 186, 246, 260, 305
redundant, xiii, 9, 10, 12, 17, 18, 21, 28, 40, 44, 45, 48, 55, 62, 75, 91, 125, 129, 134, 141, 148, 158, 167, 175, 218, 227, 231, 272, 278, 279, 281, 293, 295, 298, 300, 303
reporter, xiii, 28, 41, 48–52, 61, 117, 171, 191, 207, 219, 239, 291, 301
ribosomal RNA, 29
segmentation, hierarchy of, 76–80, 83, 103, 136, 186, 239, 247, 248, 303
segment-polarity, 79, 89, 97, 123, 131, 136, 158, 293
selector, 29, 48, 85, 89, 141, 145, 149–163, 173, 174, 188, 203, 246, 248, 262
selector, PI mode, 89, 145, 148, 163, 243
sensillar identity, 27, 75
sex determination, 30, 49, 72
subgenes, 37
switch, 9–11, 24, 28, 48, 85, 132, 148, 159–161, 165–169, 181, 184, 191, 207, 218, 240, 243, 255, 261, 271–284, 302
thin-zone vs. wide-zone, 136
ubiquitously expressed (naturally), 17, 49, 51, 59, 69, 75, 85, 92, 133, 158, 160, 161, 164, 225, 251, 263, 265, 274, 280, 282, 283, 290, 294
untranscribed, 191
upstream vs. downstream, xiv, 45

vertebrate, xiii, 71, 109, 141, 179, 181, 225, 239, 258–261, 283, 285, 289, 290, 294, 305
work loads, 247
yeast, 258, 260, 261, 265
zygotic, 77
genes, particular
14-3-3ζ, 179, 180
18 wheeler, 300
abdominal-A (abd-A), 78, 239, 242
Abdominal-B (Abd-B), 78, 239, 242
abl oncogene, 223
abnormal chemosensory jump 6 (acj6), 271
abrupt, 303
Abruptex (=Notch), 52, 195, 279
absent solo-MD neurons and alfactory sensilla (amos), 27, 28, 75, 254, 276
absent, small, or homeotic discs 1 (ash1), 242
absent, small, or homeotic discs 2 (ash2), 242, 276
achaete (ac), 17, 30–75, 91, 112, 128, 141, 142, 161, 163, 167, 177, 187–195, 225, 229, 278, 279, 292, 296, 301–303
Actin5C, 138
Additional sex combs (Asx), 242
Antennapedia (Antp), 78, 81, 83, 85, 96, 131, 239, 242, 246, 249, 251
anterior open, 169
approximated, 185
apterous (ap), 127, 141, 158–173, 248, 260, 296, 301
araucan (ara), 142, 177, 185–194, 203–208, 235, 242, 289, 292, 296, 302, 303
arc, 292
argos, 52, 147, 170, 177–188, 194, 215–218, 223, 227–229, 242, 303–305
aristaless (al), 92, 115, 117, 127–136, 141, 171, 251, 278, 292, 295, 300
aristapedia (=spineless), 80
Aristapedoid (Arp = Su(z)2), 242
armadillo (arm), 105–109, 118, 128, 134, 156, 163, 171, 173, 182, 194, 205, 242, 255, 258, 260, 294
arrow (arr), 106, 109, 205, 231, 293, 295
Arrowhead, 300
asense (ase), 17, 18, 37, 41, 48, 69, 272, 279
asteroid (ast), 185
atonal (ato), 27, 28, 73, 74, 212–235, 258, 276, 289, 296, 304, 305
baboon, 289
BarH1, 127, 131, 134–136, 193, 194, 205, 219, 220, 233, 276, 292, 296, 300
BarH2, 127, 134, 194, 219, 220, 233, 276, 300
bazooka, 11, 24, 271
Beadex, xiii, 71, 161, 260
Bearded (Brd), xi, 10, 12, 73, 272, 279
bicoid (bcd), 76–79, 84, 158, 239, 254
big brain (bib), 49, 55, 271, 279
Big brother, 223
bithorax (see Ubx), xiii, 80, 87, 237
blind spot (=poils aux pattes), 223
blistered (bs), 111, 141, 172–174, 177, 186–188, 245, 261, 289, 301, 303
bobbed, 29
Brachyury (vertebrate), 141
brahma (brm), 242
breathless, 179
bric à brac (bab), 127, 129, 135, 251, 259, 300

bride of sevenless (boss), 15, 179, 180, 213–220, 224, 305
brinker (brk), 109, 112, 141, 142, 148, 157, 265, 291, 292
Brista (=Distal-less), 251
Broad-Complex, 20, 259
Brother, 223
Brother of Brd (Bob), 73
canoe (cno), 11, 75, 180, 263, 271, 279
Casein Kinase 2 (CK2), 109, 273, 293, 294
CASK, 260
castor, 217
caudal (cad), 77–79
caupolican (caup), 142, 177, 186–194, 203–208, 242, 289, 292, 296, 303
Cf1a (=ventral veinless), 260, 263
chaoptic, 223
chickadee, 29
Chip, 45, 160–166, 191, 260
Clock, 263
clown, 180
collier (=knot), 142, 259
comb gap (cg), 111, 135, 285, 292
congested-like trachea (colt), 174
connector enhancer of ksr (cnk), 180, 182
corkscrew (csw), 179, 180, 182, 303
costal2 (cos2), 109, 154, 157, 194, 286
cousin of atonal (cato), 272
cramped (crm), 135, 242
crooked legs (crol), 127
crossveinless, 189
crossveinless 2, 189, 289
crossveinless-like 6, 189
C-terminal Binding Protein (CtBP), 21, 61, 71, 190, 193, 259, 264, 273, 280, 282, 291
cubitus interruptus (ci), 105–112, 124, 128, 141–153, 165, 171, 186, 187, 205, 233, 242, 261, 285, 287
cut, 27, 141, 142, 161–166, 219, 220, 233, 242, 243, 259, 276
cycle, 263
Cyclin (genes A-F), 44, 49, 233, 260, 280
dacapo, 11
dachs, 135, 185
dachshund (dac), 127–135, 205, 220, 223, 228, 235, 242, 251–254, 261, 287, 292, 295, 300
dachsous, 185, 209, 293
dActivin, 289
dally-like (dlp), 293
dAP-2, 127
daughter of sevenless (dos), 179, 180, 182, 303
daughterless (da), 49, 50, 69–75, 113, 173, 174, 191, 223–228, 233, 263, 276, 280
Daughters against dpp (Dad), 109, 134, 141, 148, 167, 190, 290–295
dAxin, 109, 115, 118, 292, 294
DC0, 105, 109–119, 149–157, 187, 194, 231, 287, 303
dC3G, 180, 182
dCable (dCbl), 179–181
dCdc37, 179
dCdc42, 234
DCP-1, 228
dCul-1, 189
dE2F, 295
dead ringer, 223
deadpan (dpn), 17, 72, 127, 223, 265, 274, 300
dE-cadherin, 109, 153, 294

decapentaplegic (dpp), 39, 89, 90, 104–129, 133–134, 137–158, 170–177, 185–194, 205, 228–235, 245, 252, 253, 288, 289, 295, 301, 303, 305
Deformed (Dfd), 78, 239
Delta (Dl), 9, 12, 15, 25, 47–50, 59, 61, 71, 75, 124, 127, 135, 160–169, 173–177, 203–211, 215, 216, 223–229, 242, 254, 259, 271, 272, 280, 296–300, 303–305
deltex (dx), 12, 15, 135, 175, 272, 280
Dense, 278
derailed, 179
Dfrizzled2 (Dfz2), 105, 109, 125, 146, 158, 190, 194, 293, 295
Dfrizzled3 (Dfz3), 104, 112, 125, 141, 142, 158, 295
dHSF, 254
Dichaete, 31, 45, 194, 195, 265, 280
disconnected (disco), 127, 136
discs large, 271
discs overgrown, 263
dishevelled (dsh), 15, 106, 109, 114–118, 124, 131–136, 163, 167, 193, 194, 205–209, 242, 279, 293
dispatched, 125, 185, 285
Distal-less (Dll), 29, 86–90, 96, 104, 115, 117, 127–137, 141, 142, 148, 156, 158, 205, 242, 247–254, 292, 295, 300–303
division abnormally delayed (dally), 105, 109, 146, 194, 290, 293
dJun, 179–182, 209, 215, 220, 293
dLim1, 127, 135, 240
dMyc, 92
Domina (=jumeaux), 278
dorsal, 254, 255
Dorsal switch protein 1 (Dsp1), 242
Dorsal wing (Dlw), 159–163, 166, 167
doublesex, 30
double-time (=discs overgrown), 263
downstream of receptor kinases (drk), 179–182, 303
dPax2, 28, 29, 219, 220, 233
dRac1, 180, 182, 293, 295
dRacGap, 180, 182
dRaf, 107, 153, 169, 170, 179–188, 194, 231, 242, 258
dRasGap, 304
Dredd, 228
drICE, 228
drifter (=ventral veinless), 174
Drop, 228
Drosophila CREB-binding protein (dCBP), 107, 109, 111, 288, 291, 295
Drosophila Dr1-associated protein (dDrap), 71
Drosophila Heat shock protein 90 (dHsp90), 180
Drosophila MAPK-ERK Kinase (dMEK), 169, 179–182, 304
Drosophila Mitogen-Activated Protein Kinase (dMAPK), 169, 179–183, 216, 218, 223, 227, 235, 259, 304
dSara, 289
Dscam, 191, 261
dShc, 179, 180, 182, 303
dTrk, 179
dumpy, xi, 174, 259
DWnt-4, 292
ebi, 180
ebony, 193
ecdysoneless, 278

genes, particular (*contd.*)
  echinoid, 223
  echinus, 278
  embryonic lethal, abnormal vision (elav),
    223, 273
  enabled, 300
  engrailed (en), 61, 78, 79, 87–91, 107, 111,
    112, 117, 121, 124, 137–163, 170, 182,
    187–189, 205, 209, 224, 229, 239, 246,
    248, 288, 299, 301, 303
  Enhancer of Ellipse 24D (=echinoid), 180
  Enhancer of Ellipse, 24D (=echinoid), 230
  Enhancer of split (=m8), 9, 17, 272, 281
  Enhancer of zeste, 135
  *Epidermal growth factor receptor homolog
    (Egfr)*, 29, 56, 73, 106, 153, 169, 170, 173,
    177–188, 194, 205, 212, 215–218, 227,
    228, 242, 264, 281, 288, 297, 300, 303, 305
  *Epithelial Adenomatous Polyposis Coli
    (E-APC)*, 109, 258, 294
  escargot (esg), 89, 91, 127, 131–136
  even skipped (eve), 45, 78, 262
  expanded, 209, 289, 293
  extra eye (ee), 242
  extra sex combs (esc), 80, 135, 242, 249, 265
  extradenticle (exd), 111, 131–134, 166, 195,
    205, 242, 251–254, 260–262, 300–302
  extramacrochaetae (emc), 47, 48, 71–75,
    141, 142, 161, 174, 175, 193, 194, 223,
    233, 234, 260, 281, 301–303
  eye gone (eyg), 223, 235, 242, 252, 262
  eyeless (ey), 80, 205, 228, 235, 237, 242, 249,
    252–254, 292, 296
  eyelid (=osa), 223, 292, 305
  eyes absent (eya), 205, 220, 228, 235, 242,
    252–254, 261, 287, 292
  fasciclin II (fas II), 127, 131, 134, 151, 281
  fat facets (faf), 180, 223
  fat-head, 107, 153
  flamingo (=starry night), 209, 293, 304
  forked, 29
  four jointed (fj), 127, 135, 185, 205, 301
  fringe (fng), 15, 52, 135, 159–167, 173, 194,
    203–208, 242, 279, 300
  fringe connection (frc), 292
  frizzled (fz), 125, 131, 136, 146, 158, 181,
    205, 209–211, 293, 304
  fruitless, 305
  fused (fu), 109, 149, 185, 286
  fushi tarazu (ftz), 18, 77, 78, 239, 254
  futsch, 215
  gammy legs (gam), 106, 292, 295
  Gap1, 179–182
  Geranylgeranyl transferase-1 (GGT1), 179,
    180
  giant (gt), 78
  glass, 219, 220, 233, 254, 289, 305
  glass bottom boat (gbb), 143, 185, 189, 194,
    289, 290
  glial cells missing, 271
  grain, 112, 300
  grainyhead, 217
  grim, 228
  gritz, 170, 180, 184, 218
  groucho (gro), 17, 49–52, 61, 71, 106–111,
    135, 153, 155, 164, 173, 175, 223, 142,
    254, 255, 265, 281, 291, 295, 298, 305
  gurken, 170
  gustB, 271
  Gαi, 271
  H15, 112, 114, 133, 296

hairless (H), 18, 20–28, 51, 52, 75, 175, 273,
    282, 305
hairy (h), 17, 32, 45–49, 61–63, 68–75, 112,
    127, 128, 136, 163, 177, 187, 193, 195,
    205, 223, 233, 234, 264, 265, 282, 289,
    292, 296, 302, 304, 305
head involution defective (hid), 228
heartless, 173, 179
hedgehog (hh), 79, 89, 90, 104–112, 117,
    119–129, 137–160, 165, 171, 174, 177,
    185–187, 194, 205, 208, 212, 223,
    228–235, 242, 253, 285, 288, 292, 301, 305
homothorax (hth), 112, 127–136, 141, 148,
    166, 205, 242, 251, 249–254, 261, 262,
    292, 296, 300–302, 305
huckebein, 265
hunchback (hb), 78, 79, 217, 239, 254
hyperplastic discs, 135
inscuteable (insc), 24, 271
Insulin receptor, 179
invected (inv), 141, 145, 149, 153, 161, 301
I-POU (=acj6), 71, 260
irregular chiasm C-roughest, 215
jumeaux, 271
Jun N-terminal Kinase (JNK), 180, 209,
    259, 293
kakapo, 257
karst, 180
kekkon1 (kek1), 177–183, 303
kinase suppressor of ras (ksr), 179, 180,
    182, 258
kismet (kis), 242
klingon, 218
klumpfuss (klu), 48, 127, 271, 282, 300
knirps (kni), 78, 141, 174, 177, 186, 239, 264
knirps-related, 186
knot, 139, 142, 147, 177, 183–187, 259, 289,
    301, 303
kohtalo, 223
Krüppel (Kr), 78, 239, 254, 264
kuzbanian (kuz), 49, 53, 135, 175, 258, 282
l(1)ts504, 135
l(3)1215, 300
L38, 257
labial (lab), 239, 242
leg arista wing complex (lawc), 242
lethal (2) giant discs (l(2)gd), 115, 157
lethal (2) giant larvae (lgl), 24, 273
lethal at scute (l'sc), 17, 30, 37, 41, 45, 50,
    68, 69, 73, 271
lilliputian (lilli), 180, 289
lines, 295
liquid facets (lqf), 180, 223
Lobe, 207, 292
lozenge (lz), 219, 220, 223, 233, 276
m8 and other E(spl)-C genes, 15, 17, 50,
    61, 73, 127, 142, 164–169, 175, 177,
    205–209, 219–223, 233, 264, 265, 271,
    272, 278, 281, 300, 301, 303, 305
many abnormal discs, 290
master of thickveins (mtv), 142, 148, 189,
    288, 301
mastermind (mam), 169, 175, 273, 282
maverick, 289
Medea, 109, 290
Merlin, 289
Minute (multiple genes), 21, 29, 52, 56, 201
miranda, 24, 271
mirror (mirr), 186, 190–194, 203–208, 234,
    236, 242, 296
misshapen, 209, 293

moira, 107, 242
Mothers against dpp (Mad), 109, 126, 128,
    131, 141, 143, 147, 171, 190–191,
    229–235, 253, 255, 290
multi sex combs (mxc), 135, 242
*multiple ankyrin repeats single KH
    domain (mask)*, 180
musashi (msi), 24, 264, 273
muscleblind, 220
myoglianin, 289
naked cuticle, 292
naked cuticle (nkd), 112, 134, 167, 230, 295
nanos, 76, 79
nemo, 292, 303
net, 173, 174, 177, 185, 303
neuralized (neu), 147, 156, 273, 282
NK-4, 255
Notch (N), xiv, 9, 12–17, 25, 47, 49, 61,
    71–75, 106, 114, 124, 135, 147, 157,
    160–177, 185, 188, 194, 203–211,
    215–219, 227, 229, 235, 242, 245, 259,
    264, 271, 273, 278, 283, 297–299, 301,
    304, 305
Notchless (Nle), 283
Notum, 242, 292
nubbin (nub), 127, 167, 171, 263, 301
numb, 5–27, 72, 165, 259, 264, 266, 270,
    274, 286
numb-associated kinase (nak), 24, 274
Numblike (mouse), 15
odd Oz (odz), 127, 136, 278
odd paired (opa), 262
odd skipped (odd), 78, 79, 127, 134, 136
Odorant receptor 83b (Or83b), 191
onecut, 220
ophthalmoptera (opht), 242
Ophthalmoptera (Opt), 203, 242
optix, 235, 242, 252, 253
optomotor-blind (omb), 111–113, 118, 125,
    133, 140–143, 154, 155, 158, 187, 205,
    245, 291, 292, 296, 301, 303
oroshigane, 185, 285
orthodenticle, 205, 220, 289, 296, 300, 305
osa, 242
paired (prd), 262, 264
paired box-neuro (poxn), 27, 127, 262, 277
pangolin (pan), 105, 106, 109, 128, 171,
    173, 190–194, 255, 260, 295
pannier (pnr), 41, 45, 151, 190–194, 234,
    235, 242, 292, 295, 302
partner of inscuteable, 271
partner of numb, 24, 264, 271
patched (ptc), 109, 111, 124, 125, 138–157,
    177, 185–187, 194, 208, 231, 242, 285,
    288, 301, 303
pdm-1 (=nubbin), 11, 217, 297
pdm-2, 11, 297
period, 262, 263, 302
Phospholipase Cγ, 179, 181, 182, 223, 303,
    304
phyllopod (phyl), 20, 182, 219, 220, 224,
    271
PI3 kinase, 303
pipsqueak, 220, 300
pleiohomeotic (pho), 127, 135, 242
plexus, 173, 174
poils aux pattes, 135
pointed (pnt), 153, 169, 179, 181, 182, 194,
    219, 220, 242
polychaetoid (pyd), 30, 71–75, 263, 283
Polycomb (Pc), 107, 135, 148, 160, 242, 248

polycombeotic (=Enhancer of zeste), 242
Polycomblike (Pcl), 65, 242
polyhomeotic, 107, 111, 122, 128, 149, 155, 288
porcupine (porc), 106, 242, 293
postbithorax (see Ubx), 87
Posterior sex combs (Psc), 242, 271
Presenilin (Psn), 53, 283, 295
prickle (pk), 131, 136, 209, 293, 300
proboscipedia (pb), 135, 239, 242
prospero (pros), 11, 24, 182, 219, 220, 224, 233, 254, 274
Protein Kinase A. (See DC0)
Protein Kinase C (PKC), 109, 263, 294
Protein Phosphatase 2A. (See twins)
Protein Tyrosine
  Phosphatase-ERK/Enhancer of Ras1 (PTP-ER), 179, 181, 304
puckered, 278
punt, 106, 109, 114, 115, 118, 129, 145, 185, 188, 190, 194, 229, 235, 289
radius incompletus, 177
Rap1, 179–181
Ras GTPase-activating protein (RasGAP), 181
Ras1, 29, 153, 169, 179–182, 194, 223, 228, 242, 254, 304
rasputin (rin), 181
reaper, 228
reduplicated, 123
retina aberrant in pattern (rap), 212, 223
RhoA, 209, 293
Rhodopsin (genes #1–#6), 197, 223, 224, 253, 254
rhomboid (rho), 141, 173–189, 194, 217, 223, 227, 303
roadkill, 289
rolled (rl=dMAPK), 173, 177, 180, 183, 185, 233
rotund (rn), 127
rotundRacGAP (=rotund), 29, 181, 182, 189
rough, 213, 215, 218–224, 228, 233, 234, 289, 305
Roughened (=Rap1), 181
rudimentary, 29
rugose, 223
runt, 264
sanpodo (spdo), 11, 29, 271
saxophone (sax), 143, 189, 289–291
scabrous (sca), 15, 30, 50, 55, 62, 75, 147, 157, 165, 215, 219, 223, 225–227, 233, 283, 289, 305
scalloped (sd), 142, 164, 171–173, 242, 265, 271, 296, 301, 302
schnurri (shn), 109, 188, 291, 292
scratch, 223, 274
screw, 289
Scruffy, 141
scute (sc), xiv, 17, 20, 30, 31, 37–75, 91, 113, 142, 161, 163, 191, 193, 223, 225, 245, 283, 292, 296, 301, 303
Scutoid, 278
semang, 181, 182, 223
senseless (sens), 50–52, 72, 220, 233, 271, 284, 302
Serrate (Ser), 15, 24, 25, 52, 55, 127, 135, 158–169, 173, 175, 203–209, 225, 242, 274, 296, 300, 301
seven in absentia (sina), 20, 182, 219, 220, 224, 271

sevenless (sev), 179, 180, 213–220, 227, 302, 303
seven-up (svp), 181, 215–224, 233
sex comb distal, 135
Sex comb extra (Sce), 242
Sex combs on midleg (Scm), 242
Sex combs reduced (Scr), 80, 112, 239, 242–248
shaggy (sgg), 109–118, 130, 132, 156, 157, 163, 190, 194, 205, 278, 294
shattered, 223
shaven (=dPax2), 28
shibire (shi), 26, 49, 52, 53, 59, 146, 175, 275, 284
shifted, 285
short gastrulation (sog), 185, 188, 303
shortsighted, 289, 305
shotgun (shg = dE-cadherin), 109
sine oculis (so), 205, 220, 228, 235, 242, 252–254, 261, 287, 292
singed (sn), 29, 39
single-minded (sim), 177, 262
skinhead, 169, 242, 292
sloppy paired, 84
smoothened (smo), 105, 109, 111, 128, 149–154, 185, 208, 229, 231, 286
snail (sna), 91, 264, 278
Son of sevenless, 179, 181, 182, 304
spalt, 30, 125, 140–143, 154, 155, 158, 174, 177, 185–187, 194, 205, 220, 224, 233, 245, 251, 252, 291, 292, 296, 301–303
spalt-related (salr), 141, 185, 186, 245, 301
spineless (ss), 29, 91, 127, 135, 205, 242, 249–254, 300
spiny legs (sple=prickle), 67, 131, 293
spitz (spi), 48, 52, 89, 91, 147, 169, 170, 174, 179–185, 194, 215–218, 227, 228, 242, 303, 304
split (see Notch), 15, 17, 50, 272, 281
split ends (spen), 181, 292
sprint, 181
sprouty (spry), 174, 179–185
Src42A, 181
Star, 29, 177–185, 217, 228, 303
starry night, 209, 293, 304
strabismus (=Van Gogh), 209, 293
strawberry notch, 135, 166
string, 11, 30, 44, 92, 223, 233, 271, 296, 302
Stubble, 29
sugarless (sgl), 106, 290, 293
sulfateless (sfl), 293
super sex combs (sxc), 242
supernumerary limb (slimb), 105, 109, 111, 115, 149, 154, 157, 164, 180, 242, 287, 294
Suppressor 2 of zeste, 271
Suppressor of deltex (Su(dx)), 272, 278
suppressor of forked (su(f)), 100, 119, 123, 135, 242
Suppressor of fused (Su(fu)), 109, 154, 286
Suppressor of Hairless (Su(H)), 9–25, 49–52, 135, 158, 164, 169, 171–175, 209, 219, 224, 255, 275, 284, 297
suppressor of Hairy wing (su(Hw)), 161
TACE, 258
TAF250, 254
tailless (tll), 223, 239
tam (=polychaetoid), 281, 283
tango (tgo), 29, 161, 242, 252, 263
target of poxn (tap), 258, 271
tartan (trn), 127
tartaruga (trt), 292

teashirt (tsh), 127, 134, 136, 166, 207, 242, 252, 295
tetraltera (tet), 242
thick veins (tkv), 106, 109, 114–118, 124, 125, 129–134, 141–148, 177, 185–189, 194, 229–235, 288–292, 301, 303
timeless, 263
tolkin, 185, 189
Toll-like receptor, 300, 301, 305
tolloid, 189
torso, 179, 304
tout-velu, 185
tramtrack (ttk), 11, 18, 19, 169, 182, 218, 220, 224, 242, 255, 259, 271, 275
tricornered, 29
trithorax (trx), 107, 242, 248
Tubulinaα1, 138, 154
Tufted, 278
tumorous head 1, 278
twin of eyegone (toe), 235
twin of eyeless (toy), 205, 235, 242, 253, 254
Twin of m4 (Tom), 73
twins, 11, 24, 109, 157, 179, 181, 275, 289, 294, 295
twist, 239, 254
two-faced (tfd), 242
UbcD1, 182
Ultrabithorax (Ubx), 78, 80, 87, 112, 237, 239, 242–253, 299, 305
ultraspiracle, 217, 220
u-shaped (ush), 45, 190–194, 292, 302
Van Gogh, 209, 293
vein (vn), 141, 153, 167–186, 205, 216, 227, 288, 289, 296, 302, 303, 305
ventral veinless (vvl), 141, 142, 171, 174, 177, 189, 263, 296, 303
vestigial (vg), 86, 89, 91, 139–142, 156–158, 164–173, 242, 245, 265, 292, 296, 301, 303
VP16 (herpes virus), 13, 51, 61, 254, 284
vrille, 289
white (w), 201
wingless (wg), 15, 30, 55, 78, 79, 87–91, 104–129, 133–142, 146, 156–173, 189–194, 205, 230–236, 242, 245, 279, 289, 292, 295, 301, 302
yan, 179, 180–182, 194, 220, 259
yellow (y), 1, 17, 36, 39, 41, 43, 45, 67, 193, 201
genetics. *See also* mutations; phenotypes
  analysis, xiv, 62, 202
  artifacts, xiv, xv
  background effects, 36, 38, 45, 69
  chimeric transgenes, 13, 18, 51, 68, 126, 147, 209, 216, 224, 284, 303
  complementation, 37
  constructs, constitutively active, xiii, 12, 51, 114, 124, 141, 145, 151, 170, 173, 190, 207, 216, 234
  constructs, dominant-negative, 169, 195, 216, 242
  constructs, inversion recombination, 43
  cosuppression, 298
  deletion analysis, 41
  developmental vs. physiological, 299
  dosage compensation, 71, 248
  dosage effects, 13, 21, 23, 29, 30, 31, 49, 52, 68, 69, 71–75, 113, 175, 208–211, 243, 246, 287, 297
  dosage screens, 68
  enhancer maps, 41, 239
  enhancer traps, xiii, 165, 219, 282, 296, 301

genetics (*contd.*)
  epistasis, xiii, xiv, 48, 109, 160, 161, 175,
      179, 183, 186, 188, 209, 271, 278, 287
  epistasis, paradoxical, xiv
  flp recombination, xiii
  flp-out trick, xiii, 185
  Gal4-UAS method, xiii
  genome project, xi
  gynandromorphs, 1–4, 36, 39, 52, 67, 77,
      86, 91
  haplo-insufficiency, 21, 29, 50, 52, 113,
      160, 175
  heat-shock promoters, xiii, 7, 69
  history, xv, 12–15, 32, 37–43, 57
  inhibition by RNAi, 271
  intersexuality, 63, 72, 246
  inversions, 298
  LOF vs. GOF testing, definitions, xiii
  molecular, 167
  mosaic analysis, 9, 23, 33, 35, 39, 45, 52,
      56, 69, 80, 85, 104, 114, 128, 151–155,
      209, 213, 218, 302
  mosaic analysis vs. *Gal4-UAS* method, 118
  nomenclature, xiii
  nonspecific effects, 249
  penetrance vs. expressivity, 159
  phenotypic rescue, 13, 18, 21, 47, 51, 71,
      118, 151, 154, 158, 160, 164, 165, 173,
      182, 186, 187, 188, 207, 216, 218, 223,
      228, 231, 233, 235, 253, 274, 281, 283,
      290, 291
  pleiotropy, 302
  ploidy chimeras, 55, 56
  ploidy effects, 55, 56, 62
  saturation of limiting factors, xiii, 12, 69
  sex determination, 1, 30, 36, 49, 63, 72
  somatic recombination, 33
  techniques, xiii
  temperature-dependence, 39
  transpositions, 255, 298
  transvection, 159
  ubiquitous overexpression, 28, 45, 48, 68,
      69, 75, 96, 109–119, 141, 145, 147,
      182–186, 207, 215, 218, 233, 235, 246,
      247, 276, 282
  unequal crossing over, 134, 298
  variegation, 3, 4, 248
  vs. genomics, 254, 256
Gerhart, John, 255
Ghysen, Alain, xii
Gibson, Matt, 122
Gierer, Alfred, 184
Goldschmidt, Richard, 39, 73, 237, 299
Gómez-Skarmeta, José Luis, 41, 44
gradients
  back-to-back, 89, 148, 209, 211
  biphasic (spire-ramp), 158
  bootstrapping, 153–55
  cell-size independence, 92
  centrifugal, 129
  cone-shaped, 93, 94, 101, 103, 104, 127,
      129, 131, 227
  contour lines, 31, 94, 103, 128, 139
  contour maps, 100
  curved, 101, 104, 124, 126, 128
  developmental capacity, 93
  double, reciprocal, 78, 124, 147, 291, 304
  embryo, 76, 84, 109, 113, 158
  eye disc, 205, 209, 211, 215
  heterochronic, 158, 209
  leg disc, 111, 124, 129

  leg segments, 80, 135
  linear vs. exponential, 81, 83, 158, 301
  merging, 156
  mirror-symmetric, 89, 97, 145, 148, 163,
      167, 209, 211
  notum, 104, 190–193
  nuclear import, 133
  of gene expression, 127, 129, 135, 143, 177,
      235
  overlapping, 78, 101, 104, 124, 291, 304
  parabolic, 128
  parasegmental, 84, 89
  peaks vs. valleys, 156
  perception vs. reality, 125, 134, 143, 147,
      158, 188–190, 291
  proneural, 39, 163, 164
  saddle-shaped (composite), 134
  sawtooth, 80
  seamless appearance, 153–156
  segmental, 83, 84
  shaping, 158
  sliding, 156
  slopes, 81, 92–94, 125, 147, 158, 205, 227,
      291, 301
  slopes, opposing, 78, 124, 147, 291, 304
  smoothing, 147, 158
  tent-shaped, 101, 167
  theory, 81, 83, 92, 93, 124, 158, 209, 299
  wing disc, 111, 138–146, 156–158, 186, 301
  wing veins, 185, 188
Greenwood, Simon, 231, 234
growth
  advantage, 201
  anti-apoptosis factors, 118
  asymmetric, 154
  blastemic, 96, 123
  cell-density independence, 94
  cessation, 92, 94, 99, 122, 124, 155
  compensatory, 89, 91
  control, 86, 92, 124, 155
  control by patterning system, 124, 134, 148
  disc, 57, 85, 89, 92
  disproportionate, 91, 123, 150, 155, 156,
      252
  distal, 100, 101, 103, 115, 133, 171
  even vs. uneven, 33, 35, 92, 148, 158
  extended period of, 92, 121, 155
  factors, 92, 98, 115, 118–119, 124, 196, 283,
      293
  goal of, 91
  gradients, 63
  hyperplastic, 63, 80, 85, 92, 115–119, 124,
      135, 143, 147, 157, 185
  intercalary, 67, 93–95, 103, 123, 124, 133,
      153–156
  intrinsic limits, 92, 155
  larval, 86
  maintenance, 92, 122, 148
  mitogens, 98, 118–119, 143
  need for Dpp, 92, 118–119, 144, 148, 155
  need for various pathways, 92
  orthogonal, 118
  potential, 93, 106, 124
  rate, 4, 85, 91, 105, 133, 148, 150, 151, 153,
      202, 305
  region-specific, 87, 92, 133, 134, 153
  role of nitric oxide, 92
  stunting, 158, 164
  timing, 1
  vs. differentiation, 196
  vs. patterning, 141, 158

  vs. transdetermination, 85, 169, 249
  wound-induced, 96, 98, 122
Gubb, David, 97

Hadorn, Ernst, 299
Haerry, Theodor, 143
Halder, Georg, 237, 252
Hanson, Thomas, 202
Hartenstein, Volker, 77
Heberlein, Ulrike, 229
Heitzler, Pascal, 52
histoblast nests, 1, 2, 73, 87, 299
history, xv
  bristle pattern research, 32, 57
  debate about AS-C, 37–43
  debate about eye compartments,
      202–203
  debate about Fz-Dfz2 redundancy, 293
  debate about Notch, 12–15
  debate about SOP mitoses, 5
  debate about sternopleura, 86
  embryo segmentation, 76
  embryology, 173
  epochs and eras, 57, 86, 97–99, 105, 119,
      202, 254
  GDC Model, 92–96
  genetics, 202, 254, 256
  genetics, molecular, xii, xv, 97
  ironies, 4, 33, 79, 103, 139, 156, 167, 246,
      252, 256, 287
  Morgan's team, xii, xiv, 1, 32, 33, 37, 86,
      123, 298, 299
  PC Model, 93–105, 113, 122–124, 154, 156
  PC vs. Boundary Model, 103–105
  prepattern vs. positional information,
      79–84, 251
  signal transduction pathways, 105
  studies of discs, xii, 76
  utility of studying, 254
histotypes, 9, 72, 81, 85, 169, 189, 242, 243,
      254
Holtfreter, Johannes, 304
homeosis, 84, 123, 160, 237–255. *See also*
      genes: master; mutations: homeotic
  amnesia, 249
  antenna-to-leg, 80–85, 96, 129, 249–252,
      299
  context-dependence, 252
  correlated with regeneration, 85, 246
  correlated with tissue loss, 85
  definition of, 80, 237
  etiology, 85
  eye-to-antenna, 169
  general problem, 81, 237
  haltere-to-wing, 80, 87, 243, 246
  inter-compartment, 87, 148, 149,
      158–161
  inter-disc, 84, 85, 171
  inter-leg, 65, 80, 87, 243
  inter-segmental, 243
  intra-disc, 169
  leg-to-wing, 115
  notum-to-wing, 169, 302
  of head, 141
  paradoxical, 65
  patchy, 81, 83
  single-cell. *See* cell states: switching
  vs. transdetermination, 85, 169, 237, 242,
      249, 254, 302
  wing-to-haltere, 151, 243
  wing-to-notum, 169, 302

homology
  antenna-leg, 81, 83, 249–252, 299
  bristle-scale, 67
  eye-wing, 203, 205
  inter-leg, 63, 65, 89
  shaft-socket, 23
  veins-bristle rows, 187
  wing-haltere, 89, 249
Huang, Françoise, 57

imaginal discs. *See also* imaginal discs,
    particular
  adepithelial cells, 87
  basal lamina, 87
  cell cycle. *See* mitosis
  cell death. *See* apoptosis; cell death
  co-culturing, 118
  compartments. *See* cell lineage;
      compartment boundaries
  conjoined, 98, 100, 123, 156
  culture, 84, 92, 94, 96, 169, 175, 203, 212,
      248
  cytonemes, 67, 125
  definition of, 1
  delamination, 87
  determination vs. differentiation, 86, 91
  diploidy, 86
  disc-specific markers, 86, 89, 91, 171
  dissociation-reaggregation, 65, 85, 90
  duplication vs. regeneration, 92–97, 100,
      103, 121–124, 139, 156, 169
  duplications, 86, 99, 115, 157
  embryonic origin, 1, 77, 85–91, 171
  epidermal cells, 86
  epidermal vesicles, 132, 149, 151, 153, 190
  epithelial folds, 2, 87, 124, 139, 245
  epithelial folds, extra, 115, 135
  epithelial puckering, 151
  epithelium, 46, 65, 69, 87, 197
  epithelium, columnar, 87, 121, 122, 139,
      233
  epithelium, pseudostratified, 87
  evagination, 1, 57, 65, 87, 89, 139, 174, 199
  extracellular matrix, 124, 174
  extracellular space, 124
  filopodia, 53, 65, 67, 87, 125, 174
  fusion, 123, 150, 151, 156
  geometry, 99, 121
  growth. *See* growth
  histology, 4, 86
  history of studies, xii, 76
  hollow sacs, 1, 87, 122
  hyperplasia, 85, 115, 117, 119, 135, 157,
      185
  identities, 86, 90, 149, 249, 254
  identities of compartments, 85, 89, 107,
      145, 148, 150, 153, 158–163, 173, 188,
      203, 248
  identities of subfields, 171, 240, 302
  identity codes, 84, 85, 240, 251, 254, 302
  insularity, 87
  invagination, 85, 87, 91, 99
  irradiation, 86, 95, 97, 99, 100, 106, 135,
      139
  maceration, 96
  mesoderm cells, 85–87
  mitosis, resumption of, 85
  mitotic arrest, 85
  mitotic orientations, 87, 133
  mitotic rates, 85, 87, 153, 299
  myoblasts, 87

necrosis, 113
neurons, 87
numbers of cells, 85, 86, 169
peripodial membrane, 87, 93, 97, 121–123,
    139, 205, 304
planarity, 87
prematurely metamorphosed, 133
regeneration, 86, 92–96, 100, 103, 299
regeneration, polarity of, 94, 95, 100, 103,
    304
regeneration, types of, 123
regenerative potential, 93–96, 100–105,
    123, 139
rotation, 89, 199, 205, 207, 299
size, 67, 85, 92, 245
stalk, 87, 89, 99, 199, 234
surgical fragmentation, 92, 94, 95, 99, 100,
    103, 121, 139, 169
tissue removal, 95–100, 103
tracheae, 86, 87, 139, 174
trans-lumenal extensions, 87
trans-lumenal interactions, 96, 122, 123,
    139, 212
transplantation, 92, 94, 169, 175, 203, 212,
    248
tumors, 85
wound healing, 85, 96, 100, 103, 121, 122,
    156
imaginal discs, particular, 2
  clypeolabral, 2, 77, 86, 199
  eye. *See* eye disc
  genital, xii, 1, 2, 4, 77, 87, 91, 123
  haltere, xii, 4, 77, 85, 86, 171, 239, 243, 245,
      299
  humeral, 2, 77, 87
  labial, 2, 4, 77, 86, 169, 199, 247
  leg. *See* leg disc
  wing. *See* wing disc
Ingham, Philip, 79
insects, xi, 67, 72
  cockroach, 93, 98, 119, 135, 300
  Drosophila. See Drosophila
  epidermis, 87
  grasshopper, 49, 129, 151, 300
  hemimetabola, 105, 133, 135
  hemiptera, 49, 59, 224
  holometabolous, xi
  honey bee, 223
  hymenoptera, 76, 223
  lepidoptera, 29, 31, 47, 61, 65, 67, 132, 299
  mosquito, 212
  other dipterans, 67, 69, 76, 86, 99, 193, 255
  paleoptera, 243
  vs. humans, 197
  vs. mammals, 13, 179, 297, 302
  vs. nematodes, xii, xiii, 1, 11, 50, 247
  vs. plants, 262
  vs. slime molds, 105, 151
  vs. vertebrates, xi, xiii, 86, 93, 151, 197,
      199, 224, 261, 297, 304, 305
  wasp, 302
Internet databases, xii, xiii, 112, 205,
    257, 301

Jan, Lily Yeh, 10
Jan, Yuh Nung, 10

Karch, François, 305
Karlsson, Jane, 105
Kauffman, Stuart, 84, 249
Kirschner, Marc, 255

larva, xi
  Bolwig's organ, 285
  brain, 92
  denticle belts, 79, 285
  development, 86, 87
  dimensions, 2
  discless, 86
  epidermis, 77, 86, 87, 99
  eye organ, 289
  fat body, 92
  growth, 86
  Keilin's organ, 99
  midgut, 77, 289, 291
  neurons, 99
  polyteny, 86
Lawrence, Peter, 32, 84, 89, 101, 148
Lebovitz, Richard, 212
Lecuit, Thomas, 129, 141, 143
leg
  anatomy, 99, 127, 131
  axes. *See under* axes
  branched, 99–103, 115, 123
  claws, 4, 75, 99, 103, 106, 115, 117, 118,
      127, 251, 252, 299
  clone stripes, 133
  cockroach, 93, 98, 119, 135, 300
  deformed, 106
  evolution, 132, 300
  feathery, xi
  femur-tibia detachment, 135
  fused 1st legs, 97
  fused segments, 80, 128, 135
  grasshopper, 129, 151, 300
  hemimetabolous insects, 105, 133, 135
  identity ("legness"), 240, 249, 254
  identity (vs. antenna), 83, 246, 249–254
  intersegmental membranes, 80, 300
  joints, 135, 136, 205, 299, 300
  joints, extra, 131, 135, 136
  pretarsus, 134, 151
  regeneration, 133, 135
  segmental gradients, 80, 83, 135, 300
  segmental identities, 129–135, 300
  segmentation, 80, 128, 131, 135–136, 300
  segmentation, leg vs. body, 136
  short, 65
  trochanter boundary, 132
  truncated, 118, 119, 129, 132
leg disc
  antineural stripes, 128
  bipolar duality, 105, 106, 114
  border cells, 135
  compartments, 4, 63
  dispensable regions, 119
  distalization, 115–119, 129–134
  Dpp (vs. Wg) as a major mitogen, 118–119
  Dpp and Wg as morphogens, 111–113
  Dpp-Wg circuitry transition thresholds,
      117
  Dpp-Wg cooperation in distalization,
      115–118
  Dpp-Wg mutual antagonism, 113–115
  duplications, 115, 132
  endknob, 87, 100, 103, 114, 117, 131
  extra, 115
  fate map, 96, 99, 105, 106, 111, 128
  folds, 87, 133, 134, 136
  folds, extra, 115, 135
  gene expression patterns, 89, 109,
      111–117, 121, 126, 127, 131–135, 243,
      245, 251

leg disc (*contd.*)
  geometry, 99, 105, 121, 128, 129, 137, 301
  growth control center, 119
  initiation, 77, 89, 91, 114, 117, 129, 132
  mitoses, region-specific, 133
  modes of morphogen transport, 124–125
  outgrowths, 100, 115–119, 124, 132
  outgrowths, converging vs. diverging, 119
  proximal-distal axis, 115–118, 129–135
  proximal-distal zonation, 127, 131–135
  proximal-vs.-distal cell affinities, 132
  quirks, 123, 132, 133
  regional markers, 113, 114, 115, 117, 129, 134
  role of adepithelial cells, 87
  role of *Dll*, 129–132
  role of Hh-Dpp-Wg circuit, 105–119, 128–129
  routes of morphogen transport, 125–128
  spot, Dll-sensitive, 129
  spot, peripodial Hh, 121–123
  spot, quiescent, 92, 117, 135
  sternopleura, 86, 89, 99, 106, 118, 127, 299
  tip specification, 115–118
  topology (vs. body segment), 109
  upper medial quadrant, 97, 99, 101, 122, 123, 156
  vs. wing disc, 137, 170, 171
Lewis, Edward, 243, 244, 246, 298
ligands. *See also* gradients; morphogens
  affinity for receptor, 52, 124, 125, 165, 293
  anti-apoptosis, 118
  cell-surface, 53, 135, 167
  cis interactions, 52, 61, 73, 211
  cleavage from precursor, 48, 53, 126, 146, 181, 182, 217, 259, 285, 289
  cognate receptors, 9
  diffusible, 1, 46, 49, 53, 61, 89–91, 111, 121, 142, 145, 146, 150, 153, 169, 180, 181, 211, 215, 217, 225, 285, 289, 292
  diffusion range, 55, 89–92, 107, 111, 113, 117, 119, 124–125, 140, 145–158, 183–188, 215, 218, 227, 285–288, 304
  diffusion range, modulation of, 125, 126
  diffusion rate, 125, 143, 147, 183, 193, 215, 285, 286
  effective range, 125
  endocytosis, 27, 59, 124
  epitopes, 126
  exocytosis, 124
  extinction, 50
  gradients, 147
  homodimerization, 289
  ingestion, 175, 179, 213, 215
  membrane-bound, 53, 179, 272, 275, 280
  proteolysis, 158
  redundant, 24
  regulation by receptor, 166
  regulation of receptor, 125, 158, 166, 288, 292, 295
  release, 125
  short-range inducers vs. long-range morphogens, 101, 104, 107, 109, 115, 117, 138, 187, 234, 304
  tethering to lipid rafts, 125
  transcytosis, 27, 124
  unknown, 29, 53, 119, 127, 131, 133, 134, 223, 227, 293
Lindsley, Dan, xii
links. *See also* circuitry
  definition of, xiii

fallacious, 143–148
for bipolar duality, 114, 169, 302
for cell memory, 248
for computing SOPs, 48
for creating boundaries, 107, 111, 145, 149
for disc identity, 239, 243–252
for disc initiation, 89–90
for ensuring single SOPs, 47, 50–52
for explaining regeneration, 121–123
for eye's A-P axis, 229–235
for eye's D-V axis, 205–211
for making bristle patterns, 59, 75, 190–194, 278–284
for making bristles, 11–20, 25, 271–275
for making eyes, 205
for making gradients, 142
for making leg segments, 300
for making legs, 113–119, 128–135
for making ommatidia, 219–227, 305
for making sensilla, 27, 276–277
for making wing veins, 175, 186–187, 303
for making wings, 158, 165–171, 301
for rewiring other links, 253–255, 302
for signal transduction, 109, 125, 137–148, 158, 181, 285–296
for wing's A-P axis, 145, 156
for wing's D-V axis, 156, 161–164
inconsequential, 158
wiring, *cis-trans*, 136, 246
logic. *See also* circuitry
  Boolean, xiii, 17, 25, 79, 113, 114, 117, 128, 130, 134, 157, 171, 172, 193, 213, 215, 219, 251, 252, 254, 289, 292, 295, 296, 299, 300
  disc development, 256
  double-negative, xiii, 47, 51, 78, 107, 187
  grammar, 20, 254, 300
  jigsaw-puzzle, 163, 219, 252, 254, 261, 300
  nucleosome, 239, 305
  of chemical reactions, 100
  promoter, 17, 18, 122, 254, 298
  subtractive, 59, 61, 128, 141, 163, 302
  syntax, xiii, 10, 18, 79
  Venn, 79, 112, 117, 133, 158, 166, 171–173, 251, 302

Ma, Chaoyong, 229
Maynard Smith, John, 32
mechanisms, 270. *See also* circuitry
  actomyosin motors, 26
  bootstrapping, 153, 155
  cell alignment, 151
  cell shaping, 23, 44
  clocks, intrinsic vs. extrinsic, 11, 215, 217
  clocks, oscillator, 297
  counting, 10, 11, 73, 297, 304
  dominoes, 11, 215
  error-correction, 73, 148
  gating, 10, 11, 21, 69, 233, 287
  inert decoy, 50, 68, 71, 72, 160–163, 180, 216, 260
  lateral inhibition, xiv, 39, 47, 49, 55, 59, 175, 184, 215, 224–227, 304
  licensing, 11, 44, 233, 254
  nucleosome displacement, 248
  quality control, 73
  ratcheting, 10, 11, 81, 121, 143, 147, 155, 217, 243, 244
  reaction-diffusion, 33, 69, 184, 225
  schedule, 11, 215
  servomechanisms, 153–155

shuffling of zinc fingers, 20
stripe-adding, 78, 87, 107, 131, 133, 134, 142, 145, 153–155, 159, 160, 164, 177, 208–209, 234, 302, 303
stripe-splitting, 59, 61, 62, 78, 79, 128, 142, 163, 166
time window, 10, 68, 71–74, 216, 246, 247, 299
timing, 11, 53
vesicle transport, 27, 124, 265
weird, 87
Meinhardt, Hans, 100, 101, 184
Mentzel, Christian, 298
Merriam, John, 4
metamorphosis, xi, 76, 86, 87
metaplasia, 84
Micchelli, Craig, 173
microbeam irradiation, 76
microcautery, 76, 95, 97, 106
Milán, Marco, 173
mitosis
  arrest, 24, 44, 69, 85, 92, 117, 135, 151, 160, 234, 304
  asymmetric, 7, 11, 24
  asynchronous, 9
  cell cycle, 11, 24, 44, 49, 85, 100, 223, 233, 234, 265, 304
  licensing, 44, 233
  nuclear shifts, 304
  orientation, 3, 4, 7, 24, 87, 133
  quiescent spots, 44, 69, 92, 117, 135
  quiescent zones, 92, 151, 160, 233, 234, 304
  radiosensitivity, 41
  random vs. patterned, 92, 208, 233
  rate, 4, 81, 85, 87, 91, 151, 153, 202, 299
  regulators, 92
  spindle, 3, 24
  synchrony, 44, 85, 87, 208, 211, 215, 233, 234, 305
  syncytial, 79, 86
  timing, 174
  zonation in wing, 175
Mlodzik, Marek, 225, 231
models, 266–268
  active-cell vs. passive-cell, 125
  Bateson's toy, 99
  cellular automata, 235
  computer, 79, 224, 236, 303
  hybrid, 84, 101, 103, 158
  intransigence, 147
  mathematical, 79
  overturned, 13, 38, 39, 44, 48, 84, 87, 123, 134, 145, 147, 158, 185, 202, 211, 224, 247, 253
  resurrected, 13, 79, 115, 156, 224, 299
  robustness, 99
  utility, xiv, 32, 123
Modolell, Juan, 37, 57
Mollereau, Bertrand, 224
Morata, Gines, 89, 148
Morcillo, Patrick, 161
Morgan, Thomas Hunt, xi, 31, 193, 256, 298, 302
morphogenesis, xi, 23, 87, 124, 131, 134, 135, 148, 174, 304
morphogens. *See also* gradients; ligands; morphogens, particular; positional information
  active transport, 101, 124, 125
  as mitogens, 92, 118–119, 148
  as survival factors, 92, 118, 228

binding by lipids, 124
binding by proteoglycans, 124
channeling, 101, 124, 126
circumferential, 101, 104
constraints on usage, 158, 249
constructs, membrane-tethered, 139, 145, 153, 157, 185, 285, 293
definition of, 81, 138
diffusion range. *See under* ligands
dorsalizing vs. ventralizing, 104, 109, 111–113
mistaken expression, 85
modes of movement, 104, 124–125, 146
passive diffusion, 124, 126, 130
polarized movement, 125
pumping, 129
rate of transport, 125
response modulation, 84, 125, 134, 158
retrograde transport, 125
role of actin, 125
role of cell shape, 124
role of dynamin, 125
role of receptor density, 125, 158, 188
spheres of influence, 106, 111, 124
tagging with GFP, 147
transcytosis, 27, 124, 146
transport via cytonemes, 125, 146
transport via lipid rafts, 125, 146
morphogens, particular
Bicoid, 77, 83
Decapentaplegic, 89–91, 104–129, 133–137, 140–148, 153–158, 171, 177, 185–196, 229–234, 254
Hedgehog, 84, 89, 90, 104–112, 117–129, 137–146, 150, 153, 177, 185–187, 196, 229–234, 254
Spitz, 89, 91, 196, 215, 254
tip (unknown), 83, 100, 103, 104, 122, 129, 251
Wingless, 84, 89, 90, 104–129, 133–137, 142, 146, 156–158, 161, 171, 189–190, 196, 234, 254
Wnt (unknown), 113, 132, 133, 136, 205, 231, 292, 293, 304
Muller, H. J., 32, 37, 39
mutations. *See also* phenotypes
affecting bristle patterns, 194, 278–284
affecting bristles, 271–275
affecting bristles and sensilla, 276–277
allelic series, 129
allelic specificity, 20, 24, 37, 38, 39, 44, 45, 132, 284, 291
antimorphs, xiii
atavistic, 36, 63, 67, 135, 174, 243, 246
breakpoints, 41, 43, 44, 129, 194, 195, 280
cell-lethal, 63, 97, 98, 100
deletions, 18, 23, 28, 35, 41, 43, 44, 48, 50, 51, 59, 69, 71, 134, 164, 185, 186
dominant vs. recessive, xii, 129
dominant-negative, 173, 234
duplications, 23
heterochronic, 299
homeotic, 63, 65, 80, 83, 135, 169, 237, 240, 243, 246–249, 261, 300, 302
hypomorphs, 7, 23, 37, 59, 73
idiosyncratic, 52, 75, 87, 158, 288
inversions, 41, 43, 44
lethal, 87, 141
lethal phase, 86
lethal side effects, 80
macromutations, 237

maternal effect, 278
missense, 15
Morgan's quest for, 37
neomorphs, xiii, 18, 87, 149, 271, 278, 284, 288, 302
nomenclature, xii, 37
nonsense, 44
null vs. leaky, xii, 10, 21, 23, 75, 118, 138, 154, 157, 183
null, compound, 37, 44, 69, 107, 114–119, 125, 132, 149, 153–157, 165, 188, 207, 215, 216, 218, 227, 231, 273, 275, 282, 291, 303
null, utility of, 43, 44, 117, 118, 293
P-element insertions, 49
point, 41, 44, 298
prepattern, Stern's quest for, 80, 252
rearrangements, 186
screens, 5, 18, 26, 37, 87, 141, 160, 305
screens, dosage, 68
screens, modifier, 12, 180
temperature-sensitive, xiii, 15, 23, 26, 59, 63, 73, 121–123, 138, 139, 156, 169, 173, 182, 229, 235, 283
X-ray induced, 37
mysteries, 268–270

Needham, Joseph, 301
Nellen, Denise, 141, 143
nervous system
antigens, 208, 213, 215
axon projections, 55, 97, 191–196
cell lineage, 4, 11
embryo, 49, 50, 52, 73, 149
gene expression, 27
genetics, 271, 278
grasshopper, 49
imaginal discs, 87
lobes, antennal, 191
lobes, optic, 141, 196, 198, 285, 292
midline fates, 262
mutation screens, 5
neurons, 27, 173
repressors, 20
wiring, 5, 20, 27, 55, 99, 139, 191–196, 199, 228
Neumann, Carl, 157
neurogenesis
adult, 31
axon pathfinding, 20, 55, 99, 174, 191–96
axons, retrograde transport, 125
bristle. *See* bristles
embryo, 24, 31, 166
ganglion mother cells, 11, 24
HLH genes, 72
neural competence, 49
neuroblasts, 7, 9, 11, 24, 49, 51, 52, 77, 191, 196, 217, 274, 297
Nöthiger, Rolf, xii
Nüsslein-Volhard, Christiane, 76

Oberlander, Herbert, xii
Occam's Razor, 35
O'Farrell, Patrick, 124
O'Keefe, David, 173
ommatidia
bristles, 30, 57, 199, 209, 218
cell types, 197, 199, 208–211, 215, 223, 227, 228, 233
chirality, 209–211
cone cells, 208, 212, 215–219, 228

firing center (at equator), 305
founder cell (R8), 213–218, 227, 233
lattice, 32, 201, 202, 208, 224, 227, 230
mystery cells, 208, 211, 223
number per eye, 197
photoreceptors. *See* photoreceptors
pigment cells, 209, 212, 216, 218, 228, 229
polarity, 81, 198, 207, 203–211, 236, 293
rate of creation, 208, 227
rhabdomere trapezoids, 197, 209, 211
role of Delta-Notch, 53, 211
rotation, 131, 201, 208–211, 304
spacing, 212, 304
stages of development, 208, 211, 215, 217, 225, 227, 233

Pan, Duojia, 252
pattern formation. *See also* axes; boundaries; circuitry; morphogens; prepattern; positional information; rules
alignment, 67, 148–153
amphibian limbs, 93
arthropods vs. vertebrates, 93
Bateson's Rule, 99–101
by balkanization, 131, 148, 190, 301, 302
by chain reactions, 111, 156, 167, 215, 235, 297
by determination waves, 211, 212, 235, 305
by drawing lines, 97, 173, 179
by elaboration, 63, 133, 302
by intersecting lines, 35, 45, 89, 90, 101, 157, 234
by invisible scaffolds, 156
by iteration, 29, 74, 135, 145, 217, 300
by painting sectors, 97
by painting stripes, 61–62, 69, 78
by physical forces, 33, 35, 135, 175, 189, 237
by sequential induction, 139, 211, 215
by signal relay, 139, 186
by slicing a cylinder, 136
by stratification, 87, 107, 131–134, 142, 145, 153–155, 159, 160, 164, 177, 208–211, 234, 302, 303
by symmetric annealing, 136, 167, 174, 177
by tessellation, 224
by triangulation, 56, 59, 224, 305
cell death, 87
cell lineage, irrelevance of, 3
cell-size independence, 92
checkerboard, 50, 158
community effect, 85
continuity, maintenance of, 99, 101
control of growth, 124, 134, 148
David *vs.* Goliath power, 123
development vs. regeneration, 94, 97, 123
discontinuities, 93, 95, 105, 124, 154–156
discs vs. embryo, 97
disruption, 80
distalization, 100, 115–119, 129–134, 157, 158, 166, 171
duplication vs. regeneration, 92–97, 100, 103, 121–124, 139, 156, 169
duplications, 86, 99, 105, 115, 154
duplications, mirror-image, 89, 94–99, 103–106, 113, 118, 122, 128, 129, 148–165, 169–173, 275
during growth, 86, 94, 133
edge effects, 93–96, 103, 153, 305
epimorphosis vs. morphallaxis, 96, 103, 123

pattern formation (*contd.*)
  evolution, 86
  feather lattice, 224, 305
  fingerprints, 151, 199, 299
  general problem, xi
  generic tools, 179
  homeogenetic induction, 85
  hypertrophy. *See* growth: hyperplastic
  in a syncytium, 86
  intercalation, 93–96, 133, 153–156
  local vs. global, 21, 33, 50, 55, 56, 62, 68,
      79, 81, 92, 94, 128, 139, 173, 196, 211,
      212, 224, 235, 236, 254, 301
  molecular basis, 79
  noise, 201, 202, 228, 229
  number control, 136, 304
  organizers, 100, 129, 301
  out of phase, 84, 225, 227, 233
  precision, 31, 46, 62, 69, 86, 99, 151, 156,
      201, 228, 305
  principles, 270
  quirks, 86, 299, 305
  reconstitution, 67, 90, 156
  regeneration, 81, 86, 93–95, 100, 103, 123,
      156, 246, 299, 304
  repatterning, 67
  reprograming, 63, 246
  robustness, 91
  scaffolds, 79, 80, 148
  short-range inducers, 101, 104, 107, 109,
      115, 117, 138, 187, 234, 304
  subpatterns, 20, 62, 154, 156
  top-down vs. bottom-up, 56
  topographic cues, 69
  trans-boundary cooperation, 189
  triplications, 98, 99, 100, 119, 254
  vertebrate somites, 304
  vs. growth, 141, 158
  without positional information, 133
patterns. *See also* anatomy; bristle patterns;
      phenotypes
  geometry, 31, 32, 56, 59, 93, 123, 160, 189,
      201, 211, 212, 304
  geometry, leg vs. wing, 99, 105, 128, 137,
      139, 301
  hoops, 135, 151
  ladders, 90, 189
  lattice, 32, 201, 202, 224, 227, 230, 305
  mammal coat color, 302
  orthogonal, 137, 189
  pigmentation, 32, 43, 75, 193
Peifer, Mark, 305
phenotypes
  Bateson's sports, 99
  bizarre, xii, 1, 30, 53, 63, 94, 98, 115, 117,
      119, 124, 143, 145, 153–155, 157, 203,
      237, 248, 252, 254, 265
  converging vs. diverging outgrowths, 98,
      100, 101, 103, 119
  deficient (but not duplicated), 106, 118,
      139, 169, 170
  discless, 87
  disc-specific, 87
  duplication-deficient, 97, 106, 113, 139,
      155, 169
  etiologies, 80, 96, 119, 123, 128, 131, 136,
      153, 154, 157, 163, 249, 262, 271
  hyperplasia, 63, 80, 115–119, 124, 135, 143,
      147, 157, 185
  Janus, 106, 113, 118, 128, 129, 153, 159,
      163, 169

  leakiness, 21
  mirror-image, 63, 89, 95–100, 105, 106,
      113, 118, 122, 128, 129, 148–165,
      169–173, 275
  paradoxical similarity of LOF and GOF,
      160, 163, 171, 286
  phenocopies, 80, 99, 185, 189, 237
  quantitative spectrum, 156
  segment-polarity, 136
  sensitive periods, 26, 63, 98, 123, 135, 170,
      175, 182, 189, 216, 229, 237, 252, 284
  syndromes, 49, 80, 99, 131, 136, 149, 164,
      165, 182
  triplications, 98, 99, 100, 119, 254
photoreceptors
  anatomy, 197, 199
  axon projections, 196, 198
  developmental sequence, 208–215, 218,
      227, 233
  gene expression, 220
  identity code, 219, 224, 233
  inner vs. outer, 197, 199, 233
  markers, cell-type specific, 215, 218, 220
  nuclear shifts, 304
  R1-R6, extra or missing, 216, 218, 220, 224
  R3-vs.-R4 determination, 209–211
  R7 equivalence group, 182, 215
  R7, extra or missing, 180, 213–218, 220,
      224
  R8 equivalence group, 225
  R8 spacing, 55, 62, 225
  R8, extra or missing, 223
  R8, need for *atonal*, 28, 72, 74, 217, 220
  R8, similarity to SOP, 225
  rhodopsin subtypes, 197, 223, 224, 253,
      254
  vs. bristles, patterning of, 57, 225
Plunkett, Charles, 39
Poodry, Clifton, xii, 39
positional information. *See also* axes;
      boundaries; gradients; morphogens
  abdominal segments, 84, 141
  absurd aspects, 158
  area codes, 81, 85, 191
  azimuths, 106, 124
  clockface metaphor, 103, 155
  coordinate systems, 56, 79, 81, 83, 96, 103,
      105, 193
  coordinate systems, bipolar, 106
  coordinate systems, Cartesian, 90, 93, 101,
      105, 157–158
  coordinate systems, polar, 93, 95, 97, 101,
      105, 124, 126, 128
  coordinate systems, tripolar, 101, 106
  coordinate systems, warped, 122
  coordinates, polar, 103
  crowded coordinates, 99, 103, 105, 122
  definition of, 81, 84
  down-the-slope constraint, 93
  emission vs. reception, 148, 151
  French Flag metaphor, 81, 84, 139, 299
  graininess, 83, 92, 158, 299
  individuation, 84, 89, 145, 148, 163
  interpretation, 81, 83, 84, 209, 251
  interpretation modes, 83, 84, 89, 167, 249,
      251, 300
  interpretation, minor role of, 158, 301
  intrasegmental, 83, 135
  PI-prepattern hybrid, 84, 158
  positional values, 81, 83, 84, 93, 96, 99,
      248, 251

  positional values, misleading idea of, 158,
      301
  precision, 81, 124, 158
  recording, 81, 83, 84
  reference lines, 87, 89, 95, 109, 151, 188
  reference points, 55, 59, 90, 101, 157, 234
  singularities, 93
  specification, 81, 83, 89, 251
  specified vs. determined states, 90, 111
  stages, 81
  thresholds, 84, 104, 109, 113–117, 126–131,
      138–147, 154–158, 183–193. *See also
      under* circuitry; prepattern
  thresholds, setting, 167
  thresholds, sharpening, 291
  universality, 81, 83, 84, 96, 202
  vs. prepattern, 83, 79–84
  wiring of genome, 41, 83, 84
  zonation, 78, 84, 107, 125, 127, 131,
      138–143, 163, 188–193, 252, 301
  zonation, orthogonal, 142, 158
Posthlethwait, John, xii
prepattern, xiv, 33, 43, 68, 301
  absurd aspects, 80, 83
  antenna vs. leg, 83, 251
  competence, 35, 39, 47, 83
  definition of, 33
  embryo segmentation, 79, 84
  evolution, 36
  factors, 45, 47, 69, 73, 75, 190, 193, 229, 305
  genes. *See under* genes
  gradients, 56
  hypothesis, 33, 35, 39, 79–83, 251
  landscape, 62, 69, 193
  macrochaetes, 35, 68, 190–193
  mutants, 80, 252
  nodes, 33, 79, 80
  overlapping, 83
  pair-rule stripes, 78, 79
  PI-prepattern hybrid, 84, 158
  proneural clusters, 55, 68
  proneural vs. antineural, 164, 195
  reaction-diffusion, 33, 83
  saga, 80
  sex comb, 80, 135
  singularities, 33, 35, 36, 80, 83, 84, 158, 301
  singularities, *cis*-enhancers for, 35, 92
  singularities, cryptic, 36, 41, 43, 68, 135
  thresholds, 83. *See also under* circuitry;
      positional information
  universality, implausible, 80–83
  vindication, 79, 190, 251, 299
  vs. positional information, 79–84
programming. *See under* computer
      metaphor
protein domains. *See also* protein domains,
      particular
  alterations, 72, 300
  dimerization, 15, 17, 161, 216, 248, 280,
      283
  docking, 68, 158, 300
  domain-swapping experiments, 179, 260
  idiosyncratic, 28
  interchangeability, 257
  masking vs. unmasking, 15, 21, 261, 263,
      273, 274
  overlapping, 15, 21
  propellers, 249, 265
  reference sources, 257
  repeats, definition of, 257
  repeats, extreme number, 259

scaffolding, 257, 263, 293
self-defeating, 288
size range, 257
steric effects, 72
protein domains, particular, 257–265
14–3–3, 179, 258
activator vs. repressor. *See under* gene regulation (activators or repressors)
ADAM, 258
ankyrin, 12, 13, 180, 258
arm, 258
basic, 17, 23, 161
bHLH, 17, 28, 29, 41, 48, 61, 74, 91, 177, 217, 225, 239, 242, 249, 258, 259, 272
BTB, 18–20, 135, 259
bZip, 180
COE, 186, 259
cut, 259
DEF, 180, 181, 259
DEJL, 181, 259
DNA-binding, 17–23, 28, 61, 68, 71, 85, 107, 141, 148, 161, 163, 171–174, 190, 218, 223, 239, 242, 257–265, 275, 277, 282, 288–291, 295, 303
EF hand, 295
EGF-like, 15, 52, 165, 170, 180, 181, 259
EH, 26, 259
Ets, 181, 182, 259
F, 180, 260
glycosylation sites, 15, 289, 292
GUK, 260
helix-turn-helix, 260, 263, 291
HLH, 39, 49, 50, 68, 71–73, 161, 260, 297
HMG, 180, 242, 260, 295
homeo, 11, 24, 107, 129, 135, 158, 161, 186, 190, 191, 203, 213, 217, 223, 235, 239, 242, 249–252, 260–263
Hox hexapeptide, 254, 260, 261
Ig-like, 170, 180, 181, 261, 263
KH, 180
kinase, 179
leucine zipper, 261
LID, 161
LIM, 135, 136, 158, 161, 261
LNG, 261
MADS, 174, 186, 239, 261, 290
NES, 262
NLS, 12, 13, 15, 20, 180, 262
opa, 141, 180, 262
Orange, 61, 71
paired, 252, 262
PAS, 29, 177, 249, 262
PDZ, 180, 263, 292
PEST, 12, 263
PH, 180, 181, 263
POU, 167, 217, 263, 297
PRD, 264
PTB, 24, 180, 264
PxDLSx(K/H), 71, 264
RAM23, 13
RNA-Binding, 181, 264
runt, 264
SH2, 180, 181, 264
SH3, 180, 181, 264
signal, 265
Sox, 45, 265
sterol-sensing, 125, 285
TALE, 254
TEA, 171, 265
transmembrane, 13, 15, 261, 285
VP16, 13, 51, 61, 284

WD, 24, 249, 265
WD40, 180
WRPW, 17, 51, 61, 71, 255, 265
zinc finger, 18–20, 48, 50, 52, 91, 107, 127, 132, 134, 136, 141, 180, 186, 190, 217, 223, 224, 239, 242, 252, 265
proteins. *See also* genes, particular; proteins, common; proteins, modification of
adaptors, 24, 109, 179, 180, 254, 263, 264, 294, 304
adhesive, 134, 150, 151, 153, 173, 174, 177, 261, 299
binding affinities, 259, 264, 265, 291, 294, 295
bipolar, 19, 107, 255, 288
bridging, 41, 45, 160, 161, 191, 279, 284, 287, 294
caspases, 228, 262
catalysts, 13
cellomics, 298
chimeric, 13, 18, 51, 68, 126, 147, 209, 216, 224, 284, 288, 303
chromatin-remodeling, 17, 71, 111, 207, 239, 247–249, 264, 277, 282, 295, 300
competitive binding, 17, 18, 68, 72, 79, 84, 161, 170, 215, 278, 291, 295
conformational changes in, 26, 258, 261, 264, 286, 290
constant shape vs. variable sequence, 260, 263, 264
convertases, 53
cooperativity, xiv, 79, 84, 261, 263, 265, 290, 300
cytoskeletal, 180
cytoskeletal regulators, 181
deacetylases, 293
dehydrogenases, 293
dimerization, 17, 49, 79
DNA-binding. *See under* protein domains
geranyl transferases, 180
GFP-tagged, 126
glypicans, 290, 293
GTP exchange agents, 180, 181
GTPases, 26, 180, 181, 275, 293
GTP-binding, 26
heterodimers, 12, 15, 17, 21, 49, 50, 68, 71–75, 133, 171, 172, 190, 193, 251, 252, 258–264, 272, 276–280, 284, 290, 303
homodimers, 15, 17, 71, 72, 146, 181, 258–263, 272, 273, 280, 286, 289
huge vs. tiny, xi, 174, 180, 259
immunological detection, 13, 27, 41, 43, 53, 126, 147, 303
immunological tracking, 175
isoelectric point (pI), high, 21
isoforms, 18, 19, 181, 182, 222–224, 239, 247, 258, 273–275, 281–284, 287, 294
isoforms, extreme variety of, 191, 261
kinase inhibitors, 11
kinases, 24, 180, 213, 263, 286, 287, 293, 294
ligands. *See* ligands
lipases, 181, 304
metabolic enzymes, 29, 85, 302
mimicry, 68
multi-protein, 254
multi-protein complexes, xiv, 11, 41, 182, 191, 224, 247, 248, 252, 261, 265, 271, 279, 280–289, 294–296
multi-protein machines, 300
neuronal, 191

nomenclature, xii
nuclear import, 9–11, 15, 20, 21, 131, 133, 171, 172, 251, 258, 286–295
oligmers, 258
oligomers, 181, 259, 262, 273, 275, 290
phosphatases, 11, 24, 157, 180, 181, 275, 289, 294, 304
polyadenylases, 100
proteases, 15, 53, 180, 189, 228, 258, 265, 289
proteoglycans, 55
proteolysis, 19, 20, 105, 133, 180, 182, 218, 224, 263, 271, 280, 294. *See also under* receptors
proteolysis, proteasome-dependent, 12, 107, 111, 262, 263, 271, 287, 288
receptors. *See* receptors
RNA-binding, 24, 181, 264
saturation, 9
scaffolding, 179, 180, 191, 258, 260, 283, 287, 294
schizophrenic, 255
secreted, 55, 90, 91, 105, 106, 111, 145, 146, 153, 165, 170, 182, 225, 227, 265, 285, 289, 292
self-cleaving, 146, 285
self-defeating, 259, 261, 290
sequestering, 9, 12, 15, 50, 68, 71, 72, 161, 167, 180
sheddases, 53, 258
size range, 257
stoichiometry, 13, 21, 61, 73, 163, 165, 171, 175, 286, 287, 291, 295, 298, 300, 303
sulphotransferases, 293
synthesis, 29
targeting, 20
tethering, 9, 12, 15, 24, 165, 286, 289
tetramers, 160
transcription factors, 18, 41, 48–52, 79, 105, 111, 132–136, 141, 143, 148, 156, 158, 163, 171, 180, 181, 186, 190, 213–224, 233, 239, 247–249, 254, 258–265, 272–293, 297, 300. *See also under* gene regulation
transmembrane, 9, 15, 179, 180, 182, 191, 213, 259, 261, 272, 274, 275, 279–294
turnover rates, 126, 147, 286, 287, 295
two-hybrid assays, 21, 55
two-hybrid screens, xiii, 24, 160
ubiquitin ligase adaptors, 180, 260, 287, 294, 296
ubiquitin ligases, 182, 265, 272, 273, 282, 296
ubiquitous. *See* genes: ubiquitously expressed
uncleavable constructs, 13, 287
weird, 21, 179, 213
proteins, common
actin, 125, 174
cadherins, 90, 92, 153, 191, 258, 295, 304
clathrin, 26, 124, 146
collagen, 174
connectin, 149
cyclins, 29, 44, 49, 233, 258, 260, 280
dynamin, 26, 124
importins, 258
insulin, 92
integrins, 160, 173, 174, 177
kinesin, 11, 286
laminin, 174
myosin, 11

proteins, common (*contd.*)
  neurexins, 191
  neuroglian, 149
  spectrins, 180
  tropomodulin, 11
  ubiquitin, 180
proteins, modification of, 21
  by acetylation, 248, 279, 288, 295
  by acylation, 285
  by autophosphorylation, 290
  by autoproteolysis, 285
  by cholesterol adducts, 125, 146, 285
  by cleavage, 9, 12–15, 53, 107, 111, 126,
    146, 181, 182, 213, 217, 259, 265, 274,
    285–289
  by deacetylation, 17, 248, 249, 275, 276,
    281
  by dephosphorylation, 275, 294, 304
  by geranylation, 180
  by glycosylation, 15, 124, 146, 165, 261,
    292
  by hyperphosphorylation, 294
  by phosphorylation, 9, 13, 15, 24, 107, 111,
    126, 145, 180–183, 218, 235, 260, 263,
    273, 286–296, 300, 303
  by reshaping, 26, 258, 261, 264, 286, 290
  by ubiquitination, 180, 182, 263, 280, 287,
    294
proteins, particular. *See* genes, particular

Rauskolb, Cordelia, 173
Ready, Donald, 202, 211, 212, 224
receptors
  affinity for ligand, 52, 124, 125, 165, 293
  auto-repression, 261
  capping, 211
  cassettes, 179
  cis interactions, 164, 211
  cleavage of precursor, 15, 213, 274
  constitutively active, 12, 114, 124, 141,
    145, 151, 170, 173, 190, 207, 216, 234
  co-receptors, 105, 109, 217, 286, 290, 293
  density, 125, 188
  density, effects of, 147, 158
  dimerization, 9, 181, 293
  dominant-negative, 147, 169, 173, 183,
    195, 216, 242
  endocytosis, 27, 59, 124, 146, 182, 285
  gradients, 147
  heterodimers, 289
  hypersensitized, 52, 189, 190
  hyposensitized, 143, 189, 190
  interchangeability, 216, 293
  ligand alternation, 165
  ligand-induced cleavage, 9, 12, 13, 15, 53
  maturation in Golgi, 15, 165, 274
  occupancy levels, 81
  olfactory, 191
  oligomerization, 9, 52, 261, 289
  phosphorylation, 181
  protein complexes, 179, 180, 273, 282, 286,
    289, 293
  proteolysis, 12, 13, 15, 124, 182, 286
  recycling, 124
  regulation by ligand, 125, 146, 158, 166,
    288, 292, 295
  regulation of ligand, 166
  serine-threonine kinases, 290
  trans-ingestion, 175
  tyrosine kinases, 170, 179, 182, 213, 231,
    259, 264, 290, 303

  unknown, 55, 134
  with multiple ligands, 170
Reh, Thomas, 217
Reinitz, John, 79
Renfranz, Patricia, 225
Rhyu, Michelle, 5
RNA
  antisense, 303
  binding, 264
  clocks, 297
  computation, 73
  domains, 73, 272, 279
  heteroduplexes, 73
  localization, 73, 76, 84
  maternal, 76
  processing, 24, 73, 79
  splicing, 18, 19
  translation, 73
  turnover, 73, 100
  untranslated portion, 48, 73
Roegiers, Fabrice, 53
Romani, Susana, 43
Rubin, Gerald, 225, 252
rules, xi, xii, 270, 300. *See also* cell
    instructions
  about axon wiring, 55, 191–196, 199
  about bracts, 29
  about bristle patterns, 31, 68, 229, 230
  about bristles, 20, 55
  about cell autonomy, 47
  about cell lineage, 1, 18
  about cell size, 55, 56
  about cellular automata, 236
  about disc identity, 251, 252, 305
  about distalization, 115, 119, 157, 158
  about eye's D-V axis, 203, 205, 209
  about gene regulation, 18, 91, 104, 115,
    171–173, 248, 251, 254
  about growth vs. differentiation, 59, 118
  about inhibitory fields, 55, 57, 59
  about interfaces, 148, 154, 155
  about ommatidia, 215, 227
  about regeneration, 93, 94, 95, 96, 97, 100,
    103, 156
  about regulatory hierarchies, 136
  about tissue continuity, 94, 254
  about transdetermination, 84
  about trespassing, 132
  about Turing-like models, 69
  about wing veins, 187
  Bateson's Rule, 99, 100, 101, 103, 153
  Distalization Rule, 93
  Dpp-Wg Intersection Rule, 157
  Posterior Prevalence Rule, 248
  Proximity-vs.-Pedigree Rule, 4, 5
  Reciprocity Rule, 92, 94, 95, 96, 122, 156
  Shortest Intercalation Rule, 93, 95, 101,
    154, 156
  Venn Overlap Rule, 172, 251

Sánchez-Herrero, Ernesto, 246
Schneiderman, Howard, xii, 32
Schubiger, Gerold, 122
Schultz, Jack, 38
science
  ad hoc assumptions, 99, 100, 103, 122, 131
  artifacts, xiv, xv, 48, 57, 107, 144, 147, 279
  big vs. little, 256
  cleverness, 31, 51, 97, 100, 143, 224, 239,
    246
  conventional wisdom, 246

  correlation without causation, 4, 119, 149,
    157, 234. *See also* uncoupling
  counterintuitive facts, 35, 160, 163
  debates. *See* history
  deductions, 12, 35, 37, 39, 43, 67, 83, 91,
    124, 131, 170, 251, 298
  delusions, semantic, 247
  dialectic, 12, 187
  disappointments, 33
  discoveries, xi, 2, 4, 5, 35, 37, 40, 44, 49, 53,
    85, 86, 90, 105, 121, 122, 125, 132, 149,
    158, 165, 202, 208, 224, 256, 298
  disproofs, 38, 39, 44, 48, 158, 185, 202, 211,
    224, 229
  dogmas, 53, 57, 96, 97, 246
  dramatics, 21, 75, 85, 134, 135, 143, 165,
    237
  enigmas, 268–270
  epiphanies, 270
  heresies, 15, 41, 57, 96, 119, 149, 246, 285,
    292
  heuristics, 123, 136, 266, 302
  hypotheses, 266–268
  illusions, xv, 23, 68, 72, 134, 166
  insights, xi, xv, 33, 56, 79, 83, 97, 139, 157,
    246, 254, 256, 300
  messy data, 43, 47, 183, 287
  metaphors, 266–268
  mistakes, xiv, 12, 38, 43, 147, 156
  models, 266–268
  mysteries, 268–270
  myths, 147, 202
  negative results, 147
  paradigms, xv, 43, 44, 79, 83, 84, 103, 202,
    245
  paradoxes, 10, 27, 52, 55, 56, 80, 93–96,
    113, 128, 160, 169, 183, 231, 246,
    268–270, 282, 286
  predictions, 10, 15, 33, 41, 44, 47, 49, 56,
    79, 85, 97, 153, 205, 213, 218, 265
  proofs, 1, 12, 20, 50, 61, 67, 86, 90, 113, 115,
    118, 139, 155, 157, 202, 204, 212, 217,
    231, 246, 297
  prophecies, 43, 79, 84, 97, 190, 301
  reasoning by analogy, 137, 148, 161, 163,
    203
  reasoning from first principles, xii, 31, 35,
    60, 101, 125, 158, 201, 213, 216
  red herrings, 18, 109, 134
  reductionism, 299, 302
  riddles, 268–270
  scenarios, 266–268
  skepticism, 12, 100, 114, 187, 202
  surprises, xii, 18, 33, 37, 41, 43, 44, 52, 68,
    69, 77, 87, 91, 97, 99, 105, 131, 153, 180,
    183, 209, 216–218, 243, 252, 260, 282,
    287, 303–305
  systems analysis. *See* uncoupling
  unnatural side-effects, 25, 124
self-assembly. *See also* boundaries, creation
    of; emergent properties; mechanisms;
    pattern formation; positional
    information
  general problem, xi
  of bristle rows, 65, 67, 299
  of concentric rings, 134, 224, 304
  of dynamin collars, 26
  of multi-protein complexes, 248, 249,
    303
  of ommatidia, 212, 213, 215, 227
  of stripes, 150, 151, 173, 174

sensilla
  campaniform, 28, 63, 65, 99, 131, 185, 276, 299, 304
  chordotonal, 27, 29, 74, 112
  external, 27
  extra, 73
  haltere, 246
  identity genes, 27, 75
  larval, 9, 24, 27
  nests on legs, 304
  olfactory, 27–29, 75, 191, 199, 246, 276
  role of AS-C, 44
  shifts, 63, 131
  stretch receptors, 28
  trichoid, 28, 304
  vs. bristles, 27–28
  vs. bristles on wing vein L3, 187
  wing, 174
  wing radius, 18
sensory organ precursor (SOP). *See also* bristles
  ablation, 49
  asynchrony, 55, 59, 62
  computer, 72
  delamination, 65
  eccentric, 43, 47
  filopodia, 53, 65, 67, 299
  heterochronic pairs, 53, 57
  initiation, 21, 37, 43–55, 59, 61, 68, 71–73, 163, 229, 278–284
  markers, 156, 282, 284
  mitoses, 5, 7, 41, 49, 59, 92, 229
  movements, 61, 63, 65, 67
  patterns, 39, 59, 61, 94, 139, 163, 193
  pre-SOP state, 47, 53
  regression, 53, 71
  repulsion, 65, 67
  selection, 45–74, 229, 304
  sequence, 41, 57, 59, 67, 133
  vs. PNC identity, 48, 51
  vs. R8p photoreceptor, 225
Serebrovsky, A. S., 37
Serrano, Nuria, 124
Shearn, Allen, xii
signaling pathways. *See also* signaling pathways, particular
  antagonistic, 74, 114–117, 121, 167, 179, 234, 235
  anti-mitogenic, 92
  artificial switching, 171
  as networks, 179, 300, 303
  assays for activity of, 126, 183, 218
  branched, 181, 293
  comparisons, xii, 27, 92, 105, 109, 157, 179–182, 297, 300, 303
  context-dependence, 21, 179, 181, 303
  control of AS-C, 71
  cross-talk, 15, 50, 52, 129, 136, 167, 180, 274, 279, 284, 290, 294, 300, 303
  mitogenic, 92
  PCP (planar cell polarity), 131
  shared components, 15, 109, 157, 180, 181, 275
  unorthodox variations, 11, 109, 173, 289, 292
signaling pathways, particular
  Boss-Sevenless, 179, 181, 213, 215, 218, 220, 233, 303
  dActivin, 92

Decapentaplegic, 105, 109, 124, 132, 141, 157, 167, 173, 177–180, 185–190, 194, 231, 235, 286, 289–292
EGFR, 29, 48, 52, 74, 92, 107, 153, 169, 170, 174, 177–189, 194, 216–219, 227–229, 233, 235, 242
Hedgehog, 92, 105, 109–113, 149, 154, 157, 177, 185–187, 194, 208, 231, 235, 242, 285–289
insulin receptor, 92
JAK-STAT, 92, 205, 303
JNK, 118, 263
Notch, 9–27, 49–55, 68, 69, 73–75, 92, 135, 159–169, 173–180, 188, 194, 203–211, 218, 219, 223–229, 235, 242, 263, 272–275, 279–284, 303
PCP (planar cell polarity), 293
Ras-MAPK, 92, 179, 216, 218, 224, 263, 304
RTK, xiv, 170, 179–182, 218, 231, 258, 264, 265, 303
Wingless, 15, 52, 92, 105, 109, 132, 136, 156, 157, 167, 169, 181, 182, 189–190, 194, 205, 242, 245, 279, 285–288, 292–296
Simpson, Pat, 52
Skeath, James, 41
Spemann, Hans, 36
Spencer, Susan, 227
Stern, Curt, xv, 32–43, 49, 56, 57, 75, 79–83, 86, 163, 164, 190, 246, 251, 252, 256, 298–302
Strigini, Maura, 137
Struhl, Gary, 157, 211, 224, 231, 234
Strutt, David, 231
Sturtevant, Alfred Henry, 1, 2, 4, 31, 38, 39, 86, 91, 134, 298, 299
Swammerdam, Jan, 76

Thomas, John, 173
Tokunaga, Chiyoko, 80, 83, 252
Tomlinson, Andrew, 211, 212, 224
topology
  disc-exoskeleton, 89, 99, 139, 199
  embryo-larva, 77, 89
  eye-wing, 203
  gene-anatomy, xi, 20, 31, 41, 45, 239, 242, 297, 298, 301
  gene-cell type, 83
  larva-adult, 2
  leg disc vs. body segment, 89, 109, 114
  leg segment vs. body segment, 136
  leg-antenna, 83, 199, 251, 299
  leg-wing, 137
  neuro-sensory, 5, 191
  space-time, 41, 43
  wound healing, 96
transdetermination, 84–86, 96, 123, 169, 242, 249, 254, 302
Turing, Alan, 33
typeface formats, xiii, xiv

uncoupling. *See also* circuitry
  of border from P compartment, 125, 151
  of cell size from body size, 55, 56
  of chemical reactions, 73
  of early vs. late roles of Dpp in wing, 188
  of growth from differentiation, 196
  of growth from patterning, 141, 158
  of growth in adjacent compartments, 92, 301
  of identity from affinity, 173

of identity from boundaries, 154
of identity from signaling, 149, 161
of individuation from segmentation, 84, 299
of larval from adult development, 86
of ligand binding from signaling, 285
of ligand transport from signaling, 125
of optic stalk from eye furrow, 234
of pedigrees from patterning, 3, 4
of sensilla from veins, 185, 186
of signaling from polarity, 205
of veins from crossveins, 185, 189
of veins from one another, 185, 186
Ursprung, Heinrich, xii

Waddington, C. H., 32, 224, 299, 300, 304
Weatherbee, Scott, 245
Weinstein, Alexander, 37
Weismann, August, 76
Weiss, Paul, 299, 302
White, Richard, 212
Whiting, Anna, 302
Wieschaus, Eric, 76
Wigglesworth, Vincent, 32, 49, 57, 59, 224
Wilkins, Adam, 97
wing
  airfoil, 136, 167, 174
  alula, 186
  anatomy, 139, 159, 177
  axes. *See under* axes
  blade, 49, 55, 87, 105, 153, 157, 173, 174, 185, 189, 190
  blade vs. hinge identity, 139, 158, 172, 301, 302
  butterfly, 31, 299
  corrugation, 174
  duplications, 123, 157
  evolution, 86
  expansion, 174
  extra, 141
  geometry, 137, 189
  growth, 29
  hairs, 87, 159, 189, 205
  hinge, 132, 139, 141, 172, 190, 240, 296, 301
  hinge, overgrowth of, 158
  identity ("wingness"), 171, 240, 249, 254
  margin. *See* wing margin
  missing, 137, 140, 151, 159, 160, 167, 303
  moth, 61, 65, 132
  notches, 157, 164, 165
  sensilla, 185
  shape, 164
  size, 29
  small, 140, 147, 151–154, 164, 185
  symmetry, A/P, 188
  transalar adhesivity, 174
  transalar lumen, 174
  veins. *See* wing veins
wing disc
  advantages (vs. other discs), 93, 105
  bipolar duality, 114, 167–173, 302
  compartments, 4, 87
  distalization, 157, 171
  duplications, 157
  evagination, 174
  extra, 170
  fate map, 94, 139, 156, 193, 301
  folds, 87, 126, 139
  gene expression patterns, 89, 111, 141, 142, 169, 173, 177, 193, 245

wing disc (*contd.*)
  geometry, 137, 139, 160, 301
  growth, 29
  initiation, 77, 89, 91, 171
  margin. *See* wing margin
  margin syndrome, 164, 165
  mitoses, region-specific, 92
  nomenclature, xiii
  outgrowths, 148, 153, 155, 157
  outgrowths, multiple, 157
  spot, unique distal, 157
  spot, *vg*-ON (in notum), 158
  veins. *See* wing veins
  vs. leg disc, 137, 170, 171
  wing vs. notum portion, 87, 167–173,
    190–191, 193
wing margin
  asymmetry of, 159, 160, 164
  at the eye equator (*sic!*), 203
  bracts, 28
  bristle rows, 31, 65, 67, 159, 189
  bristles, 27, 48, 139, 159, 167, 277, 279, 282,
    302
  cell affinities in, 173
  cell types in, 174
  creation of, 159, 160, 173, 302
  ectopic, 96, 159–161, 164, 165
  gene expression in, 27, 48, 141, 146, 159,
    163–167, 177, 183, 301, 303
  gene regulation in, 161, 164, 171, 179, 295,
    296
  geometry, 137, 159
  growth control by, 153, 157
  missing, 284, 302
  mitotic quiescence, 92
  mutations affecting, 160
  nomenclature, 153, 154, 155
  tip vs. remainder, 157
  vein properties of, 139, 183, 189

  vs. blade, 141, 173
  vs. haltere capitellum, 245
wing pouch
  asymmetry vs. symmetry, 167, 301
  circuitry, 163, 166
  dorsal edge, 128
  extra, 157, 158, 170
  gene expression in, 92, 141, 159, 171, 174,
    177, 301
  gene regulation in, 129, 148, 173, 177, 245,
    288, 292, 296, 301
  gradients, 142, 146
  growth, 164
  homeosis, 240
  initiation, 87, 167, 170, 171
  morphogens, 140, 142, 146, 156
  overgrowth, 157, 164, 165
  response to Dpp, 147
  response to Notch, 164, 167
  size, 139, 142
wing veins
  annealing, 137, 174, 175, 177, 245
  asymmetry, 174
  atavistic, 174
  branching, 173
  circuitry (vs. tracheae), 174
  confluent lawns, 183, 185, 186, 188
  corrugations, 160
  cross-veins, 177, 185, 188–189
  determination, 188
  displacements, 185, 186
  extra, 173, 174, 182–188, 303
  formula, 139, 154
  function, 174
  fusion, 153, 173, 185
  gene expression in, 177, 183, 186–189
  gene regulation in, 177
  inducers vs. suppressors, 183, 186, 187
  initiation, 182, 183

  initiation vs. fine-tuning, 177
  interveins, widening of, 185
  irregular, 188
  L1. *See* wing margin
  L2 and L5 (Dpp-dependent), 177,
    185–188
  L2 sprouting bristles, 187
  L3 and L4 (Hh-dependent), 177, 185–186
  L3 proneural stripe and sensilla, 185
  lateral inhibition, 175, 188
  markers, 189
  missing, 182, 183, 185–188
  mitotic zones, 175
  nomenclature, 177, 189
  pattern in disc vs. adult wing, 94, 139, 193
  patterning, 153–156, 173–191
  pigmentation, 174
  proveins, 175, 177
  regulation by Delta-Notch, 53, 175–177,
    188
  regulation by Dpp, 177, 185, 188
  regulation by EGFR, 177–185, 188
  sharpening, 184
  straightening, 173, 177
  vein vs. intervein identities. *See under* cell
    states
  vs. bristles, patterning, 175, 187
  vs. interveins, 174
  vs. mesectoderm stripes, 179
  vs. paraveins, 174
  widening, 175, 182, 188
  widening at tips, 12, 175
  widths, 167, 173, 177
Wolff, Tanya, 211
Wolpert, Lewis, 32, 79, 80, 81, 83, 84, 93, 248,
  249, 251, 299

Zecca, Myriam, 137, 157
Zimm, Georgianna, xii

Printed in the United States
30245LVS00006B/1-2